Electrical Costs with RSMeans data

Antonio D'Aulerio, Senior Editor

2021
44th annual edition

Vice President, Data
Tim Duggan

Vice President, Product
Ted Kail

Product Manager
Jason Jordan

Principal Engineer
Bob Mewis

Engineers: Architectural Divisions
1, 3, 4, 5, 6, 7, 8, 9, 10, 11, 12, 13, 41
Matthew Doheny, Manager
Sam Babbitt
Sergeleen Edouard
Richard Goldin

Scott Keller
Thomas Lane

Engineers: Civil Divisions and Wages
2, 31, 32, 33, 34, 35, 44, 46
Derrick Hale, PE, Manager
Christopher Babbitt
Stephen Bell
Michael Lynch
Elisa Mello
David Yazbek

Engineers: Mechanical, Electrical, Plumbing & Conveying Divisions
14, 21, 22, 23, 25, 26, 27, 28, 48
Joseph Kelble, Manager
Brian Adams

Michelle Curran
Antonio D'Aulerio
Thomas Lyons
Jake MacDonald

Contributing Engineers
John Melin, PE, Manager
Paul Cowan
Barry Hutchinson
Gerard Lafond, PE
Matthew Sorrentino

Technical Project Lead
Kevin Souza

Production
Debra Panarelli, Manager
Jonathan Forgit
Sharon Larsen
Sheryl Rose
Janice Thalin

Data Quality
Joseph Ingargiola, Manager
David Byars
Audrey Considine
Ellen D'Amico

Cover Design
Blaire Collins

Innovation
Ray Diwakar, Vice President
Kedar Gaikwad
Todd Glowac
Srini Narla
Joseph Woughter

RSMeans data from Gordian
Construction Publishers & Consultants
1099 Hingham Street, Suite 201
Rockland, MA 02370
United States of America
1.800.448.8182
RSMeans.com

Printed in the United States of America
ISSN 1540-6318
ISBN 978-1-950656-54-7

0031
$319.00 per copy (in United States)
Price is subject to change without prior notice.

Related Data and Services

Our engineers recommend the following products and services to complement *Electrical Costs with RSMeans data:*

Annual Cost Data Books

2021 Electrical Change Order Costs with RSMeans data
2021 Building Construction Costs with RSMeans data
2021 Mechanical Costs with RSMeans data

Reference Books

Electrical Estimating Methods
Estimating Building Costs
RSMeans Estimating Handbook
Green Building: Project Planning & Estimating
How to Estimate with RSMeans data
Plan Reading & Material Takeoff
Project Scheduling & Management for Construction

Virtual, Instructor-led & On-site Training Offerings

Unit Price Estimating
Training for our online estimation solution
Practical Project Management for Construction Professionals
Mechanical & Electrical Estimating

RSMeans data

For access to the latest cost data, an intuitive search, and an easy-to-use estimate builder, take advantage of the time savings available from our online application.

To learn more visit: **RSMeans.com/online**

Enterprise Solutions

Building owners, facility managers, building product manufacturers, and attorneys across the public and private sectors engage with RSMeans data Enterprise to solve unique challenges where trusted construction cost data is critical.

To learn more visit: **RSMeans.com/Enterprise**

Custom Built Data Sets

Building and Space Models: Quickly plan construction costs across multiple locations based on geography, project size, building system component, product options, and other variables for precise budgeting and cost control.

Predictive Analytics: Accurately plan future builds with custom graphical interactive dashboards, negotiate future costs of tenant build-outs, and identify and compare national account pricing.

Consulting

Building Product Manufacturing Analytics: Validate your claims and assist with new product launches.

Third-Party Legal Resources: Used in cases of construction cost or estimate disputes, construction product failure vs. installation failure, eminent domain, class action construction product liability, and more.

API

For resellers or internal application integration, RSMeans data is offered via API. Deliver Unit, Assembly, and Square Foot Model data within your interface. To learn more about how you can provide your customers with the latest in localized construction cost data visit:
RSMeans.com/API

Table of Contents

Foreword

The Value of RSMeans data from Gordian

Since 1942, RSMeans data has been the industry-standard materials, labor, and equipment cost information database for contractors, facility owners and managers, architects, engineers, and anyone else that requires the latest localized construction cost information. More than 75 years later, the objective remains the same: to provide facility and construction professionals with the most current and comprehensive construction cost database possible.

With the constant influx of new construction methods and materials, in addition to ever-changing labor and material costs, last year's cost data is not reliable for today's designs, estimates, or budgets. Gordian's cost engineers apply real-world construction experience to identify and quantify new building products and methodologies, adjust productivity rates, and adjust costs to local market conditions across the nation. This adds up to more than 22,000 hours in cost research annually. This unparalleled construction cost expertise is why so many facility and construction professionals rely on RSMeans data year over year.

About Gordian

Gordian originated in the spirit of innovation and a strong commitment to helping clients reach and exceed their construction goals. In 1982, Gordian's chairman and founder, Harry H. Mellon, created Job Order Contracting while serving as chief engineer at the Supreme Headquarters Allied Powers Europe. Job Order Contracting is a unique indefinite delivery/indefinite quantity (IDIQ) process, which enables facility owners to complete a substantial number of repair, maintenance, and construction projects with a single, competitively awarded contract. Realizing facility and infrastructure owners across various industries could greatly benefit from the time and cost saving advantages of this innovative construction procurement solution, he established Gordian in 1990.

Continuing the commitment to provide the most relevant and accurate facility and construction data, software, and expertise in the industry, Gordian enhanced the fortitude of its data with the acquisition of RSMeans in 2014. And in an effort to expand its facility management capabilities, Gordian acquired Sightlines, the leading provider of facilities benchmarking data and analysis, in 2015.

Our Offerings

Gordian is the leader in facility and construction cost data, software, and expertise for all phases of the building life cycle. From planning to design, procurement, construction, and operations, Gordian's solutions help clients maximize efficiency, optimize cost savings, and increase building quality with its highly specialized data engineers, software, and unique proprietary data sets.

Our Commitment

At Gordian, we do more than talk about the quality of our data and the usefulness of its application. We stand behind all of our RSMeans data—from historical cost indexes to construction materials and techniques—to craft current costs and predict future trends. If you have any questions about our products or services, please call us toll-free at 800.448.8182 or visit our website at gordian.com.

How the Cost Data Is Built: An Overview

Unit Prices*
All cost data have been divided into 50 divisions according to the MasterFormat® system of classification and numbering.

Assemblies*
The cost data in this section have been organized in an "Assemblies" format. These assemblies are the functional elements of a building and are arranged according to the 7 elements of the UNIFORMAT II classification system. For a complete explanation of a typical "Assembly", see "RSMeans data: Assemblies—How They Work."

*Residential Models**
Model buildings for four classes of construction—economy, average, custom, and luxury—are developed and shown with complete costs per square foot.

*Commercial/Industrial/ Institutional Models**
This section contains complete costs for 77 typical model buildings expressed as costs per square foot.

*Green Commercial/Industrial/ Institutional Models**
This section contains complete costs for 25 green model buildings expressed as costs per square foot.

*References**
This section includes information on Equipment Rental Costs, Crew Listings, Historical Cost Indexes, City Cost Indexes, Location Factors, Reference Tables, and Change Orders, as well as a listing of abbreviations.

- **Equipment Rental Costs:** Included are the average costs to rent and operate hundreds of pieces of construction equipment.
- **Crew Listings**: This section lists all the crews referenced in the cost data. A crew is composed of more than one trade classification and/or the addition of power equipment to any trade classification. Power equipment is included in the cost of the crew. Costs are shown both with bare labor rates and with the installing contractor's overhead and profit added. For each, the total crew cost per eight-hour day and the composite cost per labor-hour are listed.

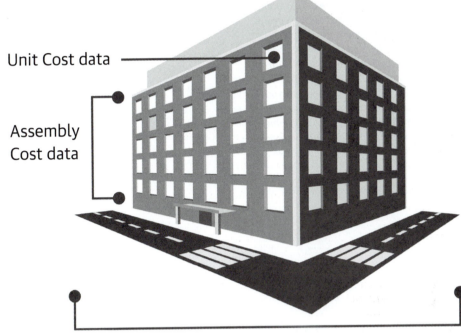

Square Foot Models

- **Historical Cost Indexes**: These indexes provide you with data to adjust construction costs over time.
- **City Cost Indexes**: All costs in this data set are U.S. national averages. Costs vary by region. You can adjust for this by CSI Division to over 730 cities in 900+ 3-digit zip codes throughout the U.S. and Canada by using this data.
- **Location Factors**: You can adjust total project costs to over 730 cities in 900+ 3-digit zip codes throughout the U.S. and Canada by using the weighted number, which applies across all divisions.
- **Reference Tables**: At the beginning of selected major classifications in the Unit Prices are reference numbers indicators. These numbers refer you to related information in the Reference Section. In this section, you'll find reference tables, explanations, and estimating information that support how we develop the unit price data, technical data, and estimating procedures.
- **Change Orders**: This section includes information on the factors that influence the pricing of change orders.

- **Abbreviations**: A listing of abbreviations used throughout this information, along with the terms they represent, is included.

Index (printed versions only)
A comprehensive listing of all terms and subjects will help you quickly find what you need when you are not sure where it occurs in MasterFormat®.

Conclusion
This information is designed to be as comprehensive and easy to use as possible.

The Construction Specifications Institute (CSI) and Construction Specifications Canada (CSC) have produced the 2018 edition of MasterFormat®, a system of titles and numbers used extensively to organize construction information.

All unit prices in the RSMeans cost data are now arranged in the 50-division MasterFormat® 2018 system.

---

*   Not all information is available in all data sets

Note: The material prices in RSMeans cost data are "contractor's prices." They are the prices that contractors can expect to pay at the lumberyards, suppliers'/distributors' warehouses, etc. Small orders of specialty items would be higher than the costs shown, while very large orders, such as truckload lots, would be less. The variation would depend on the size, timing, and negotiating power of the contractor. The labor costs are primarily for new construction or major renovation rather than repairs or minor alterations. With reasonable exercise of judgment, the figures can be used for any building work.

Estimating with RSMeans data: Unit Prices

Following these steps will allow you to complete an accurate estimate using RSMeans data: Unit Prices.

1. Scope Out the Project

- Think through the project and identify the CSI divisions needed in your estimate.
- Identify the individual work tasks that will need to be covered in your estimate.
- The Unit Price data have been divided into 50 divisions according to CSI MasterFormat® 2018.
- In printed versions, the Unit Price Section Table of Contents on page 1 may also be helpful when scoping out your project.
- Experienced estimators find it helpful to begin with Division 2 and continue through completion. Division 1 can be estimated after the full project scope is known.

2. Quantify

- Determine the number of units required for each work task that you identified.
- Experienced estimators include an allowance for waste in their quantities. (Waste is not included in our Unit Price line items unless otherwise stated.)

3. Price the Quantities

- Use the search tools available to locate individual Unit Price line items for your estimate.
- Reference Numbers indicated within a Unit Price section refer to additional information that you may find useful.
- The crew indicates who is performing the work for that task. Crew codes are expanded in the Crew Listings in the Reference Section to include all trades and equipment that comprise the crew.
- The Daily Output is the amount of work the crew is expected to complete in one day.
- The Labor-Hours value is the amount of time it will take for the crew to install one unit of work.
- The abbreviated Unit designation indicates the unit of measure upon which the crew, productivity, and prices are based.
- Bare Costs are shown for materials, labor, and equipment needed to complete the Unit Price line item. Bare costs do not include waste, project overhead, payroll insurance, payroll taxes, main office overhead, or profit.
- The Total Incl O&P cost is the billing rate or invoice amount of the installing contractor or subcontractor who performs the work for the Unit Price line item.

4. Multiply

- Multiply the total number of units needed for your project by the Total Incl O&P cost for each Unit Price line item.
- Be careful that your take off unit of measure matches the unit of measure in the Unit column.
- The price you calculate is an estimate for a completed item of work.
- Keep scoping individual tasks, determining the number of units required for those tasks, matching each task with individual Unit Price line items, and multiplying quantities by Total Incl O&P costs.
- An estimate completed in this manner is priced as if a subcontractor, or set of subcontractors, is performing the work. The estimate does not yet include Project Overhead or Estimate Summary components such as general contractor markups on subcontracted work, general contractor office overhead and profit, contingency, and location factors.

5. Project Overhead

- Include project overhead items from Division 1-General Requirements.
- These items are needed to make the job run. They are typically, but not always, provided by the general contractor. Items include, but are not limited to, field personnel, insurance, performance bond, permits, testing, temporary utilities, field office and storage facilities, temporary scaffolding and platforms, equipment mobilization and demobilization, temporary roads and sidewalks, winter protection, temporary barricades and fencing, temporary security, temporary signs, field engineering and layout, final cleaning, and commissioning.
- Each item should be quantified and matched to individual Unit Price line items in Division 1, then priced and added to your estimate.
- An alternate method of estimating project overhead costs is to apply a percentage of the total project cost—usually 5% to 15% with an average of 10% (see General Conditions).
- Include other project related expenses in your estimate such as:
  - Rented equipment not itemized in the Crew Listings
  - Rubbish handling throughout the project (see section 02 41 19.19)

6. Estimate Summary

- Include sales tax as required by laws of your state or county.
- Include the general contractor's markup on self-performed work, usually 5% to 15% with an average of 10%.
- Include the general contractor's markup on subcontracted work, usually 5% to 15% with an average of 10%.
- Include the general contractor's main office overhead and profit:
  - RSMeans data provides general guidelines on the general contractor's main office overhead (see section 01 31 13.60 and Reference Number R013113-50).
  - Markups will depend on the size of the general contractor's operations, projected annual revenue, the level of risk, and the level of competition in the local area and for this project in particular.
- Include a contingency, usually 3% to 5%, if appropriate.
- Adjust your estimate to the project's location by using the City Cost Indexes or the Location Factors in the Reference Section:
  - Look at the rules in "How to Use the City Cost Indexes" to see how to apply the Indexes for your location.
  - When the proper Index or Factor has been identified for the project's location, convert it to a multiplier by dividing it by 100, then multiply that multiplier by your estimated total cost. The original estimated total cost will now be adjusted up or down from the national average to a total that is appropriate for your location.

Editors' Note:
We urge you to spend time reading and understanding the supporting material. An accurate estimate requires experience, knowledge, and careful calculation. The more you know about how we at RSMeans developed the data, the more accurate your estimate will be. In addition, it is important to take into consideration the reference material such as Equipment Listings, Crew Listings, City Cost Indexes, Location Factors, and Reference Tables.

How to Use the Cost Data: The Details

What's Behind the Numbers? The Development of Cost Data

RSMeans data engineers continually monitor developments in the construction industry in order to ensure reliable, thorough, and up-to-date cost information. While overall construction costs may vary relative to general economic conditions, price fluctuations within the industry are dependent upon many factors. Individual price variations may, in fact, be opposite to overall economic trends. Therefore, costs are constantly tracked and complete updates are performed yearly. Also, new items are frequently added in response to changes in materials and methods.

Costs in U.S. Dollars

All costs represent U.S. national averages and are given in U.S. dollars. The City Cost Index (CCI) with RSMeans data can be used to adjust costs to a particular location. The CCI for Canada can be used to adjust U.S. national averages to local costs in Canadian dollars. No exchange rate conversion is necessary because it has already been factored in.

G The processes or products identified by the green symbol in our publications have been determined to be environmentally responsible and/or resource-efficient solely by RSMeans data engineering staff. The inclusion of the green symbol does not represent compliance with any specific industry association or standard.

Material Costs

RSMeans data engineers contact manufacturers, dealers, distributors, and contractors all across the U.S. and Canada to determine national average material costs. If you have access to current material costs for your specific location, you may wish to make adjustments to reflect differences from the national average. Included within material costs are fasteners for a normal installation. RSMeans data engineers use manufacturers' recommendations, written specifications, and/or standard construction practices for the sizing and spacing of fasteners. Adjustments to material costs may be required for your specific application or location. The manufacturer's warranty is assumed. Extended warranties are not included in the material costs. **Material costs do not include sales tax.**

Labor Costs

Labor costs are based upon a mathematical average of trade-specific wages in 30 major U.S. cities. The type of wage (union, open shop, or residential) is identified on the inside back cover of printed publications or selected by the estimator when using the electronic products. Markups for the wages can also be found on the inside back cover of printed publications and/or under the labor references found in the electronic products.

- If wage rates in your area vary from those used, or if rate increases are expected within a given year, labor costs should be adjusted accordingly.

Labor costs reflect productivity based on actual working conditions. In addition to actual installation, these figures include time spent during a normal weekday on tasks, such as material receiving and handling, mobilization at the site, site movement, breaks, and cleanup.

Productivity data is developed over an extended period so as not to be influenced by abnormal variations and reflects a typical average.

Equipment Costs

Equipment costs include not only rental but also operating costs for equipment under normal use. The operating costs include parts and labor for routine servicing, such as the repair and replacement of pumps, filters, and worn lines. Normal operating expendables, such as fuel, lubricants, tires, and electricity (where applicable), are also included. Extraordinary operating expendables with highly variable wear patterns, such as diamond bits and blades, are excluded. These costs are included under materials. Equipment rental rates are obtained from industry sources throughout North America—contractors, suppliers, dealers, manufacturers, and distributors.

Rental rates can also be treated as reimbursement costs for contractor-owned equipment. Owned equipment costs include depreciation, loan payments, interest, taxes, insurance, storage, and major repairs.

Equipment costs do not include operators' wages.

Equipment Cost/Day—The cost of equipment required for each crew is included in the Crew Listings in the Reference Section (small tools that are considered essential everyday tools are not listed out separately). The Crew Listings itemize specialized tools and heavy equipment along with labor trades. The daily cost of itemized equipment included in a crew is based on dividing the weekly bare rental rate by 5 (number of working days per week), then adding the hourly operating cost times 8 (the number of hours per day). This Equipment Cost/Day is shown in the last column of the Equipment Rental Costs in the Reference Section.

Mobilization, Demobilization—The cost to move construction equipment from an equipment yard or rental company to the job site and back again is not included in equipment costs. Mobilization (to the site) and demobilization (from the site) costs can be found in the Unit Price Section. If a piece of equipment is already at the job site, it is not appropriate to utilize mobilization or demobilization costs again in an estimate.

Overhead and Profit

Total Cost including O&P for the installing contractor is shown in the last column of the Unit Price and/or Assemblies. This figure is the sum of the bare material cost plus 10% for profit, the bare labor cost plus total overhead and profit, and the bare equipment cost plus 10% for profit. Details for the calculation of overhead and profit on labor are shown on the inside back cover of the printed product and in the Reference Section of the electronic product.

General Conditions

Cost data in this data set are presented in two ways: Bare Costs and Total Cost including O&P (Overhead and Profit). General Conditions, or General Requirements, of the contract should also be added to the Total Cost including O&P when applicable. Costs for General Conditions are listed in Division 1 of the Unit Price Section and in the Reference Section.

General Conditions for the installing contractor may range from 0% to 10% of the Total Cost including O&P. For the general or prime contractor, costs for General Conditions may range from 5% to 15% of the Total Cost including O&P, with a figure of 10% as the most typical allowance. If applicable, the Assemblies and Models sections use costs that include the installing contractor's overhead and profit (O&P).

Factors Affecting Costs

Costs can vary depending upon a number of variables. Here's a listing of some factors that affect costs and points to consider.

Quality—The prices for materials and the workmanship upon which productivity is based represent sound construction work. They are also in line with industry standard and manufacturer specifications and are frequently used by federal, state, and local governments.

Overtime—We have made no allowance for overtime. If you anticipate premium time or work beyond normal working hours, be sure to make an appropriate adjustment to your labor costs.

Productivity—The productivity, daily output, and labor-hour figures for each line item are based on an eight-hour work day in daylight hours in moderate temperatures and up to a 14' working height unless otherwise indicated. For work that extends beyond normal work hours or is performed under adverse conditions, productivity may decrease.

Size of Project—The size, scope of work, and type of construction project will have a significant impact on cost. Economies of scale can reduce costs for large projects. Unit costs can often run higher for small projects.

Location—Material prices are for metropolitan areas. However, in dense urban areas, traffic and site storage limitations may increase costs. Beyond a 20-mile radius of metropolitan areas, extra trucking or transportation charges may also increase the material costs slightly. On the other hand, lower wage rates may be in effect. Be sure to consider both of these factors when preparing an estimate, particularly if the job site is located in a central city or remote rural location. In addition, highly specialized subcontract items may require travel and per-diem expenses for mechanics.

Other Factors—

- season of year
- contractor management
- weather conditions
- local union restrictions
- building code requirements
- availability of:
  - adequate energy
  - skilled labor
  - building materials
- owner's special requirements/restrictions
- safety requirements
- environmental considerations
- access

Unpredictable Factors—General business conditions influence "in-place" costs of all items. Substitute materials and construction methods may have to be employed. These may affect the installed cost and/or life cycle costs. Such factors may be difficult to evaluate and cannot necessarily be predicted on the basis of the job's location in a particular section of the country. Thus, where these factors apply, you may find significant but unavoidable cost variations for which you will have to apply a measure of judgment to your estimate.

Rounding of Costs

In printed publications only, all unit prices in excess of $5.00 have been rounded to make them easier to use and still maintain adequate precision of the results.

How Subcontracted Items Affect Costs

A considerable portion of all large construction jobs is usually subcontracted. In fact, the percentage done by subcontractors is constantly increasing and may run over 90%. Since the workers employed by these companies do nothing else but install their particular products, they soon become experts in that line. As a result, installation by these firms is accomplished so efficiently that the total in-place cost, even with the general contractor's overhead and profit, is no more, and often less, than if the principal contractor had handled the installation. Companies that deal with construction specialties are anxious to have their products perform well and, consequently, the installation will be the best possible.

Contingencies

The allowance for contingencies generally provides for unforeseen construction difficulties. On alterations or repair jobs, 20% is not too much. If drawings are final and only field contingencies are being considered, 2% or 3% is probably sufficient and often nothing needs to be added. Contractually, changes in plans will be covered by extras. The contractor should consider inflationary price trends and possible material shortages during the course of the job. These escalation factors are dependent upon both economic conditions and the anticipated time between the estimate and actual construction. If drawings are not complete or approved, or a budget cost is wanted, it is wise to add 5% to 10%. Contingencies, then, are a matter of judgment.

Important Estimating Considerations

The productivity, or daily output, of each craftsman or crew assumes a well-managed job where tradesmen with the proper tools and equipment, along with the appropriate construction materials, are present. Included are daily set-up and cleanup time, break time, and plan layout time. Unless otherwise indicated, time for material movement on site (for items that can be transported by hand) of up to 200' into the building and to the first or second floor is also included. If material has to be transported by other means, over greater distances, or to higher floors, an additional allowance should be considered by the estimator.

While horizontal movement is typically a sole function of distances, vertical transport introduces other variables that can significantly impact productivity. In an occupied building, the use of elevators (assuming access, size, and required protective measures are acceptable) must be understood at the time of the estimate. For new construction, hoist wait and cycle times can easily be 15 minutes and may result in scheduled access extending beyond the normal work day. Finally, all vertical transport will impose strict weight limits likely to preclude the use of any motorized material handling.

The productivity, or daily output, also assumes installation that meets manufacturer/designer/ standard specifications. A time allowance for quality control checks, minor adjustments, and any task required to ensure proper function or operation is also included. For items that require connections to services, time is included for positioning, leveling, securing the unit, and making all the necessary connections (and start up where applicable) to ensure a complete installation. Estimating of the services themselves (electrical, plumbing, water, steam, hydraulics, dust collection, etc.) is separate.

In some cases, the estimator must consider the use of a crane and an appropriate crew for the installation of large or heavy items. For those situations where a crane is not included in the assigned crew and as part of the line item cost, then equipment rental costs, mobilization and demobilization costs, and operator and support personnel costs must be considered.

Labor-Hours

The labor-hours expressed in this publication are derived by dividing the total daily labor-hours for the crew by the daily output. Based on average installation time and the assumptions listed above, the labor-hours include: direct labor, indirect labor, and nonproductive time. A typical day for a craftsman might include but is not limited to:

- Direct Work
  - Measuring and layout
  - Preparing materials
  - Actual installation
  - Quality assurance/quality control
- Indirect Work
  - Reading plans or specifications
  - Preparing space
  - Receiving materials
  - Material movement
  - Giving or receiving instruction
  - Miscellaneous
- Non-Work
  - Chatting
  - Personal issues
  - Breaks
  - Interruptions (i.e., sickness, weather, material or equipment shortages, etc.)

If any of the items for a typical day do not apply to the particular work or project situation, the estimator should make any necessary adjustments.

Final Checklist

Estimating can be a straightforward process provided you remember the basics. Here's a checklist of some of the steps you should remember to complete before finalizing your estimate.

Did you remember to:

- factor in the City Cost Index for your locale?
- take into consideration which items have been marked up and by how much?
- mark up the entire estimate sufficiently for your purposes?
- read the background information on techniques and technical matters that could impact your project time span and cost?
- include all components of your project in the final estimate?
- double check your figures for accuracy?
- call RSMeans data engineers if you have any questions about your estimate or the data you've used? Remember, Gordian stands behind all of our products, including our extensive RSMeans data solutions. If you have any questions about your estimate, about the costs you've used from our data, or even about the technical aspects of the job that may affect your estimate, feel free to call the Gordian RSMeans editors at 1.800.448.8182.

Unit Price Section

Table of Contents

1

RSMeans data: Unit Prices— How They Work

All RSMeans data: Unit Prices are organized in the same way.

03 30 Cast-In-Place Concrete

03 30 53 – Miscellaneous Cast-In-Place Concrete

03 30 53.40 Concrete In Place	Crew	Daily Output	Labor-Hours	Unit	Material	2021 Bare Costs Labor	Equipment	Total	Total Incl O&P
0010 CONCRETE IN PLACE									
0020 Including forms (4 uses), Grade 60 rebar, concrete (Portland cement									
0050 Type I), placement and finishing unless otherwise indicated									
3540 Equipment pad (3000 psi), 3' x 3' x 6" thick	C-14H	45	1.067	Ea.	55.50	57	.61	113.11	147
3550 4' x 4' x 6" thick		30	1.600		84	85.50	.91	170.41	220
3560 5' x 5' x 8" thick		18	2.667		150	143	1.52	294.52	380
3570 6' x 6' x 8" thick		14	3.429		204	184	1.95	389.95	500
3580 8' x 8' x 10" thick		8	6		415	320	3.42	738.42	945
3590 10' x 10' x 12" thick		5	9.600		720	515	5.45	1,240.45	1,550
3800 Footings (3000 psi), spread under 1 C.Y.	C-14C	28	4	C.Y.	208	209	.96	417.96	540
3813 Install new concrete (3000 psi) light pole base, 24" diam. x 8'	C-1	2.66	12.030		345	625		970	1,325
3825 1 C.Y. to 5 C.Y.	C-14C	43	2.605		266	136	.63	402.63	495
3850 Over 5 C.Y.	"	75	1.493		245	78	.36	323.36	385
3900 Footings, strip (3000 psi), 18" x 9", unreinforced	C-14L	40	2.400		159	123	.67	282.67	360

It is important to understand the structure of RSMeans data: Unit Prices so that you can find information easily and use it correctly.

① Line Numbers

Line Numbers consist of 12 characters, which identify a unique location in the database for each task. The first 6 or 8 digits conform to the Construction Specifications Institute MasterFormat® 2018. The remainder of the digits are a further breakdown in order to arrange items in understandable groups of similar tasks. Line numbers are consistent across all of our publications, so a line number in any of our products will always refer to the same item of work.

② Descriptions

Descriptions are shown in a hierarchical structure to make them readable. In order to read a complete description, read up through the indents to the top of the section. Include everything that is above and to the left that is not contradicted by information below. For instance, the complete description for line 03 30 53.40 3550 is "Concrete in place, including forms (4 uses), Grade 60 rebar, concrete (Portland cement Type 1), placement and finishing unless otherwise indicated; Equipment pad (3,000 psi), 4' × 4' × 6" thick."

③ RSMeans data

When using **RSMeans data**, it is important to read through an entire section to ensure that you use the data that most closely matches your work. Note that sometimes there is additional information shown in the section that may improve your price. There are frequently lines that further describe, add to, or adjust data for specific situations.

④ Reference Information

Gordian's RSMeans engineers have created **reference** information to assist you in your estimate. **If** there is information that applies to a section, it will be indicated at the start of the section. The Reference Section is located in the back of the data set.

⑤ Crews

Crews include labor and/or equipment necessary to accomplish each task. In this case, Crew C-14H is used. Gordian's RSMeans staff selects a crew to represent the workers and equipment that are typically used for that task. In this case, Crew C-14H consists of one carpenter foreman (outside), two carpenters, one rodman, one laborer, one cement finisher, and one gas engine vibrator. Details of all crews can be found in the Reference Section.

Crews - Standard

Crew No.	Bare Costs		Incl. Subs O & P		Cost Per Labor-Hour	
Crew C-14H	Hr.	Daily	Hr.	Daily	Bare Costs	Incl. O&P
1 Carpenter Foreman (outside)	$56.70	$453.60	$84.60	$676.80	$53.53	$79.68
2 Carpenters	54.70	875.20	81.65	1306.40		
1 Rodman (reinf.)	58.90	471.20	88.05	704.40		
1 Laborer	44.40	355.20	66.25	530.00		
1 Cement Finisher	51.80	414.40	75.90	607.20		
1 Gas Engine Vibrator		27.15		29.86	.57	.62
48 L.H., Daily Totals		$2596.75		$3854.67	$54.10	$80.31

6 Daily Output

The **Daily Output** is the amount of work that the crew can do in a normal 8-hour workday, including mobilization, layout, movement of materials, and cleanup. In this case, crew C-14H can install thirty 4' × 4' × 6" thick concrete pads in a day. Daily output is variable and based on many factors, including the size of the job, location, and environmental conditions. RSMeans data represents work done in daylight (or adequate lighting) and temperate conditions.

7 Labor-Hours

The figure in the **Labor-Hours** column is the amount of labor required to perform one unit of work—in this case the amount of labor required to construct one 4' × 4' equipment pad. This figure is calculated by dividing the number of hours of labor in the crew by the daily output (48 labor-hours divided by 30 pads = 1.6 hours of labor per pad). Multiply 1.6 times 60 to see the value in minutes: 60 × 1.6 = 96

minutes. Note: the labor-hour figure is not dependent on the crew size. A change in crew size will result in a corresponding change in daily output, but the labor-hours per unit of work will not change.

8 Unit of Measure

All RSMeans data: Unit Prices include the typical **Unit of Measure** used for estimating that item. For concrete-in-place the typical unit is cubic yards (C.Y.) or each (Ea.). For installing broadloom carpet it is square yard and for gypsum board it is square foot. The estimator needs to take special care that the unit in the data matches the unit in the take-off. Unit conversions may be found in the Reference Section.

9 Bare Costs

Bare Costs are the costs of materials, labor, and equipment that the installing contractor pays. They represent the cost, in U.S. dollars, for one unit of work. They do not include any markups for profit or labor burden.

10 Bare Total

The **Total column** represents the total bare cost for the installing contractor in U.S. dollars. In this case, the sum of $84 for material + $85.50 for labor + $.91 for equipment is $170.41.

11 Total Incl O&P

The **Total Incl O&P column** is the total cost, including overhead and profit, that the installing contractor will charge the customer. This represents the cost of materials plus 10% profit, the cost of labor plus labor burden and 10% profit, and the cost of equipment plus 10% profit. It does not include the general contractor's overhead and profit. Note: See the inside back cover of the printed product or the Reference Section of the electronic product for details on how the labor burden is calculated.

National Average

The RSMeans data in our print publications represent a "national average" cost. This data should be modified to the project location using the **City Cost Indexes** or **Location Factors** tables found in the Reference Section. Use the Location Factors to adjust estimate totals if the project covers multiple trades. Use the City Cost Indexes (CCI) for single trade

projects or projects where a more detailed analysis is required. All figures in the two tables are derived from the same research. The last row of data in the CCI—the weighted average—is the same as the numbers reported for each location in the location factor table.

Project Name: Pre-Engineered Steel Building			Architect: As Shown					
Location:	Anywhere, USA							
							01/01/21	STD
Line Number	Description	Qty	Unit	Material	Labor	Equipment	SubContract	Estimate Total
03 30 53.40 3940	Strip footing, 12" x 24", reinforced	15	C.Y.	$2,625.00	$1,830.00	$8.40	$0.00	
03 30 53.40 3950	Strip footing, 12" x 36", reinforced	34	C.Y.	$5,644.00	$3,315.00	$15.30	$0.00	
03 11 13.65 3000	Concrete slab edge forms	500	L.F.	$225.00	$1,390.00	$0.00	$0.00	
03 22 11.10 0200	Welded wire fabric reinforcing	150	C.S.F.	$3,150.00	$4,575.00	$0.00	$0.00	
03 31 13.35 0300	Ready mix concrete, 4000 psi for slab on grade	278	C.Y.	$35,862.00	$0.00	$0.00	$0.00	
03 31 13.70 4300	Place, strike off & consolidate concrete slab	278	C.Y.	$0.00	$5,560.00	$136.22	$0.00	
03 35 13.30 0250	Machine float & trowel concrete slab	15,000	S.F.	$0.00	$10,350.00	$750.00	$0.00	
03 15 16.20 0140	Cut control joints in concrete slab	950	L.F.	$38.00	$437.00	$57.00	$0.00	
03 39 23.13 0300	Sprayed concrete curing membrane	150	C.S.F.	$1,830.00	$1,125.00	$0.00	$0.00	
Division 03	**Subtotal**			**$49,374.00**	**$28,582.00**	**$966.92**	**$0.00**	**$78,922.92**
08 36 13.10 2650	Manual 10' x 10' steel sectional overhead door	8	Ea.	$9,800.00	$3,880.00	$0.00	$0.00	
08 36 13.10 2860	Insulation and steel back panel for OH door	800	S.F.	$4,760.00	$0.00	$0.00	$0.00	
Division 08	**Subtotal**			**$14,560.00**	**$3,880.00**	**$0.00**	**$0.00**	**$18,440.00**
13 34 19.50 1100	Pre-Engineered Steel Building, 100' x 150' x 24'	15,000	SF Flr.	$0.00	$0.00	$0.00	$367,500.00	
13 34 19.50 6050	Framing for PESB door opening, 3' x 7'	4	Opng.	$0.00	$0.00	$0.00	$2,300.00	
13 34 19.50 6100	Framing for PESB door opening, 10' x 10'	8	Opng.	$0.00	$0.00	$0.00	$9,400.00	
13 34 19.50 6200	Framing for PESB window opening, 4' x 3'	6	Opng.	$0.00	$0.00	$0.00	$3,360.00	
13 34 19.50 5750	PESB door, 3' x 7', single leaf	4	Opng.	$2,740.00	$772.00	$0.00	$0.00	
13 34 19.50 7750	PESB sliding window, 4' x 3' with screen	6	Opng.	$2,940.00	$660.00	$68.70	$0.00	
13 34 19.50 6550	PESB gutter, eave type, 26 ga., painted	300	L.F.	$2,595.00	$906.00	$0.00	$0.00	
13 34 19.50 8650	PESB roof vent, 12" wide x 10' long	15	Ea.	$705.00	$3,615.00	$0.00	$0.00	
13 34 19.50 6900	PESB insulation, vinyl faced, 4" thick	27,400	S.F.	$14,522.00	$10,412.00	$0.00	$0.00	
Division 13	**Subtotal**			**$23,502.00**	**$16,365.00**	**$68.70**	**$382,560.00**	**$422,495.70**
			Subtotal	$87,436.00	$48,827.00	$1,035.62	$382,560.00	$519,858.62
Division 01	General Requirements @ 7%			6,120.52	3,417.89	72.49	26,779.20	
			Estimate Subtotal	$93,556.52	$52,244.89	$1,108.11	$409,339.20	$519,858.62
			Sales Tax @ 5%	4,677.83		55.41	10,233.48	
			Subtotal A	98,234.35	52,244.89	1,163.52	419,572.68	
			GC O & P	9,823.43	26,331.42	116.35	41,957.27	
			Subtotal B	108,057.78	78,576.31	1,279.87	461,529.95	$649,443.91
			Contingency @ 5%					32,472.20
			Subtotal C					$681,916.11
			Bond @ $12/1000 +10% O&P					9,001.29
			Subtotal D					$690,917.40
			Location Adjustment Factor		114.20			98,110.27
			Grand Total					**$789,027.67**

This estimate is based on an interactive spreadsheet. You are free to download it and adjust it to your methodology.
A copy of this spreadsheet is available at **RSMeans.com/2021books.**

Sample Estimate

This sample demonstrates the elements of an estimate, including a tally of the RSMeans data lines and a summary of the markups on a contractor's work to arrive at a total cost to the owner. The Location Factor with RSMeans data is added at the bottom of the estimate to adjust the cost of the work to a specific location.

❶ Work Performed

The body of the estimate shows the RSMeans data selected, including the line number, a brief description of each item, its take-off unit and quantity, and the bare costs of materials, labor, and equipment. This estimate also includes a column titled "SubContract." This data is taken from the column "Total Incl O&P" and represents the total that a subcontractor would charge a general contractor for the work, including the sub's markup for overhead and profit.

❷ Division 1, General Requirements

This is the first division numerically but the last division estimated. Division 1 includes project-wide needs provided by the general contractor. These requirements vary by project but may include temporary facilities and utilities, security, testing, project cleanup, etc. For small projects a percentage can be used—typically between 5% and 15% of project cost. For large projects the costs may be itemized and priced individually.

❸ Sales Tax

If the work is subject to state or local sales taxes, the amount must be added to the estimate. Sales tax may be added to material costs, equipment costs, and subcontracted work. In this case, sales tax was added in all three categories. It was assumed that approximately half the subcontracted work would be material cost, so the tax was applied to 50% of the subcontract total.

❹ GC O&P

This entry represents the general contractor's markup on material, labor, equipment, and subcontractor costs. Our standard markup on materials, equipment, and subcontracted work is 10%. In this estimate, the markup on the labor performed by the GC's workers uses "Skilled Workers Average" shown in Column F on the table "Installing Contractor's Overhead & Profit," which can be found on the inside back cover of the printed product or in the Reference Section of the electronic product.

❺ Contingency

A factor for contingency may be added to any estimate to represent the cost of unknowns that may occur between the time that the estimate is performed and the time the project is constructed. The amount of the allowance will depend on the stage of design at which the estimate is done and the contractor's assessment of the risk involved. Refer to section 01 21 16.50 for contingency allowances.

❻ Bonds

Bond costs should be added to the estimate. The figures here represent a typical performance bond, ensuring the owner that if the general contractor does not complete the obligations in the construction contract the bonding company will pay the cost for completion of the work.

❼ Location Adjustment

Published prices are based on national average costs. If necessary, adjust the total cost of the project using a location factor from the "Location Factor" table or the "City Cost Index" table. Use location factors if the work is general, covering multiple trades. If the work is by a single trade (e.g., masonry) use the more specific data found in the "City Cost Indexes."

Estimating Tips

01 20 00 Price and Payment Procedures

- Allowances that should be added to estimates to cover contingencies and job conditions that are not included in the national average material and labor costs are shown in Section 01 21.

- When estimating historic preservation projects (depending on the condition of the existing structure and the owner's requirements), a 15–20% contingency or allowance is recommended, regardless of the stage of the drawings.

01 30 00 Administrative Requirements

- Before determining a final cost estimate, it is good practice to review all the items listed in Subdivisions 01 31 and 01 32 to make final adjustments for items that may need customizing to specific job conditions.

- Requirements for initial and periodic submittals can represent a significant cost to the General Requirements of a job. Thoroughly check the submittal specifications when estimating a project to determine any costs that should be included.

01 40 00 Quality Requirements

- All projects will require some degree of quality control. This cost is not included in the unit cost of construction listed in each division. Depending upon the terms of the contract, the various costs of inspection and testing can be the responsibility of either the owner or the contractor. Be sure to include the required costs in your estimate.

01 50 00 Temporary Facilities and Controls

- Barricades, access roads, safety nets, scaffolding, security, and many more requirements for the execution of a safe project are elements of direct cost. These costs can easily be overlooked when preparing an estimate. When looking through the major classifications of this subdivision, determine which items apply to each division in your estimate.

- Construction equipment rental costs can be found in the Reference Section in Section 01 54 33. Operators' wages are not included in equipment rental costs.

- Equipment mobilization and demobilization costs are not included in equipment rental costs and must be considered separately.

- The cost of small tools provided by the installing contractor for his workers is covered in the "Overhead" column on the "Installing Contractor's Overhead and Profit" table that lists labor trades, base rates, and markups. Therefore, it is included in the "Total Incl. O&P" cost of any unit price line item.

01 70 00 Execution and Closeout Requirements

- When preparing an estimate, thoroughly read the specifications to determine the requirements for Contract Closeout. Final cleaning, record documentation, operation and maintenance data, warranties and bonds, and spare parts and maintenance materials can all be elements of cost for the completion of a contract. Do not overlook these in your estimate.

Reference Numbers

Reference numbers are shown at the beginning of some major classifications. These numbers refer to related items in the Reference Section. The reference information may be an estimating procedure, an alternate pricing method, or technical information.

Note: Not all subdivisions listed here necessarily appear. ■

Same Data. Simplified.

Enjoy the convenience and efficiency of accessing your costs anywhere:

- **Skip the multiplier** by setting your location
- **Quickly search**, edit, favorite and share costs
- **Stay on top of price changes** with automatic updates

Discover more at rsmeans.com/online

01 11 Summary of Work

01 11 31 – Professional Consultants

01 11 31.20 Construction Management Fees	Crew	Daily Output	Labor-Hours	Unit	Material	2021 Bare Costs Labor	Equipment	Total	Total Incl O&P
0010 CONSTRUCTION MANAGEMENT FEES									
0020 $1,000,000 job, minimum				Project				4.50%	4.50%
0050 Maximum								7.50%	7.50%
0060 For work to $100,000								10%	10%
0300 $50,000,000 job, minimum								2.50%	2.50%
0350 Maximum				↓				4%	4%

01 11 31.30 Engineering Fees

	Crew	Daily Output	Labor-Hours	Unit	Material	2021 Bare Costs Labor	Equipment	Total	Total Incl O&P
0010 ENGINEERING FEES R011110-30									
0020 Educational planning consultant, minimum				Project				.50%	.50%
0100 Maximum				"				2.50%	2.50%
0200 Electrical, minimum				Contrct				4.10%	4.10%
0300 Maximum								10.10%	10.10%
0400 Elevator & conveying systems, minimum								2.50%	2.50%
0500 Maximum								5%	5%
0600 Food service & kitchen equipment, minimum								8%	8%
0700 Maximum								12%	12%
1000 Mechanical (plumbing & HVAC), minimum								4.10%	4.10%
1100 Maximum				↓				10.10%	10.10%

01 21 Allowances

01 21 16 – Contingency Allowances

01 21 16.50 Contingencies

	Crew	Daily Output	Labor-Hours	Unit	Material	2021 Bare Costs Labor	Equipment	Total	Total Incl O&P
0010 CONTINGENCIES, Add to estimate									
0020 Conceptual stage				Project				20%	20%
0050 Schematic stage								15%	15%
0100 Preliminary working drawing stage (Design Dev.)								10%	10%
0150 Final working drawing stage				↓				3%	3%

01 21 53 – Factors Allowance

01 21 53.50 Factors

	Crew	Daily Output	Labor-Hours	Unit	Material	2021 Bare Costs Labor	Equipment	Total	Total Incl O&P
0010 FACTORS Cost adjustments									
0100 Add to construction costs for particular job requirements									
0500 Cut & patch to match existing construction, add, minimum				Costs	2%	3%			
0550 Maximum					5%	9%			
0800 Dust protection, add, minimum					1%	2%			
0850 Maximum					4%	11%			
1100 Equipment usage curtailment, add, minimum					1%	1%			
1150 Maximum					3%	10%			
1400 Material handling & storage limitation, add, minimum					1%	1%			
1450 Maximum					6%	7%			
1700 Protection of existing work, add, minimum					2%	2%			
1750 Maximum				↓	5%	7%			
2000 Shift work requirements, add, minimum				Costs		5%			
2050 Maximum						30%			
2300 Temporary shoring and bracing, add, minimum					2%	5%			
2350 Maximum				↓	5%	12%			

01 21 Allowances

01 21 53 – Factors Allowance

01 21 53.60 Security Factors

		Crew	Daily Output	Labor-Hours	Unit	Material	2021 Bare Costs Labor	Equipment	Total	Total Incl O&P
0010	**SECURITY FACTORS** R012153-60									
0100	Additional costs due to security requirements									
0110	Daily search of personnel, supplies, equipment and vehicles									
0120	Physical search, inventory and doc of assets, at entry				Costs		30%			
0130	At entry and exit						50%			
0140	Physical search, at entry						6.25%			
0150	At entry and exit						12.50%			
0160	Electronic scan search, at entry						2%			
0170	At entry and exit						4%			
0180	Visual inspection only, at entry						.25%			
0190	At entry and exit						.50%			
0200	ID card or display sticker only, at entry						.12%			
0210	At entry and exit				▼		.25%			
0220	Day 1 as described below, then visual only for up to 5 day job duration									
0230	Physical search, inventory and doc of assets, at entry				Costs		5%			
0240	At entry and exit						10%			
0250	Physical search, at entry						1.25%			
0260	At entry and exit						2.50%			
0270	Electronic scan search, at entry						.42%			
0280	At entry and exit				▼		.83%			
0290	Day 1 as described below, then visual only for 6-10 day job duration									
0300	Physical search, inventory and doc of assets, at entry				Costs		2.50%			
0310	At entry and exit						5%			
0320	Physical search, at entry						.63%			
0330	At entry and exit						1.25%			
0340	Electronic scan search, at entry						.21%			
0350	At entry and exit				▼		.42%			
0360	Day 1 as described below, then visual only for 11-20 day job duration									
0370	Physical search, inventory and doc of assets, at entry				Costs		1.25%			
0380	At entry and exit						2.50%			
0390	Physical search, at entry						.31%			
0400	At entry and exit						.63%			
0410	Electronic scan search, at entry						.10%			
0420	At entry and exit				▼		.21%			
0430	Beyond 20 days, costs are negligible									
0440	Escort required to be with tradesperson during work effort				Costs		6.25%			

01 21 53.65 Infectious Disease Precautions

		Crew	Daily Output	Labor-Hours	Unit	Material	2021 Bare Costs Labor	Equipment	Total	Total Incl O&P
0010	**INFECTIOUS DISEASE PRECAUTIONS** R012153-65									
0100	Additional costs due to infectious disease precautions									
0120	Daily temperature checks				Costs		1%			
0130	Donning & doffing masks and gloves						1%			
0140	Washing hands						1%			
0150	Informational meetings						2%			
0160	Maintaining social distance						1%			
0170	Disinfecting tools or equipment				▼		3%			
0200	N95 rated masks				Ea.	1.40			1.40	1.54
0210	Surgical masks				"	.16			.16	.18
0220	Black nitrile disposable gloves				Pair	.18			.18	.20
0250	Hand washing station & 2 x weekly service				Week				158	174

For customer support on your Electrical Costs with RSMeans Data, call 800.448.8182.

9

01 21 Allowances

01 21 55 – Job Conditions Allowance

01 21 55.50 Job Conditions

01 21 55.50 Job Conditions	Crew	Daily Output	Labor-Hours	Unit	Material	2021 Bare Costs Labor	Equipment	Total	Total Incl O&P
0010 JOB CONDITIONS Modifications to applicable									
0020 cost summaries									
0100 Economic conditions, favorable, deduct				Project				2%	2%
0200 Unfavorable, add								5%	5%
0300 Hoisting conditions, favorable, deduct								2%	2%
0400 Unfavorable, add								5%	5%
0700 Labor availability, surplus, deduct								1%	1%
0800 Shortage, add								10%	10%
0900 Material storage area, available, deduct								1%	1%
1000 Not available, add								2%	2%
1100 Subcontractor availability, surplus, deduct								5%	5%
1200 Shortage, add								12%	12%
1300 Work space, available, deduct								2%	2%
1400 Not available, add								5%	5%

01 21 57 – Overtime Allowance

01 21 57.50 Overtime

	Crew	Daily Output	Labor-Hours	Unit	Material	2021 Bare Costs Labor	Equipment	Total	Total Incl O&P
0010 OVERTIME for early completion of projects or where R012909-90									
0020 labor shortages exist, add to usual labor, up to				Costs		100%			

01 21 63 – Taxes

01 21 63.10 Taxes

	Crew	Daily Output	Labor-Hours	Unit	Material	2021 Bare Costs Labor	Equipment	Total	Total Incl O&P
0010 TAXES R012909-80									
0020 Sales tax, State, average				%	5.08%				
0050 Maximum R012909-85					7.50%				
0200 Social Security, on first $118,500 of wages						7.65%			
0300 Unemployment, combined Federal and State, minimum R012909-86						.60%			
0350 Average				%		9.60%			
0400 Maximum				"		12%			

01 31 Project Management and Coordination

01 31 13 – Project Coordination

01 31 13.20 Field Personnel

	Crew	Daily Output	Labor-Hours	Unit	Material	2021 Bare Costs Labor	Equipment	Total	Total Incl O&P
0010 FIELD PERSONNEL									
0020 Clerk, average				Week		495		495	750
0100 Field engineer, junior engineer						1,241		1,241	1,877
0120 Engineer						1,825		1,825	2,775
0140 Senior engineer						2,400		2,400	3,625
0160 General purpose laborer, average	1 Clab	.20	40			1,775		1,775	2,650
0180 Project manager, minimum						2,175		2,175	3,300
0200 Average						2,500		2,500	3,800
0220 Maximum						2,850		2,850	4,325
0240 Superintendent, minimum						2,125		2,125	3,225
0260 Average						2,325		2,325	3,525
0280 Maximum						2,650		2,650	4,025
0290 Timekeeper, average						1,350		1,350	2,050

01 31 13.30 Insurance

	Crew	Daily Output	Labor-Hours	Unit	Material	2021 Bare Costs Labor	Equipment	Total	Total Incl O&P
0010 INSURANCE R013113-40									
0020 Builders risk, standard, minimum				Job				.24%	.24%
0050 Maximum R013113-50								.80%	.80%
0200 All-risk type, minimum								.25%	.25%

01 31 Project Management and Coordination

01 31 13 – Project Coordination

		Crew	Daily Output	Labor-Hours	Unit	Material	2021 Bare Costs Labor	Equipment	Total	Total Incl O&P
01 31 13.30 Insurance										
0250	Maximum	R013113-60			Job				.62%	.62%
0400	Contractor's equipment floater, minimum				Value				.50%	.50%
0450	Maximum				"				1.50%	1.50%
0600	Public liability, average				Job				2.02%	2.02%
0800	Workers' compensation & employer's liability, average									
0850	by trade, carpentry, general				Payroll		11.97%			
1000	Electrical						4.91%			
1150	Insulation						10.07%			
1450	Plumbing						5.77%			
1550	Sheet metal work (HVAC)						7.56%			
01 31 13.50 General Contractor's Mark-Up										
0010	**GENERAL CONTRACTOR'S MARK-UP** on Change Orders				%					
0200	Extra work, by subcontractors, add								10%	10%
0250	By General Contractor, add								15%	15%
0400	Omitted work, by subcontractors, deduct all but								5%	5%
0450	By General Contractor, deduct all but								7.50%	7.50%
0600	Overtime work, by subcontractors, add								15%	15%
0650	By General Contractor, add								10%	10%
01 31 13.80 Overhead and Profit										
0010	**OVERHEAD & PROFIT** Allowance to add to items in this	R013113-50			%				25%	
0020	book that do not include Subs O&P, average									
0100	Allowance to add to items in this book that	R013113-55			%					
0110	do include Subs O&P, minimum								5%	5%
0150	Average								10%	10%
0200	Maximum								15%	15%
0300	Typical, by size of project, under $100,000								30%	
0350	$500,000 project								25%	
0400	$2,000,000 project								20%	
0450	Over $10,000,000 project								15%	
01 31 13.90 Performance Bond										
0010	**PERFORMANCE BOND**	R013113-80								
0020	For buildings, minimum				Job				.60%	.60%
0100	Maximum				"				2.50%	2.50%

01 31 14 – Facilities Services Coordination

		Crew	Daily Output	Labor-Hours	Unit	Material	2021 Bare Costs Labor	Equipment	Total	Total Incl O&P
01 31 14.20 Lock Out/Tag Out										
0010	**LOCK OUT/TAG OUT**									
0020	Miniature circuit breaker lock out device	1 Elec	220	.036	Ea.	24	2.32		26.32	30
0030	Miniature pin circuit breaker lock out device		220	.036		19.90	2.32		22.22	25.50
0040	Single circuit breaker lock out device		220	.036		21	2.32		23.32	26.50
0050	Multi-pole circuit breaker lock out device (15 to 225 Amp)		210	.038		19.90	2.43		22.33	25.50
0060	Large 3 pole circuit breaker lock out device (over 225 Amp)		210	.038		21	2.43		23.43	27
0080	Square D I-Line circuit breaker lock out device		210	.038		33	2.43		35.43	39.50
0090	Lock out disconnect switch, 30 to 100 Amp		330	.024		11	1.54		12.54	14.40
0100	100 to 400 Amp		330	.024		11	1.54		12.54	14.40
0110	Over 400 Amp		330	.024		11	1.54		12.54	14.40
0120	Lock out hasp for multiple lockout tags		200	.040		5.60	2.55		8.15	9.95
0130	Electrical cord plug lock out device		220	.036		10.50	2.32		12.82	15
0140	Electrical plug prong lock out device (3-wire grounding plug)		220	.036		6.95	2.32		9.27	11.10
0150	Wall switch lock out		200	.040		25	2.55		27.55	32
0160	Fire alarm pull station lock out	1 Stpi	200	.040		18.75	2.73		21.48	24.50
0170	Sprinkler valve tamper and flow switch lock out device	1 Skwk	220	.036		16.60	2.08		18.68	21.50

For customer support on your Electrical Costs with RSMeans Data, call 800.448.8182.

11

01 31 Project Management and Coordination

01 31 14 – Facilities Services Coordination

01 31 14.20 Lock Out/Tag Out		Crew	Daily Output	Labor-Hours	Unit	Material	2021 Bare Costs Labor	Equipment	Total	Total Incl O&P
0180	Lock out sign	1 Skwk	330	.024	Ea.	18.90	1.38		20.28	23
0190	Lock out tag	↓	440	.018	↓	5	1.04		6.04	7.05

01 32 Construction Progress Documentation

01 32 33 – Photographic Documentation

01 32 33.50 Photographs

		Crew	Daily Output	Labor-Hours	Unit	Material	2021 Bare Costs Labor	Equipment	Total	Total Incl O&P
0010	**PHOTOGRAPHS**									
0020	8" x 10", 4 shots, 2 prints ea., std. mounting				Set	545			545	600
0100	Hinged linen mounts					550			550	605
0200	8" x 10", 4 shots, 2 prints each, in color					575			575	630
0300	For I.D. slugs, add to all above				↓	5.10			5.10	5.60
1500	Time lapse equipment, camera and projector, buy				Ea.	2,600			2,600	2,850
1550	Rent per month				"	1,325			1,325	1,450
1700	Cameraman and processing, black & white				Day	1,300			1,300	1,425
1720	Color				"	1,525			1,525	1,675

01 41 Regulatory Requirements

01 41 26 – Permit Requirements

01 41 26.50 Permits

		Crew	Daily Output	Labor-Hours	Unit	Material	2021 Bare Costs Labor	Equipment	Total	Total Incl O&P
0010	**PERMITS**									
0020	Rule of thumb, most cities, minimum				Job				.50%	.50%
0100	Maximum				"				2%	2%

01 51 Temporary Utilities

01 51 13 – Temporary Electricity

01 51 13.50 Temporary Power Equip (Pro-Rated Per Job)

		Crew	Daily Output	Labor-Hours	Unit	Material	2021 Bare Costs Labor	Equipment	Total	Total Incl O&P
0010	**TEMPORARY POWER EQUIP (PRO-RATED PER JOB)** R015113-65									
0020	Service, overhead feed, 3 use									
0030	100 Amp	1 Elec	1.25	6.400	Ea.	405	410		815	1,050
0040	200 Amp		1	8		505	510		1,015	1,300
0050	400 Amp		.75	10.667		905	680		1,585	2,000
0060	600 Amp	↓	.50	16	↓	1,300	1,025		2,325	2,950
0100	Underground feed, 3 use									
0110	100 Amp	1 Elec	2	4	Ea.	415	255		670	835
0120	200 Amp		1.15	6.957		545	445		990	1,250
0130	400 Amp		1	8		925	510		1,435	1,775
0140	600 Amp		.75	10.667		1,125	680		1,805	2,250
0150	800 Amp		.50	16		1,625	1,025		2,650	3,325
0160	1000 Amp		.35	22.857		1,800	1,450		3,250	4,150
0170	1200 Amp		.25	32		2,000	2,050		4,050	5,225
0180	2000 Amp	↓	.20	40	↓	2,400	2,550		4,950	6,400
0200	Transformers, 3 use									
0210	30 kVA	1 Elec	1	8	Ea.	1,650	510		2,160	2,575
0220	45 kVA		.75	10.667		1,975	680		2,655	3,175
0230	75 kVA		.50	16		3,325	1,025		4,350	5,175
0240	112.5 kVA	↓	.40	20	↓	4,000	1,275		5,275	6,325
0250	Feeder, PVC, CU wire in trench									

12

01 51 13.50 Temporary Power Equip (Pro-Rated Per Job)	Crew	Daily Output	Labor-Hours	Unit	Material	2021 Bare Costs Labor	2021 Bare Costs Equipment	Total	Total Incl O&P
0260	60 Amp	1 Elec	96	.083	L.F.	2.42	5.30		7.72	10.55
0270	100 Amp		85	.094		4.53	6		10.53	13.90
0280	200 Amp		59	.136		10.30	8.65		18.95	24
0290	400 Amp		42	.190		27.50	12.15		39.65	48.50
0300	Feeder, PVC, aluminum wire in trench									
0310	60 Amp	1 Elec	96	.083	L.F.	1.84	5.30		7.14	9.90
0320	100 Amp		85	.094		2.01	6		8.01	11.10
0330	200 Amp		59	.136		5.50	8.65		14.15	18.90
0340	400 Amp		42	.190		8.70	12.15		20.85	27.50
0350	Feeder, EMT, CU wire									
0360	60 Amp	1 Elec	90	.089	L.F.	2.89	5.65		8.54	11.55
0370	100 Amp		80	.100		4.28	6.35		10.63	14.15
0380	200 Amp		60	.133		11.35	8.50		19.85	25
0390	400 Amp		35	.229		31.50	14.55		46.05	56.50
0400	Feeder, EMT, aluminum wire									
0410	60 Amp	1 Elec	90	.089	L.F.	3.48	5.65		9.13	12.25
0420	100 Amp		80	.100		4.51	6.35		10.86	14.40
0430	200 Amp		60	.133		7.95	8.50		16.45	21.50
0440	400 Amp		35	.229		16.40	14.55		30.95	39.50
0500	Equipment, 3 use									
0510	Spider box, 50 Amp	1 Elec	8	1	Ea.	800	63.50		863.50	970
0520	Lighting cord, 100'		8	1		118	63.50		181.50	225
0530	Light stanchion		8	1		138	63.50		201.50	247
0540	Temporary cords, 100', 3 use									
0550	Feeder cord, 50 Amp	1 Elec	16	.500	Ea.	202	32		234	271
0560	Feeder cord, 100 Amp		12	.667		1,225	42.50		1,267.50	1,425
0570	Tap cord, 50 Amp		12	.667		635	42.50		677.50	765
0580	Tap cord, 100 Amp		6	1.333		1,075	85		1,160	1,325
0590	Temporary cords, 50', 3 use									
0600	Feeder cord, 50 Amp	1 Elec	16	.500	Ea.	296	32		328	375
0610	Feeder cord, 100 Amp		12	.667		585	42.50		627.50	710
0620	Tap cord, 50 Amp		12	.667		273	42.50		315.50	365
0630	Tap cord, 100 Amp		6	1.333		590	85		675	775
0700	Connections									
0710	Compressor or pump									
0720	30 Amp	1 Elec	7	1.143	Ea.	17.30	73		90.30	127
0730	60 Amp		5.30	1.509		28	96		124	174
0740	100 Amp		4	2		73.50	127		200.50	270
0750	Tower crane									
0760	60 Amp	1 Elec	4.50	1.778	Ea.	28	113		141	199
0770	100 Amp	"	3	2.667	"	73.50	170		243.50	335
0780	Manlift									
0790	Single	1 Elec	3	2.667	Ea.	32.50	170		202.50	288
0800	Double	"	2	4	"	78	255		333	465
0810	Welder with disconnect									
0820	50 Amp	1 Elec	5	1.600	Ea.	214	102		316	385
0830	100 Amp		3.80	2.105		370	134		504	605
0840	200 Amp		2.50	3.200		680	204		884	1,050
0850	400 Amp		1	8		1,525	510		2,035	2,425
0860	Office trailer									
0870	60 Amp	1 Elec	4.50	1.778	Ea.	57.50	113		170.50	231
0880	100 Amp	1 Elec	3	2.667	Ea.	86.50	170		256.50	345
0890	200 Amp	"	2	4	"	325	255		580	735

01 51 Temporary Utilities

01 51 13 – Temporary Electricity

01 51 13.80 Temporary Utilities	Crew	Daily Output	Labor-Hours	Unit	Material	2021 Bare Costs Labor	2021 Bare Costs Equipment	Total	Total Incl O&P
0010 TEMPORARY UTILITIES									
0350 Lighting, lamps, wiring, outlets, 40,000 S.F. building, 8 strings	1 Elec	34	.235	CSF Flr	3.55	15		18.55	26.50
0360 16 strings	"	17	.471		7.10	30		37.10	52.50
0400 Power for temp lighting only, 6.6 KWH, per month								.92	1.01
0430 11.8 KWH, per month								1.65	1.82
0450 23.6 KWH, per month								3.30	3.63
0600 Power for job duration incl. elevator, etc., minimum								53	58
0650 Maximum								110	121
0675 Temporary cooling				Ea.	1,050			1,050	1,150

01 52 Construction Facilities

01 52 13 – Field Offices and Sheds

01 52 13.20 Office and Storage Space

	Crew	Daily Output	Labor-Hours	Unit	Material	2021 Bare Costs Labor	2021 Bare Costs Equipment	Total	Total Incl O&P
0010 OFFICE AND STORAGE SPACE									
0020 Office trailer, furnished, no hookups, 20' x 8', buy	2 Skwk	1	16	Ea.	10,000	915		10,915	12,400
0250 Rent per month					192			192	211
0300 32' x 8', buy	2 Skwk	.70	22.857		15,000	1,300		16,300	18,500
0350 Rent per month					243			243	267
0400 50' x 10', buy	2 Skwk	.60	26.667		30,500	1,525		32,025	35,800
0450 Rent per month					355			355	390
0500 50' x 12', buy	2 Skwk	.50	32		25,800	1,825		27,625	31,200
0550 Rent per month					450			450	500
0700 For air conditioning, rent per month, add					53			53	58.50
0800 For delivery, add per mile				Mile	11.95			11.95	13.15
0900 Bunk house trailer, 8' x 40' duplex dorm with kitchen, no hookups, buy	2 Carp	1	16	Ea.	89,000	875		89,875	99,500
0910 9 man with kitchen and bath, no hookups, buy		1	16		91,500	875		92,375	102,000
0920 18 man sleeper with bath, no hookups, buy		1	16		98,500	875		99,375	110,000
1000 Portable buildings, prefab, on skids, economy, 8' x 8'		265	.060	S.F.	29.50	3.30		32.80	37.50
1100 Deluxe, 8' x 12'		150	.107	"	33.50	5.85		39.35	45.50
1200 Storage boxes, 20' x 8', buy	2 Skwk	1.80	8.889	Ea.	3,100	510		3,610	4,200
1250 Rent per month					88			88	96.50
1300 40' x 8', buy	2 Skwk	1.40	11.429		4,600	655		5,255	6,050
1350 Rent per month					135			135	148

01 52 13.40 Field Office Expense

	Crew	Daily Output	Labor-Hours	Unit	Material	2021 Bare Costs Labor	2021 Bare Costs Equipment	Total	Total Incl O&P
0010 FIELD OFFICE EXPENSE									
0100 Office equipment rental average				Month	228			228	251
0120 Office supplies, average				"	90.50			90.50	99.50
0125 Office trailer rental, see Section 01 52 13.20									
0140 Telephone bill; avg. bill/month incl. long dist.				Month	87			87	96
0160 Lights & HVAC				Month	163			163	179

14

01 54 Construction Aids

01 54 16 – Temporary Hoists

01 54 16.50 Weekly Forklift Crew		Crew	Daily Output	Labor-Hours	Unit	Material	2021 Bare Costs Labor	Equipment	Total	Total Incl O&P
0010	**WEEKLY FORKLIFT CREW**									
0100	All-terrain forklift, 45' lift, 35' reach, 9000 lb. capacity	A-3P	.20	40	Week		2,225	1,725	3,950	5,200

01 54 19 – Temporary Cranes

01 54 19.50 Daily Crane Crews

		Crew	Daily Output	Labor-Hours	Unit	Material	Labor	Equipment	Total	Total Incl O&P
0010	**DAILY CRANE CREWS** for small jobs, portal to portal R015433-15									
0100	12-ton truck-mounted hydraulic crane	A-3H	1	8	Day		490	735	1,225	1,525
0200	25-ton	A-3I	1	8			490	810	1,300	1,625
0300	40-ton	A-3J	1	8			490	1,275	1,765	2,150
0400	55-ton	A-3K	1	16			910	1,525	2,435	3,025
0500	80-ton	A-3L	1	16			910	2,200	3,110	3,775
0900	If crane is needed on a Saturday, Sunday or Holiday									
0910	At time-and-a-half, add				Day		50%			
0920	At double time, add				"		100%			

01 54 19.60 Monthly Tower Crane Crew

		Crew	Daily Output	Labor-Hours	Unit	Material	Labor	Equipment	Total	Total Incl O&P
0010	**MONTHLY TOWER CRANE CREW**, excludes concrete footing									
0100	Static tower crane, 130' high, 106' jib, 6200 lb. capacity	A-3N	.05	176	Month	10,800	38,200		49,000	58,000

01 54 23 – Temporary Scaffolding and Platforms

01 54 23.70 Scaffolding

		Crew	Daily Output	Labor-Hours	Unit	Material	Labor	Equipment	Total	Total Incl O&P
0010	**SCAFFOLDING** R015423-10									
0015	Steel tube, regular, no plank, labor only to erect & dismantle									
0090	Building exterior, wall face, 1 to 5 stories, 6'-4" x 5' frames	3 Carp	8	3	C.S.F.		164		164	245
0200	6 to 12 stories	4 Carp	8	4			219		219	325
0301	13 to 20 stories	5 Clab	8	5			222		222	330
0460	Building interior, wall face area, up to 16' high	3 Carp	12	2			109		109	163
0560	16' to 40' high		10	2.400			131		131	196
0800	Building interior floor area, up to 30' high		150	.160	C.C.F.		8.75		8.75	13.05
0900	Over 30' high	4 Carp	160	.200	"		10.95		10.95	16.35
0906	Complete system for face of walls, no plank, material only rent/mo				C.S.F.	81.50			81.50	90
0908	Interior spaces, no plank, material only rent/mo				C.C.F.	5.60			5.60	6.15
0910	Steel tubular, heavy duty shoring, buy									
0920	Frames 5' high 2' wide				Ea.	92.50			92.50	102
0925	5' high 4' wide					118			118	130
0930	6' high 2' wide					120			120	132
0935	6' high 4' wide					119			119	131
0940	Accessories									
0945	Cross braces				Ea.	19.20			19.20	21
0950	U-head, 8" x 8"					23			23	25
0955	J-head, 4" x 8"					16.70			16.70	18.35
0960	Base plate, 8" x 8"				Ea.	18.05			18.05	19.90
0965	Leveling jack				"	36			36	39.50
1000	Steel tubular, regular, buy									
1100	Frames 3' high 5' wide				Ea.	54			54	59.50
1150	5' high 5' wide					71			71	78
1200	6'-4" high 5' wide					83.50			83.50	92
1350	7'-6" high 6' wide					156			156	171
1500	Accessories, cross braces					19.20			19.20	21
1550	Guardrail post					24			24	26.50
1600	Guardrail 7' section					8.15			8.15	9
1650	Screw jacks & plates					26			26	28.50
1700	Sidearm brackets					26			26	29
1750	8" casters					37.50			37.50	41

For customer support on your Electrical Costs with RSMeans Data, call 800.448.8182.

15

01 54 23 – Temporary Scaffolding and Platforms

01 54 23.70 Scaffolding

		Crew	Daily Output	Labor-Hours	Unit	Material	2021 Bare Costs Labor	Equipment	Total	Total Incl O&P
1800	Plank 2" x 10" x 16'-0"				Ea.	66.50			66.50	73.50
1900	Stairway section					298			298	330
1910	Stairway starter bar					32.50			32.50	35.50
1920	Stairway inside handrail					57.50			57.50	63.50
1930	Stairway outside handrail					91.50			91.50	101
1940	Walk-thru frame guardrail					42			42	46.50
2000	Steel tubular, regular, rent/mo.									
2100	Frames 3' high 5' wide				Ea.	9.90			9.90	10.90
2150	5' high 5' wide					9.90			9.90	10.90
2200	6'-4" high 5' wide					5.75			5.75	6.35
2250	7'-6" high 6' wide					9.90			9.90	10.90
2500	Accessories, cross braces					3.35			3.35	3.69
2550	Guardrail post					3.25			3.25	3.58
2600	Guardrail 7' section					11.10			11.10	12.20
2650	Screw jacks & plates					4.35			4.35	4.79
2700	Sidearm brackets					5.20			5.20	5.70
2750	8" casters					7.85			7.85	8.65
2800	Outrigger for rolling tower					9.10			9.10	10
2850	Plank 2" x 10" x 16'-0"					9.90			9.90	10.90
2900	Stairway section					32			32	35.50
2940	Walk-thru frame guardrail					8.25			8.25	9.10
3000	Steel tubular, heavy duty shoring, rent/mo.									
3250	5' high 2' & 4' wide				Ea.	8.45			8.45	9.25
3300	6' high 2' & 4' wide					8.45			8.45	9.25
3500	Accessories, cross braces					.91			.91	1
3600	U-head, 8" x 8"					2.49			2.49	2.74
3650	J-head, 4" x 8"					2.49			2.49	2.74
3700	Base plate, 8" x 8"				Ea.	.91			.91	1
3750	Leveling jack					2.47			2.47	2.72
5700	Planks, 2" x 10" x 16'-0", labor only to erect & remove to 50' H	3 Carp	72	.333			18.25		18.25	27
5800	Over 50' high	4 Carp	80	.400			22		22	32.50

01 54 23.75 Scaffolding Specialties

		Crew	Daily Output	Labor-Hours	Unit	Material	2021 Bare Costs Labor	Equipment	Total	Total Incl O&P
0010	**SCAFFOLDING SPECIALTIES**									
1500	Sidewalk bridge using tubular steel scaffold frames including									
1510	planking (material cost is rent/month)	3 Carp	45	.533	L.F.	9.80	29		38.80	54.50
1600	For 2 uses per month, deduct from all above					50%				
1700	For 1 use every 2 months, add to all above					100%				
1900	Catwalks, 20" wide, no guardrails, 7' span, buy				Ea.	148			148	163
2000	10' span, buy					207			207	228
3720	Putlog, standard, 8' span, with hangers, buy					78.50			78.50	86
3750	12' span, buy					100			100	110
3760	Trussed type, 16' span, buy					254			254	280
3790	22' span, buy					287			287	315
3795	Rent per month					32.50			32.50	35.50
3800	Rolling ladders with handrails, 30" wide, buy, 2 step					289			289	320
4000	7 step					1,125			1,125	1,225
4050	10 step					1,375			1,375	1,500
4100	Rolling towers, buy, 5' wide, 7' long, 10' high					1,250			1,250	1,375
4200	For additional 5' high sections, to buy					180			180	198
4300	Complete incl. wheels, railings, outriggers,									
4350	21' high, to buy				Ea.	2,050			2,050	2,250
4400	Rent/month = 5% of purchase cost				"	103			103	113

01 54 Construction Aids

01 54 26 – Temporary Swing Staging

01 54 26.50 Swing Staging	Crew	Daily Output	Labor-Hours	Unit	Material	2021 Bare Costs Labor	Equipment	Total	Total Incl O&P
0010 **SWING STAGING**, 500 lb. cap., 2' wide to 24' long, hand operated									
0020 Steel cable type, with 60' cables, buy				Ea.	6,425			6,425	7,075
0030 Rent per month				"	645			645	705
2200 Move swing staging (setup and remove)	E-4	2	16	Move		975	74.50	1,049.50	1,575

01 54 36 – Equipment Mobilization

01 54 36.50 Mobilization	Crew	Daily Output	Labor-Hours	Unit	Material	2021 Bare Costs Labor	Equipment	Total	Total Incl O&P
0010 **MOBILIZATION** (Use line item again for demobilization) R015436-50									
0015 Up to 25 mi. haul dist. (50 mi. RT for mob/demob crew)									
1200 Small equipment, placed in rear of, or towed by pickup truck	A-3A	4	2	Ea.		111	44	155	214
1300 Equipment hauled on 3-ton capacity towed trailer	A-3Q	2.67	3			167	93	260	350
1400 20-ton capacity	B-34U	2	8			425	224	649	885
1500 40-ton capacity	B-34N	2	8			440	345	785	1,050
1600 50-ton capacity	B-34V	1	24	↓		1,350	995	2,345	3,100
1700 Crane, truck-mounted, up to 75 ton (driver only)	1 Eqhv	4	2	Ea.		123		123	183
1800 Over 75 ton (with chase vehicle)	A-3E	2.50	6.400			360	70.50	430.50	620
2400 Crane, large lattice boom, requiring assembly	B-34W	.50	144	↓		7,775	7,075	14,850	19,400
2500 For each additional 5 miles haul distance, add						10%	10%		
3000 For large pieces of equipment, allow for assembly/knockdown									
3100 For mob/demob of micro-tunneling equip, see Section 33 05 07.36									

01 55 Vehicular Access and Parking

01 55 23 – Temporary Roads

01 55 23.50 Roads and Sidewalks	Crew	Daily Output	Labor-Hours	Unit	Material	2021 Bare Costs Labor	Equipment	Total	Total Incl O&P
0010 **ROADS AND SIDEWALKS** Temporary									
0050 Roads, gravel fill, no surfacing, 4" gravel depth	B-14	715	.067	S.Y.	5.35	3.13	.30	8.78	10.85
0100 8" gravel depth	"	615	.078	"	10.65	3.64	.35	14.64	17.55
1000 Ramp, 3/4" plywood on 2" x 6" joists, 16" OC	2 Carp	300	.053	S.F.	2.44	2.92		5.36	7.05
1100 On 2" x 10" joists, 16" OC	"	275	.058	"	3.39	3.18		6.57	8.50

01 56 Temporary Barriers and Enclosures

01 56 13 – Temporary Air Barriers

01 56 13.90 Winter Protection	Crew	Daily Output	Labor-Hours	Unit	Material	2021 Bare Costs Labor	Equipment	Total	Total Incl O&P
0010 **WINTER PROTECTION**									
0100 Framing to close openings	2 Clab	500	.032	S.F.	.77	1.42		2.19	2.97
0200 Tarpaulins hung over scaffolding, 8 uses, not incl. scaffolding		1500	.011		.25	.47		.72	.99
0300 Prefab fiberglass panels, steel frame, 8 uses	↓	1200	.013	↓	3.06	.59		3.65	4.25

01 56 16 – Temporary Dust Barriers

01 56 16.10 Dust Barriers, Temporary	Crew	Daily Output	Labor-Hours	Unit	Material	2021 Bare Costs Labor	Equipment	Total	Total Incl O&P
0010 **DUST BARRIERS, TEMPORARY**, erect and dismantle									
0020 Spring loaded telescoping pole & head, to 12', erect and dismantle	1 Clab	240	.033	Ea.		1.48		1.48	2.21
0025 Cost per day (based upon 250 days)				Day	.22			.22	.25
0030 To 21', erect and dismantle	1 Clab	240	.033	Ea.		1.48		1.48	2.21
0035 Cost per day (based upon 250 days)				Day	.73			.73	.81
0040 Accessories, caution tape reel, erect and dismantle	1 Clab	480	.017	Ea.		.74		.74	1.10
0045 Cost per day (based upon 250 days)				Day	.34			.34	.38
0060 Foam rail and connector, erect and dismantle	1 Clab	240	.033	Ea.		1.48		1.48	2.21
0065 Cost per day (based upon 250 days)				Day	.10			.10	.11
0070 Caution tape	1 Clab	384	.021	C.L.F.	2.55	.93		3.48	4.18

01 56 16.10 Dust Barriers, Temporary

		Crew	Daily Output	Labor-Hours	Unit	Material	2021 Bare Costs Labor	Equipment	Total	Total Incl O&P
0080	Zipper, standard duty	1 Clab	60	.133	Ea.	10.75	5.90		16.65	20.50
0090	Heavy duty		48	.167	"	10.85	7.40		18.25	23
0100	Polyethylene sheet, 4 mil		37	.216	Sq.	2.70	9.60		12.30	17.25
0110	6 mil		37	.216	"	7.65	9.60		17.25	23
1000	Dust partition, 6 mil polyethylene, 1" x 3" frame	2 Carp	2000	.008	S.F.	.49	.44		.93	1.19
1080	2" x 4" frame	"	2000	.008	"	.60	.44		1.04	1.31
1085	Negative air machine, 1800 CFM				Ea.	900			900	990
1090	Adhesive strip application, 2" width	1 Clab	192	.042	C.L.F.	10.65	1.85		12.50	14.50
4000	Dust & infectious control partition, adj. to 10' high, obscured, 4' panel	2 Carp	90	.178	Ea.	580	9.70		589.70	655
4010	3' panel		90	.178		550	9.70		559.70	620
4020	2' panel		90	.178		430	9.70		439.70	490
4030	1' panel	1 Carp	90	.089		285	4.86		289.86	320
4040	6" panel	"	90	.089		265	4.86		269.86	299
4050	2' panel with HEPA filtered discharge port	2 Carp	90	.178		500	9.70		509.70	565
4060	3' panel with 32" door		90	.178		895	9.70		904.70	1,000
4070	4' panel with 36" door		90	.178		995	9.70		1,004.70	1,125
4080	4'-6" panel with 44" door		90	.178		1,200	9.70		1,209.70	1,350
4090	Hinged corner		80	.200		185	10.95		195.95	220
4100	Outside corner		80	.200		150	10.95		160.95	181
4110	T post		80	.200		150	10.95		160.95	181
4120	Accessories, ceiling grid clip		360	.044		7.45	2.43		9.88	11.85
4130	Panel locking clip		360	.044		5.15	2.43		7.58	9.30
4140	Panel joint closure strip		360	.044		8	2.43		10.43	12.45
4150	Screw jack		360	.044		6.65	2.43		9.08	10.95
4160	Digital pressure difference gauge					275			275	305
4180	Combination lockset	1 Carp	13	.615		200	33.50		233.50	271
4185	Sealant tape, 2" wide	1 Clab	192	.042	C.L.F.	10.65	1.85		12.50	14.50
4190	System in place, including door and accessories									
4200	Based upon 25 uses	2 Carp	51	.314	L.F.	9.85	17.15		27	36.50
4210	Based upon 50 uses		51	.314		4.94	17.15		22.09	31
4230	Based upon 100 uses		51	.314		2.47	17.15		19.62	28

01 56 23 – Temporary Barricades

01 56 23.10 Barricades

		Crew	Daily Output	Labor-Hours	Unit	Material	2021 Bare Costs Labor	Equipment	Total	Total Incl O&P
0010	**BARRICADES**									
0020	5' high, 3 rail @ 2" x 8", fixed	2 Carp	20	.800	L.F.	9.45	44		53.45	76
0150	Movable		30	.533		8.15	29		37.15	52.50
1000	Guardrail, wooden, 3' high, 1" x 6" on 2" x 4" posts		200	.080		2.05	4.38		6.43	8.80
1100	2" x 6" on 4" x 4" posts		165	.097		3.46	5.30		8.76	11.70
1200	Portable metal with base pads, buy					14.35			14.35	15.80
1250	Typical installation, assume 10 reuses	2 Carp	600	.027		2.19	1.46		3.65	4.59
1300	Barricade tape, polyethylene, 7 mil, 3" wide x 300' long roll	1 Clab	128	.063	Ea.	7.65	2.78		10.43	12.55
3000	Detour signs, set up and remove									
3010	Reflective aluminum, MUTCD, 24" x 24", post mounted	1 Clab	20	.400	Ea.	3.46	17.75		21.21	30.50
4000	Roof edge portable barrier stands and warning flags, 50 uses	1 Rohe	9100	.001	L.F.	.08	.03		.11	.14
4010	100 uses	"	9100	.001	"	.04	.03		.07	.09

01 56 26 – Temporary Fencing

01 56 26.50 Temporary Fencing

		Crew	Daily Output	Labor-Hours	Unit	Material	2021 Bare Costs Labor	Equipment	Total	Total Incl O&P
0010	**TEMPORARY FENCING**									
0020	Chain link, 11 ga., 4' high	2 Clab	400	.040	L.F.	3.05	1.78		4.83	6
0100	6' high		300	.053		5.80	2.37		8.17	9.90
0200	Rented chain link, 6' high, to 1000' (up to 12 mo.)		400	.040		3.44	1.78		5.22	6.45
0250	Over 1000' (up to 12 mo.)		300	.053		4.07	2.37		6.44	8

For customer support on your Electrical Costs with RSMeans Data, call 800.448.8182.

01 56 Temporary Barriers and Enclosures

01 56 26 – Temporary Fencing

01 56 26.50 Temporary Fencing	Crew	Daily Output	Labor-Hours	Unit	Material	2021 Bare Costs Labor	2021 Bare Costs Equipment	Total	Total Incl O&P
0350	Plywood, painted, 2" x 4" frame, 4' high	A-4	135	.178	L.F.	9.45	9.25		18.70	24
0400	4" x 4" frame, 8' high	"	110	.218		17.05	11.35		28.40	35.50
0500	Wire mesh on 4" x 4" posts, 4' high	2 Carp	100	.160		12.25	8.75		21	26.50
0550	8' high	"	80	.200		18	10.95		28.95	36

01 56 29 – Temporary Protective Walkways

01 56 29.50 Protection

		Crew	Daily Output	Labor-Hours	Unit	Material	Labor	Equipment	Total	Total Incl O&P
0010	**PROTECTION**									
0020	Stair tread, 2" x 12" planks, 1 use	1 Carp	75	.107	Tread	7.50	5.85		13.35	16.95
0100	Exterior plywood, 1/2" thick, 1 use		65	.123		3.01	6.75		9.76	13.35
0200	3/4" thick, 1 use		60	.133		4.15	7.30		11.45	15.45
2200	Sidewalks, 2" x 12" planks, 2 uses		350	.023	S.F.	1.25	1.25		2.50	3.24
2300	Exterior plywood, 2 uses, 1/2" thick		750	.011		.50	.58		1.08	1.42
2400	5/8" thick		650	.012		.58	.67		1.25	1.64
2500	3/4" thick		600	.013		.69	.73		1.42	1.85

01 58 Project Identification

01 58 13 – Temporary Project Signage

01 58 13.50 Signs

		Crew	Daily Output	Labor-Hours	Unit	Material	Labor	Equipment	Total	Total Incl O&P
0010	**SIGNS**									
0020	High intensity reflectorized, no posts, buy				Ea.	22			22	24

01 66 Product Storage and Handling Requirements

01 66 19 – Material Handling

01 66 19.10 Material Handling

		Crew	Daily Output	Labor-Hours	Unit	Material	Labor	Equipment	Total	Total Incl O&P
0010	**MATERIAL HANDLING**									
0020	Above 2nd story, via stairs, per C.Y. of material per floor	2 Clab	145	.110	C.Y.		4.90		4.90	7.30
0030	Via elevator, per C.Y. of material		240	.067			2.96		2.96	4.42
0050	Distances greater than 200', per C.Y. of material per each addl 200'		300	.053			2.37		2.37	3.53

01 74 Cleaning and Waste Management

01 74 13 – Progress Cleaning

01 74 13.20 Cleaning Up

		Crew	Daily Output	Labor-Hours	Unit	Material	Labor	Equipment	Total	Total Incl O&P
0010	**CLEANING UP**									
0020	After job completion, allow, minimum				Job				.30%	.30%
0040	Maximum				"				1%	1%
0050	Cleanup of floor area, continuous, per day, during const.	A-5	24	.750	M.S.F.	2.77	33.50	2.07	38.34	56
0100	Final by GC at end of job	"	11.50	1.565	"	2.93	70.50	4.32	77.75	113

01 91 Commissioning

01 91 13 – General Commissioning Requirements

01 91 13.50 Building Commissioning	Crew	Daily Output	Labor-Hours	Unit	Material	2021 Bare Costs Labor	2021 Bare Costs Equipment	Total	Total Incl O&P
0010 **BUILDING COMMISSIONING**									
0100 Systems operation and verification during turnover				%				.25%	.25%
0150 Including all systems subcontractors				%				.50%	.50%
0200 Systems design assistance, operation, verification and training								.50%	.50%
0250 Including all systems subcontractors				↓				1%	1%

01 93 Facility Maintenance

01 93 13 – Facility Maintenance Procedures

01 93 13.16 Electrical Facilities Maintenance

		Crew	Daily Output	Labor-Hours	Unit	Material	2021 Bare Costs Labor	2021 Bare Costs Equipment	Total	Total Incl O&P
0010	**ELECTRICAL FACILITIES MAINTENANCE**									
0700	Cathodic protection systems									
0720	Check and adjust reading on rectifier	1 Elec	20	.400	Ea.		25.50		25.50	38
0730	Check pipe to soil potential		20	.400			25.50		25.50	38
0740	Replace lead connection		4	2			127		127	189
0800	Control device, install		5.70	1.404			89.50		89.50	133
0810	Disassemble, clean and reinstall		7	1.143			73		73	108
0820	Replace		10.70	.748			47.50		47.50	71
0830	Trouble shoot	↓	10	.800	↓		51		51	75.50
0900	Demolition, for electrical demolition see Section 26 05 05.10									
1000	Distribution systems and equipment install or repair a breaker									
1010	In power panels up to 200 amps	1 Elec	7	1.143	Ea.		73		73	108
1020	Over 200 amps		2	4			255		255	380
1030	Reset breaker or replace fuse		20	.400			25.50		25.50	38
1100	Megger test MCC (each stack)		4	2			127		127	189
1110	MCC vacuum and clean (each stack)		5.30	1.509			96		96	143
2500	Remove and replace or maint., road fixture & lamp		3	2.667		1,225	170		1,395	1,600
2510	Fluorescent fixture		7	1.143		93	73		166	210
2515	Relamp (fluor.) facility area each tube, spot	1 Clab	24	.333		7.10	14.80		21.90	30
2516	Group		100	.080		7.10	3.55		10.65	13.10
2518	Fluorescent fixture, clean (area)	↓	44	.182		.04	8.05		8.09	12.10
2520	Incandescent fixture	1 Elec	11	.727		51.50	46.50		98	126
2530	Lamp (incandescent or fluorescent)	1 Clab	60	.133		7.10	5.90		13	16.65
2535	Replace cord in socket lamp	1 Elec	13	.615		1.67	39		40.67	60.50
2540	Ballast electronic type for two tubes		8	1		37.50	63.50		101	136
2541	Starter		30	.267		3.06	17		20.06	28.50
2545	Replace other lighting parts		11	.727		13.85	46.50		60.35	84
2550	Switch		11	.727		4.72	46.50		51.22	74
2555	Receptacle		11	.727		11	46.50		57.50	81
2560	Floodlight		4	2		246	127		373	460
2570	Christmas lighting, indoor, per string	1 Clab	16	.500			22		22	33
2580	Outdoor		13	.615			27.50		27.50	41
2590	Test battery operated emergency lights	↓	40	.200	↓		8.90		8.90	13.25
2600	Repair and replace component in communication system	1 Elec	6	1.333	Ea.	53.50	85		138.50	185
2700	Repair misc. appliances (incl. clocks, vent fan, blower, etc.)	"	6	1.333			85		85	126
2710	Reset clocks & timers	1 Clab	50	.160			7.10		7.10	10.60
2720	Adjust time delay relays	1 Elec	16	.500			32		32	47.50
2730	Test specific gravity of lead-acid batteries	1 Clab	80	.100	↓		4.44		4.44	6.65
3000	Motors and generators									
3020	Disassemble, clean and reinstall motor, up to 1/4 HP	1 Elec	4	2	Ea.		127		127	189
3030	Up to 3/4 HP		3	2.667			170		170	252
3040	Up to 10 HP	↓	2	4	↓		255		255	380

01 93 Facility Maintenance

01 93 13 – Facility Maintenance Procedures

01 93 13.16 Electrical Facilities Maintenance		Crew	Daily Output	Labor-Hours	Unit	Material	2021 Bare Costs			Total Incl O&P
							Labor	Equipment	Total	
3050	Replace part, up to 1/4 HP	1 Elec	6	1.333	Ea.		85		85	126
3060	Up to 3/4 HP		4	2			127		127	189
3070	Up to 10 HP		3	2.667			170		170	252
3080	Megger test motor windings		5.33	1.501			95.50		95.50	142
3082	Motor vibration check		16	.500			32		32	47.50
3084	Oil motor bearings		25	.320			20.50		20.50	30.50
3086	Run test emergency generator for 30 minutes		11	.727			46.50		46.50	69
3090	Rewind motor, up to 1/4 HP		3	2.667			170		170	252
3100	Up to 3/4 HP		2	4			255		255	380
3110	Up to 10 HP		1.50	5.333			340		340	505
3150	Generator, repair or replace part		4	2			127		127	189
3160	Repair DC generator		2	4			255		255	380
4000	Stub pole, install or remove		3	2.667			170		170	252
4500	Transformer maintenance up to 15 kVA		2.70	2.963			189		189	280

For customer support on your Electrical Costs with RSMeans Data, call 800.448.8182.

Division Notes

		CREW	DAILY OUTPUT	LABOR-HOURS	UNIT	BARE COSTS				TOTAL INCL O&P
						MAT.	LABOR	EQUIP.	TOTAL	

Estimating Tips
02 30 00 Subsurface Investigation
In preparing estimates on structures involving earthwork or foundations, all information concerning soil characteristics should be obtained. Look particularly for hazardous waste, evidence of prior dumping of debris, and previous stream beds.

02 40 00 Demolition and Structure Moving
The costs shown for selective demolition do not include rubbish handling or disposal. These items should be estimated separately using RSMeans data or other sources.

- Historic preservation often requires that the contractor remove materials from the existing structure, rehab them, and replace them. The estimator must be aware of any related measures and precautions that must be taken when doing selective demolition and cutting and patching. Requirements may include special handling and storage, as well as security.

- In addition to Subdivision 02 41 00, you can find selective demolition items in each division. Example: Roofing demolition is in Division 7.
- Absent of any other specific reference, an approximate demolish-in-place cost can be obtained by halving the new-install labor cost. To remove for reuse, allow the entire new-install labor figure.

02 40 00 Building Deconstruction
This section provides costs for the careful dismantling and recycling of most low-rise building materials.

02 50 00 Containment of Hazardous Waste
This section addresses on-site hazardous waste disposal costs.

02 80 00 Hazardous Material Disposal/Remediation
This subdivision includes information on hazardous waste handling, asbestos remediation, lead remediation, and mold remediation. See reference numbers

R028213-20 and R028319-60 for further guidance in using these unit price lines.

02 90 00 Monitoring Chemical Sampling, Testing Analysis
This section provides costs for on-site sampling and testing hazardous waste.

Reference Numbers
Reference numbers are shown at the beginning of some major classifications. These numbers refer to related items in the Reference Section. The reference information may be an estimating procedure, an alternate pricing method, or technical information.

Note: Not all subdivisions listed here necessarily appear. ■

Same Data. Simplified.

Enjoy the convenience and efficiency of accessing your costs anywhere:

- **Skip the multiplier** by setting your location
- **Quickly search,** edit, favorite and share costs
- **Stay on top of price changes** with automatic updates

Discover more at rsmeans.com/online

02 31 Geophysical Investigations

02 31 23 – Electromagnetic Investigations

02 31 23.10 Ground Penetrating Radar	Crew	Daily Output	Labor-Hours	Unit	Material	2021 Bare Costs Labor	Equipment	Total	Total Incl O&P
0010 **GROUND PENETRATING RADAR**									
0100 Ground Penetrating Radar				Hr.		210		210	235

02 41 Demolition

02 41 13 – Selective Site Demolition

02 41 13.17 Demolish, Remove Pavement and Curb

	Crew	Daily Output	Labor-Hours	Unit	Material	2021 Bare Costs Labor	Equipment	Total	Total Incl O&P
0010 **DEMOLISH, REMOVE PAVEMENT AND CURB** R024119-10									
5010 Pavement removal, bituminous roads, up to 3" thick	B-38	690	.058	S.Y.		2.90	1.76	4.66	6.25
5050 4"-6" thick		420	.095			4.76	2.90	7.66	10.30
5100 Bituminous driveways		640	.063			3.12	1.90	5.02	6.75
5200 Concrete to 6" thick, hydraulic hammer, mesh reinforced		255	.157			7.85	4.77	12.62	16.95
5300 Rod reinforced		200	.200			10	6.10	16.10	21.50
5400 Concrete, 7"-24" thick, plain		33	1.212	C.Y.		60.50	37	97.50	131
5500 Reinforced		24	1.667	"		83	50.50	133.50	180
5600 With hand held air equipment, bituminous, to 6" thick	B-39	1900	.025	S.F.		1.18	.19	1.37	1.97
5700 Concrete to 6" thick, no reinforcing		1600	.030			1.40	.22	1.62	2.32
5800 Mesh reinforced		1400	.034			1.60	.25	1.85	2.66
5900 Rod reinforced		765	.063			2.92	.46	3.38	4.87

02 41 13.23 Utility Line Removal

	Crew	Daily Output	Labor-Hours	Unit	Material	2021 Bare Costs Labor	Equipment	Total	Total Incl O&P
0010 **UTILITY LINE REMOVAL**									
0015 No hauling, abandon catch basin or manhole	B-6	7	3.429	Ea.		165	31	196	280
0020 Remove existing catch basin or manhole, masonry		4	6			289	54	343	490
0030 Catch basin or manhole frames and covers, stored		13	1.846			89	16.65	105.65	150
0040 Remove and reset		7	3.429			165	31	196	280

02 41 13.30 Minor Site Demolition

	Crew	Daily Output	Labor-Hours	Unit	Material	2021 Bare Costs Labor	Equipment	Total	Total Incl O&P
0010 **MINOR SITE DEMOLITION** R024119-10									
4000 Sidewalk removal, bituminous, 2" thick	B-6	350	.069	S.Y.		3.30	.62	3.92	5.60
4010 2-1/2" thick		325	.074			3.55	.67	4.22	6.05
4050 Brick, set in mortar		185	.130			6.25	1.17	7.42	10.60
4100 Concrete, plain, 4"		160	.150			7.20	1.35	8.55	12.25
4110 Plain, 5"		140	.171			8.25	1.54	9.79	14
4120 Plain, 6"		120	.200			9.60	1.80	11.40	16.35
4200 Mesh reinforced, concrete, 4"		150	.160			7.70	1.44	9.14	13.10
4210 5" thick		131	.183			8.80	1.65	10.45	14.95
4220 6" thick		112	.214			10.30	1.93	12.23	17.45

02 41 19 – Selective Demolition

02 41 19.13 Selective Building Demolition

	Crew	Daily Output	Labor-Hours	Unit	Material	2021 Bare Costs Labor	Equipment	Total	Total Incl O&P
0010 **SELECTIVE BUILDING DEMOLITION**									
0020 Costs related to selective demolition of specific building components									
0025 are included under Common Work Results (XX 05)									
0030 in the component's appropriate division.									

02 41 19.16 Selective Demolition, Cutout

	Crew	Daily Output	Labor-Hours	Unit	Material	2021 Bare Costs Labor	Equipment	Total	Total Incl O&P
0010 **SELECTIVE DEMOLITION, CUTOUT** R024119-10									
0020 Concrete, elev. slab, light reinforcement, under 6 C.F.	B-9	65	.615	C.F.		27.50	5.45	32.95	47
0050 Light reinforcing, over 6 C.F.	"	75	.533	"		24	4.74	28.74	40.50
6000 Walls, interior, not including re-framing,									
6010 openings to 5 S.F.									
6100 Drywall to 5/8" thick	1 Clab	24	.333	Ea.		14.80		14.80	22
6200 Paneling to 3/4" thick		20	.400			17.75		17.75	26.50
6300 Plaster, on gypsum lath		20	.400			17.75		17.75	26.50

02 41 Demolition

02 41 19 – Selective Demolition

02 41 19.16 Selective Demolition, Cutout

		Crew	Daily Output	Labor-Hours	Unit	Material	2021 Bare Costs Labor	Equipment	Total	Total Incl O&P
6340	On wire lath	1 Clab	14	.571	Ea.		25.50		25.50	38

02 41 19.19 Selective Demolition

		Crew	Daily Output	Labor-Hours	Unit	Material	2021 Bare Costs Labor	Equipment	Total	Total Incl O&P
0010	**SELECTIVE DEMOLITION**, Rubbish Handling R024119-10									
0020	The following are to be added to the demolition prices									
0600	Dumpster, weekly rental, 1 dump/week, 6 C.Y. capacity (2 tons)				Week	415			415	455
0700	10 C.Y. capacity (3 tons)					480			480	530
0725	20 C.Y. capacity (5 tons) R024119-20					565			565	625
0800	30 C.Y. capacity (7 tons)					730			730	800
0840	40 C.Y. capacity (10 tons)					775			775	850
2000	Load, haul, dump and return, 0'-50' haul, hand carried	2 Clab	24	.667	C.Y.		29.50		29.50	44
2005	Wheeled	2 Clab	37	.432	C.Y.		19.20		19.20	28.50
2040	0'-100' haul, hand carried		16.50	.970			43		43	64
2045	Wheeled		25	.640			28.50		28.50	42.50
2050	Forklift	A-3R	25	.320			17.75	11.35	29.10	39
2080	Haul and return, add per each extra 100' haul, hand carried	2 Clab	35.50	.451			20		20	30
2085	Wheeled		54	.296			13.15		13.15	19.65
2120	For travel in elevators, up to 10 floors, add		140	.114			5.05		5.05	7.55
2130	0'-50' haul, incl. up to 5 riser stairs, hand carried		23	.696			31		31	46
2135	Wheeled		35	.457			20.50		20.50	30.50
2140	6-10 riser stairs, hand carried		22	.727			32.50		32.50	48
2145	Wheeled		34	.471			21		21	31
2150	11-20 riser stairs, hand carried		20	.800			35.50		35.50	53
2155	Wheeled		31	.516			23		23	34
2160	21-40 riser stairs, hand carried		16	1			44.50		44.50	66.50
2165	Wheeled		24	.667			29.50		29.50	44
2170	0'-100' haul, incl. 5 riser stairs, hand carried		15	1.067			47.50		47.50	70.50
2175	Wheeled		23	.696			31		31	46
2180	6-10 riser stairs, hand carried		14	1.143			50.50		50.50	75.50
2185	Wheeled		21	.762			34		34	50.50
2190	11-20 riser stairs, hand carried		12	1.333			59		59	88.50
2195	Wheeled		18	.889			39.50		39.50	59
2200	21-40 riser stairs, hand carried		8	2			89		89	133
2205	Wheeled		12	1.333			59		59	88.50
2210	Haul and return, add per each extra 100' haul, hand carried		35.50	.451			20		20	30
2215	Wheeled		54	.296			13.15		13.15	19.65
2220	For each additional flight of stairs, up to 5 risers, add		550	.029	Flight		1.29		1.29	1.93
2225	6-10 risers, add		275	.058			2.58		2.58	3.85
2230	11-20 risers, add		138	.116			5.15		5.15	7.70
2235	21-40 risers, add		69	.232			10.30		10.30	15.35
3000	Loading & trucking, including 2 mile haul, chute loaded	B-16	45	.711	C.Y.		33	12.85	45.85	63.50
3040	Hand loading truck, 50' haul	"	48	.667			31	12.05	43.05	60
3080	Machine loading truck	B-17	120	.267			13.05	5.20	18.25	25
5000	Haul, per mile, up to 8 C.Y. truck	B-34B	1165	.007			.35	.50	.85	1.08
5100	Over 8 C.Y. truck	"	1550	.005			.26	.37	.63	.81

02 41 19.20 Selective Demolition, Dump Charges

		Crew	Daily Output	Labor-Hours	Unit	Material	2021 Bare Costs Labor	Equipment	Total	Total Incl O&P
0010	**SELECTIVE DEMOLITION, DUMP CHARGES** R024119-10									
0020	Dump charges, typical urban city, tipping fees only									
0100	Building construction materials				Ton	74			74	81
0200	Trees, brush, lumber					63			63	69.50
0300	Rubbish only					63			63	69.50
0500	Reclamation station, usual charge				Ton	74			74	81

02 41 Demolition

02 41 19 – Selective Demolition

02 41 19.27 Selective Demolition, Torch Cutting	Crew	Daily Output	Labor-Hours	Unit	Material	2021 Bare Costs Labor	Equipment	Total	Total Incl O&P
0010 **SELECTIVE DEMOLITION, TORCH CUTTING** R024119-10									
0020 Steel, 1" thick plate	E-25	333	.024	L.F.	2.23	1.50	.04	3.77	4.81
0040 1" diameter bar	"	600	.013	Ea.	.37	.83	.02	1.22	1.72
1000 Oxygen lance cutting, reinforced concrete walls									
1040 12"-16" thick walls	1 Clab	10	.800	L.F.		35.50		35.50	53
1080 24" thick walls	↓	6	1.333	"		59		59	88.50
0401 For cleaning/transporting tanks (1.5 lb./100 gal. cap)	↓	500	.016	Lb.	1	.71		1.71	2.16

02 82 Asbestos Remediation

02 82 13 – Asbestos Abatement

02 82 13.47 Asbestos Waste Pkg., Handling, and Disp.	Crew	Daily Output	Labor-Hours	Unit	Material	2021 Bare Costs Labor	Equipment	Total	Total Incl O&P
0010 **ASBESTOS WASTE PACKAGING, HANDLING, AND DISPOSAL**									
0100 Collect and bag bulk material, 3 C.F. bags, by hand	A-9	400	.160	Ea.	1.23	9.75		10.98	16.30
0200 Large production vacuum loader	A-12	880	.073		1.04	4.44	.85	6.33	8.85
1000 Double bag and decontaminate	A-9	960	.067		1.23	4.07		5.30	7.55
2000 Containerize bagged material in drums, per 3 C.F. drum	"	800	.080	↓	19.80	4.88		24.68	29.50
3000 Cart bags 50' to dumpster	2 Asbe	400	.040			2.44		2.44	3.73
5000 Disposal charges, not including haul, minimum				C.Y.				61	67
5020 Maximum				"				355	395
9000 For type B (supplied air) respirator equipment, add				%				10%	10%

02 87 19 – Bio-Hazard and Infectious Disease Control Remediation

02 87 19.10 Temporary Barriers, Including Ppe	Crew	Daily Output	Labor-Hours	Unit	Material	2021 Bare Costs Labor	Equipment	Total	Total Incl O&P
0010 **TEMPORARY BARRIERS, including PPE**									
0100 Infectious control partition, adjustable to 10' high, obscured, 4' panel	2 Carp	75	.213	Ea.	580	11.65		591.65	655
0110 3' panel		75	.213		550	11.65		561.65	620
0120 2' panel	↓	75	.213		430	11.65		441.65	490
0130 1' panel	1 Carp	75	.107		285	5.85		290.85	325
0140 6" panel	"	75	.107		265	5.85		270.85	300
0150 2' panel with HEPA filtered discharge port	2 Carp	75	.213		500	11.65		511.65	565
0160 3' panel with 32" door		75	.213		895	11.65		906.65	1,000
0170 4' panel with 36" door		75	.213		995	11.65		1,006.65	1,125
0180 4'-6" panel with 44" door		75	.213		1,200	11.65		1,211.65	1,350
0190 Hinged corner		68	.235		186	12.85		198.85	223
0200 Outside corner		68	.235		151	12.85		163.85	185
0210 T post		68	.235		151	12.85		163.85	185
0220 Accessories, ceiling grid clip		300	.053		7.60	2.92		10.52	12.75
0230 Panel locking clip		300	.053		5.30	2.92		8.22	10.20
0240 Panel joint closure strip		300	.053		8.15	2.92		11.07	13.35
0250 Screw jack	↓	300	.053		6.80	2.92		9.72	11.85
0260 Digital pressure difference guage					275			275	305
0280 Combination lockset	1 Carp	11	.727	↓	202	40		242	283
0285 Sealant tape, 2" wide	1 Clab	165	.048	C.L.F.	10.85	2.15		13	15.10
0290 System in place, including door and accessories									
0300 Based upon 25 uses	2 Carp	43	.372	L.F.	11.10	20.50		31.60	42.50
0310 Based upon 50 uses		43	.372		6.15	20.50		26.65	37.50
0330 Based upon 100 uses		43	.372	↓	3.69	20.50		24.19	34.50
0400 Separation barrier, 2" x 4" @ 16" O.C., 1/2" plywood each side, 8' high		340	.047	S.F.	4.22	2.57		6.79	8.50
0410 12' high		275	.058		4.17	3.18		7.35	9.35
0420 16' high		170	.094		4.25	5.15		9.40	12.35
0430 5/8" FR GWB each side, 8' high		290	.055	↓	2.68	3.02		5.70	7.45

02 82 Asbestos Remediation

02 87 19 – Bio-Hazard and Infectious Disease Control Remediation

	Crew	Daily Output	Labor-Hours	Unit	Material	2021 Bare Costs Labor	2021 Bare Costs Equipment	Total	Total Incl O&P
02 87 19.10 Temporary Barriers, Including Ppe									
0440 12' high	2 Carp	235	.068	S.F.	2.60	3.72		6.32	8.40
0450 16' high	↓	144	.111	↓	2.55	6.10		8.65	11.85
02 87 19.20 Temporary Protection, Including Ppe									
0010 **TEMPORARY PROTECTION, including PPE**									
0100 Infectious control protection, 6 mil plastic sheeting, applied to floors	1 Skwk	1900	.004	S.F.	.08	.24		.32	.44
0110 Applied to walls		1400	.006		.08	.33		.41	.57
0120 Applied to ceilings	↓	1000	.008	↓	.08	.46		.54	.77
0130 Sealant tape, 2" wide	1 Clab	165	.048	C.L.F.	10.85	2.15		13	15.10
0140 Floor covering, 1/8" tempered hardboard	2 Carp	878	.018	S.F.	.69	1		1.69	2.25
0150 1/2" CDX		878	.018		1.18	1		2.18	2.79
0160 3/4" CDX	↓	755	.021	↓	1.56	1.16		2.72	3.45
02 87 19.30 Infectious Control Equipment and Supplies									
0010 **INFECTIOUS CONTROL EQUIPMENT AND SUPPLIES**									
0100 Entry/exit decontamination chamber, portable, 10 uses	A-10C	12	2	Ea.	228	122	16.55	366.55	455
0105 Negative air machine, 1800 CFM					900			900	990
0110 Disposable polyethylene bags, 6 mil, 3 C.F.					1.23			1.23	1.35
0120 Negative air machine, 1800 CFM					900			900	990
0130 Paper towels, carton of 12, 350' rolls					65.50			65.50	72.50
0140 Heavy duty wipes, 160 per box					38			38	42
0200 Tent, including frame, weights, roof, walls, 5' x 5'	2 Skwk	32	.500		2,150	28.50		2,178.50	2,425
0205 5' x 10'		24	.667		2,500	38		2,538	2,800
0210 10' x 10'		24	.667		3,125	38		3,163	3,475
0220 13' x 26'		12	1.333		9,350	76		9,426	10,400
0230 13' x 26', flat roof infirmary		12	1.333		7,600	76		7,676	8,475
0240 13' x 26', drive through screening	↓	12	1.333		8,800	76		8,876	9,800
0242 LED light kit for tents, 15 watt	1 Elec	48	.167		199	10.60		209.60	235
0244 Halogen heating kit, 1500 watt	"	48	.167		500	10.60		510.60	565
0250 Medical partition system, including frame, drapes, 5' x 5'	2 Skwk	48	.333	↓	1,300	19.05		1,319.05	1,450
02 87 19.40 Infectious Control Cleaning, Includes Ppe									
0010 **INFECTIOUS CONTROL CLEANING, Includes PPE**									
0100 Infectious control cleaning, including PPE and antimicrobial spray, floors	1 Skwk	3600	.002	S.F.	.01	.13		.14	.20
0110 Walls	"	3000	.003	"	.01	.15		.16	.24
0120 Ceilings	1 Skwk	2700	.003	S.F.	.01	.17		.18	.26
0130 Countertops		3300	.002		.01	.14		.15	.22
0140 Other surfaces		2600	.003	↓	.01	.18		.19	.27
0150 Plumbing fixtures		14	.571	Ea.	1.87	32.50		34.37	51
0160 Bathroom accessories		30	.267		.87	15.25		16.12	24
0170 Kitchen appliance		9	.889		2.92	51		53.92	79.50
0180 Electrical switches and receptacles		140	.057		.19	3.26		3.45	5.10
0190 Work station		9	.889		2.92	51		53.92	79.50
0200 Door and window hardware		88	.091	↓	.30	5.20		5.50	8.15
0250 Supply / return ductwork, up to 2 S.F. cross section		240	.033	L.F.	.11	1.90		2.01	2.98
0260 2 to 4 S.F. cross section		192	.042		.14	2.38		2.52	3.73
0270 4 to 8 S.F. cross section		144	.056		.19	3.17		3.36	4.97
0280 8 SF cross section		102	.078	↓	.26	4.48		4.74	7.05
0300 Duct coil		5	1.600	Ea.	5.25	91.50		96.75	143
0310 Duct damper and vane set		77	.104		.35	5.95		6.30	9.30
0320 Fan coil unit or VAV box		8	1		3.28	57		60.28	89.50
0330 Air handling unit and coil		2	4		13.10	228		241.10	360
0340 Grill / diffuser or register	↓	32	.250	↓	.82	14.30		15.12	22.50

Division Notes

		CREW	DAILY OUTPUT	LABOR-HOURS	UNIT	BARE COSTS				TOTAL INCL O&P
						MAT.	LABOR	EQUIP.	TOTAL	

Estimating Tips

General

- Carefully check all the plans and specifications. Concrete often appears on drawings other than structural drawings, including mechanical and electrical drawings for equipment pads. The cost of cutting and patching is often difficult to estimate. See Subdivision 03 81 for Concrete Cutting, Subdivision 02 41 19.16 for Cutout Demolition, Subdivision 03 05 05.10 for Concrete Demolition, and Subdivision 02 41 19.19 for Rubbish Handling (handling, loading, and hauling of debris).
- Always obtain concrete prices from suppliers near the job site. A volume discount can often be negotiated, depending upon competition in the area. Remember to add for waste, particularly for slabs and footings on grade.

03 10 00 Concrete Forming and Accessories

- A primary cost for concrete construction is forming. Most jobs today are constructed with prefabricated forms. The selection of the forms best suited for the job and the total square feet of forms required for efficient concrete forming and placing are key elements in estimating concrete construction. Enough forms must be available for erection to make efficient use of the concrete placing equipment and crew.
- Concrete accessories for forming and placing depend upon the systems used. Study the plans and specifications to ensure that all special accessory requirements have been included in the cost estimate, such as anchor bolts, inserts, and hangers.
- Included within costs for forms-in-place are all necessary bracing and shoring.

03 20 00 Concrete Reinforcing

- Ascertain that the reinforcing steel supplier has included all accessories, cutting, bending, and an allowance for lapping, splicing, and waste. A good rule of thumb is 10% for lapping, splicing, and waste. Also, 10% waste should be allowed for welded wire fabric.
- The unit price items in the subdivisions for Reinforcing In Place, Glass Fiber Reinforcing, and Welded Wire Fabric include the labor to install accessories such as beam and slab bolsters, high chairs, and bar ties and tie wire. The material cost for these accessories is not included; they may be obtained from the Accessories Subdivisions.

03 30 00 Cast-In-Place Concrete

- When estimating structural concrete, pay particular attention to requirements for concrete additives, curing methods, and surface treatments. Special consideration for climate, hot or cold, must be included in your estimate. Be sure to include requirements for concrete placing equipment and concrete finishing.
- For accurate concrete estimating, the estimator must consider each of the following major components individually: forms, reinforcing steel, ready-mix concrete, placement of the concrete, and finishing of the top surface. For faster estimating, Subdivision 03 30 53.40 for Concrete-In-Place can be used; here, various items of concrete work are presented that include the costs of all five major components (unless specifically stated otherwise).

03 40 00 Precast Concrete
03 50 00 Cast Decks and Underlayment

- The cost of hauling precast concrete structural members is often an important factor. For this reason, it is important to get a quote from the nearest supplier. It may become economically feasible to set up precasting beds on the site if the hauling costs are prohibitive.

Reference Numbers

Reference numbers are shown at the beginning of some major classifications. These numbers refer to related items in the Reference Section. The reference information may be an estimating procedure, an alternate pricing method, or technical information.

Note: Not all subdivisions listed here necessarily appear. ■

Same Data. Simplified.

Enjoy the convenience and efficiency of accessing your costs anywhere:

- **Skip the multiplier** by setting your location
- **Quickly search,** edit, favorite and share costs
- **Stay on top of price changes** with automatic updates

Discover more at rsmeans.com/online

03 01 Maintenance of Concrete

03 01 30 – Maintenance of Cast-In-Place Concrete

03 01 30.62 Concrete Patching

		Crew	Daily Output	Labor-Hours	Unit	Material	2021 Bare Costs Labor	2021 Bare Costs Equipment	Total	Total Incl O&P
0010	**CONCRETE PATCHING**									
0100	Floors, 1/4" thick, small areas, regular grout	1 Cefi	170	.047	S.F.	1.53	2.44		3.97	5.25
0150	Epoxy grout	"	100	.080	"	11.15	4.14		15.29	18.30
2000	Walls, including chipping, cleaning and epoxy grout									
2100	1/4" deep	1 Cefi	65	.123	S.F.	6.55	6.40		12.95	16.55
2150	1/2" deep		50	.160		13.10	8.30		21.40	26.50
2200	3/4" deep	↓	40	.200	↓	19.65	10.35		30	36.50

03 11 Concrete Forming

03 11 13 – Structural Cast-In-Place Concrete Forming

03 11 13.40 Forms In Place, Equipment Foundations

		Crew	Daily Output	Labor-Hours	Unit	Material	2021 Bare Costs Labor	2021 Bare Costs Equipment	Total	Total Incl O&P
0010	**FORMS IN PLACE, EQUIPMENT FOUNDATIONS**									
0020	1 use	C-2	160	.300	SFCA	3.96	16		19.96	28.50
0050	2 use		190	.253		2.18	13.45		15.63	22.50
0100	3 use		200	.240		1.59	12.80		14.39	21
0150	4 use	↓	205	.234	↓	1.30	12.50		13.80	20

03 11 13.45 Forms In Place, Footings

		Crew	Daily Output	Labor-Hours	Unit	Material	2021 Bare Costs Labor	2021 Bare Costs Equipment	Total	Total Incl O&P
0010	**FORMS IN PLACE, FOOTINGS**									
0020	Continuous wall, plywood, 1 use	C-1	375	.085	SFCA	7.85	4.45		12.30	15.30
0050	2 use		440	.073		4.32	3.79		8.11	10.40
0100	3 use		470	.068		3.14	3.55		6.69	8.75
0150	4 use		485	.066		2.56	3.44		6	7.95
5000	Spread footings, job-built lumber, 1 use	↓	305	.105	↓	3.28	5.45		8.73	11.75
5050	2 use	C-1	371	.086	SFCA	1.82	4.50		6.32	8.70
5100	3 use		401	.080		1.31	4.16		5.47	7.65
5150	4 use		414	.077		1.06	4.03		5.09	7.15

03 11 13.65 Forms In Place, Slab On Grade

		Crew	Daily Output	Labor-Hours	Unit	Material	2021 Bare Costs Labor	2021 Bare Costs Equipment	Total	Total Incl O&P
0010	**FORMS IN PLACE, SLAB ON GRADE**									
3000	Edge forms, wood, 4 use, on grade, to 6" high	C-1	600	.053	L.F.	.45	2.78		3.23	4.65
6000	Trench forms in floor, wood, 1 use		160	.200	SFCA	3.15	10.45		13.60	19
6050	2 use		175	.183		1.73	9.55		11.28	16.15
6100	3 use		180	.178		1.26	9.25		10.51	15.25
6150	4 use	↓	185	.173	↓	1.02	9		10.02	14.60

03 15 Concrete Accessories

03 15 05 – Concrete Forming Accessories

03 15 05.75 Sleeves and Chases

		Crew	Daily Output	Labor-Hours	Unit	Material	2021 Bare Costs Labor	2021 Bare Costs Equipment	Total	Total Incl O&P
0010	**SLEEVES AND CHASES**									
0100	Plastic, 1 use, 12" long, 2" diameter	1 Carp	100	.080	Ea.	1.27	4.38		5.65	7.95
0150	4" diameter		90	.089		2.99	4.86		7.85	10.55
0200	6" diameter		75	.107		11.30	5.85		17.15	21
0250	12" diameter		60	.133		18.05	7.30		25.35	31
5000	Sheet metal, 2" diameter	[G]	100	.080		1.61	4.38		5.99	8.30
5100	4" diameter	[G]	90	.089		2.01	4.86		6.87	9.45
5150	6" diameter	[G]	75	.107		1.74	5.85		7.59	10.60
5200	12" diameter	[G]	60	.133		3.67	7.30		10.97	14.95
6000	Steel pipe, 2" diameter	[G]	100	.080		4.35	4.38		8.73	11.35
6100	4" diameter	[G]	90	.089		17.20	4.86		22.06	26
6150	6" diameter	[G]	75	.107		19.45	5.85		25.30	30

03 15 Concrete Accessories

03 15 05 – Concrete Forming Accessories

03 15 05.75 Sleeves and Chases		Crew	Daily Output	Labor-Hours	Unit	Material	2021 Bare Costs Labor	Equipment	Total	Total Incl O&P
6200	12" diameter	**G** 1 Carp	60	.133	Ea.	87	7.30		94.30	107

03 15 16 – Concrete Construction Joints

03 15 16.20 Control Joints, Saw Cut

		Crew	Daily Output	Labor-Hours	Unit	Material	2021 Bare Costs Labor	Equipment	Total	Total Incl O&P
0010	**CONTROL JOINTS, SAW CUT**									
0100	Sawcut control joints in green concrete									
0120	1" depth	C-27	2000	.008	L.F.	.03	.41	.06	.50	.70
0140	1-1/2" depth		1800	.009		.04	.46	.06	.56	.79
0160	2" depth	↓	1600	.010	↓	.06	.52	.07	.65	.90
0180	Sawcut joint reservoir in cured concrete									
0182	3/8" wide x 3/4" deep, with single saw blade	C-27	1000	.016	L.F.	.04	.83	.11	.98	1.38
0184	1/2" wide x 1" deep, with double saw blades		900	.018		.08	.92	.13	1.13	1.58
0186	3/4" wide x 1-1/2" deep, with double saw blades	↓	800	.020		.17	1.04	.14	1.35	1.87
0190	Water blast joint to wash away laitance, 2 passes	C-29	2500	.003			.14	.04	.18	.25
0200	Air blast joint to blow out debris and air dry, 2 passes	C-28	2000	.004	↓		.21	.02	.23	.32
0300	For backer rod, see Section 07 91 23.10									
0342	For joint sealant, see Section 07 92 13.20									

03 15 19 – Cast-In Concrete Anchors

03 15 19.10 Anchor Bolts

		Crew	Daily Output	Labor-Hours	Unit	Material	2021 Bare Costs Labor	Equipment	Total	Total Incl O&P
0010	**ANCHOR BOLTS**									
0015	Made from recycled materials									
0025	Single bolts installed in fresh concrete, no templates									
0030	Hooked w/nut and washer, 1/2" diameter, 8" long	**G** 1 Carp	132	.061	Ea.	1.55	3.32		4.87	6.65
0040	12" long	**G**	131	.061		1.72	3.34		5.06	6.90
0050	5/8" diameter, 8" long	**G**	129	.062		3.40	3.39		6.79	8.80
0060	12" long	**G**	127	.063		4.18	3.45		7.63	9.75
0070	3/4" diameter, 8" long	**G**	127	.063		4.18	3.45		7.63	9.75
0080	12" long	**G** ↓	125	.064	↓	5.25	3.50		8.75	11

03 15 19.45 Machinery Anchors

		Crew	Daily Output	Labor-Hours	Unit	Material	2021 Bare Costs Labor	Equipment	Total	Total Incl O&P
0010	**MACHINERY ANCHORS**, heavy duty, incl. sleeve, floating base nut,									
0020	lower stud & coupling nut, fiber plug, connecting stud, washer & nut.									
0030	For flush mounted embedment in poured concrete heavy equip. pads.									
0200	Stud & bolt, 1/2" diameter	**G** E-16	40	.400	Ea.	60.50	24.50	3.72	88.72	109
0300	5/8" diameter	**G**	35	.457		67	28	4.25	99.25	122
0500	3/4" diameter	**G**	30	.533		80.50	32.50	4.96	117.96	145
0600	7/8" diameter	**G**	25	.640		95	39	5.95	139.95	171
0800	1" diameter	**G**	20	.800		105	49	7.45	161.45	199
0900	1-1/4" diameter	**G** ↓	15	1.067	↓	135	65.50	9.90	210.40	260

03 21 Reinforcement Bars

03 21 11 – Plain Steel Reinforcement Bars

03 21 11.60 Reinforcing In Place

		Crew	Daily Output	Labor-Hours	Unit	Material	2021 Bare Costs Labor	Equipment	Total	Total Incl O&P
0010	**REINFORCING IN PLACE**, 50-60 ton lots, A615 Grade 60									
0020	Includes labor, but not material cost, to install accessories									
0030	Made from recycled materials									
0502	Footings, #4 to #7	**G** 4 Rodm	4200	.008	Lb.	.62	.45		1.07	1.35
0552	#8 to #18	**G**	7200	.004		.62	.26		.88	1.07
0602	Slab on grade, #3 to #7	**G** ↓	4200	.008	↓	.62	.45		1.07	1.35
0900	For other than 50-60 ton lots									
1000	Under 10 ton job, #3 to #7, add					25%	10%			
1010	#8 to #18, add					20%	10%			

03 21 Reinforcement Bars

03 21 11 – Plain Steel Reinforcement Bars

03 21 11.60 Reinforcing In Place	Crew	Daily Output	Labor-Hours	Unit	Material	2021 Bare Costs Labor	Equipment	Total	Total Incl O&P
1050	10-50 ton job, #3 to #7, add					10%				
1060	#8 to #18, add					5%				
1100	60-100 ton job, #3 to #7, deduct					5%				
1110	#8 to #18, deduct					10%				
1150	Over 100 ton job, #3 to #7, deduct					10%				
1160	#8 to #18, deduct					15%				

03 22 Fabric and Grid Reinforcing

03 22 11 – Plain Welded Wire Fabric Reinforcing

03 22 11.10 Plain Welded Wire Fabric

		Crew	Daily Output	Labor-Hours	Unit	Material	Labor	Equipment	Total	Total Incl O&P
0010	**PLAIN WELDED WIRE FABRIC** ASTM A185									
0020	Includes labor, but not material cost, to install accessories									
0050	Sheets									
0100	6 x 6 - W1.4 x W1.4 (10 x 10) 21 lb./C.S.F.	[G] 2 Rodm	35	.457	C.S.F.	14.95	27		41.95	57

03 22 13 – Galvanized Welded Wire Fabric Reinforcing

03 22 13.10 Galvanized Welded Wire Fabric

		Crew	Daily Output	Labor-Hours	Unit	Material	Labor	Equipment	Total	Total Incl O&P
0010	**GALVANIZED WELDED WIRE FABRIC**									
0100	Add to plain welded wire pricing for galvanized welded wire				Lb.	.24			.24	.27

03 22 16 – Epoxy-Coated Welded Wire Fabric Reinforcing

03 22 16.10 Epoxy-Coated Welded Wire Fabric

		Crew	Daily Output	Labor-Hours	Unit	Material	Labor	Equipment	Total	Total Incl O&P
0010	**EPOXY-COATED WELDED WIRE FABRIC**									
0100	Add to plain welded wire pricing for epoxy-coated welded wire				Lb.	.49			.49	.53

03 30 Cast-In-Place Concrete

03 30 53 – Miscellaneous Cast-In-Place Concrete

03 30 53.40 Concrete In Place

		Crew	Daily Output	Labor-Hours	Unit	Material	Labor	Equipment	Total	Total Incl O&P
0010	**CONCRETE IN PLACE**									
0020	Including forms (4 uses), Grade 60 rebar, concrete (Portland cement									
0050	Type I), placement and finishing unless otherwise indicated									
3540	Equipment pad (3000 psi), 3' x 3' x 6" thick	C-14H	45	1.067	Ea.	55.50	57	.61	113.11	147
3550	4' x 4' x 6" thick		30	1.600		84	85.50	.91	170.41	220
3560	5' x 5' x 8" thick		18	2.667		150	143	1.52	294.52	380
3570	6' x 6' x 8" thick		14	3.429		204	184	1.95	389.95	500
3580	8' x 8' x 10" thick		8	6		415	320	3.42	738.42	945
3590	10' x 10' x 12" thick		5	9.600		720	515	5.45	1,240.45	1,550
3800	Footings (3000 psi), spread under 1 C.Y.	C-14C	28	4	C.Y.	208	209	.96	417.96	540
3813	Install new concrete (3000 psi) light pole base, 24" diam. x 8'	C-1	2.66	12.030		345	625		970	1,325
3825	1 C.Y. to 5 C.Y.	C-14C	43	2.605		266	136	.63	402.63	495
3850	Over 5 C.Y.	"	75	1.493		245	78	.36	323.36	385
3900	Footings, strip (3000 psi), 18" x 9", unreinforced	C-14L	40	2.400		159	123	.67	282.67	360
3920	18" x 9", reinforced	C-14C	35	3.200		189	167	.77	356.77	460
3925	20" x 10", unreinforced	C-14L	45	2.133		153	109	.60	262.60	335
3930	20" x 10", reinforced	C-14C	40	2.800		178	146	.67	324.67	415
3935	24" x 12", unreinforced	C-14L	55	1.745		150	89.50	.49	239.99	299
3940	24" x 12", reinforced	C-14C	48	2.333		175	122	.56	297.56	375
3945	36" x 12", unreinforced	C-14L	70	1.371		144	70	.38	214.38	263
3950	36" x 12", reinforced	C-14C	60	1.867		166	97.50	.45	263.95	330
4000	Foundation mat (3000 psi), under 10 C.Y.		38.67	2.896		259	151	.70	410.70	510

03 30 Cast-In-Place Concrete

03 30 53 – Miscellaneous Cast-In-Place Concrete

03 30 53.40 Concrete In Place

		Crew	Daily Output	Labor-Hours	Unit	Material	2021 Bare Costs Labor	Equipment	Total	Total Incl O&P
4050	Over 20 C.Y.	C-14C	56.40	1.986	C.Y.	225	104	.48	329.48	405
4650	Slab on grade (3500 psi), not including finish, 4" thick	C-14E	60.75	1.449		149	77.50	.45	226.95	279
4700	6" thick	"	92	.957	▼	144	51	.30	195.30	234
4701	Thickened slab edge (3500 psi), for slab on grade poured									
4702	monolithically with slab; depth is in addition to slab thickness;									
4703	formed vertical outside edge, earthen bottom and inside slope									
4705	8" deep x 8" wide bottom, unreinforced	C-14L	2190	.044	L.F.	4.58	2.24	.01	6.83	8.40
4710	8" x 8", reinforced	C-14C	1670	.067		7.50	3.51	.02	11.03	13.50
4715	12" deep x 12" wide bottom, unreinforced	C-14L	1800	.053		9.05	2.73	.01	11.79	14.05
4720	12" x 12", reinforced	C-14C	1310	.086		14.40	4.47	.02	18.89	22.50
4725	16" deep x 16" wide bottom, unreinforced	C-14L	1440	.067		15	3.41	.02	18.43	21.50
4730	16" x 16", reinforced	C-14C	1120	.100		21.50	5.25	.02	26.77	31.50
4735	20" deep x 20" wide bottom, unreinforced	C-14L	1150	.083		22.50	4.27	.02	26.79	31
4740	20" x 20", reinforced	C-14C	920	.122		30.50	6.35	.03	36.88	43
4745	24" deep x 24" wide bottom, unreinforced	C-14L	930	.103		31.50	5.30	.03	36.83	42.50
4750	24" x 24", reinforced	C-14C	740	.151	▼	42	7.90	.04	49.94	58

03 31 Structural Concrete

03 31 13 – Heavyweight Structural Concrete

03 31 13.35 Heavyweight Concrete, Ready Mix

		Crew	Daily Output	Labor-Hours	Unit	Material	2021 Bare Costs Labor	Equipment	Total	Total Incl O&P
0010	**HEAVYWEIGHT CONCRETE, READY MIX**, delivered									
0012	Includes local aggregate, sand, Portland cement (Type I) and water									
0015	Excludes all additives and treatments									
0020	2000 psi				C.Y.	109			109	120
0100	2500 psi					112			112	124
0150	3000 psi					124			124	137
0200	3500 psi					127			127	139
0300	4000 psi					129			129	142
1000	For high early strength (Portland cement Type III), add					10%				
1300	For winter concrete (hot water), add					5.85			5.85	6.40
1410	For mid-range water reducer, add					4.22			4.22	4.64
1420	For high-range water reducer/superplasticizer, add					6.65			6.65	7.30
1430	For retarder, add					4.22			4.22	4.64
1440	For non-Chloride accelerator, add					8.70			8.70	9.60
1450	For Chloride accelerator, per 1%, add					3.69			3.69	4.06
1460	For fiber reinforcing, synthetic (1 lb./C.Y.), add					8.30			8.30	9.15
1500	For Saturday delivery, add				▼	7.70			7.70	8.45
1510	For truck holding/waiting time past 1st hour per load, add				Hr.	114			114	125
1520	For short load (less than 4 C.Y.), add per load				Ea.	97			97	107
2000	For all lightweight aggregate, add				C.Y.	45%				

03 31 13.70 Placing Concrete

		Crew	Daily Output	Labor-Hours	Unit	Material	2021 Bare Costs Labor	Equipment	Total	Total Incl O&P
0010	**PLACING CONCRETE**									
0020	Includes labor and equipment to place, level (strike off) and consolidate									
1200	Duct bank, direct chute	C-6	155	.310	C.Y.		14.25	.35	14.60	21.50
1900	Footings, continuous, shallow, direct chute	"	120	.400			18.40	.45	18.85	28
1950	Pumped	C-20	150	.427			20	3.19	23.19	33.50
2000	With crane and bucket	C-7	90	.800			38.50	12.10	50.60	70.50
2100	Footings, continuous, deep, direct chute	C-6	140	.343			15.75	.39	16.14	24
2150	Pumped	C-20	160	.400			18.95	2.99	21.94	31.50
2200	With crane and bucket	C-7	110	.655			31.50	9.90	41.40	57.50
2400	Footings, spread, under 1 C.Y., direct chute	C-6	55	.873			40	.99	40.99	60.50

03 31 Structural Concrete

03 31 13 – Heavyweight Structural Concrete

03 31 13.70 Placing Concrete

		Crew	Daily Output	Labor-Hours	Unit	Material	2021 Bare Costs Labor	2021 Bare Costs Equipment	Total	Total Incl O&P
2450	Pumped	C-20	65	.985	C.Y.		46.50	7.35	53.85	77.50
2500	With crane and bucket	C-7	45	1.600			77	24	101	141
2600	Over 5 C.Y., direct chute	C-6	120	.400			18.40	.45	18.85	28
2650	Pumped	C-20	150	.427			20	3.19	23.19	33.50
2700	With crane and bucket	C-7	100	.720			34.50	10.90	45.40	63.50
2900	Foundation mats, over 20 C.Y., direct chute	C-6	350	.137			6.30	.15	6.45	9.55
2950	Pumped	C-20	400	.160			7.60	1.20	8.80	12.60
3000	With crane and bucket	C-7	300	.240			11.50	3.63	15.13	21

03 35 Concrete Finishing

03 35 13 – High-Tolerance Concrete Floor Finishing

03 35 13.30 Finishing Floors, High Tolerance

		Crew	Daily Output	Labor-Hours	Unit	Material	2021 Bare Costs Labor	2021 Bare Costs Equipment	Total	Total Incl O&P
0010	**FINISHING FLOORS, HIGH TOLERANCE**									
0012	Finishing of fresh concrete flatwork requires that concrete									
0013	first be placed, struck off & consolidated									
0015	Basic finishing for various unspecified flatwork									
0100	Bull float only	C-10	4000	.006	S.F.		.30		.30	.44
0125	Bull float & manual float		2000	.012			.59		.59	.87
0150	Bull float, manual float & broom finish, w/edging & joints		1850	.013			.64		.64	.94
0200	Bull float, manual float & manual steel trowel		1265	.019			.94		.94	1.38
0210	For specified Random Access Floors in ACI Classes 1, 2, 3 and 4 to achieve									
0215	Composite Overall Floor Flatness and Levelness values up to FF35/FL25									
0250	Bull float, machine float & machine trowel (walk-behind)	C-10C	1715	.014	S.F.		.69	.05	.74	1.08
0300	Power screed, bull float, machine float & trowel (walk-behind)	C-10D	2400	.010			.49	.08	.57	.81
0350	Power screed, bull float, machine float & trowel (ride-on)	C-10E	4000	.006			.30	.06	.36	.51
0352	For specified Random Access Floors in ACI Classes 5, 6, 7 and 8 to achieve									
0354	Composite Overall Floor Flatness and Levelness values up to FF50/FL50									
0356	Add for two-dimensional restraightening after power float	C-10	6000	.004	S.F.		.20		.20	.29
0358	For specified Random or Defined Access Floors in ACI Class 9 to achieve									
0360	Composite Overall Floor Flatness and Levelness values up to FF100/FL100									
0362	Add for two-dimensional restraightening after bull float & power float	C-10	3000	.008	S.F.		.39		.39	.58
0364	For specified Superflat Defined Access Floors in ACI Class 9 to achieve									
0366	Minimum Floor Flatness and Levelness values of FF100/FL100									
0368	Add for 2-dim'l restraightening after bull float, power float, power trowel	C-10	2000	.012	S.F.		.59		.59	.87

03 54 Cast Underlayment

03 54 16 – Hydraulic Cement Underlayment

03 54 16.50 Cement Underlayment

		Crew	Daily Output	Labor-Hours	Unit	Material	2021 Bare Costs Labor	2021 Bare Costs Equipment	Total	Total Incl O&P
0010	**CEMENT UNDERLAYMENT**									
2510	Underlayment, P.C. based, self-leveling, 4100 psi, pumped, 1/4" thick	C-8	20000	.003	S.F.	2.09	.14	.02	2.25	2.52
2520	1/2" thick		19000	.003		4.19	.14	.02	4.35	4.84
2530	3/4" thick		18000	.003		6.30	.15	.02	6.47	7.15
2540	1" thick		17000	.003		8.35	.16	.02	8.53	9.45
2550	1-1/2" thick		15000	.004		12.55	.18	.03	12.76	14.10
2560	Hand placed, 1/2" thick	C-18	450	.020		4.19	.89	.26	5.34	6.25
2610	Topping, P.C. based, self-leveling, 6100 psi, pumped, 1/4" thick	C-8	20000	.003		3.40	.14	.02	3.56	3.96
2620	1/2" thick		19000	.003		6.80	.14	.02	6.96	7.75
2630	3/4" thick		18000	.003		10.20	.15	.02	10.37	11.50
2660	1" thick		17000	.003		13.60	.16	.02	13.78	15.20

03 54 Cast Underlayment

03 54 16 – Hydraulic Cement Underlayment

03 54 16.50 Cement Underlayment		Crew	Daily Output	Labor-Hours	Unit	Material	2021 Bare Costs Labor	Equipment	Total	Total Incl O&P
2670	1-1/2" thick	C-8	15000	.004	S.F.	20.50	.18	.03	20.71	23
2680	Hand placed, 1/2" thick	C-18	450	.020	↓	6.80	.89	.26	7.95	9.10

03 63 Epoxy Grouting

03 63 05 – Grouting of Dowels and Fasteners

03 63 05.10 Epoxy Only

		Crew	Daily Output	Labor-Hours	Unit	Material	2021 Bare Costs Labor	Equipment	Total	Total Incl O&P
0010	**EPOXY ONLY**									
1500	Chemical anchoring, epoxy cartridge, excludes layout, drilling, fastener									
1530	For fastener 3/4" diam. x 6" embedment	2 Skwk	72	.222	Ea.	4.89	12.70		17.59	24.50
1535	1" diam. x 8" embedment		66	.242		7.35	13.85		21.20	29
1540	1-1/4" diam. x 10" embedment		60	.267		14.70	15.25		29.95	39
1545	1-3/4" diam. x 12" embedment		54	.296		24.50	16.90		41.40	52.50
1550	14" embedment		48	.333		29.50	19.05		48.55	61
1555	2" diam. x 12" embedment		42	.381		39	22		61	75.50
1560	18" embedment	↓	32	.500	↓	49	28.50		77.50	97

03 81 Concrete Cutting

03 81 13 – Flat Concrete Sawing

03 81 13.50 Concrete Floor/Slab Cutting

		Crew	Daily Output	Labor-Hours	Unit	Material	2021 Bare Costs Labor	Equipment	Total	Total Incl O&P
0010	**CONCRETE FLOOR/SLAB CUTTING**									
0050	Includes blade cost, layout and set-up time									
0300	Saw cut concrete slabs, plain, up to 3" deep	B-89	1060	.015	L.F.	.10	.79	1.01	1.90	2.40
0320	Each additional inch of depth		3180	.005		.03	.26	.34	.63	.80
0400	Mesh reinforced, up to 3" deep		980	.016		.12	.85	1.09	2.06	2.60
0420	Each additional inch of depth	↓	2940	.005	↓	.04	.28	.36	.68	.86
0500	Rod reinforced, up to 3" deep	B-89	800	.020	L.F.	.14	1.04	1.33	2.51	3.19
0520	Each additional inch of depth	"	2400	.007	"	.05	.35	.44	.84	1.06

03 81 13.75 Concrete Saw Blades

		Crew	Daily Output	Labor-Hours	Unit	Material	2021 Bare Costs Labor	Equipment	Total	Total Incl O&P
0010	**CONCRETE SAW BLADES**									
3000	Blades for saw cutting, included in cutting line items									
3020	Diamond, 12" diameter				Ea.	219			219	241
3040	18" diameter					345			345	380
3080	24" diameter					490			490	535
3120	30" diameter					865			865	950
3160	36" diameter					1,225			1,225	1,350
3200	42" diameter				↓	1,850			1,850	2,050

03 82 Concrete Boring

03 82 13 – Concrete Core Drilling

03 82 13.10 Core Drilling

		Crew	Daily Output	Labor-Hours	Unit	Material	2021 Bare Costs Labor	Equipment	Total	Total Incl O&P
0010	**CORE DRILLING**									
0015	Includes bit cost, layout and set-up time									
0020	Reinforced concrete slab, up to 6" thick									
0100	1" diameter core	B-89A	17	.941	Ea.	.20	48	7.15	55.35	79.50
0150	For each additional inch of slab thickness in same hole, add		1440	.011		.03	.56	.08	.67	.98
0200	2" diameter core		16.50	.970		.27	49	7.35	56.62	82.50
0250	For each additional inch of slab thickness in same hole, add	↓	1080	.015	↓	.04	.75	.11	.90	1.30

03 82 Concrete Boring

03 82 13 – Concrete Core Drilling

03 82 13.10 Core Drilling

		Crew	Daily Output	Labor-Hours	Unit	Material	2021 Bare Costs Labor	2021 Bare Costs Equipment	Total	Total Incl O&P
0300	3" diameter core	B-89A	16	1	Ea.	.41	51	7.60	59.01	85
0350	For each additional inch of slab thickness in same hole, add		720	.022		.07	1.13	.17	1.37	1.96
0500	4" diameter core		15	1.067		.50	54	8.10	62.60	90.50
0550	For each additional inch of slab thickness in same hole, add		480	.033		.08	1.69	.25	2.02	2.91
0700	6" diameter core		14	1.143		.80	58	8.70	67.50	97.50
0750	For each additional inch of slab thickness in same hole, add		360	.044		.13	2.26	.34	2.73	3.90
0900	8" diameter core		13	1.231		1.16	62.50	9.35	73.01	105
0950	For each additional inch of slab thickness in same hole, add		288	.056		.19	2.82	.42	3.43	4.90
1100	10" diameter core		12	1.333		1.43	67.50	10.15	79.08	114
1150	For each additional inch of slab thickness in same hole, add		240	.067		.24	3.38	.51	4.13	5.85
1300	12" diameter core		11	1.455		1.63	74	11.05	86.68	125
1350	For each additional inch of slab thickness in same hole, add		206	.078		.27	3.94	.59	4.80	6.85
1500	14" diameter core		10	1.600		2.02	81	12.15	95.17	138
1550	For each additional inch of slab thickness in same hole, add		180	.089		.34	4.51	.68	5.53	7.85
1700	18" diameter core		9	1.778		2.80	90	13.50	106.30	153
1750	For each additional inch of slab thickness in same hole, add		144	.111		.47	5.65	.84	6.96	9.90
1754	24" diameter core		8	2		4.07	102	15.20	121.27	173
1756	For each additional inch of slab thickness in same hole, add		120	.133		.68	6.75	1.01	8.44	12
1760	For horizontal holes, add to above						20%	20%		
1770	Prestressed hollow core plank, 8" thick									
1780	1" diameter core	B-89A	17.50	.914	Ea.	.26	46.50	6.95	53.71	77.50
1790	For each additional inch of plank thickness in same hole, add		3840	.004		.03	.21	.03	.27	.39
1794	2" diameter core		17.25	.928		.36	47	7.05	54.41	78.50
1796	For each additional inch of plank thickness in same hole, add		2880	.006		.04	.28	.04	.36	.52
1800	3" diameter core		17	.941		.55	48	7.15	55.70	80
1810	For each additional inch of plank thickness in same hole, add		1920	.008		.07	.42	.06	.55	.78
1820	4" diameter core		16.50	.970		.67	49	7.35	57.02	83
1830	For each additional inch of plank thickness in same hole, add		1280	.013		.08	.63	.10	.81	1.14
1840	6" diameter core		15.50	1.032		1.06	52.50	7.85	61.41	88.50
1850	For each additional inch of plank thickness in same hole, add		960	.017		.13	.85	.13	1.11	1.56
1860	8" diameter core		15	1.067		1.55	54	8.10	63.65	91.50
1870	For each additional inch of plank thickness in same hole, add		768	.021		.19	1.06	.16	1.41	1.97
1880	10" diameter core		14	1.143		1.90	58	8.70	68.60	98.50
1890	For each additional inch of plank thickness in same hole, add		640	.025		.24	1.27	.19	1.70	2.37
1900	12" diameter core		13.50	1.185		2.18	60	9	71.18	102
1910	For each additional inch of plank thickness in same hole, add		548	.029		.27	1.48	.22	1.97	2.76

03 82 16 – Concrete Drilling

03 82 16.10 Concrete Impact Drilling

		Crew	Daily Output	Labor-Hours	Unit	Material	2021 Bare Costs Labor	2021 Bare Costs Equipment	Total	Total Incl O&P
0010	**CONCRETE IMPACT DRILLING**									
0020	Includes bit cost, layout and set-up time, no anchors									
0050	Up to 4" deep in concrete/brick floors/walls									
0100	Holes, 1/4" diameter	1 Carp	75	.107	Ea.	.05	5.85		5.90	8.75
0150	For each additional inch of depth in same hole, add		430	.019		.01	1.02		1.03	1.53
0200	3/8" diameter		63	.127		.05	6.95		7	10.40
0250	For each additional inch of depth in same hole, add		340	.024		.01	1.29		1.30	1.93
0300	1/2" diameter		50	.160		.05	8.75		8.80	13.10
0350	For each additional inch of depth in same hole, add		250	.032		.01	1.75		1.76	2.62
0400	5/8" diameter		48	.167		.08	9.10		9.18	13.70
0450	For each additional inch of depth in same hole, add		240	.033		.02	1.82		1.84	2.74
0500	3/4" diameter		45	.178		.11	9.70		9.81	14.60
0550	For each additional inch of depth in same hole, add		220	.036		.03	1.99		2.02	3
0600	7/8" diameter		43	.186		.13	10.20		10.33	15.35

03 82 Concrete Boring

03 82 16 – Concrete Drilling

03 82 16.10 Concrete Impact Drilling	Crew	Daily Output	Labor-Hours	Unit	Material	2021 Bare Costs Labor	Equipment	Total	Total Incl O&P
0650	For each additional inch of depth in same hole, add	1 Carp	210	.038	Ea.	.03	2.08		2.11	3.15
0700	1" diameter		40	.200		.21	10.95		11.16	16.60
0750	For each additional inch of depth in same hole, add		190	.042		.05	2.30		2.35	3.50
0800	1-1/4" diameter		38	.211		.35	11.50		11.85	17.60
0850	For each additional inch of depth in same hole, add		180	.044		.09	2.43		2.52	3.73
0900	1-1/2" diameter	1 Carp	35	.229	Ea.	.37	12.50		12.87	19.05
0950	For each additional inch of depth in same hole, add	"	165	.048	"	.09	2.65		2.74	4.06
1000	For ceiling installations, add						40%			

Division Notes

		CREW	DAILY OUTPUT	LABOR-HOURS	UNIT	BARE COSTS				TOTAL INCL O&P
						MAT.	LABOR	EQUIP.	TOTAL	

Division Notes

Estimating Tips

05 05 00 Common Work Results for Metals

- Nuts, bolts, washers, connection angles, and plates can add a significant amount to both the tonnage of a structural steel job and the estimated cost. As a rule of thumb, add 10% to the total weight to account for these accessories.

- Type 2 steel construction, commonly referred to as "simple construction," consists generally of field-bolted connections with lateral bracing supplied by other elements of the building, such as masonry walls or x-bracing. The estimator should be aware, however, that shop connections may be accomplished by welding or bolting. The method may be particular to the fabrication shop and may have an impact on the estimated cost.

05 10 00 Structural Steel

- Steel items can be obtained from two sources: a fabrication shop or a metals service center. Fabrication shops can fabricate items under more controlled conditions than crews in the field can. They are also more efficient and can produce items more economically. Metal service centers serve as a source of long mill shapes to both fabrication shops and contractors.

- Most line items in this structural steel subdivision, and most items in 05 50 00 Metal Fabrications, are indicated as being shop fabricated. The bare material cost for these shop fabricated items is the "Invoice Cost" from the shop and includes the mill base price of steel plus mill extras, transportation to the shop, shop drawings and detailing where warranted, shop fabrication and handling, sandblasting and a shop coat of primer paint, all necessary structural bolts, and delivery to the job site. The bare labor cost and bare equipment cost for these shop fabricated items are for field installation or erection.

- Line items in Subdivision 05 12 23.40 Lightweight Framing, and other items scattered in Division 5, are indicated as being field fabricated. The bare material cost for these field fabricated items is the "Invoice Cost" from the metals service center and includes the mill base price of steel plus mill extras, transportation to the metals service center, material handling, and delivery of long lengths of mill shapes to the job site. Material costs for structural bolts and welding rods should be added to the estimate. The bare labor cost and bare equipment cost for these items are for both field fabrication and field installation or erection, and include time for cutting, welding, and drilling in the fabricated metal items. Drilling into concrete and fasteners to fasten field fabricated items to other work is not included and should be added to the estimate.

05 20 00 Steel Joist Framing

- In any given project the total weight of open web steel joists is determined by the loads to be supported and the design. However, economies can be realized in minimizing the amount of labor used to place the joists. This is done by maximizing the joist spacing and therefore minimizing the number of joists required to be installed on the job. Certain spacings and locations may be required by the design, but in other cases maximizing the spacing and keeping it as uniform as possible will keep the costs down.

05 30 00 Steel Decking

- The takeoff and estimating of a metal deck involve more than the area of the floor or roof and the type of deck specified or shown on the drawings. Many different sizes and types of openings may exist. Small openings for individual pipes or conduits may be drilled after the floor/roof is installed, but larger openings may require special deck lengths as well as reinforcing or structural support. The estimator should determine who will be supplying this reinforcing. Additionally, some deck terminations are part of the deck package, such as screed angles and pour stops, and others will be part of the steel contract, such as angles attached to structural members and cast-in-place angles and plates. The estimator must ensure that all pieces are accounted for in the complete estimate.

05 50 00 Metal Fabrications

- The most economical steel stairs are those that use common materials, standard details, and most importantly, a uniform and relatively simple method of field assembly. Commonly available A36/A992 channels and plates are very good choices for the main stringers of the stairs, as are angles and tees for the carrier members. Risers and treads are usually made by specialty shops, and it is most economical to use a typical detail in as many places as possible. The stairs should be pre-assembled and shipped directly to the site. The field connections should be simple and straightforward enough to be accomplished efficiently, and with minimum equipment and labor.

Reference Numbers

Reference numbers are shown at the beginning of some major classifications. These numbers refer to related items in the Reference Section. The reference information may be an estimating procedure, an alternate pricing method, or technical information.

Note: Not all subdivisions listed here necessarily appear. ■

Same Data. Simplified.

Enjoy the convenience and efficiency of accessing your costs anywhere:

- **Skip the multiplier** by setting your location
- **Quickly search,** edit, favorite and share costs
- **Stay on top of price changes** with automatic updates

Discover more at rsmeans.com/online

---

05 05 19 – Post-Installed Concrete Anchors

05 05 19.10 Chemical Anchors

		Crew	Daily Output	Labor-Hours	Unit	Material	2021 Bare Costs Labor	Equipment	Total	Total Incl O&P
0010	**CHEMICAL ANCHORS**									
0020	Includes layout & drilling									
1430	Chemical anchor, w/rod & epoxy cartridge, 3/4" diameter x 9-1/2" long	B-89A	27	.593	Ea.	11.90	30	4.50	46.40	63
1435	1" diameter x 11-3/4" long		24	.667		24	34	5.05	63.05	82.50
1440	1-1/4" diameter x 14" long		21	.762		38.50	38.50	5.80	82.80	106
1445	1-3/4" diameter x 15" long		20	.800		70.50	40.50	6.10	117.10	145
1450	18" long		17	.941		84.50	48	7.15	139.65	172
1455	2" diameter x 18" long		16	1		148	51	7.60	206.60	247
1460	24" long		15	1.067		194	54	8.10	256.10	305

05 05 19.20 Expansion Anchors

		Crew	Daily Output	Labor-Hours	Unit	Material	2021 Bare Costs Labor	Equipment	Total	Total Incl O&P
0010	**EXPANSION ANCHORS**									
0100	Anchors for concrete, brick or stone, no layout and drilling									
0200	Expansion shields, zinc, 1/4" diameter, 1-5/16" long, single [G]	1 Carp	90	.089	Ea.	.31	4.86		5.17	7.60
0300	1-3/8" long, double [G]		85	.094		.42	5.15		5.57	8.15
0400	3/8" diameter, 1-1/2" long, single [G]		85	.094		.51	5.15		5.66	8.25
0500	2" long, double [G]		80	.100		1.22	5.45		6.67	9.50
0600	1/2" diameter, 2-1/16" long, single [G]		80	.100		1.23	5.45		6.68	9.50
0700	2-1/2" long, double [G]		75	.107		2.25	5.85		8.10	11.20
0800	5/8" diameter, 2-5/8" long, single [G]		75	.107		1.97	5.85		7.82	10.85
0900	2-3/4" long, double [G]		70	.114		3.80	6.25		10.05	13.55
1000	3/4" diameter, 2-3/4" long, single [G]		70	.114		3.75	6.25		10	13.50
1100	3-15/16" long, double [G]		65	.123		5.55	6.75		12.30	16.15
2100	Hollow wall anchors for gypsum wall board, plaster or tile									
2300	1/8" diameter, short [G]	1 Carp	160	.050	Ea.	.31	2.74		3.05	4.42
2400	Long [G]		150	.053		.30	2.92		3.22	4.68
2500	3/16" diameter, short [G]		150	.053		.61	2.92		3.53	5
2600	Long [G]		140	.057		.71	3.13		3.84	5.45
2700	1/4" diameter, short [G]		140	.057		.58	3.13		3.71	5.30
2800	Long [G]		130	.062		1.04	3.37		4.41	6.15
3000	Toggle bolts, bright steel, 1/8" diameter, 2" long [G]		85	.094		.33	5.15		5.48	8.05
3100	4" long [G]		80	.100		.38	5.45		5.83	8.55
3400	1/4" diameter, 3" long [G]		75	.107		.44	5.85		6.29	9.20
3500	6" long [G]		70	.114		.59	6.25		6.84	10
3600	3/8" diameter, 3" long [G]		70	.114		.73	6.25		6.98	10.15
3700	6" long [G]		60	.133		1.62	7.30		8.92	12.70
3800	1/2" diameter, 4" long [G]	1 Carp	60	.133	Ea.	2.66	7.30		9.96	13.85
3900	6" long [G]	"	50	.160	"	3.05	8.75		11.80	16.40
4000	Nailing anchors									
4100	Nylon nailing anchor, 1/4" diameter, 1" long	1 Carp	3.20	2.500	C	19.40	137		156.40	226
4200	1-1/2" long		2.80	2.857		22	156		178	258
4300	2" long		2.40	3.333		24	182		206	299
4400	Metal nailing anchor, 1/4" diameter, 1" long [G]		3.20	2.500		20.50	137		157.50	227
4500	1-1/2" long [G]		2.80	2.857		21	156		177	256
4600	2" long [G]		2.40	3.333		44	182		226	320
5000	Screw anchors for concrete, masonry,									
5100	stone & tile, no layout or drilling included									
5700	Lag screw shields, 1/4" diameter, short [G]	1 Carp	90	.089	Ea.	.28	4.86		5.14	7.55
5900	3/8" diameter, short [G]		85	.094		.55	5.15		5.70	8.30
6100	1/2" diameter, short [G]		80	.100		1.02	5.45		6.47	9.25
6300	5/8" diameter, short [G]		70	.114		1.60	6.25		7.85	11.10
6600	Lead, #6 & #8, 3/4" long [G]		260	.031		.22	1.68		1.90	2.75
6700	#10 - #14, 1-1/2" long [G]		200	.040		.52	2.19		2.71	3.84

05 05 19 – Post-Installed Concrete Anchors

05 05 19.20 Expansion Anchors

		Crew	Daily Output	Labor-Hours	Unit	Material	2021 Bare Costs Labor	Equipment	Total	Total Incl O&P
6800	#16 & #18, 1-1/2" long	G 1 Carp	160	.050	Ea.	.70	2.74		3.44	4.85
6900	Plastic, #6 & #8, 3/4" long		260	.031		.05	1.68		1.73	2.57
7100	#10 & #12, 1" long	↓	220	.036	↓	.07	1.99		2.06	3.05
8000	Wedge anchors, not including layout or drilling									
8050	Carbon steel, 1/4" diameter, 1-3/4" long	G 1 Carp	150	.053	Ea.	.76	2.92		3.68	5.20
8100	3-1/4" long	G	140	.057		1	3.13		4.13	5.75
8150	3/8" diameter, 2-1/4" long	G	145	.055		.43	3.02		3.45	4.98
8200	5" long	G	140	.057		.76	3.13		3.89	5.50
8250	1/2" diameter, 2-3/4" long	G	140	.057		1.14	3.13		4.27	5.95
8300	7" long	G	125	.064		1.96	3.50		5.46	7.40
8350	5/8" diameter, 3-1/2" long	G	130	.062		2.13	3.37		5.50	7.35
8400	8-1/2" long	G	115	.070		4.54	3.81		8.35	10.70
8450	3/4" diameter, 4-1/4" long	G	115	.070		3.45	3.81		7.26	9.50
8500	10" long	G	95	.084		7.85	4.61		12.46	15.55
8550	1" diameter, 6" long	G	100	.080		5.65	4.38		10.03	12.75
8575	9" long	G	85	.094		7.35	5.15		12.50	15.75
8600	12" long	G	75	.107		7.95	5.85		13.80	17.40
8650	1-1/4" diameter, 9" long	G	70	.114		33.50	6.25		39.75	46.50
8700	12" long	G ↓	60	.133	↓	43	7.30		50.30	58.50
8750	For type 303 stainless steel, add					350%				
8800	For type 316 stainless steel, add					450%				
8950	Self-drilling concrete screw, hex washer head, 3/16" diam. x 1-3/4" long	G 1 Carp	300	.027	Ea.	.21	1.46		1.67	2.41
8960	2-1/4" long	G 1 Carp	250	.032	Ea.	.29	1.75		2.04	2.93
8970	Phillips flat head, 3/16" diam. x 1-3/4" long	G	300	.027		.20	1.46		1.66	2.40
8980	2-1/4" long	G ↓	250	.032	↓	.31	1.75		2.06	2.95

05 05 21 – Fastening Methods for Metal

05 05 21.15 Drilling Steel

		Crew	Daily Output	Labor-Hours	Unit	Material	2021 Bare Costs Labor	Equipment	Total	Total Incl O&P
0010	**DRILLING STEEL**									
1910	Drilling & layout for steel, up to 1/4" deep, no anchor									
1920	Holes, 1/4" diameter	1 Sswk	112	.071	Ea.	.06	4.31		4.37	6.70
1925	For each additional 1/4" depth, add		336	.024		.06	1.44		1.50	2.29
1930	3/8" diameter		104	.077		.08	4.64		4.72	7.30
1935	For each additional 1/4" depth, add		312	.026		.08	1.55		1.63	2.48
1940	1/2" diameter		96	.083		.08	5.05		5.13	7.90
1945	For each additional 1/4" depth, add		288	.028		.08	1.68		1.76	2.68
1950	5/8" diameter		88	.091		.12	5.50		5.62	8.65
1955	For each additional 1/4" depth, add		264	.030		.12	1.83		1.95	2.96
1960	3/4" diameter		80	.100		.16	6.05		6.21	9.55
1965	For each additional 1/4" depth, add		240	.033		.16	2.01		2.17	3.29
1970	7/8" diameter		72	.111		.19	6.70		6.89	10.55
1975	For each additional 1/4" depth, add		216	.037		.19	2.23		2.42	3.67
1980	1" diameter		64	.125		.30	7.55		7.85	12
1985	For each additional 1/4" depth, add		192	.042		.30	2.51		2.81	4.22
1990	For drilling up, add				↓		40%			

05 05 23 – Metal Fastenings

05 05 23.10 Bolts and Hex Nuts

		Crew	Daily Output	Labor-Hours	Unit	Material	2021 Bare Costs Labor	Equipment	Total	Total Incl O&P
0010	**BOLTS & HEX NUTS**, Steel, A307									
0100	1/4" diameter, 1/2" long	G 1 Sswk	140	.057	Ea.	.14	3.45		3.59	5.50
0200	1" long	G	140	.057		.17	3.45		3.62	5.55
0300	2" long	G	130	.062		.25	3.71		3.96	6.05
0400	3" long	G	130	.062		.41	3.71		4.12	6.20
0500	4" long	G ↓	120	.067		.45	4.02		4.47	6.70

41

05 05 23 – Metal Fastenings

05 05 23.10 Bolts and Hex Nuts

05 05 23.10 Bolts and Hex Nuts		Crew	Daily Output	Labor-Hours	Unit	Material	2021 Bare Costs Labor	Equipment	Total	Total Incl O&P
0600	3/8" diameter, 1" long	G 1 Sswk	130	.062	Ea.	.17	3.71		3.88	5.95
0700	2" long	G	130	.062		.20	3.71		3.91	5.95
0800	3" long	G	120	.067		.26	4.02		4.28	6.50
0900	4" long	G	120	.067		.32	4.02		4.34	6.55
1000	5" long	G	115	.070		.39	4.19		4.58	6.95
1100	1/2" diameter, 1-1/2" long	G	120	.067		.39	4.02		4.41	6.65
1200	2" long	G	120	.067		.45	4.02		4.47	6.70
1300	4" long	G	115	.070		.74	4.19		4.93	7.30
1400	6" long	G	110	.073		1.04	4.39		5.43	7.95
1500	8" long	G	105	.076		1.38	4.59		5.97	8.60
1600	5/8" diameter, 1-1/2" long	G	120	.067		.82	4.02		4.84	7.10
1700	2" long	G	120	.067		.91	4.02		4.93	7.20
1800	4" long	G	115	.070		1.29	4.19		5.48	7.90
1900	6" long	G 1 Sswk	110	.073	Ea.	1.64	4.39		6.03	8.60
2000	8" long	G	105	.076		2.42	4.59		7.01	9.75
2100	10" long	G	100	.080		3.03	4.82		7.85	10.80
2200	3/4" diameter, 2" long	G	120	.067		1.18	4.02		5.20	7.50
2300	4" long	G	110	.073		1.71	4.39		6.10	8.70
2400	6" long	G	105	.076		2.22	4.59		6.81	9.55
2500	8" long	G	95	.084		3.37	5.10		8.47	11.55
2600	10" long	G	85	.094		4.44	5.70		10.14	13.70
2700	12" long	G	80	.100		5.20	6.05		11.25	15.10
2800	1" diameter, 3" long	G	105	.076		3.11	4.59		7.70	10.50
2900	6" long	G	90	.089		4.43	5.35		9.78	13.15
3000	12" long	G	75	.107		7.75	6.45		14.20	18.50
3100	For galvanized, add					75%				
3200	For stainless, add					350%				

05 05 23.30 Lag Screws

05 05 23.30 Lag Screws		Crew	Daily Output	Labor-Hours	Unit	Material	2021 Bare Costs Labor	Equipment	Total	Total Incl O&P
0010	**LAG SCREWS**									
0020	Steel, 1/4" diameter, 2" long	G 1 Carp	200	.040	Ea.	.13	2.19		2.32	3.41
0100	3/8" diameter, 3" long	G	150	.053		.44	2.92		3.36	4.83
0200	1/2" diameter, 3" long	G	130	.062		1.08	3.37		4.45	6.20
0300	5/8" diameter, 3" long	G	120	.067		1.72	3.65		5.37	7.35

05 05 23.35 Machine Screws

05 05 23.35 Machine Screws		Crew	Daily Output	Labor-Hours	Unit	Material	2021 Bare Costs Labor	Equipment	Total	Total Incl O&P
0010	**MACHINE SCREWS**									
0020	Steel, round head, #8 x 1" long	G 1 Carp	4.80	1.667	C	3.98	91		94.98	140
0110	#8 x 2" long	G	2.40	3.333		14.70	182		196.70	288
0200	#10 x 1" long	G	4	2		6.25	109		115.25	170
0300	#10 x 2" long	G	2	4		10.90	219		229.90	335

05 05 23.50 Powder Actuated Tools and Fasteners

05 05 23.50 Powder Actuated Tools and Fasteners		Crew	Daily Output	Labor-Hours	Unit	Material	2021 Bare Costs Labor	Equipment	Total	Total Incl O&P
0010	**POWDER ACTUATED TOOLS & FASTENERS**									
0020	Stud driver, .22 caliber, single shot				Ea.	146			146	160
0100	.27 caliber, semi automatic, strip				"	625			625	685
0300	Powder load, single shot, .22 cal, power level 2, brown				C	7.85			7.85	8.60
0400	Strip, .27 cal, power level 4, red					12			12	13.20
0600	Drive pin, .300 x 3/4" long	G 1 Carp	4.80	1.667		16.95	91		107.95	155
0700	.300 x 3" long with washer	G "	4	2		22	109		131	187

05 05 23.70 Structural Blind Bolts

05 05 23.70 Structural Blind Bolts		Crew	Daily Output	Labor-Hours	Unit	Material	2021 Bare Costs Labor	Equipment	Total	Total Incl O&P
0010	**STRUCTURAL BLIND BOLTS**									
0100	1/4" diameter x 1/4" grip	G 1 Sswk	240	.033	Ea.	1.27	2.01		3.28	4.51
0150	1/2" grip	G	216	.037		1.01	2.23		3.24	4.57
0200	3/8" diameter x 1/2" grip	G	232	.034		1.69	2.08		3.77	5.10

05 05 Common Work Results for Metals

05 05 23 – Metal Fastenings

05 05 23.70 Structural Blind Bolts

			Crew	Daily Output	Labor-Hours	Unit	Material	2021 Bare Costs Labor	Equipment	Total	Total Incl O&P
0250	3/4" grip	G	1 Sswk	208	.038	Ea.	1.96	2.32		4.28	5.75
0300	1/2" diameter x 1/2" grip	G		224	.036		3.82	2.15		5.97	7.55
0350	3/4" grip	G		200	.040		3.82	2.41		6.23	7.95
0400	5/8" diameter x 3/4" grip	G		216	.037		6	2.23		8.23	10.05
0450	1" grip	G		192	.042		6	2.51		8.51	10.50

05 12 Structural Steel Framing

05 12 23 – Structural Steel for Buildings

05 12 23.60 Pipe Support Framing

			Crew	Daily Output	Labor-Hours	Unit	Material	2021 Bare Costs Labor	Equipment	Total	Total Incl O&P
0010	**PIPE SUPPORT FRAMING**										
0020	Under 10#/L.F., shop fabricated	G	E-4	3900	.008	Lb.	1.77	.50	.04	2.31	2.75
0200	10.1 to 15#/L.F.	G		4300	.007		1.74	.45	.03	2.22	2.66
0400	15.1 to 20#/L.F.	G		4800	.007		1.71	.41	.03	2.15	2.55
0600	Over 20#/L.F.	G		5400	.006		1.69	.36	.03	2.08	2.45

05 35 Raceway Decking Assemblies

05 35 13 – Steel Cellular Decking

05 35 13.50 Cellular Decking

			Crew	Daily Output	Labor-Hours	Unit	Material	2021 Bare Costs Labor	Equipment	Total	Total Incl O&P
0010	**CELLULAR DECKING**										
0015	Made from recycled materials										
0200	Cellular units, galv, 1-1/2" deep, Type BC, 20-20 ga., over 15 squares	G	E-4	1460	.022	S.F.	11.70	1.33	.10	13.13	15.05
0250	18-20 ga.	G		1420	.023		9.20	1.37	.10	10.67	12.35
0300	18-18 ga.	G		1390	.023		9.40	1.40	.11	10.91	12.65
0320	16-18 ga.	G		1360	.024		16.25	1.43	.11	17.79	20
0340	16-16 ga.	G		1330	.024		18.10	1.46	.11	19.67	22.50
0400	3" deep, Type NC, galvanized, 20-20 ga.	G		1375	.023		12.90	1.41	.11	14.42	16.45
0500	18-20 ga.	G		1350	.024		10.75	1.44	.11	12.30	14.15
0600	18-18 ga.	G		1290	.025		10.70	1.51	.12	12.33	14.20
0700	16-18 ga.	G		1230	.026		17.50	1.58	.12	19.20	22
0800	16-16 ga.	G		1150	.028		19.05	1.69	.13	20.87	24
1000	4-1/2" deep, Type JC, galvanized, 18-20 ga.	G		1100	.029		12.40	1.77	.14	14.31	16.55
1100	18-18 ga.	G		1040	.031		12.30	1.87	.14	14.31	16.60
1200	16-18 ga.	G		980	.033		20	1.99	.15	22.14	25
1300	16-16 ga.	G		935	.034		22	2.08	.16	24.24	27.50
1900	For multi-story or congested site, add							50%			

For customer support on your Electrical Costs with RSMeans Data, call 800.448.8182.

43

Division Notes

	CREW	DAILY OUTPUT	LABOR-HOURS	UNIT	BARE COSTS				TOTAL INCL O&P
					MAT.	LABOR	EQUIP.	TOTAL	

Estimating Tips
06 05 00 Common Work Results for Wood, Plastics, and Composites

- Common to any wood-framed structure are the accessory connector items such as screws, nails, adhesives, hangers, connector plates, straps, angles, and hold-downs. For typical wood-framed buildings, such as residential projects, the aggregate total for these items can be significant, especially in areas where seismic loading is a concern. For floor and wall framing, the material cost is based on 10 to 25 lbs. of accessory connectors per MBF. Hold-downs, hangers, and other connectors should be taken off by the piece.

  Included with material costs are fasteners for a normal installation. Gordian's RSMeans engineers use manufacturers' recommendations, written specifications, and/or standard construction practice for the sizing and spacing of fasteners. Prices for various fasteners are shown for informational purposes only. Adjustments should be made if unusual fastening conditions exist.

06 10 00 Carpentry

- Lumber is a traded commodity and therefore sensitive to supply and demand in the marketplace. Even with "budgetary" estimating of wood-framed projects, it is advisable to call local suppliers for the latest market pricing.

- The common quantity unit for wood-framed projects is "thousand board feet" (MBF). A board foot is a volume of wood—1" x 1' x 1' or 144 cubic inches. Board-foot quantities are generally calculated using nominal material dimensions—dressed sizes are ignored. Board foot per lineal foot of any stick of lumber can be calculated by dividing the nominal cross-sectional area by 12. As an example, 2,000 lineal feet of 2 x 12 equates to 4 MBF by dividing the nominal area, 2 x 12, by 12, which equals 2, and multiplying that by 2,000 to give 4,000 board feet. This simple rule applies to all nominal dimensioned lumber.

- Waste is an issue of concern at the quantity takeoff for any area of construction. Framing lumber is sold in even foot lengths, i.e., 8', 10', 12', 14', 16', and depending on spans, wall heights, and the grade of lumber, waste is inevitable. A rule of thumb for lumber waste is 5–10% depending on material quality and the complexity of the framing.

- Wood in various forms and shapes is used in many projects, even where the main structural framing is steel, concrete, or masonry. Plywood as a back-up partition material and 2x boards used as blocking and cant strips around roof edges are two common examples. The estimator should ensure that the costs of all wood materials are included in the final estimate.

06 20 00 Finish Carpentry

- It is necessary to consider the grade of workmanship when estimating labor costs for erecting millwork and an interior finish. In practice, there are three grades: premium, custom, and economy. The RSMeans daily output for base and case moldings is in the range of 200 to 250 L.F. per carpenter per day. This is appropriate for most average custom-grade projects. For premium projects, an adjustment to productivity of 25–50% should be made, depending on the complexity of the job.

Reference Numbers

Reference numbers are shown at the beginning of some major classifications. These numbers refer to related items in the Reference Section. The reference information may be an estimating procedure, an alternate pricing method, or technical information.

Note: Not all subdivisions listed here necessarily appear. ∎

Same Data. Simplified.

Enjoy the convenience and efficiency of accessing your costs anywhere:

- **Skip the multiplier** by setting your location
- **Quickly search,** edit, favorite and share costs
- **Stay on top of price changes** with automatic updates

Discover more at rsmeans.com/online

06 16 36 — Wood Panel Product Sheathing

06 16 36.10 Sheathing	Crew	Daily Output	Labor-Hours	Unit	Material	2021 Bare Costs Labor	Equipment	Total	Total Incl O&P
0010 **SHEATHING**									
0012 Plywood on roofs, CDX									
0030 5/16" thick	2 Carp	1600	.010	S.F.	.90	.55		1.45	1.81
0035 Pneumatic nailed		1952	.008		.90	.45		1.35	1.66
0050 3/8" thick		1525	.010		.78	.57		1.35	1.72
0055 Pneumatic nailed		1860	.009		.78	.47		1.25	1.56
0100 1/2" thick		1400	.011		1	.63		1.63	2.03
0105 Pneumatic nailed		1708	.009		1	.51		1.51	1.86
0200 5/8" thick		1300	.012		1.16	.67		1.83	2.27
0205 Pneumatic nailed		1586	.010		1.16	.55		1.71	2.09
0300 3/4" thick		1200	.013		1.38	.73		2.11	2.61
0305 Pneumatic nailed	▼	1464	.011	▼	1.38	.60		1.98	2.41
0500 Plywood on walls, with exterior CDX, 3/8" thick	2 Carp	1200	.013	S.F.	.78	.73		1.51	1.95
0505 Pneumatic nailed		1488	.011		.78	.59		1.37	1.74
0600 1/2" thick		1125	.014		1	.78		1.78	2.26
0605 Pneumatic nailed		1395	.011		1	.63		1.63	2.04
0700 5/8" thick		1050	.015		1.16	.83		1.99	2.51
0705 Pneumatic nailed		1302	.012		1.16	.67		1.83	2.27
0800 3/4" thick		975	.016		1.38	.90		2.28	2.86
0805 Pneumatic nailed	▼	1209	.013	▼	1.38	.72		2.10	2.60

Estimating Tips

07 10 00 Dampproofing and Waterproofing

- Be sure of the job specifications before pricing this subdivision. The difference in cost between waterproofing and dampproofing can be great. Waterproofing will hold back standing water. Dampproofing prevents the transmission of water vapor. Also included in this section are vapor retarding membranes.

07 20 00 Thermal Protection

- Insulation and fireproofing products are measured by area, thickness, volume, or R-value. Specifications may give only what the specific R-value should be in a certain situation. The estimator may need to choose the type of insulation to meet that R-value.

07 30 00 Steep Slope Roofing
07 40 00 Roofing and Siding Panels

- Many roofing and siding products are bought and sold by the square. One square is equal to an area that measures 100 square feet.

  This simple change in unit of measure could create a large error if the estimator is not observant. Accessories necessary for a complete installation must be figured into any calculations for both material and labor.

07 50 00 Membrane Roofing
07 60 00 Flashing and Sheet Metal
07 70 00 Roofing and Wall Specialties and Accessories

- The items in these subdivisions compose a roofing system. No one component completes the installation, and all must be estimated. Built-up or single-ply membrane roofing systems are made up of many products and installation trades. Wood blocking at roof perimeters or penetrations, parapet coverings, reglets, roof drains, gutters, downspouts, sheet metal flashing, skylights, smoke vents, and roof hatches all need to be considered along with the roofing material. Several different installation trades will need to work together on the roofing system. Inherent difficulties in the scheduling and coordination of various trades must be accounted for when estimating labor costs.

07 90 00 Joint Protection

- To complete the weather-tight shell, the sealants and caulkings must be estimated. Where different materials meet—at expansion joints, at flashing penetrations, and at hundreds of other locations throughout a construction project—caulking and sealants provide another line of defense against water penetration. Often, an entire system is based on the proper location and placement of caulking or sealants. The detailed drawings that are included as part of a set of architectural plans show typical locations for these materials. When caulking or sealants are shown at typical locations, this means the estimator must include them for all the locations where this detail is applicable. Be careful to keep different types of sealants separate, and remember to consider backer rods and primers if necessary.

Reference Numbers

Reference numbers are shown at the beginning of some major classifications. These numbers refer to related items in the Reference Section. The reference information may be an estimating procedure, an alternate pricing method, or technical information.

Note: Not all subdivisions listed here necessarily appear. ■

Same Data. Simplified.

Enjoy the convenience and efficiency of accessing your costs anywhere:

- **Skip the multiplier** by setting your location
- **Quickly search,** edit, favorite and share costs
- **Stay on top of price changes** with automatic updates

Discover more at rsmeans.com/online

07 84 13 – Penetration Firestopping

07 84 13.10 Firestopping		Crew	Daily Output	Labor-Hours	Unit	Material	2021 Bare Costs Labor	Equipment	Total	Total Incl O&P
0010	**FIRESTOPPING** R078413-30									
0100	Metallic piping, non insulated									
0110	Through walls, 2" diameter	1 Carp	16	.500	Ea.	3.70	27.50		31.20	45
0120	4" diameter		14	.571		8.30	31.50		39.80	55.50
0130	6" diameter		12	.667		12.70	36.50		49.20	68.50
0140	12" diameter		10	.800		25	44		69	93
0150	Through floors, 2" diameter		32	.250		1.88	13.70		15.58	22.50
0160	4" diameter		28	.286		3.31	15.65		18.96	27
0170	6" diameter		24	.333		6.50	18.25		24.75	34
0180	12" diameter		20	.400		12.90	22		34.90	46.50
0190	Metallic piping, insulated									
0200	Through walls, 2" diameter	1 Carp	16	.500	Ea.	8.65	27.50		36.15	50.50
0210	4" diameter		14	.571		12.65	31.50		44.15	60.50
0220	6" diameter		12	.667		16.90	36.50		53.40	73
0230	12" diameter		10	.800		28.50	44		72.50	97
0240	Through floors, 2" diameter		32	.250		3.30	13.70		17	24
0250	4" diameter		28	.286		6.35	15.65		22	30.50
0260	6" diameter		24	.333		8.75	18.25		27	36.50
0270	12" diameter		20	.400		13.05	22		35.05	47
0280	Non metallic piping, non insulated									
0290	Through walls, 2" diameter	1 Carp	12	.667	Ea.	54.50	36.50		91	115
0300	4" diameter		10	.800		109	44		153	186
0310	6" diameter		8	1		156	54.50		210.50	254
0330	Through floors, 2" diameter		16	.500		28	27.50		55.50	72
0340	4" diameter		6	1.333		64	73		137	180
0350	6" diameter		6	1.333		120	73		193	241
0370	Ductwork, insulated & non insulated, round									
0380	Through walls, 6" diameter	1 Carp	12	.667	Ea.	16.65	36.50		53.15	73
0390	12" diameter		10	.800		29	44		73	97
0400	18" diameter		8	1		34	54.50		88.50	119
0410	Through floors, 6" diameter		16	.500		8.55	27.50		36.05	50.50
0420	12" diameter	1 Carp	14	.571	Ea.	14.70	31.50		46.20	62.50
0430	18" diameter	"	12	.667	"	17.75	36.50		54.25	74
0440	Ductwork, insulated & non insulated, rectangular									
0450	With stiffener/closure angle, through walls, 6" x 12"	1 Carp	8	1	Ea.	22	54.50		76.50	106
0460	12" x 24"		6	1.333		37	73		110	150
0470	24" x 48"		4	2		93	109		202	265
0480	With stiffener/closure angle, through floors, 6" x 12"		10	.800		10.50	44		54.50	77
0490	12" x 24"		8	1		15.85	54.50		70.35	99
0500	24" x 48"		6	1.333		44	73		117	157
0510	Multi trade openings									
0520	Through walls, 6" x 12"	1 Carp	2	4	Ea.	56	219		275	385
0530	12" x 24"	"	1	8		210	440		650	885
0540	24" x 48"	2 Carp	1	16		675	875		1,550	2,050
0550	48" x 96"	"	.75	21.333		2,550	1,175		3,725	4,550
0560	Through floors, 6" x 12"	1 Carp	2	4		45	219		264	375
0570	12" x 24"	"	1	8		64	440		504	725
0580	24" x 48"	2 Carp	.75	21.333		138	1,175		1,313	1,900
0590	48" x 96"	"	.50	32		325	1,750		2,075	2,975
0600	Structural penetrations, through walls									
0610	Steel beams, W8 x 10	1 Carp	8	1	Ea.	74	54.50		128.50	163
0620	W12 x 14		6	1.333		108	73		181	228
0630	W21 x 44		5	1.600		165	87.50		252.50	315

07 84 Firestopping

07 84 13 – Penetration Firestopping

07 84 13.10 Firestopping		Crew	Daily Output	Labor-Hours	Unit	Material	2021 Bare Costs Labor	Equipment	Total	Total Incl O&P
0640	W36 x 135	1 Carp	3	2.667	Ea.	234	146		380	475
0650	Bar joists, 18" deep		6	1.333		43	73		116	157
0660	24" deep		6	1.333		61	73		134	176
0670	36" deep		5	1.600		91.50	87.50		179	232
0680	48" deep	↓	4	2	↓	132	109		241	310
0690	Construction joints, floor slab at exterior wall									
0700	Precast, brick, block or drywall exterior									
0710	2" wide joint	1 Carp	125	.064	L.F.	11.05	3.50		14.55	17.40
0720	4" wide joint	"	75	.107	"	17.05	5.85		22.90	27.50
0730	Metal panel, glass or curtain wall exterior									
0740	2" wide joint	1 Carp	40	.200	L.F.	12.55	10.95		23.50	30
0750	4" wide joint	"	25	.320	"	14.25	17.50		31.75	41.50
0760	Floor slab to drywall partition									
0770	Flat joint	1 Carp	100	.080	L.F.	14.45	4.38		18.83	22.50
0780	Fluted joint		50	.160		16.80	8.75		25.55	31.50
0790	Etched fluted joint	↓	75	.107	↓	20	5.85		25.85	30.50
0800	Floor slab to concrete/masonry partition									
0810	Flat joint	1 Carp	75	.107	L.F.	16.85	5.85		22.70	27
0820	Fluted joint	"	50	.160	"	18.35	8.75		27.10	33
0830	Concrete/CMU wall joints									
0840	1" wide	1 Carp	100	.080	L.F.	20.50	4.38		24.88	29
0850	2" wide		75	.107		32	5.85		37.85	43.50
0860	4" wide	↓	50	.160		36	8.75		44.75	52.50
0870	Concrete/CMU floor joints									
0880	1" wide	1 Carp	200	.040	L.F.	21.50	2.19		23.69	27
0890	2" wide		150	.053		24.50	2.92		27.42	31.50
0900	4" wide	↓	100	.080	↓	36	4.38		40.38	46

07 91 Preformed Joint Seals

07 91 13 – Compression Seals

07 91 13.10 Compression Seals		Crew	Daily Output	Labor-Hours	Unit	Material	2021 Bare Costs Labor	Equipment	Total	Total Incl O&P
0010	**COMPRESSION SEALS**									
4900	O-ring type cord, 1/4"	1 Bric	472	.017	L.F.	.42	.91		1.33	1.83
4910	1/2"		440	.018		1.11	.98		2.09	2.69
4920	3/4"		424	.019		2.23	1.01		3.24	3.98
4930	1"		408	.020		4.48	1.05		5.53	6.50
4940	1-1/4"		384	.021		10.80	1.12		11.92	13.55
4950	1-1/2"		368	.022		13.25	1.17		14.42	16.30
4960	1-3/4"		352	.023		15.45	1.22		16.67	18.85
4970	2"	↓	344	.023	↓	27	1.25		28.25	31.50

07 91 23 – Backer Rods

07 91 23.10 Backer Rods		Crew	Daily Output	Labor-Hours	Unit	Material	2021 Bare Costs Labor	Equipment	Total	Total Incl O&P
0010	**BACKER RODS**									
0030	Backer rod, polyethylene, 1/4" diameter	1 Bric	4.60	1.739	C.L.F.	2.77	93.50		96.27	144
0050	1/2" diameter		4.60	1.739		19.90	93.50		113.40	163
0070	3/4" diameter		4.60	1.739		7.70	93.50		101.20	149
0090	1" diameter	↓	4.60	1.739	↓	15.45	93.50		108.95	158

07 91 26.10 Joint Fillers

		Crew	Daily Output	Labor-Hours	Unit	Material	2021 Bare Costs Labor	2021 Bare Costs Equipment	Total	Total Incl O&P
0010	**JOINT FILLERS**									
4360	Butyl rubber filler, 1/4" x 1/4"	1 Bric	290	.028	L.F.	.23	1.48		1.71	2.48
4365	1/2" x 1/2"		250	.032		.92	1.72		2.64	3.60
4370	1/2" x 3/4"		210	.038		1.38	2.05		3.43	4.60
4375	3/4" x 3/4"		230	.035		2.07	1.87		3.94	5.10
4380	1" x 1"		180	.044		2.76	2.39		5.15	6.65
4390	For coloring, add					12%				
4980	Polyethylene joint backing, 1/4" x 2"	1 Bric	2.08	3.846	C.L.F.	14.85	207		221.85	325
4990	1/4" x 6"		1.28	6.250	"	33.50	335		368.50	540
5600	Silicone, room temp vulcanizing foam seal, 1/4" x 1/2"		1312	.006	L.F.	.45	.33		.78	.99
5610	1/2" x 1/2"		656	.012		.91	.65		1.56	1.99
5620	1/2" x 3/4"		442	.018		1.36	.97		2.33	2.95
5630	3/4" x 3/4"		328	.024		2.04	1.31		3.35	4.21
5640	1/8" x 1"		1312	.006		.45	.33		.78	.99
5650	1/8" x 3"		442	.018		1.36	.97		2.33	2.95
5670	1/4" x 3"		295	.027		2.72	1.46		4.18	5.20
5680	1/4" x 6"	1 Bric	148	.054	L.F.	5.45	2.90		8.35	10.35
5690	1/2" x 6"		82	.098		10.85	5.25		16.10	19.85
5700	1/2" x 9"		52.50	.152		16.30	8.20		24.50	30.50
5710	1/2" x 12"		33	.242		21.50	13		34.50	43.50

07 92 13.20 Caulking and Sealant Options

		Crew	Daily Output	Labor-Hours	Unit	Material	2021 Bare Costs Labor	2021 Bare Costs Equipment	Total	Total Incl O&P
0010	**CAULKING AND SEALANT OPTIONS**									
0050	Latex acrylic based, bulk				Gal.	30.50			30.50	33.50
0055	Bulk in place 1/4" x 1/4" bead	1 Bric	300	.027	L.F.	.10	1.43		1.53	2.27
0060	1/4" x 3/8"		294	.027		.16	1.46		1.62	2.38
0065	1/4" x 1/2"		288	.028		.21	1.49		1.70	2.48
0075	3/8" x 3/8"		284	.028		.24	1.51		1.75	2.54
0080	3/8" x 1/2"		280	.029		.32	1.53		1.85	2.66
0085	3/8" x 5/8"		276	.029		.40	1.56		1.96	2.78
0095	3/8" x 3/4"		272	.029		.48	1.58		2.06	2.90
0100	1/2" x 1/2"		275	.029		.42	1.56		1.98	2.82
0105	1/2" x 5/8"		269	.030		.53	1.60		2.13	2.99
0110	1/2" x 3/4"		263	.030		.64	1.63		2.27	3.16
0115	1/2" x 7/8"		256	.031		.74	1.68		2.42	3.35
0120	1/2" x 1"		250	.032		.85	1.72		2.57	3.52
0125	3/4" x 3/4"		244	.033		.96	1.76		2.72	3.70
0130	3/4" x 1"		225	.036		1.28	1.91		3.19	4.28
0135	1" x 1"		200	.040		1.70	2.15		3.85	5.10
0190	Cartridges				Gal.	41.50			41.50	45.50
0200	11 fl. oz. cartridge				Ea.	3.57			3.57	3.93
0600	1/2" x 1/2"	1 Bric	275	.029	L.F.	.58	1.56		2.14	2.99
3200	Polyurethane, 1 or 2 component				Gal.	54.50			54.50	60
3300	Cartridges				"	65			65	71.50
3500	Bulk, in place, 1/4" x 1/4"	1 Bric	300	.027	L.F.	.18	1.43		1.61	2.36
3655	1/2" x 1/4"		288	.028		.35	1.49		1.84	2.64
3800	3/4" x 3/8"		272	.029		.80	1.58		2.38	3.26
3900	1" x 1/2"		250	.032		1.42	1.72		3.14	4.15

07 92 Joint Sealants

07 92 19 – Acoustical Joint Sealants

07 92 19.10 Acoustical Sealant	Crew	Daily Output	Labor-Hours	Unit	Material	2021 Bare Costs Labor	Equipment	Total	Total Incl O&P
0010 **ACOUSTICAL SEALANT**									
0020 Acoustical sealant, elastomeric, cartridges				Ea.	8.05			8.05	8.85
0025 In place, 1/4" x 1/4"	1 Bric	300	.027	L.F.	.33	1.43		1.76	2.52
0030 1/4" x 1/2"		288	.028		.66	1.49		2.15	2.97
0035 1/2" x 1/2"		275	.029		1.31	1.56		2.87	3.80
0040 1/2" x 3/4"		263	.030		1.97	1.63		3.60	4.63
0045 3/4" x 3/4"		244	.033		2.96	1.76		4.72	5.90
0050 1" x 1"		200	.040		5.25	2.15		7.40	9.05

Division Notes

		CREW	DAILY OUTPUT	LABOR-HOURS	UNIT	BARE COSTS				TOTAL INCL O&P
						MAT.	LABOR	EQUIP.	TOTAL	

Estimating Tips
08 10 00 Doors and Frames

All exterior doors should be addressed for their energy conservation (insulation and seals).

- Most metal doors and frames look alike, but there may be significant differences among them. When estimating these items, be sure to choose the line item that most closely compares to the specification or door schedule requirements regarding:
  - □ type of metal
  - □ metal gauge
  - □ door core material
  - □ fire rating
  - □ finish
- Wood and plastic doors vary considerably in price. The primary determinant is the veneer material. Lauan, birch, and oak are the most common veneers. Other variables include the following:
  - □ hollow or solid core
  - □ fire rating
  - □ flush or raised panel
  - □ finish
- Door pricing includes bore for cylindrical locksets and mortise for hinges.

08 30 00 Specialty Doors and Frames

- There are many varieties of special doors, and they are usually priced per each. Add frames, hardware, or operators required for a complete installation.

08 40 00 Entrances, Storefronts, and Curtain Walls

- Glazed curtain walls consist of the metal tube framing and the glazing material. The cost data in this subdivision is presented for the metal tube framing alone or the composite wall. If your estimate requires a detailed takeoff of the framing, be sure to add the glazing cost and any tints.

08 50 00 Windows

- Steel windows are unglazed and aluminum can be glazed or unglazed. Some metal windows are priced without glass. Refer to 08 80 00 Glazing for glass pricing. The grade C indicates commercial grade windows, usually ASTM C-35.
- All wood windows and vinyl are priced preglazed. The glazing is insulating glass. Add the cost of screens and grills if required and not already included.

08 70 00 Hardware

- Hardware costs add considerably to the cost of a door. The most efficient method to determine the hardware requirements for a project is to review the door and hardware schedule together. One type of door may have different hardware, depending on the door usage.
- Door hinges are priced by the pair, with most doors requiring 1-1/2 pairs per door. The hinge prices do not include installation labor because it is included in door installation.

Hinges are classified according to the frequency of use, base material, and finish.

08 80 00 Glazing

- Different openings require different types of glass. The most common types are:
  - □ float
  - □ tempered
  - □ insulating
  - □ impact-resistant
  - □ ballistic-resistant
- Most exterior windows are glazed with insulating glass. Entrance doors and window walls, where the glass is less than 18" from the floor, are generally glazed with tempered glass. Interior windows and some residential windows are glazed with float glass.
- Coastal communities require the use of impact-resistant glass, dependent on wind speed.
- The insulation or 'u' value is a strong consideration, along with solar heat gain, to determine total energy efficiency.

Reference Numbers

Reference numbers are shown at the beginning of some major classifications. These numbers refer to related items in the Reference Section. The reference information may be an estimating procedure, an alternate pricing method, or technical information.

Note: Not all subdivisions listed here necessarily appear. ■

Same Data. Simplified.

Enjoy the convenience and efficiency of accessing your costs anywhere:

- **Skip the multiplier** by setting your location
- **Quickly search**, edit, favorite and share costs
- **Stay on top of price changes** with automatic updates

Discover more at rsmeans.com/online

08 31 13.10 Types of Framed Access Doors	Crew	Daily Output	Labor-Hours	Unit	Material	2021 Bare Costs Labor	Equipment	Total	Total Incl O&P
0010 **TYPES OF FRAMED ACCESS DOORS**									
4000 Recessed door for drywall									
4100 Metal 12" x 12"	1 Carp	6	1.333	Ea.	75.50	73		148.50	192
4150 12" x 24"		5.50	1.455		118	79.50		197.50	249
4200 24" x 36"	↓	5	1.600	↓	189	87.50		276.50	340

54

For customer support on your Electrical Costs with RSMeans Data, call 800.448.8182.

Estimating Tips
General

- Room Finish Schedule: A complete set of plans should contain a room finish schedule. If one is not available, it would be well worth the time and effort to obtain one.

09 20 00 Plaster and Gypsum Board

- Lath is estimated by the square yard plus a 5% allowance for waste. Furring, channels, and accessories are measured by the linear foot. An extra foot should be allowed for each accessory miter or stop.

- Plaster is also estimated by the square yard. Deductions for openings vary by preference, from zero deduction to 50% of all openings over 2 feet in width. The estimator should allow one extra square foot for each linear foot of horizontal interior or exterior angle located below the ceiling level. Also, double the areas of small radius work.

- Drywall accessories, studs, track, and acoustical caulking are all measured by the linear foot. Drywall taping is figured by the square foot. Gypsum wallboard is estimated by the square foot. No material deductions should be made for door or window openings under 32 S.F.

09 60 00 Flooring

- Tile and terrazzo areas are taken off on a square foot basis. Trim and base materials are measured by the linear foot. Accent tiles are listed per each. Two basic methods of installation are used. Mud set is approximately 30% more expensive than thin set.

The cost of grout is included with tile unit price lines unless otherwise noted. In terrazzo work, be sure to include the linear footage of embedded decorative strips, grounds, machine rubbing, and power cleanup.

- Wood flooring is available in strip, parquet, or block configuration. The latter two types are set in adhesives with quantities estimated by the square foot. The laying pattern will influence labor costs and material waste. In addition to the material and labor for laying wood floors, the estimator must make allowances for sanding and finishing these areas, unless the flooring is prefinished.

- Sheet flooring is measured by the square yard. Roll widths vary, so consideration should be given to use the most economical width, as waste must be figured into the total quantity. Consider also the installation methods available—direct glue down or stretched. Direct glue-down installation is assumed with sheet carpet unit price lines unless otherwise noted.

09 70 00 Wall Finishes

- Wall coverings are estimated by the square foot. The area to be covered is measured—length by height of the wall above the baseboards—to calculate the square footage of each wall. This figure is divided by the number of square feet in the single roll which is being used. Deduct, in full, the areas of openings such as doors and windows. Where a pattern match is required allow 25–30% waste.

09 80 00 Acoustic Treatment

- Acoustical systems fall into several categories. The takeoff of these materials should be by the square foot of area with a 5% allowance for waste. Do not forget about scaffolding, if applicable, when estimating these systems.

09 90 00 Painting and Coating

- New line items created for cut-ins with reference diagram.

- A major portion of the work in painting involves surface preparation. Be sure to include cleaning, sanding, filling, and masking costs in the estimate.

- Protection of adjacent surfaces is not included in painting costs. When considering the method of paint application, an important factor is the amount of protection and masking required. These must be estimated separately and may be the determining factor in choosing the method of application.

Reference Numbers

Reference numbers are shown at the beginning of some major classifications. These numbers refer to related items in the Reference Section. The reference information may be an estimating procedure, an alternate pricing method, or technical information.

Note: Not all subdivisions listed here necessarily appear. ■

Same Data. Simplified.

Enjoy the convenience and efficiency of accessing your costs anywhere:

- **Skip the multiplier** by setting your location
- **Quickly search,** edit, favorite and share costs
- **Stay on top of price changes** with automatic updates

Discover more at rsmeans.com/online

09 22 Supports for Plaster and Gypsum Board

09 22 03 – Fastening Methods for Finishes

09 22 03.20 Drilling Plaster/Drywall	Crew	Daily Output	Labor-Hours	Unit	Material	2021 Bare Costs Labor	Equipment	Total	Total Incl O&P
0010 **DRILLING PLASTER/DRYWALL**									
1100 Drilling & layout for drywall/plaster walls, up to 1" deep, no anchor									
1200 Holes, 1/4" diameter	1 Carp	150	.053	Ea.	.01	2.92		2.93	4.36
1300 3/8" diameter		140	.057		.01	3.13		3.14	4.68
1400 1/2" diameter		130	.062		.01	3.37		3.38	5
1500 3/4" diameter		120	.067		.01	3.65		3.66	5.45
1600 1" diameter		110	.073		.03	3.98		4.01	6
1700 1-1/4" diameter		100	.080		.04	4.38		4.42	6.60
1800 1-1/2" diameter		90	.089		.05	4.86		4.91	7.30
1900 For ceiling installations, add						40%			

09 57 Special Function Ceilings

09 57 53 – Security Ceiling Assemblies

09 57 53.10 Ceiling Assem., Security, Radio Freq. Shielding	Crew	Daily Output	Labor-Hours	Unit	Material	2021 Bare Costs Labor	Equipment	Total	Total Incl O&P
0010 **CEILING ASSEMBLY, SECURITY, RADIO FREQUENCY SHIELDING**									
0020 Prefabricated, galvanized steel	2 Carp	375	.043	SF Surf	5.95	2.33		8.28	10.05
0050 5 oz., copper ceiling panel		155	.103		4.84	5.65		10.49	13.75
0110 12 oz., copper ceiling panel		140	.114		10.10	6.25		16.35	20.50
0250 Ceiling hangers	E-1	45	.533	Ea.	39	31.50	3.31	73.81	95
0300 Shielding transition, 11 ga. preformed angles	"	1365	.018	L.F.	39	1.04	.11	40.15	44.50

09 69 Access Flooring

09 69 13 – Rigid-Grid Access Flooring

09 69 13.10 Access Floors	Crew	Daily Output	Labor-Hours	Unit	Material	2021 Bare Costs Labor	Equipment	Total	Total Incl O&P
0010 **ACCESS FLOORS**									
0015 Access floor pkg. including panel, pedestal, & stringers 1500 lbs. load									
0100 Package pricing, conc. fill panels, no fin., 6" ht.	4 Carp	750	.043	S.F.	15	2.33		17.33	20
0105 12" ht.		750	.043		15.20	2.33		17.53	20
0110 18" ht.		750	.043		15.40	2.33		17.73	20.50
0115 24" ht.		750	.043		15.55	2.33		17.88	20.50
0120 Package pricing, steel panels, no fin., 6" ht.		750	.043		18	2.33		20.33	23.50
0125 12" ht.		750	.043		18.05	2.33		20.38	23.50
0130 18" ht.		750	.043		18.25	2.33		20.58	23.50
0135 24" ht.		750	.043		18.45	2.33		20.78	24
0140 Package pricing, wood core panels, no fin., 6" ht.	4 Carp	750	.043	S.F.	12.20	2.33		14.53	16.90
0145 12" ht.		750	.043		12.40	2.33		14.73	17.10
0150 18" ht.		750	.043		12.55	2.33		14.88	17.30
0155 24" ht.		750	.043		12.75	2.33		15.08	17.55
0160 Pkg. pricing, conc. fill pnls, no stringers, no fin., 6" ht, 1250 lbs. load		700	.046		17.85	2.50		20.35	23.50
0165 12" ht.		700	.046		18.10	2.50		20.60	23.50
0170 18" ht.		700	.046		18.30	2.50		20.80	23.50
0175 24" ht.		700	.046		18.50	2.50		21	24
0250 Panels, 2' x 2' conc. fill, no fin.	2 Carp	500	.032		23	1.75		24.75	27.50
0255 With 1/8" high pressure laminate		500	.032		56.50	1.75		58.25	64.50
0260 Metal panels with 1/8" high pressure laminate		500	.032		66.50	1.75		68.25	76
0265 Wood core panels with 1/8" high pressure laminate		500	.032		44	1.75		45.75	50.50
0400 Aluminum panels, no fin.		500	.032		35	1.75		36.75	41
0600 For carpet covering, add					9.20			9.20	10.15
0700 For vinyl floor covering, add					9.80			9.80	10.80

09 69 Access Flooring

09 69 13 – Rigid-Grid Access Flooring

09 69 13.10 Access Floors		Crew	Daily Output	Labor-Hours	Unit	Material	2021 Bare Costs Labor	Equipment	Total	Total Incl O&P
0900	For high pressure laminate covering, add				S.F.	8.10			8.10	8.90
0910	For snap on stringer system, add	2 Carp	1000	.016	↓	1.45	.88		2.33	2.91
1000	Machine cutouts after initial installation	1 Carp	50	.160	Ea.	19.75	8.75		28.50	34.50
1050	Pedestals, 6" to 12"	2 Carp	85	.188		9.20	10.30		19.50	25.50
1100	Air conditioning grilles, 4" x 12"	1 Carp	17	.471		72.50	25.50		98	118
1150	4" x 18"	"	14	.571	↓	99	31.50		130.50	156
1200	Approach ramps, steel	2 Carp	60	.267	S.F.	29.50	14.60		44.10	54.50
1300	Aluminum	"	40	.400	"	35	22		57	71
1500	Handrail, 2 rail, aluminum	1 Carp	15	.533	L.F.	126	29		155	182

Division Notes

	CREW	DAILY OUTPUT	LABOR-HOURS	UNIT	BARE COSTS				TOTAL INCL O&P
					MAT.	LABOR	EQUIP.	TOTAL	

Estimating Tips
General
- The items in this division are usually priced per square foot or each.
- Many items in Division 10 require some type of support system or special anchors that are not usually furnished with the item. The required anchors must be added to the estimate in the appropriate division.
- Some items in Division 10, such as lockers, may require assembly before installation. Verify the amount of assembly required. Assembly can often exceed installation time.

10 20 00 Interior Specialties
- Support angles and blocking are not included in the installation of toilet compartments, shower/dressing compartments, or cubicles. Appropriate line items from Division 5 or 6 may need to be added to support the installations.
- Toilet partitions are priced by the stall. A stall consists of a side wall, pilaster, and door with hardware. Toilet tissue holders and grab bars are extra.
- The required acoustical rating of a folding partition can have a significant impact on costs. Verify the sound transmission coefficient rating of the panel priced against the specification requirements.

- Grab bar installation does not include supplemental blocking or backing to support the required load. When grab bars are installed at an existing facility, provisions must be made to attach the grab bars to a solid structure.

Reference Numbers
Reference numbers are shown at the beginning of some major classifications. These numbers refer to related items in the Reference Section. The reference information may be an estimating procedure, an alternate pricing method, or technical information.

Note: Not all subdivisions listed here necessarily appear. ■

Same Data. Simplified.
Enjoy the convenience and efficiency of accessing your costs anywhere:

- **Skip the multiplier** by setting your location
- **Quickly search**, edit, favorite and share costs
- **Stay on top of price changes** with automatic updates

Discover more at rsmeans.com/online

10 28 16.20 Medicine Cabinets	Crew	Daily Output	Labor-Hours	Unit	Material	2021 Bare Costs Labor	Equipment	Total	Total Incl O&P
0010 **MEDICINE CABINETS**									
0020 With mirror, sst frame, 16" x 22", unlighted	1 Carp	14	.571	Ea.	102	31.50		133.50	159
0100 Wood frame		14	.571		128	31.50		159.50	188
0300 Sliding mirror doors, 20" x 16" x 4-3/4", unlighted		7	1.143		136	62.50		198.50	244
0400 24" x 19" x 8-1/2", lighted		5	1.600		220	87.50		307.50	375
0600 Triple door, 30" x 32", unlighted, plywood body		7	1.143		271	62.50		333.50	390
0700 Steel body		7	1.143		355	62.50		417.50	485
0900 Oak door, wood body, beveled mirror, single door		7	1.143		119	62.50		181.50	225
1000 Double door		6	1.333		375	73		448	520
1200 Hotel cabinets, stainless, with lower shelf, unlighted		10	.800		296	44		340	390
1300 Lighted		5	1.600		310	87.50		397.50	475

Estimating Tips
General

- The items in this division are usually priced per square foot or each. Many of these items are purchased by the owner for installation by the contractor. Check the specifications for responsibilities and include time for receiving, storage, installation, and mechanical and electrical hookups in the appropriate divisions.

- Many items in Division 11 require some type of support system that is not usually furnished with the item. Examples of these systems include blocking for the attachment of casework and support angles for ceiling-hung projection screens. The required blocking or supports must be added to the estimate in the appropriate division.

- Some items in Division 11 may require assembly or electrical hookups. Verify the amount of assembly required or the need for a hard electrical connection and add the appropriate costs.

Reference Numbers

Reference numbers are shown at the beginning of some major classifications. These numbers refer to related items in the Reference Section. The reference information may be an estimating procedure, an alternate pricing method, or technical information.

Same Data. Simplified.

Enjoy the convenience and efficiency of accessing your costs anywhere:

- **Skip the multiplier** by setting your location
- **Quickly search,** edit, favorite and share costs
- **Stay on top of price changes** with automatic updates

Discover more at rsmeans.com/online

11 12 Parking Control Equipment

11 12 13 – Parking Key and Card Control Units

11 12 13.10 Parking Control Units

		Crew	Daily Output	Labor-Hours	Unit	Material	2021 Bare Costs Labor	Equipment	Total	Total Incl O&P
0010	**PARKING CONTROL UNITS**									
5100	Card reader	1 Elec	2	4	Ea.	895	255		1,150	1,375
5120	Proximity with customer display	2 Elec	1	16		6,100	1,025		7,125	8,225
6000	Parking control software, basic functionality	1 Elec	.50	16		23,000	1,025		24,025	26,800
6020	Multi-function	"	.20	40	↓	90,000	2,550		92,550	103,000

11 12 16 – Parking Ticket Dispensers

11 12 16.10 Ticket Dispensers

		Crew	Daily Output	Labor-Hours	Unit	Material	2021 Bare Costs Labor	Equipment	Total	Total Incl O&P
0010	**TICKET DISPENSERS**									
5900	Ticket spitter with time/date stamp, standard	2 Elec	2	8	Ea.	6,525	510		7,035	7,925
5920	Mag stripe encoding	"	2	8	"	18,900	510		19,410	21,600

11 12 26 – Parking Fee Collection Equipment

11 12 26.13 Parking Fee Coin Collection Equipment

		Crew	Daily Output	Labor-Hours	Unit	Material	2021 Bare Costs Labor	Equipment	Total	Total Incl O&P
0010	**PARKING FEE COIN COLLECTION EQUIPMENT**									
5200	Cashier booth, average	B-22	1	30	Ea.	10,400	1,550	287	12,237	14,100
5300	Collector station, pay on foot	2 Elec	.20	80		111,500	5,100		116,600	130,500
5320	Credit card only	"	.50	32	↓	20,600	2,050		22,650	25,700

11 12 26.23 Fee Equipment

		Crew	Daily Output	Labor-Hours	Unit	Material	2021 Bare Costs Labor	Equipment	Total	Total Incl O&P
0010	**FEE EQUIPMENT**									
5600	Fee computer	1 Elec	1.50	5.333	Ea.	13,000	340		13,340	14,800

11 12 33 – Parking Gates

11 12 33.13 Lift Arm Parking Gates

		Crew	Daily Output	Labor-Hours	Unit	Material	2021 Bare Costs Labor	Equipment	Total	Total Incl O&P
0010	**LIFT ARM PARKING GATES**									
5000	Barrier gate with programmable controller	2 Elec	3	5.333	Ea.	3,325	340		3,665	4,175
5020	Industrial		3	5.333		5,175	340		5,515	6,175
5050	Non-programmable, with reader and 12' arm		3	5.333		1,725	340		2,065	2,400
5500	Exit verifier	↓	1	16		19,800	1,025		20,825	23,300
5700	Full sign, 4" letters	1 Elec	2	4		1,225	255		1,480	1,725
5800	Inductive loop	2 Elec	4	4		208	255		463	610
5950	Vehicle detector, microprocessor based	1 Elec	3	2.667	↓	465	170		635	765

11 21 Retail and Service Equipment

11 21 73 – Commercial Laundry and Dry Cleaning Equipment

11 21 73.13 Dry Cleaning Equipment

		Crew	Daily Output	Labor-Hours	Unit	Material	2021 Bare Costs Labor	Equipment	Total	Total Incl O&P
0010	**DRY CLEANING EQUIPMENT**									
2000	Dry cleaners, electric, 20 lb. capacity, not incl. rough-in	L-1	.20	80	Ea.	35,000	5,250		40,250	46,300
2050	25 lb. capacity		.17	94.118		52,500	6,175		58,675	66,500
2100	30 lb. capacity		.15	107		55,500	7,000		62,500	71,500
2150	60 lb. capacity	↓	.09	178	↓	78,500	11,700		90,200	104,000

11 21 73.19 Finishing Equipment

		Crew	Daily Output	Labor-Hours	Unit	Material	2021 Bare Costs Labor	Equipment	Total	Total Incl O&P
0010	**FINISHING EQUIPMENT**									
3500	Folders, blankets & sheets, minimum	1 Elec	.17	47.059	Ea.	35,900	3,000		38,900	44,000
3700	King size with automatic stacker		.10	80		65,000	5,100		70,100	79,000
3800	For conveyor delivery, add	↓	.45	17.778	↓	16,500	1,125		17,625	19,900

11 21 73.23 Commercial Ironing Equipment

		Crew	Daily Output	Labor-Hours	Unit	Material	2021 Bare Costs Labor	Equipment	Total	Total Incl O&P
0010	**COMMERCIAL IRONING EQUIPMENT**									
4500	Ironers, institutional, 110", single roll	1 Elec	.20	40	Ea.	34,800	2,550		37,350	42,000

11 21 Retail and Service Equipment

11 21 73 – Commercial Laundry and Dry Cleaning Equipment

11 21 73.26 Commercial Washers and Extractors

	Crew	Daily Output	Labor-Hours	Unit	Material	2021 Bare Costs Labor	2021 Bare Costs Equipment	Total	Total Incl O&P
0010 **COMMERCIAL WASHERS AND EXTRACTORS**, not including rough-in									
6000 Combination washer/extractor, 20 lb. capacity	L-6	1.50	8	Ea.	7,175	530		7,705	8,700
6100 30 lb. capacity		.80	15		10,900	995		11,895	13,500
6200 50 lb. capacity		.68	17.647		13,800	1,175		14,975	17,000
6300 75 lb. capacity		.30	40		19,200	2,650		21,850	25,100
6350 125 lb. capacity	↓	.16	75	↓	30,100	4,975		35,075	40,500

11 21 73.33 Coin-Operated Laundry Equipment

	Crew	Daily Output	Labor-Hours	Unit	Material	2021 Bare Costs Labor	2021 Bare Costs Equipment	Total	Total Incl O&P
0010 **COIN-OPERATED LAUNDRY EQUIPMENT**									
5290 Clothes washer									
5300 Commercial, coin operated, average	1 Plum	3	2.667	Ea.	1,350	181		1,531	1,775

11 30 Residential Equipment

11 30 13 – Residential Appliances

11 30 13.15 Cooking Equipment

	Crew	Daily Output	Labor-Hours	Unit	Material	2021 Bare Costs Labor	2021 Bare Costs Equipment	Total	Total Incl O&P
0010 **COOKING EQUIPMENT**									
0020 Cooking range, 30" free standing, 1 oven, minimum	2 Clab	10	1.600	Ea.	460	71		531	610
0050 Maximum		4	4		2,300	178		2,478	2,800
0700 Free standing, 1 oven, 21" wide range, minimum		10	1.600		445	71		516	595
0750 21" wide, maximum	↓	4	4		740	178		918	1,075
0900 Countertop cooktops, 4 burner, standard, minimum	1 Elec	6	1.333		335	85		420	495
0950 Maximum		3	2.667		1,850	170		2,020	2,300
1050 As above, but with grill and griddle attachment, minimum		6	1.333		1,425	85		1,510	1,700
1100 Maximum		3	2.667		3,800	170		3,970	4,450
1200 Induction cooktop, 30" wide		3	2.667		1,475	170		1,645	1,850
1250 Microwave oven, minimum		4	2		107	127		234	305
1300 Maximum	↓	2	4	↓	485	255		740	915

11 30 13.17 Kitchen Cleaning Equipment

	Crew	Daily Output	Labor-Hours	Unit	Material	2021 Bare Costs Labor	2021 Bare Costs Equipment	Total	Total Incl O&P
0010 **KITCHEN CLEANING EQUIPMENT**									
2750 Dishwasher, built-in, 2 cycles, minimum	L-1	4	4	Ea.	310	263		573	730
2800 Maximum		2	8		445	525		970	1,275
2950 4 or more cycles, minimum		4	4		420	263		683	850
2960 Average		4	4		560	263		823	1,000
3000 Maximum	↓	2	8	↓	1,700	525		2,225	2,650

11 30 13.18 Waste Disposal Equipment

	Crew	Daily Output	Labor-Hours	Unit	Material	2021 Bare Costs Labor	2021 Bare Costs Equipment	Total	Total Incl O&P
0010 **WASTE DISPOSAL EQUIPMENT**									
3300 Garbage disposal, sink type, minimum	L-1	10	1.600	Ea.	109	105		214	277
3350 Maximum	"	10	1.600	"	242	105		347	425

11 30 13.19 Kitchen Ventilation Equipment

	Crew	Daily Output	Labor-Hours	Unit	Material	2021 Bare Costs Labor	2021 Bare Costs Equipment	Total	Total Incl O&P
0010 **KITCHEN VENTILATION EQUIPMENT**									
4150 Hood for range, 2 speed, vented, 30" wide, minimum	L-3	5	3.200	Ea.	107	191		298	400
4200 Maximum		3	5.333		1,050	320		1,370	1,625
4300 42" wide, minimum		5	3.200		152	191		343	450
4330 Custom		5	3.200		1,975	191		2,166	2,450
4350 Maximum	↓	3	5.333	↓	2,400	320		2,720	3,125
4500 For ventless hood, 2 speed, add				Ea.	18.45			18.45	20.50
4650 For vented 1 speed, deduct from maximum				"	70			70	77

11 30 13.24 Washers

	Crew	Daily Output	Labor-Hours	Unit	Material	2021 Bare Costs Labor	2021 Bare Costs Equipment	Total	Total Incl O&P
0010 **WASHERS**									
5000 Residential, 4 cycle, average	1 Plum	3	2.667	Ea.	995	181		1,176	1,375

63

For customer support on your Electrical Costs with RSMeans Data, call 800.448.8182.

11 30 Residential Equipment

11 30 15 – Miscellaneous Residential Appliances

11 30 15.23 Water Heaters

		Crew	Daily Output	Labor-Hours	Unit	Material	2021 Bare Costs Labor	Equipment	Total	Total Incl O&P
0010	**WATER HEATERS**									
6900	Electric, glass lined, 30 gallon, minimum	L-1	5	3.200	Ea.	1,100	210		1,310	1,525
6950	Maximum		3	5.333		1,525	350		1,875	2,200
7100	80 gallon, minimum		2	8		2,200	525		2,725	3,175
7150	Maximum		1	16		3,050	1,050		4,100	4,925

11 30 15.43 Air Quality

		Crew	Daily Output	Labor-Hours	Unit	Material	2021 Bare Costs Labor	Equipment	Total	Total Incl O&P
0010	**AIR QUALITY**									
2450	Dehumidifier, portable, automatic, 15 pint	1 Elec	4	2	Ea.	219	127		346	430
2550	40 pint		3.75	2.133		250	136		386	475
3550	Heater, electric, built-in, 1250 watt, ceiling type, minimum		4	2		103	127		230	300
3600	Maximum		3	2.667		196	170		366	470
3700	Wall type, minimum		4	2		235	127		362	450
3750	Maximum		3	2.667		102	170		272	365
3900	1500 watt wall type, with blower		4	2		172	127		299	380
3950	3000 watt		3	2.667		515	170		685	815
4850	Humidifier, portable, 8 gallons/day					132			132	146
5000	15 gallons/day					209			209	230

11 32 Unit Kitchens

11 32 13 – Metal Unit Kitchens

11 32 13.10 Commercial Unit Kitchens

		Crew	Daily Output	Labor-Hours	Unit	Material	2021 Bare Costs Labor	Equipment	Total	Total Incl O&P
0010	**COMMERCIAL UNIT KITCHENS**									
1500	Combination range, refrigerator and sink, 30" wide, minimum	L-1	2	8	Ea.	635	525		1,160	1,475
1550	Maximum		1	16		1,125	1,050		2,175	2,825
1570	60" wide, average		1.40	11.429		1,150	750		1,900	2,400
1590	72" wide, average		1.20	13.333		2,100	875		2,975	3,600
1600	Office model, 48" wide		2	8		1,275	525		1,800	2,175
1620	Refrigerator and sink only		2.40	6.667		2,600	440		3,040	3,500
1640	Combination range, refrigerator, sink, microwave									
1660	Oven and ice maker	L-1	.80	20	Ea.	4,875	1,325		6,200	7,300

11 41 Foodservice Storage Equipment

11 41 13 – Refrigerated Food Storage Cases

11 41 13.20 Refrigerated Food Storage Equipment

		Crew	Daily Output	Labor-Hours	Unit	Material	2021 Bare Costs Labor	Equipment	Total	Total Incl O&P
0010	**REFRIGERATED FOOD STORAGE EQUIPMENT**									
2350	Cooler, reach-in, beverage, 6' long	Q-1	6	2.667	Ea.	2,875	162		3,037	3,425

11 42 Food Preparation Equipment

11 42 10 – Commercial Food Preparation Equipment

11 42 10.10 Choppers, Mixers and Misc. Equipment	Crew	Daily Output	Labor-Hours	Unit	Material	2021 Bare Costs Labor	Equipment	Total	Total Incl O&P
0010 **CHOPPERS, MIXERS AND MISC. EQUIPMENT**									
1850 Coffee urn, twin 6 gallon urns	1 Plum	2	4	Ea.	2,625	271		2,896	3,275
3800 Food mixers, bench type, 20 quarts	L-7	7	4		3,075	212		3,287	3,700
3900 60 quarts	"	5	5.600		11,900	297		12,197	13,500

11 44 Food Cooking Equipment

11 44 13 – Commercial Ranges

11 44 13.10 Cooking Equipment

	Crew	Daily Output	Labor-Hours	Unit	Material	2021 Bare Costs Labor	Equipment	Total	Total Incl O&P
0010 **COOKING EQUIPMENT**									
0020 Bake oven, gas, one section	Q-1	8	2	Ea.	6,250	122		6,372	7,050
1300 Broiler, without oven, standard	"	8	2		3,750	122		3,872	4,300
1550 Infrared	L-7	4	7		5,300	370		5,670	6,375
6350 Kettle, w/steam jacket, tilting, w/positive lock, SS, 20 gallons		7	4		10,800	212		11,012	12,200
6600 60 gallons		6	4.667		14,600	248		14,848	16,500
8850 Steamer, electric 27 KW		7	4		16,600	212		16,812	18,600
9100 Electric, 10 KW or gas 100,000 BTU		5	5.600		7,850	297		8,147	9,075
9150 Toaster, conveyor type, 16-22 slices/minute					1,100			1,100	1,200
9200 For deluxe models of above equipment, add					75%				
9400 Rule of thumb: Equipment cost based									
9410 on kitchen work area									
9420 Office buildings, minimum	L-7	77	.364	S.F.	98.50	19.30		117.80	137
9450 Maximum		58	.483		160	25.50		185.50	214
9550 Public eating facilities, minimum		77	.364		131	19.30		150.30	173
9600 Maximum		46	.609		210	32.50		242.50	279
9750 Hospitals, minimum		58	.483		127	25.50		152.50	178
9800 Maximum		39	.718		235	38		273	315

11 46 Food Dispensing Equipment

11 46 16 – Service Line Equipment

11 46 16.10 Commercial Food Dispensing Equipment

	Crew	Daily Output	Labor-Hours	Unit	Material	2021 Bare Costs Labor	Equipment	Total	Total Incl O&P
0010 **COMMERCIAL FOOD DISPENSING EQUIPMENT**									
3300 Food warmer, counter, 1.2 KW				Ea.	630			630	695
3550 1.6 KW					2,275			2,275	2,500
5750 Ice dispenser 567 pound	Q-1	6	2.667		6,000	162		6,162	6,850

11 48 Foodservice Cleaning and Disposal Equipment

11 48 13 – Commercial Dishwashers

11 48 13.10 Dishwashers

		Crew	Daily Output	Labor-Hours	Unit	Material	2021 Bare Costs Labor	Equipment	Total	Total Incl O&P
0010 **DISHWASHERS**										
2700 Dishwasher, commercial, rack type										
2720 10 to 12 racks/hour		Q-1	3.20	5	Ea.	3,500	305		3,805	4,300
2730 Energy star rated, 35 to 40 racks/hour	G		1.30	12.308		5,650	750		6,400	7,325
2740 50 to 60 racks/hour	G		1.30	12.308		10,800	750		11,550	13,000

11 52 Audio-Visual Equipment

11 52 16 – Projectors

11 52 16.10 Movie Equipment

		Crew	Daily Output	Labor-Hours	Unit	Material	2021 Bare Costs Labor	2021 Bare Costs Equipment	Total	Total Incl O&P
0010	**MOVIE EQUIPMENT**									
0020	Changeover, minimum				Ea.	520			520	570
0100	Maximum					1,150			1,150	1,275
0800	Lamphouses, incl. rectifiers, xenon, 1,000 watt	1 Elec	2	4		8,525	255		8,780	9,750
0900	1,600 watt		2	4		9,100	255		9,355	10,400
1000	2,000 watt		1.50	5.333		8,900	340		9,240	10,300
1100	4,000 watt		1.50	5.333		11,700	340		12,040	13,400
3700	Sound systems, incl. amplifier, mono, minimum		.90	8.889		3,875	565		4,440	5,100
3800	Dolby/Super Sound, maximum		.40	20		20,800	1,275		22,075	24,800
4100	Dual system, 2 channel, front surround, minimum		.70	11.429		5,325	730		6,055	6,950
4200	Dolby/Super Sound, 4 channel, maximum		.40	20		20,000	1,275		21,275	23,900
5300	Speakers, recessed behind screen, minimum		2	4		1,250	255		1,505	1,750
5400	Maximum		1	8		3,575	510		4,085	4,675
7000	For automation, varying sophistication, minimum		1	8	System	2,750	510		3,260	3,775
7100	Maximum	2 Elec	.30	53.333	"	6,425	3,400		9,825	12,100

11 52 16.20 Movie Equipment- Digital

		Crew	Daily Output	Labor-Hours	Unit	Material	2021 Bare Costs Labor	2021 Bare Costs Equipment	Total	Total Incl O&P
0010	**MOVIE EQUIPMENT- DIGITAL**									
1000	Digital 2K projection system, 98" DMD	1 Elec	2	4	Ea.	53,500	255		53,755	59,500
1100	OEM lens		2	4		6,325	255		6,580	7,325
2000	Pedestal with power distribution		2	4		2,350	255		2,605	2,950
3000	Software		2	4		2,000	255		2,255	2,575

11 53 Laboratory Equipment

11 53 19 – Laboratory Sterilizers

11 53 19.13 Sterilizers

		Crew	Daily Output	Labor-Hours	Unit	Material	2021 Bare Costs Labor	2021 Bare Costs Equipment	Total	Total Incl O&P
0010	**STERILIZERS**									
0700	Glassware washer, undercounter, minimum	L-1	1.80	8.889	Ea.	6,950	585		7,535	8,525
0710	Maximum	"	1	16	"	15,200	1,050		16,250	18,400

11 53 33 – Emergency Safety Appliances

11 53 33.13 Emergency Equipment

		Crew	Daily Output	Labor-Hours	Unit	Material	2021 Bare Costs Labor	2021 Bare Costs Equipment	Total	Total Incl O&P
0010	**EMERGENCY EQUIPMENT**									
1400	Safety equipment, eye wash, hand held				Ea.	370			370	410
1450	Deluge shower				"	825			825	905

11 53 43 – Service Fittings and Accessories

11 53 43.13 Fittings

		Crew	Daily Output	Labor-Hours	Unit	Material	2021 Bare Costs Labor	2021 Bare Costs Equipment	Total	Total Incl O&P
0010	**FITTINGS**									
8000	Alternate pricing method: as percent of lab furniture									
8050	Installation, not incl. plumbing & duct work				% Furn.				22%	22%
8100	Plumbing, final connections, simple system								10%	10%
8110	Moderately complex system								15%	15%
8120	Complex system								20%	20%
8150	Electrical, simple system								10%	10%
8160	Moderately complex system				% Furn.				20%	20%
8170	Complex system				"				35%	35%

11 61 Broadcast, Theater, and Stage Equipment

11 61 33 – Rigging Systems and Controls

11 61 33.10 Controls

		Crew	Daily Output	Labor-Hours	Unit	Material	2021 Bare Costs Labor	Equipment	Total	Total Incl O&P
0010	**CONTROLS**									
0050	Control boards with dimmers and breakers, minimum	1 Elec	1	8	Ea.	15,500	510		16,010	17,900
0100	Average		.50	16		42,500	1,025		43,525	48,200
0150	Maximum		.20	40		124,500	2,550		127,050	141,000

11 66 Athletic Equipment

11 66 43 – Interior Scoreboards

11 66 43.10 Scoreboards

		Crew	Daily Output	Labor-Hours	Unit	Material	2021 Bare Costs Labor	Equipment	Total	Total Incl O&P
0010	**SCOREBOARDS**									
7000	Baseball, minimum	R-3	1.30	15.385	Ea.	4,200	975	147	5,322	6,225
7200	Maximum		.05	400		18,000	25,400	3,825	47,225	61,500
7300	Football, minimum		.86	23.256		5,175	1,475	222	6,872	8,150
7400	Maximum		.20	100		25,600	6,350	955	32,905	38,600
7500	Basketball (one side), minimum		2.07	9.662		2,400	615	92.50	3,107.50	3,625
7600	Maximum		.30	66.667		9,250	4,225	635	14,110	17,200
7700	Hockey-basketball (four sides), minimum		.25	80		10,300	5,075	765	16,140	19,700
7800	Maximum		.15	133		17,500	8,450	1,275	27,225	33,200

11 71 Medical Sterilizing Equipment

11 71 10 – Medical Sterilizers & Distillers

11 71 10.10 Sterilizers and Distillers

		Crew	Daily Output	Labor-Hours	Unit	Material	2021 Bare Costs Labor	Equipment	Total	Total Incl O&P
0010	**STERILIZERS AND DISTILLERS**									
6200	Steam generators, electric 10 kW to 180 kW, freestanding									
6250	Minimum	1 Elec	3	2.667	Ea.	12,800	170		12,970	14,400
6300	Maximum	"	.70	11.429	"	22,600	730		23,330	25,900

11 72 Examination and Treatment Equipment

11 72 53 – Treatment Equipment

11 72 53.13 Medical Treatment Equipment

		Crew	Daily Output	Labor-Hours	Unit	Material	2021 Bare Costs Labor	Equipment	Total	Total Incl O&P
0010	**MEDICAL TREATMENT EQUIPMENT**									
6700	Surgical lights, doctor's office, single arm	2 Elec	2	8	Ea.	3,275	510		3,785	4,350
6750	Dual arm	"	1	16	"	7,075	1,025		8,100	9,300

11 76 Operating Room Equipment

11 76 10 – Equipment for Operating Rooms

11 76 10.10 Surgical Equipment

		Crew	Daily Output	Labor-Hours	Unit	Material	2021 Bare Costs Labor	Equipment	Total	Total Incl O&P
0010	**SURGICAL EQUIPMENT**									
6800	Surgical lights, major operating room, dual head, minimum	2 Elec	1	16	Ea.	6,775	1,025		7,800	8,975
6850	Maximum	"	1	16	"	41,600	1,025		42,625	47,300

For customer support on your Electrical Costs with RSMeans Data, call 800.448.8182.

11 78 Mortuary Equipment

11 78 13 – Mortuary Refrigerators

11 78 13.10 Mortuary and Autopsy Equipment	Crew	Daily Output	Labor-Hours	Unit	Material	2021 Bare Costs Labor	Equipment	Total	Total Incl O&P
0010 **MORTUARY AND AUTOPSY EQUIPMENT**									
0015 Autopsy table, standard	1 Plum	1	8	Ea.	9,900	540		10,440	11,700

11 78 16 – Crematorium Equipment

11 78 16.10 Crematory

0010 **CREMATORY**									
1500 Crematory, not including building, 1 place	Q-3	.20	160	Ea.	81,500	10,300		91,800	105,000
1750 2 place	"	.10	320	"	116,000	20,600		136,600	159,000

11 82 Facility Solid Waste Handling Equipment

11 82 26 – Facility Waste Compactors

11 82 26.10 Compactors

	Crew	Daily Output	Labor-Hours	Unit	Material	2021 Bare Costs Labor	Equipment	Total	Total Incl O&P
0010 **COMPACTORS**									
0020 Compactors, 115 volt, 250 lbs./hr., chute fed	L-4	1	24	Ea.	12,500	1,250		13,750	15,600
1000 Heavy duty industrial compactor, 0.5 C.Y. capacity		1	24		11,100	1,250		12,350	14,100
1050 1.0 C.Y. capacity		1	24		16,900	1,250		18,150	20,500
1100 3.0 C.Y. capacity		.50	48		29,000	2,500		31,500	35,700
1150 5.0 C.Y. capacity		.50	48		36,600	2,500		39,100	44,100
1200 Combination shredder/compactor (5,000 lbs./hr.)		.50	48		71,000	2,500		73,500	82,000
1400 For handling hazardous waste materials, 55 gallon drum packer, std.					21,800			21,800	24,000
1410 55 gallon drum packer w/HEPA filter					26,200			26,200	28,800
1420 55 gallon drum packer w/charcoal & HEPA filter					34,900			34,900	38,400
1430 All of the above made explosion proof, add					2,025			2,025	2,225
5800 Shredder, industrial, minimum					29,300			29,300	32,200
5850 Maximum					147,500			147,500	162,500
5900 Baler, industrial, minimum					11,000			11,000	12,200
5950 Maximum					644,500			644,500	708,500

Estimating Tips
General

- The items in this division are usually priced per square foot or each. Most of these items are purchased by the owner and installed by the contractor. Do not assume the items in Division 12 will be purchased and installed by the contractor. Check the specifications for responsibilities and include receiving, storage, installation, and mechanical and electrical hookups in the appropriate divisions.

- Some items in this division require some type of support system that is not usually furnished with the item. Examples of these systems include blocking for the attachment of casework and heavy drapery rods. The required blocking must be added to the estimate in the appropriate division.

Reference Numbers

Reference numbers are shown at the beginning of some major classifications. These numbers refer to related items in the Reference Section. The reference information may be an estimating procedure, an alternate pricing method, or technical information.

Same Data. Simplified.

Enjoy the convenience and efficiency of accessing your costs anywhere:

- **Skip the multiplier** by setting your location
- **Quickly search,** edit, favorite and share costs
- **Stay on top of price changes** with automatic updates

Discover more at rsmeans.com/online

12 46 Furnishing Accessories

12 46 19 – Clocks

12 46 19.50 Wall Clocks		Crew	Daily Output	Labor-Hours	Unit	Material	2021 Bare Costs Labor	Equipment	Total	Total Incl O&P
0010	**WALL CLOCKS**									
0080	12" diameter, single face	1 Elec	8	1	Ea.	151	63.50		214.50	262
0100	Double face	"	6.20	1.290	"	152	82		234	289

Estimating Tips
General

- The items and systems in this division are usually estimated, purchased, supplied, and installed as a unit by one or more subcontractors. The estimator must ensure that all parties are operating from the same set of specifications and assumptions, and that all necessary items are estimated and will be provided. Many times the complex items and systems are covered, but the more common ones, such as excavation or a crane, are overlooked for the very reason that everyone assumes nobody could miss them. The estimator should be the central focus and be able to ensure that all systems are complete.

- It is important to consider factors such as site conditions, weather, shape and size of building, as well as labor availability as they may impact the overall cost of erecting special structures and systems included in this division.

- Another area where problems can develop in this division is at the interface between systems.

The estimator must ensure, for instance, that anchor bolts, nuts, and washers are estimated and included for the air-supported structures and pre-engineered buildings to be bolted to their foundations. Utility supply is a common area where essential items or pieces of equipment can be missed or overlooked because each subcontractor may feel it is another's responsibility. The estimator should also be aware of certain items which may be supplied as part of a package but installed by others, and ensure that the installing contractor's estimate includes the cost of installation. Conversely, the estimator must also ensure that items are not costed by two different subcontractors, resulting in an inflated overall estimate.

13 30 00 Special Structures

- The foundations and floor slab, as well as rough mechanical and electrical, should be estimated, as this work is required for the assembly and erection of the structure. Generally, as noted in the data set, the pre-engineered building comes

as a shell. Pricing is based on the size and structural design parameters stated in the reference section. Additional features, such as windows and doors with their related structural framing, must also be included by the estimator. Here again, the estimator must have a clear understanding of the scope of each portion of the work and all the necessary interfaces.

Reference Numbers

Reference numbers are shown at the beginning of some major classifications. These numbers refer to related items in the Reference Section. The reference information may be an estimating procedure, an alternate pricing method, or technical information.

Note: Not all subdivisions listed here necessarily appear. ∎

Same Data. Simplified.

Enjoy the convenience and efficiency of accessing your costs anywhere:

- **Skip the multiplier** by setting your location
- **Quickly search,** edit, favorite and share costs
- **Stay on top of price changes** with automatic updates

Discover more at rsmeans.com/online

13 11 Swimming Pools

13 11 13 – Below-Grade Swimming Pools

13 11 13.50 Swimming Pools	Crew	Daily Output	Labor-Hours	Unit	Material	2021 Bare Costs Labor	2021 Bare Costs Equipment	Total	Total Incl O&P
0010 **SWIMMING POOLS** Residential in-ground, vinyl lined									
0020 Concrete sides, w/equip, sand bottom	B-52	300	.187	SF Surf	29.50	9.50	1.91	40.91	48.50
0100 Metal or polystyrene sides	B-14	410	.117		24.50	5.45	.53	30.48	35.50
0200 Add for vermiculite bottom				↓	1.87			1.87	2.06
0500 Gunite bottom and sides, white plaster finish									
0600 12' x 30' pool	B-52	145	.386	SF Surf	54.50	19.60	3.95	78.05	93.50
0720 16' x 32' pool		155	.361		49	18.35	3.70	71.05	85.50
0750 20' x 40' pool	↓	250	.224	↓	44	11.35	2.29	57.64	68
0810 Concrete bottom and sides, tile finish									
0820 12' x 30' pool	B-52	80	.700	SF Surf	55	35.50	7.15	97.65	121
0830 16' x 32' pool	B-52	95	.589	SF Surf	45.50	30	6.05	81.55	101
0840 20' x 40' pool	"	130	.431	"	36.50	22	4.41	62.91	77.50
1100 Motel, gunite with plaster finish, incl. medium									
1150 capacity filtration & chlorination	B-52	115	.487	SF Surf	67.50	24.50	4.98	96.98	117
1200 Municipal, gunite with plaster finish, incl. high									
1250 capacity filtration & chlorination	B-52	100	.560	SF Surf	87	28.50	5.75	121.25	144
1350 Add for formed gutters				L.F.	128			128	141
1360 Add for stainless steel gutters				"	380			380	415
1700 Filtration and deck equipment only, as % of total				Total				20%	20%
1800 Automatic vacuum, hand tools, etc., 20' x 40' pool				SF Pool				.56	.62
1900 5,000 S.F. pool				"				.11	.12

13 11 23 – On-Grade Swimming Pools

13 11 23.50 Swimming Pools

	Crew	Daily Output	Labor-Hours	Unit	Material	2021 Bare Costs Labor	2021 Bare Costs Equipment	Total	Total Incl O&P
0010 **SWIMMING POOLS** Residential above ground, steel construction									
0100 Round, 15' diam.	B-80A	3	8	Ea.	880	355	283	1,518	1,800
0120 18' diam.		2.50	9.600		935	425	340	1,700	2,025
0140 21' diam.		2	12		1,050	535	425	2,010	2,425
0160 24' diam.		1.80	13.333		1,200	590	470	2,260	2,700
0180 27' diam.		1.50	16		1,500	710	565	2,775	3,325
0200 30' diam.		1	24		1,725	1,075	850	3,650	4,425
0220 Oval, 12' x 24'		2.30	10.435		1,525	465	370	2,360	2,775
0240 15' x 30'		1.80	13.333		1,700	590	470	2,760	3,275
0260 18' x 33'	↓	1	24	↓	1,925	1,075	850	3,850	4,650

13 11 46 – Swimming Pool Accessories

13 11 46.50 Swimming Pool Equipment

	Crew	Daily Output	Labor-Hours	Unit	Material	2021 Bare Costs Labor	2021 Bare Costs Equipment	Total	Total Incl O&P
0010 **SWIMMING POOL EQUIPMENT**									
0020 Diving stand, stainless steel, 3 meter	2 Carp	.40	40	Ea.	19,800	2,200		22,000	25,100
2100 Lights, underwater, 12 volt, with transformer, 300 watt	1 Elec	1	8		465	510		975	1,275
2200 110 volt, 500 watt, standard		1	8		291	510		801	1,075
2400 Low water cutoff type	↓	1	8	↓	315	510		825	1,100
2800 Heaters, see Section 23 52 28.10									

13 12 Fountains

13 12 13 – Exterior Fountains

13 12 13.10 Outdoor Fountains

		Crew	Daily Output	Labor-Hours	Unit	Material	2021 Bare Costs Labor	2021 Bare Costs Equipment	Total	Total Incl O&P
0010	**OUTDOOR FOUNTAINS**									
0100	Outdoor fountain, 48" high with bowl and figures	2 Clab	2	8	Ea.	385	355		740	955
0200	Commercial, concrete or cast stone, 40-60" H, simple		2	8		975	355		1,330	1,600
0220	Average		2	8		1,900	355		2,255	2,600
0240	Ornate		2	8		4,300	355		4,655	5,275
0260	Metal, 72" high		2	8		1,525	355		1,880	2,200
0280	90" high		2	8		2,200	355		2,555	2,950
0300	120" high		2	8		5,275	355		5,630	6,325
0320	Resin or fiberglass, 40-60" H, wall type	2 Clab	2	8	Ea.	690	355		1,045	1,300
0340	Waterfall type	"	2	8	"	1,175	355		1,530	1,800

13 12 23 – Interior Fountains

13 12 23.10 Indoor Fountains

		Crew	Daily Output	Labor-Hours	Unit	Material	2021 Bare Costs Labor	2021 Bare Costs Equipment	Total	Total Incl O&P
0010	**INDOOR FOUNTAINS**									
0100	Commercial, floor type, resin or fiberglass, lighted, cascade type	2 Clab	2	8	Ea.	435	355		790	1,000
0120	Tiered type		2	8		680	355		1,035	1,275
0140	Waterfall type		2	8		485	355		840	1,050

13 21 Controlled Environment Rooms

13 21 13 – Clean Rooms

13 21 13.50 Clean Room Components

		Crew	Daily Output	Labor-Hours	Unit	Material	2021 Bare Costs Labor	2021 Bare Costs Equipment	Total	Total Incl O&P
0010	**CLEAN ROOM COMPONENTS**									
1100	Clean room, soft wall, 12' x 12', Class 100	1 Carp	.18	44.444	Ea.	16,500	2,425		18,925	21,700
1110	Class 1,000		.18	44.444		17,300	2,425		19,725	22,700
1120	Class 10,000		.21	38.095		15,400	2,075		17,475	20,100
1130	Class 100,000		.21	38.095		12,600	2,075		14,675	17,000
2800	Ceiling grid support, slotted channel struts 4'-0" OC, ea. way				S.F.				5.90	6.50
3000	Ceiling panel, vinyl coated foil on mineral substrate									
3020	Sealed, non-perforated				S.F.				3.72	4.13
4000	Ceiling panel seal, silicone sealant, 150 L.F./gal.	1 Carp	150	.053	L.F.	.28	2.92		3.20	4.66
4100	Two sided adhesive tape	"	240	.033	"	.13	1.82		1.95	2.86
4200	Clips, one per panel				Ea.	1			1	1.10
6000	HEPA filter, 2' x 4', 99.97% eff., 3" dp beveled frame (silicone seal)					380			380	415
6040	6" deep skirted frame (channel seal)					515			515	565
6100	99.99% efficient, 3" deep beveled frame (silicone seal)					480			480	525
6140	6" deep skirted frame (channel seal)					555			555	610
6200	99.999% efficient, 3" deep beveled frame (silicone seal)					585			585	640
6240	6" deep skirted frame (channel seal)					660			660	730
7000	Wall panel systems, including channel strut framing									
7020	Polyester coated aluminum, particle board				S.F.				18.20	18.20
7100	Porcelain coated aluminum, particle board								32	32
7400	Wall panel support, slotted channel struts, to 12' high								16.35	16.35

13 24 Special Activity Rooms

13 24 16 – Saunas

13 24 16.50 Saunas and Heaters		Crew	Daily Output	Labor-Hours	Unit	Material	2021 Bare Costs Labor	Equipment	Total	Total Incl O&P
0010	**SAUNAS AND HEATERS**									
0020	Prefabricated, incl. heater & controls, 7' high, 6' x 4', C/C	L-7	2.20	12.727	Ea.	6,425	675		7,100	8,050
0050	6' x 4', C/P		2	14		4,300	745		5,045	5,825
0400	6' x 5', C/C		2	14		5,350	745		6,095	6,975
0450	6' x 5', C/P		2	14		5,150	745		5,895	6,775
0600	6' x 6', C/C	↓	1.80	15.556	↓	6,850	825		7,675	8,750
0650	6' x 6', C/P	L-7	1.80	15.556	Ea.	5,725	825		6,550	7,525
0800	6' x 9', C/C		1.60	17.500		7,400	930		8,330	9,525
0850	6' x 9', C/P		1.60	17.500		5,825	930		6,755	7,775
1000	8' x 12', C/C		1.10	25.455		13,900	1,350		15,250	17,300
1050	8' x 12', C/P		1.10	25.455		7,500	1,350		8,850	10,300
1200	8' x 8', C/C		1.40	20		9,750	1,050		10,800	12,300
1250	8' x 8', C/P		1.40	20		7,100	1,050		8,150	9,400
1400	8' x 10', C/C		1.20	23.333		9,050	1,250		10,300	11,800
1450	8' x 10', C/P		1.20	23.333		8,250	1,250		9,500	10,900
1600	10' x 12', C/C		1	28		14,200	1,475		15,675	17,900
1650	10' x 12', C/P	↓	1	28		14,300	1,475		15,775	17,900
2500	Heaters only (incl. above), wall mounted, to 200 C.F.					1,025			1,025	1,125
2750	To 300 C.F.					1,150			1,150	1,275
3000	Floor standing, to 720 C.F., 10,000 watts, w/controls	1 Elec	3	2.667		2,950	170		3,120	3,475
3250	To 1,000 C.F., 16,000 watts	"	3	2.667	↓	4,050	170		4,220	4,700

13 24 26 – Steam Baths

13 24 26.50 Steam Baths and Components		Crew	Daily Output	Labor-Hours	Unit	Material	2021 Bare Costs Labor	Equipment	Total	Total Incl O&P
0010	**STEAM BATHS AND COMPONENTS**									
0020	Heater, timer & head, single, to 140 C.F.	1 Plum	1.20	6.667	Ea.	2,150	450		2,600	3,025
0500	To 300 C.F.		1.10	7.273		2,425	490		2,915	3,400
1000	Commercial size, with blow-down assembly, to 800 C.F.		.90	8.889		6,125	600		6,725	7,625
1500	To 2,500 C.F.	↓	.80	10		7,225	675		7,900	8,950
2000	Multiple, motels, apts., 2 baths, w/blow-down assm., 500 C.F.	Q-1	1.30	12.308		6,750	750		7,500	8,550
2500	4 baths	"	.70	22.857	↓	10,200	1,400		11,600	13,300

13 47 Facility Protection

13 47 13 – Cathodic Protection

13 47 13.16 Cathodic Prot. for Underground Storage Tanks		Crew	Daily Output	Labor-Hours	Unit	Material	2021 Bare Costs Labor	Equipment	Total	Total Incl O&P
0010	**CATHODIC PROTECTION FOR UNDERGROUND STORAGE TANKS**									
1000	Anodes, magnesium type, 9 #	R-15	18.50	2.595	Ea.	39.50	162	15.25	216.75	300
1010	17 #		13	3.692		81.50	230	21.50	333	460
1020	32 #		10	4.800		120	300	28	448	610
1030	48 #	↓	7.20	6.667		154	415	39	608	830
1100	Graphite type w/epoxy cap, 3" x 60" (32 #)	R-22	8.40	4.438		142	259		401	540
1110	4" x 80" (68 #)		6	6.213		245	360		605	810
1120	6" x 72" (80 #)		5.20	7.169		1,500	420		1,920	2,275
1130	6" x 36" (45 #)		9.60	3.883		750	226		976	1,150
2000	Rectifiers, silicon type, air cooled, 28 V/10 A	R-19	3.50	5.714		2,350	365		2,715	3,150
2010	20 V/20 A		3.50	5.714		2,375	365		2,740	3,150
2100	Oil immersed, 28 V/10 A		3	6.667		2,400	425		2,825	3,275
2110	20 V/20 A	↓	3	6.667	↓	3,175	425		3,600	4,125
3000	Anode backfill, coke breeze	R-22	3850	.010	Lb.	.27	.56		.83	1.14
4000	Cable, HMWPE, No. 8		2.40	15.533	M.L.F.	435	905		1,340	1,825
4010	No. 6		2.40	15.533	↓	625	905		1,530	2,050

13 47 Facility Protection

13 47 13 – Cathodic Protection

13 47 13.16 Cathodic Prot. for Underground Storage Tanks	Crew	Daily Output	Labor-Hours	Unit	Material	2021 Bare Costs Labor	2021 Bare Costs Equipment	Total	Total Incl O&P
4020	No. 4	R-22	2.40	15.533	M.L.F.	950	905		1,855	2,400
4030	No. 2		2.40	15.533		1,700	905		2,605	3,225
4040	No. 1		2.20	16.945		2,000	990		2,990	3,675
4050	No. 1/0		2.20	16.945		2,925	990		3,915	4,675
4060	No. 2/0		2.20	16.945		4,800	990		5,790	6,750
4070	No. 4/0	↓	2	18.640	↓	8,500	1,075		9,575	11,000
5000	Test station, 7 terminal box, flush curb type w/lockable cover	R-19	12	1.667	Ea.	79.50	106		185.50	246
5010	Reference cell, 2" diam. PVC conduit, cplg., plug, set flush	"	4.80	4.167	"	158	266		424	570

13 49 Radiation Protection

13 49 13 – Integrated X-Ray Shielding Assemblies

13 49 13.50 Lead Sheets

13 49 13.50 Lead Sheets	Crew	Daily Output	Labor-Hours	Unit	Material	Labor	Equipment	Total	Total Incl O&P
0010	**LEAD SHEETS**									
0300	Lead sheets, 1/16" thick	2 Lath	135	.119	S.F.	11.10	6.40		17.50	21.50
0400	1/8" thick		120	.133		25	7.20		32.20	38
0500	Lead shielding, 1/4" thick		135	.119		47	6.40		53.40	61
0550	1/2" thick	↓	120	.133	↓	87.50	7.20		94.70	107
0950	Lead headed nails (average 1 lb. per sheet)				Lb.	8.85			8.85	9.70
1000	Butt joints in 1/8" lead or thicker, 2" batten strip x 7' long	2 Lath	240	.067	Ea.	34.50	3.61		38.11	43.50
1200	X-ray protection, average radiography or fluoroscopy									
1210	room, up to 300 S.F. floor, 1/16" lead, economy	2 Lath	.25	64	Total	13,300	3,450		16,750	19,700
1500	7'-0" walls, deluxe	"	.15	107	"	15,900	5,775		21,675	25,900
1600	Deep therapy X-ray room, 250 kV capacity,									
1800	up to 300 S.F. floor, 1/4" lead, economy	2 Lath	.08	200	Total	37,000	10,800		47,800	56,500
1900	7'-0" walls, deluxe	"	.06	267	"	45,700	14,400		60,100	71,000

13 49 19 – Lead-Lined Materials

13 49 19.50 Shielding Lead

13 49 19.50 Shielding Lead	Crew	Daily Output	Labor-Hours	Unit	Material	Labor	Equipment	Total	Total Incl O&P
0010	**SHIELDING LEAD**									
0100	Laminated lead in wood doors, 1/16" thick, no hardware				S.F.	52.50			52.50	57.50
0200	Lead lined door frame, not incl. hardware,									
0210	1/16" thick lead, butt prepared for hardware	1 Lath	2.40	3.333	Ea.	905	180		1,085	1,250
0850	Window frame with 1/16" lead and voice passage, 36" x 60"	2 Glaz	2	8		4,300	420		4,720	5,350
0870	24" x 36" frame		4	4	↓	2,325	211		2,536	2,875
0900	Lead gypsum board, 5/8" thick with 1/16" lead		160	.100	S.F.	16	5.25		21.25	25.50
0910	1/8" lead	↓	140	.114		27	6		33	39
0930	1/32" lead	2 Lath	200	.080	↓	12.05	4.33		16.38	19.60

13 49 21 – Lead Glazing

13 49 21.50 Lead Glazing

13 49 21.50 Lead Glazing	Crew	Daily Output	Labor-Hours	Unit	Material	Labor	Equipment	Total	Total Incl O&P
0010	**LEAD GLAZING**									
0600	Lead glass, 1/4" thick, 2.0 mm LE, 12" x 16"	2 Glaz	13	1.231	Ea.	495	65		560	635
0700	24" x 36"		8	2		1,675	105		1,780	1,975
0800	36" x 60"	↓	2	8	↓	4,025	420		4,445	5,050
2000	X-ray viewing panels, clear lead plastic									
2010	7 mm thick, 0.3 mm LE, 2.3 lb./S.F.	H-3	139	.115	S.F.	256	5.45		261.45	290
2020	12 mm thick, 0.5 mm LE, 3.9 lb./S.F.		82	.195		345	9.25		354.25	395
2030	18 mm thick, 0.8 mm LE, 5.9 lb./S.F.		54	.296		420	14.05		434.05	480
2040	22 mm thick, 1.0 mm LE, 7.2 lb./S.F.		44	.364		690	17.20		707.20	785
2050	35 mm thick, 1.5 mm LE, 11.5 lb./S.F.		28	.571		915	27		942	1,050
2060	46 mm thick, 2.0 mm LE, 15.0 lb./S.F.	↓	21	.762		1,200	36		1,236	1,350

For customer support on your Electrical Costs with RSMeans Data, call 800.448.8182.

13 49 23 – Integrated RFI/EMI Shielding Assemblies

13 49 23.50 Modular Shielding Partitions		Crew	Daily Output	Labor-Hours	Unit	Material	2021 Bare Costs Labor	Equipment	Total	Total Incl O&P
0010	**MODULAR SHIELDING PARTITIONS**									
4000	X-ray barriers, modular, panels mounted within framework for									
4002	attaching to floor, wall or ceiling, upper portion is clear lead									
4005	plastic window panels 48"H, lower portion is opaque leaded									
4008	steel panels 36"H, structural supports not incl.									
4010	1-section barrier, 36"W x 84"H overall									
4020	0.5 mm LE panels	H-3	6.40	2.500	Ea.	9,250	118		9,368	10,400
4030	0.8 mm LE panels		6.40	2.500		9,975	118		10,093	11,200
4040	1.0 mm LE panels		5.33	3.002		13,800	142		13,942	15,400
4050	1.5 mm LE panels		5.33	3.002		16,300	142		16,442	18,100
4060	2-section barrier, 72"W x 84"H overall									
4070	0.5 mm LE panels	H-3	4	4	Ea.	14,000	189		14,189	15,700
4080	0.8 mm LE panels		4	4		14,400	189		14,589	16,200
4090	1.0 mm LE panels		3.56	4.494		15,700	213		15,913	17,600
5000	1.5 mm LE panels		3.20	5		28,100	237		28,337	31,400
5010	3-section barrier, 108"W x 84"H overall									
5020	0.5 mm LE panels	H-3	3.20	5	Ea.	21,000	237		21,237	23,500
5030	0.8 mm LE panels		3.20	5		21,700	237		21,937	24,200
5040	1.0 mm LE panels		2.67	5.993		23,600	284		23,884	26,300
5050	1.5 mm LE panels		2.46	6.504		42,700	310		43,010	47,400
7000	X-ray barriers, mobile, mounted within framework w/casters on									
7005	bottom, clear lead plastic window panels on upper portion,									
7010	opaque on lower, 30"W x 75"H overall, incl. framework									
7020	24"H upper w/0.5 mm LE, 48"H lower w/0.8 mm LE	1 Carp	16	.500	Ea.	4,000	27.50		4,027.50	4,475
7030	48"W x 75"H overall, incl. framework									
7040	36"H upper w/0.5 mm LE, 36"H lower w/0.8 mm LE	1 Carp	16	.500	Ea.	6,325	27.50		6,352.50	7,000
7050	36"H upper w/1.0 mm LE, 36"H lower w/1.5 mm LE	"	16	.500	"	7,500	27.50		7,527.50	8,275
7060	72"W x 75"H overall, incl. framework									
7070	36"H upper w/0.5 mm LE, 36"H lower w/0.8 mm LE	1 Carp	16	.500	Ea.	7,500	27.50		7,527.50	8,275
7080	36"H upper w/1.0 mm LE, 36"H lower w/1.5 mm LE	"	16	.500	"	9,375	27.50		9,402.50	10,300

13 49 33 – Radio Frequency Shielding

13 49 33.50 Shielding, Radio Frequency		Crew	Daily Output	Labor-Hours	Unit	Material	2021 Bare Costs Labor	Equipment	Total	Total Incl O&P
0010	**SHIELDING, RADIO FREQUENCY**									
0020	Prefabricated, galvanized steel	2 Carp	375	.043	SF Surf	5.95	2.33		8.28	10.05
0040	5 oz., copper floor panel		480	.033		4.84	1.82		6.66	8
0050	5 oz., copper wall/ceiling panel		155	.103		4.84	5.65		10.49	13.75
0100	12 oz., copper floor panel		470	.034		10.10	1.86		11.96	13.90
0110	12 oz., copper wall/ceiling panel		140	.114		10.10	6.25		16.35	20.50
0150	Door, copper/wood laminate, 4' x 7'		1.50	10.667	Ea.	9,050	585		9,635	10,800

Estimating Tips

Pipe for fire protection and all uses is located in Subdivisions 21 11 13 and 22 11 13.

The labor adjustment factors listed in Subdivision 22 01 02.20 also apply to Division 21.

Many, but not all, areas in the U.S. require backflow protection in the fire system. Insurance underwriters may have specific requirements for the type of materials to be installed or design requirements based on the hazard to be protected. Local jurisdictions may have requirements not covered by code. It is advisable to be aware of any special conditions.

For your reference, the following is a list of the most applicable Fire Codes and Standards, which may be purchased from the NFPA, 1 Batterymarch Park, Quincy, MA 02169-7471.

- NFPA 1: Uniform Fire Code
- NFPA 10: Portable Fire Extinguishers
- NFPA 11: Low-, Medium-, and High-Expansion Foam
- NFPA 12: Carbon Dioxide Extinguishing Systems (Also companion 12A)
- NFPA 13: Installation of Sprinkler Systems (Also companion 13D, 13E, and 13R)
- NFPA 14: Installation of Standpipe and Hose Systems
- NFPA 15: Water Spray Fixed Systems for Fire Protection
- NFPA 16: Installation of Foam-Water Sprinkler and Foam-Water Spray Systems
- NFPA 17: Dry Chemical Extinguishing Systems (Also companion 17A)
- NFPA 18: Wetting Agents
- NFPA 20: Installation of Stationary Pumps for Fire Protection
- NFPA 22: Water Tanks for Private Fire Protection
- NFPA 24: Installation of Private Fire Service Mains and their Appurtenances
- NFPA 25: Inspection, Testing and Maintenance of Water-Based Fire Protection

Reference Numbers

Reference numbers are shown at the beginning of some major classifications. These numbers refer to related items in the Reference Section. The reference information may be an estimating procedure, an alternate pricing method, or technical information.

Same Data. Simplified.

Enjoy the convenience and efficiency of accessing your costs anywhere:

- **Skip the multiplier** by setting your location
- **Quickly search,** edit, favorite and share costs
- **Stay on top of price changes** with automatic updates

Discover more at rsmeans.com/online

21 21 Carbon-Dioxide Fire-Extinguishing Systems

21 21 16 – Carbon-Dioxide Fire-Extinguishing Equipment

21 21 16.50 CO2 Fire Extinguishing System	Crew	Daily Output	Labor-Hours	Unit	Material	2021 Bare Costs Labor	Equipment	Total	Total Incl O&P
0010 **CO₂ FIRE EXTINGUISHING SYSTEM**									
0100 Control panel, single zone with batteries (2 zones det., 1 suppr.)	1 Elec	1	8	Ea.	1,075	510		1,585	1,950
0150 Multizone (4) with batteries (8 zones det., 4 suppr.)	"	.50	16		3,150	1,025		4,175	5,000
1000 Dispersion nozzle, CO₂, 3" x 5"	1 Plum	18	.444		163	30		193	224
2000 Extinguisher, CO₂ system, high pressure, 75 lb. cylinder	Q-1	6	2.667		1,450	162		1,612	1,850
2100 100 lb. cylinder	"	5	3.200		1,925	195		2,120	2,400
3000 Electro/mechanical release	L-1	4	4		1,075	263		1,338	1,600
3400 Manual pull station	1 Plum	6	1.333		90.50	90.50		181	235
4000 Pneumatic damper release	"	8	1		189	67.50		256.50	310

21 22 Clean-Agent Fire-Extinguishing Systems

21 22 16 – Clean-Agent Fire-Extinguishing Equipment

21 22 16.50 Clean-Agent Extinguishing Systems	Crew	Daily Output	Labor-Hours	Unit	Material	2021 Bare Costs Labor	Equipment	Total	Total Incl O&P
0010 **CLEAN-AGENT EXTINGUISHING SYSTEMS**									
0020 FM200 fire extinguishing system									
1100 Dispersion nozzle FM200, 1-1/2"	1 Plum	14	.571	Ea.	259	38.50		297.50	345
2400 Extinguisher, FM200 system, filled, with mounting bracket									
2460 26 lb. container	Q-1	8	2	Ea.	2,150	122		2,272	2,525
2480 44 lb. container		7	2.286		2,775	139		2,914	3,250
2500 63 lb. container		6	2.667		3,950	162		4,112	4,600
2520 101 lb. container		5	3.200		5,700	195		5,895	6,575
2540 196 lb. container		4	4		6,875	244		7,119	7,950
6000 FM200 system, simple nozzle layout, with broad dispersion				C.F.	1.87			1.87	2.06
6010 Extinguisher, FM200 system, filled, with mounting bracket									
6020 Complex nozzle layout and/or including underfloor dispersion				C.F.	3.71			3.71	4.08
6100 20,000 C.F. 2 exits, 8' clng					2.08			2.08	2.29
6200 100,000 C.F. 4 exits, 8' clng					1.88			1.88	2.07
6300 250,000 C.F. 6 exits, 8' clng					1.59			1.59	1.75
7010 HFC-227ea fire extinguishing system									
7100 Cylinders with clean-agent									
7110 Does not include pallet jack/fork lift rental fees									
7120 70 lb. cyl. w/35 lb. agent, no solenoid	Q-12	14	1.143	Ea.	3,100	68.50		3,168.50	3,525
7130 70 lb. cyl w/70 lb. agent, no solenoid		10	1.600		4,100	96		4,196	4,650
7140 70 lb. cyl w/35 lb. agent, w/solenoid		14	1.143		4,650	68.50		4,718.50	5,200
7150 70 lb. cyl w/70 lb. agent, w/solenoid		10	1.600		5,875	96		5,971	6,600
7220 250 lb. cyl, w/125 lb. agent, no solenoid		8	2		7,200	120		7,320	8,100
7230 250 lb. cyl, w/250 lb. agent, no solenoid		5	3.200		7,200	192		7,392	8,200
7240 250 lb. cyl, w/125 lb. agent, w/solenoid		8	2		10,000	120		10,120	11,300
7250 250 lb. cyl, w/250 lb. agent, w/solenoid		5	3.200		14,400	192		14,592	16,200
7320 560 lb. cyl, w/300 lb. agent, no solenoid		4	4		13,800	239		14,039	15,600
7330 560 lb. cyl, w/560 lb. agent, no solenoid		2.50	6.400		21,000	385		21,385	23,700
7340 560 lb. cyl, w/300 lb. agent, w/solenoid		4	4		18,600	239		18,839	20,900
7350 560 lb. cyl, w/560 lb. agent, w/solenoid		2.50	6.400		27,700	385		28,085	31,100
7420 1,200 lb. cyl, w/600 lb. agent, no solenoid	Q-13	4	8		26,200	505		26,705	29,700
7430 1,200 lb. cyl, w/1,200 lb. agent, no solenoid		3	10.667		43,000	675		43,675	48,300
7440 1,200 lb. cyl, w/600 lb. agent, w/solenoid		4	8		64,500	505		65,005	72,000
7450 1,200 lb. cyl, w/1,200 lb. agent, w/solenoid		3	10.667		101,500	675		102,175	112,500
7500 Accessories									
7510 Dispersion nozzle	1 Spri	16	.500	Ea.	125	33.50		158.50	187
7520 Agent release panel	1 Elec	4	2		72	127		199	268
7530 Maintenance switch	"	6	1.333		39	85		124	169

For customer support on your Electrical Costs with RSMeans Data, call 800.448.8182.

21 22 Clean-Agent Fire-Extinguishing Systems

21 22 16 – Clean-Agent Fire-Extinguishing Equipment

21 22 16.50 Clean-Agent Extinguishing Systems	Crew	Daily Output	Labor-Hours	Unit	Material	2021 Bare Costs Labor	2021 Bare Costs Equipment	Total	Total Incl O&P
7540	Solenoid valve, 12v dc	1 Spri	8	1	Ea.	273	66.50		339.50	400
7550	12v ac		8	1		670	66.50		736.50	835
7560	12v dc, explosion proof		8	1		705	66.50		771.50	875

79

For customer support on your Electrical Costs with RSMeans Data, call 800.448.8182.

Division Notes

	CREW	DAILY OUTPUT	LABOR-HOURS	UNIT	BARE COSTS				TOTAL INCL O&P
					MAT.	LABOR	EQUIP.	TOTAL	

Estimating Tips
22 10 00 Plumbing
Piping and Pumps

This subdivision is primarily basic pipe and related materials. The pipe may be used by any of the mechanical disciplines, i.e., plumbing, fire protection, heating, and air conditioning.

Note: CPVC plastic piping approved for fire protection is located in 21 11 13.

- The labor adjustment factors listed in Subdivision 22 01 02.20 apply throughout Divisions 21, 22, and 23. CAUTION: the correct percentage may vary for the same items. For example, the percentage add for the basic pipe installation should be based on the maximum height that the installer must install for that particular section. If the pipe is to be located 14' above the floor but it is suspended on threaded rod from beams, the bottom flange of which is 18' high (4' rods), then the height is actually 18' and the add is 20%. The pipe cover, however, does not have to go above the 14' and so the add should be 10%.

- Most pipe is priced first as straight pipe with a joint (coupling, weld, etc.) every 10' and a hanger usually every 10'. There are exceptions with hanger spacing such as for cast iron pipe (5')

and plastic pipe (3 per 10'). Following each type of pipe there are several lines listing sizes and the amount to be subtracted to delete couplings and hangers. This is for pipe that is to be buried or supported together on trapeze hangers. The reason that the couplings are deleted is that these runs are usually long, and frequently longer lengths of pipe are used. By deleting the couplings, the estimator is expected to look up and add back the correct reduced number of couplings.

- When preparing an estimate, it may be necessary to approximate the fittings. Fittings usually run between 25% and 50% of the cost of the pipe. The lower percentage is for simpler runs, and the higher number is for complex areas, such as mechanical rooms.

- For historic restoration projects, the systems must be as invisible as possible, and pathways must be sought for pipes, conduit, and ductwork. While installations in accessible spaces (such as basements and attics) are relatively straightforward to estimate, labor costs may be more difficult to determine when delivery systems must be concealed.

22 40 00 Plumbing Fixtures

- Plumbing fixture costs usually require two lines: the fixture itself and its "rough-in, supply, and waste."

- In the Assemblies Section (Plumbing D2010) for the desired fixture, the System Components Group at the center of the page shows the fixture on the first line. The rest of the list (fittings, pipe, tubing, etc.) will total up to what we refer to in the Unit Price section as "Rough-in, supply, waste, and vent." Note that for most fixtures we allow a nominal 5' of tubing to reach from the fixture to a main or riser.

- Remember that gas- and oil-fired units need venting.

Reference Numbers

Reference numbers are shown at the beginning of some major classifications. These numbers refer to related items in the Reference Section. The reference information may be an estimating procedure, an alternate pricing method, or technical information.

Note: Not all subdivisions listed here necessarily appear. ■

Same Data. Simplified.

Enjoy the convenience and efficiency of accessing your costs anywhere:

- **Skip the multiplier** by setting your location
- **Quickly search,** edit, favorite and share costs
- **Stay on top of price changes** with automatic updates

Discover more at rsmeans.com/online

22 01 Operation and Maintenance of Plumbing

22 01 02 – Labor Adjustments

22 01 02.10 Boilers, General	Crew	Daily Output	Labor-Hours	Unit	Material	2021 Bare Costs Labor	Equipment	Total	Total Incl O&P
0010 **BOILERS, GENERAL**, Prices do not include flue piping, elec. wiring,									
0020 gas or oil piping, boiler base, pad, or tankless unless noted									
0100 Boiler H.P.: 10 KW = 34 lb./steam/hr. = 33,475 BTU/hr.									
0150 To convert SFR to BTU rating: Hot water, 150 x SFR;									
0160 Forced hot water, 180 x SFR; steam, 240 x SFR									

22 01 02.20 Labor Adjustment Factors	Crew	Daily Output	Labor-Hours	Unit	Material	2021 Bare Costs Labor	Equipment	Total	Total Incl O&P
0010 **LABOR ADJUSTMENT FACTORS** (For Div. 21, 22 and 23) R220102-20									
0100 Labor factors: The below are reasonable suggestions, but									
0110 each project must be evaluated for its own peculiarities, and									
0120 the adjustments be increased or decreased depending on the									
0130 severity of the special conditions.									
1000 Add to labor for elevated installation (Above floor level)									
1080 10' to 14.5' high						10%			
1100 15' to 19.5' high						20%			
1120 20' to 24.5' high						25%			
1140 25' to 29.5' high						35%			
1160 30' to 34.5' high						40%			
1180 35' to 39.5' high						50%			
1200 40' and higher						55%			
2000 Add to labor for crawl space									
2100 3' high						40%			
2140 4' high						30%			
3000 Add to labor for multi-story building									
3010 For new construction (No elevator available)									
3100 Add for floors 3 thru 10						5%			
3110 Add for floors 11 thru 15						10%			
3120 Add for floors 16 thru 20						15%			
3130 Add for floors 21 thru 30						20%			
3140 Add for floors 31 and up						30%			
3170 For existing structure (Elevator available)									
3180 Add for work on floor 3 and above						2%			
4000 Add to labor for working in existing occupied buildings									
4100 Hospital						35%			
4140 Office building						25%			
4180 School						20%			
4220 Factory or warehouse						15%			
4260 Multi dwelling						15%			
5000 Add to labor, miscellaneous									
5100 Cramped shaft						35%			
5140 Congested area						15%			
5180 Excessive heat or cold						30%			
9000 Labor factors: The above are reasonable suggestions, but									
9010 each project should be evaluated for its own peculiarities.									
9100 Other factors to be considered are:									
9140 Movement of material and equipment through finished areas									
9180 Equipment room									
9220 Attic space									
9260 No service road									
9300 Poor unloading/storage area									
9340 Congested site area/heavy traffic									

For customer support on your Electrical Costs with RSMeans Data, call 800.448.8182.

22 05 23 – General-Duty Valves for Plumbing Piping

22 05 23.20 Valves, Bronze	Crew	Daily Output	Labor-Hours	Unit	Material	2021 Bare Costs Labor	Equipment	Total	Total Incl O&P
0010 **VALVES, BRONZE**									
8800 Water heater water & gas safety shut off									
8810 Protection against a leaking water heater									
8814 Shut off valve	1 Plum	16	.500	Ea.	196	34		230	266
8818 Water heater dam		32	.250		52.50	16.95		69.45	83.50
8822 Gas control wiring harness		32	.250		25.50	16.95		42.45	53.50
8830 Whole house flood safety shut off									
8834 Connections									
8838 3/4" NPT	1 Plum	12	.667	Ea.	1,075	45		1,120	1,250
8842 1" NPT		11	.727		1,100	49		1,149	1,300
8846 1-1/4" NPT		10	.800		1,150	54		1,204	1,325

22 05 29 – Hangers and Supports for Plumbing Piping and Equipment

22 05 29.10 Hangers & Supp. for Plumb'g/HVAC Pipe/Equip.

	Crew	Daily Output	Labor-Hours	Unit	Material	2021 Bare Costs Labor	Equipment	Total	Total Incl O&P
0010 **HANGERS AND SUPPORTS FOR PLUMB'G/HVAC PIPE/EQUIP.**									
0011 TYPE numbers per MSS-SP58									
0050 Brackets									
0060 Beam side or wall, malleable iron, TYPE 34									
0070 3/8" threaded rod size	1 Plum	48	.167	Ea.	5.20	11.30		16.50	22.50
0080 1/2" threaded rod size		48	.167		8.85	11.30		20.15	26.50
0090 5/8" threaded rod size		48	.167		15.85	11.30		27.15	34.50
0100 3/4" threaded rod size		48	.167		21.50	11.30		32.80	40.50
0110 7/8" threaded rod size		48	.167		14.90	11.30		26.20	33.50
0120 For concrete installation, add						30%			
0150 Wall, welded steel, medium, TYPE 32									
0160 0 size, 12" wide, 18" deep	1 Plum	34	.235	Ea.	375	15.95		390.95	440
0170 1 size, 18" wide, 24" deep		34	.235		445	15.95		460.95	515
0180 2 size, 24" wide, 30" deep		34	.235		590	15.95		605.95	675
0300 Clamps									
0310 C-clamp, for mounting on steel beam flange, w/locknut, TYPE 23									
0320 3/8" threaded rod size	1 Plum	160	.050	Ea.	3.68	3.39		7.07	9.10
0330 1/2" threaded rod size		160	.050		4.57	3.39		7.96	10.10
0340 5/8" threaded rod size		160	.050		6.15	3.39		9.54	11.80
0350 3/4" threaded rod size		160	.050		8.05	3.39		11.44	13.90
0750 Riser or extension pipe, carbon steel, TYPE 8									
0760 3/4" pipe size	1 Plum	48	.167	Ea.	4.45	11.30		15.75	22
0770 1" pipe size		47	.170		4.37	11.50		15.87	22
0780 1-1/4" pipe size		46	.174		4.04	11.75		15.79	22
0790 1-1/2" pipe size		45	.178		6	12.05		18.05	24.50
0800 2" pipe size		43	.186		6	12.60		18.60	25.50
0810 2-1/2" pipe size		41	.195		6.30	13.20		19.50	26.50
0820 3" pipe size		40	.200		7.25	13.55		20.80	28
0830 3-1/2" pipe size		39	.205		8.55	13.90		22.45	30
0840 4" pipe size		38	.211		9.30	14.25		23.55	32
0850 5" pipe size		37	.216		12.40	14.65		27.05	35.50
0860 6" pipe size		36	.222		15.55	15.05		30.60	39.50
1150 Insert, concrete									
1160 Wedge type, carbon steel body, malleable iron nut, galvanized									
1170 1/4" threaded rod size	1 Plum	96	.083	Ea.	8.90	5.65		14.55	18.20
1180 3/8" threaded rod size		96	.083		2.85	5.65		8.50	11.55
1190 1/2" threaded rod size		96	.083		1.47	5.65		7.12	10
1200 5/8" threaded rod size		96	.083		2.02	5.65		7.67	10.60
1210 3/4" threaded rod size		96	.083		2.22	5.65		7.87	10.85

22 05 29.10 Hangers & Supp. for Plumb'g/HVAC Pipe/Equip.	Crew	Daily Output	Labor-Hours	Unit	Material	2021 Bare Costs Labor	Equipment	Total	Total Incl O&P
1220	7/8" threaded rod size	1 Plum	96	.083	Ea.	12.30	5.65		17.95	22
2650	Rods, carbon steel									
2660	Continuous thread									
2670	1/4" thread size	1 Plum	144	.056	L.F.	.43	3.76		4.19	6.05
2680	3/8" thread size		144	.056		.46	3.76		4.22	6.10
2690	1/2" thread size		144	.056		.72	3.76		4.48	6.40
2700	5/8" thread size		144	.056		1.03	3.76		4.79	6.75
2710	3/4" thread size		144	.056		1.81	3.76		5.57	7.60
2720	7/8" thread size		144	.056		2.27	3.76		6.03	8.10
2725	1/4" thread size, bright finish		144	.056		1.26	3.76		5.02	7
2726	1/2" thread size, bright finish		144	.056		6.50	3.76		10.26	12.75
2730	For galvanized, add					40%				
4400	U-bolt, carbon steel									
4410	Standard, with nuts, TYPE 42									
4420	1/2" pipe size	1 Plum	160	.050	Ea.	1.21	3.39		4.60	6.40
4430	3/4" pipe size		158	.051		1.73	3.43		5.16	7
4450	1" pipe size		152	.053		1.77	3.56		5.33	7.25
4460	1-1/4" pipe size		148	.054		2.09	3.66		5.75	7.75
4470	1-1/2" pipe size		143	.056		2.25	3.79		6.04	8.15
4480	2" pipe size		139	.058		2.43	3.90		6.33	8.45
4490	2-1/2" pipe size		134	.060		4.06	4.04		8.10	10.50
4500	3" pipe size		128	.063		4.27	4.23		8.50	11
4510	3-1/2" pipe size		122	.066		4.47	4.44		8.91	11.55
4520	4" pipe size		117	.068		4.59	4.63		9.22	11.95
4530	5" pipe size		114	.070		3.41	4.75		8.16	10.85
4540	6" pipe size		111	.072		7.05	4.88		11.93	15.05
4580	For plastic coating on 1/2" thru 6" size, add					150%				
8800	Wire cable support system									
8810	Cable with hook terminal and locking device									
8830	2 mm (.079") diam. cable (100 lb. cap.)									
8840	1 m (3.3') length, with hook	1 Shee	96	.083	Ea.	3.83	5.45		9.28	12.45
8850	2 m (6.6') length, with hook		84	.095		4.53	6.25		10.78	14.40
8860	3 m (9.9') length, with hook		72	.111		5.05	7.25		12.30	16.50
8870	5 m (16.4') length, with hook	Q-9	60	.267		6.20	15.70		21.90	30.50
8880	10 m (32.8') length, with hook	"	30	.533		8.90	31.50		40.40	57.50
8900	3 mm (.118") diam. cable (200 lb. cap.)									
8910	1 m (3.3') length, with hook	1 Shee	96	.083	Ea.	11.15	5.45		16.60	20.50
8920	2 m (6.6') length, with hook		84	.095		.55	6.25		6.80	10
8930	3 m (9.9') length, with hook		72	.111		13.70	7.25		20.95	26
8940	5 m (16.4') length, with hook	Q-9	60	.267		11.80	15.70		27.50	36.50
8950	10 m (32.8') length, with hook	"	30	.533		10.55	31.50		42.05	59
9000	Cable system accessories									
9010	Anchor bolt, 3/8", with nut	1 Shee	140	.057	Ea.	2.67	3.74		6.41	8.60
9020	Air duct corner protector		160	.050		.89	3.27		4.16	5.90
9030	Air duct support attachment		140	.057		2.31	3.74		6.05	8.20
9040	Flange clip, hammer-on style									
9044	For flange thickness 3/32"-9/64", 160 lb. cap.	1 Shee	180	.044	Ea.	.61	2.91		3.52	5.05
9048	For flange thickness 1/8"-1/4", 200 lb. cap.		160	.050		.69	3.27		3.96	5.70
9052	For flange thickness 5/16"-1/2", 200 lb. cap.		150	.053		1.01	3.49		4.50	6.35
9056	For flange thickness 9/16"-3/4", 200 lb. cap.		140	.057		1.22	3.74		4.96	7
9060	Wire insulation protection tube		180	.044	L.F.	.58	2.91		3.49	5.05
9070	Wire cutter				Ea.	45			45	49.50

22 33 Electric Domestic Water Heaters

22 33 13 – Instantaneous Electric Domestic Water Heaters

22 33 13.10 Hot Water Dispensers	Crew	Daily Output	Labor-Hours	Unit	Material	2021 Bare Costs Labor	Equipment	Total	Total Incl O&P
0010 **HOT WATER DISPENSERS**									
0160 Commercial, 100 cup, 11.3 amp	1 Plum	14	.571	Ea.	455	38.50		493.50	560
3180 Household, 60 cup	"	14	.571	"	239	38.50		277.50	320

22 33 30 – Residential, Electric Domestic Water Heaters

22 33 30.13 Residential, Small-Capacity Elec. Water Heaters

	Crew	Daily Output	Labor-Hours	Unit	Material	2021 Bare Costs Labor	Equipment	Total	Total Incl O&P
0010 **RESIDENTIAL, SMALL-CAPACITY ELECTRIC DOMESTIC WATER HEATERS**									
1000 Residential, electric, glass lined tank, 5 yr., 10 gal., single element R224000-10	1 Plum	2.30	3.478	Ea.	465	235		700	860
1040 20 gallon, single element R224000-20		2.20	3.636		685	246		931	1,125
1060 30 gallon, double element		2.20	3.636		1,225	246		1,471	1,725
1080 40 gallon, double element		2	4		1,400	271		1,671	1,925
1100 52 gallon, double element		2	4		1,575	271		1,846	2,150
1120 66 gallon, double element		1.80	4.444		2,150	300		2,450	2,800
1140 80 gallon, double element		1.60	5		2,425	340		2,765	3,175
1180 120 gallon, double element		1.40	5.714		3,350	385		3,735	4,275

22 33 33 – Light-Commercial Electric Domestic Water Heaters

22 33 33.10 Commercial Electric Water Heaters

	Crew	Daily Output	Labor-Hours	Unit	Material	2021 Bare Costs Labor	Equipment	Total	Total Incl O&P
0010 **COMMERCIAL ELECTRIC WATER HEATERS**									
4000 Commercial, 100° rise. NOTE: for each size tank, a range of									
4010 heaters between the ones shown is available									
4020 Electric									
4100 5 gal., 3 kW, 12 GPH, 208 volt	1 Plum	2	4	Ea.	4,200	271		4,471	5,000
4120 10 gal., 6 kW, 25 GPH, 208 volt		2	4		4,575	271		4,846	5,425
4130 30 gal., 24 kW, 98 GPH, 208 volt		1.92	4.167		7,325	282		7,607	8,475
4136 40 gal., 36 kW, 148 GPH, 208 volt		1.88	4.255		11,100	288		11,388	12,600
4140 50 gal., 9 kW, 37 GPH, 208 volt		1.80	4.444		6,225	300		6,525	7,300
4160 50 gal., 36 kW, 148 GPH, 208 volt		1.80	4.444		9,700	300		10,000	11,200
4180 80 gal., 12 kW, 49 GPH, 208 volt		1.50	5.333		7,700	360		8,060	9,025
4200 80 gal., 36 kW, 148 GPH, 208 volt		1.50	5.333		11,000	360		11,360	12,600
4220 100 gal., 36 kW, 148 GPH, 208 volt		1.20	6.667		11,800	450		12,250	13,700
4240 120 gal., 36 kW, 148 GPH, 208 volt		1.20	6.667		12,300	450		12,750	14,200
4260 150 gal., 15 kW, 61 GPH, 480 volt		1	8		34,400	540		34,940	38,600
4280 150 gal., 120 kW, 490 GPH, 480 volt		1	8		46,800	540		47,340	52,500
4300 200 gal., 15 kW, 61 GPH, 480 volt	Q-1	1.70	9.412		36,500	575		37,075	41,000
4320 200 gal., 120 kW, 490 GPH, 480 volt		1.70	9.412		49,200	575		49,775	55,000
4340 250 gal., 15 kW, 61 GPH, 480 volt		1.50	10.667		37,700	650		38,350	42,500
4360 250 gal., 150 kW, 615 GPH, 480 volt		1.50	10.667		46,200	650		46,850	52,000
4380 300 gal., 30 kW, 123 GPH, 480 volt		1.30	12.308		41,500	750		42,250	46,800
4400 300 gal., 180 kW, 738 GPH, 480 volt		1.30	12.308		71,500	750		72,250	80,000
4420 350 gal., 30 kW, 123 GPH, 480 volt		1.10	14.545		36,900	885		37,785	41,900
4440 350 gal., 180 kW, 738 GPH, 480 volt		1.10	14.545		52,000	885		52,885	58,500
4460 400 gal., 30 kW, 123 GPH, 480 volt		1	16		50,500	975		51,475	57,000
4480 400 gal., 210 kW, 860 GPH, 480 volt		1	16		84,000	975		84,975	94,000
4500 500 gal., 30 kW, 123 GPH, 480 volt		.80	20		59,000	1,225		60,225	67,000
4520 500 gal., 240 kW, 984 GPH, 480 volt		.80	20		98,500	1,225		99,725	110,000
4540 600 gal., 30 kW, 123 GPH, 480 volt	Q-2	1.20	20		66,500	1,275		67,775	75,000
4560 600 gal., 300 kW, 1,230 GPH, 480 volt		1.20	20		116,000	1,275		117,275	129,500
4580 700 gal., 30 kW, 123 GPH, 480 volt		1	24		58,500	1,525		60,025	67,000
4600 700 gal., 300 kW, 1,230 GPH, 480 volt		1	24		88,000	1,525		89,525	99,500
4620 800 gal., 60 kW, 245 GPH, 480 volt		.90	26.667		79,500	1,675		81,175	90,000
4640 800 gal., 300 kW, 1,230 GPH, 480 volt		.90	26.667		90,000	1,675		91,675	101,500
4660 1,000 gal., 60 kW, 245 GPH, 480 volt		.70	34.286		69,500	2,175		71,675	79,500
4680 1,000 gal., 480 kW, 1,970 GPH, 480 volt		.70	34.286		115,500	2,175		117,675	130,000

22 33 Electric Domestic Water Heaters

22 33 33 – Light-Commercial Electric Domestic Water Heaters

22 33 33.10 Commercial Electric Water Heaters	Crew	Daily Output	Labor-Hours	Unit	Material	2021 Bare Costs Labor	Equipment	Total	Total Incl O&P
4700	1,200 gal., 60 kW, 245 GPH, 480 volt	Q-2	.60	40	Ea.	78,500	2,525		81,025	90,500
4720	1,200 gal., 480 kW, 1,970 GPH, 480 volt		.60	40		119,500	2,525		122,025	135,000
4740	1,500 gal., 60 kW, 245 GPH, 480 volt		.50	48		103,500	3,025		106,525	118,500
4760	1,500 gal., 480 kW, 1,970 GPH, 480 volt		.50	48		143,500	3,025		146,525	162,500
5400	Modulating step control for under 90 kW, 2-5 steps	1 Elec	5.30	1.509		1,000	96		1,096	1,250
5440	For above 90 kW, 1 through 5 steps beyond standard, add		3.20	2.500		268	159		427	530
5460	For above 90 kW, 6 through 10 steps beyond standard, add		2.70	2.963		595	189		784	935
5480	For above 90 kW, 11 through 18 steps beyond standard, add		1.60	5		885	320		1,205	1,450

22 34 Fuel-Fired Domestic Water Heaters

22 34 30 – Residential Gas Domestic Water Heaters

22 34 30.13 Residential, Atmos, Gas Domestic Wtr Heaters

		Crew	Daily Output	Labor-Hours	Unit	Material	2021 Bare Costs Labor	Equipment	Total	Total Incl O&P
0010	**RESIDENTIAL, ATMOSPHERIC, GAS DOMESTIC WATER HEATERS**									
2000	Gas fired, foam lined tank, 10 yr., vent not incl.									
2040	30 gallon	1 Plum	2	4	Ea.	2,225	271		2,496	2,825
2060	40 gallon		1.90	4.211		1,850	285		2,135	2,450
2080	50 gallon		1.80	4.444		1,975	300		2,275	2,625
2090	60 gallon		1.70	4.706		1,875	320		2,195	2,550
2100	75 gallon		1.50	5.333		2,575	360		2,935	3,375
2120	100 gallon		1.30	6.154		3,475	415		3,890	4,425
2900	Water heater, safety-drain pan, 26" round		20	.400		21	27		48	63.50
3000	Tank leak safety, water & gas shut off see 22 05 23.20 8800									

22 34 46 – Oil-Fired Domestic Water Heaters

22 34 46.10 Residential Oil-Fired Water Heaters

		Crew	Daily Output	Labor-Hours	Unit	Material	2021 Bare Costs Labor	Equipment	Total	Total Incl O&P
0010	**RESIDENTIAL OIL-FIRED WATER HEATERS**									
3000	Oil fired, glass lined tank, 5 yr., vent not included, 30 gallon	1 Plum	2	4	Ea.	1,675	271		1,946	2,250
3040	50 gallon		1.80	4.444		1,525	300		1,825	2,125
3060	70 gallon		1.50	5.333		2,350	360		2,710	3,125

Estimating Tips

The labor adjustment factors listed in Subdivision 22 01 02.20 also apply to Division 23.

23 10 00 Facility Fuel Systems

- The prices in this subdivision for above- and below-ground storage tanks do not include foundations or hold-down slabs, unless noted. The estimator should refer to Divisions 3 and 31 for foundation system pricing. In addition to the foundations, required tank accessories, such as tank gauges, leak detection devices, and additional manholes and piping, must be added to the tank prices.

23 50 00 Central Heating Equipment

- When estimating the cost of an HVAC system, check to see who is responsible for providing and installing the temperature control system. It is possible to overlook controls, assuming that they would be included in the electrical estimate.

- When looking up a boiler, be careful on specified capacity. Some manufacturers rate their products on output while others use input.

- Include HVAC insulation for pipe, boiler, and duct (wrap and liner).

- Be careful when looking up mechanical items to get the correct pressure rating and connection type (thread, weld, flange).

23 70 00 Central HVAC Equipment

- Combination heating and cooling units are sized by the air conditioning requirements. (See Reference No. R236000-20 for the preliminary sizing guide.)

- A ton of air conditioning is nominally 400 CFM.

- Rectangular duct is taken off by the linear foot for each size, but its cost is usually estimated by the pound. Remember that SMACNA standards now base duct on internal pressure.

- Prefabricated duct is estimated and purchased like pipe: straight sections and fittings.

- Note that cranes or other lifting equipment are not included on any lines in Division 23. For example, if a crane is required to lift a heavy piece of pipe into place high above a gym floor, or to put a rooftop unit on the roof of a four-story building, etc., it must be added. Due to the potential for extreme variation—from nothing additional required to a major crane or helicopter—we feel that including a nominal amount for "lifting contingency" would be useless and detract from the accuracy of the estimate. When using equipment rental cost data from RSMeans, do not forget to include the cost of the operator(s).

Reference Numbers

Reference numbers are shown at the beginning of some major classifications. These numbers refer to related items in the Reference Section. The reference information may be an estimating procedure, an alternate pricing method, or technical information.

Note: Not all subdivisions listed here necessarily appear. ■

Same Data. Simplified.

Enjoy the convenience and efficiency of accessing your costs anywhere:

- **Skip the multiplier** by setting your location
- **Quickly search,** edit, favorite and share costs
- **Stay on top of price changes** with automatic updates

Discover more at rsmeans.com/online

---

23 05 Common Work Results for HVAC

23 05 02 – HVAC General

23 05 02.10 Air Conditioning, General	Crew	Daily Output	Labor-Hours	Unit	Material	2021 Bare Costs Labor	Equipment	Total	Total Incl O&P
0010 **AIR CONDITIONING, GENERAL** Prices are for standard efficiencies (SEER 13)									
0020 for upgrade to SEER 14 add					10%				

23 05 93 – Testing, Adjusting, and Balancing for HVAC

23 05 93.10 Balancing, Air

					Unit				Total	Total Incl O&P
0010	**BALANCING, AIR** (Subcontractor's quote incl. material and labor)									
0900	Heating and ventilating equipment									
1000	Centrifugal fans, utility sets				Ea.				420	420
1100	Heating and ventilating unit								630	630
1200	In-line fan								630	630
1300	Propeller and wall fan								119	119
1400	Roof exhaust fan								280	280
2000	Air conditioning equipment, central station								910	910
2100	Built-up low pressure unit								840	840
2200	Built-up high pressure unit								980	980
2300	Built-up high pressure dual duct								1,550	1,550
2400	Built-up variable volume								1,825	1,825
2500	Multi-zone A.C. and heating unit								630	630
2600	For each zone over one, add								140	140
2700	Package A.C. unit								350	350
2800	Rooftop heating and cooling unit								490	490
3000	Supply, return, exhaust, registers & diffusers, avg. height ceiling								84	84
3100	High ceiling								126	126
3200	Floor height								70	70
3300	Off mixing box								56	56
3500	Induction unit								91	91
3600	Lab fume hood								420	420
3700	Linear supply								210	210
3800	Linear supply high								245	245
4000	Linear return								70	70
4100	Light troffers								84	84
4200	Moduline - master								84	84
4300	Moduline - slaves								42	42
4400	Regenerators								560	560
4500	Taps into ceiling plenums								105	105
4600	Variable volume boxes								84	84

23 05 93.20 Balancing, Water

					Unit				Total	Total Incl O&P
0010	**BALANCING, WATER** (Subcontractor's quote incl. material and labor)									
0050	Air cooled condenser				Ea.				256	256
0080	Boiler								515	515
0100	Cabinet unit heater								88	88
0200	Chiller								620	620
0300	Convector								73	73
0400	Converter								365	365
0500	Cooling tower								475	475
0600	Fan coil unit, unit ventilator								132	132
0700	Fin tube and radiant panels								146	146
0800	Main and duct re-heat coils								135	135
0810	Heat exchanger								135	135
0900	Main balancing cocks								110	110
1000	Pumps								320	320
1100	Unit heater								102	102

23 05 93 – Testing, Adjusting, and Balancing for HVAC

23 09 23.10 Control Components/DDC Systems	Crew	Daily Output	Labor-Hours	Unit	Material	2021 Bare Costs Labor	Equipment	Total	Total Incl O&P
0010 **CONTROL COMPONENTS/DDC SYSTEMS** (Sub's quote incl. M & L)									
0100 Analog inputs									
0110 Sensors (avg. 50' run in 1/2" EMT)									
0120 Duct temperature				Ea.				465	465
0130 Space temperature								665	665
0140 Duct humidity, +/- 3%								595	595
0150 Space humidity, +/- 2%								1,150	1,150
0160 Duct static pressure								680	680
0170 CFM/transducer								920	920
0172 Water temperature								940	940
0174 Water flow								3,450	3,450
0176 Water pressure differential								975	975
0177 Steam flow								2,400	2,400
0178 Steam pressure								1,025	1,025
0180 KW/transducer								1,350	1,350
0182 KWH totalization (not incl. elec. meter pulse xmtr.)								625	625
0190 Space static pressure				▼				1,075	1,075
1000 Analog outputs (avg. 50' run in 1/2" EMT)									
1010 P/I transducer				Ea.				635	635
1020 Analog output, matl. in MUX								305	305
1030 Pneumatic (not incl. control device)								645	645
1040 Electric (not incl. control device)				▼				380	380
2000 Status (alarms)									
2100 Digital inputs (avg. 50' run in 1/2" EMT)									
2110 Freeze				Ea.				435	435
2120 Fire								395	395
2130 Differential pressure (air)	2 Elec	3	5.333		283	340		623	815
2140 Differential pressure (water)								975	975
2150 Current sensor								435	435
2160 Duct high temperature thermostat								570	570
2170 Duct smoke detector				▼				705	705
2200 Digital output (avg. 50' run in 1/2" EMT)									
2210 Start/stop				Ea.				340	340
2220 On/off (maintained contact)				"				585	585
3000 Controller MUX panel, incl. function boards									
3100 48 point				Ea.				5,275	5,275
3110 128 point				"				7,225	7,225
3200 DDC controller (avg. 50' run in conduit)									
3210 Mechanical room									
3214 16 point controller (incl. 120 volt/1 phase power supply)				Ea.				3,275	3,275
3229 32 point controller (incl. 120 volt/1 phase power supply)				"				5,425	5,425
3230 Includes software programming and checkout									
3260 Space									
3266 VAV terminal box (incl. space temp. sensor)				Ea.				840	840
3280 Host computer (avg. 50' run in conduit)									
3281 Package complete with PC, keyboard,									
3282 printer, monitor, basic software				Ea.				3,150	3,150
4000 Front end costs									
4100 Computer (P.C.) with software program				Ea.				6,350	6,350
4200 Color graphics software								3,925	3,925
4300 Color graphics slides								490	490
4350 Additional printer				▼				980	980
4400 Communications trunk cable				L.F.				3.80	3.80

For customer support on your Electrical Costs with RSMeans Data, call 800.448.8182.

89

23 05 93 – Testing, Adjusting, and Balancing for HVAC

23 09 23.10 Control Components/DDC Systems

		Crew	Daily Output	Labor-Hours	Unit	Material	2021 Bare Costs Labor	Equipment	Total	Total Incl O&P
4500	Engineering labor (not incl. dftg.)				Point				94	94
4600	Calibration labor								120	120
4700	Start-up, checkout labor				↓				120	120
4800	Programming labor, as req'd									
5000	Communications bus (data transmission cable)									
5010	#18 twisted shielded pair in 1/2" EMT conduit				C.L.F.				380	380
8000	Applications software									
8050	Basic maintenance manager software (not incl. data base entry)				Ea.				1,950	1,950
8100	Time program				Point				6.85	6.85
8120	Duty cycle								13.65	13.65
8140	Optimum start/stop								41.50	41.50
8160	Demand limiting								20.50	20.50
8180	Enthalpy program				↓				41.50	41.50
8200	Boiler optimization				Ea.				1,225	1,225
8220	Chiller optimization				"				1,625	1,625
8240	Custom applications									
8260	Cost varies with complexity									

23 09 43 – Pneumatic Control System for HVAC

23 09 43.10 Pneumatic Control Systems

			Crew	Daily Output	Labor-Hours	Unit	Material	2021 Bare Costs Labor	Equipment	Total	Total Incl O&P
0010	**PNEUMATIC CONTROL SYSTEMS**										
0011	Including a nominal 50' of tubing. Add control panelboard if req'd.										
0100	Heating and ventilating, split system										
0200	Mixed air control, economizer cycle, panel readout, tubing										
0220	Up to 10 tons	G	Q-19	.68	35.294	Ea.	4,550	2,200		6,750	8,275
0240	For 10 to 20 tons	G		.63	37.915		5,125	2,350		7,475	9,150
0260	For over 20 tons	G		.58	41.096		5,550	2,550		8,100	9,900
0270	Enthalpy cycle, up to 10 tons			.50	48.387		5,275	3,000		8,275	10,300
0280	For 10 to 20 tons			.46	52.174		5,675	3,250		8,925	11,100
0290	For over 20 tons		↓	.42	56.604	↓	5,675	3,525		9,200	11,500
0300	Heating coil, hot water, 3 way valve,										
0320	Freezestat, limit control on discharge, readout		Q-5	.69	23.088	Ea.	3,375	1,425		4,800	5,850
0500	Cooling coil, chilled water, room										
0520	Thermostat, 3 way valve		Q-5	2	8	Ea.	1,450	490		1,940	2,325
0600	Cooling tower, fan cycle, damper control,										
0620	Control system including water readout in/out at panel		Q-19	.67	35.821	Ea.	5,975	2,225		8,200	9,900
1000	Unit ventilator, day/night operation,										
1100	freezestat, ASHRAE, cycle 2		Q-19	.91	26.374	Ea.	3,200	1,650		4,850	5,975
2000	Compensated hot water from boiler, valve control,										
2100	readout and reset at panel, up to 60 GPM		Q-19	.55	43.956	Ea.	6,500	2,725		9,225	11,200
2120	For 120 GPM			.51	47.059		6,400	2,925		9,325	11,400
2140	For 240 GPM			.49	49.180		7,275	3,050		10,325	12,600
3000	Boiler room combustion air, damper to 5 S.F., controls			1.37	17.582		3,000	1,100		4,100	4,900
3500	Fan coil, heating and cooling valves, 4 pipe control system			3	8		1,400	500		1,900	2,300
3600	Heat exchanger system controls		↓	.86	27.907	↓	2,800	1,725		4,525	5,675
3900	Multizone control (one per zone), includes thermostat, damper										
3910	motor and reset of discharge temperature		Q-5	.51	31.373	Ea.	2,875	1,925		4,800	6,050
4000	Pneumatic thermostat, including controlling room radiator valve		"	2.43	6.593		870	405		1,275	1,550
4040	Program energy saving optimizer	G	Q-19	1.21	19.786		7,475	1,225		8,700	10,100
4060	Pump control system		"	3	8		1,325	500		1,825	2,225
4080	Reheat coil control system, not incl. coil		Q-5	2.43	6.593	↓	1,125	405		1,530	1,850
4500	Air supply for pneumatic control system										
4600	Tank mounted duplex compressor, starter, alternator,										

For customer support on your Electrical Costs with RSMeans Data, call 800.448.8182.

23 09 Instrumentation and Control for HVAC

23 09 43 – Pneumatic Control System for HVAC

23 09 43.10 Pneumatic Control Systems

		Crew	Daily Output	Labor-Hours	Unit	Material	2021 Bare Costs Labor	Equipment	Total	Total Incl O&P
4620	piping, dryer, PRV station and filter									
4630	1/2 HP	Q-19	.68	35.139	Ea.	11,700	2,175		13,875	16,100
4640	3/4 HP		.64	37.383		11,200	2,325		13,525	15,900
4650	1 HP		.61	39.539		12,700	2,450		15,150	17,700
4660	1-1/2 HP		.58	41.739		13,100	2,600		15,700	18,300
4680	3 HP		.55	43.956		17,800	2,725		20,525	23,700
4690	5 HP		.42	57.143		33,900	3,550		37,450	42,600
4800	Main air supply, includes 3/8" copper main and labor	Q-5	1.82	8.791	C.L.F.	395	540		935	1,225
7000	Static pressure control for air handling unit, includes pressure									
7010	sensor, receiver controller, readout and damper motors	Q-19	.64	37.383	Ea.	9,300	2,325		11,625	13,700
8600	VAV boxes, incl. thermostat, damper motor, reheat coil & tubing	Q-5	1.46	10.989	"	1,350	675		2,025	2,475

23 09 53 – Pneumatic and Electric Control System for HVAC

23 09 53.10 Control Components

		Crew	Daily Output	Labor-Hours	Unit	Material	2021 Bare Costs Labor	Equipment	Total	Total Incl O&P
0010	**CONTROL COMPONENTS**									
0700	Controller, receiver									
0850	Electric, single snap switch	1 Elec	4	2	Ea.	485	127		612	720
0860	Dual snap switches	"	3	2.667	"	650	170		820	965
3590	Sensor, electric operated									
3620	Humidity	1 Elec	8	1	Ea.	127	63.50		190.50	235
3650	Pressure		8	1		335	63.50		398.50	465
3680	Temperature		10	.800		112	51		163	199
5000	Thermostats									
5200	24 hour, automatic, clock [G]	1 Shee	8	1	Ea.	186	65.50		251.50	305
5220	Electric, low voltage, 2 wire	1 Elec	13	.615		53	39		92	117
5230	3 wire	"	10	.800		34	51		85	113
7090	Valves, motor controlled, including actuator									
7100	Electric motor actuated									
7200	Brass, two way, screwed									
7210	1/2" pipe size	L-6	36	.333	Ea.	232	22		254	288
7220	3/4" pipe size		30	.400		223	26.50		249.50	285
7230	1" pipe size		28	.429		275	28.50		303.50	345
7240	1-1/2" pipe size		19	.632		425	42		467	530
7250	2" pipe size		16	.750		495	50		545	620
7350	Brass, three way, screwed									
7360	1/2" pipe size	L-6	33	.364	Ea.	268	24		292	330
7370	3/4" pipe size		27	.444		310	29.50		339.50	390
7380	1" pipe size		25.50	.471		405	31		436	490
7384	1-1/4" pipe size		21	.571		550	38		588	660
7390	1-1/2" pipe size		17	.706		610	47		657	745
7400	2" pipe size		14	.857		955	57		1,012	1,125
7550	Iron body, two way, flanged									
7560	2-1/2" pipe size	L-6	4	3	Ea.	365	199		564	695
7570	3" pipe size		3	4		1,475	265		1,740	2,025
7580	4" pipe size		2	6		2,400	400		2,800	3,250
7850	Iron body, three way, flanged									
7860	2-1/2" pipe size	L-6	3	4	Ea.	990	265		1,255	1,500
7870	3" pipe size		2.50	4.800		1,225	320		1,545	1,825
7880	4" pipe size		2	6		1,425	400		1,825	2,175

23 34 HVAC Fans

23 34 13 – Axial HVAC Fans

23 34 13.10 Axial Flow HVAC Fans

		Crew	Daily Output	Labor-Hours	Unit	Material	2021 Bare Costs Labor	Equipment	Total	Total Incl O&P
0010	**AXIAL FLOW HVAC FANS**									
0020	Air conditioning and process air handling									
1500	Vaneaxial, low pressure, 2,000 CFM, 1/2 HP	Q-20	3.60	5.556	Ea.	2,400	335		2,735	3,125
1520	4,000 CFM, 1 HP		3.20	6.250		2,500	375		2,875	3,300
1540	8,000 CFM, 2 HP		2.80	7.143		3,375	430		3,805	4,350
1560	16,000 CFM, 5 HP		2.40	8.333		5,175	500		5,675	6,425

23 34 14 – Blower HVAC Fans

23 34 14.10 Blower Type HVAC Fans

		Crew	Daily Output	Labor-Hours	Unit	Material	2021 Bare Costs Labor	Equipment	Total	Total Incl O&P
0010	**BLOWER TYPE HVAC FANS**									
2000	Blowers, direct drive with motor, complete									
2020	1,045 CFM @ 0.5" S.P., 1/5 HP	Q-20	18	1.111	Ea.	350	66.50		416.50	490
2040	1,385 CFM @ 0.5" S.P., 1/4 HP		18	1.111		415	66.50		481.50	560
2060	1,640 CFM @ 0.5" S.P., 1/3 HP		18	1.111		395	66.50		461.50	535
2080	1,760 CFM @ 0.5" S.P., 1/2 HP		18	1.111		345	66.50		411.50	480
2090	4 speed									
2100	1,164 to 1,739 CFM @ 0.5" S.P., 1/3 HP	Q-20	16	1.250	Ea.	335	75		410	475
2120	1,467 to 2,218 CFM @ 1.0" S.P., 3/4 HP	"	14	1.429	"	380	85.50		465.50	545
2500	Ceiling fan, right angle, extra quiet, 0.10" S.P.									
2520	95 CFM	Q-20	20	1	Ea.	276	60		336	395
2540	210 CFM		19	1.053		325	63		388	455
2560	385 CFM		18	1.111		415	66.50		481.50	555
2580	885 CFM		16	1.250		815	75		890	1,000
2600	1,650 CFM		13	1.538		1,125	92		1,217	1,400
2620	2,960 CFM		11	1.818		1,500	109		1,609	1,825
2680	For speed control switch, add	1 Elec	16	.500		150	32		182	213
7500	Utility set, steel construction, pedestal, 1/4" S.P.									
7520	Direct drive, 150 CFM, 1/8 HP	Q-20	6.40	3.125	Ea.	915	187		1,102	1,275
7540	485 CFM, 1/6 HP		5.80	3.448		1,125	206		1,331	1,525
7560	1,950 CFM, 1/2 HP		4.80	4.167		1,350	249		1,599	1,850
7580	2,410 CFM, 3/4 HP		4.40	4.545		2,725	272		2,997	3,375
7600	3,328 CFM, 1-1/2 HP		3	6.667		2,675	400		3,075	3,550
7680	V-belt drive, drive cover, 3 phase									
7700	800 CFM, 1/4 HP	Q-20	6	3.333	Ea.	1,150	200		1,350	1,575
7720	1,300 CFM, 1/3 HP		5	4		1,200	239		1,439	1,675
7740	2,000 CFM, 1 HP		4.60	4.348		1,425	260		1,685	1,975
7760	2,900 CFM, 3/4 HP		4.20	4.762		1,925	285		2,210	2,550
7780	3,600 CFM, 3/4 HP		4	5		2,375	299		2,674	3,075
7800	4,800 CFM, 1 HP		3.50	5.714		2,800	340		3,140	3,625
7820	6,700 CFM, 1-1/2 HP		3	6.667		3,475	400		3,875	4,400
7830	7,500 CFM, 2 HP		2.50	8		4,700	480		5,180	5,900
7840	11,000 CFM, 3 HP		2	10		6,275	600		6,875	7,825
7860	13,000 CFM, 3 HP		1.60	12.500		6,375	750		7,125	8,150
7880	15,000 CFM, 5 HP		1	20		6,600	1,200		7,800	9,075
7900	17,000 CFM, 7-1/2 HP		.80	25		7,075	1,500		8,575	10,000
7920	20,000 CFM, 7-1/2 HP		.80	25		8,425	1,500		9,925	11,500

23 34 16 – Centrifugal HVAC Fans

23 34 16.10 Centrifugal Type HVAC Fans

		Crew	Daily Output	Labor-Hours	Unit	Material	2021 Bare Costs Labor	Equipment	Total	Total Incl O&P
0010	**CENTRIFUGAL TYPE HVAC FANS**									
0200	In-line centrifugal, supply/exhaust booster									
0220	aluminum wheel/hub, disconnect switch, 1/4" S.P.									
0240	500 CFM, 10" diameter connection	Q-20	3	6.667	Ea.	1,575	400		1,975	2,325
0260	1,380 CFM, 12" diameter connection		2	10		1,850	600		2,450	2,950

23 34 16 – Centrifugal HVAC Fans

23 34 16.10 Centrifugal Type HVAC Fans		Crew	Daily Output	Labor-Hours	Unit	Material	2021 Bare Costs Labor	Equipment	Total	Total Incl O&P
0280	1,520 CFM, 16" diameter connection	Q-20	2	10	Ea.	1,700	600		2,300	2,775
0300	2,560 CFM, 18" diameter connection		1	20		1,900	1,200		3,100	3,900
0320	3,480 CFM, 20" diameter connection		.80	25		2,175	1,500		3,675	4,650
0326	5,080 CFM, 20" diameter connection		.75	26.667		2,350	1,600		3,950	5,000
3500	Centrifugal, airfoil, motor and drive, complete									
3520	1,000 CFM, 1/2 HP	Q-20	2.50	8	Ea.	2,075	480		2,555	3,025
3540	2,000 CFM, 1 HP		2	10		2,250	600		2,850	3,375
3560	4,000 CFM, 3 HP		1.80	11.111		2,950	665		3,615	4,250
3580	8,000 CFM, 7-1/2 HP		1.40	14.286		4,475	855		5,330	6,200
3600	12,000 CFM, 10 HP		1	20		6,100	1,200		7,300	8,500
5000	Utility set, centrifugal, V belt drive, motor									
5020	1/4" S.P., 1,200 CFM, 1/4 HP	Q-20	6	3.333	Ea.	1,975	200		2,175	2,475
5040	1,520 CFM, 1/3 HP		5	4		2,025	239		2,264	2,575
5060	1,850 CFM, 1/2 HP		4	5		2,025	299		2,324	2,675
5080	2,180 CFM, 3/4 HP		3	6.667		2,400	400		2,800	3,225
5100	1/2" S.P., 3,600 CFM, 1 HP		2	10		2,400	600		3,000	3,525
5120	4,250 CFM, 1-1/2 HP		1.60	12.500		1,725	750		2,475	3,025
5140	4,800 CFM, 2 HP		1.40	14.286		3,700	855		4,555	5,350
5160	6,920 CFM, 5 HP		1.30	15.385		4,425	920		5,345	6,250
5180	7,700 CFM, 7-1/2 HP		1.20	16.667		4,725	1,000		5,725	6,700
5200	For explosion proof motor, add					15%				
5500	Fans, industrial exhauster, for air which may contain granular matl.									
5520	1,000 CFM, 1-1/2 HP	Q-20	2.50	8	Ea.	3,500	480		3,980	4,575
5540	2,000 CFM, 3 HP		2	10		4,125	600		4,725	5,425
5560	4,000 CFM, 7-1/2 HP		1.80	11.111		5,675	665		6,340	7,225
5580	8,000 CFM, 15 HP		1.40	14.286		7,325	855		8,180	9,325
5600	12,000 CFM, 30 HP		1	20		11,400	1,200		12,600	14,300
7000	Roof exhauster, centrifugal, aluminum housing, 12" galvanized									
7020	curb, bird screen, back draft damper, 1/4" S.P.									
7100	Direct drive, 320 CFM, 11" sq. damper	Q-20	7	2.857	Ea.	725	171		896	1,050
7120	600 CFM, 11" sq. damper		6	3.333		935	200		1,135	1,325
7140	815 CFM, 13" sq. damper		5	4		1,100	239		1,339	1,550
7160	1,450 CFM, 13" sq. damper		4.20	4.762		1,300	285		1,585	1,850
7180	2,050 CFM, 16" sq. damper		4	5		1,825	299		2,124	2,450
7200	V-belt drive, 1,650 CFM, 12" sq. damper		6	3.333		1,275	200		1,475	1,700
7220	2,750 CFM, 21" sq. damper		5	4		1,500	239		1,739	2,000
7230	3,500 CFM, 21" sq. damper		4.50	4.444		1,850	266		2,116	2,450
7240	4,910 CFM, 23" sq. damper		4	5		2,025	299		2,324	2,700
7260	8,525 CFM, 28" sq. damper		3	6.667		2,775	400		3,175	3,650
7280	13,760 CFM, 35" sq. damper		2	10		3,800	600		4,400	5,100
7300	20,558 CFM, 43" sq. damper		1	20		7,825	1,200		9,025	10,400
7320	For 2 speed winding, add					15%				
7340	For explosion proof motor, add					750			750	825
7360	For belt driven, top discharge, add					15%				
8500	Wall exhausters, centrifugal, auto damper, 1/8" S.P.									
8520	Direct drive, 610 CFM, 1/20 HP	Q-20	14	1.429	Ea.	460	85.50		545.50	635
8540	796 CFM, 1/12 HP		13	1.538		935	92		1,027	1,175
8560	822 CFM, 1/6 HP		12	1.667		1,175	100		1,275	1,425
8580	1,320 CFM, 1/4 HP		12	1.667		1,325	100		1,425	1,600
8600	1,756 CFM, 1/4 HP		11	1.818		1,375	109		1,484	1,675
8620	1,983 CFM, 1/4 HP		10	2		1,400	120		1,520	1,725
8640	2,900 CFM, 1/2 HP		9	2.222		1,500	133		1,633	1,850
8660	3,307 CFM, 3/4 HP		8	2.500		1,625	150		1,775	2,000

23 34 HVAC Fans

23 34 16 – Centrifugal HVAC Fans

23 34 16.10 Centrifugal Type HVAC Fans		Crew	Daily Output	Labor-Hours	Unit	Material	2021 Bare Costs Labor	Equipment	Total	Total Incl O&P
9500	V-belt drive, 3 phase									
9520	2,800 CFM, 1/4 HP	Q-20	9	2.222	Ea.	2,075	133		2,208	2,500
9540	3,740 CFM, 1/2 HP		8	2.500		2,175	150		2,325	2,600
9560	4,400 CFM, 3/4 HP		7	2.857		2,200	171		2,371	2,650
9580	5,700 CFM, 1-1/2 HP		6	3.333		2,275	200		2,475	2,800

23 34 23 – HVAC Power Ventilators

23 34 23.10 HVAC Power Circulators and Ventilators

			Crew	Daily Output	Labor-Hours	Unit	Material	2021 Bare Costs Labor	Equipment	Total	Total Incl O&P
0010	**HVAC POWER CIRCULATORS AND VENTILATORS**										
3000	Paddle blade air circulator, 3 speed switch										
3020	42", 5,000 CFM high, 3,000 CFM low	G	1 Elec	2.40	3.333	Ea.	104	212		316	430
3040	52", 6,500 CFM high, 4,000 CFM low	G	"	2.20	3.636	"	111	232		343	465
3100	For antique white motor, same cost										
3200	For brass plated motor, same cost										
3300	For light adaptor kit, add	G				Ea.	40			40	44
6000	Propeller exhaust, wall shutter										
6020	Direct drive, one speed, 0.075" S.P.										
6100	653 CFM, 1/30 HP		Q-20	10	2	Ea.	293	120		413	500
6120	1,033 CFM, 1/20 HP			9	2.222		253	133		386	480
6140	1,323 CFM, 1/15 HP			8	2.500		480	150		630	755
6160	2,444 CFM, 1/4 HP			7	2.857		335	171		506	620
6300	V-belt drive, 3 phase										
6320	6,175 CFM, 3/4 HP		Q-20	5	4	Ea.	1,075	239		1,314	1,550
6340	7,500 CFM, 3/4 HP			5	4		1,125	239		1,364	1,575
6360	10,100 CFM, 1 HP			4.50	4.444		1,300	266		1,566	1,825
6380	14,300 CFM, 1-1/2 HP			4	5		1,600	299		1,899	2,225
6400	19,800 CFM, 2 HP			3	6.667		1,775	400		2,175	2,550
6420	26,250 CFM, 3 HP			2.60	7.692		2,425	460		2,885	3,350
6440	38,500 CFM, 5 HP			2.20	9.091		3,050	545		3,595	4,175
6460	46,000 CFM, 7-1/2 HP			2	10		3,125	600		3,725	4,350
6480	51,500 CFM, 10 HP			1.80	11.111		3,175	665		3,840	4,475
6650	Residential, bath exhaust, grille, back draft damper										
6660	50 CFM		Q-20	24	.833	Ea.	48	50		98	128
6670	110 CFM			22	.909		96	54.50		150.50	188
6680	Light combination, squirrel cage, 100 watt, 70 CFM			24	.833		112	50		162	198
6700	Light/heater combination, ceiling mounted										
6710	70 CFM, 1,450 watt		Q-20	24	.833	Ea.	239	50		289	340
6800	Heater combination, recessed, 70 CFM			24	.833		62.50	50		112.50	144
6820	With 2 infrared bulbs			23	.870		94	52		146	182
6900	Kitchen exhaust, grille, complete, 160 CFM			22	.909		129	54.50		183.50	223
6910	180 CFM			20	1		103	60		163	203
6920	270 CFM			18	1.111		203	66.50		269.50	325
6930	350 CFM			16	1.250		162	75		237	290
6940	Residential roof jacks and wall caps										
6944	Wall cap with back draft damper										
6946	3" & 4" diam. round duct		1 Shee	11	.727	Ea.	18.75	47.50		66.25	92.50
6948	6" diam. round duct		"	11	.727	"	54.50	47.50		102	132
6958	Roof jack with bird screen and back draft damper										
6960	3" & 4" diam. round duct		1 Shee	11	.727	Ea.	24	47.50		71.50	98.50
6962	3-1/4" x 10" rectangular duct		"	10	.800	"	38.50	52.50		91	122
6980	Transition										
6982	3-1/4" x 10" to 6" diam. round		1 Shee	20	.400	Ea.	38	26		64	81.50
8020	Attic, roof type										

23 34 HVAC Fans

23 34 23 – HVAC Power Ventilators

23 34 23.10 HVAC Power Circulators and Ventilators

		Crew	Daily Output	Labor-Hours	Unit	Material	2021 Bare Costs Labor	Equipment	Total	Total Incl O&P
8030	Aluminum dome, damper & curb									
8080	12" diameter, 1,000 CFM (gravity)	1 Elec	10	.800	Ea.	640	51		691	780
8090	16" diameter, 1,500 CFM (gravity)		9	.889		775	56.50		831.50	935
8100	20" diameter, 2,500 CFM (gravity)		8	1		950	63.50		1,013.50	1,150
8110	26" diameter, 4,000 CFM (gravity)		7	1.143		1,150	73		1,223	1,375
8120	32" diameter, 6,500 CFM (gravity)		6	1.333		1,575	85		1,660	1,850
8130	38" diameter, 8,000 CFM (gravity)		5	1.600		2,350	102		2,452	2,725
8140	50" diameter, 13,000 CFM (gravity)	↓	4	2	↓	3,400	127		3,527	3,925
8160	Plastic, ABS dome									
8180	1,050 CFM	1 Elec	14	.571	Ea.	189	36.50		225.50	261
8200	1,600 CFM	"	12	.667	"	282	42.50		324.50	375
8240	Attic, wall type, with shutter, one speed									
8250	12" diameter, 1,000 CFM	1 Elec	14	.571	Ea.	470	36.50		506.50	570
8260	14" diameter, 1,500 CFM		12	.667		505	42.50		547.50	625
8270	16" diameter, 2,000 CFM	↓	9	.889		575	56.50		631.50	715
8290	Whole house, wall type, with shutter, one speed									
8300	30" diameter, 4,800 CFM	1 Elec	7	1.143	Ea.	1,225	73		1,298	1,450
8310	36" diameter, 7,000 CFM		6	1.333		1,325	85		1,410	1,600
8320	42" diameter, 10,000 CFM		5	1.600		1,500	102		1,602	1,800
8330	48" diameter, 16,000 CFM	↓	4	2		1,850	127		1,977	2,250
8340	For two speed, add				↓	112			112	123
8350	Whole house, lay-down type, with shutter, one speed									
8360	30" diameter, 4,500 CFM	1 Elec	8	1	Ea.	1,300	63.50		1,363.50	1,550
8370	36" diameter, 6,500 CFM		7	1.143		1,400	73		1,473	1,650
8380	42" diameter, 9,000 CFM		6	1.333		1,550	85		1,635	1,825
8390	48" diameter, 12,000 CFM	↓	5	1.600		1,750	102		1,852	2,075
8440	For two speed, add					84			84	92.50
8450	For 12 hour timer switch, add	1 Elec	32	.250		84	15.95		99.95	116

23 34 39.10 Powerfoil

		Crew	Daily Output	Labor-Hours	Unit	Material	2021 Bare Costs Labor	Equipment	Total	Total Incl O&P
0010	**POWERFOIL**									
2301	Ceiling fan, Powerfoil, 8' diameter	R-1A	2.40	6.667		8,650	380		9,030	10,100
2302	10'		2.40	6.667		8,750	380		9,130	10,200
2303	12'		2.40	6.667		8,850	380		9,230	10,300
2304	14'		2.40	6.667		8,950	380		9,330	10,400
2305	16'		2	8		9,050	460		9,510	10,600
2306	18'		2	8		9,150	460		9,610	10,800
2307	20'		2	8		9,250	460		9,710	10,900
2308	24'	↓	2	8	↓	9,450	460		9,910	11,100
2400	Powerfoil with Hybrid Airfoil									
2401	12'	R-1A	2	8	Ea.	9,350	460		9,810	11,000
2402	14'		2	8		9,450	460		9,910	11,100
2403	16'		1.75	9.143		9,550	525		10,075	11,300
2404	18'		1.75	9.143		9,650	525		10,175	11,400
2405	20'		1.75	9.143		9,750	525		10,275	11,500
2406	24'	↓	1.75	9.143	↓	9,975	525		10,500	11,800

23 52 Heating Boilers

23 52 13 – Electric Boilers

23 52 13.10 Electric Boilers, ASME	Crew	Daily Output	Labor-Hours	Unit	Material	2021 Bare Costs Labor	Equipment	Total	Total Incl O&P
0010 **ELECTRIC BOILERS, ASME**, Standard controls and trim									
1000 Steam, 6 KW, 20.5 MBH	Q-19	1.20	20	Ea.	4,500	1,250		5,750	6,800
1040 9 KW, 30.7 MBH		1.20	20		4,325	1,250		5,575	6,600
1060 18 KW, 61.4 MBH		1.20	20		4,450	1,250		5,700	6,750
1080 24 KW, 81.8 MBH		1.10	21.818		5,150	1,350		6,500	7,700
1120 36 KW, 123 MBH		1.10	21.818		5,700	1,350		7,050	8,300
1160 60 KW, 205 MBH		1	24		7,125	1,500		8,625	10,100
1220 112 KW, 382 MBH		.75	32		9,900	2,000		11,900	13,900
1240 148 KW, 505 MBH		.65	36.923		10,300	2,300		12,600	14,700
1260 168 KW, 573 MBH		.60	40		21,300	2,500		23,800	27,200
1280 222 KW, 758 MBH		.55	43.636		24,900	2,725		27,625	31,500
1300 296 KW, 1,010 MBH		.45	53.333		26,300	3,325		29,625	34,000
1320 300 KW, 1,023 MBH		.40	60		26,200	3,725		29,925	34,500
1340 370 KW, 1,263 MBH		.35	68.571		29,000	4,275		33,275	38,300
1360 444 KW, 1,515 MBH		.30	80		31,200	4,975		36,175	41,700
1380 518 KW, 1,768 MBH	Q-21	.36	88.889		32,500	5,675		38,175	44,300
1400 592 KW, 2,020 MBH		.34	94.118		35,700	6,000		41,700	48,300
1420 666 KW, 2,273 MBH		.32	100		37,100	6,375		43,475	50,500
1460 740 KW, 2,526 MBH		.28	114		38,600	7,300		45,900	53,500
1480 814 KW, 2,778 MBH		.25	128		45,700	8,175		53,875	62,500
1500 962 KW, 3,283 MBH		.22	145		45,300	9,275		54,575	63,500
1520 1,036 KW, 3,536 MBH		.20	160		47,400	10,200		57,600	67,000
1540 1,110 KW, 3,788 MBH		.19	168		50,000	10,700		60,700	71,000
1560 2,070 KW, 7,063 MBH		.18	178		69,000	11,300		80,300	92,500
1580 2,250 KW, 7,677 MBH		.17	188		80,500	12,000		92,500	106,500
1600 2,340 KW, 7,984 MBH		.16	200		103,000	12,800		115,800	132,000
2000 Hot water, 7.5 KW, 25.6 MBH	Q-19	1.30	18.462		5,150	1,150		6,300	7,375
2020 15 KW, 51.2 MBH		1.30	18.462		5,425	1,150		6,575	7,700
2040 30 KW, 102 MBH		1.20	20		5,525	1,250		6,775	7,925
2060 45 KW, 164 MBH		1.20	20		5,400	1,250		6,650	7,775
2070 60 KW, 205 MBH		1.20	20		5,775	1,250		7,025	8,200
2080 75 KW, 256 MBH		1.10	21.818		6,100	1,350		7,450	8,725
2100 90 KW, 307 MBH		1.10	21.818		6,125	1,350		7,475	8,775
2120 105 KW, 358 MBH		1	24		7,125	1,500		8,625	10,100
2140 120 KW, 410 MBH		.90	26.667		6,800	1,650		8,450	9,950
2160 135 KW, 461 MBH		.75	32		7,550	2,000		9,550	11,300
2180 150 KW, 512 MBH		.65	36.923		7,925	2,300		10,225	12,200
2200 165 KW, 563 MBH		.60	40		8,175	2,500		10,675	12,700
2220 296 KW, 1,010 MBH		.55	43.636		16,200	2,725		18,925	21,900
2280 370 KW, 1,263 MBH		.40	60		18,000	3,725		21,725	25,400
2300 444 KW, 1,515 MBH		.35	68.571		21,600	4,275		25,875	30,200
2340 518 KW, 1,768 MBH	Q-21	.44	72.727		23,900	4,650		28,550	33,200
2360 592 KW, 2,020 MBH		.43	74.419		26,200	4,750		30,950	35,900
2400 666 KW, 2,273 MBH		.40	80		28,400	5,100		33,500	38,800
2420 740 KW, 2,526 MBH		.39	82.051		28,600	5,225		33,825	39,200
2440 814 KW, 2,778 MBH		.38	84.211		30,700	5,375		36,075	41,800
2460 888 KW, 3,031 MBH		.37	86.486		33,300	5,525		38,825	44,800
2480 962 KW, 3,283 MBH		.36	88.889		35,200	5,675		40,875	47,200
2500 1,036 KW, 3,536 MBH		.34	94.118		36,400	6,000		42,400	49,000
2520 1,110 KW, 3,788 MBH		.33	96.970		39,700	6,175		45,875	53,000
2540 1,440 KW, 4,915 MBH		.32	100		44,100	6,375		50,475	58,000
2560 1,560 KW, 5,323 MBH		.31	103		50,500	6,575		57,075	65,500
2580 1,680 KW, 5,733 MBH		.30	107		53,500	6,800		60,300	69,000

23 52 Heating Boilers

23 52 13 – Electric Boilers

23 52 13.10 Electric Boilers, ASME

		Crew	Daily Output	Labor-Hours	Unit	Material	2021 Bare Costs Labor	Equipment	Total	Total Incl O&P
2600	1,800 KW, 6,143 MBH	Q-21	.29	110	Ea.	57,500	7,050		64,550	73,500
2620	1,980 KW, 6,757 MBH		.28	114		60,000	7,300		67,300	77,000
2640	2,100 KW, 7,167 MBH		.27	119		68,500	7,550		76,050	87,000
2660	2,220 KW, 7,576 MBH		.26	123		69,500	7,850		77,350	88,000
2680	2,400 KW, 8,191 MBH		.25	128		72,500	8,175		80,675	92,000
2700	2,610 KW, 8,905 MBH		.24	133		76,500	8,500		85,000	97,000
2720	2,790 KW, 9,519 MBH		.23	139		80,500	8,875		89,375	101,500
2740	2,970 KW, 10,133 MBH		.21	152		82,500	9,725		92,225	105,000
2760	3,150 KW, 10,748 MBH		.19	168		80,000	10,700		90,700	103,500
2780	3,240 KW, 11,055 MBH		.18	178		81,500	11,300		92,800	106,500
2800	3,420 KW, 11,669 MBH		.17	188		95,000	12,000		107,000	122,500
2820	3,600 KW, 12,283 MBH		.16	200		100,000	12,800		112,800	129,000

23 52 28 – Swimming Pool Boilers

23 52 28.10 Swimming Pool Heaters

		Crew	Daily Output	Labor-Hours	Unit	Material	2021 Bare Costs Labor	Equipment	Total	Total Incl O&P
0010	**SWIMMING POOL HEATERS**, Not including wiring, external									
0020	piping, base or pad									
2000	Electric, 12 KW, 4,800 gallon pool	Q-19	3	8	Ea.	2,225	500		2,725	3,200
2020	15 KW, 7,200 gallon pool		2.80	8.571		2,350	535		2,885	3,375
2040	24 KW, 9,600 gallon pool		2.40	10		2,675	625		3,300	3,875
2060	30 KW, 12,000 gallon pool		2	12		2,725	745		3,470	4,125
2080	36 KW, 14,400 gallon pool		1.60	15		3,175	935		4,110	4,900
2100	57 KW, 24,000 gallon pool		1.20	20		3,975	1,250		5,225	6,225
9000	To select pool heater: 12 BTUH x S.F. pool area									
9010	X temperature differential = required output									
9050	For electric, KW = gallons x 2.5 divided by 1,000									
9100	For family home type pool, double the									
9110	Rated gallon capacity = 1/2°F rise per hour									

23 54 Furnaces

23 54 13 – Electric-Resistance Furnaces

23 54 13.10 Electric Furnaces

		Crew	Daily Output	Labor-Hours	Unit	Material	2021 Bare Costs Labor	Equipment	Total	Total Incl O&P
0010	**ELECTRIC FURNACES**, Hot air, blowers, std. controls									
1000	Electric, UL listed									
1070	10.2 MBH	Q-20	4.40	4.545	Ea.	345	272		617	790
1080	17.1 MBH		4.60	4.348		425	260		685	855
1100	34.1 MBH		4.40	4.545		530	272		802	990
1120	51.6 MBH		4.20	4.762		585	285		870	1,075
1140	68.3 MBH		4	5		580	299		879	1,075
1160	85.3 MBH		3.80	5.263		615	315		930	1,150

23 82 16.20 Duct Heaters		Crew	Daily Output	Labor-Hours	Unit	Material	2021 Bare Costs Labor	Equipment	Total	Total Incl O&P
0010	**DUCT HEATERS**, Electric, 480 V, 3 Ph.									
0020	Finned tubular insert, 500°F									
0100	8" wide x 6" high, 4.0 kW	Q-20	16	1.250	Ea.	820	75		895	1,025
0120	12" high, 8.0 kW		15	1.333		1,350	80		1,430	1,625
0140	18" high, 12.0 kW		14	1.429		1,900	85.50		1,985.50	2,225
0160	24" high, 16.0 kW		13	1.538		2,450	92		2,542	2,850
0180	30" high, 20.0 kW		12	1.667		3,000	100		3,100	3,450
0300	12" wide x 6" high, 6.7 kW		15	1.333		870	80		950	1,075
0320	12" high, 13.3 kW		14	1.429		1,400	85.50		1,485.50	1,675
0340	18" high, 20.0 kW		13	1.538		1,975	92		2,067	2,325
0360	24" high, 26.7 kW		12	1.667		2,525	100		2,625	2,950
0380	30" high, 33.3 kW		11	1.818		3,100	109		3,209	3,575
0500	18" wide x 6" high, 13.3 kW		14	1.429		950	85.50		1,035.50	1,175
0520	12" high, 26.7 kW		13	1.538		1,625	92		1,717	1,925
0540	18" high, 40.0 kW		12	1.667		2,150	100		2,250	2,500
0560	24" high, 53.3 kW		11	1.818		2,850	109		2,959	3,300
0580	30" high, 66.7 kW		10	2		3,575	120		3,695	4,100
0700	24" wide x 6" high, 17.8 kW		13	1.538		1,050	92		1,142	1,300
0720	12" high, 35.6 kW		12	1.667		1,750	100		1,850	2,075
0740	18" high, 53.3 kW		11	1.818		2,475	109		2,584	2,900
0760	24" high, 71.1 kW		10	2		3,175	120		3,295	3,650
0780	30" high, 88.9 kW		9	2.222		3,875	133		4,008	4,475
0900	30" wide x 6" high, 22.2 kW		12	1.667		1,100	100		1,200	1,375
0920	12" high, 44.4 kW		11	1.818		1,850	109		1,959	2,200
0940	18" high, 66.7 kW		10	2		2,625	120		2,745	3,075
0960	24" high, 88.9 kW		9	2.222		3,775	133		3,908	4,350
0980	30" high, 111.0 kW		8	2.500		4,100	150		4,250	4,750
1400	Note decreased kW available for									
1410	each duct size at same cost									
1420	See line 5000 for modifications and accessories									
2000	Finned tubular flange with insulated									
2020	terminal box, 500°F									
2100	12" wide x 36" high, 54 kW	Q-20	10	2	Ea.	4,050	120		4,170	4,625
2120	40" high, 60 kW		9	2.222		4,650	133		4,783	5,325
2200	24" wide x 36" high, 118.8 kW		9	2.222		4,925	133		5,058	5,600
2220	40" high, 132 kW		8	2.500		5,100	150		5,250	5,850
2400	36" wide x 8" high, 40 kW		11	1.818		2,075	109		2,184	2,450
2420	16" high, 80 kW		10	2		2,750	120		2,870	3,200
2440	24" high, 120 kW		9	2.222		3,600	133		3,733	4,150
2460	32" high, 160 kW		8	2.500		4,600	150		4,750	5,300
2480	36" high, 180 kW		7	2.857		5,675	171		5,846	6,500
2500	40" high, 200 kW		6	3.333		6,275	200		6,475	7,200
2600	40" wide x 8" high, 45 kW		11	1.818		2,150	109		2,259	2,525
2620	16" high, 90 kW		10	2		2,950	120		3,070	3,425
2640	24" high, 135 kW		9	2.222		3,750	133		3,883	4,325
2660	32" high, 180 kW		8	2.500		5,000	150		5,150	5,725
2680	36" high, 202.5 kW		7	2.857		5,750	171		5,921	6,575
2700	40" high, 225 kW		6	3.333		6,550	200		6,750	7,525
2800	48" wide x 8" high, 54.8 kW		10	2		2,250	120		2,370	2,650
2820	16" high, 109.8 kW		9	2.222		3,125	133		3,258	3,625
2840	24" high, 164.4 kW		8	2.500		3,975	150		4,125	4,600
2860	32" high, 219.2 kW		7	2.857		5,300	171		5,471	6,075
2880	36" high, 246.6 kW		6	3.333		6,100	200		6,300	7,000

23 82 16 – Air Coils

23 82 16.20 Duct Heaters

		Crew	Daily Output	Labor-Hours	Unit	Material	2021 Bare Costs Labor	2021 Bare Costs Equipment	Total	Total Incl O&P
2900	40" high, 274 kW	Q-20	5	4	Ea.	6,900	239		7,139	7,925
3000	56" wide x 8" high, 64 kW		9	2.222		2,575	133		2,708	3,025
3020	16" high, 128 kW		8	2.500		3,550	150		3,700	4,125
3040	24" high, 192 kW		7	2.857		4,300	171		4,471	4,975
3060	32" high, 256 kW		6	3.333		6,025	200		6,225	6,925
3080	36" high, 288 kW		5	4		6,900	239		7,139	7,950
3100	40" high, 320 kW		4	5		7,625	299		7,924	8,850
3200	64" wide x 8" high, 74 kW		8	2.500		2,650	150		2,800	3,125
3220	16" high, 148 kW		7	2.857		3,650	171		3,821	4,275
3240	24" high, 222 kW		6	3.333		4,650	200		4,850	5,425
3260	32" high, 296 kW		5	4		6,225	239		6,464	7,200
3280	36" high, 333 kW		4	5		7,425	299		7,724	8,625
3300	40" high, 370 kW	▼	3	6.667	▼	8,200	400		8,600	9,625
3800	Note decreased kW available for									
3820	each duct size at same cost									
5000	Duct heater modifications and accessories									
5120	T.C.O. limit auto or manual reset	Q-20	42	.476	Ea.	135	28.50		163.50	191
5140	Thermostat		28	.714		575	43		618	695
5160	Overheat thermocouple (removable)		7	2.857		815	171		986	1,150
5180	Fan interlock relay		18	1.111		196	66.50		262.50	315
5200	Air flow switch		20	1		170	60		230	277
5220	Split terminal box cover	▼	100	.200	▼	56	11.95		67.95	79.50
8000	To obtain BTU multiply kW by 3413									

23 83 Radiant Heating Units

23 83 33 – Electric Radiant Heaters

23 83 33.10 Electric Heating

		Crew	Daily Output	Labor-Hours	Unit	Material	2021 Bare Costs Labor	2021 Bare Costs Equipment	Total	Total Incl O&P
0010	**ELECTRIC HEATING**, not incl. conduit or feed wiring									
1100	Rule of thumb: Baseboard units, including control	1 Elec	4.40	1.818	kW	119	116		235	300
1300	Baseboard heaters, 2' long, 350 watt		8	1	Ea.	31.50	63.50		95	129
1400	3' long, 750 watt		8	1		36.50	63.50		100	135
1600	4' long, 1,000 watt		6.70	1.194		38.50	76		114.50	156
1800	5' long, 935 watt		5.70	1.404		52.50	89.50		142	191
2000	6' long, 1,500 watt		5	1.600		53	102		155	210
2200	7' long, 1,310 watt		4.40	1.818		63.50	116		179.50	242
2400	8' long, 2,000 watt		4	2		73	127		200	270
2600	9' long, 1,680 watt		3.60	2.222		85.50	142		227.50	305
2800	10' long, 1,875 watt	▼	3.30	2.424	▼	97.50	154		251.50	335
2950	Wall heaters with fan, 120 to 277 volt									
3160	Recessed, residential, 750 watt	1 Elec	6	1.333	Ea.	83	85		168	218
3170	1,000 watt		6	1.333		335	85		420	490
3180	1,250 watt		5	1.600		113	102		215	276
3190	1,500 watt		4	2		360	127		487	585
3210	2,000 watt		4	2		105	127		232	305
3230	2,500 watt		3.50	2.286		365	146		511	615
3240	3,000 watt		3	2.667		325	170		495	610
3250	4,000 watt		2.70	2.963		330	189		519	640
3260	Commercial, 750 watt		6	1.333		202	85		287	350
3270	1,000 watt		6	1.333		217	85		302	365
3280	1,250 watt		5	1.600		217	102		319	390
3290	1,500 watt		4	2		200	127		327	410

23 83 33.10 Electric Heating		Crew	Daily Output	Labor-Hours	Unit	Material	2021 Bare Costs Labor	Equipment	Total	Total Incl O&P
3300	2,000 watt	1 Elec	4	2	Ea.	217	127		344	430
3310	2,500 watt		3.50	2.286		217	146		363	455
3320	3,000 watt		3	2.667		290	170		460	570
3330	4,000 watt		2.70	2.963		500	189		689	830
3600	Thermostats, integral		16	.500		29.50	32		61.50	79.50
3800	Line voltage, 1 pole		8	1		16.40	63.50		79.90	113
3810	2 pole		8	1		23	63.50		86.50	120
4000	Heat trace system, 400 degree	R238313-10								
4020	115 V, 2.5 watts/L.F.	1 Elec	530	.015	L.F.	10.05	.96		11.01	12.50
4030	5 watts/L.F.	R238313-20	530	.015		10.05	.96		11.01	12.50
4050	10 watts/L.F.		530	.015		8.40	.96		9.36	10.70
4060	208 V, 5 watts/L.F.		530	.015		10.05	.96		11.01	12.50
4080	480 V, 8 watts/L.F.		530	.015		10.05	.96		11.01	12.50
4200	Heater raceway									
4260	Heat transfer cement									
4280	1 gallon				Ea.	80.50			80.50	89
4300	5 gallon				"	244			244	268
4320	Cable tie									
4340	3/4" pipe size	1 Elec	470	.017	Ea.	.02	1.08		1.10	1.63
4360	1" pipe size		444	.018		.02	1.15		1.17	1.73
4380	1-1/4" pipe size		400	.020		.02	1.27		1.29	1.91
4400	1-1/2" pipe size		355	.023		.02	1.44		1.46	2.15
4420	2" pipe size		320	.025		.02	1.59		1.61	2.39
4440	3" pipe size		160	.050		.02	3.19		3.21	4.75
4460	4" pipe size		100	.080		.06	5.10		5.16	7.60
4480	Thermostat NEMA 3R, 22 amp, 0-150 degree, 10' cap.		8	1		190	63.50		253.50	305
4500	Thermostat NEMA 4X, 25 amp, 40 degree, 5-1/2' cap.		7	1.143		200	73		273	330
4520	Thermostat NEMA 4X, 22 amp, 25-325 degree, 10' cap.		7	1.143		525	73		598	685
4540	Thermostat NEMA 4X, 22 amp, 15-140 degree		6	1.333		440	85		525	610
4580	Thermostat NEMA 4, 7, 9, 22 amp, 25-325 degree, 10' cap.		3.60	2.222		640	142		782	915
4600	Thermostat NEMA 4, 7, 9, 22 amp, 15-140 degree		3	2.667		640	170		810	955
4720	Fiberglass application tape, 36 yard roll		11	.727		55.50	46.50		102	131
5000	Radiant heating ceiling panels, 2' x 4', 500 watt		16	.500		294	32		326	375
5050	750 watt		16	.500		257	32		289	330
5200	For recessed plaster frame, add		32	.250		70.50	15.95		86.45	101
5300	Infrared quartz heaters, 120 volts, 1,000 watt		6.70	1.194		236	76		312	370
5350	1,500 watt		5	1.600		233	102		335	405
5400	240 volts, 1,500 watt		5	1.600		239	102		341	415
5450	2,000 watt		4	2		277	127		404	495
5500	3,000 watt		3	2.667		275	170		445	550
5550	4,000 watt		2.60	3.077		315	196		511	640
5570	Modulating control		.80	10		124	635		759	1,075
5600	Unit heaters, heavy duty, with fan & mounting bracket									
5650	Single phase, 208-240-277 volt, 3 kW	1 Elec	6	1.333	Ea.	445	85		530	610
5750	5 kW		5.50	1.455		279	92.50		371.50	445
5800	7 kW		5	1.600		670	102		772	885
5850	10 kW		4	2		1,025	127		1,152	1,350
5950	15 kW		3.80	2.105		1,750	134		1,884	2,125
6000	480 volt, 3 kW		6	1.333		435	85		520	600
6020	4 kW		5.80	1.379		365	88		453	535
6040	5 kW		5.50	1.455		445	92.50		537.50	630
6060	7 kW		5	1.600		580	102		682	790
6080	10 kW		4	2		1,050	127		1,177	1,350

23 83 33 – Electric Radiant Heaters

23 83 33.10 Electric Heating		Crew	Daily Output	Labor-Hours	Unit	Material	2021 Bare Costs Labor	Equipment	Total	Total Incl O&P
6100	13 kW	1 Elec	3.80	2.105	Ea.	1,850	134		1,984	2,250
6120	15 kW		3.70	2.162		1,075	138		1,213	1,375
6140	20 kW		3.50	2.286		1,475	146		1,621	1,850
6300	3 phase, 208-240 volt, 5 kW		5.50	1.455		425	92.50		517.50	605
6320	7 kW		5	1.600		705	102		807	925
6340	10 kW		4	2		740	127		867	1,000
6360	15 kW		3.70	2.162		1,125	138		1,263	1,425
6380	20 kW		3.50	2.286		1,475	146		1,621	1,850
6400	25 kW		3.30	2.424		1,700	154		1,854	2,075
6500	480 volt, 5 kW		5.50	1.455		615	92.50		707.50	820
6520	7 kW		5	1.600		795	102		897	1,025
6540	10 kW		4	2		825	127		952	1,100
6560	13 kW		3.80	2.105		1,525	134		1,659	1,900
6580	15 kW		3.70	2.162		1,375	138		1,513	1,700
6600	20 kW		3.50	2.286		2,125	146		2,271	2,575
6620	25 kW		3.30	2.424		1,700	154		1,854	2,075
6630	30 kW		3	2.667		2,125	170		2,295	2,575
6640	40 kW		2	4		2,500	255		2,755	3,125
6650	50 kW		1.60	5		3,025	320		3,345	3,800
6800	Vertical discharge heaters, with fan									
6820	Single phase, 208-240-277 volt, 10 kW	1 Elec	4	2	Ea.	780	127		907	1,050
6840	15 kW		3.70	2.162		1,050	138		1,188	1,350
6900	3 phase, 208-240 volt, 10 kW		4	2		970	127		1,097	1,275
6920	15 kW		3.70	2.162		1,500	138		1,638	1,850
6940	20 kW		3.50	2.286		2,775	146		2,921	3,275
6960	25 kW		3.30	2.424		3,050	154		3,204	3,600
6980	30 kW		3	2.667		1,950	170		2,120	2,400
7000	40 kW		2	4		2,500	255		2,755	3,125
7020	50 kW		1.60	5		3,650	320		3,970	4,475
7100	480 volt, 10 kW		4	2		1,150	127		1,277	1,450
7120	15 kW		3.70	2.162		1,850	138		1,988	2,250
7140	20 kW		3.50	2.286		2,550	146		2,696	3,025
7160	25 kW		3.30	2.424		3,325	154		3,479	3,875
7180	30 kW		3	2.667		3,625	170		3,795	4,250
7200	40 kW		2	4		3,775	255		4,030	4,525
7410	Sill height convector heaters, 5" high x 2' long, 500 watt		6.70	1.194		360	76		436	510
7420	3' long, 750 watt		6.50	1.231		425	78.50		503.50	580
7430	4' long, 1,000 watt		6.20	1.290		490	82		572	660
7440	5' long, 1,250 watt		5.50	1.455		555	92.50		647.50	750
7450	6' long, 1,500 watt		4.80	1.667		630	106		736	855
7460	8' long, 2,000 watt		3.60	2.222		855	142		997	1,150
7470	10' long, 2,500 watt		3	2.667		1,075	170		1,245	1,425
7900	Cabinet convector heaters, 240 volt, three phase,									
7920	3' long, 2,000 watt	1 Elec	5.30	1.509	Ea.	2,300	96		2,396	2,675
7940	3,000 watt		5.30	1.509		2,400	96		2,496	2,800
7960	4,000 watt		5.30	1.509		2,475	96		2,571	2,850
7980	6,000 watt		4.60	1.739		2,675	111		2,786	3,125
8000	8,000 watt		4.60	1.739		2,750	111		2,861	3,200
8020	4' long, 4,000 watt		4.60	1.739		2,500	111		2,611	2,925
8040	6,000 watt		4	2		2,600	127		2,727	3,050
8060	8,000 watt		4	2		2,675	127		2,802	3,125
8080	10,000 watt		4	2		2,700	127		2,827	3,175
8100	Available also in 208 or 277 volt									

For customer support on your Electrical Costs with RSMeans Data, call 800.448.8182.

101

23 83 Radiant Heating Units

23 83 33 – Electric Radiant Heaters

23 83 33.10 Electric Heating

		Crew	Daily Output	Labor-Hours	Unit	Material	2021 Bare Costs Labor	Equipment	Total	Total Incl O&P
8200	Cabinet unit heaters, 120 to 277 volt, 1 pole,									
8220	wall mounted, 2 kW	1 Elec	4.60	1.739	Ea.	2,075	111		2,186	2,475
8230	3 kW		4.60	1.739		2,125	111		2,236	2,525
8240	4 kW		4.40	1.818		2,275	116		2,391	2,700
8250	5 kW		4.40	1.818		2,475	116		2,591	2,900
8260	6 kW		4.20	1.905		2,450	121		2,571	2,875
8270	8 kW		4	2		2,550	127		2,677	3,025
8280	10 kW		3.80	2.105		2,550	134		2,684	3,000
8290	12 kW		3.50	2.286		2,700	146		2,846	3,200
8300	13.5 kW		2.90	2.759		3,925	176		4,101	4,575
8310	16 kW		2.70	2.963		3,350	189		3,539	3,950
8320	20 kW		2.30	3.478		4,025	222		4,247	4,750
8330	24 kW		1.90	4.211		4,325	268		4,593	5,150
8350	Recessed, 2 kW		4.40	1.818		2,375	116		2,491	2,775
8370	3 kW		4.40	1.818		2,475	116		2,591	2,900
8380	4 kW		4.20	1.905		2,400	121		2,521	2,825
8390	5 kW		4.20	1.905		2,550	121		2,671	3,000
8400	6 kW		4	2		1,825	127		1,952	2,225
8410	8 kW		3.80	2.105		2,725	134		2,859	3,200
8420	10 kW		3.50	2.286		3,375	146		3,521	3,925
8430	12 kW		2.90	2.759		3,425	176		3,601	4,000
8440	13.5 kW		2.70	2.963		3,275	189		3,464	3,875
8450	16 kW		2.30	3.478		2,325	222		2,547	2,875
8460	20 kW		1.90	4.211		4,125	268		4,393	4,925
8470	24 kW		1.60	5		4,100	320		4,420	5,000
8490	Ceiling mounted, 2 kW		3.20	2.500		2,175	159		2,334	2,600
8510	3 kW		3.20	2.500		2,275	159		2,434	2,725
8520	4 kW		3	2.667		2,375	170		2,545	2,850
8530	5 kW		3	2.667		2,500	170		2,670	3,000
8540	6 kW		2.80	2.857		2,525	182		2,707	3,050
8550	8 kW		2.40	3.333		2,725	212		2,937	3,325
8560	10 kW		2.20	3.636		3,250	232		3,482	3,925
8570	12 kW		2	4		2,875	255		3,130	3,525
8580	13.5 kW		1.50	5.333		2,850	340		3,190	3,650
8590	16 kW		1.30	6.154		2,975	390		3,365	3,850
8600	20 kW		.90	8.889		4,025	565		4,590	5,275
8610	24 kW		.60	13.333		4,100	850		4,950	5,775
8630	208 to 480 V, 3 pole									
8650	Wall mounted, 2 kW	1 Elec	4.60	1.739	Ea.	2,025	111		2,136	2,425
8670	3 kW		4.60	1.739		2,450	111		2,561	2,875
8680	4 kW		4.40	1.818		2,550	116		2,666	3,000
8690	5 kW		4.40	1.818		2,700	116		2,816	3,150
8700	6 kW		4.20	1.905		2,725	121		2,846	3,175
8710	8 kW		4	2		2,900	127		3,027	3,375
8720	10 kW		3.80	2.105		2,975	134		3,109	3,475
8730	12 kW		3.50	2.286		3,050	146		3,196	3,600
8740	13.5 kW		2.90	2.759		3,100	176		3,276	3,650
8750	16 kW		2.70	2.963		3,250	189		3,439	3,850
8760	20 kW		2.30	3.478		4,825	222		5,047	5,625
8770	24 kW		1.90	4.211		4,500	268		4,768	5,350
8790	Recessed, 2 kW		4.40	1.818		2,250	116		2,366	2,650
8810	3 kW		4.40	1.818		2,375	116		2,491	2,800
8820	4 kW		4.20	1.905		2,525	121		2,646	2,950

23 83 Radiant Heating Units

23 83 33 – Electric Radiant Heaters

23 83 33.10 Electric Heating		Crew	Daily Output	Labor-Hours	Unit	Material	2021 Bare Costs Labor	Equipment	Total	Total Incl O&P
8830	5 kW	1 Elec	4.20	1.905	Ea.	1,800	121		1,921	2,150
8840	6 kW		4	2		2,550	127		2,677	3,000
8850	8 kW		3.80	2.105		2,700	134		2,834	3,175
8860	10 kW		3.50	2.286		1,950	146		2,096	2,375
8870	12 kW		2.90	2.759		2,975	176		3,151	3,525
8880	13.5 kW		2.70	2.963		2,650	189		2,839	3,175
8890	16 kW		2.30	3.478		3,075	222		3,297	3,725
8900	20 kW		1.90	4.211		4,475	268		4,743	5,325
8920	24 kW		1.60	5		4,650	320		4,970	5,600
8940	Ceiling mount, 2 kW		3.20	2.500		2,325	159		2,484	2,775
8950	3 kW		3.20	2.500		2,225	159		2,384	2,650
8960	4 kW		3	2.667		2,600	170		2,770	3,100
8970	5 kW		3	2.667		2,350	170		2,520	2,850
8980	6 kW		2.80	2.857		2,675	182		2,857	3,225
8990	8 kW		2.40	3.333		1,875	212		2,087	2,400
9000	10 kW		2.20	3.636		2,550	232		2,782	3,150
9020	13.5 kW		1.50	5.333		3,425	340		3,765	4,275
9030	16 kW		1.30	6.154		4,525	390		4,915	5,550
9040	20 kW		.90	8.889		4,475	565		5,040	5,775
9060	24 kW		.60	13.333		3,025	850		3,875	4,575
9230	13.5 kW, 40,956 BTU		2.20	3.636		1,925	232		2,157	2,450
9250	24 kW, 81,912 BTU	▼	2	4	▼	2,100	255		2,355	2,700

For customer support on your Electrical Costs with RSMeans Data, call 800.448.8182.

103

Division Notes

	CREW	DAILY OUTPUT	LABOR-HOURS	UNIT	BARE COSTS				TOTAL INCL O&P
					MAT.	LABOR	EQUIP.	TOTAL	

Estimating Tips

- When estimating material costs for special systems, it is always prudent to obtain manufacturers' quotations for equipment prices and special installation requirements that may affect the total cost.

- For cost modifications for elevated tray installation, add the percentages to labor according to the height of the installation and only to the quantities exceeding the different height levels, not to the total tray quantities. Refer to subdivision 26 01 02.20 for labor adjustment factors.

- Do not overlook the costs for equipment used in the installation.

Reference Numbers

Reference numbers are shown at the beginning of some major classifications. These numbers refer to related items in the Reference Section. The reference information may be an estimating procedure, an alternate pricing method, or technical information.

Same Data. Simplified.

Enjoy the convenience and efficiency of accessing your costs anywhere:

- **Skip the multiplier** by setting your location
- **Quickly search,** edit, favorite and share costs
- **Stay on top of price changes** with automatic updates

Discover more at rsmeans.com/online

25 05 Common Work Results for Integrated Automation

25 05 28 – Pathways for Integrated Automation

25 05 28.39 Surface Raceways for Integrated Automation	Crew	Daily Output	Labor-Hours	Unit	Material	2021 Bare Costs Labor	Equipment	Total	Total Incl O&P
0010 **SURFACE RACEWAYS FOR INTEGRATED AUTOMATION**									
1000 Din rail, G type, 15 mm, steel, galvanized, 3', for termination blocks	1 Elec	48	.167	Ea.	12.55	10.60		23.15	29.50
1010 32 mm, steel, galvanized, 3', for termination blocks		48	.167		16.55	10.60		27.15	34
1020 35 mm, steel, galvanized, 3', for termination blocks		48	.167		17.75	10.60		28.35	35.50
1030 Din rail, rail mounted, 35 mm, aluminum, 3', for termination blocks		48	.167		26	10.60		36.60	44.50
1040 35 mm, aluminum, 6', for termination blocks		48	.167		34.50	10.60		45.10	54
2000 Terminal block, Din mounted, 35 mm, medium screw, connector end bracket		48	.167		2.92	10.60		13.52	19
2010 35 mm, large screw, connector end bracket		48	.167		4.94	10.60		15.54	21.50
2020 Large screw, connector end bracket		48	.167		5.10	10.60		15.70	21.50
3000 Terminal block, feed through, 35 mm, Din mounted, 600V, 20A		48	.167		1.01	10.60		11.61	16.90
3010 35 mm, Din mounted, 600V, 30A		48	.167		1.66	10.60		12.26	17.65
3020 Terminal block, fuse holder, LED indicator, Din mounted, 110V, 6A		48	.167		9.15	10.60		19.75	26
3030 Din mounted, 220V, 6A		48	.167		8	10.60		18.60	24.50
3040 Din mounted, 300V, 6A		48	.167		9.45	10.60		20.05	26

25 36 Integrated Automation Instrumentation & Terminal Devices

25 36 19 – Integrated Automation Current Sensors

25 36 19.10 Integrated Automation Current Sensors

	Crew	Daily Output	Labor-Hours	Unit	Material	2021 Bare Costs Labor	Equipment	Total	Total Incl O&P
0010 **INTEGRATED AUTOMATION CURRENT SENSORS**									
1000 Current limiter inrush, 100 to 480 VAC, 10A, power supply	1 Elec	48	.167	Ea.	88.50	10.60		99.10	113

For customer support on your Electrical Costs with RSMeans Data, call 800.448.8182.

Estimating Tips
26 05 00 Common Work Results for Electrical

- Conduit should be taken off in three main categories—power distribution, branch power, and branch lighting—so the estimator can concentrate on systems and components, therefore making it easier to ensure all items have been accounted for.

- For cost modifications for elevated conduit installation, add the percentages to labor according to the height of installation and only to the quantities exceeding the different height levels, not to the total conduit quantities. Refer to subdivision 26 01 02.20 for labor adjustment factors.

- Remember that aluminum wiring of equal ampacity is larger in diameter than copper and may require larger conduit.

- If more than three wires at a time are being pulled, deduct percentages from the labor hours of that grouping of wires.

- When taking off grounding systems, identify separately the type and size of wire, and list each unique type of ground connection.

- The estimator should take the weights of materials into consideration when completing a takeoff. Topics to consider include: How will the materials be supported? What methods of support are available? How high will the support structure have to reach? Will the final support structure be able to withstand the total burden? Is the support material included or separate from the fixture, equipment, and material specified?

- Do not overlook the costs for equipment used in the installation. If scaffolding or highlifts are available in the field, contractors may use them in lieu of the proposed ladders and rolling staging.

26 20 00 Low-Voltage Electrical Transmission

- Supports and concrete pads may be shown on drawings for the larger equipment, or the support system may be only a piece of plywood for the back of a panelboard. In either case, they must be included in the costs.

26 40 00 Electrical and Cathodic Protection

- When taking off cathodic protection systems, identify the type and size of cable, and list each unique type of anode connection.

26 50 00 Lighting

- Fixtures should be taken off room by room using the fixture schedule, specifications, and the ceiling plan. For large concentrations of lighting fixtures in the same area, deduct percentages from labor hours.

Reference Numbers

Reference numbers are shown at the beginning of some major classifications. These numbers refer to related items in the Reference Section. The reference information may be an estimating procedure, an alternate pricing method, or technical information.

Note: Not all subdivisions listed here necessarily appear. ∎

Same Data. Simplified.

Enjoy the convenience and efficiency of accessing your costs anywhere:

- **Skip the multiplier** by setting your location
- **Quickly search,** edit, favorite and share costs
- **Stay on top of price changes** with automatic updates

Discover more at rsmeans.com/online

26 01 02.20 Labor Adjustment Factors	Crew	Daily Output	Labor-Hours	Unit	Material	2021 Bare Costs Labor	Equipment	Total	Total Incl O&P
0010 **LABOR ADJUSTMENT FACTORS** (For Div. 26, 27 and 28) R260519-90									
0100 Subtract from labor for Economy of Scale for Wire									
0110 4-5 wires						25%			
0120 6-10 wires						30%			
0130 11-15 wires						35%			
0140 over 15 wires						40%			
0150 Labor adjustment factors (For Div. 26, 27, 28 and 48)									
0200 Labor factors: The below are reasonable suggestions, but									
0210 each project must be evaluated for its own peculiarities, and									
0220 the adjustments be increased or decreased depending on the									
0230 severity of the special conditions.									
1000 Add to labor for elevated installation (above floor level)									
1010 10' to 14.5' high						10%			
1020 15' to 19.5' high						20%			
1030 20' to 24.5' high						25%			
1040 25' to 29.5' high						35%			
1050 30' to 34.5' high						40%			
1060 35' to 39.5' high						50%			
1070 40' and higher						55%			
2000 Add to labor for crawl space									
2010 3' high						40%			
2020 4' high						30%			
3000 Add to labor for multi-story building									
3100 For new construction (No elevator available)									
3110 Add for floors 3 thru 10						5%			
3120 Add for floors 11 thru 15						10%			
3130 Add for floors 16 thru 20						15%			
3140 Add for floors 21 thru 30						20%			
3150 Add for floors 31 and up						30%			
3200 For existing structure (Elevator available)									
3210 Add for work on floor 3 and above						2%			
4000 Add to labor for working in existing occupied buildings									
4010 Hospital						35%			
4020 Office building						25%			
4030 School						20%			
4040 Factory or warehouse						15%			
4050 Multi-dwelling						15%			
5000 Add to labor, miscellaneous									
5010 Cramped shaft						35%			
5020 Congested area						15%			
5030 Excessive heat or cold						30%			
6000 Labor factors: the above are reasonable suggestions, but									
6100 each project should be evaluated for its own peculiarities									
6200 Other factors to be considered are:									
6210 Movement of material and equipment through finished areas						10%			
6220 Equipment room min security direct access w/authorization						15%			
6230 Attic space						25%			
6240 No service road						25%			
6250 Poor unloading/storage area, no hydraulic lifts or jacks						20%			
6260 Congested site area/heavy traffic						20%			
7000 Correctional facilities (no compounding division 1 adjustment factors)									
7010 Minimum security w/facilities escort						30%			
7020 Medium security w/facilities and correctional officer escort						40%			

For customer support on your Electrical Costs with RSMeans Data, call 800.448.8182.

26 01 Operation and Maintenance of Electrical Systems

26 01 02 – Labor Adjustment

26 01 02.20 Labor Adjustment Factors	Crew	Daily Output	Labor-Hours	Unit	Material	2021 Bare Costs Labor	Equipment	Total	Total Incl O&P
7030 Max security w/facilities & correctional officer escort (no inmate contact)						50%			

26 01 40 – Operation and Maintenance of Electrical Protection Systems

26 01 40.51 Electrical Systems Repair and Replacement

0010 **ELECTRICAL SYSTEMS REPAIR AND REPLACEMENT**									
3000 Remove and replace (reinstall), switch cover	1 Elec	60	.133	Ea.		8.50		8.50	12.60
3020 Outlet cover	"	60	.133	"		8.50		8.50	12.60

26 01 50 – Operation and Maintenance of Lighting

26 01 50.81 Luminaire Replacement

0010 **LUMINAIRE REPLACEMENT**									
0100 Fluorescent fixtures, elevate from work area	1 Elec	96	.083	Ea.	84	5.30		89.30	100
0962 Remove and install new replacement lens, 2' x 2' troffer		16	.500		18.25	32		50.25	67.50
0964 2' x 4' troffer		16	.500		14.95	32		46.95	64
3200 Remove and replace (reinstall), lighting fixture	↓	4	2	↓		127		127	189

26 05 Common Work Results for Electrical

26 05 05 – Selective Demolition for Electrical

26 05 05.10 Electrical Demolition

0010 **ELECTRICAL DEMOLITION** R260105-30									
0020 Electrical demolition, conduit to 10' high, incl. fittings & hangers									
0100 Rigid galvanized steel, 1/2" to 1" diameter	1 Elec	242	.033	L.F.		2.11		2.11	3.13
0120 1-1/4" to 2"	"	200	.040			2.55		2.55	3.79
0140 2-1/2" to 3-1/2"	2 Elec	302	.053			3.37		3.37	5
0160 4" to 6"	"	160	.100			6.35		6.35	9.45
0170 PVC #40, 1/2" to 1"	1 Elec	410	.020			1.24		1.24	1.85
0172 1-1/4" to 2"		350	.023			1.46		1.46	2.16
0174 2-1/2"	↓	250	.032			2.04		2.04	3.03
0176 3" to 3-1/2"	2 Elec	340	.047			3		3	4.45
0178 4" to 6"	"	230	.070			4.43		4.43	6.60
0200 Electric metallic tubing (EMT), 1/2" to 1"	1 Elec	394	.020			1.29		1.29	1.92
0220 1-1/4" to 1-1/2"		326	.025			1.56		1.56	2.32
0240 2" to 3"	↓	236	.034	↓		2.16		2.16	3.21
0260 3-1/2" to 4"	2 Elec	310	.052	↓		3.29		3.29	4.89
0270 Armored cable (BX) avg. 50' runs									
0280 #14, 2 wire	1 Elec	690	.012	L.F.		.74		.74	1.10
0290 #14, 3 wire		571	.014			.89		.89	1.33
0300 #12, 2 wire		605	.013			.84		.84	1.25
0310 #12, 3 wire		514	.016			.99		.99	1.47
0320 #10, 2 wire		514	.016			.99		.99	1.47
0330 #10, 3 wire		425	.019			1.20		1.20	1.78
0340 #8, 3 wire	↓	342	.023	↓		1.49		1.49	2.21
0350 Non metallic sheathed cable (Romex)									
0360 #14, 2 wire	1 Elec	720	.011	L.F.		.71		.71	1.05
0370 #14, 3 wire		657	.012			.78		.78	1.15
0380 #12, 2 wire		629	.013			.81		.81	1.20
0390 #10, 3 wire	↓	450	.018			1.13		1.13	1.68
0400 Wiremold raceway, including fittings & hangers									
0420 No. 3000	1 Elec	250	.032	L.F.		2.04		2.04	3.03
0440 No. 4000		217	.037			2.35		2.35	3.49
0460 No. 6000		166	.048			3.07		3.07	4.56
0462 Plugmold with receptacle	↓	114	.070	↓		4.47		4.47	6.65

For customer support on your Electrical Costs with RSMeans Data, call 800.448.8182.

109

26 05 05.10 Electrical Demolition		Crew	Daily Output	Labor-Hours	Unit	Material	2021 Bare Costs Labor	Equipment	Total	Total Incl O&P
0465	Telephone/power pole	1 Elec	12	.667	Ea.		42.50		42.50	63
0470	Non-metallic, straight section	↓	480	.017	L.F.		1.06		1.06	1.58
0500	Channels, steel, including fittings & hangers									
0520	3/4" x 1-1/2"	1 Elec	308	.026	L.F.		1.65		1.65	2.46
0540	1-1/2" x 1-1/2"	↓	269	.030			1.89		1.89	2.81
0560	1-1/2" x 1-7/8"	↓	229	.035	↓		2.23		2.23	3.31
0600	Copper bus duct, indoor, 3 phase									
0610	Including hangers & supports									
0620	225 amp	2 Elec	135	.119	L.F.		7.55		7.55	11.20
0640	400 amp		106	.151			9.60		9.60	14.30
0660	600 amp		86	.186			11.85		11.85	17.60
0680	1,000 amp		60	.267			17		17	25
0700	1,600 amp		40	.400			25.50		25.50	38
0720	3,000 amp	↓	10	1.600	↓		102		102	151
0800	Plug-in switches, 600 V 3 ph., incl. disconnecting									
0820	wire, conduit terminations, 30 amp	1 Elec	15.50	.516	Ea.		33		33	49
0840	60 amp		13.90	.576			36.50		36.50	54.50
0850	100 amp		10.40	.769			49		49	73
0860	200 amp	↓	6.20	1.290			82		82	122
0880	400 amp	2 Elec	5.40	2.963			189		189	280
0900	600 amp		3.40	4.706			300		300	445
0920	800 amp		2.60	6.154			390		390	580
0940	1,200 amp		2	8			510		510	755
0960	1,600 amp	↓	1.70	9.412	↓		600		600	890
1010	Safety switches, 250 or 600 V, incl. disconnection									
1050	of wire & conduit terminations									
1100	30 amp	1 Elec	12.30	.650	Ea.		41.50		41.50	61.50
1120	60 amp		8.80	.909			58		58	86
1140	100 amp		7.30	1.096			70		70	104
1160	200 amp	↓	5	1.600			102		102	151
1180	400 amp	2 Elec	6.80	2.353			150		150	223
1200	600 amp	"	4.60	3.478			222		222	330
1202	Explosion proof, 60 amp	1 Elec	8.80	.909			58		58	86
1204	100 amp		5	1.600			102		102	151
1206	200 amp		3.33	2.402	↓		153		153	227
1210	Panel boards, incl. removal of all breakers,									
1220	conduit terminations & wire connections									
1230	3 wire, 120/240 V, 100A, to 20 circuits	1 Elec	2.60	3.077	Ea.		196		196	291
1240	200 amps, to 42 circuits	2 Elec	2.60	6.154			390		390	580
1241	225 amps, to 42 circuits		2.40	6.667			425		425	630
1250	400 amps, to 42 circuits	↓	2.20	7.273			465		465	690
1260	4 wire, 120/208 V, 125A, to 20 circuits	1 Elec	2.40	3.333			212		212	315
1270	200 amps, to 42 circuits	2 Elec	2.40	6.667			425		425	630
1280	400 amps, to 42 circuits	↓	1.92	8.333			530		530	790
1285	600 amps, to 42 circuits	↓	1.60	10	↓		635		635	945
1300	Transformer, dry type, 1 phase, incl. removal of									
1320	supports, wire & conduit terminations									
1340	1 kVA	1 Elec	7.70	1.039	Ea.		66		66	98.50
1360	5 kVA	↓	4.70	1.702			108		108	161
1380	10 kVA		3.60	2.222			142		142	210
1400	37.5 kVA	2 Elec	3	5.333			340		340	505
1420	75 kVA	"	2.50	6.400	↓		410		410	605
1440	3 phase to 600 V, primary									

26 05 05.10 Electrical Demolition	Crew	Daily Output	Labor-Hours	Unit	Material	2021 Bare Costs Labor	Equipment	Total	Total Incl O&P
1460	3 kVA	1 Elec	3.87	2.067	Ea.		132		132	196
1480	15 kVA	2 Elec	3.67	4.360			278		278	415
1490	25 kVA		3.51	4.558			290		290	430
1500	30 kVA		3.42	4.678			298		298	445
1510	45 kVA		3.18	5.031			320		320	475
1520	75 kVA		2.69	5.948			380		380	565
1530	112.5 kVA	R-3	2.90	6.897			440	66	506	725
1540	150 kVA		2.70	7.407			470	70.50	540.50	780
1550	300 kVA		1.80	11.111			705	106	811	1,175
1560	500 kVA		1.40	14.286			905	136	1,041	1,500
1570	750 kVA		1.10	18.182			1,150	174	1,324	1,925
1600	Pull boxes & cabinets, sheet metal, incl. removal									
1620	of supports and conduit terminations									
1640	6" x 6" x 4"	1 Elec	31.10	.257	Ea.		16.40		16.40	24.50
1660	12" x 12" x 4"		23.30	.343			22		22	32.50
1680	24" x 24" x 6"		12.30	.650			41.50		41.50	61.50
1700	36" x 36" x 8"		7.70	1.039			66		66	98.50
1720	Junction boxes, 4" sq. & oct.		80	.100			6.35		6.35	9.45
1740	Handy box		107	.075			4.76		4.76	7.10
1760	Switch box		107	.075			4.76		4.76	7.10
1780	Receptacle & switch plates		257	.031			1.98		1.98	2.95
1790	Receptacles & switches, 15 to 30 amp		135	.059			3.77		3.77	5.60
1800	Wire, THW-THWN-THHN, removed from									
1810	in place conduit, to 10' high									
1830	#14	1 Elec	65	.123	C.L.F.		7.85		7.85	11.65
1840	#12		55	.145			9.25		9.25	13.75
1850	#10		45.50	.176			11.20		11.20	16.65
1860	#8		40.40	.198			12.60		12.60	18.75
1870	#6		32.60	.245			15.65		15.65	23
1880	#4	2 Elec	53	.302			19.25		19.25	28.50
1890	#3		50	.320			20.50		20.50	30.50
1900	#2		44.60	.359			23		23	34
1910	1/0		33.20	.482			30.50		30.50	45.50
1920	2/0		29.20	.548			35		35	52
1930	3/0		25	.640			41		41	60.50
1940	4/0		22	.727			46.50		46.50	69
1950	250 kcmil		20	.800			51		51	75.50
1960	300 kcmil		19	.842			53.50		53.50	79.50
1970	350 kcmil		18	.889			56.50		56.50	84
1980	400 kcmil		17	.941			60		60	89
1990	500 kcmil		16.20	.988			63		63	93.50
2000	Interior fluorescent fixtures, incl. supports									
2010	& whips, to 10' high									
2100	Recessed drop-in 2' x 2', 2 lamp	2 Elec	35	.457	Ea.		29		29	43.50
2110	2' x 2', 4 lamp		30	.533			34		34	50.50
2120	2' x 4', 2 lamp		33	.485			31		31	46
2140	2' x 4', 4 lamp		30	.533			34		34	50.50
2160	4' x 4', 4 lamp		20	.800			51		51	75.50
2180	Surface mount, acrylic lens & hinged frame									
2200	1' x 4', 2 lamp	2 Elec	44	.364	Ea.		23		23	34.50
2210	6" x 4', 2 lamp		44	.364			23		23	34.50
2220	2' x 2', 2 lamp		44	.364			23		23	34.50
2260	2' x 4', 4 lamp		33	.485			31		31	46

26 05 05.10 Electrical Demolition

26 05 05.10 Electrical Demolition	Crew	Daily Output	Labor-Hours	Unit	Material	2021 Bare Costs Labor	2021 Bare Costs Equipment	Total	Total Incl O&P
2280	4' x 4', 4 lamp	2 Elec	23	.696	Ea.		44.50		44.50	66
2281	4' x 4', 6 lamp	↓	23	.696	↓		44.50		44.50	66
2300	Strip fixtures, surface mount									
2320	4' long, 1 lamp	2 Elec	53	.302	Ea.		19.25		19.25	28.50
2340	4' long, 2 lamp		50	.320			20.50		20.50	30.50
2360	8' long, 1 lamp		42	.381			24.50		24.50	36
2380	8' long, 2 lamp	↓	40	.400	↓		25.50		25.50	38
2400	Pendant mount, industrial, incl. removal									
2410	of chain or rod hangers, to 10' high									
2420	4' long, 2 lamp	2 Elec	35	.457	Ea.		29		29	43.50
2421	4' long, 4 lamp		35	.457			29		29	43.50
2440	8' long, 2 lamp	↓	27	.593	↓		38		38	56
2460	Interior incandescent, surface, ceiling									
2470	or wall mount, to 10' high									
2480	Metal cylinder type, 75 Watt	2 Elec	62	.258	Ea.		16.45		16.45	24.50
2500	150 Watt		62	.258			16.45		16.45	24.50
2502	300 Watt	↓	53	.302	↓		19.25		19.25	28.50
2520	Metal halide, high bay									
2540	400 Watt	2 Elec	15	1.067	Ea.		68		68	101
2560	1,000 Watt		12	1.333			85		85	126
2580	150 Watt, low bay		20	.800			51		51	75.50
2585	Globe type, ceiling mount		62	.258			16.45		16.45	24.50
2586	Can type, recessed mount	↓	62	.258	↓		16.45		16.45	24.50
2600	Exterior fixtures, incandescent, wall mount									
2620	100 Watt	2 Elec	50	.320	Ea.		20.50		20.50	30.50
2640	Quartz, 500 Watt		33	.485			31		31	46
2660	1,500 Watt	↓	27	.593	↓		38		38	56
2680	Wall pack, mercury vapor									
2700	175 Watt	2 Elec	25	.640	Ea.		41		41	60.50
2720	250 Watt	"	25	.640			41		41	60.50
7000	Weatherhead/mast, 2"	1 Elec	16	.500			32		32	47.50
7002	3"		10	.800			51		51	75.50
7004	3-1/2"		8.50	.941			60		60	89
7006	4"		8	1	↓		63.50		63.50	94.50
7100	Service entry cable, #6, +#6 neutral		420	.019	L.F.		1.21		1.21	1.80
7102	#4, +#4 neutral		360	.022			1.42		1.42	2.10
7104	#2, +#4 neutral		345	.023			1.48		1.48	2.19
7106	#2, +#2 neutral	↓	330	.024	↓		1.54		1.54	2.29
9900	Add to labor for higher elevated installation									
9905	10' to 14.5' high, add						10%			
9910	15' to 20' high, add						20%			
9920	20' to 25' high, add						25%			
9930	25' to 30' high, add						35%			
9940	30' to 35' high, add						40%			
9950	35' to 40' high, add						50%			
9960	Over 40' high, add						55%			

26 05 05.15 Electrical Demolition, Grounding

26 05 05.15 Electrical Demolition, Grounding	Crew	Daily Output	Labor-Hours	Unit	Material	Labor	Equipment	Total	Total Incl O&P
0010	**ELECTRICAL DEMOLITION, GROUNDING** Addition									
0100	Ground clamp, bronze	1 Elec	64	.125	Ea.		7.95		7.95	11.85
0140	Water pipe ground clamp, bronze, heavy duty		24	.333			21		21	31.50
0150	Ground rod, 8' to 10'	↓	13	.615	↓		39		39	58.50
0200	Ground wire, bare armored	2 Elec	900	.018	L.F.		1.13		1.13	1.68

26 05 05 – Selective Demolition for Electrical

26 05 05.15 Electrical Demolition, Grounding		Crew	Daily Output	Labor-Hours	Unit	Material	2021 Bare Costs Labor	Equipment	Total	Total Incl O&P
0240	Bare copper or aluminum	2 Elec	2800	.006	L.F.		.36		.36	.54

26 05 05.20 Electrical Demolition, Wiring Methods

		Crew	Daily Output	Labor-Hours	Unit	Material	2021 Bare Costs Labor	Equipment	Total	Total Incl O&P
0010	**ELECTRICAL DEMOLITION, WIRING METHODS** Addition									
0100	Armored cable, w/PVC jacket, in cable tray									
0110	#6	1 Elec	9.30	.860	C.L.F.		55		55	81.50
0120	#4	2 Elec	16.20	.988			63		63	93.50
0130	#2		13.80	1.159			74		74	110
0140	#1		12	1.333			85		85	126
0150	1/0		10.80	1.481			94.50		94.50	140
0160	2/0		10.20	1.569			100		100	148
0170	3/0		9.60	1.667			106		106	158
0180	4/0		9	1.778			113		113	168
0190	250 kcmil	3 Elec	10.80	2.222			142		142	210
0210	350 kcmil		9.90	2.424			154		154	229
0230	500 kcmil		9	2.667			170		170	252
0240	750 kcmil		8.10	2.963			189		189	280
1100	Control cable, 600 V or less									
1110	3 wires	1 Elec	24	.333	C.L.F.		21		21	31.50
1120	5 wires		20	.400			25.50		25.50	38
1130	7 wires		16.50	.485			31		31	46
1140	9 wires		15	.533			34		34	50.50
1150	12 wires	2 Elec	26	.615			39		39	58.50
1160	15 wires		22	.727			46.50		46.50	69
1170	19 wires		19	.842			53.50		53.50	79.50
1180	25 wires		16	1			63.50		63.50	94.50
1500	Mineral insulated (MI) cable, 600 V									
1510	#10	1 Elec	4.80	1.667	C.L.F.		106		106	158
1520	#8		4.50	1.778			113		113	168
1530	#6		4.20	1.905			121		121	180
1540	#4	2 Elec	7.20	2.222			142		142	210
1550	#2		6.60	2.424			154		154	229
1560	#1		6.30	2.540			162		162	240
1570	1/0		6	2.667			170		170	252
1580	2/0		5.70	2.807			179		179	266
1590	3/0		5.40	2.963			189		189	280
1600	4/0		4.80	3.333			212		212	315
1610	250 kcmil	3 Elec	7.20	3.333			212		212	315
1620	500 kcmil	"	5.90	4.068			259		259	385
2000	Shielded cable, XLP shielding, to 35 kV									
2010	#4	2 Elec	13.20	1.212	C.L.F.		77		77	115
2020	#1		12	1.333			85		85	126
2030	1/0		11.40	1.404			89.50		89.50	133
2040	2/0		10.80	1.481			94.50		94.50	140
2050	4/0		9.60	1.667			106		106	158
2060	250 kcmil	3 Elec	13.50	1.778			113		113	168
2070	350 kcmil		11.70	2.051			131		131	194
2080	500 kcmil		11.20	2.143			137		137	203
2090	750 kcmil		10.80	2.222			142		142	210
3000	Modular flexible wiring									
3010	Cable set	1 Elec	120	.067	Ea.		4.25		4.25	6.30
3020	Conversion module		48	.167			10.60		10.60	15.80
3030	Switching assembly		96	.083			5.30		5.30	7.90

For customer support on your Electrical Costs with RSMeans Data, call 800.448.8182.

113

26 05 05.20 Electrical Demolition, Wiring Methods	Crew	Daily Output	Labor-Hours	Unit	Material	2021 Bare Costs Labor	Equipment	Total	Total Incl O&P
3200	Undercarpet									
3210	Power or telephone, flat cable	1 Elec	1320	.006	L.F.		.39		.39	.57
3220	Transition block assemblies w/fitting		75	.107	Ea.		6.80		6.80	10.10
3230	Floor box with fitting		60	.133			8.50		8.50	12.60
3300	Data system, cable with connection		50	.160			10.20		10.20	15.15
4000	Cable tray, including fitting & support									
4010	Galvanized steel, 6" wide	2 Elec	310	.052	L.F.		3.29		3.29	4.89
4020	9" wide		295	.054			3.45		3.45	5.15
4030	12" wide		285	.056			3.58		3.58	5.30
4040	18" wide		270	.059			3.77		3.77	5.60
4050	24" wide		260	.062			3.92		3.92	5.80
4060	30" wide		240	.067			4.25		4.25	6.30
4070	36" wide		220	.073			4.63		4.63	6.90
4110	Aluminum, 6" wide		420	.038			2.43		2.43	3.61
4120	9" wide		415	.039			2.46		2.46	3.65
4130	12" wide		390	.041			2.61		2.61	3.88
4140	18" wide		370	.043			2.75		2.75	4.09
4150	24" wide		350	.046			2.91		2.91	4.33
4160	30" wide		325	.049			3.14		3.14	4.66
4170	36" wide		300	.053			3.40		3.40	5.05
4310	Cable channel, aluminum, 4" wide straight	1 Elec	240	.033			2.12		2.12	3.16
4320	Cable tray fittings, 6" to 9" wide	2 Elec	27	.593	Ea.		38		38	56
4330	12" to 18" wide	"	19	.842	"		53.50		53.50	79.50
5000	Conduit nipples, with locknuts and bushings									
5020	1/2"	1 Elec	108	.074	Ea.		4.72		4.72	7
5040	3/4"		96	.083			5.30		5.30	7.90
5060	1"		81	.099			6.30		6.30	9.35
5080	1-1/4"		69	.116			7.40		7.40	10.95
5100	1-1/2"		60	.133			8.50		8.50	12.60
5120	2"		54	.148			9.45		9.45	14
5140	2-1/2"		45	.178			11.30		11.30	16.85
5160	3"		36	.222			14.15		14.15	21
5180	3-1/2"		33	.242			15.45		15.45	23
5200	4"		27	.296			18.85		18.85	28
5220	5"		21	.381			24.50		24.50	36
5240	6"		18	.444			28.50		28.50	42
5500	Electric nonmetallic tubing (ENT), flexible, 1/2" to 1" diameter		690	.012	L.F.		.74		.74	1.10
5510	1-1/4" to 2" diameter		300	.027			1.70		1.70	2.52
5600	Flexible metallic tubing, steel, 3/8" to 3/4" diameter		600	.013			.85		.85	1.26
5610	1" to 1-1/4" diameter		260	.031			1.96		1.96	2.91
5620	1-1/2" to 2" diameter		140	.057			3.64		3.64	5.40
5630	2-1/2" diameter		100	.080			5.10		5.10	7.55
5640	3" to 3-1/2" diameter	2 Elec	150	.107			6.80		6.80	10.10
5650	4" diameter	"	100	.160			10.20		10.20	15.15
5700	Sealtite flexible conduit, 3/8" to 3/4" diameter	1 Elec	420	.019			1.21		1.21	1.80
5710	1" to 1-1/4" diameter		180	.044			2.83		2.83	4.21
5720	1-1/2" to 2" diameter		120	.067			4.25		4.25	6.30
5730	2-1/2" diameter		80	.100			6.35		6.35	9.45
5740	3" diameter	2 Elec	150	.107			6.80		6.80	10.10
5750	4" diameter	"	100	.160			10.20		10.20	15.15
5800	Wiring duct, plastic, 1-1/2" to 2-1/2" wide	1 Elec	360	.022			1.42		1.42	2.10
5810	3" wide		330	.024			1.54		1.54	2.29
5820	4" wide		300	.027			1.70		1.70	2.52

For customer support on your Electrical Costs with RSMeans Data, call 800.448.8182.

26 05 05 – Selective Demolition for Electrical

26 05 05.20 Electrical Demolition, Wiring Methods

		Crew	Daily Output	Labor-Hours	Unit	Material	2021 Bare Costs Labor	Equipment	Total	Total Incl O&P
6000	Floor box and carpet flange	1 Elec	9	.889	Ea.		56.50		56.50	84
6300	Wireway, with fittings and supports, to 10' high									
6310	2-1/2" x 2-1/2"	1 Elec	180	.044	L.F.		2.83		2.83	4.21
6320	4" x 4"	"	160	.050			3.19		3.19	4.73
6330	6" x 6"	2 Elec	240	.067			4.25		4.25	6.30
6340	8" x 8"		160	.100			6.35		6.35	9.45
6350	10" x 10"		120	.133			8.50		8.50	12.60
6360	12" x 12"		80	.200			12.75		12.75	18.95
6400	Cable reel with receptacle, 120 V/208 V		13	1.231	Ea.		78.50		78.50	116
6500	Equipment connection, to 10 HP	1 Elec	23	.348			22		22	33
6510	To 15 to 30 HP		18	.444			28.50		28.50	42
6520	To 40 to 60 HP		15	.533			34		34	50.50
6530	To 75 to 100 HP		10	.800			51		51	75.50
7008	Receptacle, explosion proof, 120 V, to 30 A		16	.500			32		32	47.50
7020	Dimmer switch, 2,000 W or less		62	.129			8.20		8.20	12.20

26 05 05.25 Electrical Demolition, Electrical Power

		Crew	Daily Output	Labor-Hours	Unit	Material	2021 Bare Costs Labor	Equipment	Total	Total Incl O&P
0010	**ELECTRICAL DEMOLITION, ELECTRICAL POWER**	R024119-10								
0100	Meter centers and sockets									
0120	Meter socket, 4 terminal	R260105-30	1 Elec	8.40	.952	Ea.		60.50		60.50	90
0140	Trans-socket, 13 terminal, 400 A	"	3.60	2.222			142		142	210
0160	800 A	2 Elec	4.40	3.636			232		232	345
0200	Meter center, 400 A		5.80	2.759			176		176	261
0210	600 A		4	4			255		255	380
0220	800 A		3.30	4.848			310		310	460
0230	1,200 A		2.80	5.714			365		365	540
0240	1,600 A		2.50	6.400			410		410	605
0300	Base meter devices, 3 meter		3.60	4.444			283		283	420
0310	4 meter		3.30	4.848			310		310	460
0320	5 meter		2.90	5.517			350		350	520
0330	6 meter		2.20	7.273			465		465	690
0340	7 meter		2	8			510		510	755
0350	8 meter		1.90	8.421			535		535	795
0400	Branch meter devices									
0410	Socket w/circuit breaker 200 A, 2 meter	2 Elec	3.30	4.848	Ea.		310		310	460
0420	3 meter		2.90	5.517			350		350	520
0430	4 meter		2.60	6.154			390		390	580
0450	Main circuit breaker, 400 A		5.80	2.759			176		176	261
0460	600 A		4	4			255		255	380
0470	800 A		3.30	4.848			310		310	460
0480	1,200 A		2.80	5.714			365		365	540
0490	1,600 A		2.50	6.400			410		410	605
0500	Main lug terminal box, 800 A		3.40	4.706			300		300	445
0510	1,200 A		2.60	6.154			390		390	580
1000	Motors, 230/460 V, 60 Hz, 3/4 HP	1 Elec	10.70	.748			47.50		47.50	71
1010	5 HP		9	.889			56.50		56.50	84
1020	10 HP		8	1			63.50		63.50	94.50
1030	15 HP		6.40	1.250			79.50		79.50	118
1040	20 HP	2 Elec	10.40	1.538			98		98	146
1050	50 HP		9.60	1.667			106		106	158
1060	75 HP		5.60	2.857			182		182	270
1070	100 HP	3 Elec	5.40	4.444			283		283	420
1080	150 HP		3.60	6.667			425		425	630

For customer support on your Electrical Costs with RSMeans Data, call 800.448.8182.

115

26 05 05.25 Electrical Demolition, Electrical Power		Crew	Daily Output	Labor-Hours	Unit	Material	2021 Bare Costs Labor	Equipment	Total	Total Incl O&P
1090	200 HP	3 Elec	3	8	Ea.		510		510	755
1200	Variable frequency drive, 460 V, for 5 HP motor size	1 Elec	3.20	2.500			159		159	237
1210	10 HP motor size	"	2.70	2.963			189		189	280
1220	20 HP motor size	2 Elec	3.60	4.444			283		283	420
1230	50 HP motor size	"	2.10	7.619			485		485	720
1240	75 HP motor size	R-3	2.20	9.091			575	87	662	955
1250	100 HP motor size		2	10			635	95.50	730.50	1,050
1260	150 HP motor size		2	10			635	95.50	730.50	1,050
1270	200 HP motor size	↓	1.70	11.765	↓		745	112	857	1,225
2000	Generator set w/accessories, 3 phase 4 wire, 277/480 V									
2010	7.5 kW	3 Elec	2	12	Ea.		765		765	1,125
2020	20 kW		1.70	14.118			900		900	1,325
2040	30 kW		1.33	18.045			1,150		1,150	1,700
2060	50 kW		1	24			1,525		1,525	2,275
2080	100 kW		.75	32			2,050		2,050	3,025
2100	150 kW		.63	38.095			2,425		2,425	3,600
2120	250 kW		.59	40.678			2,600		2,600	3,850
2140	400 kW		.50	48			3,050		3,050	4,550
2160	500 kW	↓	.44	54.545			3,475		3,475	5,175
2180	750 kW	4 Elec	.56	57.143			3,650		3,650	5,400
2200	1,000 kW	"	.46	69.565	↓		4,425		4,425	6,575
3000	Uninterruptible power supply system (UPS)									
3010	Single phase, 120 V, 1 kVA	1 Elec	3.20	2.500	Ea.		159		159	237
3020	2 kVA	2 Elec	3.60	4.444			283		283	420
3030	5 kVA	3 Elec	2.50	9.600			610		610	910
3040	10 kVA		2.30	10.435			665		665	990
3050	15 kVA	↓	1.80	13.333	↓		850		850	1,250
4000	Transformer, incl. support, wire & conduit termination									
4010	Buck-boost, single phase, 120/240 V, 0.1 kVA	1 Elec	25	.320	Ea.		20.50		20.50	30.50
4020	0.5 kVA		12.50	.640			41		41	60.50
4030	1 kVA		6.30	1.270			81		81	120
4040	5 kVA	↓	3.80	2.105			134		134	199
4800	5 kV or 15 kV primary, 277/480 V second, 112.5 kVA	R-3	3.50	5.714	↓		365	54.50	419.50	600
4801	5 kV or 15 kV primary, 277/480 V second, 112.5 kVA		400	.050	kVA		3.17	.48	3.65	5.25
4810	150 kVA		2.70	7.407	Ea.		470	70.50	540.50	780
4811	150 kVA		412	.049	kVA		3.08	.46	3.54	5.10
4820	225 kVA		2.30	8.696	Ea.		550	83	633	910
4821	225 kVA		512	.039	kVA		2.48	.37	2.85	4.09
4830	500 kVA		1.45	13.793	Ea.		875	132	1,007	1,450
4831	500 kVA		700	.029	kVA		1.81	.27	2.08	3
4840	750 kVA		1.33	15.038	Ea.		955	144	1,099	1,575
4841	750 kVA		1065	.019	kVA		1.19	.18	1.37	1.97
4850	1,000 kVA		1.25	16	Ea.		1,025	153	1,178	1,675
4851	1,000 kVA		1250	.016	kVA		1.02	.15	1.17	1.68
4860	2,000 kVA		1.05	19.048	Ea.		1,200	182	1,382	2,000
4861	2,000 kVA		20000	.001	kVA		.06	.01	.07	.10
4870	3,000 kVA		.75	26.667	Ea.		1,700	255	1,955	2,800
4871	3,000 kVA		2200	.009	kVA		.58	.09	.67	.96
6010	Isolation panel, 3 kVA	1 Elec	2.30	3.478	Ea.		222		222	330
6020	5 kVA		2.20	3.636			232		232	345
6030	7.5 kVA		2.10	3.810			243		243	360
6040	10 kVA		1.80	4.444			283		283	420
6050	15 kVA	↓	1.40	5.714	↓		365		365	540

26 05 05 – Selective Demolition for Electrical

26 05 05.25 Electrical Demolition, Electrical Power

		Crew	Daily Output	Labor-Hours	Unit	Material	2021 Bare Costs Labor	2021 Bare Costs Equipment	Total	Total Incl O&P
7000	Power filters & conditioners									
7100	Automatic voltage regulator	2 Elec	6	2.667	Ea.		170		170	252
7200	Capacitor, 1 kVAR	1 Elec	8.60	.930			59.50		59.50	88
7210	5 kVAR		5.80	1.379			88		88	131
7220	10 kVAR		4.80	1.667			106		106	158
7230	15 kVAR		4.20	1.905			121		121	180
7240	20 kVAR		3.50	2.286			146		146	216
7250	30 kVAR		3.40	2.353			150		150	223
7260	50 kVAR		3.20	2.500			159		159	237
7400	Computer isolator transformer									
7410	Single phase, 120/240 V, 0.5 kVAR	1 Elec	12.80	.625	Ea.		40		40	59
7420	1 kVAR		8.50	.941			60		60	89
7430	2.5 kVAR		6.40	1.250			79.50		79.50	118
7440	5 kVAR		3.70	2.162			138		138	205
7500	Computer regulator transformer									
7510	Single phase, 240 V, 0.5 kVAR	1 Elec	8.50	.941	Ea.		60		60	89
7520	1 kVAR		6.40	1.250			79.50		79.50	118
7530	2 kVAR		3.20	2.500			159		159	237
7540	Single phase, plug-in unit 120 V, 0.5 kVAR		26	.308			19.60		19.60	29
7550	1 kVAR		17	.471			30		30	44.50
7600	Power conditioner transformer									
7610	Single phase 115 V - 240 V, 3 kVA	2 Elec	5.10	3.137	Ea.		200		200	297
7620	5 kVA	"	3.70	4.324			275		275	410
7630	7.5 kVA	3 Elec	4.80	5			320		320	475
7640	10 kVA	"	4.30	5.581			355		355	530
7700	Transient suppressor/voltage regulator									
7710	Single phase 115 or 220 V, 1 kVA	1 Elec	8.50	.941	Ea.		60		60	89
7720	2 kVA		7.30	1.096			70		70	104
7730	4 kVA		6.80	1.176			75		75	111
7800	Transient voltage suppressor transformer									
7810	Single phase, 115 or 220 V, 3.6 kVA	1 Elec	12.80	.625	Ea.		40		40	59
7820	7.2 kVA		11.50	.696			44.50		44.50	66
7830	14.4 kVA		10.20	.784			50		50	74
7840	Single phase, plug-in, 120 V, 1.8 kVA		26	.308			19.60		19.60	29
8000	Power measurement & control									
8010	Switchboard instruments, 3 phase 4 wire, indicating unit	1 Elec	25	.320	Ea.		20.50		20.50	30.50
8020	Recording unit		12	.667			42.50		42.50	63
8100	3 current transformers, 3 phase 4 wire, 5 to 800 A		6.40	1.250			79.50		79.50	118
8110	1,000 to 1,500 A		4.20	1.905			121		121	180
8120	2,000 to 4,000 A		3.20	2.500			159		159	237

26 05 05.30 Electrical Demolition, Transmission and Distribution

		Crew	Daily Output	Labor-Hours	Unit	Material	2021 Bare Costs Labor	2021 Bare Costs Equipment	Total	Total Incl O&P
0010	**ELECTRICAL DEMOLITION, TRANSMISSION & DISTRIBUTION**									
0100	Load interrupter switch, 600 A, NEMA 1, 4.8 kV	R-3	1.33	15.038	Ea.		955	144	1,099	1,575
0120	13.8 kV (R024119-10)	"	1.27	15.748			1,000	150	1,150	1,650
0200	Lightning arrester, 4.8 kV	1 Elec	9	.889			56.50		56.50	84
0220	13.8 kV (R260105-30)		6.67	1.199			76.50		76.50	114
0300	Alarm or option items		3.33	2.402			153		153	227

26 05 05.35 Electrical Demolition, L.V. Distribution

		Crew	Daily Output	Labor-Hours	Unit	Material	2021 Bare Costs Labor	2021 Bare Costs Equipment	Total	Total Incl O&P
0010	**ELECTRICAL DEMOLITION, L.V. DISTRIBUTION** Addition (R024119-10)									
0100	Circuit breakers in enclosure									
0120	Enclosed (NEMA 1), 600 V, 3 pole, 30 A (R260105-30)	1 Elec	12.30	.650	Ea.		41.50		41.50	61.50
0140	60 A		8.80	.909			58		58	86

For customer support on your Electrical Costs with RSMeans Data, call 800.448.8182.

117

26 05 05.35 Electrical Demolition, L.V. Distribution	Crew	Daily Output	Labor-Hours	Unit	Material	2021 Bare Costs Labor	Equipment	Total	Total Incl O&P
0160	100 A	1 Elec	7.30	1.096	Ea.		70		70	104
0180	225 A	↓	5	1.600			102		102	151
0200	400 A	2 Elec	6.80	2.353			150		150	223
0220	600 A		4.60	3.478			222		222	330
0240	800 A		3.60	4.444			283		283	420
0260	1,200 A		3.10	5.161			330		330	490
0280	1,600 A		2.80	5.714			365		365	540
0300	2,000 A	↓	2.50	6.400			410		410	605
0400	Enclosed (NEMA 7), 600 V, 3 pole, 50 A	1 Elec	8.80	.909			58		58	86
0410	100 A		5.80	1.379			88		88	131
0420	150 A	↓	3.80	2.105			134		134	199
0430	250 A	2 Elec	6.20	2.581			164		164	244
0440	400 A	"	4.60	3.478			222		222	330
0460	Manual motor starter, NEMA 1	1 Elec	24	.333			21		21	31.50
0480	NEMA 4 or NEMA 7		15	.533			34		34	50.50
0500	Time switches, single pole single throw		15.40	.519			33		33	49
0520	Photo cell		30	.267			17		17	25
0540	Load management device, 4 loads		7.70	1.039			66		66	98.50
0550	8 loads	↓	3.80	2.105			134		134	199
0560	Master light control panel	2 Elec	2	8			510		510	755
0600	Transfer switches, enclosed, 30 A	1 Elec	12	.667			42.50		42.50	63
0610	60 A		9.50	.842			53.50		53.50	79.50
0620	100 A	↓	6.50	1.231			78.50		78.50	116
0630	150 A	2 Elec	12	1.333			85		85	126
0640	260 A		10	1.600			102		102	151
0660	400 A		8	2			127		127	189
0670	600 A		5	3.200			204		204	305
0680	800 A		4	4			255		255	380
0690	1,200 A		3.50	4.571			291		291	435
0700	1,600 A		3	5.333			340		340	505
0710	2,000 A	↓	2.50	6.400			410		410	605
0800	Air terminal with base and connector	1 Elec	24	.333			21		21	31.50
1000	Enclosed controller, NEMA 1, 30 A		13.80	.580			37		37	55
1020	60 A		11.50	.696			44.50		44.50	66
1040	100 A		9.60	.833			53		53	79
1060	150 A		7.70	1.039			66		66	98.50
1080	200 A	↓	5.40	1.481			94.50		94.50	140
1100	400 A	2 Elec	6.90	2.319			148		148	219
1120	600 A		4.60	3.478			222		222	330
1140	800 A		3.80	4.211			268		268	400
1160	1,200 A	↓	3.10	5.161			330		330	490
1200	Control station, NEMA 1	1 Elec	30	.267			17		17	25
1220	NEMA 7		23	.348			22		22	33
1230	Control switches, push button		69	.116			7.40		7.40	10.95
1240	Indicating light unit		120	.067			4.25		4.25	6.30
1250	Relay	↓	15	.533	↓		34		34	50.50
2000	Motor control center components									
2020	Starter, NEMA 1, size 1	1 Elec	9	.889	Ea.		56.50		56.50	84
2040	Size 2	2 Elec	13.30	1.203			76.50		76.50	114
2060	Size 3		6.70	2.388			152		152	226
2080	Size 4		5.30	3.019			192		192	286
2100	Size 5	↓	3.30	4.848			310		310	460
2120	NEMA 7, size 1	1 Elec	8.70	.920			58.50		58.50	87

26 05 05 – Selective Demolition for Electrical

26 05 05.35 Electrical Demolition, L.V. Distribution	Crew	Daily Output	Labor-Hours	Unit	Material	2021 Bare Costs Labor	Equipment	Total	Total Incl O&P
2140	Size 2	2 Elec	12.70	1.260	Ea.		80.50		80.50	119
2160	Size 3		6.30	2.540			162		162	240
2180	Size 4		5	3.200			204		204	305
2200	Size 5		3.20	5			320		320	475
2300	Fuse, light contactor, NEMA 1, 30 A	1 Elec	9	.889			56.50		56.50	84
2310	60 A		6.70	1.194			76		76	113
2320	100 A		3.30	2.424			154		154	229
2330	200 A		2.70	2.963			189		189	280
2350	Motor control center, incoming section	2 Elec	4	4			255		255	380
2352	Structure per section	"	4.20	3.810			243		243	360
2400	Starter & structure, 10 HP	1 Elec	9	.889			56.50		56.50	84
2410	25 HP	2 Elec	13.30	1.203			76.50		76.50	114
2420	50 HP		6.70	2.388			152		152	226
2430	75 HP		5.30	3.019			192		192	286
2440	100 HP		4.70	3.404			217		217	320
2450	200 HP		3.30	4.848			310		310	460
2460	400 HP		2.70	5.926			375		375	560
2500	Motor starter & control									
2510	Motor starter, NEMA 1, 5 HP	1 Elec	8.90	.899	Ea.		57.50		57.50	85
2520	10 HP	"	6.20	1.290			82		82	122
2530	25 HP	2 Elec	8.40	1.905			121		121	180
2540	50 HP		6.90	2.319			148		148	219
2550	100 HP		4.60	3.478			222		222	330
2560	200 HP		3.50	4.571			291		291	435
2570	400 HP		3.10	5.161			330		330	490
2610	Motor starter, NEMA 7, 5 HP	1 Elec	6.20	1.290			82		82	122
2620	10 HP	"	4.20	1.905			121		121	180
2630	25 HP	2 Elec	6.90	2.319			148		148	219
2640	50 HP		4.60	3.478			222		222	330
2650	100 HP		3.50	4.571			291		291	435
2660	200 HP		1.90	8.421			535		535	795
2710	Combination control unit, NEMA 1, 5 HP	1 Elec	6.90	1.159			74		74	110
2720	10 HP	"	5	1.600			102		102	151
2730	25 HP	2 Elec	7.70	2.078			132		132	197
2740	50 HP		5.10	3.137			200		200	297
2750	100 HP		3.10	5.161			330		330	490
2810	NEMA 7, 5 HP	1 Elec	5	1.600			102		102	151
2820	10 HP	"	3.90	2.051			131		131	194
2830	25 HP	2 Elec	5.10	3.137			200		200	297
2840	50 HP		3.10	5.161			330		330	490
2850	100 HP		2.30	6.957			445		445	660
2860	200 HP		1.50	10.667			680		680	1,000
3000	Panelboard or load center circuit breaker									
3010	Bolt-on or plug in, 15 A to 50 A	1 Elec	20	.400	Ea.		25.50		25.50	38
3020	60 A to 70 A		16	.500			32		32	47.50
3030	Bolt-on, 80 A to 100 A		14	.571			36.50		36.50	54
3040	Up to 250 A		7.30	1.096			70		70	104
3050	Motor operated, 30 A		13	.615			39		39	58.50
3060	60 A		10	.800			51		51	75.50
3070	100 A		8	1			63.50		63.50	94.50
3200	Switchboard circuit breaker									
3210	15 A to 60 A	1 Elec	19	.421	Ea.		27		27	40
3220	70 A to 100 A		14	.571			36.50		36.50	54

For customer support on your Electrical Costs with RSMeans Data, call 800.448.8182.

119

26 05 05.35 Electrical Demolition, L.V. Distribution

		Crew	Daily Output	Labor-Hours	Unit	Material	2021 Bare Costs Labor	2021 Bare Costs Equipment	Total	Total Incl O&P
3230	125 A to 400 A	1 Elec	10	.800	Ea.		51		51	75.50
3240	450 A to 600 A		5.30	1.509			96		96	143
3250	700 A to 800 A		4.30	1.860			119		119	176
3260	1,000 A		3.30	2.424			154		154	229
3270	1,200 A		2.70	2.963			189		189	280
3500	Switchboard, incoming section, 400 A	2 Elec	3.70	4.324			275		275	410
3510	600 A		3.30	4.848			310		310	460
3520	800 A		2.90	5.517			350		350	520
3530	1,200 A		2.40	6.667			425		425	630
3540	1,600 A		2.20	7.273			465		465	690
3550	2,000 A		2.10	7.619			485		485	720
3560	3,000 A		1.90	8.421			535		535	795
3570	4,000 A		1.70	9.412			600		600	890
3610	Distribution section, 600 A		4	4			255		255	380
3620	800 A		3.60	4.444			283		283	420
3630	1,200 A		3.10	5.161			330		330	490
3640	1,600 A		2.90	5.517			350		350	520
3650	2,000 A		2.70	5.926			375		375	560
3710	Transition section, 600 A		3.80	4.211			268		268	400
3720	800 A		3.30	4.848			310		310	460
3730	1,200 A		2.70	5.926			375		375	560
3740	1,600 A		2.40	6.667			425		425	630
3750	2,000 A		2.20	7.273			465		465	690
3760	2,500 A		2.10	7.619			485		485	720
3770	3,000 A		1.90	8.421			535		535	795
4000	Bus duct, aluminum or copper, 30 A	1 Elec	200	.040	L.F.		2.55		2.55	3.79
4020	60 A		160	.050			3.19		3.19	4.73
4040	100 A		140	.057			3.64		3.64	5.40
5000	Feedrail, trolley busway, up to 60 A		160	.050			3.19		3.19	4.73
5020	100 A		120	.067			4.25		4.25	6.30
5040	Busway, 50 A		200	.040			2.55		2.55	3.79
6000	Fuse, 30 A		133	.060	Ea.		3.83		3.83	5.70
6010	60 A		133	.060			3.83		3.83	5.70
6020	100 A		105	.076			4.85		4.85	7.20
6030	200 A		95	.084			5.35		5.35	7.95
6040	400 A		80	.100			6.35		6.35	9.45
6050	600 A		53	.151			9.60		9.60	14.30
6060	601 A to 1,200 A		42	.190			12.15		12.15	18.05
6070	1,500 A to 1,600 A		34	.235			15		15	22.50
6080	1,800 A to 2,500 A		26	.308			19.60		19.60	29
6090	4,000 A		21	.381			24.50		24.50	36
6100	4,500 A to 5,000 A		18	.444			28.50		28.50	42
6120	6,000 A		15	.533			34		34	50.50
6200	Fuse plug or fustat		100	.080			5.10		5.10	7.55

26 05 05.50 Electrical Demolition, Lighting

		Crew	Daily Output	Labor-Hours	Unit	Material	2021 Bare Costs Labor	2021 Bare Costs Equipment	Total	Total Incl O&P
0010	**ELECTRICAL DEMOLITION, LIGHTING** Addition R024119-10									
0100	Fixture hanger, flexible, 1/2" diameter	1 Elec	36	.222	Ea.		14.15		14.15	21
0120	3/4" diameter R260105-30	"	30	.267	"		17		17	25
3000	Light pole, anchor base, excl. concrete bases									
3010	Metal light pole, 10'	2 Elec	24	.667	Ea.		42.50		42.50	63
3020	16'	"	18	.889			56.50		56.50	84
3030	20'	R-3	8.70	2.299			146	22	168	241

26 05 05 – Selective Demolition for Electrical

26 05 05.50 Electrical Demolition, Lighting

		Crew	Daily Output	Labor-Hours	Unit	Material	2021 Bare Costs Labor	Equipment	Total	Total Incl O&P
3035	25'	R-3	8.60	2.326	Ea.		148	22	170	244
3038	30'		6	3.333			212	32	244	350
3039	35'		6	3.333			212	32	244	350
3040	40'		6	3.333			212	32	244	350
3100	Wood light pole, 10'	2 Elec	36	.444			28.50		28.50	42
3120	20'	"	24	.667			42.50		42.50	63
3140	Bollard light, 42"	1 Elec	9	.889			56.50		56.50	84
3160	Walkway luminaire	"	8.10	.988			63		63	93.50
4000	Explosion proof									
4010	Metal halide, 175 W	1 Elec	8.70	.920	Ea.		58.50		58.50	87
4020	250 W	"	8.10	.988			63		63	93.50
4030	400 W	2 Elec	14.40	1.111			71		71	105
4050	High pressure sodium, 70 W	1 Elec	9	.889			56.50		56.50	84
4060	100 W		9	.889			56.50		56.50	84
4070	150 W		8.10	.988			63		63	93.50
4100	Incandescent		8.70	.920			58.50		58.50	87
4200	Fluorescent		8.10	.988			63		63	93.50
5000	Ballast, fluorescent fixture		24	.333			21		21	31.50
5040	High intensity discharge fixture		24	.333			21		21	31.50
5300	Exit and emergency lighting									
5310	Exit light	1 Elec	24	.333	Ea.		21		21	31.50
5320	Emergency battery pack lighting unit		12	.667			42.50		42.50	63
5330	Remote lamp only		80	.100			6.35		6.35	9.45
5340	Self-contained fluorescent lamp pack		30	.267			17		17	25
5500	Track lighting, 8' section	2 Elec	40	.400			25.50		25.50	38
5510	Track lighting fixture	1 Elec	64	.125			7.95		7.95	11.85
5800	Energy saving devices									
5810	Occupancy sensor	1 Elec	21	.381	Ea.		24.50		24.50	36
5820	Automatic wall switch		72	.111			7.10		7.10	10.50
5830	Remote power pack		30	.267			17		17	25
5840	Photoelectric control		24	.333			21		21	31.50
5850	Fixture whip		100	.080			5.10		5.10	7.55
6000	Lamps									
6010	Fluorescent	1 Elec	200	.040	Ea.		2.55		2.55	3.79
6030	High intensity discharge lamp, up to 400 W		68	.118			7.50		7.50	11.15
6040	Up to 1,000 W		45	.178			11.30		11.30	16.85
6050	Quartz		90	.089			5.65		5.65	8.40
6070	Incandescent		360	.022			1.42		1.42	2.10
6080	Exterior, PAR		290	.028			1.76		1.76	2.61
6090	Guards for fluorescent lamp		24	.333			21		21	31.50

26 05 13 – Medium-Voltage Cables

26 05 13.10 Cable Terminations

		Crew	Daily Output	Labor-Hours	Unit	Material	2021 Bare Costs Labor	Equipment	Total	Total Incl O&P
0010	**CABLE TERMINATIONS**, 5 kV to 35 kV									
0100	Indoor, insulation diameter range 0.64" to 1.08"									
0300	Padmount, 5 kV	1 Elec	8	1	Ea.	86	63.50		149.50	190
0400	10 kV		6.40	1.250		114	79.50		193.50	243
0500	15 kV		6	1.333		153	85		238	294
0600	25 kV		5.60	1.429		205	91		296	360
0700	Insulation diameter range 1.05" to 1.8"									
0800	Padmount, 5 kV	1 Elec	8	1	Ea.	37	63.50		100.50	136
0900	10 kV		6	1.333		170	85		255	315
1000	15 kV		5.60	1.429		128	91		219	275

For customer support on your Electrical Costs with RSMeans Data, call 800.448.8182.

121

26 05 13.10 Cable Terminations

		Crew	Daily Output	Labor-Hours	Unit	Material	2021 Bare Costs Labor	Equipment	Total	Total Incl O&P
1100	25 kV	1 Elec	5.30	1.509	Ea.	256	96		352	425
1200	Insulation diameter range 1.53" to 2.32"									
1300	Padmount, 5 kV	1 Elec	7.40	1.081	Ea.	155	69		224	273
1400	10 kV		5.60	1.429		146	91		237	295
1500	15 kV		5.30	1.509		194	96		290	355
1600	25 kV		5	1.600		286	102		388	465
1700	Outdoor systems, #4 stranded to 1/0 stranded									
1800	5 kV	1 Elec	7.40	1.081	Ea.	90	69		159	201
1900	15 kV		5.30	1.509		485	96		581	675
2000	25 kV		5	1.600		555	102		657	760
2100	35 kV		4.80	1.667		330	106		436	520
2200	#1 solid to 4/0 stranded, 5 kV		6.90	1.159		143	74		217	267
2300	15 kV		5	1.600		555	102		657	760
2400	25 kV		4.80	1.667		610	106		716	830
2500	35 kV		4.60	1.739		1,150	111		1,261	1,450
2600	2/0 solid to 350 kcmil stranded, 5 kV		6.40	1.250		197	79.50		276.50	335
2700	15 kV		4.80	1.667		665	106		771	890
2800	25 kV		4.60	1.739		820	111		931	1,075
2900	35 kV		4.40	1.818		450	116		566	665
3000	400 kcmil compact to 750 kcmil stranded, 5 kV		6	1.333		665	85		750	855
3100	15 kV		4.60	1.739		820	111		931	1,075
3200	25 kV		4.40	1.818		1,250	116		1,366	1,550
3300	35 kV		4.20	1.905		1,425	121		1,546	1,725
3400	1,000 kcmil, 5 kV		5.60	1.429		250	91		341	410
3500	15 kV		4.40	1.818		820	116		936	1,075
3600	25 kV		4.20	1.905		1,025	121		1,146	1,325
3700	35 kV		4	2		1,525	127		1,652	1,875

26 05 13.16 Medium-Voltage, Single Cable

		Crew	Daily Output	Labor-Hours	Unit	Material	2021 Bare Costs Labor	Equipment	Total	Total Incl O&P
0010	**MEDIUM-VOLTAGE, SINGLE CABLE** Splicing & terminations not included									
0040	Copper, XLP shielding, 5 kV, #6	2 Elec	4.40	3.636	C.L.F.	168	232		400	530
0050	#4		4.40	3.636		218	232		450	585
0100	#2		4	4		232	255		487	635
0200	#1		4	4		330	255		585	745
0400	1/0		3.80	4.211		370	268		638	810
0600	2/0		3.60	4.444		400	283		683	860
0800	4/0		3.20	5		610	320		930	1,150
1000	250 kcmil	3 Elec	4.50	5.333		725	340		1,065	1,300
1200	350 kcmil		3.90	6.154		945	390		1,335	1,625
1400	500 kcmil		3.60	6.667		1,000	425		1,425	1,725
1600	15 kV, ungrounded neutral, #1	2 Elec	4	4		360	255		615	775
1800	1/0		3.80	4.211		430	268		698	875
2000	2/0		3.60	4.444		490	283		773	960
2200	4/0		3.20	5		650	320		970	1,200
2400	250 kcmil	3 Elec	4.50	5.333		720	340		1,060	1,300
2600	350 kcmil		3.90	6.154		910	390		1,300	1,575
2800	500 kcmil		3.60	6.667		1,025	425		1,450	1,750
3000	25 kV, grounded neutral, 1/0	2 Elec	3.60	4.444		595	283		878	1,075
3200	2/0		3.40	4.706		655	300		955	1,175
3400	4/0		3	5.333		825	340		1,165	1,400
3600	250 kcmil	3 Elec	4.20	5.714		1,025	365		1,390	1,675
3800	350 kcmil		3.60	6.667		1,200	425		1,625	1,950
3900	500 kcmil		3.30	7.273		1,400	465		1,865	2,250

26 05 13 – Medium-Voltage Cables

26 05 13.16 Medium-Voltage, Single Cable

		Crew	Daily Output	Labor-Hours	Unit	Material	2021 Bare Costs Labor	Equipment	Total	Total Incl O&P
4000	35 kV, grounded neutral, 1/0	2 Elec	3.40	4.706	C.L.F.	610	300		910	1,125
4200	2/0		3.20	5		745	320		1,065	1,300
4400	4/0		2.80	5.714		935	365		1,300	1,575
4600	250 kcmil	3 Elec	3.90	6.154		1,100	390		1,490	1,775
4800	350 kcmil		3.30	7.273		1,325	465		1,790	2,150
5000	500 kcmil		3	8		1,550	510		2,060	2,450
5050	Aluminum, XLP shielding, 5 kV, #2	2 Elec	5	3.200		263	204		467	595
5070	#1		4.40	3.636		272	232		504	645
5090	1/0		4	4		315	255		570	725
5100	2/0		3.80	4.211		350	268		618	790
5150	4/0		3.60	4.444		420	283		703	880
5200	250 kcmil	3 Elec	4.80	5		505	320		825	1,025
5220	350 kcmil		4.50	5.333		595	340		935	1,150
5240	500 kcmil		3.90	6.154		765	390		1,155	1,425
5260	750 kcmil		3.60	6.667		1,000	425		1,425	1,725
5300	15 kV aluminum, XLP, #1	2 Elec	4.40	3.636		335	232		567	715
5320	1/0		4	4		350	255		605	765
5340	2/0		3.80	4.211		375	268		643	815
5360	4/0		3.60	4.444		460	283		743	925
5380	250 kcmil	3 Elec	4.80	5		550	320		870	1,075
5400	350 kcmil		4.50	5.333		615	340		955	1,175
5420	500 kcmil		3.90	6.154		855	390		1,245	1,525
5440	750 kcmil		3.60	6.667		1,125	425		1,550	1,875

26 05 19 – Low-Voltage Electrical Power Conductors and Cables

26 05 19.13 Undercarpet Electrical Power Cables

		Crew	Daily Output	Labor-Hours	Unit	Material	2021 Bare Costs Labor	Equipment	Total	Total Incl O&P
0010	**UNDERCARPET ELECTRICAL POWER CABLES** R260519-80									
0020	Power System									
0100	Cable flat, 3 conductor, #12, w/attached bottom shield	1 Elec	982	.008	L.F.	5.20	.52		5.72	6.50
0200	Shield, top, steel		1768	.005	"	4.71	.29		5	5.65
0250	Splice, 3 conductor		48	.167	Ea.	14.30	10.60		24.90	31.50
0300	Top shield		96	.083		1.34	5.30		6.64	9.35
0350	Tap		40	.200		18.30	12.75		31.05	39
0400	Insulating patch, splice, tap & end		48	.167		46	10.60		56.60	66.50
0450	Fold		230	.035			2.22		2.22	3.29
0500	Top shield, tap & fold		96	.083		1.34	5.30		6.64	9.35
0700	Transition, block assembly		77	.104		67	6.60		73.60	83.50
0750	Receptacle frame & base		32	.250		36.50	15.95		52.45	64
0800	Cover receptacle		120	.067		3.09	4.25		7.34	9.70
0850	Cover blank		160	.050		3.65	3.19		6.84	8.75
0860	Receptacle, direct connected, single		25	.320		78.50	20.50		99	117
0870	Dual		16	.500		129	32		161	190
0880	Combination high & low, tension		21	.381		95	24.50		119.50	140
0900	Box, floor with cover		20	.400		79.50	25.50		105	126
0920	Floor service w/barrier		4	2		225	127		352	435
1000	Wall, surface, with cover		20	.400		52.50	25.50		78	95.50
1100	Wall, flush, with cover		20	.400		36.50	25.50		62	78.50
1450	Cable flat, 5 conductor #12, w/attached bottom shield		800	.010	L.F.	8.55	.64		9.19	10.40
1550	Shield, top, steel		1768	.005	"	8.50	.29		8.79	9.80
1600	Splice, 5 conductor		48	.167	Ea.	23	10.60		33.60	41.50
1650	Top shield		96	.083		1.34	5.30		6.64	9.35
1700	Tap		48	.167		30.50	10.60		41.10	50
1750	Insulating patch, splice tap & end		83	.096		54	6.15		60.15	68

26 05 19.13 Undercarpet Electrical Power Cables		Crew	Daily Output	Labor-Hours	Unit	Material	2021 Bare Costs Labor	2021 Bare Costs Equipment	Total	Total Incl O&P
1800	Transition, block assembly	1 Elec	77	.104	Ea.	48	6.60		54.60	63
1850	Box, wall, flush with cover		20	.400		49	25.50		74.50	91.50
1900	Cable flat, 4 conductor #12		933	.009	L.F.	6.10	.55		6.65	7.50
1950	3 conductor #10		982	.008		6	.52		6.52	7.35
1960	4 conductor #10		933	.009		7.85	.55		8.40	9.45
1970	5 conductor #10		884	.009		9.60	.58		10.18	11.40
2500	Telephone System									
2510	Transition fitting wall box, surface	1 Elec	24	.333	Ea.	65	21		86	103
2520	Flush		24	.333		65	21		86	103
2530	Flush, for PC board		24	.333		65	21		86	103
2540	Floor service box		4	2		240	127		367	455
2550	Cover, surface					21.50			21.50	23.50
2560	Flush					21.50			21.50	23.50
2570	Flush for PC board					21.50			21.50	23.50
2700	Floor fitting w/duplex jack & cover	1 Elec	21	.381		49.50	24.50		74	90.50
2720	Low profile		53	.151		18.30	9.60		27.90	34.50
2740	Miniature w/duplex jack		53	.151		33.50	9.60		43.10	51.50
2760	25 pair kit		21	.381		46	24.50		70.50	86.50
2780	Low profile		53	.151		18.70	9.60		28.30	35
2800	Call director kit for 5 cable		19	.421		86	27		113	135
2820	4 pair kit		19	.421		87.50	27		114.50	136
2840	3 pair kit		19	.421		108	27		135	159
2860	Comb. 25 pair & 3 conductor power		21	.381		86	24.50		110.50	131
2880	5 conductor power		21	.381		84	24.50		108.50	129
2900	PC board, 8 per 3 pair		161	.050		63.50	3.17		66.67	74.50
2920	6 per 4 pair		161	.050		63.50	3.17		66.67	74.50
2940	3 pair adapter		161	.050		58	3.17		61.17	68
2950	Plug		77	.104		3.23	6.60		9.83	13.40
2960	Couplers		321	.025		9.10	1.59		10.69	12.35
3000	Bottom shield for 25 pair cable		4420	.002	L.F.	.86	.12		.98	1.12
3020	4 pair		4420	.002		.41	.12		.53	.62
3040	Top shield for 25 pair cable		4420	.002		.86	.12		.98	1.12
3100	Cable assembly, double-end, 50', 25 pair		11.80	.678	Ea.	224	43		267	310
3110	3 pair		23.60	.339		79	21.50		100.50	119
3120	4 pair		23.60	.339		86	21.50		107.50	127
3140	Bulk 3 pair		1473	.005	L.F.	1.35	.35		1.70	2
3160	4 pair		1473	.005	"	1.67	.35		2.02	2.35
3500	Data System									
3520	Cable 25 conductor w/connection 40', 75 ohm	1 Elec	14.50	.552	Ea.	77.50	35		112.50	137
3530	Single lead		22	.364		211	23		234	267
3540	Dual lead		22	.364		225	23		248	282
3560	Shields same for 25 conductor as 25 pair telephone									
3570	Single & dual, none required									
3590	BNC coax connectors, plug	1 Elec	40	.200	Ea.	15	12.75		27.75	35.50
3600	TNC coax connectors, plug	"	40	.200	"	16.05	12.75		28.80	36.50
3700	Cable-bulk									
3710	Single lead	1 Elec	1473	.005	L.F.	3.63	.35		3.98	4.50
3720	Dual lead	"	1473	.005	"	5.25	.35		5.60	6.30
3730	Hand tool crimp				Ea.	515			515	570
3740	Hand tool notch				"	21			21	23
3750	Boxes & floor fitting same as telephone									
3790	Data cable notching, 90°	1 Elec	97	.082	Ea.		5.25		5.25	7.80
3800	180°		60	.133			8.50		8.50	12.60

26 05 19 – Low-Voltage Electrical Power Conductors and Cables

26 05 19.13 Undercarpet Electrical Power Cables	Crew	Daily Output	Labor-Hours	Unit	Material	2021 Bare Costs Labor	Equipment	Total	Total Incl O&P
8100	Drill floor	1 Elec	160	.050	Ea.	4.60	3.19		7.79	9.80
8200	Marking floor		1600	.005	L.F.		.32		.32	.47
8300	Tape, hold down		6400	.001	"	.18	.08		.26	.32
8350	Tape primer, 500' per can	↓	96	.083	Ea.	104	5.30		109.30	123
8400	Tool, splicing				"	163			163	179

26 05 19.20 Armored Cable

		Crew	Daily Output	Labor-Hours	Unit	Material	2021 Bare Costs Labor	Equipment	Total	Total Incl O&P
0010	**ARMORED CABLE** R260519-20									
0050	600 volt, copper (BX), #14, 2 conductor, solid	1 Elec	2.40	3.333	C.L.F.	38.50	212		250.50	360
0100	3 conductor, solid		2.20	3.636		63.50	232		295.50	415
0120	4 conductor, solid		2	4		87	255		342	475
0150	#12, 2 conductor, solid		2.30	3.478		40.50	222		262.50	375
0200	3 conductor, solid		2	4		65.50	255		320.50	450
0220	4 conductor, solid		1.80	4.444		93	283		376	520
0250	#10, 2 conductor, solid		2	4		79.50	255		334.50	470
0300	3 conductor, solid		1.60	5		107	320		427	595
0320	4 conductor, solid		1.40	5.714		166	365		531	725
0340	#8, 2 conductor, stranded		1.50	5.333		156	340		496	675
0350	3 conductor, stranded		1.30	6.154		156	390		546	750
0370	4 conductor, stranded		1.10	7.273		385	465		850	1,125
0380	#6, 2 conductor, stranded		1.30	6.154		255	390		645	860
0390	#4, 3 conductor, stranded		1.40	5.714		475	365		840	1,050
0400	3 conductor with PVC jacket, in cable tray, #6	↓	3.10	2.581		297	164		461	570
0450	#4	2 Elec	5.40	2.963		555	189		744	890
0500	#2		4.60	3.478		640	222		862	1,025
0550	#1		4	4		845	255		1,100	1,300
0600	1/0		3.60	4.444		1,100	283		1,383	1,625
0650	2/0		3.40	4.706		1,025	300		1,325	1,600
0700	3/0		3.20	5		1,775	320		2,095	2,425
0750	4/0	↓	3	5.333		2,125	340		2,465	2,825
0800	250 kcmil	3 Elec	3.60	6.667		2,625	425		3,050	3,525
0850	350 kcmil		3.30	7.273		3,300	465		3,765	4,350
0900	500 kcmil	↓	3	8		4,575	510		5,085	5,775
0910	4 conductor with PVC jacket, in cable tray, #6	1 Elec	2.70	2.963		1,375	189		1,564	1,800
0920	#4	2 Elec	4.60	3.478		1,275	222		1,497	1,725
0930	#2		4	4		1,275	255		1,530	1,775
0940	#1		3.60	4.444		1,550	283		1,833	2,125
0950	1/0		3.40	4.706		1,425	300		1,725	2,000
0960	2/0		3.20	5		1,750	320		2,070	2,400
0970	3/0		3	5.333		2,300	340		2,640	3,050
0980	4/0	↓	2.40	6.667		2,100	425		2,525	2,925
0990	250 kcmil	3 Elec	3.30	7.273		3,225	465		3,690	4,250
1000	350 kcmil		3	8		4,325	510		4,835	5,525
1010	500 kcmil	↓	2.70	8.889	↓	6,000	565		6,565	7,450
1050	5 kV, copper, 3 conductor with PVC jacket,									
1060	non-shielded, in cable tray, #4	2 Elec	380	.042	L.F.	6.30	2.68		8.98	10.90
1100	#2		360	.044		12.35	2.83		15.18	17.80
1200	#1		300	.053		15.35	3.40		18.75	22
1400	1/0		290	.055		18.15	3.51		21.66	25
1600	2/0		260	.062		20	3.92		23.92	28
2000	4/0		240	.067		28	4.25		32.25	37.50
2100	250 kcmil	3 Elec	330	.073		32.50	4.63		37.13	43
2150	350 kcmil	↓	315	.076	↓	47.50	4.85		52.35	59

26 05 Common Work Results for Electrical

26 05 19 – Low-Voltage Electrical Power Conductors and Cables

26 05 19.20 Armored Cable		Crew	Daily Output	Labor-Hours	Unit	Material	2021 Bare Costs Labor	Equipment	Total	Total Incl O&P
2200	500 kcmil	3 Elec	270	.089	L.F.	44.50	5.65		50.15	57.50
2400	15 kV, copper, 3 conductor with PVC jacket galv., steel armored									
2500	grounded neutral, in cable tray, #2	2 Elec	300	.053	L.F.	19.90	3.40		23.30	27
2600	#1		280	.057		21.50	3.64		25.14	29
2800	1/0		260	.062		24	3.92		27.92	32.50
2900	2/0		220	.073		32.50	4.63		37.13	42.50
3000	4/0		190	.084		36.50	5.35		41.85	48.50
3100	250 kcmil	3 Elec	270	.089		41	5.65		46.65	53.50
3150	350 kcmil		240	.100		48	6.35		54.35	62.50
3200	500 kcmil		210	.114		64.50	7.30		71.80	82
3400	15 kV, copper, 3 conductor with PVC jacket,									
3450	ungrounded neutral, in cable tray, #2	2 Elec	260	.062	L.F.	21.50	3.92		25.42	29.50
3500	#1		230	.070		24	4.43		28.43	32.50
3600	1/0		200	.080		27.50	5.10		32.60	37.50
3700	2/0		190	.084		33.50	5.35		38.85	45
3800	4/0		160	.100		40.50	6.35		46.85	54
4000	250 kcmil	3 Elec	210	.114		47	7.30		54.30	63
4050	350 kcmil		195	.123		62	7.85		69.85	79.50
4100	500 kcmil		180	.133		75.50	8.50		84	95.50
4200	600 volt, aluminum, 3 conductor in cable tray with PVC jacket									
4300	#2	2 Elec	540	.030	L.F.	5.10	1.89		6.99	8.40
4400	#1		460	.035		4.72	2.22		6.94	8.50
4500	1/0		400	.040		5.65	2.55		8.20	10
4600	2/0		360	.044		5.95	2.83		8.78	10.75
4700	3/0		340	.047		6.70	3		9.70	11.80
4800	4/0		320	.050		8.10	3.19		11.29	13.65
4900	250 kcmil	3 Elec	450	.053		9.75	3.40		13.15	15.75
5000	350 kcmil		360	.067		12.10	4.25		16.35	19.60
5200	500 kcmil		330	.073		15.05	4.63		19.68	23.50
5300	750 kcmil		285	.084		18.60	5.35		23.95	28.50
5400	600 volt, aluminum, 4 conductor in cable tray with PVC jacket									
5410	#2	2 Elec	520	.031	L.F.	4.72	1.96		6.68	8.10
5430	#1		440	.036		5.75	2.32		8.07	9.80
5450	1/0		380	.042		6.75	2.68		9.43	11.40
5470	2/0		340	.047		6.90	3		9.90	12.05
5480	3/0		320	.050		8.10	3.19		11.29	13.65
5500	4/0		300	.053		9.50	3.40		12.90	15.50
5520	250 kcmil	3 Elec	420	.057		10.20	3.64		13.84	16.65
5540	350 kcmil		330	.073		13.20	4.63		17.83	21.50
5560	500 kcmil		300	.080		16.55	5.10		21.65	26
5580	750 kcmil		270	.089		22.50	5.65		28.15	33.50
5600	5 kV, aluminum, unshielded in cable tray, #2 with PVC jacket	2 Elec	380	.042		5.95	2.68		8.63	10.55
5700	#1 with PVC jacket		360	.044		6.40	2.83		9.23	11.25
5800	1/0 with PVC jacket		300	.053		6.85	3.40		10.25	12.55
6000	2/0 with PVC jacket		290	.055		6.70	3.51		10.21	12.55
6200	3/0 with PVC jacket		260	.062		8.40	3.92		12.32	15.05
6300	4/0 with PVC jacket		240	.067		9.45	4.25		13.70	16.70
6400	250 kcmil with PVC jacket	3 Elec	330	.073		10.35	4.63		14.98	18.30
6500	350 kcmil with PVC jacket		315	.076		12.15	4.85		17	20.50
6600	500 kcmil with PVC jacket		300	.080		14.35	5.10		19.45	23.50
6800	750 kcmil with PVC jacket		270	.089		17.70	5.65		23.35	28
6900	15 kV, aluminum, shielded-grounded, #2 with PVC jacket	2 Elec	320	.050		13.60	3.19		16.79	19.70
7000	#1 with PVC jacket		300	.053		13.40	3.40		16.80	19.75

26 05 19.20 Armored Cable	Crew	Daily Output	Labor-Hours	Unit	Material	2021 Bare Costs Labor	2021 Bare Costs Equipment	Total	Total Incl O&P
7200	1/0 with PVC jacket	2 Elec	280	.057	L.F.	15.05	3.64		18.69	22
7300	2/0 with PVC jacket		260	.062		14.65	3.92		18.57	22
7400	3/0 with PVC jacket		240	.067		16.45	4.25		20.70	24.50
7500	4/0 with PVC jacket		220	.073		16.90	4.63		21.53	25.50
7600	250 kcmil with PVC jacket	3 Elec	300	.080		19.35	5.10		24.45	29
7700	350 kcmil with PVC jacket		270	.089		21.50	5.65		27.15	32.50
7800	500 kcmil with PVC jacket		240	.100		26	6.35		32.35	38
8000	750 kcmil with PVC jacket		204	.118		31	7.50		38.50	45
8200	15 kV, aluminum, shielded-ungrounded, #1 with PVC jacket	2 Elec	250	.064		16.30	4.08		20.38	24
8300	1/0 with PVC jacket		230	.070		16.90	4.43		21.33	25
8400	2/0 with PVC jacket		210	.076		19.35	4.85		24.20	28.50
8500	3/0 with PVC jacket		200	.080		18.85	5.10		23.95	28
8600	4/0 with PVC jacket		190	.084		20.50	5.35		25.85	30.50
8700	250 kcmil with PVC jacket	3 Elec	270	.089		22	5.65		27.65	32.50
8800	350 kcmil with PVC jacket		240	.100		25	6.35		31.35	37
8900	500 kcmil with PVC jacket		210	.114		30.50	7.30		37.80	44.50
8950	750 kcmil with PVC jacket		174	.138		37.50	8.80		46.30	54.50
9010	600 volt, copper (MC) steel clad, #14, 2 wire	R-1A	8.16	1.961	C.L.F.	64.50	112		176.50	238
9020	3 wire		7.21	2.219		71.50	127		198.50	268
9030	4 wire		6.69	2.392		185	137		322	405
9040	#12, 2 wire		7.21	2.219		45.50	127		172.50	240
9050	3 wire		6.69	2.392		110	137		247	325
9060	4 wire		6.04	2.649		148	152		300	390
9070	#10, 2 wire		6.04	2.649		155	152		307	395
9080	3 wire		5.65	2.832		190	162		352	450
9090	4 wire		5.23	3.059		192	175		367	470
9100	#8, 2 wire, stranded		3.94	4.061		146	233		379	505
9110	3 wire, stranded		3.40	4.706		194	270		464	615
9120	4 wire, stranded		2.80	5.714		315	330		645	835
9130	#6, 2 wire, stranded		3.45	4.638		214	266		480	630
9200	600 volt, copper (MC) aluminum clad, #14, 2 wire		8.16	1.961		40.50	112		152.50	212
9210	3 wire		7.21	2.219		67.50	127		194.50	263
9220	4 wire		6.69	2.392		118	137		255	335
9230	#12, 2 wire		7.21	2.219		41	127		168	234
9240	3 wire		6.69	2.392		74.50	137		211.50	286
9250	4 wire		6.04	2.649		153	152		305	395
9260	#10, 2 wire		6.04	2.649		93.50	152		245.50	330
9270	3 wire		5.65	2.832		106	162		268	355
9280	4 wire		5.23	3.059		276	175		451	565
9600	Alum (MC) aluminum clad, #6, 3 conductor w/#6 grnd		2.75	5.818		291	335		626	815
9610	4 conductor w/#6 grnd		2.29	6.987		445	400		845	1,075
9620	#4, 3 conductor w/#6 grnd	R-1B	3.13	7.668		264	425		689	920
9630	4 conductor w/#6 grnd		2.67	8.989		545	495		1,040	1,325
9640	#2, 3 conductor w/#4 grnd		2.59	9.266		465	510		975	1,275
9650	4 conductor w/#4 grnd		2.21	10.860		520	600		1,120	1,475
9660	#1, 3 conductor w/#4 grnd		2.15	11.163		320	615		935	1,275
9670	4 conductor w/#4 grnd		1.85	12.973		705	715		1,420	1,850
9680	1/0, 3 conductor w/#4 grnd		2.16	11.111		405	615		1,020	1,350
9690	4 conductor w/#4 grnd		1.73	13.873		1,025	765		1,790	2,275
9700	2/0, 3 conductor w/#4 grnd		1.91	12.565		715	695		1,410	1,825
9710	4 conductor w/#4 grnd		1.53	15.686		1,150	865		2,015	2,550
9720	3/0, 3 conductor w/#4 grnd		1.75	13.714		915	755		1,670	2,125
9730	4 conductor w/#4 grnd		1.40	17.143		1,425	945		2,370	2,975

For customer support on your Electrical Costs with RSMeans Data, call 800.448.8182.

127

26 05 19 – Low-Voltage Electrical Power Conductors and Cables

26 05 19.20 Armored Cable

		Crew	Daily Output	Labor-Hours	Unit	Material	2021 Bare Costs Labor	Equipment	Total	Total Incl O&P
9740	4/0, 3 conductor w/#2 grnd	R-1B	1.54	15.584	C.L.F.	1,025	860		1,885	2,400
9750	4 conductor w/#2 grnd		1.23	19.512		1,600	1,075		2,675	3,375
9760	250 kcmil, 3 conductor w/#1 grnd	R-1C	1.90	16.842		1,325	965	54	2,344	2,925
9770	4 conductor w/#1 grnd		1.52	21.053		1,800	1,200	67.50	3,067.50	3,875
9780	350 kcmil, 3 conductor w/1/0 grnd		1.76	18.182		1,750	1,050	58.50	2,858.50	3,550
9790	4 conductor w/1/0 grnd		1.41	22.695		2,725	1,300	73	4,098	5,000
9800	500 kcmil, 3 conductor w/#1 grnd		1.64	19.512		1,975	1,125	62.50	3,162.50	3,900
9810	4 conductor w/2/0 grnd		1.31	24.427		890	1,400	78.50	2,368.50	3,150
9840	750 kcmil, 3 conductor w/1/0 grnd		1.31	24.427		3,150	1,400	78.50	4,628.50	5,600
9850	4 conductor w/3/0 grnd		1.04	30.769		1,600	1,775	99	3,474	4,500

26 05 19.25 Cable Connectors

		Crew	Daily Output	Labor-Hours	Unit	Material	2021 Bare Costs Labor	Equipment	Total	Total Incl O&P
0010	**CABLE CONNECTORS**									
0100	600 volt, nonmetallic, #14-2 wire	1 Elec	160	.050	Ea.	1.92	3.19		5.11	6.85
0200	#14-3 wire to #12-2 wire		133	.060		1.92	3.83		5.75	7.80
0300	#12-3 wire to #10-2 wire		114	.070		1.92	4.47		6.39	8.75
0400	#10-3 wire to #14-4 and #12-4 wire		100	.080		1.92	5.10		7.02	9.65
0500	#8-3 wire to #10-4 wire		80	.100		3.61	6.35		9.96	13.40
0600	#6-3 wire		40	.200		4.93	12.75		17.68	24.50
0800	SER, 3 #8 insulated + 1 #8 ground		32	.250		4.39	15.95		20.34	28.50
0900	3 #6 + 1 #6 ground		24	.333		5.50	21		26.50	37.50
1000	3 #4 + 1 #6 ground		22	.364		5.30	23		28.30	40.50
1100	3 #2 + 1 #4 ground		20	.400		13.05	25.50		38.55	52.50
1200	3 1/0 + 1 #2 ground		18	.444		26.50	28.50		55	71
1400	3 2/0 + 1 #1 ground		16	.500		24	32		56	74
1600	3 4/0 + 1 #2/0 ground		14	.571		28	36.50		64.50	85
1800	600 volt, armored, #14-2 wire		80	.100		1.80	6.35		8.15	11.45
2200	#14-4, #12-3 and #10-2 wire		40	.200		.61	12.75		13.36	19.60
2400	#12-4, #10-3 and #8-2 wire		32	.250		3.65	15.95		19.60	27.50
2600	#8-3 and #10-4 wire		26	.308		4.19	19.60		23.79	33.50
2650	#8-4 wire		22	.364		6.35	23		29.35	41.50
2652	Non-PVC jacket connector, #6-3 wire, #6-4 wire		22	.364		52.50	23		75.50	92.50
2660	1/0-3 wire		11	.727		72	46.50		118.50	149
2670	300 kcmil-3 wire		6	1.333		109	85		194	246
2680	700 kcmil-3 wire		4	2		162	127		289	365
2700	PVC jacket connector, #6-3 wire, #6-4 wire		16	.500		5.85	32		37.85	54
2800	#4-3 wire, #4-4 wire		16	.500		5.85	32		37.85	54
2900	#2-3 wire		12	.667		5.85	42.50		48.35	69.50
3000	#1-3 wire, #2-4 wire		12	.667		10.60	42.50		53.10	74.50
3200	1/0-3 wire		11	.727		10.60	46.50		57.10	80.50
3400	2/0-3 wire, 1/0-4 wire		10	.800		10.60	51		61.60	87
3500	3/0-3 wire, 2/0-4 wire		9	.889		14.40	56.50		70.90	100
3600	4/0-3 wire, 3/0-4 wire		7	1.143		14.40	73		87.40	124
3800	250 kcmil-3 wire, 4/0-4 wire		6	1.333		28	85		113	157
4000	350 kcmil-3 wire, 250 kcmil-4 wire		5	1.600		28	102		130	182
4100	350 kcmil-4 wire		4	2		187	127		314	395
4200	500 kcmil-3 wire		4	2		187	127		314	395
4250	500 kcmil-4 wire, 750 kcmil-3 wire		3.50	2.286		227	146		373	465
4300	750 kcmil-4 wire		3	2.667		227	170		397	500
4400	5 kV, armored, #4		8	1		28	63.50		91.50	126
4600	#2		8	1		32.50	63.50		96	131
4800	#1		8	1		49.50	63.50		113	149
5000	1/0		6.40	1.250		91	79.50		170.50	218

26 05 19 – Low-Voltage Electrical Power Conductors and Cables

26 05 19.25 Cable Connectors

		Crew	Daily Output	Labor-Hours	Unit	Material	2021 Bare Costs Labor	Equipment	Total	Total Incl O&P
5200	2/0	1 Elec	5.30	1.509	Ea.	91	96		187	243
5500	4/0		4	2		91	127		218	289
5600	250 kcmil		3.60	2.222		91	142		233	310
5650	350 kcmil		3.20	2.500		102	159		261	350
5700	500 kcmil		2.50	3.200		198	204		402	525
5720	750 kcmil		2.20	3.636		260	232		492	630
5750	1,000 kcmil		2	4		395	255		650	815
5800	15 kV, armored, #1		4	2		121	127		248	320
5900	1/0		4	2		151	127		278	355
6000	3/0		3.60	2.222		267	142		409	505
6100	4/0		3.40	2.353		299	150		449	555
6200	250 kcmil		3.20	2.500		289	159		448	555
6300	350 kcmil		2.70	2.963		305	189		494	615
6400	500 kcmil		2	4		460	255		715	885

26 05 19.30 Cable Splicing

		Crew	Daily Output	Labor-Hours	Unit	Material	2021 Bare Costs Labor	Equipment	Total	Total Incl O&P
0010	**CABLE SPLICING** URD or similar, ideal conditions									
0100	#6 stranded to #1 stranded, 5 kV	1 Elec	4	2	Ea.	174	127		301	380
0120	15 kV		3.60	2.222		214	142		356	445
0140	25 kV		3.20	2.500		257	159		416	520
0200	#1 stranded to 4/0 stranded, 5 kV		3.60	2.222		180	142		322	410
0210	15 kV		3.20	2.500		243	159		402	505
0220	25 kV		2.80	2.857		257	182		439	555
0300	4/0 stranded to 500 kcmil stranded, 5 kV		3.30	2.424		180	154		334	425
0310	15 kV		2.90	2.759		425	176		601	725
0320	25 kV		2.50	3.200		485	204		689	840
0400	500 kcmil, 5 kV		3.20	2.500		220	159		379	480
0410	15 kV		2.80	2.857		410	182		592	720
0420	25 kV		2.30	3.478		485	222		707	865
0500	600 kcmil, 5 kV		2.90	2.759		220	176		396	505
0510	15 kV		2.40	3.333		435	212		647	795
0520	25 kV		2	4		525	255		780	955
0600	750 kcmil, 5 kV		2.60	3.077		220	196		416	535
0610	15 kV		2.20	3.636		435	232		667	825
0620	25 kV		1.90	4.211		525	268		793	975
0700	1,000 kcmil, 5 kV		2.30	3.478		220	222		442	570
0710	15 kV		1.90	4.211		475	268		743	920
0720	25 kV		1.60	5		455	320		775	975

26 05 19.35 Cable Terminations

		Crew	Daily Output	Labor-Hours	Unit	Material	2021 Bare Costs Labor	Equipment	Total	Total Incl O&P
0010	**CABLE TERMINATIONS**									
0015	Wire connectors, screw type, #22 to #14	1 Elec	260	.031	Ea.	.05	1.96		2.01	2.97
0020	#18 to #12		240	.033		.14	2.12		2.26	3.31
0035	#16 to #10		230	.035		.19	2.22		2.41	3.50
0040	#14 to #8		210	.038		.47	2.43		2.90	4.13
0045	#12 to #6		180	.044		.47	2.83		3.30	4.73
0050	Terminal lugs, solderless, #16 to #10		50	.160		.35	10.20		10.55	15.55
0100	#8 to #4		30	.267		.69	17		17.69	26
0150	#2 to #1		22	.364		.86	23		23.86	35.50
0200	1/0 to 2/0		16	.500		1.61	32		33.61	49.50
0250	3/0		12	.667		3.72	42.50		46.22	67
0300	4/0		11	.727		22	46.50		68.50	93
0350	250 kcmil		9	.889		2.92	56.50		59.42	87
0400	350 kcmil		7	1.143		4.06	73		77.06	112

129

26 05 19.35 Cable Terminations	Crew	Daily Output	Labor-Hours	Unit	Material	2021 Bare Costs Labor	Equipment	Total	Total Incl O&P
0450	500 kcmil	1 Elec	6	1.333	Ea.	7.90	85		92.90	135
0500	600 kcmil		5.80	1.379		8.60	88		96.60	140
0550	750 kcmil		5.20	1.538		10.85	98		108.85	158
0600	Split bolt connectors, tapped, #6		16	.500		3	32		35	51
0650	#4		14	.571		3.59	36.50		40.09	58
0700	#2		12	.667		5.30	42.50		47.80	69
0750	#1		11	.727		6.85	46.50		53.35	76.50
0800	1/0		10	.800		6.85	51		57.85	83
0850	2/0		9	.889		11.05	56.50		67.55	96
0900	3/0		7.20	1.111		15.30	71		86.30	122
1000	4/0		6.40	1.250		18.30	79.50		97.80	138
1100	250 kcmil		5.70	1.404		18.30	89.50		107.80	153
1200	300 kcmil		5.30	1.509		34.50	96		130.50	181
1400	350 kcmil		4.60	1.739		34.50	111		145.50	203
1500	500 kcmil		4	2		45	127		172	239
1600	Crimp 1 hole lugs, copper or aluminum, 600 volt									
1620	#14	1 Elec	60	.133	Ea.	.59	8.50		9.09	13.25
1630	#12		50	.160		1.19	10.20		11.39	16.45
1640	#10		45	.178		1.19	11.30		12.49	18.15
1780	#8		36	.222		1.65	14.15		15.80	23
1800	#6		30	.267		1.94	17		18.94	27
2000	#4		27	.296		2.61	18.85		21.46	31
2200	#2		24	.333		3.03	21		24.03	35
2400	#1		20	.400		4.28	25.50		29.78	42.50
2500	1/0		17.50	.457		5.05	29		34.05	49
2600	2/0		15	.533		9.15	34		43.15	60.50
2800	3/0		12	.667		6.45	42.50		48.95	70
3000	4/0		11	.727		7.40	46.50		53.90	77
3200	250 kcmil		9	.889		8	56.50		64.50	93
3400	300 kcmil		8	1		8.90	63.50		72.40	104
3500	350 kcmil		7	1.143		10	73		83	119
3600	400 kcmil		6.50	1.231		12.55	78.50		91.05	130
3800	500 kcmil		6	1.333		14.80	85		99.80	142
4000	600 kcmil		5.80	1.379		29.50	88		117.50	164
4200	700 kcmil		5.50	1.455		39.50	92.50		132	182
4400	750 kcmil		5.20	1.538		39.50	98		137.50	190
4500	Crimp 2-way connectors, copper or alum., 600 volt									
4510	#14	1 Elec	60	.133	Ea.	2.08	8.50		10.58	14.90
4520	#12		50	.160		2.21	10.20		12.41	17.60
4530	#10		45	.178		2.32	11.30		13.62	19.40
4540	#8		27	.296		2.35	18.85		21.20	30.50
4600	#6		25	.320		4.05	20.50		24.55	35
4800	#4		23	.348		4.30	22		26.30	37.50
5000	#2		20	.400		5.25	25.50		30.75	44
5200	#1		16	.500		5.90	32		37.90	54
5400	1/0		13	.615		6.20	39		45.20	65.50
5420	2/0		12	.667		6.65	42.50		49.15	70.50
5440	3/0		11	.727		7.30	46.50		53.80	77
5460	4/0		10	.800		8.85	51		59.85	85
5480	250 kcmil		9	.889		12.45	56.50		68.95	97.50
5500	300 kcmil		8.50	.941		17.70	60		77.70	108
5520	350 kcmil		8	1		16.95	63.50		80.45	113
5540	400 kcmil		7.30	1.096		21	70		91	127

26 05 19.35 Cable Terminations

		Crew	Daily Output	Labor-Hours	Unit	Material	2021 Bare Costs Labor	Equipment	Total	Total Incl O&P
5560	500 kcmil	1 Elec	6.20	1.290	Ea.	27.50	82		109.50	153
5580	600 kcmil		5.50	1.455		40	92.50		132.50	182
5600	700 kcmil		4.50	1.778		54	113		167	228
5620	750 kcmil		4	2		60.50	127		187.50	256
7000	Compression equipment adapter, aluminum wire, #6		30	.267		5.80	17		22.80	31.50
7020	#4		27	.296		5.90	18.85		24.75	34.50
7040	#2		24	.333		10.70	21		31.70	43.50
7060	#1		20	.400		14.05	25.50		39.55	53.50
7080	1/0		18	.444		13.90	28.50		42.40	57.50
7100	2/0		15	.533		13.25	34		47.25	65
7140	4/0		11	.727		15.35	46.50		61.85	86
7160	250 kcmil		9	.889		15.55	56.50		72.05	101
7180	300 kcmil		8	1		17.15	63.50		80.65	113
7200	350 kcmil		7	1.143		20.50	73		93.50	131
7220	400 kcmil		6.50	1.231		23.50	78.50		102	142
7240	500 kcmil		6	1.333		25	85		110	154
7260	600 kcmil		5.80	1.379		30	88		118	165
7280	750 kcmil		5.20	1.538		32.50	98		130.50	182
8000	Compression tool, hand					1,175			1,175	1,275
8100	Hydraulic					2,650			2,650	2,925
8500	Hydraulic dies					360			360	395

26 05 19.50 Mineral Insulated Cable

		Crew	Daily Output	Labor-Hours	Unit	Material	2021 Bare Costs Labor	Equipment	Total	Total Incl O&P
0010	**MINERAL INSULATED CABLE** 600 volt									
0100	1 conductor, #12	1 Elec	1.60	5	C.L.F.	410	320		730	925
0200	#10		1.60	5		615	320		935	1,150
0400	#8		1.50	5.333		690	340		1,030	1,275
0500	#6		1.40	5.714		825	365		1,190	1,450
0600	#4	2 Elec	2.40	6.667		1,100	425		1,525	1,850
0800	#2		2.20	7.273		1,575	465		2,040	2,425
0900	#1		2.10	7.619		1,825	485		2,310	2,725
1000	1/0		2	8		2,125	510		2,635	3,075
1100	2/0		1.90	8.421		2,525	535		3,060	3,600
1200	3/0		1.80	8.889		3,050	565		3,615	4,200
1400	4/0		1.60	10		3,500	635		4,135	4,800
1410	250 kcmil	3 Elec	2.40	10		3,950	635		4,585	5,300
1420	350 kcmil		1.95	12.308		4,650	785		5,435	6,275
1430	500 kcmil		1.95	12.308		5,800	785		6,585	7,575
1500	2 conductor, #12	1 Elec	1.40	5.714		1,000	365		1,365	1,650
1600	#10		1.20	6.667		1,250	425		1,675	2,000
1800	#8		1.10	7.273		1,325	465		1,790	2,175
2000	#6		1.05	7.619		1,975	485		2,460	2,875
2100	#4	2 Elec	2	8		2,650	510		3,160	3,675
2200	3 conductor, #12	1 Elec	1.20	6.667		1,250	425		1,675	2,000
2400	#10		1.10	7.273		1,425	465		1,890	2,275
2600	#8		1.05	7.619		1,800	485		2,285	2,700
2800	#6		1	8		2,375	510		2,885	3,350
3000	#4	2 Elec	1.80	8.889		2,950	565		3,515	4,100
3100	4 conductor, #12	1 Elec	1.20	6.667		1,325	425		1,750	2,075
3200	#10		1.10	7.273		1,600	465		2,065	2,450
3400	#8		1	8		2,100	510		2,610	3,050
3600	#6		.90	8.889		2,725	565		3,290	3,825
3620	7 conductor, #12		1.10	7.273		1,700	465		2,165	2,575

26 05 19.50 Mineral Insulated Cable

		Crew	Daily Output	Labor-Hours	Unit	Material	2021 Bare Costs Labor	Equipment	Total	Total Incl O&P
3640	#10	1 Elec	1	8	C.L.F.	2,250	510		2,760	3,225
3800	Terminations, 600 volt, 1 conductor, #12		8	1	Ea.	21.50	63.50		85	118
4000	#10		7.60	1.053		21.50	67		88.50	123
4100	#8		7.30	1.096		21.50	70		91.50	128
4200	#6		6.70	1.194		21.50	76		97.50	137
4400	#4		6.20	1.290		21.50	82		103.50	146
4600	#2		5.70	1.404		32	89.50		121.50	168
4800	#1		5.30	1.509		32	96		128	178
5000	1/0		5	1.600		32	102		134	186
5100	2/0		4.70	1.702		32	108		140	196
5200	3/0		4.30	1.860		32	119		151	211
5400	4/0		4	2		71	127		198	267
5410	250 kcmil		4	2		71	127		198	267
5420	350 kcmil		4	2		99.50	127		226.50	299
5430	500 kcmil		4	2		99.50	127		226.50	299
5500	2 conductor, #12		6.70	1.194		21.50	76		97.50	137
5600	#10		6.40	1.250		32	79.50		111.50	153
5800	#8		6.20	1.290		32	82		114	157
6000	#6		5.70	1.404		32	89.50		121.50	168
6200	#4		5.30	1.509		71	96		167	221
6400	3 conductor, #12		5.70	1.404		32	89.50		121.50	168
6500	#10		5.50	1.455		32	92.50		124.50	173
6600	#8		5.20	1.538		32	98		130	181
6800	#6		4.80	1.667		32	106		138	193
7200	#4		4.60	1.739		71	111		182	243
7400	4 conductor, #12		4.60	1.739		35.50	111		146.50	204
7500	#10		4.40	1.818		35.50	116		151.50	211
7600	#8		4.20	1.905		35.50	121		156.50	219
8400	#6		4	2		74.50	127		201.50	271
8500	7 conductor, #12		3.50	2.286		35.50	146		181.50	255
8600	#10	▼	3	2.667		77	170		247	335
8800	Crimping tool, plier type					78.50			78.50	86
9000	Stripping tool					296			296	325
9200	Hand vise				▼	88			88	97

26 05 19.55 Non-Metallic Sheathed Cable

		Crew	Daily Output	Labor-Hours	Unit	Material	2021 Bare Costs Labor	Equipment	Total	Total Incl O&P
0010	**NON-METALLIC SHEATHED CABLE** 600 volt									
0100	Copper with ground wire (Romex)									
0150	#14, 2 conductor	1 Elec	2.70	2.963	C.L.F.	17.40	189		206.40	299
0200	3 conductor		2.40	3.333		24.50	212		236.50	340
0220	4 conductor		2.20	3.636		41	232		273	390
0250	#12, 2 conductor		2.50	3.200		23.50	204		227.50	330
0300	3 conductor		2.20	3.636		37	232		269	385
0320	4 conductor		2	4		60.50	255		315.50	445
0350	#10, 2 conductor		2.20	3.636		45.50	232		277.50	395
0400	3 conductor		1.80	4.444		57.50	283		340.50	485
0420	4 conductor		1.60	5		96	320		416	580
0430	#8, 2 conductor		1.60	5		77.50	320		397.50	560
0450	3 conductor		1.50	5.333		110	340		450	625
0500	#6, 3 conductor	▼	1.40	5.714		385	365		750	960
0520	#4, 3 conductor	2 Elec	2.40	6.667		405	425		830	1,075
0540	#2, 3 conductor	"	2.20	7.273	▼	630	465		1,095	1,375
0550	SE type SER aluminum cable, 3 RHW and									

26 05 19.55 Non-Metallic Sheathed Cable		Crew	Daily Output	Labor-Hours	Unit	Material	2021 Bare Costs Labor	Equipment	Total	Total Incl O&P
0600	1 bare neutral, 3 #8 & 1 #8	1 Elec	1.60	5	C.L.F.	71	320		391	555
0650	3 #6 & 1 #6	"	1.40	5.714		74.50	365		439.50	620
0700	3 #4 & 1 #6	2 Elec	2.40	6.667		82	425		507	720
0750	3 #2 & 1 #4		2.20	7.273		128	465		593	830
0800	3 #1/0 & 1 #2		2	8		173	510		683	945
0850	3 #2/0 & 1 #1		1.80	8.889		202	565		767	1,050
0900	3 #4/0 & 1 #2/0		1.60	10		325	635		960	1,300
1000	URD - triplex underground distribution cable, alum. 2 #4 + #4 neutral		2.80	5.714		83.50	365		448.50	630
1010	2 #2 + #4 neutral		2.65	6.038		102	385		487	680
1020	2 #2 + #2 neutral		2.55	6.275		103	400		503	710
1030	2 1/0 + #2 neutral		2.40	6.667		124	425		549	765
1040	2 1/0 + 1/0 neutral		2.30	6.957		140	445		585	815
1050	2 2/0 + #1 neutral		2.20	7.273		169	465		634	875
1060	2 2/0 + 2/0 neutral		2.10	7.619		171	485		656	910
1070	2 3/0 + 1/0 neutral		1.95	8.205		187	525		712	980
1080	2 3/0 + 3/0 neutral		1.95	8.205		203	525		728	1,000
1090	2 4/0 + 2/0 neutral		1.85	8.649		192	550		742	1,025
1100	2 4/0 + 4/0 neutral		1.85	8.649		241	550		791	1,075
1450	UF underground feeder cable, copper with ground, #14, 2 conductor	1 Elec	4	2		19.70	127		146.70	211
1500	#12, 2 conductor		3.50	2.286		34	146		180	254
1550	#10, 2 conductor		3	2.667		53	170		223	310
1600	#14, 3 conductor		3.50	2.286		36.50	146		182.50	256
1650	#12, 3 conductor		3	2.667		47	170		217	305
1700	#10, 3 conductor		2.50	3.200		78.50	204		282.50	390
1710	#8, 3 conductor		2	4		144	255		399	540
1720	#6, 3 conductor		1.80	4.444		212	283		495	655
2400	SEU service entrance cable, copper 2 conductors, #8 + #8 neutral		1.50	5.333		97	340		437	610
2600	#6 + #8 neutral		1.30	6.154		135	390		525	730
2800	#6 + #6 neutral		1.30	6.154		155	390		545	750
3000	#4 + #6 neutral	2 Elec	2.20	7.273		215	465		680	925
3200	#4 + #4 neutral		2.20	7.273		249	465		714	965
3400	#3 + #5 neutral		2.10	7.619		283	485		768	1,025
3600	#3 + #3 neutral		2.10	7.619		335	485		820	1,100
3800	#2 + #4 neutral		2	8		355	510		865	1,150
4000	#1 + #1 neutral		1.90	8.421		395	535		930	1,225
4200	1/0 + 1/0 neutral		1.80	8.889		870	565		1,435	1,800
4400	2/0 + 2/0 neutral		1.70	9.412		840	600		1,440	1,825
4600	3/0 + 3/0 neutral		1.60	10		1,100	635		1,735	2,175
4620	4/0 + 4/0 neutral		1.45	11.034		785	705		1,490	1,925
4800	Aluminum 2 conductors, #8 + #8 neutral	1 Elec	1.60	5		54	320		374	535
5000	#6 + #6 neutral	"	1.40	5.714		58.50	365		423.50	605
5100	#4 + #6 neutral	2 Elec	2.50	6.400		74	410		484	685
5200	#4 + #4 neutral		2.40	6.667		78	425		503	715
5300	#2 + #4 neutral		2.30	6.957		107	445		552	780
5400	#2 + #2 neutral		2.20	7.273		107	465		572	810
5450	1/0 + #2 neutral		2.10	7.619		179	485		664	915
5500	1/0 + 1/0 neutral		2	8		167	510		677	940
5550	2/0 + #1 neutral		1.90	8.421		200	535		735	1,025
5600	2/0 + 2/0 neutral		1.80	8.889		187	565		752	1,050
5800	3/0 + 1/0 neutral		1.70	9.412		288	600		888	1,200
6000	3/0 + 3/0 neutral		1.70	9.412		237	600		837	1,150
6200	4/0 + 2/0 neutral		1.60	10		221	635		856	1,200
6400	4/0 + 4/0 neutral		1.60	10		237	635		872	1,200

133

26 05 19.55 Non-Metallic Sheathed Cable

		Crew	Daily Output	Labor-Hours	Unit	Material	2021 Bare Costs Labor	Equipment	Total	Total Incl O&P
6500	Service entrance cap for copper SEU									
6600	100 amp	1 Elec	12	.667	Ea.	7	42.50		49.50	70.50
6700	150 amp		10	.800		12.95	51		63.95	90
6800	200 amp		8	1		19.65	63.50		83.15	116

26 05 19.70 Portable Cord

		Crew	Daily Output	Labor-Hours	Unit	Material	2021 Bare Costs Labor	Equipment	Total	Total Incl O&P
0010	**PORTABLE CORD** 600 volt									
0100	Type SO, #18, 2 conductor	1 Elec	980	.008	L.F.	.29	.52		.81	1.09
0110	3 conductor		980	.008		.35	.52		.87	1.16
0120	#16, 2 conductor		840	.010		.35	.61		.96	1.29
0130	3 conductor		840	.010		3.48	.61		4.09	4.73
0140	4 conductor		840	.010		.54	.61		1.15	1.49
0240	#14, 2 conductor		840	.010		.58	.61		1.19	1.54
0250	3 conductor		840	.010		.63	.61		1.24	1.59
0260	4 conductor		840	.010		.80	.61		1.41	1.78
0280	#12, 2 conductor		840	.010		.74	.61		1.35	1.71
0290	3 conductor		840	.010		.89	.61		1.50	1.88
0300	4 conductor		840	.010		1.06	.61		1.67	2.07
0320	#10, 2 conductor		765	.010		1.03	.67		1.70	2.12
0330	3 conductor		765	.010		1.21	.67		1.88	2.32
0340	4 conductor		765	.010		1.33	.67		2	2.45
0360	#8, 2 conductor		555	.014		1.74	.92		2.66	3.27
0370	3 conductor		540	.015		1.76	.94		2.70	3.34
0380	4 conductor		525	.015		2.13	.97		3.10	3.78
0400	#6, 2 conductor		525	.015		2.64	.97		3.61	4.34
0410	3 conductor		490	.016		2.52	1.04		3.56	4.32
0420	4 conductor		415	.019		3.16	1.23		4.39	5.30
0440	#4, 2 conductor	2 Elec	830	.019		3.75	1.23		4.98	5.95
0450	3 conductor		700	.023		4.08	1.46		5.54	6.65
0460	4 conductor		660	.024		5.15	1.54		6.69	8
0480	#2, 2 conductor		450	.036		4.96	2.26		7.22	8.80
0490	3 conductor		350	.046		6.70	2.91		9.61	11.70
0500	4 conductor		280	.057		8.20	3.64		11.84	14.40
2000	See 26 27 26.20 for Wiring Devices Elements									

26 05 19.75 Modular Flexible Wiring System

		Crew	Daily Output	Labor-Hours	Unit	Material	2021 Bare Costs Labor	Equipment	Total	Total Incl O&P
0010	**MODULAR FLEXIBLE WIRING SYSTEM**									
0020	Commercial system grid ceiling									
0100	Conversion Module	1 Elec	32	.250	Ea.	8.35	15.95		24.30	32.50
0120	Fixture cable, for fixture to fixture, 3 conductor, 15' long		40	.200		33.50	12.75		46.25	56
0150	Extender cable, 3 conductor, 15' long		48	.167		33.50	10.60		44.10	53
0200	Switch drop, 1 level, 9' long		32	.250		35.50	15.95		51.45	62.50
0220	2 level, 9' long		32	.250		24	15.95		39.95	50
0250	Power tee, 9' long		32	.250		52	15.95		67.95	80.50
1020	Industrial system open ceiling									
1100	Converter, interface between hardwiring and modular wiring	1 Elec	32	.250	Ea.	20.50	15.95		36.45	46
1120	Fixture cable, for fixture to fixture, 3 conductor, 21' long		32	.250		117	15.95		132.95	153
1125	3 conductor, 25' long		24	.333		150	21		171	197
1130	3 conductor, 31' long		19	.421		133	27		160	186
1150	Fixture cord drop, 10' long		40	.200		36	12.75		48.75	58.50

26 05 19.90 Wire

			Crew	Daily Output	Labor-Hours	Unit	Material	2021 Bare Costs Labor	Equipment	Total	Total Incl O&P
0010	**WIRE**, normal installation conditions in wireway, conduit, cable tray										
0020	600 volt, copper type THW, solid, #14	R260519-90	1 Elec	13	.615	C.L.F.	8.30	39		47.30	67.50
0030	#12			11	.727		10.25	46.50		56.75	80.50

26 05 19 – Low-Voltage Electrical Power Conductors and Cables

26 05 19.90 Wire			Crew	Daily Output	Labor-Hours	Unit	Material	2021 Bare Costs Labor	Equipment	Total	Total Incl O&P
0040	#10		1 Elec	10	.800	C.L.F.	16.55	51		67.55	93.50
0050	Stranded, #14			13	.615		7.55	39		46.55	67
0100	#12			11	.727		17.45	46.50		63.95	88
0120	#10	R260519-92		10	.800		28	51		79	107
0140	#8			8	1		30.50	63.50		94	128
0160	#6	R260519-93		6.50	1.231		73	78.50		151.50	197
0180	#4		2 Elec	10.60	1.509		112	96		208	266
0200	#3	R260519-94		10	1.600		152	102		254	320
0220	#2			9	1.778		176	113		289	360
0240	#1	R260533-22		8	2		219	127		346	430
0260	1/0			6.60	2.424		285	154		439	545
0280	2/0			5.80	2.759		495	176		671	805
0300	3/0			5	3.200		350	204		554	690
0350	4/0			4.40	3.636		570	232		802	975
0400	250 kcmil		3 Elec	6	4		620	255		875	1,075
0420	300 kcmil			5.70	4.211		890	268		1,158	1,375
0450	350 kcmil			5.40	4.444		585	283		868	1,075
0480	400 kcmil			5.10	4.706		1,525	300		1,825	2,125
0490	500 kcmil			4.80	5		1,250	320		1,570	1,850
0500	600 kcmil			3.90	6.154		1,625	390		2,015	2,375
0510	750 kcmil			3.30	7.273		1,050	465		1,515	1,875
0520	1,000 kcmil			2.70	8.889		1,900	565		2,465	2,925
0540	600 volt, aluminum type THHN, stranded, #6		1 Elec	8	1		16.25	63.50		79.75	112
0560	#4		2 Elec	13	1.231		20.50	78.50		99	139
0580	#2			10.60	1.509		28	96		124	174
0600	#1			9	1.778		39	113		152	211
0620	1/0			8	2		43.50	127		170.50	237
0640	2/0			7.20	2.222		53	142		195	269
0680	3/0			6.60	2.424		65.50	154		219.50	300
0700	4/0			6.20	2.581		70	164		234	320
0720	250 kcmil		3 Elec	8.70	2.759		84	176		260	355
0740	300 kcmil			8.10	2.963		122	189		311	415
0760	350 kcmil			7.50	3.200		121	204		325	440
0780	400 kcmil			6.90	3.478		146	222		368	490
0800	500 kcmil			6	4		157	255		412	555
0850	600 kcmil			5.70	4.211		209	268		477	630
0880	700 kcmil			5.10	4.706		238	300		538	705
0900	750 kcmil			4.80	5		240	320		560	740
0910	1,000 kcmil			3.78	6.349		370	405		775	1,000
0920	600 volt, copper type THWN-THHN, solid, #14		1 Elec	13	.615		6.90	39		45.90	66
0940	#12			11	.727		9.60	46.50		56.10	79.50
0960	#10			10	.800		15.55	51		66.55	92.50
1000	Stranded, #14			13	.615		7.15	39		46.15	66.50
1200	#12			11	.727		17.95	46.50		64.45	89
1250	#10			10	.800		25	51		76	103
1300	#8			8	1		26	63.50		89.50	123
1350	#6			6.50	1.231		42	78.50		120.50	162
1400	#4		2 Elec	10.60	1.509		65	96		161	215
1450	#3			10	1.600		79.50	102		181.50	239
1500	#2			9	1.778		99.50	113		212.50	277
1550	#1			8	2		122	127		249	325
1600	1/0			6.60	2.424		151	154		305	395
1650	2/0			5.80	2.759		186	176		362	465

135

26 05 19 – Low-Voltage Electrical Power Conductors and Cables

26 05 19.90 Wire		Crew	Daily Output	Labor-Hours	Unit	Material	2021 Bare Costs Labor	Equipment	Total	Total Incl O&P
1700	3/0	2 Elec	5	3.200	C.L.F.	232	204		436	560
2000	4/0		4.40	3.636		290	232		522	665
2200	250 kcmil	3 Elec	6	4		335	255		590	750
2400	300 kcmil		5.70	4.211		405	268		673	845
2600	350 kcmil		5.40	4.444		470	283		753	940
2700	400 kcmil		5.10	4.706		515	300		815	1,000
2800	500 kcmil		4.80	5		675	320		995	1,225
2802	600 kcmil		3.90	6.154		645	390		1,035	1,300
2804	750 kcmil		3.30	7.273		1,550	465		2,015	2,400
2805	1,000 kcmil		1.94	12.371		2,300	790		3,090	3,700
2900	600 volt, copper type XHHW, solid, #14	1 Elec	13	.615		23.50	39		62.50	84.50
2920	#12		11	.727		37.50	46.50		84	111
2940	#10		10	.800		58	51		109	140
3000	Stranded, #14		13	.615		9.80	39		48.80	69.50
3020	#12		11	.727		13.55	46.50		60.05	84
3040	#10		10	.800		20.50	51		71.50	98
3060	#8		8	1		34.50	63.50		98	133
3080	#6		6.50	1.231		54.50	78.50		133	176
3100	#4	2 Elec	10.60	1.509		71	96		167	222
3110	#3		10	1.600		90	102		192	250
3120	#2		9	1.778		113	113		226	292
3140	#1		8	2		145	127		272	350
3160	1/0		6.60	2.424		179	154		333	425
3180	2/0		5.80	2.759		218	176		394	500
3200	3/0		5	3.200		274	204		478	605
3220	4/0		4.40	3.636		345	232		577	725
3240	250 kcmil	3 Elec	6	4		390	255		645	810
3260	300 kcmil		5.70	4.211		455	268		723	900
3280	350 kcmil		5.40	4.444		530	283		813	1,000
3300	400 kcmil		5.10	4.706		665	300		965	1,175
3320	500 kcmil		4.80	5		755	320		1,075	1,300
3340	600 kcmil		3.90	6.154		940	390		1,330	1,600
3360	750 kcmil		3.30	7.273		1,600	465		2,065	2,450
3380	1,000 kcmil		2.40	10		2,300	635		2,935	3,500
5020	600 volt, aluminum type XHHW, stranded, #6	1 Elec	8	1		15.55	63.50		79.05	112
5040	#4	2 Elec	13	1.231		18.85	78.50		97.35	137
5060	#2		10.60	1.509		25.50	96		121.50	171
5080	#1		9	1.778		35.50	113		148.50	207
5100	1/0		8	2		40.50	127		167.50	234
5120	2/0		7.20	2.222		49	142		191	264
5140	3/0		6.60	2.424		57.50	154		211.50	293
5160	4/0		6.20	2.581		65	164		229	315
5180	250 kcmil	3 Elec	8.70	2.759		83	176		259	350
5200	300 kcmil		8.10	2.963		110	189		299	400
5220	350 kcmil		7.50	3.200		110	204		314	425
5240	400 kcmil		6.90	3.478		131	222		353	475
5260	500 kcmil		6	4		139	255		394	535
5280	600 kcmil		5.70	4.211		180	268		448	600
5300	700 kcmil		5.40	4.444		227	283		510	670
5320	750 kcmil		5.10	4.706		206	300		506	670
5340	1,000 kcmil		3.60	6.667		350	425		775	1,025
5390	600 volt, copper type XLPE-USE (RHW), solid, #14	1 Elec	12	.667		19.55	42.50		62.05	84.50
5400	#12		11	.727		40.50	46.50		87	114

26 05 19 – Low-Voltage Electrical Power Conductors and Cables

26 05 19.90 Wire		Crew	Daily Output	Labor-Hours	Unit	Material	2021 Bare Costs Labor	Equipment	Total	Total Incl O&P
5420	#10	1 Elec	10	.800	C.L.F.	28.50	51		79.50	107
5440	Stranded, #14		13	.615		21	39		60	81.50
5460	#12		11	.727		23	46.50		69.50	94
5480	#10		10	.800		27	51		78	106
5500	#8		8	1		42	63.50		105.50	141
5520	#6		6.50	1.231		59	78.50		137.50	181
5540	#4	2 Elec	10.60	1.509		93	96		189	245
5560	#2		9	1.778		167	113		280	350
5580	#1		8	2		218	127		345	430
5600	1/0		6.60	2.424		278	154		432	535
5620	2/0		5.80	2.759		292	176		468	580
5640	3/0		5	3.200		360	204		564	705
5660	4/0		4.40	3.636		510	232		742	905
5680	250 kcmil	3 Elec	6	4		480	255		735	905
5700	300 kcmil		5.70	4.211		515	268		783	965
5720	350 kcmil		5.40	4.444		615	283		898	1,100
5740	400 kcmil		5.10	4.706		695	300		995	1,200
5760	500 kcmil		4.80	5		1,025	320		1,345	1,600
5780	600 kcmil		3.90	6.154		1,050	390		1,440	1,725
5800	750 kcmil		3.30	7.273		1,775	465		2,240	2,650
5820	1,000 kcmil		2.70	8.889		2,450	565		3,015	3,550
5840	600 volt, aluminum type XLPE-USE (RHW), stranded, #6	1 Elec	8	1		19.85	63.50		83.35	117
5860	#4	2 Elec	13	1.231		22	78.50		100.50	141
5880	#2		10.60	1.509		31.50	96		127.50	178
5900	#1		9	1.778		45	113		158	218
5920	1/0		8	2		52.50	127		179.50	247
5940	2/0		7.20	2.222		57.50	142		199.50	274
5960	3/0		6.60	2.424		74.50	154		228.50	310
5980	4/0		6.20	2.581		77.50	164		241.50	330
6000	250 kcmil	3 Elec	8.70	2.759		110	176		286	380
6020	300 kcmil		8.10	2.963		145	189		334	440
6040	350 kcmil		7.50	3.200		149	204		353	470
6060	400 kcmil		6.90	3.478		405	222		627	775
6080	500 kcmil		6	4		202	255		457	600
6100	600 kcmil		5.70	4.211		263	268		531	690
6110	700 kcmil		5.40	4.444		278	283		561	725
6120	750 kcmil		5.10	4.706		300	300		600	775

26 05 23 – Control-Voltage Electrical Power Cables

26 05 23.10 Control Cable		Crew	Daily Output	Labor-Hours	Unit	Material	2021 Bare Costs Labor	Equipment	Total	Total Incl O&P
0010	**CONTROL CABLE**									
0020	600 volt, copper, #14 THWN wire with PVC jacket, 2 wires	1 Elec	9	.889	C.L.F.	48	56.50		104.50	137
0030	3 wires		8	1		57	63.50		120.50	157
0100	4 wires		7	1.143		139	73		212	261
0150	5 wires		6.50	1.231		186	78.50		264.50	320
0200	6 wires		6	1.333		85.50	85		170.50	220
0300	8 wires		5.30	1.509		211	96		307	375
0400	10 wires		4.80	1.667		172	106		278	345
0500	12 wires		4.30	1.860		183	119		302	380
0600	14 wires		3.80	2.105		233	134		367	455
0700	16 wires		3.50	2.286		292	146		438	535
0800	18 wires		3.30	2.424		325	154		479	590
0810	19 wires		3.10	2.581		235	164		399	505

137

26 05 23.10 Control Cable		Crew	Daily Output	Labor-Hours	Unit	Material	2021 Bare Costs Labor	2021 Bare Costs Equipment	Total	Total Incl O&P
0900	20 wires	1 Elec	3	2.667	C.L.F.	250	170		420	525
1000	22 wires	↓	2.80	2.857	↓	370	182		552	675

26 05 23.20 Special Wires and Fittings

		Crew	Daily Output	Labor-Hours	Unit	Material	2021 Bare Costs Labor	2021 Bare Costs Equipment	Total	Total Incl O&P
0010	**SPECIAL WIRES & FITTINGS**									
0100	Fixture TFFN 600 volt 90°C stranded, #18	1 Elec	13	.615	C.L.F.	10.90	39		49.90	70.50
0150	#16		13	.615		25.50	39		64.50	86.50
0500	Thermostat, jacket non-plenum, twisted, #18-2 conductor		8	1		13.10	63.50		76.60	109
0550	#18-3 conductor		7	1.143		14.25	73		87.25	124
0600	#18-4 conductor		6.50	1.231		24.50	78.50		103	143
0650	#18-5 conductor		6	1.333		29	85		114	158
0700	#18-6 conductor		5.50	1.455		39	92.50		131.50	181
0750	#18-7 conductor		5	1.600		38	102		140	193
0800	#18-8 conductor		4.80	1.667		39.50	106		145.50	202
2460	Tray cable, type TC, copper, #16-2 conductor		9.40	.851		32	54		86	116
2464	#16-3 conductor		8.40	.952		56.50	60.50		117	152
2468	#16-4 conductor		7.30	1.096		55.50	70		125.50	165
2472	#16-5 conductor		6.70	1.194		42	76		118	160
2476	#16-7 conductor		5.50	1.455		58	92.50		150.50	202
2480	#16-9 conductor		5	1.600		72.50	102		174.50	231
2484	#16-12 conductor	2 Elec	8.80	1.818		131	116		247	315
2488	#16-15 conductor		7.20	2.222		126	142		268	350
2492	#16-19 conductor		6.60	2.424		156	154		310	400
2496	#16-25 conductor	↓	6.40	2.500		219	159		378	480
2500	#14-2 conductor	1 Elec	9	.889		19.35	56.50		75.85	106
2520	#14-3 conductor		8	1		31	63.50		94.50	129
2540	#14-4 conductor		7	1.143		42.50	73		115.50	155
2560	#14-5 conductor		6.50	1.231		47	78.50		125.50	168
2564	#14-7 conductor	↓	5.30	1.509		91.50	96		187.50	244
2568	#14-9 conductor	2 Elec	9.60	1.667		91	106		197	258
2572	#14-12 conductor		8.60	1.860		125	119		244	315
2576	#14-15 conductor		7.20	2.222		173	142		315	400
2578	#14-19 conductor		6.20	2.581		218	164		382	485
2582	#14-25 conductor	↓	5.30	3.019		288	192		480	600
2590	#12-2 conductor	1 Elec	8.40	.952		53.50	60.50		114	149
2592	#12-3 conductor		7.60	1.053		42.50	67		109.50	146
2594	#12-4 conductor		6.60	1.212		105	77		182	230
2596	#12-5 conductor	↓	6.20	1.290		119	82		201	252
2598	#12-7 conductor	2 Elec	10.40	1.538		107	98		205	264
2602	#12-9 conductor		8.80	1.818		219	116		335	415
2604	#12-12 conductor		7.80	2.051		190	131		321	405
2606	#12-15 conductor		6.80	2.353		260	150		410	510
2608	#12-19 conductor	↓	6	2.667		320	170		490	600
2610	#12-25 conductor	3 Elec	7.70	3.117		435	199		634	770
2618	#10-2 conductor	1 Elec	8	1		135	63.50		198.50	243
2622	#10-3 conductor	"	7.30	1.096		64.50	70		134.50	175
2624	#10-4 conductor	2 Elec	12.80	1.250		170	79.50		249.50	305
2626	#10-5 conductor		11.80	1.356		103	86.50		189.50	242
2628	#10-7 conductor		9.40	1.702		195	108		303	375
2630	#10-9 conductor		8.40	1.905		189	121		310	390
2632	#10-12 conductor	↓	7.20	2.222	↓	212	142		354	445
2640	300 V, copper braided shield, PVC jacket									
2650	2 conductor #18 stranded	1 Elec	7	1.143	C.L.F.	14.20	73		87.20	124

For customer support on your Electrical Costs with RSMeans Data, call 800.448.8182.

26 05 23 – Control-Voltage Electrical Power Cables

26 05 23.20 Special Wires and Fittings

		Crew	Daily Output	Labor-Hours	Unit	Material	2021 Bare Costs Labor	Equipment	Total	Total Incl O&P
2660	3 conductor #18 stranded	1 Elec	6	1.333	C.L.F.	15.60	85		100.60	143
2670	4 conductor #18 stranded	↓	6	1.333	↓	160	85		245	300
3000	Strain relief grip for cable									
3050	Cord, top, #12-3	1 Elec	40	.200	Ea.	11.40	12.75		24.15	31.50
3060	#12-4		40	.200		11.40	12.75		24.15	31.50
3070	#12-5		39	.205		13.10	13.05		26.15	34
3100	#10-3		39	.205		13.10	13.05		26.15	34
3110	#10-4		38	.211		13.10	13.40		26.50	34.50
3120	#10-5		38	.211		16	13.40		29.40	37.50
3200	Bottom, #12-3		40	.200		15.35	12.75		28.10	36
3210	#12-4		40	.200		15.35	12.75		28.10	36
3220	#12-5		39	.205		16.95	13.05		30	38
3230	#10-3		39	.205		16.95	13.05		30	38
3300	#10-4		38	.211		16.95	13.40		30.35	38.50
3310	#10-5		38	.211		16.55	13.40		29.95	38
3400	Cable ties, standard, 4" length		190	.042		.15	2.68		2.83	4.16
3410	7" length		160	.050		.18	3.19		3.37	4.93
3420	14.5" length		90	.089		.42	5.65		6.07	8.85
3430	Heavy, 14.5" length		80	.100		.58	6.35		6.93	10.10
3500	Cable gland, nylon, 1/4" NPT thread		60	.133		2.38	8.50		10.88	15.20
3510	Cable gland, nylon, 3/8" NPT thread		60	.133		2.73	8.50		11.23	15.60
3520	Cable gland, nylon, 1/2" NPT thread		60	.133		3.41	8.50		11.91	16.35
3530	Cable gland, nylon, 3/4" NPT thread		60	.133		5.10	8.50		13.60	18.20
3540	Cable gland, nylon, 1" NPT thread		60	.133		7.40	8.50		15.90	21
3550	Cable gland, nylon, 1-1/4" NPT thread	↓	60	.133	↓	9.50	8.50		18	23
3970	U-cable guards									
3980	.75" x 5' U-cable guard, 14 gauge	1 Elec	24	.333	Ea.	24	21		45	57.50
3990	.75" x 8' U-cable guard, 14 gauge		24	.333		25	21		46	59
4000	1-1/8" x 5' U-cable guard		24	.333		18.35	21		39.35	51.50
4010	1-1/8" x 8' U-cable guard		24	.333		30	21		51	64.50
4020	2-3/16" x 5' U-cable guard		24	.333		35.50	21		56.50	70.50
4030	2-3/16" x 8' U-cable guard		24	.333		48.50	21		69.50	85
4040	3-3/16" x 5' U-cable guard		24	.333		44	21		65	80
4050	3-3/16" x 8' U-cable guard		24	.333		68	21		89	107
4060	4" x 8' U-cable guard		24	.333		42.50	21		63.50	78.50
4070	2" x 9' flanged-cable guard		20	.400		100	25.50		125.50	148
4080	3" x 9' flanged-cable guard		20	.400		128	25.50		153.50	179
4090	3.5" x 9' flanged-cable guard		20	.400		155	25.50		180.50	208
5000	2" x 5' flanged-cable guard, extension		20	.400		98.50	25.50		124	146
5010	3" x 5' flanged-cable guard, extension		20	.400		86	25.50		111.50	133
5020	.5" x 8' U-cable guard, black plastic		26	.308		5.65	19.60		25.25	35.50
5030	.75" x 8' U-cable guard, black plastic		26	.308		7.55	19.60		27.15	37.50
5040	1" x 8' U-cable guard, black plastic		26	.308		12.90	19.60		32.50	43

26 05 26 – Grounding and Bonding for Electrical Systems

26 05 26.80 Grounding

			Crew	Daily Output	Labor-Hours	Unit	Material	2021 Bare Costs Labor	Equipment	Total	Total Incl O&P
0010	**GROUNDING**	R260526-80									
0030	Rod, copper clad, 8' long, 1/2" diameter		1 Elec	5.50	1.455	Ea.	17	92.50		109.50	157
0040	5/8" diameter			5.50	1.455		15	92.50		107.50	155
0050	3/4" diameter			5.30	1.509		28.50	96		124.50	175
0080	10' long, 1/2" diameter			4.80	1.667		21	106		127	181
0090	5/8" diameter			4.60	1.739		20	111		131	187
0100	3/4" diameter		↓	4.40	1.818	↓	52	116		168	229

For customer support on your Electrical Costs with RSMeans Data, call 800.448.8182.

139

26 05 26.80 Grounding		Crew	Daily Output	Labor-Hours	Unit	Material	2021 Bare Costs Labor	Equipment	Total	Total Incl O&P
0130	15' long, 3/4" diameter	1 Elec	4	2	Ea.	63	127		190	259
0150	Coupling, bronze, 1/2" diameter					9.30			9.30	10.25
0160	5/8" diameter					6.85			6.85	7.55
0170	3/4" diameter					19.20			19.20	21
0190	Drive studs, 1/2" diameter					16.60			16.60	18.30
0210	5/8" diameter					26			26	28.50
0220	3/4" diameter					29			29	32
0230	Clamp, bronze, 1/2" diameter	1 Elec	32	.250		5.30	15.95		21.25	29.50
0240	5/8" diameter		32	.250		2.46	15.95		18.41	26
0250	3/4" diameter		32	.250		3.52	15.95		19.47	27.50
0260	Wire ground bare armored, #8-1 conductor		2	4	C.L.F.	77.50	255		332.50	465
0270	#6-1 conductor		1.80	4.444		150	283		433	585
0280	#4-1 conductor		1.60	5		148	320		468	640
0320	Bare copper wire, #14 solid		14	.571		6.90	36.50		43.40	61.50
0330	#12		13	.615		12.10	39		51.10	72
0340	#10		12	.667		16.55	42.50		59.05	81
0350	#8		11	.727		25	46.50		71.50	96.50
0360	#6		10	.800		39.50	51		90.50	119
0370	#4		8	1		80	63.50		143.50	183
0380	#2		5	1.600		126	102		228	290
0390	Bare copper wire, stranded, #8		11	.727		27.50	46.50		74	99.50
0400	#6		10	.800		43.50	51		94.50	124
0450	#4	2 Elec	16	1		64	63.50		127.50	165
0600	#2		10	1.600		112	102		214	274
0650	#1		9	1.778		122	113		235	305
0700	1/0		8	2		145	127		272	350
0750	2/0		7.20	2.222		192	142		334	420
0800	3/0		6.60	2.424		275	154		429	535
1000	4/0		5.70	2.807		305	179		484	600
1200	250 kcmil	3 Elec	7.20	3.333		375	212		587	730
1210	300 kcmil		6.60	3.636		760	232		992	1,175
1220	350 kcmil		6	4		540	255		795	975
1230	400 kcmil		5.70	4.211		715	268		983	1,200
1240	500 kcmil		5.10	4.706		710	300		1,010	1,225
1260	750 kcmil		3.60	6.667		1,250	425		1,675	2,000
1270	1,000 kcmil		3	8		1,200	510		1,710	2,050
1360	Bare aluminum, stranded, #6	1 Elec	9	.889		13.70	56.50		70.20	99
1370	#4	2 Elec	16	1		33.50	63.50		97	131
1380	#2		13	1.231		55.50	78.50		134	177
1390	#1		10.60	1.509		40	96		136	187
1400	1/0		9	1.778		80.50	113		193.50	257
1410	2/0		8	2		95.50	127		222.50	294
1420	3/0		7.20	2.222		136	142		278	360
1430	4/0		6.60	2.424		134	154		288	375
1440	250 kcmil	3 Elec	9.30	2.581		64	164		228	315
1450	300 kcmil		8.70	2.759		106	176		282	380
1460	400 kcmil		7.50	3.200		92	204		296	405
1470	500 kcmil		6.90	3.478		145	222		367	490
1480	600 kcmil		6	4		182	255		437	580
1490	700 kcmil		5.70	4.211		201	268		469	620
1500	750 Kcmil		5.10	4.706		223	300		523	690
1510	1,000 kcmil		4.80	5		227	320		547	725
1800	Water pipe ground clamps, heavy duty									

26 05 26 – Grounding and Bonding for Electrical Systems

26 05 26.80 Grounding	Crew	Daily Output	Labor-Hours	Unit	Material	2021 Bare Costs Labor	Equipment	Total	Total Incl O&P
2000	Bronze, 1/2" to 1" diameter	1 Elec	8	1	Ea.	24	63.50		87.50	121
2100	1-1/4" to 2" diameter		8	1		44.50	63.50		108	144
2200	2-1/2" to 3" diameter		6	1.333		66	85		151	199
2400	Grounding, exothermic welding									
2500	Exothermic welding reusable mold, cable to cable, parallel, vertical	1 Elec	8	1	Ea.	425	63.50		488.50	565
2505	Splice single		8	1		360	63.50		423.50	495
2510	Splice single to double		8	1		425	63.50		488.50	565
2520	Cable to cable, termination, Tee		8	1		420	63.50		483.50	555
2530	Cable to rod, termination, Tee		8	1		330	63.50		393.50	455
2535	Cable to rebar over, termination, Tee		8	1		97.50	63.50		161	202
2540	Cable to rod, termination, 90 Deg		8	1		355	63.50		418.50	485
2550	Cable to vertical flat steel		8	1		420	63.50		483.50	560
2555	Cable to 45 Deg		8	1		430	63.50		493.50	565
2730	Exothermic weld, 4/0 wire to 1" ground rod		7	1.143		2.74	73		75.74	111
2740	4/0 wire to building steel		7	1.143		2.74	73		75.74	111
2750	4/0 wire to motor frame		7	1.143		2.74	73		75.74	111
2760	4/0 wire to 4/0 wire		7	1.143		2.74	73		75.74	111
2770	4/0 wire to #4 wire		7	1.143		2.74	73		75.74	111
2780	4/0 wire to #8 wire		7	1.143		2.74	73		75.74	111
2790	Mold, reusable, for above					142			142	156
2800	Brazed connections, #6 wire	1 Elec	12	.667		16	42.50		58.50	80.50
3000	#2 wire		10	.800		21.50	51		72.50	99
3100	3/0 wire		8	1		32	63.50		95.50	130
3200	4/0 wire		7	1.143		36.50	73		109.50	149
3400	250 kcmil wire		5	1.600		42.50	102		144.50	198
3600	500 kcmil wire		4	2		52.50	127		179.50	247
3700	Insulated ground wire, copper #14		13	.615	C.L.F.	7.15	39		46.15	66.50
3710	#12		11	.727		17.95	46.50		64.45	89
3720	#10		10	.800		25	51		76	103
3730	#8		8	1		26	63.50		89.50	123
3740	#6		6.50	1.231		42	78.50		120.50	162
3750	#4	2 Elec	10.60	1.509		65	96		161	215
3770	#2		9	1.778		99.50	113		212.50	277
3780	#1		8	2		122	127		249	325
3790	1/0		6.60	2.424		151	154		305	395
3800	2/0		5.80	2.759		186	176		362	465
3810	3/0		5	3.200		232	204		436	560
3820	4/0		4.40	3.636		290	232		522	665
3830	250 kcmil	3 Elec	6	4		335	255		590	750
3840	300 kcmil		5.70	4.211		405	268		673	845
3850	350 kcmil		5.40	4.444		470	283		753	940
3860	400 kcmil		5.10	4.706		515	300		815	1,000
3870	500 kcmil		4.80	5		675	320		995	1,225
3880	600 kcmil		3.90	6.154		1,625	390		2,015	2,375
3890	750 kcmil		3.30	7.273		1,050	465		1,515	1,875
3900	1,000 kcmil		2.70	8.889		1,900	565		2,465	2,925
3960	Insulated ground wire, aluminum, #6	1 Elec	8	1		16.25	63.50		79.75	112
3970	#4	2 Elec	13	1.231		20.50	78.50		99	139
3980	#2		10.60	1.509		28	96		124	174
3990	#1		9	1.778		39	113		152	211
4000	1/0		8	2		43.50	127		170.50	237
4010	2/0		7.20	2.222		53	142		195	269
4020	3/0		6.60	2.424		65.50	154		219.50	300

26 05 26 – Grounding and Bonding for Electrical Systems

26 05 26.80 Grounding

		Crew	Daily Output	Labor-Hours	Unit	Material	2021 Bare Costs Labor	2021 Bare Costs Equipment	Total	Total Incl O&P
4030	4/0	2 Elec	6.20	2.581	C.L.F.	70	164		234	320
4040	250 kcmil	3 Elec	8.70	2.759		84	176		260	355
4050	300 kcmil		8.10	2.963		122	189		311	415
4060	350 kcmil		7.50	3.200		121	204		325	440
4070	400 kcmil		6.90	3.478		146	222		368	490
4080	500 kcmil		6	4		157	255		412	555
4090	600 kcmil		5.70	4.211		209	268		477	630
4100	700 kcmil		5.10	4.706		238	300		538	705
4110	750 kcmil	↓	4.80	5	↓	240	320		560	740
5000	Copper electrolytic ground rod system									
5010	Includes augering hole, mixing bentonite clay,									
5020	Installing rod, and terminating ground wire									
5100	Straight vertical type, 2" diam.									
5120	8.5' long, clamp connection	1 Elec	2.67	2.996	Ea.	755	191		946	1,125
5130	With exothermic weld connection		1.95	4.103		725	261		986	1,200
5140	10' long		2.35	3.404		740	217		957	1,125
5150	With exothermic weld connection		1.78	4.494		925	286		1,211	1,450
5160	12' long		2.16	3.704		1,000	236		1,236	1,450
5170	With exothermic weld connection		1.67	4.790		1,100	305		1,405	1,650
5180	20' long		1.74	4.598		1,450	293		1,743	2,025
5190	With exothermic weld connection	↓	1.40	5.714		1,375	365		1,740	2,050
5195	40' long with exothermic weld connection	2 Elec	2	8	↓	3,025	510		3,535	4,100
5200	L-shaped, 2" diam.									
5220	4' vert. x 10' horiz., clamp connection	1 Elec	5.33	1.501	Ea.	1,300	95.50		1,395.50	1,575
5230	With exothermic weld connection	"	3.08	2.597	"	1,300	165		1,465	1,700
5300	Protective box at grade level, with breather slots									
5320	Round 12" long, fiberlyte	1 Elec	32	.250	Ea.	150	15.95		165.95	189
5330	Concrete	"	16	.500		113	32		145	172
5400	Bentonite clay, 50# bag, 1 per 10' of rod					52			52	57
5500	Equipotential earthing bar	1 Elec	2	4		136	255		391	530
7000	Exothermic welding kit, multi vertical	"	8	1		715	63.50		778.50	880

26 05 29 – Hangers and Supports for Electrical Systems

26 05 29.20 Hangers

		Crew	Daily Output	Labor-Hours	Unit	Material	2021 Bare Costs Labor	2021 Bare Costs Equipment	Total	Total Incl O&P
0010	**HANGERS** R260533-21									
0015	See section 22 05 29.10 for additional items									
0030	Conduit supports									
0050	Strap w/2 holes, rigid steel conduit									
0100	1/2" diameter	1 Elec	470	.017	Ea.	.09	1.08		1.17	1.71
0150	3/4" diameter		440	.018		.12	1.16		1.28	1.85
0200	1" diameter		400	.020		.19	1.27		1.46	2.10
0300	1-1/4" diameter		355	.023		.28	1.44		1.72	2.44
0350	1-1/2" diameter		320	.025		.33	1.59		1.92	2.73
0400	2" diameter		266	.030		.43	1.92		2.35	3.32
0500	2-1/2" diameter		160	.050		.69	3.19		3.88	5.50
0550	3" diameter		133	.060		.93	3.83		4.76	6.70
0600	3-1/2" diameter		100	.080		2.15	5.10		7.25	9.90
0650	4" diameter		80	.100		1.30	6.35		7.65	10.90
0700	EMT, 1/2" diameter		470	.017		.17	1.08		1.25	1.80
0800	3/4" diameter		440	.018		.16	1.16		1.32	1.90
0850	1" diameter		400	.020		.28	1.27		1.55	2.20
0900	1-1/4" diameter		355	.023		.35	1.44		1.79	2.52
0950	1-1/2" diameter		320	.025		.43	1.59		2.02	2.84

26 05 29 – Hangers and Supports for Electrical Systems

26 05 29.20 Hangers	Crew	Daily Output	Labor-Hours	Unit	Material	2021 Bare Costs Labor	Equipment	Total	Total Incl O&P
1000	2" diameter	1 Elec	266	.030	Ea.	.50	1.92		2.42	3.40
1100	2-1/2" diameter		160	.050		1.41	3.19		4.60	6.30
1150	3" diameter		133	.060		1.84	3.83		5.67	7.70
1200	3-1/2" diameter		100	.080		2.11	5.10		7.21	9.85
1250	4" diameter		80	.100		2.67	6.35		9.02	12.40
1400	Hanger, with bolt, 1/2" diameter		200	.040		.55	2.55		3.10	4.40
1450	3/4" diameter		190	.042		.43	2.68		3.11	4.46
1500	1" diameter		176	.045		.53	2.90		3.43	4.88
1550	1-1/4" diameter		160	.050		.63	3.19		3.82	5.40
1600	1-1/2" diameter		140	.057		.90	3.64		4.54	6.40
1650	2" diameter		130	.062		1.17	3.92		5.09	7.10
1700	2-1/2" diameter		100	.080		1.49	5.10		6.59	9.20
1750	3" diameter		64	.125		1.75	7.95		9.70	13.80
1800	3-1/2" diameter		50	.160		1.99	10.20		12.19	17.35
1850	4" diameter		40	.200		3.67	12.75		16.42	23
1900	Riser clamps, conduit, 1/2" diameter		40	.200		5.70	12.75		18.45	25.50
1950	3/4" diameter		36	.222		6	14.15		20.15	27.50
2000	1" diameter		30	.267		6.05	17		23.05	31.50
2100	1-1/4" diameter		27	.296		7.55	18.85		26.40	36.50
2150	1-1/2" diameter		27	.296		7.25	18.85		26.10	36
2200	2" diameter		20	.400		7.70	25.50		33.20	46.50
2250	2-1/2" diameter		20	.400		8.75	25.50		34.25	47.50
2300	3" diameter		18	.444		8.65	28.50		37.15	51.50
2350	3-1/2" diameter		18	.444		11	28.50		39.50	54
2400	4" diameter		14	.571		12.45	36.50		48.95	67.50
2500	Threaded rod, painted, 1/4" diameter		260	.031	L.F.	1.59	1.96		3.55	4.66
2600	3/8" diameter		200	.040		7.30	2.55		9.85	11.85
2700	1/2" diameter		140	.057		8.30	3.64		11.94	14.55
2800	5/8" diameter		100	.080		10	5.10		15.10	18.55
2900	3/4" diameter		60	.133		6.10	8.50		14.60	19.30
2940	Couplings painted, 1/4" diameter				C	475			475	520
2960	3/8" diameter					985			985	1,075
2970	1/2" diameter					186			186	204
2980	5/8" diameter					1,000			1,000	1,100
2990	3/4" diameter					615			615	680
3000	Nuts, galvanized, 1/4" diameter					10.35			10.35	11.35
3050	3/8" diameter					21.50			21.50	23.50
3100	1/2" diameter					42			42	46
3150	5/8" diameter					67.50			67.50	74
3200	3/4" diameter					121			121	134
3250	Washers, galvanized, 1/4" diameter					11.20			11.20	12.35
3300	3/8" diameter					14.75			14.75	16.25
3350	1/2" diameter					24			24	26.50
3400	5/8" diameter					61			61	67
3450	3/4" diameter					117			117	129
3500	Lock washers, galvanized, 1/4" diameter					9.60			9.60	10.55
3550	3/8" diameter					13.45			13.45	14.80
3600	1/2" diameter					17.80			17.80	19.55
3650	5/8" diameter					40			40	44
3700	3/4" diameter					73.50			73.50	80.50
3705	Metal channel slotted, 304 stainless steel or aluminum									
3710	304 stainless steel, 1' long, 12GA, 1-5/8" x 1-5/8"		50	.160	Ea.	19.25	10.20		29.45	36

For customer support on your Electrical Costs with RSMeans Data, call 800.448.8182.

143

26 05 29 – Hangers and Supports for Electrical Systems

26 05 29.20 Hangers		Crew	Daily Output	Labor-Hours	Unit	Material	2021 Bare Costs Labor	Equipment	Total	Total Incl O&P
3712	Per L.F.	1 Elec	50	.160	L.F.	19.25	10.20		29.45	36
3715	304 stainless steel, 1'6" long, 12GA, 1-5/8" x 1-5/8"		33.33	.240	Ea.	26	15.30		41.30	51
3717	Per L.F.		50	.160		17.15	10.20		27.35	34
3720	304 stainless steel, 2' long, 12GA, 1-5/8" x 1-5/8"		25	.320	↓	38.50	20.50		59	73
3722	Per L.F.		50	.160	L.F.	19.25	10.20		29.45	36
3725	304 stainless steel, 3' long, 12GA, 1-5/8" x 1-5/8"		16.67	.480	Ea.	51.50	30.50		82	102
3730	304 stainless steel, 4' long, 12GA, 1-5/8" x 1-5/8"		12.50	.640		61	41		102	128
3735	304 stainless steel, 5' long, 12GA, 1-5/8" x 1-5/8"		10	.800		92.50	51		143.50	178
3740	304 stainless steel, 10' long, 12GA, 1-5/8" x 1-5/8"		5	1.600	↓	147	102		249	310
3742	Aluminum, 12GA, 1-5/8" x 1-5/8"		50	.160	L.F.	5.55	10.20		15.75	21.50
3745	Aluminum, 1' length, 12GA, 1-5/8" x 1-5/8"		50	.160	Ea.	5.55	10.20		15.75	21.50
3750	Aluminum, 1'6" length, 12GA, 1-5/8" x 1-5/8"		33.33	.240		8.35	15.30		23.65	31.50
3755	Aluminum, 2' length, 12GA, 1-5/8" x 1-5/8"		25	.320		11.15	20.50		31.65	43
3760	Aluminum, 3' length, 12GA, 1-5/8" x 1-5/8"		16.67	.480		16.70	30.50		47.20	64
3765	Aluminum, 4' length, 12GA, 1-5/8" x 1-5/8"		12.50	.640		22.50	41		63.50	85
3770	Aluminum, 5' length, 12GA, 1-5/8" x 1-5/8"		10	.800		26.50	51		77.50	105
3775	Aluminum, 10' length, 12GA, 1-5/8" x 1-5/8"		5	1.600	↓	47.50	102		149.50	203
3800	Channels, steel, 3/4" x 1-1/2", 14 ga.		80	.100	L.F.	3.74	6.35		10.09	13.55
3900	1-1/2" x 1-1/2", 12 ga.		70	.114		4.25	7.30		11.55	15.50
4000	1-7/8" x 1-1/2"		60	.133		24.50	8.50		33	39.50
4100	3" x 1-1/2"		50	.160		23.50	10.20		33.70	41
4110	1-5/8" x 13/16", 16 ga.		80	.100	↓	4.96	6.35		11.31	14.90
4114	Trapeze channel support, 12" wide, steel, 12 ga.		8.80	.909	Ea.	36.50	58		94.50	126
4116	18" wide, steel		8.30	.964	"	38.50	61.50		100	134
4120	1-5/8" x 1-5/8", 14 ga.		70	.114	L.F.	6.60	7.30		13.90	18.05
4130	1-5/8" x 7/8", 12 ga.		70	.114		5.75	7.30		13.05	17.15
4140	1-5/8" x 1-3/8", 12 ga.		60	.133		7.25	8.50		15.75	20.50
4150	1-5/8" x 1-5/8", 12 ga.		50	.160	↓	6.45	10.20		16.65	22.50
4160	Flat plate fitting, 2 hole, 3-1/2", material only				Ea.	2.65			2.65	2.92
4170	3 hole, 1-7/16" x 4-1/8"					3.55			3.55	3.91
4200	Spring nuts, long, 1/4"	1 Elec	120	.067		.50	4.25		4.75	6.85
4250	3/8"		100	.080		.64	5.10		5.74	8.25
4300	1/2"		80	.100		.72	6.35		7.07	10.25
4350	Spring nuts, short, 1/4"		120	.067		1.27	4.25		5.52	7.70
4400	3/8"		100	.080		1.40	5.10		6.50	9.10
4450	1/2"		80	.100	↓	1.67	6.35		8.02	11.30
4500	Closure strip		200	.040	L.F.	5.75	2.55		8.30	10.15
4550	End cap		60	.133	Ea.	1.72	8.50		10.22	14.50
4600	End connector 3/4" conduit		40	.200		1.28	12.75		14.03	20.50
4650	Junction box, 1 channel		16	.500		48.50	32		80.50	101
4700	2 channel		14	.571		61.50	36.50		98	122
4750	3 channel		12	.667		83.50	42.50		126	155
4800	4 channel		10	.800		80.50	51		131.50	164
4850	Splice plate		40	.200		9.45	12.75		22.20	29.50
4900	Continuous concrete insert, 1-1/2" deep, 1' long		16	.500		18.85	32		50.85	68
4950	2' long		14	.571		22.50	36.50		59	79
5000	3' long		12	.667		24.50	42.50		67	89.50
5050	4' long		10	.800		35	51		86	114
5100	6' long		8	1		52.50	63.50		116	153
5150	3/4" deep, 1' long		16	.500		21	32		53	70.50
5200	2' long		14	.571		21	36.50		57.50	77
5250	3' long		12	.667		21.50	42.50		64	86.50
5300	4' long		10	.800		35.50	51		86.50	115

26 05 29.20 Hangers		Crew	Daily Output	Labor-Hours	Unit	Material	2021 Bare Costs Labor	Equipment	Total	Total Incl O&P
5350	6' long	1 Elec	8	1	Ea.	53.50	63.50		117	154
5400	90° angle fitting 2-1/8" x 2-1/8"		60	.133		3.86	8.50		12.36	16.85
5450	Supports, suspension rod type, small		60	.133		68.50	8.50		77	87.50
5500	Large		40	.200		69	12.75		81.75	95
5550	Beam clamp, small		60	.133		32.50	8.50		41	48.50
5600	Large		40	.200		32.50	12.75		45.25	55
5650	U-support, small		60	.133		5.75	8.50		14.25	18.90
5700	Large		40	.200		11.35	12.75		24.10	31.50
5750	Concrete insert, cast, for up to 1/2" threaded rod		16	.500		14.75	32		46.75	63.50
5800	Beam clamp, 1/4" clamp, for 1/4" threaded drop rod		32	.250		3.03	15.95		18.98	27
5900	3/8" clamp, for 3/8" threaded drop rod		32	.250		6.80	15.95		22.75	31
6000	Strap, rigid conduit, 1/2" diameter		540	.015		1.37	.94		2.31	2.91
6050	3/4" diameter		440	.018		1.49	1.16		2.65	3.36
6100	1" diameter		420	.019		1.74	1.21		2.95	3.71
6150	1-1/4" diameter		400	.020		2.01	1.27		3.28	4.10
6200	1-1/2" diameter		400	.020		2.40	1.27		3.67	4.53
6250	2" diameter		267	.030		2.59	1.91		4.50	5.70
6300	2-1/2" diameter		267	.030		2.59	1.91		4.50	5.70
6350	3" diameter		160	.050		3.03	3.19		6.22	8.05
6400	3-1/2" diameter		133	.060		3.60	3.83		7.43	9.65
6450	4" diameter		100	.080		3.75	5.10		8.85	11.70
6500	5" diameter		80	.100		12.35	6.35		18.70	23
6550	6" diameter		60	.133		21	8.50		29.50	35.50
6600	EMT, 1/2" diameter		540	.015		.70	.94		1.64	2.17
6650	3/4" diameter		440	.018		.76	1.16		1.92	2.56
6700	1" diameter		420	.019		.83	1.21		2.04	2.71
6750	1-1/4" diameter		400	.020		1.19	1.27		2.46	3.20
6800	1-1/2" diameter		400	.020		1.49	1.27		2.76	3.53
6850	2" diameter		267	.030		1.59	1.91		3.50	4.59
6900	2-1/2" diameter		267	.030		1.30	1.91		3.21	4.27
6950	3" diameter		160	.050		1.42	3.19		4.61	6.30
6970	3-1/2" diameter		133	.060		1.69	3.83		5.52	7.55
6990	4" diameter		100	.080		1.93	5.10		7.03	9.65
7000	Clip, 1 hole for rigid conduit, 1/2" diameter		500	.016		.27	1.02		1.29	1.81
7050	3/4" diameter		470	.017		.38	1.08		1.46	2.03
7100	1" diameter		440	.018		1.82	1.16		2.98	3.72
7150	1-1/4" diameter		400	.020		.94	1.27		2.21	2.92
7200	1-1/2" diameter		355	.023		1.11	1.44		2.55	3.35
7250	2" diameter		320	.025		1.94	1.59		3.53	4.50
7300	2-1/2" diameter		266	.030		5.55	1.92		7.47	8.95
7350	3" diameter		160	.050		5.50	3.19		8.69	10.80
7400	3-1/2" diameter		133	.060		9.45	3.83		13.28	16.10
7450	4" diameter		100	.080		4.01	5.10		9.11	11.95
7500	5" diameter		80	.100		144	6.35		150.35	167
7550	6" diameter		60	.133		219	8.50		227.50	254
7600	Hammer on purlin clip, 1/4" hole, 1/16" flange		60	.133		.64	8.50		9.14	13.30
7610	1/4" hole threaded, 1/4" flange, rod hanger		60	.133		1.42	8.50		9.92	14.15
7620	3/8" hole threaded, 1/4" flange, rod hanger		60	.133		1.46	8.50		9.96	14.20
7630	Z purlin clip, 1/4" & 3/8" flange threaded rod hanger		60	.133		1.47	8.50		9.97	14.20
7640	1/2" rigid, 3/4" conduit, 1" flange		60	.133		1.47	8.50		9.97	14.20
7650	1" conduit, 1" flange		60	.133		1.64	8.50		10.14	14.40
7660	1-1/4" conduit, 1" flange		60	.133		1.97	8.50		10.47	14.75
7670	Push-in type conduit clip 3/4" EMT, with 9/32" hole		60	.133		.49	8.50		8.99	13.15

For customer support on your Electrical Costs with RSMeans Data, call 800.448.8182.

145

26 05 29.20 Hangers		Crew	Daily Output	Labor-Hours	Unit	Material	2021 Bare Costs Labor	Equipment	Total	Total Incl O&P
7680	3/4" EMT, 1/2" rigid	1 Elec	60	.133	Ea.	1.35	8.50		9.85	14.10
7690	Push-in type tandem conduit clip 3/4" - 1" EMT, 1/2" - 3/4" rigid		60	.133		1.38	8.50		9.88	14.10
7820	Conduit hangers, with bolt & 12" rod, 1/2" diameter		150	.053		7.85	3.40		11.25	13.70
7830	3/4" diameter		145	.055		7.75	3.51		11.26	13.70
7840	1" diameter		135	.059		7.85	3.77		11.62	14.20
7850	1-1/4" diameter		120	.067		7.95	4.25		12.20	15
7860	1-1/2" diameter		110	.073		8.20	4.63		12.83	15.90
7870	2" diameter		100	.080		9.50	5.10		14.60	18
7880	2-1/2" diameter		80	.100		9.80	6.35		16.15	20.50
7890	3" diameter		60	.133		10.05	8.50		18.55	23.50
7900	3-1/2" diameter		45	.178		10.30	11.30		21.60	28
7910	4" diameter		35	.229		12	14.55		26.55	34.50
7920	5" diameter		30	.267		13.70	17		30.70	40
7930	6" diameter		25	.320		14.15	20.50		34.65	46
7950	Jay clamp, 1/2" diameter		32	.250		5.55	15.95		21.50	29.50
7960	3/4" diameter		32	.250		6.85	15.95		22.80	31
7970	1" diameter		32	.250		6.85	15.95		22.80	31
7980	1-1/4" diameter		30	.267		8.85	17		25.85	34.50
7990	1-1/2" diameter		30	.267		11	17		28	37
8000	2" diameter		30	.267		13.85	17		30.85	40.50
8010	2-1/2" diameter		28	.286		23	18.20		41.20	52
8020	3" diameter		28	.286		23	18.20		41.20	52
8030	3-1/2" diameter		25	.320		68.50	20.50		89	106
8040	4" diameter		25	.320		68.50	20.50		89	106
8050	5" diameter		20	.400		140	25.50		165.50	192
8060	6" diameter		16	.500		267	32		299	340
8070	Channels, 3/4" x 1-1/2" w/12" rods for 1/2" to 1" conduit		30	.267		12.90	17		29.90	39
8080	1-1/2" x 1-1/2" w/12" rods for 1-1/4" to 2" conduit		28	.286		12.60	18.20		30.80	41
8090	1-1/2" x 1-1/2" w/12" rods for 2-1/2" to 4" conduit		26	.308		15.35	19.60		34.95	46
8100	1-1/2" x 1-7/8" w/12" rods for 5" to 6" conduit		24	.333		55	21		76	92
8110	Beam clamp, conduit, plastic coated steel, 1/2" diam.		30	.267		34	17		51	62.50
8120	3/4" diameter		30	.267		35	17		52	63.50
8130	1" diameter		30	.267		36	17		53	64.50
8140	1-1/4" diameter		28	.286		49	18.20		67.20	81
8150	1-1/2" diameter		28	.286		59.50	18.20		77.70	92.50
8160	2" diameter		28	.286		79	18.20		97.20	114
8170	2-1/2" diameter		26	.308		88.50	19.60		108.10	127
8180	3" diameter		26	.308		101	19.60		120.60	140
8190	3-1/2" diameter		23	.348		101	22		123	144
8200	4" diameter		23	.348		106	22		128	149
8210	5" diameter		18	.444		219	28.50		247.50	283
8220	Channels, plastic coated									
8250	3/4" x 1-1/2", w/12" rods for 1/2" to 1" conduit	1 Elec	28	.286	Ea.	42	18.20		60.20	73.50
8260	1-1/2" x 1-1/2", w/12" rods for 1-1/4" to 2" conduit		26	.308		52.50	19.60		72.10	87
8270	1-1/2" x 1-1/2", w/12" rods for 2-1/2" to 3-1/2" conduit		24	.333		57	21		78	94.50
8280	1-1/2" x 1-7/8", w/12" rods for 4" to 5" conduit		22	.364		95	23		118	139
8290	1-1/2" x 1-7/8", w/12" rods for 6" conduit		20	.400		104	25.50		129.50	153
8320	Conduit hangers, plastic coated steel, with bolt & 12" rod, 1/2" diam.		140	.057		27	3.64		30.64	35
8330	3/4" diameter		135	.059		28	3.77		31.77	36
8340	1" diameter		125	.064		28	4.08		32.08	37
8350	1-1/4" diameter		110	.073		30	4.63		34.63	40
8360	1-1/2" diameter		100	.080		33.50	5.10		38.60	44.50
8370	2" diameter		90	.089		37	5.65		42.65	49.50

146

26 05 29.20 Hangers

		Crew	Daily Output	Labor-Hours	Unit	Material	2021 Bare Costs Labor	Equipment	Total	Total Incl O&P
8380	2-1/2" diameter	1 Elec	70	.114	Ea.	44	7.30		51.30	59.50
8390	3" diameter		50	.160		53	10.20		63.20	73.50
8400	3-1/2" diameter		35	.229		39	14.55		53.55	64.50
8410	4" diameter		25	.320		79	20.50		99.50	118
8420	5" diameter		20	.400		86.50	25.50		112	133
9000	Parallel type, conduit beam clamp, 1/2"		32	.250		7.95	15.95		23.90	32.50
9010	3/4"		32	.250		4.81	15.95		20.76	29
9020	1"		32	.250		5.35	15.95		21.30	29.50
9030	1-1/4"		30	.267		10.85	17		27.85	37
9040	1-1/2"		30	.267		12.25	17		29.25	38.50
9050	2"		30	.267		10.30	17		27.30	36.50
9060	2-1/2"		28	.286		17.05	18.20		35.25	46
9070	3"		28	.286		19.75	18.20		37.95	49
9090	4"		25	.320		15.75	20.50		36.25	48
9110	Right angle, conduit beam clamp, 1/2"		32	.250		5.30	15.95		21.25	29.50
9120	3/4"		32	.250		5.60	15.95		21.55	29.50
9130	1"		32	.250		10.25	15.95		26.20	35
9140	1-1/4"		30	.267		6.70	17		23.70	32.50
9150	1-1/2"		30	.267		7.15	17		24.15	33
9160	2"		30	.267		9.50	17		26.50	35.50
9170	2-1/2"		28	.286		8.05	18.20		26.25	36
9180	3"		28	.286		9.45	18.20		27.65	37.50
9190	3-1/2"		25	.320		10.70	20.50		31.20	42.50
9200	4"		25	.320		12.40	20.50		32.90	44
9230	Adjustable, conduit hanger, 1/2"		32	.250		7	15.95		22.95	31
9240	3/4"		32	.250		6.55	15.95		22.50	30.50
9250	1"		32	.250		6.50	15.95		22.45	30.50
9260	1-1/4"		30	.267		6.75	17		23.75	32.50
9270	1-1/2"		30	.267		7.60	17		24.60	33.50
9280	2"		30	.267		8.50	17		25.50	34.50
9290	2-1/2"		28	.286		16.10	18.20		34.30	44.50
9300	3"		28	.286		17.35	18.20		35.55	46
9310	3-1/2"		25	.320		20.50	20.50		41	53
9320	4"		25	.320		25	20.50		45.50	58
9330	5"		20	.400		30	25.50		55.50	71
9340	6"		16	.500		41.50	32		73.50	93.50
9350	Combination conduit hanger, 3/8"		32	.250		15	15.95		30.95	40
9360	Adjustable flange, 3/8"	▼	32	.250	▼	17.55	15.95		33.50	43

26 05 29.30 Fittings and Channel Support

		Crew	Daily Output	Labor-Hours	Unit	Material	2021 Bare Costs Labor	Equipment	Total	Total Incl O&P
0010	**FITTINGS & CHANNEL SUPPORT**									
0020	Rooftop channel support									
0200	2-7/8" L x 1-5/8" W, 12 ga. pre galv., dbl. base	1 Elec	46	.174	Ea.	3.76	11.10		14.86	20.50
0210	2-7/8" L x 1-5/8" W, 12 ga. hot dip galv., dbl. base		46	.174		2.19	11.10		13.29	18.85
0220	2-7/16" L x 1-5/8" W, 12 ga. pre galv., sngl base		48	.167		1.80	10.60		12.40	17.80
0230	2-7/16" L x 1-5/8" W, 12 ga. hot dip galv., sngl base		48	.167		6.40	10.60		17	23
0240	13/16" L x 1-5/8" W, 14 ga. pre galv., sngl base		56	.143		1.53	9.10		10.63	15.20
0250	13/16" L x 1-5/8" W, 14 ga. hot dip galv., sngl base		56	.143		6.30	9.10		15.40	20.50
0260	1-5/8" L x 1-5/8" W, 12 ga. pre galv., sngl base		54	.148		1.66	9.45		11.11	15.85
0270	1-5/8" L x 1-5/8" W, 12 ga. hot dip galv., sngl base		54	.148		1.76	9.45		11.21	15.95
0280	1-5/8" x 28", 12 ga. pre galv., dbl. H block base		42	.190		3.81	12.15		15.96	22
0290	1-5/8" x 36", 12 ga. pre galv., dbl. H block base		40	.200		4.03	12.75		16.78	23.50
0300	1-5/8" x 42", 12 ga. pre galv., dbl. H block base	▼	34	.235	▼	4.53	15		19.53	27.50

147

26 05 29 – Hangers and Supports for Electrical Systems

26 05 29.30 Fittings and Channel Support

		Crew	Daily Output	Labor-Hours	Unit	Material	2021 Bare Costs Labor	2021 Bare Costs Equipment	Total	Total Incl O&P
0310	1-5/8" x 50", 12 ga. pre galv., dbl. H block base	1 Elec	38	.211	Ea.	4.83	13.40		18.23	25.50
0320	1-5/8" x 60", 12 ga. pre galv., dbl. H block base		36	.222		5.30	14.15		19.45	27
0330	13/16" x 8" H, threaded rod pre galv., H block base		35	.229		2.56	14.55		17.11	24.50
0340	13/16" x 12" H, threaded rod pre galv., H block base		34.60	.231		3.20	14.75		17.95	25.50
0350	1-5/8" x 16" H, threaded rod dbl. H block base		34	.235		4.77	15		19.77	28
0360	1-5/8" x 12" H, pre galv., dbl. H block base		34.60	.231		9.15	14.75		23.90	32
0370	1-5/8" x 24"H, pre galv., dbl. H block base		34	.235		9.65	15		24.65	33
0380	3-1/4" x 24" H, pre galv., dbl. H block base		32	.250		22	15.95		37.95	47.50
0390	3-1/4" x 36" H, pre galv., dbl. H block base		30	.267		24	17		41	51.50

26 05 33 – Raceway and Boxes for Electrical Systems

26 05 33.13 Conduit

		Crew	Daily Output	Labor-Hours	Unit	Material	2021 Bare Costs Labor	2021 Bare Costs Equipment	Total	Total Incl O&P
0010	**CONDUIT** To 10' high, includes 2 terminations, 2 elbows,	R260533-20								
0020	11 beam clamps, and 11 couplings per 100 L.F.									
0300	Aluminum, 1/2" diameter	1 Elec	100	.080	L.F.	1.97	5.10		7.07	9.70
0500	3/4" diameter	R260533-25	90	.089		2.75	5.65		8.40	11.40
0700	1" diameter		80	.100		3.10	6.35		9.45	12.85
1000	1-1/4" diameter		70	.114		4.24	7.30		11.54	15.45
1030	1-1/2" diameter		65	.123		5.10	7.85		12.95	17.25
1050	2" diameter		60	.133		6.80	8.50		15.30	20
1070	2-1/2" diameter		50	.160		10.25	10.20		20.45	26.50
1100	3" diameter	2 Elec	90	.178		13.15	11.30		24.45	31.50
1130	3-1/2" diameter		80	.200		18.55	12.75		31.30	39.50
1140	4" diameter		70	.229		19.80	14.55		34.35	43.50
1150	5" diameter		50	.320		43.50	20.50		64	78.50
1160	6" diameter		40	.400		62.50	25.50		88	107
1161	Field bends, 45° to 90°, 1/2" diameter	1 Elec	53	.151	Ea.		9.60		9.60	14.30
1162	3/4" diameter		47	.170			10.85		10.85	16.10
1163	1" diameter		44	.182			11.60		11.60	17.20
1164	1-1/4" diameter		23	.348			22		22	33
1165	1-1/2" diameter		21	.381			24.50		24.50	36
1166	2" diameter		16	.500			32		32	47.50
1170	Elbows, 1/2" diameter		40	.200		17.90	12.75		30.65	38.50
1200	3/4" diameter		32	.250		24	15.95		39.95	50
1230	1" diameter		28	.286		10.40	18.20		28.60	38.50
1250	1-1/4" diameter		24	.333		16.65	21		37.65	50
1270	1-1/2" diameter		20	.400		22	25.50		47.50	62.50
1300	2" diameter		16	.500		32.50	32		64.50	83.50
1330	2-1/2" diameter		12	.667		55.50	42.50		98	124
1350	3" diameter		8	1		85	63.50		148.50	188
1370	3-1/2" diameter		6	1.333		146	85		231	286
1400	4" diameter		5	1.600		156	102		258	325
1410	5" diameter		4	2		505	127		632	750
1420	6" diameter		2.50	3.200		700	204		904	1,075
1430	Couplings, 1/2" diameter		320	.025		1.81	1.59		3.40	4.36
1450	3/4" diameter		192	.042		2.76	2.65		5.41	7
1470	1" diameter		160	.050		3.52	3.19		6.71	8.60
1500	1-1/4" diameter		120	.067		5.55	4.25		9.80	12.40
1530	1-1/2" diameter		96	.083		4.98	5.30		10.28	13.40
1550	2" diameter		80	.100		7.05	6.35		13.40	17.20
1570	2-1/2" diameter		69	.116		22.50	7.40		29.90	35.50
1600	3" diameter		64	.125		21.50	7.95		29.45	35.50
1630	3-1/2" diameter		56	.143		41	9.10		50.10	59

26 05 33 – Raceway and Boxes for Electrical Systems

26 05 33.13 Conduit		Crew	Daily Output	Labor-Hours	Unit	Material	2021 Bare Costs Labor	Equipment	Total	Total Incl O&P
1650	4" diameter	1 Elec	53	.151	Ea.	34.50	9.60		44.10	52.50
1670	5" diameter		48	.167		143	10.60		153.60	174
1690	6" diameter		46	.174		198	11.10		209.10	234
1691	See note on line 26 05 33.13 9995 R260533-30									
1750	Rigid galvanized steel, 1/2" diameter	1 Elec	90	.089	L.F.	2.62	5.65		8.27	11.30
1770	3/4" diameter		80	.100		2.92	6.35		9.27	12.65
1800	1" diameter		65	.123		4.18	7.85		12.03	16.25
1830	1-1/4" diameter		60	.133		5.95	8.50		14.45	19.10
1850	1-1/2" diameter		55	.145		6.80	9.25		16.05	21.50
1870	2" diameter		45	.178		8.45	11.30		19.75	26
1900	2-1/2" diameter		35	.229		15.40	14.55		29.95	38.50
1930	3" diameter	2 Elec	50	.320		17.30	20.50		37.80	49.50
1950	3-1/2" diameter		44	.364		22	23		45	58.50
1970	4" diameter		40	.400		24	25.50		49.50	64.50
1980	5" diameter		30	.533		50.50	34		84.50	106
1990	6" diameter		20	.800		63.50	51		114.50	146
1991	Field bends, 45° to 90°, 1/2" diameter	1 Elec	44	.182	Ea.		11.60		11.60	17.20
1992	3/4" diameter		40	.200			12.75		12.75	18.95
1993	1" diameter		36	.222			14.15		14.15	21
1994	1-1/4" diameter		19	.421			27		27	40
1995	1-1/2" diameter		18	.444			28.50		28.50	42
1996	2" diameter		13	.615			39		39	58.50
2000	Elbows, 1/2" diameter		32	.250		4.10	15.95		20.05	28
2030	3/4" diameter		28	.286		4.69	18.20		22.89	32
2050	1" diameter		24	.333		6.30	21		27.30	38.50
2070	1-1/4" diameter		18	.444		9.25	28.50		37.75	52
2100	1-1/2" diameter		16	.500		12.55	32		44.55	61.50
2130	2" diameter		12	.667		17.25	42.50		59.75	82
2150	2-1/2" diameter		8	1		28.50	63.50		92	126
2170	3" diameter		6	1.333		40.50	85		125.50	171
2200	3-1/2" diameter		4.20	1.905		63	121		184	249
2220	4" diameter		4	2		74	127		201	270
2230	5" diameter		3.50	2.286		249	146		395	490
2240	6" diameter		2	4		283	255		538	690
2250	Couplings, 1/2" diameter		267	.030		1.12	1.91		3.03	4.07
2270	3/4" diameter		160	.050		1.52	3.19		4.71	6.40
2300	1" diameter		133	.060		2.38	3.83		6.21	8.30
2330	1-1/4" diameter		100	.080		2.84	5.10		7.94	10.65
2350	1-1/2" diameter		80	.100		3.85	6.35		10.20	13.70
2370	2" diameter		67	.119		4.58	7.60		12.18	16.35
2400	2-1/2" diameter		57	.140		11.45	8.95		20.40	26
2430	3" diameter		53	.151		14.50	9.60		24.10	30.50
2450	3-1/2" diameter		47	.170		19.70	10.85		30.55	37.50
2470	4" diameter		44	.182		21	11.60		32.60	40
2480	5" diameter		40	.200		50	12.75		62.75	74
2490	6" diameter		38	.211		58.50	13.40		71.90	84.50
2491	See note on line 26 05 33.13 9995 R260533-30									
2500	Steel, intermediate conduit (IMC), 1/2" diameter	1 Elec	100	.080	L.F.	1.97	5.10		7.07	9.70
2530	3/4" diameter		90	.089		2.92	5.65		8.57	11.60
2550	1" diameter		70	.114		3.36	7.30		10.66	14.50
2570	1-1/4" diameter		65	.123		4.49	7.85		12.34	16.60
2600	1-1/2" diameter		60	.133		5.60	8.50		14.10	18.75
2630	2" diameter		50	.160		6.75	10.20		16.95	22.50

26 05 33.13 Conduit

		Crew	Daily Output	Labor-Hours	Unit	Material	2021 Bare Costs Labor	Equipment	Total	Total Incl O&P
2650	2-1/2" diameter	1 Elec	40	.200	L.F.	12.10	12.75		24.85	32.50
2670	3" diameter	2 Elec	60	.267		16.10	17		33.10	42.50
2700	3-1/2" diameter		54	.296		20	18.85		38.85	50
2730	4" diameter		50	.320		23.50	20.50		44	56.50
2731	Field bends, 45° to 90°, 1/2" diameter	1 Elec	44	.182	Ea.		11.60		11.60	17.20
2732	3/4" diameter		40	.200			12.75		12.75	18.95
2733	1" diameter		36	.222			14.15		14.15	21
2734	1-1/4" diameter		19	.421			27		27	40
2735	1-1/2" diameter		18	.444			28.50		28.50	42
2736	2" diameter		13	.615			39		39	58.50
2750	Elbows, 1/2" diameter		32	.250		5.70	15.95		21.65	30
2770	3/4" diameter		28	.286		6	18.20		24.20	33.50
2800	1" diameter		24	.333		9.10	21		30.10	41.50
2830	1-1/4" diameter		18	.444		12.70	28.50		41.20	56
2850	1-1/2" diameter		16	.500		15.45	32		47.45	64.50
2870	2" diameter		12	.667		23.50	42.50		66	88.50
2900	2-1/2" diameter		8	1		45	63.50		108.50	144
2930	3" diameter		6	1.333		61.50	85		146.50	194
2950	3-1/2" diameter		4.20	1.905		92	121		213	281
2970	4" diameter		4	2		113	127		240	315
3000	Couplings, 1/2" diameter		293	.027		1.12	1.74		2.86	3.81
3030	3/4" diameter		176	.045		1.52	2.90		4.42	5.95
3050	1" diameter		147	.054		2.38	3.47		5.85	7.75
3070	1-1/4" diameter		110	.073		2.84	4.63		7.47	10
3100	1-1/2" diameter		88	.091		3.85	5.80		9.65	12.85
3130	2" diameter		73	.110		4.58	7		11.58	15.40
3150	2-1/2" diameter		63	.127		11.45	8.10		19.55	24.50
3170	3" diameter		59	.136		14.50	8.65		23.15	29
3200	3-1/2" diameter		52	.154		19.70	9.80		29.50	36
3230	4" diameter		49	.163		21	10.40		31.40	38.50
3231	See note on line 26 05 33.13 9995	R260533-30								
4100	Rigid steel, plastic coated, 40 mil thick									
4130	1/2" diameter	1 Elec	80	.100	L.F.	11	6.35		17.35	21.50
4150	3/4" diameter		70	.114		11.50	7.30		18.80	23.50
4170	1" diameter		55	.145		13.55	9.25		22.80	28.50
4200	1-1/4" diameter		50	.160		17.85	10.20		28.05	35
4230	1-1/2" diameter		45	.178		21.50	11.30		32.80	40.50
4250	2" diameter		35	.229		28.50	14.55		43.05	52.50
4270	2-1/2" diameter		25	.320		42.50	20.50		63	77.50
4300	3" diameter	2 Elec	44	.364		51.50	23		74.50	91
4330	3-1/2" diameter		40	.400		63.50	25.50		89	108
4350	4" diameter		36	.444		70	28.50		98.50	119
4370	5" diameter		30	.533		154	34		188	220
4400	Elbows, 1/2" diameter	1 Elec	28	.286	Ea.	24	18.20		42.20	53.50
4430	3/4" diameter		24	.333		22	21		43	55.50
4450	1" diameter		18	.444		25.50	28.50		54	70
4470	1-1/4" diameter		16	.500		36	32		68	87
4500	1-1/2" diameter		12	.667		39	42.50		81.50	106
4530	2" diameter		8	1		53.50	63.50		117	153
4550	2-1/2" diameter		6	1.333		117	85		202	255
4570	3" diameter		4.20	1.905		163	121		284	360
4600	3-1/2" diameter		4	2		242	127		369	455
4630	4" diameter		3.80	2.105		229	134		363	450

For customer support on your Electrical Costs with RSMeans Data, call 800.448.8182.

26 05 33 – Raceway and Boxes for Electrical Systems

26 05 33.13 Conduit

		Crew	Daily Output	Labor-Hours	Unit	Material	2021 Bare Costs Labor	Equipment	Total	Total Incl O&P
4650	5" diameter	1 Elec	3.50	2.286	Ea.	645	146		791	925
4680	Couplings, 1/2" diameter		213	.038		6.85	2.39		9.24	11.10
4700	3/4" diameter		128	.063		7.55	3.98		11.53	14.20
4730	1" diameter		107	.075		8.60	4.76		13.36	16.55
4750	1-1/4" diameter		80	.100		10.85	6.35		17.20	21.50
4770	1-1/2" diameter		64	.125		12.25	7.95		20.20	25.50
4800	2" diameter		53	.151		17.80	9.60		27.40	34
4830	2-1/2" diameter		46	.174		46.50	11.10		57.60	68
4850	3" diameter		43	.186		54.50	11.85		66.35	77.50
4870	3-1/2" diameter		38	.211		66.50	13.40		79.90	93
4900	4" diameter		36	.222		79.50	14.15		93.65	109
4950	5" diameter		32	.250		291	15.95		306.95	345
4951	See note on line 26 05 33.13 9995	R260533-30								
5000	Electric metallic tubing (EMT), 1/2" diameter	1 Elec	170	.047	L.F.	.72	3		3.72	5.25
5020	3/4" diameter		130	.062		1.01	3.92		4.93	6.90
5040	1" diameter		115	.070		1.68	4.43		6.11	8.45
5060	1-1/4" diameter		100	.080		2.69	5.10		7.79	10.50
5080	1-1/2" diameter		90	.089		3.55	5.65		9.20	12.30
5100	2" diameter		80	.100		3.82	6.35		10.17	13.65
5120	2-1/2" diameter		60	.133		5.50	8.50		14	18.65
5140	3" diameter	2 Elec	100	.160		8.85	10.20		19.05	25
5160	3-1/2" diameter		90	.178		10.25	11.30		21.55	28
5180	4" diameter		80	.200		11.65	12.75		24.40	32
5200	Field bends, 45° to 90°, 1/2" diameter	1 Elec	89	.090	Ea.		5.75		5.75	8.50
5220	3/4" diameter		80	.100			6.35		6.35	9.45
5240	1" diameter		73	.110			7		7	10.35
5260	1-1/4" diameter		38	.211			13.40		13.40	19.95
5280	1-1/2" diameter		36	.222			14.15		14.15	21
5300	2" diameter		26	.308			19.60		19.60	29
5320	Offsets, 1/2" diameter		65	.123			7.85		7.85	11.65
5340	3/4" diameter		62	.129			8.20		8.20	12.20
5360	1" diameter		53	.151			9.60		9.60	14.30
5380	1-1/4" diameter		30	.267			17		17	25
5400	1-1/2" diameter		28	.286			18.20		18.20	27
5420	2" diameter		20	.400			25.50		25.50	38
5700	Elbows, 1" diameter		40	.200		3.73	12.75		16.48	23
5720	1-1/4" diameter		32	.250		4.79	15.95		20.74	29
5740	1-1/2" diameter		24	.333		5.50	21		26.50	37.50
5760	2" diameter		20	.400		7.90	25.50		33.40	46.50
5780	2-1/2" diameter		12	.667		17.20	42.50		59.70	82
5800	3" diameter		9	.889		25	56.50		81.50	112
5820	3-1/2" diameter		7	1.143		34.50	73		107.50	146
5840	4" diameter		6	1.333		39	85		124	169
6200	Couplings, set screw, steel, 1/2" diameter		235	.034		.55	2.17		2.72	3.83
6220	3/4" diameter		188	.043		.75	2.71		3.46	4.86
6240	1" diameter		157	.051		1.25	3.25		4.50	6.20
6260	1-1/4" diameter		118	.068		2.79	4.32		7.11	9.45
6280	1-1/2" diameter		94	.085		3.35	5.40		8.75	11.75
6300	2" diameter		78	.103		4.62	6.55		11.17	14.80
6320	2-1/2" diameter		67	.119		3.72	7.60		11.32	15.40
6340	3" diameter		59	.136		16.15	8.65		24.80	30.50
6360	3-1/2" diameter		47	.170		18.25	10.85		29.10	36
6380	4" diameter		43	.186		26.50	11.85		38.35	47

For customer support on your Electrical Costs with RSMeans Data, call 800.448.8182.

151

26 05 33.13 Conduit		Crew	Daily Output	Labor-Hours	Unit	Material	2021 Bare Costs Labor	Equipment	Total	Total Incl O&P
6381	See note on line 26 05 33.13 9995 R260533-30									
6500	Box connectors, set screw, steel, 1/2" diameter	1 Elec	120	.067	Ea.	.62	4.25		4.87	7
6520	3/4" diameter		110	.073		1.05	4.63		5.68	8.05
6540	1" diameter		90	.089		1.82	5.65		7.47	10.40
6560	1-1/4" diameter		70	.114		3.75	7.30		11.05	14.95
6580	1-1/2" diameter		60	.133		5.20	8.50		13.70	18.35
6600	2" diameter		50	.160		6.90	10.20		17.10	23
6620	2-1/2" diameter		36	.222		10.10	14.15		24.25	32
6640	3" diameter		27	.296		38.50	18.85		57.35	70
6680	3-1/2" diameter		21	.381		8.90	24.50		33.40	46
6700	4" diameter		16	.500		16.40	32		48.40	65.50
6740	Insulated box connectors, set screw, steel, 1/2" diameter		120	.067		1.05	4.25		5.30	7.45
6760	3/4" diameter		110	.073		1.54	4.63		6.17	8.60
6780	1" diameter		90	.089		3.05	5.65		8.70	11.75
6800	1-1/4" diameter		70	.114		5.55	7.30		12.85	16.90
6820	1-1/2" diameter		60	.133		7.85	8.50		16.35	21.50
6840	2" diameter		50	.160		11.40	10.20		21.60	27.50
6860	2-1/2" diameter		36	.222		58.50	14.15		72.65	85
6880	3" diameter		27	.296		69	18.85		87.85	104
6900	3-1/2" diameter		21	.381		96	24.50		120.50	142
6920	4" diameter		16	.500		105	32		137	163
7000	EMT to conduit adapters, 1/2" diameter (compression)		70	.114		4.27	7.30		11.57	15.50
7020	3/4" diameter		60	.133		5.65	8.50		14.15	18.85
7040	1" diameter		50	.160		9.05	10.20		19.25	25
7060	1-1/4" diameter		40	.200		14.60	12.75		27.35	35
7080	1-1/2" diameter		30	.267		17.35	17		34.35	44
7100	2" diameter		25	.320		26	20.50		46.50	59.50
7200	EMT to Greenfield adapters, 1/2" to 3/8" diameter (compression)		90	.089		2.52	5.65		8.17	11.15
7220	1/2" diameter		90	.089		4.86	5.65		10.51	13.75
7240	3/4" diameter		80	.100		6.10	6.35		12.45	16.15
7260	1" diameter		70	.114		17	7.30		24.30	29.50
7270	1-1/4" diameter		60	.133		18.75	8.50		27.25	33
7280	1-1/2" diameter		50	.160		21.50	10.20		31.70	39
7290	2" diameter		40	.200		31.50	12.75		44.25	53.50
7400	EMT, LB, LR or LL fittings with covers, 1/2" diameter, set screw		24	.333		6.05	21		27.05	38
7420	3/4" diameter		20	.400		7.40	25.50		32.90	46
7440	1" diameter		16	.500		11.10	32		43.10	59.50
7450	1-1/4" diameter		13	.615		16.25	39		55.25	76.50
7460	1-1/2" diameter		11	.727		20.50	46.50		67	91.50
7470	2" diameter		9	.889		35.50	56.50		92	123
7600	EMT, "T" fittings with covers, 1/2" diameter, set screw		16	.500		5.70	32		37.70	54
7620	3/4" diameter		15	.533		7.10	34		41.10	58.50
7640	1" diameter		12	.667		9.35	42.50		51.85	73.50
7650	1-1/4" diameter		11	.727		17.10	46.50		63.60	88
7660	1-1/2" diameter		10	.800		19.35	51		70.35	97
7670	2" diameter		8	1		25.50	63.50		89	123
8000	EMT, expansion fittings, no jumper, 1/2" diameter		24	.333		79.50	21		100.50	119
8020	3/4" diameter		20	.400		99.50	25.50		125	147
8040	1" diameter		16	.500		127	32		159	187
8060	1-1/4" diameter		13	.615		164	39		203	239
8080	1-1/2" diameter		11	.727		217	46.50		263.50	310
8100	2" diameter		9	.889		296	56.50		352.50	410
8110	2-1/2" diameter		7	1.143		440	73		513	595

152

26 05 33 – Raceway and Boxes for Electrical Systems

26 05 33.13 Conduit		Crew	Daily Output	Labor-Hours	Unit	Material	2021 Bare Costs Labor	Equipment	Total	Total Incl O&P
8120	3" diameter	1 Elec	6	1.333	Ea.	560	85		645	740
8140	4" diameter		5	1.600		870	102		972	1,100
8200	Split adapter, 1/2" diameter		110	.073		3.47	4.63		8.10	10.70
8210	3/4" diameter		90	.089		3.15	5.65		8.80	11.85
8220	1" diameter		70	.114		5.10	7.30		12.40	16.40
8230	1-1/4" diameter		60	.133		8.20	8.50		16.70	21.50
8240	1-1/2" diameter		50	.160		9.80	10.20		20	26
8250	2" diameter		36	.222		34	14.15		48.15	58
8300	1 hole clips, 1/2" diameter		500	.016		.11	1.02		1.13	1.63
8320	3/4" diameter		470	.017		.26	1.08		1.34	1.90
8340	1" diameter		444	.018		.51	1.15		1.66	2.27
8360	1-1/4" diameter		400	.020		.68	1.27		1.95	2.64
8380	1-1/2" diameter		355	.023		1.04	1.44		2.48	3.27
8400	2" diameter		320	.025		1.63	1.59		3.22	4.16
8420	2-1/2" diameter		266	.030		1.86	1.92		3.78	4.90
8440	3" diameter		160	.050		2.47	3.19		5.66	7.45
8460	3-1/2" diameter		133	.060		3.34	3.83		7.17	9.35
8480	4" diameter		100	.080		4.15	5.10		9.25	12.10
8500	Clamp back spacers, 1/2" diameter		500	.016		1.16	1.02		2.18	2.79
8510	3/4" diameter		470	.017		1.23	1.08		2.31	2.96
8520	1" diameter		444	.018		2.23	1.15		3.38	4.16
8530	1-1/4" diameter		400	.020		3.29	1.27		4.56	5.50
8540	1-1/2" diameter		355	.023		4.23	1.44		5.67	6.80
8550	2" diameter		320	.025		5.65	1.59		7.24	8.60
8560	2-1/2" diameter		266	.030		13.25	1.92		15.17	17.45
8570	3" diameter		160	.050		43.50	3.19		46.69	52
8580	3-1/2" diameter		133	.060		55.50	3.83		59.33	66.50
8590	4" diameter		100	.080		126	5.10		131.10	147
8600	Offset connectors, 1/2" diameter		40	.200		2.42	12.75		15.17	21.50
8610	3/4" diameter		32	.250		3.92	15.95		19.87	28
8620	1" diameter		24	.333		4.83	21		25.83	37
8650	90° pulling elbows, female, 1/2" diameter, with gasket		24	.333		9.05	21		30.05	41.50
8660	3/4" diameter		20	.400		15.60	25.50		41.10	55
8700	Couplings, compression, 1/2" diameter, steel		78	.103		1.52	6.55		8.07	11.35
8710	3/4" diameter		67	.119		1.82	7.60		9.42	13.30
8720	1" diameter		59	.136		3.06	8.65		11.71	16.20
8730	1-1/4" diameter		47	.170		6.15	10.85		17	23
8740	1-1/2" diameter		38	.211		9.10	13.40		22.50	30
8750	2" diameter		31	.258		12.10	16.45		28.55	38
8760	2-1/2" diameter		24	.333		68	21		89	106
8770	3" diameter		21	.381		84.50	24.50		109	129
8780	3-1/2" diameter		19	.421		138	27		165	192
8790	4" diameter		16	.500		141	32		173	203
8791	See note on line 26 05 33.13 9995 R260533-30									
8800	Box connectors, compression, 1/2" diam., steel	1 Elec	120	.067	Ea.	.41	4.25		4.66	6.75
8810	3/4" diameter		110	.073		.49	4.63		5.12	7.45
8820	1" diameter		90	.089		.95	5.65		6.60	9.45
8830	1-1/4" diameter		70	.114		1.55	7.30		8.85	12.50
8840	1-1/2" diameter		60	.133		2.56	8.50		11.06	15.40
8850	2" diameter		50	.160		3.57	10.20		13.77	19.10
8860	2-1/2" diameter		36	.222		9.55	14.15		23.70	31.50
8870	3" diameter		27	.296		14.15	18.85		33	43.50
8880	3-1/2" diameter		21	.381		27.50	24.50		52	66.50

For customer support on your Electrical Costs with RSMeans Data, call 800.448.8182.

153

26 05 33.13 Conduit		Crew	Daily Output	Labor-Hours	Unit	Material	2021 Bare Costs Labor	2021 Bare Costs Equipment	Total	Total Incl O&P
8890	4" diameter	1 Elec	16	.500	Ea.	20	32		52	70
8900	Box connectors, insulated compression, 1/2" diam., steel		120	.067		.46	4.25		4.71	6.80
8910	3/4" diameter		110	.073		.62	4.63		5.25	7.60
8920	1" diameter		90	.089		1.02	5.65		6.67	9.50
8930	1-1/4" diameter		70	.114		2.23	7.30		9.53	13.25
8940	1-1/2" diameter		60	.133		3.09	8.50		11.59	16
8950	2" diameter		50	.160		4.51	10.20		14.71	20
8960	2-1/2" diameter		36	.222		14.20	14.15		28.35	36.50
8970	3" diameter		27	.296		20.50	18.85		39.35	50.50
8980	3-1/2" diameter		21	.381		29	24.50		53.50	67.50
8990	4" diameter		16	.500		31	32		63	81.50
9100	PVC, schedule 40, 1/2" diameter		190	.042	L.F.	.90	2.68		3.58	4.98
9110	3/4" diameter		145	.055		.95	3.51		4.46	6.25
9120	1" diameter		125	.064		1.37	4.08		5.45	7.55
9130	1-1/4" diameter		110	.073		1.77	4.63		6.40	8.85
9140	1-1/2" diameter		100	.080		1.88	5.10		6.98	9.60
9150	2" diameter		90	.089		2.40	5.65		8.05	11.05
9160	2-1/2" diameter		65	.123		4.61	7.85		12.46	16.70
9170	3" diameter	2 Elec	110	.145		3.94	9.25		13.19	18.10
9180	3-1/2" diameter		100	.160		5.70	10.20		15.90	21.50
9190	4" diameter		90	.178		5.35	11.30		16.65	23
9200	5" diameter		70	.229		9.25	14.55		23.80	31.50
9210	6" diameter		60	.267		9.35	17		26.35	35.50
9220	Elbows, 1/2" diameter	1 Elec	50	.160	Ea.	.49	10.20		10.69	15.70
9225	3/4" diameter		42	.190		.55	12.15		12.70	18.65
9230	1" diameter		35	.229		.86	14.55		15.41	22.50
9235	1-1/4" diameter		28	.286		1.30	18.20		19.50	28.50
9240	1-1/2" diameter		20	.400		2.27	25.50		27.77	40.50
9245	2" diameter	R-1A	36.40	.440		3.08	25		28.08	41
9250	2-1/2" diameter		26.70	.599		5.15	34.50		39.65	56.50
9255	3" diameter		22.90	.699		8.40	40		48.40	69
9260	3-1/2" diameter		22.20	.721		9.70	41.50		51.20	72
9265	4" diameter		18.20	.879		17.05	50.50		67.55	94
9270	5" diameter		12.10	1.322		18.35	76		94.35	133
9275	6" diameter		11.10	1.441		36	82.50		118.50	163
9312	Couplings, 1/2" diameter	1 Elec	50	.160		.11	10.20		10.31	15.25
9314	3/4" diameter		42	.190		.20	12.15		12.35	18.25
9316	1" diameter		35	.229		.31	14.55		14.86	22
9318	1-1/4" diameter		28	.286		.41	18.20		18.61	27.50
9320	1-1/2" diameter		20	.400		.63	25.50		26.13	38.50
9322	2" diameter	R-1A	36.40	.440		.80	25		25.80	38.50
9324	2-1/2" diameter		26.70	.599		2.10	34.50		36.60	53.50
9326	3" diameter		22.90	.699		2.04	40		42.04	61.50
9328	3-1/2" diameter		22.20	.721		2.35	41.50		43.85	64
9330	4" diameter		18.20	.879		3.06	50.50		53.56	78.50
9332	5" diameter		12.10	1.322		8.80	76		84.80	123
9334	6" diameter		11.10	1.441		12.35	82.50		94.85	137
9335	See note on line 26 05 33.13 9995 R260533-30									
9340	Field bends, 45° & 90°, 1/2" diameter	1 Elec	45	.178	Ea.		11.30		11.30	16.85
9350	3/4" diameter		40	.200			12.75		12.75	18.95
9360	1" diameter		35	.229			14.55		14.55	21.50
9370	1-1/4" diameter		32	.250			15.95		15.95	23.50
9380	1-1/2" diameter		27	.296			18.85		18.85	28

26 05 33 – Raceway and Boxes for Electrical Systems

26 05 33.13 Conduit

		Crew	Daily Output	Labor-Hours	Unit	Material	2021 Bare Costs Labor	Equipment	Total	Total Incl O&P
9390	2" diameter	1 Elec	20	.400	Ea.		25.50		25.50	38
9400	2-1/2" diameter		16	.500			32		32	47.50
9410	3" diameter		13	.615			39		39	58.50
9420	3-1/2" diameter		12	.667			42.50		42.50	63
9430	4" diameter		10	.800			51		51	75.50
9440	5" diameter		9	.889			56.50		56.50	84
9450	6" diameter		8	1			63.50		63.50	94.50
9460	PVC adapters, 1/2" diameter		50	.160		.21	10.20		10.41	15.40
9470	3/4" diameter		42	.190		.35	12.15		12.50	18.45
9480	1" diameter		38	.211		.38	13.40		13.78	20.50
9490	1-1/4" diameter		35	.229		.61	14.55		15.16	22
9500	1-1/2" diameter		32	.250		.81	15.95		16.76	24.50
9510	2" diameter		27	.296		1.02	18.85		19.87	29
9520	2-1/2" diameter		23	.348		1.67	22		23.67	35
9530	3" diameter		18	.444		2.13	28.50		30.63	44.50
9540	3-1/2" diameter		13	.615		2.80	39		41.80	61.50
9550	4" diameter		11	.727		4.33	46.50		50.83	74
9560	5" diameter		8	1		8.65	63.50		72.15	104
9570	6" diameter		6	1.333		10.30	85		95.30	137
9580	PVC-LB, LR or LL fittings & covers									
9590	1/2" diameter	1 Elec	20	.400	Ea.	2.96	25.50		28.46	41.50
9600	3/4" diameter		16	.500		4.25	32		36.25	52
9610	1" diameter		12	.667		4.65	42.50		47.15	68
9620	1-1/4" diameter		9	.889		7.40	56.50		63.90	92
9630	1-1/2" diameter		7	1.143		8.65	73		81.65	118
9640	2" diameter		6	1.333		16.25	85		101.25	144
9650	2-1/2" diameter		6	1.333		47	85		132	178
9660	3" diameter		5	1.600		56.50	102		158.50	214
9670	3-1/2" diameter		4	2		45.50	127		172.50	240
9680	4" diameter		3	2.667		63.50	170		233.50	320
9690	PVC-tee fitting & cover									
9700	1/2"	1 Elec	14	.571	Ea.	4.99	36.50		41.49	59.50
9710	3/4"		13	.615		5.65	39		44.65	64.50
9720	1"		10	.800		6.20	51		57.20	82.50
9730	1-1/4"		9	.889		10.65	56.50		67.15	96
9740	1-1/2"		8	1		11.35	63.50		74.85	107
9750	2"		7	1.143		19.35	73		92.35	130
9760	PVC-reducers, 3/4" x 1/2" diameter					1.30			1.30	1.43
9770	1" x 1/2" diameter					2.45			2.45	2.70
9780	1" x 3/4" diameter					2.48			2.48	2.73
9790	1-1/4" x 3/4" diameter					2.49			2.49	2.74
9800	1-1/4" x 1" diameter					3.80			3.80	4.18
9810	1-1/2" x 1-1/4" diameter					4.63			4.63	5.10
9820	2" x 1-1/4" diameter					5.15			5.15	5.65
9830	2-1/2" x 2" diameter					15			15	16.50
9840	3" x 2" diameter					14.90			14.90	16.35
9850	4" x 3" diameter					17.95			17.95	19.75
9860	Cement, quart					17.50			17.50	19.25
9870	Gallon					111			111	123
9880	Heat bender, to 6" diameter					1,625			1,625	1,800
9900	Add to labor for higher elevated installation									
9905	10' to 14.5' high, add						10%			
9910	15' to 20' high, add						20%			

26 05 33.13 Conduit

		Crew	Daily Output	Labor-Hours	Unit	Material	2021 Bare Costs Labor	Equipment	Total	Total Incl O&P
9920	20' to 25' high, add						25%			
9930	25' to 30' high, add						35%			
9940	30' to 35' high, add						40%			
9950	35' to 40' high, add						50%			
9960	Over 40' high, add						55%			
9995	Do not include labor when adding couplings to a fitting installation R260533-30									

26 05 33.14 Conduit

		Crew	Daily Output	Labor-Hours	Unit	Material	2021 Bare Costs Labor	Equipment	Total	Total Incl O&P
0010	**CONDUIT** To 10' high, includes 11 couplings per 100'									
0200	Electric metallic tubing, 1/2" diameter	1 Elec	435	.018	L.F.	.48	1.17		1.65	2.27
0220	3/4" diameter		253	.032		.76	2.01		2.77	3.83
0240	1" diameter		207	.039		1.31	2.46		3.77	5.10
0260	1-1/4" diameter		173	.046		2.25	2.95		5.20	6.85
0280	1-1/2" diameter		153	.052		3.07	3.33		6.40	8.35
0300	2" diameter		130	.062		3.22	3.92		7.14	9.35
0320	2-1/2" diameter		92	.087		4.54	5.55		10.09	13.25
0340	3" diameter	2 Elec	148	.108		7.15	6.90		14.05	18.10
0360	3-1/2" diameter		134	.119		8.95	7.60		16.55	21
0380	4" diameter		114	.140		10	8.95		18.95	24.50
0500	Steel rigid galvanized, 1/2" diameter	1 Elec	146	.055		1.92	3.49		5.41	7.30
0520	3/4" diameter		125	.064		2.06	4.08		6.14	8.30
0540	1" diameter		93	.086		3.27	5.50		8.77	11.75
0560	1-1/4" diameter		88	.091		4.73	5.80		10.53	13.80
0580	1-1/2" diameter		80	.100		5.30	6.35		11.65	15.25
0600	2" diameter		65	.123		6.50	7.85		14.35	18.80
0620	2-1/2" diameter		48	.167		12.80	10.60		23.40	30
0640	3" diameter	2 Elec	64	.250		14.40	15.95		30.35	39.50
0660	3-1/2" diameter		60	.267		18.85	17		35.85	45.50
0680	4" diameter		52	.308		20	19.60		39.60	51
0700	5" diameter		50	.320		42	20.50		62.50	76.50
0720	6" diameter		48	.333		51	21		72	88
1000	Steel intermediate conduit (IMC), 1/2" diameter R260533-25	1 Elec	155	.052		1.34	3.29		4.63	6.35
1010	3/4" diameter		130	.062		2.08	3.92		6	8.10
1020	1" diameter		100	.080		2.44	5.10		7.54	10.25
1030	1-1/4" diameter		93	.086		3.29	5.50		8.79	11.75
1040	1-1/2" diameter		85	.094		4.14	6		10.14	13.45
1050	2" diameter		70	.114		5.15	7.30		12.45	16.45
1060	2-1/2" diameter		53	.151		10.50	9.60		20.10	26
1070	3" diameter	2 Elec	80	.200		13.70	12.75		26.45	34
1080	3-1/2" diameter		70	.229		16.90	14.55		31.45	40
1090	4" diameter		60	.267		19.60	17		36.60	46.50

26 05 33.15 Conduit Nipples

		Crew	Daily Output	Labor-Hours	Unit	Material	2021 Bare Costs Labor	Equipment	Total	Total Incl O&P
0010	**CONDUIT NIPPLES** With locknuts and bushings									
0100	Aluminum, 1/2" diameter, close	1 Elec	36	.222	Ea.	7.15	14.15		21.30	29
0120	1-1/2" long		36	.222		10.40	14.15		24.55	32.50
0140	2" long		36	.222		7.20	14.15		21.35	29
0160	2-1/2" long		36	.222		7.90	14.15		22.05	29.50
0180	3" long		36	.222		8.30	14.15		22.45	30
0200	3-1/2" long		36	.222		12.70	14.15		26.85	35
0220	4" long		36	.222		8.70	14.15		22.85	30.50
0240	5" long		36	.222		14.70	14.15		28.85	37
0260	6" long		36	.222		10.25	14.15		24.40	32.50
0280	8" long		36	.222		11.70	14.15		25.85	34

26 05 33.15 Conduit Nipples		Crew	Daily Output	Labor-Hours	Unit	Material	2021 Bare Costs Labor	Equipment	Total	Total Incl O&P
0300	10" long	1 Elec	36	.222	Ea.	12.60	14.15		26.75	35
0320	12" long		36	.222		13.50	14.15		27.65	36
0340	3/4" diameter, close		32	.250		9.80	15.95		25.75	34.50
0360	1-1/2" long		32	.250		9.95	15.95		25.90	34.50
0380	2" long		32	.250		10.05	15.95		26	34.50
0400	2-1/2" long		32	.250		10.55	15.95		26.50	35
0420	3" long		32	.250		10.80	15.95		26.75	35.50
0440	3-1/2" long		32	.250		10.70	15.95		26.65	35.50
0460	4" long		32	.250		16.65	15.95		32.60	42
0480	5" long		32	.250		12.65	15.95		28.60	37.50
0500	6" long		32	.250		20.50	15.95		36.45	46
0520	8" long		32	.250		15.65	15.95		31.60	41
0540	10" long		32	.250		16.60	15.95		32.55	42
0560	12" long		32	.250		18.15	15.95		34.10	43.50
0580	1" diameter, close		27	.296		15.75	18.85		34.60	45.50
0600	2" long		27	.296		16.25	18.85		35.10	46
0620	2-1/2" long		27	.296		16.55	18.85		35.40	46
0640	3" long		27	.296		17.30	18.85		36.15	47
0660	3-1/2" long		27	.296		17.80	18.85		36.65	47.50
0680	4" long		27	.296		26.50	18.85		45.35	57
0700	5" long		27	.296		19.70	18.85		38.55	49.50
0720	6" long		27	.296		21.50	18.85		40.35	51.50
0740	8" long		27	.296		23	18.85		41.85	53
0760	10" long		27	.296		27.50	18.85		46.35	58.50
0780	12" long		27	.296		28.50	18.85		47.35	59.50
0800	1-1/4" diameter, close		23	.348		22.50	22		44.50	57.50
0820	2" long		23	.348		29.50	22		51.50	65.50
0840	2-1/2" long		23	.348		23	22		45	58
0860	3" long		23	.348		24.50	22		46.50	59.50
0880	3-1/2" long		23	.348		24	22		46	59.50
0900	4" long		23	.348		25	22		47	60.50
0920	5" long		23	.348		26.50	22		48.50	62.50
0940	6" long		23	.348		29.50	22		51.50	65.50
0960	8" long		23	.348		31.50	22		53.50	67.50
0980	10" long		23	.348		35	22		57	71.50
1000	12" long		23	.348		37.50	22		59.50	74.50
1020	1-1/2" diameter, close		20	.400		30.50	25.50		56	71.50
1040	2" long		20	.400		30.50	25.50		56	71.50
1060	2-1/2" long		20	.400		30.50	25.50		56	72
1080	3" long		20	.400		31.50	25.50		57	72.50
1100	3-1/2" long		20	.400		33.50	25.50		59	75
1120	4" long		20	.400		34.50	25.50		60	76
1140	5" long		20	.400		38	25.50		63.50	80
1160	6" long		20	.400		38	25.50		63.50	80
1180	8" long		20	.400		43	25.50		68.50	85.50
1200	10" long		20	.400		46	25.50		71.50	89
1220	12" long		20	.400		50	25.50		75.50	93
1240	2" diameter, close		18	.444		38	28.50		66.50	84
1260	2-1/2" long		18	.444		38.50	28.50		67	84.50
1280	3" long		18	.444		39.50	28.50		68	85.50
1300	3-1/2" long		18	.444		40	28.50		68.50	86
1320	4" long		18	.444		43	28.50		71.50	89.50
1340	5" long		18	.444		44.50	28.50		73	91

26 05 33.15 Conduit Nipples		Crew	Daily Output	Labor-Hours	Unit	Material	2021 Bare Costs Labor	Equipment	Total	Total Incl O&P
1360	6" long	1 Elec	18	.444	Ea.	48	28.50		76.50	94.50
1380	8" long		18	.444		54.50	28.50		83	102
1400	10" long		18	.444		56.50	28.50		85	104
1420	12" long		18	.444		62.50	28.50		91	111
1440	2-1/2" diameter, close		15	.533		97	34		131	157
1460	3" long		15	.533		95.50	34		129.50	156
1480	3-1/2" long		15	.533		96.50	34		130.50	157
1500	4" long		15	.533		99.50	34		133.50	160
1520	5" long		15	.533		100	34		134	161
1540	6" long		15	.533		104	34		138	166
1560	8" long		15	.533		113	34		147	176
1580	10" long		15	.533		144	34		178	209
1600	12" long		15	.533		127	34		161	190
1620	3" diameter, close		12	.667		119	42.50		161.50	194
1640	3" long		12	.667		124	42.50		166.50	199
1660	3-1/2" long		12	.667		140	42.50		182.50	217
1680	4" long		12	.667		127	42.50		169.50	203
1700	5" long		12	.667		129	42.50		171.50	205
1720	6" long		12	.667		136	42.50		178.50	212
1740	8" long		12	.667		143	42.50		185.50	220
1760	10" long		12	.667		152	42.50		194.50	230
1780	12" long		12	.667		164	42.50		206.50	244
1800	3-1/2" diameter, close		11	.727		241	46.50		287.50	335
1820	4" long		11	.727		247	46.50		293.50	340
1840	5" long		11	.727		249	46.50		295.50	345
1860	6" long		11	.727		261	46.50		307.50	355
1880	8" long		11	.727		263	46.50		309.50	360
1900	10" long		11	.727		315	46.50		361.50	415
1920	12" long		11	.727		297	46.50		343.50	395
1940	4" diameter, close		9	.889		263	56.50		319.50	375
1960	4" long		9	.889		270	56.50		326.50	380
1980	5" long		9	.889		281	56.50		337.50	395
2000	6" long		9	.889		286	56.50		342.50	400
2020	8" long		9	.889		299	56.50		355.50	415
2040	10" long		9	.889		310	56.50		366.50	425
2060	12" long		9	.889		325	56.50		381.50	445
2080	5" diameter, close		7	1.143		510	73		583	670
2100	5" long		7	1.143		590	73		663	755
2120	6" long		7	1.143		570	73		643	735
2140	8" long		7	1.143		640	73		713	815
2160	10" long		7	1.143		560	73		633	730
2180	12" long		7	1.143		590	73		663	760
2200	6" diameter, close		6	1.333		810	85		895	1,025
2220	5" long		6	1.333		925	85		1,010	1,150
2240	6" long		6	1.333		840	85		925	1,050
2260	8" long		6	1.333		1,100	85		1,185	1,325
2280	10" long		6	1.333		1,100	85		1,185	1,325
2300	12" long		6	1.333		920	85		1,005	1,125
2320	Rigid galvanized steel, 1/2" diameter, close		32	.250		2.52	15.95		18.47	26.50
2340	1-1/2" long		32	.250		2.61	15.95		18.56	26.50
2360	2" long		32	.250		2.82	15.95		18.77	26.50
2380	2-1/2" long		32	.250		2.82	15.95		18.77	26.50
2400	3" long		32	.250		2.88	15.95		18.83	26.50

For customer support on your Electrical Costs with RSMeans Data, call 800.448.8182.

26 05 33 – Raceway and Boxes for Electrical Systems

26 05 33.15 Conduit Nipples		Crew	Daily Output	Labor-Hours	Unit	Material	2021 Bare Costs Labor	Equipment	Total	Total Incl O&P
2420	3-1/2" long	1 Elec	32	.250	Ea.	3	15.95		18.95	27
2440	4" long		32	.250		3.10	15.95		19.05	27
2460	5" long		32	.250		3.26	15.95		19.21	27
2480	6" long		32	.250		3.52	15.95		19.47	27.50
2500	8" long		32	.250		4.89	15.95		20.84	29
2520	10" long		32	.250		5.35	15.95		21.30	29.50
2540	12" long		32	.250		5.80	15.95		21.75	30
2560	3/4" diameter, close		27	.296		3.67	18.85		22.52	32
2580	2" long		27	.296		3.88	18.85		22.73	32.50
2600	2-1/2" long		27	.296		3.96	18.85		22.81	32.50
2620	3" long		27	.296		4.05	18.85		22.90	32.50
2640	3-1/2" long		27	.296		4.13	18.85		22.98	32.50
2660	4" long		27	.296		4.27	18.85		23.12	32.50
2680	5" long		27	.296		4.53	18.85		23.38	33
2700	6" long		27	.296		4.78	18.85		23.63	33.50
2720	8" long		27	.296		6.10	18.85		24.95	34.50
2740	10" long		27	.296		6.95	18.85		25.80	35.50
2760	12" long		27	.296		7.25	18.85		26.10	36
2780	1" diameter, close		23	.348		5.65	22		27.65	39.50
2800	2" long		23	.348		5.80	22		27.80	39.50
2820	2-1/2" long		23	.348		5.95	22		27.95	39.50
2840	3" long		23	.348		6.15	22		28.15	40
2860	3-1/2" long		23	.348		6.40	22		28.40	40
2880	4" long		23	.348		6.55	22		28.55	40
2900	5" long		23	.348		6.85	22		28.85	40.50
2920	6" long		23	.348		7	22		29	40.50
2940	8" long		23	.348		8.65	22		30.65	42.50
2960	10" long		23	.348		10.05	22		32.05	44
2980	12" long		23	.348		10.50	22		32.50	44.50
3000	1-1/4" diameter, close		20	.400		7.65	25.50		33.15	46.50
3020	2" long		20	.400		7.75	25.50		33.25	46.50
3040	3" long		20	.400		8.15	25.50		33.65	47
3060	3-1/2" long		20	.400		8.50	25.50		34	47.50
3080	4" long		20	.400		8.60	25.50		34.10	47.50
3100	5" long		20	.400		9.05	25.50		34.55	48
3120	6" long		20	.400		9.40	25.50		34.90	48.50
3140	8" long		20	.400		11.75	25.50		37.25	51
3160	10" long		20	.400		13.25	25.50		38.75	52.50
3180	12" long		20	.400		14.25	25.50		39.75	53.50
3200	1-1/2" diameter, close		18	.444		11.25	28.50		39.75	54.50
3220	2" long		18	.444		11.40	28.50		39.90	54.50
3240	2-1/2" long		18	.444		11.70	28.50		40.20	55
3260	3" long		18	.444		11.90	28.50		40.40	55
3280	3-1/2" long		18	.444		12.40	28.50		40.90	55.50
3300	4" long		18	.444		12.60	28.50		41.10	56
3320	5" long		18	.444		13.05	28.50		41.55	56.50
3340	6" long		18	.444		13.85	28.50		42.35	57.50
3360	8" long		18	.444		16.40	28.50		44.90	60
3380	10" long		18	.444		17.95	28.50		46.45	61.50
3400	12" long		18	.444		18.50	28.50		47	62.50
3420	2" diameter, close		16	.500		14.70	32		46.70	63.50
3440	2-1/2" long		16	.500		15.30	32		47.30	64.50
3460	3" long		16	.500		15.65	32		47.65	64.50

26 05 33.15 Conduit Nipples

	26 05 33.15 Conduit Nipples	Crew	Daily Output	Labor-Hours	Unit	Material	2021 Bare Costs Labor	Equipment	Total	Total Incl O&P
3480	3-1/2" long	1 Elec	16	.500	Ea.	16.25	32		48.25	65.50
3500	4" long		16	.500		16.55	32		48.55	65.50
3520	5" long		16	.500		17.35	32		49.35	66.50
3540	6" long		16	.500		18.30	32		50.30	67.50
3560	8" long		16	.500		21	32		53	70.50
3580	10" long		16	.500		22.50	32		54.50	72
3600	12" long		16	.500		24	32		56	74
3620	2-1/2" diameter, close		13	.615		50	39		89	114
3640	3" long		13	.615		50	39		89	114
3660	3-1/2" long		13	.615		52	39		91	116
3680	4" long		13	.615		51.50	39		90.50	116
3700	5" long		13	.615		53.50	39		92.50	118
3720	6" long		13	.615		54.50	39		93.50	119
3740	8" long		13	.615		58.50	39		97.50	123
3760	10" long		13	.615		62	39		101	127
3780	12" long		13	.615		63.50	39		102.50	129
3800	3" diameter, close		12	.667		49.50	42.50		92	118
3820	3" long		12	.667		50.50	42.50		93	119
3900	3-1/2" long		12	.667		52	42.50		94.50	120
3920	4" long		12	.667		52	42.50		94.50	120
3940	5" long		12	.667		54	42.50		96.50	123
3960	6" long		12	.667		55.50	42.50		98	124
3980	8" long		12	.667		60	42.50		102.50	129
4000	10" long		12	.667		63.50	42.50		106	133
4020	12" long		12	.667		68	42.50		110.50	138
4040	3-1/2" diameter, close		10	.800		82.50	51		133.50	166
4060	4" long		10	.800		85.50	51		136.50	170
4080	5" long		10	.800		87	51		138	172
4100	6" long		10	.800		89	51		140	174
4120	8" long		10	.800		95	51		146	180
4140	10" long		10	.800		99	51		150	185
4160	12" long		10	.800		103	51		154	190
4180	4" diameter, close		8	1		152	63.50		215.50	262
4200	4" long		8	1		155	63.50		218.50	265
4220	5" long		8	1		158	63.50		221.50	269
4240	6" long		8	1		159	63.50		222.50	270
4260	8" long		8	1		164	63.50		227.50	275
4280	10" long		8	1		170	63.50		233.50	282
4300	12" long		8	1		174	63.50		237.50	287
4320	5" diameter, close		6	1.333		204	85		289	350
4340	5" long		6	1.333		216	85		301	365
4360	6" long		6	1.333		219	85		304	365
4380	8" long		6	1.333		232	85		317	380
4400	10" long		6	1.333		235	85		320	385
4420	12" long		6	1.333		244	85		329	395
4440	6" diameter, close		5	1.600		360	102		462	545
4460	5" long		5	1.600		370	102		472	560
4480	6" long		5	1.600		375	102		477	565
4500	8" long		5	1.600		385	102		487	575
4520	10" long		5	1.600		400	102		502	590
4540	12" long		5	1.600		405	102		507	600
4560	Plastic coated, 40 mil thick, 1/2" diameter, 2" long		32	.250		25.50	15.95		41.45	51.50
4580	2-1/2" long		32	.250		47	15.95		62.95	75

26 05 33 – Raceway and Boxes for Electrical Systems

26 05 33.15 Conduit Nipples	Crew	Daily Output	Labor-Hours	Unit	Material	2021 Bare Costs Labor	Equipment	Total	Total Incl O&P
4600	3" long	1 Elec	32	.250	Ea.	27	15.95		42.95	53.50
4680	3-1/2" long		32	.250		28	15.95		43.95	54.50
4700	4" long		32	.250		32	15.95		47.95	58.50
4720	5" long		32	.250		30	15.95		45.95	56.50
4740	6" long		32	.250		33	15.95		48.95	59.50
4760	8" long		32	.250		30.50	15.95		46.45	57
4780	10" long		32	.250		32	15.95		47.95	58.50
4800	12" long		32	.250		36	15.95		51.95	63
4820	3/4" diameter, 2" long		26	.308		26	19.60		45.60	57.50
4840	2-1/2" long		26	.308		27	19.60		46.60	59
4860	3" long		26	.308		28.50	19.60		48.10	60.50
4880	3-1/2" long		26	.308		29.50	19.60		49.10	61.50
4900	4" long		26	.308		31.50	19.60		51.10	63.50
4920	5" long		26	.308		31	19.60		50.60	63.50
4940	6" long		26	.308		34	19.60		53.60	66.50
4960	8" long		26	.308		32.50	19.60		52.10	65
4980	10" long		26	.308		34	19.60		53.60	66.50
5000	12" long		26	.308		38.50	19.60		58.10	71.50
5020	1" diameter, 2" long		22	.364		29	23		52	66.50
5040	2-1/2" long		22	.364		30.50	23		53.50	68
5060	3" long		22	.364		31	23		54	69
5080	3-1/2" long		22	.364		32.50	23		55.50	70.50
5100	4" long		22	.364		36	23		59	74
5120	5" long		22	.364		34.50	23		57.50	72.50
5140	6" long		22	.364		36.50	23		59.50	74.50
5160	8" long		22	.364		36	23		59	74
5180	10" long		22	.364		38.50	23		61.50	77
5200	12" long		22	.364		45.50	23		68.50	84.50
5220	1-1/4" diameter, 2" long		18	.444		34	28.50		62.50	79.50
5240	2-1/2" long		18	.444		36	28.50		64.50	82
5260	3" long		18	.444		39	28.50		67.50	85
5280	3-1/2" long		18	.444		38.50	28.50		67	84
5300	4" long		18	.444		43.50	28.50		72	90
5320	5" long		18	.444		42.50	28.50		71	88.50
5340	6" long		18	.444		43	28.50		71.50	89
5360	8" long		18	.444		46	28.50		74.50	92.50
5380	10" long		18	.444		49.50	28.50		78	96.50
5400	12" long		18	.444		63.50	28.50		92	112
5420	1-1/2" diameter, 2" long		16	.500		40	32		72	91.50
5440	2-1/2" long		16	.500		42	32		74	94
5460	3" long		16	.500		40.50	32		72.50	92
5480	3-1/2" long		16	.500		44.50	32		76.50	96.50
5500	4" long		16	.500		49	32		81	102
5520	5" long		16	.500		46.50	32		78.50	99
5540	6" long		16	.500		52.50	32		84.50	105
5560	8" long		16	.500		53	32		85	106
5580	10" long		16	.500		67	32		99	121
5600	12" long		16	.500		80.50	32		112.50	136
5620	2" diameter, 2-1/2" long		14	.571		46.50	36.50		83	106
5640	3" long		14	.571		49.50	36.50		86	109
5660	3-1/2" long		14	.571		52.50	36.50		89	112
5680	4" long		14	.571		58	36.50		94.50	118
5700	5" long		14	.571		59	36.50		95.50	119

For customer support on your Electrical Costs with RSMeans Data, call 800.448.8182.

161

26 05 33.15 Conduit Nipples

		Crew	Daily Output	Labor-Hours	Unit	Material	2021 Bare Costs Labor	Equipment	Total	Total Incl O&P
5720	6" long	1 Elec	14	.571	Ea.	64.50	36.50		101	125
5740	8" long		14	.571		66.50	36.50		103	127
5760	10" long		14	.571		83.50	36.50		120	146
5780	12" long		14	.571		102	36.50		138.50	167
5800	2-1/2" diameter, 3-1/2" long		12	.667		107	42.50		149.50	181
5820	4" long		12	.667		111	42.50		153.50	185
5840	5" long		12	.667		123	42.50		165.50	199
5860	6" long		12	.667		129	42.50		171.50	205
5880	8" long		12	.667		145	42.50		187.50	222
5900	10" long		12	.667		155	42.50		197.50	233
5920	12" long		12	.667		187	42.50		229.50	269
5940	3" diameter, 3-1/2" long		11	.727		132	46.50		178.50	214
5960	4" long		11	.727		122	46.50		168.50	203
5980	5" long		11	.727		132	46.50		178.50	215
6000	6" long		11	.727		140	46.50		186.50	223
6020	8" long		11	.727		159	46.50		205.50	244
6040	10" long		11	.727		202	46.50		248.50	291
6060	12" long		11	.727		228	46.50		274.50	320
6080	3-1/2" diameter, 4" long		9	.889		169	56.50		225.50	270
6100	5" long		9	.889		171	56.50		227.50	272
6120	6" long		9	.889		198	56.50		254.50	300
6140	8" long		9	.889		208	56.50		264.50	315
6160	10" long		9	.889		241	56.50		297.50	350
6180	12" long		9	.889		275	56.50		331.50	390
6200	4" diameter, 4" long		7.50	1.067		260	68		328	385
6220	5" long		7.50	1.067		252	68		320	380
6240	6" long		7.50	1.067		267	68		335	395
6260	8" long		7.50	1.067		292	68		360	420
6280	10" long		7.50	1.067		345	68		413	480
6300	12" long		7.50	1.067		385	68		453	520
6320	5" diameter, 5" long		5.50	1.455		224	92.50		316.50	385
6340	6" long		5.50	1.455		254	92.50		346.50	415
6360	8" long		5.50	1.455		261	92.50		353.50	425
6380	10" long		5.50	1.455		271	92.50		363.50	435
6400	12" long		5.50	1.455		295	92.50		387.50	465
6420	6" diameter, 5" long		4.50	1.778		410	113		523	620
6440	6" long		4.50	1.778		420	113		533	630
6460	8" long		4.50	1.778		435	113		548	645
6480	10" long		4.50	1.778		445	113		558	660
6500	12" long		4.50	1.778		465	113		578	685

26 05 33.16 Boxes for Electrical Systems

		Crew	Daily Output	Labor-Hours	Unit	Material	2021 Bare Costs Labor	Equipment	Total	Total Incl O&P
0010	**BOXES FOR ELECTRICAL SYSTEMS** R260533-65									
0020	Pressed steel, octagon, 4"	1 Elec	20	.400	Ea.	1.31	25.50		26.81	39.50
0040	For Romex or BX		20	.400		2.47	25.50		27.97	40.50
0050	For Romex or BX, with bracket		20	.400		3.36	25.50		28.86	41.50
0060	Covers, blank		64	.125		.67	7.95		8.62	12.60
0100	Extension rings		40	.200		2.11	12.75		14.86	21.50
0150	Square, 4"		20	.400		2.20	25.50		27.70	40.50
0160	For Romex or BX		20	.400		6.40	25.50		31.90	45
0170	For Romex or BX, with bracket		20	.400		9.05	25.50		34.55	48
0200	Extension rings		40	.200		2.68	12.75		15.43	22
0220	2-1/8" deep, 1" KO		20	.400		3.07	25.50		28.57	41.50

26 05 33.16 Boxes for Electrical Systems		Crew	Daily Output	Labor-Hours	Unit	Material	2021 Bare Costs Labor	Equipment	Total	Total Incl O&P
0250	Covers, blank	1 Elec	64	.125	Ea.	.98	7.95		8.93	12.95
0260	Raised device		64	.125		1.54	7.95		9.49	13.55
0300	Plaster rings		64	.125		1.66	7.95		9.61	13.70
0350	Square, 4-11/16"		20	.400		4.23	25.50		29.73	42.50
0370	2-1/8" deep, 3/4" to 1-1/4" KO		20	.400		5.10	25.50		30.60	43.50
0400	Extension rings		40	.200		4.99	12.75		17.74	24.50
0450	Covers, blank		53	.151		1.53	9.60		11.13	16
0460	Raised device		53	.151		3.40	9.60		13	18.05
0500	Plaster rings		53	.151		3.91	9.60		13.51	18.60
0550	Handy box		27	.296		2.84	18.85		21.69	31
0560	Covers, device		64	.125		.69	7.95		8.64	12.60
0600	Extension rings		54	.148		2.58	9.45		12.03	16.85
0650	Switchbox		27	.296		4.70	18.85		23.55	33
0660	Romex or BX		27	.296		3.20	18.85		22.05	31.50
0670	with bracket		27	.296		7.90	18.85		26.75	36.50
0680	Partition, metal		27	.296		3.23	18.85		22.08	31.50
0700	Masonry, 1 gang, 2-1/2" deep		27	.296		3.71	18.85		22.56	32
0710	3-1/2" deep		27	.296		4.35	18.85		23.20	33
0750	2 gang, 2-1/2" deep		20	.400		10.35	25.50		35.85	49.50
0760	3-1/2" deep		20	.400		6.20	25.50		31.70	45
0800	3 gang, 2-1/2" deep		13	.615		12.40	39		51.40	72
0850	4 gang, 2-1/2" deep		10	.800		13.45	51		64.45	90.50
0860	5 gang, 2-1/2" deep		9	.889		20	56.50		76.50	106
0870	6 gang, 2-1/2" deep		8	1		56	63.50		119.50	156
0880	Masonry thru-the-wall, 1 gang, 4" block		16	.500		38.50	32		70.50	89.50
0890	6" block		16	.500		60.50	32		92.50	114
0900	8" block		16	.500		79	32		111	135
0920	2 gang, 6" block		16	.500		90.50	32		122.50	147
0940	Bar hanger with 3/8" stud, for wood and masonry boxes		53	.151		5.50	9.60		15.10	20.50
0950	Concrete, set flush, 4" deep		20	.400		15.85	25.50		41.35	55.50
1000	Plate with 3/8" stud		80	.100		6.55	6.35		12.90	16.65
1100	Concrete, floor, 1 gang		5.30	1.509		118	96		214	273
1150	2 gang		4	2		180	127		307	385
1200	3 gang		2.70	2.963		252	189		441	555
1250	For duplex receptacle, pedestal mounted, add		24	.333		119	21		140	163
1270	Flush mounted, add		27	.296		30.50	18.85		49.35	61.50
1300	For telephone, pedestal mounted, add		30	.267		110	17		127	146
1350	Carpet flange, 1 gang		53	.151		62	9.60		71.60	83
1400	Cast, 1 gang, FS (2" deep), 1/2" hub		12	.667		27	42.50		69.50	93
1410	3/4" hub		12	.667		27.50	42.50		70	93.50
1420	FD (2-11/16" deep), 1/2" hub		12	.667		19.40	42.50		61.90	84.50
1430	3/4" hub		12	.667		21.50	42.50		64	87
1450	2 gang, FS, 1/2" hub		10	.800		50	51		101	131
1460	3/4" hub		10	.800		52	51		103	133
1470	FD, 1/2" hub		10	.800		59.50	51		110.50	141
1480	3/4" hub		10	.800		57.50	51		108.50	139
1500	3 gang, FS, 3/4" hub		9	.889		76.50	56.50		133	168
1510	Switch cover, 1 gang, FS		64	.125		7.10	7.95		15.05	19.65
1520	2 gang		53	.151		11.15	9.60		20.75	26.50
1530	Duplex receptacle cover, 1 gang, FS		64	.125		7	7.95		14.95	19.55
1540	2 gang, FS		53	.151		10.95	9.60		20.55	26.50
1542	Weatherproof blank cover, 1 gang		64	.125		1.27	7.95		9.22	13.25
1544	2 gang		53	.151		1.27	9.60		10.87	15.70

For customer support on your Electrical Costs with RSMeans Data, call 800.448.8182.

163

26 05 33.16 Boxes for Electrical Systems

	26 05 33.16 Boxes for Electrical Systems	Crew	Daily Output	Labor-Hours	Unit	Material	2021 Bare Costs Labor	Equipment	Total	Total Incl O&P
1550	Weatherproof switch cover, 1 gang	1 Elec	64	.125	Ea.	5.95	7.95		13.90	18.40
1554	2 gang		53	.151		10.15	9.60		19.75	25.50
1600	Weatherproof receptacle cover, 1 gang		64	.125		2.61	7.95		10.56	14.70
1604	2 gang		53	.151		10.05	9.60		19.65	25.50
1620	Weatherproof receptacle cover, tamper resistant, 1 gang		58	.138		8.90	8.80		17.70	23
1624	2 gang		48	.167		27	10.60		37.60	45.50
1750	FSC, 1 gang, 1/2" hub		11	.727		29	46.50		75.50	101
1760	3/4" hub		11	.727		32.50	46.50		79	105
1770	2 gang, 1/2" hub		9	.889		57	56.50		113.50	147
1780	3/4" hub		9	.889		58.50	56.50		115	149
1790	FDC, 1 gang, 1/2" hub		11	.727		32.50	46.50		79	105
1800	3/4" hub		11	.727		35	46.50		81.50	108
1810	2 gang, 1/2" hub		9	.889		70	56.50		126.50	161
1820	3/4" hub		9	.889		61	56.50		117.50	151
1850	Weatherproof in-use cover, 1 gang		64	.125		33.50	7.95		41.45	48.50
1870	2 gang		53	.151		31	9.60		40.60	48.50
2000	Poke-thru fitting, fire rated, for 3-3/4" floor		6.80	1.176		180	75		255	310
2040	For 7" floor		6.80	1.176		231	75		306	365
2100	Pedestal, 15 amp, duplex receptacle & blank plate		5.25	1.524		180	97		277	340
2120	Duplex receptacle and telephone plate		5.25	1.524		180	97		277	340
2140	Pedestal, 20 amp, duplex recept. & phone plate		5	1.600		181	102		283	350
2160	Telephone plate, both sides		5.25	1.524		167	97		264	325
2200	Abandonment plate		32	.250		46	15.95		61.95	74

26 05 33.17 Outlet Boxes, Plastic

	26 05 33.17 Outlet Boxes, Plastic	Crew	Daily Output	Labor-Hours	Unit	Material	2021 Bare Costs Labor	Equipment	Total	Total Incl O&P
0010	**OUTLET BOXES, PLASTIC**									
0050	4" diameter, round with 2 mounting nails	1 Elec	25	.320	Ea.	2.49	20.50		22.99	33
0100	Bar hanger mounted		25	.320		4.83	20.50		25.33	36
0200	4", square with 2 mounting nails		25	.320		5.35	20.50		25.85	36.50
0300	Plaster ring		64	.125		2.38	7.95		10.33	14.45
0400	Switch box with 2 mounting nails, 1 gang		30	.267		2.04	17		19.04	27
0500	2 gang		25	.320		3.38	20.50		23.88	34
0600	3 gang		20	.400		5.70	25.50		31.20	44.50
0700	Old work box		30	.267		2.19	17		19.19	27.50
1400	PVC, FSS, 1 gang, 1/2" hub		14	.571		10.60	36.50		47.10	65.50
1410	3/4" hub		14	.571		10.40	36.50		46.90	65.50
1420	FD, 1 gang for variable terminations		14	.571		9.55	36.50		46.05	64.50
1450	FS, 2 gang for variable terminations		12	.667		9.50	42.50		52	73.50
1480	Weatherproof blank cover, FS, 1 gang		64	.125		4.72	7.95		12.67	17.05
1500	2 gang		53	.151		5.20	9.60		14.80	20
1510	Weatherproof switch cover, FS, 1 gang		64	.125		10.75	7.95		18.70	23.50
1520	2 gang		53	.151		18.85	9.60		28.45	35
1530	Weatherproof duplex receptacle cover, FS, 1 gang		64	.125		16.60	7.95		24.55	30
1540	2 gang		53	.151		18.35	9.60		27.95	34.50
1750	FSC, 1 gang, 1/2" hub		13	.615		9.50	39		48.50	69
1760	3/4" hub		13	.615		10.10	39		49.10	69.50
1770	FSC, 2 gang, 1/2" hub		11	.727		13.10	46.50		59.60	83.50
1780	3/4" hub		11	.727		12.90	46.50		59.40	83
1790	FDC, 1 gang, 1/2" hub		13	.615		8.80	39		47.80	68
1800	3/4" hub		13	.615		12.80	39		51.80	72.50
1810	Weatherproof, T box w/3 holes		14	.571		14.10	36.50		50.60	69.50
1820	4" diameter round w/5 holes		14	.571		13.75	36.50		50.25	69
1850	In-use cover, 1 gang		64	.125		8.90	7.95		16.85	21.50

26 05 Common Work Results for Electrical

26 05 33 – Raceway and Boxes for Electrical Systems

26 05 33.17 Outlet Boxes, Plastic

		Crew	Daily Output	Labor-Hours	Unit	Material	2021 Bare Costs Labor	Equipment	Total	Total Incl O&P
1870	2 gang	1 Elec	53	.151	Ea.	12.20	9.60		21.80	27.50

26 05 33.18 Pull Boxes

		Crew	Daily Output	Labor-Hours	Unit	Material	2021 Bare Costs Labor	Equipment	Total	Total Incl O&P
0010	**PULL BOXES** R260533-70									
0100	Steel, pull box, NEMA 1, type SC, 6" W x 6" H x 4" D	1 Elec	8	1	Ea.	9.50	63.50		73	105
0180	8" W x 6" H x 4" D		8	1		13.30	63.50		76.80	109
0200	8" W x 8" H x 4" D		8	1		13.05	63.50		76.55	109
0210	10" W x 10" H x 4" D		7	1.143		17.10	73		90.10	127
0220	12" W x 12" H x 4" D		6.50	1.231		22	78.50		100.50	140
0230	15" W x 15" H x 4" D		5.20	1.538		40.50	98		138.50	191
0240	18" W x 18" H x 4" D		4.40	1.818		40	116		156	216
0250	6" W x 6" H x 6" D		8	1		11.70	63.50		75.20	107
0260	8" W x 8" H x 6" D		7.50	1.067		15.55	68		83.55	118
0270	10" W x 10" H x 6" D		5.50	1.455		20.50	92.50		113	161
0300	10" W x 12" H x 6" D		5.30	1.509		32	96		128	179
0310	12" W x 12" H x 6" D		5.20	1.538		26	98		124	175
0320	15" W x 15" H x 6" D		4.60	1.739		101	111		212	276
0330	18" W x 18" H x 6" D		4.20	1.905		46.50	121		167.50	231
0340	24" W x 24" H x 6" D		3.20	2.500		92.50	159		251.50	340
0350	12" W x 12" H x 8" D		5	1.600		29.50	102		131.50	184
0360	15" W x 15" H x 8" D		4.50	1.778		45.50	113		158.50	218
0370	18" W x 18" H x 8" D		4	2		61.50	127		188.50	257
0380	24" W x 18" H x 6" D		3.70	2.162		89.50	138		227.50	305
0400	16" W x 20" H x 8" D		4	2		83	127		210	280
0500	20" W x 24" H x 8" D		3.20	2.500		97	159		256	345
0510	24" W x 24" H x 8" D		3	2.667		104	170		274	365
0600	24" W x 36" H x 8" D		2.70	2.963		151	189		340	445
0610	30" W x 30" H x 8" D		2.70	2.963		197	189		386	495
0620	36" W x 36" H x 8" D		2	4		210	255		465	610
0630	24" W x 24" H x 10" D		2.50	3.200		169	204		373	490
0650	Pull box, hinged, NEMA 1, 6" W x 6" H x 4" D		8	1		12.75	63.50		76.25	109
0660	8" W x 8" H x 4" D		8	1		17.75	63.50		81.25	114
0670	10" W x 10" H x 4" D		7	1.143		24	73		97	135
0680	12" W x 12" H x 4" D		6	1.333		28.50	85		113.50	158
0690	15" W x 15" H x 4" D		5.20	1.538		42.50	98		140.50	193
0700	18" W x 18" H x 4" D		4.40	1.818		58	116		174	236
0710	6" W x 6" H x 6" D		8	1		16.30	63.50		79.80	112
0720	8" W x 8" H x 6" D		7.50	1.067		21.50	68		89.50	125
0730	10" W x 10" H x 6" D		5.50	1.455		33	92.50		125.50	175
0740	12" W x 12" H x 6" D		5.20	1.538		34	98		132	184
0800	12" W x 16" H x 6" D		4.70	1.702		80	108		188	250
0820	18" W x 18" H x 6" D		4.20	1.905		59.50	121		180.50	246
1000	20" W x 20" H x 6" D		3.60	2.222		216	142		358	445
1010	24" W x 24" H x 6" D		3.20	2.500		119	159		278	370
1020	12" W x 12" H x 8" D		5	1.600		74.50	102		176.50	233
1030	15" W x 15" H x 8" D		4.50	1.778		90	113		203	267
1040	18" W x 18" H x 8" D		4	2		126	127		253	330
1200	20" W x 20" H x 8" D		3.20	2.500		170	159		329	425
1210	24" W x 24" H x 8" D		3	2.667		178	170		348	450
1220	30" W x 30" H x 8" D		2.70	2.963		300	189		489	610
1400	24" W x 36" H x 8" D		2.70	2.963		250	189		439	555
1600	24" W x 42" H x 8" D		2	4		370	255		625	785
1610	36" W x 36" H x 8" D		2	4		310	255		565	720

For customer support on your Electrical Costs with RSMeans Data, call 800.448.8182.

165

26 05 33.18 Pull Boxes

		Crew	Daily Output	Labor-Hours	Unit	Material	2021 Bare Costs Labor	2021 Bare Costs Equipment	Total	Total Incl O&P
2100	Pull box, NEMA 3R, type SC, raintight & weatherproof									
2150	6" L x 6" W x 6" D	1 Elec	10	.800	Ea.	19.50	51		70.50	97
2200	8" L x 6" W x 6" D		8	1		40.50	63.50		104	139
2250	10" L x 6" W x 6" D		7	1.143		32.50	73		105.50	144
2300	12" L x 12" W x 6" D		5	1.600		46.50	102		148.50	202
2350	16" L x 16" W x 6" D		4.50	1.778		92.50	113		205.50	270
2400	20" L x 20" W x 6" D		4	2		89	127		216	287
2450	24" L x 18" W x 8" D		3	2.667		148	170		318	415
2500	24" L x 24" W x 10" D		2.50	3.200		190	204		394	515
2550	30" L x 24" W x 12" D		2	4		860	255		1,115	1,325
2600	36" L x 36" W x 12" D	↓	1.50	5.333	↓	455	340		795	1,000
2800	Cast iron, pull boxes for surface mounting									
3000	NEMA 4, watertight & dust tight									
3050	6" L x 6" W x 6" D	1 Elec	4	2	Ea.	283	127		410	500
3100	8" L x 6" W x 6" D		3.20	2.500		470	159		629	755
3150	10" L x 6" W x 6" D		2.50	3.200		490	204		694	845
3200	12" L x 12" W x 6" D		2.30	3.478		715	222		937	1,125
3250	16" L x 16" W x 6" D		1.30	6.154		805	390		1,195	1,475
3300	20" L x 20" W x 6" D		.80	10		980	635		1,615	2,025
3350	24" L x 18" W x 8" D		.70	11.429		2,575	730		3,305	3,900
3400	24" L x 24" W x 10" D		.50	16		4,825	1,025		5,850	6,825
3450	30" L x 24" W x 12" D		.40	20		6,350	1,275		7,625	8,875
3500	36" L x 36" W x 12" D		.20	40		2,375	2,550		4,925	6,375
3510	NEMA 4 clamp cover, 6" L x 6" W x 4" D		4	2		203	127		330	410
3520	8" L x 6" W x 4" D	↓	4	2	↓	335	127		462	560
4000	NEMA 7, explosion proof									
4050	6" L x 6" W x 6" D	1 Elec	2	4	Ea.	565	255		820	1,000
4100	8" L x 6" W x 6" D		1.80	4.444		990	283		1,273	1,500
4150	10" L x 6" W x 6" D		1.60	5		1,425	320		1,745	2,025
4200	12" L x 12" W x 6" D		1	8		2,600	510		3,110	3,600
4250	16" L x 14" W x 6" D		.60	13.333		3,300	850		4,150	4,875
4300	18" L x 18" W x 8" D		.50	16		6,525	1,025		7,550	8,700
4350	24" L x 18" W x 8" D		.40	20		7,800	1,275		9,075	10,500
4400	24" L x 24" W x 10" D		.30	26.667		11,300	1,700		13,000	14,900
4450	30" L x 24" W x 12" D		.20	40		14,800	2,550		17,350	20,100
5000	NEMA 9, dust tight 6" L x 6" W x 6" D		3.20	2.500		425	159		584	705
5050	8" L x 6" W x 6" D		2.70	2.963		510	189		699	840
5100	10" L x 6" W x 6" D		2	4		675	255		930	1,125
5150	12" L x 12" W x 6" D		1.60	5		1,300	320		1,620	1,900
5200	16" L x 16" W x 6" D		1	8		2,150	510		2,660	3,125
5250	18" L x 18" W x 8" D		.70	11.429		3,375	730		4,105	4,775
5300	24" L x 18" W x 8" D		.60	13.333		4,775	850		5,625	6,500
5350	24" L x 24" W x 10" D		.40	20		6,400	1,275		7,675	8,950
5400	30" L x 24" W x 12" D	↓	.30	26.667	↓	9,675	1,700		11,375	13,100
6000	J.I.C. wiring boxes, NEMA 12, dust tight & drip tight									
6050	6" L x 8" W x 4" D	1 Elec	10	.800	Ea.	77	51		128	161
6100	8" L x 10" W x 4" D		8	1		96	63.50		159.50	201
6150	12" L x 14" W x 6" D		5.30	1.509		161	96		257	320
6200	14" L x 16" W x 6" D		4.70	1.702		187	108		295	365
6250	16" L x 20" W x 6" D		4.40	1.818		251	116		367	450
6300	24" L x 30" W x 6" D		3.20	2.500		370	159		529	640
6350	24" L x 30" W x 8" D		2.90	2.759		375	176		551	675
6400	24" L x 36" W x 8" D	↓	2.70	2.963		420	189		609	740

26 05 33 – Raceway and Boxes for Electrical Systems

26 05 33.18 Pull Boxes		Crew	Daily Output	Labor-Hours	Unit	Material	2021 Bare Costs Labor	Equipment	Total	Total Incl O&P
6450	24" L x 42" W x 8" D	1 Elec	2.30	3.478	Ea.	480	222		702	855
6500	24" L x 48" W x 8" D		2	4		525	255		780	955

26 05 33.23 Wireway		Crew	Daily Output	Labor-Hours	Unit	Material	2021 Bare Costs Labor	Equipment	Total	Total Incl O&P
0010	**WIREWAY** to 10' high R260533-60									
0020	For higher elevations, see Section 26 05 36.40									
0100	NEMA 1, screw cover w/fittings and supports, 2-1/2" x 2-1/2"	1 Elec	45	.178	L.F.	15.85	11.30		27.15	34.50
0200	4" x 4"	"	40	.200		23	12.75		35.75	44
0400	6" x 6"	2 Elec	60	.267		31	17		48	59.50
0600	8" x 8"		40	.400		22	25.50		47.50	62
0620	10" x 10"		30	.533		66.50	34		100.50	124
0640	12" x 12"		20	.800		57	51		108	139
0800	Elbows, 90°, 2-1/2"	1 Elec	24	.333	Ea.	46	21		67	82.50
1000	4"		20	.400		47.50	25.50		73	90
1200	6"		18	.444		59	28.50		87.50	107
1400	8"		16	.500		66.50	32		98.50	121
1420	10"		12	.667		96	42.50		138.50	168
1440	12"		10	.800		122	51		173	210
1500	Elbows, 45°, 2-1/2"		24	.333		33	21		54	67.50
1510	4"		20	.400		39.50	25.50		65	81.50
1520	6"		18	.444		42	28.50		70.50	88.50
1530	8"		16	.500		66.50	32		98.50	121
1540	10"		12	.667		97	42.50		139.50	169
1550	12"		10	.800		169	51		220	261
1600	"T" box, 2-1/2"		18	.444		37	28.50		65.50	83
1800	4"		16	.500		54.50	32		86.50	108
2000	6"		14	.571		53.50	36.50		90	113
2200	8"		12	.667		93	42.50		135.50	165
2220	10"		10	.800		115	51		166	202
2240	12"		8	1		165	63.50		228.50	276
2300	Cross, 2-1/2"		16	.500		41.50	32		73.50	93
2310	4"		14	.571		50.50	36.50		87	110
2320	6"		12	.667		62.50	42.50		105	132
2400	Panel adapter, 2-1/2"		24	.333		6.95	21		27.95	39
2600	4"		20	.400		7.55	25.50		33.05	46.50
2800	6"		18	.444		14.25	28.50		42.75	57.50
3000	8"		16	.500		20.50	32		52.50	70
3020	10"		14	.571		33	36.50		69.50	90
3040	12"		12	.667		34	42.50		76.50	101
3200	Reducer, 4" to 2-1/2"		24	.333		12.90	21		33.90	45.50
3400	6" to 4"		20	.400		29.50	25.50		55	70.50
3600	8" to 6"		18	.444		36	28.50		64.50	81.50
3620	10" to 8"		16	.500		61.50	32		93.50	116
3640	12" to 10"		14	.571		56.50	36.50		93	117
3780	End cap, 2-1/2"		24	.333		6.30	21		27.30	38.50
3800	4"		20	.400		7.25	25.50		32.75	46
4000	6"		18	.444		7.30	28.50		35.80	50
4200	8"		16	.500		11.10	32		43.10	59.50
4220	10"		14	.571		20	36.50		56.50	76
4240	12"		12	.667		25	42.50		67.50	90.50
4300	U-connector, 2-1/2"		200	.040		4.64	2.55		7.19	8.90
4320	4"		200	.040		5.85	2.55		8.40	10.25
4340	6"		180	.044		7.65	2.83		10.48	12.60

For customer support on your Electrical Costs with RSMeans Data, call 800.448.8182.

167

26 05 33.23 Wireway		Crew	Daily Output	Labor-Hours	Unit	Material	2021 Bare Costs Labor	Equipment	Total	Total Incl O&P
4360	8"	1 Elec	170	.047	Ea.	12.70	3		15.70	18.40
4380	10"		150	.053		23.50	3.40		26.90	31
4400	12"		130	.062		23	3.92		26.92	31.50
4420	Hanger, 2-1/2"		100	.080		11.60	5.10		16.70	20.50
4430	4"		100	.080		12.10	5.10		17.20	21
4440	6"		80	.100		12.45	6.35		18.80	23
4450	8"		65	.123		24.50	7.85		32.35	38.50
4460	10"		50	.160		44.50	10.20		54.70	64
4470	12"		40	.200		65	12.75		77.75	90.50
4475	NEMA 3R, screw cover w/fittings and supports, 4" x 4"		36	.222	L.F.	17	14.15		31.15	39.50
4480	6" x 6"	2 Elec	55	.291		18.85	18.55		37.40	48.50
4485	8" x 8"		36	.444		33.50	28.50		62	78.50
4490	12" x 12"		18	.889		61	56.50		117.50	151
4500	Hinged cover, with fittings and supports, 2-1/2" x 2-1/2"	1 Elec	60	.133		18.95	8.50		27.45	33.50
4520	4" x 4"	"	45	.178		14.20	11.30		25.50	32.50
4540	6" x 6"	2 Elec	80	.200		21.50	12.75		34.25	43
4560	8" x 8"		60	.267		38	17		55	66.50
4580	10" x 10"		50	.320		49	20.50		69.50	84.50
4600	12" x 12"		24	.667		87	42.50		129.50	159
4700	Elbows 90°, 2-1/2" x 2-1/2"	1 Elec	32	.250	Ea.	37	15.95		52.95	64.50
4720	4"		27	.296		157	18.85		175.85	201
4730	6"		23	.348		186	22		208	237
4740	8"		18	.444		62.50	28.50		91	111
4750	10"		14	.571		136	36.50		172.50	204
4760	12"		12	.667		142	42.50		184.50	219
4800	Tee box, hinged cover, 2-1/2" x 2-1/2"		23	.348		55.50	22		77.50	94
4810	4"		20	.400		189	25.50		214.50	246
4820	6"		18	.444		71.50	28.50		100	121
4830	8"		16	.500		285	32		317	365
4840	10"		12	.667		129	42.50		171.50	205
4860	12"		10	.800		217	51		268	315
4880	Cross box, hinged cover, 2-1/2" x 2-1/2"		18	.444		62	28.50		90.50	110
4900	4"		16	.500		253	32		285	325
4920	6"		13	.615		305	39		344	395
4940	8"		11	.727		263	46.50		309.50	360
4960	10"		10	.800		192	51		243	287
4980	12"		9	.889		185	56.50		241.50	287
5000	NEMA 12, hinged cover, 2-1/2" x 2-1/2"		40	.200	L.F.	26.50	12.75		39.25	48.50
5020	4" x 4"		35	.229		30.50	14.55		45.05	55
5040	6" x 6"	2 Elec	60	.267		45	17		62	74.50
5060	8" x 8"	"	50	.320		60	20.50		80.50	96.50
5120	Elbows 90°, flanged, 2-1/2" x 2-1/2"	1 Elec	23	.348	Ea.	100	22		122	143
5140	4"		20	.400		126	25.50		151.50	176
5160	6"		18	.444		158	28.50		186.50	215
5180	8"		15	.533		230	34		264	305
5240	Tee box, flanged, 2-1/2" x 2-1/2"		18	.444		136	28.50		164.50	191
5260	4"		16	.500		157	32		189	221
5280	6"		15	.533		211	34		245	284
5300	8"		13	.615		305	39		344	400
5360	Cross box, flanged, 2-1/2" x 2-1/2"		15	.533		169	34		203	237
5380	4"		13	.615		220	39		259	300
5400	6"		12	.667		278	42.50		320.50	370
5420	8"		10	.800		380	51		431	495

26 05 33 – Raceway and Boxes for Electrical Systems

		Crew	Daily Output	Labor-Hours	Unit	Material	2021 Bare Costs Labor	Equipment	Total	Total Incl O&P
26 05 33.23 Wireway										
5480	Flange gasket, 2-1/2"	1 Elec	160	.050	Ea.	5.05	3.19		8.24	10.35
5500	4"		80	.100		6.85	6.35		13.20	16.95
5520	6"		53	.151		9.05	9.60		18.65	24.50
5530	8"		40	.200		12.05	12.75		24.80	32
26 05 33.25 Conduit Fittings for Rigid Galvanized Steel										
0010	**CONDUIT FITTINGS FOR RIGID GALVANIZED STEEL**									
0050	Standard, locknuts, 1/2" diameter				Ea.	.20			.20	.22
0100	3/4" diameter					.42			.42	.46
0300	1" diameter					.58			.58	.64
0500	1-1/4" diameter					.76			.76	.84
0700	1-1/2" diameter					1.24			1.24	1.36
1000	2" diameter					1.76			1.76	1.94
1030	2-1/2" diameter					4.56			4.56	5
1050	3" diameter					5.90			5.90	6.50
1070	3-1/2" diameter					9.70			9.70	10.65
1100	4" diameter					12.30			12.30	13.55
1110	5" diameter					26			26	28.50
1120	6" diameter					45.50			45.50	50
1130	Bushings, plastic, 1/2" diameter	1 Elec	40	.200		.12	12.75		12.87	19.10
1150	3/4" diameter		32	.250		.13	15.95		16.08	23.50
1170	1" diameter		28	.286		.23	18.20		18.43	27.50
1200	1-1/4" diameter		24	.333		.28	21		21.28	32
1230	1-1/2" diameter		18	.444		.43	28.50		28.93	42.50
1250	2" diameter		15	.533		.91	34		34.91	51.50
1270	2-1/2" diameter		13	.615		1.31	39		40.31	60
1300	3" diameter		12	.667		1.51	42.50		44.01	64.50
1330	3-1/2" diameter		11	.727		2.02	46.50		48.52	71
1350	4" diameter		9	.889		2.20	56.50		58.70	86.50
1360	5" diameter		7	1.143		16.75	73		89.75	126
1370	6" diameter		5	1.600		49	102		151	205
1390	Steel, 1/2" diameter		40	.200		.68	12.75		13.43	19.70
1400	3/4" diameter		32	.250		.95	15.95		16.90	24.50
1430	1" diameter		28	.286		1.57	18.20		19.77	28.50
1450	Steel insulated, 1-1/4" diameter		24	.333		4.38	21		25.38	36.50
1470	1-1/2" diameter		18	.444		5.70	28.50		34.20	48.50
1500	2" diameter		15	.533		7.85	34		41.85	59
1530	2-1/2" diameter		13	.615		15.90	39		54.90	76
1550	3" diameter		12	.667		20.50	42.50		63	86
1570	3-1/2" diameter		11	.727		26.50	46.50		73	98
1600	4" diameter		9	.889		34	56.50		90.50	122
1610	5" diameter		7	1.143		69	73		142	184
1620	6" diameter		5	1.600		106	102		208	267
1630	Sealing locknuts, 1/2" diameter		40	.200		1.37	12.75		14.12	20.50
1650	3/4" diameter		32	.250		1.62	15.95		17.57	25.50
1670	1" diameter		28	.286		2.41	18.20		20.61	29.50
1700	1-1/4" diameter		24	.333		3.71	21		24.71	35.50
1730	1-1/2" diameter		18	.444		4.72	28.50		33.22	47
1750	2" diameter		15	.533		5.95	34		39.95	57
1760	Grounding bushing, insulated, 1/2" diameter		32	.250		6.50	15.95		22.45	30.50
1770	3/4" diameter		28	.286		6.10	18.20		24.30	34
1780	1" diameter		20	.400		9.30	25.50		34.80	48.50
1800	1-1/4" diameter		18	.444		11.25	28.50		39.75	54.50

26 05 33.25 Conduit Fittings for Rigid Galvanized Steel	Crew	Daily Output	Labor-Hours	Unit	Material	2021 Bare Costs Labor	Equipment	Total	Total Incl O&P
1830	1-1/2" diameter	1 Elec	16	.500	Ea.	13.15	32		45.15	62
1850	2" diameter		13	.615		15.45	39		54.45	75.50
1870	2-1/2" diameter		12	.667		28	42.50		70.50	93.50
1900	3" diameter		11	.727		29	46.50		75.50	101
1930	3-1/2" diameter		9	.889		44	56.50		100.50	132
1950	4" diameter		8	1		47.50	63.50		111	147
1960	5" diameter		6	1.333		106	85		191	243
1970	6" diameter		4	2		155	127		282	360
1990	Coupling, with set screw, 1/2" diameter		50	.160		3.34	10.20		13.54	18.80
2000	3/4" diameter		40	.200		4.18	12.75		16.93	23.50
2030	1" diameter		35	.229		7.20	14.55		21.75	29.50
2050	1-1/4" diameter		28	.286		10.55	18.20		28.75	38.50
2070	1-1/2" diameter		23	.348		13.70	22		35.70	48
2090	2" diameter		20	.400		30.50	25.50		56	71.50
2100	2-1/2" diameter		18	.444		72.50	28.50		101	122
2110	3" diameter		15	.533		84	34		118	143
2120	3-1/2" diameter		12	.667		126	42.50		168.50	202
2130	4" diameter		10	.800		161	51		212	253
2140	5" diameter		9	.889		285	56.50		341.50	400
2150	6" diameter		8	1		370	63.50		433.50	500
2160	Box connector with set screw, plain, 1/2" diameter		70	.114		2.20	7.30		9.50	13.20
2170	3/4" diameter		60	.133		3.23	8.50		11.73	16.15
2180	1" diameter		50	.160		4.99	10.20		15.19	20.50
2190	Insulated, 1-1/4" diameter		40	.200		9.50	12.75		22.25	29.50
2200	1-1/2" diameter		30	.267		13.65	17		30.65	40
2210	2" diameter		20	.400		29.50	25.50		55	70
2220	2-1/2" diameter		18	.444		77.50	28.50		106	128
2230	3" diameter		15	.533		101	34		135	162
2240	3-1/2" diameter		12	.667		143	42.50		185.50	220
2250	4" diameter		10	.800		188	51		239	283
2260	5" diameter		9	.889		315	56.50		371.50	430
2270	6" diameter		8	1		435	63.50		498.50	575
2280	LB, LR or LL fittings & covers, 1/2" diameter		16	.500		8	32		40	56.50
2290	3/4" diameter		13	.615		18.10	39		57.10	78.50
2300	1" diameter		11	.727		14.45	46.50		60.95	85
2330	1-1/4" diameter		8	1		22.50	63.50		86	119
2350	1-1/2" diameter		6	1.333		29.50	85		114.50	159
2370	2" diameter		5	1.600		42.50	102		144.50	198
2380	2-1/2" diameter		4	2		102	127		229	300
2390	3" diameter		3.50	2.286		138	146		284	370
2400	3-1/2" diameter		3	2.667		218	170		388	490
2410	4" diameter		2.50	3.200		251	204		455	580
2420	T fittings, with cover, 1/2" diameter		12	.667		10.35	42.50		52.85	74.50
2430	3/4" diameter		11	.727		12.20	46.50		58.70	82.50
2440	1" diameter		9	.889		17.90	56.50		74.40	104
2450	1-1/4" diameter		6	1.333		23.50	85		108.50	152
2470	1-1/2" diameter		5	1.600		35	102		137	190
2500	2" diameter		4	2		41.50	127		168.50	235
2510	2-1/2" diameter		3.50	2.286		98.50	146		244.50	325
2520	3" diameter		3	2.667		123	170		293	390
2530	3-1/2" diameter		2.50	3.200		214	204		418	540
2540	4" diameter		2	4		245	255		500	650
2550	Nipples chase, plain, 1/2" diameter		40	.200		.86	12.75		13.61	19.90

26 05 33 – Raceway and Boxes for Electrical Systems

26 05 33.25 Conduit Fittings for Rigid Galvanized Steel		Crew	Daily Output	Labor-Hours	Unit	Material	2021 Bare Costs Labor	Equipment	Total	Total Incl O&P
2560	3/4" diameter	1 Elec	32	.250	Ea.	1.48	15.95		17.43	25
2570	1" diameter		28	.286		2.49	18.20		20.69	29.50
2600	Insulated, 1-1/4" diameter		24	.333		8.10	21		29.10	40.50
2630	1-1/2" diameter		18	.444		10.50	28.50		39	53.50
2650	2" diameter		15	.533		15.45	34		49.45	67.50
2660	2-1/2" diameter		12	.667		37	42.50		79.50	104
2670	3" diameter		10	.800		46.50	51		97.50	127
2680	3-1/2" diameter		9	.889		60.50	56.50		117	151
2690	4" diameter		8	1		97	63.50		160.50	202
2700	5" diameter		7	1.143		310	73		383	450
2710	6" diameter		6	1.333		505	85		590	680
2720	Nipples offset, plain, 1/2" diameter		40	.200		4.25	12.75		17	23.50
2730	3/4" diameter		32	.250		5.10	15.95		21.05	29
2740	1" diameter		24	.333		5.40	21		26.40	37.50
2750	Insulated, 1-1/4" diameter		20	.400		6.85	25.50		32.35	45.50
2760	1-1/2" diameter		18	.444		8.30	28.50		36.80	51
2770	2" diameter		16	.500		11.65	32		43.65	60.50
2780	3" diameter		14	.571		62.50	36.50		99	123
2850	Coupling, expansion, 1/2" diameter		12	.667		54	42.50		96.50	123
2880	3/4" diameter		10	.800		64	51		115	146
2900	1" diameter		8	1		75.50	63.50		139	178
2920	1-1/4" diameter		6.40	1.250		100	79.50		179.50	228
2940	1-1/2" diameter		5.30	1.509		151	96		247	310
2960	2" diameter		4.60	1.739		205	111		316	390
2980	2-1/2" diameter		3.60	2.222		360	142		502	610
3000	3" diameter		3	2.667		400	170		570	690
3020	3-1/2" diameter		2.80	2.857		670	182		852	1,000
3040	4" diameter		2.40	3.333		785	212		997	1,175
3060	5" diameter		2	4		1,075	255		1,330	1,550
3080	6" diameter		1.80	4.444		1,925	283		2,208	2,550
3100	Expansion deflection, 1/2" diameter		12	.667		271	42.50		313.50	360
3120	3/4" diameter		12	.667		305	42.50		347.50	405
3140	1" diameter		10	.800		360	51		411	475
3160	1-1/4" diameter		6.40	1.250		435	79.50		514.50	595
3180	1-1/2" diameter		5.30	1.509		670	96		766	885
3200	2" diameter		4.60	1.739		625	111		736	855
3220	2-1/2" diameter		3.60	2.222		920	142		1,062	1,225
3240	3" diameter		3	2.667		1,675	170		1,845	2,100
3260	3-1/2" diameter		2.80	2.857		1,350	182		1,532	1,750
3280	4" diameter		2.40	3.333		1,675	212		1,887	2,150
3300	5" diameter		2	4		2,525	255		2,780	3,150
3320	6" diameter		1.80	4.444		4,400	283		4,683	5,250
3340	Ericson, 1/2" diameter		16	.500		3.98	32		35.98	52
3360	3/4" diameter		14	.571		5.15	36.50		41.65	59.50
3380	1" diameter		11	.727		8.15	46.50		54.65	78
3400	1-1/4" diameter		8	1		14.85	63.50		78.35	111
3420	1-1/2" diameter		7	1.143		20.50	73		93.50	131
3440	2" diameter		5	1.600		38.50	102		140.50	193
3460	2-1/2" diameter		4	2		77.50	127		204.50	274
3480	3" diameter		3.50	2.286		120	146		266	350
3500	3-1/2" diameter		3	2.667		173	170		343	440
3520	4" diameter		2.70	2.963		200	189		389	500
3540	5" diameter		2.50	3.200		350	204		554	690

For customer support on your Electrical Costs with RSMeans Data, call 800.448.8182.

171

26 05 33.25 Conduit Fittings for Rigid Galvanized Steel	Crew	Daily Output	Labor-Hours	Unit	Material	2021 Bare Costs Labor	Equipment	Total	Total Incl O&P
3560	6" diameter	1 Elec	2.30	3.478	Ea.	440	222		662	815
3580	Split, 1/2" diameter		32	.250		4.04	15.95		19.99	28
3600	3/4" diameter		27	.296		5.35	18.85		24.20	34
3620	1" diameter		20	.400		8.80	25.50		34.30	47.50
3640	1-1/4" diameter		16	.500		13.10	32		45.10	62
3660	1-1/2" diameter		14	.571		13.95	36.50		50.45	69.50
3680	2" diameter		12	.667		29	42.50		71.50	95
3700	2-1/2" diameter		10	.800		73	51		124	156
3720	3" diameter		9	.889		95	56.50		151.50	188
3740	3-1/2" diameter		8	1		159	63.50		222.50	270
3760	4" diameter		7	1.143		188	73		261	315
3780	5" diameter		6	1.333		325	85		410	480
3800	6" diameter		5	1.600		455	102		557	650
4600	Reducing bushings, 3/4" to 1/2" diameter		54	.148		2.47	9.45		11.92	16.70
4620	1" to 3/4" diameter		46	.174		2.47	11.10		13.57	19.15
4640	1-1/4" to 1" diameter		40	.200		5.10	12.75		17.85	24.50
4660	1-1/2" to 1-1/4" diameter		36	.222		6.30	14.15		20.45	28
4680	2" to 1-1/2" diameter		32	.250		14.50	15.95		30.45	39.50
4740	2-1/2" to 2" diameter		30	.267		17.75	17		34.75	44.50
4760	3" to 2-1/2" diameter		28	.286		19.25	18.20		37.45	48
4800	Through-wall seal, 1/2" diameter		8	1		212	63.50		275.50	330
4820	3/4" diameter		7.50	1.067		273	68		341	400
4840	1" diameter		6.50	1.231		274	78.50		352.50	415
4860	1-1/4" diameter		5.50	1.455		350	92.50		442.50	525
4880	1-1/2" diameter		5	1.600		420	102		522	615
4900	2" diameter		4.20	1.905		305	121		426	515
4920	2-1/2" diameter		3.50	2.286		515	146		661	780
4940	3" diameter		3	2.667		545	170		715	850
4960	3-1/2" diameter		2.50	3.200		830	204		1,034	1,225
4980	4" diameter		2	4		840	255		1,095	1,300
5000	5" diameter		1.50	5.333		865	340		1,205	1,450
5020	6" diameter		1	8		865	510		1,375	1,700
5100	Cable supports, 2 or more wires									
5120	1-1/2" diameter	1 Elec	8	1	Ea.	156	63.50		219.50	266
5140	2" diameter		6	1.333		212	85		297	360
5160	2-1/2" diameter		4	2		229	127		356	440
5180	3" diameter		3.50	2.286		293	146		439	535
5200	3-1/2" diameter		2.60	3.077		380	196		576	705
5220	4" diameter		2	4		480	255		735	910
5240	5" diameter		1.50	5.333		655	340		995	1,225
5260	6" diameter		1	8		1,050	510		1,560	1,925
5280	Service entrance cap, 1/2" diameter		16	.500		5.25	32		37.25	53.50
5300	3/4" diameter		13	.615		4.18	39		43.18	63
5320	1" diameter		10	.800		7.10	51		58.10	83.50
5340	1-1/4" diameter		8	1		5.20	63.50		68.70	100
5360	1-1/2" diameter		6.50	1.231		12.20	78.50		90.70	129
5380	2" diameter		5.50	1.455		23.50	92.50		116	164
5400	2-1/2" diameter		4	2		61	127		188	256
5420	3" diameter		3.40	2.353		89.50	150		239.50	320
5440	3-1/2" diameter		3	2.667		128	170		298	395
5460	4" diameter		2.70	2.963		128	189		317	420
5750	90° pull elbows steel, female, 1/2" diameter		16	.500		7.40	32		39.40	55.50
5760	3/4" diameter		13	.615		8.45	39		47.45	68

26 05 33 – Raceway and Boxes for Electrical Systems

26 05 33.25 Conduit Fittings for Rigid Galvanized Steel		Crew	Daily Output	Labor-Hours	Unit	Material	2021 Bare Costs Labor	Equipment	Total	Total Incl O&P
5780	1" diameter	1 Elec	11	.727	Ea.	13.80	46.50		60.30	84
5800	1-1/4" diameter		8	1		19.90	63.50		83.40	117
5820	1-1/2" diameter		6	1.333		28	85		113	157
5840	2" diameter	▼	5	1.600	▼	44	102		146	200
6000	Explosion proof, flexible coupling									
6010	1/2" diameter, 4" long	1 Elec	12	.667	Ea.	98.50	42.50		141	172
6020	6" long		12	.667		110	42.50		152.50	184
6050	12" long		12	.667		156	42.50		198.50	235
6070	18" long		12	.667		175	42.50		217.50	256
6090	24" long		12	.667		204	42.50		246.50	288
6110	30" long		12	.667		237	42.50		279.50	325
6130	36" long		12	.667		261	42.50		303.50	350
6140	3/4" diameter, 4" long		10	.800		119	51		170	206
6150	6" long		10	.800		136	51		187	226
6180	12" long		10	.800		182	51		233	276
6200	18" long		10	.800		237	51		288	335
6220	24" long		10	.800		269	51		320	370
6240	30" long		10	.800		315	51		366	425
6260	36" long		10	.800		350	51		401	460
6270	1" diameter, 6" long		8	1		245	63.50		308.50	365
6300	12" long		8	1		305	63.50		368.50	430
6320	18" long		8	1		365	63.50		428.50	495
6340	24" long		8	1		425	63.50		488.50	565
6360	30" long		8	1		490	63.50		553.50	635
6380	36" long		8	1		550	63.50		613.50	695
6390	1-1/4" diameter, 12" long		6.40	1.250		450	79.50		529.50	615
6410	18" long		6.40	1.250		530	79.50		609.50	705
6430	24" long		6.40	1.250		650	79.50		729.50	835
6450	30" long		6.40	1.250		715	79.50		794.50	905
6470	36" long		6.40	1.250		1,025	79.50		1,104.50	1,250
6480	1-1/2" diameter, 12" long		5.30	1.509		605	96		701	810
6500	18" long		5.30	1.509		710	96		806	925
6520	24" long		5.30	1.509		825	96		921	1,050
6540	30" long		5.30	1.509		1,000	96		1,096	1,250
6560	36" long		5.30	1.509		1,050	96		1,146	1,300
6570	2" diameter, 12" long		4.60	1.739		775	111		886	1,025
6590	18" long		4.60	1.739		980	111		1,091	1,250
6610	24" long		4.60	1.739		1,075	111		1,186	1,375
6630	30" long		4.60	1.739		1,300	111		1,411	1,600
6650	36" long		4.60	1.739		1,475	111		1,586	1,800
7000	Close up plug, 1/2" diameter, explosion proof		40	.200		2.49	12.75		15.24	21.50
7010	3/4" diameter		32	.250		2.93	15.95		18.88	26.50
7020	1" diameter		28	.286		3.38	18.20		21.58	30.50
7030	1-1/4" diameter		24	.333		3.58	21		24.58	35.50
7040	1-1/2" diameter		18	.444		5.15	28.50		33.65	47.50
7050	2" diameter		15	.533		8.50	34		42.50	60
7060	2-1/2" diameter		13	.615		13.10	39		52.10	73
7070	3" diameter		12	.667		20.50	42.50		63	85.50
7080	3-1/2" diameter		11	.727		23.50	46.50		70	95
7090	4" diameter		9	.889		33	56.50		89.50	121
7091	Elbow female, 45°, 1/2"		16	.500		17.55	32		49.55	67
7092	3/4"		13	.615		20	39		59	80.50
7093	1"		11	.727		22.50	46.50		69	94

For customer support on your Electrical Costs with RSMeans Data, call 800.448.8182.

173

26 05 33.25 Conduit Fittings for Rigid Galvanized Steel	Crew	Daily Output	Labor-Hours	Unit	Material	2021 Bare Costs Labor	Equipment	Total	Total Incl O&P
7094	1-1/4"	1 Elec	8	1	Ea.	34	63.50		97.50	132
7095	1-1/2"		6	1.333		39	85		124	169
7096	2"		5	1.600		42.50	102		144.50	198
7097	2-1/2"		4.50	1.778		125	113		238	305
7098	3"		4.20	1.905		156	121		277	350
7099	3-1/2"		4	2		221	127		348	430
7100	4"		3.80	2.105		270	134		404	495
7101	90°, 1/2"		16	.500		15.35	32		47.35	64.50
7102	3/4"		13	.615		18.30	39		57.30	78.50
7103	1"		11	.727		25	46.50		71.50	96.50
7104	1-1/4"		8	1		36	63.50		99.50	134
7105	1-1/2"		6	1.333		55.50	85		140.50	187
7106	2"		5	1.600		91.50	102		193.50	252
7107	2-1/2"		4.50	1.778		215	113		328	405
7110	Elbows 90°, long male & female, 1/2" diameter, explosion proof		16	.500		24.50	32		56.50	74.50
7120	3/4" diameter		13	.615		27	39		66	88
7130	1" diameter		11	.727		33.50	46.50		80	106
7140	1-1/4" diameter		8	1		39.50	63.50		103	138
7150	1-1/2" diameter		6	1.333		44.50	85		129.50	175
7160	2" diameter		5	1.600		56.50	102		158.50	213
7170	Capped elbow, 1/2" diameter, explosion proof		11	.727		25.50	46.50		72	97
7180	3/4" diameter		8	1		32	63.50		95.50	130
7190	1" diameter		6	1.333		34.50	85		119.50	164
7200	1-1/4" diameter		5	1.600		73.50	102		175.50	232
7210	Pulling elbow, 1/2" diameter, explosion proof		11	.727		96.50	46.50		143	175
7220	3/4" diameter		8	1		186	63.50		249.50	300
7230	1" diameter		6	1.333		276	85		361	430
7240	1-1/4" diameter		5	1.600		355	102		457	545
7250	1-1/2" diameter		5	1.600		440	102		542	630
7260	2" diameter		4	2		455	127		582	690
7270	2-1/2" diameter		3.50	2.286		1,100	146		1,246	1,450
7280	3" diameter		3	2.667		1,375	170		1,545	1,775
7290	3-1/2" diameter		2.50	3.200		1,650	204		1,854	2,125
7300	4" diameter		2.20	3.636		2,175	232		2,407	2,750
7310	LB conduit body, 1/2" diameter		11	.727		64	46.50		110.50	140
7320	3/4" diameter		8	1		76.50	63.50		140	179
7330	T conduit body, 1/2" diameter		9	.889		66.50	56.50		123	157
7340	3/4" diameter		6	1.333		81	85		166	215
7350	Explosion proof, round box w/cover, 3 threaded hubs, 1/2" diameter		8	1		58.50	63.50		122	159
7351	3/4" diameter		8	1		84	63.50		147.50	187
7352	1" diameter		7.50	1.067		89.50	68		157.50	200
7353	1-1/4" diameter		7	1.143		89	73		162	206
7354	1-1/2" diameter		7	1.143		238	73		311	370
7355	2" diameter		6	1.333		244	85		329	395
7356	Round box w/cover & mtng flange, 3 threaded hubs, 1/2" diameter		8	1		54.50	63.50		118	155
7357	3/4" diameter		8	1		60	63.50		123.50	161
7358	4 threaded hubs, 1" diameter		7	1.143		68	73		141	183
7400	Unions, 1/2" diameter		20	.400		16.25	25.50		41.75	56
7410	3/4" - 1/2" diameter		16	.500		18.20	32		50.20	67.50
7420	3/4" diameter		16	.500		22.50	32		54.50	72.50
7430	1" diameter		14	.571		40.50	36.50		77	98.50
7440	1-1/4" diameter		12	.667		62	42.50		104.50	131
7450	1-1/2" diameter		10	.800		76.50	51		127.50	160

26 05 33 – Raceway and Boxes for Electrical Systems

26 05 33.25 Conduit Fittings for Rigid Galvanized Steel		Crew	Daily Output	Labor-Hours	Unit	Material	2021 Bare Costs Labor	Equipment	Total	Total Incl O&P
7460	2" diameter	1 Elec	8.50	.941	Ea.	96.50	60		156.50	195
7480	2-1/2" diameter		8	1		143	63.50		206.50	252
7490	3" diameter		7	1.143		186	73		259	315
7500	3-1/2" diameter		6	1.333		425	85		510	595
7510	4" diameter		5	1.600		350	102		452	535
7680	Reducer, 3/4" to 1/2"		54	.148		3.86	9.45		13.31	18.25
7690	1" to 1/2"		46	.174		4.49	11.10		15.59	21.50
7700	1" to 3/4"		46	.174		4.34	11.10		15.44	21
7710	1-1/4" to 3/4"		40	.200		6.50	12.75		19.25	26
7720	1-1/4" to 1"		40	.200		6.60	12.75		19.35	26
7730	1-1/2" to 1"		36	.222		10.80	14.15		24.95	33
7740	1-1/2" to 1-1/4"		36	.222		10.30	14.15		24.45	32.50
7750	2" to 3/4"		32	.250		16.65	15.95		32.60	42
7760	2" to 1-1/4"		32	.250		14.15	15.95		30.10	39
7770	2" to 1-1/2"		32	.250		15.90	15.95		31.85	41
7780	2-1/2" to 1-1/2"		30	.267		22	17		39	49
7790	3" to 2"		30	.267		26	17		43	53.50
7800	3-1/2" to 2-1/2"		28	.286		43	18.20		61.20	74.50
7810	4" to 3"		28	.286		53	18.20		71.20	85
7820	Sealing fitting, vertical/horizontal, 1/2" diameter		14.50	.552		14.85	35		49.85	68.50
7830	3/4" diameter		13.30	.602		17.65	38.50		56.15	76.50
7840	1" diameter		11.40	.702		22.50	44.50		67	91
7850	1-1/4" diameter		10	.800		27.50	51		78.50	106
7860	1-1/2" diameter		8.80	.909		42	58		100	132
7870	2" diameter		8	1		53.50	63.50		117	153
7880	2-1/2" diameter		6.70	1.194		84.50	76		160.50	206
7890	3" diameter		5.70	1.404		114	89.50		203.50	258
7900	3-1/2" diameter		4.70	1.702		299	108		407	490
7910	4" diameter		4	2		470	127		597	710
7920	Sealing hubs, 1" by 1-1/2"		12	.667		43	42.50		85.50	111
7930	1-1/4" by 2"		10	.800		49.50	51		100.50	130
7940	1-1/2" by 2"		9	.889		93	56.50		149.50	186
7950	2" by 2-1/2"		8	1		111	63.50		174.50	218
7960	3" by 4"		7	1.143		208	73		281	335
7970	4" by 5"		6	1.333		390	85		475	555
7980	Drain, 1/2"		32	.250		120	15.95		135.95	156
7990	Breather, 1/2"		32	.250		123	15.95		138.95	159
7991	Explosion proof sealant compound, hub, fittings, 60 min set time, 1lb. pail		48	.167		22	10.60		32.60	40
7992	5 lb. pail		48	.167		56	10.60		66.60	77.50
7993	2 oz. tube		50	.160		67	10.20		77.20	89
7994	6 oz. tube		50	.160		136	10.20		146.20	165
7995	2.0 oz. cartridge		50	.160		39	10.20		49.20	57.50
7996	6.0 oz. cartridge		50	.160		79.50	10.20		89.70	103
7997	2.0 oz. box		50	.160		28.50	10.20		38.70	46.50
7998	8.0 oz. box		50	.160		95	10.20		105.20	119
7999	1.0 lb. box		50	.160		153	10.20		163.20	183
8000	Plastic coated 40 mil thick									
8010	LB, LR or LL conduit body w/cover, 1/2" diameter	1 Elec	13	.615	Ea.	61.50	39		100.50	126
8020	3/4" diameter		11	.727		67	46.50		113.50	143
8030	1" diameter		8	1		89.50	63.50		153	193
8040	1-1/4" diameter		6	1.333		140	85		225	279
8050	1-1/2" diameter		5	1.600		160	102		262	330
8060	2" diameter		4.50	1.778		241	113		354	435

For customer support on your Electrical Costs with RSMeans Data, call 800.448.8182.

175

26 05 33.25 Conduit Fittings for Rigid Galvanized Steel	Crew	Daily Output	Labor-Hours	Unit	Material	2021 Bare Costs Labor	Equipment	Total	Total Incl O&P
8070	2-1/2" diameter	1 Elec	4	2	Ea.	445	127		572	680
8080	3" diameter		3.50	2.286		590	146		736	860
8090	3-1/2" diameter		3	2.667		810	170		980	1,150
8100	4" diameter		2.50	3.200		970	204		1,174	1,375
8150	T conduit body with cover, 1/2" diameter		11	.727		72	46.50		118.50	149
8160	3/4" diameter		9	.889		87.50	56.50		144	181
8170	1" diameter		6	1.333		105	85		190	241
8180	1-1/4" diameter		5	1.600		164	102		266	330
8190	1-1/2" diameter		4.50	1.778		181	113		294	365
8200	2" diameter		4	2		273	127		400	490
8210	2-1/2" diameter		3.50	2.286		440	146		586	700
8220	3" diameter		3	2.667		605	170		775	915
8230	3-1/2" diameter		2.50	3.200		1,050	204		1,254	1,450
8240	4" diameter		2	4		980	255		1,235	1,450
8300	FS conduit body, 1 gang, 3/4" diameter		11	.727		74	46.50		120.50	151
8310	1" diameter		10	.800		91	51		142	176
8350	2 gang, 3/4" diameter		9	.889		140	56.50		196.50	238
8360	1" diameter		8	1		165	63.50		228.50	276
8400	Duplex receptacle cover		64	.125		55.50	7.95		63.45	73
8410	Switch cover		64	.125		91.50	7.95		99.45	113
8420	Switch, vaportight cover		53	.151		320	9.60		329.60	365
8430	Blank, cover		64	.125		48.50	7.95		56.45	65.50
8520	FSC conduit body, 1 gang, 3/4" diameter		10	.800		90.50	51		141.50	176
8530	1" diameter		9	.889		110	56.50		166.50	205
8550	2 gang, 3/4" diameter		8	1		160	63.50		223.50	271
8560	1" diameter		7	1.143		172	73		245	297
8590	Conduit hubs, 1/2" diameter		18	.444		45.50	28.50		74	92
8600	3/4" diameter		16	.500		51	32		83	104
8610	1" diameter		14	.571		63.50	36.50		100	124
8620	1-1/4" diameter		12	.667		77.50	42.50		120	149
8630	1-1/2" diameter		10	.800		84	51		135	168
8640	2" diameter		8.80	.909		120	58		178	218
8650	2-1/2" diameter		8.50	.941		216	60		276	325
8660	3" diameter		8	1		259	63.50		322.50	380
8670	3-1/2" diameter		7.50	1.067		355	68		423	490
8680	4" diameter		7	1.143		425	73		498	575
8690	5" diameter		6	1.333		635	85		720	820
8700	Plastic coated 40 mil thick									
8710	Pipe strap, stamped 1 hole, 1/2" diameter	1 Elec	470	.017	Ea.	15.90	1.08		16.98	19.10
8720	3/4" diameter		440	.018		14.95	1.16		16.11	18.15
8730	1" diameter		400	.020		15.45	1.27		16.72	18.90
8740	1-1/4" diameter		355	.023		24	1.44		25.44	28
8750	1-1/2" diameter		320	.025		24.50	1.59		26.09	29.50
8760	2" diameter		266	.030		34.50	1.92		36.42	41
8770	2-1/2" diameter		200	.040		50	2.55		52.55	59
8780	3" diameter		133	.060		63	3.83		66.83	75
8790	3-1/2" diameter		110	.073		109	4.63		113.63	126
8800	4" diameter		90	.089		123	5.65		128.65	144
8810	5" diameter		70	.114		135	7.30		142.30	159
8840	Clamp back spacers, 3/4" diameter		440	.018		21	1.16		22.16	25
8850	1" diameter		400	.020		29	1.27		30.27	34
8860	1-1/4" diameter		355	.023		40	1.44		41.44	46
8870	1-1/2" diameter		320	.025		52.50	1.59		54.09	60.50

26 05 33 – Raceway and Boxes for Electrical Systems

26 05 33.25 Conduit Fittings for Rigid Galvanized Steel

		Crew	Daily Output	Labor-Hours	Unit	Material	2021 Bare Costs Labor	2021 Bare Costs Equipment	Total	Total Incl O&P
8880	2" diameter	1 Elec	266	.030	Ea.	73	1.92		74.92	83.50
8900	3" diameter		133	.060		184	3.83		187.83	208
8920	4" diameter	↓	90	.089		72	5.65		77.65	88
8950	Touch-up plastic coating, spray, 12 oz.					48			48	52.50
8960	Sealing fittings, 1/2" diameter	1 Elec	11	.727		110	46.50		156.50	190
8970	3/4" diameter		9	.889		102	56.50		158.50	197
8980	1" diameter		7.50	1.067		127	68		195	241
8990	1-1/4" diameter		6.50	1.231		132	78.50		210.50	261
9000	1-1/2" diameter		5.50	1.455		210	92.50		302.50	370
9010	2" diameter		4.80	1.667		255	106		361	440
9020	2-1/2" diameter		4	2		315	127		442	535
9030	3" diameter		3.50	2.286		410	146		556	665
9040	3-1/2" diameter		3	2.667		1,000	170		1,170	1,350
9050	4" diameter		2.50	3.200		1,825	204		2,029	2,325
9060	5" diameter		1.70	4.706		1,775	300		2,075	2,400
9070	Unions, 1/2" diameter		18	.444		74.50	28.50		103	124
9080	3/4" diameter		15	.533		77	34		111	135
9090	1" diameter		13	.615		101	39		140	170
9100	1-1/4" diameter		11	.727		152	46.50		198.50	237
9110	1-1/2" diameter		9.50	.842		216	53.50		269.50	320
9120	2" diameter		8	1		266	63.50		329.50	385
9130	2-1/2" diameter		7.50	1.067		425	68		493	570
9140	3" diameter		6.80	1.176		470	75		545	625
9150	3-1/2" diameter		5.80	1.379		675	88		763	875
9160	4" diameter		4.80	1.667		945	106		1,051	1,200
9170	5" diameter	↓	4	2	↓	930	127		1,057	1,225

26 05 33.30 Electrical Nonmetallic Tubing (ENT)

		Crew	Daily Output	Labor-Hours	Unit	Material	2021 Bare Costs Labor	2021 Bare Costs Equipment	Total	Total Incl O&P
0010	**ELECTRICAL NONMETALLIC TUBING (ENT)**									
0050	Flexible, 1/2" diameter	1 Elec	270	.030	L.F.	.63	1.89		2.52	3.49
0100	3/4" diameter		230	.035		.98	2.22		3.20	4.37
0200	1" diameter		145	.055		1.84	3.51		5.35	7.20
0210	1-1/4" diameter		125	.064		1.70	4.08		5.78	7.90
0220	1-1/2" diameter		100	.080		2.40	5.10		7.50	10.20
0230	2" diameter		75	.107	↓	3.11	6.80		9.91	13.50
0300	Connectors, to outlet box, 1/2" diameter		230	.035	Ea.	1.02	2.22		3.24	4.41
0310	3/4" diameter		210	.038		1.79	2.43		4.22	5.60
0320	1" diameter		200	.040		2.64	2.55		5.19	6.70
0400	Couplings, to conduit, 1/2" diameter		145	.055		.95	3.51		4.46	6.25
0410	3/4" diameter		130	.062		1.25	3.92		5.17	7.20
0420	1" diameter	↓	125	.064	↓	2.37	4.08		6.45	8.65

26 05 33.33 Raceway/Boxes for Utility Substations

		Crew	Daily Output	Labor-Hours	Unit	Material	2021 Bare Costs Labor	2021 Bare Costs Equipment	Total	Total Incl O&P
0010	**RACEWAY/BOXES FOR UTILITY SUBSTATIONS**									
7000	Conduit, conductors, and insulators									
7100	Conduit, metallic	R-11	560	.100	Lb.	3.74	6.05	1.15	10.94	14.45
7110	Non-metallic	"	800	.070	"	9.05	4.25	.81	14.11	17.15
7190	See Section 26 05 33									
7200	Wire and cable	R-11	700	.080	Lb.	11.45	4.86	.92	17.23	21
7290	See Section 26 05 19									

26 05 33.35 Flexible Metallic Conduit

		Crew	Daily Output	Labor-Hours	Unit	Material	2021 Bare Costs Labor	2021 Bare Costs Equipment	Total	Total Incl O&P
0010	**FLEXIBLE METALLIC CONDUIT**									
0050	Steel, 3/8" diameter	1 Elec	200	.040	L.F.	.39	2.55		2.94	4.22
0100	1/2" diameter	↓	200	.040		.44	2.55		2.99	4.27

26 05 33.35 Flexible Metallic Conduit		Crew	Daily Output	Labor-Hours	Unit	Material	2021 Bare Costs Labor	Equipment	Total	Total Incl O&P
0200	3/4" diameter	1 Elec	160	.050	L.F.	.63	3.19		3.82	5.40
0250	1" diameter		100	.080		1.10	5.10		6.20	8.75
0300	1-1/4" diameter		70	.114		1.42	7.30		8.72	12.35
0350	1-1/2" diameter		50	.160		2.30	10.20		12.50	17.70
0370	2" diameter		40	.200		3.05	12.75		15.80	22.50
0380	2-1/2" diameter		30	.267		3.49	17		20.49	29
0390	3" diameter	2 Elec	50	.320		5.95	20.50		26.45	37
0400	3-1/2" diameter		40	.400		6.75	25.50		32.25	45.50
0410	4" diameter		30	.533		7.60	34		41.60	59
0420	Connectors, plain, 3/8" diameter	1 Elec	100	.080	Ea.	1.61	5.10		6.71	9.30
0430	1/2" diameter		80	.100		2.02	6.35		8.37	11.65
0440	3/4" diameter		70	.114		1.90	7.30		9.20	12.90
0450	1" diameter		50	.160		3.94	10.20		14.14	19.50
0452	1-1/4" diameter		45	.178		5.30	11.30		16.60	22.50
0454	1-1/2" diameter		40	.200		7.30	12.75		20.05	27
0456	2" diameter		28	.286		10.95	18.20		29.15	39
0458	2-1/2" diameter		25	.320		18.25	20.50		38.75	50.50
0460	3" diameter		20	.400		27.50	25.50		53	68
0462	3-1/2" diameter		16	.500		96	32		128	154
0464	4" diameter		13	.615		110	39		149	180
0490	Insulated, 1" diameter		40	.200		7.25	12.75		20	27
0500	1-1/4" diameter		40	.200		11.90	12.75		24.65	32
0550	1-1/2" diameter		32	.250		21.50	15.95		37.45	47
0600	2" diameter		23	.348		30	22		52	66
0610	2-1/2" diameter		20	.400		48	25.50		73.50	91
0620	3" diameter		17	.471		69.50	30		99.50	121
0630	3-1/2" diameter		13	.615		280	39		319	370
0640	4" diameter		10	.800		315	51		366	420
0650	Connectors 90°, plain, 3/8" diameter		80	.100		3.34	6.35		9.69	13.10
0660	1/2" diameter		60	.133		6.90	8.50		15.40	20
0700	3/4" diameter		50	.160		8.85	10.20		19.05	25
0750	1" diameter		40	.200		14.40	12.75		27.15	35
0790	Insulated, 1" diameter		40	.200		13	12.75		25.75	33.50
0800	1-1/4" diameter		30	.267		23	17		40	50
0850	1-1/2" diameter		23	.348		37.50	22		59.50	74.50
0900	2" diameter		18	.444		55.50	28.50		84	103
0910	2-1/2" diameter		16	.500		116	32		148	176
0920	3" diameter		14	.571		155	36.50		191.50	224
0930	3-1/2" diameter		11	.727		450	46.50		496.50	565
0940	4" diameter		8	1		805	63.50		868.50	980
0960	Couplings, to flexible conduit, 1/2" diameter		50	.160		.94	10.20		11.14	16.20
0970	3/4" diameter		40	.200		1.52	12.75		14.27	20.50
0980	1" diameter		35	.229		2.53	14.55		17.08	24.50
0990	1-1/4" diameter		28	.286		5.75	18.20		23.95	33.50
1000	1-1/2" diameter		23	.348		6.65	22		28.65	40.50
1010	2" diameter		20	.400		16.70	25.50		42.20	56.50
1020	2-1/2" diameter		18	.444		38	28.50		66.50	84
1030	3" diameter		15	.533		96	34		130	156
1032	Aluminum, 3/8" diameter		210	.038	L.F.	.37	2.43		2.80	4.02
1034	1/2" diameter		210	.038		.43	2.43		2.86	4.08
1036	3/4" diameter		165	.048		.60	3.09		3.69	5.25
1038	1" diameter		105	.076		1.11	4.85		5.96	8.40
1040	1-1/4" diameter		75	.107		1.55	6.80		8.35	11.80

26 05 33 – Raceway and Boxes for Electrical Systems

26 05 33.35 Flexible Metallic Conduit		Crew	Daily Output	Labor-Hours	Unit	Material	2021 Bare Costs Labor	Equipment	Total	Total Incl O&P
1042	1-1/2" diameter	1 Elec	53	.151	L.F.	2.21	9.60		11.81	16.75
1044	2" diameter		42	.190		2.86	12.15		15.01	21
1046	2-1/2" diameter		32	.250		3.69	15.95		19.64	27.50
1048	3" diameter	2 Elec	53	.302		8.40	19.25		27.65	38
1050	3-1/2" diameter		42	.381		9.70	24.50		34.20	46.50
1052	4" diameter		32	.500		7.70	32		39.70	56
1070	Sealtite, 3/8" diameter	1 Elec	140	.057		.66	3.64		4.30	6.15
1080	1/2" diameter		140	.057		1.30	3.64		4.94	6.85
1090	3/4" diameter		100	.080		1.71	5.10		6.81	9.45
1100	1" diameter		70	.114		3.07	7.30		10.37	14.20
1200	1-1/4" diameter		50	.160		2.97	10.20		13.17	18.40
1300	1-1/2" diameter		40	.200		3.32	12.75		16.07	22.50
1400	2" diameter		30	.267		4.25	17		21.25	29.50
1410	2-1/2" diameter		27	.296		7.10	18.85		25.95	36
1420	3" diameter	2 Elec	50	.320		11.35	20.50		31.85	43
1440	4" diameter	"	30	.533		9.35	34		43.35	61
1490	Connectors, plain, 3/8" diameter	1 Elec	70	.114	Ea.	2.70	7.30		10	13.75
1500	1/2" diameter		70	.114		3.26	7.30		10.56	14.40
1700	3/4" diameter		50	.160		4.21	10.20		14.41	19.80
1900	1" diameter		40	.200		13.40	12.75		26.15	33.50
1910	Insulated, 1" diameter		40	.200		7.10	12.75		19.85	27
2000	1-1/4" diameter		32	.250		12.10	15.95		28.05	37
2100	1-1/2" diameter		27	.296		14.75	18.85		33.60	44
2200	2" diameter		20	.400		28	25.50		53.50	69
2210	2-1/2" diameter		15	.533		88.50	34		122.50	148
2220	3" diameter		12	.667		134	42.50		176.50	210
2240	4" diameter		8	1		187	63.50		250.50	300
2290	Connectors, 90°, 3/8" diameter		70	.114		6.15	7.30		13.45	17.60
2300	1/2" diameter		70	.114		9.25	7.30		16.55	21
2400	3/4" diameter		50	.160		8.45	10.20		18.65	24.50
2600	1" diameter		40	.200		17.20	12.75		29.95	38
2790	Insulated, 1" diameter		40	.200		13.35	12.75		26.10	33.50
2800	1-1/4" diameter		32	.250		20.50	15.95		36.45	46.50
3000	1-1/2" diameter		27	.296		26	18.85		44.85	56.50
3100	2" diameter		20	.400		53.50	25.50		79	97
3110	2-1/2" diameter		14	.571		170	36.50		206.50	240
3120	3" diameter		11	.727		390	46.50		436.50	500
3140	4" diameter		7	1.143		230	73		303	360
4300	Coupling sealtite to rigid, 1/2" diameter		20	.400		4.57	25.50		30.07	43
4500	3/4" diameter		18	.444		6.25	28.50		34.75	49
4800	1" diameter		14	.571		7.05	36.50		43.55	62
4900	1-1/4" diameter		12	.667		10.60	42.50		53.10	74.50
5000	1-1/2" diameter		11	.727		20.50	46.50		67	91.50
5100	2" diameter		10	.800		28	51		79	106
5110	2-1/2" diameter		9.50	.842		130	53.50		183.50	223
5120	3" diameter		9	.889		153	56.50		209.50	252
5130	3-1/2" diameter		9	.889		168	56.50		224.50	269
5140	4" diameter		8.50	.941		184	60		244	291

For customer support on your Electrical Costs with RSMeans Data, call 800.448.8182.

179

26 05 33.95 Cutting and Drilling	Crew	Daily Output	Labor-Hours	Unit	Material	2021 Bare Costs Labor	Equipment	Total	Total Incl O&P
0010	**CUTTING AND DRILLING**								
0100	Hole drilling to 10' high, concrete wall								
0110	8" thick, 1/2" pipe size	R-31	12	.667	Ea.	.26	42.50	5.20	47.96	69
0120	3/4" pipe size		12	.667		.26	42.50	5.20	47.96	69
0130	1" pipe size		9.50	.842		.36	53.50	6.60	60.46	87
0140	1-1/4" pipe size		9.50	.842		.36	53.50	6.60	60.46	87
0150	1-1/2" pipe size	R-31	9.50	.842	Ea.	.36	53.50	6.60	60.46	87
0160	2" pipe size		4.40	1.818		.55	116	14.25	130.80	188
0170	2-1/2" pipe size		4.40	1.818		.55	116	14.25	130.80	188
0180	3" pipe size		4.40	1.818		.55	116	14.25	130.80	188
0190	3-1/2" pipe size		3.30	2.424		.67	154	19	173.67	251
0200	4" pipe size		3.30	2.424		.67	154	19	173.67	251
0500	12" thick, 1/2" pipe size		9.40	.851		.39	54	6.65	61.04	88.50
0520	3/4" pipe size		9.40	.851		.39	54	6.65	61.04	88.50
0540	1" pipe size		7.30	1.096		.53	70	8.60	79.13	114
0560	1-1/4" pipe size		7.30	1.096		.53	70	8.60	79.13	114
0570	1-1/2" pipe size		7.30	1.096		.53	70	8.60	79.13	114
0580	2" pipe size		3.60	2.222		.83	142	17.40	160.23	230
0590	2-1/2" pipe size		3.60	2.222		.83	142	17.40	160.23	230
0600	3" pipe size		3.60	2.222		.83	142	17.40	160.23	230
0610	3-1/2" pipe size		2.80	2.857		1.01	182	22.50	205.51	296
0630	4" pipe size		2.50	3.200		1.01	204	25	230.01	335
0650	16" thick, 1/2" pipe size		7.60	1.053		.52	67	8.25	75.77	109
0670	3/4" pipe size		7	1.143		.52	73	8.95	82.47	118
0690	1" pipe size		6	1.333		.71	85	10.45	96.16	138
0710	1-1/4" pipe size		5.50	1.455		.71	92.50	11.40	104.61	151
0730	1-1/2" pipe size		5.50	1.455		.71	92.50	11.40	104.61	151
0750	2" pipe size		3	2.667		1.10	170	21	192.10	276
0770	2-1/2" pipe size		2.70	2.963		1.10	189	23	213.10	305
0790	3" pipe size		2.50	3.200		1.10	204	25	230.10	335
0810	3-1/2" pipe size		2.30	3.478		1.34	222	27	250.34	360
0830	4" pipe size		2	4		1.34	255	31.50	287.84	415
0850	20" thick, 1/2" pipe size		6.40	1.250		.65	79.50	9.80	89.95	129
0870	3/4" pipe size		6	1.333		.65	85	10.45	96.10	138
0890	1" pipe size		5	1.600		.89	102	12.55	115.44	166
0910	1-1/4" pipe size		4.80	1.667		.89	106	13.05	119.94	173
0930	1-1/2" pipe size		4.60	1.739		.89	111	13.60	125.49	181
0950	2" pipe size		2.70	2.963		1.38	189	23	213.38	305
0970	2-1/2" pipe size		2.40	3.333		1.38	212	26	239.38	345
0990	3" pipe size		2.20	3.636		1.38	232	28.50	261.88	380
1010	3-1/2" pipe size		2	4		1.68	255	31.50	288.18	415
1030	4" pipe size		1.70	4.706		1.68	300	37	338.68	485
1050	24" thick, 1/2" pipe size		5.50	1.455		.78	92.50	11.40	104.68	151
1070	3/4" pipe size		5.10	1.569		.78	100	12.30	113.08	162
1090	1" pipe size		4.30	1.860		1.07	119	14.55	134.62	193
1110	1-1/4" pipe size		4	2		1.07	127	15.65	143.72	207
1130	1-1/2" pipe size		4	2		1.07	127	15.65	143.72	207
1150	2" pipe size		2.40	3.333		1.65	212	26	239.65	345
1170	2-1/2" pipe size		2.20	3.636		1.65	232	28.50	262.15	380
1190	3" pipe size		2	4		1.65	255	31.50	288.15	415
1210	3-1/2" pipe size		1.80	4.444		2.02	283	35	320.02	460
1230	4" pipe size		1.50	5.333		2.02	340	42	384.02	555

180

For customer support on your Electrical Costs with RSMeans Data, call 800.448.8182.

26 05 33 – Raceway and Boxes for Electrical Systems

		Crew	Daily Output	Labor-Hours	Unit	Material	2021 Bare Costs Labor	Equipment	Total	Total Incl O&P
26 05 33.95 Cutting and Drilling										
1500	Brick wall, 8" thick, 1/2" pipe size	R-31	18	.444	Ea.	.26	28.50	3.48	32.24	46
1520	3/4" pipe size		18	.444		.26	28.50	3.48	32.24	46
1540	1" pipe size		13.30	.602		.36	38.50	4.71	43.57	62.50
1560	1-1/4" pipe size		13.30	.602		.36	38.50	4.71	43.57	62.50
1580	1-1/2" pipe size		13.30	.602		.36	38.50	4.71	43.57	62.50
1600	2" pipe size		5.70	1.404		.55	89.50	11	101.05	146
1620	2-1/2" pipe size		5.70	1.404		.55	89.50	11	101.05	146
1640	3" pipe size		5.70	1.404		.55	89.50	11	101.05	146
1660	3-1/2" pipe size		4.40	1.818		.67	116	14.25	130.92	188
1680	4" pipe size		4	2		.67	127	15.65	143.32	207
1700	12" thick, 1/2" pipe size		14.50	.552		.39	35	4.32	39.71	57
1720	3/4" pipe size		14.50	.552		.39	35	4.32	39.71	57
1740	1" pipe size		11	.727		.53	46.50	5.70	52.73	76
1760	1-1/4" pipe size		11	.727		.53	46.50	5.70	52.73	76
1780	1-1/2" pipe size		11	.727		.53	46.50	5.70	52.73	76
1800	2" pipe size		5	1.600		.83	102	12.55	115.38	166
1820	2-1/2" pipe size		5	1.600		.83	102	12.55	115.38	166
1840	3" pipe size		5	1.600		.83	102	12.55	115.38	166
1860	3-1/2" pipe size		3.80	2.105		1.01	134	16.50	151.51	218
1880	4" pipe size		3.30	2.424		1.01	154	19	174.01	251
1900	16" thick, 1/2" pipe size		12.30	.650		.52	41.50	5.10	47.12	67.50
1920	3/4" pipe size		12.30	.650		.52	41.50	5.10	47.12	67.50
1940	1" pipe size		9.30	.860		.71	55	6.75	62.46	89.50
1960	1-1/4" pipe size		9.30	.860		.71	55	6.75	62.46	89.50
1980	1-1/2" pipe size		9.30	.860		.71	55	6.75	62.46	89.50
2000	2" pipe size		4.40	1.818		1.10	116	14.25	131.35	189
2010	2-1/2" pipe size		4.40	1.818		1.10	116	14.25	131.35	189
2030	3" pipe size		4.40	1.818		1.10	116	14.25	131.35	189
2050	3-1/2" pipe size		3.30	2.424		1.34	154	19	174.34	251
2070	4" pipe size		3	2.667		1.34	170	21	192.34	276
2090	20" thick, 1/2" pipe size		10.70	.748		.65	47.50	5.85	54	78
2110	3/4" pipe size		10.70	.748		.65	47.50	5.85	54	78
2130	1" pipe size		8	1		.89	63.50	7.85	72.24	104
2150	1-1/4" pipe size		8	1		.89	63.50	7.85	72.24	104
2170	1-1/2" pipe size		8	1		.89	63.50	7.85	72.24	104
2190	2" pipe size		4	2		1.38	127	15.65	144.03	208
2210	2-1/2" pipe size		4	2		1.38	127	15.65	144.03	208
2230	3" pipe size		4	2		1.38	127	15.65	144.03	208
2250	3-1/2" pipe size		3	2.667		1.68	170	21	192.68	277
2270	4" pipe size		2.70	2.963		1.68	189	23	213.68	305
2290	24" thick, 1/2" pipe size		9.40	.851		.78	54	6.65	61.43	88.50
2310	3/4" pipe size		9.40	.851		.78	54	6.65	61.43	88.50
2330	1" pipe size		7.10	1.127		1.07	72	8.80	81.87	118
2350	1-1/4" pipe size		7.10	1.127		1.07	72	8.80	81.87	118
2370	1-1/2" pipe size		7.10	1.127		1.07	72	8.80	81.87	118
2390	2" pipe size		3.60	2.222		1.65	142	17.40	161.05	231
2410	2-1/2" pipe size		3.60	2.222		1.65	142	17.40	161.05	231
2430	3" pipe size		3.60	2.222		1.65	142	17.40	161.05	231
2450	3-1/2" pipe size		2.80	2.857		2.02	182	22.50	206.52	297
2470	4" pipe size		2.50	3.200		2.02	204	25	231.02	335
3000	Knockouts to 8' high, metal boxes & enclosures									
3020	With hole saw, 1/2" pipe size	1 Elec	53	.151	Ea.		9.60		9.60	14.30
3040	3/4" pipe size		47	.170			10.85		10.85	16.10

181

26 05 33 – Raceway and Boxes for Electrical Systems

26 05 33.95 Cutting and Drilling

		Crew	Daily Output	Labor-Hours	Unit	Material	2021 Bare Costs Labor	Equipment	Total	Total Incl O&P
3050	1" pipe size	1 Elec	40	.200	Ea.		12.75		12.75	18.95
3060	1-1/4" pipe size		36	.222			14.15		14.15	21
3070	1-1/2" pipe size		32	.250			15.95		15.95	23.50
3080	2" pipe size		27	.296			18.85		18.85	28
3090	2-1/2" pipe size		20	.400			25.50		25.50	38
4010	3" pipe size		16	.500			32		32	47.50
4030	3-1/2" pipe size		13	.615			39		39	58.50
4050	4" pipe size		11	.727			46.50		46.50	69

26 05 36 – Cable Trays for Electrical Systems

26 05 36.10 Cable Tray Ladder Type

		Crew	Daily Output	Labor-Hours	Unit	Material	2021 Bare Costs Labor	Equipment	Total	Total Incl O&P
0010	**CABLE TRAY LADDER TYPE** w/ftngs. & supports, 4" dp., to 15' elev.									
0100	For higher elevations, see Section 26 05 36.40									
0160	Galvanized steel tray									
0170	4" rung spacing, 6" wide	2 Elec	98	.163	L.F.	15.75	10.40		26.15	33
0180	9" wide		92	.174		17.20	11.10		28.30	35.50
0200	12" wide		86	.186		19	11.85		30.85	38.50
0400	18" wide		82	.195		22	12.45		34.45	42.50
0600	24" wide		78	.205		25.50	13.05		38.55	47.50
0650	30" wide		68	.235		32	15		47	58
0800	6" rung spacing, 6" wide		100	.160		14.35	10.20		24.55	31
0850	9" wide		94	.170		15.75	10.85		26.60	33.50
0860	12" wide		88	.182		16.55	11.60		28.15	35.50
0870	18" wide		84	.190		18.90	12.15		31.05	39
0880	24" wide		80	.200		21	12.75		33.75	42
0890	30" wide		70	.229		25	14.55		39.55	49
0910	9" rung spacing, 6" wide	2 Elec	102	.157	L.F.	13.40	10		23.40	29.50
0920	9" wide		98	.163		13.70	10.40		24.10	30.50
0930	12" wide		94	.170		15.75	10.85		26.60	33.50
0940	18" wide		90	.178		17.80	11.30		29.10	36.50
0950	24" wide		86	.186		20	11.85		31.85	40
0960	30" wide		80	.200		22	12.75		34.75	43.50
0980	12" rung spacing, 6" wide		106	.151		13.15	9.60		22.75	29
0990	9" wide		104	.154		13.50	9.80		23.30	29.50
1000	12" wide		100	.160		14.25	10.20		24.45	31
1010	18" wide		96	.167		15.30	10.60		25.90	32.50
1020	24" wide		94	.170		16.40	10.85		27.25	34
1030	30" wide		88	.182		18.65	11.60		30.25	37.50
1041	18" rung spacing, 6" wide		108	.148		13.10	9.45		22.55	28.50
1042	9" wide		106	.151		13.25	9.60		22.85	29
1043	12" wide		102	.157		14.15	10		24.15	30.50
1044	18" wide		98	.163		14.85	10.40		25.25	32
1045	24" wide		96	.167		15.30	10.60		25.90	32.50
1046	30" wide		90	.178		16.15	11.30		27.45	34.50
1050	Elbows horiz. 9" rung spacing, 90°, 12" radius, 6" wide		9.60	1.667	Ea.	61	106		167	225
1060	9" wide		8.40	1.905		64.50	121		185.50	251
1070	12" wide		7.60	2.105		70	134		204	276
1080	18" wide		6.20	2.581		92	164		256	345
1090	24" wide		5.40	2.963		103	189		292	395
1100	30" wide		4.80	3.333		136	212		348	465
1120	90°, 24" radius, 6" wide		9.20	1.739		125	111		236	305
1130	9" wide		8	2		129	127		256	330
1140	12" wide		7.20	2.222		133	142		275	355

R260536-10

R260536-11

182

For customer support on your Electrical Costs with RSMeans Data, call 800.448.8182.

26 05 36 – Cable Trays for Electrical Systems

26 05 36.10 Cable Tray Ladder Type	Crew	Daily Output	Labor-Hours	Unit	Material	2021 Bare Costs Labor	Equipment	Total	Total Incl O&P
1150	18" wide	2 Elec	5.80	2.759	Ea.	144	176		320	420
1160	24" wide		5	3.200		155	204		359	475
1170	30" wide		4.40	3.636		168	232		400	530
1190	90°, 36" radius, 6" wide		8.80	1.818		138	116		254	325
1200	9" wide		7.60	2.105		146	134		280	360
1210	12" wide		6.80	2.353		153	150		303	390
1220	18" wide		5.40	2.963		175	189		364	470
1230	24" wide		4.60	3.478		192	222		414	540
1240	30" wide		4	4		230	255		485	635
1260	45°, 12" radius, 6" wide		13.20	1.212		47	77		124	167
1270	9" wide		11	1.455		49	92.50		141.50	192
1280	12" wide		9.60	1.667		51	106		157	214
1290	18" wide		7.60	2.105		53.50	134		187.50	258
1300	24" wide		6.20	2.581		69	164		233	320
1310	30" wide		5.40	2.963		88	189		277	375
1330	45°, 24" radius, 6" wide		12.80	1.250		59.50	79.50		139	184
1340	9" wide		10.60	1.509		63	96		159	213
1350	12" wide		9.20	1.739		71	111		182	243
1360	18" wide		7.20	2.222		72.50	142		214.50	290
1370	24" wide		5.80	2.759		88	176		264	360
1380	30" wide		5	3.200		105	204		309	420
1400	45°, 36" radius, 6" wide		12.40	1.290		85.50	82		167.50	216
1410	9" wide		10.20	1.569		90.50	100		190.50	248
1420	12" wide		8.80	1.818		92	116		208	273
1430	18" wide		6.80	2.353		99.50	150		249.50	330
1440	24" wide		5.40	2.963		105	189		294	395
1450	30" wide	▼	4.60	3.478	▼	136	222		358	480
1470	Elbows horizontal, 4" rung spacing, use 9" rung x 1.50									
1480	6" rung spacing, use 9" rung x 1.20									
1490	12" rung spacing, use 9" rung x 0.93									
1500	Elbows vertical, 9" rung spacing, 90°, 12" radius, 6" wide	2 Elec	9.60	1.667	Ea.	83.50	106		189.50	250
1510	9" wide		8.40	1.905		85.50	121		206.50	274
1520	12" wide		7.60	2.105		88	134		222	296
1530	18" wide		6.20	2.581		92	164		256	345
1540	24" wide		5.40	2.963		96	189		285	385
1550	30" wide		4.80	3.333		103	212		315	430
1570	24" radius, 6" wide		9.20	1.739		118	111		229	295
1580	9" wide		8	2		122	127		249	325
1590	12" wide		7.20	2.222		125	142		267	350
1600	18" wide		5.80	2.759		133	176		309	405
1610	24" wide		5	3.200		140	204		344	460
1620	30" wide		4.40	3.636		146	232		378	505
1640	36" radius, 6" wide		8.80	1.818		158	116		274	345
1650	9" wide		7.60	2.105		164	134		298	380
1660	12" wide		6.80	2.353		166	150		316	405
1670	18" wide		5.40	2.963		181	189		370	480
1680	24" wide		4.60	3.478		182	222		404	530
1690	30" wide	▼	4	4	▼	206	255		461	605
1710	Elbows vertical, 4" rung spacing, use 9" rung x 1.25									
1720	6" rung spacing, use 9" rung x 1.15									
1730	12" rung spacing, use 9" rung x 0.90									
1740	Tee horizontal, 9" rung spacing, 12" radius, 6" wide	2 Elec	5	3.200	Ea.	129	204		333	445
1750	9" wide	↓	4.60	3.478	↓	138	222		360	480

183

26 05 36.10 Cable Tray Ladder Type		Crew	Daily Output	Labor-Hours	Unit	Material	2021 Bare Costs Labor	2021 Bare Costs Equipment	Total	Total Incl O&P
1760	12" wide	2 Elec	4.40	3.636	Ea.	146	232		378	505
1770	18" wide		4	4		166	255		421	560
1780	24" wide		3.60	4.444		195	283		478	635
1790	30" wide		3.40	4.706		234	300		534	700
1810	24" radius, 6" wide		4.60	3.478		201	222		423	550
1820	9" wide		4.20	3.810		210	243		453	590
1830	12" wide		4	4		217	255		472	620
1840	18" wide		3.60	4.444		236	283		519	680
1850	24" wide		3.20	5		256	320		576	755
1860	30" wide		3	5.333		278	340		618	810
1880	36" radius, 6" wide		4.20	3.810		300	243		543	690
1890	9" wide		3.80	4.211		315	268		583	745
1900	12" wide		3.60	4.444		335	283		618	785
1910	18" wide		3.20	5		355	320		675	870
1920	24" wide		2.80	5.714		405	365		770	985
1930	30" wide		2.60	6.154		440	390		830	1,075
1980	Tee vertical, 9" rung spacing, 12" radius, 6" wide		5.40	2.963		219	189		408	520
1990	9" wide		5.20	3.077		221	196		417	535
2000	12" wide		5	3.200		223	204		427	550
2010	18" wide		4.60	3.478		225	222		447	575
2020	24" wide		4.40	3.636		230	232		462	600
2030	30" wide		4	4		245	255		500	650
2050	24" radius, 6" wide		5	3.200		395	204		599	740
2060	9" wide		4.80	3.333		400	212		612	755
2070	12" wide		4.60	3.478		405	222		627	775
2080	18" wide		4.20	3.810		420	243		663	820
2090	24" wide		4	4		430	255		685	850
2100	30" wide		3.60	4.444		440	283		723	905
2120	36" radius, 6" wide		4.60	3.478		735	222		957	1,150
2130	9" wide		4.40	3.636		750	232		982	1,175
2140	12" wide		4.20	3.810		755	243		998	1,200
2150	18" wide		3.80	4.211		765	268		1,033	1,250
2160	24" wide		3.60	4.444		775	283		1,058	1,275
2170	30" wide		3.20	5		805	320		1,125	1,350
2190	Tee, 4" rung spacing, use 9" rung x 1.30									
2200	6" rung spacing, use 9" rung x 1.20									
2210	12" rung spacing, use 9" rung x 0.90									
2220	Cross horizontal, 9" rung spacing, 12" radius, 6" wide	2 Elec	4	4	Ea.	159	255		414	555
2230	9" wide		3.80	4.211		167	268		435	585
2240	12" wide		3.60	4.444		182	283		465	620
2250	18" wide		3.40	4.706		199	300		499	665
2260	24" wide		3	5.333		221	340		561	750
2270	30" wide		2.80	5.714		267	365		632	835
2290	24" radius, 6" wide		3.60	4.444		310	283		593	760
2300	9" wide		3.40	4.706		320	300		620	795
2310	12" wide		3.20	5		330	320		650	835
2320	18" wide		3	5.333		350	340		690	890
2330	24" wide		2.60	6.154		370	390		760	990
2340	30" wide		2.40	6.667		395	425		820	1,075
2360	36" radius, 6" wide		3.20	5		430	320		750	950
2370	9" wide		3	5.333		450	340		790	1,000
2380	12" wide		2.80	5.714		470	365		835	1,050
2390	18" wide		2.60	6.154		520	390		910	1,150

26 05 36 – Cable Trays for Electrical Systems

26 05 36.10 Cable Tray Ladder Type		Crew	Daily Output	Labor-Hours	Unit	Material	2021 Bare Costs Labor	2021 Bare Costs Equipment	Total	Total Incl O&P
2400	24" wide	2 Elec	2.20	7.273	Ea.	575	465		1,040	1,325
2410	30" wide	↓	2	8	↓	625	510		1,135	1,450
2430	Cross horizontal, 4" rung spacing, use 9" rung x 1.30									
2440	6" rung spacing, use 9" rung x 1.20									
2450	12" rung spacing, use 9" rung x 0.90									
2460	Reducer, 9" to 6" wide tray	2 Elec	13	1.231	Ea.	88.50	78.50		167	214
2470	12" to 9" wide tray		12	1.333		90.50	85		175.50	226
2480	18" to 12" wide tray		10.40	1.538		90.50	98		188.50	246
2490	24" to 18" wide tray		9	1.778		92	113		205	269
2500	30" to 24" wide tray		8	2		94	127		221	292
2510	36" to 30" wide tray		7	2.286		105	146		251	330
2511	Reducer, 18" to 6" wide tray		10.40	1.538		94	98		192	249
2512	24" to 12" wide tray		9	1.778		96	113		209	273
2513	30" to 18" wide tray		8	2		97.50	127		224.50	296
2514	30" to 12" wide tray		8	2		104	127		231	305
2515	36" to 24" wide tray		7	2.286		107	146		253	335
2516	36" to 18" wide tray		7	2.286		107	146		253	335
2517	36" to 12" wide tray		7	2.286		107	146		253	335
2520	Dropout or end plate, 6" wide		32	.500		9.85	32		41.85	58.50
2530	9" wide		28	.571		11.60	36.50		48.10	67
2540	12" wide		26	.615		12.75	39		51.75	72.50
2550	18" wide		22	.727		14.90	46.50		61.40	85.50
2560	24" wide		20	.800		18.05	51		69.05	95.50
2570	30" wide		18	.889		19.75	56.50		76.25	106
2590	Tray connector		48	.333	↓	17.50	21		38.50	51
3200	Aluminum tray, 4" deep, 6" rung spacing, 6" wide		134	.119	L.F.	17.60	7.60		25.20	30.50
3210	9" wide		128	.125		18.70	7.95		26.65	32.50
3220	12" wide		124	.129		19.70	8.20		27.90	33.50
3230	18" wide		114	.140		22	8.95		30.95	37.50
3240	24" wide		106	.151		25.50	9.60		35.10	42.50
3250	30" wide		100	.160		28	10.20		38.20	46
3270	9" rung spacing, 6" wide		140	.114		14.25	7.30		21.55	26.50
3280	9" wide	↓	134	.119	↓	15.05	7.60		22.65	28
3290	12" wide	2 Elec	130	.123	L.F.	15.30	7.85		23.15	28.50
3300	18" wide		122	.131		17.90	8.35		26.25	32
3310	24" wide		116	.138		21	8.80		29.80	36
3320	30" wide		108	.148		23	9.45		32.45	39
3340	12" rung spacing, 6" wide		146	.110		15.25	7		22.25	27
3350	9" wide		140	.114		15.40	7.30		22.70	27.50
3360	12" wide		134	.119		15.80	7.60		23.40	28.50
3370	18" wide		128	.125		16.85	7.95		24.80	30.50
3380	24" wide		124	.129		18.70	8.20		26.90	32.50
3390	30" wide		114	.140		20	8.95		28.95	35.50
3401	18" rung spacing, 6" wide		150	.107		15.15	6.80		21.95	27
3402	9" wide tray		144	.111		15.30	7.10		22.40	27.50
3403	12" wide tray		140	.114		15.80	7.30		23.10	28
3404	18" wide tray		134	.119		16.75	7.60		24.35	30
3405	24" wide tray		130	.123		18.05	7.85		25.90	31.50
3406	30" wide tray		120	.133	↓	18.70	8.50		27.20	33
3410	Elbows horiz., 9" rung spacing, 90°, 12" radius, 6" wide		9.60	1.667	Ea.	68	106		174	233
3420	9" wide		8.40	1.905		69.50	121		190.50	257
3430	12" wide		7.60	2.105		82.50	134		216.50	290
3440	18" wide	↓	6.20	2.581	↓	93	164		257	345

185

26 05 36.10 Cable Tray Ladder Type	Crew	Daily Output	Labor-Hours	Unit	Material	2021 Bare Costs Labor	Equipment	Total	Total Incl O&P
3450	24" wide	2 Elec	5.40	2.963	Ea.	115	189		304	405
3460	30" wide		4.80	3.333		122	212		334	450
3480	24" radius, 6" wide		9.20	1.739		121	111		232	298
3490	9" wide		8	2		126	127		253	325
3500	12" wide		7.20	2.222		131	142		273	355
3510	18" wide		5.80	2.759		143	176		319	420
3520	24" wide		5	3.200		157	204		361	480
3530	30" wide		4.40	3.636		170	232		402	530
3550	90°, 36" radius, 6" wide		8.80	1.818		132	116		248	320
3560	9" wide		7.60	2.105		136	134		270	350
3570	12" wide		6.80	2.353		152	150		302	390
3580	18" wide		5.40	2.963		170	189		359	470
3590	24" wide		4.60	3.478		192	222		414	540
3600	30" wide		4	4		212	255		467	615
3620	45°, 12" radius, 6" wide		13.20	1.212		44.50	77		121.50	164
3630	9" wide		11	1.455		46.50	92.50		139	189
3640	12" wide		9.60	1.667		48	106		154	211
3650	18" wide		7.60	2.105		57	134		191	262
3660	24" wide		6.20	2.581		70.50	164		234.50	320
3670	30" wide		5.40	2.963		81	189		270	370
3690	45°, 24" radius, 6" wide		12.80	1.250		59.50	79.50		139	184
3700	9" wide		10.60	1.509		68	96		164	218
3710	12" wide		9.20	1.739		69.50	111		180.50	242
3720	18" wide		7.20	2.222		74.50	142		216.50	292
3730	24" wide		5.80	2.759		86	176		262	355
3740	30" wide		5	3.200		103	204		307	420
3760	45°, 36" radius, 6" wide		12.40	1.290		81	82		163	211
3770	9" wide		10.20	1.569		82.50	100		182.50	239
3780	12" wide		8.80	1.818		84.50	116		200.50	265
3790	18" wide		6.80	2.353		92.50	150		242.50	325
3800	24" wide		5.40	2.963		114	189		303	405
3810	30" wide	↓	4.60	3.478	↓	121	222		343	465
3830	Elbows horizontal, 4" rung spacing, use 9" rung x 1.50									
3840	6" rung spacing, use 9" rung x 1.20									
3850	12" rung spacing, use 9" rung x 0.93									
3860	Elbows vertical, 9" rung spacing, 90°, 12" radius, 6" wide	2 Elec	9.60	1.667	Ea.	94.50	106		200.50	262
3870	9" wide		8.40	1.905		99	121		220	289
3880	12" wide		7.60	2.105		101	134		235	310
3890	18" wide		6.20	2.581		104	164		268	360
3900	24" wide		5.40	2.963		111	189		300	400
3910	30" wide		4.80	3.333		116	212		328	440
3930	24" radius, 6" wide		9.20	1.739		103	111		214	278
3940	9" wide		8	2		106	127		233	305
3950	12" wide		7.20	2.222		109	142		251	330
3960	18" wide		5.80	2.759		116	176		292	390
3970	24" wide		5	3.200		122	204		326	440
3980	30" wide		4.40	3.636		129	232		361	485
4000	36" radius, 6" wide		8.80	1.818		159	116		275	345
4010	9" wide		7.60	2.105		162	134		296	375
4020	12" wide		6.80	2.353		172	150		322	410
4030	18" wide		5.40	2.963		179	189		368	475
4040	24" wide		4.60	3.478		185	222		407	535
4050	30" wide	↓	4	4	↓	200	255		455	600

26 05 36 – Cable Trays for Electrical Systems

26 05 36.10 Cable Tray Ladder Type		Crew	Daily Output	Labor-Hours	Unit	Material	2021 Bare Costs Labor	Equipment	Total	Total Incl O&P
4070	Elbows vertical, 4" rung spacing, use 9" rung x 1.25									
4080	6" rung spacing, use 9" rung x 1.15									
4090	12" rung spacing, use 9" rung x 0.90									
4100	Tee horizontal, 9" rung spacing, 12" radius, 6" wide	2 Elec	5	3.200	Ea.	105	204		309	420
4110	9" wide		4.60	3.478		113	222		335	455
4120	12" wide		4.40	3.636		95.50	232		327.50	450
4130	18" wide		4.20	3.810		136	243		379	510
4140	24" wide		4	4		149	255		404	545
4150	30" wide		3.60	4.444		174	283		457	610
4170	24" radius, 6" wide		4.60	3.478		179	222		401	525
4180	9" wide		4.20	3.810		189	243		432	565
4190	12" wide		4	4		194	255		449	595
4200	18" wide		3.80	4.211		212	268		480	635
4210	24" wide		3.60	4.444		232	283		515	675
4220	30" wide		3.20	5		251	320		571	750
4240	36" radius, 6" wide		4.20	3.810		253	243		496	640
4250	9" wide		3.80	4.211		260	268		528	685
4260	12" wide		3.60	4.444		285	283		568	735
4270	18" wide		3.40	4.706		299	300		599	775
4280	24" wide		3.20	5		330	320		650	840
4290	30" wide		2.80	5.714		420	365		785	1,000
4310	Tee vertical, 9" rung spacing, 12" radius, 6" wide		5.40	2.963		235	189		424	540
4320	9" wide		5.20	3.077		243	196		439	560
4330	12" wide		5	3.200		243	204		447	575
4340	18" wide		4.60	3.478		250	222		472	605
4350	24" wide		4.40	3.636		270	232		502	640
4360	30" wide		4.20	3.810		275	243		518	660
4380	24" radius, 6" wide		5	3.200		400	204		604	745
4390	9" wide		4.80	3.333		405	212		617	760
4400	12" wide		4.60	3.478		410	222		632	780
4410	18" wide		4.20	3.810		420	243		663	820
4420	24" wide		4	4		430	255		685	855
4430	30" wide		3.80	4.211		445	268		713	890
4450	36" radius, 6" wide		4.60	3.478		875	222		1,097	1,300
4460	9" wide		4.40	3.636		895	232		1,127	1,325
4470	12" wide		4.20	3.810		930	243		1,173	1,375
4480	18" wide		3.80	4.211		940	268		1,208	1,425
4490	24" wide		3.60	4.444		965	283		1,248	1,475
4500	30" wide		3.40	4.706		975	300		1,275	1,525
4520	Tees, 4" rung spacing, use 9" rung x 1.30									
4530	6" rung spacing, use 9" rung x 1.20									
4540	12" rung spacing, use 9" rung x 0.90									
4550	Cross horizontal, 9" rung spacing, 12" radius, 6" wide	2 Elec	4.40	3.636	Ea.	146	232		378	505
4560	9" wide		4.20	3.810		162	243		405	540
4570	12" wide		4	4		138	255		393	530
4580	18" wide		3.60	4.444		199	283		482	640
4590	24" wide		3.40	4.706		205	300		505	670
4600	30" wide		3	5.333		242	340		582	770
4620	24" radius, 6" wide		4	4		280	255		535	690
4630	9" wide		3.80	4.211		290	268		558	720
4640	12" wide		3.60	4.444		298	283		581	750
4650	18" wide		3.20	5		315	320		635	820
4660	24" wide		3	5.333		340	340		680	880

187

26 05 36.10 Cable Tray Ladder Type		Crew	Daily Output	Labor-Hours	Unit	Material	2021 Bare Costs Labor	Equipment	Total	Total Incl O&P
4670	30" wide	2 Elec	2.60	6.154	Ea.	365	390		755	980
4690	36" radius, 6" wide		3.60	4.444		330	283		613	785
4700	9" wide		3.40	4.706		345	300		645	825
4710	12" wide		3.20	5		365	320		685	875
4720	18" wide		2.80	5.714		390	365		755	970
4730	24" wide		2.60	6.154		480	390		870	1,100
4740	30" wide		2.20	7.273		535	465		1,000	1,275
4760	Cross horizontal, 4" rung spacing, use 9" rung x 1.30									
4770	6" rung spacing, use 9" rung x 1.20									
4780	12" rung spacing, use 9" rung x 0.90									
4790	Reducer, 9" to 6" wide tray	2 Elec	16	1	Ea.	77.50	63.50		141	180
4800	12" to 9" wide tray		14	1.143		81	73		154	197
4810	18" to 12" wide tray		12.40	1.290		81	82		163	211
4820	24" to 18" wide tray		10.60	1.509		82.50	96		178.50	234
4830	30" to 24" wide tray		9.20	1.739		84.50	111		195.50	258
4840	36" to 30" wide tray		8	2		89.50	127		216.50	288
4841	Reducer, 18" to 6" wide tray		12.40	1.290		81	82		163	211
4842	24" to 12" wide tray		10.60	1.509		82.50	96		178.50	234
4843	30" to 18" wide tray		9.20	1.739		84.50	111		195.50	258
4844	30" to 12" wide tray		9.20	1.739		84.50	111		195.50	258
4845	36" to 24" wide tray		8	2		89.50	127		216.50	288
4846	36" to 18" wide tray		8	2		89.50	127		216.50	288
4847	36" to 12" wide tray		8	2		89.50	127		216.50	288
4850	Dropout or end plate, 6" wide		32	.500		10.15	32		42.15	58.50
4860	9" wide tray		28	.571		11	36.50		47.50	66
4870	12" wide tray		26	.615		11.80	39		50.80	71.50
4880	18" wide tray		22	.727		15.95	46.50		62.45	86.50
4890	24" wide tray		20	.800		18.70	51		69.70	96
4900	30" wide tray		18	.889		22	56.50		78.50	109
4920	Tray connector		48	.333		14.85	21		35.85	48
8000	Elbow 36" radius horiz., 60°, 6" wide tray		10.60	1.509		113	96		209	267
8010	9" wide tray		9	1.778		122	113		235	305
8020	12" wide tray		7.80	2.051		126	131		257	330
8030	18" wide tray		6.20	2.581		134	164		298	390
8040	24" wide tray		5	3.200		157	204		361	480
8050	30" wide tray		4.40	3.636		161	232		393	520
8060	30°, 6" wide tray		14	1.143		86	73		159	203
8070	9" wide tray		11.40	1.404		91	89.50		180.50	233
8080	12" wide tray		9.80	1.633		92.50	104		196.50	257
8090	18" wide tray		7.40	2.162		97.50	138		235.50	310
8100	24" wide tray		5.80	2.759		119	176		295	390
8110	30" wide tray		4.80	3.333		122	212		334	450
8120	Adjustable, 6" wide tray		12.40	1.290		114	82		196	248
8130	9" wide tray		10.20	1.569		121	100		221	281
8140	12" wide tray		8.80	1.818		122	116		238	305
8150	18" wide tray		6.80	2.353		134	150		284	370
8160	24" wide tray		5.40	2.963		142	189		331	435
8170	30" wide tray		4.60	3.478		159	222		381	505
8180	Wye 36" radius horiz., 45°, 6" wide tray		4.60	3.478		124	222		346	465
8190	9" wide tray		4.40	3.636		131	232		363	490
8200	12" wide tray		4.20	3.810		136	243		379	510
8210	18" wide tray		3.80	4.211		154	268		422	570
8220	24" wide tray		3.60	4.444		172	283		455	610

26 05 36 – Cable Trays for Electrical Systems

26 05 36.10 Cable Tray Ladder Type		Crew	Daily Output	Labor-Hours	Unit	Material	2021 Bare Costs Labor	Equipment	Total	Total Incl O&P
8230	30" wide tray	2 Elec	3.40	4.706	Ea.	192	300		492	655
8240	Elbow 36" radius vert. in/outside, 60°, 6" wide tray		10.60	1.509		124	96		220	280
8250	9" wide tray		9	1.778		129	113		242	310
8260	12" wide tray		7.80	2.051		131	131		262	340
8270	18" wide tray		6.20	2.581		136	164		300	395
8280	24" wide tray		5	3.200		144	204		348	465
8290	30" wide tray		4.40	3.636		154	232		386	515
8300	45°, 6" wide tray		12.40	1.290		112	82		194	245
8310	9" wide tray		10.20	1.569		119	100		219	279
8320	12" wide tray		8.80	1.818		120	116		236	305
8330	18" wide tray		6.80	2.353		124	150		274	360
8340	24" wide tray		5.40	2.963		131	189		320	425
8350	30" wide tray		4.60	3.478		134	222		356	475
8360	30°, 6" wide tray		14	1.143		98	73		171	216
8370	9" wide tray		11.40	1.404		100	89.50		189.50	243
8380	12" wide tray		9.80	1.633		107	104		211	272
8390	18" wide tray		7.40	2.162		108	138		246	325
8400	24" wide tray		5.80	2.759		111	176		287	385
8410	30" wide tray		4.80	3.333		117	212		329	445
8660	Adjustable, 6" wide tray		12.40	1.290		112	82		194	245
8670	9" wide tray		10.20	1.569		119	100		219	279
8680	12" wide tray		8.80	1.818		121	116		237	305
8690	18" wide tray		6.80	2.353		134	150		284	370
8700	24" wide tray		5.40	2.963		142	189		331	435
8710	30" wide tray		4.60	3.478		157	222		379	505
8720	Cross, vertical, 24" radius, 6" wide tray		3.60	4.444		815	283		1,098	1,325
8730	9" wide tray		3.40	4.706		830	300		1,130	1,350
8740	12" wide tray		3.20	5		850	320		1,170	1,400
8750	18" wide tray		2.80	5.714		890	365		1,255	1,525
8760	24" wide tray		2.60	6.154		920	390		1,310	1,600
8770	30" wide tray		2.20	7.273		975	465		1,440	1,775
9200	Splice plate	1 Elec	48	.167	Pr.	15.55	10.60		26.15	33
9210	Expansion joint		48	.167		18.70	10.60		29.30	36.50
9220	Horizontal hinged		48	.167		15.55	10.60		26.15	33
9230	Vertical hinged		48	.167		18.70	10.60		29.30	36.50
9240	Ladder hanger, vertical		28	.286	Ea.	4.67	18.20		22.87	32
9250	Ladder to channel connector		24	.333		54.50	21		75.50	91
9260	Ladder to box connector, 30" wide		19	.421		54.50	27		81.50	99.50
9270	24" wide		20	.400		54.50	25.50		80	97.50
9280	18" wide		21	.381		52.50	24.50		77	93.50
9290	12" wide		22	.364		49	23		72	88.50
9300	9" wide		23	.348		44	22		66	81
9310	6" wide		24	.333		30	21		51	64.50
9320	Ladder floor flange		24	.333		35	21		56	70
9330	Cable roller for tray, 30" wide		10	.800		310	51		361	415
9340	24" wide		11	.727		261	46.50		307.50	355
9350	18" wide		12	.667		249	42.50		291.50	335
9360	12" wide		13	.615		196	39		235	274
9370	9" wide		14	.571		174	36.50		210.50	245
9380	6" wide		15	.533		141	34		175	206
9390	Pulley, single wheel		12	.667		320	42.50		362.50	420
9400	Triple wheel		10	.800		640	51		691	780
9440	Nylon cable tie, 14" long		80	.100		.57	6.35		6.92	10.10

26 05 Common Work Results for Electrical

26 05 36 – Cable Trays for Electrical Systems

26 05 36.10 Cable Tray Ladder Type

		Crew	Daily Output	Labor-Hours	Unit	Material	2021 Bare Costs Labor	Equipment	Total	Total Incl O&P
9450	Ladder, hold down clamp	1 Elec	60	.133	Ea.	3.98	8.50		12.48	17
9460	Cable clamp		60	.133		10.05	8.50		18.55	23.50
9470	Wall bracket, 30" wide tray		19	.421		56.50	27		83.50	103
9480	24" wide tray		20	.400		44	25.50		69.50	86.50
9490	18" wide tray		21	.381		39.50	24.50		64	79.50
9500	12" wide tray		22	.364		37	23		60	75
9510	9" wide tray		23	.348		32.50	22		54.50	68.50
9520	6" wide tray		24	.333		31.50	21		52.50	66.50

26 05 36.20 Cable Tray Solid Bottom

			Crew	Daily Output	Labor-Hours	Unit	Material	2021 Bare Costs Labor	Equipment	Total	Total Incl O&P
0010	**CABLE TRAY SOLID BOTTOM** w/ftngs. & supports, 3" deep, to 15' high										
0200	For higher elevations, see Section 26 05 36.40										
0220	Galvanized steel, tray, 6" wide	R260536-10	2 Elec	120	.133	L.F.	12.45	8.50		20.95	26.50
0240	12" wide			100	.160		16.05	10.20		26.25	33
0260	18" wide			70	.229		19.45	14.55		34	43
0280	24" wide	R260536-11		60	.267		23	17		40	50
0300	30" wide			50	.320		27	20.50		47.50	60.50
0340	Elbow horizontal, 90°, 12" radius, 6" wide			9.60	1.667	Ea.	95	106		201	262
0360	12" wide			6.80	2.353		106	150		256	340
0370	18" wide			5.40	2.963		121	189		310	415
0380	24" wide			4.40	3.636		148	232		380	505
0390	30" wide			3.80	4.211		177	268		445	595
0420	24" radius, 6" wide			9.20	1.739		135	111		246	315
0440	12" wide			6.40	2.500		157	159		316	410
0450	18" wide			5	3.200		182	204		386	505
0460	24" wide			4	4		211	255		466	615
0470	30" wide			3.40	4.706		248	300		548	720
0500	36" radius, 6" wide			8.80	1.818		197	116		313	390
0520	12" wide			6	2.667		222	170		392	495
0530	18" wide			4.60	3.478		272	222		494	630
0540	24" wide			3.60	4.444		293	283		576	745
0550	30" wide			3	5.333		355	340		695	895
0580	Elbow vertical, 90°, 12" radius, 6" wide			9.60	1.667		115	106		221	284
0600	12" wide			6.80	2.353		126	150		276	360
0610	18" wide			5.40	2.963		159	189		348	455
0620	24" wide			4.40	3.636		144	232		376	505
0630	30" wide			3.80	4.211		153	268		421	570
0670	24" radius, 6" wide			9.20	1.739		162	111		273	345
0690	12" wide			6.40	2.500		179	159		338	435
0700	18" wide			5	3.200		193	204		397	515
0710	24" wide			4	4		210	255		465	610
0720	30" wide			3.40	4.706		222	300		522	690
0750	36" radius, 6" wide			8.80	1.818		222	116		338	415
0770	12" wide			6.60	2.424		248	154		402	500
0780	18" wide			4.60	3.478		272	222		494	630
0790	24" wide			3.60	4.444		297	283		580	745
0800	30" wide			3	5.333		325	340		665	860
0840	Tee horizontal, 12" radius, 6" wide			5	3.200		133	204		337	450
0860	12" wide			4	4		144	255		399	540
0870	18" wide			3.40	4.706		171	300		471	635
0880	24" wide			2.80	5.714		191	365		556	750
0890	30" wide			2.60	6.154		219	390		609	820
0940	24" radius, 6" wide			4.60	3.478		206	222		428	555

26 05 36 – Cable Trays for Electrical Systems

26 05 36.20 Cable Tray Solid Bottom		Crew	Daily Output	Labor-Hours	Unit	Material	2021 Bare Costs Labor	Equipment	Total	Total Incl O&P
0960	12" wide	2 Elec	3.60	4.444	Ea.	241	283		524	685
0970	18" wide		3	5.333		268	340		608	800
0980	24" wide		2.40	6.667		365	425		790	1,025
0990	30" wide		2.20	7.273		395	465		860	1,125
1020	36" radius, 6" wide		4.20	3.810		325	243		568	715
1040	12" wide		3.20	5		365	320		685	875
1050	18" wide		2.60	6.154		395	390		785	1,025
1060	24" wide		2.20	7.273		505	465		970	1,250
1070	30" wide		2	8		550	510		1,060	1,350
1100	Tee vertical, 12" radius, 6" wide		5	3.200		219	204		423	545
1120	12" wide		4	4		228	255		483	630
1130	18" wide		3.60	4.444		231	283		514	675
1140	24" wide		3.40	4.706		253	300		553	725
1150	30" wide		3	5.333		272	340		612	805
1180	24" radius, 6" wide		4.60	3.478		325	222		547	685
1200	12" wide		3.60	4.444		335	283		618	790
1210	18" wide		3.20	5		355	320		675	865
1220	24" wide		3	5.333		370	340		710	915
1230	30" wide		2.60	6.154		410	390		800	1,025
1260	36" radius, 6" wide		4.20	3.810		505	243		748	915
1280	12" wide		3.20	5		510	320		830	1,050
1290	18" wide		2.80	5.714		550	365		915	1,150
1300	24" wide		2.60	6.154		570	390		960	1,200
1310	30" wide		2.20	7.273		665	465		1,130	1,425
1340	Cross horizontal, 12" radius, 6" wide		4	4		159	255		414	555
1360	12" wide		3.40	4.706		173	300		473	635
1370	18" wide		2.80	5.714		202	365		567	765
1380	24" wide		2.40	6.667		228	425		653	880
1390	30" wide		2	8		253	510		763	1,025
1420	24" radius, 6" wide		3.60	4.444		290	283		573	740
1440	12" wide		3	5.333		325	340		665	860
1450	18" wide		2.40	6.667		365	425		790	1,025
1460	24" wide		2	8		465	510		975	1,275
1470	30" wide		1.80	8.889		505	565		1,070	1,400
1500	36" radius, 6" wide		3.20	5		475	320		795	995
1520	12" wide		2.60	6.154		520	390		910	1,150
1530	18" wide		2	8		570	510		1,080	1,375
1540	24" wide		1.80	8.889		670	565		1,235	1,575
1550	30" wide		1.60	10		730	635		1,365	1,750
1580	Drop out or end plate, 6" wide		32	.500		19.35	32		51.35	69
1600	12" wide		26	.615		24	39		63	85
1610	18" wide		22	.727		26	46.50		72.50	97.50
1620	24" wide		20	.800		30.50	51		81.50	109
1630	30" wide		18	.889		34	56.50		90.50	122
1660	Reducer, 12" to 6" wide		12	1.333		93	85		178	228
1680	18" to 12" wide		10.60	1.509		95	96		191	247
1700	18" to 6" wide		10.60	1.509		95	96		191	247
1720	24" to 18" wide		9.20	1.739		98.50	111		209.50	273
1740	24" to 12" wide		9.20	1.739		98.50	111		209.50	273
1760	30" to 24" wide		8	2		106	127		233	305
1780	30" to 18" wide		8	2		106	127		233	305
1800	30" to 12" wide		8	2		106	127		233	305
1820	36" to 30" wide		7.20	2.222		109	142		251	330

26 05 36.20 Cable Tray Solid Bottom		Crew	Daily Output	Labor-Hours	Unit	Material	2021 Bare Costs Labor	Equipment	Total	Total Incl O&P
1840	36" to 24" wide	2 Elec	7.20	2.222	Ea.	109	142		251	330
1860	36" to 18" wide		7.20	2.222		109	142		251	330
1880	36" to 12" wide		7.20	2.222		109	142		251	330
2000	Aluminum tray, 6" wide		150	.107	L.F.	12.25	6.80		19.05	23.50
2020	12" wide		130	.123		16.05	7.85		23.90	29.50
2030	18" wide		100	.160		20	10.20		30.20	37
2040	24" wide		90	.178		25.50	11.30		36.80	45
2050	30" wide		70	.229		30	14.55		44.55	54.50
2080	Elbow horizontal, 90°, 12" radius, 6" wide		9.60	1.667	Ea.	98.50	106		204.50	266
2100	12" wide		7.60	2.105		111	134		245	320
2110	18" wide		6.80	2.353		136	150		286	375
2120	24" wide		5.80	2.759		155	176		331	430
2130	30" wide		5	3.200		189	204		393	515
2160	24" radius, 6" wide		9.20	1.739		140	111		251	320
2180	12" wide		7.20	2.222		166	142		308	390
2190	18" wide		6.40	2.500		189	159		348	445
2200	24" wide		5.40	2.963		233	189		422	535
2210	30" wide		4.60	3.478		267	222		489	625
2240	36" radius, 6" wide		8.80	1.818		207	116		323	400
2260	12" wide		6.80	2.353		248	150		398	495
2270	18" wide		6	2.667		296	170		466	575
2280	24" wide		5	3.200		315	204		519	655
2290	30" wide		4.20	3.810		360	243		603	755
2320	Elbow vertical, 90°, 12" radius, 6" wide		9.60	1.667		117	106		223	286
2340	12" wide		7.60	2.105		121	134		255	330
2350	18" wide		6.80	2.353		137	150		287	375
2360	24" wide		5.80	2.759		145	176		321	420
2370	30" wide		5	3.200		154	204		358	475
2400	24" radius, 6" wide		9.20	1.739		166	111		277	345
2420	12" wide		7.20	2.222		180	142		322	410
2430	18" wide		6.40	2.500		194	159		353	450
2440	24" wide		5.40	2.963		210	189		399	510
2450	30" wide		4.60	3.478		223	222		445	575
2480	36" radius, 6" wide		8.80	1.818		219	116		335	415
2500	12" wide		6.80	2.353		245	150		395	495
2510	18" wide		6	2.667		267	170		437	545
2520	24" wide		5	3.200		278	204		482	610
2530	30" wide		4.20	3.810		305	243		548	700
2560	Tee horizontal, 12" radius, 6" wide		5	3.200		152	204		356	470
2580	12" wide		4.40	3.636		185	232		417	550
2590	18" wide		4	4		207	255		462	610
2600	24" wide		3.60	4.444		232	283		515	675
2610	30" wide		3	5.333		267	340		607	800
2640	24" radius, 6" wide		4.60	3.478		251	222		473	605
2660	12" wide		4	4		282	255		537	690
2670	18" wide		3.60	4.444		320	283		603	770
2680	24" wide		3	5.333		390	340		730	935
2690	30" wide		2.40	6.667		445	425		870	1,125
2720	36" radius, 6" wide		4.20	3.810		410	243		653	810
2740	12" wide		3.60	4.444		445	283		728	910
2750	18" wide		3.20	5		515	320		835	1,050
2760	24" wide		2.60	6.154		600	390		990	1,250
2770	30" wide		2	8		660	510		1,170	1,475

26 05 36 – Cable Trays for Electrical Systems

26 05 36.20 Cable Tray Solid Bottom		Crew	Daily Output	Labor-Hours	Unit	Material	2021 Bare Costs Labor	Equipment	Total	Total Incl O&P
2800	Tee vertical, 12" radius, 6" wide	2 Elec	5	3.200	Ea.	217	204		421	545
2820	12" wide		4.40	3.636		222	232		454	590
2830	18" wide		4.20	3.810		231	243		474	615
2840	24" wide		4	4		245	255		500	650
2850	30" wide		3.60	4.444		267	283		550	715
2880	24" radius, 6" wide		4.60	3.478		320	222		542	680
2900	12" wide		4	4		340	255		595	750
2910	18" wide		3.80	4.211		360	268		628	795
2920	24" wide		3.60	4.444		390	283		673	850
2930	30" wide		3.20	5		430	320		750	950
2960	36" radius, 6" wide		4.20	3.810		500	243		743	910
2980	12" wide		3.40	4.706		525	300		825	1,025
2990	18" wide		3.40	4.706		555	300		855	1,050
3000	24" wide		3.20	5		570	320		890	1,100
3010	30" wide		2.80	5.714		615	365		980	1,225
3040	Cross horizontal, 12" radius, 6" wide		4.40	3.636		194	232		426	560
3060	12" wide		4	4		219	255		474	620
3070	18" wide		3.40	4.706		248	300		548	720
3080	24" wide		2.80	5.714		282	365		647	850
3090	30" wide		2.60	6.154		320	390		710	930
3120	24" radius, 6" wide		4	4		340	255		595	750
3140	12" wide		3.60	4.444		390	283		673	850
3150	18" wide		3	5.333		425	340		765	975
3160	24" wide		2.40	6.667		515	425		940	1,200
3170	30" wide		2.20	7.273		570	465		1,035	1,325
3200	36" radius, 6" wide		3.60	4.444		590	283		873	1,075
3220	12" wide		3.20	5		620	320		940	1,150
3230	18" wide		2.60	6.154		690	390		1,080	1,350
3240	24" wide		2	8		755	510		1,265	1,575
3250	30" wide		1.80	8.889		815	565		1,380	1,725
3280	Dropout, or end plate, 6" wide		32	.500		19.45	32		51.45	69
3300	12" wide		26	.615		22	39		61	82.50
3310	18" wide		22	.727		26.50	46.50		73	98
3320	24" wide		20	.800		29	51		80	108
3330	30" wide		18	.889		34	56.50		90.50	121
3380	Reducer, 12" to 6" wide		14	1.143		95.50	73		168.50	213
3400	18" to 12" wide		12	1.333		98.50	85		183.50	234
3420	18" to 6" wide		12	1.333		98.50	85		183.50	234
3440	24" to 18" wide		10.60	1.509		104	96		200	258
3460	24" to 12" wide		10.60	1.509		104	96		200	258
3480	30" to 24" wide		9.20	1.739		114	111		225	290
3500	30" to 18" wide		9.20	1.739		114	111		225	290
3520	30" to 12" wide		9.20	1.739		114	111		225	290
3540	36" to 30" wide		8	2		120	127		247	320
3560	36" to 24" wide		8	2		121	127		248	320
3580	36" to 18" wide		8	2		121	127		248	320
3600	36" to 12" wide		8	2		124	127		251	325

26 05 36.30 Cable Tray Trough

		Crew	Daily Output	Labor-Hours	Unit	Material	2021 Bare Costs Labor	Equipment	Total	Total Incl O&P
0010	**CABLE TRAY TROUGH** vented, w/ftngs. & supports, 6" deep, to 10' high									
0020	For higher elevations, see Section 26 05 36.40									
0200	Galvanized steel, tray, 6" wide	R260536-10	2 Elec	90	.178	L.F.	16.15	11.30		27.45	34.50
0240	12" wide			80	.200		15.15	12.75		27.90	35.50

26 05 36.30 Cable Tray Trough		Crew	Daily Output	Labor-Hours	Unit	Material	2021 Bare Costs Labor	Equipment	Total	Total Incl O&P
0260	18" wide	2 Elec	70	.229	L.F.	23.50	14.55		38.05	47.50
0280	24" wide		60	.267		50	17		67	80
0300	30" wide		50	.320		47	20.50		67.50	82.50
0340	Elbow horizontal, 90°, 12" radius, 6" wide		7.60	2.105	Ea.	132	134		266	345
0360	12" wide		5.60	2.857		156	182		338	440
0370	18" wide		4.40	3.636		180	232		412	545
0380	24" wide		3.60	4.444		223	283		506	665
0390	30" wide		3.20	5		251	320		571	750
0420	24" radius, 6" wide		7.20	2.222		194	142		336	425
0440	12" wide		5.20	3.077		225	196		421	540
0450	18" wide		4	4		270	255		525	675
0460	24" wide		3.20	5		305	320		625	810
0470	30" wide		2.80	5.714		355	365		720	930
0500	36" radius, 6" wide		6.80	2.353		252	150		402	500
0520	12" wide		4.80	3.333		287	212		499	630
0530	18" wide		3.60	4.444		345	283		628	800
0540	24" wide		2.80	5.714		350	365		715	930
0550	30" wide		2.40	6.667		395	425		820	1,075
0580	Elbow vertical, 90°, 12" radius, 6" wide		7.60	2.105		143	134		277	355
0600	12" wide		5.60	2.857		158	182		340	445
0610	18" wide		4.40	3.636		160	232		392	520
0620	24" wide		3.60	4.444		184	283		467	620
0630	30" wide		3.20	5		188	320		508	680
0660	24" radius, 6" wide		7.20	2.222		204	142		346	435
0680	12" wide		5.20	3.077		216	196		412	530
0690	18" wide		4	4		239	255		494	645
0700	24" wide		3.20	5		247	320		567	745
0710	30" wide		2.80	5.714		277	365		642	845
0740	36" radius, 6" wide		6.80	2.353		275	150		425	530
0760	12" wide		4.80	3.333		293	212		505	635
0770	18" wide		3.60	4.444		325	283		608	775
0780	24" wide		2.80	5.714		350	365		715	925
0790	30" wide		2.40	6.667		355	425		780	1,025
0820	Tee horizontal, 12" radius, 6" wide		4	4		166	255		421	565
0840	12" wide		3.20	5		186	320		506	680
0850	18" wide		2.80	5.714		210	365		575	770
0860	24" wide		2.40	6.667		243	425		668	900
0870	30" wide		2.20	7.273		277	465		742	995
0900	24" radius, 6" wide		3.60	4.444		275	283		558	725
0920	12" wide		2.80	5.714		300	365		665	870
0930	18" wide		2.40	6.667		335	425		760	1,000
0940	24" wide		2	8		435	510		945	1,225
0950	30" wide		1.80	8.889		475	565		1,040	1,350
0980	36" radius, 6" wide		3.20	5		395	320		715	910
1000	12" wide		2.40	6.667		450	425		875	1,125
1010	18" wide		2	8		495	510		1,005	1,300
1020	24" wide		1.60	10		595	635		1,230	1,600
1030	30" wide		1.40	11.429		630	730		1,360	1,775
1060	Tee vertical, 12" radius, 6" wide		4	4		263	255		518	670
1080	12" wide		3.20	5		267	320		587	770
1090	18" wide		3	5.333		275	340		615	805
1100	24" wide		2.80	5.714		290	365		655	860
1110	30" wide		2.60	6.154		305	390		695	915

R260536-11

194

26 05 36 – Cable Trays for Electrical Systems

26 05 36.30 Cable Tray Trough	Crew	Daily Output	Labor-Hours	Unit	Material	2021 Bare Costs Labor	Equipment	Total	Total Incl O&P
1140	24" radius, 6" wide	2 Elec	3.60	4.444	Ea.	355	283		638	810
1160	12" wide		2.80	5.714		370	365		735	950
1170	18" wide		2.60	6.154		390	390		780	1,000
1180	24" wide		2.40	6.667		420	425		845	1,100
1190	30" wide		2.20	7.273		440	465		905	1,175
1220	36" radius, 6" wide		3.20	5		575	320		895	1,100
1240	12" wide		2.40	6.667		590	425		1,015	1,275
1250	18" wide		2.20	7.273		610	465		1,075	1,350
1260	24" wide		2	8		645	510		1,155	1,475
1270	30" wide		1.80	8.889		750	565		1,315	1,675
1300	Cross horizontal, 12" radius, 6" wide		3.20	5		207	320		527	705
1320	12" wide		2.80	5.714		209	365		574	770
1330	18" wide		2.40	6.667		234	425		659	890
1340	24" wide		2	8		246	510		756	1,025
1350	30" wide		1.80	8.889		277	565		842	1,150
1380	24" radius, 6" wide		2.80	5.714		325	365		690	900
1400	12" wide		2.40	6.667		340	425		765	1,000
1410	18" wide		2	8		375	510		885	1,175
1420	24" wide		1.60	10		480	635		1,115	1,475
1430	30" wide		1.40	11.429		530	730		1,260	1,650
1460	36" radius, 6" wide		2.40	6.667		560	425		985	1,250
1480	12" wide		2	8		580	510		1,090	1,400
1490	18" wide		1.60	10		590	635		1,225	1,600
1500	24" wide		1.20	13.333		710	850		1,560	2,025
1510	30" wide		1	16		780	1,025		1,805	2,375
1540	Dropout or end plate, 6" wide		26	.615		25	39		64	85.50
1560	12" wide		22	.727		30.50	46.50		77	103
1580	18" wide		20	.800		33	51		84	112
1600	24" wide		18	.889		38	56.50		94.50	126
1620	30" wide		16	1		41	63.50		104.50	140
1660	Reducer, 12" to 6" wide		9.40	1.702		101	108		209	272
1680	18" to 12" wide		8.40	1.905		105	121		226	295
1700	18" to 6" wide		8.40	1.905		105	121		226	295
1720	24" to 18" Trough		7.20	2.222		112	142		254	335
1740	24" to 12" wide		7.20	2.222		112	142		254	335
1760	30" to 24" wide		6.40	2.500		116	159		275	365
1780	30" to 18" wide		6.40	2.500		116	159		275	365
1800	30" to 12" wide		6.40	2.500		116	159		275	365
1820	36" to 30" wide		5.80	2.759		124	176		300	395
1840	36" to 24" wide		5.80	2.759		124	176		300	395
1860	36" to 18" wide		5.80	2.759		126	176		302	400
1880	36" to 12" wide		5.80	2.759		126	176		302	400
2000	Aluminum, tray, vented, 6" wide		120	.133	L.F.	18.15	8.50		26.65	32.50
2010	9" wide		110	.145		20.50	9.25		29.75	36.50
2020	12" wide		100	.160		30.50	10.20		40.70	48.50
2030	18" wide		90	.178		27.50	11.30		38.80	47
2040	24" wide		80	.200		32.50	12.75		45.25	54.50
2050	30" wide		70	.229		43	14.55		57.55	69
2080	Elbow horiz., 90°, 12" radius, 6" wide		7.60	2.105	Ea.	114	134		248	325
2090	9" wide		7	2.286		124	146		270	350
2100	12" wide		6.20	2.581		131	164		295	390
2110	18" wide		5.60	2.857		154	182		336	440
2120	24" wide		4.60	3.478		177	222		399	525

For customer support on your Electrical Costs with RSMeans Data, call 800.448.8182.

195

26 05 36 – Cable Trays for Electrical Systems

26 05 36.30 Cable Tray Trough		Crew	Daily Output	Labor-Hours	Unit	Material	2021 Bare Costs Labor	Equipment	Total	Total Incl O&P
2130	30" wide	2 Elec	4	4	Ea.	218	255		473	620
2160	24" radius, 6" wide		7.20	2.222		165	142		307	390
2180	12" wide		5.80	2.759		193	176		369	475
2190	18" wide		5.20	3.077		222	196		418	535
2200	24" wide		4.20	3.810		253	243		496	640
2210	30" wide		3.60	4.444		290	283		573	740
2240	36" radius, 6" wide		6.80	2.353		238	150		388	485
2260	12" wide		5.40	2.963		271	189		460	580
2270	18" wide		4.80	3.333		310	212		522	655
2280	24" wide		3.80	4.211		335	268		603	770
2290	30" wide		3.40	4.706		395	300		695	880
2320	Elbow vertical, 90°, 12" radius, 6" wide		7.60	2.105		138	134		272	350
2330	9" wide		7	2.286		147	146		293	380
2340	12" wide		6.20	2.581		149	164		313	410
2350	18" wide		5.60	2.857		154	182		336	440
2360	24" wide		4.60	3.478		169	222		391	515
2370	30" wide		4	4		177	255		432	575
2400	24" radius, 6" wide		7.20	2.222		187	142		329	415
2420	12" wide		5.80	2.759		203	176		379	485
2430	18" wide		5.20	3.077		220	196		416	535
2440	24" wide		4.20	3.810		223	243		466	605
2450	30" wide		3.60	4.444		236	283		519	680
2480	36" radius, 6" wide		6.80	2.353		237	150		387	485
2500	12" wide		5.40	2.963		253	189		442	560
2510	18" wide		4.80	3.333		284	212		496	625
2520	24" wide		3.80	4.211		305	268		573	735
2530	30" wide		3.40	4.706		335	300		635	815
2560	Tee horizontal, 12" radius, 6" wide		4	4		173	255		428	570
2570	9" wide		3.80	4.211		177	268		445	595
2580	12" wide		3.60	4.444		193	283		476	630
2590	18" wide		3.20	5		223	320		543	720
2600	24" wide		2.80	5.714		256	365		621	820
2610	30" wide		2.40	6.667		275	425		700	935
2640	24" radius, 6" wide		3.60	4.444		263	283		546	710
2660	12" wide		3.20	5		299	320		619	805
2670	18" wide		2.80	5.714		330	365		695	900
2680	24" wide		2.40	6.667		410	425		835	1,075
2690	30" wide		2	8		445	510		955	1,250
2720	36" radius, 6" wide		3.20	5		425	320		745	940
2740	12" wide		2.80	5.714		485	365		850	1,075
2750	18" wide		2.40	6.667		550	425		975	1,225
2760	24" wide		2	8		645	510		1,155	1,450
2770	30" wide		1.60	10		685	635		1,320	1,700
2800	Tee vertical, 12" radius, 6" wide		4	4		243	255		498	650
2810	9" wide		3.80	4.211		245	268		513	670
2820	12" wide		3.60	4.444		263	283		546	710
2830	18" wide		3.40	4.706		267	300		567	740
2840	24" wide		3.20	5		271	320		591	775
2850	30" wide		3	5.333		284	340		624	815
2880	24" radius, 6" wide		3.60	4.444		345	283		628	800
2900	12" wide		3.20	5		370	320		690	885
2910	18" wide		3	5.333		395	340		735	940
2920	24" wide		2.80	5.714		420	365		785	1,000

For customer support on your Electrical Costs with RSMeans Data, call 800.448.8182.

26 05 36 – Cable Trays for Electrical Systems

26 05 36.30 Cable Tray Trough	Crew	Daily Output	Labor-Hours	Unit	Material	2021 Bare Costs Labor	Equipment	Total	Total Incl O&P
2930	30" wide	2 Elec	2.60	6.154	Ea.	455	390		845	1,075
2960	36" radius, 6" wide		3.20	5		520	320		840	1,050
2980	12" wide		2.80	5.714		550	365		915	1,150
2990	18" wide		2.60	6.154		570	390		960	1,200
3000	24" wide		2.40	6.667		610	425		1,035	1,300
3010	30" wide		2.20	7.273		660	465		1,125	1,425
3040	Cross horizontal, 12" radius, 6" wide		3.60	4.444		215	283		498	655
3050	9" wide		3.40	4.706		231	300		531	700
3060	12" wide		3.20	5		237	320		557	735
3070	18" wide		2.80	5.714		249	365		614	815
3080	24" wide		2.40	6.667		277	425		702	935
3090	30" wide		2.20	7.273		355	465		820	1,075
3120	24" radius, 6" wide		3.20	5		385	320		705	900
3140	12" wide		2.80	5.714		425	365		790	1,000
3150	18" wide		2.40	6.667		455	425		880	1,125
3160	24" wide		2	8		525	510		1,035	1,325
3170	30" wide		1.80	8.889		570	565		1,135	1,475
3200	36" radius, 6" wide		2.80	5.714		670	365		1,035	1,275
3220	12" wide		2.40	6.667		705	425		1,130	1,400
3230	18" wide		2	8		765	510		1,275	1,600
3240	24" wide		1.60	10		885	635		1,520	1,925
3250	30" wide		1.40	11.429		965	730		1,695	2,125
3280	Dropout, or end plate, 6" wide		26	.615		22	39		61	82.50
3300	12" wide		22	.727		26	46.50		72.50	97.50
3310	18" wide		20	.800		31.50	51		82.50	110
3320	24" wide		18	.889		38	56.50		94.50	126
3330	30" wide		16	1		40.50	63.50		104	139
3370	Reducer, 9" to 6" wide		12	1.333		108	85		193	245
3380	12" to 6" wide		11.40	1.404		113	89.50		202.50	258
3390	12" to 9" wide		11.40	1.404		113	89.50		202.50	258
3400	18" to 12" wide		9.60	1.667		122	106		228	292
3420	18" to 6" wide		9.60	1.667		122	106		228	292
3430	18" to 9" wide		9.60	1.667		122	106		228	292
3440	24" to 18" wide		8.40	1.905		130	121		251	325
3460	24" to 12" wide		8.40	1.905		130	121		251	325
3470	24" to 9" wide		8.40	1.905		131	121		252	325
3475	24" to 6" wide		8.40	1.905		131	121		252	325
3480	30" to 24" wide		7.20	2.222		133	142		275	355
3500	30" to 18" wide		7.20	2.222		133	142		275	355
3520	30" to 12" wide		7.20	2.222		136	142		278	360
3540	36" to 30" wide		6.40	2.500		138	159		297	390
3560	36" to 24" wide		6.40	2.500		138	159		297	390
3580	36" to 18" wide		6.40	2.500		138	159		297	390
3600	36" to 12" wide		6.40	2.500		138	159		297	390
3610	Elbow horizontal, 60°, 12" radius, 6" wide		7.80	2.051		94	131		225	297
3620	9" wide		7.20	2.222		103	142		245	325
3630	12" wide		6.40	2.500		113	159		272	360
3640	18" wide		5.80	2.759		127	176		303	400
3650	24" wide		4.80	3.333		149	212		361	480
3680	Elbow horizontal, 45°, 12" radius, 6" wide		8	2		64	127		191	260
3690	9" wide		7.40	2.162		86	138		224	300
3700	12" wide		6.60	2.424		91	154		245	330
3710	18" wide		6	2.667		102	170		272	365

26 05 36.30 Cable Tray Trough

	26 05 36.30 Cable Tray Trough	Crew	Daily Output	Labor-Hours	Unit	Material	2021 Bare Costs Labor	Equipment	Total	Total Incl O&P
3720	24" wide	2 Elec	5	3.200	Ea.	119	204		323	435
3750	Elbow horizontal, 30°, 12" radius, 6" wide		8.20	1.951		71	124		195	263
3760	9" wide		7.60	2.105		74.50	134		208.50	281
3770	12" wide		6.80	2.353		81	150		231	315
3780	18" wide		6.20	2.581		86	164		250	340
3790	24" wide		5.20	3.077		94	196		290	395
3820	Elbow vertical, 60° in/outside, 12" radius, 6" wide		7.80	2.051		117	131		248	320
3830	9" wide		7.20	2.222		118	142		260	340
3840	12" wide		6.40	2.500		120	159		279	370
3850	18" wide		5.80	2.759		125	176		301	400
3860	24" wide		4.80	3.333		130	212		342	460
3890	Elbow vertical, 45° in/outside, 12" radius, 6" wide		8	2		75	127		202	272
3900	9" wide		7.40	2.162		95.50	138		233.50	310
3910	12" wide		6.60	2.424		103	154		257	340
3920	18" wide		6	2.667		106	170		276	370
3930	24" wide		5	3.200		116	204		320	435
3960	Elbow vertical, 30° in/outside, 12" radius, 6" wide		8.20	1.951		81	124		205	274
3970	9" wide		7.60	2.105		86	134		220	294
3980	12" wide		6.80	2.353		87.50	150		237.50	320
3990	18" wide		6.20	2.581		91	164		255	345
4000	24" wide		5.20	3.077		92.50	196		288.50	395
4250	Reducer, left or right hand, 24" to 18" wide		8.40	1.905		123	121		244	315
4260	24" to 12" wide		8.40	1.905		123	121		244	315
4270	24" to 9" wide		8.40	1.905		123	121		244	315
4280	24" to 6" wide		8.40	1.905		125	121		246	315
4290	18" to 12" wide		9.60	1.667		115	106		221	285
4300	18" to 9" wide		9.60	1.667		115	106		221	285
4310	18" to 6" wide		9.60	1.667		115	106		221	285
4320	12" to 9" wide		11.40	1.404		110	89.50		199.50	254
4330	12" to 6" wide		11.40	1.404		110	89.50		199.50	254
4340	9" to 6" wide	↓	12	1.333		107	85		192	244
4350	Splice plate	1 Elec	48	.167		10.20	10.60		20.80	27
4360	Splice plate, expansion joint		48	.167		10.50	10.60		21.10	27.50
4370	Splice plate, hinged, horizontal		48	.167		8.40	10.60		19	25
4380	Vertical		48	.167		12	10.60		22.60	29
4390	Trough, hanger, vertical		28	.286		33.50	18.20		51.70	64
4400	Box connector, 24" wide		20	.400		45	25.50		70.50	87.50
4410	18" wide		21	.381		38	24.50		62.50	77.50
4420	12" wide		22	.364		36.50	23		59.50	74.50
4430	9" wide		23	.348		34.50	22		56.50	71
4440	6" wide		24	.333		33	21		54	67.50
4450	Floor flange		24	.333		34.50	21		55.50	69.50
4460	Hold down clamp		60	.133		3.96	8.50		12.46	16.95
4520	Wall bracket, 24" wide tray		20	.400		39.50	25.50		65	81.50
4530	18" wide tray		21	.381		38.50	24.50		63	78.50
4540	12" wide tray		22	.364		20.50	23		43.50	57
4550	9" wide tray		23	.348		18.35	22		40.35	53
4560	6" wide tray	↓	24	.333	↓	16.70	21		37.70	50
5000	Cable channel aluminum, vented, 1-1/4" deep, 4" wide, straight		80	.100	L.F.	8.95	6.35		15.30	19.30
5010	Elbow horizontal, 36" radius, 90°		5	1.600	Ea.	162	102		264	330
5020	60°		5.50	1.455		124	92.50		216.50	274
5030	45°		6	1.333		101	85		186	237
5040	30°	↓	6.50	1.231	↓	86.50	78.50		165	211

26 05 36.30 Cable Tray Trough

		Crew	Daily Output	Labor-Hours	Unit	Material	2021 Bare Costs Labor	Equipment	Total	Total Incl O&P
5050	Adjustable	1 Elec	6	1.333	Ea.	82.50	85		167.50	217
5060	Elbow vertical, 36" radius, 90°		5	1.600		174	102		276	340
5070	60°		5.50	1.455		134	92.50		226.50	286
5080	45°		6	1.333		110	85		195	247
5090	30°		6.50	1.231		97.50	78.50		176	223
5100	Adjustable		6	1.333		82.50	85		167.50	217
5110	Splice plate, hinged, horizontal		48	.167		10.40	10.60		21	27
5120	Splice plate, hinged, vertical		48	.167		14.70	10.60		25.30	32
5130	Hanger, vertical		28	.286		16.20	18.20		34.40	45
5140	Single		28	.286		24.50	18.20		42.70	54
5150	Double		20	.400		25	25.50		50.50	66
5160	Channel to box connector		24	.333		33.50	21		54.50	68.50
5170	Hold down clip		80	.100		4.68	6.35		11.03	14.60
5180	Wall bracket, single		28	.286		13.25	18.20		31.45	41.50
5190	Double		20	.400		17	25.50		42.50	56.50
5200	Cable roller		16	.500		208	32		240	277
5210	Splice plate		48	.167		6.40	10.60		17	23

26 05 36.36 Cable Trays for Utility Substations

		Crew	Daily Output	Labor-Hours	Unit	Material	2021 Bare Costs Labor	Equipment	Total	Total Incl O&P
0010	**CABLE TRAYS FOR UTILITY SUBSTATIONS**									
7700	Cable tray	R-11	40	1.400	L.F.	20	85	16.10	121.10	166
7790	See Section 26 05 36									

26 05 36.40 Cable Tray, Covers and Dividers

		Crew	Daily Output	Labor-Hours	Unit	Material	2021 Bare Costs Labor	Equipment	Total	Total Incl O&P
0010	**CABLE TRAY, COVERS AND DIVIDERS** To 10' high R260536-10									
0011	For higher elevations, see lines 9900 – 9960									
0100	Covers, ventilated galv. steel, straight, 6" wide tray size	2 Elec	520	.031	L.F.	8.70	1.96		10.66	12.45
0200	9" wide tray size		460	.035		8.55	2.22		10.77	12.75
0300	12" wide tray size R260536-11		400	.040		16.55	2.55		19.10	22
0400	18" wide tray size		300	.053		31	3.40		34.40	39
0500	24" wide tray size		220	.073		41	4.63		45.63	52
0600	30" wide tray size		180	.089		26	5.65		31.65	37
1000	Elbow horizontal, 90°, 12" radius, 6" wide tray size		150	.107	Ea.	46	6.80		52.80	60.50
1020	9" wide tray size		128	.125		51	7.95		58.95	68
1040	12" wide tray size		108	.148		54	9.45		63.45	73
1060	18" wide tray size		84	.190		74.50	12.15		86.65	99.50
1080	24" wide tray size		66	.242		90	15.45		105.45	122
1100	30" wide tray size		60	.267		114	17		131	151
1160	24" radius, 6" wide tray size		136	.118		76.50	7.50		84	95
1180	9" wide tray size		116	.138		78.50	8.80		87.30	99
1200	12" wide tray size		96	.167		87	10.60		97.60	112
1220	18" wide tray size		76	.211		108	13.40		121.40	138
1240	24" wide tray size		60	.267		131	17		148	169
1260	30" wide tray size		52	.308		172	19.60		191.60	218
1320	36" radius, 6" wide tray size		120	.133		114	8.50		122.50	139
1340	9" wide tray size		104	.154		126	9.80		135.80	153
1360	12" wide tray size		84	.190		134	12.15		146.15	166
1380	18" wide tray size		72	.222		171	14.15		185.15	209
1400	24" wide tray size	2 Elec	52	.308	Ea.	204	19.60		223.60	253
1420	30" wide tray size		46	.348		242	22		264	299
1480	Elbow horizontal, 45°, 12" radius, 6" wide tray size		150	.107		34.50	6.80		41.30	48
1500	9" wide tray size		128	.125		41	7.95		48.95	57.50
1520	12" wide tray size		108	.148		45	9.45		54.45	63
1540	18" wide tray size		88	.182		54.50	11.60		66.10	77

For customer support on your Electrical Costs with RSMeans Data, call 800.448.8182.

199

26 05 36.40 Cable Tray, Covers and Dividers		Crew	Daily Output	Labor-Hours	Unit	Material	2021 Bare Costs Labor	Equipment	Total	Total Incl O&P
1560	24" wide tray size	2 Elec	76	.211	Ea.	64	13.40		77.40	90
1580	30" wide tray size		66	.242		75	15.45		90.45	106
1640	24" radius, 6" wide tray size		136	.118		48	7.50		55.50	64
1660	9" wide tray size		116	.138		54	8.80		62.80	72
1680	12" wide tray size		96	.167		60.50	10.60		71.10	82.50
1700	18" wide tray size		80	.200		70.50	12.75		83.25	96.50
1720	24" wide tray size		70	.229		80	14.55		94.55	110
1740	30" wide tray size		60	.267		102	17		119	137
1800	36" radius, 6" wide tray size		120	.133		71	8.50		79.50	90.50
1820	9" wide tray size		104	.154		80	9.80		89.80	103
1840	12" wide tray size		84	.190		88	12.15		100.15	115
1860	18" wide tray size		76	.211		101	13.40		114.40	131
1880	24" wide tray size		62	.258		123	16.45		139.45	160
1900	30" wide tray size		52	.308		134	19.60		153.60	176
1960	Elbow vertical, 90°, 12" radius, 6" wide tray size		150	.107		40.50	6.80		47.30	54.50
1980	9" wide tray size		128	.125		41	7.95		48.95	57.50
2000	12" wide tray size		108	.148		44.50	9.45		53.95	63
2020	18" wide tray size		88	.182		50.50	11.60		62.10	72.50
2040	24" wide tray size		68	.235		51.50	15		66.50	79
2060	30" wide tray size		60	.267		56.50	17		73.50	87.50
2120	24" radius, 6" wide tray size		136	.118		49.50	7.50		57	65.50
2140	9" wide tray size		116	.138		54.50	8.80		63.30	73
2160	12" wide tray size		96	.167		56.50	10.60		67.10	78.50
2180	18" wide tray size		80	.200		75	12.75		87.75	102
2200	24" wide tray size		62	.258		84.50	16.45		100.95	118
2220	30" wide tray size		52	.308		93	19.60		112.60	131
2280	36" radius, 6" wide tray size		120	.133		56.50	8.50		65	74.50
2300	9" wide tray size		104	.154		72	9.80		81.80	93.50
2320	12" wide tray size		84	.190		80	12.15		92.15	106
2340	18" wide tray size		76	.211		101	13.40		114.40	131
2350	24" wide tray size		54	.296		112	18.85		130.85	151
2360	30" wide tray size		46	.348		134	22		156	180
2400	Tee horizontal, 12" radius, 6" wide tray size		92	.174		73	11.10		84.10	97
2410	9" wide tray size		80	.200		75	12.75		87.75	102
2420	12" wide tray size	2 Elec	68	.235	Ea.	85	15		100	116
2430	18" wide tray size		60	.267		103	17		120	138
2440	24" wide tray size		52	.308		130	19.60		149.60	172
2460	30" wide tray size		36	.444		152	28.50		180.50	210
2500	24" radius, 6" wide tray size		88	.182		118	11.60		129.60	147
2510	9" wide tray size		76	.211		125	13.40		138.40	158
2520	12" wide tray size		64	.250		129	15.95		144.95	166
2530	18" wide tray size		56	.286		165	18.20		183.20	209
2540	24" wide tray size		48	.333		254	21		275	310
2560	30" wide tray size		32	.500		290	32		322	370
2600	36" radius, 6" wide tray size		84	.190		202	12.15		214.15	240
2610	9" wide tray size		72	.222		206	14.15		220.15	247
2620	12" wide tray size		60	.267		229	17		246	276
2630	18" wide tray size		52	.308		267	19.60		286.60	325
2640	24" wide tray size		44	.364		350	23		373	420
2660	30" wide tray size		28	.571		380	36.50		416.50	475
2700	Cross horizontal, 12" radius, 6" wide tray size		68	.235		109	15		124	143
2710	9" wide tray size		64	.250		115	15.95		130.95	151
2720	12" wide tray size		60	.267		129	17		146	167

For customer support on your Electrical Costs with RSMeans Data, call 800.448.8182.

26 05 36 – Cable Trays for Electrical Systems

26 05 36.40 Cable Tray, Covers and Dividers	Crew	Daily Output	Labor-Hours	Unit	Material	2021 Bare Costs Labor	Equipment	Total	Total Incl O&P
2730	18" wide tray size	2 Elec	52	.308	Ea.	152	19.60		171.60	197
2740	24" wide tray size		36	.444		181	28.50		209.50	241
2760	30" wide tray size		30	.533		206	34		240	277
2800	24" radius, 6" wide tray size		64	.250		198	15.95		213.95	241
2810	9" wide tray size		60	.267		214	17		231	261
2820	12" wide tray size		56	.286		231	18.20		249.20	282
2830	18" wide tray size		48	.333		277	21		298	335
2840	24" wide tray size		32	.500		335	32		367	420
2860	30" wide tray size		26	.615		370	39		409	470
2900	36" radius, 6" wide tray size		60	.267		340	17		357	395
2910	9" wide tray size		56	.286		335	18.20		353.20	395
2920	12" wide tray size		52	.308		365	19.60		384.60	435
2930	18" wide tray size		44	.364		415	23		438	490
2940	24" wide tray size		28	.571		520	36.50		556.50	625
2960	30" wide tray size		22	.727		555	46.50		601.50	680
3000	Reducer, 9" to 6" wide tray size		128	.125		47	7.95		54.95	63.50
3010	12" to 6" wide tray size		108	.148		49	9.45		58.45	67.50
3020	12" to 9" wide tray size		108	.148		49	9.45		58.45	67.50
3030	18" to 12" wide tray size		88	.182		52	11.60		63.60	74.50
3050	18" to 6" wide tray size		88	.182		52	11.60		63.60	74.50
3060	24" to 18" wide tray size		80	.200		73	12.75		85.75	99
3070	24" to 12" wide tray size		80	.200		66.50	12.75		79.25	92.50
3090	30" to 24" wide tray size		70	.229		77	14.55		91.55	107
3100	30" to 18" wide tray size		70	.229		77	14.55		91.55	107
3110	30" to 12" wide tray size		70	.229		66.50	14.55		81.05	95
3140	36" to 30" wide tray size		64	.250		84	15.95		99.95	116
3150	36" to 24" wide tray size		64	.250		84	15.95		99.95	116
3160	36" to 18" wide tray size		64	.250		84	15.95		99.95	116
3170	36" to 12" wide tray size		64	.250		84	15.95		99.95	116
3250	Covers, aluminum, straight, 6" wide tray size		520	.031	L.F.	4.08	1.96		6.04	7.40
3270	9" wide tray size		460	.035		4.92	2.22		7.14	8.70
3290	12" wide tray size		400	.040		5.90	2.55		8.45	10.25
3310	18" wide tray size		320	.050		7.75	3.19		10.94	13.30
3330	24" wide tray size		260	.062		9.70	3.92		13.62	16.45
3350	30" wide tray size		200	.080		10.70	5.10		15.80	19.30
3400	Elbow horizontal, 90°, 12" radius, 6" wide tray size		150	.107	Ea.	30.50	6.80		37.30	43.50
3410	9" wide tray size		128	.125		32	7.95		39.95	47
3420	12" wide tray size		108	.148		34.50	9.45		43.95	52
3430	18" wide tray size		88	.182		59	11.60		70.60	82
3440	24" wide tray size		70	.229		75	14.55		89.55	104
3460	30" wide tray size		64	.250		89.50	15.95		105.45	122
3500	24" radius, 6" wide tray size		136	.118		52	7.50		59.50	68
3510	9" wide tray size		116	.138		64.50	8.80		73.30	83.50
3520	12" wide tray size		96	.167		71	10.60		81.60	94
3530	18" wide tray size		80	.200		84	12.75		96.75	111
3540	24" wide tray size		64	.250		104	15.95		119.95	138
3560	30" wide tray size		56	.286		127	18.20		145.20	166
3600	36" radius, 6" wide tray size		120	.133		91	8.50		99.50	113
3610	9" wide tray size		104	.154		99.50	9.80		109.30	125
3620	12" wide tray size		84	.190		113	12.15		125.15	142
3630	18" wide tray size		76	.211		135	13.40		148.40	168
3640	24" wide tray size		56	.286		164	18.20		182.20	207
3660	30" wide tray size		50	.320		188	20.50		208.50	238

26 05 36.40 Cable Tray, Covers and Dividers		Crew	Daily Output	Labor-Hours	Unit	Material	2021 Bare Costs Labor	Equipment	Total	Total Incl O&P
3700	Elbow horizontal, 45°, 12" radius, 6" wide tray size	2 Elec	150	.107	Ea.	29.50	6.80		36.30	42.50
3710	9" wide tray size		128	.125		30.50	7.95		38.45	45.50
3720	12" wide tray size		108	.148		34	9.45		43.45	51.50
3730	18" wide tray size		88	.182		39	11.60		50.60	60
3740	24" wide tray size		80	.200		43.50	12.75		56.25	67
3760	30" wide tray size		70	.229		55	14.55		69.55	82
3800	24" radius, 6" wide tray size		136	.118		32.50	7.50		40	46.50
3810	9" wide tray size		116	.138		41	8.80		49.80	58.50
3820	12" wide tray size		96	.167		43	10.60		53.60	63
3830	18" wide tray size		80	.200		52	12.75		64.75	76
3840	24" wide tray size		72	.222		64.50	14.15		78.65	91.50
3860	30" wide tray size		64	.250		74	15.95		89.95	105
3900	36" radius, 6" wide tray size		120	.133		57.50	8.50		66	75.50
3910	9" wide tray size		104	.154		62.50	9.80		72.30	83
3920	12" wide tray size		84	.190		66	12.15		78.15	90.50
3930	18" wide tray size		76	.211		79.50	13.40		92.90	107
3940	24" wide tray size		64	.250		96	15.95		111.95	130
3960	30" wide tray size		56	.286		110	18.20		128.20	148
3970	36" wide tray size		50	.320		125	20.50		145.50	168
4000	Elbow vertical, 90°, 12" radius, 6" wide tray size		150	.107		34	6.80		40.80	47.50
4010	9" wide tray size		128	.125		34.50	7.95		42.45	49.50
4020	12" wide tray size		108	.148		37.50	9.45		46.95	55.50
4030	18" wide tray size		88	.182		43.50	11.60		55.10	65
4040	24" wide tray size		70	.229		45	14.55		59.55	71.50
4060	30" wide tray size		64	.250		46	15.95		61.95	74.50
4070	36" wide tray size		54	.296		56	18.85		74.85	89.50
4100	24" radius, 6" wide tray size		136	.118		39	7.50		46.50	54
4110	9" wide tray size		116	.138		42.50	8.80		51.30	60
4120	12" wide tray size		96	.167		46	10.60		56.60	67
4130	18" wide tray size		80	.200		55.50	12.75		68.25	80.50
4140	24" wide tray size		64	.250		61	15.95		76.95	90.50
4160	30" wide tray size		56	.286		74	18.20		92.20	109
4170	36" wide tray size		48	.333		80.50	21		101.50	121
4200	36" radius, 6" wide tray size		120	.133		44	8.50		52.50	60.50
4210	9" wide tray size		104	.154		54	9.80		63.80	74
4220	12" wide tray size		84	.190		62.50	12.15		74.65	86.50
4230	18" wide tray size		76	.211		76	13.40		89.40	103
4240	24" wide tray size		56	.286		89.50	18.20		107.70	126
4260	30" wide tray size		50	.320		113	20.50		133.50	155
4270	36" wide tray size		44	.364		121	23		144	169
4300	Tee horizontal, 12" radius, 6" wide tray size		108	.148		55.50	9.45		64.95	75.50
4310	9" wide tray size		88	.182		59.50	11.60		71.10	82
4320	12" wide tray size		80	.200		66	12.75		78.75	91.50
4330	18" wide tray size		68	.235		78.50	15		93.50	109
4340	24" wide tray size		56	.286		99	18.20		117.20	136
4360	30" wide tray size		44	.364		117	23		140	163
4370	36" wide tray size		36	.444		141	28.50		169.50	197
4400	24" radius, 6" wide tray size		96	.167		90.50	10.60		101.10	115
4410	9" wide tray size		80	.200		95.50	12.75		108.25	124
4420	12" wide tray size		72	.222		107	14.15		121.15	139
4430	18" wide tray size		60	.267		125	17		142	163
4440	24" wide tray size		48	.333		199	21		220	251
4460	30" wide tray size		40	.400		220	25.50		245.50	280

26 05 36 – Cable Trays for Electrical Systems

26 05 36.40 Cable Tray, Covers and Dividers	Crew	Daily Output	Labor-Hours	Unit	Material	2021 Bare Costs Labor	Equipment	Total	Total Incl O&P
4470	36" wide tray size	2 Elec	32	.500	Ea.	245	32		277	320
4500	36" radius, 6" wide tray size		88	.182		161	11.60		172.60	194
4510	9" wide tray size		72	.222		164	14.15		178.15	201
4520	12" wide tray size		64	.250		180	15.95		195.95	222
4530	18" wide tray size		56	.286		204	18.20		222.20	252
4540	24" wide tray size		44	.364		259	23		282	320
4560	30" wide tray size		36	.444		292	28.50		320.50	360
4570	36" wide tray size		28	.571		330	36.50		366.50	415
4600	Cross horizontal, 12" radius, 6" wide tray size		80	.200		85.50	12.75		98.25	113
4610	9" wide tray size		72	.222		90.50	14.15		104.65	121
4620	12" wide tray size		64	.250		100	15.95		115.95	134
4630	18" wide tray size		56	.286		122	18.20		140.20	161
4640	24" wide tray size		48	.333		141	21		162	187
4660	30" wide tray size		40	.400		167	25.50		192.50	222
4670	36" wide tray size		32	.500		192	32		224	260
4700	24" radius, 6" wide tray size		72	.222		158	14.15		172.15	195
4710	9" wide tray size		64	.250		169	15.95		184.95	210
4720	12" wide tray size		56	.286		182	18.20		200.20	227
4730	18" wide tray size		48	.333		212	21		233	265
4740	24" wide tray size		40	.400		253	25.50		278.50	315
4760	30" wide tray size		32	.500		296	32		328	375
4770	36" wide tray size		24	.667		320	42.50		362.50	420
4800	36" radius, 6" wide tray size		64	.250		259	15.95		274.95	310
4810	9" wide tray size		56	.286		271	18.20		289.20	325
4820	12" wide tray size		50	.320		298	20.50		318.50	355
4830	18" wide tray size		44	.364		330	23		353	395
4840	24" wide tray size		36	.444		405	28.50		433.50	485
4860	30" wide tray size		28	.571		465	36.50		501.50	565
4870	36" wide tray size		22	.727		510	46.50		556.50	630
4900	Reducer, 9" to 6" wide tray size		128	.125		32.50	7.95		40.45	47.50
4910	12" to 6" wide tray size		108	.148		34	9.45		43.45	51
4920	12" to 9" wide tray size		108	.148		34	9.45		43.45	51
4930	18" to 12" wide tray size		88	.182		37	11.60		48.60	58
4950	18" to 6" wide tray size		88	.182		37	11.60		48.60	58
4960	24" to 18" wide tray size		80	.200		47.50	12.75		60.25	71.50
4970	24" to 12" wide tray size		80	.200		41	12.75		53.75	64
4990	30" to 24" wide tray size		70	.229		49.50	14.55		64.05	76
5000	30" to 18" wide tray size		70	.229		49.50	14.55		64.05	76
5010	30" to 12" wide tray size		70	.229		49.50	14.55		64.05	76
5040	36" to 30" wide tray size		64	.250		55	15.95		70.95	84
5050	36" to 24" wide tray size		64	.250		55	15.95		70.95	84
5060	36" to 18" wide tray size		64	.250		55	15.95		70.95	84
5070	36" to 12" wide tray size		64	.250		55	15.95		70.95	84
5710	Tray cover hold down clamp	1 Elec	60	.133		8.80	8.50		17.30	22.50
8000	Divider strip, straight, galvanized, 3" deep		200	.040	L.F.	5.85	2.55		8.40	10.20
8020	4" deep		180	.044		7.10	2.83		9.93	12
8040	6" deep		160	.050		9.30	3.19		12.49	15
8060	Aluminum, straight, 3" deep		210	.038		5.85	2.43		8.28	10
8080	4" deep		190	.042		7.20	2.68		9.88	11.95
8100	6" deep		170	.047		9.20	3		12.20	14.60
8110	Divider strip, vertical fitting, 3" deep									
8120	12" radius, galvanized, 30°	1 Elec	28	.286	Ea.	28	18.20		46.20	58
8140	45°		27	.296		35	18.85		53.85	67

For customer support on your Electrical Costs with RSMeans Data, call 800.448.8182.

203

26 05 36.40 Cable Tray, Covers and Dividers		Crew	Daily Output	Labor-Hours	Unit	Material	2021 Bare Costs Labor	Equipment	Total	Total Incl O&P
8160	60°	1 Elec	26	.308	Ea.	37.50	19.60		57.10	70
8180	90°		25	.320		47	20.50		67.50	82
8200	Aluminum, 30°		29	.276		15.15	17.55		32.70	42.50
8220	45°		28	.286		17.90	18.20		36.10	46.50
8240	60°		27	.296		27.50	18.85		46.35	58
8260	90°		26	.308		34.50	19.60		54.10	67
8280	24" radius, galvanized, 30°		25	.320		42	20.50		62.50	77
8300	45°		24	.333		45	21		66	81
8320	60°		23	.348		57	22		79	95.50
8340	90°		22	.364		77.50	23		100.50	120
8360	Aluminum, 30°		26	.308		31.50	19.60		51.10	63.50
8380	45°		25	.320		36.50	20.50		57	70.50
8400	60°		24	.333		45.50	21		66.50	81.50
8420	90°		23	.348		61.50	22		83.50	101
8440	36" radius, galvanized, 30°		22	.364		54	23		77	93.50
8460	45°		21	.381		62.50	24.50		87	105
8480	60°		20	.400		73.50	25.50		99	119
8500	90°		19	.421		102	27		129	152
8520	Aluminum, 30°		23	.348		48	22		70	85.50
8540	45°		22	.364		61.50	23		84.50	103
8560	60°		21	.381		80	24.50		104.50	124
8570	90°		20	.400		105	25.50		130.50	153
8590	Divider strip, vertical fitting, 4" deep									
8600	12" radius, galvanized, 30°	1 Elec	27	.296	Ea.	33.50	18.85		52.35	65
8610	45°	"	26	.308	"	39	19.60		58.60	72
8620	60°	1 Elec	25	.320	Ea.	44	20.50		64.50	79
8630	90°		24	.333		54	21		75	90.50
8640	Aluminum, 30°		28	.286		27	18.20		45.20	56.50
8650	45°		27	.296		31	18.85		49.85	62
8660	60°		26	.308		35.50	19.60		55.10	68
8670	90°		25	.320		42	20.50		62.50	76.50
8680	24" radius, galvanized, 30°		24	.333		54	21		75	90.50
8690	45°		23	.348		67.50	22		89.50	108
8700	60°		22	.364		76.50	23		99.50	119
8710	90°		21	.381		102	24.50		126.50	148
8720	Aluminum, 30°		25	.320		42	20.50		62.50	76.50
8730	45°		24	.333		51	21		72	87.50
8740	60°		23	.348		60	22		82	99
8750	90°		22	.364		82.50	23		105.50	125
8760	36" radius, galvanized, 30°		23	.348		64	22		86	104
8770	45°		22	.364		73	23		96	115
8780	60°		21	.381		92	24.50		116.50	137
8790	90°		20	.400		122	25.50		147.50	173
8800	Aluminum, 30°		24	.333		66	21		87	104
8810	45°		23	.348		84.50	22		106.50	126
8820	60°		22	.364		103	23		126	148
8830	90°		21	.381		130	24.50		154.50	179
8840	Divider strip, vertical fitting, 6" deep									
8850	12" radius, galvanized, 30°	1 Elec	24	.333	Ea.	38.50	21		59.50	74
8860	45°		23	.348		43	22		65	80.50
8870	60°		22	.364		49.50	23		72.50	89
8880	90°		21	.381		62	24.50		86.50	104
8890	Aluminum, 30°		25	.320		31	20.50		51.50	64.50

26 05 36 – Cable Trays for Electrical Systems

26 05 36.40 Cable Tray, Covers and Dividers

		Crew	Daily Output	Labor-Hours	Unit	Material	2021 Bare Costs Labor	Equipment	Total	Total Incl O&P
8900	45°	1 Elec	24	.333	Ea.	36.50	21		57.50	71.50
8910	60°		23	.348		38.50	22		60.50	75
8920	90°		22	.364		45	23		68	84.50
8930	24" radius, galvanized, 30°		23	.348		54	22		76	92
8940	45°		22	.364		67.50	23		90.50	109
8950	60°		21	.381		77.50	24.50		102	121
8960	90°		20	.400		102	25.50		127.50	150
8970	Aluminum, 30°		24	.333		43	21		64	78.50
8980	45°		23	.348		58.50	22		80.50	97
8990	60°		22	.364		64.50	23		87.50	105
9000	90°		21	.381		87	24.50		111.50	132
9010	36" radius, galvanized, 30°		22	.364		62.50	23		85.50	104
9020	45°		21	.381		77.50	24.50		102	121
9030	60°		20	.400		102	25.50		127.50	150
9040	90°		19	.421		134	27		161	187
9050	Aluminum, 30°	1 Elec	23	.348	Ea.	66.50	22		88.50	107
9060	45°		22	.364		89.50	23		112.50	133
9070	60°		21	.381		105	24.50		129.50	152
9080	90°		20	.400		126	25.50		151.50	177
9120	Divider strip, horizontal fitting, galvanized, 3" deep		33	.242		38.50	15.45		53.95	65.50
9130	4" deep		30	.267		43	17		60	72.50
9140	6" deep		27	.296		55	18.85		73.85	88.50
9150	Aluminum, 3" deep		35	.229		30.50	14.55		45.05	55
9160	4" deep		32	.250		33.50	15.95		49.45	60.50
9170	6" deep		29	.276		45	17.55		62.55	75
9300	Divider strip protector		300	.027	L.F.	2.94	1.70		4.64	5.75
9310	Fastener, ladder tray				Ea.	.52			.52	.57
9320	Trough or solid bottom tray				"	.41			.41	.45
9900	Add to labor for higher elevated installation									
9905	10' to 14.5' high, add						10%			
9910	15' to 20' high, add						20%			
9920	20' to 25' high, add						25%			
9930	25' to 30' high, add						35%			
9940	30' to 35' high, add						40%			
9950	35' to 40' high, add						50%			
9960	Over 40' high, add						55%			

26 05 39 – Underfloor Raceways for Electrical Systems

26 05 39.30 Conduit In Concrete Slab

		Crew	Daily Output	Labor-Hours	Unit	Material	2021 Bare Costs Labor	Equipment	Total	Total Incl O&P
0010	**CONDUIT IN CONCRETE SLAB** Including terminations, R260539-30									
0020	fittings and supports									
3230	PVC, schedule 40, 1/2" diameter	1 Elec	270	.030	L.F.	.36	1.89		2.25	3.19
3250	3/4" diameter		230	.035		.37	2.22		2.59	3.70
3270	1" diameter		200	.040		.42	2.55		2.97	4.25
3300	1-1/4" diameter		170	.047		.73	3		3.73	5.25
3330	1-1/2" diameter		140	.057		.76	3.64		4.40	6.25
3350	2" diameter		120	.067		.90	4.25		5.15	7.30
3370	2-1/2" diameter		90	.089		1.67	5.65		7.32	10.25
3400	3" diameter	2 Elec	160	.100		2.14	6.35		8.49	11.80
3430	3-1/2" diameter		120	.133		2.65	8.50		11.15	15.50
3440	4" diameter		100	.160		3.07	10.20		13.27	18.55
3450	5" diameter		80	.200		7.10	12.75		19.85	27
3460	6" diameter		60	.267		6.10	17		23.10	31.50

For customer support on your Electrical Costs with RSMeans Data, call 800.448.8182.

205

26 05 39.30 Conduit In Concrete Slab

		Crew	Daily Output	Labor-Hours	Unit	Material	2021 Bare Costs Labor	Equipment	Total	Total Incl O&P
3530	Sweeps, 1" diameter, 30" radius	1 Elec	32	.250	Ea.	11.50	15.95		27.45	36
3550	1-1/4" diameter		24	.333		10.30	21		31.30	43
3570	1-1/2" diameter		21	.381		11.20	24.50		35.70	48.50
3600	2" diameter		18	.444		13.45	28.50		41.95	57
3630	2-1/2" diameter		14	.571		88.50	36.50		125	151
3650	3" diameter		10	.800		50	51		101	131
3670	3-1/2" diameter		8	1		49.50	63.50		113	149
3700	4" diameter		7	1.143		38.50	73		111.50	150
3710	5" diameter		6	1.333		58	85		143	190
3730	Couplings, 1/2" diameter					.17			.17	.19
3750	3/4" diameter					.16			.16	.18
3770	1" diameter					.24			.24	.26
3800	1-1/4" diameter					.47			.47	.52
3830	1-1/2" diameter					.46			.46	.51
3850	2" diameter					.60			.60	.66
3870	2-1/2" diameter					1.27			1.27	1.40
3900	3" diameter					2.13			2.13	2.34
3930	3-1/2" diameter					2.08			2.08	2.29
3950	4" diameter					2.58			2.58	2.84
3960	5" diameter					6.15			6.15	6.75
3970	6" diameter					8.65			8.65	9.50
4030	End bells, 1" diameter, PVC	1 Elec	60	.133		3.19	8.50		11.69	16.10
4050	1-1/4" diameter		53	.151		4.98	9.60		14.58	19.80
4100	1-1/2" diameter		48	.167		3.38	10.60		13.98	19.50
4150	2" diameter		34	.235		4.70	15		19.70	27.50
4170	2-1/2" diameter		27	.296		3.69	18.85		22.54	32
4200	3" diameter		20	.400		5.45	25.50		30.95	44
4250	3-1/2" diameter		16	.500		7.35	32		39.35	55.50
4300	4" diameter		14	.571		6.40	36.50		42.90	61
4310	5" diameter		12	.667		10.05	42.50		52.55	74
4320	6" diameter		9	.889		12.10	56.50		68.60	97.50
4350	Rigid galvanized steel, 1/2" diameter		200	.040	L.F.	2.32	2.55		4.87	6.35
4400	3/4" diameter		170	.047		2.48	3		5.48	7.15
4450	1" diameter		130	.062		3.73	3.92		7.65	9.90
4500	1-1/4" diameter		110	.073		5.25	4.63		9.88	12.70
4600	1-1/2" diameter		100	.080		5.90	5.10		11	14.05
4800	2" diameter		90	.089		7.20	5.65		12.85	16.35

26 05 39.40 Conduit In Trench

		Crew	Daily Output	Labor-Hours	Unit	Material	2021 Bare Costs Labor	Equipment	Total	Total Incl O&P
0010	**CONDUIT IN TRENCH** Includes terminations and fittings	R260539-40								
0020	Does not include excavation or backfill, see Section 31 23 16									
0200	Rigid galvanized steel, 2" diameter	1 Elec	150	.053	L.F.	6.90	3.40		10.30	12.65
0400	2-1/2" diameter	"	100	.080		13.60	5.10		18.70	22.50
0600	3" diameter	2 Elec	160	.100		15.45	6.35		21.80	26.50
0800	3-1/2" diameter		140	.114		20.50	7.30		27.80	33.50
1000	4" diameter		100	.160		22	10.20		32.20	39
1200	5" diameter		80	.200		48	12.75		60.75	72
1400	6" diameter		60	.267		59.50	17		76.50	90.50

26 05 43.10 Trench Duct

		Crew	Daily Output	Labor-Hours	Unit	Material	2021 Bare Costs Labor	Equipment	Total	Total Incl O&P
0010	**TRENCH DUCT** Steel with cover									
0020	Standard adjustable, depths to 4"									
0100	Straight, single compartment, 9" wide	2 Elec	40	.400	L.F.	400	25.50		425.50	480
0200	12" wide		32	.500		530	32		562	630
0400	18" wide		26	.615		500	39		539	610
0600	24" wide		22	.727		595	46.50		641.50	725
0700	27" wide		21	.762		214	48.50		262.50	310
0800	30" wide		20	.800		231	51		282	330
1000	36" wide		16	1		256	63.50		319.50	375
1020	Two compartment, 9" wide		38	.421		123	27		150	175
1030	12" wide		30	.533		144	34		178	210
1040	18" wide		24	.667		205	42.50		247.50	288
1050	24" wide		20	.800		253	51		304	355
1060	30" wide		18	.889		325	56.50		381.50	445
1070	36" wide		14	1.143		360	73		433	505
1090	Three compartment, 9" wide		36	.444		155	28.50		183.50	213
1100	12" wide		28	.571		180	36.50		216.50	252
1110	18" wide		22	.727		219	46.50		265.50	310
1120	24" wide		18	.889		273	56.50		329.50	385
1130	30" wide		16	1		340	63.50		403.50	470
1140	36" wide		12	1.333	▼	390	85		475	555
1200	Horizontal elbow, 9" wide		5.40	2.963	Ea.	415	189		604	740
1400	12" wide		4.60	3.478		440	222		662	810
1600	18" wide		4	4		590	255		845	1,025
1800	24" wide		3.20	5		835	320		1,155	1,400
1900	27" wide		3	5.333		995	340		1,335	1,600
2000	30" wide		2.60	6.154		1,225	390		1,615	1,925
2200	36" wide		2.40	6.667		1,450	425		1,875	2,225
2220	Two compartment, 9" wide		3.80	4.211		675	268		943	1,150
2230	12" wide		3	5.333		745	340		1,085	1,325
2240	18" wide		2.40	6.667		895	425		1,320	1,625
2250	24" wide		2	8		1,200	510		1,710	2,075
2260	30" wide		1.80	8.889		1,600	565		2,165	2,625
2270	36" wide		1.60	10		1,950	635		2,585	3,100
2290	Three compartment, 9" wide		3.60	4.444		745	283		1,028	1,250
2300	12" wide		2.80	5.714		835	365		1,200	1,450
2310	18" wide		2.20	7.273		1,025	465		1,490	1,825
2320	24" wide		1.80	8.889		1,275	565		1,840	2,275
2330	30" wide		1.60	10		1,675	635		2,310	2,800
2350	36" wide		1.40	11.429		2,050	730		2,780	3,325
2400	Vertical elbow, 9" wide		5.40	2.963		186	189		375	485
2600	12" wide		4.60	3.478		158	222		380	505
2800	18" wide		4	4		181	255		436	580
3000	24" wide		3.20	5		225	320		545	725
3100	27" wide		3	5.333		235	340		575	765
3200	30" wide		2.60	6.154		249	390		639	855
3400	36" wide		2.40	6.667		274	425		699	930
3600	Cross, 9" wide		4	4		685	255		940	1,125
3800	12" wide		3.20	5		725	320		1,045	1,275
4000	18" wide		2.60	6.154		865	390		1,255	1,525
4200	24" wide		2.20	7.273		1,075	465		1,540	1,875
4300	27" wide		2.20	7.273		1,375	465		1,840	2,200

For customer support on your Electrical Costs with RSMeans Data, call 800.448.8182.

207

26 05 43.10 Trench Duct		Crew	Daily Output	Labor-Hours	Unit	Material	2021 Bare Costs Labor	Equipment	Total	Total Incl O&P
4400	30" wide	2 Elec	2	8	Ea.	1,500	510		2,010	2,400
4600	36" wide		1.80	8.889		1,875	565		2,440	2,925
4620	Two compartment, 9" wide		3.80	4.211		755	268		1,023	1,225
4630	12" wide		3	5.333		720	340		1,060	1,300
4640	18" wide		2.40	6.667		960	425		1,385	1,675
4650	24" wide		2	8		1,225	510		1,735	2,100
4660	30" wide		1.80	8.889		1,600	565		2,165	2,625
4670	36" wide		1.60	10		1,950	635		2,585	3,100
4690	Three compartment, 9" wide		3.60	4.444		765	283		1,048	1,250
4700	12" wide		2.80	5.714		875	365		1,240	1,500
4710	18" wide		2.20	7.273		1,050	465		1,515	1,850
4720	24" wide		1.80	8.889		1,300	565		1,865	2,275
4730	30" wide		1.60	10		1,525	635		2,160	2,650
4740	36" wide		1.40	11.429		1,900	730		2,630	3,175
4800	End closure, 9" wide		14.40	1.111		45.50	71		116.50	155
5000	12" wide		12	1.333		52	85		137	183
5200	18" wide		10	1.600		79.50	102		181.50	239
5400	24" wide		8	2		105	127		232	305
5500	27" wide		7	2.286		121	146		267	350
5600	30" wide		6.60	2.424		162	154		316	410
5800	36" wide		5.80	2.759		156	176		332	430
6000	Tees, 9" wide		4	4		440	255		695	865
6200	12" wide		3.60	4.444		510	283		793	980
6400	18" wide		3.20	5		655	320		975	1,200
6600	24" wide		3	5.333		940	340		1,280	1,525
6700	27" wide		2.80	5.714		1,050	365		1,415	1,700
6800	30" wide		2.60	6.154		1,225	390		1,615	1,925
7000	36" wide		2	8		1,600	510		2,110	2,525
7020	Two compartment, 9" wide		3.80	4.211		510	268		778	960
7030	12" wide		3.40	4.706		550	300		850	1,050
7040	18" wide		3	5.333		735	340		1,075	1,300
7050	24" wide		2.80	5.714		990	365		1,355	1,650
7060	30" wide		2.40	6.667		1,350	425		1,775	2,100
7070	36" wide		1.90	8.421		1,675	535		2,210	2,650
7090	Three compartment, 9" wide		3.60	4.444		585	283		868	1,050
7100	12" wide		3.20	5		610	320		930	1,150
7110	18" wide		2.80	5.714		775	365		1,140	1,400
7120	24" wide		2.60	6.154		1,075	390		1,465	1,750
7130	30" wide		2.20	7.273		1,400	465		1,865	2,225
7140	36" wide		1.80	8.889		1,775	565		2,340	2,825
7200	Riser, and cabinet connector, 9" wide		5.40	2.963		192	189		381	490
7400	12" wide		4.60	3.478		224	222		446	575
7600	18" wide		4	4		238	255		493	640
7800	24" wide		3.20	5		335	320		655	840
7900	27" wide		3	5.333		283	340		623	815
8000	30" wide		2.60	6.154		385	390		775	1,000
8200	36" wide		2	8		445	510		955	1,250
8400	Insert assembly, cell to conduit adapter, 1-1/4"	1 Elec	16	.500		69.50	32		101.50	124
8500	Adjustable partition	"	320	.025	L.F.	27	1.59		28.59	32.50
8600	Depth of duct over 4", per 1", add					11.15			11.15	12.25
8700	Support post	1 Elec	240	.033		25.50	2.12		27.62	31
8800	Cover double tile trim, 2 sides					39.50			39.50	43.50
8900	4 sides					124			124	136

26 05 43 – Underground Ducts and Raceways for Electrical Systems

26 05 43.10 Trench Duct

		Crew	Daily Output	Labor-Hours	Unit	Material	2021 Bare Costs Labor	Equipment	Total	Total Incl O&P
9160	Trench duct 3-1/2" x 4-1/2", add				L.F.	10.55			10.55	11.65
9170	Trench duct 4" x 5", add					10.55			10.55	11.65
9200	For carpet trim, add					36			36	39.50
9210	For double carpet trim, add				▼	109			109	120

26 05 43.20 Underfloor Duct

		Crew	Daily Output	Labor-Hours	Unit	Material	2021 Bare Costs Labor	Equipment	Total	Total Incl O&P
0010	**UNDERFLOOR DUCT** R260543-50									
0100	Duct, 1-3/8" x 3-1/8" blank, standard	2 Elec	160	.100	L.F.	12.95	6.35		19.30	23.50
0200	1-3/8" x 7-1/4" blank, super duct		120	.133		30.50	8.50		39	46
0400	7/8" or 1-3/8" insert type, 24" OC, 1-3/8" x 3-1/8", std.		140	.114		20.50	7.30		27.80	33.50
0600	1-3/8" x 7-1/4", super duct	▼	100	.160	▼	35.50	10.20		45.70	54
0800	Junction box, single duct, 1 level, 3-1/8"	1 Elec	4	2	Ea.	450	127		577	685
0820	3-1/8" x 7-1/4"		4	2		585	127		712	830
0840	2 level, 3-1/8" upper & lower		3.20	2.500		530	159		689	820
0860	3-1/8" upper, 7-1/4" lower		2.70	2.963		515	189		704	845
0880	Carpet pan for above		80	.100		375	6.35		381.35	420
0900	Terrazzo pan for above		67	.119		895	7.60		902.60	995
1000	Junction box, single duct, 1 level, 7-1/4"		2.70	2.963		525	189		714	855
1020	2 level, 7-1/4" upper & lower		2.70	2.963		600	189		789	940
1040	2 duct, two 3-1/8" upper & lower		3.20	2.500		815	159		974	1,125
1200	1 level, 2 duct, 3-1/8"		3.20	2.500		605	159		764	900
1220	Carpet pan for above boxes		80	.100		365	6.35		371.35	410
1240	Terrazzo pan for above boxes		67	.119		855	7.60		862.60	950
1260	Junction box, 1 level, two 3-1/8" x one 3-1/8" + one 7-1/4"		2.30	3.478		1,000	222		1,222	1,425
1280	2 level, two 3-1/8" upper, one 3-1/8" + one 7-1/4" lower		2	4		1,100	255		1,355	1,575
1300	Carpet pan for above boxes		80	.100		365	6.35		371.35	410
1320	Terrazzo pan for above boxes		67	.119		855	7.60		862.60	950
1400	Junction box, 1 level, 2 duct, 7-1/4"		2.30	3.478		1,475	222		1,697	1,950
1420	Two 3-1/8" + one 7-1/4"		2	4		1,525	255		1,780	2,050
1440	Carpet pan for above		80	.100		365	6.35		371.35	410
1460	Terrazzo pan for above		67	.119		855	7.60		862.60	950
1580	Junction box, 1 level, one 3-1/8" + one 7-1/4" x same		2.30	3.478		1,025	222		1,247	1,450
1600	Triple duct, 3-1/8"		2.30	3.478		1,025	222		1,247	1,450
1700	Junction box, 1 level, one 3-1/8" + two 7-1/4"		2	4		1,675	255		1,930	2,225
1720	Carpet pan for above		80	.100		365	6.35		371.35	410
1740	Terrazzo pan for above		67	.119		855	7.60		862.60	950
1800	Insert to conduit adapter, 3/4" & 1"		32	.250		37	15.95		52.95	64
2000	Support, single cell		27	.296		55	18.85		73.85	88.50
2200	Super duct		16	.500		55.50	32		87.50	109
2400	Double cell		16	.500		55.50	32		87.50	109
2600	Triple cell		11	.727		63	46.50		109.50	139
2800	Vertical elbow, standard duct		10	.800		99.50	51		150.50	185
3000	Super duct		8	1		99.50	63.50		163	204
3200	Cabinet connector, standard duct		32	.250		79.50	15.95		95.45	111
3400	Super duct		27	.296		77.50	18.85		96.35	114
3600	Conduit adapter, 1" to 1-1/4"		32	.250		74.50	15.95		90.45	105
3800	2" to 1-1/4"		27	.296		89	18.85		107.85	126
4000	Outlet, low tension (tele, computer, etc.)		8	1		105	63.50		168.50	210
4200	High tension, receptacle (120 volt)		8	1		107	63.50		170.50	212
4300	End closure, standard duct		160	.050		4.49	3.19		7.68	9.65
4310	Super duct		160	.050		8.50	3.19		11.69	14.10
4350	Elbow, horiz., standard duct		26	.308		255	19.60		274.60	310
4360	Super duct	▼	26	.308	▼	237	19.60		256.60	289

For customer support on your Electrical Costs with RSMeans Data, call 800.448.8182.

209

26 05 43.20 Underfloor Duct

		Crew	Daily Output	Labor-Hours	Unit	Material	2021 Bare Costs Labor	2021 Bare Costs Equipment	Total	Total Incl O&P
4380	Elbow, offset, standard duct	1 Elec	26	.308	Ea.	99.50	19.60		119.10	138
4390	Super duct		26	.308		101	19.60		120.60	140
4400	Marker screw assembly for inserts		50	.160		19.60	10.20		29.80	36.50
4410	Y take off, standard duct		26	.308		155	19.60		174.60	200
4420	Super duct		26	.308		155	19.60		174.60	200
4430	Box opening plug, standard duct		160	.050		18.80	3.19		21.99	25
4440	Super duct		160	.050		18.80	3.19		21.99	25
4450	Sleeve coupling, standard duct	1 Elec	160	.050	Ea.	50	3.19		53.19	59.50
4460	Super duct		160	.050		50	3.19		53.19	59.50
4470	Conduit adapter, standard duct, 3/4"		32	.250		80.50	15.95		96.45	112
4480	1" or 1-1/4"		32	.250		76.50	15.95		92.45	108
4500	1-1/2"		32	.250		74.50	15.95		90.45	106

26 05 83.10 Motor Connections

		Crew	Daily Output	Labor-Hours	Unit	Material	2021 Bare Costs Labor	2021 Bare Costs Equipment	Total	Total Incl O&P
0010	**MOTOR CONNECTIONS**									
0020	Flexible conduit and fittings, 115 volt, 1 phase, up to 1 HP motor	1 Elec	8	1	Ea.	6.20	63.50		69.70	101
0050	2 HP motor		6.50	1.231		10.30	78.50		88.80	127
0100	3 HP motor		5.50	1.455		8.95	92.50		101.45	148
0110	230 volt, 3 phase, 3 HP motor		6.78	1.180		7.15	75		82.15	120
0112	5 HP motor		5.47	1.463		6.35	93		99.35	145
0114	7-1/2 HP motor		4.61	1.735		10.35	111		121.35	175
0120	10 HP motor		4.20	1.905		16.80	121		137.80	198
0150	15 HP motor		3.30	2.424		16.80	154		170.80	247
0200	25 HP motor		2.70	2.963		25	189		214	310
0400	50 HP motor		2.20	3.636		51.50	232		283.50	400
0600	100 HP motor		1.50	5.333		119	340		459	635
1500	460 volt, 5 HP motor, 3 phase		8	1		6.70	63.50		70.20	102
1520	10 HP motor		8	1		6.70	63.50		70.20	102
1530	25 HP motor		6	1.333		11.05	85		96.05	138
1540	30 HP motor		6	1.333		11.05	85		96.05	138
1550	40 HP motor		5	1.600		15.85	102		117.85	168
1560	50 HP motor		5	1.600		19.70	102		121.70	173
1570	60 HP motor		3.80	2.105		22	134		156	223
1580	75 HP motor		3.50	2.286		31	146		177	250
1590	100 HP motor		2.50	3.200		57	204		261	370
1600	125 HP motor		2	4		56.50	255		311.50	440
1610	150 HP motor		1.80	4.444		64	283		347	490
1620	200 HP motor		1.50	5.333		89.50	340		429.50	605
2005	460 volt, 5 HP motor, 3 phase, w/sealtite		8	1		10.45	63.50		73.95	106
2010	10 HP motor		8	1		10.45	63.50		73.95	106
2015	25 HP motor		6	1.333		17.30	85		102.30	145
2020	30 HP motor		6	1.333		17.30	85		102.30	145
2025	40 HP motor		5	1.600		28	102		130	182
2030	50 HP motor		5	1.600		29	102		131	183
2035	60 HP motor		3.80	2.105		41.50	134		175.50	245
2040	75 HP motor		3.50	2.286		47	146		193	268
2045	100 HP motor		2.50	3.200		73.50	204		277.50	385
2055	150 HP motor		1.80	4.444		73	283		356	500
2060	200 HP motor		1.50	5.333		237	340		577	765

Note: Row 0050 references R260580-75

210

For customer support on your Electrical Costs with RSMeans Data, call 800.448.8182.

26 05 90.10 Residential Wiring

	Crew	Daily Output	Labor-Hours	Unit	Material	2021 Bare Costs Labor	Equipment	Total	Total Incl O&P
0010 RESIDENTIAL WIRING									
0020 20' avg. runs and #14/2 wiring incl. unless otherwise noted									
1000 Service & panel, includes 24' SE-AL cable, service eye, meter,									
1010 Socket, panel board, main bkr., ground rod, 15 or 20 amp									
1020 1-pole circuit breakers, and misc. hardware									
1100 100 amp, with 10 branch breakers	1 Elec	1.19	6.723	Ea.	400	430		830	1,075
1110 With PVC conduit and wire		.92	8.696		425	555		980	1,300
1120 With RGS conduit and wire		.73	10.959		600	700		1,300	1,675
1150 150 amp, with 14 branch breakers		1.03	7.767		650	495		1,145	1,450
1170 With PVC conduit and wire		.82	9.756		690	620		1,310	1,675
1180 With RGS conduit and wire		.67	11.940		1,050	760		1,810	2,275
1200 200 amp, with 18 branch breakers	2 Elec	1.80	8.889		1,075	565		1,640	2,050
1220 With PVC conduit and wire		1.46	10.959		1,125	700		1,825	2,275
1230 With RGS conduit and wire		1.24	12.903		1,550	820		2,370	2,950
1800 Lightning surge suppressor	1 Elec	32	.250		50.50	15.95		66.45	79.50
2000 Switch devices									
2100 Single pole, 15 amp, ivory, with a 1-gang box, cover plate,									
2110 Type NM (Romex) cable	1 Elec	17.10	.468	Ea.	6.85	30		36.85	52
2120 Type MC cable		14.30	.559		20.50	35.50		56	75.50
2130 EMT & wire		5.71	1.401		29	89.50		118.50	165
2150 3-way, #14/3, type NM cable		14.55	.550		8.50	35		43.50	61.50
2170 Type MC cable		12.31	.650		26	41.50		67.50	90
2180 EMT & wire		5	1.600		31	102		133	185
2200 4-way, #14/3, type NM cable		14.55	.550		15.40	35		50.40	69
2220 Type MC cable		12.31	.650		32.50	41.50		74	97.50
2230 EMT & wire		5	1.600		38	102		140	193
2250 S.P., 20 amp, #12/2, type NM cable		13.33	.600		11.85	38		49.85	70
2270 Type MC cable		11.43	.700		24.50	44.50		69	93.50
2280 EMT & wire		4.85	1.649		35.50	105		140.50	195
2290 S.P. rotary dimmer, 600 W, no wiring		17	.471		27.50	30		57.50	74.50
2300 S.P. rotary dimmer, 600 W, type NM cable		14.55	.550		31	35		66	86
2320 Type MC cable		12.31	.650		44.50	41.50		86	111
2330 EMT & wire		5	1.600		55	102		157	212
2350 3-way rotary dimmer, type NM cable		13.33	.600		22	38		60	81.50
2370 Type MC cable		11.43	.700		36	44.50		80.50	106
2380 EMT & wire		4.85	1.649		46	105		151	207
2400 Interval timer wall switch, 20 amp, 1-30 min., #12/2									
2410 Type NM cable	1 Elec	14.55	.550	Ea.	57.50	35		92.50	116
2420 Type MC cable		12.31	.650		64.50	41.50		106	133
2430 EMT & wire		5	1.600		81.50	102		183.50	241
2500 Decorator style									
2510 S.P., 15 amp, type NM cable	1 Elec	17.10	.468	Ea.	11.60	30		41.60	57.50
2520 Type MC cable		14.30	.559		25.50	35.50		61	81
2530 EMT & wire		5.71	1.401		34	89.50		123.50	170
2550 3-way, #14/3, type NM cable		14.55	.550		13.30	35		48.30	66.50
2570 Type MC cable		12.31	.650		30.50	41.50		72	95
2580 EMT & wire		5	1.600		36	102		138	191
2600 4-way, #14/3, type NM cable		14.55	.550		20	35		55	74
2620 Type MC cable		12.31	.650		37.50	41.50		79	103
2630 EMT & wire		5	1.600		42.50	102		144.50	198
2650 S.P., 20 amp, #12/2, type NM cable		13.33	.600		16.60	38		54.60	75.50
2670 Type MC cable		11.43	.700		29.50	44.50		74	99

26 05 90.10 Residential Wiring	Crew	Daily Output	Labor-Hours	Unit	Material	2021 Bare Costs Labor	Equipment	Total	Total Incl O&P
2680	EMT & wire	1 Elec	4.85	1.649	Ea.	40.50	105		145.50	201
2700	S.P., slide dimmer, type NM cable		17.10	.468		37	30		67	85
2720	Type MC cable		14.30	.559		50.50	35.50		86	109
2730	EMT & wire		5.71	1.401		60.50	89.50		150	200
2750	S.P., touch dimmer, type NM cable		17.10	.468		24.50	30		54.50	71.50
2770	Type MC cable		14.30	.559		38	35.50		73.50	95
2780	EMT & wire		5.71	1.401		48.50	89.50		138	186
2800	3-way touch dimmer, type NM cable		13.33	.600		48.50	38		86.50	111
2820	Type MC cable		11.43	.700		62.50	44.50		107	135
2830	EMT & wire		4.85	1.649		72.50	105		177.50	236
3000	Combination devices									
3100	S.P. switch/15 amp recpt., ivory, 1-gang box, plate									
3110	Type NM cable	1 Elec	11.43	.700	Ea.	19.50	44.50		64	88
3120	Type MC cable		10	.800		33	51		84	112
3130	EMT & wire		4.40	1.818		43.50	116		159.50	220
3150	S.P. switch/pilot light, type NM cable		11.43	.700		20.50	44.50		65	89
3170	Type MC cable		10	.800		34	51		85	113
3180	EMT & wire		4.43	1.806		44	115		159	220
3190	2-S.P. switches, 2-#14/2, no wiring		14	.571		5.20	36.50		41.70	59.50
3200	2-S.P. switches, 2-#14/2, type NM cables		10	.800		21	51		72	98.50
3220	Type MC cable		8.89	.900		39	57.50		96.50	128
3230	EMT & wire		4.10	1.951		46.50	124		170.50	236
3250	3-way switch/15 amp recpt., #14/3, type NM cable		10	.800		26	51		77	104
3270	Type MC cable		8.89	.900		43.50	57.50		101	133
3280	EMT & wire		4.10	1.951		48.50	124		172.50	239
3300	2-3 way switches, 2-#14/3, type NM cables	1 Elec	8.89	.900	Ea.	33	57.50		90.50	122
3320	Type MC cable		8	1		58	63.50		121.50	159
3330	EMT & wire		4	2		55.50	127		182.50	250
3350	S.P. switch/20 amp recpt., #12/2, type NM cable		10	.800		27	51		78	105
3370	Type MC cable		8.89	.900		34	57.50		91.50	122
3380	EMT & wire		4.10	1.951		50.50	124		174.50	241
3400	Decorator style									
3410	S.P. switch/15 amp recpt., type NM cable	1 Elec	11.43	.700	Ea.	24.50	44.50		69	93
3420	Type MC cable		10	.800		38	51		89	118
3430	EMT & wire		4.40	1.818		48	116		164	225
3450	S.P. switch/pilot light, type NM cable		11.43	.700		25	44.50		69.50	94
3470	Type MC cable		10	.800		39	51		90	118
3480	EMT & wire		4.40	1.818		49	116		165	226
3500	2-S.P. switches, 2-#14/2, type NM cables		10	.800		26	51		77	104
3520	Type MC cable		8.89	.900		44	57.50		101.50	133
3530	EMT & wire		4.10	1.951		51	124		175	241
3550	3-way/15 amp recpt., #14/3, type NM cable		10	.800		31	51		82	110
3570	Type MC cable		8.89	.900		48	57.50		105.50	138
3580	EMT & wire		4.10	1.951		53.50	124		177.50	244
3650	2-3 way switches, 2-#14/3, type NM cables		8.89	.900		38	57.50		95.50	127
3670	Type MC cable		8	1		63	63.50		126.50	164
3680	EMT & wire		4	2		60.50	127		187.50	256
3700	S.P. switch/20 amp recpt., #12/2, type NM cable		10	.800		31.50	51		82.50	111
3720	Type MC cable		8.89	.900		38.50	57.50		96	128
3730	EMT & wire		4.10	1.951		55.50	124		179.50	246
4000	Receptacle devices									
4010	Duplex outlet, 15 amp recpt., ivory, 1-gang box, plate									
4015	Type NM cable	1 Elec	14.55	.550	Ea.	7.30	35		42.30	60

212

26 05 90 – Residential Applications

26 05 90.10 Residential Wiring		Crew	Daily Output	Labor-Hours	Unit	Material	2021 Bare Costs Labor	2021 Bare Costs Equipment	Total	Total Incl O&P
4020	Type MC cable	1 Elec	12.31	.650	Ea.	21	41.50		62.50	84.50
4030	EMT & wire		5.33	1.501		29.50	95.50		125	175
4050	With #12/2, type NM cable		12.31	.650		8.50	41.50		50	71
4070	Type MC cable		10.67	.750		21.50	48		69.50	94.50
4080	EMT & wire		4.71	1.699		32.50	108		140.50	197
4100	20 amp recpt., #12/2, type NM cable		12.31	.650		18.10	41.50		59.60	81.50
4120	Type MC cable		10.67	.750		31	48		79	105
4130	EMT & wire	▼	4.71	1.699	▼	42	108		150	207
4140	For GFI see Section 26 05 90.10 line 4300 below									
4150	Decorator style, 15 amp recpt., type NM cable	1 Elec	14.55	.550	Ea.	12.05	35		47.05	65.50
4170	Type MC cable		12.31	.650		26	41.50		67.50	90
4180	EMT & wire	▼	5.33	1.501	▼	34	95.50		129.50	180
4200	With #12/2, type NM cable	1 Elec	12.31	.650	Ea.	13.25	41.50		54.75	76
4220	Type MC cable		10.67	.750		26	48		74	99.50
4230	EMT & wire		4.71	1.699		37	108		145	202
4250	20 amp recpt., #12/2, type NM cable		12.31	.650		23	41.50		64.50	86.50
4270	Type MC cable		10.67	.750		35.50	48		83.50	111
4280	EMT & wire		4.71	1.699		46.50	108		154.50	213
4300	GFI, 15 amp recpt., type NM cable		12.31	.650		21	41.50		62.50	84.50
4320	Type MC cable		10.67	.750		34.50	48		82.50	109
4330	EMT & wire		4.71	1.699		43	108		151	209
4350	GFI with #12/2, type NM cable		10.67	.750		22	48		70	95.50
4370	Type MC cable		9.20	.870		35	55.50		90.50	121
4380	EMT & wire		4.21	1.900		46	121		167	231
4400	20 amp recpt., #12/2, type NM cable		10.67	.750		62.50	48		110.50	140
4420	Type MC cable		9.20	.870		75.50	55.50		131	166
4430	EMT & wire		4.21	1.900		86.50	121		207.50	275
4500	Weather-proof cover for above receptacles, add	▼	32	.250	▼	3.29	15.95		19.24	27
4550	Air conditioner outlet, 20 amp-240 volt recpt.									
4560	30' of #12/2, 2 pole circuit breaker									
4570	Type NM cable	1 Elec	10	.800	Ea.	59.50	51		110.50	141
4580	Type MC cable		9	.889		74	56.50		130.50	166
4590	EMT & wire		4	2		84	127		211	282
4600	Decorator style, type NM cable		10	.800		64.50	51		115.50	147
4620	Type MC cable		9	.889		79	56.50		135.50	171
4630	EMT & wire	▼	4	2	▼	89	127		216	287
4650	Dryer outlet, 30 amp-240 volt recpt., 20' of #10/3									
4660	2 pole circuit breaker									
4670	Type NM cable	1 Elec	6.41	1.248	Ea.	54	79.50		133.50	178
4680	Type MC cable		5.71	1.401		64	89.50		153.50	203
4690	EMT & wire	▼	3.48	2.299	▼	75	146		221	300
4700	Range outlet, 50 amp-240 volt recpt., 30' of #8/3									
4710	Type NM cable	1 Elec	4.21	1.900	Ea.	85.50	121		206.50	274
4720	Type MC cable		4	2		108	127		235	310
4730	EMT & wire		2.96	2.703		104	172		276	370
4750	Central vacuum outlet, type NM cable		6.40	1.250		61.50	79.50		141	186
4770	Type MC cable		5.71	1.401		74	89.50		163.50	215
4780	EMT & wire	▼	3.48	2.299	▼	90.50	146		236.50	320
4800	30 amp-110 volt locking recpt., #10/2 circ. bkr.									
4810	Type NM cable	1 Elec	6.20	1.290	Ea.	74.50	82		156.50	204
4820	Type MC cable		5.40	1.481		89.50	94.50		184	239
4830	EMT & wire	▼	3.20	2.500	▼	106	159		265	355
4900	Low voltage outlets									

213

26 05 90.10 Residential Wiring		Crew	Daily Output	Labor-Hours	Unit	Material	2021 Bare Costs Labor	2021 Bare Costs Equipment	Total	Total Incl O&P
4910	Telephone recpt., 20' of 4/C phone wire	1 Elec	26	.308	Ea.	7.60	19.60		27.20	37.50
4920	TV recpt., 20' of RG59U coax wire, F type connector	"	16	.500	"	15.40	32		47.40	64.50
4950	Door bell chime, transformer, 2 buttons, 60' of bellwire									
4970	Economy model	1 Elec	11.50	.696	Ea.	55.50	44.50		100	127
4980	Custom model		11.50	.696		94	44.50		138.50	170
4990	Luxury model, 3 buttons		9.50	.842		193	53.50		246.50	292
6000	Lighting outlets									
6050	Wire only (for fixture), type NM cable	1 Elec	32	.250	Ea.	5.90	15.95		21.85	30
6070	Type MC cable		24	.333		12.25	21		33.25	45
6080	EMT & wire		10	.800		19.95	51		70.95	97.50
6100	Box (4") and wire (for fixture), type NM cable		25	.320		14.95	20.50		35.45	47
6120	Type MC cable		20	.400		21.50	25.50		47	61.50
6130	EMT & wire		11	.727		29	46.50		75.50	101
6200	Fixtures (use with line 6050 or 6100 above)									
6210	Canopy style, economy grade	1 Elec	40	.200	Ea.	22.50	12.75		35.25	44
6220	Custom grade		40	.200		53	12.75		65.75	77.50
6250	Dining room chandelier, economy grade		19	.421		80.50	27		107.50	129
6260	Custom grade		19	.421		340	27		367	415
6270	Luxury grade		15	.533		3,050	34		3,084	3,400
6310	Kitchen fixture (fluorescent), economy grade		30	.267		76	17		93	109
6320	Custom grade		25	.320		220	20.50		240.50	273
6350	Outdoor, wall mounted, economy grade		30	.267		38.50	17		55.50	67.50
6360	Custom grade		30	.267		116	17		133	153
6370	Luxury grade		25	.320		248	20.50		268.50	305
6410	Outdoor PAR floodlights, 1 lamp, 150 watt		20	.400		15.55	25.50		41.05	55
6420	2 lamp, 150 watt each		20	.400		28	25.50		53.50	69
6425	Motion sensing, 2 lamp, 150 watt each		20	.400		107	25.50		132.50	155
6430	For infrared security sensor, add		32	.250		102	15.95		117.95	136
6450	Outdoor, quartz-halogen, 300 watt flood		20	.400		40	25.50		65.50	82
6600	Recessed downlight, round, pre-wired, 50 or 75 watt trim		30	.267		46.50	17		63.50	76
6610	With shower light trim		30	.267		71.50	17		88.50	104
6620	With wall washer trim		28	.286		72	18.20		90.20	106
6630	With eye-ball trim		28	.286		31	18.20		49.20	61
6700	Porcelain lamp holder		40	.200		2.75	12.75		15.50	22
6710	With pull switch		40	.200		11.30	12.75		24.05	31.50
6750	Fluorescent strip, 2-20 watt tube, wrap around diffuser, 24"		24	.333		43.50	21		64.50	79.50
6760	1-34 watt tube, 48"		24	.333		113	21		134	156
6770	2-34 watt tubes, 48"		20	.400		148	25.50		173.50	200
6800	Bathroom heat lamp, 1-250 watt		28	.286		33.50	18.20		51.70	64
6810	2-250 watt lamps		28	.286		60	18.20		78.20	93
6820	For timer switch, see Section 26 05 90.10 line 2400									
6900	Outdoor post lamp, incl. post, fixture, 35' of #14/2									
6910	Type NM cable	1 Elec	3.50	2.286	Ea.	310	146		456	555
6920	Photo-eye, add		27	.296		15.20	18.85		34.05	44.50
6950	Clock dial time switch, 24 hr., w/enclosure, type NM cable		11.43	.700		71	44.50		115.50	145
6970	Type MC cable		11	.727		84.50	46.50		131	162
6980	EMT & wire		4.85	1.649		93	105		198	258
7000	Alarm systems									
7050	Smoke detectors, box, #14/3, type NM cable	1 Elec	14.55	.550	Ea.	31	35		66	86
7070	Type MC cable		12.31	.650		42.50	41.50		84	108
7080	EMT & wire		5	1.600		47.50	102		149.50	203
7090	For relay output to security system, add					9.80			9.80	10.80
8000	Residential equipment									

For customer support on your Electrical Costs with RSMeans Data, call 800.448.8182.

26 05 90 – Residential Applications

26 05 90.10 Residential Wiring	Crew	Daily Output	Labor-Hours	Unit	Material	2021 Bare Costs Labor	Equipment	Total	Total Incl O&P
8050 Disposal hook-up, incl. switch, outlet box, 3' of flex									
8060 20 amp-1 pole circ. bkr., and 25' of #12/2									
8070 Type NM cable	1 Elec	10	.800	Ea.	25	51		76	103
8080 Type MC cable		8	1		38.50	63.50		102	137
8090 EMT & wire	↓	5	1.600	↓	52.50	102		154.50	209
8100 Trash compactor or dishwasher hook-up, incl. outlet box,									
8110 3' of flex, 15 amp-1 pole circ. bkr., and 25' of #14/2									
8120 Type NM cable	1 Elec	10	.800	Ea.	21.50	51		72.50	99
8130 Type MC cable		8	1		36.50	63.50		100	135
8140 EMT & wire	↓	5	1.600		48	102		150	204
8150 Hot water sink dispenser hook-up, use line 8100									
8200 Vent/exhaust fan hook-up, type NM cable	1 Elec	32	.250	Ea.	5.90	15.95		21.85	30
8220 Type MC cable		24	.333		12.25	21		33.25	45
8230 EMT & wire	↓	10	.800	↓	19.95	51		70.95	97.50
8250 Bathroom vent fan, 50 CFM (use with above hook-up)									
8260 Economy model	1 Elec	15	.533	Ea.	18.60	34		52.60	71
8270 Low noise model		15	.533		50.50	34		84.50	107
8280 Custom model	↓	12	.667	↓	116	42.50		158.50	191
8300 Bathroom or kitchen vent fan, 110 CFM									
8310 Economy model	1 Elec	15	.533	Ea.	63	34		97	120
8320 Low noise model	"	15	.533	"	101	34		135	162
8350 Paddle fan, variable speed (w/o lights)									
8360 Economy model (AC motor)	1 Elec	10	.800	Ea.	132	51		183	221
8362 With light kit	"	10	.800	"	171	51		222	265
8370 Custom model (AC motor)	1 Elec	10	.800	Ea.	320	51		371	430
8372 With light kit		10	.800		360	51		411	470
8380 Luxury model (DC motor)		8	1		279	63.50		342.50	400
8382 With light kit		8	1		320	63.50		383.50	445
8390 Remote speed switch for above, add	↓	12	.667	↓	40.50	42.50		83	108
8500 Whole house exhaust fan, ceiling mount, 36", variable speed									
8510 Remote switch, incl. shutters, 20 amp-1 pole circ. bkr.									
8520 30' of #12/2, type NM cable	1 Elec	4	2	Ea.	1,500	127		1,627	1,875
8530 Type MC cable		3.50	2.286		1,525	146		1,671	1,900
8540 EMT & wire	↓	3	2.667	↓	1,550	170		1,720	1,950
8600 Whirlpool tub hook-up, incl. timer switch, outlet box									
8610 3' of flex, 20 amp-1 pole GFI circ. bkr.									
8620 30' of #12/2, type NM cable	1 Elec	5	1.600	Ea.	241	102		343	415
8630 Type MC cable		4.20	1.905		252	121		373	455
8640 EMT & wire	↓	3.40	2.353	↓	263	150		413	510
8650 Hot water heater hook-up, incl. 1-2 pole circ. bkr., box;									
8660 3' of flex, 20' of #10/2, type NM cable	1 Elec	5	1.600	Ea.	29	102		131	183
8670 Type MC cable		4.20	1.905		43	121		164	228
8680 EMT & wire	↓	3.40	2.353	↓	48.50	150		198.50	277
9000 Heating/air conditioning									
9050 Furnace/boiler hook-up, incl. firestat, local on-off switch									
9060 Emergency switch, and 40' of type NM cable	1 Elec	4	2	Ea.	51	127		178	245
9070 Type MC cable		3.50	2.286		69	146		215	292
9080 EMT & wire	↓	1.50	5.333	↓	88	340		428	600
9100 Air conditioner hook-up, incl. local 60 amp disc. switch									
9110 3' sealtite, 40 amp, 2 pole circuit breaker									
9130 40' of #8/2, type NM cable	1 Elec	3.50	2.286	Ea.	149	146		295	380
9140 Type MC cable		3	2.667		180	170		350	450
9150 EMT & wire	↓	1.30	6.154	↓	186	390		576	785

26 05 90.10 Residential Wiring

		Crew	Daily Output	Labor-Hours	Unit	Material	2021 Bare Costs Labor	Equipment	Total	Total Incl O&P
9200	Heat pump hook-up, 1-40 & 1-100 amp 2 pole circ. bkr.									
9210	Local disconnect switch, 3' sealtite									
9220	40' of #8/2 & 30' of #3/2									
9230	Type NM cable	1 Elec	1.30	6.154	Ea.	425	390		815	1,050
9240	Type MC cable		1.08	7.407		510	470		980	1,250
9250	EMT & wire	↓	.94	8.511	↓	555	540		1,095	1,425
9500	Thermostat hook-up, using low voltage wire									
9520	Heating only, 25' of #18-3	1 Elec	24	.333	Ea.	6.15	21		27.15	38.50
9530	Heating/cooling, 25' of #18-4	"	20	.400	"	8.75	25.50		34.25	47.50

26 09 Instrumentation and Control for Electrical Systems
26 09 13 – Electrical Power Monitoring

26 09 13.10 Switchboard Instruments

			Crew	Daily Output	Labor-Hours	Unit	Material	2021 Bare Costs Labor	Equipment	Total	Total Incl O&P
0010	**SWITCHBOARD INSTRUMENTS** 3 phase, 4 wire	R260913-80									
0100	AC indicating, ammeter & switch		1 Elec	8	1	Ea.	2,700	63.50		2,763.50	3,075
0200	Voltmeter & switch			8	1		3,650	63.50		3,713.50	4,125
0300	Wattmeter			8	1		4,150	63.50		4,213.50	4,650
0400	AC recording, ammeter			4	2		7,375	127		7,502	8,325
0500	Voltmeter			4	2		7,375	127		7,502	8,325
0600	Ground fault protection, zero sequence			2.70	2.963		6,525	189		6,714	7,450
0700	Ground return path			2.70	2.963		6,525	189		6,714	7,450
0800	3 current transformers, 5 to 800 amp			2	4		3,025	255		3,280	3,700
0900	1,000 to 1,500 amp			1.30	6.154		4,350	390		4,740	5,375
1200	2,000 to 4,000 amp			1	8		5,150	510		5,660	6,400
1300	Fused potential transformer, maximum 600 volt		↓	8	1	↓	1,150	63.50		1,213.50	1,350

26 09 13.20 Voltage Monitor Systems

					Unit	Material			Total	Total Incl O&P
0010	**VOLTAGE MONITOR SYSTEMS** (test equipment)									
0100	AC voltage monitor system, 120/240 V, one-channel				Ea.	2,975			2,975	3,250
0110	Modem adapter					370			370	405
0120	Add-on detector only					1,550			1,550	1,700
0150	AC voltage remote monitor sys., 3 channel, 120, 230, or 480 V					5,375			5,375	5,925
0160	With internal modem					5,675			5,675	6,250
0170	Combination temperature and humidity probe					835			835	920
0180	Add-on detector only					3,900			3,900	4,300
0190	With internal modem					4,250			4,250	4,675

26 09 13.30 Smart Metering

			Crew	Daily Output	Labor-Hours	Unit	Material	2021 Bare Costs Labor	Equipment	Total	Total Incl O&P
0010	**SMART METERING**, In panel										
0100	Single phase, 120/208 volt, 100 amp	G	1 Elec	8.78	.911	Ea.	455	58		513	585
0120	200 amp	G		8.78	.911		435	58		493	560
0200	277 volt, 100 amp	G		8.78	.911		400	58		458	525
0220	200 amp	G		8.78	.911		460	58		518	595
1100	Three phase, 120/208 volt, 100 amp	G		4.69	1.706		800	109		909	1,050
1120	200 amp	G		4.69	1.706		975	109		1,084	1,225
1130	400 amp	G		4.69	1.706		2,225	109		2,334	2,600
1140	800 amp	G		4.69	1.706		2,625	109		2,734	3,050
1150	1,600 amp	G		4.69	1.706		2,625	109		2,734	3,050
1200	277/480 volt, 100 amp	G		4.69	1.706		1,000	109		1,109	1,250
1220	200 amp	G		4.69	1.706		2,425	109		2,534	2,825
1230	400 amp	G		4.69	1.706		2,425	109		2,534	2,825
1240	800 amp	G		4.69	1.706		2,875	109		2,984	3,300

26 09 Instrumentation and Control for Electrical Systems

26 09 13 – Electrical Power Monitoring

26 09 13.30 Smart Metering

		Crew	Daily Output	Labor-Hours	Unit	Material	2021 Bare Costs Labor	Equipment	Total	Total Incl O&P
1250	1,600 amp	G	1 Elec	4.69	1.706	Ea.	2,875	109		2,984	3,325
2000	Data recorder, 8 meters	G	↓	10.97	.729	↓	2,275	46.50		2,321.50	2,575
2100	16 meters	G	1 Elec	8.53	.938	Ea.	3,025	59.50		3,084.50	3,425
3000	Software package, per meter, basic	G					264			264	290
3100	Premium	G					695			695	765

26 09 23 – Lighting Control Devices

26 09 23.10 Energy Saving Lighting Devices

		Crew	Daily Output	Labor-Hours	Unit	Material	2021 Bare Costs Labor	Equipment	Total	Total Incl O&P
0010	**ENERGY SAVING LIGHTING DEVICES**										
0100	Occupancy sensors, passive infrared ceiling mounted	G	1 Elec	7	1.143	Ea.	77	73		150	193
0110	Ultrasonic ceiling mounted	G		7	1.143		90.50	73		163.50	208
0120	Dual technology ceiling mounted	G		6.50	1.231		127	78.50		205.50	256
0150	Automatic wall switches	G		24	.333		75	21		96	114
0160	Daylighting sensor, manual control, ceiling mounted	G		7	1.143		137	73		210	258
0170	Remote and dimming control with remote controller	G		6.50	1.231		201	78.50		279.50	335
0200	Passive infrared ceiling mounted	G		6.50	1.231		37	78.50		115.50	157
0400	Remote power pack	G		10	.800		37	51		88	117
0450	Photoelectric control, S.P.S.T. 120 V	G		8	1		22.50	63.50		86	120
0500	S.P.S.T. 208 V/277 V	G		8	1		22	63.50		85.50	119
0550	D.P.S.T. 120 V	G		6	1.333		209	85		294	355
0600	D.P.S.T. 208 V/277 V	G		6	1.333		237	85		322	385
0650	S.P.D.T. 208 V/277 V	G		6	1.333		230	85		315	380
0660	Daylight level sensor, wall mounted, on/off or dimming	G	↓	8	1	↓	135	63.50		198.50	244

26 09 26 – Lighting Control Panelboards

26 09 26.10 Lighting Control Relay Panel

		Crew	Daily Output	Labor-Hours	Unit	Material	2021 Bare Costs Labor	Equipment	Total	Total Incl O&P
0010	**LIGHTING CONTROL RELAY PANEL** with timeclock										
0100	4 Relay	G	1 Elec	2.50	3.200	Ea.	1,300	204		1,504	1,725
0110	8 Relay	G		2.30	3.478		2,475	222		2,697	3,050
0120	16 Relay	G		1.80	4.444		1,825	283		2,108	2,425
0130	24 Relay	G		1.50	5.333		2,475	340		2,815	3,225
0140	48 Relay	G	↓	1	8	↓	1,600	510		2,110	2,500
0200	Room Controller, switching only										
0210	1 Relay	G	1 Elec	3	2.667	Ea.	113	170		283	375
0220	2 Relay	G		3	2.667		162	170		332	430
0230	3 Relay	G	↓	3	2.667	↓	1,600	170		1,770	2,000
0240	Dimming										
0250	1 Relay	G	1 Elec	3	2.667	Ea.	1,300	170		1,470	1,675
0260	2 Relay	G		3	2.667		1,350	170		1,520	1,750
0270	3 Relay	G	↓	3	2.667	↓	1,600	170		1,770	2,000

26 09 36 – Modular Dimming Controls

26 09 36.13 Manual Modular Dimming Controls

		Crew	Daily Output	Labor-Hours	Unit	Material	2021 Bare Costs Labor	Equipment	Total	Total Incl O&P
0010	**MANUAL MODULAR DIMMING CONTROLS**										
2000	Lighting control module	G	1 Elec	2	4	Ea.	315	255		570	730

26 12 Medium-Voltage Transformers

26 12 19 – Pad-Mounted, Liquid-Filled, Medium-Voltage Transformers

26 12 19.10 Transformer, Oil-Filled

	Crew	Daily Output	Labor-Hours	Unit	Material	2021 Bare Costs Labor	2021 Bare Costs Equipment	Total	Total Incl O&P
0010 **TRANSFORMER, OIL-FILLED** primary delta or Y, R262213-60									
0050 Pad mounted 5 kV or 15 kV, with taps, 277/480 V secondary, 3 phase									
0100 150 kVA	R-3	.65	30.769	Ea.	9,925	1,950	294	12,169	14,100
0110 225 kVA		.55	36.364		20,400	2,300	345	23,045	26,200
0200 300 kVA		.45	44.444		17,100	2,825	425	20,350	23,500
0300 500 kVA		.40	50		21,100	3,175	480	24,755	28,500
0400 750 kVA		.38	52.632		30,800	3,350	505	34,655	39,400
0500 1,000 kVA		.26	76.923		36,500	4,875	735	42,110	48,200
0600 1,500 kVA		.23	86.957		43,300	5,525	830	49,655	57,000
0700 2,000 kVA		.20	100		54,500	6,350	955	61,805	70,500
0710 2,500 kVA		.19	105		66,000	6,675	1,000	73,675	83,500
0720 3,000 kVA		.17	118		80,000	7,475	1,125	88,600	100,500
0800 3,750 kVA	▼	.16	125	▼	105,500	7,925	1,200	114,625	129,000
1990 Pole mounted distribution type, single phase									
2000 13.8 kV primary, 120/240 V secondary, 10 kVA	R-15	7.45	6.443	Ea.	1,100	400	38	1,538	1,850
2010 50 kVA		3.70	12.973		2,425	810	76	3,311	3,950
2020 100 kVA		2.75	17.455		9,050	1,100	102	10,252	11,700
2030 167 kVA		2.15	22.326		5,575	1,400	131	7,106	8,375
2900 2,400 V primary, 120/240 V secondary, 10 kVA		7.45	6.443		1,075	400	38	1,513	1,850
2910 15 kVA		6.70	7.164		1,325	445	42	1,812	2,175
2920 25 kVA		6	8		1,700	500	47	2,247	2,675
2930 37.5 kVA		4.30	11.163		2,100	695	65.50	2,860.50	3,400
2940 50 kVA		3.70	12.973		2,400	810	76	3,286	3,900
2950 75 kVA		3	16		3,625	1,000	94	4,719	5,550
2960 100 kVA	▼	2.75	17.455	▼	3,575	1,100	102	4,777	5,675

26 12 19.20 Transformer, Liquid-Filled

	Crew	Daily Output	Labor-Hours	Unit	Material	2021 Bare Costs Labor	2021 Bare Costs Equipment	Total	Total Incl O&P
0010 **TRANSFORMER, LIQUID-FILLED** Pad mounted									
0020 5 kV or 15 kV primary, 277/480 volt secondary, 3 phase									
0050 225 kVA	R-3	.55	36.364	Ea.	12,700	2,300	345	15,345	17,700
0100 300 kVA		.45	44.444		15,000	2,825	425	18,250	21,200
0200 500 kVA		.40	50		19,000	3,175	480	22,655	26,200
0250 750 kVA		.38	52.632		24,500	3,350	505	28,355	32,500
0300 1,000 kVA		.26	76.923		28,500	4,875	735	34,110	39,400
0350 1,500 kVA		.23	86.957		33,200	5,525	830	39,555	45,700
0400 2,000 kVA		.20	100		41,200	6,350	955	48,505	56,000
0450 2,500 kVA	▼	.19	105	▼	75,500	6,675	1,000	83,175	94,000

26 13 Medium-Voltage Switchgear

26 13 16 – Medium-Voltage Fusible Interrupter Switchgear

26 13 16.10 Switchgear

	Crew	Daily Output	Labor-Hours	Unit	Material	2021 Bare Costs Labor	2021 Bare Costs Equipment	Total	Total Incl O&P
0010 **SWITCHGEAR**, Incorporate switch with cable connections, transformer,									
0100 & low voltage section									
0200 Load interrupter switch, 600 amp, 2 position									
0300 NEMA 1, 4.8 kV, 300 kVA & below w/CLF fuses	R-3	.40	50	Ea.	19,400	3,175	480	23,055	26,600
0400 400 kVA & above w/CLF fuses		.38	52.632		22,600	3,350	505	26,455	30,300
0500 Non fusible		.41	48.780		17,500	3,100	465	21,065	24,400
0600 13.8 kV, 300 kVA & below w/CLF fuses		.38	52.632		29,800	3,350	505	33,655	38,300
0700 400 kVA & above w/CLF fuses		.36	55.556		29,800	3,525	530	33,855	38,600
0800 Non fusible	▼	.40	50	▼	19,200	3,175	480	22,855	26,500
0900 Cable lugs for 2 feeders 4.8 kV or 13.8 kV	1 Elec	8	1		735	63.50		798.50	900
1000 Pothead, one 3 conductor or three 1 conductor		4	2		3,525	127		3,652	4,075

26 13 16 – Medium-Voltage Fusible Interrupter Switchgear

26 13 16.10 Switchgear		Crew	Daily Output	Labor-Hours	Unit	Material	2021 Bare Costs Labor	2021 Bare Costs Equipment	Total	Total Incl O&P
1100	Two 3 conductor or six 1 conductor	1 Elec	2	4	Ea.	6,950	255		7,205	8,025
1200	Key interlocks	↓	8	1	↓	815	63.50		878.50	990
1300	Lightning arresters, distribution class (no charge)									
1400	Intermediate class or line type 4.8 kV	1 Elec	2.70	2.963	Ea.	3,950	189		4,139	4,625
1500	13.8 kV		2	4		5,225	255		5,480	6,125
1600	Station class, 4.8 kV		2.70	2.963		6,775	189		6,964	7,725
1700	13.8 kV	↓	2	4		11,600	255		11,855	13,200
1800	Transformers, 4,800 volts to 480/277 volts, 75 kVA	R-3	.68	29.412		20,600	1,875	281	22,756	25,700
1900	112.5 kVA		.65	30.769		25,100	1,950	294	27,344	30,800
2000	150 kVA		.57	35.088		28,600	2,225	335	31,160	35,200
2100	225 kVA		.48	41.667		32,900	2,650	400	35,950	40,500
2200	300 kVA		.41	48.780		36,700	3,100	465	40,265	45,500
2300	500 kVA		.36	55.556		48,300	3,525	530	52,355	59,000
2400	750 kVA		.29	68.966		55,000	4,375	660	60,035	67,500
2500	13,800 volts to 480/277 volts, 75 kVA		.61	32.787		29,000	2,075	315	31,390	35,300
2600	112.5 kVA		.55	36.364		38,500	2,300	345	41,145	46,100
2700	150 kVA		.49	40.816		38,800	2,600	390	41,790	47,000
2800	225 kVA		.41	48.780		44,800	3,100	465	48,365	54,500
2900	300 kVA		.37	54.054		45,800	3,425	515	49,740	56,000
3000	500 kVA		.31	64.516		50,500	4,100	615	55,215	62,500
3100	750 kVA	↓	.26	76.923	↓	55,500	4,875	735	61,110	69,500
3200	Forced air cooling & temperature alarm	1 Elec	1	8		4,475	510		4,985	5,675
3300	Low voltage components									
3400	Maximum panel height 49-1/2", single or twin row									
3500	Breaker heights, type FA or FH, 6"									
3600	type KA or KH, 8"									
3700	type LA, 11"									
3800	type MA, 14"									
3900	Breakers, 2 pole, 15 to 60 amp, type FA	1 Elec	5.60	1.429	Ea.	635	91		726	830
4000	70 to 100 amp, type FA		4.20	1.905		1,425	121		1,546	1,750
4100	15 to 60 amp, type FH	↓	5.60	1.429	↓	955	91		1,046	1,175
4200	70 to 100 amp, type FH	1 Elec	4.20	1.905	Ea.	2,100	121		2,221	2,500
4300	125 to 225 amp, type KA		3.40	2.353		1,400	150		1,550	1,775
4400	125 to 225 amp, type KH		3.40	2.353		2,450	150		2,600	2,925
4500	125 to 400 amp, type LA		2.50	3.200		1,900	204		2,104	2,400
4600	125 to 600 amp, type MA		1.80	4.444		6,850	283		7,133	7,975
4700	700 & 800 amp, type MA		1.50	5.333		8,625	340		8,965	9,975
4800	3 pole, 15 to 60 amp, type FA		5.30	1.509		1,425	96		1,521	1,725
4900	70 to 100 amp, type FA		4	2		1,775	127		1,902	2,150
5000	15 to 60 amp, type FH		5.30	1.509		2,075	96		2,171	2,450
5100	70 to 100 amp, type FH		4	2		2,550	127		2,677	3,000
5200	125 to 225 amp, type KA		3.20	2.500		2,150	159		2,309	2,600
5300	125 to 225 amp, type KH		3.20	2.500		3,900	159		4,059	4,525
5400	125 to 400 amp, type LA		2.30	3.478		2,050	222		2,272	2,575
5500	125 to 600 amp, type MA		1.60	5		5,175	320		5,495	6,175
5600	700 & 800 amp, type MA	↓	1.30	6.154	↓	7,625	390		8,015	8,975

219

For customer support on your Electrical Costs with RSMeans Data, call 800.448.8182.

26 22 13.10 Transformer, Dry-Type		Crew	Daily Output	Labor-Hours	Unit	Material	2021 Bare Costs		Total	Total Incl O&P
							Labor	Equipment		
0010	**TRANSFORMER, DRY-TYPE**	R262213-10								
0050	Single phase, 240/480 volt primary, 120/240 volt secondary									
0100	1 kVA	R262213-60	1 Elec	2	4	Ea.	305	255		560	720
0300	2 kVA			1.60	5		610	320		930	1,150
0500	3 kVA			1.40	5.714		690	365		1,055	1,300
0700	5 kVA			1.20	6.667		840	425		1,265	1,550
0900	7.5 kVA		2 Elec	2.20	7.273		1,200	465		1,665	2,025
1100	10 kVA			1.60	10		1,550	635		2,185	2,650
1300	15 kVA			1.20	13.333		2,225	850		3,075	3,700
1500	25 kVA			1	16		3,950	1,025		4,975	5,875
1700	37.5 kVA			.80	20		3,000	1,275		4,275	5,200
1900	50 kVA			.70	22.857		4,725	1,450		6,175	7,375
2100	75 kVA			.65	24.615		4,700	1,575		6,275	7,500
2110	100 kVA		R-3	.90	22.222		6,125	1,400	212	7,737	9,075
2120	167 kVA		"	.80	25		10,200	1,575	239	12,014	13,800
2190	480 V primary, 120/240 V secondary, nonvent., 15 kVA		2 Elec	1.20	13.333		5,850	850		6,700	7,700
2200	25 kVA			.90	17.778		2,650	1,125		3,775	4,600
2210	37 kVA			.75	21.333		11,200	1,350		12,550	14,400
2220	50 kVA			.65	24.615		14,400	1,575		15,975	18,100
2230	75 kVA			.60	26.667		17,600	1,700		19,300	21,900
2240	100 kVA			.50	32		22,300	2,050		24,350	27,500
2250	Low operating temperature (80°C), 25 kVA			1	16		4,625	1,025		5,650	6,625
2260	37 kVA			.80	20		5,000	1,275		6,275	7,400
2270	50 kVA			.70	22.857		6,525	1,450		7,975	9,350
2280	75 kVA			.65	24.615		10,300	1,575		11,875	13,700
2290	100 kVA			.55	29.091		10,800	1,850		12,650	14,700
2300	3 phase, 480 volt primary, 120/208 volt secondary									
2310	Ventilated, 3 kVA		1 Elec	1	8	Ea.	1,250	510		1,760	2,125
2700	6 kVA			.80	10		1,325	635		1,960	2,400
2900	9 kVA			.70	11.429		1,550	730		2,280	2,775
3100	15 kVA		2 Elec	1.10	14.545		1,625	925		2,550	3,150
3300	30 kVA			.90	17.778		5,175	1,125		6,300	7,375
3500	45 kVA			.80	20		6,775	1,275		8,050	9,350
3700	75 kVA			.70	22.857		4,625	1,450		6,075	7,275
3900	112.5 kVA		R-3	.90	22.222		5,575	1,400	212	7,187	8,450
4100	150 kVA			.85	23.529		7,175	1,500	225	8,900	10,300
4300	225 kVA			.65	30.769		10,500	1,950	294	12,744	14,800
4500	300 kVA			.55	36.364		34,900	2,300	345	37,545	42,100
4700	500 kVA			.45	44.444		19,800	2,825	425	23,050	26,400
4800	750 kVA			.35	57.143		18,800	3,625	545	22,970	26,700
4820	1,000 kVA			.32	62.500		25,100	3,975	595	29,670	34,200
4850	K-4 rated, 15 kVA		2 Elec	1.10	14.545		3,325	925		4,250	5,025
4855	30 kVA			.90	17.778		5,000	1,125		6,125	7,150
4860	45 kVA			.80	20		6,000	1,275		7,275	8,500
4865	75 kVA			.70	22.857		9,050	1,450		10,500	12,100
4870	112.5 kVA		R-3	.90	22.222		12,000	1,400	212	13,612	15,600
4875	150 kVA			.85	23.529		15,700	1,500	225	17,425	19,800
4880	225 kVA			.65	30.769		21,900	1,950	294	24,144	27,300
4885	300 kVA			.55	36.364		13,600	2,300	345	16,245	18,700
4890	500 kVA			.45	44.444		34,200	2,825	425	37,450	42,400
4900	K-13 rated, 15 kVA		2 Elec	1.10	14.545		5,250	925		6,175	7,150
4905	30 kVA			.90	17.778		5,675	1,125		6,800	7,925
4910	45 kVA			.80	20		6,825	1,275		8,100	9,400

26 22 13 – Low-Voltage Distribution Transformers

26 22 13.10 Transformer, Dry-Type		Crew	Daily Output	Labor-Hours	Unit	Material	2021 Bare Costs Labor	Equipment	Total	Total Incl O&P
4915	75 kVA	2 Elec	.70	22.857	Ea.	11,700	1,450		13,150	15,100
4920	112.5 kVA	R-3	.90	22.222		13,700	1,400	212	15,312	17,400
4925	150 kVA		.85	23.529		17,900	1,500	225	19,625	22,200
4930	225 kVA		.65	30.769		24,400	1,950	294	26,644	30,000
4935	300 kVA		.55	36.364		32,500	2,300	345	35,145	39,500
4940	500 kVA		.45	44.444		25,100	2,825	425	28,350	32,300
5020	480 volt primary, 120/208 volt secondary									
5030	Nonventilated, 15 kVA	2 Elec	1.10	14.545	Ea.	2,275	925		3,200	3,875
5040	30 kVA		.80	20		4,050	1,275		5,325	6,350
5050	45 kVA		.70	22.857		5,675	1,450		7,125	8,425
5060	75 kVA		.65	24.615		9,100	1,575		10,675	12,300
5070	112.5 kVA	R-3	.85	23.529		14,700	1,500	225	16,425	18,700
5081	150 kVA		.85	23.529		19,000	1,500	225	20,725	23,400
5090	225 kVA		.60	33.333		28,600	2,125	320	31,045	35,000
5100	300 kVA	R-3	.50	40	Ea.	25,000	2,550	380	27,930	31,700
5200	Low operating temperature (80°C), 30 kVA	2 Elec	.90	17.778		4,525	1,125		5,650	6,675
5210	45 kVA		.80	20		5,450	1,275		6,725	7,900
5220	75 kVA		.70	22.857		8,225	1,450		9,675	11,200
5230	112.5 kVA	R-3	.90	22.222		11,000	1,400	212	12,612	14,300
5240	150 kVA		.85	23.529		14,000	1,500	225	15,725	17,900
5250	225 kVA		.65	30.769		19,100	1,950	294	21,344	24,200
5260	300 kVA		.55	36.364		31,400	2,300	345	34,045	38,300
5270	500 kVA		.45	44.444		34,500	2,825	425	37,750	42,700
5380	3 phase, 5 kV primary, 277/480 volt secondary									
5400	High voltage, 112.5 kVA	R-3	.85	23.529	Ea.	7,250	1,500	225	8,975	10,400
5410	150 kVA		.65	30.769		9,150	1,950	294	11,394	13,300
5420	225 kVA		.55	36.364		10,200	2,300	345	12,845	15,000
5430	300 kVA		.45	44.444		11,600	2,825	425	14,850	17,400
5440	500 kVA		.35	57.143		15,500	3,625	545	19,670	23,100
5450	750 kVA		.32	62.500		22,100	3,975	595	26,670	31,000
5460	1,000 kVA		.30	66.667		29,100	4,225	635	33,960	39,100
5470	1,500 kVA		.27	74.074		39,600	4,700	705	45,005	51,500
5480	2,000 kVA		.25	80		92,500	5,075	765	98,340	110,000
5490	2,500 kVA		.20	100		104,000	6,350	955	111,305	125,000
5500	3,000 kVA		.18	111		136,000	7,050	1,050	144,100	161,000
5590	15 kV primary, 277/480 volt secondary									
5600	High voltage, 112.5 kVA	R-3	.85	23.529	Ea.	16,000	1,500	225	17,725	20,100
5610	150 kVA		.65	30.769		22,300	1,950	294	24,544	27,700
5620	225 kVA		.55	36.364		27,000	2,300	345	29,645	33,500
5630	300 kVA		.45	44.444		29,600	2,825	425	32,850	37,300
5640	500 kVA		.35	57.143		32,400	3,625	545	36,570	41,700
5650	750 kVA		.32	62.500		34,700	3,975	595	39,270	44,800
5660	1,000 kVA		.30	66.667		40,500	4,225	635	45,360	51,500
5670	1,500 kVA		.27	74.074		48,500	4,700	705	53,905	61,500
5680	2,000 kVA		.25	80		56,000	5,075	765	61,840	70,000
5690	2,500 kVA		.20	100		61,500	6,350	955	68,805	78,000
5700	3,000 kVA		.18	111		64,500	7,050	1,050	72,600	82,500
6000	2400 volt primary, 480 volt secondary, 300 kVA		.45	44.444		9,000	2,825	425	12,250	14,600
6010	500 kVA		.35	57.143		14,200	3,625	545	18,370	21,600
6020	750 kVA		.32	62.500		18,600	3,975	595	23,170	27,000
9300	Energy efficient transformer 3 ph									
9400	15 kVA, 480VAC delta, 208Y/120VAC	R-3	1.50	13.333	Ea.	2,425	845	127	3,397	4,075
9405	30 kVA, 480VAC delta, 208Y/120VAC		1.50	13.333		2,550	845	127	3,522	4,200

26 22 13 – Low-Voltage Distribution Transformers

26 22 13.10 Transformer, Dry-Type		Crew	Daily Output	Labor-Hours	Unit	Material	2021 Bare Costs Labor	2021 Bare Costs Equipment	Total	Total Incl O&P
9410	45 kVA, 480VAC delta, 208Y/120VAC	R-3	1.50	13.333	Ea.	2,000	845	127	2,972	3,600
9415	75 kVA, 480VAC delta, 208Y/120VAC	↓	1.25	16	↓	2,425	1,025	153	3,603	4,350
9420	112 kVA, 480VAC delta, 208Y/120VAC	R-3	1.25	16	Ea.	7,125	1,025	153	8,303	9,525
9425	150 kVA, 480VAC delta, 208Y/120VAC		1.25	16		4,500	1,025	153	5,678	6,625
9430	225 kVA, 480VAC delta, 208Y/120VAC		1.25	16		5,775	1,025	153	6,953	8,025
9435	300 kVA, 480VAC delta, 208Y/120VAC	↓	1.25	16	↓	8,050	1,025	153	9,228	10,500

26 22 13.20 Isolating Panels

		Crew	Daily Output	Labor-Hours	Unit	Material	2021 Bare Costs Labor	2021 Bare Costs Equipment	Total	Total Incl O&P
0010	**ISOLATING PANELS** used with isolating transformers									
0020	For hospital applications									
0100	Critical care area, 8 circuit, 3 kVA	1 Elec	.58	13.793	Ea.	7,150	880		8,030	9,175
0200	5 kVA		.54	14.815		6,850	945		7,795	8,925
0400	7.5 kVA		.52	15.385		3,525	980		4,505	5,325
0600	10 kVA		.44	18.182		8,525	1,150		9,675	11,100
0800	Operating room power & lighting, 8 circuit, 3 kVA		.58	13.793		5,450	880		6,330	7,300
1000	5 kVA		.54	14.815		5,925	945		6,870	7,900
1200	7.5 kVA		.52	15.385		3,400	980		4,380	5,200
1400	10 kVA		.44	18.182		3,200	1,150		4,350	5,250
1600	X-ray systems, 15 kVA, 90 amp		.44	18.182		10,100	1,150		11,250	12,800
1800	25 kVA, 125 amp		.36	22.222		8,950	1,425		10,375	12,000

26 22 13.30 Isolating Transformer

			Crew	Daily Output	Labor-Hours	Unit	Material	2021 Bare Costs Labor	2021 Bare Costs Equipment	Total	Total Incl O&P
0010	**ISOLATING TRANSFORMER**										
0100	Single phase, 120/240 volt primary, 120/240 volt secondary										
0200	0.50 kVA	R262213-60	1 Elec	4	2	Ea.	435	127		562	665
0400	1 kVA			2	4		1,100	255		1,355	1,575
0600	2 kVA			1.60	5		1,750	320		2,070	2,400
0800	3 kVA			1.40	5.714		1,900	365		2,265	2,625
1000	5 kVA			1.20	6.667		2,425	425		2,850	3,275
1200	7.5 kVA			1.10	7.273		3,225	465		3,690	4,250
1400	10 kVA			.80	10		4,000	635		4,635	5,350
1600	15 kVA			.60	13.333		3,100	850		3,950	4,650
1800	25 kVA			.50	16		7,875	1,025		8,900	10,200
1810	37.5 kVA		2 Elec	.80	20		5,900	1,275		7,175	8,400
1820	75 kVA		"	.65	24.615		8,875	1,575		10,450	12,100
1830	3 phase, 120/240 V primary, 120/240 V secondary, 112.5 kVA		R-3	.90	22.222		9,550	1,400	212	11,162	12,800
1840	150 kVA			.85	23.529		12,100	1,500	225	13,825	15,800
1850	225 kVA			.65	30.769		16,900	1,950	294	19,144	21,800
1860	300 kVA			.55	36.364		22,500	2,300	345	25,145	28,500
1870	500 kVA			.45	44.444		37,400	2,825	425	40,650	45,900
1880	750 kVA		↓	.35	57.143		37,800	3,625	545	41,970	47,600

26 22 13.90 Transformer Handling

		Crew	Daily Output	Labor-Hours	Unit	Material	2021 Bare Costs Labor	2021 Bare Costs Equipment	Total	Total Incl O&P
0010	**TRANSFORMER HANDLING** Add to normal labor cost in restricted areas									
5000	Transformers									
5150	15 kVA, approximately 200 pounds	2 Elec	2.70	5.926	Ea.		375		375	560
5160	25 kVA, approximately 300 pounds	"	2.50	6.400	"		410		410	605
5170	37.5 kVA, approximately 400 pounds	2 Elec	2.30	6.957	Ea.		445		445	660
5180	50 kVA, approximately 500 pounds		2	8			510		510	755
5190	75 kVA, approximately 600 pounds		1.80	8.889			565		565	840
5200	100 kVA, approximately 700 pounds		1.60	10			635		635	945
5210	112.5 kVA, approximately 800 pounds	3 Elec	2.20	10.909			695		695	1,025
5220	125 kVA, approximately 900 pounds		2	12			765		765	1,125
5230	150 kVA, approximately 1,000 pounds		1.80	13.333			850		850	1,250
5240	167 kVA, approximately 1,200 pounds	↓	1.60	15			955		955	1,425

26 22 Low-Voltage Transformers

26 22 13 – Low-Voltage Distribution Transformers

26 22 13.90 Transformer Handling

		Crew	Daily Output	Labor-Hours	Unit	Material	2021 Bare Costs Labor	Equipment	Total	Total Incl O&P
5250	200 kVA, approximately 1,400 pounds	3 Elec	1.40	17.143	Ea.		1,100		1,100	1,625
5260	225 kVA, approximately 1,600 pounds		1.30	18.462			1,175		1,175	1,750
5270	250 kVA, approximately 1,800 pounds		1.10	21.818			1,400		1,400	2,075
5280	300 kVA, approximately 2,000 pounds		1	24			1,525		1,525	2,275
5290	500 kVA, approximately 3,000 pounds		.75	32			2,050		2,050	3,025
5300	600 kVA, approximately 3,500 pounds		.67	35.821			2,275		2,275	3,400
5310	750 kVA, approximately 4,000 pounds		.60	40			2,550		2,550	3,775
5320	1,000 kVA, approximately 5,000 pounds		.50	48			3,050		3,050	4,550

26 22 16 – Low-Voltage Buck-Boost Transformers

26 22 16.10 Buck-Boost Transformer

		Crew	Daily Output	Labor-Hours	Unit	Material	2021 Bare Costs Labor	Equipment	Total	Total Incl O&P
0010	**BUCK-BOOST TRANSFORMER** R262213-60									
0100	Single phase, 120/240 V primary, 12/24 V secondary									
0200	0.10 kVA	1 Elec	8	1	Ea.	130	63.50		193.50	238
0400	0.25 kVA		5.70	1.404		205	89.50		294.50	360
0600	0.50 kVA		4	2		355	127		482	580
0800	0.75 kVA		3.10	2.581		370	164		534	650
1000	1.0 kVA		2	4		435	255		690	860
1200	1.5 kVA		1.80	4.444		465	283		748	935
1400	2.0 kVA		1.60	5		620	320		940	1,150
1600	3.0 kVA		1.40	5.714		840	365		1,205	1,450
1800	5.0 kVA		1.20	6.667		1,625	425		2,050	2,425
2000	3 phase, 240 V primary, 208/120 V secondary, 15 kVA	2 Elec	2.40	6.667		735	425		1,160	1,450
2200	30 kVA		1.60	10		2,900	635		3,535	4,150
2400	45 kVA		1.40	11.429		3,500	730		4,230	4,900
2600	75 kVA		1.20	13.333		4,225	850		5,075	5,900
2800	112.5 kVA	R-3	1.40	14.286		5,250	905	136	6,291	7,275
3000	150 kVA		1.10	18.182		6,975	1,150	174	8,299	9,600
3200	225 kVA		1	20		9,100	1,275	191	10,566	12,100
3400	300 kVA		.90	22.222		12,300	1,400	212	13,912	15,900

26 24 Switchboards and Panelboards

26 24 13 – Switchboards

26 24 13.10 Incoming Switchboards

		Crew	Daily Output	Labor-Hours	Unit	Material	2021 Bare Costs Labor	Equipment	Total	Total Incl O&P
0010	**INCOMING SWITCHBOARDS** main service section R262419-84									
0100	Aluminum bus bars, not including CT's or PT's									
0200	No main disconnect, includes CT compartment									
0300	120/208 volt, 4 wire, 600 amp	2 Elec	1	16	Ea.	4,900	1,025		5,925	6,925
0400	800 amp		.88	18.182		4,900	1,150		6,050	7,125
0500	1,000 amp		.80	20		5,900	1,275		7,175	8,375
0600	1,200 amp		.72	22.222		5,900	1,425		7,325	8,575
0700	1,600 amp		.66	24.242		5,900	1,550		7,450	8,775
0800	2,000 amp		.62	25.806		6,450	1,650		8,100	9,550
1000	3,000 amp		.56	28.571		8,500	1,825		10,325	12,100
1200	277/480 volt, 4 wire, 600 amp		1	16		4,900	1,025		5,925	6,925
1300	800 amp		.88	18.182		4,900	1,150		6,050	7,125
1400	1,000 amp		.80	20		6,225	1,275		7,500	8,750
1500	1,200 amp		.72	22.222		6,225	1,425		7,650	8,950
1600	1,600 amp		.66	24.242		6,225	1,550		7,775	9,150
1700	2,000 amp		.62	25.806		6,450	1,650		8,100	9,550
1800	3,000 amp		.56	28.571		8,500	1,825		10,325	12,100

26 24 13 – Switchboards

26 24 13.10 Incoming Switchboards

		Crew	Daily Output	Labor-Hours	Unit	Material	2021 Bare Costs Labor	Equipment	Total	Total Incl O&P
1900	4,000 amp	2 Elec	.52	30.769	Ea.	10,400	1,950		12,350	14,300
2000	Fused switch & CT compartment									
2100	120/208 volt, 4 wire, 400 amp	2 Elec	1.12	14.286	Ea.	2,875	910		3,785	4,525
2200	600 amp		.94	17.021		3,425	1,075		4,500	5,375
2300	800 amp		.84	19.048		13,300	1,225		14,525	16,400
2400	1,200 amp		.68	23.529		17,300	1,500		18,800	21,200
2500	277/480 volt, 4 wire, 400 amp		1.14	14.035		4,025	895		4,920	5,750
2600	600 amp		.94	17.021		6,275	1,075		7,350	8,500
2700	800 amp		.84	19.048		13,300	1,225		14,525	16,400
2800	1,200 amp		.68	23.529		17,300	1,500		18,800	21,200
2900	Pressure switch & CT compartment									
3000	120/208 volt, 4 wire, 800 amp	2 Elec	.80	20	Ea.	11,900	1,275		13,175	15,000
3100	1,200 amp		.66	24.242		23,400	1,550		24,950	28,100
3200	1,600 amp		.62	25.806		24,900	1,650		26,550	29,900
3300	2,000 amp		.56	28.571		26,100	1,825		27,925	31,400
3310	2,500 amp		.50	32		32,600	2,050		34,650	38,800
3320	3,000 amp		.44	36.364		43,200	2,325		45,525	51,000
3330	4,000 amp		.40	40		56,500	2,550		59,050	66,000
3340	120/208 volt, 4 wire, 800 amp, with ground fault		.80	20		19,700	1,275		20,975	23,600
3350	1,200 amp, with ground fault		.66	24.242		25,400	1,550		26,950	30,300
3360	1,600 amp, with ground fault		.62	25.806		28,100	1,650		29,750	33,400
3370	2,000 amp, with ground fault		.56	28.571		30,300	1,825		32,125	36,000
3400	277/480 volt, 4 wire, 800 amp, with ground fault		.80	20		19,700	1,275		20,975	23,600
3600	1,200 amp, with ground fault		.66	24.242		25,400	1,550		26,950	30,300
4000	1,600 amp, with ground fault		.62	25.806		28,100	1,650		29,750	33,400
4200	2,000 amp, with ground fault		.56	28.571		30,300	1,825		32,125	36,000
4400	Circuit breaker, molded case & CT compartment									
4600	3 pole, 4 wire, 600 amp	2 Elec	.94	17.021	Ea.	10,300	1,075		11,375	12,900
4800	800 amp		.84	19.048		12,500	1,225		13,725	15,500
5000	1,200 amp		.68	23.529		16,700	1,500		18,200	20,600
5100	Copper bus bars, not incl. CT's or PT's, add, minimum					15%				

26 24 13.20 In Plant Distribution Switchboards

		Crew	Daily Output	Labor-Hours	Unit	Material	2021 Bare Costs Labor	Equipment	Total	Total Incl O&P
0010	**IN PLANT DISTRIBUTION SWITCHBOARDS**									
0100	Main lugs only, to 600 volt, 3 pole, 3 wire, 200 amp	2 Elec	1.20	13.333	Ea.	1,250	850		2,100	2,625
0110	400 amp		1.20	13.333		1,325	850		2,175	2,725
0120	600 amp		1.20	13.333		1,275	850		2,125	2,675
0130	800 amp		1.08	14.815		1,400	945		2,345	2,925
0140	1,200 amp		.92	17.391		1,700	1,100		2,800	3,500
0150	1,600 amp		.86	18.605		2,225	1,175		3,400	4,200
0160	2,000 amp		.82	19.512		2,450	1,250		3,700	4,550
0250	To 480 volt, 3 pole, 4 wire, 200 amp		1.20	13.333		1,075	850		1,925	2,450
0260	400 amp		1.20	13.333		1,250	850		2,100	2,625
0270	600 amp		1.20	13.333		1,350	850		2,200	2,750
0280	800 amp		1.08	14.815		1,500	945		2,445	3,025
0290	1,200 amp		.92	17.391		1,825	1,100		2,925	3,650
0300	1,600 amp		.86	18.605		2,100	1,175		3,275	4,050
0310	2,000 amp		.82	19.512		2,450	1,250		3,700	4,550
0400	Main circuit breaker, to 600 volt, 3 pole, 3 wire, 200 amp		1.20	13.333		3,625	850		4,475	5,250
0410	400 amp		1.14	14.035		2,950	895		3,845	4,575
0420	600 amp		1.10	14.545		4,600	925		5,525	6,425
0430	800 amp		1.04	15.385		7,700	980		8,680	9,900
0440	1,200 amp		.88	18.182		10,100	1,150		11,250	12,800

26 24 13 – Switchboards

26 24 13.20 In Plant Distribution Switchboards

		Crew	Daily Output	Labor-Hours	Unit	Material	2021 Bare Costs Labor	Equipment	Total	Total Incl O&P
0450	1,600 amp	2 Elec	.84	19.048	Ea.	16,100	1,225		17,325	19,600
0460	2,000 amp		.80	20		17,200	1,275		18,475	20,900
0550	277/480 volt, 3 pole, 4 wire, 200 amp		1.20	13.333		3,875	850		4,725	5,500
0560	400 amp		1.14	14.035		3,875	895		4,770	5,575
0570	600 amp		1.10	14.545		6,050	925		6,975	8,025
0580	800 amp		1.04	15.385		10,800	980		11,780	13,300
0590	1,200 amp		.88	18.182		10,500	1,150		11,650	13,200
0600	1,600 amp		.84	19.048		16,200	1,225		17,425	19,700
0610	2,000 amp		.80	20		17,300	1,275		18,575	21,000
0700	Main fusible switch w/fuse, 208/240 volt, 3 pole, 3 wire, 200 amp		1.20	13.333		4,000	850		4,850	5,650
0710	400 amp		1.14	14.035		3,275	895		4,170	4,925
0720	600 amp		1.10	14.545		4,650	925		5,575	6,475
0730	800 amp		1.04	15.385		9,975	980		10,955	12,500
0740	1,200 amp		.88	18.182		11,600	1,150		12,750	14,400
0800	120/208, 120/240 volt, 3 pole, 4 wire, 200 amp		1.20	13.333		3,600	850		4,450	5,225
0810	400 amp		1.14	14.035		3,275	895		4,170	4,925
0820	600 amp		1.10	14.545		4,650	925		5,575	6,475
0830	800 amp		1.04	15.385		7,375	980		8,355	9,575
0840	1,200 amp		.88	18.182		8,575	1,150		9,725	11,200
0900	480 or 600 volt, 3 pole, 3 wire, 200 amp		1.20	13.333		3,850	850		4,700	5,475
0910	400 amp		1.14	14.035		3,150	895		4,045	4,775
0920	600 amp		1.10	14.545		4,625	925		5,550	6,475
0930	800 amp		1.04	15.385		7,150	980		8,130	9,325
0940	1,200 amp		.88	18.182		8,425	1,150		9,575	11,000
1000	277 or 480 volt, 3 pole, 4 wire, 200 amp		1.20	13.333		4,050	850		4,900	5,725
1010	400 amp		1.14	14.035		3,875	895		4,770	5,575
1020	600 amp		1.10	14.545		6,050	925		6,975	8,025
1030	800 amp		1.04	15.385		7,400	980		8,380	9,575
1040	1,200 amp		.88	18.182		8,575	1,150		9,725	11,200
1120	1,600 amp		.76	21.053		15,700	1,350		17,050	19,200
1130	2,000 amp		.68	23.529		20,600	1,500		22,100	24,900
1150	Pressure switch, bolted, 3 pole, 208/240 volt, 3 wire, 800 amp		.96	16.667		11,600	1,050		12,650	14,400
1160	1,200 amp		.80	20		14,800	1,275		16,075	18,200
1170	1,600 amp		.76	21.053		17,200	1,350		18,550	20,900
1180	2,000 amp		.68	23.529		19,300	1,500		20,800	23,400
1200	120/208 or 120/240 volt, 3 pole, 4 wire, 800 amp		.96	16.667		9,175	1,050		10,225	11,700
1210	1,200 amp		.80	20		10,700	1,275		11,975	13,700
1220	1,600 amp		.76	21.053		17,200	1,350		18,550	20,900
1230	2,000 amp		.68	23.529		19,300	1,500		20,800	23,400
1300	480 or 600 volt, 3 wire, 800 amp		.96	16.667		11,600	1,050		12,650	14,400
1310	1,200 amp		.80	20		16,500	1,275		17,775	20,000
1320	1,600 amp		.76	21.053		18,300	1,350		19,650	22,100
1330	2,000 amp		.68	23.529		20,600	1,500		22,100	24,900
1400	277/480 volt, 4 wire, 800 amp		.96	16.667		11,600	1,050		12,650	14,400
1410	1,200 amp		.80	20		16,500	1,275		17,775	20,000
1420	1,600 amp		.76	21.053		18,300	1,350		19,650	22,100
1430	2,000 amp		.68	23.529		20,600	1,500		22,100	24,900
1500	Main ground fault protector, 1,200-2,000 amp		5.40	2.963		3,375	189		3,564	3,975
1600	Busway connection, 200 amp		5.40	2.963		540	189		729	875
1610	400 amp		4.60	3.478		540	222		762	925
1620	600 amp		4	4		540	255		795	975
1630	800 amp		3.20	5		540	320		860	1,075
1640	1,200 amp		2.60	6.154		540	390		930	1,175

26 24 13.20 In Plant Distribution Switchboards

		Crew	Daily Output	Labor-Hours	Unit	Material	2021 Bare Costs Labor	Equipment	Total	Total Incl O&P
1650	1,600 amp	2 Elec	2.40	6.667	Ea.	1,125	425		1,550	1,850
1660	2,000 amp		2	8		1,125	510		1,635	1,975
1700	Shunt trip for remote operation, 200 amp		8	2		650	127		777	910
1710	400 amp		8	2		1,050	127		1,177	1,350
1720	600 amp		8	2		1,225	127		1,352	1,550
1730	800 amp		8	2		1,575	127		1,702	1,950
1740	1,200-2,000 amp		8	2		3,450	127		3,577	3,975
1800	Motor operated main breaker, 200 amp		8	2		3,875	127		4,002	4,475
1810	400 amp		8	2		3,925	127		4,052	4,525
1820	600 amp		8	2		3,650	127		3,777	4,200
1830	800 amp		8	2		3,100	127		3,227	3,625
1840	1,200-2,000 amp		8	2		3,400	127		3,527	3,950
1900	Current/potential transformer metering compartment, 200-800 amp		5.40	2.963		2,125	189		2,314	2,600
1940	1,200 amp		5.40	2.963		6,050	189		6,239	6,950
1950	1,600-2,000 amp		5.40	2.963		9,700	189		9,889	11,000
2000	With watt meter, 200-800 amp		4	4		9,950	255		10,205	11,300
2040	1,200 amp		4	4		10,000	255		10,255	11,400
2050	1,600-2,000 amp		4	4		12,100	255		12,355	13,700
2100	Split bus, 60-200 amp	1 Elec	5.30	1.509		202	96		298	365
2130	400 amp	2 Elec	4.60	3.478		340	222		562	705
2140	600 amp		3.60	4.444		415	283		698	880
2150	800 amp		2.60	6.154		530	390		920	1,175
2170	1,200 amp		2	8		605	510		1,115	1,425
2250	Contactor control, 60 amp	1 Elec	2	4		1,325	255		1,580	1,825
2260	100 amp		1.50	5.333		1,450	340		1,790	2,100
2270	200 amp		1	8		2,200	510		2,710	3,150
2280	400 amp	2 Elec	1	16		6,825	1,025		7,850	9,025
2290	600 amp		.84	19.048		7,625	1,225		8,850	10,200
2300	800 amp		.72	22.222		8,825	1,425		10,250	11,800
2500	Modifier, two distribution sections, add		.80	20		2,900	1,275		4,175	5,100
2520	Three distribution sections, add		.40	40		4,400	2,550		6,950	8,600
2560	Auxiliary pull section, 20", add		2	8		1,100	510		1,610	1,975
2580	24", add		1.80	8.889		1,100	565		1,665	2,075
2600	30", add		1.60	10		1,100	635		1,735	2,175
2620	36", add		1.40	11.429		1,325	730		2,055	2,525
2640	Dog house, 12", add		2.40	6.667		228	425		653	880
2660	18", add		2	8		455	510		965	1,250
3000	Transition section between switchboard and transformer									
3050	or motor control center, 4 wire alum. bus, 600 amp	2 Elec	1.14	14.035	Ea.	2,175	895		3,070	3,725
3100	800 amp		1	16		2,425	1,025		3,450	4,200
3150	1,000 amp		.88	18.182		2,750	1,150		3,900	4,750
3200	1,200 amp		.80	20		3,025	1,275		4,300	5,225
3250	1,600 amp		.72	22.222		3,575	1,425		5,000	6,025
3300	2,000 amp		.66	24.242		4,100	1,550		5,650	6,800
3350	2,500 amp		.62	25.806		4,775	1,650		6,425	7,700
3400	3,000 amp		.56	28.571		5,500	1,825		7,325	8,750
4000	Weatherproof construction, per vertical section		1.76	9.091		2,675	580		3,255	3,800

26 24 13.30 Distribution Switchboards Section

		Crew	Daily Output	Labor-Hours	Unit	Material	2021 Bare Costs Labor	Equipment	Total	Total Incl O&P
0010	**DISTRIBUTION SWITCHBOARDS SECTION** R262419-80									
0100	Aluminum bus bars, not including breakers									
0160	Subfeed lug-rated at 60 amp	2 Elec	1.30	12.308	Ea.	1,200	785		1,985	2,500
0170	100 amp		1.26	12.698		1,250	810		2,060	2,575

For customer support on your Electrical Costs with RSMeans Data, call 800.448.8182.

26 24 13 – Switchboards

26 24 13.30 Distribution Switchboards Section

		Crew	Daily Output	Labor-Hours	Unit	Material	2021 Bare Costs Labor	2021 Bare Costs Equipment	Total	Total Incl O&P
0180	200 amp	2 Elec	1.20	13.333	Ea.	1,600	850		2,450	3,025
0190	400 amp		1.10	14.545		1,175	925		2,100	2,675
0195	120/208 or 277/480 volt, 4 wire, 400 amp		1.10	14.545		1,650	925		2,575	3,200
0200	600 amp		1	16		1,575	1,025		2,600	3,250
0300	800 amp		.88	18.182		1,975	1,150		3,125	3,900
0400	1,000 amp		.80	20		2,525	1,275		3,800	4,675
0500	1,200 amp		.72	22.222		2,975	1,425		4,400	5,375
0600	1,600 amp		.66	24.242		4,650	1,550		6,200	7,425
0700	2,000 amp		.62	25.806		5,900	1,650		7,550	8,950
0800	2,500 amp		.60	26.667		6,250	1,700		7,950	9,400
0900	3,000 amp		.56	28.571		8,475	1,825		10,300	12,000
0950	4,000 amp		.52	30.769		7,950	1,950		9,900	11,700

26 24 13.40 Switchboards Feeder Section

		Crew	Daily Output	Labor-Hours	Unit	Material	2021 Bare Costs Labor	2021 Bare Costs Equipment	Total	Total Incl O&P
0010	**SWITCHBOARDS FEEDER SECTION** group mounted devices R262419-82									
0030	Circuit breakers									
0160	FA frame, 15 to 60 amp, 240 volt, 1 pole	1 Elec	8	1	Ea.	135	63.50		198.50	243
0170	2 pole		7	1.143		239	73		312	370
0180	3 pole		5.30	1.509		575	96		671	775
0210	480 volt, 1 pole		8	1		197	63.50		260.50	310
0220	2 pole		7	1.143		515	73		588	675
0230	3 pole		5.30	1.509		870	96		966	1,100
0260	600 volt, 2 pole		7	1.143		800	73		873	990
0270	3 pole		5.30	1.509		940	96		1,036	1,175
0280	FA frame, 70 to 100 amp, 240 volt, 1 pole		7	1.143		237	73		310	370
0310	2 pole		5	1.600		640	102		742	855
0320	3 pole		4	2		790	127		917	1,050
0330	480 volt, 1 pole		7	1.143		365	73		438	510
0360	2 pole		5	1.600		950	102		1,052	1,200
0370	3 pole		4	2		1,025	127		1,152	1,325
0380	600 volt, 2 pole		5	1.600		655	102		757	870
0410	3 pole		4	2		1,025	127		1,152	1,325
0420	KA frame, 70 to 225 amp		3.20	2.500		1,425	159		1,584	1,800
0430	LA frame, 125 to 400 amp		2.30	3.478		4,250	222		4,472	5,000
0460	MA frame, 450 to 600 amp		1.60	5		5,925	320		6,245	7,000
0470	700 to 800 amp	1 Elec	1.30	6.154	Ea.	6,000	390		6,390	7,175
0480	MAL frame, 1,000 amp		1	8		8,000	510		8,510	9,550
0490	PA frame, 1,200 amp		.80	10		16,300	635		16,935	18,800
0500	Branch circuit, fusible switch, 600 volt, double 30/30 amp		4	2		885	127		1,012	1,150
0550	60/60 amp		3.20	2.500		905	159		1,064	1,225
0600	100/100 amp		2.70	2.963		1,150	189		1,339	1,550
0650	Single, 30 amp		5.30	1.509		830	96		926	1,050
0700	60 amp		4.70	1.702		880	108		988	1,125
0750	100 amp		4	2		1,550	127		1,677	1,900
0800	200 amp		2.70	2.963		1,275	189		1,464	1,675
0850	400 amp		2.30	3.478		2,300	222		2,522	2,850
0900	600 amp		1.80	4.444		2,850	283		3,133	3,550
0950	800 amp		1.30	6.154		4,800	390		5,190	5,850
1000	1,200 amp		.80	10		5,450	635		6,085	6,950
1080	Branch circuit, circuit breakers, high interrupting capacity									
1100	60 amp, 240, 480 or 600 volt, 1 pole	1 Elec	8	1	Ea.	340	63.50		403.50	470
1120	2 pole		7	1.143		805	73		878	1,000
1140	3 pole		5.30	1.509		945	96		1,041	1,200

For customer support on your Electrical Costs with RSMeans Data, call 800.448.8182.

227

26 24 13.40 Switchboards Feeder Section	Crew	Daily Output	Labor-Hours	Unit	Material	2021 Bare Costs Labor	Equipment	Total	Total Incl O&P
1150	100 amp, 240, 480 or 600 volt, 1 pole	1 Elec	7	1.143	Ea.	395	73		468	540
1160	2 pole		5	1.600		1,300	102		1,402	1,575
1180	3 pole		4	2		1,150	127		1,277	1,475
1200	225 amp, 240, 480 or 600 volt, 2 pole		3.50	2.286		2,975	146		3,121	3,500
1220	3 pole		3.20	2.500		3,500	159		3,659	4,075
1240	400 amp, 240, 480 or 600 volt, 2 pole		2.50	3.200		5,000	204		5,204	5,800
1260	3 pole		2.30	3.478		2,900	222		3,122	3,525
1280	600 amp, 240, 480 or 600 volt, 2 pole		1.80	4.444		3,450	283		3,733	4,225
1300	3 pole		1.60	5		3,450	320		3,770	4,275
1320	800 amp, 240, 480 or 600 volt, 2 pole		1.50	5.333		4,400	340		4,740	5,350
1340	3 pole		1.30	6.154		5,400	390		5,790	6,500
1360	1,000 amp, 240, 480 or 600 volt, 2 pole		1.10	7.273		5,525	465		5,990	6,775
1380	3 pole		1	8		5,925	510		6,435	7,275
1400	1,200 amp, 240, 480 or 600 volt, 2 pole		.90	8.889		7,825	565		8,390	9,450
1420	3 pole		.80	10		7,475	635		8,110	9,175
1700	Fusible switch, 240 V, 60 amp, 2 pole		3.20	2.500		495	159		654	780
1720	3 pole		3	2.667		680	170		850	1,000
1740	100 amp, 2 pole		2.70	2.963		505	189		694	835
1760	3 pole		2.50	3.200		765	204		969	1,150
1780	200 amp, 2 pole		2	4		1,050	255		1,305	1,525
1800	3 pole		1.90	4.211		1,225	268		1,493	1,750
1820	400 amp, 2 pole		1.50	5.333		1,975	340		2,315	2,675
1840	3 pole		1.30	6.154		2,475	390		2,865	3,300
1860	600 amp, 2 pole		1	8		2,800	510		3,310	3,850
1880	3 pole		.90	8.889		3,450	565		4,015	4,650
1900	240-600 V, 800 amp, 2 pole		.70	11.429		6,250	730		6,980	7,950
1920	3 pole		.60	13.333		7,650	850		8,500	9,675
2000	600 V, 60 amp, 2 pole		3.20	2.500		750	159		909	1,050
2040	100 amp, 2 pole		2.70	2.963		775	189		964	1,125
2080	200 amp, 2 pole		2	4		1,350	255		1,605	1,850
2120	400 amp, 2 pole		1.50	5.333		2,650	340		2,990	3,425
2160	600 amp, 2 pole		1	8		3,225	510		3,735	4,300
2500	Branch circuit, circuit breakers, 60 amp, 600 volt, 3 pole		5.30	1.509		900	96		996	1,125
2520	240, 480 or 600 volt, 1 pole		8	1		229	63.50		292.50	345
2540	240 volt, 2 pole		7	1.143		390	73		463	540
2560	480 or 600 volt, 2 pole		7	1.143		600	73		673	770
2580	240 volt, 3 pole		5.30	1.509		600	96		696	805
2600	480 volt, 3 pole		5.30	1.509		750	96		846	970
2620	100 amp, 600 volt, 2 pole		5	1.600		725	102		827	950
2640	3 pole		4	2		905	127		1,032	1,175
2660	480 volt, 2 pole		5	1.600		725	102		827	950
2680	240 volt, 2 pole		5	1.600		620	102		722	830
2700	3 pole		4	2		775	127		902	1,050
2720	480 volt, 3 pole		4	2		775	127		902	1,050
2740	225 amp, 240, 480 or 600 volt, 2 pole		3.50	2.286		1,600	146		1,746	2,000
2760	3 pole		3.20	2.500		815	159		974	1,125
2780	400 amp, 240, 480 or 600 volt, 2 pole		2.50	3.200		2,825	204		3,029	3,400
2800	3 pole		2.30	3.478		3,700	222		3,922	4,400
2820	600 amp, 240 or 480 volt, 2 pole		1.80	4.444		4,400	283		4,683	5,275
2840	3 pole		1.60	5		5,600	320		5,920	6,625
2860	800 amp, 240, 480 or 600 volt, 2 pole		1.50	5.333		5,725	340		6,065	6,775
2880	3 pole		1.30	6.154		7,350	390		7,740	8,650
2900	1,000 amp, 240, 480 or 600 volt, 2 pole		1.10	7.273		13,300	465		13,765	15,400

26 24 13 – Switchboards

26 24 13.40 Switchboards Feeder Section	Crew	Daily Output	Labor-Hours	Unit	Material	2021 Bare Costs Labor	Equipment	Total	Total Incl O&P
2920 480 or 600 volt, 3 pole	1 Elec	1	8	Ea.	14,600	510		15,110	16,900
2940 1,200 amp, 240, 480 or 600 volt, 2 pole		.90	8.889		13,300	565		13,865	15,500
2960 3 pole		.80	10		14,600	635		15,235	17,000
2980 600 volt, 3 pole		.80	10		8,050	635		8,685	9,800

26 24 16 – Panelboards

26 24 16.10 Load Centers

	Crew	Daily Output	Labor-Hours	Unit	Material	2021 Bare Costs Labor	Equipment	Total	Total Incl O&P
0010 **LOAD CENTERS** (residential type) R262416-50									
0100 3 wire, 120/240 V, 1 phase, including 1 pole plug-in breakers									
0200 100 amp main lugs, indoor, 8 circuits	1 Elec	1.40	5.714	Ea.	164	365		529	720
0300 12 circuits		1.20	6.667		216	425		641	870
0400 Rainproof, 8 circuits		1.40	5.714		196	365		561	755
0500 12 circuits		1.20	6.667		248	425		673	905
0600 200 amp main lugs, indoor, 16 circuits	R-1A	1.80	8.889		335	510		845	1,125
0700 20 circuits		1.50	10.667		360	610		970	1,300
0800 24 circuits		1.30	12.308		440	705		1,145	1,525
0900 30 circuits		1.20	13.333		530	765		1,295	1,700
1000 40 circuits		.80	20		740	1,150		1,890	2,525
1200 Rainproof, 16 circuits		1.80	8.889		385	510		895	1,175
1300 20 circuits		1.50	10.667		475	610		1,085	1,425
1400 24 circuits		1.30	12.308		540	705		1,245	1,650
1500 30 circuits		1.20	13.333		525	765		1,290	1,700
1600 40 circuits		.80	20		690	1,150		1,840	2,450
1800 400 amp main lugs, indoor, 42 circuits		.72	22.222		1,550	1,275		2,825	3,600
1900 Rainproof, 42 circuits		.72	22.222		1,975	1,275		3,250	4,075
2200 Plug in breakers, 20 amp, 1 pole, 4 wire, 120/208 volts									
2210 125 amp main lugs, indoor, 12 circuits	1 Elec	1.20	6.667	Ea.	222	425		647	875
2300 18 circuits		.80	10		410	635		1,045	1,400
2400 Rainproof, 12 circuits		1.20	6.667		250	425		675	905
2500 18 circuits		.80	10		480	635		1,115	1,475
2600 200 amp main lugs, indoor, 24 circuits	R-1A	1.30	12.308		425	705		1,130	1,525
2700 30 circuits		1.20	13.333		515	765		1,280	1,700
2800 36 circuits		1	16		650	915		1,565	2,100
2900 42 circuits		.80	20		700	1,150		1,850	2,475
3000 Rainproof, 24 circuits		1.30	12.308		460	705		1,165	1,550
3100 30 circuits		1.20	13.333		605	765		1,370	1,800
3200 36 circuits		1	16		755	915		1,670	2,200
3300 42 circuits		.80	20		970	1,150		2,120	2,775
3500 400 amp main lugs, indoor, 42 circuits		.72	22.222		1,575	1,275		2,850	3,650
3600 Rainproof, 42 circuits		.72	22.222		2,025	1,275		3,300	4,125
3700 Plug-in breakers, 20 amp, 1 pole, 3 wire, 120/240 volts									
3800 100 amp main breaker, indoor, 12 circuits	1 Elec	1.20	6.667	Ea.	257	425		682	915
3900 18 circuits	"	.80	10		335	635		970	1,325
4000 200 amp main breaker, indoor, 20 circuits	R-1A	1.50	10.667		470	610		1,080	1,425
4200 24 circuits		1.30	12.308		525	705		1,230	1,625
4300 30 circuits		1.20	13.333		640	765		1,405	1,825
4400 40 circuits		.90	17.778		770	1,025		1,795	2,375
4500 Rainproof, 20 circuits		1.50	10.667		375	610		985	1,325
4600 24 circuits		1.30	12.308		605	705		1,310	1,725
4700 30 circuits		1.20	13.333		535	765		1,300	1,725
4800 40 circuits		.90	17.778		690	1,025		1,715	2,275
5000 400 amp main breaker, indoor, 42 circuits		.72	22.222		2,950	1,275		4,225	5,125
5100 Rainproof, 42 circuits		.72	22.222		3,500	1,275		4,775	5,725

26 24 16.10 Load Centers

		Crew	Daily Output	Labor-Hours	Unit	Material	2021 Bare Costs Labor	Equipment	Total	Total Incl O&P
5300	Plug in breakers, 20 amp, 1 pole, 4 wire, 120/208 volts									
5400	200 amp main breaker, indoor, 30 circuits	R-1A	1.20	13.333	Ea.	640	765		1,405	1,825
5500	42 circuits		.80	20		1,375	1,150		2,525	3,225
5600	Rainproof, 30 circuits		1.20	13.333		1,200	765		1,965	2,425
5700	42 circuits		.80	20		885	1,150		2,035	2,675

26 24 16.20 Panelboard and Load Center Circuit Breakers

		Crew	Daily Output	Labor-Hours	Unit	Material	2021 Bare Costs Labor	Equipment	Total	Total Incl O&P
0010	**PANELBOARD AND LOAD CENTER CIRCUIT BREAKERS** R262416-50									
0050	Bolt-on, 10,000 amp I.C., 120 volt, 1 pole									
0100	15-50 amp	1 Elec	10	.800	Ea.	26.50	51		77.50	105
0200	60 amp		8	1		23	63.50		86.50	120
0300	70 amp		8	1		62	63.50		125.50	163
0350	240 volt, 2 pole									
0400	15-50 amp	1 Elec	8	1	Ea.	63.50	63.50		127	165
0500	60 amp		7.50	1.067		62	68		130	170
0600	80-100 amp		5	1.600		156	102		258	320
0700	3 pole, 15-60 amp		6.20	1.290		195	82		277	335
0800	70 amp		5	1.600		240	102		342	415
0900	80-100 amp		3.60	2.222		289	142		431	530
1000	22,000 amp I.C., 240 volt, 2 pole, 70-225 amp		2.70	2.963		175	189		364	470
1100	3 pole, 70-225 amp		2.30	3.478		430	222		652	800
1200	14,000 amp I.C., 277 volts, 1 pole, 15-30 amp		8	1		108	63.50		171.50	214
1300	22,000 amp I.C., 480 volts, 2 pole, 70-225 amp		2.70	2.963		134	189		323	425
1400	3 pole, 70-225 amp		2.30	3.478		1,100	222		1,322	1,550
2000	Plug-in panel or load center, 120/240 volt, to 60 amp, 1 pole		12	.667		13	42.50		55.50	77.50
2004	Circuit breaker, 120/240 volt, 20 A, 1 pole with NM cable		6.50	1.231		17.70	78.50		96.20	135
2006	30 A, 1 pole with NM cable		6.50	1.231		17.70	78.50		96.20	135
2010	2 pole		9	.889		26.50	56.50		83	114
2014	50 A, 2 pole with NM cable		5.50	1.455		31.50	92.50		124	173
2020	3 pole		7.50	1.067		110	68		178	222
2030	100 amp, 2 pole		6	1.333		86.50	85		171.50	221
2040	3 pole		4.50	1.778		160	113		273	345
2050	150-200 amp, 2 pole		3	2.667		126	170		296	390
2060	Plug-in tandem, 120/240 V, 2-15 A, 1 pole		11	.727		21.50	46.50		68	92.50
2070	1-15 A & 1-20 A		11	.727		24	46.50		70.50	95.50
2080	2-20 A		11	.727		22	46.50		68.50	93
2082	Arc fault circuit interrupter, 120/240 V, 1-15 A & 1-20 A, 1 pole		11	.727		50.50	46.50		97	125
2100	High interrupting capacity, 120/240 volt, plug-in, 30 amp, 1 pole		12	.667		34	42.50		76.50	100
2110	60 amp, 2 pole		9	.889		62.50	56.50		119	153
2120	3 pole		7.50	1.067		161	68		229	278
2130	100 amp, 2 pole		6	1.333		121	85		206	260
2140	3 pole		4.50	1.778		236	113		349	430
2150	125 amp, 2 pole		3	2.667		360	170		530	650
2200	Bolt-on, 30 amp, 1 pole		10	.800		60.50	51		111.50	142
2210	60 amp, 2 pole		7.50	1.067		147	68		215	263
2220	3 pole		6.20	1.290		370	82		452	525
2230	100 amp, 2 pole		5	1.600		320	102		422	505
2240	3 pole		3.60	2.222		525	142		667	790
2300	Ground fault, 240 volt, 30 amp, 1 pole		7	1.143		207	73		280	335
2310	2 pole		6	1.333		375	85		460	535
2350	Key operated, 240 volt, 1 pole, 30 amp		7	1.143		141	73		214	263
2360	Switched neutral, 240 volt, 30 amp, 2 pole		6	1.333		107	85		192	244
2370	3 pole		5.50	1.455		151	92.50		243.50	305

26 24 16 – Panelboards

26 24 16.20 Panelboard and Load Center Circuit Breakers

		Crew	Daily Output	Labor-Hours	Unit	Material	2021 Bare Costs Labor	Equipment	Total	Total Incl O&P
2400	Shunt trip, for 240 volt breaker, 60 amp, 1 pole	1 Elec	4	2	Ea.	86	127		213	284
2410	2 pole		3.50	2.286		86	146		232	310
2420	3 pole		3	2.667		86	170		256	345
2430	100 amp, 2 pole		3	2.667		86	170		256	345
2440	3 pole		2.50	3.200		86	204		290	400
2450	150 amp, 2 pole		2	4		248	255		503	655
2500	Auxiliary switch, for 240 volt breaker, 60 amp, 1 pole		4	2		95	127		222	293
2510	2 pole		3.50	2.286		95	146		241	320
2520	3 pole		3	2.667		95	170		265	355
2530	100 amp, 2 pole		3	2.667		95	170		265	355
2540	3 pole		2.50	3.200		95	204		299	410
2550	150 amp, 2 pole		2	4		130	255		385	525
2600	Panel or load center, 277/480 volt, plug-in, 30 amp, 1 pole		12	.667		75	42.50		117.50	146
2610	60 amp, 2 pole		9	.889		230	56.50		286.50	335
2620	3 pole		7.50	1.067		335	68		403	470
2650	Bolt-on, 60 amp, 2 pole		7.50	1.067		230	68		298	355
2660	3 pole		6.20	1.290		335	82		417	490
2700	I-line, 277/480 volt, 30 amp, 1 pole		8	1		72.50	63.50		136	174
2710	60 amp, 2 pole		7.50	1.067		254	68		322	380
2720	3 pole		6.20	1.290		390	82		472	550
2730	100 amp, 1 pole		7.50	1.067		169	68		237	287
2740	2 pole		5	1.600		390	102		492	580
2750	3 pole		3.50	2.286		465	146		611	725
2800	High interrupting capacity, 277/480 volt, plug-in, 30 amp, 1 pole		12	.667		445	42.50		487.50	550
2810	60 amp, 2 pole		9	.889		685	56.50		741.50	835
2820	3 pole		7	1.143		795	73		868	985
2830	Bolt-on, 30 amp, 1 pole		8	1		445	63.50		508.50	580
2840	60 amp, 2 pole		7.50	1.067		615	68		683	780
2850	3 pole		6.20	1.290		680	82		762	870
2900	I-line, 30 amp, 1 pole		8	1		445	63.50		508.50	580
2910	60 amp, 2 pole		7.50	1.067		615	68		683	780
2920	3 pole		6.20	1.290		700	82		782	890
2930	100 amp, 1 pole		7.50	1.067		715	68		783	885
2940	2 pole		5	1.600		490	102		592	685
2950	3 pole		3.60	2.222		770	142		912	1,050
2960	Shunt trip, 277/480 volt breaker, remote oper., 30 amp, 1 pole		4	2		465	127		592	700
2970	60 amp, 2 pole		3.50	2.286		680	146		826	960
2980	3 pole		3	2.667		770	170		940	1,100
2990	100 amp, 1 pole		3.50	2.286		555	146		701	825
3000	2 pole		3	2.667		770	170		940	1,100
3010	3 pole		2.50	3.200		710	204		914	1,075
3050	Under voltage trip, 277/480 volt breaker, 30 amp, 1 pole		4	2		465	127		592	700
3060	60 amp, 2 pole		3.50	2.286		680	146		826	960
3070	3 pole		3	2.667		770	170		940	1,100
3080	100 amp, 1 pole		3.50	2.286		555	146		701	825
3090	2 pole		3	2.667		770	170		940	1,100
3100	3 pole		2.50	3.200		710	204		914	1,075
3150	Motor operated, 277/480 volt breaker, 30 amp, 1 pole		4	2		825	127		952	1,100
3160	60 amp, 2 pole		3.50	2.286		865	146		1,011	1,175
3170	3 pole		3	2.667		1,150	170		1,320	1,525
3180	100 amp, 1 pole		3.50	2.286		925	146		1,071	1,250
3190	2 pole		3	2.667		1,150	170		1,320	1,525
3200	3 pole		2.50	3.200		995	204		1,199	1,400

26 24 16 – Panelboards

26 24 16.20 Panelboard and Load Center Circuit Breakers	Crew	Daily Output	Labor-Hours	Unit	Material	2021 Bare Costs Labor	Equipment	Total	Total Incl O&P
3250	Panelboard spacers, per pole	1 Elec	40	.200	Ea.	1.79	12.75		14.54	21
5110	NEMA 1 enclosure only, 600V, 3 p, 14k AIC, 100A	↓	2.20	3.636	↓	147	232		379	505

26 24 16.30 Panelboards Commercial Applications

	26 24 16.30 Panelboards Commercial Applications	Crew	Daily Output	Labor-Hours	Unit	Material	Labor	Equipment	Total	Total Incl O&P
0010	**PANELBOARDS COMMERCIAL APPLICATIONS** R262416-50									
0050	NQOD, w/20 amp 1 pole bolt-on circuit breakers									
0100	3 wire, 120/240 volts, 100 amp main lugs									
0150	10 circuits	1 Elec	1	8	Ea.	950	510		1,460	1,800
0200	14 circuits		.88	9.091		1,075	580		1,655	2,025
0250	18 circuits		.75	10.667		1,150	680		1,830	2,275
0300	20 circuits	↓	.65	12.308		1,250	785		2,035	2,550
0350	225 amp main lugs, 24 circuits	2 Elec	1.20	13.333		1,425	850		2,275	2,825
0400	30 circuits		.90	17.778		1,650	1,125		2,775	3,475
0450	36 circuits		.80	20		1,900	1,275		3,175	4,000
0500	38 circuits		.72	22.222		2,050	1,425		3,475	4,350
0550	42 circuits		.66	24.242		2,125	1,550		3,675	4,650
0600	4 wire, 120/208 volts, 100 amp main lugs, 12 circuits	1 Elec	1	8		1,025	510		1,535	1,875
0650	16 circuits		.75	10.667		1,150	680		1,830	2,250
0700	20 circuits		.65	12.308		1,325	785		2,110	2,625
0750	24 circuits		.60	13.333		415	850		1,265	1,700
0800	30 circuits	↓	.53	15.094		1,650	960		2,610	3,225
0850	225 amp main lugs, 32 circuits	2 Elec	.90	17.778		1,875	1,125		3,000	3,750
0900	34 circuits		.84	19.048		1,925	1,225		3,150	3,900
0950	36 circuits		.80	20		1,950	1,275		3,225	4,050
1000	42 circuits		.68	23.529		2,200	1,500		3,700	4,650
1010	400 amp main lugs, 42 circs		.68	23.529		2,200	1,500		3,700	4,650
1040	225 amp main lugs, NEMA 7, 12 circuits		1	16		5,825	1,025		6,850	7,925
1100	24 circuits	↓	.40	40	↓	7,025	2,550		9,575	11,500
1200	NEHB, w/20 amp, 1 pole bolt-on circuit breakers									
1250	4 wire, 277/480 volts, 100 amp main lugs, 12 circuits	1 Elec	.88	9.091	Ea.	1,500	580		2,080	2,500
1300	20 circuits	"	.60	13.333		2,150	850		3,000	3,625
1350	225 amp main lugs, 24 circuits	2 Elec	.90	17.778		2,525	1,125		3,650	4,450
1400	30 circuits		.80	20		3,025	1,275		4,300	5,225
1448	32 circuits		4.90	3.265		3,500	208		3,708	4,150
1450	36 circuits		.72	22.222		3,500	1,425		4,925	5,950
1500	42 circuits		.60	26.667		4,000	1,700		5,700	6,925
1510	225 amp main lugs, NEMA 7, 12 circuits		.90	17.778		5,975	1,125		7,100	8,250
1590	24 circuits	↓	.30	53.333	↓	8,550	3,400		11,950	14,500
1600	NQOD panel, w/20 amp, 1 pole, circuit breakers									
1650	3 wire, 120/240 volt with main circuit breaker									
1700	100 amp main, 12 circuits	1 Elec	.80	10	Ea.	1,225	635		1,860	2,275
1750	20 circuits	"	.60	13.333		1,550	850		2,400	2,950
1800	225 amp main, 30 circuits	2 Elec	.68	23.529		2,925	1,500		4,425	5,450
1801	225 amp main, 32 circuits		5	3.200		2,925	204		3,129	3,525
1850	42 circuits		.52	30.769		3,375	1,950		5,325	6,625
1900	400 amp main, 30 circuits		.54	29.630		4,050	1,875		5,925	7,250
1950	42 circuits	↓	.50	32		4,525	2,050		6,575	8,000
2000	4 wire, 120/208 volts with main circuit breaker									
2050	100 amp main, 24 circuits	1 Elec	.47	17.021	Ea.	1,750	1,075		2,825	3,525
2100	30 circuits	"	.40	20		2,050	1,275		3,325	4,150
2200	225 amp main, 32 circuits	2 Elec	.72	22.222		3,400	1,425		4,825	5,850
2250	42 circuits		.56	28.571		4,000	1,825		5,825	7,100
2300	400 amp main, 42 circuits	↓	.48	33.333		5,025	2,125		7,150	8,675

26 24 16 – Panelboards

26 24 16.30 Panelboards Commercial Applications

		Crew	Daily Output	Labor-Hours	Unit	Material	2021 Bare Costs Labor	Equipment	Total	Total Incl O&P
2350	600 amp main, 42 circuits	2 Elec	.40	40	Ea.	7,400	2,550		9,950	11,900
2400	NEHB, with 20 amp, 1 pole circuit breaker									
2450	4 wire, 277/480 volts with main circuit breaker									
2500	100 amp main, 24 circuits	1 Elec	.42	19.048	Ea.	2,900	1,225		4,125	5,000
2550	30 circuits	"	.38	21.053		3,400	1,350		4,750	5,725
2600	225 amp main, 30 circuits	2 Elec	.72	22.222		4,250	1,425		5,675	6,800
2650	42 circuits		.56	28.571		5,250	1,825		7,075	8,475
2700	400 amp main, 42 circuits		.46	34.783		6,300	2,225		8,525	10,200
2750	600 amp main, 42 circuits		.38	42.105		8,625	2,675		11,300	13,500
2900	Note: the following line items don't include branch circuit breakers									
2910	For branch circuit breakers information, see Section 26 24 16.20									
3010	Main lug, no main breaker, 240 volt, 1 pole, 3 wire, 100 amp	1 Elec	2.30	3.478	Ea.	610	222		832	1,000
3020	225 amp	2 Elec	2.40	6.667		645	425		1,070	1,350
3030	400 amp	"	1.80	8.889		1,150	565		1,715	2,125
3060	3 pole, 3 wire, 100 amp	1 Elec	2.30	3.478		660	222		882	1,050
3070	225 amp	2 Elec	2.40	6.667		685	425		1,110	1,375
3080	400 amp		1.80	8.889		1,200	565		1,765	2,175
3090	600 amp		1.60	10		1,450	635		2,085	2,550
3110	3 pole, 4 wire, 100 amp	1 Elec	2.30	3.478		725	222		947	1,125
3120	225 amp	2 Elec	2.40	6.667		810	425		1,235	1,525
3130	400 amp		1.80	8.889		1,200	565		1,765	2,175
3140	600 amp		1.60	10		1,450	635		2,085	2,550
3160	480 volt, 3 pole, 3 wire, 100 amp	1 Elec	2.30	3.478		770	222		992	1,175
3170	225 amp	2 Elec	2.40	6.667		1,100	425		1,525	1,825
3180	400 amp		1.80	8.889		1,575	565		2,140	2,575
3190	600 amp		1.60	10		1,750	635		2,385	2,875
3210	277/480 volt, 3 pole, 4 wire, 100 amp	1 Elec	2.30	3.478		750	222		972	1,150
3220	225 amp	2 Elec	2.40	6.667		910	425		1,335	1,625
3230	400 amp		1.80	8.889		1,525	565		2,090	2,525
3240	600 amp		1.60	10		1,700	635		2,335	2,825
3260	Main circuit breaker, 240 volt, 1 pole, 3 wire, 100 amp	1 Elec	2	4		725	255		980	1,175
3270	225 amp	2 Elec	2	8		1,775	510		2,285	2,725
3280	400 amp	"	1.60	10		2,800	635		3,435	4,050
3310	3 pole, 3 wire, 100 amp	1 Elec	2	4		930	255		1,185	1,400
3320	225 amp	2 Elec	2	8		2,050	510		2,560	3,000
3330	400 amp		1.60	10		3,200	635		3,835	4,475
3360	120/208 volt, 3 pole, 4 wire, 100 amp		4	4		835	255		1,090	1,300
3370	225 amp		2	8		2,050	510		2,560	3,000
3380	400 amp		1.60	10		3,200	635		3,835	4,475
3410	480 volt, 3 pole, 3 wire, 100 amp	1 Elec	2	4		1,400	255		1,655	1,925
3420	225 amp	2 Elec	2	8		2,400	510		2,910	3,400
3430	400 amp		1.60	10		3,600	635		4,235	4,900
3460	277/480 volt, 3 pole, 4 wire, 100 amp		4	4		1,375	255		1,630	1,875
3470	225 amp		2	8		2,350	510		2,860	3,325
3480	400 amp		1.60	10		3,625	635		4,260	4,925
3510	Main circuit breaker, HIC, 240 volt, 1 pole, 3 wire, 100 amp	1 Elec	2	4		1,325	255		1,580	1,850
3520	225 amp	2 Elec	2	8		3,500	510		4,010	4,600
3530	400 amp	"	1.60	10		4,675	635		5,310	6,100
3560	3 pole, 3 wire, 100 amp	1 Elec	2	4		1,500	255		1,755	2,025
3570	225 amp	2 Elec	2	8		3,950	510		4,460	5,075
3580	400 amp	"	1.60	10		5,200	635		5,835	6,675
3610	120/208 volt, 3 pole, 4 wire, 100 amp	1 Elec	2	4		1,500	255		1,755	2,025
3620	225 amp	2 Elec	2	8		3,950	510		4,460	5,075

26 24 16 – Panelboards

26 24 16.30 Panelboards Commercial Applications

		Crew	Daily Output	Labor-Hours	Unit	Material	2021 Bare Costs Labor	Equipment	Total	Total Incl O&P
3630	400 amp	2 Elec	1.60	10	Ea.	5,200	635		5,835	6,675
3660	480 volt, 3 pole, 3 wire, 100 amp	1 Elec	2	4		2,225	255		2,480	2,825
3670	225 amp	2 Elec	2	8		4,375	510		4,885	5,550
3680	400 amp	"	1.60	10		5,575	635		6,210	7,075
3710	277/480 volt, 3 pole, 4 wire, 100 amp	1 Elec	2	4		2,125	255		2,380	2,700
3720	225 amp	2 Elec	2	8		4,175	510		4,685	5,325
3730	400 amp	"	1.60	10		5,525	635		6,160	7,025
3760	Main circuit breaker, shunt trip, 100 amp	1 Elec	1.20	6.667		1,100	425		1,525	1,850
3770	225 amp	2 Elec	1.60	10		2,550	635		3,185	3,750
3780	400 amp	"	1.40	11.429		3,675	730		4,405	5,125

26 24 19 – Motor-Control Centers

26 24 19.20 Motor Control Center Components

		Crew	Daily Output	Labor-Hours	Unit	Material	2021 Bare Costs Labor	Equipment	Total	Total Incl O&P
0010	**MOTOR CONTROL CENTER COMPONENTS** R262419-60									
0100	Starter, size 1, FVNR, NEMA 1, type A, fusible	1 Elec	2.70	2.963	Ea.	1,600	189		1,789	2,025
0120	Circuit breaker		2.70	2.963		1,725	189		1,914	2,175
0140	Type B, fusible		2.70	2.963		1,750	189		1,939	2,200
0160	Circuit breaker		2.70	2.963		1,900	189		2,089	2,375
0180	NEMA 12, type A, fusible		2.60	3.077		1,625	196		1,821	2,075
0200	Circuit breaker		2.60	3.077		1,750	196		1,946	2,225
0220	Type B, fusible		2.60	3.077		1,775	196		1,971	2,250
0240	Circuit breaker		2.60	3.077		1,925	196		2,121	2,425
0300	Starter, size 1, FVR, NEMA 1, type A, fusible		2	4		2,275	255		2,530	2,875
0320	Circuit breaker		2	4		2,275	255		2,530	2,875
0340	Type B, fusible		2	4		2,500	255		2,755	3,150
0360	Circuit breaker		2	4		2,500	255		2,755	3,150
0380	NEMA 12, type A, fusible		1.90	4.211		2,300	268		2,568	2,950
0400	Circuit breaker		1.90	4.211		2,300	268		2,568	2,950
0420	Type B, fusible		1.90	4.211		2,550	268		2,818	3,200
0440	Circuit breaker		1.90	4.211		2,550	268		2,818	3,200
0490	Starter size 1, 2 speed, separate winding									
0500	NEMA 1, type A, fusible	1 Elec	2.60	3.077	Ea.	3,000	196		3,196	3,600
0520	Circuit breaker		2.60	3.077		3,000	196		3,196	3,600
0540	Type B, fusible		2.60	3.077		3,300	196		3,496	3,950
0560	Circuit breaker		2.60	3.077		3,300	196		3,496	3,950
0580	NEMA 12, type A, fusible		2.50	3.200		3,050	204		3,254	3,675
0600	Circuit breaker		2.50	3.200		3,050	204		3,254	3,675
0620	Type B, fusible		2.50	3.200		3,375	204		3,579	4,000
0640	Circuit breaker		2.50	3.200		3,375	204		3,579	4,000
0650	Starter size 1, 2 speed, consequent pole									
0660	NEMA 1, type A, fusible	1 Elec	2.60	3.077	Ea.	3,000	196		3,196	3,600
0680	Circuit breaker		2.60	3.077		3,000	196		3,196	3,600
0700	Type B, fusible		2.60	3.077		3,300	196		3,496	3,950
0720	Circuit breaker		2.60	3.077		3,300	196		3,496	3,950
0740	NEMA 12, type A, fusible		2.50	3.200		3,050	204		3,254	3,675
0760	Circuit breaker		2.50	3.200		3,050	204		3,254	3,675
0780	Type B, fusible		2.50	3.200		3,350	204		3,554	3,975
0800	Circuit breaker		2.50	3.200		3,375	204		3,579	4,000
0810	Starter size 1, 2 speed, space only									
0820	NEMA 1, type A, fusible	1 Elec	16	.500	Ea.	675	32		707	790
0840	Circuit breaker		16	.500		675	32		707	790
0860	Type B, fusible		16	.500		675	32		707	790
0880	Circuit breaker		16	.500		675	32		707	790

26 24 19 – Motor-Control Centers

26 24 19.20 Motor Control Center Components		Crew	Daily Output	Labor-Hours	Unit	Material	2021 Bare Costs Labor	Equipment	Total	Total Incl O&P
0900	NEMA 12, type A, fusible	1 Elec	15	.533	Ea.	700	34		734	825
0920	Circuit breaker		15	.533		700	34		734	825
0940	Type B, fusible		15	.533		700	34		734	825
0960	Circuit breaker		15	.533		700	34		734	825
1100	Starter size 2, FVNR, NEMA 1, type A, fusible	2 Elec	4	4		1,800	255		2,055	2,350
1120	Circuit breaker		4	4		1,975	255		2,230	2,550
1140	Type B, fusible		4	4		1,975	255		2,230	2,550
1160	Circuit breaker		4	4		2,150	255		2,405	2,750
1180	NEMA 12, type A, fusible		3.80	4.211		1,825	268		2,093	2,400
1200	Circuit breaker		3.80	4.211		2,000	268		2,268	2,600
1220	Type B, fusible		3.80	4.211		2,000	268		2,268	2,600
1240	Circuit breaker		3.80	4.211		2,175	268		2,443	2,800
1300	FVR, NEMA 1, type A, fusible		3.20	5		3,050	320		3,370	3,825
1320	Circuit breaker		3.20	5		3,050	320		3,370	3,825
1340	Type B, fusible		3.20	5		3,350	320		3,670	4,150
1360	Circuit breaker		3.20	5		3,350	320		3,670	4,150
1380	NEMA type 12, type A, fusible		3	5.333		3,100	340		3,440	3,900
1400	Circuit breaker		3	5.333		3,100	340		3,440	3,900
1420	Type B, fusible		3	5.333		3,400	340		3,740	4,250
1440	Circuit breaker		3	5.333		3,400	340		3,740	4,250
1490	Starter size 2, 2 speed, separate winding									
1500	NEMA 1, type A, fusible	2 Elec	3.80	4.211	Ea.	3,400	268		3,668	4,150
1520	Circuit breaker		3.80	4.211		3,400	268		3,668	4,150
1540	Type B, fusible		3.80	4.211		3,725	268		3,993	4,500
1560	Circuit breaker		3.80	4.211		3,725	268		3,993	4,500
1570	NEMA 12, type A, fusible		3.60	4.444		3,450	283		3,733	4,225
1580	Circuit breaker		3.60	4.444		3,450	283		3,733	4,225
1600	Type B, fusible		3.60	4.444		3,800	283		4,083	4,600
1620	Circuit breaker		3.60	4.444		3,800	283		4,083	4,600
1630	Starter size 2, 2 speed, consequent pole									
1640	NEMA 1, type A, fusible	2 Elec	3.80	4.211	Ea.	3,850	268		4,118	4,650
1660	Circuit breaker		3.80	4.211		3,850	268		4,118	4,650
1680	Type B, fusible		3.80	4.211		4,075	268		4,343	4,900
1700	Circuit breaker		3.80	4.211		4,075	268		4,343	4,900
1720	NEMA 12, type A, fusible		3.80	4.211		3,875	268		4,143	4,650
1740	Circuit breaker		3.60	4.444		3,925	283		4,208	4,725
1760	Type B, fusible		3.60	4.444		4,150	283		4,433	4,975
1780	Circuit breaker		3.60	4.444		4,150	283		4,433	4,975
1830	Starter size 2, autotransformer									
1840	NEMA 1, type A, fusible	2 Elec	3.40	4.706	Ea.	6,025	300		6,325	7,075
1860	Circuit breaker		3.40	4.706		6,150	300		6,450	7,225
1880	Type B, fusible		3.40	4.706		6,575	300		6,875	7,675
1900	Circuit breaker		3.40	4.706		6,575	300		6,875	7,675
1920	NEMA 12, type A, fusible		3.20	5		6,125	320		6,445	7,225
1940	Circuit breaker		3.20	5		6,125	320		6,445	7,225
1960	Type B, fusible		3.20	5		6,675	320		6,995	7,825
1980	Circuit breaker		3.20	5		6,675	320		6,995	7,825
2030	Starter size 2, space only									
2040	NEMA 1, type A, fusible	1 Elec	16	.500	Ea.	675	32		707	790
2060	Circuit breaker		16	.500		675	32		707	790
2080	Type B, fusible		16	.500		675	32		707	790
2100	Circuit breaker		16	.500		675	32		707	790
2120	NEMA 12, type A, fusible		15	.533		700	34		734	825

235

26 24 19.20 Motor Control Center Components		Crew	Daily Output	Labor-Hours	Unit	Material	2021 Bare Costs Labor	Equipment	Total	Total Incl O&P
2140	Circuit breaker	1 Elec	15	.533	Ea.	700	34		734	825
2160	Type B, fusible		15	.533		700	34		734	825
2180	Circuit breaker		15	.533		700	34		734	825
2300	Starter size 3, FVNR, NEMA 1, type A, fusible	2 Elec	2	8		3,425	510		3,935	4,525
2320	Circuit breaker		2	8		3,050	510		3,560	4,100
2340	Type B, fusible		2	8		3,775	510		4,285	4,900
2360	Circuit breaker		2	8		3,350	510		3,860	4,450
2380	NEMA 12, type A, fusible		1.90	8.421		3,500	535		4,035	4,650
2400	Circuit breaker		1.90	8.421		3,100	535		3,635	4,200
2420	Type B, fusible		1.90	8.421		3,900	535		4,435	5,075
2440	Circuit breaker		1.90	8.421		3,400	535		3,935	4,550
2500	Starter size 3, FVR, NEMA 1, type A, fusible		1.60	10		4,675	635		5,310	6,100
2520	Circuit breaker		1.60	10		4,475	635		5,110	5,850
2540	Type B, fusible		1.60	10		5,075	635		5,710	6,550
2560	Circuit breaker		1.60	10		4,875	635		5,510	6,325
2580	NEMA 12, type A, fusible		1.50	10.667		4,775	680		5,455	6,250
2600	Circuit breaker		1.50	10.667		4,550	680		5,230	6,000
2620	Type B, fusible		1.50	10.667		5,175	680		5,855	6,700
2640	Circuit breaker		1.50	10.667		5,375	680		6,055	6,925
2690	Starter size 3, 2 speed, separate winding									
2700	NEMA 1, type A, fusible	2 Elec	2	8	Ea.	5,375	510		5,885	6,650
2720	Circuit breaker		2	8		4,750	510		5,260	5,975
2740	Type B, fusible		2	8		5,875	510		6,385	7,225
2760	Circuit breaker		2	8		5,200	510		5,710	6,450
2780	NEMA 12, type A, fusible		1.90	8.421		5,475	535		6,010	6,825
2800	Circuit breaker		1.90	8.421		4,850	535		5,385	6,125
2820	Type B, fusible		1.90	8.421		6,000	535		6,535	7,400
2840	Circuit breaker		1.90	8.421		5,275	535		5,810	6,625
2850	Starter size 3, 2 speed, consequent pole									
2860	NEMA 1, type A, fusible	2 Elec	2	8	Ea.	6,000	510		6,510	7,350
2880	Circuit breaker		2	8		5,375	510		5,885	6,650
2900	Type B, fusible		2	8		6,500	510		7,010	7,900
2920	Circuit breaker		2	8		5,825	510		6,335	7,150
2940	NEMA 12, type A, fusible		1.90	8.421		6,100	535		6,635	7,525
2960	Circuit breaker		1.90	8.421		5,475	535		6,010	6,800
2980	Type B, fusible		1.90	8.421		6,625	535		7,160	8,075
3000	Circuit breaker		1.90	8.421		5,900	535		6,435	7,300
3100	Starter size 3, autotransformer, NEMA 1, type A, fusible		1.60	10		7,925	635		8,560	9,675
3120	Circuit breaker		1.60	10		7,450	635		8,085	9,150
3140	Type B, fusible		1.60	10		8,150	635		8,785	9,925
3160	Circuit breaker		1.60	10		8,150	635		8,785	9,925
3180	NEMA 12, type A, fusible		1.50	10.667		7,600	680		8,280	9,350
3200	Circuit breaker		1.50	10.667		7,600	680		8,280	9,350
3220	Type B, fusible		1.50	10.667		8,300	680		8,980	10,100
3240	Circuit breaker		1.50	10.667		8,300	680		8,980	10,100
3260	Starter size 3, space only, NEMA 1, type A, fusible	1 Elec	15	.533		1,150	34		1,184	1,325
3280	Circuit breaker		15	.533		900	34		934	1,050
3300	Type B, fusible		15	.533		1,150	34		1,184	1,325
3320	Circuit breaker		15	.533		900	34		934	1,050
3340	NEMA 12, type A, fusible		14	.571		1,225	36.50		1,261.50	1,400
3360	Circuit breaker		14	.571		955	36.50		991.50	1,100
3380	Type B, fusible		14	.571		1,225	36.50		1,261.50	1,400
3400	Circuit breaker		14	.571		955	36.50		991.50	1,100

236

26 24 19 – Motor-Control Centers

26 24 19.20 Motor Control Center Components	Crew	Daily Output	Labor-Hours	Unit	Material	2021 Bare Costs Labor	Equipment	Total	Total Incl O&P
3500	Starter size 4, FVNR, NEMA 1, type A, fusible	2 Elec	1.60	10	Ea.	4,475	635		5,110	5,850
3520	Circuit breaker		1.60	10		4,050	635		4,685	5,400
3540	Type B, fusible		1.60	10		4,875	635		5,510	6,325
3560	Circuit breaker		1.60	10		4,475	635		5,110	5,850
3580	NEMA 12, type A, fusible		1.50	10.667		4,575	680		5,255	6,025
3600	Circuit breaker		1.50	10.667		4,125	680		4,805	5,525
3620	Type B, fusible		1.50	10.667		5,000	680		5,680	6,500
3640	Circuit breaker		1.50	10.667		4,525	680		5,205	5,975
3700	Starter size 4, FVR, NEMA 1, type A, fusible		1.20	13.333		6,100	850		6,950	7,975
3720	Circuit breaker		1.20	13.333		5,500	850		6,350	7,300
3740	Type B, fusible		1.20	13.333		6,650	850		7,500	8,575
3760	Circuit breaker		1.20	13.333		6,050	850		6,900	7,900
3780	NEMA 12, type A, fusible		1.16	13.793		6,250	880		7,130	8,175
3800	Circuit breaker		1.16	13.793		5,575	880		6,455	7,450
3820	Type B, fusible		1.16	13.793		5,650	880		6,530	7,500
3840	Circuit breaker		1.16	13.793		6,150	880		7,030	8,050
3890	Starter size 4, 2 speed, separate windings									
3900	NEMA 1, type A, fusible	2 Elec	1.60	10	Ea.	7,700	635		8,335	9,425
3920	Circuit breaker		1.60	10		5,775	635		6,410	7,325
3940	Type B, fusible		1.60	10		8,475	635		9,110	10,200
3960	Circuit breaker		1.60	10		6,350	635		6,985	7,925
3980	NEMA 12, type A, fusible		1.50	10.667		7,850	680		8,530	9,625
4000	Circuit breaker		1.50	10.667		5,875	680		6,555	7,475
4020	Type B, fusible		1.50	10.667		8,600	680		9,280	10,500
4040	Circuit breaker		1.50	10.667		6,425	680		7,105	8,075
4050	Starter size 4, 2 speed, consequent pole									
4060	NEMA 1, type A, fusible	2 Elec	1.60	10	Ea.	9,000	635		9,635	10,800
4080	Circuit breaker		1.60	10		6,700	635		7,335	8,300
4100	Type B, fusible		1.60	10		9,900	635		10,535	11,800
4120	Circuit breaker		1.60	10		7,325	635		7,960	9,025
4140	NEMA 12, type A, fusible		1.50	10.667		9,125	680		9,805	11,000
4160	Circuit breaker		1.50	10.667		6,800	680		7,480	8,475
4180	Type B, fusible		1.50	10.667		10,000	680		10,680	12,000
4200	Circuit breaker		1.50	10.667		7,425	680		8,105	9,175
4300	Starter size 4, autotransformer, NEMA 1, type A, fusible		1.30	12.308		8,850	785		9,635	10,900
4320	Circuit breaker		1.30	12.308		8,925	785		9,710	11,000
4340	Type B, fusible		1.30	12.308		9,750	785		10,535	11,900
4360	Circuit breaker		1.30	12.308		9,750	785		10,535	11,900
4380	NEMA 12, type A, fusible		1.24	12.903		9,000	820		9,820	11,100
4400	Circuit breaker		1.24	12.903		9,050	820		9,870	11,200
4420	Type B, fusible		1.24	12.903		9,900	820		10,720	12,100
4440	Circuit breaker		1.24	12.903		9,875	820		10,695	12,100
4500	Starter size 4, space only, NEMA 1, type A, fusible	1 Elec	14	.571		1,600	36.50		1,636.50	1,825
4520	Circuit breaker		14	.571		1,150	36.50		1,186.50	1,325
4540	Type B, fusible		14	.571		1,600	36.50		1,636.50	1,825
4560	Circuit breaker		14	.571		1,150	36.50		1,186.50	1,325
4580	NEMA 12, type A, fusible		13	.615		1,700	39		1,739	1,925
4600	Circuit breaker		13	.615		1,225	39		1,264	1,400
4620	Type B, fusible		13	.615		1,700	39		1,739	1,925
4640	Circuit breaker		13	.615		1,225	39		1,264	1,400
4800	Starter size 5, FVNR, NEMA 1, type A, fusible	2 Elec	1	16		9,225	1,025		10,250	11,600
4820	Circuit breaker		1	16		6,550	1,025		7,575	8,725
4840	Type B, fusible		1	16		10,100	1,025		11,125	12,600

For customer support on your Electrical Costs with RSMeans Data, call 800.448.8182.

237

26 24 19.20 Motor Control Center Components		Crew	Daily Output	Labor-Hours	Unit	Material	2021 Bare Costs Labor	Equipment	Total	Total Incl O&P
4860	Circuit breaker	2 Elec	1	16	Ea.	7,200	1,025		8,225	9,450
4880	NEMA 12, type A, fusible		.96	16.667		9,350	1,050		10,400	11,900
4900	Circuit breaker		.96	16.667		6,675	1,050		7,725	8,900
4920	Type B, fusible		.96	16.667		10,200	1,050		11,250	12,900
4940	Circuit breaker		.96	16.667		7,325	1,050		8,375	9,625
5000	Starter size 5, FVR, NEMA 1, type A, fusible		.80	20		13,600	1,275		14,875	16,800
5020	Circuit breaker		.80	20		10,700	1,275		11,975	13,600
5040	Type B, fusible		.80	20		14,900	1,275		16,175	18,300
5060	Circuit breaker		.80	20		11,700	1,275		12,975	14,800
5080	NEMA 12, type A, fusible		.76	21.053		13,800	1,350		15,150	17,200
5100	Circuit breaker		.76	21.053		10,800	1,350		12,150	13,900
5120	Type B, fusible		.76	21.053		15,100	1,350		16,450	18,700
5140	Circuit breaker		.76	21.053		11,900	1,350		13,250	15,000
5190	Starter size 5, 2 speed, separate windings									
5200	NEMA 1, type A, fusible	2 Elec	1	16	Ea.	18,700	1,025		19,725	22,100
5220	Circuit breaker		1	16		14,000	1,025		15,025	16,900
5240	Type B, fusible		1	16		20,600	1,025		21,625	24,100
5260	Circuit breaker		1	16		15,300	1,025		16,325	18,400
5280	NEMA 12, type A, fusible		.96	16.667		19,000	1,050		20,050	22,500
5300	Circuit breaker		.96	16.667		14,100	1,050		15,150	17,100
5320	Type B, fusible		.96	16.667		20,800	1,050		21,850	24,500
5340	Circuit breaker		.96	16.667		15,500	1,050		16,550	18,700
5400	Starter size 5, autotransformer, NEMA 1, type A, fusible	2 Elec	.70	22.857	Ea.	15,300	1,450		16,750	19,100
5420	Circuit breaker		.70	22.857		12,700	1,450		14,150	16,200
5440	Type B, fusible		.70	22.857		16,800	1,450		18,250	20,700
5460	Circuit breaker		.70	22.857		14,000	1,450		15,450	17,600
5480	NEMA 12, type A, fusible		.68	23.529		15,500	1,500		17,000	19,200
5500	Circuit breaker		.68	23.529		12,900	1,500		14,400	16,300
5520	Type B, fusible		.68	23.529		17,000	1,500		18,500	20,900
5540	Circuit breakers		.68	23.529		14,100	1,500		15,600	17,700
5600	Starter size 5, space only, NEMA 1, type A, fusible	1 Elec	12	.667		2,050	42.50		2,092.50	2,350
5620	Circuit breaker		12	.667		2,050	42.50		2,092.50	2,350
5640	Type B, fusible		12	.667		2,050	42.50		2,092.50	2,350
5660	Circuit breaker		12	.667		1,350	42.50		1,392.50	1,550
5680	NEMA 12, type A, fusible		11	.727		2,175	46.50		2,221.50	2,475
5700	Circuit breaker		11	.727		1,425	46.50		1,471.50	1,650
5720	Type B, fusible		11	.727		2,175	46.50		2,221.50	2,475
5740	Circuit breaker		11	.727		1,425	46.50		1,471.50	1,650
5800	Fuse, light contactor NEMA 1, type A, 30 amp		2.70	2.963		1,550	189		1,739	1,975
5820	60 amp		2	4		1,750	255		2,005	2,300
5840	100 amp		1	8		3,325	510		3,835	4,400
5860	200 amp		.80	10		7,725	635		8,360	9,450
5880	Type B, 30 amp		2.70	2.963		1,675	189		1,864	2,125
5900	60 amp		2	4		1,875	255		2,130	2,450
5920	100 amp		1	8		3,650	510		4,160	4,775
5940	200 amp	2 Elec	1.60	10		8,450	635		9,085	10,200
5960	NEMA 12, type A, 30 amp	1 Elec	2.60	3.077		1,575	196		1,771	2,025
5980	60 amp		1.90	4.211		1,775	268		2,043	2,350
6000	100 amp		.95	8.421		3,400	535		3,935	4,525
6020	200 amp	2 Elec	1.50	10.667		7,850	680		8,530	9,625
6040	Type B, 30 amp	1 Elec	2.60	3.077		1,700	196		1,896	2,175
6060	60 amp		1.90	4.211		1,900	268		2,168	2,500
6080	100 amp		.95	8.421		3,725	535		4,260	4,900

26 24 19 – Motor-Control Centers

26 24 19.20 Motor Control Center Components

26 24 19.20 Motor Control Center Components		Crew	Daily Output	Labor-Hours	Unit	Material	2021 Bare Costs Labor	Equipment	Total	Total Incl O&P
6100	200 amp	2 Elec	1.50	10.667	Ea.	8,600	680		9,280	10,500
6200	Circuit breaker, light contactor NEMA 1, type A, 30 amp	1 Elec	2.70	2.963		1,675	189		1,864	2,125
6220	60 amp		2	4		1,925	255		2,180	2,475
6240	100 amp		1	8		2,950	510		3,460	4,000
6260	200 amp	2 Elec	1.60	10		6,400	635		7,035	8,000
6280	Type B, 30 amp	1 Elec	2.70	2.963		1,850	189		2,039	2,300
6300	60 amp		2	4		2,125	255		2,380	2,700
6320	100 amp		1	8		3,225	510		3,735	4,300
6340	200 amp	2 Elec	1.60	10		6,975	635		7,610	8,625
6360	NEMA 12, type A, 30 amp	1 Elec	2.60	3.077		1,700	196		1,896	2,175
6380	60 amp		1.90	4.211		1,950	268		2,218	2,525
6400	100 amp		.95	8.421		3,000	535		3,535	4,100
6420	200 amp	2 Elec	1.50	10.667		6,425	680		7,105	8,075
6440	Type B, 30 amp	1 Elec	2.60	3.077		2,000	196		2,196	2,500
6460	60 amp		1.90	4.211		2,150	268		2,418	2,750
6480	100 amp		.95	8.421		3,275	535		3,810	4,400
6500	200 amp	2 Elec	1.50	10.667		7,075	680		7,755	8,775
6600	Fusible switch, NEMA 1, type A, 30 amp	1 Elec	5.30	1.509		1,050	96		1,146	1,300
6620	60 amp		5	1.600		1,125	102		1,227	1,375
6640	100 amp		4	2		1,250	127		1,377	1,575
6660	200 amp		3.20	2.500		2,150	159		2,309	2,575
6680	400 amp	2 Elec	4.60	3.478		5,050	222		5,272	5,875
6700	600 amp		3.20	5		5,475	320		5,795	6,500
6720	800 amp		2.60	6.154		15,000	390		15,390	17,100
6740	NEMA 12, type A, 30 amp	1 Elec	5.20	1.538		1,075	98		1,173	1,350
6760	60 amp		4.90	1.633		1,150	104		1,254	1,425
6780	100 amp		3.90	2.051		1,275	131		1,406	1,600
6800	200 amp		3.10	2.581		2,200	164		2,364	2,675
6820	400 amp	2 Elec	4.40	3.636		5,150	232		5,382	6,025
6840	600 amp		3	5.333		5,625	340		5,965	6,675
6860	800 amp		2.40	6.667		15,200	425		15,625	17,300
6900	Circuit breaker, NEMA 1, type A, 30 amp	1 Elec	5.30	1.509		970	96		1,066	1,225
6920	60 amp		5	1.600		970	102		1,072	1,225
6940	100 amp		4	2		970	127		1,097	1,275
6960	225 amp		3.20	2.500		1,725	159		1,884	2,125
6980	400 amp	2 Elec	4.60	3.478		3,275	222		3,497	3,950
7000	600 amp		3.20	5		3,850	320		4,170	4,700
7020	800 amp		2.60	6.154		8,025	390		8,415	9,400
7040	NEMA 12, type A, 30 amp	1 Elec	5.20	1.538		995	98		1,093	1,250
7060	60 amp		4.90	1.633		995	104		1,099	1,250
7080	100 amp		3.90	2.051		995	131		1,126	1,300
7100	225 amp		3.10	2.581		1,750	164		1,914	2,175
7120	400 amp	2 Elec	4.40	3.636		3,350	232		3,582	4,025
7140	600 amp		3	5.333		3,900	340		4,240	4,775
7160	800 amp		2.40	6.667		8,175	425		8,600	9,625
7300	Incoming line, main lug only, 600 amp, alum., NEMA 1		1.60	10		1,225	635		1,860	2,300
7320	NEMA 12		1.50	10.667		1,250	680		1,930	2,375
7340	Copper, NEMA 1		1.60	10		1,300	635		1,935	2,375
7360	800 amp, alum., NEMA 1		1.50	10.667		3,075	680		3,755	4,375
7380	NEMA 12		1.40	11.429		3,125	730		3,855	4,500
7400	Copper, NEMA 1		1.50	10.667		3,200	680		3,880	4,525
7420	1,200 amp, copper, NEMA 1		1.40	11.429		3,300	730		4,030	4,700
7440	Incoming line, fusible switch, 400 amp, alum., NEMA 1		1.20	13.333		4,225	850		5,075	5,900

26 24 19.20 Motor Control Center Components

		Crew	Daily Output	Labor-Hours	Unit	Material	2021 Bare Costs Labor	2021 Bare Costs Equipment	Total	Total Incl O&P
7460	NEMA 12	2 Elec	1.10	14.545	Ea.	4,300	925		5,225	6,125
7480	Copper, NEMA 1		1.20	13.333		4,300	850		5,150	5,975
7500	600 amp, alum., NEMA 1		1.10	14.545		5,150	925		6,075	7,050
7520	NEMA 12		1	16		5,250	1,025		6,275	7,300
7540	Copper, NEMA 1		1.10	14.545		5,225	925		6,150	7,125
7560	Incoming line, circuit breaker, 225 amp, alum., NEMA 1		1.20	13.333		2,200	850		3,050	3,675
7580	NEMA 12		1.10	14.545		2,250	925		3,175	3,850
7600	Copper, NEMA 1		1.20	13.333		2,275	850		3,125	3,750
7620	400 amp, alum., NEMA 1		1.20	13.333		3,275	850		4,125	4,875
7640	NEMA 12		1.10	14.545		3,350	925		4,275	5,050
7660	Copper, NEMA 1		1.20	13.333		3,350	850		4,200	4,950
7680	600 amp, alum., NEMA 1		1.10	14.545		3,850	925		4,775	5,600
7700	NEMA 12		1	16		3,900	1,025		4,925	5,800
7720	Copper, NEMA 1		1.10	14.545		3,900	925		4,825	5,675
7740	800 amp, copper, NEMA 1		.90	17.778		8,025	1,125		9,150	10,500
7760	Incoming line, for copper bus, add					121			121	133
7780	For 65,000 amp bus bracing, add					180			180	198
7800	For NEMA 3R enclosure, add					5,650			5,650	6,225
7820	For NEMA 12 enclosure, add					156			156	171
7840	For 1/4" x 1" ground bus, add	1 Elec	16	.500		100	32		132	158
7860	For 1/4" x 2" ground bus, add	"	12	.667		100	42.50		142.50	173
7900	Main rating basic section, alum., NEMA 1, 800 amp	2 Elec	1.40	11.429		330	730		1,060	1,425
7920	1,200 amp	"	1.20	13.333		645	850		1,495	1,950
7940	For copper bus, add					485			485	535
7960	For 65,000 amp bus bracing, add					325			325	360
7980	For NEMA 3R enclosure, add					5,650			5,650	6,225
8000	For NEMA 12, enclosure, add					156			156	171
8020	For 1/4" x 1" ground bus, add	1 Elec	16	.500		100	32		132	158
8040	For 1/4" x 2" ground bus, add		12	.667		100	42.50		142.50	173
8060	Unit devices, pilot light, standard		16	.500		86.50	32		118.50	143
8080	Pilot light, push to test		16	.500		121	32		153	181
8100	Pilot light, standard, and push button		12	.667		208	42.50		250.50	291
8120	Pilot light, push to test, and push button		12	.667		242	42.50		284.50	330
8140	Pilot light, standard, and select switch		12	.667		208	42.50		250.50	291
8160	Pilot light, push to test, and select switch		12	.667		242	42.50		284.50	330

26 24 19.30 Motor Control Center

		Crew	Daily Output	Labor-Hours	Unit	Material	2021 Bare Costs Labor	2021 Bare Costs Equipment	Total	Total Incl O&P
0010	**MOTOR CONTROL CENTER** Consists of starters & structures R262419-60									
0050	Starters, class 1, type B, comb. MCP, FVNR, with									
0100	control transformer, 10 HP, size 1, 12" high	1 Elec	2.70	2.963	Ea.	1,725	189		1,914	2,175
0200	25 HP, size 2, 18" high	2 Elec	4	4		1,975	255		2,230	2,550
0300	50 HP, size 3, 24" high		2	8		3,050	510		3,560	4,100
0350	75 HP, size 4, 24" high		1.60	10		4,050	635		4,685	5,400
0400	100 HP, size 4, 30" high		1.40	11.429		5,350	730		6,080	6,950
0500	200 HP, size 5, 48" high		1	16		7,975	1,025		9,000	10,300
0600	400 HP, size 6, 72" high		.80	20		16,400	1,275		17,675	19,900
0800	Structures, 600 amp, 22,000 rms, takes any									
0900	combination of starters up to 72" high	2 Elec	1.60	10	Ea.	1,875	635		2,510	3,025
1000	Back to back, 72" front & 66" back	"	1.20	13.333		2,575	850		3,425	4,100
1100	For copper bus, add per structure					266			266	293
1200	For NEMA 12, add per structure					154			154	169
1300	For 42,000 rms, add per structure					214			214	236
1400	For 100,000 rms, size 1 & 2, add					685			685	750

26 24 Switchboards and Panelboards

26 24 19 – Motor-Control Centers

26 24 19.30 Motor Control Center		Crew	Daily Output	Labor-Hours	Unit	Material	2021 Bare Costs Labor	Equipment	Total	Total Incl O&P
1500	Size 3, add				Ea.	1,100			1,100	1,200
1600	Size 4, add					890			890	975
1700	For pilot lights, add per starter	1 Elec	16	.500		120	32		152	180
1800	For push button, add per starter		16	.500		120	32		152	180
1900	For auxiliary contacts, add per starter	↓	16	.500	↓	248	32		280	320

26 24 19.40 Motor Starters and Controls

26 24 19.40 Motor Starters and Controls		Crew	Daily Output	Labor-Hours	Unit	Material	2021 Bare Costs Labor	Equipment	Total	Total Incl O&P
0010	**MOTOR STARTERS AND CONTROLS** R262419-65									
0050	Magnetic, FVNR, with enclosure and heaters, 480 volt									
0080	2 HP, size 00	1 Elec	3.50	2.286	Ea.	120	146		266	345
0100	5 HP, size 0		2.30	3.478		385	222		607	750
0200	10 HP, size 1	↓	1.60	5		169	320		489	660
0300	25 HP, size 2 R263413-33	2 Elec	2.20	7.273		335	465		800	1,050
0400	50 HP, size 3		1.80	8.889		555	565		1,120	1,450
0500	100 HP, size 4		1.20	13.333		1,250	850		2,100	2,625
0600	200 HP, size 5		.90	17.778		2,975	1,125		4,100	4,925
0610	400 HP, size 6	↓	.80	20		8,350	1,275		9,625	11,100
0620	NEMA 7, 5 HP, size 0	1 Elec	1.60	5		2,550	320		2,870	3,275
0630	10 HP, size 1	"	1.10	7.273		2,650	465		3,115	3,625
0640	25 HP, size 2	2 Elec	1.80	8.889		4,300	565		4,865	5,575
0650	50 HP, size 3		1.20	13.333		6,450	850		7,300	8,350
0660	100 HP, size 4		.90	17.778		8,325	1,125		9,450	10,800
0670	200 HP, size 5	↓	.50	32		13,500	2,050		15,550	17,900
0700	Combination, with motor circuit protectors, 5 HP, size 0	1 Elec	1.80	4.444		1,450	283		1,733	2,000
0800	10 HP, size 1	"	1.30	6.154		1,500	390		1,890	2,225
0900	25 HP, size 2	2 Elec	2	8		2,075	510		2,585	3,050
1000	50 HP, size 3		1.32	12.121		3,025	770		3,795	4,475
1200	100 HP, size 4	↓	.80	20		6,650	1,275		7,925	9,225
1220	NEMA 7, 5 HP, size 0	1 Elec	1.30	6.154		3,400	390		3,790	4,325
1230	10 HP, size 1	"	1	8		3,475	510		3,985	4,575
1240	25 HP, size 2	2 Elec	1.32	12.121		4,650	770		5,420	6,275
1250	50 HP, size 3		.80	20		7,700	1,275		8,975	10,400
1260	100 HP, size 4		.60	26.667		12,000	1,700		13,700	15,700
1270	200 HP, size 5	↓	.40	40		26,000	2,550		28,550	32,400
1400	Combination, with fused switch, 5 HP, size 0	1 Elec	1.80	4.444		1,250	283		1,533	1,800
1600	10 HP, size 1	"	1.30	6.154		1,325	390		1,715	2,025
1800	25 HP, size 2	2 Elec	2	8		2,050	510		2,560	3,025
2000	50 HP, size 3		1.32	12.121		3,675	770		4,445	5,175
2200	100 HP, size 4	↓	.80	20		7,300	1,275		8,575	9,925
2610	NEMA 4, with start-stop push button, size 1	1 Elec	1.30	6.154		1,550	390		1,940	2,275
2620	Size 2	2 Elec	2	8		2,075	510		2,585	3,025
2630	Size 3		1.32	12.121		3,300	770		4,070	4,775
2640	Size 4	↓	.80	20	↓	5,050	1,275		6,325	7,475
2650	NEMA 4, FVNR, including control transformer									
2660	Size 1	2 Elec	2.60	6.154	Ea.	1,650	390		2,040	2,400
2670	Size 2		2	8		2,350	510		2,860	3,350
2680	Size 3		1.32	12.121		3,750	770		4,520	5,275
2690	Size 4	↓	.80	20		5,800	1,275		7,075	8,275
2710	Magnetic, FVR, control circuit transformer, NEMA 1, size 1	1 Elec	1.30	6.154		975	390		1,365	1,650
2720	Size 2	2 Elec	2	8		1,550	510		2,060	2,475
2730	Size 3		1.32	12.121		2,375	770		3,145	3,750
2740	Size 4	↓	.80	20		5,150	1,275		6,425	7,575
2760	NEMA 4, size 1	1 Elec	1.10	7.273		1,400	465		1,865	2,250

26 24 19.40 Motor Starters and Controls		Crew	Daily Output	Labor-Hours	Unit	Material	2021 Bare Costs Labor	Equipment	Total	Total Incl O&P
2770	Size 2	2 Elec	1.60	10	Ea.	2,250	635		2,885	3,425
2780	Size 3		1.20	13.333		3,400	850		4,250	4,975
2790	Size 4		.70	22.857		6,975	1,450		8,425	9,850
2820	NEMA 12, size 1	1 Elec	1.10	7.273		1,150	465		1,615	1,975
2830	Size 2	2 Elec	1.60	10		1,825	635		2,460	2,950
2840	Size 3		1.20	13.333		2,850	850		3,700	4,400
2850	Size 4		.70	22.857		5,900	1,450		7,350	8,650
2870	Combination FVR, fused, w/control XFMR & PB, NEMA 1, size 1	1 Elec	1	8		1,675	510		2,185	2,600
2880	Size 2	2 Elec	1.50	10.667		2,400	680		3,080	3,625
2890	Size 3		1.10	14.545		3,550	925		4,475	5,300
2900	Size 4		.70	22.857		7,225	1,450		8,675	10,100
2910	NEMA 4, size 1	1 Elec	.90	8.889		2,500	565		3,065	3,600
2920	Size 2	2 Elec	1.40	11.429		3,625	730		4,355	5,075
2930	Size 3		1	16		5,725	1,025		6,750	7,825
2940	Size 4		.60	26.667		9,525	1,700		11,225	13,000
2950	NEMA 12, size 1	1 Elec	1	8		1,900	510		2,410	2,850
2960	Size 2	2 Elec	1.40	11.429		2,675	730		3,405	4,025
2970	Size 3		1	16		3,950	1,025		4,975	5,875
2980	Size 4		.60	26.667		7,825	1,700		9,525	11,200
3010	Manual, single phase, w/pilot, 1 pole, 120 V, NEMA 1	1 Elec	6.40	1.250		102	79.50		181.50	230
3020	NEMA 4		4	2		395	127		522	620
3030	2 pole, 120/240 V, NEMA 1		6.40	1.250		138	79.50		217.50	270
3040	NEMA 4		4	2		375	127		502	605
3041	3 phase, 3 pole, 600 V, NEMA 1		5.50	1.455		96.50	92.50		189	244
3042	NEMA 4		3.50	2.286		320	146		466	565
3043	NEMA 12		3.50	2.286		300	146		446	545
3070	Auxiliary contact, normally open					100			100	110
3500	Magnetic FVNR with NEMA 12, enclosure & heaters, 480 volt									
3600	5 HP, size 0	1 Elec	2.20	3.636	Ea.	256	232		488	625
3700	10 HP, size 1	"	1.50	5.333		385	340		725	930
3800	25 HP, size 2	2 Elec	2	8		720	510		1,230	1,550
3900	50 HP, size 3		1.60	10		1,100	635		1,735	2,175
4000	100 HP, size 4		1	16		2,650	1,025		3,675	4,450
4100	200 HP, size 5		.80	20		6,375	1,275		7,650	8,925
4200	Combination, with motor circuit protectors, 5 HP, size 0	1 Elec	1.70	4.706		850	300		1,150	1,375
4300	10 HP, size 1	"	1.20	6.667		885	425		1,310	1,600
4400	25 HP, size 2	2 Elec	1.80	8.889		1,325	565		1,890	2,300
4500	50 HP, size 3		1.20	13.333		2,150	850		3,000	3,625
4600	100 HP, size 4		.74	21.622		4,850	1,375		6,225	7,400
4700	Combination, with fused switch, 5 HP, size 0	1 Elec	1.70	4.706		700	300		1,000	1,225
4800	10 HP, size 1	"	1.20	6.667		725	425		1,150	1,425
4900	25 HP, size 2	2 Elec	1.80	8.889		1,100	565		1,665	2,075
5000	50 HP, size 3		1.20	13.333		1,775	850		2,625	3,200
5100	100 HP, size 4		.74	21.622		3,600	1,375		4,975	6,000
5200	Factory installed controls, adders to size 0 thru 5									
5300	Start-stop push button	1 Elec	32	.250	Ea.	54	15.95		69.95	82.50
5400	Hand-off-auto-selector switch		32	.250		54	15.95		69.95	82.50
5500	Pilot light		32	.250		101	15.95		116.95	135
5600	Start-stop-pilot		32	.250		155	15.95		170.95	194
5700	Auxiliary contact, NO or NC		32	.250		74	15.95		89.95	105
5800	NO-NC		32	.250		148	15.95		163.95	187
5810	Magnetic FVR, NEMA 7, w/heaters, size 1		.66	12.121		3,975	770		4,745	5,525
5830	Size 2	2 Elec	1.10	14.545		5,450	925		6,375	7,375

For customer support on your Electrical Costs with RSMeans Data, call 800.448.8182.

26 24 19 – Motor-Control Centers

26 24 19.40 Motor Starters and Controls		Crew	Daily Output	Labor-Hours	Unit	Material	2021 Bare Costs Labor	Equipment	Total	Total Incl O&P
5840	Size 3	2 Elec	.70	22.857	Ea.	8,750	1,450		10,200	11,800
5850	Size 4	▼	.60	26.667		11,500	1,700		13,200	15,100
5860	Combination w/circuit breakers, heaters, control XFMR PB, size 1	1 Elec	.60	13.333		1,600	850		2,450	3,025
5870	Size 2	2 Elec	.80	20	▼	2,075	1,275		3,350	4,175
5880	Size 3	2 Elec	.50	32	Ea.	2,750	2,050		4,800	6,050
5890	Size 4	"	.40	40		5,475	2,550		8,025	9,775
5900	Manual, 240 volt, 0.75 HP motor	1 Elec	4	2		57	127		184	252
5910	2 HP motor		4	2		155	127		282	360
6000	Magnetic, 240 volt, 1 or 2 pole, 0.75 HP motor		4	2		194	127		321	400
6020	2 HP motor		4	2		214	127		341	425
6040	5 HP motor		3	2.667		305	170		475	585
6060	10 HP motor		2.30	3.478		760	222		982	1,175
6100	3 pole, 0.75 HP motor		3	2.667		193	170		363	465
6120	5 HP motor		2.30	3.478		262	222		484	620
6140	10 HP motor		1.60	5		490	320		810	1,025
6160	15 HP motor		1.60	5		490	320		810	1,025
6180	20 HP motor	▼	1.10	7.273		800	465		1,265	1,575
6200	25 HP motor	2 Elec	2.20	7.273		770	465		1,235	1,525
6210	30 HP motor		1.80	8.889		770	565		1,335	1,675
6220	40 HP motor		1.80	8.889		1,700	565		2,265	2,725
6230	50 HP motor		1.80	8.889		1,700	565		2,265	2,725
6240	60 HP motor		1.20	13.333		4,000	850		4,850	5,650
6250	75 HP motor		1.20	13.333		4,000	850		4,850	5,650
6260	100 HP motor		1.20	13.333		4,000	850		4,850	5,650
6270	125 HP motor		.90	17.778		11,200	1,125		12,325	14,000
6280	150 HP motor		.90	17.778		11,200	1,125		12,325	14,000
6290	200 HP motor	▼	.90	17.778		11,200	1,125		12,325	14,000
6400	Starter & nonfused disconnect, 240 volt, 1-2 pole, 0.75 HP motor	1 Elec	2	4		247	255		502	650
6410	2 HP motor		2	4		267	255		522	675
6420	5 HP motor		1.80	4.444		360	283		643	815
6430	10 HP motor		1.40	5.714		830	365		1,195	1,450
6440	3 pole, 0.75 HP motor		1.60	5		246	320		566	745
6450	5 HP motor		1.40	5.714		315	365		680	885
6460	10 HP motor		1.10	7.273		560	465		1,025	1,300
6470	15 HP motor	▼	1	8		560	510		1,070	1,375
6480	20 HP motor	2 Elec	1.50	10.667		965	680		1,645	2,050
6490	25 HP motor		1.50	10.667		930	680		1,610	2,025
6500	30 HP motor		1.30	12.308		930	785		1,715	2,200
6510	40 HP motor		1.24	12.903		2,000	820		2,820	3,425
6520	50 HP motor		1.12	14.286		2,000	910		2,910	3,550
6530	60 HP motor		.90	17.778		4,300	1,125		5,425	6,400
6540	75 HP motor		.76	21.053		4,300	1,350		5,650	6,725
6550	100 HP motor		.70	22.857		4,300	1,450		5,750	6,900
6560	125 HP motor		.60	26.667		12,000	1,700		13,700	15,700
6570	150 HP motor		.52	30.769		12,000	1,950		13,950	16,100
6580	200 HP motor	▼	.50	32		12,000	2,050		14,050	16,200
6600	Starter & fused disconnect, 240 volt, 1-2 pole, 0.75 HP motor	1 Elec	2	4		259	255		514	665
6610	2 HP motor		2	4		279	255		534	685
6620	5 HP motor		1.80	4.444		370	283		653	830
6630	10 HP motor		1.40	5.714		870	365		1,235	1,500
6640	3 pole, 0.75 HP motor		1.60	5		258	320		578	760
6650	5 HP motor		1.40	5.714		325	365		690	900
6660	10 HP motor	▼	1.10	7.273	▼	605	465		1,070	1,350

For customer support on your Electrical Costs with RSMeans Data, call 800.448.8182.

243

26 24 19.40 Motor Starters and Controls		Crew	Daily Output	Labor-Hours	Unit	Material	2021 Bare Costs Labor	Equipment	Total	Total Incl O&P
6690	15 HP motor	1 Elec	1	8	Ea.	605	510		1,115	1,425
6700	20 HP motor	2 Elec	1.60	10		995	635		1,630	2,050
6710	25 HP motor		1.60	10		960	635		1,595	2,000
6720	30 HP motor		1.40	11.429		960	730		1,690	2,125
6730	40 HP motor		1.20	13.333		2,125	850		2,975	3,575
6740	50 HP motor		1.20	13.333		2,125	850		2,975	3,575
6750	60 HP motor		.90	17.778		4,400	1,125		5,525	6,525
6760	75 HP motor		.90	17.778		4,400	1,125		5,525	6,525
6770	100 HP motor		.70	22.857		4,400	1,450		5,850	7,025
6780	125 HP motor		.54	29.630		12,200	1,875		14,075	16,300
6790	Combination starter & nonfusible disconnect									
6800	240 volt, 1-2 pole, 0.75 HP motor	1 Elec	2	4	Ea.	605	255		860	1,050
6810	2 HP motor		2	4		645	255		900	1,100
6820	5 HP motor		1.50	5.333		645	340		985	1,225
6830	10 HP motor		1.20	6.667		1,000	425		1,425	1,725
6840	3 pole, 0.75 HP motor		1.80	4.444		615	283		898	1,100
6850	5 HP motor		1.30	6.154		645	390		1,035	1,300
6860	10 HP motor		1	8		1,000	510		1,510	1,850
6870	15 HP motor		1	8		1,000	510		1,510	1,850
6880	20 HP motor	2 Elec	1.32	12.121		1,650	770		2,420	2,975
6890	25 HP motor		1.32	12.121		1,650	770		2,420	2,975
6900	30 HP motor		1.32	12.121		1,650	770		2,420	2,975
6910	40 HP motor		.80	20		3,150	1,275		4,425	5,375
6920	50 HP motor		.80	20		3,150	1,275		4,425	5,375
6930	60 HP motor		.70	22.857		7,050	1,450		8,500	9,925
6940	75 HP motor		.70	22.857		7,050	1,450		8,500	9,925
6950	100 HP motor		.70	22.857		7,050	1,450		8,500	9,925
6960	125 HP motor		.60	26.667		18,600	1,700		20,300	22,900
6970	150 HP motor		.60	26.667		18,600	1,700		20,300	22,900
6980	200 HP motor		.60	26.667		18,600	1,700		20,300	22,900
6990	Combination starter and fused disconnect									
7000	240 volt, 1-2 pole, 0.75 HP motor	1 Elec	2	4	Ea.	615	255		870	1,050
7010	2 HP motor		2	4		615	255		870	1,050
7020	5 HP motor		1.50	5.333		645	340		985	1,225
7030	10 HP motor		1.20	6.667		1,000	425		1,425	1,725
7040	3 pole, 0.75 HP motor		1.80	4.444		615	283		898	1,100
7050	5 HP motor		1.30	6.154		645	390		1,035	1,300
7060	10 HP motor		1	8		1,000	510		1,510	1,850
7070	15 HP motor		1	8		1,000	510		1,510	1,850
7080	20 HP motor	2 Elec	1.32	12.121		1,675	770		2,445	2,975
7090	25 HP motor		1.32	12.121		1,675	770		2,445	2,975
7100	30 HP motor		1.32	12.121		1,675	770		2,445	2,975
7110	40 HP motor		.80	20		3,175	1,275		4,450	5,400
7120	50 HP motor		.80	20		3,175	1,275		4,450	5,400
7130	60 HP motor		.80	20		7,075	1,275		8,350	9,700
7140	75 HP motor		.70	22.857		7,075	1,450		8,525	9,975
7150	100 HP motor		.70	22.857		7,075	1,450		8,525	9,975
7160	125 HP motor		.70	22.857		18,700	1,450		20,150	22,800
7170	150 HP motor		.60	26.667		18,700	1,700		20,400	23,100
7180	200 HP motor		.60	26.667		18,700	1,700		20,400	23,100
7190	Combination starter & circuit breaker disconnect									
7200	240 volt, 1-2 pole, 0.75 HP motor	1 Elec	2	4	Ea.	640	255		895	1,075
7210	2 HP motor		2	4		640	255		895	1,075

For customer support on your Electrical Costs with RSMeans Data, call 800.448.8182.

26 24 19.40 Motor Starters and Controls		Crew	Daily Output	Labor-Hours	Unit	Material	2021 Bare Costs Labor	Equipment	Total	Total Incl O&P
7220	5 HP motor	1 Elec	1.50	5.333	Ea.	670	340		1,010	1,250
7230	10 HP motor		1.20	6.667		1,025	425		1,450	1,750
7240	3 pole, 0.75 HP motor		1.80	4.444		655	283		938	1,150
7250	5 HP motor		1.30	6.154		690	390		1,080	1,350
7260	10 HP motor		1	8		1,050	510		1,560	1,900
7270	15 HP motor		1	8		1,050	510		1,560	1,900
7280	20 HP motor	2 Elec	1.32	12.121		1,775	770		2,545	3,100
7290	25 HP motor		1.32	12.121		1,775	770		2,545	3,100
7300	30 HP motor		1.32	12.121		1,775	770		2,545	3,100
7310	40 HP motor		.80	20		3,850	1,275		5,125	6,150
7320	50 HP motor		.80	20		3,850	1,275		5,125	6,150
7330	60 HP motor		.80	20		8,900	1,275		10,175	11,700
7340	75 HP motor		.70	22.857		8,900	1,450		10,350	12,000
7350	100 HP motor		.70	22.857		8,900	1,450		10,350	12,000
7360	125 HP motor		.70	22.857		19,300	1,450		20,750	23,400
7370	150 HP motor		.60	26.667		19,300	1,700		21,000	23,700
7380	200 HP motor		.60	26.667		19,300	1,700		21,000	23,700
7400	Magnetic FVNR with enclosure & heaters, 2 pole,									
7410	230 volt, 1 HP, size 00	1 Elec	4	2	Ea.	175	127		302	380
7420	2 HP, size 0		4	2		196	127		323	405
7430	3 HP, size 1		3	2.667		224	170		394	500
7440	5 HP, size 1P		3	2.667		288	170		458	565
7450	115 volt, 1/3 HP, size 00		4	2		175	127		302	380
7460	1 HP, size 0		4	2		196	127		323	405
7470	2 HP, size 1		3	2.667		224	170		394	500
7480	3 HP, size 1P		3	2.667		279	170		449	555
7500	3 pole, 480 volt, 600 HP, size 7	2 Elec	.70	22.857		14,900	1,450		16,350	18,600
7590	Magnetic FVNR with heater, NEMA 1									
7600	600 volt, 3 pole, 5 HP motor	1 Elec	2.30	3.478	Ea.	233	222		455	585
7610	10 HP motor	"	1.60	5		262	320		582	765
7620	25 HP motor	2 Elec	2.20	7.273		490	465		955	1,225
7630	30 HP motor		1.80	8.889		800	565		1,365	1,725
7640	40 HP motor		1.80	8.889		800	565		1,365	1,725
7650	50 HP motor		1.80	8.889		800	565		1,365	1,725
7660	60 HP motor		1.20	13.333		1,775	850		2,625	3,200
7670	75 HP motor		1.20	13.333		2,375	850		3,225	3,850
7680	100 HP motor		1.20	13.333		2,625	850		3,475	4,150
7690	125 HP motor		.90	17.778		4,075	1,125		5,200	6,150
7700	150 HP motor		.90	17.778		4,250	1,125		5,375	6,350
7710	200 HP motor		.90	17.778		4,675	1,125		5,800	6,825
7750	Starter & nonfused disconnect, 600 volt, 3 pole, 5 HP motor	1 Elec	1.40	5.714		325	365		690	900
7760	10 HP motor	"	1.10	7.273		355	465		820	1,075
7770	25 HP motor	2 Elec	1.50	10.667		585	680		1,265	1,650
7780	30 HP motor		1.30	12.308		965	785		1,750	2,225
7790	40 HP motor		1.30	12.308		965	785		1,750	2,225
7800	50 HP motor		1.30	12.308		965	785		1,750	2,225
7810	60 HP motor		.92	17.391		2,050	1,100		3,150	3,900
7820	75 HP motor		.92	17.391		2,625	1,100		3,725	4,550
7830	100 HP motor		.84	19.048		2,900	1,225		4,125	4,975
7840	125 HP motor		.70	22.857		4,475	1,450		5,925	7,100
7850	150 HP motor		.70	22.857		4,650	1,450		6,100	7,300
7860	200 HP motor		.60	26.667		5,075	1,700		6,775	8,125
7870	Starter & fused disconnect, 600 volt, 3 pole, 5 HP motor	1 Elec	1.40	5.714		410	365		775	990

245

26 24 Switchboards and Panelboards

26 24 19 – Motor-Control Centers

26 24 19.40 Motor Starters and Controls

		Crew	Daily Output	Labor-Hours	Unit	Material	2021 Bare Costs Labor	Equipment	Total	Total Incl O&P
7880	10 HP motor	1 Elec	1.10	7.273	Ea.	435	465		900	1,175
7890	25 HP motor	2 Elec	1.50	10.667		665	680		1,345	1,725
7900	30 HP motor		1.30	12.308		1,025	785		1,810	2,300
7910	40 HP motor		1.30	12.308		1,025	785		1,810	2,300
7920	50 HP motor		1.30	12.308		1,025	785		1,810	2,300
7930	60 HP motor		.92	17.391		2,175	1,100		3,275	4,050
7940	75 HP motor		.92	17.391		2,775	1,100		3,875	4,700
7950	100 HP motor		.84	19.048		3,025	1,225		4,250	5,125
7960	125 HP motor		.70	22.857		4,650	1,450		6,100	7,300
7970	150 HP motor		.70	22.857		4,825	1,450		6,275	7,475
7980	200 HP motor	↓	.60	26.667	↓	5,250	1,700		6,950	8,300
7990	Combination starter and nonfusible disconnect									
8000	600 volt, 3 pole, 5 HP motor	1 Elec	1.80	4.444	Ea.	610	283		893	1,100
8010	10 HP motor	"	1.30	6.154		820	390		1,210	1,475
8020	25 HP motor	2 Elec	2	8		985	510		1,495	1,825
8030	30 HP motor		1.32	12.121		1,725	770		2,495	3,025
8040	40 HP motor		1.32	12.121		1,725	770		2,495	3,025
8050	50 HP motor		1.32	12.121		1,975	770		2,745	3,325
8060	60 HP motor		.80	20		3,275	1,275		4,550	5,500
8070	75 HP motor		.80	20		3,275	1,275		4,550	5,500
8080	100 HP motor		.80	20		3,900	1,275		5,175	6,175
8090	125 HP motor		.70	22.857		7,450	1,450		8,900	10,400
8100	150 HP motor		.70	22.857		8,550	1,450		10,000	11,600
8110	200 HP motor	↓	.70	22.857	↓	7,525	1,450		8,975	10,500
8140	Combination starter and fused disconnect									
8150	600 volt, 3 pole, 5 HP motor	1 Elec	1.80	4.444	Ea.	615	283		898	1,100
8160	10 HP motor	"	1.30	6.154		670	390		1,060	1,325
8170	25 HP motor	2 Elec	2	8		1,100	510		1,610	1,950
8180	30 HP motor		1.32	12.121		1,700	770		2,470	3,025
8190	40 HP motor		1.32	12.121		1,825	770		2,595	3,175
8200	50 HP motor		1.32	12.121		1,825	770		2,595	3,175
8210	60 HP motor		.80	20		3,200	1,275		4,475	5,425
8220	75 HP motor		.80	20		3,200	1,275		4,475	5,425
8230	100 HP motor		.80	20		3,200	1,275		4,475	5,425
8240	125 HP motor		.70	22.857		7,075	1,450		8,525	9,975
8250	150 HP motor		.70	22.857		7,075	1,450		8,525	9,975
8260	200 HP motor	↓	.70	22.857	↓	7,075	1,450		8,525	9,975
8290	Combination starter & circuit breaker disconnect									
8300	600 volt, 3 pole, 5 HP motor	1 Elec	1.80	4.444	Ea.	915	283		1,198	1,425
8310	10 HP motor	"	1.30	6.154		895	390		1,285	1,575
8320	25 HP motor	2 Elec	2	8		1,325	510		1,835	2,225
8330	30 HP motor		1.32	12.121		1,775	770		2,545	3,100
8340	40 HP motor		1.32	12.121		1,775	770		2,545	3,100
8350	50 HP motor		1.32	12.121		1,750	770		2,520	3,075
8360	60 HP motor		.80	20		3,850	1,275		5,125	6,150
8370	75 HP motor		.80	20		3,850	1,275		5,125	6,150
8380	100 HP motor		.80	20		3,800	1,275		5,075	6,100
8390	125 HP motor		.70	22.857		8,900	1,450		10,350	12,000
8400	150 HP motor		.70	22.857		8,725	1,450		10,175	11,800
8410	200 HP motor	↓	.70	22.857	↓	8,800	1,450		10,250	11,900
8430	Starter & circuit breaker disconnect									
8440	600 volt, 3 pole, 5 HP motor	1 Elec	1.40	5.714	Ea.	785	365		1,150	1,400
8450	10 HP motor	"	1.10	7.273	↓	815	465		1,280	1,575

246

26 24 19 – Motor-Control Centers

26 24 19.40 Motor Starters and Controls

		Crew	Daily Output	Labor-Hours	Unit	Material	2021 Bare Costs Labor	2021 Bare Costs Equipment	Total	Total Incl O&P
8460	25 HP motor	2 Elec	1.50	10.667	Ea.	1,050	680		1,730	2,150
8470	30 HP motor		1.30	12.308		1,500	785		2,285	2,825
8480	40 HP motor		1.30	12.308		1,500	785		2,285	2,825
8490	50 HP motor		1.30	12.308		1,500	785		2,285	2,825
8500	60 HP motor		.92	17.391		3,625	1,100		4,725	5,650
8510	75 HP motor		.92	17.391		4,225	1,100		5,325	6,300
8520	100 HP motor		.84	19.048		4,475	1,225		5,700	6,725
8530	125 HP motor		.70	22.857		5,925	1,450		7,375	8,700
8540	150 HP motor		.70	22.857		6,100	1,450		7,550	8,900
8550	200 HP motor	▼	.60	26.667		7,850	1,700		9,550	11,200
8900	240 volt, 1-2 pole, 0.75 HP motor	1 Elec	2	4		745	255		1,000	1,200
8910	2 HP motor		2	4		765	255		1,020	1,225
8920	5 HP motor		1.80	4.444		860	283		1,143	1,375
8930	10 HP motor		1.40	5.714		1,450	365		1,815	2,150
8950	3 pole, 0.75 HP motor		1.60	5		745	320		1,065	1,300
8970	5 HP motor		1.40	5.714		815	365		1,180	1,425
8980	10 HP motor		1.10	7.273		1,200	465		1,665	2,000
8990	15 HP motor	▼	1	8		1,200	510		1,710	2,050
9100	20 HP motor	2 Elec	1.50	10.667		1,600	680		2,280	2,750
9110	25 HP motor		1.50	10.667		1,575	680		2,255	2,725
9120	30 HP motor		1.30	12.308		1,575	785		2,360	2,900
9130	40 HP motor		1.24	12.903		3,550	820		4,370	5,125
9140	50 HP motor		1.12	14.286		3,550	910		4,460	5,250
9150	60 HP motor		.90	17.778		7,150	1,125		8,275	9,525
9160	75 HP motor		.76	21.053		7,150	1,350		8,500	9,850
9170	100 HP motor		.70	22.857		7,150	1,450		8,600	10,000
9180	125 HP motor		.60	26.667		15,800	1,700		17,500	19,800
9190	150 HP motor		.52	30.769		15,800	1,950		17,750	20,200
9200	200 HP motor	▼	.50	32		15,800	2,050		17,850	20,300

26 25 Low-Voltage Enclosed Bus Assemblies

26 25 13 – Low-Voltage Busways

26 25 13.10 Aluminum Bus Duct

		Crew	Daily Output	Labor-Hours	Unit	Material	2021 Bare Costs Labor	2021 Bare Costs Equipment	Total	Total Incl O&P
0010	**ALUMINUM BUS DUCT** 10 ft. long R262513-10									
0050	Indoor 3 pole 4 wire, plug-in, straight section, 225 amp	2 Elec	44	.364	L.F.	171	23		194	223
0100	400 amp		36	.444		215	28.50		243.50	278
0150	600 amp		32	.500		297	32		329	375
0200	800 amp		26	.615		345	39		384	440
0250	1,000 amp		24	.667		430	42.50		472.50	540
0270	1,200 amp		23	.696		550	44.50		594.50	670
0300	1,350 amp		22	.727		310	46.50		356.50	410
0310	1,600 amp		18	.889		1,000	56.50		1,056.50	1,175
0320	2,000 amp		16	1		1,225	63.50		1,288.50	1,450
0330	2,500 amp		14	1.143		1,500	73		1,573	1,750
0340	3,000 amp		12	1.333		1,750	85		1,835	2,050
0350	Feeder, 600 amp		34	.471		116	30		146	172
0400	800 amp		28	.571		430	36.50		466.50	530
0450	1,000 amp		26	.615		550	39		589	665
0455	1,200 amp		25	.640		650	41		691	775
0500	1,350 amp		24	.667		241	42.50		283.50	330
0550	1,600 amp	▼	20	.800		900	51		951	1,075

For customer support on your Electrical Costs with RSMeans Data, call 800.448.8182.

247

26 25 13.10 Aluminum Bus Duct		Crew	Daily Output	Labor-Hours	Unit	Material	2021 Bare Costs Labor	Equipment	Total	Total Incl O&P
0600	2,000 amp	2 Elec	18	.889	L.F.	1,100	56.50		1,156.50	1,275
0620	2,500 amp		14	1.143		1,350	73		1,423	1,575
0630	3,000 amp		12	1.333		1,575	85		1,660	1,850
0640	4,000 amp		10	1.600	▼	695	102		797	915
0650	Elbow, 225 amp		4.40	3.636	Ea.	1,900	232		2,132	2,425
0700	400 amp		3.80	4.211		2,025	268		2,293	2,625
0750	600 amp		3.40	4.706		2,275	300		2,575	2,950
0800	800 amp		3	5.333		2,425	340		2,765	3,150
0850	1,000 amp		2.80	5.714		3,200	365		3,565	4,075
0870	1,200 amp		2.70	5.926		3,550	375		3,925	4,450
0900	1,350 amp		2.60	6.154		1,375	390		1,765	2,075
0950	1,600 amp		2.40	6.667		4,650	425		5,075	5,725
1000	2,000 amp		2	8		4,475	510		4,985	5,675
1020	2,500 amp		1.80	8.889		6,725	565		7,290	8,250
1030	3,000 amp		1.60	10		7,700	635		8,335	9,425
1040	4,000 amp		1.40	11.429		3,525	730		4,255	4,950
1100	Cable tap box end, 225 amp		3.60	4.444		1,825	283		2,108	2,425
1150	400 amp		3.20	5		2,275	320		2,595	2,975
1200	600 amp		2.60	6.154		3,325	390		3,715	4,250
1250	800 amp		2.20	7.273		3,375	465		3,840	4,400
1300	1,000 amp		2	8		3,400	510		3,910	4,500
1320	1,200 amp		2	8		6,050	510		6,560	7,400
1350	1,350 amp		1.60	10		1,675	635		2,310	2,800
1400	1,600 amp		1.40	11.429		6,675	730		7,405	8,425
1450	2,000 amp		1.20	13.333		7,400	850		8,250	9,400
1460	2,500 amp		1	16		9,000	1,025		10,025	11,400
1470	3,000 amp		.80	20		9,925	1,275		11,200	12,800
1480	4,000 amp		.60	26.667		11,600	1,700		13,300	15,300
1500	Switchboard stub, 225 amp		5.80	2.759		1,650	176		1,826	2,075
1550	400 amp		5.40	2.963		1,700	189		1,889	2,125
1600	600 amp		4.60	3.478		1,875	222		2,097	2,400
1650	800 amp		4	4		1,925	255		2,180	2,500
1700	1,000 amp		3.20	5		1,700	320		2,020	2,325
1720	1,200 amp		3.10	5.161		1,725	330		2,055	2,400
1750	1,350 amp		3	5.333		1,700	340		2,040	2,375
1800	1,600 amp		2.60	6.154		2,000	390		2,390	2,775
1850	2,000 amp		2.40	6.667		1,975	425		2,400	2,800
1860	2,500 amp		2.20	7.273		2,250	465		2,715	3,175
1870	3,000 amp		2	8		2,950	510		3,460	3,975
1880	4,000 amp		1.80	8.889		1,900	565		2,465	2,950
1890	Tee fittings, 225 amp		3.20	5		3,725	320		4,045	4,575
1900	400 amp		2.80	5.714		3,725	365		4,090	4,650
1950	600 amp		2.60	6.154		3,725	390		4,115	4,675
2000	800 amp		2.40	6.667		970	425		1,395	1,700
2050	1,000 amp		2.20	7.273		1,025	465		1,490	1,825
2070	1,200 amp		2.10	7.619		1,200	485		1,685	2,050
2100	1,350 amp		2	8		1,675	510		2,185	2,575
2150	1,600 amp		1.60	10		2,000	635		2,635	3,150
2200	2,000 amp		1.20	13.333		2,225	850		3,075	3,700
2220	2,500 amp		1	16		2,650	1,025		3,675	4,450
2230	3,000 amp		.80	20		3,025	1,275		4,300	5,225
2240	4,000 amp		.60	26.667		5,025	1,700		6,725	8,075
2300	Wall flange, 600 amp		20	.800		420	51		471	535

248

26 25 13 – Low-Voltage Busways

26 25 13.10 Aluminum Bus Duct		Crew	Daily Output	Labor-Hours	Unit	Material	2021 Bare Costs Labor	Equipment	Total	Total Incl O&P
2310	800 amp	2 Elec	16	1	Ea.	420	63.50		483.50	555
2320	1,000 amp		13	1.231		420	78.50		498.50	575
2325	1,200 amp		12	1.333		420	85		505	585
2330	1,350 amp		10.80	1.481		420	94.50		514.50	600
2340	1,600 amp		9	1.778		420	113		533	630
2350	2,000 amp		8	2		420	127		547	650
2360	2,500 amp		6.60	2.424		420	154		574	690
2370	3,000 amp		5.40	2.963		605	189		794	945
2380	4,000 amp		4	4		605	255		860	1,050
2390	5,000 amp		3	5.333		360	340		700	900
2400	Vapor barrier		8	2		475	127		602	710
2420	Roof flange kit		4	4		980	255		1,235	1,450
2600	Expansion fitting, 225 amp		10	1.600		1,525	102		1,627	1,825
2610	400 amp		8	2		1,525	127		1,652	1,875
2620	600 amp		6	2.667		1,525	170		1,695	1,925
2630	800 amp		4.60	3.478		1,700	222		1,922	2,200
2640	1,000 amp		4	4		1,950	255		2,205	2,525
2650	1,350 amp		3.60	4.444		2,125	283		2,408	2,775
2660	1,600 amp		3.20	5		2,550	320		2,870	3,275
2670	2,000 amp		2.80	5.714		2,825	365		3,190	3,675
2680	2,500 amp		2.40	6.667		3,425	425		3,850	4,400
2690	3,000 amp		2	8		3,950	510		4,460	5,100
2700	4,000 amp		1.60	10		5,200	635		5,835	6,650
2800	Reducer nonfused, 400 amp		8	2		910	127		1,037	1,200
2810	600 amp		6	2.667		910	170		1,080	1,250
2820	800 amp		4.60	3.478		1,100	222		1,322	1,525
2830	1,000 amp		4	4		1,275	255		1,530	1,775
2840	1,350 amp		3.60	4.444		1,675	283		1,958	2,250
2850	1,600 amp		3.20	5		2,250	320		2,570	2,950
2860	2,000 amp		2.80	5.714		2,575	365		2,940	3,400
2870	2,500 amp		2.40	6.667		3,250	425		3,675	4,200
2880	3,000 amp		2	8		3,775	510		4,285	4,900
2890	4,000 amp		1.60	10		5,025	635		5,660	6,475
2950	Reducer fuse included, 225 amp		4.40	3.636		2,650	232		2,882	3,275
2960	400 amp		4.20	3.810		2,700	243		2,943	3,325
2970	600 amp		3.60	4.444		3,200	283		3,483	3,950
2980	800 amp		3.20	5		5,075	320		5,395	6,075
2990	1,000 amp		3	5.333		5,800	340		6,140	6,900
3000	1,200 amp		2.80	5.714		5,800	365		6,165	6,950
3010	1,600 amp		2.20	7.273		13,300	465		13,765	15,300
3020	2,000 amp		1.80	8.889		14,700	565		15,265	17,000
3100	Reducer circuit breaker, 225 amp		4.40	3.636		2,625	232		2,857	3,225
3110	400 amp		4.20	3.810		3,200	243		3,443	3,850
3120	600 amp		3.60	4.444		4,525	283		4,808	5,400
3130	800 amp		3.20	5		5,300	320		5,620	6,325
3140	1,000 amp		3	5.333		6,025	340		6,365	7,125
3150	1,200 amp		2.80	5.714		7,250	365		7,615	8,525
3160	1,600 amp		2.20	7.273		10,700	465		11,165	12,500
3170	2,000 amp		1.80	8.889		11,700	565		12,265	13,700
3250	Reducer circuit breaker, 75,000 AIC, 225 amp		4.40	3.636		4,100	232		4,332	4,850
3260	400 amp		4.20	3.810		4,100	243		4,343	4,850
3270	600 amp		3.60	4.444		5,475	283		5,758	6,450
3280	800 amp		3.20	5		6,025	320		6,345	7,100

For customer support on your Electrical Costs with RSMeans Data, call 800.448.8182.

249

26 25 13.10 Aluminum Bus Duct		Crew	Daily Output	Labor-Hours	Unit	Material	2021 Bare Costs Labor	Equipment	Total	Total Incl O&P
3290	1,000 amp	2 Elec	3	5.333	Ea.	9,675	340		10,015	11,200
3300	1,200 amp		2.80	5.714		9,675	365		10,040	11,200
3310	1,600 amp		2.20	7.273		10,700	465		11,165	12,500
3320	2,000 amp		1.80	8.889		11,700	565		12,265	13,700
3400	Reducer circuit breaker CLF 225 amp		4.40	3.636		4,200	232		4,432	4,975
3410	400 amp		4.20	3.810		5,000	243		5,243	5,850
3420	600 amp		3.60	4.444		7,500	283		7,783	8,675
3430	800 amp		3.20	5		7,825	320		8,145	9,075
3440	1,000 amp		3	5.333		8,150	340		8,490	9,475
3450	1,200 amp		2.80	5.714		10,700	365		11,065	12,200
3460	1,600 amp		2.20	7.273		10,700	465		11,165	12,500
3470	2,000 amp		1.80	8.889		11,700	565		12,265	13,700
3550	Ground bus added to bus duct, 225 amp		320	.050	L.F.	34	3.19		37.19	42
3560	400 amp		320	.050		34	3.19		37.19	42
3570	600 amp		280	.057		34	3.64		37.64	43
3580	800 amp		240	.067		34	4.25		38.25	44
3590	1,000 amp		200	.080		34	5.10		39.10	45
3600	1,350 amp		180	.089		34	5.65		39.65	46
3610	1,600 amp		160	.100		34	6.35		40.35	47
3620	2,000 amp		160	.100		34	6.35		40.35	47
3630	2,500 amp		140	.114		34	7.30		41.30	48.50
3640	3,000 amp		120	.133		34	8.50		42.50	50
3650	4,000 amp		100	.160		34	10.20		44.20	52.50
3810	High short circuit, 400 amp		36	.444		205	28.50		233.50	268
3820	600 amp		32	.500		355	32		387	440
3830	800 amp		26	.615		415	39		454	515
3840	1,000 amp		24	.667		520	42.50		562.50	635
3850	1,350 amp		22	.727		193	46.50		239.50	281
3860	1,600 amp		18	.889		222	56.50		278.50	330
3870	2,000 amp		16	1		260	63.50		323.50	380
3880	2,500 amp		14	1.143		435	73		508	585
3890	3,000 amp		12	1.333		490	85		575	665
3920	Cross, 225 amp		5.60	2.857	Ea.	1,375	182		1,557	1,775
3930	400 amp		4.60	3.478		1,375	222		1,597	1,825
3940	600 amp		4	4		1,375	255		1,630	1,875
3950	800 amp		3.40	4.706		1,450	300		1,750	2,025
3960	1,000 amp		3	5.333		1,525	340		1,865	2,175
3970	1,350 amp		2.80	5.714		2,450	365		2,815	3,250
3980	1,600 amp		2.20	7.273		2,900	465		3,365	3,900
3990	2,000 amp		1.80	8.889		3,200	565		3,765	4,375
4000	2,500 amp		1.60	10		3,800	635		4,435	5,125
4010	3,000 amp		1.20	13.333		4,350	850		5,200	6,050
4020	4,000 amp		1	16		6,650	1,025		7,675	8,850
4040	Cable tap box center, 225 amp		3.60	4.444		3,025	283		3,308	3,750
4050	400 amp		3.20	5		3,475	320		3,795	4,300
4060	600 amp		2.60	6.154		4,625	390		5,015	5,675
4070	800 amp		2.20	7.273		5,125	465		5,590	6,325
4080	1,000 amp		2	8		6,500	510		7,010	7,900
4090	1,350 amp		1.60	10		5,325	635		5,960	6,800
4100	1,600 amp		1.40	11.429		5,325	730		6,055	6,925
4110	2,000 amp		1.20	13.333		1,900	850		2,750	3,350
4120	2,500 amp		1	16		2,325	1,025		3,350	4,075
4130	3,000 amp		.80	20		2,575	1,275		3,850	4,725

26 25 13 – Low-Voltage Busways

26 25 13.10 Aluminum Bus Duct

		Crew	Daily Output	Labor-Hours	Unit	Material	2021 Bare Costs Labor	Equipment	Total	Total Incl O&P
4140	4,000 amp	2 Elec	.60	26.667	Ea.	3,400	1,700		5,100	6,275
4500	Weatherproof 3 pole 4 wire, feeder, 600 amp		30	.533	L.F.	138	34		172	203
4520	800 amp		24	.667		162	42.50		204.50	241
4540	1,000 amp		22	.727		185	46.50		231.50	272
4550	1,200 amp		21	.762		256	48.50		304.50	355
4560	1,350 amp		20	.800		289	51		340	395
4580	1,600 amp		17	.941		335	60		395	460
4600	2,000 amp		16	1		395	63.50		458.50	530
4620	2,500 amp		12	1.333		510	85		595	685
4640	3,000 amp		10	1.600		580	102		682	785
4660	4,000 amp		8	2		830	127		957	1,100
5000	Indoor 3 pole, 3 wire, feeder, 600 amp		40	.400		106	25.50		131.50	155
5010	800 amp		32	.500		355	32		387	440
5020	1,000 amp		30	.533		395	34		429	485
5025	1,200 amp		29	.552		840	35		875	975
5030	1,350 amp		28	.571		183	36.50		219.50	255
5040	1,600 amp		24	.667		755	42.50		797.50	895
5050	2,000 amp		20	.800		905	51		956	1,075
5060	2,500 amp		16	1		1,100	63.50		1,163.50	1,300
5070	3,000 amp		14	1.143		1,250	73		1,323	1,475
5080	4,000 amp		12	1.333		500	85		585	675
5200	Plug-in type, 225 amp		50	.320		138	20.50		158.50	183
5210	400 amp		42	.381		170	24.50		194.50	223
5220	600 amp		36	.444		211	28.50		239.50	274
5230	800 amp		30	.533		296	34		330	375
5240	1,000 amp		28	.571		335	36.50		371.50	425
5245	1,200 amp		27	.593		450	38		488	550
5250	1,350 amp		26	.615		193	39		232	271
5260	1,600 amp		20	.800		840	51		891	995
5270	2,000 amp		18	.889		1,000	56.50		1,056.50	1,175
5280	2,500 amp		16	1		1,225	63.50		1,288.50	1,450
5290	3,000 amp		14	1.143		1,400	73		1,473	1,650
5300	4,000 amp		12	1.333		510	85		595	685
5330	High short circuit, 400 amp		42	.381		204	24.50		228.50	260
5340	600 amp		36	.444		253	28.50		281.50	320
5350	800 amp		30	.533		355	34		389	440
5360	1,000 amp		28	.571		400	36.50		436.50	495
5370	1,350 amp		26	.615		193	39		232	271
5380	1,600 amp		20	.800		222	51		273	320
5390	2,000 amp		18	.889		260	56.50		316.50	370
5400	2,500 amp		16	1		410	63.50		473.50	545
5410	3,000 amp		14	1.143		425	73		498	575
5440	Elbow, 225 amp		5	3.200	Ea.	1,800	204		2,004	2,275
5450	400 amp		4.40	3.636		1,650	232		1,882	2,175
5460	600 amp		4	4		1,775	255		2,030	2,325
5470	800 amp		3.40	4.706		2,025	300		2,325	2,675
5480	1,000 amp		3.20	5		735	320		1,055	1,275
5485	1,200 amp	2 Elec	3.10	5.161	Ea.	785	330		1,115	1,350
5490	1,350 amp		3	5.333		745	340		1,085	1,325
5500	1,600 amp		2.80	5.714		3,775	365		4,140	4,700
5510	2,000 amp		2.40	6.667		4,475	425		4,900	5,550
5520	2,500 amp		2	8		5,425	510		5,935	6,725
5530	3,000 amp		1.80	8.889		6,150	565		6,715	7,625

For customer support on your Electrical Costs with RSMeans Data, call 800.448.8182.

251

26 25 13.10 Aluminum Bus Duct		Crew	Daily Output	Labor-Hours	Unit	Material	2021 Bare Costs Labor	Equipment	Total	Total Incl O&P
5540	4,000 amp	2 Elec	1.60	10	Ea.	9,500	635		10,135	11,300
5560	Tee fittings, 225 amp		3.60	4.444		3,050	283		3,333	3,800
5570	400 amp		3.20	5		3,050	320		3,370	3,850
5580	600 amp		3	5.333		3,050	340		3,390	3,875
5590	800 amp		2.80	5.714		875	365		1,240	1,500
5600	1,000 amp		2.60	6.154		905	390		1,295	1,575
5605	1,200 amp		2.50	6.400		965	410		1,375	1,650
5610	1,350 amp		2.40	6.667		1,350	425		1,775	2,100
5620	1,600 amp		1.80	8.889		1,575	565		2,140	2,575
5630	2,000 amp		1.40	11.429		1,750	730		2,480	3,000
5640	2,500 amp		1.20	13.333		2,200	850		3,050	3,650
5650	3,000 amp		1	16		2,575	1,025		3,600	4,350
5660	4,000 amp		.70	22.857		3,775	1,450		5,225	6,325
5680	Cross, 225 amp		6.40	2.500		1,325	159		1,484	1,675
5690	400 amp		5.40	2.963		1,325	189		1,514	1,725
5700	600 amp		4.60	3.478		1,325	222		1,547	1,775
5710	800 amp		4	4		1,400	255		1,655	1,925
5720	1,000 amp		3.60	4.444		1,450	283		1,733	2,000
5730	1,350 amp		3.20	5		2,150	320		2,470	2,825
5740	1,600 amp		2.60	6.154		2,500	390		2,890	3,325
5750	2,000 amp		2.20	7.273		2,750	465		3,215	3,700
5760	2,500 amp		1.80	8.889		3,300	565		3,865	4,500
5770	3,000 amp		1.40	11.429		3,950	730		4,680	5,425
5780	4,000 amp		1.20	13.333		5,500	850		6,350	7,300
5800	Expansion fitting, 225 amp		11.60	1.379		1,150	88		1,238	1,400
5810	400 amp		9.20	1.739		1,150	111		1,261	1,450
5820	600 amp		7	2.286		1,150	146		1,296	1,500
5830	800 amp		5.20	3.077		1,375	196		1,571	1,825
5840	1,000 amp		4.60	3.478		1,500	222		1,722	1,975
5850	1,350 amp		4.20	3.810		1,500	243		1,743	2,000
5860	1,600 amp		3.60	4.444		1,875	283		2,158	2,475
5870	2,000 amp		3.20	5		2,225	320		2,545	2,925
5880	2,500 amp		2.80	5.714		2,500	365		2,865	3,300
5890	3,000 amp		2.40	6.667		2,925	425		3,350	3,850
5900	4,000 amp		1.80	8.889		4,025	565		4,590	5,275
5940	Reducer, nonfused, 400 amp		9.20	1.739		780	111		891	1,025
5950	600 amp		7	2.286		780	146		926	1,075
5960	800 amp		5.20	3.077		820	196		1,016	1,200
5970	1,000 amp		4.60	3.478		1,000	222		1,222	1,425
5980	1,350 amp		4.20	3.810		1,500	243		1,743	2,000
5990	1,600 amp		3.60	4.444		1,700	283		1,983	2,300
6000	2,000 amp		3.20	5		1,975	320		2,295	2,650
6010	2,500 amp		2.80	5.714		2,525	365		2,890	3,350
6020	3,000 amp		2.20	7.273		2,975	465		3,440	3,975
6030	4,000 amp		1.80	8.889		4,075	565		4,640	5,325
6050	Reducer, fuse included, 225 amp		5	3.200		2,050	204		2,254	2,550
6060	400 amp		4.80	3.333		2,700	212		2,912	3,300
6070	600 amp		4.20	3.810		3,475	243		3,718	4,175
6080	800 amp		3.60	4.444		5,375	283		5,658	6,325
6090	1,000 amp		3.40	4.706		5,850	300		6,150	6,900
6100	1,350 amp		3.20	5		10,700	320		11,020	12,300
6110	1,600 amp		2.60	6.154		12,700	390		13,090	14,500
6120	2,000 amp		2	8		14,700	510		15,210	17,000

For customer support on your Electrical Costs with RSMeans Data, call 800.448.8182.

26 25 13.10 Aluminum Bus Duct

		Crew	Daily Output	Labor-Hours	Unit	Material	2021 Bare Costs Labor	Equipment	Total	Total Incl O&P
6160	Reducer, circuit breaker, 225 amp	2 Elec	5	3.200	Ea.	2,500	204		2,704	3,050
6170	400 amp		4.80	3.333		3,075	212		3,287	3,700
6180	600 amp		4.20	3.810		4,400	243		4,643	5,200
6190	800 amp		3.60	4.444		5,150	283		5,433	6,100
6200	1,000 amp		3.40	4.706		5,850	300		6,150	6,900
6210	1,350 amp		3.20	5		7,075	320		7,395	8,250
6220	1,600 amp		2.60	6.154		10,600	390		10,990	12,200
6230	2,000 amp		2	8		11,500	510		12,010	13,500
6270	Cable tap box center, 225 amp		4.20	3.810		3,050	243		3,293	3,700
6280	400 amp		3.60	4.444		1,050	283		1,333	1,575
6290	600 amp		3	5.333		1,050	340		1,390	1,650
6300	800 amp		2.60	6.154		1,150	390		1,540	1,850
6310	1,000 amp		2.40	6.667		1,225	425		1,650	1,975
6320	1,350 amp		1.80	8.889		4,575	565		5,140	5,875
6330	1,600 amp		1.60	10		4,675	635		5,310	6,075
6340	2,000 amp		1.40	11.429		1,650	730		2,380	2,900
6350	2,500 amp		1.20	13.333		2,050	850		2,900	3,525
6360	3,000 amp		1	16		2,350	1,025		3,375	4,100
6370	4,000 amp		.70	22.857		2,775	1,450		4,225	5,225
6390	Cable tap box end, 225 amp		4.20	3.810		1,650	243		1,893	2,150
6400	400 amp		3.60	4.444		1,975	283		2,258	2,600
6410	600 amp		3	5.333		645	340		985	1,225
6420	800 amp		2.60	6.154		705	390		1,095	1,350
6430	1,000 amp		2.40	6.667		760	425		1,185	1,475
6435	1,200 amp		2.10	7.619		825	485		1,310	1,625
6440	1,350 amp		1.80	8.889		800	565		1,365	1,725
6450	1,600 amp		1.60	10		5,775	635		6,410	7,325
6460	2,000 amp		1.40	11.429		6,250	730		6,980	7,950
6470	2,500 amp		1.20	13.333		7,425	850		8,275	9,425
6480	3,000 amp		1	16		8,000	1,025		9,025	10,300
6490	4,000 amp		.70	22.857		1,700	1,450		3,150	4,025
7000	Weatherproof 3 pole 3 wire, feeder, 600 amp		34	.471	L.F.	137	30		167	195
7020	800 amp		28	.571		151	36.50		187.50	220
7040	1,000 amp		26	.615		162	39		201	237
7050	1,200 amp		25	.640		189	41		230	269
7060	1,350 amp		24	.667		220	42.50		262.50	305
7080	1,600 amp		20	.800		254	51		305	355
7100	2,000 amp		18	.889		300	56.50		356.50	415
7120	2,500 amp		14	1.143		415	73		488	570
7140	3,000 amp		12	1.333		500	85		585	670
7160	4,000 amp		10	1.600		600	102		702	810

26 25 13.20 Bus Duct

		Crew	Daily Output	Labor-Hours	Unit	Material	2021 Bare Costs Labor	Equipment	Total	Total Incl O&P
0010	**BUS DUCT** 100 amp and less, aluminum or copper, plug-in									
0080	Bus duct, 3 pole 3 wire, 100 amp	1 Elec	42	.190	L.F.	81.50	12.15		93.65	108
0110	Elbow		4	2	Ea.	131	127		258	335
0120	Tee		2	4		187	255		442	585
0130	Wall flange		8	1		28.50	63.50		92	126
0140	Ground kit		16	.500		61	32		93	115
0180	3 pole 4 wire, 100 amp		40	.200	L.F.	89.50	12.75		102.25	117
0200	Cable tap box		3.10	2.581	Ea.	268	164		432	540
0300	End closure		16	.500		36	32		68	87
0400	Elbow		4	2		224	127		351	435

For customer support on your Electrical Costs with RSMeans Data, call 800.448.8182.

253

26 25 13.20 Bus Duct		Crew	Daily Output	Labor-Hours	Unit	Material	2021 Bare Costs Labor	2021 Bare Costs Equipment	Total	Total Incl O&P
0500	Tee	1 Elec	2	4	Ea.	325	255		580	740
0600	Hangers		10	.800		59.50	51		110.50	141
0700	Circuit breakers, 15 to 50 amp, 1 pole		8	1		670	63.50		733.50	830
0800	15 to 60 amp, 2 pole		6.70	1.194		380	76		456	535
0900	3 pole		5.30	1.509		470	96		566	665
1000	60 to 100 amp, 1 pole		6.70	1.194		785	76		861	980
1100	70 to 100 amp, 2 pole		5.30	1.509		1,350	96		1,446	1,650
1200	3 pole		4.50	1.778		1,850	113		1,963	2,200
1220	Switch, nonfused, 3 pole, 4 wire		8	1		227	63.50		290.50	345
1240	Fused, 3 fuses, 4 wire, 30 amp		8	1		360	63.50		423.50	490
1260	60 amp		5.30	1.509		420	96		516	610
1280	100 amp		4.50	1.778		565	113		678	795
1300	Plug, fusible, 3 pole 250 volt, 30 amp		5.30	1.509		555	96		651	755
1310	60 amp		5.30	1.509		625	96		721	830
1320	100 amp		4.50	1.778		840	113		953	1,100
1330	3 pole 480 volt, 30 amp		5.30	1.509		560	96		656	760
1340	60 amp		5.30	1.509		600	96		696	805
1350	100 amp		4.50	1.778		870	113		983	1,125
1360	Circuit breaker, 3 pole 250 volt, 60 amp		5.30	1.509		990	96		1,086	1,250
1370	3 pole 480 volt, 100 amp		4.50	1.778		990	113		1,103	1,275
2000	Bus duct, 2 wire, 250 volt, 30 amp		60	.133	L.F.	8.65	8.50		17.15	22
2100	60 amp		50	.160		8.65	10.20		18.85	24.50
2200	300 volt, 30 amp		60	.133		8.65	8.50		17.15	22
2300	60 amp		50	.160		8.65	10.20		18.85	24.50
2400	3 wire, 250 volt, 30 amp		60	.133		11.15	8.50		19.65	25
2500	60 amp		50	.160		10.90	10.20		21.10	27
2600	480/277 volt, 30 amp		60	.133		11.15	8.50		19.65	25
2700	60 amp		50	.160		11.15	10.20		21.35	27.50
2750	End feed, 300 volt 2 wire max. 30 amp		6	1.333	Ea.	78	85		163	212
2800	60 amp		5.50	1.455		78	92.50		170.50	224
2850	30 amp miniature		6	1.333		78	85		163	212
2900	3 wire, 30 amp		6	1.333		98	85		183	234
2950	60 amp		5.50	1.455		98	92.50		190.50	246
3000	30 amp miniature		6	1.333		98	85		183	234
3050	Center feed, 300 volt 2 wire, 30 amp		6	1.333		107	85		192	244
3100	60 amp		5.50	1.455		107	92.50		199.50	256
3150	3 wire, 30 amp		6	1.333		122	85		207	261
3200	60 amp		5.50	1.455		122	92.50		214.50	273
3220	Elbow, 30 amp		6	1.333		42.50	85		127.50	173
3240	60 amp		5.50	1.455		42.50	92.50		135	185
3260	End cap		40	.200		10.25	12.75		23	30.50
3280	Strength beam, 10'		15	.533		25	34		59	78
3300	Hanger		24	.333		5.35	21		26.35	37.50
3320	Tap box, nonfusible		6.30	1.270		87.50	81		168.50	216
3340	Fusible switch 30 amp, 1 fuse		6	1.333		460	85		545	630
3360	2 fuse		6	1.333		485	85		570	655
3380	3 fuse		6	1.333		545	85		630	725
3400	Circuit breaker handle on cover, 1 pole		6	1.333		62	85		147	194
3420	2 pole		6	1.333		630	85		715	815
3440	3 pole		6	1.333		775	85		860	975
3460	Circuit breaker external operhandle, 1 pole		6	1.333		74	85		159	207
3480	2 pole		6	1.333		815	85		900	1,025
3500	3 pole		6	1.333		890	85		975	1,100

26 25 13.20 Bus Duct

		Crew	Daily Output	Labor-Hours	Unit	Material	2021 Bare Costs Labor	Equipment	Total	Total Incl O&P
3520	Terminal plug only	1 Elec	16	.500	Ea.	98.50	32		130.50	157
3540	Terminal with receptacle		16	.500		108	32		140	167
3560	Fixture plug		16	.500	▼	73.50	32		105.50	128
4000	Copper bus duct, lighting, 2 wire 300 volt, 20 amp		70	.114	L.F.	7.80	7.30		15.10	19.40
4020	35 amp		60	.133		7.80	8.50		16.30	21
4040	50 amp		55	.145		7.80	9.25		17.05	22.50
4060	60 amp		50	.160		7.80	10.20		18	24
4080	3 wire 300 volt, 20 amp		70	.114		7.65	7.30		14.95	19.20
4100	35 amp		60	.133		7.40	8.50		15.90	21
4120	50 amp		55	.145		7.65	9.25		16.90	22
4140	60 amp		50	.160	▼	7.65	10.20		17.85	23.50
4160	Feeder in box, end, 1 circuit		6	1.333	Ea.	96	85		181	231
4180	2 circuit		5.50	1.455		100	92.50		192.50	249
4200	Center, 1 circuit		6	1.333		133	85		218	272
4220	2 circuit		5.50	1.455		138	92.50		230.50	290
4240	End cap		40	.200		16.25	12.75		29	37
4260	Hanger, surface mount		24	.333		9.70	21		30.70	42
4280	Coupling	▼	40	.200	▼	12.35	12.75		25.10	32.50

26 25 13.30 Copper Bus Duct

		Crew	Daily Output	Labor-Hours	Unit	Material	2021 Bare Costs Labor	Equipment	Total	Total Incl O&P
0010	**COPPER BUS DUCT**									
0100	Weatherproof 3 pole 4 wire, feeder duct, 600 amp	2 Elec	24	.667	L.F.	310	42.50		352.50	405
0110	800 amp		18	.889		375	56.50		431.50	500
0120	1,000 amp		17	.941		500	60		560	640
0125	1,200 amp		16.50	.970		545	62		607	690
0130	1,350 amp		16	1		605	63.50		668.50	760
0140	1,600 amp		12	1.333		640	85		725	830
0150	2,000 amp		10	1.600		780	102		882	1,000
0160	2,500 amp		7	2.286		885	146		1,031	1,200
0170	3,000 amp		5	3.200		1,175	204		1,379	1,600
0180	4,000 amp		3.60	4.444		1,800	283		2,083	2,400
0200	Indoor 3 pole 4 wire, plug-in, bus duct high short circuit, 400 amp		32	.500		450	32		482	545
0210	600 amp		26	.615		510	39		549	620
0220	800 amp		20	.800		740	51		791	890
0230	1,000 amp		18	.889		870	56.50		926.50	1,050
0240	1,350 amp		16	1		555	63.50		618.50	710
0250	1,600 amp		12	1.333		630	85		715	820
0260	2,000 amp		10	1.600		800	102		902	1,025
0270	2,500 amp		8	2		985	127		1,112	1,275
0280	3,000 amp		6	2.667	▼	1,050	170		1,220	1,400
0310	Cross, 225 amp		3	5.333	Ea.	3,650	340		3,990	4,500
0320	400 amp		2.80	5.714		3,650	365		4,015	4,550
0330	600 amp		2.60	6.154		3,650	390		4,040	4,575
0340	800 amp		2.20	7.273		3,950	465		4,415	5,050
0350	1,000 amp		2	8		4,425	510		4,935	5,600
0360	1,350 amp		1.80	8.889		4,825	565		5,390	6,150
0370	1,600 amp		1.70	9.412	▼	5,350	600		5,950	6,800
0380	2,000 amp	2 Elec	1.60	10	Ea.	8,800	635		9,435	10,600
0390	2,500 amp		1.40	11.429		10,700	730		11,430	12,800
0400	3,000 amp		1.20	13.333		10,900	850		11,750	13,300
0410	4,000 amp		1	16		14,200	1,025		15,225	17,100
0430	Expansion fitting, 225 amp		5.40	2.963		2,250	189		2,439	2,750
0440	400 amp	▼	4.60	3.478	▼	2,550	222		2,772	3,125

255

26 25 13.30 Copper Bus Duct		Crew	Daily Output	Labor-Hours	Unit	Material	2021 Bare Costs Labor	Equipment	Total	Total Incl O&P
0450	600 amp	2 Elec	4	4	Ea.	3,175	255		3,430	3,850
0460	800 amp		3.40	4.706		3,725	300		4,025	4,550
0470	1,000 amp		3	5.333		4,275	340		4,615	5,200
0480	1,350 amp		2.80	5.714		4,400	365		4,765	5,375
0490	1,600 amp		2.60	6.154		6,125	390		6,515	7,300
0500	2,000 amp		2.20	7.273		7,050	465		7,515	8,450
0510	2,500 amp		1.80	8.889		8,575	565		9,140	10,300
0520	3,000 amp		1.60	10		8,975	635		9,610	10,800
0530	4,000 amp		1.20	13.333		11,600	850		12,450	14,000
0550	Reducer nonfused, 225 amp		5.40	2.963		2,150	189		2,339	2,650
0560	400 amp		4.60	3.478		2,150	222		2,372	2,700
0570	600 amp		4	4		2,150	255		2,405	2,750
0580	800 amp		3.40	4.706		2,600	300		2,900	3,325
0590	1,000 amp		3	5.333		3,125	340		3,465	3,950
0600	1,350 amp		2.80	5.714		4,725	365		5,090	5,750
0610	1,600 amp		2.60	6.154		5,425	390		5,815	6,550
0620	2,000 amp		2.20	7.273		6,550	465		7,015	7,900
0630	2,500 amp		1.80	8.889		8,250	565		8,815	9,925
0640	3,000 amp		1.60	10		8,475	635		9,110	10,300
0650	4,000 amp		1.20	13.333		11,000	850		11,850	13,400
0670	Reducer fuse included, 225 amp		4.40	3.636		4,600	232		4,832	5,425
0680	400 amp		4.20	3.810		5,775	243		6,018	6,700
0690	600 amp		3.60	4.444		7,100	283		7,383	8,250
0700	800 amp		3.20	5		10,000	320		10,320	11,500
0710	1,000 amp		3	5.333		12,600	340		12,940	14,300
0720	1,350 amp		2.80	5.714		15,600	365		15,965	17,600
0730	1,600 amp		2.20	7.273		27,800	465		28,265	31,200
0740	2,000 amp		1.80	8.889		31,300	565		31,865	35,200
0790	Reducer, circuit breaker, 225 amp		4.40	3.636		5,825	232		6,057	6,775
0800	400 amp		4.20	3.810		6,800	243		7,043	7,850
0810	600 amp		3.60	4.444		9,675	283		9,958	11,000
0820	800 amp		3.20	5		11,300	320		11,620	13,000
0830	1,000 amp		3	5.333		12,900	340		13,240	14,600
0840	1,350 amp		2.80	5.714		15,400	365		15,765	17,400
0850	1,600 amp		2.20	7.273		22,700	465		23,165	25,700
0860	2,000 amp		1.80	8.889		24,900	565		25,465	28,200
0910	Cable tap box, center, 225 amp		3.20	5		3,575	320		3,895	4,425
0920	400 amp		2.60	6.154		5,100	390		5,490	6,200
0930	600 amp		2.20	7.273		5,950	465		6,415	7,250
0940	800 amp		2	8		2,325	510		2,835	3,325
0950	1,000 amp		1.60	10		2,525	635		3,160	3,725
0960	1,350 amp		1.40	11.429		3,225	730		3,955	4,625
0970	1,600 amp		1.20	13.333		3,600	850		4,450	5,200
0980	2,000 amp		1	16		4,325	1,025		5,350	6,300
1040	2,500 amp		.80	20		5,125	1,275		6,400	7,550
1060	3,000 amp		.60	26.667		5,250	1,700		6,950	8,300
1080	4,000 amp		.40	40		6,600	2,550		9,150	11,100
1800	Weatherproof 3 pole 3 wire, feeder duct, 600 amp		28	.571	L.F.	325	36.50		361.50	410
1820	800 amp		22	.727		390	46.50		436.50	500
1840	1,000 amp		20	.800		440	51		491	555
1850	1,200 amp		19	.842		455	53.50		508.50	580
1860	1,350 amp		18	.889		625	56.50		681.50	770
1880	1,600 amp		14	1.143		715	73		788	895

26 25 13 – Low-Voltage Busways

26 25 13.30 Copper Bus Duct

		Crew	Daily Output	Labor-Hours	Unit	Material	2021 Bare Costs		Total	Total Incl O&P
							Labor	Equipment		
1900	2,000 amp	2 Elec	12	1.333	L.F.	915	85		1,000	1,125
1920	2,500 amp		8	2		1,125	127		1,252	1,450
1940	3,000 amp		6	2.667		1,225	170		1,395	1,600
1960	4,000 amp		4	4		1,625	255		1,880	2,150
2000	Indoor 3 pole 3 wire, feeder duct, 600 amp		32	.500		269	32		301	345
2010	800 amp		26	.615		580	39		619	700
2020	1,000 amp		24	.667		365	42.50		407.50	465
2025	1,200 amp		22	.727		480	46.50		526.50	600
2030	1,350 amp		20	.800		520	51		571	645
2040	1,600 amp		16	1		1,075	63.50		1,138.50	1,275
2050	2,000 amp		14	1.143		1,375	73		1,448	1,600
2060	2,500 amp		10	1.600		945	102		1,047	1,200
2070	3,000 amp		8	2		990	127		1,117	1,275
2080	4,000 amp		6	2.667		1,350	170		1,520	1,725
2090	5,000 amp		5	3.200		1,650	204		1,854	2,100
2200	Indoor 3 pole 3 wire, bus duct plug-in, 225 amp		46	.348		171	22		193	221
2210	400 amp		36	.444		259	28.50		287.50	325
2220	600 amp		30	.533		330	34		364	415
2230	800 amp		24	.667		485	42.50		527.50	600
2240	1,000 amp		20	.800		450	51		501	570
2250	1,350 amp		18	.889		790	56.50		846.50	955
2260	1,600 amp		14	1.143		845	73		918	1,050
2270	2,000 amp		12	1.333		800	85		885	1,000
2280	2,500 amp		10	1.600		985	102		1,087	1,225
2290	3,000 amp		8	2		1,050	127		1,177	1,350
2330	High short circuit, 400 amp		36	.444		310	28.50		338.50	380
2340	600 amp		30	.533		400	34		434	485
2350	800 amp		24	.667		580	42.50		622.50	705
2360	1,000 amp		20	.800		630	51		681	765
2370	1,350 amp		18	.889		950	56.50		1,006.50	1,125
2380	1,600 amp		14	1.143		1,025	73		1,098	1,225
2390	2,000 amp		12	1.333		800	85		885	1,000
2400	2,500 amp		10	1.600		985	102		1,087	1,225
2410	3,000 amp		8	2		1,050	127		1,177	1,350
2440	Elbows, 225 amp		4.60	3.478	Ea.	1,675	222		1,897	2,175
2450	400 amp		4.20	3.810		1,950	243		2,193	2,500
2460	600 amp		3.60	4.444		2,175	283		2,458	2,800
2470	800 amp		3.20	5		2,650	320		2,970	3,375
2480	1,000 amp		3	5.333		2,850	340		3,190	3,650
2485	1,200 amp		2.90	5.517		1,875	350		2,225	2,575
2490	1,350 amp		2.80	5.714		3,950	365		4,315	4,875
2500	1,600 amp		2.60	6.154		4,100	390		4,490	5,075
2510	2,000 amp		2	8		2,625	510		3,135	3,625
2520	2,500 amp		1.80	8.889		3,975	565		4,540	5,225
2530	3,000 amp		1.60	10		4,050	635		4,685	5,425
2540	4,000 amp		1.40	11.429		5,200	730		5,930	6,775
2560	Tee fittings, 225 amp		2.80	5.714		1,825	365		2,190	2,550
2570	400 amp		2.40	6.667		1,825	425		2,250	2,625
2580	600 amp		2	8		1,825	510		2,335	2,750
2590	800 amp		1.80	8.889		1,975	565		2,540	3,025
2600	1,000 amp		1.60	10		2,200	635		2,835	3,375
2605	1,200 amp		1.50	10.667		2,350	680		3,030	3,600
2610	1,350 amp		1.40	11.429		2,500	730		3,230	3,825

For customer support on your Electrical Costs with RSMeans Data, call 800.448.8182.

257

26 25 13.30 Copper Bus Duct		Crew	Daily Output	Labor-Hours	Unit	Material	2021 Bare Costs Labor	Equipment	Total	Total Incl O&P
2620	1,600 amp	2 Elec	1.20	13.333	Ea.	2,900	850		3,750	4,450
2630	2,000 amp		1	16		4,850	1,025		5,875	6,850
2640	2,500 amp		.70	22.857		5,675	1,450		7,125	8,425
2650	3,000 amp		.60	26.667		5,825	1,700		7,525	8,950
2660	4,000 amp		.50	32		7,525	2,050		9,575	11,300
2680	Cross, 225 amp		3.60	4.444		2,775	283		3,058	3,475
2690	400 amp		3.20	5		2,775	320		3,095	3,525
2700	600 amp		3	5.333		2,775	340		3,115	3,550
2710	800 amp		2.60	6.154		3,300	390		3,690	4,225
2720	1,000 amp		2.40	6.667		3,450	425		3,875	4,425
2730	1,350 amp		2.20	7.273		4,000	465		4,465	5,100
2740	1,600 amp		2	8		4,300	510		4,810	5,500
2750	2,000 amp		1.80	8.889		6,825	565		7,390	8,350
2760	2,500 amp		1.60	10		7,950	635		8,585	9,675
2770	3,000 amp		1.40	11.429		8,125	730		8,855	10,000
2780	4,000 amp		1	16		10,400	1,025		11,425	12,900
2800	Expansion fitting, 225 amp		6.40	2.500		2,375	159		2,534	2,850
2810	400 amp		5.40	2.963		4,825	189		5,014	5,575
2820	600 amp		4.60	3.478		2,375	222		2,597	2,950
2830	800 amp		4	4		2,825	255		3,080	3,500
2840	1,000 amp		3.60	4.444		3,100	283		3,383	3,850
2850	1,350 amp		3.20	5		3,400	320		3,720	4,200
2860	1,600 amp		3	5.333		3,725	340		4,065	4,600
2870	2,000 amp		2.60	6.154		4,625	390		5,015	5,675
2880	2,500 amp		2.20	7.273		6,475	465		6,940	7,825
2890	3,000 amp		1.80	8.889		6,750	565		7,315	8,275
2900	4,000 amp		1.40	11.429		8,625	730		9,355	10,600
2920	Reducer nonfused, 225 amp		6.40	2.500		1,775	159		1,934	2,175
2930	400 amp		5.40	2.963		1,775	189		1,964	2,225
2940	600 amp		4.60	3.478		1,775	222		1,997	2,275
2950	800 amp		4	4		2,050	255		2,305	2,650
2960	1,000 amp		3.60	4.444		2,325	283		2,608	2,975
2970	1,350 amp		3.20	5		3,050	320		3,370	3,825
2980	1,600 amp		3	5.333		3,450	340		3,790	4,300
2990	2,000 amp		2.60	6.154		4,150	390		4,540	5,150
3000	2,500 amp		2.20	7.273		6,075	465		6,540	7,400
3010	3,000 amp		1.80	8.889		6,200	565		6,765	7,675
3020	4,000 amp		1.40	11.429		8,075	730		8,805	9,950
3040	Reducer fuse included, 225 amp		5	3.200		4,175	204		4,379	4,900
3050	400 amp		4.80	3.333		5,575	212		5,787	6,450
3060	600 amp		4.20	3.810		6,625	243		6,868	7,650
3070	800 amp		3.60	4.444		9,450	283		9,733	10,800
3080	1,000 amp		3.40	4.706		11,200	300		11,500	12,700
3090	1,350 amp		3.20	5		11,400	320		11,720	13,000
3100	1,600 amp		2.60	6.154		26,500	390		26,890	29,700
3110	2,000 amp		2	8		30,000	510		30,510	33,800
3160	Reducer circuit breaker, 225 amp		5	3.200		5,350	204		5,554	6,175
3170	400 amp		4.80	3.333		6,550	212		6,762	7,525
3180	600 amp		4.20	3.810		9,400	243		9,643	10,700
3190	800 amp		3.60	4.444		11,000	283		11,283	12,500
3200	1,000 amp		3.40	4.706		12,500	300		12,800	14,100
3210	1,350 amp		3.20	5		15,000	320		15,320	17,000
3220	1,600 amp		2.60	6.154		22,400	390		22,790	25,300

26 25 13.30 Copper Bus Duct		Crew	Daily Output	Labor-Hours	Unit	Material	2021 Bare Costs Labor	Equipment	Total	Total Incl O&P
3230	2,000 amp	2 Elec	2	8	Ea.	24,500	510		25,010	27,800
3280	3 pole, 3 wire, cable tap box center, 225 amp		3.60	4.444		3,075	283		3,358	3,800
3290	400 amp		3	5.333		3,950	340		4,290	4,850
3300	600 amp		2.60	6.154		5,025	390		5,415	6,125
3310	800 amp		2.40	6.667		6,600	425		7,025	7,900
3320	1,000 amp		1.80	8.889		6,975	565		7,540	8,525
3330	1,350 amp		1.60	10		10,200	635		10,835	12,200
3340	1,600 amp		1.40	11.429		10,800	730		11,530	13,000
3350	2,000 amp		1.20	13.333		5,125	850		5,975	6,900
3360	2,500 amp		1	16		6,100	1,025		7,125	8,250
3370	3,000 amp		.70	22.857		6,300	1,450		7,750	9,100
3380	4,000 amp		.50	32		7,975	2,050		10,025	11,800
3400	Cable tap box end, 225 amp		3.60	4.444		2,100	283		2,383	2,750
3410	400 amp		3	5.333		2,525	340		2,865	3,275
3420	600 amp		2.60	6.154		3,725	390		4,115	4,675
3430	800 amp		2.40	6.667		2,900	425		3,325	3,800
3440	1,000 amp		1.80	8.889		3,800	565		4,365	5,025
3445	1,200 amp		1.70	9.412		1,900	600		2,500	2,975
3450	1,350 amp		1.60	10		1,900	635		2,535	3,050
3460	1,600 amp		1.40	11.429		3,850	730		4,580	5,300
3470	2,000 amp		1.20	13.333		2,650	850		3,500	4,175
3480	2,500 amp		1	16		3,175	1,025		4,200	5,025
3490	3,000 amp		.70	22.857		3,250	1,450		4,700	5,750
3500	4,000 amp		.50	32		4,125	2,050		6,175	7,575
4600	Plug-in, fusible switch w/3 fuses, 3 pole, 250 volt, 30 amp	1 Elec	4	2		525	127		652	765
4610	60 amp		3.60	2.222		685	142		827	960
4620	100 amp		2.70	2.963		980	189		1,169	1,350
4630	200 amp	2 Elec	3.20	5		1,650	320		1,970	2,275
4640	400 amp		1.40	11.429		4,300	730		5,030	5,800
4650	600 amp		.90	17.778		5,925	1,125		7,050	8,200
4700	4 pole, 120/208 volt, 30 amp	1 Elec	3.90	2.051		715	131		846	980
4710	60 amp		3.50	2.286		770	146		916	1,075
4720	100 amp		2.60	3.077		1,075	196		1,271	1,475
4730	200 amp	2 Elec	3	5.333		1,800	340		2,140	2,500
4740	400 amp		1.30	12.308		4,225	785		5,010	5,825
4750	600 amp		.80	20		5,950	1,275		7,225	8,425
4800	3 pole, 480 volt, 30 amp	1 Elec	4	2		540	127		667	780
4810	60 amp		3.60	2.222		570	142		712	835
4820	100 amp		2.70	2.963		960	189		1,149	1,325
4830	200 amp	2 Elec	3.20	5		1,650	320		1,970	2,275
4840	400 amp		1.40	11.429		3,825	730		4,555	5,300
4850	600 amp		.90	17.778		5,450	1,125		6,575	7,675
4860	800 amp		.66	24.242		22,900	1,550		24,450	27,500
4870	1,000 amp		.60	26.667		22,900	1,700		24,600	27,700
4880	1,200 amp		.50	32		25,600	2,050		27,650	31,100
4890	1,600 amp		.44	36.364		26,200	2,325		28,525	32,300
4900	4 pole, 277/480 volt, 30 amp	1 Elec	3.90	2.051		775	131		906	1,050
4910	60 amp		3.50	2.286		825	146		971	1,125
4920	100 amp		2.60	3.077		1,200	196		1,396	1,625
4930	200 amp	2 Elec	3	5.333		2,400	340		2,740	3,150
4940	400 amp		1.30	12.308		4,525	785		5,310	6,150
4950	600 amp		.80	20		6,200	1,275		7,475	8,700
5050	800 amp		.60	26.667		20,400	1,700		22,100	24,900

26 25 13.30 Copper Bus Duct		Crew	Daily Output	Labor-Hours	Unit	Material	2021 Bare Costs Labor	Equipment	Total	Total Incl O&P
5060	1,000 amp	2 Elec	.56	28.571	Ea.	23,500	1,825		25,325	28,500
5070	1,200 amp		.48	33.333		23,700	2,125		25,825	29,200
5080	1,600 amp		.42	38.095		27,400	2,425		29,825	33,700
5150	Fusible with starter, 3 pole 250 volt, 30 amp	1 Elec	3.50	2.286		3,600	146		3,746	4,175
5160	60 amp		3.20	2.500		3,825	159		3,984	4,425
5170	100 amp		2.50	3.200		4,300	204		4,504	5,050
5180	200 amp	2 Elec	2.80	5.714		7,125	365		7,490	8,375
5200	3 pole 480 volt, 30 amp	1 Elec	3.50	2.286		3,600	146		3,746	4,175
5210	60 amp		3.20	2.500		3,825	159		3,984	4,425
5220	100 amp		2.50	3.200		4,300	204		4,504	5,050
5230	200 amp	2 Elec	2.80	5.714		7,125	365		7,490	8,375
5300	Fusible with contactor, 3 pole 250 volt, 30 amp	1 Elec	3.50	2.286		3,500	146		3,646	4,100
5310	60 amp		3.20	2.500		4,475	159		4,634	5,150
5320	100 amp		2.50	3.200		6,250	204		6,454	7,175
5330	200 amp	2 Elec	2.80	5.714		7,150	365		7,515	8,425
5400	3 pole 480 volt, 30 amp	1 Elec	3.50	2.286		3,775	146		3,921	4,375
5410	60 amp		3.20	2.500		5,325	159		5,484	6,100
5420	100 amp		2.50	3.200		7,325	204		7,529	8,350
5430	200 amp	2 Elec	2.80	5.714		7,500	365		7,865	8,800
5450	Fusible with capacitor, 3 pole 250 volt, 30 amp	1 Elec	3	2.667		9,125	170		9,295	10,300
5460	60 amp		2	4		10,600	255		10,855	12,100
5500	3 pole 480 volt, 30 amp		3	2.667		7,650	170		7,820	8,650
5510	60 amp		2	4		9,550	255		9,805	10,900
5600	Circuit breaker, 3 pole, 250 volt, 60 amp		4.50	1.778		735	113		848	980
5610	100 amp		3.20	2.500		905	159		1,064	1,225
5650	4 pole, 120/208 volt, 60 amp		4.40	1.818		830	116		946	1,075
5660	100 amp		3.10	2.581		985	164		1,149	1,325
5700	3 pole, 4 wire 277/480 volt, 60 amp		4.30	1.860		1,125	119		1,244	1,400
5710	100 amp		3	2.667		1,250	170		1,420	1,625
5720	225 amp	2 Elec	3.20	5		2,750	320		3,070	3,525
5730	400 amp		1.20	13.333		5,725	850		6,575	7,550
5740	600 amp		.96	16.667		7,700	1,050		8,750	10,100
5750	700 amp		.60	26.667		9,750	1,700		11,450	13,200
5760	800 amp		.60	26.667		9,750	1,700		11,450	13,200
5770	900 amp		.54	29.630		12,900	1,875		14,775	16,900
5780	1,000 amp		.54	29.630		12,900	1,875		14,775	16,900
5790	1,200 amp		.42	38.095		15,500	2,425		17,925	20,600
5810	Circuit breaker w/HIC fuses, 3 pole 480 volt, 60 amp	1 Elec	4.40	1.818		1,400	116		1,516	1,725
5820	100 amp	"	3.10	2.581		1,550	164		1,714	1,950
5830	225 amp	2 Elec	3.40	4.706		4,925	300		5,225	5,875
5840	400 amp		1.40	11.429		7,825	730		8,555	9,675
5850	600 amp		1	16		7,925	1,025		8,950	10,300
5860	700 amp		.64	25		10,500	1,600		12,100	14,000
5870	800 amp		.64	25		10,500	1,600		12,100	14,000
5880	900 amp		.56	28.571		22,700	1,825		24,525	27,700
5890	1,000 amp		.56	28.571		22,700	1,825		24,525	27,700
5950	3 pole 4 wire 277/480 volt, 60 amp	1 Elec	4.30	1.860		1,400	119		1,519	1,725
5960	100 amp	"	3	2.667		1,550	170		1,720	1,950
5970	225 amp	2 Elec	3	5.333		4,925	340		5,265	5,925
5980	400 amp		1.10	14.545		7,825	925		8,750	9,975
5990	600 amp		.94	17.021		7,925	1,075		9,000	10,300
6000	700 amp		.58	27.586		10,500	1,750		12,250	14,200
6010	800 amp		.58	27.586		10,500	1,750		12,250	14,200

26 25 13 – Low-Voltage Busways

26 25 13.30 Copper Bus Duct		Crew	Daily Output	Labor-Hours	Unit	Material	2021 Bare Costs Labor	Equipment	Total	Total Incl O&P
6020	900 amp	2 Elec	.52	30.769	Ea.	22,700	1,950		24,650	27,900
6030	1,000 amp		.52	30.769		22,700	1,950		24,650	27,900
6040	1,200 amp	↓	.40	40		22,700	2,550		25,250	28,800
6100	Circuit breaker with starter, 3 pole 250 volt, 60 amp	1 Elec	3.20	2.500		1,850	159		2,009	2,250
6110	100 amp	"	2.50	3.200		3,050	204		3,254	3,675
6120	225 amp	2 Elec	3	5.333		4,525	340		4,865	5,500
6130	3 pole 480 volt, 60 amp	1 Elec	3.20	2.500		2,200	159		2,359	2,650
6140	100 amp	"	2.50	3.200		3,050	204		3,254	3,675
6150	225 amp	2 Elec	3	5.333		3,825	340		4,165	4,700
6200	Circuit breaker with contactor, 3 pole 250 volt, 60 amp	1 Elec	3.20	2.500		2,075	159		2,234	2,525
6210	100 amp	"	2.50	3.200		2,825	204		3,029	3,425
6220	225 amp	2 Elec	3	5.333		4,200	340		4,540	5,125
6250	3 pole 480 volt, 60 amp	1 Elec	3.20	2.500		2,075	159		2,234	2,525
6260	100 amp	"	2.50	3.200		2,825	204		3,029	3,425
6270	225 amp	2 Elec	3	5.333		3,650	340		3,990	4,525
6300	Circuit breaker with capacitor, 3 pole 250 volt, 60 amp	1 Elec	2	4		11,900	255		12,155	13,500
6310	3 pole 480 volt, 60 amp		2	4		11,500	255		11,755	13,000
6400	Add control transformer with pilot light to above starter		16	.500		625	32		657	740
6410	Switch, fusible, mechanically held contactor optional		16	.500		1,650	32		1,682	1,875
6430	Circuit breaker, mechanically held contactor optional		16	.500		1,650	32		1,682	1,875
6450	Ground neutralizer, 3 pole	↓	16	.500	↓	76.50	32		108.50	132

26 25 13.40 Copper Bus Duct		Crew	Daily Output	Labor-Hours	Unit	Material	2021 Bare Costs Labor	Equipment	Total	Total Incl O&P
0010	**COPPER BUS DUCT** 10' long									
0050	Indoor 3 pole 4 wire, plug-in, straight section, 225 amp	2 Elec	40	.400	L.F.	222	25.50		247.50	282
1000	400 amp		32	.500		181	32		213	247
1500	600 amp		26	.615		425	39		464	525
2400	800 amp		20	.800		620	51		671	755
2450	1,000 amp		18	.889		305	56.50		361.50	420
2470	1,200 amp		17	.941		1,175	60		1,235	1,400
2500	1,350 amp		16	1		405	63.50		468.50	540
2510	1,600 amp		12	1.333		1,600	85		1,685	1,875
2520	2,000 amp		10	1.600		1,950	102		2,052	2,300
2530	2,500 amp		8	2		715	127		842	975
2540	3,000 amp		6	2.667		3,050	170		3,220	3,625
2550	Feeder, 600 amp		28	.571		202	36.50		238.50	276
2600	800 amp		22	.727		750	46.50		796.50	895
2700	1,000 amp		20	.800		885	51		936	1,050
2750	1,200 amp		19	.842		1,050	53.50		1,103.50	1,250
2800	1,350 amp		18	.889		375	56.50		431.50	500
2900	1,600 amp		14	1.143		1,425	73		1,498	1,675
3000	2,000 amp		12	1.333		1,775	85		1,860	2,075
3010	2,500 amp		8	2		685	127		812	945
3020	3,000 amp		6	2.667		830	170		1,000	1,150
3030	4,000 amp		4	4		1,100	255		1,355	1,575
3040	5,000 amp		2	8	↓	1,325	510		1,835	2,225
3100	Elbows, 225 amp		4	4	Ea.	2,075	255		2,330	2,650
3200	400 amp		3.60	4.444		2,525	283		2,808	3,200
3300	600 amp		3.20	5		2,675	320		2,995	3,425
3400	800 amp		2.80	5.714		3,575	365		3,940	4,475
3500	1,000 amp		2.60	6.154		3,825	390		4,215	4,775
3550	1,200 amp		2.50	6.400		4,850	410		5,260	5,925
3600	1,350 amp	↓	2.40	6.667	↓	5,125	425		5,550	6,250

For customer support on your Electrical Costs with RSMeans Data, call 800.448.8182.

261

26 25 13.40 Copper Bus Duct		Crew	Daily Output	Labor-Hours	Unit	Material	2021 Bare Costs Labor	Equipment	Total	Total Incl O&P
3700	1,600 amp	2 Elec	2.20	7.273	Ea.	5,675	465		6,140	6,925
3800	2,000 amp		1.80	8.889		6,725	565		7,290	8,250
3810	2,500 amp		1.60	10		8,500	635		9,135	10,300
3820	3,000 amp		1.40	11.429		10,600	730		11,330	12,800
3830	4,000 amp		1.20	13.333		16,500	850		17,350	19,400
3840	5,000 amp		1	16		19,300	1,025		20,325	22,800
4000	End box, 225 amp		34	.471		75	30		105	127
4100	400 amp		32	.500		187	32		219	254
4200	600 amp		28	.571		187	36.50		223.50	260
4300	800 amp		26	.615		187	39		226	265
4400	1,000 amp		24	.667		110	42.50		152.50	184
4410	1,200 amp		23	.696		187	44.50		231.50	272
4500	1,350 amp		22	.727		175	46.50		221.50	261
4600	1,600 amp		20	.800		175	51		226	268
4700	2,000 amp		18	.889		214	56.50		270.50	320
4710	2,500 amp		16	1		214	63.50		277.50	330
4720	3,000 amp		14	1.143		214	73		287	345
4730	4,000 amp		12	1.333		259	85		344	410
4740	5,000 amp		10	1.600		259	102		361	435
4800	Cable tap box end, 225 amp		3.20	5		2,350	320		2,670	3,075
5000	400 amp		2.60	6.154		2,825	390		3,215	3,700
5100	600 amp		2.20	7.273		4,150	465		4,615	5,275
5200	800 amp		2	8		4,350	510		4,860	5,550
5300	1,000 amp		1.60	10		4,400	635		5,035	5,800
5350	1,200 amp		1.50	10.667		1,925	680		2,605	3,125
5400	1,350 amp		1.40	11.429		2,275	730		3,005	3,575
5500	1,600 amp		1.20	13.333		2,550	850		3,400	4,050
5600	2,000 amp		1	16		2,775	1,025		3,800	4,575
5610	2,500 amp		.80	20		3,075	1,275		4,350	5,275
5620	3,000 amp		.60	26.667		3,825	1,700		5,525	6,725
5630	4,000 amp		.40	40		4,275	2,550		6,825	8,475
5640	5,000 amp		.20	80		5,300	5,100		10,400	13,400
5700	Switchboard stub, 225 amp		5.40	2.963		1,175	189		1,364	1,550
5800	400 amp		4.60	3.478		1,300	222		1,522	1,750
5900	600 amp		4	4		1,375	255		1,630	1,900
6000	800 amp		3.20	5		1,650	320		1,970	2,300
6100	1,000 amp		3	5.333		1,925	340		2,265	2,625
6150	1,200 amp		2.80	5.714		2,250	365		2,615	3,025
6200	1,350 amp		2.60	6.154		2,325	390		2,715	3,150
6300	1,600 amp		2.40	6.667		2,625	425		3,050	3,525
6400	2,000 amp		2	8		3,175	510		3,685	4,250
6410	2,500 amp		1.80	8.889		3,875	565		4,440	5,100
6420	3,000 amp		1.60	10		4,500	635		5,135	5,900
6430	4,000 amp		1.40	11.429		5,875	730		6,605	7,525
6440	5,000 amp		1.20	13.333		7,200	850		8,050	9,175
6490	Tee fittings, 225 amp		2.40	6.667		3,350	425		3,775	4,325
6500	400 amp		2	8		3,350	510		3,860	4,450
6600	600 amp		1.80	8.889		3,350	565		3,915	4,550
6700	800 amp		1.60	10		2,325	635		2,960	3,500
6750	1,000 amp		1.40	11.429		2,550	730		3,280	3,875
6770	1,200 amp		1.30	12.308		2,875	785		3,660	4,325
6800	1,350 amp		1.20	13.333		2,975	850		3,825	4,525
7000	1,600 amp		1	16		3,375	1,025		4,400	5,250

26 25 13 – Low-Voltage Busways

26 25 13.40 Copper Bus Duct		Crew	Daily Output	Labor-Hours	Unit	Material	2021 Bare Costs Labor	Equipment	Total	Total Incl O&P
7100	2,000 amp	2 Elec	.80	20	Ea.	4,025	1,275		5,300	6,325
7110	2,500 amp		.60	26.667		4,925	1,700		6,625	7,950
7120	3,000 amp		.50	32		5,725	2,050		7,775	9,325
7130	4,000 amp		.40	40		7,325	2,550		9,875	11,800
7140	5,000 amp		.20	80		8,650	5,100		13,750	17,100
7200	Plug-in fusible switches w/3 fuses, 600 volt, 3 pole, 30 amp	1 Elec	4	2		820	127		947	1,100
7300	60 amp		3.60	2.222		920	142		1,062	1,200
7400	100 amp		2.70	2.963		1,200	189		1,389	1,600
7500	200 amp	2 Elec	3.20	5		2,125	320		2,445	2,825
7600	400 amp		1.40	11.429		7,350	730		8,080	9,150
7700	600 amp		.90	17.778		7,175	1,125		8,300	9,575
7800	800 amp		.66	24.242		11,900	1,550		13,450	15,400
7900	1,200 amp		.50	32		22,300	2,050		24,350	27,500
7910	1,600 amp		.44	36.364		21,900	2,325		24,225	27,500
8000	Plug-in circuit breakers, molded case, 15 to 50 amp	1 Elec	4.40	1.818		775	116		891	1,025
8100	70 to 100 amp	"	3.10	2.581		860	164		1,024	1,200
8200	150 to 225 amp	2 Elec	3.40	4.706		2,325	300		2,625	3,025
8300	250 to 400 amp		1.40	11.429		4,100	730		4,830	5,575
8400	500 to 600 amp		1	16		5,525	1,025		6,550	7,600
8500	700 to 800 amp		.64	25		6,800	1,600		8,400	9,875
8600	900 to 1,000 amp		.56	28.571		9,750	1,825		11,575	13,400
8700	1,200 amp		.44	36.364		11,700	2,325		14,025	16,400
8720	1,400 amp		.40	40		16,800	2,550		19,350	22,300
8730	1,600 amp		.40	40		18,400	2,550		20,950	24,000
8750	Circuit breakers, with current limiting fuse, 15 to 50 amp	1 Elec	4.40	1.818		1,550	116		1,666	1,875
8760	70 to 100 amp	"	3.10	2.581		1,825	164		1,989	2,275
8770	150 to 225 amp	2 Elec	3.40	4.706		3,950	300		4,250	4,800
8780	250 to 400 amp		1.40	11.429		6,100	730		6,830	7,775
8790	500 to 600 amp		1	16		7,025	1,025		8,050	9,250
8800	700 to 800 amp		.64	25		11,500	1,600		13,100	15,100
8810	900 to 1,000 amp		.56	28.571		13,200	1,825		15,025	17,200
8850	Combination starter FVNR, fusible switch, NEMA size 0, 30 amp	1 Elec	2	4		2,075	255		2,330	2,675
8860	NEMA size 1, 60 amp		1.80	4.444		2,200	283		2,483	2,850
8870	NEMA size 2, 100 amp		1.30	6.154		2,775	390		3,165	3,625
8880	NEMA size 3, 200 amp	2 Elec	2	8		4,450	510		4,960	5,650
8900	Circuit breaker, NEMA size 0, 30 amp	1 Elec	2	4		2,150	255		2,405	2,725
8910	NEMA size 1, 60 amp		1.80	4.444		2,225	283		2,508	2,875
8920	NEMA size 2, 100 amp		1.30	6.154		3,200	390		3,590	4,075
8930	NEMA size 3, 200 amp	2 Elec	2	8		4,050	510		4,560	5,200
8950	Combination contactor, fusible switch, NEMA size 0, 30 amp	1 Elec	2	4		1,225	255		1,480	1,725
8960	NEMA size 1, 60 amp		1.80	4.444		1,250	283		1,533	1,800
8970	NEMA size 2, 100 amp		1.30	6.154		1,875	390		2,265	2,625
8980	NEMA size 3, 200 amp	2 Elec	2	8		2,150	510		2,660	3,125
9000	Circuit breaker, NEMA size 0, 30 amp	1 Elec	2	4		1,400	255		1,655	1,925
9010	NEMA size 1, 60 amp		1.80	4.444		1,425	283		1,708	2,000
9020	NEMA size 2, 100 amp		1.30	6.154		2,225	390		2,615	3,025
9030	NEMA size 3, 200 amp	2 Elec	2	8		2,700	510		3,210	3,700
9050	Control transformer for above, NEMA size 0, 30 amp	1 Elec	8	1		237	63.50		300.50	355
9060	NEMA size 1, 60 amp		8	1		237	63.50		300.50	355
9070	NEMA size 2, 100 amp		7	1.143		330	73		403	475
9080	NEMA size 3, 200 amp	2 Elec	14	1.143		460	73		533	615
9100	Comb. fusible switch & lighting control, electrically held, 30 amp	1 Elec	2	4		970	255		1,225	1,450
9110	60 amp		1.80	4.444		1,400	283		1,683	1,950

For customer support on your Electrical Costs with RSMeans Data, call 800.448.8182.

263

26 25 13.40 Copper Bus Duct

		Crew	Daily Output	Labor-Hours	Unit	Material	2021 Bare Costs Labor	Equipment	Total	Total Incl O&P
9120	100 amp	1 Elec	1.30	6.154	Ea.	1,800	390		2,190	2,550
9130	200 amp	2 Elec	2	8		4,400	510		4,910	5,600
9150	Mechanically held, 30 amp	1 Elec	2	4		1,200	255		1,455	1,700
9160	60 amp		1.80	4.444		1,800	283		2,083	2,425
9170	100 amp		1.30	6.154		2,325	390		2,715	3,125
9180	200 amp	2 Elec	2	8		4,725	510		5,235	5,950
9200	Ground bus added to bus duct, 225 amp		320	.050	L.F.	44	3.19		47.19	53
9210	400 amp		240	.067		44	4.25		48.25	55
9220	600 amp		240	.067		44	4.25		48.25	55
9230	800 amp		160	.100		51.50	6.35		57.85	66
9240	1,000 amp		160	.100		59	6.35		65.35	74
9250	1,350 amp		140	.114		82.50	7.30		89.80	101
9260	1,600 amp		120	.133		89.50	8.50		98	111
9270	2,000 amp		110	.145		117	9.25		126.25	142
9280	2,500 amp		100	.160		144	10.20		154.20	174
9290	3,000 amp		90	.178		171	11.30		182.30	205
9300	4,000 amp		80	.200		226	12.75		238.75	267
9310	5,000 amp		70	.229		271	14.55		285.55	320
9320	High short circuit bracing, add					18.15			18.15	19.95

26 25 13.60 Copper or Aluminum Bus Duct Fittings

		Crew	Daily Output	Labor-Hours	Unit	Material	2021 Bare Costs Labor	Equipment	Total	Total Incl O&P
0010	**COPPER OR ALUMINUM BUS DUCT FITTINGS**									
0100	Flange, wall, with vapor barrier, 225 amp	2 Elec	6.20	2.581	Ea.	860	164		1,024	1,200
0110	400 amp		6	2.667		765	170		935	1,100
0120	600 amp		5.80	2.759		860	176		1,036	1,200
0130	800 amp		5.40	2.963		860	189		1,049	1,225
0140	1,000 amp		5	3.200		860	204		1,064	1,250
0145	1,200 amp		4.80	3.333		860	212		1,072	1,250
0150	1,350 amp		4.60	3.478		860	222		1,082	1,275
0160	1,600 amp		4.20	3.810		860	243		1,103	1,300
0170	2,000 amp		4	4		860	255		1,115	1,325
0180	2,500 amp		3.60	4.444		860	283		1,143	1,375
0190	3,000 amp		3.20	5		860	320		1,180	1,425
0200	4,000 amp		2.60	6.154		830	390		1,220	1,500
0300	Roof, 225 amp		6.20	2.581		1,075	164		1,239	1,450
0310	400 amp		6	2.667		1,075	170		1,245	1,450
0320	600 amp		5.80	2.759		1,075	176		1,251	1,450
0330	800 amp		5.40	2.963		1,075	189		1,264	1,475
0340	1,000 amp		5	3.200		1,075	204		1,279	1,500
0345	1,200 amp		4.80	3.333		1,075	212		1,287	1,525
0350	1,350 amp		4.60	3.478		1,075	222		1,297	1,525
0360	1,600 amp		4.20	3.810		1,075	243		1,318	1,550
0370	2,000 amp		4	4		1,075	255		1,330	1,575
0380	2,500 amp		3.60	4.444		1,075	283		1,358	1,625
0390	3,000 amp		3.20	5		1,075	320		1,395	1,675
0400	4,000 amp		2.60	6.154		1,075	390		1,465	1,775
0420	Support, floor mounted, 225 amp		20	.800		172	51		223	265
0430	400 amp		20	.800		172	51		223	265
0440	600 amp		18	.889		172	56.50		228.50	273
0450	800 amp		16	1		172	63.50		235.50	284
0460	1,000 amp		13	1.231		172	78.50		250.50	305
0465	1,200 amp		11.80	1.356		172	86.50		258.50	315
0470	1,350 amp		10.60	1.509		172	96		268	330

26 25 13.60 Copper or Aluminum Bus Duct Fittings		Crew	Daily Output	Labor-Hours	Unit	Material	2021 Bare Costs Labor	Equipment	Total	Total Incl O&P
0480	1,600 amp	2 Elec	9.20	1.739	Ea.	172	111		283	355
0490	2,000 amp		8	2		172	127		299	380
0500	2,500 amp		6.40	2.500		172	159		331	425
0510	3,000 amp		5.40	2.963		172	189		361	470
0520	4,000 amp		4	4		172	255		427	570
0540	Weather stop, 225 amp		12	1.333		510	85		595	685
0550	400 amp		10	1.600		510	102		612	710
0560	600 amp		9	1.778		510	113		623	730
0570	800 amp		8	2		510	127		637	750
0580	1,000 amp		6.40	2.500		510	159		669	795
0585	1,200 amp		5.90	2.712		510	173		683	815
0590	1,350 amp		5.40	2.963		510	189		699	840
0600	1,600 amp		4.60	3.478		510	222		732	890
0610	2,000 amp		4	4		510	255		765	940
0620	2,500 amp		3.20	5		510	320		830	1,025
0630	3,000 amp		2.60	6.154		510	390		900	1,150
0640	4,000 amp		2	8		510	510		1,020	1,325
0660	End closure, 225 amp		34	.471		400	30		430	485
0670	400 amp		32	.500		400	32		432	490
0680	600 amp		28	.571		400	36.50		436.50	495
0690	800 amp		26	.615		175	39		214	251
0700	1,000 amp		24	.667		400	42.50		442.50	505
0705	1,200 amp		23	.696		171	44.50		215.50	254
0710	1,350 amp		22	.727		171	46.50		217.50	257
0720	1,600 amp		20	.800		400	51		451	515
0730	2,000 amp		18	.889		520	56.50		576.50	655
0740	2,500 amp		16	1		286	63.50		349.50	410
0750	3,000 amp		14	1.143		286	73		359	425
0760	4,000 amp		12	1.333		255	85		340	405
0780	Switchboard stub, 3 pole 3 wire, 225 amp		6	2.667		1,025	170		1,195	1,400
0790	400 amp		5.20	3.077		1,025	196		1,221	1,450
0800	600 amp		4.60	3.478		1,025	222		1,247	1,475
0810	800 amp		3.60	4.444		1,250	283		1,533	1,800
0820	1,000 amp		3.40	4.706		1,450	300		1,750	2,050
0825	1,200 amp		3.20	5		1,450	320		1,770	2,075
0830	1,350 amp		3	5.333		1,750	340		2,090	2,425
0840	1,600 amp		2.80	5.714		2,075	365		2,440	2,825
0850	2,000 amp		2.40	6.667		2,400	425		2,825	3,250
0860	2,500 amp		2	8		2,925	510		3,435	3,975
0870	3,000 amp		1.80	8.889		3,325	565		3,890	4,500
0880	4,000 amp		1.60	10		4,300	635		4,935	5,675
0890	5,000 amp		1.40	11.429		5,325	730		6,055	6,925
0900	3 pole 4 wire, 225 amp		5.40	2.963		1,275	189		1,464	1,675
0910	400 amp		4.60	3.478		1,275	222		1,497	1,725
0920	600 amp		4	4		1,275	255		1,530	1,775
0930	800 amp		3.20	5		1,550	320		1,870	2,175
0940	1,000 amp		3	5.333		1,800	340		2,140	2,500
0950	1,350 amp		2.60	6.154		2,300	390		2,690	3,100
0960	1,600 amp		2.40	6.667		2,625	425		3,050	3,525
0970	2,000 amp		2	8		3,150	510		3,660	4,225
0980	2,500 amp		1.80	8.889		3,875	565		4,440	5,125
0990	3,000 amp		1.60	10		4,500	635		5,135	5,900
1000	4,000 amp		1.40	11.429		5,875	730		6,605	7,550

For customer support on your Electrical Costs with RSMeans Data, call 800.448.8182.

265

26 25 13.60 Copper or Aluminum Bus Duct Fittings

		Crew	Daily Output	Labor-Hours	Unit	Material	2021 Bare Costs Labor	Equipment	Total	Total Incl O&P
1050	Service head, weatherproof, 3 pole 3 wire, 225 amp	2 Elec	3	5.333	Ea.	1,750	340		2,090	2,425
1060	400 amp		2.80	5.714		1,750	365		2,115	2,475
1070	600 amp		2.60	6.154		1,750	390		2,140	2,500
1080	800 amp		2.40	6.667		1,950	425		2,375	2,775
1090	1,000 amp		2	8		2,125	510		2,635	3,075
1100	1,350 amp		1.80	8.889		2,750	565		3,315	3,875
1110	1,600 amp		1.60	10		3,075	635		3,710	4,325
1120	2,000 amp		1.40	11.429		3,775	730		4,505	5,225
1130	2,500 amp		1.20	13.333		4,500	850		5,350	6,200
1140	3,000 amp		.90	17.778		5,200	1,125		6,325	7,400
1150	4,000 amp		.70	22.857		6,600	1,450		8,050	9,450
1200	3 pole 4 wire, 225 amp		2.60	6.154		1,950	390		2,340	2,725
1210	400 amp		2.40	6.667		1,950	425		2,375	2,775
1220	600 amp		2.20	7.273		1,950	465		2,415	2,850
1230	800 amp		2	8		2,275	510		2,785	3,250
1240	1,000 amp		1.70	9.412		2,600	600		3,200	3,775
1250	1,350 amp		1.50	10.667		3,100	680		3,780	4,400
1260	1,600 amp		1.40	11.429		3,400	730		4,130	4,825
1270	2,000 amp		1.20	13.333		4,525	850		5,375	6,225
1280	2,500 amp		1	16		5,550	1,025		6,575	7,650
1290	3,000 amp		.80	20		6,575	1,275		7,850	9,125
1300	4,000 amp		.60	26.667		8,450	1,700		10,150	11,800
1350	Flanged end, 3 pole 3 wire, 225 amp		6	2.667		950	170		1,120	1,300
1360	400 amp		5.20	3.077		950	196		1,146	1,350
1370	600 amp		4.60	3.478		950	222		1,172	1,375
1380	800 amp		3.60	4.444		1,050	283		1,333	1,600
1390	1,000 amp		3.40	4.706		1,200	300		1,500	1,750
1395	1,200 amp		3.20	5		1,325	320		1,645	1,925
1400	1,350 amp		3	5.333		1,425	340		1,765	2,050
1410	1,600 amp		2.80	5.714		1,600	365		1,965	2,325
1420	2,000 amp		2.40	6.667		1,900	425		2,325	2,725
1430	2,500 amp		2	8		2,200	510		2,710	3,175
1440	3,000 amp		1.80	8.889		2,525	565		3,090	3,625
1450	4,000 amp		1.60	10		3,125	635		3,760	4,400
1500	3 pole 4 wire, 225 amp		5.40	2.963		1,100	189		1,289	1,475
1510	400 amp		4.60	3.478		1,100	222		1,322	1,525
1520	600 amp		4	4		1,100	255		1,355	1,575
1530	800 amp		3.20	5		1,300	320		1,620	1,900
1540	1,000 amp		3	5.333		1,425	340		1,765	2,075
1545	1,200 amp		2.80	5.714		1,650	365		2,015	2,350
1550	1,350 amp		2.60	6.154		1,750	390		2,140	2,500
1560	1,600 amp		2.40	6.667		2,000	425		2,425	2,825
1570	2,000 amp		2	8		2,375	510		2,885	3,375
1580	2,500 amp		1.80	8.889		2,825	565		3,390	3,950
1590	3,000 amp		1.60	10		3,250	635		3,885	4,525
1600	4,000 amp		1.40	11.429		4,200	730		4,930	5,700
1650	Hanger, standard, 225 amp		64	.250		57.50	15.95		73.45	87
1660	400 amp		48	.333		57.50	21		78.50	95
1670	600 amp		40	.400		57.50	25.50		83	102
1680	800 amp		32	.500		57.50	32		89.50	111
1690	1,000 amp		24	.667		57.50	42.50		100	127
1695	1,200 amp		22	.727		57.50	46.50		104	133
1700	1,350 amp		20	.800		57.50	51		108.50	139

26 25 13 – Low-Voltage Busways

26 25 13.60 Copper or Aluminum Bus Duct Fittings

		Crew	Daily Output	Labor-Hours	Unit	Material	2021 Bare Costs Labor	Equipment	Total	Total Incl O&P
1710	1,600 amp	2 Elec	20	.800	Ea.	57.50	51		108.50	139
1720	2,000 amp		18	.889		57.50	56.50		114	148
1730	2,500 amp		16	1		57.50	63.50		121	158
1740	3,000 amp		16	1		57.50	63.50		121	158
1750	4,000 amp		16	1		57.50	63.50		121	158
1800	Spring type, 225 amp		16	1		89	63.50		152.50	193
1810	400 amp		14	1.143		89	73		162	206
1820	600 amp		14	1.143		89	73		162	206
1830	800 amp		14	1.143		89	73		162	206
1840	1,000 amp		14	1.143		89	73		162	206
1845	1,200 amp		14	1.143		89	73		162	206
1850	1,350 amp		14	1.143		89	73		162	206
1860	1,600 amp		12	1.333		89	85		174	224
1870	2,000 amp		12	1.333		89	85		174	224
1880	2,500 amp		12	1.333		89	85		174	224
1890	3,000 amp		10	1.600		89	102		191	249
1900	4,000 amp		10	1.600		89	102		191	249

26 25 13.70 Feedrail

		Crew	Daily Output	Labor-Hours	Unit	Material	2021 Bare Costs Labor	Equipment	Total	Total Incl O&P
0010	**FEEDRAIL**, 12' mounting									
0050	Trolley busway, 3 pole									
0100	300 volt 60 amp, plain, 10' lengths	1 Elec	50	.160	L.F.	22	10.20		32.20	39.50
0300	Door track		50	.160		43	10.20		53.20	62.50
0500	Curved track		30	.267		20.50	17		37.50	47.50
0700	Coupling				Ea.	13.70			13.70	15.05
0900	Center feed	1 Elec	5.30	1.509		46.50	96		142.50	194
1100	End feed		5.30	1.509		41	96		137	188
1300	Hanger set		24	.333		2.84	21		23.84	34.50
3000	600 volt 100 amp, plain, 10' lengths		35	.229	L.F.	60.50	14.55		75.05	88
3300	Door track		35	.229	"	93.50	14.55		108.05	125
3700	Coupling				Ea.	48.50			48.50	53.50
4000	End cap	1 Elec	40	.200		65.50	12.75		78.25	91
4200	End feed		4	2		221	127		348	430
4500	Trolley, 600 volt, 20 amp		5.30	1.509		267	96		363	435
4700	30 amp		5.30	1.509		267	96		363	435
4900	Duplex, 40 amp		4	2		820	127		947	1,100
5000	60 amp		4	2		815	127		942	1,075
5300	Fusible, 20 amp		4	2		610	127		737	860
5500	30 amp		4	2		610	127		737	860
5900	300 volt, 20 amp		5.30	1.509		252	96		348	420
6000	30 amp		5.30	1.509		335	96		431	515
6300	Fusible, 20 amp		4.70	1.702		350	108		458	545
6500	30 amp		4.70	1.702		485	108		593	695
7300	Busway, 250 volt 50 amp, 2 wire		70	.114	L.F.	19.90	7.30		27.20	33
7330	Coupling				Ea.	43.50			43.50	48
7340	Center feed	1 Elec	6	1.333		515	85		600	690
7350	End feed		6	1.333		119	85		204	257
7360	End cap		40	.200		29	12.75		41.75	51
7370	Hanger set		24	.333		3.36	21		24.36	35
7400	125/250 volt 50 amp, 3 wire		60	.133	L.F.	20.50	8.50		29	35
7430	Coupling		6	1.333	Ea.	60.50	85		145.50	193
7440	Center feed		6	1.333		510	85		595	685
7450	End feed		6	1.333		114	85		199	252

For customer support on your Electrical Costs with RSMeans Data, call 800.448.8182.

267

26 25 13.70 Feedrail		Crew	Daily Output	Labor-Hours	Unit	Material	2021 Bare Costs Labor	2021 Bare Costs Equipment	Total	Total Incl O&P
7460	End cap	1 Elec	40	.200	Ea.	29	12.75		41.75	51
7470	Hanger set		24	.333		4.80	21		25.80	37
7480	Trolley, 250 volt, 2 pole, 20 amp		6	1.333		40.50	85		125.50	171
7490	30 amp		6	1.333		40.50	85		125.50	171
7500	125/250 volt, 3 pole, 20 amp		6	1.333		36.50	85		121.50	166
7510	30 amp		6	1.333		36.50	85		121.50	166
8000	Cleaning tools, 300 volt, dust remover					106			106	117
8100	Bus bar cleaner					199			199	219
8300	600 volt, dust remover, 60 amp					305			305	335
8400	100 amp					795			795	875
8600	Bus bar cleaner, 60 amp					755			755	830
8700	100 amp					975			975	1,075

26 27 13.10 Meter Centers and Sockets		Crew	Daily Output	Labor-Hours	Unit	Material	2021 Bare Costs Labor	2021 Bare Costs Equipment	Total	Total Incl O&P
0010	**METER CENTERS AND SOCKETS**									
0100	Sockets, single position, 4 terminal, 100 amp	1 Elec	3.20	2.500	Ea.	73	159		232	320
0200	150 amp		2.30	3.478		55	222		277	390
0300	200 amp		1.90	4.211		106	268		374	515
0400	Transformer rated, 20 amp		3.20	2.500		174	159		333	430
0500	Double position, 4 terminal, 100 amp		2.80	2.857		234	182		416	525
0600	150 amp		2.10	3.810		275	243		518	660
0700	200 amp		1.70	4.706		515	300		815	1,025
0800	Trans-socket, 13 terminal, 3 CT mounts, 400 amp		1	8		1,225	510		1,735	2,100
0900	800 amp	2 Elec	1.20	13.333		1,625	850		2,475	3,025
1100	Meter centers and sockets, three phase, single pos, 7 terminal, 100 amp	1 Elec	2.80	2.857		157	182		339	440
1200	200 amp		2.10	3.810		264	243		507	650
1400	400 amp		1.70	4.706		840	300		1,140	1,375
2000	Meter center, main fusible switch, 1P 3W 120/240 V									
2030	400 amp	2 Elec	1.60	10	Ea.	1,000	635		1,635	2,050
2040	600 amp		1.10	14.545		1,600	925		2,525	3,125
2050	800 amp		.90	17.778		5,775	1,125		6,900	8,025
2060	Rainproof 1P 3W 120/240 V, 400 A		1.60	10		2,000	635		2,635	3,150
2070	600 amp		1.10	14.545		3,475	925		4,400	5,200
2080	800 amp		.90	17.778		5,450	1,125		6,575	7,650
2100	3P 4W 120/208 V, 400 amp		1.60	10		1,075	635		1,710	2,125
2110	600 amp	2 Elec	1.10	14.545	Ea.	1,800	925		2,725	3,350
2120	800 amp		.90	17.778		2,450	1,125		3,575	4,375
2130	Rainproof 3P 4W 120/208 V, 400 amp		1.60	10		2,275	635		2,910	3,475
2140	600 amp		1.10	14.545		3,675	925		4,600	5,425
2150	800 amp		.90	17.778		7,925	1,125		9,050	10,400
2170	Main circuit breaker, 1P 3W 120/240 V									
2180	400 amp	2 Elec	1.60	10	Ea.	1,475	635		2,110	2,550
2190	600 amp		1.10	14.545		1,900	925		2,825	3,450
2200	800 amp		.90	17.778		3,450	1,125		4,575	5,450
2210	1,000 amp		.80	20		4,850	1,275		6,125	7,250
2220	1,200 amp		.76	21.053		4,875	1,350		6,225	7,350
2230	1,600 amp		.68	23.529		8,500	1,500		10,000	11,600
2240	Rainproof 1P 3W 120/240 V, 400 amp		1.60	10		3,075	635		3,710	4,325
2250	600 amp		1.10	14.545		4,800	925		5,725	6,675

26 27 13.10 Meter Centers and Sockets

		Crew	Daily Output	Labor-Hours	Unit	Material	2021 Bare Costs Labor	Equipment	Total	Total Incl O&P
2260	800 amp	2 Elec	.90	17.778	Ea.	5,600	1,125		6,725	7,850
2270	1,000 amp		.80	20		7,725	1,275		9,000	10,400
2280	1,200 amp		.76	21.053		10,400	1,350		11,750	13,500
2300	3P 4W 120/208 V, 400 amp		1.60	10		3,750	635		4,385	5,075
2310	600 amp		1.10	14.545		5,725	925		6,650	7,675
2320	800 amp		.90	17.778		6,800	1,125		7,925	9,175
2330	1,000 amp		.80	20		8,950	1,275		10,225	11,800
2340	1,200 amp		.76	21.053		11,400	1,350		12,750	14,600
2350	1,600 amp		.68	23.529		10,300	1,500		11,800	13,500
2360	Rainproof 3P 4W 120/208 V, 400 amp		1.60	10		4,075	635		4,710	5,425
2370	600 amp		1.10	14.545		5,725	925		6,650	7,675
2380	800 amp		.90	17.778		6,800	1,125		7,925	9,175
2390	1,000 amp		.76	21.053		8,950	1,350		10,300	11,900
2400	1,200 amp	▼	.68	23.529	▼	11,400	1,500		12,900	14,800
2420	Main lugs terminal box, 1P 3W 120/240 V									
2430	800 amp	2 Elec	.94	17.021	Ea.	470	1,075		1,545	2,125
2440	1,200 amp		.72	22.222		660	1,425		2,085	2,825
2450	Rainproof 1P 3W 120/240 V, 225 amp		2.40	6.667		455	425		880	1,125
2460	800 amp		.94	17.021		670	1,075		1,745	2,350
2470	1,200 amp		.72	22.222		1,350	1,425		2,775	3,575
2500	3P 4W 120/208 V, 800 amp		.94	17.021		745	1,075		1,820	2,425
2510	1,200 amp		.72	22.222		1,500	1,425		2,925	3,750
2520	Rainproof 3P 4W 120/208 V, 225 amp		2.40	6.667		455	425		880	1,125
2530	800 amp		.94	17.021		745	1,075		1,820	2,425
2540	1,200 amp	▼	.72	22.222	▼	1,500	1,425		2,925	3,750
2590	Basic meter device									
2600	1P 3W 120/240 V 4 jaw 125A sockets, 3 meter	2 Elec	1	16	Ea.	485	1,025		1,510	2,050
2610	4 meter		.90	17.778		515	1,125		1,640	2,250
2620	5 meter		.80	20		615	1,275		1,890	2,575
2630	6 meter		.60	26.667		745	1,700		2,445	3,350
2640	7 meter		.56	28.571		1,700	1,825		3,525	4,550
2650	8 meter		.52	30.769		1,850	1,950		3,800	4,925
2660	10 meter	▼	.48	33.333	▼	2,300	2,125		4,425	5,675
2680	Rainproof 1P 3W 120/240 V 4 jaw 125A sockets									
2690	3 meter	2 Elec	1	16	Ea.	770	1,025		1,795	2,375
2700	4 meter		.90	17.778		920	1,125		2,045	2,675
2710	6 meter		.60	26.667		1,325	1,700		3,025	3,975
2720	7 meter		.56	28.571		1,700	1,825		3,525	4,550
2730	8 meter	▼	.52	30.769	▼	1,850	1,950		3,800	4,925
2750	1P 3W 120/240 V 4 jaw sockets									
2760	with 125A circuit breaker, 3 meter	2 Elec	1	16	Ea.	1,450	1,025		2,475	3,100
2770	4 meter		.90	17.778		1,825	1,125		2,950	3,675
2780	5 meter		.80	20		2,275	1,275		3,550	4,400
2790	6 meter		.60	26.667		2,675	1,700		4,375	5,450
2800	7 meter		.56	28.571		3,250	1,825		5,075	6,275
2810	8 meter		.52	30.769		3,625	1,950		5,575	6,900
2820	10 meter	▼	.48	33.333	▼	4,525	2,125		6,650	8,150
2830	Rainproof 1P 3W 120/240 V 4 jaw sockets									
2840	with 125A circuit breaker, 3 meter	2 Elec	1	16	Ea.	1,450	1,025		2,475	3,100
2850	4 meter		.90	17.778		1,825	1,125		2,950	3,675
2870	6 meter		.60	26.667		2,675	1,700		4,375	5,450
2880	7 meter		.56	28.571		3,250	1,825		5,075	6,275
2890	8 meter	▼	.52	30.769	▼	3,625	1,950		5,575	6,900

26 27 13.10 Meter Centers and Sockets

		Crew	Daily Output	Labor-Hours	Unit	Material	2021 Bare Costs Labor	Equipment	Total	Total Incl O&P
2920	1P 3W on 3P 4W 120/208 V system 5 jaw									
2930	125A sockets, 3 meter	2 Elec	1	16	Ea.	770	1,025		1,795	2,375
2940	4 meter		.90	17.778		920	1,125		2,045	2,675
2950	5 meter		.80	20		1,150	1,275		2,425	3,175
2960	6 meter		.60	26.667		1,325	1,700		3,025	3,975
2970	7 meter		.56	28.571		1,700	1,825		3,525	4,550
2980	8 meter		.52	30.769		1,850	1,950		3,800	4,925
2990	10 meter		.48	33.333		2,300	2,125		4,425	5,675
3000	Rainproof 1P 3W on 3P 4W 120/208 V system									
3020	5 jaw 125A sockets, 3 meter	2 Elec	1	16	Ea.	770	1,025		1,795	2,375
3030	4 meter		.90	17.778		920	1,125		2,045	2,675
3050	6 meter		.60	26.667		1,325	1,700		3,025	3,975
3060	7 meter		.56	28.571		1,700	1,825		3,525	4,550
3070	8 meter		.52	30.769		1,850	1,950		3,800	4,925
3090	1P 3W on 3P 4W 120/208 V system 5 jaw sockets									
3100	With 125A circuit breaker, 3 meter	2 Elec	1	16	Ea.	1,450	1,025		2,475	3,100
3110	4 meter	"	.90	17.778	"	1,825	1,125		2,950	3,675
3120	5 meter	2 Elec	.80	20	Ea.	2,275	1,275		3,550	4,400
3130	6 meter		.60	26.667		2,675	1,700		4,375	5,450
3140	7 meter		.56	28.571		3,250	1,825		5,075	6,275
3150	8 meter		.52	30.769		3,625	1,950		5,575	6,900
3160	10 meter		.48	33.333		4,525	2,125		6,650	8,150
3170	Rainproof 1P 3W on 3P 4W 120/208 V system									
3180	5 jaw sockets w/125A circuit breaker, 3 meter	2 Elec	1	16	Ea.	1,450	1,025		2,475	3,100
3190	4 meter		.90	17.778		1,825	1,125		2,950	3,675
3210	6 meter		.60	26.667		2,675	1,700		4,375	5,450
3220	7 meter		.56	28.571		3,250	1,825		5,075	6,275
3230	8 meter		.52	30.769		3,625	1,950		5,575	6,900
3250	1P 3W 120/240 V 4 jaw sockets									
3260	with 200A circuit breaker, 3 meter	2 Elec	1	16	Ea.	2,150	1,025		3,175	3,900
3270	4 meter		.90	17.778		2,925	1,125		4,050	4,875
3290	6 meter		.60	26.667		4,300	1,700		6,000	7,275
3300	7 meter		.56	28.571		5,075	1,825		6,900	8,275
3310	8 meter		.56	28.571		5,825	1,825		7,650	9,100
3330	Rainproof 1P 3W 120/240 V 4 jaw sockets									
3350	with 200A circuit breaker, 3 meter	2 Elec	1	16	Ea.	2,150	1,025		3,175	3,900
3360	4 meter		.90	17.778		2,925	1,125		4,050	4,875
3380	6 meter		.60	26.667		4,300	1,700		6,000	7,275
3390	7 meter		.56	28.571		5,075	1,825		6,900	8,275
3400	8 meter		.52	30.769		5,825	1,950		7,775	9,300
3420	1P 3W on 3P 4W 120/208 V 5 jaw sockets									
3430	with 200A circuit breaker, 3 meter	2 Elec	1	16	Ea.	2,150	1,025		3,175	3,900
3440	4 meter		.90	17.778		2,925	1,125		4,050	4,875
3460	6 meter		.60	26.667		4,325	1,700		6,025	7,275
3470	7 meter		.56	28.571		5,075	1,825		6,900	8,275
3480	8 meter		.52	30.769		5,825	1,950		7,775	9,300
3500	Rainproof 1P 3W on 3P 4W 120/208 V 5 jaw socket									
3510	with 200A circuit breaker, 3 meter	2 Elec	1	16	Ea.	2,150	1,025		3,175	3,900
3520	4 meter		.90	17.778		2,925	1,125		4,050	4,875
3540	6 meter		.60	26.667		4,300	1,700		6,000	7,275
3550	7 meter		.56	28.571		5,075	1,825		6,900	8,275
3560	8 meter		.52	30.769		5,825	1,950		7,775	9,300
3600	Automatic circuit closing, add					83.50			83.50	91.50

26 27 Low-Voltage Distribution Equipment

26 27 13 – Electricity Metering

26 27 13.10 Meter Centers and Sockets

		Crew	Daily Output	Labor-Hours	Unit	Material	2021 Bare Costs Labor	Equipment	Total	Total Incl O&P
3610	Manual circuit closing, add				Ea.	95			95	105
3650	Branch meter device									
3660	3P 4W 208/120 or 240/120 V 7 jaw sockets									
3670	with 200A circuit breaker, 2 meter	2 Elec	.90	17.778	Ea.	3,600	1,125		4,725	5,625
3680	3 meter	"	.80	20	"	5,400	1,275		6,675	7,850
3690	4 meter	2 Elec	.70	22.857	Ea.	7,200	1,450		8,650	10,100
3700	Main circuit breaker 42,000 rms, 400 amp		1.60	10		1,975	635		2,610	3,125
3710	600 amp		1.10	14.545		2,725	925		3,650	4,375
3720	800 amp		.90	17.778		4,150	1,125		5,275	6,250
3730	Rainproof main circ. breaker 42,000 rms, 400 amp		1.60	10		3,150	635		3,785	4,400
3740	600 amp		1.10	14.545		5,075	925		6,000	6,950
3750	800 amp		.90	17.778		6,800	1,125		7,925	9,175
3760	Main circuit breaker 65,000 rms, 400 amp		1.60	10		4,275	635		4,910	5,650
3770	600 amp		1.10	14.545		6,000	925		6,925	8,000
3780	800 amp		.90	17.778		6,800	1,125		7,925	9,175
3790	1,000 amp		.80	20		8,950	1,275		10,225	11,800
3800	1,200 amp		.76	21.053		11,400	1,350		12,750	14,600
3810	1,600 amp		.68	23.529		23,500	1,500		25,000	28,000
3820	Rainproof main circ. breaker 65,000 rms, 400 amp		1.60	10		4,275	635		4,910	5,650
3830	600 amp		1.10	14.545		6,000	925		6,925	8,000
3840	800 amp		.90	17.778		6,800	1,125		7,925	9,150
3850	1,000 amp		.80	20		8,950	1,275		10,225	11,800
3860	1,200 amp		.76	21.053		11,400	1,350		12,750	14,600
3880	Main circuit breaker 100,000 rms, 400 amp		1.60	10		4,275	635		4,910	5,650
3890	600 amp		1.10	14.545		6,000	925		6,925	8,000
3900	800 amp		.90	17.778		7,050	1,125		8,175	9,450
3910	Rainproof main circ. breaker 100,000 rms, 400 amp		1.60	10		4,275	635		4,910	5,650
3920	600 amp		1.10	14.545		6,000	925		6,925	8,000
3930	800 amp		.90	17.778		7,050	1,125		8,175	9,450
3940	Main lugs terminal box, 800 amp		.94	17.021		745	1,075		1,820	2,425
3950	1,600 amp		.72	22.222		3,525	1,425		4,950	5,975
3960	Rainproof, 800 amp		.94	17.021		745	1,075		1,820	2,425
3970	1,600 amp		.72	22.222		3,550	1,425		4,975	6,000

26 27 16 – Electrical Cabinets and Enclosures

26 27 16.10 Cabinets

			Crew	Daily Output	Labor-Hours	Unit	Material	2021 Bare Costs Labor	Equipment	Total	Total Incl O&P
0010	**CABINETS**	R260533-70									
7000	Cabinets, current transformer										
7050	Single door, 24" H x 24" W x 10" D	R262716-40	1 Elec	1.60	5	Ea.	143	320		463	635
7100	30" H x 24" W x 10" D			1.30	6.154		159	390		549	755
7150	36" H x 24" W x 10" D			1.10	7.273		405	465		870	1,125
7200	30" H x 30" W x 10" D			1	8		265	510		775	1,050
7250	36" H x 30" W x 10" D			.90	8.889		330	565		895	1,200
7300	36" H x 36" W x 10" D			.80	10		340	635		975	1,325
7500	Double door, 48" H x 36" W x 10" D			.60	13.333		630	850		1,480	1,950
7550	24" H x 24" W x 12" D			1	8		251	510		761	1,025
8000	NEMA 12, double door, floor mounted										
8020	54" H x 42" W x 8" D		2 Elec	6	2.667	Ea.	1,575	170		1,745	1,975
8040	60" H x 48" W x 8" D			5.40	2.963		2,125	189		2,314	2,625
8060	60" H x 48" W x 10" D			5.40	2.963		2,125	189		2,314	2,625
8080	60" H x 60" W x 10" D			5	3.200		2,475	204		2,679	3,000
8100	72" H x 60" W x 10" D			4	4		2,825	255		3,080	3,475
8120	72" H x 72" W x 10" D			3.40	4.706		3,125	300		3,425	3,875

For customer support on your Electrical Costs with RSMeans Data, call 800.448.8182.

271

26 27 16.10 Cabinets		Crew	Daily Output	Labor-Hours	Unit	Material	2021 Bare Costs Labor	Equipment	Total	Total Incl O&P
8140	60" H x 48" W x 12" D	2 Elec	3.40	4.706	Ea.	2,150	300		2,450	2,825
8160	60" H x 60" W x 12" D		3.20	5		2,450	320		2,770	3,175
8180	72" H x 60" W x 12" D		3	5.333		2,800	340		3,140	3,575
8200	72" H x 72" W x 12" D		3	5.333		3,150	340		3,490	3,950
8220	60" H x 48" W x 16" D		3.20	5		2,325	320		2,645	3,050
8240	72" H x 72" W x 16" D		2.60	6.154		3,325	390		3,715	4,225
8260	60" H x 48" W x 20" D		3	5.333		2,525	340		2,865	3,275
8280	72" H x 72" W x 20" D		2.20	7.273		3,600	465		4,065	4,650
8300	60" H x 48" W x 24" D		2.60	6.154		2,650	390		3,040	3,500
8320	72" H x 72" W x 24" D	▼	2	8	▼	3,800	510		4,310	4,925
8340	Pushbutton enclosure, oiltight									
8360	3-1/2" H x 3-1/4" W x 2-3/4" D, for 1 P.B.	1 Elec	12	.667	Ea.	41.50	42.50		84	109
8380	5-3/4" H x 3-1/4" W x 2-3/4" D, for 2 P.B.		11	.727		47	46.50		93.50	121
8400	8" H x 3-1/4" W x 2-3/4" D, for 3 P.B.		10.50	.762		88.50	48.50		137	169
8420	10-1/4" H x 3-1/4" W x 2-3/4" D, for 4 P.B.		10.50	.762		99	48.50		147.50	181
8460	12-1/2" H x 3-1/4" W x 3" D, for 5 P.B.		9	.889		109	56.50		165.50	204
8480	9-1/2" H x 6-1/4" W x 3" D, for 6 P.B.		8.50	.941		127	60		187	229
8500	9-1/2" H x 8-1/2" W x 3" D, for 9 P.B.		8	1		149	63.50		212.50	259
8510	11-3/4" H x 8-1/2" W x 3" D, for 12 P.B.		7	1.143		166	73		239	291
8520	11-3/4" H x 10-3/4" W x 3" D, for 16 P.B.		6.50	1.231		192	78.50		270.50	330
8540	14" H x 10-3/4" W x 3" D, for 20 P.B.		5	1.600		161	102		263	330
8560	14" H x 13" W x 3" D, for 25 P.B.	▼	4.50	1.778	▼	175	113		288	360
8580	Sloping front pushbutton enclosures									
8600	3-1/2" H x 7-3/4" W x 4-7/8" D, for 3 P.B.	1 Elec	10	.800	Ea.	121	51		172	209
8620	7-1/4" H x 8-1/2" W x 6-3/4" D, for 6 P.B.		8	1		174	63.50		237.50	287
8640	9-1/2" H x 8-1/2" W x 7-7/8" D, for 9 P.B.		7	1.143		210	73		283	340
8660	11-1/4" H x 8-1/2" W x 9" D, for 12 P.B.		5	1.600		243	102		345	420
8680	11-3/4" H x 10" W x 9" D, for 16 P.B.		5	1.600		259	102		361	435
8700	11-3/4" H x 13" W x 9" D, for 20 P.B.		5	1.600		298	102		400	480
8720	14" H x 13" W x 10-1/8" D, for 25 P.B.	▼	4.50	1.778	▼	340	113		453	545
8740	Pedestals, not including P.B. enclosure or base									
8760	Straight column 4" x 4"	1 Elec	4.50	1.778	Ea.	305	113		418	505
8780	6" x 6"		4	2		460	127		587	695
8800	Angled column 4" x 4"		4.50	1.778		360	113		473	570
8820	6" x 6"		4	2		540	127		667	780
8840	Pedestal, base 18" x 18"		10	.800		189	51		240	283
8860	24" x 24"	▼	9	.889	▼	420	56.50		476.50	550
8900	Electronic rack enclosures									
8920	72" H x 19" W x 24" D	1 Elec	1.50	5.333	Ea.	2,350	340		2,690	3,075
8940	72" H x 23" W x 24" D		1.50	5.333		2,500	340		2,840	3,250
8960	72" H x 19" W x 30" D		1.30	6.154		2,675	390		3,065	3,500
8980	72" H x 19" W x 36" D		1.20	6.667		2,725	425		3,150	3,625
9000	72" H x 23" W x 36" D	▼	1.20	6.667	▼	3,025	425		3,450	3,950
9020	NEMA 12 & 4 enclosure panels									
9040	12" x 24"	1 Elec	20	.400	Ea.	44	25.50		69.50	86.50
9060	16" x 12"		20	.400		31.50	25.50		57	73
9080	20" x 16"		20	.400		46.50	25.50		72	89
9100	20" x 20"		19	.421		56	27		83	102
9120	24" x 20"		18	.444		73	28.50		101.50	122
9140	24" x 24"		17	.471		83.50	30		113.50	137
9160	30" x 20"		16	.500		143	32		175	205
9180	30" x 24"		16	.500		102	32		134	160
9200	36" x 24"	▼	15	.533	▼	121	34		155	184

For customer support on your Electrical Costs with RSMeans Data, call 800.448.8182.

26 27 16 – Electrical Cabinets and Enclosures

26 27 16.10 Cabinets

		Crew	Daily Output	Labor-Hours	Unit	Material	2021 Bare Costs Labor	Equipment	Total	Total Incl O&P
9220	36" x 30"	1 Elec	15	.533	Ea.	161	34		195	228
9240	42" x 24"		15	.533		146	34		180	212
9260	42" x 30"		14	.571		183	36.50		219.50	255
9280	42" x 36"		14	.571		217	36.50		253.50	293
9300	48" x 24"		14	.571		169	36.50		205.50	240
9320	48" x 30"		14	.571		207	36.50		243.50	282
9340	48" x 36"		13	.615		243	39		282	325
9360	60" x 36"		12	.667		300	42.50		342.50	395
9400	Wiring trough steel JIC, clamp cover									
9490	4" x 4", 12" long	1 Elec	12	.667	Ea.	112	42.50		154.50	186
9510	24" long		10	.800		143	51		194	233
9530	36" long		8	1		174	63.50		237.50	286
9540	48" long		7	1.143		205	73		278	335
9550	60" long		6	1.333		236	85		321	385
9560	6" x 6", 12" long		11	.727		146	46.50		192.50	230
9580	24" long		9	.889		191	56.50		247.50	294
9600	36" long		7	1.143		242	73		315	375
9610	48" long		6	1.333		289	85		374	445
9620	60" long		5	1.600		335	102		437	520

26 27 16.20 Cabinets and Enclosures

		Crew	Daily Output	Labor-Hours	Unit	Material	2021 Bare Costs Labor	Equipment	Total	Total Incl O&P
0010	**CABINETS AND ENCLOSURES** Nonmetallic									
0080	Enclosures fiberglass NEMA 4X									
0100	Wall mount, quick release latch door, 20" H x 16" W x 6" D	1 Elec	4.80	1.667	Ea.	695	106		801	925
0110	20" H x 20" W x 6" D		4.50	1.778		840	113		953	1,100
0120	24" H x 20" W x 6" D		4.20	1.905		900	121		1,021	1,175
0130	20" H x 16" W x 8" D		4.50	1.778		790	113		903	1,025
0140	20" H x 20" W x 8" D		4.20	1.905		885	121		1,006	1,150
0150	24" H x 24" W x 8" D		3.80	2.105		985	134		1,119	1,275
0160	30" H x 24" W x 8" D		3.20	2.500		1,100	159		1,259	1,425
0170	36" H x 30" W x 8" D		3	2.667		1,525	170		1,695	1,925
0180	20" H x 16" W x 10" D		3.50	2.286		920	146		1,066	1,250
0190	20" H x 20" W x 10" D		3.20	2.500		980	159		1,139	1,300
0200	24" H x 20" W x 10" D		3	2.667		1,050	170		1,220	1,400
0210	30" H x 24" W x 10" D		2.80	2.857		1,200	182		1,382	1,575
0220	20" H x 16" W x 12" D		3	2.667		1,025	170		1,195	1,375
0230	20" H x 20" W x 12" D		2.80	2.857		1,050	182		1,232	1,425
0240	24" H x 24" W x 12" D		2.60	3.077		1,150	196		1,346	1,550
0250	30" H x 24" W x 12" D		2.40	3.333		1,275	212		1,487	1,725
0260	36" H x 30" W x 12" D		2.20	3.636		1,750	232		1,982	2,275
0270	36" H x 36" W x 12" D		2.10	3.810		1,925	243		2,168	2,475
0280	48" H x 36" W x 12" D		2	4		1,375	255		1,630	1,900
0290	60" H x 36" W x 12" D		1.80	4.444		2,550	283		2,833	3,225
0300	30" H x 24" W x 16" D		1.40	5.714		1,525	365		1,890	2,225
0310	48" H x 36" W x 16" D		1.20	6.667		2,575	425		3,000	3,450
0320	60" H x 36" W x 16" D		1	8		2,900	510		3,410	3,925
0480	Freestanding, one door, 72" H x 25" W x 25" D		.80	10		5,025	635		5,660	6,475
0490	Two doors with two panels, 72" H x 49" W x 24" D		.50	16		11,900	1,025		12,925	14,600
0500	Floor stand kits, for NEMA 4 & 12, 20" W or more, 6" H x 8" D		24	.333		190	21		211	241
0510	6" H x 10" D		24	.333		209	21		230	261
0520	6" H x 12" D		24	.333		237	21		258	293
0530	6" H x 18" D		24	.333		282	21		303	340
0540	12" H x 8" D		22	.364		245	23		268	305

273

26 27 16.20 Cabinets and Enclosures	Crew	Daily Output	Labor-Hours	Unit	Material	2021 Bare Costs Labor	Equipment	Total	Total Incl O&P
0550	12" H x 10" D	1 Elec	22	.364	Ea.	263	23		286	325
0560	12" H x 12" D		22	.364		278	23		301	340
0570	12" H x 16" D		22	.364		315	23		338	385
0580	12" H x 18" D		22	.364		340	23		363	405
0590	12" H x 20" D		22	.364		355	23		378	425
0600	18" H x 8" D		20	.400		300	25.50		325.50	370
0610	18" H x 10" D		20	.400		320	25.50		345.50	390
0620	18" H x 12" D		20	.400		335	25.50		360.50	410
0630	18" H x 16" D		20	.400		375	25.50		400.50	450
0640	24" H x 8" D		16	.500		355	32		387	440
0650	24" H x 10" D		16	.500		375	32		407	465
0660	24" H x 12" D		16	.500		390	32		422	480
0670	24" H x 16" D		16	.500		430	32		462	525
0680	Small, screw cover, 5-1/2" H x 4" W x 4-15/16" D		12	.667		120	42.50		162.50	195
0690	7-1/2" H x 4" W x 4-15/16" D		12	.667		126	42.50		168.50	202
0700	7-1/2" H x 6" W x 5-3/16" D		10	.800		145	51		196	235
0710	9-1/2" H x 6" W x 5-11/16" D		10	.800		151	51		202	242
0720	11-1/2" H x 8" W x 6-11/16" D		8	1		222	63.50		285.50	340
0730	13-1/2" H x 10" W x 7-3/16" D		7	1.143		272	73		345	405
0740	15-1/2" H x 12" W x 8-3/16" D		6	1.333		345	85		430	505
0750	17-1/2" H x 14" W x 8-11/16" D		5	1.600		415	102		517	605
0760	Screw cover with window, 6" H x 4" W x 5" D		12	.667		155	42.50		197.50	233
0770	8" H x 4" W x 5" D		11	.727		205	46.50		251.50	294
0780	8" H x 6" W x 5" D		11	.727		292	46.50		338.50	390
0790	10" H x 6" W x 6" D		10	.800		310	51		361	415
0800	12" H x 8" W x 7" D		8	1		395	63.50		458.50	530
0810	14" H x 10" W x 7" D		7	1.143		470	73		543	630
0820	16" H x 12" W x 8" D		6	1.333		570	85		655	750
0830	18" H x 14" W x 9" D		5	1.600		680	102		782	895
0840	Quick-release latch cover, 5-1/2" H x 4" W x 5" D		12	.667		155	42.50		197.50	234
0850	7-1/2" H x 4" W x 5" D		12	.667		163	42.50		205.50	242
0860	7-1/2" H x 6" W x 5-1/4" D		10	.800		186	51		237	281
0870	9-1/2" H x 6" W x 5-3/4" D		10	.800		194	51		245	289
0880	11-1/2" H x 8" W x 6-3/4" D		8	1		283	63.50		346.50	405
0890	13-1/2" H x 10" W x 7-1/4" D		7	1.143		350	73		423	495
0900	15-1/2" H x 12" W x 8-1/4" D		6	1.333		445	85		530	615
0910	17-1/2" H x 14" W x 8-3/4" D		5	1.600		535	102		637	740
0920	Pushbutton, 1 hole 5-1/2" H x 4" W x 4-15/16" D		12	.667		435	42.50		477.50	545
0930	2 hole 7-1/2" H x 4" W x 4-15/16" D		11	.727		600	46.50		646.50	730
0940	4 hole 7-1/2" H x 6" W x 5-3/16" D		10.50	.762		425	48.50		473.50	535
0950	6 hole 9-1/2" H x 6" W x 5-11/16" D		9	.889		585	56.50		641.50	730
0960	8 hole 11-1/2" H x 8" W x 6-11/16" D		8.50	.941		745	60		805	910
0970	12 hole 13-1/2" H x 10" W x 7-3/16" D		8	1		865	63.50		928.50	1,050
0980	20 hole 15-1/2" H x 12" W x 8-3/16" D		5	1.600		795	102		897	1,025
0990	30 hole 17-1/2" H x 14" W x 8-11/16" D	▼	4.50	1.778	▼	940	113		1,053	1,200
1450	Enclosures polyester NEMA 4X									
1460	Small, screw cover,									
1500	3-15/16" H x 3-15/16" W x 3-1/16" D	1 Elec	12	.667	Ea.	65.50	42.50		108	136
1510	5-3/16" H x 3-5/16" W x 3-1/16" D		12	.667		64.50	42.50		107	134
1520	5-7/8" H x 3-7/8" W x 4-3/16" D		12	.667		68	42.50		110.50	138
1530	5-7/8" H x 5-7/8" W x 4-3/16" D		12	.667		76.50	42.50		119	147
1540	7-5/8" H x 3-5/16" W x 3-1/16" D		12	.667		70.50	42.50		113	141
1550	10-3/16" H x 3-5/16" W x 3-1/16" D	▼	10	.800	▼	88	51		139	172

26 27 16 – Electrical Cabinets and Enclosures

26 27 16.20 Cabinets and Enclosures		Crew	Daily Output	Labor-Hours	Unit	Material	2021 Bare Costs Labor	2021 Bare Costs Equipment	Total	Total Incl O&P
1560	Clear cover, 3-15/16" H x 3-15/16" W x 2-7/8" D	1 Elec	12	.667	Ea.	74	42.50		116.50	145
1570	5-3/16" H x 3-5/16" W x 2-7/8" D		12	.667		79	42.50		121.50	150
1580	5-7/8" H x 3-7/8" W x 4" D		12	.667		98	42.50		140.50	171
1590	5-7/8" H x 5-7/8" W x 4" D		12	.667		121	42.50		163.50	196
1600	7-5/8" H x 3-5/16" W x 2-7/8" D		12	.667		88.50	42.50		131	161
1610	10-3/16" H x 3-5/16" W x 2-7/8" D		10	.800		111	51		162	198
1620	Pushbutton, 1 hole, 5-5/16" H x 3-5/16" W x 3-1/16" D		12	.667		60	42.50		102.50	130
1630	2 hole, 7-5/8" H x 3-5/16" W x 3-1/8" D		11	.727		65	46.50		111.50	141
1640	3 hole, 10-3/16" H x 3-5/16" W x 3-1/16" D		10.50	.762		77	48.50		125.50	157
8000	Wireway fiberglass, straight sect. screwcover, 12" L, 4" W x 4" D		40	.200		220	12.75		232.75	261
8010	6" W x 6" D		30	.267		295	17		312	350
8020	24" L, 4" W x 4" D		20	.400		280	25.50		305.50	350
8030	6" W x 6" D		15	.533		430	34		464	520
8040	36" L, 4" W x 4" D		13.30	.602		335	38.50		373.50	425
8050	6" W x 6" D		10	.800		540	51		591	670
8060	48" L, 4" W x 4" D		10	.800		400	51		451	515
8070	6" W x 6" D		7.50	1.067		650	68		718	820
8080	60" L, 4" W x 4" D		8	1		460	63.50		523.50	600
8090	6" W x 6" D		6	1.333		760	85		845	960
8100	Elbow, 90°, 4" W x 4" D		20	.400		219	25.50		244.50	279
8110	6" W x 6" D		18	.444		440	28.50		468.50	525
8120	Elbow, 45°, 4" W x 4" D		20	.400		216	25.50		241.50	276
8130	6" W x 6" D		18	.444		430	28.50		458.50	510
8140	Tee, 4" W x 4" D		16	.500		273	32		305	350
8150	6" W x 6" D		14	.571		495	36.50		531.50	595
8160	Cross, 4" W x 4" D		14	.571		251	36.50		287.50	330
8170	6" W x 6" D		12	.667		735	42.50		777.50	875
8180	Cut-off fitting, w/flange & adhesive, 4" W x 4" D		18	.444		169	28.50		197.50	228
8190	6" W x 6" D		16	.500		345	32		377	425
8200	Flexible ftng., hvy. neoprene coated nylon, 4" W x 4" D		20	.400		305	25.50		330.50	380
8210	6" W x 6" D		18	.444		460	28.50		488.50	550
8220	Closure plate, fiberglass, 4" W x 4" D		20	.400		59	25.50		84.50	103
8230	6" W x 6" D		18	.444		68	28.50		96.50	117
8240	Box connector, stainless steel type 304, 4" W x 4" D	1 Elec	20	.400	Ea.	106	25.50		131.50	154
8250	6" W x 6" D		18	.444		117	28.50		145.50	170
8260	Hanger, 4" W x 4" D		100	.080		25.50	5.10		30.60	35.50
8270	6" W x 6" D		80	.100		34	6.35		40.35	47
8280	Straight tube section fiberglass, 4" W x 4" D, 12" long		40	.200		192	12.75		204.75	230
8290	24" long		20	.400		224	25.50		249.50	284
8300	36" long		13.30	.602		177	38.50		215.50	252
8310	48" long		10	.800		315	51		366	420
8320	60" long		8	1		355	63.50		418.50	485
8330	120" long		4	2		525	127		652	770

26 27 19 – Multi-Outlet Assemblies

26 27 19.10 Wiring Duct

		Crew	Daily Output	Labor-Hours	Unit	Material	2021 Bare Costs Labor	2021 Bare Costs Equipment	Total	Total Incl O&P
0010	**WIRING DUCT** Plastic									
1250	PVC, snap-in slots, adhesive backed									
1270	1-1/2" W x 2" H	2 Elec	120	.133	L.F.	4.55	8.50		13.05	17.60
1280	1-1/2" W x 3" H		120	.133		5.45	8.50		13.95	18.60
1290	1-1/2" W x 4" H		120	.133		6.50	8.50		15	19.75
1300	2" W x 1" H		120	.133		4.08	8.50		12.58	17.10
1310	2" W x 1-1/2" H		120	.133		5	8.50		13.50	18.10

275

For customer support on your Electrical Costs with RSMeans Data, call 800.448.8182.

26 27 19 – Multi-Outlet Assemblies

26 27 19.10 Wiring Duct		Crew	Daily Output	Labor-Hours	Unit	Material	2021 Bare Costs Labor	Equipment	Total	Total Incl O&P
1320	2" W x 2" H	2 Elec	120	.133	L.F.	5.30	8.50		13.80	18.40
1340	2" W x 3" H		120	.133		5.55	8.50		14.05	18.70
1350	2" W x 4" H		120	.133		6.70	8.50		15.20	19.95
1360	2-1/2" W x 3" H		120	.133		6.25	8.50		14.75	19.50
1370	3" W x 1" H		110	.145		4.66	9.25		13.91	18.90
1390	3" W x 2" H		110	.145		5.55	9.25		14.80	19.85
1400	3" W x 3" H		110	.145		6.35	9.25		15.60	21
1410	3" W x 4" H		110	.145		8.40	9.25		17.65	23
1420	3" W x 5" H		110	.145		11	9.25		20.25	26
1430	4" W x 1-1/2" H		100	.160		5.80	10.20		16	21.50
1440	4" W x 2" H		100	.160		6.50	10.20		16.70	22.50
1450	4" W x 3" H		100	.160		7.30	10.20		17.50	23
1460	4" W x 4" H		100	.160		8.85	10.20		19.05	25
1470	4" W x 5" H		100	.160		12.45	10.20		22.65	29
1550	Cover, 1-1/2" W		200	.080		.89	5.10		5.99	8.55
1560	2" W		200	.080		.95	5.10		6.05	8.60
1570	2-1/2" W		200	.080		1.37	5.10		6.47	9.05
1580	3" W		200	.080		1.45	5.10		6.55	9.15
1590	4" W		200	.080		1.75	5.10		6.85	9.50

26 27 23 – Indoor Service Poles

26 27 23.40 Surface Raceway		Crew	Daily Output	Labor-Hours	Unit	Material	2021 Bare Costs Labor	Equipment	Total	Total Incl O&P
0010	**SURFACE RACEWAY**									
0090	Metal, straight section									
0100	No. 500	1 Elec	100	.080	L.F.	1.13	5.10		6.23	8.80
0110	No. 700	"	100	.080	"	1.26	5.10		6.36	8.95
0200	No. 1000	1 Elec	90	.089	L.F.	1.85	5.65		7.50	10.45
0400	No. 1500, small pancake		90	.089		2.44	5.65		8.09	11.10
0600	No. 2000, base & cover, blank		90	.089		2.49	5.65		8.14	11.15
0610	Receptacle, 6" OC		40	.200		41.50	12.75		54.25	64.50
0620	12" OC		44	.182		12.80	11.60		24.40	31.50
0630	18" OC		46	.174		8.40	11.10		19.50	25.50
0650	30" OC		50	.160		4.76	10.20		14.96	20.50
0660	60" OC		50	.160		4.41	10.20		14.61	20
0670	No. 2400, base & cover, blank		80	.100		2.16	6.35		8.51	11.85
0680	Receptacle, 6" OC		42	.190		39	12.15		51.15	60.50
0690	12" OC		53	.151		48	9.60		57.60	67.50
0700	18" OC		55	.145		48	9.25		57.25	66.50
0710	24" OC		57	.140		10.05	8.95		19	24.50
0720	30" OC		59	.136		7.20	8.65		15.85	21
0730	60" OC		61	.131		6.65	8.35		15	19.75
0800	No. 3000, base & cover, blank		75	.107		4.74	6.80		11.54	15.30
0810	Receptacle, 6" OC		45	.178		47.50	11.30		58.80	69
0820	12" OC		62	.129		27	8.20		35.20	41.50
0830	18" OC		64	.125		21.50	7.95		29.45	36
0840	24" OC		66	.121		16.25	7.70		23.95	29.50
0850	30" OC		68	.118		14.60	7.50		22.10	27.50
0860	60" OC		70	.114		10.70	7.30		18	22.50
1000	No. 4000, base & cover, blank		65	.123		7.65	7.85		15.50	20
1010	Receptacle, 6" OC		41	.195		50	12.45		62.45	73
1020	12" OC		52	.154		39.50	9.80		49.30	58
1030	18" OC		54	.148		32.50	9.45		41.95	49.50
1040	24" OC		56	.143		26.50	9.10		35.60	42.50

276

For customer support on your Electrical Costs with RSMeans Data, call 800.448.8182.

26 27 23.40 Surface Raceway

26 27 23.40 Surface Raceway	Crew	Daily Output	Labor-Hours	Unit	Material	2021 Bare Costs Labor	Equipment	Total	Total Incl O&P
1050	30" OC	1 Elec	58	.138	L.F.	25	8.80		33.80	40.50
1060	60" OC		60	.133		18.45	8.50		26.95	33
1200	No. 6000, base & cover, blank		50	.160		12.80	10.20		23	29.50
1210	Receptacle, 6" OC		30	.267		61	17		78	92
1220	12" OC		37	.216		51.50	13.75		65.25	77
1230	18" OC		39	.205		43.50	13.05		56.55	67.50
1240	24" OC		41	.195		35.50	12.45		47.95	58
1250	30" OC		43	.186		34	11.85		45.85	55
1260	60" OC		45	.178	↓	26.50	11.30		37.80	46
2400	Fittings, elbows, No. 500		40	.200	Ea.	2.06	12.75		14.81	21
2800	Elbow cover, No. 2000		40	.200		4.02	12.75		16.77	23.50
2880	Tee, No. 500		42	.190		3.98	12.15		16.13	22.50
2900	No. 2000		27	.296		13.40	18.85		32.25	43
3000	Switch box, No. 500		16	.500		11.95	32		43.95	60.50
3400	Telephone outlet, No. 1500		16	.500	↓	16.10	32		48.10	65
3600	Junction box, No. 1500	1 Elec	16	.500	Ea.	11.05	32		43.05	59.50
3800	Plugmold wired sections, No. 2000									
4000	1 circuit, 6 outlets, 3' long	1 Elec	8	1	Ea.	40.50	63.50		104	139
4100	2 circuits, 8 outlets, 6' long		5.30	1.509		60	96		156	209
4110	Tele-power pole, alum, w/2 recept, 10'		4	2		231	127		358	445
4120	12'		3.85	2.078		221	132		353	440
4130	15'		3.70	2.162		350	138		488	590
4140	Steel, w/2 recept, 10'		4	2		146	127		273	350
4150	One phone fitting, 10'		4	2		157	127		284	360
4160	Alum, 4 outlets, 10'		3.70	2.162	↓	365	138		503	605
4300	Overhead distribution systems, 125 volt									
4800	No. 2000, entrance end fitting	1 Elec	20	.400	Ea.	6.20	25.50		31.70	45
5000	Blank end fitting		40	.200		2.47	12.75		15.22	21.50
5200	Supporting clip		40	.200		1.35	12.75		14.10	20.50
5800	No. 3000, entrance end fitting		20	.400		11.65	25.50		37.15	51
6000	Blank end fitting		40	.200		2.85	12.75		15.60	22
6020	Internal elbow		20	.400		15.30	25.50		40.80	55
6030	External elbow		20	.400		21.50	25.50		47	62
6040	Device bracket		53	.151		5.20	9.60		14.80	20
6400	Hanger clamp		32	.250	↓	7.90	15.95		23.85	32
7000	No. 4000 base		90	.089	L.F.	5.55	5.65		11.20	14.55
7200	Divider		100	.080	"	.93	5.10		6.03	8.55
7400	Entrance end fitting		16	.500	Ea.	25.50	32		57.50	75.50
7600	Blank end fitting		40	.200		6.70	12.75		19.45	26.50
7610	Recpt. & tele. cover		53	.151		13.30	9.60		22.90	29
7620	External elbow		16	.500		43.50	32		75.50	95.50
7630	Coupling		53	.151		6.65	9.60		16.25	21.50
7640	Divider clip & coupling		80	.100		1.32	6.35		7.67	10.90
7650	Panel connector		16	.500		27.50	32		59.50	78
7800	Take off connector		16	.500		92	32		124	149
8000	No. 6000, take off connector		16	.500		109	32		141	168
8100	Take off fitting		16	.500		81	32		113	137
8200	Hanger clamp		32	.250	↓	20.50	15.95		36.45	46
8230	Coupling					9.70			9.70	10.70
8240	One gang device plate	1 Elec	53	.151		10.50	9.60		20.10	26
8250	Two gang device plate		40	.200		12.80	12.75		25.55	33
8260	Blank end fitting		40	.200		11.60	12.75		24.35	31.50
8270	Combination elbow		14	.571	↓	45.50	36.50		82	104

26 27 23 – Indoor Service Poles

26 27 23.40 Surface Raceway

		Crew	Daily Output	Labor-Hours	Unit	Material	2021 Bare Costs Labor	Equipment	Total	Total Incl O&P
8300	Panel connector	1 Elec	16	.500	Ea.	19.95	32		51.95	69.50
8500	Chan-L-Wire system installed in 1-5/8" x 1-5/8" strut. Strut									
8600	not incl., 30 amp, 4 wire, 3 phase	1 Elec	200	.040	L.F.	5.05	2.55		7.60	9.35
8700	Junction box		8	1	Ea.	33.50	63.50		97	132
8800	Insulating end cap		40	.200		8.95	12.75		21.70	29
8900	Strut splice plate		40	.200		11.85	12.75		24.60	32
9000	Tap		40	.200		24	12.75		36.75	45.50
9100	Fixture hanger		60	.133		10.50	8.50		19	24
9200	Pulling tool					90.50			90.50	99.50
9300	Non-metallic, straight section									
9310	7/16" x 7/8", base & cover, blank	1 Elec	160	.050	L.F.	2.17	3.19		5.36	7.10
9320	Base & cover w/adhesive		160	.050		1.49	3.19		4.68	6.35
9340	7/16" x 1-5/16", base & cover, blank		145	.055		2.38	3.51		5.89	7.80
9350	Base & cover w/adhesive		145	.055		2.27	3.51		5.78	7.70
9370	11/16" x 2-1/4", base & cover, blank		130	.062		2.88	3.92		6.80	8.95
9380	Base & cover w/adhesive		130	.062		3.16	3.92		7.08	9.30
9385	1-11/16" x 5-1/4", two compartment base & cover w/screws		80	.100		10.50	6.35		16.85	21
9400	Fittings, elbows, 7/16" x 7/8"		50	.160	Ea.	1.98	10.20		12.18	17.35
9410	7/16" x 1-5/16"		45	.178		2.05	11.30		13.35	19.10
9420	11/16" x 2-1/4"		40	.200		2.22	12.75		14.97	21.50
9425	1-11/16" x 5-1/4"		28	.286		11.65	18.20		29.85	40
9430	Tees, 7/16" x 7/8"		35	.229		2.56	14.55		17.11	24.50
9440	7/16" x 1-5/16"		32	.250		2.62	15.95		18.57	26.50
9450	11/16" x 2-1/4"		30	.267		2.70	17		19.70	28
9455	1-11/16" x 5-1/4"		24	.333		18.70	21		39.70	52
9460	Cover clip, 7/16" x 7/8"		80	.100		.52	6.35		6.87	10
9470	7/16" x 1-5/16"		72	.111		.47	7.10		7.57	11
9480	11/16" x 2-1/4"		64	.125		.79	7.95		8.74	12.70
9484	1-11/16" x 5-1/4"		42	.190		2.76	12.15		14.91	21
9486	Wire clip, 1-11/16" x 5-1/4"		68	.118		.48	7.50		7.98	11.70
9490	Blank end, 7/16" x 7/8"		50	.160		.72	10.20		10.92	15.95
9500	7/16" x 1-5/16"		45	.178		.83	11.30		12.13	17.75
9510	11/16" x 2-1/4"		40	.200		1.23	12.75		13.98	20.50
9515	1-11/16" x 5-1/4"		38	.211		5.75	13.40		19.15	26.50
9520	Round fixture box, 5.5" diam. x 1"		25	.320		11.70	20.50		32.20	43.50
9530	Device box, 1 gang		30	.267		5.30	17		22.30	31
9540	2 gang		25	.320		7.65	20.50		28.15	39

26 27 26 – Wiring Devices

26 27 26.10 Low Voltage Switching

		Crew	Daily Output	Labor-Hours	Unit	Material	2021 Bare Costs Labor	Equipment	Total	Total Incl O&P
0010	**LOW VOLTAGE SWITCHING**									
3600	Relays, 120 V or 277 V standard	1 Elec	12	.667	Ea.	40	42.50		82.50	107
3800	Flush switch, standard		40	.200		14.55	12.75		27.30	35
4000	Interchangeable		40	.200		19	12.75		31.75	40
4100	Surface switch, standard	1 Elec	40	.200	Ea.	8.45	12.75		21.20	28.50
4200	Transformer 115 V to 25 V		12	.667		127	42.50		169.50	202
4400	Master control, 12 circuit, manual		4	2		129	127		256	330
4500	25 circuit, motorized		4	2		164	127		291	370
4600	Rectifier, silicon		12	.667		51.50	42.50		94	120
4800	Switchplates, 1 gang, 1, 2 or 3 switch, plastic		80	.100		5.05	6.35		11.40	15
5000	Stainless steel		80	.100		11.20	6.35		17.55	22
5400	2 gang, 3 switch, stainless steel		53	.151		23.50	9.60		33.10	40
5500	4 switch, plastic		53	.151		10.35	9.60		19.95	25.50

26 27 Low-Voltage Distribution Equipment

26 27 26 – Wiring Devices

26 27 26.10 Low Voltage Switching

		Crew	Daily Output	Labor-Hours	Unit	Material	2021 Bare Costs Labor	2021 Bare Costs Equipment	Total	Total Incl O&P
5600	2 gang, 4 switch, stainless steel	1 Elec	53	.151	Ea.	23	9.60		32.60	40
5700	6 switch, stainless steel		53	.151		44.50	9.60		54.10	63.50
5800	3 gang, 9 switch, stainless steel		32	.250		70	15.95		85.95	101
5900	Receptacle, triple, 1 return, 1 feed		26	.308		43.50	19.60		63.10	77
6000	2 feed		20	.400		43.50	25.50		69	86
6100	Relay gang boxes, flush or surface, 6 gang		5.30	1.509		95.50	96		191.50	248
6200	12 gang		4.70	1.702		114	108		222	287
6400	18 gang		4	2		117	127		244	315
6500	Frame, to hold up to 6 relays		12	.667		94	42.50		136.50	166
7200	Control wire, 2 conductor		6.30	1.270	C.L.F.	19.35	81		100.35	142
7400	3 conductor		5	1.600		31	102		133	185
7600	19 conductor		2.50	3.200		320	204		524	655
7800	26 conductor		2	4		435	255		690	855
8000	Weatherproof, 3 conductor		5	1.600		65	102		167	223

26 27 26.20 Wiring Devices Elements

		Crew	Daily Output	Labor-Hours	Unit	Material	2021 Bare Costs Labor	2021 Bare Costs Equipment	Total	Total Incl O&P
0010	**WIRING DEVICES ELEMENTS** R262726-90									
0200	Toggle switch, quiet type, single pole, 15 amp	1 Elec	40	.200	Ea.	.49	12.75		13.24	19.50
0500	20 amp		27	.296		4.72	18.85		23.57	33
0510	30 amp		23	.348		15.65	22		37.65	50
0530	Lock handle, 20 amp		27	.296		33	18.85		51.85	64
0540	Security key, 20 amp		26	.308		93.50	19.60		113.10	132
0550	Rocker, 15 amp		40	.200		2.39	12.75		15.14	21.50
0560	20 amp		27	.296		6.35	18.85		25.20	35
0600	3 way, 15 amp		23	.348		1.49	22		23.49	34.50
0800	20 amp		18	.444		5.45	28.50		33.95	48
0810	30 amp		9	.889		19.40	56.50		75.90	106
0830	Lock handle, 20 amp		18	.444		31	28.50		59.50	76
0840	Security key, 20 amp		17	.471		67.50	30		97.50	119
0850	Rocker, 15 amp		23	.348		8.50	22		30.50	42.50
0860	20 amp		18	.444		8.30	28.50		36.80	51
0900	4 way, 15 amp		15	.533		11.35	34		45.35	63
1000	20 amp	1 Elec	11	.727	Ea.	52	46.50		98.50	127
1020	Lock handle, 20 amp		11	.727		50	46.50		96.50	124
1030	Rocker, 15 amp		15	.533		9.95	34		43.95	61.50
1040	20 amp		11	.727		17.90	46.50		64.40	88.50
1100	Toggle switch, quiet type, double pole, 15 amp		15	.533		14.70	34		48.70	66.50
1200	20 amp		11	.727		21	46.50		67.50	92
1210	30 amp		9	.889		35.50	56.50		92	123
1230	Lock handle, 20 amp		11	.727		26.50	46.50		73	98
1250	Security key, 20 amp		10	.800		80	51		131	163
1420	Toggle switch quiet type, 1 pole, 2 throw center off, 15 amp		23	.348		48	22		70	86
1440	20 amp		18	.444		71	28.50		99.50	120
1460	2 pole, 2 throw center off, lock handle, 20 amp		11	.727		76	46.50		122.50	153
1480	1 pole, momentary contact, 15 amp		23	.348		21	22		43	56.50
1500	20 amp		18	.444		16.85	28.50		45.35	60.50
1520	Momentary contact, lock handle, 20 amp		18	.444		32	28.50		60.50	77.50
1650	Dimmer switch, 120 volt, incandescent, 600 watt, 1 pole G		16	.500		20.50	32		52.50	70
1700	600 watt, 3 way G		12	.667		11.45	42.50		53.95	75.50
1750	1,000 watt, 1 pole G		16	.500		34	32		66	85
1800	1,000 watt, 3 way G		12	.667		74	42.50		116.50	144
2000	1,500 watt, 1 pole G		11	.727		109	46.50		155.50	189
2100	2,000 watt, 1 pole G		8	1		159	63.50		222.50	269

279

26 27 26.20 Wiring Devices Elements		Crew	Daily Output	Labor-Hours	Unit	Material	2021 Bare Costs Labor	2021 Bare Costs Equipment	Total	Total Incl O&P
2110	Fluorescent, 600 watt	G	1 Elec	15	.533	Ea.	111	34		145	173
2120	1,000 watt	G		15	.533		155	34		189	221
2130	1,500 watt	G		10	.800		286	51		337	390
2160	Explosion proof, toggle switch, wall, single pole 20 amp			5.30	1.509		224	96		320	390
2180	Receptacle, single outlet, 20 amp			5.30	1.509		440	96		536	625
2190	30 amp			4	2		925	127		1,052	1,225
2290	60 amp			2.50	3.200		875	204		1,079	1,275
2360	Plug, 20 amp			16	.500		243	32		275	315
2370	30 amp			12	.667		410	42.50		452.50	515
2380	60 amp			8	1		525	63.50		588.50	670
2410	Furnace, thermal cutoff switch with plate			26	.308		17.25	19.60		36.85	48
2460	Receptacle, duplex, 120 volt, grounded, 15 amp			40	.200		1.38	12.75		14.13	20.50
2470	20 amp			27	.296		11	18.85		29.85	40
2480	Ground fault interrupting, 15 amp			27	.296		15.05	18.85		33.90	44.50
2482	20 amp			27	.296		55.50	18.85		74.35	89
2486	Clock receptacle, 15 amp			40	.200		29.50	12.75		42.25	51.50
2490	Dryer, 30 amp			15	.533		5.35	34		39.35	56.50
2500	Range, 50 amp			11	.727		14.55	46.50		61.05	85
2530	Surge suppressor receptacle, duplex, 20 amp		1 Elec	27	.296	Ea.	51.50	18.85		70.35	84.50
2532	Quad, 20 amp			20	.400		112	25.50		137.50	161
2540	Isolated ground receptacle, duplex, 20 amp			27	.296		31.50	18.85		50.35	62.50
2542	Quad, 20 amp			20	.400		35	25.50		60.50	76.50
2550	Simplex, 20 amp			27	.296		30.50	18.85		49.35	61.50
2560	Simplex, 30 amp			15	.533		39.50	34		73.50	94
2570	Cable reel w/receptacle 50' w/3#12, 120 V, 20 A		2 Elec	2.67	5.999		1,125	380		1,505	1,800
2600	Wall plates, stainless steel, 1 gang		1 Elec	80	.100		1.41	6.35		7.76	11
2800	2 gang			53	.151		4.39	9.60		13.99	19.15
3000	3 gang			32	.250		10.20	15.95		26.15	34.50
3100	4 gang			27	.296		11.10	18.85		29.95	40
3110	Brown plastic, 1 gang			80	.100		.38	6.35		6.73	9.85
3120	2 gang			53	.151		.89	9.60		10.49	15.30
3130	3 gang			32	.250		1.44	15.95		17.39	25
3140	4 gang			27	.296		2.94	18.85		21.79	31
3150	Brushed brass, 1 gang			80	.100		5.15	6.35		11.50	15.10
3160	Anodized aluminum, 1 gang			80	.100		3.47	6.35		9.82	13.25
3170	Switch cover, weatherproof, 1 gang			60	.133		6.05	8.50		14.55	19.25
3180	Vandal proof lock, 1 gang			60	.133		14.70	8.50		23.20	29
3200	Lampholder, keyless			26	.308		20	19.60		39.60	51
3400	Pullchain with receptacle			22	.364		21.50	23		44.50	58
3500	Pilot light, neon with jewel			27	.296		10.45	18.85		29.30	39.50
3600	Receptacle, 20 amp, 250 volt, NEMA 6			27	.296		23.50	18.85		42.35	53.50
3620	277 volt NEMA 7			27	.296		24	18.85		42.85	54.50
3640	125/250 volt NEMA 10			27	.296		25	18.85		43.85	55.50
3680	125/250 volt NEMA 14			25	.320		31.50	20.50		52	65
3700	3 pole, 250 volt NEMA 15			25	.320		32	20.50		52.50	65.50
3720	120/208 volt NEMA 18			25	.320		35	20.50		55.50	69
3740	30 amp, 125 volt NEMA 5			15	.533		25.50	34		59.50	78.50
3760	250 volt NEMA 6			15	.533		29	34		63	82.50
3780	277 volt NEMA 7			15	.533		33.50	34		67.50	87
3820	125/250 volt NEMA 14			14	.571		68	36.50		104.50	129
3840	3 pole, 250 volt NEMA 15			14	.571		63	36.50		99.50	123
3880	50 amp, 125 volt NEMA 5			11	.727		32.50	46.50		79	105
3900	250 volt NEMA 6			11	.727		33.50	46.50		80	106

26 27 Low-Voltage Distribution Equipment

26 27 26 – Wiring Devices

26 27 26.20 Wiring Devices Elements	Crew	Daily Output	Labor-Hours	Unit	Material	2021 Bare Costs Labor	Equipment	Total	Total Incl O&P
3920	277 volt NEMA 7	1 Elec	11	.727	Ea.	37.50	46.50		84	111
3960	125/250 volt NEMA 14		10	.800		88.50	51		139.50	173
3980	3 pole, 250 volt NEMA 15		10	.800		88	51		139	173
4020	60 amp, 125/250 volt, NEMA 14		8	1		97	63.50		160.50	202
4040	3 pole, 250 volt NEMA 15	1 Elec	8	1	Ea.	103	63.50		166.50	208
4060	120/208 volt NEMA 18		8	1		85	63.50		148.50	188
4100	Receptacle locking, 20 amp, 125 volt NEMA L5		27	.296		24.50	18.85		43.35	55
4120	250 volt NEMA L6		27	.296		24.50	18.85		43.35	55
4140	277 volt NEMA L7		27	.296		23.50	18.85		42.35	54
4150	3 pole, 250 volt, NEMA L11		27	.296		20	18.85		38.85	50
4160	20 amp, 480 volt NEMA L8		27	.296		26.50	18.85		45.35	57
4180	600 volt NEMA L9		27	.296		33.50	18.85		52.35	65
4200	125/250 volt NEMA L10		27	.296		29	18.85		47.85	60
4230	125/250 volt NEMA L14		25	.320		32.50	20.50		53	66.50
4280	250 volt NEMA L15		25	.320		35	20.50		55.50	69
4300	480 volt NEMA L16		25	.320		28.50	20.50		49	61.50
4320	3 phase, 120/208 volt NEMA L18		25	.320		36.50	20.50		57	70.50
4340	277/480 volt NEMA L19		25	.320		37	20.50		57.50	71.50
4360	347/600 volt NEMA L20		25	.320		38.50	20.50		59	73
4380	120/208 volt NEMA L21		23	.348		41	22		63	78
4400	277/480 volt NEMA L22		23	.348		42.50	22		64.50	80
4420	347/600 volt NEMA L23		23	.348		48.50	22		70.50	86
4440	30 amp, 125 volt NEMA L5		15	.533		34	34		68	88
4460	250 volt NEMA L6		15	.533		36.50	34		70.50	90.50
4480	277 volt NEMA L7		15	.533		37	34		71	91
4500	480 volt NEMA L8		15	.533		38.50	34		72.50	93
4520	600 volt NEMA L9		15	.533		41.50	34		75.50	96
4540	125/250 volt NEMA L10		15	.533		44.50	34		78.50	99.50
4560	3 phase, 250 volt NEMA L11		15	.533		25	34		59	78
4620	125/250 volt NEMA L14		14	.571		53	36.50		89.50	113
4640	250 volt NEMA L15		14	.571		54	36.50		90.50	114
4660	480 volt NEMA L16		14	.571		55	36.50		91.50	115
4680	600 volt NEMA L17		14	.571		51.50	36.50		88	111
4700	120/208 volt NEMA L18		14	.571		53.50	36.50		90	113
4720	277/480 volt NEMA L19		14	.571		54.50	36.50		91	114
4740	347/600 volt NEMA L20		14	.571		55	36.50		91.50	115
4760	120/208 volt NEMA L21		13	.615		56.50	39		95.50	121
4780	277/480 volt NEMA L22		13	.615		55.50	39		94.50	120
4800	347/600 volt NEMA L23		13	.615		61.50	39		100.50	126
4840	Receptacle, corrosion resistant, 15 or 20 amp, 125 volt NEMA L5		27	.296		30	18.85		48.85	61
4860	250 volt NEMA L6		27	.296		23	18.85		41.85	53.50
4870	Receptacle box assembly, cast aluminum, 60A, 4P5W, 3P, 120/208V, NEMA PR4		22	.364		710	23		733	815
4875	100A, 4P5W, 3P, 347/600V, NEMA PR4		22	.364		815	23		838	930
4880	200A, 4P5W, 3P, 347/600V, NEMA PR4		22	.364		1,775	23		1,798	1,975
4900	Receptacle, cover plate, phenolic plastic, NEMA 5 & 6	1 Elec	80	.100	Ea.	.67	6.35		7.02	10.20
4910	NEMA 7-23		80	.100		.64	6.35		6.99	10.15
4920	Stainless steel, NEMA 5 & 6		80	.100		1.25	6.35		7.60	10.85
4930	NEMA 7-23		80	.100		2.80	6.35		9.15	12.55
4940	Brushed brass NEMA 5 & 6		80	.100		5.50	6.35		11.85	15.50
4950	NEMA 7-23		80	.100		5.95	6.35		12.30	16
4960	Anodized aluminum, NEMA 5 & 6		80	.100		3.58	6.35		9.93	13.40
4970	NEMA 7-23		80	.100		5.90	6.35		12.25	15.95
4980	Weatherproof NEMA 7-23		60	.133		43.50	8.50		52	60.50

281

26 27 26.20 Wiring Devices Elements	Crew	Daily Output	Labor-Hours	Unit	Material	2021 Bare Costs Labor	2021 Bare Costs Equipment	Total	Total Incl O&P
5000	Duplex receptacle, combo 15A/125V, 3 wire w/2-5V 0.7A, port USB, AL	1 Elec	27	.296	Ea.	16.05	18.85		34.90	45.50
5002	15A/125V, 3 wire w/2-5V 0.7A, port USB, BK		27	.296		17.85	18.85		36.70	47.50
5004	15A/125V, 3 wire w/2-5V 0.7A, port USB, LA		27	.296		16.15	18.85		35	46
5006	15A/125V, 3 wire w/2-5V 0.7A, port USB, IV		27	.296		15.80	18.85		34.65	45.50
5008	15A/125V, 3 wire w/2-5V 0.7A, port USB, WH		27	.296		18.85	18.85		37.70	48.50
5010	Duplex receptacle, combo 15A/125V, 3 wire w/2-5V 2.1A, port USB, AL		27	.296		71.50	18.85		90.35	107
5012	15A/125V, 3 wire w/2-5V 2.1A, port USB, BK		27	.296		71.50	18.85		90.35	107
5014	15A/125V, 3 wire w/2-5V 2.1A, port USB, LA		27	.296		28	18.85		46.85	58.50
5016	15A/125V, 3 wire w/2-5V 2.1A, port USB, IV		27	.296		71.50	18.85		90.35	107
5018	15A/125V, 3 wire w/2-5V 2.1A, port USB, WH		27	.296		71.50	18.85		90.35	107
5020	Duplex receptacle, combo 20A/125V, 3 wire w/2-5V 2.1A, port USB, AL		27	.296		86.50	18.85		105.35	123
5022	20A/125V, 3 wire w/2-5V 2.1A, port USB, BK		27	.296		77.50	18.85		96.35	113
5024	20A/125V, 3 wire w/2-5V 2.1A, port USB, LA		27	.296		77.50	18.85		96.35	113
5026	20A/125V, 3 wire w/2-5V 2.1A, port USB, IV		27	.296		34.50	18.85		53.35	66
5028	20A/125V, 3 wire w/2-5V 2.1A, port USB, WH		27	.296		77.50	18.85		96.35	113
5100	Plug, 20 amp, 250 volt NEMA 6		30	.267		21.50	17		38.50	49
5110	277 volt NEMA 7		30	.267		22.50	17		39.50	49.50
5120	3 pole, 120/250 volt NEMA 10		26	.308		26	19.60		45.60	57.50
5130	125/250 volt NEMA 14		26	.308		50	19.60		69.60	84
5140	250 volt NEMA 15		26	.308		54	19.60		73.60	88.50
5150	120/208 volt NEMA 8		26	.308		56	19.60		75.60	91
5160	30 amp, 125 volt NEMA 5		13	.615		50.50	39		89.50	114
5170	250 volt NEMA 6		13	.615		48	39		87	111
5180	277 volt NEMA 7		13	.615		47.50	39		86.50	111
5190	125/250 volt NEMA 14		13	.615		57	39		96	121
5200	3 pole, 250 volt NEMA 15		12	.667		63	42.50		105.50	133
5210	50 amp, 125 volt NEMA 5		9	.889		78	56.50		134.50	170
5220	250 volt NEMA 6		9	.889		81.50	56.50		138	174
5230	277 volt NEMA 7		9	.889		87	56.50		143.50	180
5240	125/250 volt NEMA 14	1 Elec	9	.889	Ea.	74.50	56.50		131	166
5250	3 pole, 250 volt NEMA 15		8	1		75.50	63.50		139	178
5260	60 amp, 125/250 volt NEMA 14		7	1.143		84	73		157	201
5270	3 pole, 250 volt NEMA 15		7	1.143		88	73		161	205
5280	120/208 volt NEMA 18		7	1.143		104	73		177	223
5300	Plug angle, 20 amp, 250 volt NEMA 6		30	.267		30.50	17		47.50	58.50
5310	30 amp, 125 volt NEMA 5		13	.615		62	39		101	127
5320	250 volt NEMA 6		13	.615		63	39		102	128
5330	277 volt NEMA 7		13	.615		71.50	39		110.50	137
5340	125/250 volt NEMA 14		13	.615		67.50	39		106.50	133
5350	3 pole, 250 volt NEMA 15		12	.667		70	42.50		112.50	140
5360	50 amp, 125 volt NEMA 5		9	.889		67.50	56.50		124	158
5370	250 volt NEMA 6		9	.889		66.50	56.50		123	157
5380	277 volt NEMA 7		9	.889		78	56.50		134.50	170
5390	125/250 volt NEMA 14		9	.889		80	56.50		136.50	172
5400	3 pole, 250 volt NEMA 15		8	1		83.50	63.50		147	186
5410	60 amp, 125/250 volt NEMA 14		7	1.143		93	73		166	210
5420	3 pole, 250 volt NEMA 15		7	1.143		95.50	73		168.50	213
5430	120/208 volt NEMA 18		7	1.143		102	73		175	220
5500	Plug, locking, 20 amp, 125 volt NEMA L5		30	.267		19.10	17		36.10	46
5510	250 volt NEMA L6		30	.267		19.10	17		36.10	46
5520	277 volt NEMA L7		30	.267		18.85	17		35.85	46
5530	480 volt NEMA L8		30	.267		20.50	17		37.50	47.50
5540	600 volt NEMA L9		30	.267		21	17		38	48

For customer support on your Electrical Costs with RSMeans Data, call 800.448.8182.

26 27 26.20 Wiring Devices Elements	Crew	Daily Output	Labor-Hours	Unit	Material	2021 Bare Costs Labor	Equipment	Total	Total Incl O&P
5550	3 pole, 125/250 volt NEMA L10	1 Elec	26	.308	Ea.	23	19.60		42.60	54.50
5560	250 volt NEMA L11		26	.308		23	19.60		42.60	54.50
5570	480 volt NEMA L12		26	.308		27	19.60		46.60	59
5580	125/250 volt NEMA L14		26	.308		29	19.60		48.60	61
5590	250 volt NEMA L15		26	.308		31	19.60		50.60	63
5600	480 volt NEMA L16		26	.308		31	19.60		50.60	63
5610	4 pole, 120/208 volt NEMA L18		24	.333		32	21		53	66.50
5620	277/480 volt NEMA L19		24	.333		32.50	21		53.50	67.50
5630	347/600 volt NEMA L20		24	.333		34	21		55	68.50
5640	120/208 volt NEMA L21		24	.333		37	21		58	72.50
5650	277/480 volt NEMA L22		24	.333		38	21		59	73.50
5660	347/600 volt NEMA L23		24	.333		42.50	21		63.50	78
5670	30 amp, 125 volt NEMA L5		13	.615		29.50	39		68.50	91
5680	250 volt NEMA L6		13	.615		30	39		69	91.50
5690	277 volt NEMA L7		13	.615		28.50	39		67.50	89.50
5700	480 volt NEMA L8		13	.615		30.50	39		69.50	92.50
5710	600 volt NEMA L9	1 Elec	13	.615	Ea.	31.50	39		70.50	93
5720	3 pole, 125/250 volt NEMA L10		11	.727		25.50	46.50		72	97
5730	250 volt NEMA L11		11	.727		25	46.50		71.50	97
5760	125/250 volt NEMA L14		11	.727		39.50	46.50		86	113
5770	250 volt NEMA L15		11	.727		39.50	46.50		86	113
5780	480 volt NEMA L16		11	.727		40	46.50		86.50	113
5790	600 volt NEMA L17		11	.727		40	46.50		86.50	113
5800	4 pole, 120/208 volt NEMA L18		10	.800		40	51		91	120
5810	120/208 volt NEMA L19		10	.800		40	51		91	120
5820	347/600 volt NEMA L20		10	.800		42	51		93	122
5830	120/208 volt NEMA L21		10	.800		43.50	51		94.50	124
5840	277/480 volt NEMA L22		10	.800		44.50	51		95.50	125
5850	347/600 volt NEMA L23		10	.800		46	51		97	126
6000	Connector, 20 amp, 250 volt NEMA 6		30	.267		33.50	17		50.50	62
6010	277 volt NEMA 7		30	.267		31	17		48	59.50
6020	3 pole, 120/250 volt NEMA 10		26	.308		35	19.60		54.60	67.50
6030	125/250 volt NEMA 14		26	.308		35	19.60		54.60	67.50
6040	250 volt NEMA 15		26	.308		35.50	19.60		55.10	68
6050	120/208 volt NEMA 18		26	.308		38.50	19.60		58.10	71.50
6060	30 amp, 125 volt NEMA 5		13	.615		61	39		100	126
6070	250 volt NEMA 6		13	.615		61	39		100	126
6080	277 volt NEMA 7		13	.615		61	39		100	126
6110	50 amp, 125 volt NEMA 5		9	.889		84	56.50		140.50	177
6120	250 volt NEMA 6		9	.889		84	56.50		140.50	177
6130	277 volt NEMA 7		9	.889		84	56.50		140.50	177
6200	Connector, locking, 20 amp, 125 volt NEMA L5		30	.267		29.50	17		46.50	57.50
6210	250 volt NEMA L6		30	.267		29.50	17		46.50	57.50
6220	277 volt NEMA L7		30	.267		28	17		45	56
6230	480 volt NEMA L8		30	.267		30.50	17		47.50	58.50
6240	600 volt NEMA L9		30	.267		32.50	17		49.50	60.50
6250	3 pole, 125/250 volt NEMA L10		26	.308		36	19.60		55.60	69
6260	250 volt NEMA L11		26	.308		36	19.60		55.60	69
6280	125/250 volt NEMA L14		26	.308		40.50	19.60		60.10	73.50
6290	250 volt NEMA L15		26	.308		40.50	19.60		60.10	73.50
6300	480 volt NEMA L16		26	.308		42	19.60		61.60	75
6310	4 pole, 120/208 volt NEMA L18		24	.333		44	21		65	80
6320	277/480 volt NEMA L19		24	.333		47.50	21		68.50	84

283

26 27 26.20 Wiring Devices Elements	Crew	Daily Output	Labor-Hours	Unit	Material	2021 Bare Costs Labor	Equipment	Total	Total Incl O&P
6330	347/600 volt NEMA L20	1 Elec	24	.333	Ea.	48	21		69	84
6340	120/208 volt NEMA L21		24	.333		60.50	21		81.50	98.50
6350	277/480 volt NEMA L22		24	.333		68	21		89	106
6360	347/600 volt NEMA L23	1 Elec	24	.333	Ea.	73.50	21		94.50	113
6370	30 amp, 125 volt NEMA L5		13	.615		58.50	39		97.50	123
6380	250 volt NEMA L6		13	.615		59.50	39		98.50	124
6390	277 volt NEMA L7		13	.615		49	39		88	112
6400	480 volt NEMA L8		13	.615		60.50	39		99.50	125
6410	600 volt NEMA L9		13	.615		66.50	39		105.50	132
6420	3 pole, 125/250 volt NEMA L10		11	.727		69.50	46.50		116	146
6430	250 volt NEMA L11		11	.727		69.50	46.50		116	146
6460	125/250 volt NEMA L14		11	.727		81	46.50		127.50	158
6470	250 volt NEMA L15		11	.727		81.50	46.50		128	159
6480	480 volt NEMA L16		11	.727		84.50	46.50		131	162
6490	600 volt NEMA L17		11	.727		84.50	46.50		131	162
6500	4 pole, 120/208 volt NEMA L18		10	.800		84	51		135	168
6510	120/208 volt NEMA L19		10	.800		85.50	51		136.50	170
6520	347/600 volt NEMA L20		10	.800		85.50	51		136.50	170
6530	120/208 volt NEMA L21		10	.800		78	51		129	162
6540	277/480 volt NEMA L22		10	.800		83.50	51		134.50	168
6550	347/600 volt NEMA L23		10	.800		84	51		135	168
7000	Receptacle computer, 250 volt, 15 amp, 3 pole 4 wire		8	1		98	63.50		161.50	203
7010	20 amp, 2 pole 3 wire		8	1		113	63.50		176.50	219
7020	30 amp, 2 pole 3 wire		6.50	1.231		181	78.50		259.50	315
7030	30 amp, 3 pole 4 wire		6.50	1.231		193	78.50		271.50	330
7040	60 amp, 3 pole 4 wire		4.50	1.778		320	113		433	520
7050	100 amp, 3 pole 4 wire		3	2.667		395	170		565	685
7100	Connector computer, 250 volt, 15 amp, 3 pole 4 wire		27	.296		151	18.85		169.85	194
7110	20 amp, 2 pole 3 wire		27	.296		157	18.85		175.85	201
7120	30 amp, 2 pole 3 wire		15	.533		220	34		254	293
7130	30 amp, 3 pole 4 wire		15	.533		242	34		276	320
7140	60 amp, 3 pole 4 wire		8	1		400	63.50		463.50	535
7150	100 amp, 3 pole 4 wire		4	2		550	127		677	795
7200	Plug, computer, 250 volt, 15 amp, 3 pole 4 wire		27	.296		135	18.85		153.85	176
7210	20 amp, 2 pole 3 wire		27	.296		143	18.85		161.85	185
7220	30 amp, 2 pole 3 wire		15	.533		219	34		253	292
7230	30 amp, 3 pole 4 wire		15	.533		213	34		247	286
7240	60 amp, 3 pole 4 wire		8	1		305	63.50		368.50	430
7250	100 amp, 3 pole 4 wire		4	2		380	127		507	610
7300	Connector adapter to flexible conduit, 1/2"		60	.133		3.70	8.50		12.20	16.65
7310	3/4"		50	.160		5.35	10.20		15.55	21
7320	1-1/4"		30	.267		15.60	17		32.60	42
7330	1-1/2"		23	.348		22	22		44	57
8100	Pin/sleeve, 20A, 480V, DSN1, male inlet, 3 pole, 3 phase, 4 wire	1 Elec	16	.500	Ea.	45.50	32		77.50	97.50
8110	20A, 125V, DSN20, male inlet, 3 pole, 3 phase, 3 wire		16	.500		41.50	32		73.50	93.50
8120	60A, 480V, DS60, male inlet, 3 pole, 3 phase, 3 wire		16	.500		345	32		377	430
8130	Pin/sleeve, 20A, 480V, DSN10, female RCPT, 3P, 3 phase, 4 wire		16	.500		91.50	32		123.50	149
8140	30A, 480V, DS30, female RCPT, 3P, 3 phase, 4 wire		16	.500		320	32		352	400
8500	Wiring device terminal strip, 2 pole, 12-2AWG, 300VAC, CSA, 600VAC, 10A		36	.222		3.48	14.15		17.63	25
8505	4 pole, 12-2AWG, 300VAC, CSA, 600VAC, 10A		36	.222		4.92	14.15		19.07	26.50
8510	6 pole, 12-2AWG, 300VAC, CSA, 600VAC, 10A		36	.222		5.55	14.15		19.70	27
8515	8 pole, 12-2AWG, 300VAC, CSA, 600VAC, 10A		30	.267		6.20	17		23.20	32
8520	10 pole, 12-2AWG, 300VAC, CSA, 600VAC, 10A		30	.267		6.85	17		23.85	32.50

26 27 26 – Wiring Devices

26 27 26.20 Wiring Devices Elements

		Crew	Daily Output	Labor-Hours	Unit	Material	2021 Bare Costs Labor	2021 Bare Costs Equipment	Total	Total Incl O&P
8525	12 pole, 12-2AWG, 300VAC, CSA, 600VAC, 10A	1 Elec	30	.267	Ea.	7.55	17		24.55	33.50
8530	Wiring device terminal strip jumper, 10A		36	.222		.86	14.15		15.01	22
8535	Wiring device terminal strip, 2 pole, 12-22AWG, 300VAC, CSA, 600VAC, 20A		34	.235		3.78	15		18.78	26.50
8540	4 pole, 12-22AWG, 300VAC, CSA, 600VAC, 20A		34	.235		5.15	15		20.15	28
8545	6 pole, 12-22AWG, 300VAC, CSA, 600VAC, 20A		34	.235		5.90	15		20.90	29
8550	8 pole, 12-22AWG, 300VAC, CSA, 600VAC, 20A		30	.267		6.55	17		23.55	32
8555	10 pole, 12-22AWG, 300VAC, CSA, 600VAC, 20A		30	.267		7.25	17		24.25	33
8560	12 pole, 12-22AWG, 300VAC, CSA, 600VAC, 20A		30	.267		7.95	17		24.95	34
8565	Wiring device terminal strip jumper, 20A		34	.235		1.06	15		16.06	23.50
8570	Wiring device terminal shorting type no cover, 4 pole, 6-10AWG, 600VAC, 50A		30	.267		20	17		37	47
8575	6 pole, 6-10AWG, 600VAC, 50A		30	.267		26	17		43	53.50
8580	8 pole, 6-10AWG, 600VAC, 50A		30	.267		29.50	17		46.50	57.50
8585	12 pole, 6-10AWG, 600VAC, 50A		30	.267		33	17		50	61.50
8590	Wiring device terminal 2 shorting pin 1/2 cover, 2 pole, 8-4AWG, 600V, 45A		30	.267		21	17		38	48
8595	4 pole, 8-4AWG, 600V, 45A		30	.267		26	17		43	54
8600	6 pole, 8-4AWG, 600V, 45A		30	.267		32.50	17		49.50	60.50
8605	8 pole, 8-4AWG, 600V, 45A		30	.267		47	17		64	76.50
8610	12 pole, 8-4AWG, 600V, 45A		30	.267		56	17		73	86.50
8615	Wiring device terminal 2 shorting pin W/cover, 4 pole, 8-4AWG, 600V, 45A		30	.267		35.50	17		52.50	64
8620	6 pole, 8-4AWG, 600V, 45A		30	.267		42	17		59	71.50
8625	8 pole, 8-4AWG, 600V, 45A		30	.267		56.50	17		73.50	87
8630	12 pole, 8-4AWG, 600V, 45A		30	.267		67.50	17		84.50	99

26 27 33 – Power Distribution Units

26 27 33.10 Power Distribution Unit Cabinet

		Crew	Daily Output	Labor-Hours	Unit	Material	2021 Bare Costs Labor	2021 Bare Costs Equipment	Total	Total Incl O&P
0010	**POWER DISTRIBUTION UNIT CABINET**									
0100	Power distribution unit, single cabinet, 50 kVA output, 480/208 input	1 Elec	2	4	Ea.	9,250	255		9,505	10,600
0110	75 kVA output, 480/208 input	"	2	4	"	10,000	255		10,255	11,400
0120	100 kVA output, 480/208 input	1 Elec	2	4	Ea.	10,900	255		11,155	12,400
0130	125 kVA output, 480/208 input		1.50	5.333		12,000	340		12,340	13,700
0140	150 kVA output, 480/208 input		1.50	5.333		14,600	340		14,940	16,600
0150	200 kVA output, 480/208 input		1	8		16,100	510		16,610	18,500
0160	225 kVA output, 480/208 input		1	8		17,000	510		17,510	19,500

26 27 33.20 Power Distribution Unit

		Crew	Daily Output	Labor-Hours	Unit	Material	2021 Bare Costs Labor	2021 Bare Costs Equipment	Total	Total Incl O&P
0010	**POWER DISTRIBUTION UNIT**									
0050	3PH, 60 kVA, PDU									
0100	208V-208V/120V	2 Elec	2	8	Ea.	14,400	510		14,910	16,600
0110	480V-208V/120V		2	8		14,400	510		14,910	16,600
0120	600V-208V/120V		2	8		14,400	510		14,910	16,600
0125	3PH, 80 kVA, PDU									
0130	208V-208V/120V	2 Elec	2	8	Ea.	26,100	510		26,610	29,500
0140	480V-208V/120V		2	8		27,800	510		28,310	31,400
0150	600V-208V/120V		2	8		27,500	510		28,010	31,100
0200	3PH, 15 kVA, 480-208/120V, 208-208/120V		3	5.333		11,500	340		11,840	13,100
0210	30 kVA, 400-208/120V, 208-208/120V		3	5.333		18,900	340		19,240	21,300
0250	50 kVA, 480-208/120V, 208-208/120V		2.50	6.400		10,400	410		10,810	12,000
0260	75 kVA, 480-208/120V, 208-208/120V		2.50	6.400		11,700	410		12,110	13,500
0270	100 kVA, 480-208/120V, 208-208/120V		2	8		12,600	510		13,110	14,600
0280	125 kVA, 480-208/120V, 208-208/120V		2	8		13,500	510		14,010	15,700
0290	150 kVA, 480-208/120V, 208-208/120V		2	8		12,400	510		12,910	14,400
0300	200 kVA, 480-208/120V, 208-208/120V		1.50	10.667		13,600	680		14,280	16,000
0310	225 kVA, 480-208/120V, 208-208/120V		1.50	10.667		14,400	680		15,080	16,800

For customer support on your Electrical Costs with RSMeans Data, call 800.448.8182.

285

26 27 Low-Voltage Distribution Equipment

26 27 73 – Door Chimes

26 27 73.10 Doorbell System

		Crew	Daily Output	Labor-Hours	Unit	Material	2021 Bare Costs Labor	Equipment	Total	Total Incl O&P
0010	**DOORBELL SYSTEM**, incl. transformer, button & signal									
0100	6" bell	1 Elec	4	2	Ea.	153	127		280	355
0200	Buzzer		4	2		123	127		250	325
1000	Door chimes, 2 notes		16	.500		32.50	32		64.50	83
1020	with ambient light		12	.667		120	42.50		162.50	195
1100	Tube type, 3 tube system		12	.667		236	42.50		278.50	325
1180	4 tube system		10	.800		475	51		526	600
1900	For transformer & button, add		5	1.600		16	102		118	169
3000	For push button only		24	.333		.92	21		21.92	32.50
3200	Bell transformer		16	.500		26	32		58	76

26 28 Low-Voltage Circuit Protective Devices

26 28 13 – Fuses

26 28 13.10 Fuse Elements

		Crew	Daily Output	Labor-Hours	Unit	Material	2021 Bare Costs Labor	Equipment	Total	Total Incl O&P
0010	**FUSE ELEMENTS**									
0020	Cartridge, nonrenewable									
0050	250 volt, 30 amp	1 Elec	50	.160	Ea.	2.28	10.20		12.48	17.65
0100	60 amp	"	50	.160	"	3.99	10.20		14.19	19.55
0150	100 amp	1 Elec	40	.200	Ea.	16.55	12.75		29.30	37
0200	200 amp		36	.222		37	14.15		51.15	61.50
0250	400 amp		30	.267		75.50	17		92.50	108
0300	600 amp		24	.333		118	21		139	162
0400	600 volt, 30 amp		40	.200		9	12.75		21.75	29
0450	60 amp		40	.200		13.90	12.75		26.65	34.50
0500	100 amp		36	.222		28	14.15		42.15	52
0550	200 amp		30	.267		55.50	17		72.50	86
0600	400 amp		24	.333		120	21		141	164
0650	600 amp		20	.400		200	25.50		225.50	258
0800	Dual element, time delay, 250 volt, 30 amp		50	.160		9.95	10.20		20.15	26
0840	50 amp		50	.160		11.30	10.20		21.50	27.50
0850	60 amp		50	.160		11.60	10.20		21.80	28
0900	100 amp		40	.200		28	12.75		40.75	50
0950	200 amp		36	.222		61.50	14.15		75.65	88.50
1000	400 amp		30	.267		110	17		127	146
1050	600 amp		24	.333		181	21		202	231
1300	600 volt, 15 to 30 amp		40	.200		21.50	12.75		34.25	42.50
1350	35 to 60 amp		40	.200		31.50	12.75		44.25	53.50
1400	70 to 100 amp		36	.222		66.50	14.15		80.65	94
1450	110 to 200 amp		30	.267		129	17		146	167
1500	225 to 400 amp		24	.333		284	21		305	340
1550	600 amp		20	.400		355	25.50		380.50	435
1800	Class RK1, high capacity, 250 volt, 30 amp		50	.160		7.40	10.20		17.60	23.50
1850	60 amp		50	.160		13.35	10.20		23.55	30
1900	100 amp		40	.200		26	12.75		38.75	47.50
1950	200 amp		36	.222		66	14.15		80.15	93.50
2000	400 amp		30	.267		110	17		127	146
2050	600 amp		24	.333		195	21		216	246
2200	600 volt, 30 amp		40	.200		14.65	12.75		27.40	35
2250	60 amp		40	.200		21.50	12.75		34.25	43
2300	100 amp		36	.222		46.50	14.15		60.65	72

26 28 13 – Fuses

26 28 13.10 Fuse Elements

		Crew	Daily Output	Labor-Hours	Unit	Material	2021 Bare Costs Labor	Equipment	Total	Total Incl O&P
2350	200 amp	1 Elec	30	.267	Ea.	93.50	17		110.50	128
2400	400 amp		24	.333		220	21		241	274
2450	600 amp		20	.400		375	25.50		400.50	455
2700	Class J, current limiting, 250 or 600 volt, 30 amp		40	.200		19.60	12.75		32.35	40.50
2750	60 amp		40	.200		32.50	12.75		45.25	55
2800	100 amp		36	.222		46	14.15		60.15	71.50
2850	200 amp		30	.267		90.50	17		107.50	125
2900	400 amp		24	.333		201	21		222	253
2950	600 amp		20	.400		325	25.50		350.50	395
3100	Class L, current limiting, 250 or 600 volt, 601 to 1,200 amp		16	.500		470	32		502	570
3150	1,500 to 1,600 amp	1 Elec	13	.615	Ea.	610	39		649	730
3200	1,800 to 2,000 amp		10	.800		770	51		821	920
3250	2,500 amp		10	.800		1,175	51		1,226	1,375
3300	3,000 amp		8	1		1,400	63.50		1,463.50	1,625
3350	3,500 to 4,000 amp		8	1		2,050	63.50		2,113.50	2,350
3400	4,500 to 5,000 amp		6.70	1.194		2,350	76		2,426	2,700
3450	6,000 amp		5.70	1.404		2,850	89.50		2,939.50	3,275
3600	Plug, 120 volt, 1 to 10 amp		50	.160		3.94	10.20		14.14	19.50
3650	15 to 30 amp		50	.160		3.97	10.20		14.17	19.50
3700	Dual element 0.3 to 14 amp		50	.160		7.50	10.20		17.70	23.50
3750	15 to 30 amp		50	.160		7.60	10.20		17.80	23.50
3800	Fustat, 120 volt, 15 to 30 amp		50	.160		5.75	10.20		15.95	21.50
3850	0.3 to 14 amp		50	.160		8.75	10.20		18.95	25
3900	Adapters 0.3 to 10 amp		50	.160		6.65	10.20		16.85	22.50
3950	15 to 30 amp		50	.160		8.70	10.20		18.90	24.50
4000	F-frame current limiting fuse, 14 to 2 AWG, 3 ampere, 3P, aluminum terminal		40	.200		1,975	12.75		1,987.75	2,200
4010	7 ampere, 3P, aluminum terminal		40	.200		1,975	12.75		1,987.75	2,200
4020	15 AMP, 3P, aluminum terminal		40	.200		1,825	12.75		1,837.75	2,025
4030	30 AMP, 3P, aluminum terminal		40	.200		1,875	12.75		1,887.75	2,075
4040	50 AMP, 3P, aluminum terminal		40	.200		1,575	12.75		1,587.75	1,750
4050	F-frame current limiting fuse, 1 to 4/0 AWG, 100 AMP, 3P, aluminum terminal		40	.200		2,275	12.75		2,287.75	2,550
4060	150 AMP, 3P, aluminum terminal		40	.200		2,600	12.75		2,612.75	2,900

26 28 16 – Enclosed Switches and Circuit Breakers

26 28 16.10 Circuit Breakers

		Crew	Daily Output	Labor-Hours	Unit	Material	2021 Bare Costs Labor	Equipment	Total	Total Incl O&P
0010	**CIRCUIT BREAKERS** (in enclosure)									
0100	Enclosed (NEMA 1), 600 volt, 3 pole, 30 amp	1 Elec	3.20	2.500	Ea.	550	159		709	840
0200	60 amp		2.80	2.857		700	182		882	1,050
0400	100 amp		2.30	3.478		800	222		1,022	1,200
0500	200 amp		1.50	5.333		1,900	340		2,240	2,575
0600	225 amp		1.50	5.333		1,850	340		2,190	2,525
0700	400 amp	2 Elec	1.60	10		3,150	635		3,785	4,425
0800	600 amp		1.20	13.333		4,575	850		5,425	6,275
1000	800 amp		.94	17.021		5,950	1,075		7,025	8,150
1200	1,000 amp		.84	19.048		7,525	1,225		8,750	10,100
1220	1,200 amp		.80	20		9,625	1,275		10,900	12,500
1240	1,600 amp		.72	22.222		16,100	1,425		17,525	19,900
1260	2,000 amp		.64	25		17,500	1,600		19,100	21,600
1400	1,200 amp with ground fault		.80	20		14,600	1,275		15,875	17,900
1600	1,600 amp with ground fault		.72	22.222		18,300	1,425		19,725	22,200
1800	2,000 amp with ground fault		.64	25		19,600	1,600		21,200	24,000
2000	Disconnect, 240 volt 3 pole, 5 HP motor	1 Elec	3.20	2.500		490	159		649	775
2020	10 HP motor		3.20	2.500		490	159		649	775

26 28 16.10 Circuit Breakers

		Crew	Daily Output	Labor-Hours	Unit	Material	2021 Bare Costs Labor	Equipment	Total	Total Incl O&P
2040	15 HP motor	1 Elec	2.80	2.857	Ea.	490	182		672	810
2060	20 HP motor		2.30	3.478		595	222		817	985
2080	25 HP motor		2.30	3.478		595	222		817	985
2100	30 HP motor		2.30	3.478		595	222		817	985
2120	40 HP motor		2	4		1,025	255		1,280	1,500
2140	50 HP motor		1.50	5.333		1,025	340		1,365	1,625
2160	60 HP motor		1.50	5.333		2,350	340		2,690	3,075
2180	75 HP motor	2 Elec	2	8		2,350	510		2,860	3,325
2200	100 HP motor		1.60	10		2,350	635		2,985	3,525
2220	125 HP motor		1.60	10		2,350	635		2,985	3,525
2240	150 HP motor		1.20	13.333		4,575	850		5,425	6,275
2260	200 HP motor		1.20	13.333		5,950	850		6,800	7,800
2300	Enclosed (NEMA 7), explosion proof, 600 volt 3 pole, 50 amp	1 Elec	2.30	3.478		1,975	222		2,197	2,500
2350	100 amp		1.50	5.333		2,050	340		2,390	2,750
2400	150 amp		1	8		4,925	510		5,435	6,175
2450	250 amp	2 Elec	1.60	10		6,150	635		6,785	7,725
2500	400 amp	"	1.20	13.333		6,200	850		7,050	8,075

26 28 16.13 Circuit Breakers

		Crew	Daily Output	Labor-Hours	Unit	Material	2021 Bare Costs Labor	Equipment	Total	Total Incl O&P
0010	**CIRCUIT BREAKERS**									
0100	Circuit breaker current limiter, 225 to 400, ampere, 1 pole	1 Elec	48	.167	Ea.	810	10.60		820.60	905
0200	Circuit breaker current limiter, 200 to 500, ampere, 1 pole, thermal magnet		46	.174		490	11.10		501.10	555
0300	300 to 500, ampere, 1 pole, thermal magnet		46	.174		1,375	11.10		1,386.10	1,550
0400	Circuit breaker current limiter, 600 to 1000, ampere, 1 pole, tri pac		46	.174		1,775	11.10		1,786.10	1,975

26 28 16.20 Safety Switches

		Crew	Daily Output	Labor-Hours	Unit	Material	2021 Bare Costs Labor	Equipment	Total	Total Incl O&P
0010	**SAFETY SWITCHES** R262816-80									
0100	General duty 240 volt, 3 pole NEMA 1, fusible, 30 amp	1 Elec	3.20	2.500	Ea.	65.50	159		224.50	310
0200	60 amp		2.30	3.478		111	222		333	450
0300	100 amp		1.90	4.211		193	268		461	610
0400	200 amp		1.30	6.154		415	390		805	1,025
0500	400 amp	2 Elec	1.80	8.889		1,050	565		1,615	2,000
0600	600 amp	"	1.20	13.333		1,950	850		2,800	3,400
0610	Nonfusible, 30 amp	1 Elec	3.20	2.500		53	159		212	296
0650	60 amp		2.30	3.478		69.50	222		291.50	405
0700	100 amp		1.90	4.211		163	268		431	580
0750	200 amp		1.30	6.154		305	390		695	915
0800	400 amp	2 Elec	1.80	8.889		775	565		1,340	1,700
0850	600 amp	"	1.20	13.333		1,350	850		2,200	2,750
1100	Heavy duty, 600 volt, 3 pole NEMA 1 nonfused									
1110	30 amp	1 Elec	3.20	2.500	Ea.	93.50	159		252.50	340
1500	60 amp		2.30	3.478		164	222		386	510
1700	100 amp		1.90	4.211		260	268		528	685
1900	200 amp		1.30	6.154		400	390		790	1,025
2100	400 amp	2 Elec	1.80	8.889		920	565		1,485	1,850
2300	600 amp		1.20	13.333		1,800	850		2,650	3,225
2500	800 amp		.94	17.021		3,175	1,075		4,250	5,100
2700	1,200 amp		.80	20		4,500	1,275		5,775	6,850
2900	Heavy duty, 240 volt, 3 pole NEMA 1 fusible									
2910	30 amp	1 Elec	3.20	2.500	Ea.	111	159		270	360
3000	60 amp		2.30	3.478		179	222		401	525
3300	100 amp		1.90	4.211		284	268		552	710
3500	200 amp		1.30	6.154		495	390		885	1,125
3700	400 amp	2 Elec	1.80	8.889		1,200	565		1,765	2,175

26 28 16.20 Safety Switches		Crew	Daily Output	Labor-Hours	Unit	Material	2021 Bare Costs Labor	Equipment	Total	Total Incl O&P
3900	600 amp	2 Elec	1.20	13.333	Ea.	2,375	850		3,225	3,850
4100	800 amp		.94	17.021		5,075	1,075		6,150	7,200
4300	1,200 amp	↓	.80	20		4,200	1,275		5,475	6,525
4340	2 pole fusible, 30 amp	1 Elec	3.50	2.286		80.50	146		226.50	305
4350	600 volt, 3 pole, fusible, 30 amp		3.20	2.500		175	159		334	430
4380	60 amp		2.30	3.478		214	222		436	565
4400	100 amp		1.90	4.211		395	268		663	830
4420	200 amp		1.30	6.154		570	390		960	1,200
4440	400 amp	2 Elec	1.80	8.889		1,475	565		2,040	2,475
4450	600 amp		1.20	13.333		2,600	850		3,450	4,100
4460	800 amp		.94	17.021		5,075	1,075		6,150	7,200
4480	1,200 amp	↓	.80	20	↓	4,200	1,275		5,475	6,525
4500	240 volt 3 pole NEMA 3R (no hubs), fusible									
4510	30 amp	1 Elec	3.10	2.581	Ea.	184	164		348	445
4700	60 amp		2.20	3.636		296	232		528	670
4900	100 amp		1.80	4.444		460	283		743	925
5100	200 amp	↓	1.20	6.667		580	425		1,005	1,275
5300	400 amp	2 Elec	1.60	10		1,325	635		1,960	2,425
5500	600 amp	"	1	16		2,700	1,025		3,725	4,500
5510	Heavy duty, 600 volt, 3 pole 3 ph. NEMA 3R fusible, 30 amp	1 Elec	3.10	2.581		295	164		459	570
5520	60 amp		2.20	3.636		365	232		597	750
5530	100 amp		1.80	4.444		530	283		813	1,000
5540	200 amp	↓	1.20	6.667		725	425		1,150	1,425
5550	400 amp	2 Elec	1.60	10	↓	1,800	635		2,435	2,925
5700	600 volt, 3 pole NEMA 3R nonfused									
5710	30 amp	1 Elec	3.10	2.581	Ea.	160	164		324	420
5900	60 amp		2.20	3.636		283	232		515	655
6100	100 amp		1.80	4.444		405	283		688	865
6300	200 amp	↓	1.20	6.667		490	425		915	1,175
6500	400 amp	2 Elec	1.60	10		1,325	635		1,960	2,400
6700	600 amp	"	1	16		2,550	1,025		3,575	4,325
6900	600 volt, 6 pole NEMA 3R nonfused, 30 amp	1 Elec	2.70	2.963		1,150	189		1,339	1,550
7100	60 amp		2	4		1,450	255		1,705	1,975
7300	100 amp		1.50	5.333		1,900	340		2,240	2,600
7500	200 amp	↓	1.20	6.667	↓	4,050	425		4,475	5,075
7600	600 volt, 3 pole NEMA 7 explosion proof nonfused									
7610	30 amp	1 Elec	2.20	3.636	Ea.	1,425	232		1,657	1,925
7620	60 amp		1.80	4.444		1,725	283		2,008	2,325
7630	100 amp		1.20	6.667		2,875	425		3,300	3,775
7640	200 amp		.80	10		6,175	635		6,810	7,725
7710	600 volt 6 pole, NEMA 3R fusible, 30 amp		2.70	2.963		1,475	189		1,664	1,900
7900	60 amp		2	4		1,900	255		2,155	2,450
8100	100 amp		1.50	5.333		2,300	340		2,640	3,025
8110	240 volt 3 pole, NEMA 12 fusible, 30 amp		3.10	2.581		256	164		420	525
8120	60 amp		2.20	3.636		425	232		657	810
8130	100 amp		1.80	4.444		555	283		838	1,025
8140	200 amp	↓	1.20	6.667		745	425		1,170	1,450
8150	400 amp	2 Elec	1.60	10		1,775	635		2,410	2,900
8160	600 amp	"	1	16		3,000	1,025		4,025	4,825
8180	600 volt 3 pole, NEMA 12 fusible, 30 amp	1 Elec	3.10	2.581		375	164		539	655
8190	60 amp		2.20	3.636		435	232		667	825
8200	100 amp		1.80	4.444		635	283		918	1,125
8210	200 amp	↓	1.20	6.667		900	425		1,325	1,625

For customer support on your Electrical Costs with RSMeans Data, call 800.448.8182.

289

26 28 16.20 Safety Switches

	26 28 16.20 Safety Switches	Crew	Daily Output	Labor-Hours	Unit	Material	2021 Bare Costs Labor	Equipment	Total	Total Incl O&P
8220	400 amp	2 Elec	1.60	10	Ea.	2,075	635		2,710	3,225
8230	600 amp	"	1	16		3,775	1,025		4,800	5,675
8240	600 volt 3 pole, NEMA 12 nonfused, 30 amp	1 Elec	3.10	2.581		235	164		399	500
8250	60 amp		2.20	3.636		315	232		547	695
8260	100 amp		1.80	4.444		460	283		743	925
8270	200 amp		1.20	6.667		570	425		995	1,250
8280	400 amp	2 Elec	1.60	10		1,425	635		2,060	2,500
8290	600 amp	"	1	16		2,975	1,025		4,000	4,800
8310	600 volt, 3 pole NEMA 4 fusible, 30 amp	1 Elec	3	2.667		885	170		1,055	1,225
8320	60 amp		2.20	3.636		975	232		1,207	1,425
8330	100 amp		1.80	4.444		1,950	283		2,233	2,575
8340	200 amp		1.20	6.667		2,650	425		3,075	3,525
8350	400 amp	2 Elec	1.60	10		5,375	635		6,010	6,850
8360	600 volt 3 pole NEMA 4 nonfused, 30 amp	1 Elec	3	2.667		780	170		950	1,100
8370	60 amp		2.20	3.636		895	232		1,127	1,325
8380	100 amp		1.80	4.444		1,800	283		2,083	2,425
8390	200 amp		1.20	6.667		2,475	425		2,900	3,350
8400	400 amp	2 Elec	1.60	10		4,500	635		5,135	5,900
8490	Motor starters, manual, single phase, NEMA 1	1 Elec	6.40	1.250		102	79.50		181.50	230
8500	NEMA 4		4	2		276	127		403	495
8700	NEMA 7		4	2		350	127		477	575
8900	NEMA 1 with pilot		6.40	1.250		129	79.50		208.50	260
8920	3 pole, NEMA 1, 230/460 volt, 5 HP, size 0		3.50	2.286		202	146		348	440
8940	10 HP, size 1		2	4		240	255		495	645
9010	Disc. switch, 600 volt 3 pole fusible, 30 amp, to 10 HP motor		3.20	2.500		360	159		519	630
9050	60 amp, to 30 HP motor		2.30	3.478		825	222		1,047	1,225
9070	100 amp, to 60 HP motor		1.90	4.211		825	268		1,093	1,300
9100	200 amp, to 125 HP motor		1.30	6.154		1,225	390		1,615	1,925
9110	400 amp, to 200 HP motor	2 Elec	1.80	8.889		3,125	565		3,690	4,275

26 28 16.40 Time Switches

	26 28 16.40 Time Switches	Crew	Daily Output	Labor-Hours	Unit	Material	2021 Bare Costs Labor	Equipment	Total	Total Incl O&P
0010	**TIME SWITCHES**									
0100	Single pole, single throw, 24 hour dial	1 Elec	4	2	Ea.	144	127		271	350
0200	24 hour dial with reserve power		3.60	2.222		740	142		882	1,025
0300	Astronomic dial		3.60	2.222		257	142		399	495
0400	Astronomic dial with reserve power		3.30	2.424		960	154		1,114	1,275
0500	7 day calendar dial		3.30	2.424		214	154		368	465
0600	7 day calendar dial with reserve power		3.20	2.500		193	159		352	450
0700	Photo cell 2,000 watt		8	1		22	63.50		85.50	119
1080	Load management device, 4 loads		2	4		1,150	255		1,405	1,650
1100	8 loads		1	8		2,600	510		3,110	3,600

26 28 16.50 Meter Socket Entry Hub

	26 28 16.50 Meter Socket Entry Hub	Crew	Daily Output	Labor-Hours	Unit	Material	2021 Bare Costs Labor	Equipment	Total	Total Incl O&P
0010	**METER SOCKET ENTRY HUB**									
0100	Meter socket entry hub closing cap, 3/4 to 4 inch	1 Elec	40	.200	Ea.	2.63	12.75		15.38	22
0110	Meter socket entry hub, 3/4 conduit		40	.200		16.35	12.75		29.10	37
0120	1 inch conduit		40	.200		16.95	12.75		29.70	37.50
0130	1.25 inch conduit		40	.200		16.15	12.75		28.90	37
0140	1.50 inch conduit		40	.200		16.85	12.75		29.60	37.50
0150	2 inch conduit		40	.200		26	12.75		38.75	47.50
0160	2.50 inch conduit		40	.200		45.50	12.75		58.25	69
0170	3 inch conduit		40	.200		79	12.75		91.75	106

26 29 Low-Voltage Controllers

26 29 13 – Enclosed Controllers

26 29 13.10 Contactors, AC

		Crew	Daily Output	Labor-Hours	Unit	Material	2021 Bare Costs Labor	Equipment	Total	Total Incl O&P
0010	**CONTACTORS, AC** Enclosed (NEMA 1)									
0050	Lighting, 600 volt 3 pole, electrically held									
0100	20 amp	1 Elec	4	2	Ea.	395	127		522	620
0200	30 amp		3.60	2.222		305	142		447	545
0300	60 amp		3	2.667		655	170		825	970
0400	100 amp		2.50	3.200		1,050	204		1,254	1,450
0500	200 amp		1.40	5.714		2,525	365		2,890	3,325
0600	300 amp	2 Elec	1.60	10		5,250	635		5,885	6,725
0800	600 volt 3 pole, mechanically held, 30 amp	1 Elec	3.60	2.222		445	142		587	700
0900	60 amp	"	3	2.667		880	170		1,050	1,225
1000	75 amp	1 Elec	2.80	2.857	Ea.	1,300	182		1,482	1,725
1100	100 amp		2.50	3.200		1,750	204		1,954	2,225
1200	150 amp		2	4		3,200	255		3,455	3,900
1300	200 amp		1.40	5.714		3,425	365		3,790	4,325
1500	Magnetic with auxiliary contact, size 00, 9 amp		4	2		192	127		319	400
1600	Size 0, 18 amp		4	2		229	127		356	440
1700	Size 1, 27 amp		3.60	2.222		260	142		402	495
1800	Size 2, 45 amp		3	2.667		485	170		655	780
1900	Size 3, 90 amp		2.50	3.200		780	204		984	1,175
2000	Size 4, 135 amp		2.30	3.478		1,775	222		1,997	2,300
2100	Size 5, 270 amp	2 Elec	1.80	8.889		3,750	565		4,315	4,975
2200	Size 6, 540 amp		1.20	13.333		11,000	850		11,850	13,400
2300	Size 7, 810 amp		1	16		14,700	1,025		15,725	17,700
2310	Size 8, 1,215 amp		.80	20		23,000	1,275		24,275	27,200
2500	Magnetic, 240 volt, 1-2 pole, 0.75 HP motor	1 Elec	4	2		155	127		282	360
2520	2 HP motor		3.60	2.222		380	142		522	630
2540	5 HP motor		2.50	3.200		420	204		624	770
2560	10 HP motor		1.40	5.714		695	365		1,060	1,300
2600	240 volt or less, 3 pole, 0.75 HP motor		4	2		155	127		282	360
2620	5 HP motor		3.60	2.222		192	142		334	420
2640	10 HP motor		3.60	2.222		223	142		365	455
2660	15 HP motor		2.50	3.200		445	204		649	795
2700	25 HP motor		2.50	3.200		445	204		649	795
2720	30 HP motor	2 Elec	2.80	5.714		740	365		1,105	1,350
2740	40 HP motor		2.80	5.714		740	365		1,105	1,350
2760	50 HP motor		1.60	10		740	635		1,375	1,750
2800	75 HP motor		1.60	10		1,750	635		2,385	2,875
2820	100 HP motor		1	16		1,750	1,025		2,775	3,450
2860	150 HP motor		1	16		3,700	1,025		4,725	5,600
2880	200 HP motor		1	16		3,700	1,025		4,725	5,600
3000	600 volt, 3 pole, 5 HP motor	1 Elec	4	2		192	127		319	400
3020	10 HP motor		3.60	2.222		223	142		365	455
3040	25 HP motor		3	2.667		445	170		615	740
3100	50 HP motor		2.50	3.200		740	204		944	1,125
3160	100 HP motor	2 Elec	2.80	5.714		1,750	365		2,115	2,475
3220	200 HP motor	"	1.60	10		3,700	635		4,335	5,025

26 29 13.20 Control Stations

		Crew	Daily Output	Labor-Hours	Unit	Material	2021 Bare Costs Labor	Equipment	Total	Total Incl O&P
0010	**CONTROL STATIONS**									
0050	NEMA 1, heavy duty, stop/start	1 Elec	8	1	Ea.	155	63.50		218.50	265
0100	Stop/start, pilot light		6.20	1.290		211	82		293	355
0200	Hand/off/automatic		6.20	1.290		114	82		196	248
0400	Stop/start/reverse		5.30	1.509		209	96		305	375

For customer support on your Electrical Costs with RSMeans Data, call 800.448.8182.

291

26 29 13.20 Control Stations

		Crew	Daily Output	Labor-Hours	Unit	Material	2021 Bare Costs Labor	Equipment	Total	Total Incl O&P
0500	NEMA 7, heavy duty, stop/start	1 Elec	6	1.333	Ea.	575	85		660	755
0600	Stop/start, pilot light		4	2		700	127		827	960
0700	NEMA 7 or 9, 1 element		6	1.333		465	85		550	640
0800	2 element		6	1.333		600	85		685	785
0900	3 element		4	2		1,275	127		1,402	1,625
0910	Selector switch, 2 position		6	1.333		470	85		555	640
0920	3 position		4	2		470	127		597	705
0930	Oiltight, 1 element		8	1		148	63.50		211.50	258
0940	2 element		6.20	1.290		260	82		342	410
0950	3 element		5.30	1.509		315	96		411	490
0960	Selector switch, 2 position		6.20	1.290		158	82		240	296
0970	3 position		5.30	1.509		200	96		296	365

26 29 13.30 Control Switches

		Crew	Daily Output	Labor-Hours	Unit	Material	2021 Bare Costs Labor	Equipment	Total	Total Incl O&P
0010	**CONTROL SWITCHES** Field installed									
6000	Push button 600 V 10A, momentary contact									
6150	Standard operator with colored button	1 Elec	34	.235	Ea.	50	15		65	77.50
6160	With single block 1NO 1NC		18	.444		69	28.50		97.50	118
6170	With double block 2NO 2NC		15	.533		106	34		140	168
6180	Std operator w/mushroom button 1-9/16" diam.		34	.235		80.50	15		95.50	111
6190	Std operator w/mushroom button 2-1/4" diam.									
6200	With single block 1NO 1NC	1 Elec	18	.444	Ea.	58	28.50		86.50	106
6210	With double block 2NO 2NC		15	.533		78	34		112	136
6500	Maintained contact, selector operator		34	.235		58	15		73	86
6510	With single block 1NO 1NC		18	.444		78	28.50		106.50	128
6520	With double block 2NO 2NC		15	.533		98	34		132	159
6560	Spring-return selector operator		34	.235		58	15		73	86
6570	With single block 1NO 1NC		18	.444		78	28.50		106.50	128
6580	With double block 2NO 2NC		15	.533		98	34		132	159
6620	Transformer operator w/illuminated									
6630	button 6 V #12 lamp	1 Elec	32	.250	Ea.	98	15.95		113.95	132
6640	With single block 1NO 1NC w/guard		16	.500		118	32		150	178
6650	With double block 2NO 2NC w/guard		13	.615		138	39		177	211
6690	Combination operator		34	.235		58	15		73	86
6700	With single block 1NO 1NC		18	.444		78	28.50		106.50	128
6710	With double block 2NO 2NC		15	.533		98	34		132	159
9000	Indicating light unit, full voltage									
9010	110-125 V front mount	1 Elec	32	.250	Ea.	78	15.95		93.95	109
9020	130 V resistor type		32	.250		63	15.95		78.95	92.50
9030	6 V transformer type		32	.250		71.50	15.95		87.45	102

26 29 13.40 Relays

		Crew	Daily Output	Labor-Hours	Unit	Material	2021 Bare Costs Labor	Equipment	Total	Total Incl O&P
0010	**RELAYS** Enclosed (NEMA 1)									
0050	600 volt AC, 1 pole, 12 amp	1 Elec	5.30	1.509	Ea.	90.50	96		186.50	243
0100	2 pole, 12 amp		5	1.600		90.50	102		192.50	251
0200	4 pole, 10 amp		4.50	1.778		121	113		234	300
0500	250 volt DC, 1 pole, 15 amp		5.30	1.509		119	96		215	273
0600	2 pole, 10 amp		5	1.600		111	102		213	273
0700	4 pole, 4 amp		4.50	1.778		150	113		263	335

26 29 23.10 Variable Frequency Drives/Adj. Frequency Drives

			Crew	Daily Output	Labor-Hours	Unit	Material	2021 Bare Costs Labor	Equipment	Total	Total Incl O&P
0010	**VARIABLE FREQUENCY DRIVES/ADJ. FREQUENCY DRIVES**										
0100	Enclosed (NEMA 1), 460 volt, for 3 HP motor size	G	1 Elec	.80	10	Ea.	1,750	635		2,385	2,875
0110	5 HP motor size	G		.80	10		1,975	635		2,610	3,125

26 29 Low-Voltage Controllers

26 29 23 – Variable-Frequency Motor Controllers

26 29 23.10 Variable Frequency Drives/Adj. Frequency Drives		Crew	Daily Output	Labor-Hours	Unit	Material	2021 Bare Costs Labor	Equipment	Total	Total Incl O&P
0120	7.5 HP motor size	G	1 Elec	.67	11.940	Ea.	2,400	760		3,160	3,775
0130	10 HP motor size	G		.67	11.940		2,125	760		2,885	3,450
0140	15 HP motor size	G	2 Elec	.89	17.978		3,400	1,150		4,550	5,425
0150	20 HP motor size	G		.89	17.978		4,275	1,150		5,425	6,400
0160	25 HP motor size	G		.67	23.881		5,075	1,525		6,600	7,825
0170	30 HP motor size	G		.67	23.881		6,150	1,525		7,675	9,025
0180	40 HP motor size	G		.67	23.881		7,125	1,525		8,650	10,100
0190	50 HP motor size	G		.53	30.189		9,375	1,925		11,300	13,200
0200	60 HP motor size	G	R-3	.56	35.714		11,800	2,275	340	14,415	16,700
0210	75 HP motor size	G		.56	35.714		13,600	2,275	340	16,215	18,700
0220	100 HP motor size	G		.50	40		16,400	2,550	380	19,330	22,300
0230	125 HP motor size	G		.50	40		17,800	2,550	380	20,730	23,800
0240	150 HP motor size	G		.50	40		23,300	2,550	380	26,230	29,800
0250	200 HP motor size	G		.42	47.619		32,900	3,025	455	36,380	41,200
1100	Custom-engineered, 460 volt, for 3 HP motor size	G	1 Elec	.56	14.286		3,575	910		4,485	5,275
1110	5 HP motor size	G		.56	14.286		3,625	910		4,535	5,350
1120	7.5 HP motor size	G		.47	17.021		3,075	1,075		4,150	4,975
1130	10 HP motor size	G		.47	17.021		3,250	1,075		4,325	5,200
1140	15 HP motor size	G	2 Elec	.62	25.806		4,925	1,650		6,575	7,875
1150	20 HP motor size	G		.62	25.806		2,650	1,650		4,300	5,375
1160	25 HP motor size	G		.47	34.043		5,675	2,175		7,850	9,450
1170	30 HP motor size	G		.47	34.043		6,575	2,175		8,750	10,500
1180	40 HP motor size	G		.47	34.043		7,425	2,175		9,600	11,400
1190	50 HP motor size	G		.37	43.243		11,000	2,750		13,750	16,200
1200	60 HP motor size	G	R-3	.39	51.282		16,500	3,250	490	20,240	23,500
1210	75 HP motor size	G		.39	51.282		13,800	3,250	490	17,540	20,600
1220	100 HP motor size	G		.35	57.143		19,300	3,625	545	23,470	27,300
1230	125 HP motor size	G		.35	57.143		20,700	3,625	545	24,870	28,800
1240	150 HP motor size	G		.35	57.143		23,500	3,625	545	27,670	31,800
1250	200 HP motor size	G		.29	68.966		27,000	4,375	660	32,035	36,900
2000	For complex & special design systems to meet specific										
2010	requirements, obtain quote from vendor.										

26 31 Photovoltaic Collectors

26 31 13 – Photovoltaics

26 31 13.50 Solar Energy - Photovoltaics

			Crew	Daily Output	Labor-Hours	Unit	Material	Labor	Equipment	Total	Total Incl O&P
0010	**SOLAR ENERGY - PHOTOVOLTAICS**										
0220	Alt. energy source, photovoltaic module, 6 watt, 15 V	G	1 Elec	8	1	Ea.	58	63.50		121.50	159
0230	10 watt, 16.3 V	G		8	1		116	63.50		179.50	222
0240	20 watt, 14.5 V	G		8	1		184	63.50		247.50	297
0250	36 watt, 17 V	G		8	1		184	63.50		247.50	297
0260	55 watt, 17 V	G		8	1		241	63.50		304.50	360
0270	75 watt, 17 V	G		8	1		425	63.50		488.50	565
0280	130 watt, 33 V	G		8	1		615	63.50		678.50	775
0290	140 watt, 33 V	G		8	1		490	63.50		553.50	630
0300	150 watt, 33 V	G		8	1		505	63.50		568.50	650
0310	DC to AC inverter for, 12 V, 2,000 watt	G		4	2		1,525	127		1,652	1,875
0320	12 V, 2,500 watt	G		4	2		1,150	127		1,277	1,475
0330	24 V, 2,500 watt	G		4	2		1,675	127		1,802	2,050
0340	12 V, 3,000 watt	G		3	2.667		1,300	170		1,470	1,675
0350	24 V, 3,000 watt	G		3	2.667		1,100	170		1,270	1,475

For customer support on your Electrical Costs with RSMeans Data, call 800.448.8182.

293

26 31 Photovoltaic Collectors

26 31 13 – Photovoltaics

26 31 13.50 Solar Energy - Photovoltaics

			Crew	Daily Output	Labor-Hours	Unit	Material	2021 Bare Costs Labor	Equipment	Total	Total Incl O&P
0360	24 V, 4,000 watt	G	1 Elec	2	4	Ea.	3,750	255		4,005	4,500
0370	48 V, 4,000 watt	G		2	4		3,125	255		3,380	3,825
0380	48 V, 5,500 watt	G		2	4		2,775	255		3,030	3,450
0390	PV components, combiner box, 10 lug, NEMA 3R enclosure	G		4	2		269	127		396	485
0400	Fuse, 15 A for combiner box	G		40	.200		21.50	12.75		34.25	42.50
0410	Battery charger controller w/temperature sensor	G		4	2		405	127		532	635
0420	Digital readout panel, displays hours, volts, amps, etc.	G		4	2		211	127		338	420
0430	Deep cycle solar battery, 6 V, 180 Ah (C/20)	G		8	1		340	63.50		403.50	465
0440	Battery interconn, 15" AWG #2/0, sealed w/copper ring lugs	G		16	.500		17.05	32		49.05	66.50
0442	Battery interconn, 24" AWG #2/0, sealed w/copper ring lugs	G		16	.500		26.50	32		58.50	76.50
0444	Battery interconn, 60" AWG #2/0, sealed w/copper ring lugs	G		16	.500		47.50	32		79.50	100
0446	Batt temp computer probe, RJ11 jack, 15' cord	G		16	.500		22	32		54	72
0450	System disconnect, DC 175 amp circuit breaker	G		8	1		244	63.50		307.50	365
0460	Conduit box for inverter	G		8	1		63	63.50		126.50	164
0470	Low voltage disconnect	G		8	1		73	63.50		136.50	175
0480	Vented battery enclosure, wood	G	1 Carp	2	4		305	219		524	660
0490	PV rack system, roof, non-penetrating ballast, 1 panel	G	R-1A	30.50	.525		760	30		790	885
0500	Penetrating surface mount, on steel framing, 1 panel			5.13	3.122		58.50	179		237.50	330
0510	On wood framing, 1 panel			11	1.455		57.50	83.50		141	187
0520	With standoff, 1 panel		R-1A	11	1.455	Ea.	66.50	83.50		150	198
0530	Ground, ballast, fixed, 3 panel			20.50	.780		1,100	45		1,145	1,300
0540	4 panel			20.50	.780		1,500	45		1,545	1,750
0550	5 panel			20.50	.780		2,000	45		2,045	2,275
0560	6 panel			20.50	.780		2,300	45		2,345	2,600
0570	Adjustable, 3 panel			20.50	.780		1,175	45		1,220	1,375
0580	4 panel			20.50	.780		1,600	45		1,645	1,850
0590	5 panel			20.50	.780		2,125	45		2,170	2,425
0600	6 panel			20.50	.780		2,450	45		2,495	2,775
0605	Top of Pole, see 26 56 13.10 3800+, 5400+ & 33 71 16.33 7400+ for poles										
0610	Passive tracking, 1 panel		R-1A	20.50	.780	Ea.	540	45		585	660
0620	2 panel			20.50	.780		1,375	45		1,420	1,575
0630	3 panel			20.50	.780		565	45		610	685
0640	4 panel			20.50	.780		975	45		1,020	1,150
0650	6 panel			20.50	.780		1,475	45		1,520	1,675
0660	8 panel			20.50	.780		3,450	45		3,495	3,875
1020	Photovoltaic module, Thin film		1 Elec	8	1		184	63.50		247.50	297
1040	Photovoltaic module, Cadium telluride			8	1		241	63.50		304.50	360
1060	Photovoltaic module, Polycrystalline			8	1		184	63.50		247.50	297
1080	Photovoltaic module, Monocrystalline			8	1		615	63.50		678.50	775

26 32 Packaged Generator Assemblies

26 32 13 – Engine Generators

26 32 13.13 Diesel-Engine-Driven Generator Sets

		Crew	Daily Output	Labor-Hours	Unit	Material	2021 Bare Costs Labor	Equipment	Total	Total Incl O&P
0010	**DIESEL-ENGINE-DRIVEN GENERATOR SETS**									
2000	Diesel engine, including battery, charger,									
2010	muffler & day tank, 30 kW	R-3	.55	36.364	Ea.	11,800	2,300	345	14,445	16,800
2100	50 kW		.42	47.619		25,200	3,025	455	28,680	32,800
2110	60 kW		.39	51.282		24,500	3,250	490	28,240	32,400
2200	75 kW		.35	57.143		19,500	3,625	545	23,670	27,500
2300	100 kW		.31	64.516		33,600	4,100	615	38,315	43,800
2400	125 kW		.29	68.966		33,300	4,375	660	38,335	43,800

26 32 13 – Engine Generators

26 32 13.13 Diesel-Engine-Driven Generator Sets

		Crew	Daily Output	Labor-Hours	Unit	Material	2021 Bare Costs Labor	Equipment	Total	Total Incl O&P
2500	150 kW	R-3	.26	76.923	Ea.	24,300	4,875	735	29,910	34,800
2501	Generator set, dsl eng in alum encl, incl btry, chgr, muf & day tank, 150 kW		.26	76.923		41,500	4,875	735	47,110	53,500
2600	175 kW		.25	80		44,800	5,075	765	50,640	57,500
2700	200 kW		.24	83.333		45,400	5,300	795	51,495	58,500
2800	250 kW		.23	86.957		60,500	5,525	830	66,855	76,000
2850	275 kW		.22	90.909		67,500	5,775	870	74,145	83,500
2900	300 kW		.22	90.909		55,500	5,775	870	62,145	70,500
3000	350 kW		.20	100		70,000	6,350	955	77,305	87,500
3100	400 kW		.19	105		60,500	6,675	1,000	68,175	78,000
3200	500 kW		.18	111		97,500	7,050	1,050	105,600	119,000
3220	600 kW	▼	.17	118	▼	124,500	7,475	1,125	133,100	149,500
3230	650 kW	R-13	.38	111	Ea.	197,000	6,800	645	204,445	228,000
3240	750 kW		.38	111		185,500	6,800	645	192,945	215,000
3250	800 kW		.36	117		180,500	7,175	685	188,360	210,000
3260	900 kW		.31	135		221,000	8,350	795	230,145	256,500
3270	1,000 kW	▼	.27	156	▼	205,000	9,575	910	215,485	240,500

26 32 13.16 Gas-Engine-Driven Generator Sets

		Crew	Daily Output	Labor-Hours	Unit	Material	2021 Bare Costs Labor	Equipment	Total	Total Incl O&P
0010	**GAS-ENGINE-DRIVEN GENERATOR SETS**									
0020	Gas or gasoline operated, includes battery,									
0050	charger & muffler									
0200	7.5 kW	R-3	.83	24.096	Ea.	7,525	1,525	230	9,280	10,800
0300	11.5 kW		.71	28.169		10,700	1,775	269	12,744	14,700
0400	20 kW		.63	31.746		10,600	2,025	305	12,930	15,000
0460	25 kW, steel enclosure		.63	31.746		15,600	2,025	305	17,930	20,400
0470	27 kW, aluminum enclosure		.63	31.746		21,700	2,025	305	24,030	27,200
0480	30 kW, steel enclosure		.63	31.746		17,600	2,025	305	19,930	22,700
0485	120/208 V, 30kW		.63	31.746		19,100	2,025	305	21,430	24,300
0500	35 kW		.55	36.364		15,000	2,300	345	17,645	20,300
0510	45 kW, steel enclosure		.50	40		34,000	2,550	380	36,930	41,700
0515	48 kW, aluminum enclosure		.50	40		37,000	2,550	380	39,930	44,900
0520	60 kW	▼	.50	40		14,900	2,550	380	17,830	20,600
0550	70 kW, aluminum enclosure	R-13	.46	91.304		42,400	5,625	535	48,560	55,500
0600	80 kW		.40	105		23,600	6,475	615	30,690	36,200
0700	100 kW		.33	127		25,100	7,825	745	33,670	40,000
0800	125 kW		.28	150		27,800	9,225	880	37,905	45,200
0850	130 kW, aluminum enclosure		.28	150		65,000	9,225	880	75,105	86,000
0875	150 kW, aluminum enclosure		.25	168		80,000	10,300	985	91,285	104,500
0900	185 kW	▼	.25	168	▼	73,000	10,300	985	84,285	97,000

26 33 19 – Battery Units

26 33 19.10 Battery Units

		Crew	Daily Output	Labor-Hours	Unit	Material	2021 Bare Costs Labor	Equipment	Total	Total Incl O&P
0010	**BATTERY UNITS**									
0500	Salt water battery, 2.1 kWA, 24V, 750W, 30A, wired in crate, IP22 rated	1 Elec	16.25	.492	Ea.	940	31.50		971.50	1,075
0510	2.2 kWh, 48V, 800W, 17A, wired in crate, IP22 rated		16.25	.492		1,175	31.50		1,206.50	1,350
0520	25.9 kWh, 48V, 11700W, 240A, wired in crate, IP2X rated		16.25	.492		14,100	31.50		14,131.50	15,500
0700	Nickel iron nife battery, deep cycle, 100Ah, 12V, 10 cells		14.25	.561		1,000	36		1,036	1,150
0710	24V, 10 cells		14.25	.561		2,150	36		2,186	2,425
0720	48V, 10 cells		14.25	.561		4,025	36		4,061	4,500
0730	Nickel iron nife battery, deep cycle, 300Ah, 12V, 10 cells		12.25	.653		2,975	41.50		3,016.50	3,325
0740	24V, 10 cells	▼	12.25	.653	▼	5,950	41.50		5,991.50	6,600

For customer support on your Electrical Costs with RSMeans Data, call 800.448.8182.

295

26 33 Battery Equipment

26 33 19 – Battery Units

26 33 19.10 Battery Units		Crew	Daily Output	Labor-Hours	Unit	Material	2021 Bare Costs Labor	2021 Bare Costs Equipment	Total	Total Incl O&P
0750	48V, 10 cells	1 Elec	12.25	.653	Ea.	12,500	41.50		12,541.50	13,800

26 33 43 – Battery Chargers

26 33 43.55 Electric Vehicle Charging

			Crew	Daily Output	Labor-Hours	Unit	Material	Labor	Equipment	Total	Total Incl O&P
0010	**ELECTRIC VEHICLE CHARGING**										
0020	Level 2, wall mounted										
2200	Heavy duty	G	R-1A	15.36	1.042	Ea.	800	59.50		859.50	970
2210	with RFID	G		12.29	1.302		3,375	74.50		3,449.50	3,800
2300	Free standing, single connector	G		10.24	1.563		1,950	89.50		2,039.50	2,250
2310	with RFID	G		8.78	1.822		4,475	104		4,579	5,050
2320	Double connector	G		7.68	2.083		3,600	119		3,719	4,125
2330	with RFID	G		6.83	2.343		8,750	134		8,884	9,825

26 33 53 – Static Uninterruptible Power Supply

26 33 53.10 Uninterruptible Power Supply/Conditioner Trans.

			Crew	Daily Output	Labor-Hours	Unit	Material	Labor	Equipment	Total	Total Incl O&P
0010	**UNINTERRUPTIBLE POWER SUPPLY/CONDITIONER TRANSFORMERS**										
0100	Volt. regulating, isolating transf., w/invert. & 10 min. battery pack										
0110	Single-phase, 120 V, 0.35 kVA	R263353-80	1 Elec	2.29	3.493	Ea.	1,225	223		1,448	1,650
0120	0.5 kVA			2	4		1,275	255		1,530	1,775
0130	For additional 55 min. battery, add to 0.35 kVA			2.29	3.493		715	223		938	1,125
0140	Add to 0.5 kVA			1.14	7.018		750	445		1,195	1,500
0150	Single-phase, 120 V, 0.75 kVA			.80	10		1,600	635		2,235	2,725
0160	1.0 kVA			.80	10		2,300	635		2,935	3,475
0170	1.5 kVA		2 Elec	1.14	14.035		3,950	895		4,845	5,675
0180	2 kVA		"	.89	17.978		4,300	1,150		5,450	6,425
0190	3 kVA		R-3	.63	31.746		5,300	2,025	305	7,630	9,150
0200	5 kVA			.42	47.619		7,550	3,025	455	11,030	13,300
0210	7.5 kVA			.33	60.606		9,625	3,850	580	14,055	17,000
0220	10 kVA			.28	71.429		10,700	4,525	680	15,905	19,300
0230	15 kVA			.22	90.909		15,300	5,775	870	21,945	26,300
0240	3 phase, 120/208 V input 120/208 V output, 20 kVA, incl 17 min. battery			.21	95.238		29,100	6,050	910	36,060	42,000
0242	30 kVA, incl 11 min. battery			.20	100		32,500	6,350	955	39,805	46,300
0250	40 kVA, incl 15 min. battery			.20	100		43,400	6,350	955	50,705	58,500
0260	480 V input 277/480 V output, 60 kVA, incl 6 min. battery			.19	105		48,800	6,675	1,000	56,475	64,500
0262	80 kVA, incl 4 min. battery			.19	105		58,000	6,675	1,000	65,675	75,000
0400	For additional 34 min./15 min. battery, add to 40 kVA			.71	28.011		15,100	1,775	268	17,143	19,600
0600	For complex & special design systems to meet specific										
0610	requirements, obtain quote from vendor										

26 35 Power Filters and Conditioners

26 35 13 – Capacitors

26 35 13.10 Capacitors Indoor

			Crew	Daily Output	Labor-Hours	Unit	Material	Labor	Equipment	Total	Total Incl O&P
0010	**CAPACITORS INDOOR**										
0020	240 volts, single & 3 phase, 0.5 kVAR		1 Elec	2.70	2.963	Ea.	605	189		794	945
0100	1.0 kVAR			2.70	2.963		725	189		914	1,075
0150	2.5 kVAR			2	4		820	255		1,075	1,275
0200	5.0 kVAR			1.80	4.444		790	283		1,073	1,300
0250	7.5 kVAR			1.60	5		1,500	320		1,820	2,125
0300	10 kVAR		1 Elec	1.50	5.333	Ea.	1,750	340		2,090	2,425
0350	15 kVAR			1.30	6.154		1,325	390		1,715	2,025
0400	20 kVAR			1.10	7.273		2,725	465		3,190	3,700
0450	25 kVAR			1	8		3,125	510		3,635	4,200

For customer support on your Electrical Costs with RSMeans Data, call 800.448.8182.

26 35 13 – Capacitors

		Crew	Daily Output	Labor-Hours	Unit	Material	2021 Bare Costs Labor	Equipment	Total	Total Incl O&P
26 35 13.10 Capacitors Indoor										
1000	480 volts, single & 3 phase, 1 kVAR	1 Elec	2.70	2.963	Ea.	550	189		739	885
1050	2 kVAR		2.70	2.963		630	189		819	975
1100	5 kVAR		2	4		795	255		1,050	1,250
1150	7.5 kVAR		2	4		860	255		1,115	1,325
1200	10 kVAR		2	4		1,200	255		1,455	1,700
1250	15 kVAR		2	4		1,425	255		1,680	1,950
1300	20 kVAR		1.60	5		1,575	320		1,895	2,200
1350	30 kVAR		1.50	5.333		1,400	340		1,740	2,025
1400	40 kVAR		1.20	6.667		2,350	425		2,775	3,200
1450	50 kVAR		1.10	7.273		2,750	465		3,215	3,725
2000	600 volts, single & 3 phase, 1 kVAR		2.70	2.963		565	189		754	905
2050	2 kVAR		2.70	2.963		650	189		839	995
2100	5 kVAR		2	4		795	255		1,050	1,250
2150	7.5 kVAR		2	4		860	255		1,115	1,325
2200	10 kVAR		2	4		1,225	255		1,480	1,700
2250	15 kVAR		1.60	5		1,425	320		1,745	2,050
2300	20 kVAR		1.60	5		1,575	320		1,895	2,200
2350	25 kVAR		1.50	5.333		1,725	340		2,065	2,400
2400	35 kVAR		1.40	5.714		2,125	365		2,490	2,875
2450	50 kVAR		1.30	6.154		2,750	390		3,140	3,625

26 35 26 – Harmonic Filters

26 35 26.10 Computer Isolation Transformer

		Crew	Daily Output	Labor-Hours	Unit	Material	2021 Bare Costs Labor	Equipment	Total	Total Incl O&P
0010	**COMPUTER ISOLATION TRANSFORMER**									
0100	Computer grade									
0110	Single-phase, 120/240 V, 0.5 kVA	1 Elec	4	2	Ea.	465	127		592	700
0120	1.0 kVA		2.67	2.996		660	191		851	1,000
0130	2.5 kVA		2	4		1,000	255		1,255	1,475
0140	5 kVA		1.14	7.018		1,150	445		1,595	1,925

26 35 26.20 Computer Regulator Transformer

		Crew	Daily Output	Labor-Hours	Unit	Material	2021 Bare Costs Labor	Equipment	Total	Total Incl O&P
0010	**COMPUTER REGULATOR TRANSFORMER**									
0100	Ferro-resonant, constant voltage, variable transformer									
0110	Single-phase, 240 V, 0.5 kVA	1 Elec	2.67	2.996	Ea.	575	191		766	915
0120	1.0 kVA		2	4		785	255		1,040	1,250
0130	2.0 kVA		1	8		1,350	510		1,860	2,225
0210	Plug-in unit 120 V, 0.14 kVA		8	1		330	63.50		393.50	460
0220	0.25 kVA		8	1		385	63.50		448.50	520
0230	0.5 kVA		8	1		575	63.50		638.50	725
0240	1.0 kVA		5.33	1.501		785	95.50		880.50	1,000
0250	2.0 kVA		4	2		1,350	127		1,477	1,675

26 35 26.30 Power Conditioner Transformer

		Crew	Daily Output	Labor-Hours	Unit	Material	2021 Bare Costs Labor	Equipment	Total	Total Incl O&P
0010	**POWER CONDITIONER TRANSFORMER**									
0100	Electronic solid state, buck-boost, transformer, w/tap switch									
0110	Single-phase, 115 V, 3.0 kVA, + or - 3% accuracy	2 Elec	1.60	10	Ea.	3,100	635		3,735	4,350
0120	208, 220, 230, or 240 V, 5.0 kVA, + or - 1.5% accuracy	3 Elec	1.60	15		4,000	955		4,955	5,825
0130	5.0 kVA, + or - 6% accuracy	2 Elec	1.14	14.035		3,575	895		4,470	5,275
0140	7.5 kVA, + or - 1.5% accuracy	3 Elec	1.50	16		5,075	1,025		6,100	7,125
0150	7.5 kVA, + or - 6% accuracy		1.60	15		4,225	955		5,180	6,075
0160	10.0 kVA, + or - 1.5% accuracy		1.33	18.045		6,775	1,150		7,925	9,150
0170	10.0 kVA, + or - 6% accuracy		1.41	17.021		5,775	1,075		6,850	7,950

26 35 26 – Harmonic Filters

26 35 26.40 Transient Voltage Suppressor Transformer	Crew	Daily Output	Labor-Hours	Unit	Material	2021 Bare Costs Labor	Equipment	Total	Total Incl O&P
0010 **TRANSIENT VOLTAGE SUPPRESSOR TRANSFORMER**									
0110 Single-phase, 120 V, 1.8 kVA	1 Elec	4	2	Ea.	1,525	127		1,652	1,875
0120 3.6 kVA		4	2		2,175	127		2,302	2,600
0130 7.2 kVA		3.20	2.500		3,175	159		3,334	3,725
0150 240 V, 3.6 kVA		4	2		3,500	127		3,627	4,050
0160 7.2 kVA		4	2		3,750	127		3,877	4,325
0170 14.4 kVA		3.20	2.500		6,025	159		6,184	6,850
0210 Plug-in unit, 120 V, 1.8 kVA		8	1		1,250	63.50		1,313.50	1,475

26 35 53 – Voltage Regulators

26 35 53.10 Automatic Voltage Regulators

	Crew	Daily Output	Labor-Hours	Unit	Material	2021 Bare Costs Labor	Equipment	Total	Total Incl O&P
0010 **AUTOMATIC VOLTAGE REGULATORS**									
0100 Computer grade, solid state, variable transf. volt. regulator									
0110 Single-phase, 120 V, 8.6 kVA	2 Elec	1.33	12.030	Ea.	6,475	765		7,240	8,275
0120 17.3 kVA		1.14	14.035		7,650	895		8,545	9,750
0130 208/240 V, 7.5/8.6 kVA		1.33	12.030		6,475	765		7,240	8,275
0140 13.5/15.6 kVA		1.33	12.030		7,650	765		8,415	9,575
0150 27.0/31.2 kVA		1.14	14.035		9,650	895		10,545	11,900
0210 Two-phase, single control, 208/240 V, 15.0/17.3 kVA		1.14	14.035		7,650	895		8,545	9,750
0220 Individual phase control, 15.0/17.3 kVA		1.14	14.035		7,650	895		8,545	9,750
0230 30.0/34.6 kVA	3 Elec	1.33	18.045		9,650	1,150		10,800	12,300
0310 Three-phase single control, 208/240 V, 26/30 kVA	2 Elec	1	16		7,650	1,025		8,675	9,950
0320 380/480 V, 24/30 kVA	"	1	16		7,650	1,025		8,675	9,950
0330 43/54 kVA	3 Elec	1.33	18.045		13,900	1,150		15,050	17,000
0340 Individual phase control, 208 V, 26 kVA	"	1.33	18.045		7,650	1,150		8,800	10,100
0350 52 kVA	R-3	.91	21.978		9,650	1,400	210	11,260	12,900
0360 340/480 V, 24/30 kVA	2 Elec	1	16		7,650	1,025		8,675	9,950
0370 43/54 kVA	"	1	16		9,650	1,025		10,675	12,100
0380 48/60 kVA	3 Elec	1.33	18.045		14,000	1,150		15,150	17,100
0390 86/108 kVA	R-3	.91	21.978		15,600	1,400	210	17,210	19,400
0500 Standard grade, solid state, variable transformer volt. regulator									
0510 Single-phase, 115 V, 2.3 kVA	1 Elec	2.29	3.493	Ea.	2,425	223		2,648	2,975
0520 4.2 kVA		2	4		3,875	255		4,130	4,625
0530 6.6 kVA		1.14	7.018		4,750	445		5,195	5,900
0540 13.0 kVA		1.14	7.018		7,625	445		8,070	9,075
0550 16.6 kVA	2 Elec	1.23	13.008		9,025	830		9,855	11,200
0610 230 V, 8.3 kVA		1.33	12.030		7,625	765		8,390	9,550
0620 21.4 kVA		1.23	13.008		9,025	830		9,855	11,200
0630 29.9 kVA		1.23	13.008		9,025	830		9,855	11,200
0710 460 V, 9.2 kVA		1.33	12.030		7,625	765		8,390	9,550
0720 20.7 kVA		1.23	13.008		9,025	830		9,855	11,200
0810 Three-phase, 230 V, 13.1 kVA	3 Elec	1.41	17.021		7,625	1,075		8,700	10,000
0820 19.1 kVA		1.41	17.021		9,025	1,075		10,100	11,500
0830 25.1 kVA		1.23	19.512		9,025	1,250		10,275	11,800
0840 57.8 kVA	R-3	.95	21.053		16,400	1,325	201	17,926	20,200
0850 74.9 kVA	"	.91	21.978		16,400	1,400	210	18,010	20,300
0910 460 V, 14.3 kVA	3 Elec	1.41	17.021		7,625	1,075		8,700	10,000
0920 19.1 kVA		1.41	17.021		9,025	1,075		10,100	11,500
0930 27.9 kVA		1.23	19.512		9,025	1,250		10,275	11,800
0940 59.8 kVA	R-3	1	20		16,400	1,275	191	17,866	20,100
0950 79.7 kVA		.95	21.053		18,300	1,325	201	19,826	22,300
0960 118 kVA		.95	21.053		19,300	1,325	201	20,826	23,400
1000 Laboratory grade, precision, electronic voltage regulator									

26 35 53 – Voltage Regulators

26 35 53.10 Automatic Voltage Regulators		Crew	Daily Output	Labor-Hours	Unit	Material	2021 Bare Costs Labor	Equipment	Total	Total Incl O&P
1110	Single-phase, 115 V, 0.5 kVA	1 Elec	2.29	3.493	Ea.	1,600	223		1,823	2,075
1120	1.0 kVA		2	4		1,700	255		1,955	2,250
1130	3.0 kVA		.80	10		2,375	635		3,010	3,550
1140	6.0 kVA	2 Elec	1.46	10.959		4,375	700		5,075	5,850
1150	10.0 kVA	3 Elec	1	24		5,700	1,525		7,225	8,550
1160	15.0 kVA	"	1.50	16		6,500	1,025		7,525	8,675
1210	230 V, 3.0 kVA	1 Elec	.80	10		2,675	635		3,310	3,900
1220	6.0 kVA	2 Elec	1.46	10.959		4,475	700		5,175	5,950
1230	10.0 kVA	3 Elec	1.71	14.035		5,950	895		6,845	7,875
1240	15.0 kVA	"	1.60	15		6,725	955		7,680	8,825

26 35 53.30 Transient Suppressor/Voltage Regulator

26 35 53.30		Crew	Daily Output	Labor-Hours	Unit	Material	Labor	Equipment	Total	Total Incl O&P
0010	**TRANSIENT SUPPRESSOR/VOLTAGE REGULATOR** (without isolation)									
0110	Single-phase, 115 V, 1.0 kVA	1 Elec	2.67	2.996	Ea.	1,125	191		1,316	1,525
0120	2.0 kVA		2.29	3.493		1,550	223		1,773	2,025
0130	4.0 kVA		2.13	3.756		1,900	239		2,139	2,450
0140	220 V, 1.0 kVA		2.67	2.996		1,125	191		1,316	1,525
0150	2.0 kVA		2.29	3.493		2,175	223		2,398	2,725
0160	4.0 kVA		2.13	3.756		2,000	239		2,239	2,550
0210	Plug-in unit, 120 V, 1.0 kVA		8	1		1,075	63.50		1,138.50	1,300
0220	2.0 kVA		8	1		1,525	63.50		1,588.50	1,775

26 36 Transfer Switches

26 36 13 – Manual Transfer Switches

26 36 13.10 Non-Automatic Transfer Switches		Crew	Daily Output	Labor-Hours	Unit	Material	Labor	Equipment	Total	Total Incl O&P
0010	**NON-AUTOMATIC TRANSFER SWITCHES** enclosed									
0100	Manual operated, 480 volt 3 pole, 30 amp	1 Elec	2.30	3.478	Ea.	1,525	222		1,747	2,000
0150	60 amp		1.90	4.211		2,100	268		2,368	2,700
0200	100 amp		1.30	6.154		3,475	390		3,865	4,400
0250	200 amp	2 Elec	2	8		4,600	510		5,110	5,800
0300	400 amp		1.60	10		4,525	635		5,160	5,950
0350	600 amp		1	16		5,525	1,025		6,550	7,625
1000	250 volt 3 pole, 30 amp	1 Elec	2.30	3.478		1,000	222		1,222	1,425
1100	60 amp		1.90	4.211		1,675	268		1,943	2,250
1150	100 amp		1.30	6.154		2,850	390		3,240	3,700
1200	200 amp	2 Elec	2	8		4,300	510		4,810	5,475
1300	600 amp	"	1	16		11,800	1,025		12,825	14,500
1500	Electrically operated, 480 volt 3 pole, 60 amp	1 Elec	1.90	4.211		1,600	268		1,868	2,175
1600	100 amp	"	1.30	6.154		1,600	390		1,990	2,350
1650	200 amp	2 Elec	2	8		2,650	510		3,160	3,675
1700	400 amp		1.60	10		3,700	635		4,335	5,025
1750	600 amp		1	16		5,350	1,025		6,375	7,400
2000	250 volt 3 pole, 30 amp	1 Elec	2.30	3.478		545	222		767	930
2050	60 amp	"	1.90	4.211		680	268		948	1,150
2150	200 amp	2 Elec	2	8		1,275	510		1,785	2,150
2200	400 amp		1.60	10		3,375	635		4,010	4,650
2250	600 amp		1	16		5,050	1,025		6,075	7,075
2500	NEMA 3R, 480 volt 3 pole, 60 amp	1 Elec	1.80	4.444		2,575	283		2,858	3,250
2550	100 amp	"	1.20	6.667		3,150	425		3,575	4,075
2600	200 amp	2 Elec	1.80	8.889		5,050	565		5,615	6,425
2650	400 amp	"	1.40	11.429		4,050	730		4,780	5,550
2800	NEMA 3R, 250 volt 3 pole solid state, 100 amp	1 Elec	1.20	6.667		3,625	425		4,050	4,600

For customer support on your Electrical Costs with RSMeans Data, call 800.448.8182.

299

26 36 Transfer Switches

26 36 13 – Manual Transfer Switches

26 36 13.10 Non-Automatic Transfer Switches		Crew	Daily Output	Labor-Hours	Unit	Material	2021 Bare Costs Labor	Equipment	Total	Total Incl O&P
2850	150 amp	2 Elec	1.80	8.889	Ea.	4,825	565		5,390	6,150
2900	250 volt 2 pole solid state, 100 amp	1 Elec	1.30	6.154		3,525	390		3,915	4,450
2950	150 amp	2 Elec	2	8	↓	4,675	510		5,185	5,900

26 36 23 – Automatic Transfer Switches

26 36 23.10 Automatic Transfer Switch Devices

	26 36 23.10 Automatic Transfer Switch Devices		Crew	Daily Output	Labor-Hours	Unit	Material	2021 Bare Costs Labor	Equipment	Total	Total Incl O&P
0010	**AUTOMATIC TRANSFER SWITCH DEVICES**	R263623-60									
0015	Switches, enclosed 120/240 volt, 2 pole, 30 amp		1 Elec	2.40	3.333	Ea.	1,975	212		2,187	2,500
0020	70 amp			2	4		1,975	255		2,230	2,550
0030	100 amp			1.35	5.926		1,250	375		1,625	1,925
0040	225 amp		2 Elec	2.10	7.619		1,925	485		2,410	2,850
0050	400 amp			1.70	9.412		2,600	600		3,200	3,775
0060	600 amp			1.06	15.094		4,700	960		5,660	6,600
0070	800 amp		↓	.84	19.048		4,800	1,225		6,025	7,100
0100	Switches, enclosed 480 volt, 3 pole, 30 amp		1 Elec	2.30	3.478		3,425	222		3,647	4,100
0200	60 amp			1.90	4.211		3,425	268		3,693	4,175
0300	100 amp			1.30	6.154		1,675	390		2,065	2,400
0400	150 amp		2 Elec	2.40	6.667		1,800	425		2,225	2,600
0500	225 amp			2	8		2,075	510		2,585	3,050
0600	260 amp			2	8		2,250	510		2,760	3,225
0700	400 amp			1.60	10		2,800	635		3,435	4,025
0800	600 amp			1	16		4,850	1,025		5,875	6,850
0900	800 amp			.80	20		5,125	1,275		6,400	7,525
1000	1,000 amp			.76	21.053		16,400	1,350		17,750	20,000
1100	1,200 amp			.70	22.857		21,500	1,450		22,950	25,900
1200	1,600 amp			.60	26.667		24,200	1,700		25,900	29,100
1300	2,000 amp			.50	32		26,600	2,050		28,650	32,300
1500	Switches, enclosed 480 V, 4 pole, Bypass/Iso. SW 150 A			2.40	6.667		18,000	425		18,425	20,400
1510	260 A			2	8		19,000	510		19,510	21,700
1520	400 A			1.60	10		24,000	635		24,635	27,300
1530	600 A		↓	1	16		31,000	1,025		32,025	35,600
1600	Accessories, time delay on engine starting						251			251	276
1700	Adjustable time delay on retransfer						251			251	276
1800	Shunt trips for customer connections						450			450	495
1900	Maintenance select switch						102			102	112
2000	Auxiliary contact when normal fails						101			101	111
2100	Pilot light-emergency						102			102	112
2200	Pilot light-normal						102			102	112
2300	Auxiliary contact-closed on normal						118			118	130
2400	Auxiliary contact-closed on emergency						118			118	130
2500	Emergency source sensing, frequency relay					↓	520			520	570

26 41 13 – Lightning Protection for Structures

26 41 13.13 Lightning Protection for Buildings	Crew	Daily Output	Labor-Hours	Unit	Material	2021 Bare Costs Labor	Equipment	Total	Total Incl O&P
0010 **LIGHTNING PROTECTION FOR BUILDINGS**									
0200 Air terminals & base, copper									
0400 3/8" diameter x 10" (to 75' high)	1 Elec	8	1	Ea.	27	63.50		90.50	125
0500 1/2" diameter x 12" (over 75' high)		8	1		31.50	63.50		95	130
0520 1/2" diameter x 24"		7.30	1.096		42.50	70		112.50	151
0540 1/2" diameter x 60"		6.70	1.194		83.50	76		159.50	205
1000 Aluminum, 1/2" diameter x 12" (to 75' high)		8	1		17.45	63.50		80.95	114
1020 1/2" diameter x 24"		7.30	1.096		18.45	70		88.45	125
1040 1/2" diameter x 60"		6.70	1.194		24	76		100	140
1100 5/8" diameter x 12" (over 75' high)		8	1	↓	17.95	63.50		81.45	114
2000 Cable, copper, 220 lb. per thousand ft. (to 75' high)		320	.025	L.F.	4.78	1.59		6.37	7.60
2100 375 lb. per thousand ft. (over 75' high)	↓	230	.035	"	4.63	2.22		6.85	8.40
2500 Aluminum, 101 lb. per thousand ft. (to 75' high)	1 Elec	280	.029	L.F.	1.71	1.82		3.53	4.58
2600 199 lb. per thousand ft. (over 75' high)		240	.033	"	1.63	2.12		3.75	4.95
3000 Arrester, 175 volt AC to ground		8	1	Ea.	72.50	63.50		136	175
3100 650 volt AC to ground	↓	6.70	1.194	"	47	76		123	165
4000 Air terminals, copper									
4010 3/8" x 10"	1 Elec	50	.160	Ea.	7.70	10.20		17.90	23.50
4020 1/2" x 12"		38	.211		12.20	13.40		25.60	33.50
4030 5/8" x 12"		33	.242		21	15.45		36.45	46
4040 1/2" x 24"		30	.267		23	17		40	50.50
4050 1/2" x 60"	↓	19	.421	↓	64	27		91	111
4060 Air terminals, aluminum									
4070 3/8" x 10"	1 Elec	50	.160	Ea.	3.20	10.20		13.40	18.65
4080 1/2" x 12"		38	.211		4.47	13.40		17.87	25
4090 5/8" x 12"		33	.242		4.96	15.45		20.41	28.50
4100 1/2" x 24"		30	.267		5.45	17		22.45	31
4110 1/2" x 60"	↓	19	.421	↓	11.20	27		38.20	52.50
4200 Air terminal bases, copper									
4210 Adhesive bases, 1/2"	1 Elec	15	.533	Ea.	25	34		59	78
4215 Adhesive base, lightning air terminal					25			25	27.50
4220 Bolted bases	1 Elec	9	.889		19.45	56.50		75.95	106
4230 Hinged bases		7	1.143		35	73		108	147
4240 Side mounted bases		9	.889		20.50	56.50		77	107
4250 Offset point support		9	.889		33	56.50		89.50	121
4260 Concealed base assembly		5	1.600		16.85	102		118.85	170
4270 Tee connector base		13	.615		97	39		136	166
4280 Tripod base, 36" for 60" air terminal		7	1.143		42.50	73		115.50	155
4290 Intermediate base, 1/2"	↓	15	.533	↓	63	34		97	120
4300 Air terminal bases, aluminum									
4310 Adhesive bases, 1/2"	1 Elec	15	.533	Ea.	8.95	34		42.95	60.50
4320 Bolted bases		9	.889		13	56.50		69.50	98.50
4330 Hinged bases		7	1.143		24	73		97	135
4340 Side mounted bases		9	.889		12.55	56.50		69.05	98
4350 Offset point support		9	.889		21	56.50		77.50	107
4360 Concealed base assembly		5	1.600		9.05	102		111.05	161
4370 Tee connector base		13	.615		12.65	39		51.65	72.50
4380 Tripod base, 36" for 60" air terminal		7	1.143		33.50	73		106.50	145
4390 Intermediate base, 1/2"	↓	15	.533	↓	14.80	34		48.80	67
4400 Connector cable, copper									
4410 Through wall connector	1 Elec	7	1.143	Ea.	124	73		197	244
4420 Through roof connector	"	5	1.600	"	100	102		202	261
4430 Beam connector	1 Elec	7	1.143	Ea.	38.50	73		111.50	151

For customer support on your Electrical Costs with RSMeans Data, call 800.448.8182.

301

26 41 Facility Lightning Protection

26 41 13 – Lightning Protection for Structures

26 41 13.13 Lightning Protection for Buildings

26 41 13.13 Lightning Protection for Buildings		Crew	Daily Output	Labor-Hours	Unit	Material	2021 Bare Costs Labor	Equipment	Total	Total Incl O&P
4440	Double bolt connector	1 Elec	38	.211	Ea.	10.25	13.40		23.65	31
4450	Bar, copper	↓	40	.200	↓	14.30	12.75		27.05	34.50
4500	Connector cable, aluminum									
4510	Through wall connector	1 Elec	7	1.143	Ea.	34	73		107	146
4520	Through roof connector		5	1.600		25	102		127	179
4530	Beam connector		7	1.143		17.85	73		90.85	128
4540	Double bolt connector	↓	38	.211	↓	5.45	13.40		18.85	26
4600	Bonding plates, copper									
4610	I-beam, 8" square	1 Elec	7	1.143	Ea.	22	73		95	132
4620	Purlin, 8" square		7	1.143		22.50	73		95.50	133
4630	Large heavy duty, 16" square		5	1.600		35	102		137	190
4640	Aluminum, I-beam, 8" square		7	1.143		17.90	73		90.90	128
4650	Purlin, 8" square		7	1.143		18.25	73		91.25	128
4660	Large heavy duty, 16" square		5	1.600		18.55	102		120.55	172
4700	Cable support, copper, loop fastener		38	.211		1.19	13.40		14.59	21.50
4710	Adhesive cable holder		38	.211		1.39	13.40		14.79	21.50
4720	Aluminum, loop fastener		38	.211		.81	13.40		14.21	21
4730	Adhesive cable holder		38	.211		.62	13.40		14.02	20.50
4740	Bonding strap, copper, 3/4" x 9-1/2"		15	.533		9.70	34		43.70	61
4750	Aluminum, 3/4" x 9-1/2"		15	.533		4.90	34		38.90	56
4760	Swivel adapter, copper, 1/2"		73	.110		15.60	7		22.60	27.50
4770	Aluminum, 1/2"	↓	73	.110	↓	8.25	7		15.25	19.40

26 51 Interior Lighting

26 51 13 – Interior Lighting Fixtures, Lamps, and Ballasts

26 51 13.10 Fixture Hangers

		Crew	Daily Output	Labor-Hours	Unit	Material	2021 Bare Costs Labor	Equipment	Total	Total Incl O&P
0010	**FIXTURE HANGERS**									
0220	Box hub cover	1 Elec	32	.250	Ea.	3.97	15.95		19.92	28
0240	Canopy		12	.667		9.15	42.50		51.65	73
0260	Connecting block		40	.200		2.82	12.75		15.57	22
0280	Cushion hanger		16	.500		26	32		58	76
0300	Box hanger, with mounting strap		8	1		9.35	63.50		72.85	105
0320	Connecting block		40	.200		2.68	12.75		15.43	22
0340	Flexible, 1/2" diameter, 4" long		12	.667		16.70	42.50		59.20	81.50
0360	6" long		12	.667		18.20	42.50		60.70	83
0380	8" long		12	.667		20	42.50		62.50	85
0400	10" long		12	.667		21.50	42.50		64	86.50
0420	12" long		12	.667		23	42.50		65.50	88
0440	15" long		12	.667		23.50	42.50		66	89
0460	18" long		12	.667		28	42.50		70.50	93.50
0480	3/4" diameter, 4" long	↓	10	.800	↓	19.85	51		70.85	97.50
0500	6" long	1 Elec	10	.800	Ea.	22	51		73	99.50
0520	8" long		10	.800		23	51		74	101
0540	10" long		10	.800		23	51		74	101
0560	12" long		10	.800		25	51		76	103
0580	15" long		10	.800		28	51		79	106
0600	18" long	↓	10	.800		30.50	51		81.50	109

26 51 13.40 Interior HID Fixtures

		Crew	Daily Output	Labor-Hours	Unit	Material	2021 Bare Costs Labor	Equipment	Total	Total Incl O&P
0010	**INTERIOR HID FIXTURES** Incl. lamps and mounting hardware									
0700	High pressure sodium, recessed, round, 70 watt	1 Elec	3.50	2.286	Ea.	545	146		691	810
0720	100 watt	↓	3.50	2.286	↓	700	146		846	985

26 51 Interior Lighting

26 51 13 – Interior Lighting Fixtures, Lamps, and Ballasts

26 51 13.40 Interior HID Fixtures		Crew	Daily Output	Labor-Hours	Unit	Material	2021 Bare Costs Labor	Equipment	Total	Total Incl O&P
0740	150 watt	1 Elec	3.20	2.500	Ea.	730	159		889	1,050
0760	Square, 70 watt		3.60	2.222		665	142		807	945
0780	100 watt		3.60	2.222		700	142		842	980
0820	250 watt	↓	3	2.667		820	170		990	1,150
0840	1,000 watt	2 Elec	4.80	3.333		1,650	212		1,862	2,150
0860	Surface, round, 70 watt	1 Elec	3	2.667		835	170		1,005	1,175
0880	100 watt		3	2.667		855	170		1,025	1,200
0900	150 watt		2.70	2.963		1,025	189		1,214	1,400
0920	Square, 70 watt		3	2.667		845	170		1,015	1,175
0940	100 watt		3	2.667		825	170		995	1,150
0980	250 watt		2.50	3.200		735	204		939	1,100
1040	Pendent, round, 70 watt		3	2.667		930	170		1,100	1,275
1060	100 watt		3	2.667		825	170		995	1,150
1080	150 watt		2.70	2.963		910	189		1,099	1,275
1100	Square, 70 watt		3	2.667		905	170		1,075	1,250
1120	100 watt		3	2.667		915	170		1,085	1,250
1140	150 watt		2.70	2.963		940	189		1,129	1,300
1160	250 watt		2.50	3.200		1,275	204		1,479	1,700
1180	400 watt		2.40	3.333		1,350	212		1,562	1,800
1220	Wall, round, 70 watt		3	2.667		855	170		1,025	1,200
1240	100 watt		3	2.667		855	170		1,025	1,200
1260	150 watt		2.70	2.963		865	189		1,054	1,225
1300	Square, 70 watt		3	2.667		715	170		885	1,025
1320	100 watt		3	2.667		845	170		1,015	1,175
1340	150 watt		2.70	2.963		885	189		1,074	1,250
1360	250 watt	↓	2.50	3.200		800	204		1,004	1,175
1380	400 watt	2 Elec	4.80	3.333		985	212		1,197	1,400
1400	1,000 watt	"	3.60	4.444		1,325	283		1,608	1,900
1500	Metal halide, recessed, round, 175 watt	1 Elec	3.40	2.353		450	150		600	720
1520	250 watt	"	3.20	2.500		620	159		779	915
1540	400 watt	2 Elec	5.80	2.759		995	176		1,171	1,350
1580	Square, 175 watt	1 Elec	3.40	2.353		460	150		610	730
1640	Surface, round, 175 watt		2.90	2.759		605	176		781	925
1660	250 watt	↓	2.70	2.963		1,225	189		1,414	1,625
1680	400 watt	2 Elec	4.80	3.333		1,500	212		1,712	1,975
1720	Square, 175 watt	1 Elec	2.90	2.759		655	176		831	985
1800	Pendent, round, 175 watt		2.90	2.759		995	176		1,171	1,350
1820	250 watt	↓	2.70	2.963		1,350	189		1,539	1,750
1840	400 watt	2 Elec	4.80	3.333		1,225	212		1,437	1,650
1880	Square, 175 watt	1 Elec	2.90	2.759		515	176		691	825
1900	250 watt	"	2.70	2.963		745	189		934	1,100
1920	400 watt	2 Elec	4.80	3.333		1,425	212		1,637	1,875
1980	Wall, round, 175 watt	1 Elec	2.90	2.759		1,125	176		1,301	1,475
2000	250 watt	"	2.70	2.963		1,175	189		1,364	1,550
2020	400 watt	2 Elec	4.80	3.333		1,050	212		1,262	1,475
2060	Square, 175 watt	1 Elec	2.90	2.759		520	176		696	835
2080	250 watt	"	2.70	2.963		720	189		909	1,075
2100	400 watt	2 Elec	4.80	3.333		1,025	212		1,237	1,450
2800	High pressure sodium, vaporproof, recessed, 70 watt	1 Elec	3.50	2.286		760	146		906	1,050
2820	100 watt		3.50	2.286		780	146		926	1,075
2840	150 watt		3.20	2.500		800	159		959	1,125
2900	Surface, 70 watt		3	2.667		860	170		1,030	1,200
2920	100 watt	↓	3	2.667		900	170		1,070	1,250

26 51 13.40 Interior HID Fixtures

		Crew	Daily Output	Labor-Hours	Unit	Material	2021 Bare Costs Labor	2021 Bare Costs Equipment	Total	Total Incl O&P
2940	150 watt	1 Elec	2.70	2.963	Ea.	935	189		1,124	1,300
3000	Pendent, 70 watt		3	2.667		850	170		1,020	1,175
3020	100 watt		3	2.667		875	170		1,045	1,225
3040	150 watt		2.70	2.963		925	189		1,114	1,300
3100	Wall, 70 watt		3	2.667		915	170		1,085	1,250
3120	100 watt		3	2.667		955	170		1,125	1,300
3140	150 watt		2.70	2.963		995	189		1,184	1,375
3200	Metal halide, vaporproof, recessed, 175 watt		3.40	2.353		755	150		905	1,050
3220	250 watt	▼	3.20	2.500		715	159		874	1,025
3240	400 watt	2 Elec	5.80	2.759		900	176		1,076	1,250
3260	1,000 watt	"	4.80	3.333		1,650	212		1,862	2,125
3280	Surface, 175 watt	1 Elec	2.90	2.759		1,200	176		1,376	1,575
3300	250 watt	"	2.70	2.963		1,025	189		1,214	1,400
3320	400 watt	2 Elec	4.80	3.333		1,275	212		1,487	1,725
3340	1,000 watt	"	3.60	4.444		1,850	283		2,133	2,450
3360	Pendent, 175 watt	1 Elec	2.90	2.759		1,300	176		1,476	1,675
3380	250 watt	"	2.70	2.963		1,050	189		1,239	1,425
3400	400 watt	2 Elec	4.80	3.333		1,250	212		1,462	1,700
3420	1,000 watt	"	3.60	4.444		2,025	283		2,308	2,650
3440	Wall, 175 watt	1 Elec	2.90	2.759		1,375	176		1,551	1,750
3460	250 watt	"	2.70	2.963		1,100	189		1,289	1,500
3480	400 watt	2 Elec	4.80	3.333		1,325	212		1,537	1,775
3500	1,000 watt	"	3.60	4.444		2,100	283		2,383	2,750

26 51 13.50 Interior Lighting Fixtures

		Crew	Daily Output	Labor-Hours	Unit	Material	2021 Bare Costs Labor	2021 Bare Costs Equipment	Total	Total Incl O&P
0010	**INTERIOR LIGHTING FIXTURES** Including lamps, mounting	R265113-40								
0030	hardware and connections									
0100	Fluorescent, C.W. lamps, troffer, recess mounted in grid, RS									
0130	Grid ceiling mount									
0200	Acrylic lens, 1' W x 4' L, two 40 watt	1 Elec	5.70	1.404	Ea.	54	89.50		143.50	193
0210	1' W x 4' L, three 40 watt		5.40	1.481		61.50	94.50		156	208
0300	2' W x 2' L, two U40 watt		5.70	1.404		58	89.50		147.50	197
0400	2' W x 4' L, two 40 watt		5.30	1.509		57	96		153	206
0500	2' W x 4' L, three 40 watt		5	1.600		63	102		165	220
0600	2' W x 4' L, four 40 watt	▼	4.70	1.702		65.50	108		173.50	234
0700	4' W x 4' L, four 40 watt	2 Elec	6.40	2.500		300	159		459	565
0800	4' W x 4' L, six 40 watt		6.20	2.581		315	164		479	590
0900	4' W x 4' L, eight 40 watt	▼	5.80	2.759		355	176		531	650
0910	Acrylic lens, 1' W x 4' L, two 32 watt T8 [G]	1 Elec	5.70	1.404		78	89.50		167.50	219
0930	2' W x 2' L, two U32 watt T8 [G]		5.70	1.404		131	89.50		220.50	277
0940	2' W x 4' L, two 32 watt T8 [G]		5.30	1.509		93	96		189	245
0950	2' W x 4' L, three 32 watt T8 [G]		5	1.600		82.50	102		184.50	242
0960	2' W x 4' L, four 32 watt T8 [G]	▼	4.70	1.702	▼	75.50	108		183.50	244
1000	Surface mounted, RS									
1030	Acrylic lens with hinged & latched door frame									
1100	1' W x 4' L, two 40 watt	1 Elec	7	1.143	Ea.	71	73		144	186
1110	1' W x 4' L, three 40 watt		6.70	1.194		73.50	76		149.50	194
1200	2' W x 2' L, two U40 watt		7	1.143		76	73		149	192
1300	2' W x 4' L, two 40 watt		6.20	1.290		86.50	82		168.50	217
1400	2' W x 4' L, three 40 watt		5.70	1.404		86	89.50		175.50	228
1500	2' W x 4' L, four 40 watt		5.30	1.509		88	96		184	240
1501	2' W x 4' L, six 40 watt T8		5.20	1.538		88	98		186	243
1600	4' W x 4' L, four 40 watt	2 Elec	7.20	2.222	▼	435	142		577	690

26 51 13.50 Interior Lighting Fixtures		Crew	Daily Output	Labor-Hours	Unit	Material	2021 Bare Costs Labor	Equipment	Total	Total Incl O&P
1700	4' W x 4' L, six 40 watt	2 Elec	6.60	2.424	Ea.	470	154		624	750
1800	4' W x 4' L, eight 40 watt		6.20	2.581		490	164		654	785
1900	2' W x 8' L, four 40 watt		6.40	2.500		174	159		333	430
2000	2' W x 8' L, eight 40 watt		6.20	2.581		187	164		351	450
2010	Acrylic wrap around lens									
2020	6" W x 4' L, one 40 watt	1 Elec	8	1	Ea.	70.50	63.50		134	172
2030	6" W x 8' L, two 40 watt	2 Elec	8	2		78	127		205	275
2040	11" W x 4' L, two 40 watt	1 Elec	7	1.143		47.50	73		120.50	160
2050	11" W x 8' L, four 40 watt	2 Elec	6.60	2.424		78.50	154		232.50	315
2060	16" W x 4' L, four 40 watt	1 Elec	5.30	1.509		78.50	96		174.50	230
2070	16" W x 8' L, eight 40 watt	2 Elec	6.40	2.500		171	159		330	425
2080	2' W x 2' L, two U40 watt	1 Elec	7	1.143		96	73		169	213
2100	Strip fixture									
2130	Surface mounted									
2200	4' long, one 40 watt, RS	1 Elec	8.50	.941	Ea.	28	60		88	120
2300	4' long, two 40 watt, RS		8	1		50	63.50		113.50	150
2310	4' long, two 32 watt T8, RS [G]		8	1		75.50	63.50		139	178
2400	4' long, one 40 watt, SL		8	1		54	63.50		117.50	154
2500	4' long, two 40 watt, SL		7	1.143		73.50	73		146.50	189
2580	8' long, one 60 watt T8, SL [G]	2 Elec	13.40	1.194		113	76		189	237
2590	8' long, two 60 watt T8, SL [G]		12.40	1.290		101	82		183	233
2600	8' long, one 75 watt, SL		13.40	1.194		57.50	76		133.50	177
2700	8' long, two 75 watt, SL		12.40	1.290		72.50	82		154.50	202
2800	4' long, two 60 watt, HO	1 Elec	6.70	1.194		112	76		188	236
2810	4' long, two 54 watt, T5HO [G]	"	6.70	1.194		177	76		253	310
2900	8' long, two 110 watt, HO	2 Elec	10.60	1.509		109	96		205	263
2910	4' long, two 115 watt, VHO	1 Elec	6.50	1.231		144	78.50		222.50	274
2920	8' long, two 215 watt, VHO	2 Elec	10.40	1.538		155	98		253	315
2950	High bay pendent mounted, 16" W x 4' L, four 54 watt, T5HO [G]		8.90	1.798		280	115		395	480
2952	2' W x 4' L, six 54 watt, T5HO [G]		8.50	1.882		305	120		425	520
2954	2' W x 4' L, six 32 watt, T8 [G]		8.50	1.882		178	120		298	375
3000	Strip, pendent mounted, industrial, white porcelain enamel									
3100	4' long, two 40 watt, RS	1 Elec	5.70	1.404	Ea.	53.50	89.50		143	192
3110	4' long, two 32 watt T8, RS [G]		5.70	1.404		108	89.50		197.50	252
3200	4' long, two 60 watt, HO		5	1.600		83	102		185	242
3290	8' long, two 60 watt T8, SL [G]	2 Elec	8.80	1.818		122	116		238	305
3300	8' long, two 75 watt, SL		8.80	1.818		101	116		217	283
3400	8' long, two 110 watt, HO		8	2		129	127		256	330
3410	Acrylic finish, 4' long, two 40 watt, RS	1 Elec	5.70	1.404		87	89.50		176.50	229
3420	4' long, two 60 watt, HO		5	1.600		162	102		264	330
3430	4' long, two 115 watt, VHO		4.80	1.667		213	106		319	390
3440	8' long, two 75 watt, SL	2 Elec	8.80	1.818		174	116		290	365
3450	8' long, two 110 watt, HO		8	2		201	127		328	410
3460	8' long, two 215 watt, VHO		7.60	2.105		285	134		419	515
3470	Troffer, air handling, 2' W x 4' L with four 32 watt T8 [G]	1 Elec	4	2		119	127		246	320
3480	2' W x 2' L with two U32 watt T8 [G]		5.50	1.455		103	92.50		195.50	251
3490	Air connector insulated, 5" diameter		20	.400		69.50	25.50		95	115
3500	6" diameter		20	.400		69	25.50		94.50	114
3502	Troffer, direct/indirect, 2' W x 4' L with two 32 W T8 [G]		5.30	1.509		206	96		302	370
3510	Troffer parabolic lay-in, 1' W x 4' L with one 32 W T8 [G]		5.70	1.404		118	89.50		207.50	263
3520	1' W x 4' L with two 32 W T8 [G]		5.30	1.509		86.50	96		182.50	238
3525	2' W x 2' L with two U32 W T8 [G]		5.70	1.404		123	89.50		212.50	269
3530	2' W x 4' L with three 32 W T8 [G]		5	1.600		132	102		234	296

For customer support on your Electrical Costs with RSMeans Data, call 800.448.8182.

305

26 51 13 – Interior Lighting Fixtures, Lamps, and Ballasts

26 51 13.50 Interior Lighting Fixtures		Crew	Daily Output	Labor-Hours	Unit	Material	2021 Bare Costs Labor	Equipment	Total	Total Incl O&P
3531	Intr fxtr, fluor, troffer prismatic lay-in, 2' W x 4'l w/three 32 W T8	1 Elec	5	1.600	Ea.	150	102		252	315
3535	Downlight, recess mounted **G**		8	1		149	63.50		212.50	259
3540	Wall wash reflector, recess mounted **G**		8	1		118	63.50		181.50	224
3550	Direct/indirect, 4' long, stl., pendent mtd. **G**		5	1.600		162	102		264	330
3560	4' long, alum., pendent mtd. **G**		5	1.600		335	102		437	520
3565	Prefabricated cove, 4' long, stl. continuous row **G**		5	1.600		209	102		311	380
3570	4' long, alum. continuous row **G**		5	1.600		345	102		447	530
3580	Wet location, recess mounted, 2' W x 4' L with two 32 watt T8 **G**		5.30	1.509		241	96		337	410
3590	Pendent mounted, 2' W x 4' L with two 32 watt T8 **G**	▼	5.70	1.404	▼	390	89.50		479.50	560
4000	Induction lamp, integral ballast, ceiling mounted									
4110	High bay, aluminum reflector, 160 watt	1 Elec	3.20	2.500	Ea.	935	159		1,094	1,250
4120	320 watt	"	3	2.667		1,675	170		1,845	2,100
4130	480 watt	2 Elec	5.80	2.759		2,550	176		2,726	3,050
4150	Low bay, aluminum reflector, 250 watt	1 Elec	3.20	2.500		945	159		1,104	1,275
4170	Garage, aluminum reflector, 80 watt		3.60	2.222		845	142		987	1,150
4180	Vandalproof, aluminum reflector, 100 watt	▼	3.20	2.500	▼	625	159		784	925
4220	Metal halide, integral ballast, ceiling, recess mounted									
4230	prismatic glass lens, floating door									
4240	2' W x 2' L, 250 watt	1 Elec	3.20	2.500	Ea.	370	159		529	640
4250	2' W x 2' L, 400 watt	2 Elec	5.80	2.759		375	176		551	675
4260	Surface mounted, 2' W x 2' L, 250 watt	1 Elec	2.70	2.963		340	189		529	655
4270	400 watt	2 Elec	4.80	3.333	▼	405	212		617	760
4280	High bay, aluminum reflector,									
4290	Single unit, 400 watt	2 Elec	4.60	3.478	Ea.	425	222		647	795
4300	Single unit, 1,000 watt		4	4		610	255		865	1,050
4310	Twin unit, 400 watt	▼	3.20	5		825	320		1,145	1,375
4320	Low bay, aluminum reflector, 250W DX lamp	1 Elec	3.20	2.500		365	159		524	635
4330	400 watt lamp	2 Elec	5	3.200	▼	540	204		744	900
4340	High pressure sodium integral ballast ceiling, recess mounted									
4350	prismatic glass lens, floating door									
4360	2' W x 2' L, 150 watt lamp	1 Elec	3.20	2.500	Ea.	400	159		559	675
4370	2' W x 2' L, 400 watt lamp	2 Elec	5.80	2.759		475	176		651	785
4380	Surface mounted, 2' W x 2' L, 150 watt lamp	1 Elec	2.70	2.963		485	189		674	815
4390	400 watt lamp	2 Elec	4.80	3.333	▼	545	212		757	910
4400	High bay, aluminum reflector,									
4410	Single unit, 400 watt lamp	2 Elec	4.60	3.478	Ea.	390	222		612	760
4430	Single unit, 1,000 watt lamp	"	4	4		545	255		800	980
4440	Low bay, aluminum reflector, 150 watt lamp	1 Elec	3.20	2.500	▼	335	159		494	605
4445	High bay H.I.D. quartz restrike	"	16	.500	▼	168	32		200	233
4450	Incandescent, high hat can, round alzak reflector, prewired									
4470	100 watt	1 Elec	8	1	Ea.	68.50	63.50		132	170
4480	150 watt		8	1		104	63.50		167.50	210
4500	300 watt		6.70	1.194		241	76		317	380
4520	Round with reflector and baffles, 150 watt		8	1		53.50	63.50		117	154
4540	Round with concentric louver, 150 watt PAR	▼	8	1	▼	78.50	63.50		142	181
4600	Square glass lens with metal trim, prewired									
4630	100 watt	1 Elec	6.70	1.194	Ea.	55	76		131	174
4680	150 watt		6.70	1.194		98	76		174	221
4700	200 watt		6.70	1.194		98	76		174	221
4800	300 watt		5.70	1.404		146	89.50		235.50	294
4810	500 watt		5	1.600		287	102		389	465
4900	Ceiling/wall, surface mounted, metal cylinder, 75 watt		10	.800		47	51		98	127
4920	150 watt	▼	10	.800	▼	86	51		137	170

26 51 13.50 Interior Lighting Fixtures		Crew	Daily Output	Labor-Hours	Unit	Material	2021 Bare Costs Labor	Equipment	Total	Total Incl O&P
4930	300 watt	1 Elec	8	1	Ea.	172	63.50		235.50	285
5000	500 watt		6.70	1.194		365	76		441	515
5010	Square, 100 watt		8	1		120	63.50		183.50	227
5020	150 watt		8	1		123	63.50		186.50	230
5030	300 watt		7	1.143		330	73		403	475
5040	500 watt		6	1.333		360	85		445	520
5200	Ceiling, surface mounted, opal glass drum									
5300	8", one 60 watt lamp	1 Elec	10	.800	Ea.	65	51		116	147
5400	10", two 60 watt lamps		8	1		72.50	63.50		136	175
5500	12", four 60 watt lamps		6.70	1.194		101	76		177	224
5510	Pendent, round, 100 watt		8	1		120	63.50		183.50	227
5520	150 watt		8	1		120	63.50		183.50	228
5530	300 watt		6.70	1.194		172	76		248	300
5540	500 watt		5.50	1.455		330	92.50		422.50	505
5550	Square, 100 watt		6.70	1.194		152	76		228	281
5560	150 watt		6.70	1.194		161	76		237	290
5570	300 watt		5.70	1.404		236	89.50		325.50	395
5580	500 watt		5	1.600		320	102		422	500
5600	Wall, round, 100 watt		8	1		65	63.50		128.50	166
5620	300 watt		8	1		134	63.50		197.50	242
5630	500 watt		6.70	1.194		390	76		466	545
5640	Square, 100 watt		8	1		123	63.50		186.50	230
5650	150 watt		8	1		108	63.50		171.50	214
5660	300 watt		7	1.143		170	73		243	295
5670	500 watt		6	1.333		279	85		364	430
6010	Vapor tight, incandescent, ceiling mounted, 200 watt		6.20	1.290		70	82		152	199
6020	Recessed, 200 watt		6.70	1.194		125	76		201	250
6030	Pendent, 200 watt		6.70	1.194		70	76		146	190
6040	Wall, 200 watt		8	1		83.50	63.50		147	187
6100	Fluorescent, surface mounted, 2 lamps, 4' L, RS, 40 watt		3.20	2.500		119	159		278	370
6110	Industrial, 2 lamps, 4' L in tandem, 430 MA		2.20	3.636		212	232		444	580
6130	2 lamps, 4' L, 800 MA		1.90	4.211		181	268		449	600
6160	Pendent, indust, 2 lamps, 4' L in tandem, 430 MA		1.90	4.211		247	268		515	670
6170	2 lamps, 4' L, 430 MA		2.30	3.478		166	222		388	510
6180	2 lamps, 4' L, 800 MA		1.70	4.706		209	300		509	675
6850	Vandalproof, surface mounted, fluorescent, two 32 watt T8 [G]		3.20	2.500		268	159		427	530
6860	Incandescent, one 150 watt		8	1		90	63.50		153.50	194
6900	Mirror light, fluorescent, RS, acrylic enclosure, two 40 watt		8	1		117	63.50		180.50	223
6910	One 40 watt		8	1		100	63.50		163.50	206
6920	One 20 watt		12	.667		87.50	42.50		130	159
7000	Low bay, aluminum reflector, 70 watt, high pressure sodium		4	2		270	127		397	485
7010	250 watt		3.20	2.500		370	159		529	640
7020	400 watt	2 Elec	5	3.200		410	204		614	760
7500	Ballast replacement, by weight of ballast, to 15' high									
7520	Indoor fluorescent, less than 2 lb.	1 Elec	10	.800	Ea.	27	51		78	106
7540	Two 40W, watt reducer, 2 to 5 lb.		9.40	.851		68.50	54		122.50	156
7560	Two F96 slimline, over 5 lb.		8	1		109	63.50		172.50	214
7580	Vaportite ballast, less than 2 lb.		9.40	.851		27	54		81	111
7600	2 lb. to 5 lb.		8.90	.899		68.50	57.50		126	161
7620	Over 5 lb.		7.60	1.053		109	67		176	219
7630	Electronic ballast for two tubes		8	1		37.50	63.50		101	136
7640	Dimmable ballast one-lamp [G]		8	1		105	63.50		168.50	211
7650	Dimmable ballast two-lamp [G]		7.60	1.053		110	67		177	221

For customer support on your Electrical Costs with RSMeans Data, call 800.448.8182.

307

26 51 13.50 Interior Lighting Fixtures		Crew	Daily Output	Labor-Hours	Unit	Material	2021 Bare Costs Labor	Equipment	Total	Total Incl O&P
7660	Dimmable ballast three-lamp	1 Elec	6.90	1.159	Ea.	126	74		200	249
7690	Emergency ballast (factory installed in fixture)				↓	151			151	166
7990	Decorator									
8000	Pendent RLM in colors, shallow dome, 12" diam., 100 watt	1 Elec	8	1	Ea.	80	63.50		143.50	183
8010	Regular dome, 12" diam., 100 watt	1 Elec	8	1	Ea.	83	63.50		146.50	186
8020	16" diam., 200 watt		7	1.143		84.50	73		157.50	201
8030	18" diam., 300 watt		6	1.333		92	85		177	227
8100	Picture framing light		16	.500		89	32		121	146
8150	Miniature low voltage, recessed, pinhole		8	1		134	63.50		197.50	242
8160	Star		8	1		133	63.50		196.50	242
8170	Adjustable cone		8	1		169	63.50		232.50	281
8180	Eyeball		8	1		106	63.50		169.50	211
8190	Cone		8	1		129	63.50		192.50	237
8200	Coilex baffle		8	1		182	63.50		245.50	295
8210	Surface mounted, adjustable cylinder	↓	8	1	↓	139	63.50		202.50	248
8250	Chandeliers, incandescent									
8260	24" diam. x 42" high, 6 light candle	1 Elec	6	1.333	Ea.	470	85		555	645
8270	24" diam. x 42" high, 6 light candle w/glass shade		6	1.333		430	85		515	600
8280	17" diam. x 12" high, 8 light w/glass panels		8	1		274	63.50		337.50	395
8300	27" diam. x 29" high, 10 light bohemian lead crystal		4	2		470	127		597	710
8310	21" diam. x 9" high, 6 light sculptured ice crystal	↓	8	1	↓	365	63.50		428.50	495
8500	Accent lights, on floor or edge, 0.5 W low volt incandescent									
8520	incl. transformer & fastenings, based on 100' lengths									
8550	Lights in clear tubing, 12" OC	1 Elec	230	.035	L.F.	8.50	2.22		10.72	12.65
8560	6" OC		160	.050		11.10	3.19		14.29	16.95
8570	4" OC		130	.062		17	3.92		20.92	24.50
8580	3" OC		125	.064		18.85	4.08		22.93	27
8590	2" OC		100	.080		27.50	5.10		32.60	37.50
8600	Carpet, lights both sides 6" OC, in alum. extrusion		270	.030		25.50	1.89		27.39	31.50
8610	In bronze extrusion		270	.030		29	1.89		30.89	35
8620	Carpet-bare floor, lights 18" OC, in alum. extrusion		270	.030		20.50	1.89		22.39	25.50
8630	In bronze extrusion		270	.030		24	1.89		25.89	29.50
8640	Carpet edge-wall, lights 6" OC in alum. extrusion		270	.030		25.50	1.89		27.39	31.50
8650	In bronze extrusion		270	.030		29	1.89		30.89	35
8660	Bare floor, lights 18" OC, in alum. extrusion		300	.027		20.50	1.70		22.20	25
8670	In bronze extrusion		300	.027		24	1.70		25.70	29
8680	Bare floor conduit, alum. extrusion		300	.027		6.80	1.70		8.50	10
8690	In bronze extrusion		300	.027	↓	13.60	1.70		15.30	17.45
8700	Step edge to 36", lights 6" OC, in alum. extrusion		100	.080	Ea.	68.50	5.10		73.60	83
8710	In bronze extrusion		100	.080		71.50	5.10		76.60	86
8720	Step edge to 54", lights 6" OC, in alum. extrusion		100	.080		103	5.10		108.10	121
8730	In bronze extrusion	↓	100	.080	↓	109	5.10		114.10	127
8740	Step edge to 72", lights 6" OC, in alum. extrusion	1 Elec	100	.080	Ea.	137	5.10		142.10	158
8750	In bronze extrusion		100	.080		149	5.10		154.10	172
8760	Connector, male		32	.250		2.54	15.95		18.49	26.50
8770	Female with pigtail		32	.250		5.30	15.95		21.25	29.50
8780	Clamps		400	.020		.51	1.27		1.78	2.45
8790	Transformers, 50 watt		8	1		102	63.50		165.50	207
8800	250 watt		4	2		315	127		442	540
8810	1,000 watt	↓	2.70	2.963	↓	355	189		544	670

26 51 13 – Interior Lighting Fixtures, Lamps, and Ballasts

26 51 13.55 Interior LED Fixtures		Crew	Daily Output	Labor-Hours	Unit	Material	2021 Bare Costs Labor	Equipment	Total	Total Incl O&P
0010	**INTERIOR LED FIXTURES** Incl. lamps and mounting hardware									
0100	Downlight, recess mounted, 7.5" diameter, 25 watt	G	1 Elec	8	1	Ea.	345	63.50		408.50	475
0120	10" diameter, 36 watt	G		8	1		365	63.50		428.50	495
0160	cylinder, 10 watts	G		8	1		63	63.50		126.50	164
0180	20 watts	G		8	1		89	63.50		152.50	193
0900	Interior LED fixts, troffer, recess mounted, 2' x 2', 3,500K			8.50	.941		130	60		190	232
0910	2' x 2', 4,000K			8.50	.941		132	60		192	234
1000	Troffer, recess mounted, 2' x 4', 3,200 lumens	G		5.30	1.509		141	96		237	298
1010	4,800 lumens	G		5	1.600		140	102		242	305
1020	6,400 lumens	G		4.70	1.702		190	108		298	370
1100	Troffer retrofit lamp, 38 watt	G		21	.381		52.50	24.50		77	93.50
1110	60 watt	G		20	.400		97.50	25.50		123	145
1120	100 watt	G		18	.444		142	28.50		170.50	199
1200	Troffer, volumetric recess mounted, 2' x 2'	G		5.70	1.404		365	89.50		454.50	535
2000	Strip, surface mounted, one light bar 4' long, 3,500 K	G		8.50	.941		310	60		370	435
2010	5,000 K	G		8	1		269	63.50		332.50	390
2020	Two light bar 4' long, 5,000 K	G		7	1.143		420	73		493	575
3000	Linear, suspended mounted, one light bar 4' long, 37 watt	G		6.70	1.194		171	76		247	300
3010	One light bar 8' long, 74 watt	G	2 Elec	12.20	1.311		315	83.50		398.50	470
3020	Two light bar 4' long, 74 watt	G	1 Elec	5.70	1.404		335	89.50		424.50	505
3030	Two light bar 8' long, 148 watt	G	2 Elec	8.80	1.818		365	116		481	575
4000	High bay, surface mounted, round, 150 watts	G		5.41	2.959		470	188		658	795
4010	2 bars, 164 watts	G		5.41	2.959		385	188		573	700
4020	3 bars, 246 watts	G		5.01	3.197		515	204		719	870
4030	4 bars, 328 watts	G		4.60	3.478		735	222		957	1,150
4040	5 bars, 410 watts	G	3 Elec	4.20	5.716		830	365		1,195	1,450
4050	6 bars, 492 watts	G		3.80	6.324		930	405		1,335	1,625
4060	7 bars, 574 watts	G		3.39	7.075		1,000	450		1,450	1,775
4070	8 bars, 656 watts	G		2.99	8.029		1,050	510		1,560	1,900
5000	Track, lighthead, 6 watt	G	1 Elec	32	.250		55.50	15.95		71.45	84.50
5010	9 watt	G	"	32	.250		62.50	15.95		78.45	92.50
6000	Garage, surface mounted, 103 watts	G	2 Elec	6.50	2.462		990	157		1,147	1,300
6100	pendent mounted, 80 watts	G		6.50	2.462		705	157		862	1,000
6200	95 watts	G		6.50	2.462		825	157		982	1,150
6300	125 watts	G		6.50	2.462		840	157		997	1,150
7000	Downlight, recess mtd., low profile, 4" diam., 9W, 3,000K		1 Elec	8.50	.941		24	60		84	116
7010	4,000K			8.50	.941		23	60		83	115
7020	5,000K			8.50	.941		17.75	60		77.75	109

26 51 13.70 Residential Fixtures

		Crew	Daily Output	Labor-Hours	Unit	Material	2021 Bare Costs Labor	Equipment	Total	Total Incl O&P
0010	**RESIDENTIAL FIXTURES**									
0400	Fluorescent, interior, surface, circline, 32 watt & 40 watt	1 Elec	20	.400	Ea.	178	25.50		203.50	234
0500	2' x 2', two U-tube 32 watt T8		8	1		138	63.50		201.50	247
0700	Shallow under cabinet, two 20 watt		16	.500		65.50	32		97.50	120
0900	Wall mounted, 4' L, two 32 watt T8, with baffle		10	.800		163	51		214	255
2000	Incandescent, exterior lantern, wall mounted, 60 watt		16	.500		64.50	32		96.50	119
2100	Post light, 150 W, with 7' post		4	2		254	127		381	470
2500	Lamp holder, weatherproof with 150 W PAR		16	.500		35.50	32		67.50	86.50
2550	With reflector and guard		12	.667		63	42.50		105.50	132
2600	Interior pendent, globe with shade, 150 W		20	.400		180	25.50		205.50	236

For customer support on your Electrical Costs with RSMeans Data, call 800.448.8182.

309

26 51 Interior Lighting

26 51 13 – Interior Lighting Fixtures, Lamps, and Ballasts

26 51 13.90 Ballast, Replacement HID	Crew	Daily Output	Labor-Hours	Unit	Material	2021 Bare Costs Labor	Equipment	Total	Total Incl O&P
0010 BALLAST, REPLACEMENT HID									
7510 Multi-tap 120/208/240/277 V									
7550 High pressure sodium, 70 watt	1 Elec	10	.800	Ea.	56	51		107	138
7560 100 watt		9.40	.851		63	54		117	150
7570 150 watt		9	.889		67	56.50		123.50	158
7580 250 watt		8.50	.941		98.50	60		158.50	197
7590 400 watt		7	1.143		110	73		183	229
7600 1,000 watt		6	1.333		197	85		282	340
7610 Metal halide, 175 watt		8	1		58.50	63.50		122	159
7620 250 watt		8	1		73	63.50		136.50	175
7630 400 watt		7	1.143		76	73		149	192
7640 1,000 watt		6	1.333		141	85		226	281
7650 1,500 watt		5	1.600		228	102		330	400

26 51 19 – LED Interior Lighting

26 51 19.10 LED Interior Lighting

	Crew	Daily Output	Labor-Hours	Unit	Material	2021 Bare Costs Labor	Equipment	Total	Total Incl O&P
0010 LED INTERIOR LIGHTING									
7000 Interior LED fixts, tape rope kit, 120V, 3' strip, 4W, w/5' cord, 3,000K	1 Elec	24	.333	Ea.	20.50	21		41.50	54
7010 5,000K		24	.333		11.10	21		32.10	43.50
7020 Interior LED fixts, tape rope kit, 120V, 6' strip, 8W, w/5' cord, 3,000K		23	.348		15.20	22		37.20	49.50
7030 5,000K		23	.348		22.50	22		44.50	58
7040 Interior LED fixts, tape rope kit, 120V, 13' strip, 15W, w/5' cord, 3,000K		22	.364		26.50	23		49.50	63.50
7050 5,000K		22	.364		28	23		51	65.50
7060 Interior LED fixts, tape rope kit, 120V, 20' strip, 22W, w/5' cord, 3,000K	1 Elec	21	.381	Ea.	40.50	24.50		65	80.50
7070 5,000K		21	.381		40.50	24.50		65	80.50
7080 Interior LED fixts, tape rope kit, 120V, 33' strip, 37W, w/5' cord, 3,000K		20	.400		60	25.50		85.50	104
7090 5,000K		20	.400		60	25.50		85.50	104
7100 Interior LED fixts, tape rope kit, 120V, 50' strip, 55W, w/5' cord, 3,000K		19	.421		90.50	27		117.50	140
7110 5,000K		19	.421		129	27		156	182
7120 Interior LED tape light, starstrand, dimmable, 12V, super star, 2", 3,000K		25	.320		11.15	20.50		31.65	43
7130 Ultra star, 2", 3,000K		25	.320		14.10	20.50		34.60	46
7140 Rainbow star, 2", 3,000K		25	.320		11.30	20.50		31.80	43
7150 Interior LED tape light, starstrand, dimmable, 24V, elite star, 4", 35K		25	.320		5.10	20.50		25.60	36
7160 12", 35K		25	.320		19.35	20.50		39.85	52
7170 60", 35K		23	.348		52.50	22		74.50	90.50
7180 120", 35K		21	.381		101	24.50		125.50	147
7190 240", 35K		18	.444		310	28.50		338.50	380
7200 Interior LED tape lighting channel , 36", AL flex, starstrand, elite star		48	.167		10.55	10.60		21.15	27.50

26 52 Safety Lighting

26 52 13 – Emergency and Exit Lighting

26 52 13.10 Emergency Lighting and Battery Units

	Crew	Daily Output	Labor-Hours	Unit	Material	2021 Bare Costs Labor	Equipment	Total	Total Incl O&P
0010 EMERGENCY LIGHTING AND BATTERY UNITS									
0300 Emergency light units, battery operated									
0350 Twin sealed beam light, 25 W, 6 V each									
0500 Lead battery operated	1 Elec	4	2	Ea.	139	127		266	340
0700 Nickel cadmium battery operated		4	2		340	127		467	565
0780 Additional remote mount, sealed beam, 25 W 6 V		26.70	.300		20	19.10		39.10	50.50
0790 Twin sealed beam light, 25 W 6 V each		26.70	.300		62.50	19.10		81.60	97.50
0900 Self-contained fluorescent lamp pack		10	.800		242	51		293	340

26 52 Safety Lighting

26 52 13 – Emergency and Exit Lighting

26 52 13.16 Exit Signs

		Crew	Daily Output	Labor-Hours	Unit	Material	2021 Bare Costs Labor	Equipment	Total	Total Incl O&P
0010	**EXIT SIGNS**									
0080	Exit light ceiling or wall mount, incandescent, single face	1 Elec	8	1	Ea.	72.50	63.50		136	175
0100	Double face		6.70	1.194		50.50	76		126.50	169
0120	Explosion proof		3.80	2.105		570	134		704	825
0150	Fluorescent, single face		8	1		96	63.50		159.50	200
0160	Double face		6.70	1.194		76.50	76		152.50	198
0200	LED standard, single face G		8	1		47.50	63.50		111	147
0220	Double face G		6.70	1.194		51	76		127	169
0230	LED vandal-resistant, single face G		7.27	1.100		310	70		380	445
0240	LED w/battery unit, single face G		4.40	1.818		335	116		451	535
0260	Double face G		4	2		405	127		532	635
0262	LED w/battery unit, vandal-resistant, single face G		4.40	1.818		350	116		466	555
0270	Combination emergency light units and exit sign	↓	4	2	↓	160	127		287	365
0290	LED retrofit kits G	1 Elec	60	.133	Ea.	52.50	8.50		61	70.50
1500	Exit sign, 12 V, 1 face, remote end mounted (Type 602A1)	R-19	18	1.111		72.50	71		143.50	185
1780	With emergency battery, explosion proof	"	7.70	2.597	↓	3,600	166		3,766	4,200

26 54 Classified Location Lighting

26 54 13 – Incandescent Classified Location Lighting

26 54 13.20 Explosion Proof

		Crew	Daily Output	Labor-Hours	Unit	Material	2021 Bare Costs Labor	Equipment	Total	Total Incl O&P
0010	**EXPLOSION PROOF**, incl. lamps, mounting hardware and connections									
6310	Metal halide with ballast, ceiling, surface mounted, 175 watt	1 Elec	2.90	2.759	Ea.	1,425	176		1,601	1,825
6320	250 watt	"	2.70	2.963		1,725	189		1,914	2,175
6330	400 watt	2 Elec	4.80	3.333		1,850	212		2,062	2,350
6340	Ceiling, pendent mounted, 175 watt	1 Elec	2.60	3.077		1,350	196		1,546	1,800
6350	250 watt	"	2.40	3.333		1,650	212		1,862	2,150
6360	400 watt	2 Elec	4.20	3.810		1,775	243		2,018	2,300
6370	Wall, surface mounted, 175 watt	1 Elec	2.90	2.759		1,525	176		1,701	1,950
6380	250 watt	"	2.70	2.963		1,825	189		2,014	2,275
6390	400 watt	2 Elec	4.80	3.333		1,950	212		2,162	2,475
6400	High pressure sodium, ceiling surface mounted, 70 watt	1 Elec	3	2.667		2,250	170		2,420	2,725
6410	100 watt		3	2.667		2,325	170		2,495	2,825
6420	150 watt		2.70	2.963		2,575	189		2,764	3,125
6430	Pendent mounted, 70 watt		2.70	2.963		2,150	189		2,339	2,625
6440	100 watt		2.70	2.963		2,225	189		2,414	2,725
6450	150 watt		2.40	3.333		2,075	212		2,287	2,600
6460	Wall mounted, 70 watt		3	2.667		2,400	170		2,570	2,900
6470	100 watt		3	2.667		2,675	170		2,845	3,175
6480	150 watt		2.70	2.963		2,550	189		2,739	3,100
6510	Incandescent, ceiling mounted, 200 watt		4	2		1,775	127		1,902	2,150
6520	Pendent mounted, 200 watt		3.50	2.286		1,525	146		1,671	1,900
6530	Wall mounted, 200 watt		4	2		1,750	127		1,877	2,125
6600	Fluorescent, RS, 4' long, ceiling mounted, two 40 watt		2.70	2.963		5,150	189		5,339	5,925
6610	Three 40 watt		2.20	3.636		7,450	232		7,682	8,550
6620	Four 40 watt		1.90	4.211		9,575	268		9,843	10,900
6630	Pendent mounted, two 40 watt		2.30	3.478		5,950	222		6,172	6,875
6640	Three 40 watt		1.90	4.211		8,450	268		8,718	9,700
6650	Four 40 watt	↓	1.70	4.706	↓	11,100	300		11,400	12,700

311

26 55 Special Purpose Lighting

26 55 33 – Hazard Warning Lighting

26 55 33.10 Warning Beacons

		Crew	Daily Output	Labor-Hours	Unit	Material	2021 Bare Costs Labor	Equipment	Total	Total Incl O&P
0010	**WARNING BEACONS**									
0015	Surface mount with colored or clear lens									
0100	Rotating beacon									
0110	120V, 40 watt halogen	1 Elec	3.50	2.286	Ea.	233	146		379	475
0120	24V, 20 watt halogen	"	3.50	2.286	"	239	146		385	480
0200	Steady beacon									
0210	120V, 40 watt halogen	1 Elec	3.50	2.286	Ea.	236	146		382	475
0220	24V, 20 watt		3.50	2.286		236	146		382	475
0230	12V DC, incandescent		3.50	2.286		119	146		265	345
0300	Flashing beacon									
0310	120V, 40 watt halogen	1 Elec	3.50	2.286	Ea.	237	146		383	475
0320	24V, 20 watt halogen		3.50	2.286		222	146		368	460
0410	12V DC with two 6V lantern batteries		7	1.143		221	73		294	350

26 55 59 – Display Lighting

26 55 59.10 Track Lighting

		Crew	Daily Output	Labor-Hours	Unit	Material	2021 Bare Costs Labor	Equipment	Total	Total Incl O&P
0010	**TRACK LIGHTING**									
0080	Track, 1 circuit, 4' section	1 Elec	6.70	1.194	Ea.	50.50	76		126.50	169
0100	8' section	2 Elec	10.60	1.509		66	96		162	216
0200	12' section	"	8.80	1.818		124	116		240	310
0300	3 circuits, 4' section	1 Elec	6.70	1.194		105	76		181	229
0400	8' section	2 Elec	10.60	1.509		131	96		227	287
0500	12' section	"	8.80	1.818		176	116		292	365
1000	Feed kit, surface mounting	1 Elec	16	.500		17.65	32		49.65	67
1100	End cover		24	.333		8.25	21		29.25	40.50
1200	Feed kit, stem mounting, 1 circuit		16	.500		60.50	32		92.50	114
1300	3 circuit		16	.500		60.50	32		92.50	114
2000	Electrical joiner, for continuous runs, 1 circuit		32	.250		34.50	15.95		50.45	61.50
2100	3 circuit		32	.250		92.50	15.95		108.45	126
2200	Fixtures, spotlight, 75 W PAR halogen		16	.500		51.50	32		83.50	105
2210	50 W MR16 halogen		16	.500		212	32		244	281
3000	Wall washer, 250 W tungsten halogen		16	.500		135	32		167	196
3100	Low voltage, 25/50 W, 1 circuit		16	.500		136	32		168	197
3120	3 circuit		16	.500		193	32		225	261

26 55 61 – Theatrical Lighting

26 55 61.10 Lights

		Crew	Daily Output	Labor-Hours	Unit	Material	2021 Bare Costs Labor	Equipment	Total	Total Incl O&P
0010	**LIGHTS**									
2000	Lights, border, quartz, reflector, vented,									
2100	colored or white	1 Elec	20	.400	L.F.	190	25.50		215.50	247
2500	Spotlight, follow spot, with transformer, 2,100 watt	"	4	2	Ea.	3,675	127		3,802	4,225
2600	For no transformer, deduct					1,000			1,000	1,100
3000	Stationary spot, fresnel quartz, 6" lens	1 Elec	4	2		245	127		372	460
3100	8" lens		4	2		254	127		381	470
3500	Ellipsoidal quartz, 1,000 watt, 6" lens		4	2		375	127		502	600
3600	12" lens		4	2		665	127		792	920
4000	Strobe light, 1 to 15 flashes per second, quartz	1 Elec	3	2.667	Ea.	905	170		1,075	1,250
4500	Color wheel, portable, five hole, motorized	"	4	2	"	232	127		359	445

26 55 63 – Detention Lighting

26 55 63.10 Detention Lighting Fixtures

		Crew	Daily Output	Labor-Hours	Unit	Material	2021 Bare Costs Labor	Equipment	Total	Total Incl O&P
0010	**DETENTION LIGHTING FIXTURES**									
1000	Surface mounted, cold rolled steel, 14 ga., 1' x 4'	1 Elec	5.40	1.481	Ea.	855	94.50		949.50	1,075
1010	2' x 2'		5.40	1.481		825	94.50		919.50	1,050

26 55 Special Purpose Lighting

26 55 63 – Detention Lighting

26 55 63.10 Detention Lighting Fixtures	Crew	Daily Output	Labor-Hours	Unit	Material	2021 Bare Costs Labor	Equipment	Total	Total Incl O&P
Detention Lighting Fixtures	1 Elec	5	1.600	Ea.	1,000	102		1,102	1,250
1020 2' x 4'		5.40	1.481		1,075	94.50		1,169.50	1,325
1100 12 ga., 1' x 4'		5.40	1.481		825	94.50		919.50	1,050
1110 2' x 2'		5	1.600		1,075	102		1,177	1,325
1120 2' x 4'		5.40	1.481		222	94.50		316.50	385
3000 Fluorescent vandal resistant light fixture, high abuse troffer									
3010 50" L x 8" W x 4" H		5.40	1.481		192	94.50		286.50	350
3020 32W		5.40	1.481		280	94.50		374.50	450
3030 T8/32W		5.40	1.481		203	94.50		297.50	365
3040 No lamp		5.60	1.429		335	91		426	505
3050 Lamp included		5.40	1.481		385	94.50		479.50	565
3060 16 ga. cold rolled steel		5.40	1.481		385	94.50		479.50	565
3070 16 ga. cold rolled steel, ceiling/wall mount		5.40	1.481		320	94.50		414.50	495
3080 w/lamp, white		5.40	1.481		251	94.50		345.50	415
3090 wo/lamp, bronze		5.60	1.429		223	91		314	380
3100 w/lamp 6" L x 4-1/2" W x 8-1/2" H		5.40	1.481		38	94.50		132.50	182
3110 w/lamp 5-7/8" W x 4-1/2" D x 8-1/2" H clear		5.40	1.481		65.50	94.50		160	212
3120 w/lamp white		5.40	1.481		320	94.50		414.50	490
3130 Tall wallpack, w/lamp, bronze		5.40	1.481		111	94.50		205.50	262
3140 Tall wallpack, w/lamp, white		5.40	1.481		110	94.50		204.50	261

26 56 Exterior Lighting

26 56 13 – Lighting Poles and Standards

26 56 13.10 Lighting Poles

No.	Description	Crew	Daily Output	Labor-Hours	Unit	Material	Labor	Equipment	Total	Total Incl O&P
0010	**LIGHTING POLES**									
0100	Exterior, light poles, concrete, 30' above 5' below, 13.5" Base, 5.5" Tip	2 Elec	4.60	3.478	Ea.	1,600	222		1,822	2,075
0110	39' above 6' below, 15.5" Base, 5.25" Tip		4.60	3.478		1,875	222		2,097	2,400
0120	43' above 7' below, 17.25" Base, 6.5" Tip		4.60	3.478		1,900	222		2,122	2,425
0130	43' above 7' below, 19.5" Base, 8.25" Tip		4.60	3.478		1,925	222		2,147	2,450
2800	Light poles, anchor base									
2820	not including concrete bases									
2840	Aluminum pole, 8' high	1 Elec	4	2	Ea.	365	127		492	595
2850	10' high		4	2		400	127		527	630
2860	12' high		3.80	2.105		430	134		564	670
2870	14' high		3.40	2.353		445	150		595	715
2880	16' high		3	2.667		635	170		805	950
3000	20' high	R-3	2.90	6.897		605	440	66	1,111	1,400
3200	30' high		2.60	7.692		1,550	490	73.50	2,113.50	2,525
3400	35' high		2.30	8.696		1,575	550	83	2,208	2,650
3600	40' high		2	10		2,000	635	95.50	2,730.50	3,250
3800	Bracket arms, 1 arm	1 Elec	8	1		137	63.50		200.50	246
4000	2 arms		8	1		273	63.50		336.50	395
4200	3 arms		5.30	1.509		410	96		506	595
4400	4 arms		4.80	1.667		550	106		656	765
4500	Steel pole, galvanized, 8' high		3.80	2.105		720	134		854	995
4510	10' high		3.70	2.162		325	138		463	565
4520	12' high		3.40	2.353		810	150		960	1,125
4530	14' high		3.10	2.581		865	164		1,029	1,200
4540	16' high		2.90	2.759		915	176		1,091	1,250
4550	18' high		2.70	2.963		965	189		1,154	1,350
4600	20' high	R-3	2.60	7.692		1,150	490	73.50	1,713.50	2,050
4800	30' high		2.30	8.696		1,275	550	83	1,908	2,300

313

26 56 13 – Lighting Poles and Standards

26 56 13.10 Lighting Poles

		Crew	Daily Output	Labor-Hours	Unit	Material	2021 Bare Costs Labor	Equipment	Total	Total Incl O&P
5000	35' high	R-3	2.20	9.091	Ea.	1,375	575	87	2,037	2,450
5200	40' high	↓	1.70	11.765		1,500	745	112	2,357	2,875
5400	Bracket arms, 1 arm	1 Elec	8	1		219	63.50		282.50	335
5600	2 arms		8	1		295	63.50		358.50	420
5800	3 arms		5.30	1.509		244	96		340	410
6000	4 arms		5.30	1.509		335	96		431	515
6100	Fiberglass pole, 1 or 2 fixtures, 20' high	R-3	4	5		970	315	48	1,333	1,600
6200	30' high		3.60	5.556		1,200	355	53	1,608	1,900
6300	35' high		3.20	6.250		3,325	395	59.50	3,779.50	4,325
6400	40' high	↓	2.80	7.143		2,250	455	68	2,773	3,225
6420	Wood pole, 4-1/2" x 5-1/8", 8' high	1 Elec	6	1.333		390	85		475	550
6430	10' high		6	1.333		460	85		545	630
6440	12' high		5.70	1.404		570	89.50		659.50	765
6450	15' high		5	1.600		665	102		767	880
6460	20' high	↓	4	2		810	127		937	1,075
6461	Light poles, anchor base, w/o conc base, pwdr ct stl, 16' H	2 Elec	3.10	5.161		915	330		1,245	1,500
6462	20' high	R-3	2.90	6.897		1,150	440	66	1,656	1,975
6463	30' high		2.30	8.696		1,275	550	83	1,908	2,300
6464	35' high	↓	2.40	8.333		1,375	530	79.50	1,984.50	2,375
6470	Light pole conc base, max 6' buried, 2' exposed, 18" diam., average cost	C-6	6	8	↓	200	370	9.05	579.05	775
7300	Transformer bases, not including concrete bases									
7320	Maximum pole size, steel, 40' high	1 Elec	2	4	Ea.	1,450	255		1,705	1,950
7340	Cast aluminum, 30' high		3	2.667		765	170		935	1,100
7350	40' high	↓	2.50	3.200	↓	1,150	204		1,354	1,575
8000	Line cover protective devices									
8010	line cover									
8015	Refer to 26 01 02.20 for labor adjustment factors as they apply									
8100	MVLC 1" W x 1-1/2" H, 14/5'	1 Elec	60	.133	Ea.	4.84	8.50		13.34	17.90
8110	8'		58	.138		5.40	8.80		14.20	19
8120	MVLC 1" W x 1-1/2" H, 18/5'		60	.133		6.15	8.50		14.65	19.35
8130	8'		58	.138		6.90	8.80		15.70	20.50
8140	MVLC 1" W x 1-1/2" H, 38/5'		60	.133		10.70	8.50		19.20	24.50
8150	8'	↓	58	.138		13.25	8.80		22.05	27.50

26 56 19 – LED Exterior Lighting

26 56 19.55 Roadway LED Luminaire

			Crew	Daily Output	Labor-Hours	Unit	Material	2021 Bare Costs Labor	Equipment	Total	Total Incl O&P
0010	**ROADWAY LED LUMINAIRE**										
0100	LED fixture, 72 LEDs, 120 V AC or 12 V DC, equal to 60 watt	G	1 Elec	2.70	2.963	Ea.	575	189		764	910
0110	108 LEDs, 120 V AC or 12 V DC, equal to 90 watt	G		2.70	2.963		765	189		954	1,125
0120	144 LEDs, 120 V AC or 12 V DC, equal to 120 watt	G	↓	2.70	2.963		940	189		1,129	1,300
0130	252 LEDs, 120 V AC or 12 V DC, equal to 210 watt	G	2 Elec	4.40	3.636		1,300	232		1,532	1,775
0140	Replaces high pressure sodium fixture, 75 watt	G	1 Elec	2.70	2.963		410	189		599	730
0150	125 watt	G		2.70	2.963		470	189		659	795
0160	150 watt	G		2.70	2.963		545	189		734	880
0170	175 watt	G		2.70	2.963		805	189		994	1,175
0180	200 watt	G		2.70	2.963		770	189		959	1,125
0190	250 watt	G	2 Elec	4.40	3.636		880	232		1,112	1,325
0200	320 watt	G		4.40	3.636		960	232		1,192	1,400
2000	Exterior fixtures, LED roadway, type 2, 50W, neutral slipfitter bronze			5.20	3.077		755	196		951	1,125
2010	3, 50W, neutral slipfitter bronze			5.20	3.077		755	196		951	1,125
2020	4, 50W, neutral slipfitter bronze			5.20	3.077		755	196		951	1,125
2030	Exterior fixtures, LED roadway, type 2, 78W, neutral slipfitter bronze			5.20	3.077		725	196		921	1,075
2040	3, 78W, neutral slipfitter bronze			5.20	3.077		725	196		921	1,075

26 56 Exterior Lighting

26 56 19 – LED Exterior Lighting

26 56 19.55 Roadway LED Luminaire

		Crew	Daily Output	Labor-Hours	Unit	Material	2021 Bare Costs Labor	Equipment	Total	Total Incl O&P
2050	4, 78W, neutral slipfitter bronze	2 Elec	5.20	3.077	Ea.	740	196		936	1,100
2060	Exterior fixtures, LED roadway, type 2, 105W, neutral slipfitter bronze		5.20	3.077		865	196		1,061	1,250
2070	3, 105W, neutral slipfitter bronze		5.20	3.077		845	196		1,041	1,225
2080	4, 105W, neutral slipfitter bronze		5.20	3.077		865	196		1,061	1,250
2090	Exterior fixtures, LED roadway, type 2, 125W, neutral slipfitter bronze	↓	5.20	3.077	↓	840	196		1,036	1,225

26 56 19.60 Parking LED Lighting

		Crew	Daily Output	Labor-Hours	Unit	Material	2021 Bare Costs Labor	Equipment	Total	Total Incl O&P
0010	**PARKING LED LIGHTING**									
0100	Round pole mounting, 88 lamp watts [G]	1 Elec	2	4	Ea.	1,275	255		1,530	1,775

26 56 21 – HID Exterior Lighting

26 56 21.20 Roadway Luminaire

		Crew	Daily Output	Labor-Hours	Unit	Material	2021 Bare Costs Labor	Equipment	Total	Total Incl O&P
0010	**ROADWAY LUMINAIRE**									
2650	Roadway area luminaire, low pressure sodium, 135 watt	1 Elec	2	4	Ea.	730	255		985	1,175
2700	180 watt	"	2	4		880	255		1,135	1,350
2750	Metal halide, 400 watt	2 Elec	4.40	3.636		640	232		872	1,050
2760	1,000 watt		4	4		775	255		1,030	1,225
2780	High pressure sodium, 400 watt		4.40	3.636		770	232		1,002	1,200
2790	1,000 watt	↓	4	4	↓	875	255		1,130	1,350

26 56 23 – Area Lighting

26 56 23.10 Exterior Fixtures

		Crew	Daily Output	Labor-Hours	Unit	Material	2021 Bare Costs Labor	Equipment	Total	Total Incl O&P
0010	**EXTERIOR FIXTURES** With lamps									
0200	Wall mounted, incandescent, 100 watt	1 Elec	8	1	Ea.	168	63.50		231.50	280
0400	Quartz, 500 watt		5.30	1.509		61.50	96		157.50	211
0420	1,500 watt		4.20	1.905		100	121		221	290
1100	Wall pack, low pressure sodium, 35 watt		4	2		207	127		334	415
1150	55 watt		4	2		243	127		370	455
1160	High pressure sodium, 70 watt		4	2		208	127		335	420
1170	150 watt		4	2		219	127		346	430
1175	High pressure sodium, 250 watt		4	2		219	127		346	430
1180	Metal halide, 175 watt		4	2		223	127		350	435
1190	250 watt		4	2		253	127		380	470
1195	400 watt		4	2		410	127		537	645
1250	Induction lamp, 40 watt		4	2		520	127		647	765
1260	80 watt		4	2		600	127		727	850
1278	LED, poly lens, 26 watt		4	2		1,675	127		1,802	2,050
1280	110 watt		4	2		820	127		947	1,100
1500	LED, glass lens, 13 watt	↓	4	2		1,675	127		1,802	2,050

26 56 23.55 Exterior LED Fixtures

		Crew	Daily Output	Labor-Hours	Unit	Material	2021 Bare Costs Labor	Equipment	Total	Total Incl O&P
0010	**EXTERIOR LED FIXTURES**									
0100	Wall mounted, indoor/outdoor, 12 watt [G]	1 Elec	10	.800	Ea.	203	51		254	300
0110	32 watt [G]		10	.800		460	51		511	580
0120	66 watt [G]		10	.800		345	51		396	455
0200	outdoor, 110 watt [G]		10	.800		1,200	51		1,251	1,375
0210	220 watt [G]		10	.800		305	51		356	410
0300	modular, type IV, 120 V, 50 lamp watts [G]		9	.889		1,425	56.50		1,481.50	1,650
0310	101 lamp watts [G]		9	.889		1,325	56.50		1,381.50	1,525
0320	126 lamp watts [G]		9	.889		1,650	56.50		1,706.50	1,875
0330	202 lamp watts [G]		9	.889		1,900	56.50		1,956.50	2,150
0340	240 V, 50 lamp watts [G]		8	1		1,225	63.50		1,288.50	1,450
0350	101 lamp watts [G]		8	1		1,375	63.50		1,438.50	1,600
0360	126 lamp watts [G]		8	1		1,300	63.50		1,363.50	1,525
0370	202 lamp watts [G]	↓	8	1		1,925	63.50		1,988.50	2,200

315

26 56 Exterior Lighting

26 56 23 - Area Lighting

26 56 23.55 Exterior LED Fixtures

		Crew	Daily Output	Labor-Hours	Unit	Material	2021 Bare Costs Labor	Equipment	Total	Total Incl O&P
0400	wall pack, glass, 13 lamp watts **G**	1 Elec	4	2	Ea.	450	127		577	685
0410	poly w/photocell, 26 lamp watts **G**		4	2		335	127		462	560
0420	50 lamp watts **G**		4	2		785	127		912	1,050
0430	replacement, 40 watts **G**		4	2		122	127		249	325
0440	60 watts **G**		4	2		154	127		281	360

26 56 26 - Landscape Lighting

26 56 26.20 Landscape Fixtures

		Crew	Daily Output	Labor-Hours	Unit	Material	Labor	Equipment	Total	Total Incl O&P
0010	**LANDSCAPE FIXTURES**									
7380	Landscape recessed uplight, incl. housing, ballast, transformer									
7390	& reflector, not incl. conduit, wire, trench									
7420	Incandescent, 250 watt	1 Elec	5	1.600	Ea.	685	102		787	905
7440	Quartz, 250 watt		5	1.600		650	102		752	865
7460	500 watt		4	2		670	127		797	925

26 56 26.50 Landscape LED Fixtures

		Crew	Daily Output	Labor-Hours	Unit	Material	Labor	Equipment	Total	Total Incl O&P
0010	**LANDSCAPE LED FIXTURES**									
0100	12 volt alum bullet hooded-BLK	1 Elec	5	1.600	Ea.	35.50	102		137.50	190
0200	12 volt alum bullet hooded-BRZ		5	1.600		98.50	102		200.50	260
0300	12 volt alum bullet hooded-GRN		5	1.600		98.50	102		200.50	260
1000	12 volt alum large bullet hooded-BLK		5	1.600		72.50	102		174.50	231
1100	12 volt alum large bullet hooded-BRZ		5	1.600		72.50	102		174.50	231
1200	12 volt alum large bullet hooded-GRN		5	1.600		72.50	102		174.50	231
2000	12 volt large bullet landscape light fixture		5	1.600		72.50	102		174.50	231
2100	12 volt alum light large bullet		5	1.600		72.50	102		174.50	231
2200	12 volt alum bullet light		5	1.600		72.50	102		174.50	231

26 56 33 - Walkway Lighting

26 56 33.10 Walkway Luminaire

		Crew	Daily Output	Labor-Hours	Unit	Material	Labor	Equipment	Total	Total Incl O&P
0010	**WALKWAY LUMINAIRE**									
6500	Bollard light, lamp & ballast, 42" high with polycarbonate lens									
6800	Metal halide, 175 watt	1 Elec	3	2.667	Ea.	980	170		1,150	1,325
6900	High pressure sodium, 70 watt		3	2.667		1,050	170		1,220	1,400
7000	100 watt		3	2.667		1,050	170		1,220	1,400
7100	150 watt		3	2.667		1,025	170		1,195	1,375
7200	Incandescent, 150 watt		3	2.667		605	170		775	915
7810	Walkway luminaire, square 16", metal halide 250 watt		2.70	2.963		780	189		969	1,125
7820	High pressure sodium, 70 watt		3	2.667		890	170		1,060	1,225
7830	100 watt		3	2.667		910	170		1,080	1,250
7840	150 watt		3	2.667		910	170		1,080	1,250
7850	200 watt		3	2.667		915	170		1,085	1,250
7910	Round 19", metal halide, 250 watt		2.70	2.963		1,250	189		1,439	1,650
7920	High pressure sodium, 70 watt		3	2.667		1,375	170		1,545	1,750
7930	100 watt		3	2.667		1,375	170		1,545	1,750
7940	150 watt		3	2.667		1,375	170		1,545	1,775
7950	250 watt		2.70	2.963		1,325	189		1,514	1,725
8000	Sphere 14" opal, incandescent, 200 watt		4	2		395	127		522	625
8020	Sphere 18" opal, incandescent, 300 watt		3.50	2.286		480	146		626	740
8040	Sphere 16" clear, high pressure sodium, 70 watt		3	2.667		830	170		1,000	1,175
8050	100 watt		3	2.667		885	170		1,055	1,225
8100	Cube 16" opal, incandescent, 300 watt		3.50	2.286		525	146		671	790
8120	High pressure sodium, 70 watt		3	2.667		765	170		935	1,100
8130	100 watt		3	2.667		785	170		955	1,125
8230	Lantern, high pressure sodium, 70 watt		3	2.667		690	170		860	1,000

26 56 33 – Walkway Lighting

26 56 33.10 Walkway Luminaire

		Crew	Daily Output	Labor-Hours	Unit	Material	2021 Bare Costs Labor	Equipment	Total	Total Incl O&P
8240	100 watt	1 Elec	3	2.667	Ea.	740	170		910	1,075
8250	150 watt		3	2.667		695	170		865	1,025
8260	250 watt		2.70	2.963		970	189		1,159	1,350
8270	Incandescent, 300 watt		3.50	2.286		510	146		656	780
8330	Reflector 22" w/globe, high pressure sodium, 70 watt		3	2.667		505	170		675	805
8340	100 watt		3	2.667		510	170		680	810
8350	150 watt		3	2.667		515	170		685	820
8360	250 watt	▼	2.70	2.963	▼	660	189		849	1,000

26 56 33.55 Walkway LED Luminaire

			Crew	Daily Output	Labor-Hours	Unit	Material	2021 Bare Costs Labor	Equipment	Total	Total Incl O&P
0010	**WALKWAY LED LUMINAIRE**										
0100	Pole mounted, 86 watts, 4,350 lumens	G	1 Elec	3	2.667	Ea.	1,350	170		1,520	1,725
0110	4,630 lumens	G		3	2.667		2,225	170		2,395	2,700
0120	80 watts, 4,000 lumens	G		3	2.667		2,050	170		2,220	2,500
0600	LED bollard pole, 6" diam. x 36" H, cast alum, surf. mtd., 8W, 6,000K, 120V			8.50	.941		390	60		450	520
0610	6" diam. x 42" H, cast alum, surf. mtd., 12W, 5,100K,120V			8.50	.941		625	60		685	775
0620	6" diam. x 42" H, cast alum, surf. mtd., 18W, 5,100K,120V			8.50	.941		645	60		705	795
0630	6" diam. x 42" H, cast alum, surf. mtd., 24W, 5,100K,120V			8.50	.941		660	60		720	815
0640	9.5" square x 42" H, cast resin, surf. mtd., A19, 9W, 120V			8.50	.941		615	60		675	770
0650	9.75" diam. x 42" H, cast resin, side mount, dome, 9W, 120V			8.50	.941		700	60		760	860
0660	9.75" diam. x 43" H, cast resin, burial base, dome, 9W, 120V			8.50	.941		700	60		760	860
0670	8" diam. x 36" H, concrete security base, w/rebar, 8W, HID			8	1		1,050	63.50		1,113.50	1,250
0680	10" diam. x 49" H, concrete base, w/Sch 40 Steel pipe, 8W			6.75	1.185		1,175	75.50		1,250.50	1,400
0690	12" diam. x 48" H, concrete base, w/Sch 40 Steel pipe, 8W			6.75	1.185		1,350	75.50		1,425.50	1,575
0700	8" diam. x 36" H, concrete, cast aluminum dome, 90 degree, 4.1W			7	1.143		1,050	73		1,123	1,250
0710	8" diam. x 36" H, concrete, cast aluminum dome, 180 degree, 4.1W			7	1.143		1,275	73		1,348	1,525
0720	8" diam. x 48" H, concrete, cast aluminum dome, 360 degree, 4.1W		▼	7	1.143		1,575	73		1,648	1,850

26 56 36 – Flood Lighting

26 56 36.20 Floodlights

			Crew	Daily Output	Labor-Hours	Unit	Material	2021 Bare Costs Labor	Equipment	Total	Total Incl O&P
0010	**FLOODLIGHTS** with ballast and lamp,										
1290	floor mtd, mount with swivel bracket										
1300	Induction lamp, 40 watt		1 Elec	3	2.667	Ea.	380	170		550	670
1310	80 watt			3	2.667		770	170		940	1,100
1320	150 watt		▼	3	2.667	▼	1,375	170		1,545	1,775
1400	Pole mounted, pole not included										
1950	Metal halide, 175 watt		1 Elec	2.70	2.963	Ea.	188	189		377	485
2000	400 watt		2 Elec	4.40	3.636	Ea.	186	232		418	550
2200	1,000 watt			4	4		1,000	255		1,255	1,475
2210	1,500 watt		▼	3.70	4.324		450	275		725	905
2250	Low pressure sodium, 55 watt		1 Elec	2.70	2.963		420	189		609	745
2270	90 watt			2	4		600	255		855	1,050
2290	180 watt			2	4		625	255		880	1,075
2340	High pressure sodium, 70 watt			2.70	2.963		255	189		444	560
2360	100 watt			2.70	2.963		262	189		451	570
2380	150 watt		▼	2.70	2.963		295	189		484	605
2400	400 watt		2 Elec	4.40	3.636		300	232		532	675
2600	1,000 watt		"	4	4		605	255		860	1,050
9005	Solar powered floodlight, w/motion det, incl. batt pack for cloudy days	G	1 Elec	8	1		74	63.50		137.50	176
9020	Replacement battery pack	G	"	8	1	▼	20	63.50		83.50	117

317

26 56 Exterior Lighting

26 56 36 – Flood Lighting

26 56 36.55 LED Floodlights

		Crew	Daily Output	Labor-Hours	Unit	Material	2021 Bare Costs Labor	Equipment	Total	Total Incl O&P
0010	**LED FLOODLIGHTS** with ballast and lamp,									
0020	Pole mounted, pole not included									
0100	11 watt	G 1 Elec	4	2	Ea.	320	127		447	540
0110	46 watt	G	4	2		267	127		394	480
0120	90 watt	G	4	2		2,225	127		2,352	2,650
0130	288 watt	G	4	2		705	127		832	965

26 61 Lighting Systems and Accessories

26 61 13 – Lighting Accessories

26 61 13.30 Fixture Whips

		Crew	Daily Output	Labor-Hours	Unit	Material	2021 Bare Costs Labor	Equipment	Total	Total Incl O&P
0010	**FIXTURE WHIPS**									
0080	3/8" Greenfield, 2 connectors, 6' long									
0100	TFFN wire, three #18	1 Elec	32	.250	Ea.	6.80	15.95		22.75	31
0150	Four #18		28	.286		8.45	18.20		26.65	36.50
0200	Three #16		32	.250		6.35	15.95		22.30	30.50
0250	Four #16		28	.286		8.80	18.20		27	36.50
0300	THHN wire, three #14		32	.250		4.87	15.95		20.82	29
0350	Four #14		28	.286		6.75	18.20		24.95	34.50
0360	Three #12		32	.250		9.30	15.95		25.25	34

26 61 23 – Lamps Applications

26 61 23.10 Lamps

		Crew	Daily Output	Labor-Hours	Unit	Material	2021 Bare Costs Labor	Equipment	Total	Total Incl O&P
0010	**LAMPS**									
0080	Fluorescent, rapid start, cool white, 2' long, 20 watt	1 Elec	1	8	C	283	510		793	1,075
0100	4' long, 40 watt		.90	8.889		297	565		862	1,175
0120	3' long, 30 watt		.90	8.889		355	565		920	1,225
0125	3' long, 25 watt energy saver	R265723-10	.90	8.889		850	565		1,415	1,775
0150	U-40 watt		.80	10		1,550	635		2,185	2,650
0155	U-34 watt energy saver	G	.80	10		11.80	635		646.80	960
0170	4' long, 34 watt energy saver	G	.90	8.889		3.99	565		568.99	845
0176	2' long, T8, 17 watt energy saver	G	1	8		365	510		875	1,150
0178	3' long, T8, 25 watt energy saver	G 1 Elec	.90	8.889	C	390	565		955	1,275
0180	4' long, T8, 32 watt energy saver	G	.90	8.889		710	565		1,275	1,625
0200	Slimline, 4' long, 40 watt		.90	8.889		745	565		1,310	1,650
0210	4' long, 30 watt energy saver	G	.90	8.889		745	565		1,310	1,650
0300	8' long, 75 watt		.80	10		1,475	635		2,110	2,575
0350	8' long, 60 watt energy saver	G	.80	10		395	635		1,030	1,375
0400	High output, 4' long, 60 watt		.90	8.889		645	565		1,210	1,550
0410	8' long, 95 watt energy saver	G	.80	10		905	635		1,540	1,950
0500	8' long, 110 watt		.80	10		905	635		1,540	1,950
0512	2' long, T5, 14 watt energy saver	G	1	8		235	510		745	1,025
0514	3' long, T5, 21 watt energy saver	G	.90	8.889		232	565		797	1,100
0516	4' long, T5, 28 watt energy saver	G	.90	8.889		232	565		797	1,100
0517	4' long, T5, 54 watt energy saver	G	.90	8.889		450	565		1,015	1,325
0520	Very high output, 4' long, 110 watt		.90	8.889		2,375	565		2,940	3,475
0525	8' long, 195 watt energy saver	G	.70	11.429		1,875	730		2,605	3,125
0550	8' long, 215 watt		.70	11.429		1,550	730		2,280	2,775
0554	Full spectrum, 4' long, 60 watt		.90	8.889		750	565		1,315	1,675
0556	6' long, 85 watt		.90	8.889		640	565		1,205	1,550
0558	8' long, 110 watt		.80	10		430	635		1,065	1,425
0560	Twin tube compact lamp	G	.90	8.889		294	565		859	1,175

26 61 23.10 Lamps

		Crew	Daily Output	Labor-Hours	Unit	Material	2021 Bare Costs Labor	2021 Bare Costs Equipment	Total	Total Incl O&P
0570	Double twin tube compact lamp **G**	1 Elec	.80	10	C	705	635		1,340	1,725
0600	Mercury vapor, mogul base, deluxe white, 100 watt		.30	26.667		14,600	1,700		16,300	18,600
0650	175 watt		.30	26.667		2,575	1,700		4,275	5,375
0700	250 watt		.30	26.667		11,100	1,700		12,800	14,700
0800	400 watt		.30	26.667		10,700	1,700		12,400	14,300
0900	1,000 watt		.20	40		10,800	2,550		13,350	15,700
1000	Metal halide, mogul base, 175 watt		.30	26.667		1,125	1,700		2,825	3,750
1100	250 watt		.30	26.667		2,825	1,700		4,525	5,625
1200	400 watt		.30	26.667		2,550	1,700		4,250	5,325
1300	1,000 watt		.20	40		3,725	2,550		6,275	7,875
1320	1,000 watt, 125,000 initial lumens		.20	40		6,875	2,550		9,425	11,400
1330	1,500 watt		.20	40		8,450	2,550		11,000	13,100
1350	High pressure sodium, 70 watt		.30	26.667		1,800	1,700		3,500	4,500
1360	100 watt		.30	26.667		1,875	1,700		3,575	4,575
1370	150 watt		.30	26.667		1,650	1,700		3,350	4,325
1380	250 watt		.30	26.667		1,625	1,700		3,325	4,300
1400	400 watt		.30	26.667		1,625	1,700		3,325	4,300
1450	1,000 watt		.20	40		4,575	2,550		7,125	8,800
1500	Low pressure sodium, 35 watt		.30	26.667		9,075	1,700		10,775	12,500
1550	55 watt	1 Elec	.30	26.667	C	4,975	1,700		6,675	8,000
1600	90 watt		.30	26.667		14,100	1,700		15,800	18,000
1650	135 watt		.20	40		9,175	2,550		11,725	13,900
1700	180 watt		.20	40		17,000	2,550		19,550	22,500
1750	Quartz line, clear, 500 watt		1.10	7.273		750	465		1,215	1,525
1760	1,500 watt		.20	40		2,275	2,550		4,825	6,275
1762	Spot, MR 16, 50 watt		1.30	6.154		2,125	390		2,515	2,925
1770	Tungsten halogen, T4, 400 watt		1.10	7.273		900	465		1,365	1,675
1775	T3, 1,200 watt		.30	26.667		2,900	1,700		4,600	5,725
1778	PAR 30, 50 watt		1.30	6.154		1,025	390		1,415	1,725
1780	PAR 38, 90 watt		1.30	6.154		3,525	390		3,915	4,450
1800	Incandescent, interior, A21, 100 watt		1.60	5		390	320		710	905
1900	A21, 150 watt		1.60	5		150	320		470	640
2000	A23, 200 watt		1.60	5		234	320		554	730
2200	PS 35, 300 watt		1.60	5		515	320		835	1,050
2210	PS 35, 500 watt		1.60	5		475	320		795	1,000
2230	PS 52, 1,000 watt		1.30	6.154		2,300	390		2,690	3,100
2240	PS 52, 1,500 watt		1.30	6.154		1,850	390		2,240	2,600
2300	R30, 75 watt		1.30	6.154		745	390		1,135	1,400
2400	R40, 100 watt		1.30	6.154		1,800	390		2,190	2,550
2500	Exterior, PAR 38, 75 watt		1.30	6.154		1,400	390		1,790	2,125
2600	PAR 38, 150 watt		1.30	6.154		2,150	390		2,540	2,925
2700	PAR 46, 200 watt		1.10	7.273		1,750	465		2,215	2,625
2800	PAR 56, 300 watt		1.10	7.273		1,575	465		2,040	2,450
3000	Guards, fluorescent lamp, 4' long		100	.080	Ea.	18.05	5.10		23.15	27.50
3200	8' long		90	.089	"	36	5.65		41.65	48

26 61 23.55 LED Lamps

		Crew	Daily Output	Labor-Hours	Unit	Material	2021 Bare Costs Labor	2021 Bare Costs Equipment	Total	Total Incl O&P
0010	**LED LAMPS**									
0100	LED lamp, interior, shape A60, equal to 60 W **G**	1 Elec	160	.050	Ea.	18.65	3.19		21.84	25
0110	7 W LED decorative c, ca, f, g shape		160	.050		7.65	3.19		10.84	13.15
0120	12V mini lamp LED		160	.050		5.30	3.19		8.49	10.60
0200	Globe frosted A60, equal to 60 W **G**		160	.050		5.75	3.19		8.94	11.10
0205	LED lamp, interior, globe		160	.050		10.20	3.19		13.39	15.95

For customer support on your Electrical Costs with RSMeans Data, call 800.448.8182.

319

26 61 23.55 LED Lamps

		Crew	Daily Output	Labor-Hours	Unit	Material	2021 Bare Costs Labor	Equipment	Total	Total Incl O&P
0210	2.2 W LED LMP	1 Elec	160	.050	Ea.	64	3.19		67.19	75
0220	2.2 W LED replacement decorative lamp		160	.050		8.40	3.19		11.59	14
0230	3.5 W LED replacement decorative lamp		160	.050		8.95	3.19		12.14	14.55
0240	4.5 W 120V LED replacement decorative lamp		160	.050		11.20	3.19		14.39	17.10
0250	4.5 W 120V LED, 2700k replacement decorative lamp		160	.050		10.85	3.19		14.04	16.65
0260	4.9 W 120V LED, 2700k replacement decorative lamp candelabra base		160	.050		9.70	3.19		12.89	15.40
0270	4.9 W 120V LED, 3000k replacement decorative lamp candelabra base		160	.050		4.97	3.19		8.16	10.20
0280	5 W LED PAR 20 parabolic reflector lamp FL		140	.057		58	3.64		61.64	69.50
0300	Globe earth, equal to 100 W [G]		140	.057		4.99	3.64		8.63	10.90
0305	7 W LED reflector lamp 3000k DIM		140	.057		4.65	3.64		8.29	10.50
0310	10 W omni LED warm white light bulb E26 medium base 120 volt card		140	.057		17.10	3.64		20.74	24
0315	7 W LED reflector lamp WFL 2700k DIM		140	.057		4.68	3.64		8.32	10.55
0320	9 W omni LED warm white light bulb E26 medium base 120 volt card		140	.057		9.05	3.64		12.69	15.35
0500	8 W LED, A19 lamp, equal to 40 W		140	.057		5.70	3.64		9.34	11.65
0505	9 W LED, A19 lamp, 5000K dimmable, equal to 60 W		140	.057		7.10	3.64		10.74	13.25
0510	10.5 W LED, A19 lamp, dimmable, equal to 60 W		140	.057		8.10	3.64		11.74	14.30
0515	10 W LED, A19 lamp, frosted, dimmable, equal to 60 W		140	.057		3.39	3.64		7.03	9.15
0520	10 W LED, A19 lamp, omni-directional, dimmable, equal to 60 W		140	.057		7.50	3.64		11.14	13.65
0525	12 W LED, A19 lamp, dimmable, equal to 75 W		140	.057		6.90	3.64		10.54	13
1100	MR16, 3 W, replacement of halogen lamp 25 W [G]		130	.062		18.70	3.92		22.62	26.50
1200	6 W replacement of halogen lamp 45 W [G]		130	.062		19.65	3.92		23.57	27.50
2100	10 W, PAR20, equal to 60 W [G]		130	.062		26	3.92		29.92	34.50
2200	15 W, PAR30, equal to 100 W [G]		130	.062		49	3.92		52.92	60
2210	50 W LED lamp		130	.062		735	3.92		738.92	815
2220	11 Watt reflector dimmable warm white LED light bulb with medium base		130	.062		40	3.92		43.92	50
2221	12 W A-Line LED lamp DIM		130	.062		75.50	3.92		79.42	89
2225	13 Watt reflector LED warm white e26 with medium base 120 volt box		130	.062		38	3.92		41.92	47.50
2226	13 W br30 LED lamp		130	.062		65.50	3.92		69.42	78
2227	15 W 120V br30 inc LED lamp, 2700k		130	.062		61	3.92		64.92	73.50
2228	15 W 120V br30 inc LED lamp, 4000k		130	.062		58	3.92		61.92	70
2230	3 Watt dimmable warm white decorative LED lamp with medium base		160	.050		11.85	3.19		15.04	17.75
2240	.43 W night light LED daylight bulb E12 candelabra base 120 volt 2 pack		160	.050		4.56	3.19		7.75	9.75
2250	15 W omni-directional LED warm white e26 medium base 120 volt box		160	.050		39	3.19		42.19	47
2251	11 W omni-directional LED warm white e26 medium base 120 volt box		160	.050		27.50	3.19		30.69	35
2252	10 W omni-directional LED warm white e26 medium base 120 volt box		160	.050		27	3.19		30.19	34
2253	7 W omni A19 LED warm white e26 medium base 120 volt box		160	.050		20.50	3.19		23.69	27
2255	8 PAR 20 parabolic reflector 2700 LED lamp		160	.050		18.20	3.19		21.39	24.50
2256	8 PAR 20 parabolic reflector 3000 LED lamp		160	.050		18.05	3.19		21.24	24.50
2260	12 W PAR 38 120V LED 15 degree directional lamp		160	.050		196	3.19		199.19	220
2270	12 W PAR 38 120V LED 25 degree directional lamp		160	.050		30.50	3.19		33.69	38
2280	12 W PAR 38 120V LED 40 degree directional lamp		160	.050		196	3.19		199.19	220
2285	16 W PAR 38 120V 2700k LED parabolic reflector lamp		160	.050		29.50	3.19		32.69	37
2290	16 W PAR 38 120V 3000k LED parabolic reflector lamp		160	.050		22	3.19		25.19	29
3000	15 W PAR 30 LED daylight E26 medium base 120V box	1 Elec	160	.050	Ea.	58	3.19		61.19	68.50
3100	17 W LED 3000k PAR 38 100 W replacement		160	.050		45	3.19		48.19	54
3110	17 W LED PAR 38 100 W replacement		160	.050		34	3.19		37.19	42
3120	17 W LED PAR 38 100 W replacement parabolic reflector		160	.050		35.50	3.19		38.69	43.50
3130	24 W LED T8 PW straight fluorescent lamp		8.89	.900		279	57.50		336.50	390
3135	Linear fluorescent LED lamp 120V		8.89	.900		128	57.50		185.50	226
3200	30 W LED 2700K recessed 8055E PAR 38 high power		160	.050		86.50	3.19		89.69	99.50
3210	30 W LED 4200K 120 degree 8055E PAR 38 high power		160	.050		81	3.19		84.19	93.50
3220	30 W LED 5700K 8055E PAR 38 high power		160	.050		81	3.19		84.19	93.50
3230	50 W LED 2700K 8045M PAR 38 277V high power		160	.050		330	3.19		333.19	365

For customer support on your Electrical Costs with RSMeans Data, call 800.448.8182.

26 61 Lighting Systems and Accessories

26 61 23 – Lamps Applications

26 61 23.55 LED Lamps

		Crew	Daily Output	Labor-Hours	Unit	Material	2021 Bare Costs Labor	Equipment	Total	Total Incl O&P
3240	50 W LED 4200K 8045M PAR 38 277V high power retro fit	1 Elec	160	.050	Ea.	85	3.19		88.19	98
3250	50 W LED 5700K 8045M PAR 38 277V high power retro fit		160	.050		330	3.19		333.19	365
8000	10 W LED PAR 30/fl 10 pk		160	.050		93	3.19		96.19	107
8010	Gen 3 PAR 30 15 W short neck power LED 120 VAC E26 80 +cri 300k dimm		160	.050		12.50	3.19		15.69	18.50
8020	12 PAR 30 2700K parabolic reflector LED lamp		160	.050		40.50	3.19		43.69	49
8030	12 PAR 30 3000K parabolic reflector LED lamp		160	.050		31	3.19		34.19	38.50
8040	3500K LED advantage T8 9 W 800LM 2ft linear 2 BD frosted		69	.116		9.60	7.40		17	21.50
8050	3500K LED litespan T8 9 W 900LM 2ft linear frosted		69	.116		12.75	7.40		20.15	25
8060	4000K LED advantage T8 9 W 800LM 2ft linear 2BD frosted		69	.116		12.15	7.40		19.55	24.50
8070	4000K LED litespan T8 9 W 900LM 2ft linear frosted		69	.116		13.35	7.40		20.75	25.50
8080	5000K LED advantage T8 9 W 800LM 2ft linear 2BD frosted		69	.116		9.60	7.40		17	21.50
8090	5000K LED litespan T8 9 W 900LM 2ft linear frosted		69	.116		10.60	7.40		18	22.50
8100	3500K LED advantage T8 18 W 1600LM 4ft linear 2 BD frosted		69	.116		10	7.40		17.40	22
8105	18 W LED 4ft T8 4000K frost 1600L linear lamp		69	.116		17.85	7.40		25.25	30.50
8108	18 W LED 48 inch T8 4100K 1890LM linear lamp		69	.116		55	7.40		62.40	71.50
8110	3500K LED advantage T8 18 W 1800LM 4ft linear 2 BD frosted		69	.116		28.50	7.40		35.90	42.50
8120	4000K LED advantage T8 18 W 1600LM 4ft linear 2 BD frosted		69	.116		19.55	7.40		26.95	32.50
8130	4000K LED litespan T8 18 W 1800LM 4ft linear frosted		69	.116		27	7.40		34.40	40.50
8140	5000K LED advantage T8 18 W 1600LM 4ft linear 2BD frosted		69	.116		12.75	7.40		20.15	25
8150	5000K LED litespan T8 18 W 1800LM 4ft linear 2BD frosted		69	.116		20.50	7.40		27.90	33.50
8200	16.5 T8 3000 IF-6U U-shape fluorescent LED lamp		65	.123		20.50	7.85		28.35	34.50
8210	16.5 T8 3500 IF-6U U-shape fluorescent LED lamp		65	.123		21	7.85		28.85	35
8220	16.5 T8 4000 IF-6U U-shape fluorescent LED lamp		65	.123		21	7.85		28.85	34.50
8230	16.5 T8 5000 IF-6U U-shape fluorescent LED lamp		65	.123		20.50	7.85		28.35	34.50
8240	18 W 6 inch T8 4100K U-shape frosted fluorescent LED lamp		60	.133		11.60	8.50		20.10	25.50
8250	18 W 6 inch T8 5000K U-shape frosted fluorescent LED lamp	1 Elec	60	.133	Ea.	11.60	8.50		20.10	25.50
8260	Circular 12 W linear fluorescent LED 2700K MOD lamp		62.40	.128		291	8.15		299.15	330
8270	Circular 12 W linear fluorescent LED 3000K MOD lamp		62.40	.128		291	8.15		299.15	330
8280	Circular 12 W linear fluorescent LED 3500K MOD lamp		62.40	.128		264	8.15		272.15	300
8290	Circular 18 W linear fluorescent LED 3000K MOD lamp		62.40	.128		360	8.15		368.15	405
8300	Circular 18 W linear fluorescent LED 3500K MOD lamp		62.40	.128		395	8.15		403.15	445
8310	Circular 18 W linear fluorescent LED 5000K MOD lamp		62.40	.128		360	8.15		368.15	405
8320	Circular 18 W linear fluorescent LED 4100K MOD lamp		62.40	.128		385	8.15		393.15	430

26 71 Electrical Machines

26 71 13 – Motors Applications

26 71 13.10 Handling

		Crew	Daily Output	Labor-Hours	Unit	Material	2021 Bare Costs Labor	Equipment	Total	Total Incl O&P
0010	**HANDLING** Add to normal labor cost for restricted areas									
5000	Motors									
5100	1/2 HP, 23 pounds	1 Elec	4	2	Ea.		127		127	189
5110	3/4 HP, 28 pounds		4	2			127		127	189
5120	1 HP, 33 pounds		4	2			127		127	189
5130	1-1/2 HP, 44 pounds		3.20	2.500			159		159	237
5140	2 HP, 56 pounds		3	2.667			170		170	252
5150	3 HP, 71 pounds		2.30	3.478			222		222	330
5160	5 HP, 82 pounds		1.90	4.211			268		268	400
5170	7-1/2 HP, 124 pounds		1.50	5.333			340		340	505
5180	10 HP, 144 pounds		1.20	6.667			425		425	630
5190	15 HP, 185 pounds		1	8			510		510	755
5200	20 HP, 214 pounds	2 Elec	1.50	10.667			680		680	1,000
5210	25 HP, 266 pounds		1.40	11.429			730		730	1,075

26 71 Electrical Machines

26 71 13 – Motors Applications

26 71 13.10 Handling

		Crew	Daily Output	Labor-Hours	Unit	Material	2021 Bare Costs Labor	Equipment	Total	Total Incl O&P
5220	30 HP, 310 pounds	2 Elec	1.20	13.333	Ea.		850		850	1,250
5230	40 HP, 400 pounds		1	16			1,025		1,025	1,525
5240	50 HP, 450 pounds		.90	17.778			1,125		1,125	1,675
5250	75 HP, 680 pounds		.80	20			1,275		1,275	1,900
5260	100 HP, 870 pounds	3 Elec	1	24			1,525		1,525	2,275
5270	125 HP, 940 pounds		.80	30			1,900		1,900	2,850
5280	150 HP, 1,200 pounds		.70	34.286			2,175		2,175	3,250
5290	175 HP, 1,300 pounds		.60	40			2,550		2,550	3,775
5300	200 HP, 1,400 pounds		.50	48			3,050		3,050	4,550

26 71 13.20 Motors

		Crew	Daily Output	Labor-Hours	Unit	Material	2021 Bare Costs Labor	Equipment	Total	Total Incl O&P
0010	**MOTORS** 230/460 V, 60 HZ R263413-30									
0050	Dripproof, premium efficiency, 1.15 service factor									
0060	1,800 RPM, 1/4 HP	1 Elec	5.33	1.501	Ea.	179	95.50		274.50	340
0070	1/3 HP		5.33	1.501		145	95.50		240.50	300
0080	1/2 HP		5.33	1.501		147	95.50		242.50	305
0090	3/4 HP		5.33	1.501		227	95.50		322.50	390
0100	1 HP	1 Elec	4.50	1.778	Ea.	254	113		367	445
0150	2 HP		4.50	1.778		227	113		340	420
0200	3 HP		4.50	1.778		500	113		613	720
0250	5 HP		4.50	1.778		595	113		708	825
0300	7.5 HP		4.20	1.905		890	121		1,011	1,150
0350	10 HP		4	2		1,050	127		1,177	1,375
0400	15 HP		3.20	2.500		1,475	159		1,634	1,850
0450	20 HP	2 Elec	5.20	3.077		1,875	196		2,071	2,350
0500	25 HP		5	3.200		2,425	204		2,629	2,975
0550	30 HP		4.80	3.333		2,650	212		2,862	3,250
0600	40 HP		4	4		3,250	255		3,505	3,950
0650	50 HP		3.20	5		3,825	320		4,145	4,675
0700	60 HP		2.80	5.714		4,750	365		5,115	5,775
0750	75 HP		2.40	6.667		5,825	425		6,250	7,025
0800	100 HP	3 Elec	2.70	8.889		5,500	565		6,065	6,900
0850	125 HP		2.10	11.429		7,575	730		8,305	9,425
0900	150 HP		1.80	13.333		9,250	850		10,100	11,500
0950	200 HP		1.50	16		9,650	1,025		10,675	12,100
1000	1,200 RPM, 1 HP	1 Elec	4.50	1.778		475	113		588	690
1050	2 HP		4.50	1.778		565	113		678	790
1100	3 HP		4.50	1.778		720	113		833	960
1150	5 HP		4.50	1.778		945	113		1,058	1,200
1200	3,600 RPM, 2 HP		4.50	1.778		465	113		578	685
1250	3 HP		4.50	1.778		550	113		663	775
1300	5 HP		4.50	1.778		585	113		698	815
1350	Totally enclosed, premium efficiency 1.15 service factor									
1360	1,800 RPM, 1/4 HP	1 Elec	5.33	1.501	Ea.	201	95.50		296.50	365
1370	1/3 HP		5.33	1.501		291	95.50		386.50	460
1380	1/2 HP		5.33	1.501		370	95.50		465.50	545
1390	3/4 HP		5.33	1.501		360	95.50		455.50	535
1400	1 HP		4.50	1.778		415	113		528	625
1450	2 HP		4.50	1.778		615	113		728	845
1500	3 HP		4.50	1.778		775	113		888	1,025
1550	5 HP		4.50	1.778		800	113		913	1,050
1600	7.5 HP		4.20	1.905		960	121		1,081	1,225
1650	10 HP		4	2		1,175	127		1,302	1,500

26 71 13 – Motors Applications

26 71 13.20 Motors

		Crew	Daily Output	Labor-Hours	Unit	Material	2021 Bare Costs Labor	2021 Bare Costs Equipment	Total	Total Incl O&P
1700	15 HP	1 Elec	3.20	2.500	Ea.	1,625	159		1,784	2,025
1750	20 HP	2 Elec	5.20	3.077		2,575	196		2,771	3,125
1800	25 HP		5	3.200		2,575	204		2,779	3,125
1850	30 HP		4.80	3.333		3,375	212		3,587	4,050
1900	40 HP		4	4		4,000	255		4,255	4,775
1950	50 HP		3.20	5		4,875	320		5,195	5,825
2000	60 HP		2.80	5.714		5,100	365		5,465	6,150
2050	75 HP		2.40	6.667		6,675	425		7,100	7,975
2100	100 HP	3 Elec	2.70	8.889		7,525	565		8,090	9,125
2150	125 HP		2.10	11.429		9,250	730		9,980	11,300
2200	150 HP		1.80	13.333		10,500	850		11,350	12,800
2250	200 HP		1.50	16		14,400	1,025		15,425	17,300
2300	1,200 RPM, 1 HP	1 Elec	4.50	1.778		505	113		618	725
2350	2 HP		4.50	1.778		490	113		603	710
2400	3 HP		4.50	1.778		635	113		748	870
2450	5 HP		4.50	1.778		935	113		1,048	1,200
2500	3,600 RPM, 2 HP		4.50	1.778		495	113		608	715
2550	3 HP		4.50	1.778		500	113		613	720
2600	5 HP		4.50	1.778		650	113		763	885

26 71 13.40 Motors Explosion Proof

		Crew	Daily Output	Labor-Hours	Unit	Material	2021 Bare Costs Labor	2021 Bare Costs Equipment	Total	Total Incl O&P
0010	**MOTORS EXPLOSION PROOF**, 208-230/460 V, 60 HZ									
0020	1,800 RPM, 1/4 HP	1 Elec	5	1.600	Ea.	565	102		667	775
0030	1/3 HP		5	1.600		450	102		552	645
0040	1/2 HP		5	1.600		610	102		712	820
0050	3/4 HP		5.33	1.501		825	95.50		920.50	1,050
0060	1 HP		4.20	1.905		785	121		906	1,050
0070	2 HP		4.20	1.905		645	121		766	890
0080	3 HP		4.20	1.905		895	121		1,016	1,175
0090	5 HP		4.20	1.905		890	121		1,011	1,150
0100	7.5 HP		4	2		1,100	127		1,227	1,400
0110	10 HP		3.70	2.162		1,300	138		1,438	1,625
0120	15 HP		3.20	2.500		1,700	159		1,859	2,100
0130	20 HP	2 Elec	5	3.200		2,125	204		2,329	2,625
0140	25 HP		4.80	3.333		2,550	212		2,762	3,125
0150	30 HP		4.40	3.636		2,950	232		3,182	3,600
0160	40 HP		3.80	4.211		4,125	268		4,393	4,925
0170	50 HP		3.20	5		5,200	320		5,520	6,200
0180	60 HP		2.60	6.154		7,625	390		8,015	8,975
0190	75 HP		2	8		9,375	510		9,885	11,100
0200	100 HP	3 Elec	2.50	9.600		12,300	610		12,910	14,400
0210	125 HP		2.20	10.909		15,600	695		16,295	18,100
0220	150 HP		1.60	15		18,500	955		19,455	21,700
0230	200 HP		1.20	20		22,600	1,275		23,875	26,800
1000	1,200 RPM, 1 HP	1 Elec	4.20	1.905		835	121		956	1,100
1010	2 HP		4.20	1.905		1,025	121		1,146	1,300
1020	3 HP		4.20	1.905		1,250	121		1,371	1,550
1030	5 HP		4.20	1.905		1,800	121		1,921	2,150
2000	3,600 RPM, 1/4 HP		5	1.600		425	102		527	615
2010	1/3 HP		5	1.600		445	102		547	640
2020	1/2 HP		5	1.600		565	102		667	770
2030	3/4 HP		5	1.600		580	102		682	785
2040	1 HP		4.20	1.905		610	121		731	850

26 71 13.40 Motors Explosion Proof		Crew	Daily Output	Labor-Hours	Unit	Material	2021 Bare Costs Labor	Equipment	Total	Total Incl O&P
2050	2 HP	1 Elec	4.20	1.905	Ea.	795	121		916	1,050
2060	3 HP		4.20	1.905		865	121		986	1,125
2070	5 HP		4.20	1.905		1,150	121		1,271	1,450
2080	7.5 HP		4	2		1,300	127		1,427	1,625
2090	10 HP		3.70	2.162		1,500	138		1,638	1,850
2100	15 HP		3.20	2.500		2,025	159		2,184	2,450
2110	20 HP	2 Elec	5	3.200		2,500	204		2,704	3,050
2120	25 HP		4.80	3.333		3,075	212		3,287	3,725
2130	30 HP		4.40	3.636		3,600	232		3,832	4,300
2140	40 HP		3.80	4.211		4,575	268		4,843	5,425
2150	50 HP		3.20	5		5,825	320		6,145	6,875
2160	60 HP		2.60	6.154		7,375	390		7,765	8,700
2170	75 HP		2	8		8,950	510		9,460	10,600
2180	100 HP	3 Elec	2.50	9.600		12,500	610		13,110	14,700
2190	125 HP		2.20	10.909		15,200	695		15,895	17,800
2200	150 HP		1.60	15		18,700	955		19,655	22,000
2210	200 HP		1.20	20		23,700	1,275		24,975	27,900
2220	250 HP		1.20	20		28,400	1,275		29,675	33,200

Estimating Tips

27 20 00 Data Communications
27 30 00 Voice Communications
27 40 00 Audio-Video Communications

- When estimating material costs for special systems, it is always prudent to obtain manufacturers' quotations for equipment prices and special installation requirements that may affect the total cost.

- For cost modifications for elevated tray installation, add the percentages to labor according to the height of the installation and only to the quantities exceeding the different height levels, not to the total tray quantities. Refer to subdivision 26 01 02.20 for labor adjustment factors.

- Do not overlook the costs for equipment used in the installation. If scissor lifts and boom lifts are available in the field, contractors may use them in lieu of the proposed ladders and rolling staging.

Reference Numbers

Reference numbers are shown at the beginning of some major classifications. These numbers refer to related items in the Reference Section. The reference information may be an estimating procedure, an alternate pricing method, or technical information.

Note: Not all subdivisions listed here necessarily appear. ■

Same Data. Simplified.

Enjoy the convenience and efficiency of accessing your costs anywhere:

- **Skip the multiplier** by setting your location
- **Quickly search,** edit, favorite and share costs
- **Stay on top of price changes** with automatic updates

Discover more at rsmeans.com/online

27 01 30 – Operation and Maintenance of Voice Communications

27 01 30.51 Operation and Maintenance of Voice Equipment	Crew	Daily Output	Labor-Hours	Unit	Material	2021 Bare Costs Labor	Equipment	Total	Total Incl O&P
0010	**OPERATION AND MAINTENANCE OF VOICE EQUIPMENT**								
3400	Remove and replace (reinstall), speaker	1 Elec	6	1.333	Ea.		85		85	126

27 05 Common Work Results for Communications

27 05 05 – Selective Demolition for Communications

27 05 05.20 Electrical Demolition, Communications

		Crew	Daily Output	Labor-Hours	Unit	Material	2021 Bare Costs Labor	Equipment	Total	Total Incl O&P
0010	**ELECTRICAL DEMOLITION, COMMUNICATIONS** R024119-10									
0100	Fiber optics									
0120	Cable R260105-30	1 Elec	2400	.003	L.F.		.21		.21	.32
0160	Multi-channel rack enclosure		6	1.333	Ea.		85		85	126
0180	Patch panel		18	.444	"		28.50		28.50	42
0200	Communication cables & fittings									
0220	Voice/data outlet	1 Elec	140	.057	Ea.		3.64		3.64	5.40
0240	Telephone cable		2800	.003	L.F.		.18		.18	.27
0260	Phone jack		135	.059	Ea.		3.77		3.77	5.60
0280	Data jack		135	.059	"		3.77		3.77	5.60
0300	High performance cable, 2 pair		3000	.003	L.F.		.17		.17	.25
0320	4 pair		2100	.004			.24		.24	.36
0340	25 pair		900	.009			.57		.57	.84
0400	Terminal cabinet		5	1.600	Ea.		102		102	151
1000	Nurse call system									
1020	Station	1 Elec	24	.333	Ea.		21		21	31.50
1040	Standard call button		24	.333			21		21	31.50
1060	Corridor dome light or zone indicator		24	.333			21		21	31.50
1080	Master control station	2 Elec	2	8			510		510	755

27 05 05.30 Electrical Demolition, Sound and Video

		Crew	Daily Output	Labor-Hours	Unit	Material	2021 Bare Costs Labor	Equipment	Total	Total Incl O&P
0010	**ELECTRICAL DEMOLITION, SOUND & VIDEO** R024119-10									
0100	Cables									
0120	TV antenna lead-in cable R260105-30	1 Elec	2100	.004	L.F.		.24		.24	.36
0140	Sound cable		2400	.003			.21		.21	.32
0160	Microphone cable		2400	.003			.21		.21	.32
0180	Coaxial cable		2400	.003			.21		.21	.32
0200	Doorbell system, not including wires, cables, and conduit		16	.500	Ea.		32		32	47.50
0220	Door chime or devices		36	.222	"		14.15		14.15	21
0300	Public address system, not including wires, cables, and conduit									
0320	Conventional office	1 Elec	16	.500	Speaker		32		32	47.50
0340	Conventional industrial	"	8	1	"		63.50		63.50	94.50
0352	PA cabinet/panel	2 Elec	5	3.200	Ea.		204		204	305
0400	Sound system, not including wires, cables, and conduit									
0410	Components	1 Elec	24	.333	Ea.		21		21	31.50
0412	Speaker		24	.333			21		21	31.50
0416	Volume control		24	.333			21		21	31.50
0420	Intercom, master station		6	1.333			85		85	126
0440	Remote station		24	.333			21		21	31.50
0460	Intercom outlets		24	.333			21		21	31.50
0480	Handset		12	.667			42.50		42.50	63
0500	Emergency call system, not including wires, cables, and conduit									
0520	Annunciator	1 Elec	4	2	Ea.		127		127	189
0540	Devices	"	16	.500			32		32	47.50
0600	Master door, buzzer type unit	2 Elec	1.60	10			635		635	945
0800	TV system, not including wires, cables, and conduit									

27 05 Common Work Results for Communications

27 05 05 – Selective Demolition for Communications

27 05 05.30 Electrical Demolition, Sound and Video	Crew	Daily Output	Labor-Hours	Unit	Material	2021 Bare Costs Labor	Equipment	Total	Total Incl O&P
0820 Master TV antenna system, per outlet	1 Elec	39	.205	Outlet		13.05		13.05	19.40
0840 School application, per outlet		16	.500	"		32		32	47.50
0860 Amplifier		12	.667	Ea.		42.50		42.50	63
0880 Antenna	↓	6	1.333	"		85		85	126
0900 One camera & one monitor	2 Elec	7.80	2.051	Total		131		131	194
0920 One camera	1 Elec	8	1	Ea.		63.50		63.50	94.50

27 05 29 – Hangers and Supports for Communications Systems

27 05 29.10 Cable Support

	Crew	Daily Output	Labor-Hours	Unit	Material	2021 Bare Costs Labor	Equipment	Total	Total Incl O&P
0010 **CABLE SUPPORT**									
0110 J-hook, single tier, single sided, 1" diam.	1 Elec	68.57	.117	Ea.	2.37	7.45		9.82	13.65
0120 1-1/2" diam.		67.80	.118		2.96	7.50		10.46	14.40
0130 2" diam.		67.30	.119		3.49	7.55		11.04	15.10
0140 4" diam.		66.80	.120		6.90	7.65		14.55	18.95
0330 Double tier, single sided, 2" diam.		67	.119		14.75	7.60		22.35	27.50
0340 4" diam.		66.20	.121		18.40	7.70		26.10	31.50
0430 Double sided, 2" diam.		65.80	.122		23.50	7.75		31.25	37
0440 4" diam.		65.40	.122		32.50	7.80		40.30	47
0530 Triple tier, single sided, 2" diam.		66.40	.120		19.45	7.65		27.10	33
0540 4" diam.		65.85	.121		25	7.75		32.75	39
0630 Double sided, 2" diam.		65.10	.123		30	7.85		37.85	45
0640 4" diam.	↓	64.70	.124	↓	39.50	7.90		47.40	55

27 11 Communications Equipment Room Fittings

27 11 16 – Communications Cabinets, Racks, Frames and Enclosures

27 11 16.10 Public Phone

	Crew	Daily Output	Labor-Hours	Unit	Material	2021 Bare Costs Labor	Equipment	Total	Total Incl O&P
0010 **PUBLIC PHONE**									
7600 Telephone with wood backboard									
7620 Single door, 12" H x 12" W x 4" D	1 Elec	5.30	1.509	Ea.	125	96		221	281
7650 18" H x 12" W x 4" D		4.70	1.702		146	108		254	320
7700 24" H x 12" W x 4" D		4.20	1.905		161	121		282	355
7720 18" H x 18" W x 4" D		4.20	1.905		171	121		292	370
7750 24" H x 18" W x 4" D		4	2		219	127		346	430
7780 36" H x 36" W x 4" D		3.60	2.222		279	142		421	515
7800 24" H x 24" W x 6" D		3.60	2.222		320	142		462	560
7820 30" H x 24" W x 6" D		3.20	2.500		296	159		455	560
7850 30" H x 30" W x 6" D		2.70	2.963		450	189		639	775
7880 36" H x 30" W x 6" D		2.50	3.200		480	204		684	830
7900 48" H x 36" W x 6" D		2.20	3.636		735	232		967	1,150
7920 Double door, 48" H x 36" W x 6" D	↓	2	4	↓	990	255		1,245	1,475

27 11 16.20 Rack Mount Cabinet

	Crew	Daily Output	Labor-Hours	Unit	Material	2021 Bare Costs Labor	Equipment	Total	Total Incl O&P
0010 **RACK MOUNT CABINET**									
0100 80" H x 24" W x 40" D									
0110 No sides	1 Elec	9.60	.833	Ea.	1,425	53		1,478	1,650
0120 One side		8	1		1,500	63.50		1,563.50	1,750
0130 Two sides	↓	7.86	1.018	↓	1,650	65		1,715	1,900
0200 80" H x 24" W x 42" D									
0210 No sides	1 Elec	9.60	.833	Ea.	1,450	53		1,503	1,675
0220 One side		8	1		1,500	63.50		1,563.50	1,750
0230 Two sides	↓	7.86	1.018	↓	1,625	65		1,690	1,875
0300 80" H x 24" W x 48" D									

For customer support on your Electrical Costs with RSMeans Data, call 800.448.8182.

327

27 11 16 – Communications Cabinets, Racks, Frames and Enclosures

27 11 16.20 Rack Mount Cabinet		Crew	Daily Output	Labor-Hours	Unit	Material	2021 Bare Costs Labor	Equipment	Total	Total Incl O&P
0310	No sides	1 Elec	9.60	.833	Ea.	1,675	53		1,728	1,900
0320	One side		8	1		1,500	63.50		1,563.50	1,750
0330	Two sides	↓	7.86	1.018	↓	1,625	65		1,690	1,900
0400	80" H x 30" W x 40" D									
0410	No sides	1 Elec	9.60	.833	Ea.	1,825	53		1,878	2,075
0420	One side		8	1		1,900	63.50		1,963.50	2,200
0430	Two sides	↓	7.86	1.018	↓	1,975	65		2,040	2,275
0500	80" H x 30" W x 42" D									
0510	No sides	1 Elec	9.60	.833	Ea.	1,900	53		1,953	2,175
0520	One side		8	1		2,025	63.50		2,088.50	2,325
0530	Two sides	↓	7.86	1.018	↓	1,975	65		2,040	2,275
0600	80" H x 30" W x 48" D									
0610	No sides	1 Elec	9.60	.833	Ea.	2,050	53		2,103	2,350
0620	One side		8	1		2,175	63.50		2,238.50	2,500
0630	Two sides	↓	7.86	1.018	↓	2,325	65		2,390	2,650
0700	80" H x 32" W x 32" D									
0710	No sides	1 Elec	9.60	.833	Ea.	1,400	53		1,453	1,600
0720	One side		8	1		1,425	63.50		1,488.50	1,675
0730	Two sides	↓	7.86	1.018	↓	1,450	65		1,515	1,700
0800	80" H x 32" W x 40" D									
0810	No sides	1 Elec	9.60	.833	Ea.	1,575	53		1,628	1,800
0820	One side		8	1		1,625	63.50		1,688.50	1,900
0830	Two sides	↓	7.86	1.018	↓	1,700	65		1,765	1,975
0900	80" H x 32" W x 42" D									
0910	No sides	1 Elec	9.60	.833	Ea.	1,525	53		1,578	1,750
0920	One side		8	1		1,550	63.50		1,613.50	1,800
0930	Two sides	↓	7.86	1.018	↓	1,575	65		1,640	1,850
1000	80" H x 32" W x 48" D									
1010	No sides	1 Elec	9.60	.833	Ea.	1,775	53		1,828	2,025
1020	One side		8	1		1,875	63.50		1,938.50	2,150
1030	Two sides	↓	7.86	1.018	↓	2,375	65		2,440	2,700
1200	Freestanding cabinet with front and rear doors									
1300	84" H x 19" W x 24" D, gray									
1310	No sides	1 Elec	9.60	.833	Ea.	1,225	53		1,278	1,425
1320	One side		9.60	.833		2,575	53		2,628	2,900
1330	Two sides	↓	9.60	.833	↓	2,575	53		2,628	2,900
1400	84" H x 19" W x 24" D, white									
1410	No sides	1 Elec	9.60	.833	Ea.	2,175	53		2,228	2,450
1420	One side		9.60	.833		2,575	53		2,628	2,900
1430	Two sides	↓	9.60	.833	↓	2,575	53		2,628	2,900
1500	84" H x 19" W x 24" D, black									
1510	No sides	1 Elec	9.60	.833	Ea.	2,475	53		2,528	2,800
1520	One side		9.60	.833		2,450	53		2,503	2,775
1530	Two sides		9.60	.833		1,775	53		1,828	2,025
1600	84" H x 28" W x 34" D front glass, rear louvered door	↓	9.60	.833	↓	1,575	53		1,628	1,800

27 11 19 – Communications Termination Blocks and Patch Panels

27 11 19.10 Termination Blocks and Patch Panels

		Crew	Daily Output	Labor-Hours	Unit	Material	2021 Bare Costs Labor	Equipment	Total	Total Incl O&P
0010	**TERMINATION BLOCKS AND PATCH PANELS**									
2960	Patch panel, RJ-45/110 type, 24 ports	2 Elec	6	2.667	Ea.	151	170		321	420
3000	48 ports	3 Elec	6	4		297	255		552	705
3040	96 ports	"	4	6		515	380		895	1,125
3100	Punch down termination per port	1 Elec	107	.075	↓		4.76		4.76	7.10

27 13 23 – Communications Optical Fiber Backbone Cabling

27 13 23.13 Communications Optical Fiber	Crew	Daily Output	Labor-Hours	Unit	Material	2021 Bare Costs Labor	Equipment	Total	Total Incl O&P
0010 **COMMUNICATIONS OPTICAL FIBER**									
0040 Specialized tools & techniques cause installation costs to vary.									
0070 Fiber optic, cable, bulk simplex, single mode	R271323-40	1 Elec	8	1	C.L.F.	22.50	63.50		86	120
0080 Multi mode		8	1		29.50	63.50		93	127
0090 4 strand, single mode		7.34	1.090		38	69.50		107.50	145
0095 Multi mode		7.34	1.090		50.50	69.50		120	159
0100 12 strand, single mode		6.67	1.199		72	76.50		148.50	193
0105 Multi mode		6.67	1.199		96.50	76.50		173	220
0150 Jumper				Ea.	31.50			31.50	34.50
0200 Pigtail					38			38	41.50
0300 Connector	1 Elec	24	.333		25	21		46	59.50
0350 Finger splice		32	.250		37.50	15.95		53.45	64.50
0400 Transceiver (low cost bi-directional)		8	1		440	63.50		503.50	580
0450 Rack housing, 4 rack spaces, 12 panels (144 fibers)		2	4		1,075	255		1,330	1,550
0500 Patch panel, 12 ports		6	1.333		315	85		400	470
0600 Cable connector panel, 6 fiber		40	.200		55.50	12.75		68.25	80
1000 Cable, 62.5 microns, direct burial, 4 fiber	R-15	1200	.040	L.F.	.77	2.50	.23	3.50	4.82
1020 Indoor, 2 fiber	R-19	1000	.020		.49	1.28		1.77	2.44
1040 Outdoor, aerial/duct	"	1670	.012		.64	.76		1.40	1.84
1060 50 microns, direct burial, 8 fiber	R-22	4000	.009		1.28	.54		1.82	2.22
1080 12 fiber		4000	.009		1.52	.54		2.06	2.48
1100 Indoor, 12 fiber		759	.049		1.38	2.86		4.24	5.75
1120 Connectors, 62.5 micron cable, transmission	R-19	40	.500	Ea.	14.90	32		46.90	64
1140 Cable splice		40	.500		17.35	32		49.35	66.50
1160 125 micron cable, transmission		16	1.250		15.65	80		95.65	136
1180 Receiver, 1.2 mile range		20	1		340	64		404	470
1200 1.9 mile range		20	1		226	64		290	345
1220 6.2 mile range		5	4		375	255		630	795
1240 Transmitter, 1.2 mile range		20	1		375	64		439	510
1260 1.9 mile range		20	1		284	64		348	405
1280 6.2 mile range		5	4		1,175	255		1,430	1,675
1300 Modem, 1.2 mile range		5	4		200	255		455	600
1320 6.2 mile range		5	4		375	255		630	795
1340 1.9 mile range, 12 channel		5	4		2,100	255		2,355	2,700
1360 Repeater, 1.2 mile range		10	2		340	128		468	565
1380 1.9 mile range		10	2		500	128		628	740
1400 6.2 mile range		5	4		905	255		1,160	1,375
1420 1.2 mile range, digital		5	4		460	255		715	885
1440 Transceiver, 1.9 mile range		5	4		430	255		685	850
1460 1.2 mile range, digital		5	4		415	255		670	840
1480 Cable enclosure, interior NEMA 13		7	2.857		155	182		337	440
1500 Splice w/enclosure encapsulant		16	1.250		230	80		310	370
1510 1/4", black, non-metallic, flexible, tube, liquid tight		8	2.500	C.L.F.	221	160		381	480
1520 3/8"		8	2.500		130	160		290	380
1530 1/2"		7.34	2.725		129	174		303	400
1540 3/4"		7.34	2.725		234	174		408	515
1550 1"		7.34	2.725		365	174		539	660
1560 1-1/4"		6.67	2.999		600	191		791	945
1570 1-1/2"		6.67	2.999		695	191		886	1,050
1580 2"		6.35	3.150		1,100	201		1,301	1,525
2000 Fiber optic cable, 48 strand, single mode, indoor/outdoor		1.67	11.976	M.L.F.	3,450	765		4,215	4,900
2002 72 strand		1.50	13.307		5,525	850		6,375	7,325
2004 96 strand		1.42	14.085		7,375	900		8,275	9,425

For customer support on your Electrical Costs with RSMeans Data, call 800.448.8182.

329

27 13 Communications Backbone Cabling

27 13 23 – Communications Optical Fiber Backbone Cabling

27 13 23.13 Communications Optical Fiber

		Crew	Daily Output	Labor-Hours	Unit	Material	2021 Bare Costs Labor	Equipment	Total	Total Incl O&P
2006	144 strand	R-19	1.34	14.970	M.L.F.	11,100	955		12,055	13,600
2010	Fiber optic cable, 48 strand, Single mode, Indoor, Plenum		1.67	11.976		4,325	765		5,090	5,900
2012	72 strand		1.50	13.307		6,725	850		7,575	8,650
2014	96 strand		1.42	14.085		9,225	900		10,125	11,500
2020	Fiber optic cable, 48 strand, Single mode, Indoor, Riser		1.67	11.976		2,750	765		3,515	4,150
2030	Armored		1.67	11.976		5,400	765		6,165	7,075
2040	Fiber optic cable, 48 strand, single mode, armored, Plenum		1.67	11.976		5,050	765		5,815	6,675
2042	72 strand		1.50	13.307		7,050	850		7,900	9,000
2044	96 strand		1.42	14.085		12,500	900		13,400	15,000
2046	144 strand		1.50	13.307		18,800	850		19,650	22,000
2060	Fiber optic cable 48 strand Single mode Armored Gel Filled Dry Block Burial		1.35	14.815		805	945		1,750	2,275
2062	72 strand		1.22	16.461		2,250	1,050		3,300	4,050
2064	96 strand		1.15	17.422		2,550	1,100		3,650	4,475
2066	144 strand		1.08	18.519		3,375	1,175		4,550	5,450
2070	Fiber optic cable 48 strand single mode gel filled dry block outdoor duct		1.25	16		1,150	1,025		2,175	2,800
2090	Armored		1	20		700	1,275		1,975	2,675
2100	Aerial, Self-Supported		.95	21.053		920	1,350		2,270	3,000

27 15 Communications Horizontal Cabling

27 15 01 – Communications Horizontal Cabling Applications

27 15 01.19 Fire Alarm Communications Conductors & Cables

		Crew	Daily Output	Labor-Hours	Unit	Material	2021 Bare Costs Labor	Equipment	Total	Total Incl O&P
0010	**FIRE ALARM COMMUNICATIONS CONDUCTORS AND CABLES**									
1500	Fire alarm FEP teflon 150 V to 200°C									
1550	#22, 1 pair	1 Elec	10	.800	C.L.F.	96.50	51		147.50	182
1600	2 pair		8	1		97.50	63.50		161	202
1650	4 pair		7	1.143		515	73		588	675
1700	6 pair		6	1.333		510	85		595	690
1750	8 pair		5.50	1.455		875	92.50		967.50	1,100
1800	10 pair		5	1.600		805	102		907	1,025
1850	#18, 1 pair	2 Elec	16	1		63	63.50		126.50	164
1900	2 pair		13	1.231		267	78.50		345.50	410
1950	4 pair		9.60	1.667		210	106		316	390
2000	6 pair		8	2		450	127		577	685
2050	8 pair		7	2.286		330	146		476	575
2100	10 pair		6	2.667		490	170		660	790

27 15 01.23 Audio-Video Communications Horizontal Cabling

		Crew	Daily Output	Labor-Hours	Unit	Material	2021 Bare Costs Labor	Equipment	Total	Total Incl O&P
0010	**AUDIO-VIDEO COMMUNICATIONS HORIZONTAL CABLING**									
3000	S-video cable, MD4, 1M Pro interconnects, 3'	1 Elec	38	.211	Ea.	21	13.40		34.40	43
3010	6'		38	.211		29.50	13.40		42.90	52.50
3020	9'		38	.211		30	13.40		43.40	53
3030	16'		36	.222		50	14.15		64.15	76
3040	22'		36	.222		58.50	14.15		72.65	85.50
3300	HDMI cable, coaxial, HDMI/HDMI, Non plenum, 3', 24 kt terminations		40	.200		13.55	12.75		26.30	34
3310	6', 24 kt terminations		40	.200		15.80	12.75		28.55	36.50
3320	HDMI cable, coaxial, HDMI/HDMI metal, Non plenum, 3', 24 kt terminations		40	.200		16.40	12.75		29.15	37
3330	6', 24 kt terminations		40	.200		19.05	12.75		31.80	40
3340	15', 24 kt terminations		38	.211		26.50	13.40		39.90	49
3350	25', 24 kt terminations		38	.211		34.50	13.40		47.90	58

27 15 10 – Special Communications Cabling

27 15 10.23 Sound and Video Cables and Fittings

	Crew	Daily Output	Labor-Hours	Unit	Material	2021 Bare Costs Labor	2021 Bare Costs Equipment	Total	Total Incl O&P
0010 **SOUND AND VIDEO CABLES & FITTINGS**									
0900 TV antenna lead-in, 300 ohm, #20-2 conductor	1 Elec	7	1.143	C.L.F.	23.50	73		96.50	134
0950 Coaxial, feeder outlet		7	1.143		20.50	73		93.50	131
1000 Coaxial, main riser		6	1.333		10.30	85		95.30	137
1100 Sound, shielded with drain, #22-2 conductor		8	1		9.80	63.50		73.30	105
1150 #22-3 conductor		7.50	1.067		12.15	68		80.15	114
1200 #22-4 conductor		6.50	1.231		16.50	78.50		95	134
1250 Nonshielded, #22-2 conductor		10	.800		14.25	51		65.25	91
1300 #22-3 conductor		9	.889		20.50	56.50		77	107
1350 #22-4 conductor		8	1		23.50	63.50		87	121
1400 Microphone cable		8	1		28.50	63.50		92	126

27 15 13 – Communications Copper Horizontal Cabling

27 15 13.13 Communication Cables and Fittings

	Crew	Daily Output	Labor-Hours	Unit	Material	2021 Bare Costs Labor	2021 Bare Costs Equipment	Total	Total Incl O&P
0010 **COMMUNICATION CABLES AND FITTINGS**									
2200 Telephone twisted, PVC insulation, #22-2 conductor	1 Elec	10	.800	C.L.F.	10.40	51		61.40	87
2250 #22-3 conductor		9	.889		13.10	56.50		69.60	98.50
2300 #22-4 conductor		8	1		14.85	63.50		78.35	111
2350 #18-2 conductor		9	.889		20.50	56.50		77	107
2370 Telephone jack, eight pins		32	.250	Ea.	3.54	15.95		19.49	27.50
5000 High performance unshielded twisted pair (UTP)									
5100 Cable, category 3, #24, 2 pair solid, PVC jacket R271513-75	1 Elec	10	.800	C.L.F.	15.90	51		66.90	93
5200 4 pair solid, PVC jacket		7	1.143		24.50	73		97.50	135
5300 25 pair solid, PVC jacket		3	2.667		143	170		313	410
5400 2 pair solid, plenum		10	.800		10.60	51		61.60	87
5500 4 pair solid, plenum		7	1.143		14.15	73		87.15	124
5600 25 pair solid, plenum		3	2.667		101	170		271	365
5700 4 pair stranded, PVC jacket		7	1.143		14.40	73		87.40	124
7000 Category 5, #24, 4 pair solid, PVC jacket		7	1.143		57	73		130	171
7100 4 pair solid, plenum		7	1.143		17.05	73		90.05	127
7200 4 pair stranded, PVC jacket		7	1.143		27.50	73		100.50	138
7210 Category 5e, #24, 4 pair solid, PVC jacket		7	1.143		25	73		98	136
7212 4 pair solid, plenum		7	1.143		45.50	73		118.50	158
7214 4 pair stranded, PVC jacket		7	1.143		19.30	73		92.30	130
7240 Category 6, #24, 4 pair solid, PVC jacket		7	1.143		52.50	73		125.50	166
7242 4 pair solid, plenum		7	1.143		12.05	73		85.05	121
7244 4 pair stranded, PVC jacket		7	1.143		19.90	73		92.90	130
7300 Connector, RJ45, category 5		80	.100	Ea.	.75	6.35		7.10	10.30
7302 Shielded RJ45, category 5		72	.111		1.59	7.10		8.69	12.25
7310 Jack, UTP RJ45, category 3		72	.111		.54	7.10		7.64	11.10
7312 Category 5		65	.123		6.45	7.85		14.30	18.75
7314 Category 5e		65	.123		5.15	7.85		13	17.30
7316 Category 6		65	.123		3.80	7.85		11.65	15.85
7322 Jack, shielded RJ45, category 5		60	.133		5.65	8.50		14.15	18.85
7324 Category 5e		60	.133		5.65	8.50		14.15	18.85
7326 Category 6		60	.133		6.70	8.50		15.20	20
7400 Voice/data expansion module, category 5e		8	1		31	63.50		94.50	129
7401 Modular jack, cat 6 keystone, RJ45, office white		16.50	.485		6.20	31		37.20	53
7402 RJ45, green		16.50	.485		6.40	31		37.40	53
7404 RJ45, grey		16.50	.485		7.45	31		38.45	54
7408 RJ45, orange		16.50	.485		6.50	31		37.50	53
7420 5M, RJ45, 568B, cord set		16.50	.485		46.50	31		77.50	97.50
7422 3M, RJ45, 568B, cord set		16.50	.485		31.50	31		62.50	81

For customer support on your Electrical Costs with RSMeans Data, call 800.448.8182.

331

27 15 13.13 Communication Cables and Fittings		Crew	Daily Output	Labor-Hours	Unit	Material	2021 Bare Costs Labor	Equipment	Total	Total Incl O&P
7424	1M, RJ45, 568B, cord set	1 Elec	16.50	.485	Ea.	29.50	31		60.50	78.50
7450	M12 TO RJ45, 2M ultra-loch D-mode, cord set	↓	15.50	.516	↓	45	33		78	98.50
8000	Multipair unshielded non-plenum cable, 150 V PVC jacket									
8002	#22, 2 pair	1 Elec	8.40	.952	C.L.F.	87.50	60.50		148	186
8003	3 pair		8	1		126	63.50		189.50	234
8004	4 pair		7.30	1.096		47	70		117	156
8006	6 pair		6.20	1.290		234	82		316	380
8008	8 pair		5.70	1.404		500	89.50		589.50	685
8010	10 pair		5.30	1.509		63	96		159	213
8012	12 pair		4.20	1.905		134	121		255	330
8015	15 pair	↓	3.80	2.105		580	134		714	835
8020	20 pair	2 Elec	6.60	2.424		1,125	154		1,279	1,450
8025	25 pair		6	2.667		1,375	170		1,545	1,750
8030	30 pair		5.60	2.857		490	182		672	810
8040	40 pair		5.20	3.077		420	196		616	755
8050	50 pair	↓	4.90	3.265	↓	600	208		808	970
8100	Multipair unshielded non-plenum cable, 300 V PVC jacket									
8102	#20, 2 pair	1 Elec	7.30	1.096	C.L.F.	228	70		298	355
8103	3 pair		6.70	1.194		296	76		372	440
8104	4 pair		6	1.333		365	85		450	530
8106	6 pair		5.30	1.509		510	96		606	705
8108	8 pair	↓	4.40	1.818		770	116		886	1,025
8110	10 pair	2 Elec	7.30	2.192		99.50	140		239.50	315
8112	12 pair		6.20	2.581		124	164		288	380
8115	15 pair	↓	5.90	2.712		155	173		328	430
8201	#18, 1 pair	1 Elec	8	1		92	63.50		155.50	196
8202	2 pair		6.50	1.231		98	78.50		176.50	224
8203	3 pair		5.60	1.429		147	91		238	296
8204	4 pair		5	1.600		167	102		269	335
8206	6 pair		4.40	1.818		390	116		506	600
8208	8 pair	↓	4	2		245	127		372	460
8215	15 pair	2 Elec	5	3.200	↓	650	204		854	1,025
8300	Multipair shielded non-plenum cable, 300 V PVC jacket									
8303	#22, 3 pair	1 Elec	6.70	1.194	C.L.F.	164	76		240	293
8306	6 pair		5.70	1.404		385	89.50		474.50	560
8309	9 pair	↓	5	1.600		247	102		349	425
8312	12 pair	2 Elec	8	2		500	127		627	740
8315	15 pair		7.30	2.192		550	140		690	810
8317	17 pair		6.80	2.353		915	150		1,065	1,225
8319	19 pair		6.40	2.500		730	159		889	1,050
8327	27 pair	↓	5.90	2.712		1,325	173		1,498	1,725
8402	#20, 2 pair	1 Elec	6.70	1.194		95	76		171	217
8403	3 pair		6.20	1.290		121	82		203	256
8406	6 pair		4.40	1.818		305	116		421	505
8409	9 pair	↓	3.60	2.222		278	142		420	515
8412	12 pair	2 Elec	5.90	2.712		615	173		788	930
8415	15 pair	"	5.70	2.807		555	179		734	875
8502	#18, 2 pair	1 Elec	6.20	1.290		116	82		198	250
8503	3 pair		5.30	1.509		233	96		329	400
8504	4 pair		4.70	1.702		186	108		294	365
8506	6 pair		4	2		505	127		632	745
8509	9 pair	↓	3.40	2.353		585	150		735	870
8515	15 pair	2 Elec	4.80	3.333	↓	1,375	212		1,587	1,850

27 15 Communications Horizontal Cabling

27 15 33 – Communications Coaxial Horizontal Cabling

27 15 33.10 Coaxial Cable and Fittings

27 15 33.10 Coaxial Cable and Fittings	Crew	Daily Output	Labor-Hours	Unit	Material	2021 Bare Costs Labor	Equipment	Total	Total Incl O&P
0010 **COAXIAL CABLE & FITTINGS**									
3500 Coaxial connectors, 50 ohm impedance quick disconnect									
3540 BNC plug, for RG A/U #58 cable	1 Elec	42	.190	Ea.	2.88	12.15		15.03	21
3550 RG A/U #59 cable		42	.190		2.78	12.15		14.93	21
3560 RG A/U #62 cable		42	.190		2.82	12.15		14.97	21
3600 BNC jack, for RG A/U #58 cable		42	.190		2.94	12.15		15.09	21.50
3610 RG A/U #59 cable		42	.190		3.53	12.15		15.68	22
3620 RG A/U #62 cable		42	.190		3.43	12.15		15.58	22
3660 BNC panel jack, for RG A/U #58 cable		40	.200		3.04	12.75		15.79	22.50
3670 RG A/U #59 cable		40	.200		6.05	12.75		18.80	25.50
3680 RG A/U #62 cable		40	.200		6.05	12.75		18.80	25.50
3720 BNC bulkhead jack, for RG A/U #58 cable		40	.200		6.05	12.75		18.80	25.50
3730 RG A/U #59 cable		40	.200		6.05	12.75		18.80	25.50
3740 RG A/U #62 cable		40	.200		6.05	12.75		18.80	25.50
3850 Coaxial cable, RG A/U #58, 50 ohm		8	1	C.L.F.	49	63.50		112.50	149
3860 RG A/U #59, 75 ohm		8	1		39.50	63.50		103	138
3870 RG A/U #62, 93 ohm		8	1		45.50	63.50		109	145
3875 RG 6/U, 75 ohm		8	1		34.50	63.50		98	132
3950 Fire rated, RG A/U #58, 50 ohm		8	1		94.50	63.50		158	199
3960 RG A/U #59, 75 ohm		8	1		124	63.50		187.50	231
3970 RG A/U #62, 93 ohm		8	1		109	63.50		172.50	215

27 15 43 – Communications Faceplates and Connectors

27 15 43.13 Communication Outlets

27 15 43.13 Communication Outlets	Crew	Daily Output	Labor-Hours	Unit	Material	2021 Bare Costs Labor	Equipment	Total	Total Incl O&P
0010 **COMMUNICATION OUTLETS**									
0100 Voice/data devices not included									
0120 Voice/data outlets, single opening	1 Elec	48	.167	Ea.	6.90	10.60		17.50	23.50
0140 Two jack openings		48	.167		2.56	10.60		13.16	18.60
0160 One jack & one 3/4" round opening		48	.167		7.45	10.60		18.05	24
0180 One jack & one twinaxial opening		48	.167		7.45	10.60		18.05	24
0200 One jack & one connector cabling opening		48	.167		7.45	10.60		18.05	24
0220 Two 3/8" coaxial openings	1 Elec	48	.167	Ea.	7.45	10.60		18.05	24
0300 Data outlets, single opening		48	.167		7.45	10.60		18.05	24
0320 One 25-pin subminiature opening		48	.167		7.45	10.60		18.05	24
1000 Voice/data wall plate, plastic, 1 gang, 1-port		72	.111		2.17	7.10		9.27	12.90
1020 2-port		72	.111		2.28	7.10		9.38	13
1040 3-port		72	.111		2.28	7.10		9.38	13
1060 4-port		72	.111		2.71	7.10		9.81	13.50
1080 6-port		72	.111		2.16	7.10		9.26	12.90
1100 2 gang, 6-port		48	.167		5.80	10.60		16.40	22
1120 Voice/data wall plate, stainless steel, 1 gang, 1-port		72	.111		8.45	7.10		15.55	19.80
1140 2-port		72	.111		8.40	7.10		15.50	19.75
1160 3-port		72	.111		8.45	7.10		15.55	19.80
1180 4-port		72	.111		8.35	7.10		15.45	19.70
1200 2 gang, 6-port		48	.167		13.75	10.60		24.35	31

27 21 Data Communications Network Equipment

27 21 29 – Data Communications Switches and Hubs

27 21 29.10 Switching and Routing Equipment	Crew	Daily Output	Labor-Hours	Unit	Material	2021 Bare Costs Labor	Equipment	Total	Total Incl O&P
0010 SWITCHING AND ROUTING EQUIPMENT									
1100 Network hub, dual speed, 24 ports, includes cabinet	3 Elec	.66	36.364	Ea.	1,000	2,325		3,325	4,550
1300 Network switch, 50/60 HZ, 8 port, multi-platform, analog KVM	1 Elec	6.85	1.168		1,225	74.50		1,299.50	1,450
1310 16 port, multi-platform, analog KVM		6	1.333		1,400	85		1,485	1,675
1320 Network switch, 0x2x16,CAT5, analog KVM		6	1.333		1,225	85		1,310	1,475
1330 2x1x16, digital KVM, w/VM		16	.500		3,875	32		3,907	4,325
1340 2x1x32, digital KVM, w/VM		16	.500		4,625	32		4,657	5,125
1350 8x1x32, digital KVM, w/VM		16	.500		4,925	32		4,957	5,475
1500 KVM,1 RMU 16 port, no keyboard		16	.500		1,675	32		1,707	1,875
1510 KVM,1 RMU, 17" LCD, 16 port, US keyboard		15.15	.528		3,200	33.50		3,233.50	3,575
1520 KVM,1 RMU, 17" LCD, 16 port, UK keyboard		15.15	.528		3,200	33.50		3,233.50	3,575
2000 10/100/1000 Mbps, 24 ports		14	.571		1,925	36.50		1,961.50	2,150
2040 10/100/1000 Mbps, 48 ports		12	.667		600	42.50		642.50	725
2050 10/100/1000 Mbps, 5 port, industrial ethernet type		32	.250		281	15.95		296.95	335
2060 10/100/1000 Mbps, 6 port, industrial ethernet type		32	.250		1,400	15.95		1,415.95	1,550
2070 10/100/1000 Mbps, 10 port, X307-3		28	.286		2,700	18.20		2,718.20	3,000
2080 10/100/1000 Mbps, 10 port, X308-2		28	.286		3,025	18.20		3,043.20	3,375
3000 10/100/1000 Mbps, 20 port, front ports		20	.400		1,900	25.50		1,925.50	2,150
3010 10/100/1000 Mbps, 20 port, rear ports		20	.400		1,850	25.50		1,875.50	2,100
3040 KVM, 10/100/1000/10000 Mbps, 28 port, rear ports		16	.500		8,975	32		9,007	9,925
3050 KVM, 10/100/1000/10000 Mbps, 28 port, front ports	▼	16	.500		10,500	32		10,532	11,600
3070 KVM, 10/100/1000/10000 Mbps, 52 port, front ports	2 Elec	12	1.333	▼	11,100	85		11,185	12,300
3080 KVM, 10/100/1000/10000 Mbps, 52 port, rear ports	2 Elec	12	1.333	Ea.	11,100	85		11,185	12,300

27 32 Voice Communications Terminal Equipment

27 32 26 – Ring-Down Emergency Telephones

27 32 26.10 Emergency Phone Stations

	Crew	Daily Output	Labor-Hours	Unit	Material	2021 Bare Costs Labor	Equipment	Total	Total Incl O&P
0010 EMERGENCY PHONE STATIONS									
0100 Hands free emergency speaker phone w/LED strobe	1 Elec	7.22	1.108	Ea.	860	70.50		930.50	1,050
0110 Wall/pole mounted emergency phone station w/LED strobe	"	4.60	1.739		1,525	111		1,636	1,850
2000 Call station, 30", pole mounted emergency phone, 120V/ VoIP, blue beacon	2 Elec	5	3.200		1,100	204		1,304	1,500
2010 Call station, 36"		5	3.200		1,575	204		1,779	2,025
2020 Call station, 5'-8"		4.80	3.333		2,650	212		2,862	3,250
2030 Call station, 9'	▼	4.80	3.333	▼	2,675	212		2,887	3,275

27 32 36 – TTY Equipment

27 32 36.10 TTY Telephone Equipment

	Crew	Daily Output	Labor-Hours	Unit	Material	2021 Bare Costs Labor	Equipment	Total	Total Incl O&P
0010 TTY TELEPHONE EQUIPMENT									
1620 Telephone, TTY, compact, pocket type				Ea.	320			320	355
1630 Advanced, desk type	2 Elec	20	.800		500	51		551	625
1640 Full-featured public, wall type	"	4	4	▼	750	255		1,005	1,200

27 41 Audio-Video Systems

27 41 33 – Master Antenna Television Systems

27 41 33.10 TV Systems

		Crew	Daily Output	Labor-Hours	Unit	Material	2021 Bare Costs Labor	Equipment	Total	Total Incl O&P
0010	**TV SYSTEMS,** not including rough-in wires, cables & conduits									
0100	Master TV antenna system									
0200	VHF reception & distribution, 12 outlets	1 Elec	6	1.333	Outlet	136	85		221	275
0400	30 outlets		10	.800		260	51		311	360
0600	100 outlets		13	.615		300	39		339	390
0800	VHF & UHF reception & distribution, 12 outlets		6	1.333		248	85		333	400
1000	30 outlets		10	.800		168	51		219	261
1200	100 outlets		13	.615		173	39		212	249
1400	School and deluxe systems, 12 outlets		2.40	3.333		325	212		537	675
1600	30 outlets		4	2		287	127		414	505
1800	80 outlets		5.30	1.509	↓	276	96		372	450
1900	Amplifier		4	2	Ea.	820	127		947	1,100
5000	Antenna, small		6	1.333		55	85		140	187
5100	Large		4	2		230	127		357	440
5110	Rotor unit		8	1		57.50	63.50		121	158
5120	Single booster		8	1		28	63.50		91.50	125
5130	Antenna pole, 10'	↓	3.20	2.500		21.50	159		180.50	261
6100	Satellite TV system	2 Elec	1	16		2,050	1,025		3,075	3,800
6110	Dish, mesh, 10' diam.	"	2.40	6.667		1,600	425		2,025	2,375
6111	Two way RF/IF tapeoff	1 Elec	36	.222		3.57	14.15		17.72	25
6112	Two way RF/IF splitter		24	.333		13.75	21		34.75	46.50
6113	Line amplifier	↓	24	.333	↓	12.45	21		33.45	45
6114	Line splitters	1 Elec	36	.222	Ea.	4.58	14.15		18.73	26
6115	Line multi switches		8	1		655	63.50		718.50	820
6120	Motor unit		2.40	3.333		232	212		444	570
7000	Home theater, widescreen, 42", high definition, TV		10.24	.781		455	50		505	575
7050	Flat wall mount bracket		10.24	.781		57	50		107	137
7100	7 channel home theater receiver		10.24	.781	↓	415	50		465	535
7200	Home theater speakers		10.24	.781	Set	215	50		265	310
7300	Home theater programmable remote		10.24	.781	Ea.	276	50		326	380
8000	Main video splitter		4	2		3,250	127		3,377	3,775
8010	Video distribution units	↓	4	2	↓	140	127		267	345

27 42 Electronic Digital Systems

27 42 13 – Point of Sale Systems

27 42 13.10 Bar Code Scanner

		Crew	Daily Output	Labor-Hours	Unit	Material	2021 Bare Costs Labor	Equipment	Total	Total Incl O&P
0010	**BAR CODE SCANNER**									
0100	CR3600, w/B4 battery, handle	1 Elec	6	1.333	Outlet	1,025	85		1,110	1,250
0110	Palm		12	.667		790	42.50		832.50	935
0120	Batch, B4 batt, handle, chgstn, 3' USB		9	.889		990	56.50		1,046.50	1,175
0130	Palm, chgstn, 3' USB		12	.667		760	42.50		802.50	900
0140	CR2500, w/H2 handle, no cable		10	.800		615	51		666	750
0150	Core unit, palm, no cable	↓	16	.500	↓	585	32		617	695

27 51 Distributed Audio-Video Communications Systems

27 51 16 – Public Address Systems

27 51 16.10 Public Address System

		Crew	Daily Output	Labor-Hours	Unit	Material	2021 Bare Costs Labor	Equipment	Total	Total Incl O&P
0010	**PUBLIC ADDRESS SYSTEM**									
0100	Conventional, office	1 Elec	5.33	1.501	Speaker	167	95.50		262.50	325
0200	Industrial	"	2.70	2.963	"	325	189		514	635

27 51 19 – Sound Masking Systems

27 51 19.10 Sound System

		Crew	Daily Output	Labor-Hours	Unit	Material	2021 Bare Costs Labor	Equipment	Total	Total Incl O&P
0010	**SOUND SYSTEM**, not including rough-in wires, cables & conduits									
0100	Components, projector outlet	1 Elec	8	1	Ea.	55	63.50		118.50	155
0200	Microphone		4	2		121	127		248	320
0400	Speakers, ceiling or wall		8	1		138	63.50		201.50	246
0600	Trumpets		4	2		291	127		418	510
0800	Privacy switch		8	1		102	63.50		165.50	207
1000	Monitor panel		4	2		455	127		582	690
1200	Antenna, AM/FM		4	2		147	127		274	350
1400	Volume control		8	1		66	63.50		129.50	167
1600	Amplifier, 250 W		1	8		1,425	510		1,935	2,300
1800	Cabinets		1	8		1,125	510		1,635	2,000
2000	Intercom, 30 station capacity, master station	2 Elec	2	8		2,275	510		2,785	3,250
2020	10 station capacity	"	4	4		1,425	255		1,680	1,950
2200	Remote station	1 Elec	8	1		217	63.50		280.50	335
2400	Intercom outlets	1 Elec	8	1	Ea.	128	63.50		191.50	235
2600	Handset		4	2		425	127		552	655
2800	Emergency call system, 12 zones, annunciator		1.30	6.154		1,275	390		1,665	1,975
3000	Bell		5.30	1.509		116	96		212	270
3200	Light or relay		8	1		66	63.50		129.50	167
3400	Transformer		4	2		255	127		382	470
3600	House telephone, talking station		1.60	5		545	320		865	1,075
3800	Press to talk, release to listen		5.30	1.509		145	96		241	300
4000	System-on button					76			76	84
4200	Door release	1 Elec	4	2		136	127		263	340
4400	Combination speaker and microphone		8	1		232	63.50		295.50	350
4600	Termination box		3.20	2.500		73	159		232	320
4800	Amplifier or power supply		5.30	1.509		840	96		936	1,075
5000	Vestibule door unit		16	.500	Name	161	32		193	225
5200	Strip cabinet		27	.296	Ea.	291	18.85		309.85	350
5400	Directory		16	.500		156	32		188	219
6000	Master door, button buzzer type, 100 unit	2 Elec	.54	29.630		1,400	1,875		3,275	4,325
6020	200 unit		.30	53.333		2,625	3,400		6,025	7,950
6040	300 unit		.20	80		4,000	5,100		9,100	12,000
6060	Transformer	1 Elec	8	1		40.50	63.50		104	139
6080	Door opener		5.30	1.509		51	96		147	199
6100	Buzzer with door release and plate		4	2		51	127		178	245
6200	Intercom type, 100 unit	2 Elec	.54	29.630		1,725	1,875		3,600	4,700
6220	200 unit		.30	53.333		3,375	3,400		6,775	8,775
6240	300 unit		.20	80		5,100	5,100		10,200	13,200
6260	Amplifier	1 Elec	2	4		255	255		510	660
6280	Speaker with door release	"	4	2		76	127		203	273

27 52 Healthcare Communications and Monitoring Systems

27 52 23 – Nurse Call/Code Blue Systems

27 52 23.10 Nurse Call Systems

		Crew	Daily Output	Labor-Hours	Unit	Material	2021 Bare Costs Labor	Equipment	Total	Total Incl O&P
0010	**NURSE CALL SYSTEMS**									
0100	Single bedside call station	1 Elec	8	1	Ea.	167	63.50		230.50	279
0200	Ceiling speaker station		8	1		96.50	63.50		160	201
0400	Emergency call station		8	1		71	63.50		134.50	173
0600	Pillow speaker		8	1		209	63.50		272.50	325
0800	Double bedside call station		4	2		183	127		310	390
1000	Duty station		4	2		166	127		293	370
1200	Standard call button		8	1		132	63.50		195.50	240
1400	Lights, corridor, dome or zone indicator		8	1		55.50	63.50		119	156
1600	Master control station for 20 stations	2 Elec	.65	24.615	Total	3,725	1,575		5,300	6,425

27 53 Distributed Systems

27 53 13 – Clock Systems

27 53 13.50 Clock Equipments

		Crew	Daily Output	Labor-Hours	Unit	Material	2021 Bare Costs Labor	Equipment	Total	Total Incl O&P
0010	**CLOCK EQUIPMENTS**, not including wires & conduits									
0100	Time system components, master controller	1 Elec	.33	24.242	Ea.	1,525	1,550		3,075	4,000
0200	Program bell		8	1		88	63.50		151.50	191
0400	Combination clock & speaker		3.20	2.500		203	159		362	460
0600	Frequency generator		2	4		2,425	255		2,680	3,050
0800	Job time automatic stamp recorder		4	2		560	127		687	805
1600	Master time clock system, clocks & bells, 20 room	4 Elec	.20	160		6,325	10,200		16,525	22,100
1800	50 room	"	.08	400		13,500	25,500		39,000	52,500
1900	Time clock	1 Elec	3.20	2.500		370	159		529	645
2000	100 cards in & out, 1 color					9.45			9.45	10.40
2200	2 colors					11.25			11.25	12.40
2800	Metal rack for 25 cards	1 Elec	7	1.143		59.50	73		132.50	174
4000	Wireless time systems component, master controller	"	2	4		630	255		885	1,075
4010	For transceiver and antenna, see Section 28 47 12.10									
4100	Wireless analog clock w/battery operated	1 Elec	8	1	Ea.	110	63.50		173.50	216
4200	Wireless digital clock w/battery operated	"	8	1	"	420	63.50		483.50	555

Division Notes

	CREW	DAILY OUTPUT	LABOR-HOURS	UNIT	BARE COSTS				TOTAL INCL O&P
					MAT.	LABOR	EQUIP.	TOTAL	

Estimating Tips

- When estimating material costs for electronic safety and security systems, it is always prudent to obtain manufacturers' quotations for equipment prices and special installation requirements that may affect the total cost.

- Fire alarm systems consist of control panels, annunciator panels, batteries with rack, charger, and fire alarm actuating and indicating devices. Some fire alarm systems include speakers, telephone lines, door closer controls, and other components. Be careful not to overlook the costs related to installation for these items. Also be aware of costs for integrated automation instrumentation and terminal devices, control equipment, control wiring, and programming. Insurance underwriters may have specific requirements for the type of materials to be installed or design requirements based on the hazard to be protected. Local jurisdictions may have requirements not covered by code. It is advisable to be aware of any special conditions.

- Security equipment includes items such as CCTV, access control, and other detection and identification systems to perform alert and alarm functions. Be sure to consider the costs related to installation for this security equipment, such as for integrated automation instrumentation and terminal devices, control equipment, control wiring, and programming.

Reference Numbers

Reference numbers are shown at the beginning of some major classifications. These numbers refer to related items in the Reference Section. The reference information may be an estimating procedure, an alternate pricing method, or technical information.

Same Data. Simplified.

Enjoy the convenience and efficiency of accessing your costs anywhere:

- **Skip the multiplier** by setting your location
- **Quickly search,** edit, favorite and share costs
- **Stay on top of price changes** with automatic updates

Discover more at rsmeans.com/online

Note: Trade Service, in part, has been used as a reference source for some of the material prices used in Division 28.

28 01 Operation and Maint. of Electronic Safety and Security

28 01 80 – Operation and Maintenance of Fire Detection and Alarm

28 01 80.51 Maintenance & Admin. of Fire Detection & Alarm	Crew	Daily Output	Labor-Hours	Unit	Material	2021 Bare Costs Labor	Equipment	Total	Total Incl O&P
0010 **MAINTENANCE AND ADMINISTRATION OF FIRE DETECTION AND ALARM**									
3300 Remove and replace (reinstall), fire alarm device	1 Elec	5.33	1.501	Ea.		95.50		95.50	142

28 05 Common Work Results for Electronic Safety and Security

28 05 05 – Selective Demolition for Electronic Safety and Security

28 05 05.10 Safety and Security Demolition

		Crew	Daily Output	Labor-Hours	Unit	Material	Labor	Equipment	Total	Total Incl O&P
0010	**SAFETY AND SECURITY DEMOLITION**									
1050	Finger print/card reader	1 Elec	8	1	Ea.		63.50		63.50	94.50
1051	Electrical dml, CA 3000 card reader sys w/25 users		.31	25.600			1,625		1,625	2,425
1060	Video camera		8	1			63.50		63.50	94.50
1070	Motion detector, multi-channel		7	1.143			73		73	108
1090	Infrared detector		7	1.143			73		73	108
1130	Video monitor, 19"		8	1			63.50		63.50	94.50
1210	Fire alarm horn and strobe light		16	.500			32		32	47.50
1215	Remote fire alarm indicator light		23	.348			22		22	33
1220	Flame detector		23	.348			22		22	33
1230	Duct detector		10	.800			51		51	75.50
1240	Smoke detector		23	.348			22		22	33
1241	Fire alarm duct smoke remote test station		21.60	.370			23.50		23.50	35
1250	Fire alarm pull station		23	.348			22		22	33
1260	Fire alarm control panel, 4 to 8 zone	2 Elec	4	4	Ea.		255		255	380
1270	12 to 16 zone	"	2.67	5.993			380		380	565
1280	Fire alarm annunciation panel, 4 to 8 zone	1 Elec	6	1.333			85		85	126
1290	12 to 16 zone	2 Elec	6	2.667			170		170	252
1300	Electrical dml, tamper switch	1 Elec	38	.211			13.40		13.40	19.95
1310	Electrical dml, flow switch	"	38	.211			13.40		13.40	19.95

28 05 19 – Storage Appliances for Electronic Safety and Security

28 05 19.11 Digital Video Recorder (DVR)

		Crew	Daily Output	Labor-Hours	Unit	Material	Labor	Equipment	Total	Total Incl O&P
0010	**DIGITAL VIDEO RECORDERS**									
0100	Pentaplex hybrid, internet protocol, and hard drive									
0200	4 channel	1 Elec	1.33	6.015	Ea.	2,000	385		2,385	2,775
0300	8 channel		1	8		3,875	510		4,385	5,025
0400	16 channel		1	8		3,325	510		3,835	4,400

28 15 Access Control Hardware Devices

28 15 11 – Integrated Credential Readers and Field Entry Management

28 15 11.11 Standard Card Readers

		Crew	Daily Output	Labor-Hours	Unit	Material	Labor	Equipment	Total	Total Incl O&P
0010	**STANDARD CARD READERS**									
0015	Card key access									
0020	Computerized system, processor, proximity reader and cards									
0030	Does not include door hardware, lockset or wiring									
0040	Card key system for 1 door				Ea.	1,675			1,675	1,825
0060	Card key system for 2 doors					2,400			2,400	2,625
0080	Card key system for 4 doors					2,600			2,600	2,875
0100	Processor for card key access system					1,225			1,225	1,350
0160	Magnetic lock for electric access, 600 pound holding force					192			192	212
0170	Magnetic lock for electric access, 1200 pound holding force					192			192	212
0200	Proximity card reader					197			197	217
0210	Entrance card systems									

For customer support on your Electrical Costs with RSMeans Data, call 800.448.8182.

28 15 Access Control Hardware Devices

28 15 11 – Integrated Credential Readers and Field Entry Management

28 15 11.11 Standard Card Readers

		Crew	Daily Output	Labor-Hours	Unit	Material	2021 Bare Costs Labor	Equipment	Total	Total Incl O&P
0300	Entrance card, barium ferrite				Ea.	4.93			4.93	5.40
0320	Credential					6.65			6.65	7.30
0340	Proximity					3.69			3.69	4.06
0360	Weigand					10.15			10.15	11.20
0700	Entrance card reader, barium ferrite	R-19	4	5		265	320		585	765
0720	Credential		4	5		212	320		532	710
0740	Proximity		4	5		150	320		470	640
0760	Weigand		4	5		540	320		860	1,075
0850	Scanner, eye retina		4	5		2,650	320		2,970	3,375
0900	Gate opener, cantilever	R-18	3	8.667		2,300	485		2,785	3,250
0910	Light duty		3	8.667		2,050	485		2,535	2,975
0912	Custom		3	8.667		1,700	485		2,185	2,600
0950	Switch, tamper	R-19	6	3.333		34	213		247	355
1100	Accessories, electric door strike/bolt	R-19	5	4	Ea.	83	255		338	470
1120	Electromagnetic lock		5	4		164	255		419	560
1140	Keypad for card reader		4	5		167	320		487	660

28 15 11.13 Keypads

		Crew	Daily Output	Labor-Hours	Unit	Material	2021 Bare Costs Labor	Equipment	Total	Total Incl O&P
0010	**KEYPADS**									
0340	Digital keypad, int./ext., basic, excl. striker/power/wiring	1 Elec	3	2.667	Ea.	113	170		283	375
0350	Lockset, mechanical push-button type, complete, incl. hardware	1 Carp	4	2	"	365	109		474	565

28 15 11.15 Biometric Identity Devices

		Crew	Daily Output	Labor-Hours	Unit	Material	2021 Bare Costs Labor	Equipment	Total	Total Incl O&P
0010	**BIOMETRIC IDENTITY DEVICES**									
0200	Fingerprint scanner unit, excl. striker/power supply	1 Elec	4	2	Ea.	1,600	127		1,727	1,975
0210	Fingerprint scanner unit, for computer keyboard access		8	1		1,375	63.50		1,438.50	1,625
0220	Hand geometry scanner, mem of 512 users, excl. striker/power		3	2.667		2,450	170		2,620	2,950
0230	Memory upgrade for, adds 9,700 user profiles		8	1		440	63.50		503.50	575
0240	Adds 32,500 user profiles		8	1		580	63.50		643.50	735
0250	Prison type, memory of 256 users, excl. striker/power		3	2.667		2,800	170		2,970	3,325
0260	Memory upgrade for, adds 3,300 user profiles		8	1		410	63.50		473.50	545
0270	Adds 9,700 user profiles		8	1		505	63.50		568.50	650
0280	Adds 27,900 user profiles		8	1		770	63.50		833.50	945
0290	All weather, mem of 512 users, excl. striker/power		3	2.667		4,425	170		4,595	5,100
0300	Facial & fingerprint scanner, combination unit, excl. striker/power		3	2.667		940	170		1,110	1,275
0310	Access for, for initial setup, excl. striker/power		3	2.667		1,375	170		1,545	1,775

28 15 11.19 Security Access Control Accessories

		Crew	Daily Output	Labor-Hours	Unit	Material	2021 Bare Costs Labor	Equipment	Total	Total Incl O&P
0010	**SECURITY ACCESS CONTROL ACCESSORIES**									
0360	Scanner/reader access, power supply/transf, 110V to 12/24V	1 Elec	4	2	Ea.	225	127		352	435
0370	Elec/mag strikers, 12/24V		4	2		92	127		219	290
0380	1 hour battery backup power supply		4	2		45.50	127		172.50	239
0390	Deadbolt, digital, batt-operated, indoor/outdoor, complete, incl. hardware	1 Carp	4	2		405	109		514	615

28 18 Security Access Detection Equipment

28 18 11 – Security Access Metal Detectors

28 18 11.13 Security Access Metal Detectors

28 18 11.13 Security Access Metal Detectors	Crew	Daily Output	Labor-Hours	Unit	Material	2021 Bare Costs Labor	Equipment	Total	Total Incl O&P
0010 **SECURITY ACCESS METAL DETECTORS**									
0240 Metal detector, hand-held, wand type, unit only				Ea.	129			129	138
0250 Metal detector, walk through portal type, single zone	1 Elec	2	4		3,850	255		4,105	4,600
0260 Multi-zone	"	2	4		4,975	255		5,230	5,850

28 18 13 – Security Access X-Ray Equipment

28 18 13.16 Security Access X-Ray Equipment

	Crew	Daily Output	Labor-Hours	Unit	Material	Labor	Equipment	Total	Total Incl O&P
0010 **SECURITY ACCESS X-RAY EQUIPMENT**									
0290 X-ray machine, desk top, for mail/small packages/letters	1 Elec	4	2	Ea.	3,375	127		3,502	3,925
0300 Conveyor type, incl. monitor		2	4		15,600	255		15,855	17,600
0310 Includes additional features		2	4		27,800	255		28,055	31,000
0320 X-ray machine, large unit, for airports, incl. monitor	2 Elec	1	16		39,000	1,025		40,025	44,400
0330 Full console	2 Elec	.50	32	Ea.	67,000	2,050		69,050	76,500

28 18 15 – Security Access Explosive Detection Equipment

28 18 15.23 Security Access Explosive Detection Equipment

	Crew	Daily Output	Labor-Hours	Unit	Material	Labor	Equipment	Total	Total Incl O&P
0010 **SECURITY ACCESS EXPLOSIVE DETECTION EQUIPMENT**									
0270 Explosives detector, walk through portal type	1 Elec	2	4	Ea.	57,500	255		57,755	64,000
0280 Hand-held, battery operated				"	25,500				28,100

28 23 Video Management System

28 23 13 – Video Management System Interfaces

28 23 13.10 Closed Circuit Television System

	Crew	Daily Output	Labor-Hours	Unit	Material	Labor	Equipment	Total	Total Incl O&P
0010 **CLOSED CIRCUIT TELEVISION SYSTEM**									
2000 Surveillance, one station (camera & monitor)	2 Elec	2.60	6.154	Total	655	390		1,045	1,300
2200 For additional camera stations, add	1 Elec	2.70	2.963	Ea.	315	189		504	625
2400 Industrial quality, one station (camera & monitor)	2 Elec	2.60	6.154	Total	1,625	390		2,015	2,375
2600 For additional camera stations, add	1 Elec	2.70	2.963	Ea.	680	189		869	1,025
2610 For low light, add		2.70	2.963		515	189		704	845
2620 For very low light, add		2.70	2.963		3,400	189		3,589	4,000
2800 For weatherproof camera station, add		1.30	6.154		530	390		920	1,175
3000 For pan and tilt, add		1.30	6.154		1,550	390		1,940	2,275
3200 For zoom lens - remote control, add		2	4		1,775	255		2,030	2,325
3400 Extended zoom lens		2	4		4,925	255		5,180	5,800
3410 For automatic iris for low light, add		2	4		1,000	255		1,255	1,475
3600 Educational TV studio, basic 3 camera system, black & white,									
3800 electrical & electronic equip. only	4 Elec	.80	40	Total	8,125	2,550		10,675	12,700
4000 Full console		.28	114		24,900	7,275		32,175	38,200
4100 As above, but color system		.28	114		49,500	7,275		56,775	65,500
4120 Full console		.12	267		186,500	17,000		203,500	230,000
4200 For film chain, black & white, add	1 Elec	1	8	Ea.	12,200	510		12,710	14,300
4250 Color, add		.25	32		8,750	2,050		10,800	12,700
4400 For video recorders, add		1	8		2,400	510		2,910	3,375
4600 Premium	4 Elec	.40	80		15,800	5,100		20,900	24,900

28 23 23 – Video Surveillance Systems Infrastructure

28 23 23.50 Video Surveillance Equipments

	Crew	Daily Output	Labor-Hours	Unit	Material	Labor	Equipment	Total	Total Incl O&P
0010 **VIDEO SURVEILLANCE EQUIPMENTS**									
0200 Video cameras, wireless, hidden in exit signs, clocks, etc., incl. receiver	1 Elec	3	2.667	Ea.	176	170		346	445
0210 Accessories for video recorder, single camera		3	2.667		177	170		347	445
0220 For multiple cameras		3	2.667		1,700	170		1,870	2,125
0230 Video cameras, wireless, for under vehicle searching, complete		2	4		15,200	255		15,455	17,200

28 23 Video Management System

28 23 23 – Video Surveillance Systems Infrastructure

28 23 23.50 Video Surveillance Equipments	Crew	Daily Output	Labor-Hours	Unit	Material	2021 Bare Costs Labor	Equipment	Total	Total Incl O&P
0234	Master monitor station, 3 doors x 5 color monitor with tilt feature	1 Elec	2	4	Ea.	1,100	255		1,355	1,575
0400	Internet protocol network camera, day/night, color & power supply		2.60	3.077		930	196		1,126	1,325
0500	Monitor, color flat screen, liquid crystal display (LCD), 15"		2.70	2.963		223	189		412	525
0520	17"		2.70	2.963		345	189		534	660
0540	19"		2.70	2.963		435	189		624	760

28 31 Intrusion Detection

28 31 16 – Intrusion Detection Systems Infrastructure

28 31 16.50 Intrusion Detection

		Crew	Daily Output	Labor-Hours	Unit	Material	2021 Bare Costs Labor	Equipment	Total	Total Incl O&P
0010	**INTRUSION DETECTION,** not including wires & conduits									
0100	Burglar alarm, battery operated, mechanical trigger	1 Elec	4	2	Ea.	289	127		416	510
0200	Electrical trigger		4	2		345	127		472	570
0400	For outside key control, add		8	1		89	63.50		152.50	193
0600	For remote signaling circuitry, add		8	1		147	63.50		210.50	257
0800	Card reader, flush type, standard		2.70	2.963		680	189		869	1,025
1000	Multi-code		2.70	2.963		1,250	189		1,439	1,650
1010	Card reader, proximity type		2.70	2.963		150	189		339	445
1200	Door switches, hinge switch		5.30	1.509		65.50	96		161.50	216
1400	Magnetic switch		5.30	1.509		109	96		205	263
1600	Exit control locks, horn alarm		4	2		219	127		346	430
1800	Flashing light alarm		4	2		248	127		375	460
2000	Indicating panels, 1 channel		2.70	2.963		222	189		411	525
2200	10 channel	2 Elec	3.20	5		1,175	320		1,495	1,775
2400	20 channel		2	8		2,500	510		3,010	3,500
2600	40 channel		1.14	14.035		4,650	895		5,545	6,425
2800	Ultrasonic motion detector, 12 V	1 Elec	2.30	3.478		193	222		415	540
3000	Infrared photoelectric detector		4	2		122	127		249	325
3200	Passive infrared detector		4	2		55	127		182	250
3400	Glass break alarm switch		8	1		81	63.50		144.50	184
3420	Switchmats, 30" x 5'		5.30	1.509		120	96		216	275
3440	30" x 25'		4	2		211	127		338	420
3460	Police connect panel		4	2		292	127		419	510
3480	Telephone dialer		5.30	1.509		400	96		496	585
3500	Alarm bell		4	2		101	127		228	300
3520	Siren		4	2		152	127		279	355
3540	Microwave detector, 10' to 200'		2	4		555	255		810	990
3560	10' to 350'		2	4		1,700	255		1,955	2,250

28 33 Security Monitoring and Control

28 33 11 – Electronic Structural Monitoring Systems

28 33 11.10 Load Cell Units

		Crew	Daily Output	Labor-Hours	Unit	Material	2021 Bare Costs Labor	Equipment	Total	Total Incl O&P
0010	**LOAD CELL UNITS**									
0100	Load cell, stainless steel, 0.5/1 ton rated, mounting unit	2 Elec	1.33	12.030	Ea.	655	765		1,420	1,875
0110	2/3 to 5/5 ton rated, mounting unit		1.20	13.333		695	850		1,545	2,025
0120	0.5/1/2/3.5/5 ton, self aligning bottom base		1.66	9.639		192	615		807	1,125
0130	Combo mounting unit, 10/25 ton rated		.86	18.605		1,800	1,175		2,975	3,725
0140	40/60 ton rated		.66	24.242		2,900	1,550		4,450	5,500
0150	60kg/3m, rated		1.84	8.696		1,525	555		2,080	2,500
0160	130kg/3m, rated		1.55	10.323		1,450	660		2,110	2,575

For customer support on your Electrical Costs with RSMeans Data, call 800.448.8182.

343

28 33 Security Monitoring and Control

28 33 11 – Electronic Structural Monitoring Systems

28 33 11.10 Load Cell Units	Crew	Daily Output	Labor-Hours	Unit	Material	2021 Bare Costs Labor	Equipment	Total	Total Incl O&P
0170 280 kg, rated	2 Elec	1.22	13.115	Ea.	1,450	835		2,285	2,850

28 46 Fire Detection and Alarm

28 46 11 – Fire Sensors and Detectors

28 46 11.21 Carbon-Monoxide Detection Sensors

	Crew	Daily Output	Labor-Hours	Unit	Material	2021 Bare Costs Labor	Equipment	Total	Total Incl O&P
0010 **CARBON-MONOXIDE DETECTION SENSORS**									
8400 Smoke and carbon monoxide alarm battery operated photoelectric low profile	1 Elec	24	.333	Ea.	54.50	21		75.50	91.50
8410 low profile photoelectric battery powered		24	.333		42.50	21		63.50	78.50
8420 photoelectric low profile sealed lithium		24	.333		74.50	21		95.50	114
8430 Photoelectric low profile sealed lithium smoke and CO with voice combo		24	.333		52	21		73	88.50
8500 Carbon monoxide sensor, wall mount 1 Mod 1 relay output smoke & heat		24	.333		455	21		476	530
8510 1Mod, 1relay no display, smoke & heat		24	.333		445	21		466	520
8530 1Mod 1 relay no HW logo smoke & heat		24	.333		480	21		501	555
8700 Carbon monoxide detector, battery operated, wall mounted		16	.500		44	32		76	96
8710 Hardwired, wall and ceiling mounted		8	1		73	63.50		136.50	175
8720 Duct mounted		8	1		320	63.50		383.50	450
8730 Continuous air monitoring system, PC based, incl. remote sensors, min.	2 Elec	.25	64		31,600	4,075		35,675	40,800
8740 Maximum	"	.10	160		57,000	10,200		67,200	77,500

28 46 11.27 Other Sensors

	Crew	Daily Output	Labor-Hours	Unit	Material	2021 Bare Costs Labor	Equipment	Total	Total Incl O&P
0010 **OTHER SENSORS**									
5200 Smoke detector, ceiling type	1 Elec	6.20	1.290	Ea.	126	82		208	260
5240 Smoke detector, addressable type		6	1.333		214	85		299	360
5400 Duct type		3.20	2.500		295	159		454	560
5420 Duct addressable type		3.20	2.500		180	159		339	435
8300 Smoke alarm with integrated strobe light 120 V, 16DB 60 fpm flash rate		16	.500		126	32		158	186
8310 Photoelectric smoke detector with strobe 120 V, 90 DB ceiling mount		12	.667		240	42.50		282.50	325
8320 120 V, 90 DB wall mount		12	.667		214	42.50		256.50	298
8330 120 V, 9 VDC, 90 DB ceiling mount		12	.667		213	42.50		255.50	298
8340 120 V, 9 VDC backup 90 DB wall mount		12	.667		214	42.50		256.50	298
8440 Fire alarm beam detector, motorized reflective, infrared optical beam	2 Elec	1.60	10		685	635		1,320	1,700
8450 Fire alarm beam detector, motorized reflective, IR/UV optical beam	"	1.60	10		970	635		1,605	2,025

28 46 11.50 Fire and Heat Detectors

	Crew	Daily Output	Labor-Hours	Unit	Material	2021 Bare Costs Labor	Equipment	Total	Total Incl O&P
0010 **FIRE & HEAT DETECTORS**									
5000 Detector, rate of rise	1 Elec	8	1	Ea.	29.50	63.50		93	127
5010 Heat addressable type		7.25	1.103		208	70.50		278.50	335
5100 Fixed temp fire alarm		7	1.143		39	73		112	151
5200 125 V SRF MT-135F		7.25	1.103		35.50	70.50		106	143
5300 MT-195F	1 Elec	7.25	1.103	Ea.	36	70.50		106.50	144
5400 104/85 DB, indoor/outdoor, ceiling/wall, red		10.25	.780		36	49.50		85.50	114
5410 Ceiling/wall, white		10.25	.780		35	49.50		84.50	113
5420 102/98 DB, lumin 110cd, indoor/wall, red		10.25	.780		109	49.50		158.50	193
5430 Lumin15/75cd, indoor/wall, red		10.25	.780		129	49.50		178.50	216
5440 HI/LO DB, 24 V, fire marking, red		10.25	.780		85.50	49.50		135	168

28 46 20 – Fire Alarm

28 46 20.50 Alarm Panels and Devices

	Crew	Daily Output	Labor-Hours	Unit	Material	2021 Bare Costs Labor	Equipment	Total	Total Incl O&P
0010 **ALARM PANELS AND DEVICES**, not including wires & conduits									
2200 Intercom remote station	1 Elec	8	1	Ea.	94.50	63.50		158	199
2400 Intercom outlet		8	1		66	63.50		129.50	167
2600 Sound system, intercom handset		8	1		515	63.50		578.50	660
3590 Signal device, beacon		8	1		207	63.50		270.50	325

28 46 20 – Fire Alarm

28 46 20.50 Alarm Panels and Devices

		Crew	Daily Output	Labor-Hours	Unit	Material	2021 Bare Costs Labor	Equipment	Total	Total Incl O&P
3594	2 zone	1 Elec	1.35	5.926	Ea.	385	375		760	985
3598	3 zone	2 Elec	2	8		465	510		975	1,275
3600	4 zone		2	8		435	510		945	1,225
3700	5 zone		1.50	10.667		740	680		1,420	1,825
3800	8 zone		1	16		420	1,025		1,445	1,975
3900	10 zone		1.25	12.800		890	815		1,705	2,175
4000	12 zone		.67	23.988		2,550	1,525		4,075	5,100
4020	Alarm device, tamper, flow	1 Elec	8	1		206	63.50		269.50	320
4025	Fire alarm, loop expander card		16	.500		810	32		842	940
4050	Actuating device		8	1		330	63.50		393.50	455
4160	Alarm control panel, addressable w/o voice, up to 200 points	2 Elec	1.14	13.998		1,975	890		2,865	3,475
4170	addressable w/voice, up to 400 points	"	.73	22.008		10,100	1,400		11,500	13,200
4175	Addressable interface device	1 Elec	7.25	1.103		149	70.50		219.50	268
4200	Battery and rack		4	2		465	127		592	700
4400	Automatic charger		8	1		525	63.50		588.50	675
4600	Signal bell		8	1		68	63.50		131.50	170
4610	Fire alarm signal bell 10" red 20-24 V P		8	1		153	63.50		216.50	264
4800	Trouble buzzer or manual station		8	1		75.50	63.50		139	178
5425	Duct smoke and heat detector 2 wire		8	1		155	63.50		218.50	265
5430	Fire alarm duct detector controller		3	2.667		183	170		353	455
5435	Fire alarm duct detector sensor kit		8	1		77.50	63.50		141	180
5440	Remote test station for smoke detector duct type		5.30	1.509		54.50	96		150.50	203
5460	Remote fire alarm indicator light		5.30	1.509		28	96		124	174
5600	Strobe and horn		5.30	1.509		129	96		225	285
5610	Strobe and horn (ADA type)		5.30	1.509		155	96		251	315
5620	Visual alarm (ADA type)		6.70	1.194		103	76		179	226
5800	electric bell		6.70	1.194		50.50	76		126.50	169
6000	Door holder, electro-magnetic		4	2		92	127		219	290
6200	Combination holder and closer		3.20	2.500		157	159		316	410
6600	Drill switch		8	1		485	63.50		548.50	630
6800	Master box		2.70	2.963		5,850	189		6,039	6,725
7000	Break glass station		8	1		62.50	63.50		126	163
7010	Break glass station, addressable		7.25	1.103		148	70.50		218.50	267
7800	Remote annunciator, 8 zone lamp		1.80	4.444		219	283		502	660
8000	12 zone lamp	2 Elec	2.60	6.154		430	390		820	1,050
8200	16 zone lamp	"	2.20	7.273		385	465		850	1,100

28 47 Mass Notification

28 47 12 – Notification Systems

28 47 12.10 Mass Notification System

		Crew	Daily Output	Labor-Hours	Unit	Material	2021 Bare Costs Labor	Equipment	Total	Total Incl O&P
0010	**MASS NOTIFICATION SYSTEM**									
0100	Wireless command center, 10,000 devices	2 Elec	1.33	12.030	Ea.	3,050	765		3,815	4,500
0200	Option, email notification					2,125			2,125	2,325
0210	Remote device supervision & monitor					2,650			2,650	2,925
0300	Antenna VHF or UHF, for medium range	1 Elec	4	2		93.50	127		220.50	292
0310	For high-power transmitter		2	4		1,125	255		1,380	1,625
0400	Transmitter, 25 watt		4	2		1,750	127		1,877	2,125
0410	40 watt		2.66	3.008		2,025	192		2,217	2,500
0420	100 watt		1.33	6.015		6,625	385		7,010	7,850
0500	Wireless receiver/control module for speaker		8	1		287	63.50		350.50	410
0600	Desktop paging controller, stand alone		4	2		855	127		982	1,125

345

28 47 Mass Notification

28 47 12 – Notification Systems

28 47 12.10 Mass Notification System	Crew	Daily Output	Labor-Hours	Unit	Material	2021 Bare Costs Labor	Equipment	Total	Total Incl O&P
0700	Auxiliary level inputs	1 Elec	8.36	.957	Ea.	103	61		164	204

28 52 Detention Security Systems

28 52 11 – Detention Monitoring and Control Systems

28 52 11.10 Detention Control Systems

		Crew	Daily Output	Labor-Hours	Unit	Material	2021 Bare Costs Labor	Equipment	Total	Total Incl O&P
0010	**DETENTION CONTROL SYSTEMS**									
1000	Desk top control systems for 10 doors, 10 intercoms, and 10 lights	2 Elec	.30	53.333	Ea.	64,500	3,400		67,900	75,500
1020	Push button control panel systems for 10 doors, 10 intercoms, and 10 lights	"	.32	50	"	58,000	3,175		61,175	68,000

For customer support on your Electrical Costs with RSMeans Data, call 800.448.8182.

Estimating Tips
31 05 00 Common Work Results for Earthwork

- Estimating the actual cost of performing earthwork requires careful consideration of the variables involved. This includes items such as type of soil, whether water will be encountered, dewatering, whether banks need bracing, disposal of excavated earth, and length of haul to fill or spoil sites, etc. If the project has large quantities of cut or fill, consider raising or lowering the site to reduce costs, while paying close attention to the effect on site drainage and utilities.

- If the project has large quantities of fill, creating a borrow pit on the site can significantly lower the costs.

- It is very important to consider what time of year the project is scheduled for completion. Bad weather can create large cost overruns from dewatering, site repair, and lost productivity from cold weather.

Reference Numbers

Reference numbers are shown at the beginning of some major classifications. These numbers refer to related items in the Reference Section. The reference information may be an estimating procedure, an alternate pricing method, or technical information.

Note: Not all subdivisions listed here necessarily appear. ■

Same Data. Simplified.

Enjoy the convenience and efficiency of accessing your costs anywhere:

- **Skip the multiplier** by setting your location
- **Quickly search,** edit, favorite and share costs
- **Stay on top of price changes** with automatic updates

Discover more at rsmeans.com/online

31 23 Excavation and Fill

31 23 16 – Excavation

31 23 16.13 Excavating, Trench

		Crew	Daily Output	Labor-Hours	Unit	Material	2021 Bare Costs Labor	2021 Bare Costs Equipment	Total	Total Incl O&P
0010	**EXCAVATING, TRENCH**									
0011	Or continuous footing									
0020	Common earth with no sheeting or dewatering included									
0050	1' to 4' deep, 3/8 C.Y. excavator	B-11C	150	.107	B.C.Y.		5.50	1.44	6.94	9.80
0060	1/2 C.Y. excavator	B-11M	200	.080			4.14	1.18	5.32	7.45
0062	3/4 C.Y. excavator	B-12F	270	.059			3.14	2.60	5.74	7.55
1400	By hand with pick and shovel 2' to 6' deep, light soil	1 Clab	8	1			44.50		44.50	66.50
1500	Heavy soil	"	4	2			89		89	133
3000	Backfill trench, F.E. loader, wheel mtd., 1 C.Y. bucket									
3020	Minimal haul	B-10R	400	.030	L.C.Y.		1.62	.76	2.38	3.26
5020	Loam & sandy clay with no sheeting or dewatering included									
5050	1' to 4' deep, 3/8 C.Y. tractor loader/backhoe	B-11C	162	.099	B.C.Y.		5.10	1.33	6.43	9.05
5060	1/2 C.Y. excavator	B-11M	216	.074			3.83	1.09	4.92	6.90
5070	3/4 C.Y. excavator	B-12F	292	.055			2.90	2.40	5.30	6.95
6020	Sand & gravel with no sheeting or dewatering included									
6050	1' to 4' deep, 3/8 C.Y. excavator	B-11C	165	.097	B.C.Y.		5	1.31	6.31	8.90
6060	1/2 C.Y. excavator	B-11M	220	.073			3.76	1.07	4.83	6.80
6070	3/4 C.Y. excavator	B-12F	297	.054			2.85	2.36	5.21	6.85
7020	Dense hard clay with no sheeting or dewatering included									
7050	1' to 4' deep, 3/8 C.Y. excavator	B-11C	132	.121	B.C.Y.		6.25	1.64	7.89	11.15
7060	1/2 C.Y. excavator	B-11M	176	.091			4.70	1.34	6.04	8.45
7070	3/4 C.Y. excavator	B-12F	238	.067			3.56	2.95	6.51	8.55

31 23 16.14 Excavating, Utility Trench

		Crew	Daily Output	Labor-Hours	Unit	Material	2021 Bare Costs Labor	2021 Bare Costs Equipment	Total	Total Incl O&P
0010	**EXCAVATING, UTILITY TRENCH**									
0011	Common earth									
0050	Trenching with chain trencher, 12 HP, operator walking									
0100	4" wide trench, 12" deep	B-53	800	.010	L.F.		.56	.20	.76	1.05
0150	18" deep		750	.011			.59	.21	.80	1.11
0200	24" deep		700	.011			.63	.23	.86	1.20
0300	6" wide trench, 12" deep		650	.012			.68	.24	.92	1.29
0350	18" deep		600	.013			.74	.26	1	1.39
0400	24" deep		550	.015			.81	.29	1.10	1.52
0450	36" deep		450	.018			.99	.35	1.34	1.86
0600	8" wide trench, 12" deep		475	.017			.93	.33	1.26	1.76
0650	18" deep		400	.020			1.11	.40	1.51	2.09
0700	24" deep		350	.023			1.27	.45	1.72	2.39
0750	36" deep		300	.027			1.48	.53	2.01	2.79
1000	Backfill by hand including compaction, add									
1050	4" wide trench, 12" deep	A-1G	800	.010	L.F.		.44	.07	.51	.74
1100	18" deep		530	.015			.67	.10	.77	1.11
1150	24" deep		400	.020			.89	.14	1.03	1.48
1300	6" wide trench, 12" deep		540	.015			.66	.10	.76	1.09
1350	18" deep		405	.020			.88	.13	1.01	1.46
1400	24" deep		270	.030			1.32	.20	1.52	2.18
1450	36" deep		180	.044			1.97	.30	2.27	3.27
1600	8" wide trench, 12" deep		400	.020			.89	.14	1.03	1.48
1650	18" deep		265	.030			1.34	.21	1.55	2.23
1700	24" deep		200	.040			1.78	.27	2.05	2.95
1750	36" deep		135	.059			2.63	.40	3.03	4.38
2000	Chain trencher, 40 HP operator riding									
2050	6" wide trench and backfill, 12" deep	B-54	1200	.007	L.F.		.37	.38	.75	.96
2100	18" deep		1000	.008			.44	.45	.89	1.16

31 23 16 – Excavation

31 23 16.14 Excavating, Utility Trench

		Crew	Daily Output	Labor-Hours	Unit	Material	2021 Bare Costs Labor	Equipment	Total	Total Incl O&P
2150	24" deep	B-54	975	.008	L.F.		.46	.46	.92	1.19
2200	36" deep		900	.009			.49	.50	.99	1.29
2250	48" deep		750	.011			.59	.60	1.19	1.54
2300	60" deep		650	.012			.68	.69	1.37	1.78
2400	8" wide trench and backfill, 12" deep		1000	.008			.44	.45	.89	1.16
2450	18" deep		950	.008			.47	.47	.94	1.22
2500	24" deep		900	.009			.49	.50	.99	1.29
2550	36" deep		800	.010			.56	.56	1.12	1.45
2600	48" deep		650	.012			.68	.69	1.37	1.78
2700	12" wide trench and backfill, 12" deep		975	.008			.46	.46	.92	1.19
2750	18" deep		860	.009			.52	.52	1.04	1.35
2800	24" deep		800	.010			.56	.56	1.12	1.45
2850	36" deep		725	.011			.61	.62	1.23	1.59
3000	16" wide trench and backfill, 12" deep		835	.010			.53	.54	1.07	1.38
3050	18" deep		750	.011			.59	.60	1.19	1.54
3100	24" deep	▼	700	.011	▼		.63	.64	1.27	1.66
3200	Compaction with vibratory plate, add								35%	35%
5100	Hand excavate and trim for pipe bells after trench excavation									
5200	8" pipe	1 Clab	155	.052	L.F.		2.29		2.29	3.42
5300	18" pipe	"	130	.062	"		2.73		2.73	4.08

31 23 16.15 Excavating, Utility Trench, Plow

		Crew	Daily Output	Labor-Hours	Unit	Material	2021 Bare Costs Labor	Equipment	Total	Total Incl O&P
0010	**EXCAVATING, UTILITY TRENCH, PLOW**									
0100	Single cable, plowed into fine material	B-54C	3800	.004	L.F.		.22	.30	.52	.65
0200	Two cable		3200	.005			.26	.36	.62	.78
0300	Single cable, plowed into coarse material	▼	2000	.008	▼		.41	.57	.98	1.25

31 23 16.16 Structural Excavation for Minor Structures

		Crew	Daily Output	Labor-Hours	Unit	Material	2021 Bare Costs Labor	Equipment	Total	Total Incl O&P
0010	**STRUCTURAL EXCAVATION FOR MINOR STRUCTURES**									
0015	Hand, pits to 6' deep, sandy soil	1 Clab	8	1	B.C.Y.		44.50		44.50	66.50
0100	Heavy soil or clay		4	2			89		89	133
0300	Pits 6' to 12' deep, sandy soil		5	1.600			71		71	106
0500	Heavy soil or clay		3	2.667			118		118	177
0700	Pits 12' to 18' deep, sandy soil		4	2			89		89	133
0900	Heavy soil or clay	▼	2	4			178		178	265
6030	Common earth, hydraulic backhoe, 1/2 C.Y. bucket	B-12E	55	.291			15.40	8.30	23.70	32
6035	3/4 C.Y. bucket	B-12F	90	.178			9.40	7.80	17.20	22.50
6040	1 C.Y. bucket	B-12A	108	.148			7.85	7.70	15.55	20
6050	1-1/2 C.Y. bucket	B-12B	144	.111			5.90	4.83	10.73	14.05
6060	2 C.Y. bucket	B-12C	200	.080			4.23	4.71	8.94	11.50
6070	Sand and gravel, 3/4 C.Y. bucket	B-12F	100	.160	▼		8.45	7	15.45	20.50
6080	1 C.Y. bucket	B-12A	120	.133	B.C.Y.		7.05	6.95	14	18.15
6090	1-1/2 C.Y. bucket	B-12B	160	.100			5.30	4.35	9.65	12.70
6100	2 C.Y. bucket	B-12C	220	.073			3.85	4.29	8.14	10.45
6110	Clay, till, or blasted rock, 3/4 C.Y. bucket	B-12F	80	.200			10.60	8.75	19.35	25.50
6120	1 C.Y. bucket	B-12A	95	.168			8.90	8.75	17.65	23
6130	1-1/2 C.Y. bucket	B-12B	130	.123			6.50	5.35	11.85	15.60
6140	2 C.Y. bucket	B-12C	175	.091			4.84	5.40	10.24	13.15
6230	Sandy clay & loam, hydraulic backhoe, 1/2 C.Y. bucket	B-12E	60	.267			14.10	7.60	21.70	29.50
6235	3/4 C.Y. bucket	B-12F	98	.163			8.65	7.15	15.80	21
6240	1 C.Y. bucket	B-12A	116	.138			7.30	7.20	14.50	18.80
6250	1-1/2 C.Y. bucket	B-12B	156	.103	▼		5.45	4.46	9.91	13

31 23 Excavation and Fill

31 23 19 – Dewatering

31 23 19.20 Dewatering Systems

		Crew	Daily Output	Labor-Hours	Unit	Material	2021 Bare Costs Labor	2021 Bare Costs Equipment	Total	Total Incl O&P
0010	**DEWATERING SYSTEMS**									
0020	Excavate drainage trench, 2' wide, 2' deep	B-11C	90	.178	C.Y.		9.20	2.40	11.60	16.35
0100	2' wide, 3' deep, with backhoe loader	"	135	.119			6.15	1.60	7.75	10.90
0200	Excavate sump pits by hand, light soil	1 Clab	7.10	1.127			50		50	74.50
0300	Heavy soil	"	3.50	2.286	↓		101		101	151
0500	Pumping 8 hrs., attended 2 hrs./day, incl. 20 L.F.									
0550	of suction hose & 100 L.F. discharge hose									
0600	2" diaphragm pump used for 8 hrs.	B-10H	4	3	Day		162	25	187	270
0650	4" diaphragm pump used for 8 hrs.	B-10I	4	3			162	37.50	199.50	283
0800	8 hrs. attended, 2" diaphragm pump	B-10H	1	12			650	99	749	1,075
0900	3" centrifugal pump	B-10J	1	12			650	92	742	1,075
1000	4" diaphragm pump	B-10I	1	12			650	149	799	1,125
1100	6" centrifugal pump	B-10K	1	12	↓		650	297	947	1,300

31 23 23 – Fill

31 23 23.13 Backfill

		Crew	Daily Output	Labor-Hours	Unit	Material	2021 Bare Costs Labor	2021 Bare Costs Equipment	Total	Total Incl O&P
0010	**BACKFILL**									
0015	By hand, no compaction, light soil	1 Clab	14	.571	L.C.Y.		25.50		25.50	38
0100	Heavy soil		11	.727	"		32.50		32.50	48
0300	Compaction in 6" layers, hand tamp, add to above	↓	20.60	.388	E.C.Y.		17.25		17.25	25.50
0400	Roller compaction operator walking, add	B-10A	100	.120			6.50	1.67	8.17	11.55
0500	Air tamp, add	B-9D	190	.211			9.45	1.73	11.18	15.95
0600	Vibrating plate, add	A-1D	60	.133			5.90	.53	6.43	9.45
0800	Compaction in 12" layers, hand tamp, add to above	1 Clab	34	.235			10.45		10.45	15.60
0900	Roller compaction operator walking, add	B-10A	150	.080			4.33	1.11	5.44	7.65
1000	Air tamp, add	B-9	285	.140			6.30	1.25	7.55	10.75
1100	Vibrating plate, add	A-1E	90	.089	↓		3.95	1.84	5.79	7.90
1300	Dozer backfilling, bulk, up to 300' haul, no compaction	B-10B	1200	.010	L.C.Y.		.54	1.27	1.81	2.20
1400	Air tamped, add	B-11B	80	.200	E.C.Y.		10	5.35	15.35	21
1600	Compacting backfill, 6" to 12" lifts, vibrating roller	B-10C	800	.015			.81	2.55	3.36	4.02
1700	Sheepsfoot roller	B-10D	750	.016	↓		.87	2.60	3.47	4.15
1900	Dozer backfilling, trench, up to 300' haul, no compaction	B-10B	900	.013	L.C.Y.		.72	1.69	2.41	2.94
2000	Air tamped, add	B-11B	80	.200	E.C.Y.		10	5.35	15.35	21
2200	Compacting backfill, 6" to 12" lifts, vibrating roller	B-10C	700	.017			.93	2.91	3.84	4.59
2300	Sheepsfoot roller	B-10D	650	.018	↓		1	3	4	4.78

31 23 23.16 Fill By Borrow and Utility Bedding

		Crew	Daily Output	Labor-Hours	Unit	Material	2021 Bare Costs Labor	2021 Bare Costs Equipment	Total	Total Incl O&P
0010	**FILL BY BORROW AND UTILITY BEDDING**									
0049	Utility bedding, for pipe & conduit, not incl. compaction									
0050	Crushed or screened bank run gravel	B-6	150	.160	L.C.Y.	33.50	7.70	1.44	42.64	49.50
0100	Crushed stone 3/4" to 1/2"		150	.160		25.50	7.70	1.44	34.64	41.50
0200	Sand, dead or bank	↓	150	.160	↓	18.75	7.70	1.44	27.89	33.50
0500	Compacting bedding in trench	A-1D	90	.089	E.C.Y.		3.95	.35	4.30	6.30
0610	See Section 31 23 23.20 for hauling mileage add.									

31 23 23.20 Hauling

		Crew	Daily Output	Labor-Hours	Unit	Material	2021 Bare Costs Labor	2021 Bare Costs Equipment	Total	Total Incl O&P
0010	**HAULING**									
0011	Excavated or borrow, loose cubic yards									
0012	no loading equipment, including hauling, waiting, loading/dumping									
0013	time per cycle (wait, load, travel, unload or dump & return)									
0014	8 C.Y. truck, 15 MPH avg., cycle 0.5 miles, 10 min. wait/ld./uld.	B-34A	320	.025	L.C.Y.		1.28	1.27	2.55	3.32
0016	cycle 1 mile		272	.029			1.51	1.50	3.01	3.91
0018	cycle 2 miles		208	.038			1.97	1.96	3.93	5.10
0020	cycle 4 miles	↓	144	.056	↓		2.85	2.83	5.68	7.35

31 23 23 – Fill

31 23 23.20 Hauling		Crew	Daily Output	Labor-Hours	Unit	Material	2021 Bare Costs Labor	Equipment	Total	Total Incl O&P
0022	cycle 6 miles	B-34A	112	.071	L.C.Y.		3.66	3.64	7.30	9.50
0024	cycle 8 miles		88	.091			4.66	4.63	9.29	12.05
0026	20 MPH avg., cycle 0.5 mile		336	.024			1.22	1.21	2.43	3.16
0028	cycle 1 mile		296	.027			1.39	1.38	2.77	3.58
0030	cycle 2 miles		240	.033			1.71	1.70	3.41	4.43
0032	cycle 4 miles		176	.045			2.33	2.32	4.65	6.05
0034	cycle 6 miles		136	.059			3.02	3	6.02	7.80
0036	cycle 8 miles		112	.071			3.66	3.64	7.30	9.50
0044	25 MPH avg., cycle 4 miles		192	.042			2.14	2.12	4.26	5.55
0046	cycle 6 miles		160	.050			2.57	2.55	5.12	6.65
0048	cycle 8 miles		128	.063			3.21	3.18	6.39	8.30
0050	30 MPH avg., cycle 4 miles		216	.037			1.90	1.89	3.79	4.92
0052	cycle 6 miles		176	.045			2.33	2.32	4.65	6.05
0054	cycle 8 miles		144	.056			2.85	2.83	5.68	7.35
0114	15 MPH avg., cycle 0.5 mile, 15 min. wait/ld./uld.		224	.036			1.83	1.82	3.65	4.74
0116	cycle 1 mile		200	.040			2.05	2.04	4.09	5.30
0118	cycle 2 miles		168	.048			2.44	2.43	4.87	6.30
0120	cycle 4 miles	B-34A	120	.067	L.C.Y.		3.42	3.40	6.82	8.85
0122	cycle 6 miles		96	.083			4.28	4.25	8.53	11.05
0124	cycle 8 miles		80	.100			5.15	5.10	10.25	13.25
0126	20 MPH avg., cycle 0.5 mile		232	.034			1.77	1.76	3.53	4.57
0128	cycle 1 mile		208	.038			1.97	1.96	3.93	5.10
0130	cycle 2 miles		184	.043			2.23	2.22	4.45	5.75
0132	cycle 4 miles		144	.056			2.85	2.83	5.68	7.35
0134	cycle 6 miles		112	.071			3.66	3.64	7.30	9.50
0136	cycle 8 miles		96	.083			4.28	4.25	8.53	11.05
0144	25 MPH avg., cycle 4 miles		152	.053			2.70	2.68	5.38	7
0146	cycle 6 miles		128	.063			3.21	3.18	6.39	8.30
0148	cycle 8 miles		112	.071			3.66	3.64	7.30	9.50
0150	30 MPH avg., cycle 4 miles		168	.048			2.44	2.43	4.87	6.30
0152	cycle 6 miles		144	.056			2.85	2.83	5.68	7.35
0154	cycle 8 miles		120	.067			3.42	3.40	6.82	8.85
0214	15 MPH avg., cycle 0.5 mile, 20 min. wait/ld./uld.		176	.045			2.33	2.32	4.65	6.05
0216	cycle 1 mile		160	.050			2.57	2.55	5.12	6.65
0218	cycle 2 miles		136	.059			3.02	3	6.02	7.80
0220	cycle 4 miles		104	.077			3.95	3.92	7.87	10.20
0222	cycle 6 miles		88	.091			4.66	4.63	9.29	12.05
0224	cycle 8 miles		72	.111			5.70	5.65	11.35	14.75
0226	20 MPH avg., cycle 0.5 mile		176	.045			2.33	2.32	4.65	6.05
0228	cycle 1 mile		168	.048			2.44	2.43	4.87	6.30
0230	cycle 2 miles		144	.056			2.85	2.83	5.68	7.35
0232	cycle 4 miles		120	.067			3.42	3.40	6.82	8.85
0234	cycle 6 miles		96	.083			4.28	4.25	8.53	11.05
0236	cycle 8 miles		88	.091			4.66	4.63	9.29	12.05
0244	25 MPH avg., cycle 4 miles		128	.063			3.21	3.18	6.39	8.30
0246	cycle 6 miles		112	.071			3.66	3.64	7.30	9.50
0248	cycle 8 miles		96	.083			4.28	4.25	8.53	11.05
0250	30 MPH avg., cycle 4 miles		136	.059			3.02	3	6.02	7.80
0252	cycle 6 miles		120	.067			3.42	3.40	6.82	8.85
0254	cycle 8 miles		104	.077			3.95	3.92	7.87	10.20
0314	15 MPH avg., cycle 0.5 mile, 25 min. wait/ld./uld.		144	.056			2.85	2.83	5.68	7.35
0316	cycle 1 mile		128	.063			3.21	3.18	6.39	8.30
0318	cycle 2 miles		112	.071			3.66	3.64	7.30	9.50

351

31 23 23.20 Hauling		Crew	Daily Output	Labor-Hours	Unit	Material	2021 Bare Costs		Total	Total Incl O&P
							Labor	Equipment		
0320	cycle 4 miles	B-34A	96	.083	L.C.Y.		4.28	4.25	8.53	11.05
0322	cycle 6 miles		80	.100			5.15	5.10	10.25	13.25
0324	cycle 8 miles		64	.125			6.40	6.35	12.75	16.60
0326	20 MPH avg., cycle 0.5 mile		144	.056			2.85	2.83	5.68	7.35
0328	cycle 1 mile		136	.059			3.02	3	6.02	7.80
0330	cycle 2 miles	B-34A	120	.067	L.C.Y.		3.42	3.40	6.82	8.85
0332	cycle 4 miles		104	.077			3.95	3.92	7.87	10.20
0334	cycle 6 miles		88	.091			4.66	4.63	9.29	12.05
0336	cycle 8 miles		80	.100			5.15	5.10	10.25	13.25
0344	25 MPH avg., cycle 4 miles		112	.071			3.66	3.64	7.30	9.50
0346	cycle 6 miles		96	.083			4.28	4.25	8.53	11.05
0348	cycle 8 miles		88	.091			4.66	4.63	9.29	12.05
0350	30 MPH avg., cycle 4 miles		112	.071			3.66	3.64	7.30	9.50
0352	cycle 6 miles		104	.077			3.95	3.92	7.87	10.20
0354	cycle 8 miles		96	.083			4.28	4.25	8.53	11.05
0414	15 MPH avg., cycle 0.5 mile, 30 min. wait/ld./uld.		120	.067			3.42	3.40	6.82	8.85
0416	cycle 1 mile		112	.071			3.66	3.64	7.30	9.50
0418	cycle 2 miles		96	.083			4.28	4.25	8.53	11.05
0420	cycle 4 miles		80	.100			5.15	5.10	10.25	13.25
0422	cycle 6 miles		72	.111			5.70	5.65	11.35	14.75
0424	cycle 8 miles		64	.125			6.40	6.35	12.75	16.60
0426	20 MPH avg., cycle 0.5 mile		120	.067			3.42	3.40	6.82	8.85
0428	cycle 1 mile		112	.071			3.66	3.64	7.30	9.50
0430	cycle 2 miles		104	.077			3.95	3.92	7.87	10.20
0432	cycle 4 miles		88	.091			4.66	4.63	9.29	12.05
0434	cycle 6 miles		80	.100			5.15	5.10	10.25	13.25
0436	cycle 8 miles		72	.111			5.70	5.65	11.35	14.75
0444	25 MPH avg., cycle 4 miles		96	.083			4.28	4.25	8.53	11.05
0446	cycle 6 miles		88	.091			4.66	4.63	9.29	12.05
0448	cycle 8 miles		80	.100			5.15	5.10	10.25	13.25
0450	30 MPH avg., cycle 4 miles		96	.083			4.28	4.25	8.53	11.05
0452	cycle 6 miles		88	.091			4.66	4.63	9.29	12.05
0454	cycle 8 miles		80	.100			5.15	5.10	10.25	13.25
0514	15 MPH avg., cycle 0.5 mile, 35 min. wait/ld./uld.		104	.077			3.95	3.92	7.87	10.20
0516	cycle 1 mile		96	.083			4.28	4.25	8.53	11.05
0518	cycle 2 miles		88	.091			4.66	4.63	9.29	12.05
0520	cycle 4 miles		72	.111			5.70	5.65	11.35	14.75
0522	cycle 6 miles		64	.125			6.40	6.35	12.75	16.60
0524	cycle 8 miles		56	.143			7.35	7.30	14.65	18.95
0526	20 MPH avg., cycle 0.5 mile		104	.077			3.95	3.92	7.87	10.20
0528	cycle 1 mile		96	.083			4.28	4.25	8.53	11.05
0530	cycle 2 miles		96	.083			4.28	4.25	8.53	11.05
0532	cycle 4 miles		80	.100			5.15	5.10	10.25	13.25
0534	cycle 6 miles		72	.111			5.70	5.65	11.35	14.75
0536	cycle 8 miles		64	.125			6.40	6.35	12.75	16.60
0544	25 MPH avg., cycle 4 miles		88	.091			4.66	4.63	9.29	12.05
0546	cycle 6 miles	B-34A	80	.100	L.C.Y.		5.15	5.10	10.25	13.25
0548	cycle 8 miles		72	.111			5.70	5.65	11.35	14.75
0550	30 MPH avg., cycle 4 miles		88	.091			4.66	4.63	9.29	12.05
0552	cycle 6 miles		80	.100			5.15	5.10	10.25	13.25
0554	cycle 8 miles		72	.111			5.70	5.65	11.35	14.75
1014	12 C.Y. truck, cycle 0.5 mile, 15 MPH avg., 15 min. wait/ld./uld.	B-34B	336	.024			1.22	1.72	2.94	3.73
1016	cycle 1 mile		300	.027			1.37	1.93	3.30	4.17

352

31 23 23.20 Hauling		Crew	Daily Output	Labor-Hours	Unit	Material	2021 Bare Costs Labor	Equipment	Total	Total Incl O&P
1018	cycle 2 miles	B-34B	252	.032	L.C.Y.		1.63	2.30	3.93	4.96
1020	cycle 4 miles		180	.044			2.28	3.22	5.50	6.95
1022	cycle 6 miles		144	.056			2.85	4.02	6.87	8.70
1024	cycle 8 miles		120	.067			3.42	4.83	8.25	10.40
1025	cycle 10 miles		96	.083			4.28	6.05	10.33	13.05
1026	20 MPH avg., cycle 0.5 mile		348	.023			1.18	1.66	2.84	3.59
1028	cycle 1 mile		312	.026			1.32	1.86	3.18	4.01
1030	cycle 2 miles		276	.029			1.49	2.10	3.59	4.53
1032	cycle 4 miles		216	.037			1.90	2.68	4.58	5.80
1034	cycle 6 miles		168	.048			2.44	3.45	5.89	7.45
1036	cycle 8 miles		144	.056			2.85	4.02	6.87	8.70
1038	cycle 10 miles		120	.067			3.42	4.83	8.25	10.40
1040	25 MPH avg., cycle 4 miles		228	.035			1.80	2.54	4.34	5.50
1042	cycle 6 miles		192	.042			2.14	3.02	5.16	6.50
1044	cycle 8 miles		168	.048			2.44	3.45	5.89	7.45
1046	cycle 10 miles		144	.056			2.85	4.02	6.87	8.70
1050	30 MPH avg., cycle 4 miles		252	.032			1.63	2.30	3.93	4.96
1052	cycle 6 miles		216	.037			1.90	2.68	4.58	5.80
1054	cycle 8 miles		180	.044			2.28	3.22	5.50	6.95
1056	cycle 10 miles		156	.051			2.63	3.71	6.34	8
1060	35 MPH avg., cycle 4 miles		264	.030			1.55	2.19	3.74	4.73
1062	cycle 6 miles		228	.035			1.80	2.54	4.34	5.50
1064	cycle 8 miles		204	.039			2.01	2.84	4.85	6.15
1066	cycle 10 miles		180	.044			2.28	3.22	5.50	6.95
1068	cycle 20 miles		120	.067			3.42	4.83	8.25	10.40
1069	cycle 30 miles		84	.095			4.89	6.90	11.79	14.90
1070	cycle 40 miles		72	.111			5.70	8.05	13.75	17.35
1072	40 MPH avg., cycle 6 miles		240	.033			1.71	2.41	4.12	5.20
1074	cycle 8 miles		216	.037			1.90	2.68	4.58	5.80
1076	cycle 10 miles		192	.042			2.14	3.02	5.16	6.50
1078	cycle 20 miles		120	.067			3.42	4.83	8.25	10.40
1080	cycle 30 miles		96	.083			4.28	6.05	10.33	13.05
1082	cycle 40 miles		72	.111			5.70	8.05	13.75	17.35
1084	cycle 50 miles		60	.133			6.85	9.65	16.50	21
1094	45 MPH avg., cycle 8 miles	B-34B	216	.037	L.C.Y.		1.90	2.68	4.58	5.80
1096	cycle 10 miles		204	.039			2.01	2.84	4.85	6.15
1098	cycle 20 miles		132	.061			3.11	4.39	7.50	9.50
1100	cycle 30 miles		108	.074			3.80	5.35	9.15	11.60
1102	cycle 40 miles		84	.095			4.89	6.90	11.79	14.90
1104	cycle 50 miles		72	.111			5.70	8.05	13.75	17.35
1106	50 MPH avg., cycle 10 miles		216	.037			1.90	2.68	4.58	5.80
1108	cycle 20 miles		144	.056			2.85	4.02	6.87	8.70
1110	cycle 30 miles		108	.074			3.80	5.35	9.15	11.60
1112	cycle 40 miles		84	.095			4.89	6.90	11.79	14.90
1114	cycle 50 miles		72	.111			5.70	8.05	13.75	17.35
1214	15 MPH avg., cycle 0.5 mile, 20 min. wait/ld./uld.		264	.030			1.55	2.19	3.74	4.73
1216	cycle 1 mile		240	.033			1.71	2.41	4.12	5.20
1218	cycle 2 miles		204	.039			2.01	2.84	4.85	6.15
1220	cycle 4 miles		156	.051			2.63	3.71	6.34	8
1222	cycle 6 miles		132	.061			3.11	4.39	7.50	9.50
1224	cycle 8 miles		108	.074			3.80	5.35	9.15	11.60
1225	cycle 10 miles		96	.083			4.28	6.05	10.33	13.05
1226	20 MPH avg., cycle 0.5 mile		264	.030			1.55	2.19	3.74	4.73

31 23 23.20 Hauling

		Crew	Daily Output	Labor-Hours	Unit	Material	2021 Bare Costs Labor	2021 Bare Costs Equipment	Total	Total Incl O&P
1228	cycle 1 mile	B-34B	252	.032	L.C.Y.		1.63	2.30	3.93	4.96
1230	cycle 2 miles		216	.037			1.90	2.68	4.58	5.80
1232	cycle 4 miles		180	.044			2.28	3.22	5.50	6.95
1234	cycle 6 miles		144	.056			2.85	4.02	6.87	8.70
1236	cycle 8 miles		132	.061			3.11	4.39	7.50	9.50
1238	cycle 10 miles		108	.074			3.80	5.35	9.15	11.60
1240	25 MPH avg., cycle 4 miles		192	.042			2.14	3.02	5.16	6.50
1242	cycle 6 miles		168	.048			2.44	3.45	5.89	7.45
1244	cycle 8 miles		144	.056			2.85	4.02	6.87	8.70
1246	cycle 10 miles		132	.061			3.11	4.39	7.50	9.50
1250	30 MPH avg., cycle 4 miles		204	.039			2.01	2.84	4.85	6.15
1252	cycle 6 miles		180	.044			2.28	3.22	5.50	6.95
1254	cycle 8 miles		156	.051			2.63	3.71	6.34	8
1256	cycle 10 miles		144	.056			2.85	4.02	6.87	8.70
1260	35 MPH avg., cycle 4 miles		216	.037			1.90	2.68	4.58	5.80
1262	cycle 6 miles		192	.042			2.14	3.02	5.16	6.50
1264	cycle 8 miles		168	.048			2.44	3.45	5.89	7.45
1266	cycle 10 miles		156	.051			2.63	3.71	6.34	8
1268	cycle 20 miles		108	.074			3.80	5.35	9.15	11.60
1269	cycle 30 miles		72	.111			5.70	8.05	13.75	17.35
1270	cycle 40 miles		60	.133			6.85	9.65	16.50	21
1272	40 MPH avg., cycle 6 miles	▼	192	.042	▼		2.14	3.02	5.16	6.50
1274	cycle 8 miles	B-34B	180	.044	L.C.Y.		2.28	3.22	5.50	6.95
1276	cycle 10 miles		156	.051			2.63	3.71	6.34	8
1278	cycle 20 miles		108	.074			3.80	5.35	9.15	11.60
1280	cycle 30 miles		84	.095			4.89	6.90	11.79	14.90
1282	cycle 40 miles		72	.111			5.70	8.05	13.75	17.35
1284	cycle 50 miles		60	.133			6.85	9.65	16.50	21
1294	45 MPH avg., cycle 8 miles		180	.044			2.28	3.22	5.50	6.95
1296	cycle 10 miles		168	.048			2.44	3.45	5.89	7.45
1298	cycle 20 miles		120	.067			3.42	4.83	8.25	10.40
1300	cycle 30 miles		96	.083			4.28	6.05	10.33	13.05
1302	cycle 40 miles		72	.111			5.70	8.05	13.75	17.35
1304	cycle 50 miles		60	.133			6.85	9.65	16.50	21
1306	50 MPH avg., cycle 10 miles		180	.044			2.28	3.22	5.50	6.95
1308	cycle 20 miles		132	.061			3.11	4.39	7.50	9.50
1310	cycle 30 miles		96	.083			4.28	6.05	10.33	13.05
1312	cycle 40 miles		84	.095			4.89	6.90	11.79	14.90
1314	cycle 50 miles		72	.111			5.70	8.05	13.75	17.35
1414	15 MPH avg., cycle 0.5 mile, 25 min. wait/ld./uld.		204	.039			2.01	2.84	4.85	6.15
1416	cycle 1 mile		192	.042			2.14	3.02	5.16	6.50
1418	cycle 2 miles		168	.048			2.44	3.45	5.89	7.45
1420	cycle 4 miles		132	.061			3.11	4.39	7.50	9.50
1422	cycle 6 miles		120	.067			3.42	4.83	8.25	10.40
1424	cycle 8 miles		96	.083			4.28	6.05	10.33	13.05
1425	cycle 10 miles		84	.095			4.89	6.90	11.79	14.90
1426	20 MPH avg., cycle 0.5 mile		216	.037			1.90	2.68	4.58	5.80
1428	cycle 1 mile		204	.039			2.01	2.84	4.85	6.15
1430	cycle 2 miles		180	.044			2.28	3.22	5.50	6.95
1432	cycle 4 miles		156	.051			2.63	3.71	6.34	8
1434	cycle 6 miles		132	.061			3.11	4.39	7.50	9.50
1436	cycle 8 miles		120	.067			3.42	4.83	8.25	10.40
1438	cycle 10 miles	▼	96	.083	▼		4.28	6.05	10.33	13.05

31 23 23 – Fill

31 23 23.20 Hauling

		Crew	Daily Output	Labor-Hours	Unit	Material	2021 Bare Costs Labor	2021 Bare Costs Equipment	Total	Total Incl O&P
1440	25 MPH avg., cycle 4 miles	B-34B	168	.048	L.C.Y.		2.44	3.45	5.89	7.45
1442	cycle 6 miles		144	.056			2.85	4.02	6.87	8.70
1444	cycle 8 miles		132	.061			3.11	4.39	7.50	9.50
1446	cycle 10 miles		108	.074			3.80	5.35	9.15	11.60
1450	30 MPH avg., cycle 4 miles		168	.048			2.44	3.45	5.89	7.45
1452	cycle 6 miles		156	.051			2.63	3.71	6.34	8
1454	cycle 8 miles		132	.061			3.11	4.39	7.50	9.50
1456	cycle 10 miles		120	.067			3.42	4.83	8.25	10.40
1460	35 MPH avg., cycle 4 miles		180	.044			2.28	3.22	5.50	6.95
1462	cycle 6 miles		156	.051			2.63	3.71	6.34	8
1464	cycle 8 miles	B-34B	144	.056	L.C.Y.		2.85	4.02	6.87	8.70
1466	cycle 10 miles		132	.061			3.11	4.39	7.50	9.50
1468	cycle 20 miles		96	.083			4.28	6.05	10.33	13.05
1469	cycle 30 miles		72	.111			5.70	8.05	13.75	17.35
1470	cycle 40 miles		60	.133			6.85	9.65	16.50	21
1472	40 MPH avg., cycle 6 miles		168	.048			2.44	3.45	5.89	7.45
1474	cycle 8 miles		156	.051			2.63	3.71	6.34	8
1476	cycle 10 miles		144	.056			2.85	4.02	6.87	8.70
1478	cycle 20 miles		96	.083			4.28	6.05	10.33	13.05
1480	cycle 30 miles		84	.095			4.89	6.90	11.79	14.90
1482	cycle 40 miles		60	.133			6.85	9.65	16.50	21
1484	cycle 50 miles		60	.133			6.85	9.65	16.50	21
1494	45 MPH avg., cycle 8 miles		156	.051			2.63	3.71	6.34	8
1496	cycle 10 miles		144	.056			2.85	4.02	6.87	8.70
1498	cycle 20 miles		108	.074			3.80	5.35	9.15	11.60
1500	cycle 30 miles		84	.095			4.89	6.90	11.79	14.90
1502	cycle 40 miles		72	.111			5.70	8.05	13.75	17.35
1504	cycle 50 miles		60	.133			6.85	9.65	16.50	21
1506	50 MPH avg., cycle 10 miles		156	.051			2.63	3.71	6.34	8
1508	cycle 20 miles		120	.067			3.42	4.83	8.25	10.40
1510	cycle 30 miles		96	.083			4.28	6.05	10.33	13.05
1512	cycle 40 miles		72	.111			5.70	8.05	13.75	17.35
1514	cycle 50 miles		60	.133			6.85	9.65	16.50	21
1614	15 MPH avg., cycle 0.5 mile, 30 min. wait/ld./uld.		180	.044			2.28	3.22	5.50	6.95
1616	cycle 1 mile		168	.048			2.44	3.45	5.89	7.45
1618	cycle 2 miles		144	.056			2.85	4.02	6.87	8.70
1620	cycle 4 miles		120	.067			3.42	4.83	8.25	10.40
1622	cycle 6 miles		108	.074			3.80	5.35	9.15	11.60
1624	cycle 8 miles		84	.095			4.89	6.90	11.79	14.90
1625	cycle 10 miles		84	.095			4.89	6.90	11.79	14.90
1626	20 MPH avg., cycle 0.5 mile		180	.044			2.28	3.22	5.50	6.95
1628	cycle 1 mile		168	.048			2.44	3.45	5.89	7.45
1630	cycle 2 miles		156	.051			2.63	3.71	6.34	8
1632	cycle 4 miles		132	.061			3.11	4.39	7.50	9.50
1634	cycle 6 miles		120	.067			3.42	4.83	8.25	10.40
1636	cycle 8 miles		108	.074			3.80	5.35	9.15	11.60
1638	cycle 10 miles		96	.083			4.28	6.05	10.33	13.05
1640	25 MPH avg., cycle 4 miles		144	.056			2.85	4.02	6.87	8.70
1642	cycle 6 miles		132	.061			3.11	4.39	7.50	9.50
1644	cycle 8 miles		108	.074			3.80	5.35	9.15	11.60
1646	cycle 10 miles		108	.074			3.80	5.35	9.15	11.60
1650	30 MPH avg., cycle 4 miles	B-34B	144	.056	L.C.Y.		2.85	4.02	6.87	8.70
1652	cycle 6 miles		132	.061			3.11	4.39	7.50	9.50

355

31 23 23.20 Hauling		Crew	Daily Output	Labor-Hours	Unit	Material	2021 Bare Costs Labor	2021 Bare Costs Equipment	Total	Total Incl O&P
1654	cycle 8 miles	B-34B	120	.067	L.C.Y.		3.42	4.83	8.25	10.40
1656	cycle 10 miles		108	.074			3.80	5.35	9.15	11.60
1660	35 MPH avg., cycle 4 miles		156	.051			2.63	3.71	6.34	8
1662	cycle 6 miles		144	.056			2.85	4.02	6.87	8.70
1664	cycle 8 miles		132	.061			3.11	4.39	7.50	9.50
1666	cycle 10 miles		120	.067			3.42	4.83	8.25	10.40
1668	cycle 20 miles		84	.095			4.89	6.90	11.79	14.90
1669	cycle 30 miles		72	.111			5.70	8.05	13.75	17.35
1670	cycle 40 miles		60	.133			6.85	9.65	16.50	21
1672	40 MPH avg., cycle 6 miles		144	.056			2.85	4.02	6.87	8.70
1674	cycle 8 miles		132	.061			3.11	4.39	7.50	9.50
1676	cycle 10 miles		120	.067			3.42	4.83	8.25	10.40
1678	cycle 20 miles		96	.083			4.28	6.05	10.33	13.05
1680	cycle 30 miles		72	.111			5.70	8.05	13.75	17.35
1682	cycle 40 miles		60	.133			6.85	9.65	16.50	21
1684	cycle 50 miles		48	.167			8.55	12.05	20.60	26
1694	45 MPH avg., cycle 8 miles		144	.056			2.85	4.02	6.87	8.70
1696	cycle 10 miles		132	.061			3.11	4.39	7.50	9.50
1698	cycle 20 miles		96	.083			4.28	6.05	10.33	13.05
1700	cycle 30 miles		84	.095			4.89	6.90	11.79	14.90
1702	cycle 40 miles		60	.133			6.85	9.65	16.50	21
1704	cycle 50 miles		60	.133			6.85	9.65	16.50	21
1706	50 MPH avg., cycle 10 miles		132	.061			3.11	4.39	7.50	9.50
1708	cycle 20 miles		108	.074			3.80	5.35	9.15	11.60
1710	cycle 30 miles		84	.095			4.89	6.90	11.79	14.90
1712	cycle 40 miles		72	.111			5.70	8.05	13.75	17.35
1714	cycle 50 miles		60	.133			6.85	9.65	16.50	21
2000	Hauling, 8 C.Y. truck, small project cost per hour	B-34A	8	1	Hr.		51.50	51	102.50	133
2100	12 C.Y. truck	B-34B	8	1			51.50	72.50	124	156
2150	16.5 C.Y. truck	B-34C	8	1			51.50	79.50	131	164
2175	18 C.Y. 8 wheel truck	B-34I	8	1			51.50	94	145.50	181
2200	20 C.Y. truck	B-34D	8	1			51.50	81.50	133	167
2300	Grading at dump, or embankment if required, by dozer	B-10B	1000	.012	L.C.Y.		.65	1.52	2.17	2.64
2310	Spotter at fill or cut, if required	1 Clab	8	1	Hr.		44.50		44.50	66.50
9014	18 C.Y. truck, 8 wheels,15 min. wait/ld./uld.,15 MPH, cycle 0.5 mi.	B-34I	504	.016	L.C.Y.		.81	1.50	2.31	2.86
9016	cycle 1 mile		450	.018			.91	1.67	2.58	3.20
9018	cycle 2 miles		378	.021			1.09	1.99	3.08	3.81
9020	cycle 4 miles		270	.030			1.52	2.79	4.31	5.35
9022	cycle 6 miles		216	.037			1.90	3.49	5.39	6.70
9024	cycle 8 miles	B-34I	180	.044	L.C.Y.		2.28	4.19	6.47	8
9025	cycle 10 miles		144	.056			2.85	5.25	8.10	10
9026	20 MPH avg., cycle 0.5 mile		522	.015			.79	1.44	2.23	2.77
9028	cycle 1 mile		468	.017			.88	1.61	2.49	3.08
9030	cycle 2 miles		414	.019			.99	1.82	2.81	3.48
9032	cycle 4 miles		324	.025			1.27	2.33	3.60	4.45
9034	cycle 6 miles		252	.032			1.63	2.99	4.62	5.70
9036	cycle 8 miles		216	.037			1.90	3.49	5.39	6.70
9038	cycle 10 miles		180	.044			2.28	4.19	6.47	8
9040	25 MPH avg., cycle 4 miles		342	.023			1.20	2.20	3.40	4.21
9042	cycle 6 miles		288	.028			1.43	2.62	4.05	5
9044	cycle 8 miles		252	.032			1.63	2.99	4.62	5.70
9046	cycle 10 miles		216	.037			1.90	3.49	5.39	6.70
9050	30 MPH avg., cycle 4 miles		378	.021			1.09	1.99	3.08	3.81

31 23 23.20 Hauling	Crew	Daily Output	Labor-Hours	Unit	Material	Labor	Equipment	Total	Total Incl O&P
						2021 Bare Costs			
9052　cycle 6 miles	B-34I	324	.025	L.C.Y.		1.27	2.33	3.60	4.45
9054　cycle 8 miles		270	.030			1.52	2.79	4.31	5.35
9056　cycle 10 miles		234	.034			1.75	3.22	4.97	6.15
9060　35 MPH avg., cycle 4 miles		396	.020			1.04	1.90	2.94	3.64
9062　cycle 6 miles		342	.023			1.20	2.20	3.40	4.21
9064　cycle 8 miles		288	.028			1.43	2.62	4.05	5
9066　cycle 10 miles		270	.030			1.52	2.79	4.31	5.35
9068　cycle 20 miles		162	.049			2.53	4.65	7.18	8.90
9070　cycle 30 miles		126	.063			3.26	6	9.26	11.45
9072　cycle 40 miles		90	.089			4.56	8.35	12.91	16
9074　40 MPH avg., cycle 6 miles		360	.022			1.14	2.09	3.23	4
9076　cycle 8 miles		324	.025			1.27	2.33	3.60	4.45
9078　cycle 10 miles		288	.028			1.43	2.62	4.05	5
9080　cycle 20 miles		180	.044			2.28	4.19	6.47	8
9082　cycle 30 miles		144	.056			2.85	5.25	8.10	10
9084　cycle 40 miles		108	.074			3.80	7	10.80	13.40
9086　cycle 50 miles		90	.089			4.56	8.35	12.91	16
9094　45 MPH avg., cycle 8 miles		324	.025			1.27	2.33	3.60	4.45
9096　cycle 10 miles		306	.026			1.34	2.46	3.80	4.72
9098　cycle 20 miles		198	.040			2.07	3.81	5.88	7.30
9100　cycle 30 miles		144	.056			2.85	5.25	8.10	10
9102　cycle 40 miles		126	.063			3.26	6	9.26	11.45
9104　cycle 50 miles		108	.074			3.80	7	10.80	13.40
9106　50 MPH avg., cycle 10 miles		324	.025			1.27	2.33	3.60	4.45
9108　cycle 20 miles		216	.037			1.90	3.49	5.39	6.70
9110　cycle 30 miles		162	.049			2.53	4.65	7.18	8.90
9112　cycle 40 miles		126	.063			3.26	6	9.26	11.45
9114　cycle 50 miles		108	.074			3.80	7	10.80	13.40
9214　20 min. wait/ld./uld.,15 MPH, cycle 0.5 mi.	B-34I	396	.020	L.C.Y.		1.04	1.90	2.94	3.64
9216　cycle 1 mile		360	.022			1.14	2.09	3.23	4
9218　cycle 2 miles		306	.026			1.34	2.46	3.80	4.72
9220　cycle 4 miles		234	.034			1.75	3.22	4.97	6.15
9222　cycle 6 miles		198	.040			2.07	3.81	5.88	7.30
9224　cycle 8 miles		162	.049			2.53	4.65	7.18	8.90
9225　cycle 10 miles		144	.056			2.85	5.25	8.10	10
9226　20 MPH avg., cycle 0.5 mile		396	.020			1.04	1.90	2.94	3.64
9228　cycle 1 mile		378	.021			1.09	1.99	3.08	3.81
9230　cycle 2 miles		324	.025			1.27	2.33	3.60	4.45
9232　cycle 4 miles		270	.030			1.52	2.79	4.31	5.35
9234　cycle 6 miles		216	.037			1.90	3.49	5.39	6.70
9236　cycle 8 miles		198	.040			2.07	3.81	5.88	7.30
9238　cycle 10 miles		162	.049			2.53	4.65	7.18	8.90
9240　25 MPH avg., cycle 4 miles		288	.028			1.43	2.62	4.05	5
9242　cycle 6 miles		252	.032			1.63	2.99	4.62	5.70
9244　cycle 8 miles		216	.037			1.90	3.49	5.39	6.70
9246　cycle 10 miles		198	.040			2.07	3.81	5.88	7.30
9250　30 MPH avg., cycle 4 miles		306	.026			1.34	2.46	3.80	4.72
9252　cycle 6 miles		270	.030			1.52	2.79	4.31	5.35
9254　cycle 8 miles		234	.034			1.75	3.22	4.97	6.15
9256　cycle 10 miles		216	.037			1.90	3.49	5.39	6.70
9260　35 MPH avg., cycle 4 miles		324	.025			1.27	2.33	3.60	4.45
9262　cycle 6 miles		288	.028			1.43	2.62	4.05	5
9264　cycle 8 miles		252	.032			1.63	2.99	4.62	5.70

For customer support on your Electrical Costs with RSMeans Data, call 800.448.8182.

357

31 23 23.20 Hauling		Crew	Daily Output	Labor-Hours	Unit	Material	2021 Bare Costs Labor	2021 Bare Costs Equipment	Total	Total Incl O&P
9266	cycle 10 miles	B-34I	234	.034	L.C.Y.		1.75	3.22	4.97	6.15
9268	cycle 20 miles		162	.049			2.53	4.65	7.18	8.90
9270	cycle 30 miles		108	.074			3.80	7	10.80	13.40
9272	cycle 40 miles		90	.089			4.56	8.35	12.91	16
9274	40 MPH avg., cycle 6 miles		288	.028			1.43	2.62	4.05	5
9276	cycle 8 miles		270	.030			1.52	2.79	4.31	5.35
9278	cycle 10 miles		234	.034			1.75	3.22	4.97	6.15
9280	cycle 20 miles		162	.049			2.53	4.65	7.18	8.90
9282	cycle 30 miles		126	.063			3.26	6	9.26	11.45
9284	cycle 40 miles		108	.074			3.80	7	10.80	13.40
9286	cycle 50 miles		90	.089			4.56	8.35	12.91	16
9294	45 MPH avg., cycle 8 miles		270	.030			1.52	2.79	4.31	5.35
9296	cycle 10 miles		252	.032			1.63	2.99	4.62	5.70
9298	cycle 20 miles		180	.044			2.28	4.19	6.47	8
9300	cycle 30 miles		144	.056			2.85	5.25	8.10	10
9302	cycle 40 miles		108	.074			3.80	7	10.80	13.40
9304	cycle 50 miles	B-34I	90	.089	L.C.Y.		4.56	8.35	12.91	16
9306	50 MPH avg., cycle 10 miles		270	.030			1.52	2.79	4.31	5.35
9308	cycle 20 miles		198	.040			2.07	3.81	5.88	7.30
9310	cycle 30 miles		144	.056			2.85	5.25	8.10	10
9312	cycle 40 miles		126	.063			3.26	6	9.26	11.45
9314	cycle 50 miles		108	.074			3.80	7	10.80	13.40
9414	25 min. wait/ld./uld.,15 MPH, cycle 0.5 mi.		306	.026			1.34	2.46	3.80	4.72
9416	cycle 1 mile		288	.028			1.43	2.62	4.05	5
9418	cycle 2 miles		252	.032			1.63	2.99	4.62	5.70
9420	cycle 4 miles		198	.040			2.07	3.81	5.88	7.30
9422	cycle 6 miles		180	.044			2.28	4.19	6.47	8
9424	cycle 8 miles		144	.056			2.85	5.25	8.10	10
9425	cycle 10 miles		126	.063			3.26	6	9.26	11.45
9426	20 MPH avg., cycle 0.5 mile		324	.025			1.27	2.33	3.60	4.45
9428	cycle 1 mile		306	.026			1.34	2.46	3.80	4.72
9430	cycle 2 miles		270	.030			1.52	2.79	4.31	5.35
9432	cycle 4 miles		234	.034			1.75	3.22	4.97	6.15
9434	cycle 6 miles		198	.040			2.07	3.81	5.88	7.30
9436	cycle 8 miles		180	.044			2.28	4.19	6.47	8
9438	cycle 10 miles		144	.056			2.85	5.25	8.10	10
9440	25 MPH avg., cycle 4 miles		252	.032			1.63	2.99	4.62	5.70
9442	cycle 6 miles		216	.037			1.90	3.49	5.39	6.70
9444	cycle 8 miles		198	.040			2.07	3.81	5.88	7.30
9446	cycle 10 miles		180	.044			2.28	4.19	6.47	8
9450	30 MPH avg., cycle 4 miles		252	.032			1.63	2.99	4.62	5.70
9452	cycle 6 miles		234	.034			1.75	3.22	4.97	6.15
9454	cycle 8 miles		198	.040			2.07	3.81	5.88	7.30
9456	cycle 10 miles		180	.044			2.28	4.19	6.47	8
9460	35 MPH avg., cycle 4 miles		270	.030			1.52	2.79	4.31	5.35
9462	cycle 6 miles		234	.034			1.75	3.22	4.97	6.15
9464	cycle 8 miles		216	.037			1.90	3.49	5.39	6.70
9466	cycle 10 miles		198	.040			2.07	3.81	5.88	7.30
9468	cycle 20 miles		144	.056			2.85	5.25	8.10	10
9470	cycle 30 miles		108	.074			3.80	7	10.80	13.40
9472	cycle 40 miles		90	.089			4.56	8.35	12.91	16
9474	40 MPH avg., cycle 6 miles		252	.032			1.63	2.99	4.62	5.70
9476	cycle 8 miles		234	.034			1.75	3.22	4.97	6.15

For customer support on your Electrical Costs with RSMeans Data, call 800.448.8182.

31 23 23 – Fill

31 23 23.20 Hauling	Crew	Daily Output	Labor-Hours	Unit	Material	2021 Bare Costs Labor	2021 Bare Costs Equipment	Total	Total Incl O&P
9478 cycle 10 miles	B-34I	216	.037	L.C.Y.		1.90	3.49	5.39	6.70
9480 cycle 20 miles		144	.056			2.85	5.25	8.10	10
9482 cycle 30 miles		126	.063			3.26	6	9.26	11.45
9484 cycle 40 miles		90	.089			4.56	8.35	12.91	16
9486 cycle 50 miles	B-34I	90	.089	L.C.Y.		4.56	8.35	12.91	16
9494 45 MPH avg., cycle 8 miles		234	.034			1.75	3.22	4.97	6.15
9496 cycle 10 miles		216	.037			1.90	3.49	5.39	6.70
9498 cycle 20 miles		162	.049			2.53	4.65	7.18	8.90
9500 cycle 30 miles		126	.063			3.26	6	9.26	11.45
9502 cycle 40 miles		108	.074			3.80	7	10.80	13.40
9504 cycle 50 miles		90	.089			4.56	8.35	12.91	16
9506 50 MPH avg., cycle 10 miles		234	.034			1.75	3.22	4.97	6.15
9508 cycle 20 miles		180	.044			2.28	4.19	6.47	8
9510 cycle 30 miles		144	.056			2.85	5.25	8.10	10
9512 cycle 40 miles		108	.074			3.80	7	10.80	13.40
9514 cycle 50 miles		90	.089			4.56	8.35	12.91	16
9614 30 min. wait/ld./uld.,15 MPH, cycle 0.5 mi.		270	.030			1.52	2.79	4.31	5.35
9616 cycle 1 mile		252	.032			1.63	2.99	4.62	5.70
9618 cycle 2 miles		216	.037			1.90	3.49	5.39	6.70
9620 cycle 4 miles		180	.044			2.28	4.19	6.47	8
9622 cycle 6 miles		162	.049			2.53	4.65	7.18	8.90
9624 cycle 8 miles		126	.063			3.26	6	9.26	11.45
9625 cycle 10 miles		126	.063			3.26	6	9.26	11.45
9626 20 MPH avg., cycle 0.5 mile		270	.030			1.52	2.79	4.31	5.35
9628 cycle 1 mile		252	.032			1.63	2.99	4.62	5.70
9630 cycle 2 miles		234	.034			1.75	3.22	4.97	6.15
9632 cycle 4 miles		198	.040			2.07	3.81	5.88	7.30
9634 cycle 6 miles		180	.044			2.28	4.19	6.47	8
9636 cycle 8 miles		162	.049			2.53	4.65	7.18	8.90
9638 cycle 10 miles		144	.056			2.85	5.25	8.10	10
9640 25 MPH avg., cycle 4 miles		216	.037			1.90	3.49	5.39	6.70
9642 cycle 6 miles		198	.040			2.07	3.81	5.88	7.30
9644 cycle 8 miles		180	.044			2.28	4.19	6.47	8
9646 cycle 10 miles		162	.049			2.53	4.65	7.18	8.90
9650 30 MPH avg., cycle 4 miles		216	.037			1.90	3.49	5.39	6.70
9652 cycle 6 miles		198	.040			2.07	3.81	5.88	7.30
9654 cycle 8 miles		180	.044			2.28	4.19	6.47	8
9656 cycle 10 miles		162	.049			2.53	4.65	7.18	8.90
9660 35 MPH avg., cycle 4 miles		234	.034			1.75	3.22	4.97	6.15
9662 cycle 6 miles		216	.037			1.90	3.49	5.39	6.70
9664 cycle 8 miles		198	.040			2.07	3.81	5.88	7.30
9666 cycle 10 miles		180	.044			2.28	4.19	6.47	8
9668 cycle 20 miles		126	.063			3.26	6	9.26	11.45
9670 cycle 30 miles		108	.074			3.80	7	10.80	13.40
9672 cycle 40 miles		90	.089			4.56	8.35	12.91	16
9674 40 MPH avg., cycle 6 miles	B-34I	216	.037	L.C.Y.		1.90	3.49	5.39	6.70
9676 cycle 8 miles		198	.040			2.07	3.81	5.88	7.30
9678 cycle 10 miles		180	.044			2.28	4.19	6.47	8
9680 cycle 20 miles		144	.056			2.85	5.25	8.10	10
9682 cycle 30 miles		108	.074			3.80	7	10.80	13.40
9684 cycle 40 miles		90	.089			4.56	8.35	12.91	16
9686 cycle 50 miles		72	.111			5.70	10.45	16.15	20
9694 45 MPH avg., cycle 8 miles		216	.037			1.90	3.49	5.39	6.70

359

31 23 Excavation and Fill

31 23 23 – Fill

31 23 23.20 Hauling

		Crew	Daily Output	Labor-Hours	Unit	Material	2021 Bare Costs Labor	Equipment	Total	Total Incl O&P
9696	cycle 10 miles	B-34I	198	.040	L.C.Y.		2.07	3.81	5.88	7.30
9698	cycle 20 miles		144	.056			2.85	5.25	8.10	10
9700	cycle 30 miles		126	.063			3.26	6	9.26	11.45
9702	cycle 40 miles		108	.074			3.80	7	10.80	13.40
9704	cycle 50 miles		90	.089			4.56	8.35	12.91	16
9706	50 MPH avg., cycle 10 miles		198	.040			2.07	3.81	5.88	7.30
9708	cycle 20 miles		162	.049			2.53	4.65	7.18	8.90
9710	cycle 30 miles		126	.063			3.26	6	9.26	11.45
9712	cycle 40 miles		108	.074			3.80	7	10.80	13.40
9714	cycle 50 miles		90	.089			4.56	8.35	12.91	16

31 23 23.23 Compaction

		Crew	Daily Output	Labor-Hours	Unit	Material	2021 Bare Costs Labor	Equipment	Total	Total Incl O&P
0010	**COMPACTION**									
7000	Walk behind, vibrating plate 18" wide, 6" lifts, 2 passes	A-1D	200	.040	E.C.Y.		1.78	.16	1.94	2.83
7020	3 passes		185	.043			1.92	.17	2.09	3.05
7040	4 passes		140	.057			2.54	.23	2.77	4.04

Estimating Tips

32 01 00 Operations and Maintenance of Exterior Improvements

- Recycling of asphalt pavement is becoming very popular and is an alternative to removal and replacement. It can be a good value engineering proposal if removed pavement can be recycled, either at the project site or at another site that is reasonably close to the project site. Sections on repair of flexible and rigid pavement are included.

32 10 00 Bases, Ballasts, and Paving

- When estimating paving, keep in mind the project schedule. Also note that prices for asphalt and concrete are generally higher in the cold seasons. Lines for pavement markings, including tactile warning systems and fence lines, are included.

32 90 00 Planting

- The timing of planting and guarantee specifications often dictate the costs for establishing tree and shrub growth and a stand of grass or ground cover. Establish the work performance schedule to coincide with the local planting season. Maintenance and growth guarantees can add 20–100% to the total landscaping cost and can be contractually cumbersome. The cost to replace trees and shrubs can be as high as 5% of the total cost, depending on the planting zone, soil conditions, and time of year.

Reference Numbers

Reference numbers are shown at the beginning of some major classifications. These numbers refer to related items in the Reference Section. The reference information may be an estimating procedure, an alternate pricing method, or technical information.

Note: Not all subdivisions listed here necessarily appear. ∎

Same Data. Simplified.

Enjoy the convenience and efficiency of accessing your costs anywhere:

- **Skip the multiplier** by setting your location
- **Quickly search,** edit, favorite and share costs
- **Stay on top of price changes** with automatic updates

Discover more at rsmeans.com/online

32 31 13.30 Fence, Chain Link, Gates and Posts		Crew	Daily Output	Labor-Hours	Unit	Material	2021 Bare Costs Labor	2021 Bare Costs Equipment	Total	Total Incl O&P
0010	**FENCE, CHAIN LINK, GATES & POSTS**									
0011	(1/3 post length in ground)									
0013	For Concrete, See Section 03 31									
6710	Corner post, galv. steel, 4" OD, set in conc., 4'	B-80	65	.492	Ea.	91.50	24	16.20	131.70	154
6715	6'		63	.508		121	25	16.70	162.70	188
6720	7'		61	.525		120	25.50	17.25	162.75	189
6725	8'		65	.492		185	24	16.20	225.20	258
7900	Auger fence post hole, 3' deep, medium soil, by hand	1 Clab	30	.267			11.85		11.85	17.65
7925	By machine	B-80	175	.183			8.90	6	14.90	19.90
7950	Rock, with jackhammer	B-9	32	1.250			56	11.10	67.10	96
7975	With rock drill	B-47C	65	.246			12.30	27.50	39.80	49

Estimating Tips
33 10 00 Water Utilities
33 30 00 Sanitary Sewerage Utilities
33 40 00 Storm Drainage Utilities

- Never assume that the water, sewer, and drainage lines will go in at the early stages of the project. Consider the site access needs before dividing the site in half with open trenches, loose pipe, and machinery obstructions. Always inspect the site to establish that the site drawings are complete. Check off all existing utilities on your drawings as you locate them. Be especially careful with underground utilities because appurtenances are sometimes buried during regrading or repaving operations. If you find any discrepancies, mark up the site plan for further research. Differing site conditions can be very costly if discovered later in the project.

- See also Section 33 01 00 for restoration of pipe where removal/replacement may be undesirable. Use of new types of piping materials can reduce the overall project cost. Owners/design engineers should consider the installing contractor as a valuable source of current information on utility products and local conditions that could lead to significant cost savings.

Reference Numbers

Reference numbers are shown at the beginning of some major classifications. These numbers refer to related items in the Reference Section. The reference information may be an estimating procedure, an alternate pricing method, or technical information.

Note: Not all subdivisions listed here necessarily appear. ■

Same Data. Simplified.

Enjoy the convenience and efficiency of accessing your costs anywhere:

- **Skip the multiplier** by setting your location
- **Quickly search,** edit, favorite and share costs
- **Stay on top of price changes** with automatic updates

Discover more at rsmeans.com/online

---

33 05 07.13 Utility Directional Drilling

		Daily Output	Labor-Hours	Unit	Material	2021 Bare Costs Labor	2021 Bare Costs Equipment	Total	Total Incl O&P
0010	**UTILITY DIRECTIONAL DRILLING**								
0011	Excluding access and splice pits (if required) and conduit (required)								
0012	Drilled hole diameters shown should be 50% larger than conduit								
0013	Assume access to H2O & removal of spoil/drilling mud as non-hazard material								
0014	Actual production rates adj. to account for risk factors for each soil type								
0100	Sand, silt, clay, common earth								
0110	Mobilization or demobilization	B-82A	1	32	Ea.		1,600	2,500	4,100	5,125
0120	6" diameter		180	.178	L.F.		8.90	13.90	22.80	28.50
0130	12" diameter		70	.457			23	36	59	73.50
0140	18" diameter		50	.640			32	50	82	103
0150	24" diameter		35	.914			45.50	71.50	117	147
0210	Mobilization or demobilization	B-82B	1	32	Ea.		1,600	2,600	4,200	5,250
0220	6" diameter		125	.256	L.F.		12.80	21	33.80	42
0230	12" diameter		50	.640			32	52	84	105
0240	18" diameter		30	1.067			53.50	87	140.50	175
0250	24" diameter		25	1.280			64	104	168	211
0260	30" diameter		20	1.600			80	130	210	262
0270	36" diameter		15	2.133			107	174	281	350
0300	Hard rock (solid bed, 24,000+ psi)								
0310	Mobilization or demobilization	B-82C	1	32	Ea.		1,600	2,775	4,375	5,425
0320	6" diameter		30	1.067	L.F.		53.50	93	146.50	182
0330	12" diameter		12	2.667			133	232	365	455
0340	18" diameter		8	4			200	350	550	685
0344	20" diameter		7.50	4.267			213	370	583	730
0350	24" diameter		7	4.571			228	400	628	775
0360	30" diameter		6	5.333			266	465	731	905
0370	36" diameter		4.50	7.111			355	620	975	1,200

33 05 07.23 Utility Boring and Jacking

		Crew	Daily Output	Labor-Hours	Unit	Material	2021 Bare Costs Labor	2021 Bare Costs Equipment	Total	Total Incl O&P
0010	**UTILITY BORING AND JACKING**									
0011	Casing only, 100' minimum,									
0020	not incl. jacking pits or dewatering									
0100	Roadwork, 1/2" thick wall, 24" diameter casing	B-42	20	3.200	L.F.	142	159	53.50	354.50	455
0200	36" diameter		16	4		219	199	66.50	484.50	615
0300	48" diameter		15	4.267		305	212	71	588	735
0500	Railroad work, 24" diameter		15	4.267		142	212	71	425	555
0600	36" diameter		14	4.571		219	228	76	523	665
0700	48" diameter		12	5.333		305	265	89	659	835
0900	For ledge, add								20%	20%
1000	Small diameter boring, 3", sandy soil	B-82	900	.018		23	.89	.20	24.09	27
1040	Rocky soil	"	500	.032		23	1.60	.37	24.97	28.50
1100	Prepare jacking pits, incl. mobilization & demobilization, minimum				Ea.				3,225	3,700
1101	Maximum				"				22,000	25,500

33 05 07.36 Microtunneling

		Crew	Daily Output	Labor-Hours	Unit	Material	2021 Bare Costs Labor	2021 Bare Costs Equipment	Total	Total Incl O&P
0010	**MICROTUNNELING**									
0011	Not including excavation, backfill, shoring,									
0020	or dewatering, average 50'/day, slurry method									
0100	24" to 48" outside diameter, minimum				L.F.				965	965
0110	Adverse conditions, add				"				500	500
1000	Rent microtunneling machine, average monthly lease				Month				97,500	107,000
1010	Operating technician				Day				630	705
1100	Mobilization and demobilization, minimum				Job				41,200	45,900
1110	Maximum				"				445,500	490,500

33 05 Common Work Results for Utilities

33 05 61 – Concrete Manholes

33 05 61.10 Storm Drainage Manholes, Frames and Covers

33 05 61.10 Storm Drainage Manholes, Frames and Covers	Crew	Daily Output	Labor-Hours	Unit	Material	2021 Bare Costs Labor	2021 Bare Costs Equipment	Total	Total Incl O&P
0010	**STORM DRAINAGE MANHOLES, FRAMES & COVERS**									
0020	Excludes footing, excavation, backfill (See line items for frame & cover)									
0050	Brick, 4' inside diameter, 4' deep	D-1	1	16	Ea.	700	780		1,480	1,950
0100	6' deep		.70	22.857		1,000	1,125		2,125	2,775
0150	8' deep		.50	32	▼	1,300	1,550		2,850	3,775
0200	Add for 1' depth increase		4	4	V.L.F.	176	195		371	485
0400	Concrete blocks (radial), 4' ID, 4' deep		1.50	10.667	Ea.	430	520		950	1,250
0500	6' deep		1	16		570	780		1,350	1,800
0600	8' deep		.70	22.857	▼	710	1,125		1,835	2,450
0700	For depths over 8', add		5.50	2.909	V.L.F.	73	142		215	293
0800	Concrete, cast in place, 4' x 4', 8" thick, 4' deep	C-14H	2	24	Ea.	680	1,275	13.70	1,968.70	2,650
0900	6' deep		1.50	32		985	1,725	18.25	2,728.25	3,650
1000	8' deep		1	48	▼	1,400	2,575	27.50	4,002.50	5,400
1100	For depths over 8', add	▼	8	6	V.L.F.	166	320	3.42	489.42	665
1110	Precast, 4' ID, 4' deep	B-22	4.10	7.317	Ea.	905	380	70	1,355	1,625
1120	6' deep		3	10		1,075	515	95.50	1,685.50	2,050
1130	8' deep		2	15	▼	1,900	775	143	2,818	3,400
1140	For depths over 8', add	▼	16	1.875	V.L.F.	107	97	17.90	221.90	283
1150	5' ID, 4' deep	B-6	3	8	Ea.	1,875	385	72	2,332	2,725
1160	6' deep		2	12		3,250	575	108	3,933	4,550
1170	8' deep		1.50	16		2,750	770	144	3,664	4,325
1180	For depths over 8', add		12	2	V.L.F.	470	96	18	584	680
1190	6' ID, 4' deep		2	12	Ea.	2,100	575	108	2,783	3,275
1200	6' deep		1.50	16		4,600	770	144	5,514	6,350
1210	8' deep		1	24		5,375	1,150	216	6,741	7,900
1220	For depths over 8', add	▼	8	3	V.L.F.	425	144	27	596	715
1250	Slab tops, precast, 8" thick									
1300	4' diameter manhole	B-6	8	3	Ea.	288	144	27	459	560
1400	5' diameter manhole		7.50	3.200		685	154	29	868	1,000
1500	6' diameter manhole	▼	7	3.429	▼	715	165	31	911	1,075
3800	Steps, heavyweight cast iron, 7" x 9"	1 Bric	40	.200	Ea.	17.25	10.75		28	35
3900	8" x 9"		40	.200		20.50	10.75		31.25	39
3928	12" x 10-1/2"		40	.200		31	10.75		41.75	50
4000	Standard sizes, galvanized steel		40	.200		24	10.75		34.75	42.50
4100	Aluminum		40	.200		31	10.75		41.75	50
4150	Polyethylene	▼	40	.200	▼	32.50	10.75		43.25	52

33 05 63 – Concrete Vaults and Chambers

33 05 63.13 Precast Concrete Utility Structures

		Crew	Daily Output	Labor-Hours	Unit	Material	Labor	Equipment	Total	Total Incl O&P
0010	**PRECAST CONCRETE UTILITY STRUCTURES**, 6" thick									
0050	5' x 10' x 6' high, ID	B-13	2	28	Ea.	1,825	1,350	293	3,468	4,350
0100	6' x 10' x 6' high, ID		2	28		1,900	1,350	293	3,543	4,450
0150	5' x 12' x 6' high, ID		2	28		2,000	1,350	293	3,643	4,550
0200	6' x 12' x 6' high, ID		1.80	31.111		2,250	1,500	325	4,075	5,075
0250	6' x 13' x 6' high, ID		1.50	37.333		2,950	1,800	390	5,140	6,375
0300	8' x 14' x 7' high, ID	▼	1	56	▼	3,200	2,700	585	6,485	8,175

33 05 97 – Identification and Signage for Utilities

33 05 97.05 Utility Connection

		Crew	Daily Output	Labor-Hours	Unit	Material	Labor	Equipment	Total	Total Incl O&P
0010	**UTILITY CONNECTION**									
0020	Water, sanitary, stormwater, gas, single connection	B-14	1	48	Ea.	4,925	2,225	216	7,366	9,000
0030	Telecommunication	"	3	16	"	420	745	72	1,237	1,650

33 05 Common Work Results for Utilities

33 05 97 – Identification and Signage for Utilities

33 05 97.10 Utility Accessories	Crew	Daily Output	Labor-Hours	Unit	Material	2021 Bare Costs Labor	2021 Bare Costs Equipment	Total	Total Incl O&P
0010 UTILITY ACCESSORIES									
0400　Underground tape, detectable, reinforced, alum. foil core, 2"	1 Clab	150	.053	C.L.F.	3	2.37		5.37	6.85

33 42 Stormwater Conveyance

33 42 33 – Stormwater Curbside Drains and Inlets

33 42 33.13 Catch Basins

		Crew	Daily Output	Labor-Hours	Unit	Material	Labor	Equipment	Total	Total Incl O&P
0010	**CATCH BASINS**									
0011	Not including footing & excavation									
1600	Frames & grates, C.I., 24" square, 500 lb.	B-6	7.80	3.077	Ea.	395	148	27.50	570.50	685
1700	26" D shape, 600 lb.		7	3.429		735	165	31	931	1,075
1800	Light traffic, 18" diameter, 100 lb.		10	2.400		216	115	21.50	352.50	435
1900	24" diameter, 300 lb.		8.70	2.759		209	133	25	367	455
2000	36" diameter, 900 lb.		5.80	4.138		720	199	37.50	956.50	1,125
2100	Heavy traffic, 24" diameter, 400 lb.		7.80	3.077		291	148	27.50	466.50	570
2200	36" diameter, 1,150 lb.		3	8		955	385	72	1,412	1,700
2300	Mass. State standard, 26" diameter, 475 lb.		7	3.429		695	165	31	891	1,050
2400	30" diameter, 620 lb.		7	3.429		380	165	31	576	695
2500	Watertight, 24" diameter, 350 lb.		7.80	3.077		430	148	27.50	605.50	725
2600	26" diameter, 500 lb.		7	3.429		445	165	31	641	770
2700	32" diameter, 575 lb.		6	4		915	192	36	1,143	1,325
2800	3 piece cover & frame, 10" deep,	▼			▼					
2900	1,200 lb., for heavy equipment	B-6	3	8	Ea.	1,075	385	72	1,532	1,850
3000	Raised for paving 1-1/4" to 2" high									
3100	4 piece expansion ring									
3200	20" to 26" diameter	1 Clab	3	2.667	Ea.	223	118		341	425
3300	30" to 36" diameter	"	3	2.667	"	310	118		428	515
3320	Frames and covers, existing, raised for paving, 2", including									
3340	row of brick, concrete collar, up to 12" wide frame	B-6	18	1.333	Ea.	52.50	64	12	128.50	166
3360	20" to 26" wide frame		11	2.182		79	105	19.65	203.65	266
3380	30" to 36" wide frame	▼	9	2.667		98	128	24	250	325
3400	Inverts, single channel brick	D-1	3	5.333		113	260		373	515
3500	Concrete		5	3.200		137	156		293	385
3600	Triple channel, brick		2	8		186	390		576	790
3700	Concrete	▼	3	5.333	▼	156	260		416	560

33 71 Electrical Utility Transmission and Distribution

33 71 13 – Electrical Utility Towers

33 71 13.23 Steel Electrical Utility Towers

		Crew	Daily Output	Labor-Hours	Unit	Material	Labor	Equipment	Total	Total Incl O&P
0010	**STEEL ELECTRICAL UTILITY TOWERS**									
0500	Towers: material handling and spotting	R-7	22.56	2.128	Ton		97.50	7.45	104.95	155
0540	Steel tower erection	R-5	7.65	11.503		2,100	645	156	2,901	3,450
0550	Lace and box	"	7.10	12.394		2,100	690	168	2,958	3,525
0600	Special towers: material handling and spotting	R-7	12.31	3.899			178	13.65	191.65	284
0640	Special steel structure erection	R-6	6.52	13.497		2,650	755	269	3,674	4,325
0650	Special steel lace and box	"	6.29	13.990	▼	2,650	780	279	3,709	4,375

33 71 Electrical Utility Transmission and Distribution

33 71 13 – Electrical Utility Towers

33 71 13.80 Transmission Line Right of Way	Crew	Daily Output	Labor-Hours	Unit	Material	2021 Bare Costs Labor	Equipment	Total	Total Incl O&P
0010 **TRANSMISSION LINE RIGHT OF WAY**									
0100 Clearing right of way	B-87	6.67	5.997	Acre		335	375	710	910
0200 Restoration & seeding	B-10D	4	3	"	420	162	485	1,067	1,225

33 71 16 – Electrical Utility Poles

33 71 16.20 Line Poles and Fixtures

	Crew	Daily Output	Labor-Hours	Unit	Material	Labor	Equipment	Total	Total Incl O&P
0010 **LINE POLES & FIXTURES**									
0200 Wood poles, material handling and spotting	R-7	6.49	7.396	Ea.		340	26	366	540
0220 Erect wood poles & backfill holes, in earth	R-5	6.77	12.999	"	1,850	725	177	2,752	3,325
0300 Crossarms for wood pole structure									
0310 Material handling and spotting	R-7	14.55	3.299	Ea.		151	11.55	162.55	240
0320 Install crossarms	R-5	11	8	"	610	445	109	1,164	1,450

33 71 16.23 Steel Electrical Utility Poles

	Crew	Daily Output	Labor-Hours	Unit	Material	Labor	Equipment	Total	Total Incl O&P
0010 **STEEL ELECTRICAL UTILITY POLES**									
0100 Steel, square, tapered shaft, 20'	R-15A	6	8	Ea.	465	450	47	962	1,225
0120 25'		5.35	8.972		2,325	500	52.50	2,877.50	3,350
0140 30'		4.80	10		2,500	560	58.50	3,118.50	3,650
0160 35'		4.35	11.034		545	620	65	1,230	1,600
0180 40'		4	12		1,750	670	70.50	2,490.50	3,000
0300 Galvanized, round, tapered shaft, 40'		4	12		1,900	670	70.50	2,640.50	3,175
0320 45'	R-15A	3.70	12.973	Ea.	2,100	725	76	2,901	3,450
0340 50'		3.45	13.913		2,250	780	81.50	3,111.50	3,750
0360 55'		3	16		3,350	895	94	4,339	5,125
0380 60'		2.65	18.113		3,550	1,025	106	4,681	5,525
0400 65'	R-13	2.20	19.091		3,775	1,175	112	5,062	6,025
0420 70'		1.90	22.105		4,050	1,350	129	5,529	6,625
0440 75'		1.70	24.706		4,325	1,525	145	5,995	7,150
0460 80'		1.50	28		4,625	1,725	164	6,514	7,825
0480 85'		1.40	30		4,850	1,850	176	6,876	8,275
0500 90'		1.25	33.600		5,175	2,075	197	7,447	9,000
0520 95'		1.15	36.522		5,400	2,250	214	7,864	9,525
0540 100'		1	42		6,600	2,575	246	9,421	11,400
0700 12 sided, tapered shaft, 40'	R-15A	4	12		2,425	670	70.50	3,165.50	3,750
0720 45'		3.70	12.973		2,725	725	76	3,526	4,150
0740 50'		3.45	13.913		2,950	780	81.50	3,811.50	4,500
0760 55'		3	16		3,275	895	94	4,264	5,025
0780 60'		2.65	18.113		4,225	1,025	106	5,356	6,275
0800 65'	R-13	2.20	19.091		4,475	1,175	112	5,762	6,800
0820 70'		1.90	22.105		4,775	1,350	129	6,254	7,425
0840 75'		1.70	24.706		5,075	1,525	145	6,745	8,000
0860 80'		1.50	28		5,400	1,725	164	7,289	8,675
0880 85'		1.40	30		5,750	1,850	176	7,776	9,275
0900 90'		1.25	33.600		6,025	2,075	197	8,297	9,925
0920 95'		1.15	36.522		6,325	2,250	214	8,789	10,500
0940 100'		1	42		6,700	2,575	246	9,521	11,500
0960 105'		.90	46.667		7,275	2,875	273	10,423	12,600
0980 Ladder clips					335			335	370
1000 Galvanized steel, round, tapered, w/one 6' arm, 20'	R-15A	9.20	5.217		1,500	292	30.50	1,822.50	2,125
1020 25'		7.65	6.275		1,700	350	37	2,087	2,450
1040 30'		6.55	7.328		1,925	410	43	2,378	2,775
1060 35'		5.75	8.348		2,100	465	49	2,614	3,050
1080 40'		5.15	9.320		2,350	520	54.50	2,924.50	3,425

For customer support on your Electrical Costs with RSMeans Data, call 800.448.8182.

367

33 71 16 – Electrical Utility Poles

33 71 16.23 Steel Electrical Utility Poles

		Crew	Daily Output	Labor-Hours	Unit	Material	2021 Bare Costs Labor	Equipment	Total	Total Incl O&P
1200	Two 6' arms, 20'	R-15A	6.55	7.328	Ea.	1,700	410	43	2,153	2,525
1220	25'		5.75	8.348		1,900	465	49	2,414	2,850
1240	30'		5.15	9.320		2,125	520	54.50	2,699.50	3,175
1260	35'		4.60	10.435		2,300	585	61.50	2,946.50	3,475
1280	40'		4.20	11.429		2,550	640	67	3,257	3,850
1400	One 12' truss arm, 25'		7.65	6.275		1,225	350	37	1,612	1,925
1420	30'		6.55	7.328		1,250	410	43	1,703	2,050
1440	35'		5.75	8.348		1,425	465	49	1,939	2,325
1460	40'		5.15	9.320		1,950	520	54.50	2,524.50	2,975
1480	45'		4.60	10.435		2,125	585	61.50	2,771.50	3,300
1600	Two 12' truss arms, 25'		5.75	8.348		1,650	465	49	2,164	2,575
1620	30'		5.15	9.320		1,625	520	54.50	2,199.50	2,600
1640	35'		4.60	10.435		1,850	585	61.50	2,496.50	2,975
1660	40'		4.20	11.429		2,250	640	67	2,957	3,500
1680	45'		3.85	12.468		2,075	700	73	2,848	3,425
3400	Galvanized steel, tapered, 10'		15.50	3.097		440	173	18.20	631.20	765
3420	12'		11.50	4.174		655	234	24.50	913.50	1,100
3440	14'		9.20	5.217		390	292	30.50	712.50	900
3460	16'		8.45	5.680		405	320	33.50	758.50	955
3480	18'		7.65	6.275		415	350	37	802	1,025
3500	20'		7.10	6.761		405	380	39.50	824.50	1,050
5200	Galvanized steel, bracket type, 2", 6' long		18.50	2.595		166	145	15.25	326.25	415
5220	8' long		15.50	3.097		213	173	18.20	404.20	510
5240	10' long		12.50	3.840		355	215	22.50	592.50	735
6000	Digging holes in earth, average	R-5	25.14	3.500			196	47.50	243.50	345
6010	In rock, average	"	4.51	19.512			1,100	265	1,365	1,925
6020	Formed plate pole structure									
6030	Material handling and spotting	R-7	2.40	20	Ea.		915	70	985	1,450
6040	Erect steel plate pole	R-5	1.95	45.128		11,500	2,525	615	14,640	17,200
6050	Guys, anchors and hardware for pole, in earth		7.04	12.500		700	700	170	1,570	2,000
6060	In rock		17.96	4.900		835	274	66.50	1,175.50	1,400
6070	Foundations for line poles									
6080	Excavation, in earth	R-5	135.38	.650	C.Y.		36.50	8.85	45.35	63.50
6090	In rock		20	4.400			246	60	306	430
6110	Concrete foundations		11	8		164	445	109	718	970

33 71 16.33 Wood Electrical Utility Poles

		Crew	Daily Output	Labor-Hours	Unit	Material	2021 Bare Costs Labor	Equipment	Total	Total Incl O&P
0010	**WOOD ELECTRICAL UTILITY POLES**									
0011	Excludes excavation, backfill and cast-in-place concrete									
1001	Pole, wood, class 2 yellow pine, CCA-treated, 30'	R-15A	7.70	6.234	Ea.	480	350	36.50	866.50	1,075
1002	35'		5.80	8.276		575	465	48.50	1,088.50	1,375
1003	40'		5.30	9.057		700	505	53	1,258	1,575
1004	45'		4.70	10.213		805	570	60	1,435	1,800
1005	50'		4.20	11.429		920	640	67	1,627	2,025
1006	55'		3.80	12.632		1,050	705	74	1,829	2,275
1007	Electrical utility pole, wood pole, parametric ,CCA/ACA-treated, 30', class		3.50	13.714		1,550	770	80.50	2,400.50	2,975
1020	12" Ponderosa Pine Poles treated 0.40 ACQ, 16'	R-3	3.20	6.250		1,200	395	59.50	1,654.50	1,950
2001	Pole, wood, class 4 yellow pine, CCA-treated, 30'	R-15A	7.70	6.234		385	350	36.50	771.50	985
2002	35'		5.80	8.276		440	465	48.50	953.50	1,225
2003	40'		5.30	9.057		650	505	53	1,208	1,525
2004	45'		4.70	10.213		805	570	60	1,435	1,800
2005	50'		4.20	11.429		920	640	67	1,627	2,025
3000	Pole, wood, class 5 yellow pine, CCA-treated, 25'		8.60	5.581		230	310	33	573	755

33 71 16 – Electrical Utility Poles

33 71 16.33 Wood Electrical Utility Poles	Crew	Daily Output	Labor-Hours	Unit	Material	2021 Bare Costs Labor	2021 Bare Costs Equipment	Total	Total Incl O&P
3001	30'	R-15A	7.70	6.234	Ea.	299	350	36.50	685.50	890
3002	35'		5.80	8.276		380	465	48.50	893.50	1,175
3003	40'		5.30	9.057		525	505	53	1,083	1,400
3004	45'		4.70	10.213		600	570	60	1,230	1,575
3005	50'		4.20	11.429		920	640	67	1,627	2,025
3101	Pole, wood, class 6 yellow pine, CCA-treated, 30'		7.70	6.234		164	350	36.50	550.50	740
3102	35'		5.80	8.276		208	465	48.50	721.50	975
3103	40'		5.30	9.057		258	505	53	816	1,100
3104	45'		4.70	10.213		315	570	60	945	1,250
5000	Wood, class 3 yellow pine, penta-treated, 25'		8.60	5.581		252	310	33	595	780
5005	50'		4.20	11.429		920	640	67	1,627	2,025
5020	30'		7.70	6.234		355	350	36.50	741.50	950
5040	35'		5.80	8.276		460	465	48.50	973.50	1,250
5060	40'		5.30	9.057		565	505	53	1,123	1,425
5080	45'		4.70	10.213		645	570	60	1,275	1,625
5100	50'		4.20	11.429		770	640	67	1,477	1,875
5120	55'		3.80	12.632		900	705	74	1,679	2,125
5140	60'		3.50	13.714		1,025	770	80.50	1,875.50	2,375
5160	65'		3.20	15		1,400	840	88	2,328	2,875
5180	70'		3	16		1,575	895	94	2,564	3,150
5301	Wood, class 3 yellow pine, CCA-treated, 30'		7.70	6.234		430	350	36.50	816.50	1,025
5305	50'		4.20	11.429		950	640	67	1,657	2,075
5306	55'		3.80	12.632		950	705	74	1,729	2,175
5307	60'		3.50	13.714		1,525	770	80.50	2,375.50	2,925
6000	Wood, class 1 type C, CCA/ACA-treated, 25'		8.60	5.581		350	310	33	693	885
6020	30'		7.70	6.234		490	350	36.50	876.50	1,100
6040	35'		5.80	8.276		630	465	48.50	1,143.50	1,450
6060	40'		5.30	9.057		780	505	53	1,338	1,675
6080	45'		4.70	10.213		735	570	60	1,365	1,725
6100	50'		4.20	11.429		1,050	640	67	1,757	2,175
6120	55'		3.80	12.632		1,200	705	74	1,979	2,450
6200	Wood, class 3 Douglas Fir, penta-treated, 20'	R-3	3.10	6.452		250	410	61.50	721.50	955
6600	30'		2.60	7.692		325	490	73.50	888.50	1,150
7000	40'		2.30	8.696		595	550	83	1,228	1,575
7200	45'		1.70	11.765		870	745	112	1,727	2,175
7400	Cross arms with hardware & insulators									
7600	4' long	1 Elec	2.50	3.200	Ea.	151	204		355	470
7800	5' long		2.40	3.333		166	212		378	495
8000	6' long		2.20	3.636		200	232		432	565
9000	Disposal of pole & hardware surplus material	R-7	20.87	2.300	Mile		105	8.05	113.05	168
9100	Disposal of crossarms & hardware surplus material	"	40	1.200	"		55	4.20	59.20	87

33 71 19 – Electrical Underground Ducts and Manholes

33 71 19.15 Underground Ducts and Manholes

		Crew	Daily Output	Labor-Hours	Unit	Material	2021 Bare Costs Labor	2021 Bare Costs Equipment	Total	Total Incl O&P
0010	**UNDERGROUND DUCTS AND MANHOLES**									
0011	Not incl. excavation, backfill, or concrete in slab and duct bank									
1000	Direct burial R337119-30									
1010	PVC, schedule 40, w/coupling, 1/2" diameter	1 Elec	340	.024	L.F.	.18	1.50		1.68	2.43
1020	3/4" diameter		290	.028		.22	1.76		1.98	2.85
1030	1" diameter		260	.031		.32	1.96		2.28	3.26
1040	1-1/2" diameter		210	.038		.53	2.43		2.96	4.19
1050	2" diameter		180	.044		.64	2.83		3.47	4.91
1060	3" diameter	2 Elec	240	.067		1.24	4.25		5.49	7.65

For customer support on your Electrical Costs with RSMeans Data, call 800.448.8182.

369

33 71 19.15 Underground Ducts and Manholes	Crew	Daily Output	Labor-Hours	Unit	Material	2021 Bare Costs Labor	2021 Bare Costs Equipment	Total	Total Incl O&P
1070	4" diameter	2 Elec	160	.100	L.F.	1.76	6.35		8.11	11.40
1080	5" diameter		120	.133		4.79	8.50		13.29	17.85
1090	6" diameter		90	.178		3.15	11.30		14.45	20.50
1110	Elbows, 1/2" diameter	1 Elec	48	.167	Ea.	.49	10.60		11.09	16.35
1120	3/4" diameter		38	.211		.55	13.40		13.95	20.50
1130	1" diameter		32	.250		.86	15.95		16.81	24.50
1140	1-1/2" diameter		21	.381		2.27	24.50		26.77	38.50
1150	2" diameter		16	.500		3.08	32		35.08	51
1160	3" diameter		12	.667		8.40	42.50		50.90	72.50
1170	4" diameter		9	.889		17.05	56.50		73.55	103
1180	5" diameter		8	1		18.35	63.50		81.85	115
1190	6" diameter		5	1.600		36	102		138	191
1210	Adapters, 1/2" diameter		52	.154		.21	9.80		10.01	14.80
1220	3/4" diameter		43	.186		.35	11.85		12.20	18
1230	1" diameter		39	.205		.38	13.05		13.43	19.80
1240	1-1/2" diameter		35	.229		.81	14.55		15.36	22.50
1250	2" diameter		26	.308		1.02	19.60		20.62	30
1260	3" diameter		20	.400		2.13	25.50		27.63	40.50
1270	4" diameter		14	.571		4.33	36.50		40.83	59
1280	5" diameter		12	.667		8.65	42.50		51.15	72.50
1290	6" diameter		9	.889		10.30	56.50		66.80	95.50
1340	Bell end & cap, 1-1/2" diameter		35	.229		5.25	14.55		19.80	27.50
1350	Bell end & plug, 2" diameter		26	.308		6.50	19.60		26.10	36
1360	3" diameter		20	.400		7.95	25.50		33.45	47
1370	4" diameter		14	.571		9.35	36.50		45.85	64.50
1380	5" diameter		12	.667		14.20	42.50		56.70	78.50
1390	6" diameter		9	.889		16.15	56.50		72.65	102
1450	Base spacer, 2" diameter		56	.143		1.37	9.10		10.47	15
1460	3" diameter		46	.174		1.77	11.10		12.87	18.40
1470	4" diameter		41	.195		1.78	12.45		14.23	20.50
1480	5" diameter		37	.216		1.99	13.75		15.74	22.50
1490	6" diameter		34	.235		2.48	15		17.48	25
1550	Intermediate spacer, 2" diameter		60	.133		1.25	8.50		9.75	14
1560	3" diameter	1 Elec	46	.174	Ea.	1.76	11.10		12.86	18.40
1570	4" diameter		41	.195		1.59	12.45		14.04	20
1580	5" diameter		37	.216		1.94	13.75		15.69	22.50
1590	6" diameter		34	.235		2.58	15		17.58	25.50
4010	PVC, schedule 80, w/coupling, 1/2" diameter		215	.037	L.F.	.97	2.37		3.34	4.59
4020	3/4" diameter		180	.044		1.20	2.83		4.03	5.55
4030	1" diameter		145	.055		1.79	3.51		5.30	7.15
4040	1-1/2" diameter		120	.067		2.94	4.25		7.19	9.55
4050	2" diameter		100	.080		3.23	5.10		8.33	11.10
4060	3" diameter	2 Elec	130	.123		8.40	7.85		16.25	21
4070	4" diameter		90	.178		12	11.30		23.30	30
4080	5" diameter		70	.229		16.70	14.55		31.25	40
4090	6" diameter		50	.320		18.45	20.50		38.95	51
4110	Elbows, 1/2" diameter	1 Elec	29	.276	Ea.	2.11	17.55		19.66	28.50
4120	3/4" diameter		23	.348		5.70	22		27.70	39.50
4130	1" diameter		20	.400		4.60	25.50		30.10	43
4140	1-1/2" diameter		16	.500		6.60	32		38.60	55
4150	2" diameter		12	.667		7.95	42.50		50.45	72
4160	3" diameter		9	.889		21	56.50		77.50	107
4170	4" diameter		7	1.143		32	73		105	143

33 71 19 – Electrical Underground Ducts and Manholes

33 71 19.15 Underground Ducts and Manholes

		Crew	Daily Output	Labor-Hours	Unit	Material	2021 Bare Costs Labor	Equipment	Total	Total Incl O&P
4180	5" diameter	1 Elec	6	1.333	Ea.	81.50	85		166.50	216
4190	6" diameter		4	2		90.50	127		217.50	289
4210	Adapter, 1/2" diameter		39	.205		.21	13.05		13.26	19.65
4220	3/4" diameter		33	.242		.35	15.45		15.80	23.50
4230	1" diameter		29	.276		.38	17.55		17.93	26.50
4240	1-1/2" diameter		26	.308		.81	19.60		20.41	30
4250	2" diameter		23	.348		1.02	22		23.02	34
4260	3" diameter		18	.444		2.13	28.50		30.63	44.50
4270	4" diameter		13	.615		4.33	39		43.33	63.50
4280	5" diameter		11	.727		8.65	46.50		55.15	78.50
4290	6" diameter		8	1		10.30	63.50		73.80	106
4310	Bell end & cap, 1-1/2" diameter		26	.308		5.25	19.60		24.85	35
4320	Bell end & plug, 2" diameter		23	.348		6.50	22		28.50	40
4330	3" diameter		18	.444		7.95	28.50		36.45	51
4340	4" diameter		13	.615		9.35	39		48.35	69
4350	5" diameter		11	.727		14.20	46.50		60.70	84.50
4360	6" diameter		8	1		16.15	63.50		79.65	112
4370	Base spacer, 2" diameter		42	.190		1.37	12.15		13.52	19.55
4380	3" diameter		33	.242		1.77	15.45		17.22	25
4390	4" diameter		29	.276		1.78	17.55		19.33	28
4400	5" diameter		26	.308		1.99	19.60		21.59	31
4410	6" diameter		25	.320		2.48	20.50		22.98	33
4420	Intermediate spacer, 2" diameter	1 Elec	45	.178	Ea.	1.25	11.30		12.55	18.25
4430	3" diameter		34	.235		1.76	15		16.76	24.50
4440	4" diameter		31	.258		1.59	16.45		18.04	26.50
4450	5" diameter		28	.286		1.94	18.20		20.14	29
4460	6" diameter		25	.320		2.58	20.50		23.08	33.50

33 71 19.17 Electric and Telephone Underground

		Crew	Daily Output	Labor-Hours	Unit	Material	2021 Bare Costs Labor	Equipment	Total	Total Incl O&P
0010	**ELECTRIC AND TELEPHONE UNDERGROUND**									
0011	Not including excavation									
0200	backfill and cast in place concrete									
0250	For bedding, see Section 31 23 23.16 R337119-30									
0400	Hand holes, precast concrete, with concrete cover									
0600	2' x 2' x 3' deep	R-3	2.40	8.333	Ea.	530	530	79.50	1,139.50	1,450
0800	3' x 3' x 3' deep		1.90	10.526		690	670	101	1,461	1,850
1000	4' x 4' x 4' deep		1.40	14.286		1,550	905	136	2,591	3,200
1200	Manholes, precast with iron racks & pulling irons, C.I. frame									
1400	and cover, 4' x 6' x 7' deep	B-13	2	28	Ea.	2,950	1,350	293	4,593	5,600
1600	6' x 8' x 7' deep		1.90	29.474		3,325	1,425	310	5,060	6,125
1800	6' x 10' x 7' deep		1.80	31.111		3,725	1,500	325	5,550	6,700
4200	Underground duct, banks ready for concrete fill, min. of 7.5"									
4400	between conduits, center to center									
4580	PVC, type EB, 1 @ 2" diameter	2 Elec	480	.033	L.F.	.96	2.12		3.08	4.21
4600	2 @ 2" diameter		240	.067		1.91	4.25		6.16	8.40
4800	4 @ 2" diameter		120	.133		3.83	8.50		12.33	16.80
4900	1 @ 3" diameter		400	.040		1.62	2.55		4.17	5.55
5000	2 @ 3" diameter		200	.080		3.23	5.10		8.33	11.10
5200	4 @ 3" diameter		100	.160		6.45	10.20		16.65	22.50
5300	1 @ 4" diameter		320	.050		1.58	3.19		4.77	6.45
5400	2 @ 4" diameter		160	.100		3.16	6.35		9.51	12.90
5600	4 @ 4" diameter		80	.200		6.30	12.75		19.05	26
5800	6 @ 4" diameter		54	.296		9.45	18.85		28.30	38.50

For customer support on your Electrical Costs with RSMeans Data, call 800.448.8182.

371

33 71 19.17 Electric and Telephone Underground	Crew	Daily Output	Labor-Hours	Unit	Material	2021 Bare Costs Labor	Equipment	Total	Total Incl O&P
5810	1 @ 5" diameter	2 Elec	260	.062	L.F.	1.78	3.92		5.70	7.75
5820	2 @ 5" diameter		130	.123		3.56	7.85		11.41	15.55
5840	4 @ 5" diameter		70	.229		7.10	14.55		21.65	29.50
5860	6 @ 5" diameter		50	.320		10.65	20.50		31.15	42.50
5870	1 @ 6" diameter		200	.080		3.03	5.10		8.13	10.90
5880	2 @ 6" diameter		100	.160		6.05	10.20		16.25	22
5900	4 @ 6" diameter		50	.320		12.10	20.50		32.60	44
5920	6 @ 6" diameter		30	.533		18.15	34		52.15	70.50
6200	Rigid galvanized steel, 2 @ 2" diameter		180	.089		13.95	5.65		19.60	24
6400	4 @ 2" diameter		90	.178		28	11.30		39.30	47.50
6800	2 @ 3" diameter	2 Elec	100	.160	L.F.	30.50	10.20		40.70	49
7000	4 @ 3" diameter		50	.320		61.50	20.50		82	98
7200	2 @ 4" diameter		70	.229		43.50	14.55		58.05	69
7400	4 @ 4" diameter		34	.471		86.50	30		116.50	140
7600	6 @ 4" diameter		22	.727		130	46.50		176.50	212
7620	2 @ 5" diameter		60	.267		94	17		111	128
7640	4 @ 5" diameter		30	.533		188	34		222	258
7660	6 @ 5" diameter		18	.889		282	56.50		338.50	395
7680	2 @ 6" diameter		40	.400		114	25.50		139.50	164
7700	4 @ 6" diameter		20	.800		229	51		280	325
7720	6 @ 6" diameter		14	1.143		345	73		418	485
7800	For cast-in-place concrete, add									
7810	Under 1 C.Y.	C-6	16	3	C.Y.	199	138	3.39	340.39	430
7820	1 C.Y. to 5 C.Y.		19.20	2.500		180	115	2.83	297.83	375
7830	Over 5 C.Y.		24	2		149	92	2.26	243.26	305
7850	For reinforcing rods, add									
7860	#4 to #7	2 Rodm	1.10	14.545	Ton	1,250	855		2,105	2,650
7870	#8 to #14	"	1.50	10.667	"	1,250	630		1,880	2,325
8000	Fittings, PVC type EB, elbow, 2" diameter	1 Elec	16	.500	Ea.	15.45	32		47.45	64.50
8200	3" diameter		14	.571		17.80	36.50		54.30	73.50
8400	4" diameter		12	.667		36.50	42.50		79	104
8420	5" diameter		10	.800		113	51		164	200
8440	6" diameter		9	.889		217	56.50		273.50	325
8500	Coupling, 2" diameter					.90			.90	.99
8600	3" diameter					4.48			4.48	4.93
8700	4" diameter					4.79			4.79	5.25
8720	5" diameter					9.15			9.15	10.05
8740	6" diameter					25.50			25.50	28
8800	Adapter, 2" diameter	1 Elec	26	.308		.63	19.60		20.23	29.50
9000	3" diameter		20	.400		3.01	25.50		28.51	41.50
9200	4" diameter		16	.500		3.58	32		35.58	51.50
9220	5" diameter		13	.615		6.15	39		45.15	65.50
9240	6" diameter		10	.800		13.30	51		64.30	90
9400	End bell, 2" diameter		16	.500		2.30	32		34.30	50
9600	3" diameter		14	.571		4.07	36.50		40.57	58.50
9800	4" diameter		12	.667		4.62	42.50		47.12	68
9810	5" diameter		10	.800		8.10	51		59.10	84.50
9820	6" diameter		8	1		10.10	63.50		73.60	106
9830	5° angle coupling, 2" diameter		26	.308		6.95	19.60		26.55	36.50
9840	3" diameter		20	.400		25	25.50		50.50	65.50
9850	4" diameter		16	.500		20.50	32		52.50	70
9860	5" diameter		13	.615		17.90	39		56.90	78
9870	6" diameter	1 Elec	10	.800	Ea.	27.50	51		78.50	106

33 71 Electrical Utility Transmission and Distribution

33 71 19 – Electrical Underground Ducts and Manholes

33 71 19.17 Electric and Telephone Underground

		Crew	Daily Output	Labor-Hours	Unit	Material	2021 Bare Costs Labor	2021 Bare Costs Equipment	Total	Total Incl O&P
9880	Expansion joint, 2" diameter	1 Elec	16	.500	Ea.	35.50	32		67.50	86.50
9890	3" diameter		18	.444		74.50	28.50		103	124
9900	4" diameter		12	.667		100	42.50		142.50	173
9910	5" diameter		10	.800		181	51		232	275
9920	6" diameter		8	1		172	63.50		235.50	284
9930	Heat bender, 2" diameter					515			515	565
9940	6" diameter					1,625			1,625	1,800
9950	Cement, quart					17.50			17.50	19.25
9960	Nylon polyethylene pull rope, 1/4"	2 Elec	2000	.008	L.F.	.18	.51		.69	.96

33 71 23 – Insulators and Fittings

33 71 23.16 Post Insulators

		Crew	Daily Output	Labor-Hours	Unit	Material	2021 Bare Costs Labor	2021 Bare Costs Equipment	Total	Total Incl O&P
0010	**POST INSULATORS**									
7400	Insulators, pedestal type	R-11	112	.500	Ea.		30.50	5.75	36.25	51.50
7490	See also line 33 71 39.13 1000									

33 71 26 – Transmission and Distribution Equipment

33 71 26.13 Capacitor Banks

		Crew	Daily Output	Labor-Hours	Unit	Material	2021 Bare Costs Labor	2021 Bare Costs Equipment	Total	Total Incl O&P
0010	**CAPACITOR BANKS**									
1300	Station capacitors									
1350	Synchronous, 13 to 26 kV	R-11	3.11	18.006	MVAR	7,625	1,100	207	8,932	10,300
1360	46 kV		3.33	16.817		9,725	1,025	193	10,943	12,400
1370	69 kV		3.81	14.698		9,575	890	169	10,634	12,000
1380	161 kV		6.51	8.602		8,950	520	99	9,569	10,700
1390	500 kV		10.37	5.400		7,775	330	62	8,167	9,100
1450	Static, 13 to 26 kV		3.11	18.006		6,475	1,100	207	7,782	8,975
1460	46 kV		3.01	18.605		8,175	1,125	214	9,514	10,900
1470	69 kV		3.81	14.698		7,925	890	169	8,984	10,200
1480	161 kV		6.51	8.602		7,375	520	99	7,994	8,975
1490	500 kV		10.37	5.400		6,725	330	62	7,117	7,950
1600	Voltage regulators, 13 to 26 kV		.75	74.667	Ea.	293,000	4,525	860	298,385	329,500

33 71 26.23 Current Transformers

		Crew	Daily Output	Labor-Hours	Unit	Material	2021 Bare Costs Labor	2021 Bare Costs Equipment	Total	Total Incl O&P
0010	**CURRENT TRANSFORMERS**									
4050	Current transformers, 13 to 26 kV	R-11	14	4	Ea.	950	243	46	1,239	1,450
4060	46 kV		9.33	6.002		3,850	365	69	4,284	4,850
4070	69 kV		7	8		4,600	485	92	5,177	5,875
4080	161 kV		1.87	29.947		35,300	1,825	345	37,470	42,000

33 71 26.26 Potential Transformers

		Crew	Daily Output	Labor-Hours	Unit	Material	2021 Bare Costs Labor	2021 Bare Costs Equipment	Total	Total Incl O&P
0010	**POTENTIAL TRANSFORMERS**									
4100	Potential transformers, 13 to 26 kV	R-11	11.20	5	Ea.	1,200	305	57.50	1,562.50	1,850
4110	46 kV		8	7		4,000	425	80.50	4,505.50	5,125
4120	69 kV		6.22	9.003		4,900	545	104	5,549	6,325
4130	161 kV		2.24	25		24,200	1,525	288	26,013	29,300
4140	500 kV		1.40	40		72,500	2,425	460	75,385	83,500

33 71 26.28 Overhead Fuse Cutouts

		Crew	Daily Output	Labor-Hours	Unit	Material	2021 Bare Costs Labor	2021 Bare Costs Equipment	Total	Total Incl O&P
0010	**OVERHEAD FUSE CUTOUTS**									
1000	Cutout Fuse Assemblies									
1100	15.0 kV, 110 kV BIL, 100 amp, type C, silicone	R-8	6	8	Ea.	121	455	46.50	622.50	860
1200	200 amp		6	8		197	455	46.50	698.50	945
1300	300 amp		6	8		235	455	46.50	736.50	985
2000	Cutout Fuse Links									
2100	Removeable button head, type K, 100 amp	R-3	10	2	Ea.	11.55	127	19.10	157.65	223
2140	140 amp	"	10	2	"	26	127	19.10	172.10	239

373

33 71 26 – Transmission and Distribution Equipment

33 71 26.28 Overhead Fuse Cutouts	Crew	Daily Output	Labor-Hours	Unit	Material	2021 Bare Costs Labor	Equipment	Total	Total Incl O&P
3000 Cutout Fuse Hardware									
3100 Tri-mount bracket, aluminum	R-3	10	2	Ea.	230	127	19.10	376.10	465
3200 Arrester, surge, 12.7 kV		10	2		61.50	127	19.10	207.60	278
3300 Clamp, hot-line tap, bronze, 5"		10	2		23	127	19.10	169.10	236
3400 Clamp, stirrup, bronze, 1-13/16"		10	2		16.85	127	19.10	162.95	229
3500 Insulator, dead-end, clevis tongue, 12-1/2"	↓	10	2	↓	29	127	19.10	175.10	242

33 71 39 – High-Voltage Wiring

33 71 39.13 Overhead High-Voltage Wiring

	Crew	Daily Output	Labor-Hours	Unit	Material	2021 Bare Costs Labor	Equipment	Total	Total Incl O&P
0010 **OVERHEAD HIGH-VOLTAGE WIRING** R337116-60									
0100 Conductors, primary circuits									
0110 Material handling and spotting	R-5	9.78	8.998	W.Mile		505	122	627	885
0120 For river crossing, add		11	8			445	109	554	790
0150 Conductors, per wire, 210 to 636 kcmil		1.96	44.898		7,825	2,500	610	10,935	13,000
0160 795 to 954 kcmil		1.87	47.059		14,300	2,625	640	17,565	20,300
0170 1,000 to 1,600 kcmil		1.47	59.864		22,800	3,350	815	26,965	30,900
0180 Over 1,600 kcmil		1.35	65.185		31,000	3,650	885	35,535	40,500
0200 For river crossing, add, 210 to 636 kcmil		1.24	70.968			3,975	965	4,940	6,975
0220 795 to 954 kcmil		1.09	80.734			4,500	1,100	5,600	7,925
0230 1,000 to 1,600 kcmil		.97	90.722			5,075	1,225	6,300	8,925
0240 Over 1,600 kcmil	↓	.87	101	↓		5,650	1,375	7,025	9,950
0300 Joints and dead ends	R-8	6	8	Ea.	2,250	455	46.50	2,751.50	3,200
0400 Sagging	R-6	7.33	12.001	W.Mile		670	239	909	1,275
0500 Clipping, per structure, 69 kV	R-10	9.60	5	Ea.		300	40.50	340.50	495
0510 161 kV		5.33	9.006			540	73	613	885
0520 345 to 500 kV	↓	2.53	18.972			1,150	153	1,303	1,875
0600 Make and install jumpers, per structure, 69 kV	R-8	3.20	15		610	850	87.50	1,547.50	2,050
0620 161 kV		1.20	40		1,225	2,275	234	3,734	4,950
0640 345 to 500 kV	↓	.32	150		2,050	8,475	875	11,400	15,900
0700 Spacers	R-10	68.57	.700		121	42	5.65	168.65	202
0720 For river crossings, add	"	60	.800	↓		48	6.45	54.45	78.50
0800 Installing pulling line (500 kV only)	R-9	1.45	44.138	W.Mile	900	2,325	193	3,418	4,700
0810 Disposal of surplus material, high voltage conductors	R-7	6.96	6.897	Mile		315	24	339	500
0820 With trailer mounted reel stands	"	13.71	3.501	"		160	12.25	172.25	255
0900 Insulators and hardware, primary circuits									
0920 Material handling and spotting, 69 kV	R-7	480	.100	Ea.		4.57	.35	4.92	7.30
0930 161 kV		685.71	.070			3.20	.25	3.45	5.10
0950 345 to 500 kV	↓	960	.050			2.29	.18	2.47	3.64
1000 Disk insulators, 69 kV	R-5	880	.100		124	5.60	1.36	130.96	146
1020 161 kV		977.78	.090		142	5.05	1.22	148.27	165
1040 345 to 500 kV	↓	1100	.080	↓	142	4.47	1.09	147.56	164
1060 See Section 33 71 23.16 for pin or pedestal insulator									
1100 Install disk insulator at river crossing, add									
1110 69 kV	R-5	586.67	.150	Ea.		8.40	2.04	10.44	14.75
1120 161 kV		880	.100			5.60	1.36	6.96	9.85
1140 345 to 500 kV	↓	880	.100	↓		5.60	1.36	6.96	9.85
1150 Disposal of surplus material, high voltage insulators	R-7	41.74	1.150	Mile		52.50	4.02	56.52	84
1300 Overhead ground wire installation									
1320 Material handling and spotting	R-7	5.65	8.496	W.Mile		390	29.50	419.50	620
1340 Overhead ground wire	R-5	1.76	50		4,625	2,800	680	8,105	10,000
1350 At river crossing, add		1.17	75.214	↓		4,200	1,025	5,225	7,400
1360 Disposal of surplus material, grounding wire	↓	41.74	2.108	Mile		118	28.50	146.50	208
1400 Installing conductors, underbuilt circuits									

For customer support on your Electrical Costs with RSMeans Data, call 800.448.8182.

33 71 Electrical Utility Transmission and Distribution

33 71 39 – High-Voltage Wiring

33 71 39.13 Overhead High-Voltage Wiring

		Crew	Daily Output	Labor-Hours	Unit	Material	2021 Bare Costs Labor	2021 Bare Costs Equipment	Total	Total Incl O&P
1420	Material handling and spotting	R-7	5.65	8.496	W.Mile		390	29.50	419.50	620
1440	Conductors, per wire, 210 to 636 kcmil	R-5	1.96	44.898		7,825	2,500	610	10,935	13,000
1450	795 to 954 kcmil		1.87	47.059		14,300	2,625	640	17,565	20,300
1460	1,000 to 1,600 kcmil		1.47	59.864		22,800	3,350	815	26,965	30,900
1470	Over 1,600 kcmil		1.35	65.185		31,000	3,650	885	35,535	40,500
1500	Joints and dead ends	R-8	6	8	Ea.	2,250	455	46.50	2,751.50	3,200
1550	Sagging	R-6	8.80	10	W.Mile		560	199	759	1,050
1600	Clipping, per structure, 69 kV	R-10	9.60	5	Ea.		300	40.50	340.50	495
1620	161 kV		5.33	9.006			540	73	613	885
1640	345 to 500 kV		2.53	18.972			1,150	153	1,303	1,875
1700	Making and installing jumpers, per structure, 69 kV	R-8	5.87	8.177		610	465	48	1,123	1,425
1720	161 kV		.96	50		1,225	2,825	292	4,342	5,875
1740	345 to 500 kV		.32	150		2,050	8,475	875	11,400	15,900
1800	Spacers	R-10	96	.500		121	30	4.04	155.04	182
1810	Disposal of surplus material, conductors & hardware	R-7	6.96	6.897	Mile		315	24	339	500
2000	Insulators and hardware for underbuilt circuits									
2100	Material handling and spotting	R-7	1200	.040	Ea.		1.83	.14	1.97	2.91
2150	Disk insulators, 69 kV	R-8	600	.080		124	4.53	.47	129	143
2160	161 kV		686	.070		142	3.96	.41	146.37	162
2170	345 to 500 kV		800	.060		142	3.39	.35	145.74	161
2180	Disposal of surplus material, insulators & hardware	R-7	41.74	1.150	Mile		52.50	4.02	56.52	84
2300	Sectionalizing switches, 69 kV	R-5	1.26	69.841	Ea.	31,900	3,900	950	36,750	42,000
2310	161 kV		.80	110		36,100	6,150	1,500	43,750	50,500
2500	Protective devices		5.50	16		10,400	895	217	11,512	13,100
2600	Clearance poles, 8 poles per mile									
2650	In earth, 69 kV	R-5	1.16	75.862	Mile	8,575	4,250	1,025	13,850	16,900
2660	161 kV	"	.64	138		14,100	7,675	1,875	23,650	29,100
2670	345 to 500 kV	R-6	.48	183		16,900	10,200	3,650	30,750	37,900
2800	In rock, 69 kV	R-5	.69	128		8,575	7,125	1,725	17,425	22,000
2820	161 kV	"	.35	251		14,100	14,000	3,425	31,525	40,300
2840	345 to 500 kV	R-6	.24	367		16,900	20,500	7,300	44,700	57,000

33 72 Utility Substations

33 72 26 – Substation Bus Assemblies

33 72 26.13 Aluminum Substation Bus Assemblies

		Crew	Daily Output	Labor-Hours	Unit	Material	Labor	Equipment	Total	Total Incl O&P
0010	**ALUMINUM SUBSTATION BUS ASSEMBLIES**									
7300	Bus	R-11	590	.095	Lb.	3.07	5.75	1.09	9.91	13.15

33 72 33 – Control House Equipment

33 72 33.43 Substation Backup Batteries

		Crew	Daily Output	Labor-Hours	Unit	Material	Labor	Equipment	Total	Total Incl O&P
0010	**SUBSTATION BACKUP BATTERIES**									
9120	Battery chargers	R-11	11.20	5	Ea.	4,375	305	57.50	4,737.50	5,350
9200	Control batteries	"	14	4	K.A.H.	540	243	46	829	1,000

33 72 53 – Shunt Reactors

33 72 53.13 Reactors

		Crew	Daily Output	Labor-Hours	Unit	Material	Labor	Equipment	Total	Total Incl O&P
0010	**REACTORS**									
8000	Protective equipment									
8150	Reactors and resistors, 13 to 26 kV	R-11	28	2	Ea.	4,000	121	23	4,144	4,600
8160	46 kV		4.31	12.993		14,100	790	149	15,039	16,800
8170	69 kV		2.80	20		22,300	1,225	230	23,755	26,600
8180	161 kV		2.24	25		24,800	1,525	288	26,613	29,900

33 72 Utility Substations

33 72 53 – Shunt Reactors

33 72 53.13 Reactors		Crew	Daily Output	Labor-Hours	Unit	Material	2021 Bare Costs Labor	Equipment	Total	Total Incl O&P
8190	500 kV	R-11	.08	700	Ea.	95,500	42,500	8,050	146,050	177,500

33 73 Utility Transformers

33 73 23 – Dry-Type Utility Transformers

33 73 23.20 Primary Transformers

		Crew	Daily Output	Labor-Hours	Unit	Material	2021 Bare Costs Labor	Equipment	Total	Total Incl O&P
0010	**PRIMARY TRANSFORMERS**									
1000	Main conversion equipment									
1050	Power transformers, 13 to 26 kV	R-11	1.72	32.558	MVA	27,300	1,975	375	29,650	33,500
1060	46 kV		3.50	16		25,600	970	184	26,754	29,900
1070	69 kV		3.11	18.006		21,900	1,100	207	23,207	25,900
1080	110 kV		3.29	17.021		20,800	1,025	196	22,021	24,600
1090	161 kV		4.31	12.993		19,300	790	149	20,239	22,500
1100	500 kV		7	8		19,200	485	92	19,777	21,900
1200	Grounding transformers	R-11	3.11	18.006	Ea.	122,000	1,100	207	123,307	136,500

33 75 High-Voltage Switchgear and Protection Devices

33 75 13 – Air High-Voltage Circuit Breaker

33 75 13.13 High-Voltage Circuit Breaker, Air

		Crew	Daily Output	Labor-Hours	Unit	Material	2021 Bare Costs Labor	Equipment	Total	Total Incl O&P
0010	**HIGH-VOLTAGE CIRCUIT BREAKER, AIR**									
2100	Air circuit breakers, 13 to 26 kV	R-11	.56	100	Ea.	68,500	6,075	1,150	75,725	86,000
2110	161 kV	"	.24	233	"	293,500	14,200	2,675	310,375	347,000

33 75 16 – Oil High-Voltage Circuit Breaker

33 75 16.13 Oil Circuit Breaker

		Crew	Daily Output	Labor-Hours	Unit	Material	2021 Bare Costs Labor	Equipment	Total	Total Incl O&P
0010	**OIL CIRCUIT BREAKER**									
2000	Power circuit breakers									
2050	Oil circuit breakers, 13 to 26 kV	R-11	1.12	50	Ea.	70,000	3,025	575	73,600	82,000
2060	46 kV		.75	74.667		101,000	4,525	860	106,385	119,000
2070	69 kV		.45	124		235,000	7,550	1,425	243,975	271,500
2080	161 kV		.16	350		339,000	21,200	4,025	364,225	409,000
2090	500 kV		.06	933		1,266,500	56,500	10,700	1,333,700	1,489,000

33 75 19 – Gas High-Voltage Circuit Breaker

33 75 19.13 Gas Circuit Breaker

		Crew	Daily Output	Labor-Hours	Unit	Material	2021 Bare Costs Labor	Equipment	Total	Total Incl O&P
0010	**GAS CIRCUIT BREAKER**									
2000	Power circuit breakers									
2150	Gas circuit breakers, 13 to 26 kV	R-11	.56	100	Ea.	55,000	6,075	1,150	62,225	71,000
2160	161 kV		.08	700		93,000	42,500	8,050	143,550	174,500
2170	500 kV		.04	1400		432,000	85,000	16,100	533,100	619,000

33 75 23 – Vacuum High-Voltage Circuit Breaker

33 75 23.13 Vacuum Circuit Breaker

		Crew	Daily Output	Labor-Hours	Unit	Material	2021 Bare Costs Labor	Equipment	Total	Total Incl O&P
0010	**VACUUM CIRCUIT BREAKER**									
2000	Power circuit breakers									
2200	Vacuum circuit breakers, 13 to 26 kV	R-11	.56	100	Ea.	27,100	6,075	1,150	34,325	40,100

33 75 High-Voltage Switchgear and Protection Devices

33 75 36 – High-Voltage Utility Fuses

33 75 36.13 Fuses

		Crew	Daily Output	Labor-Hours	Unit	Material	2021 Bare Costs Labor	Equipment	Total	Total Incl O&P
0010	**FUSES**									
8000	Protective equipment									
8250	Fuses, 13 to 26 kV	R-11	18.67	2.999	Ea.	2,525	182	34.50	2,741.50	3,100
8260	46 kV		11.20	5		2,925	305	57.50	3,287.50	3,725
8270	69 kV		8	7		3,150	425	80.50	3,655.50	4,200
8280	161 kV	↓	4.67	11.991	↓	3,800	730	138	4,668	5,425

33 75 39 – High-Voltage Surge Arresters

33 75 39.13 Surge Arresters

		Crew	Daily Output	Labor-Hours	Unit	Material	2021 Bare Costs Labor	Equipment	Total	Total Incl O&P
0010	**SURGE ARRESTERS**									
8000	Protective equipment									
8050	Lightning arresters, 13 to 26 kV	R-11	18.67	2.999	Ea.	340	182	34.50	556.50	685
8060	46 kV		14	4		380	243	46	669	830
8070	69 kV		11.20	5		1,100	305	57.50	1,462.50	1,725
8080	161 kV		5.60	10		1,775	605	115	2,495	2,975
8090	500 kV	↓	1.40	40	↓	4,150	2,425	460	7,035	8,675

33 75 53 – High-Voltage Switches

33 75 53.13 Switches

		Crew	Daily Output	Labor-Hours	Unit	Material	2021 Bare Costs Labor	Equipment	Total	Total Incl O&P
0010	**SWITCHES**									
3000	Disconnecting switches									
3050	Gang operated switches									
3060	Manual operation, 13 to 26 kV	R-11	1.65	33.939	Ea.	5,075	2,050	390	7,515	9,100
3070	46 kV		1.12	50		7,625	3,025	575	11,225	13,500
3080	69 kV		.80	70		10,200	4,250	805	15,255	18,400
3090	161 kV		.56	100		15,200	6,075	1,150	22,425	27,100
3100	500 kV		.14	400		33,400	24,300	4,600	62,300	78,000
3110	Motor operation, 161 kV		.51	110		20,300	6,675	1,275	28,250	33,700
3120	500 kV		.28	200		38,100	12,100	2,300	52,500	62,500
3250	Circuit switches, 161 kV	↓	.41	137	↓	106,000	8,300	1,575	115,875	131,000
3300	Single pole switches									
3350	Disconnecting switches, 13 to 26 kV	R-11	28	2	Ea.	3,800	121	23	3,944	4,400
3360	46 kV		8	7		4,650	425	80.50	5,155.50	5,850
3370	69 kV		5.60	10		7,625	605	115	8,345	9,400
3380	161 kV		2.80	20		109,500	1,225	230	110,955	122,500
3390	500 kV		.22	255		310,500	15,400	2,925	328,825	367,500
3450	Grounding switches, 46 kV		5.60	10		41,800	605	115	42,520	47,000
3460	69 kV		3.73	15.013		42,100	910	173	43,183	47,800
3470	161 kV		2.24	25		45,000	1,525	288	46,813	52,000
3480	500 kV	↓	.62	90.323	↓	57,000	5,475	1,050	63,525	72,500

33 78 Substation Converter Stations

33 78 33 – Converter Stations

33 78 33.46 Substation Converter Stations

		Crew	Daily Output	Labor-Hours	Unit	Material	2021 Bare Costs Labor	Equipment	Total	Total Incl O&P
0010	**SUBSTATION CONVERTER STATIONS**									
9000	Station service equipment									
9100	Conversion equipment									
9110	Station service transformers	R-11	5.60	10	Ea.	106,000	605	115	106,720	117,500

33 79 Site Grounding

33 79 83 – Site Grounding Conductors

33 79 83.13 Grounding Wire, Bar, and Rod	Crew	Daily Output	Labor-Hours	Unit	Material	2021 Bare Costs Labor	2021 Bare Costs Equipment	Total	Total Incl O&P
0010 **GROUNDING WIRE, BAR, AND ROD**									
7500 Grounding systems	R-11	280	.200	Lb.	11.75	12.15	2.30	26.20	33.50

33 81 Communications Structures

33 81 13 – Communications Transmission Towers

33 81 13.10 Radio Towers

	Crew	Daily Output	Labor-Hours	Unit	Material	2021 Bare Costs Labor	2021 Bare Costs Equipment	Total	Total Incl O&P
0010 **RADIO TOWERS**									
0020 Guyed, 50' H, 40 lb. section, 70 MPH basic wind speed	2 Sswk	1	16	Ea.	2,775	965		3,740	4,550
0100 Wind load 90 MPH basic wind speed	"	1	16		3,650	965		4,615	5,500
0300 190' high, 40 lb. section, wind load 70 MPH basic wind speed	K-2	.33	72.727		8,950	4,150	2,575	15,675	19,000
0400 200' high, 70 lb. section, wind load 90 MPH basic wind speed		.33	72.727		15,300	4,150	2,575	22,025	26,000
0600 300' high, 70 lb. section, wind load 70 MPH basic wind speed		.20	120		18,600	6,850	4,250	29,700	35,600
0700 270' high, 90 lb. section, wind load 90 MPH basic wind speed	K-2	.20	120	Ea.	21,600	6,850	4,250	32,700	38,900
0800 400' high, 100 lb. section, wind load 70 MPH basic wind speed		.14	171		30,600	9,800	6,075	46,475	55,500
0900 Self-supporting, 60' high, wind load 70 MPH basic wind speed		.80	30		4,575	1,725	1,075	7,375	8,825
0910 60' high, wind load 90 MPH basic wind speed		.45	53.333		5,175	3,050	1,900	10,125	12,500
1000 120' high, wind load 70 MPH basic wind speed		.40	60		9,650	3,425	2,125	15,200	18,200
1200 190' high, wind load 90 MPH basic wind speed		.20	120		28,600	6,850	4,250	39,700	46,700
2000 For states west of Rocky Mountains, add for shipping					10%				

Estimating Tips
34 11 00 Rail Tracks
This subdivision includes items that may involve either repair of existing or construction of new railroad tracks. Additional preparation work, such as the roadbed earthwork, would be found in Division 31. Additional new construction siding and turnouts are found in Subdivision 34 72. Maintenance of railroads is found under 34 01 23 Operation and Maintenance of Railways.

34 40 00 Traffic Signals
This subdivision includes traffic signal systems. Other traffic control devices such as traffic signs are found in Subdivision 10 14 53 Traffic Signage.

34 70 00 Vehicle Barriers
This subdivision includes security vehicle barriers, guide and guard rails, crash barriers, and delineators. The actual maintenance and construction of concrete and asphalt pavement are found in Division 32.

Reference Numbers
Reference numbers are shown at the beginning of some major classifications. These numbers refer to related items in the Reference Section. The reference information may be an estimating procedure, an alternate pricing method, or technical information.

Note: Not all subdivisions listed here necessarily appear. ■

34 43 Airfield Signaling and Control Equipment

34 43 13 – Airfield Signals and Lighting

34 43 13.16 Airfield Runway and Taxiway Inset Lighting

	34 43 13.16 Airfield Runway and Taxiway Inset Lighting	Crew	Daily Output	Labor-Hours	Unit	Material	2021 Bare Costs Labor	2021 Bare Costs Equipment	Total	Total Incl O&P
0010	**AIRFIELD RUNWAY AND TAXIWAY INSET LIGHTING**									
0100	Runway centerline, bidir., semi-flush, 200 W, w/shallow insert base	R-22	12.40	3.006	Ea.	2,350	175		2,525	2,850
0110	for mounting in base housing		18.60	2.004		1,150	117		1,267	1,425
0120	Flush, 200 W, w/shallow insert base		12.40	3.006		2,150	175		2,325	2,625
0130	for mounting in base housing		18.64	2		1,075	117		1,192	1,375
0140	45 W, for mounting in base housing		18.64	2		1,075	117		1,192	1,375
0150	Touchdown zone light, unidirectional, 200 W, w/shallow insert base		12.40	3.006		1,750	175		1,925	2,175
0160	115 W		12.40	3.006		1,850	175		2,025	2,300
0170	Bidirectional, 62 W		12.40	3.006		1,775	175		1,950	2,200
0180	Unidirectional, 200 W, for mounting in base housing		18.60	2.004		845	117		962	1,100
0190	115 W		18.60	2.004		905	117		1,022	1,175
0200	Bidirectional, 62 W		18.60	2.004		1,000	117		1,117	1,275
0210	Runway edge & threshold light, bidir., 200 W, for base housing		9.36	3.983		1,125	232		1,357	1,600
0220	300 W		9.36	3.983		1,575	232		1,807	2,100
0230	499 W		7.44	5.011		2,650	292		2,942	3,325
0240	Threshold & approach light, unidir., 200 W, for base housing		9.36	3.983		645	232		877	1,050
0250	499 W		7.44	5.011		925	292		1,217	1,450
0260	Runway edge, bidirectional, 2-115 W, for base housing		12.40	3.006		1,700	175		1,875	2,125
0270	2-185 W		12.40	3.006		1,900	175		2,075	2,350
0280	Runway threshold & end, bidir., 2-115 W, for base housing		12.40	3.006		1,875	175		2,050	2,325
0370	45 W, flush, for mounting in base housing		18.64	2		920	117		1,037	1,200
0380	115 W		18.64	2		945	117		1,062	1,200
0400	Transformer, 45 W		37.20	1.002		202	58.50		260.50	310
0410	65 W		37.20	1.002		207	58.50		265.50	315
0420	100 W		37.20	1.002		254	58.50		312.50	365
0430	200 W, 6.6/6.6 A		37.20	1.002		281	58.50		339.50	395
0440	20/6.6 A		24.80	1.503		268	87.50		355.50	425
0450	500 W		24.80	1.503		385	87.50		472.50	550
0460	Base housing, 24" deep, 12" diameter		18.64	2		570	117		687	800
0470	15" diameter		18.64	2		1,250	117		1,367	1,550
0480	Connection kit for 1 conductor cable		24.80	1.503		40.50	87.50		128	175
0500	Elevated runway marker, high intensity quartz, edge, 115 W		14.88	2.505		420	146		566	675
0510	175 W		14.88	2.505		440	146		586	700
0520	Threshold, 115 W		14.88	2.505		480	146		626	740
0530	Edge, 200 W, w/base and transformer		11.85	3.146		1,275	183		1,458	1,675
0540	Elevated runway light, quartz, runway edge, 45 W		14.88	2.505		315	146		461	560
0550	Threshold, 115 W		14.88	2.505		450	146		596	710
0560	Taxiway, 30 W		14.88	2.505		249	146		395	490
0570	Fixture, 200 W, w/base and transformer		11.85	3.146		1,275	183		1,458	1,675
0580	Elevated threshold marker, quartz, 115 W		14.88	2.505		560	146		706	830
0590	175 W		14.88	2.505		560	146		706	830

34 43 23 – Weather Observation Equipment

34 43 23.16 Airfield Wind Cones

		Crew	Daily Output	Labor-Hours	Unit	Material	2021 Bare Costs Labor	2021 Bare Costs Equipment	Total	Total Incl O&P
0010	**AIRFIELD WIND CONES**									
1200	Wind cone, 12' lighted assembly, rigid, w/obstruction light	R-21	1.36	24.118	Ea.	12,200	1,525	85	13,810	15,800
1210	Without obstruction light		1.52	21.579		10,200	1,375	76	11,651	13,400
1220	Unlighted assembly, w/obstruction light		1.68	19.524		11,200	1,250	68.50	12,518.50	14,200
1230	Without obstruction light		1.84	17.826		10,200	1,125	63	11,388	13,100
1240	Wind cone slip fitter, 2-1/2" pipe		21.84	1.502		136	95.50	5.30	236.80	298
1250	Wind cone sock, 12' x 3', cotton		6.56	5		740	320	17.60	1,077.60	1,300
1260	Nylon		6.56	5		690	320	17.60	1,027.60	1,250

Estimating Tips

- When estimating costs for the installation of electrical power generation equipment, factors to review include access to the job site, access and setting up at the installation site, required connections, uncrating pads, anchors, leveling, final assembly of the components, and temporary protection from physical damage, such as environmental exposure.

- Be aware of the costs of equipment supports, concrete pads, and vibration isolators. Cross-reference them against other trades' specifications. Also, review site and structural drawings for items that must be included in the estimates.

- It is important to include items that are not documented in the plans and specifications but must be priced. These items include, but are not limited to, testing, dust protection, roof penetration, core drilling concrete floors and walls, patching, cleanup, and final adjustments. Add a contingency or allowance for utility company fees for power hookups, if needed.

- The project size and scope of electrical power generation equipment will have a significant impact on cost. The intent of RSMeans cost data is to provide a benchmark cost so that owners, engineers, and electrical contractors will have a comfortable number with which to start a project. Additionally, there are many websites available to use for research and to obtain a vendor's quote to finalize costs.

Reference Numbers

Reference numbers are shown at the beginning of some major classifications. These numbers refer to related items in the Reference Section. The reference information may be an estimating procedure, an alternate pricing method, or technical information.

Same Data. Simplified.

Enjoy the convenience and efficiency of accessing your costs anywhere:

- **Skip the multiplier** by setting your location
- **Quickly search,** edit, favorite and share costs
- **Stay on top of price changes** with automatic updates

Discover more at rsmeans.com/online

48 13 Hydroelectric Plant Electrical Power Generation Equipment

48 13 13 – Hydroelectric Power Plant Water Turbines

48 13 13.10 Water Turbines and Components	Crew	Daily Output	Labor-Hours	Unit	Material	2021 Bare Costs Labor	2021 Bare Costs Equipment	Total	Total Incl O&P
0010 **WATER TURBINES & COMPONENTS**									
0500 Water turbine, 100-500 watt, 3P, 48VDC, 54"W x 24"H, 260LBS	2 Elec	2	8	Ea.	10,300	510		10,810	12,100
0600 Water turbine, 1500-2000 watt, 3P, 130-200VAC to DC, 95LBS	"	3	5.333	"	3,900	340		4,240	4,800

48 15 Wind Energy Electrical Power Generation Equipment

48 15 13 – Wind Turbines

48 15 13.50 Wind Turbines and Components

	Crew	Daily Output	Labor-Hours	Unit	Material	2021 Bare Costs Labor	2021 Bare Costs Equipment	Total	Total Incl O&P
0010 **WIND TURBINES & COMPONENTS**									
0500 Complete system, grid connected									
0501 Wind turbine marine rated, 160 W, 12 V, DC	2 Elec	2.20	7.273	Ea.	1,200	465		1,665	2,000
0510 160 W, 24 V, DC		2.20	7.273		1,200	465		1,665	2,000
0520 160 W, 48 V, DC	↓	2.20	7.273	↓	1,200	465		1,665	2,000
1000 20 kW, 31' diam., incl. labor & material [G]				System				49,900	49,900
1001 Wind turbine, 300 W, 12 V, DC	2 Elec	2.20	7.273	Ea.	950	465		1,415	1,750
1011 300 W, 24 V, DC		2.20	7.273		950	465		1,415	1,750
1020 300 W, 48 V, DC		2.20	7.273		950	465		1,415	1,750
1030 Wind turbine, 400 W, 12 V, DC	↓	2.20	7.273	↓	950	465		1,415	1,750
1040 400 W, 24 V, DC	2 Elec	2.20	7.273	Ea.	950	465		1,415	1,750
1050 400 W, 48 V, DC	"	2.20	7.273	"	950	465		1,415	1,750
2000 2.4 kW, 12' diam., incl. labor & material [G]				System				18,000	18,000
2900 Component system									
3200 1,000 W, 9' diam. [G]	1 Elec	2.05	3.902	Ea.	1,500	249		1,749	2,025
3300 Wind turbine/charge control, 2,900-7,900 W, 5.5" H x 11' W, 180 lb.	2 Elec	2	8		16,700	510		17,210	19,100
3320 Wind turbine/charge control, 4,000-10,800 W, 5.5" H x 13' W, 240 lb.	"	2	8	↓	27,700	510		28,210	31,200
3400 Mounting hardware									
3401 Wind turbine tower, 2 stage steel, 8' H x 6' W	2 Elec	2	8	Ea.	2,100	510		2,610	3,050
3500 30' guyed tower kit [G]	2 Clab	5.12	3.125		415	139		554	665
3505 3' galvanized helical earth screw [G]	1 Clab	8	1		58.50	44.50		103	131
3510 Attic mount kit [G]	1 Rofc	2.56	3.125		247	151		398	515
3520 Roof mount kit [G]	1 Clab	3.41	2.346		248	104		352	430
4000 Anemometer, standard AC or 9V, 2-digit LCD, wood case, +/-2% of Input	2 Elec	2	8	↓	179	510		689	950
8900 Equipment									
9100 DC to AC inverter for, 48 V, 4,000 W [G]	1 Elec	2	4	Ea.	2,125	255		2,380	2,725

48 18 Fuel Cell Electrical Power Generation Equipment

48 18 13 – Electrical Power Generation Fuel Cells

48 18 13.10 Fuel Cells

	Crew	Daily Output	Labor-Hours	Unit	Material	2021 Bare Costs Labor	2021 Bare Costs Equipment	Total	Total Incl O&P
0010 **FUEL CELLS**									
2001 Complete system, natural gas, installed, 200 kW [G]				System				1,150,000	1,150,000
2002 400 kW [G]								2,168,000	2,168,000
2003 600 kW [G]								3,060,000	3,060,000
2004 800 kW [G]								4,000,000	4,000,000
2005 1,000 kW [G]				↓				5,005,000	5,005,000
2010 Comm type, for battery charging, uses hydrogen, 100 W, 12 V [G]				Ea.	1,575			1,575	1,725
2020 500 W, 12 V [G]					9,475			9,475	10,400
2030 1,000 W, 48 V [G]					12,300			12,300	13,500
2040 Small, for demonstration or battery charging units, 12 V [G]					179			179	197
2050 Spare anode set for unit [G]					41.50			41.50	46
2060 3 W, uses hydrogen gas [G]				↓	315			315	345

48 18 13.10 Fuel Cells		Crew	Daily Output	Labor-Hours	Unit	Material	2021 Bare Costs Labor	Equipment	Total	Total Incl O&P
2070	6 W, uses hydrogen gas	G				Ea.	595			595	655
2080	10 W, uses hydrogen gas	G				↓	725			725	800

For customer support on your Electrical Costs with RSMeans Data, call 800.448.8182.

383

Assemblies Section

Table of Contents

RSMeans data: Assemblies— How They Work

Assemblies estimating provides a fast and reasonably accurate way to develop construction costs. An assembly is the grouping of individual work items—with appropriate quantities— to provide a cost for a major construction component in a convenient unit of measure.

An assemblies estimate is often used during early stages of design development to compare the cost impact of various design alternatives on total building cost.

Assemblies estimates are also used as an efficient tool to verify construction estimates.

Assemblies estimates do not require a completed design or detailed drawings. Instead, they are based on the general size of the structure and other known parameters of the project. The degree of accuracy of an assemblies estimate is generally within +/- 15%.

Most assemblies consist of three major elements: a graphic, the system components, and the cost data itself. The **Graphic** is a visual representation showing the typical appearance of the assembly

① Unique 12-character Identifier

Our assemblies are identified by a **unique 12-character identifier**. The assemblies are numbered using UNIFORMAT II, ASTM Standard E1557. The first 5 characters represent this system to Level 3. The last 7 characters represent further breakdown in order to arrange items in understandable groups of similar tasks. Line numbers are consistent across all of our publications, so a line number in any assemblies data set will always refer to the same item.

② Reference Box

Information is available in the Reference Section to assist the estimator with estimating procedures, alternate pricing methods, and additional technical information.

The **Reference Box** indicates the exact location of this information in the Reference Section. The "R" stands for "reference," and the remaining characters are the line numbers.

③ Narrative Descriptions

Our assemblies descriptions appear in two formats: narrative and table. **Narrative descriptions** are shown in a hierarchical structure to make them readable. In order to read a complete description, read up through the indents to the top of the section. Include everything that is above and to the left that is not contradicted by information below.

D20 Plumbing

D2020 Domestic Water Distribution

Installation includes piping and fittings within 10' of heater. Electric water heaters do not require venting.

1 Kilowatt hour will raise:			
Gallons of Water	Degrees F	Gallons of Water	Degrees F
4.1	100°	6.8	60°
4.5	90°	8.2	50°
5.1	80°	10.0	40°
5.9	70°		

System Components			COST EACH		
	QUANTITY	UNIT	MAT.	INST.	TOTAL
① SYSTEM D2020 210 1780 **ELECTRIC WATER HEATER, RESIDENTIAL, 100°F RISE** **10 GALLON TANK, 7 GPH**					
Water heater, residential electric, glass lined tank, 10 gal.	1.000	Ea.	510	350	860
Copper tubing, type L, solder joint, hanger 10' OC 1/2" diam.	30.000	L.F.	82.80	300	382.80
Wrought copper 90° elbow for solder joints 1/2" diam.	4.000	Ea.	7.80	162	169.80
Wrought copper Tee for solder joints, 1/2" diam.	2.000	Ea.	6.52	124	130.52
Union, wrought copper, 1/2" diam.	2.000	Ea.	77	85	162
Valve, gate, bronze, 125 lb, NRS, soldered 1/2" diam.	2.000	Ea.	168	67	235
Relief valve, bronze, press & temp, self-close, 3/4" IPS	1.000	Ea.	293	29	322
Wrought copper adapter, CTS to MPT 3/4" IPS	1.000	Ea.	9.80	47.50	57.30
Copper tubing, type L, solder joints, 3/4" diam.	1.000	L.F.	3.35	10.65	14
Wrought copper 90° elbow for solder joints 3/4" diam.	1.000	Ea.	4.28	42.50	46.78
TOTAL			1,162.55	1,217.65	2,380.20

D2020 210	Electric Water Heaters - Residential Systems		COST EACH		
			MAT.	INST.	TOTAL
1760	Electric water heater, residential, 100°F rise				
1780	10 gallon tank, 7 GPH		1,175	1,225	2,400
1820	20 gallon tank, 7 GPH		1,475	1,300	2,775
1860	30 gallon tank, 7 GPH	② R224000 -20	2,275	1,375	3,650
1900	40 gallon tank, 8 GPH		2,650	1,500	4,150
1940	52 gallon tank, 10 GPH		2,875	1,525	4,400
1980	66 gallon tank, 13 GPH		4,575	1,725	6,300
2020	80 gallon tank, 16 GPH		4,900	1,800	6,700
2060	120 gallon tank, 23 GPH		6,825	2,100	8,925

For supplemental customizable square foot estimating forms, visit: RSMeans.com/2021books in question. It is frequently accompanied by additional explanatory technical information describing the class of items. The **System Components** is a listing of the individual tasks that make up the assembly, including the quantity and unit of measure for each item, along with the cost of material and installation. The **Assemblies**

data below lists prices for other similar systems with dimensional and/or size variations.

All of our assemblies costs represent the cost for the installing contractor. An allowance for profit has been added to all material, labor, and equipment rental costs. A markup for labor burdens, including workers' compensation, fixed

overhead, and business overhead, is included with installation costs.

The information in RSMeans cost data represents a "national average" cost. This data should be modified to the project location using the **City Cost Indexes** or **Location Factors** tables found in the Reference Section.

D20 Plumbing

D2020 Domestic Water Distribution

Installation includes piping and fittings within 10' of heater. Electric water heaters do not require venting.

1 Kilowatt hour will raise:			
Gallons of Water	Degrees F	Gallons of Water	Degrees F
4.1	100°	6.8	60°
4.5	90°	8.2	50°
5.1	80°	10.0	40°
5.9	70°		

④ Unit of Measure
All RSMeans data: Assemblies include a typical **Unit of Measure** used for estimating that item. For instance, for continuous footings or foundation walls the unit is linear feet (L.F.). For spread footings the unit is each (Ea.). The estimator needs to take special care that the unit in the data matches the unit in the takeoff. Abbreviations and unit conversions can be found in the Reference Section.

⑤ System Components
System components are listed separately to detail what is included in the development of the total system price.

System Components ⑤	QUANTITY	UNIT	COST EACH MAT.	COST EACH INST.	COST EACH TOTAL
SYSTEM D2020 210 1780					
ELECTRIC WATER HEATER, RESIDENTIAL, 100°F RISE					
10 GALLON TANK, 7 GPH					
Water heater, residential electric, glass lined tank, 10 gal.	1.000	Ea.	510	350	860
Copper tubing, type L, solder joint, hanger 10' OC 1/2" diam.	30.000	L.F.	82.80	300	382.80
Wrought copper 90° elbow for solder joints 1/2" diam.	4.000	Ea.	7.80	162	169.80
Wrought copper Tee for solder joints, 1/2" diam.	2.000	Ea.	6.52	124	130.52
Union, wrought copper, 1/2" diam.	2.000	Ea.	77	85	162
Valve, gate, bronze, 125 lb, NRS, soldered 1/2" diam.	2.000	Ea.	168	67	235
Relief valve, bronze, press & temp, self-close, 3/4" IPS	1.000	Ea.	293	29	322
Wrought copper adapter, CTS to MPT 3/4" IPS	1.000	Ea.	9.80	47.50	57.30
Copper tubing, type L, solder joints, 3/4" diam.	1.000	L.F.	3.35	10.65	14
Wrought copper 90° elbow for solder joints 3/4" diam.	1.000	Ea.	4.28	42.50	46.78
TOTAL			1,162.55	1,217.65	2,380.20

D2020 210	Electric Water Heaters - Residential Systems		COST EACH MAT.	COST EACH INST.	COST EACH TOTAL
1760	Electric water heater, residential, 100°F rise		1,175	1,225	2,400
1780	10 gallon tank, 7 GPH		1,475	1,300	2,775
1820	20 gallon tank, 7 GPH	R224000 -20	2,275	1,375	3,650
1860	30 gallon tank, 7 GPH		2,650	1,500	4,150
1900	40 gallon tank, 8 GPH		2,875	1,525	4,400
1940	52 gallon tank, 10 GPH		4,575	1,725	6,300
1980	66 gallon tank, 13 GPH		4,900	1,800	6,700
2020	80 gallon tank, 16 GPH		6,825	2,100	8,925
2060	120 gallon tank, 23 GPH				

RSMeans data: Assemblies—How They Work (Continued)

Sample Estimate

This sample demonstrates the elements of an estimate, including a tally of the RSMeans data lines. Published assemblies costs include all markups for labor burden and profit for the installing contractor. This estimate adds a summary of the markups applied by a general contractor on the installing contractor's work. These figures represent the total cost to the owner. The location factor with RSMeans data is applied at the bottom of the estimate to adjust the cost of the work to a specific location.

Project Name:	Interior Fit-out, ABC Office			
Location:	Anywhere, USA	Date: 1/1/2021		STD
Assembly Number	**Description**	Qty.	Unit	Subtotal
① C1010 124 1200	Wood partition, 2 x 4 @ 16" OC w/5/8" FR gypsum board	560	S.F.	$3,169.60
C1020 114 1800	Metal door & frame, flush hollow core, 3'-0" x 7'-0"	2	Ea.	$2,580.00
C3010 230 0080	Painting, brushwork, primer & 2 coats	1,120	S.F.	$1,512.00
C3020 410 0140	Carpet, tufted, nylon, roll goods, 12' wide, 26 oz	240	S.F.	$684.00
C3030 210 6000	Acoustic ceilings, 24" x 48" tile, tee grid suspension	200	S.F.	$1,812.00
D5020 125 0560	Receptacles incl plate, box, conduit, wire, 20 A duplex	8	Ea.	$2,448.00
D5020 125 0720	Light switch incl plate, box, conduit, wire, 20 A single pole	2	Ea.	$598.00
D5020 210 0560	Fluorescent fixtures, recess mounted, 20 per 1000 SF	200	S.F.	$2,414.00
Assembly Subtotal				**$15,217.60**
Sales Tax @ ②		5 %		$ 380.44
General Requirements @ ③		7 %		$ 1,065.23
Subtotal A				**$16,663.27**
GC Overhead @ ④		5 %		$ 833.16
Subtotal B				**$17,496.44**
GC Profit @ ⑤		5 %		$ 874.82
Subtotal C				**$18,371.26**
Adjusted by Location Factor ⑥		114.2		$ 20,979.98
Architects Fee @ ⑦		8 %		$ 1,678.40
Contingency @ ⑧		15 %		$ 3,147.00
Project Total Cost				**$ 25,805.38**

This estimate is based on an interactive spreadsheet. You are free to download it and adjust it to your methodology. A copy of this spreadsheet is available at **RSMeans.com/2021books**.

1 Work Performed

The body of the estimate shows the RSMeans data selected, including line numbers, a brief description of each item, its takeoff quantity and unit, and the total installed cost, including the installing contractor's overhead and profit.

2 Sales Tax

If the work is subject to state or local sales taxes, the amount must be added to the estimate. In a conceptual estimate it can be assumed that one half of the total represents material costs. Therefore, apply the sales tax rate to 50% of the assembly subtotal.

3 General Requirements

This item covers project-wide needs provided by the general contractor. These items vary by project but may include temporary facilities and utilities, security, testing, project cleanup, etc. In assemblies estimates a percentage is used—typically between 5% and 15% of project cost.

4 General Contractor Overhead

This entry represents the general contractor's markup on all work to cover project administration costs.

5 General Contractor Profit

This entry represents the GC's profit on all work performed. The value included here can vary widely by project and is influenced by the GC's perception of the project's financial risk and market conditions.

6 Location Factor

RSMeans published data are based on national average costs. If necessary, adjust the total cost of the project using a location factor from the "Location Factor" table or the "City Cost Indexes" table found in the Reference Section. Use location factors if the work is general, covering the work of multiple trades. If the work is by a single trade (e.g., masonry) use the more specific data found in the City Cost Indexes.

To adjust costs by location factors, multiply the base cost by the factor and divide by 100.

7 Architect's Fee

If appropriate, add the design cost to the project estimate. These fees vary based on project complexity and size. Typical design and engineering fees can be found in the Reference Section.

8 Contingency

A factor for contingency may be added to any estimate to represent the cost of unknowns that may occur between the time that the estimate is performed and the time the project is constructed. The amount of the allowance will depend on the stage of design at which the estimate is done, as well as the contractor's assessment of the risk involved.

Same Data. Simplified.

Enjoy the convenience and efficiency of accessing your costs anywhere:

- **Skip the multiplier** by setting your location
- **Quickly search,** edit, favorite and share costs
- **Stay on top of price changes** with automatic updates

Discover more at rsmeans.com/online

D2020 Domestic Water Distribution

Installation includes piping and fittings within 10' of heater. Electric water heaters do not require venting.

1 Kilowatt hour will raise:			
Gallons of Water	Degrees F	Gallons of Water	Degrees F
4.1	100°	6.8	60°
4.5	90°	8.2	50°
5.1	80°	10.0	40°
5.9	70°		

System Components	QUANTITY	UNIT	COST EACH		
			MAT.	INST.	TOTAL
SYSTEM D2020 210 1780					
ELECTRIC WATER HEATER, RESIDENTIAL, 100°F RISE					
10 GALLON TANK, 7 GPH					
Water heater, residential electric, glass lined tank, 10 gal.	1.000	Ea.	510	350	860
Copper tubing, type L, solder joint, hanger 10' OC 1/2" diam.	30.000	L.F.	82.80	300	382.80
Wrought copper 90° elbow for solder joints 1/2" diam.	4.000	Ea.	7.80	162	169.80
Wrought copper Tee for solder joints, 1/2" diam.	2.000	Ea.	6.52	124	130.52
Union, wrought copper, 1/2" diam.	2.000	Ea.	6.52	124	130.52
Valve, gate, bronze, 125 lb, NRS, soldered 1/2" diam.	2.000	Ea.	77	85	162
Relief valve, bronze, press & temp, self-close, 3/4" IPS	1.000	Ea.	168	67	235
Wrought copper adapter, CTS to MPT 3/4" IPS	1.000	Ea.	293	29	322
Copper tubing, type L, solder joints, 3/4" diam.	1.000	Ea.	9.80	47.50	57.30
Wrought copper 90° elbow for solder joints 3/4" diam.	1.000	L.F.	3.35	10.65	14
	1.000	Ea.	4.28	42.50	46.78
TOTAL			1,162.55	1,217.65	2,380.20

D2020 210	Electric Water Heaters - Residential Systems	COST EACH		
		MAT.	INST.	TOTAL
1760	Electric water heater, residential, 100°F rise			
1780	10 gallon tank, 7 GPH	1,175	1,225	2,400
1820	20 gallon tank, 7 GPH	1,475	1,300	2,775
1860	30 gallon tank, 7 GPH	2,275	1,375	3,650
1900	40 gallon tank, 8 GPH	2,650	1,500	4,150
1940	52 gallon tank, 10 GPH	2,875	1,525	4,400
1980	66 gallon tank, 13 GPH	4,575	1,725	6,300
2020	80 gallon tank, 16 GPH	4,900	1,800	6,700
2060	120 gallon tank, 23 GPH	6,825	2,100	8,925

R224000 -20

D2020 Domestic Water Distribution

Systems below include piping and fittings within 10' of heater. Electric water heaters do not require venting.

System Components	QUANTITY	UNIT	COST EACH MAT.	COST EACH INST.	COST EACH TOTAL
SYSTEM D2020 240 1820					
ELECTRIC WATER HEATER, COMMERCIAL, 100°F RISE					
50 GALLON TANK, 9 KW, 37 GPH					
Water heater, commercial, electric, 50 Gal, 9 KW, 37 GPH	1.000	Ea.	6,850	450	7,300
Copper tubing, type L, solder joint, hanger 10' OC, 3/4" diam	34.000	L.F.	113.90	362.10	476
Wrought copper 90° elbow for solder joints 3/4" diam	5.000	Ea.	21.40	212.50	233.90
Wrought copper Tee for solder joints, 3/4" diam	2.000	Ea.	15.70	135	150.70
Wrought copper union for soldered joints, 3/4" diam.	2.000	Ea.	96	90	186
Valve, gate, bronze, 125 lb, NRS, soldered 3/4" diam	2.000	Ea.	186	81	267
Relief valve, bronze, press & temp, self-close, 3/4" IPS	1.000	Ea.	293	29	322
Wrought copper adapter, copper tubing to male, 3/4" IPS	1.000	Ea.	9.80	47.50	57.30
TOTAL			7,585.80	1,407.10	8,992.90

D2020 240	Electric Water Heaters - Commercial Systems		COST EACH MAT.	COST EACH INST.	COST EACH TOTAL
1800	Electric water heater, commercial, 100°F rise				
1820	50 gallon tank, 9 KW 37 GPH		7,575	1,400	8,975
1860	80 gal, 12 KW 49 GPH		9,950	1,750	11,700
1900	36 KW 147 GPH	R224000 -10	13,600	1,875	15,475
1940	120 gal, 36 KW 147 GPH		15,000	2,025	17,025
1980	150 gal, 120 KW 490 GPH	R224000 -20	53,000	2,175	55,175
2020	200 gal, 120 KW 490 GPH		55,500	2,225	57,725
2060	250 gal, 150 KW 615 GPH		53,500	2,600	56,100
2100	300 gal, 180 KW 738 GPH		81,500	2,750	84,250
2140	350 gal, 30 KW 123 GPH		43,000	2,950	45,950
2180	180 KW 738 GPH		59,500	2,950	62,450
2220	500 gal, 30 KW 123 GPH		67,500	3,475	70,975
2260	240 KW 984 GPH		110,500	3,475	113,975
2300	700 gal, 30 KW 123 GPH		67,500	3,975	71,475
2340	300 KW 1230 GPH		100,000	3,975	103,975
2380	1000 gal, 60 KW 245 GPH		82,000	5,525	87,525
2420	480 KW 1970 GPH		132,500	5,550	138,050
2460	1500 gal, 60 KW 245 GPH		119,500	6,850	126,350
2500	480 KW 1970 GPH		163,500	6,850	170,350

393

D3020 Heat Generating Systems

Boiler

Baseboard Radiation

Small Electric Boiler
System Considerations:
1. Terminal units are fin tube baseboard radiation rated at 720 BTU/hr with 200° water temperature or 820 BTU/hr steam.
2. Primary use being for residential or smaller supplementary areas, the floor levels are based on 7-1/2' ceiling heights.
3. All distribution piping is copper for boilers through 205 MBH. All piping for larger systems is steel pipe.

System Components	QUANTITY	UNIT	COST EACH		
			MAT.	INST.	TOTAL
SYSTEM D3020 102 1120					
SMALL HEATING SYSTEM, HYDRONIC, ELECTRIC BOILER					
1,480 S.F., 61 MBH, STEAM, 1 FLOOR					
Boiler, electric steam, std cntrls, trim, ftngs and valves, 18 KW, 61.4 MBH	1.000	Ea.	5,390	2,035	7,425
Copper tubing type L, solder joint, hanger 10'OC, 1-1/4" diam	160.000	L.F.	1,888	2,232	4,120
Radiation, 3/4" copper tube w/alum fin baseboard pkg 7" high	60.000	L.F.	516	1,530	2,046
Rough in baseboard panel or fin tube with valves & traps	10.000	Set	3,250	7,700	10,950
Pipe covering, calcium silicate w/cover, 1" wall 1-1/4" diam	160.000	L.F.	832	1,304	2,136
Low water cut-off, quick hookup, in gage glass tappings	1.000	Ea.	400	51	451
TOTAL			12,276	14,852	27,128
COST PER S.F.			8.29	10.04	18.33

D3020 102	Small Heating Systems, Hydronic, Electric Boilers	COST PER S.F.		
		MAT.	INST.	TOTAL
1100	Small heating systems, hydronic, electric boilers			
1120	Steam, 1 floor, 1480 S.F., 61 M.B.H.	8.31	10.07	18.38
1160	3,000 S.F., 123 M.B.H.	6.40	8.80	15.20
1200	5,000 S.F., 205 M.B.H.	5.05	8.10	13.15
1240	2 floors, 12,400 S.F., 512 M.B.H.	4.39	8.05	12.44
1280	3 floors, 24,800 S.F., 1023 M.B.H.	4.80	8	12.80
1320	34,750 S.F., 1,433 M.B.H.	4.44	7.75	12.19
1360	Hot water, 1 floor, 1,000 S.F., 41 M.B.H.	13.75	5.60	19.35
1400	2,500 S.F., 103 M.B.H.	9.15	10.05	19.20
1440	2 floors, 4,850 S.F., 205 M.B.H.	8.25	12.05	20.30
1480	3 floors, 9,700 S.F., 410 M.B.H.	8.35	12.45	20.80

D3020 Heat Generating Systems

Boiler

Unit Heater

Large Electric Boiler System Considerations:

1. Terminal units are all unit heaters of the same size. Quantities are varied to accommodate total requirements.
2. All air is circulated through the heaters a minimum of three times per hour.
3. As the capacities are adequate for commercial use, floor levels are based on 10' ceiling heights.
4. All distribution piping is black steel pipe.

System Components	QUANTITY	UNIT	COST EACH MAT.	COST EACH INST.	COST EACH TOTAL
SYSTEM D3020 104 1240					
LARGE HEATING SYSTEM, HYDRONIC, ELECTRIC BOILER					
9,280 S.F., 135 KW, 461 MBH, 1 FLOOR					
Boiler, electric hot water, std ctrls, trim, ftngs, valves, 135 KW, 461 MBH	1.000	Ea.	12,450	4,462.50	16,912.50
Expansion tank, painted steel, 60 Gal capacity ASME	1.000	Ea.	5,975	245	6,220
Circulating pump, CI, close cpld, 50 GPM, 2 HP, 2" pipe conn	1.000	Ea.	3,250	485	3,735
Unit heater, 1 speed propeller, horizontal, 200° EWT, 72.7 MBH	7.000	Ea.	7,350	1,869	9,219
Unit heater piping hookup with controls	7.000	Set	6,405	12,075	18,480
Pipe, steel, black, schedule 40, welded, 2-1/2" diam	380.000	L.F.	5,282	12,737.60	18,019.60
Pipe covering, calcium silicate w/cover, 1" wall, 2-1/2" diam	380.000	L.F.	2,432	3,192	5,624
TOTAL			43,144	35,066.10	78,210.10
COST PER S.F.			4.65	3.78	8.43

D3020 104	Large Heating Systems, Hydronic, Electric Boilers	COST PER S.F. MAT.	COST PER S.F. INST.	COST PER S.F. TOTAL
1230	Large heating systems, hydronic, electric boilers			
1240	9,280 S.F., 135 K.W., 461 M.B.H., 1 floor	4.64	3.78	8.42
1280	14,900 S.F., 240 K.W., 820 M.B.H., 2 floors	6	6.15	12.15
1320	18,600 S.F., 296 K.W., 1,010 M.B.H., 3 floors	5.90	6.70	12.60
1360	26,100 S.F., 420 K.W., 1,432 M.B.H., 4 floors	5.85	6.55	12.40
1400	39,100 S.F., 666 K.W., 2,273 M.B.H., 4 floors	5	5.50	10.50
1440	57,700 S.F., 900 K.W., 3,071 M.B.H., 5 floors	4.82	5.40	10.22
1480	111,700 S.F., 1,800 K.W., 6,148 M.B.H., 6 floors	4.31	4.64	8.95
1520	149,000 S.F., 2,400 K.W., 8,191 M.B.H., 8 floors	4.27	4.63	8.90
1560	223,300 S.F., 3,600 K.W., 12,283 M.B.H., 14 floors	4.53	5.25	9.78

395

D40 Fire Protection

D4090 Other Fire Protection Systems

General: Automatic fire protection (suppression) systems other than water sprinklers may be desired for special environments, high risk areas, isolated locations or unusual hazards. Some typical applications would include:

Paint dip tanks
Securities vaults
Electronic data processing
Tape and data storage
Transformer rooms
Spray booths
Petroleum storage
High rack storage

Piping and wiring costs are dependent on the individual application and must be added to the component costs shown below.

All areas are assumed to be open.

D4090 910	Fire Suppression Unit Components	COST EACH		
		MAT.	INST.	TOTAL
0020	Detectors with brackets			
0040	Fixed temperature heat detector	43	108	151
0060	Rate of temperature rise detector	32.50	94.50	127
0080	Ion detector (smoke) detector	138	122	260
0200	Extinguisher agent			
0240	200 lb FM200, container	7,575	365	7,940
0280	75 lb carbon dioxide cylinder	1,600	242	1,842
0320	Dispersion nozzle			
0340	FM200 1-1/2" dispersion nozzle	285	57.50	342.50
0380	Carbon dioxide 3" x 5" dispersion nozzle	179	45	224
0420	Control station			
0440	Single zone control station with batteries	1,200	755	1,955
0470	Multizone (4) control station with batteries	3,475	1,525	5,000
0490				
0500	Electric mechanical release			
0520		1,200	390	1,590
0550	Manual pull station			
0570		100	135	235
0640	Battery standby power 10" x 10" x 17"			
0700		510	189	699
0740	Bell signalling device	75	94.50	169.50

D4090 920	FM200 Systems	COST PER C.F.		
		MAT.	INST.	TOTAL
0820	Average FM200 system, minimum			2.06
0840	Maximum			4.08

D5010 Electrical Service/Distribution

System Components	QUANTITY	UNIT	COST PER L.F.		
			MAT.	INST.	TOTAL
SYSTEM D5010 110 0200					
HIGH VOLTAGE CABLE, NEUTRAL AND CONDUIT INCLUDED, COPPER #2, 5 kV					
Shielded cable, no splice/termn, copper, XLP shielding, 5 kV, #2	.030	C.L.F.	7.68	11.40	19.08
Wire, 600 volt, type THW, copper, stranded, #8	.010	C.L.F.	.34	.95	1.29
Rigid galv steel conduit to 15' H, 2" diam, w/term, ftng & support	1.000	L.F.	9.30	16.85	26.15
TOTAL			17.32	29.20	46.52

D5010 110	High Voltage Shielded Conductors	COST PER L.F.		
		MAT.	INST.	TOTAL
0200	High voltage cable, neutral & conduit included, copper #2, 5 kV	17.30	29	46.30
0240	Copper #1, 5 kV	28.50	34	62.50
0280	15 kV	29.50	34	63.50
0320	Copper 1/0, 5 kV	32	43.50	75.50
0360	15 kV	34	43.50	77.50
0400	25 kV	39.50	44.50	84
0440	35 kV	40	45	85
0480	Copper 2/0, 5 kV	33.50	44.50	78
0520	15 kV	36.50	44.50	81
0560	25 kV	42	45.50	87.50
0600	35 kV	50	50	100
0640	Copper 4/0, 5 kV	46	50.50	96.50
0680	15 kV	47.50	50.50	98
0720	25 kV	53	51.50	104.50
0760	35 kV	56.50	52.50	109
0800	Copper 250 kcmil, 5 kV	50	51.50	101.50
0840	15 kV	50	51.50	101.50
0880	25 kV	59.50	52.50	112
0920	35 kV	64.50	57	121.50
0960	Copper 350 kcmil, 5 kV	60	57	117
1000	15 kV	58.50	57	115.50
1040	25 kV	68	58.50	126.50
1080	35 kV	72	60.50	132.50
1120	Copper 500 kcmil, 5 kV	91.50	71.50	163
1160	15 kV	92.50	71.50	164
1200	25 kV	105	73.50	178.50
1240	35 kV	110	75.50	185.50

D5010 Electrical Service/Distribution

System Components

System Components	QUANTITY	UNIT	COST EACH		
			MAT.	INST.	TOTAL
SYSTEM D5010 120 0220					
SERVICE INSTALLATION, INCLUDES BREAKERS, METERING, 20' CONDUIT & WIRE					
3 PHASE, 4 WIRE, 60 A					
Wire, 600 volt, copper type XHHW, stranded #6	.600	C.L.F.	36	69.60	105.60
Rigid galvanized steel conduit, 3/4" including fittings	20.000	L.F.	64.40	189	253.40
Service entrance cap, 3/4" diameter	1.000	Ea.	4.60	58.50	63.10
Conduit LB fitting with cover, 3/4" diameter	1.000	Ea.	19.90	58.50	78.40
Meter socket, three phase, 100 A	1.000	Ea.	172	270	442
Safety switches, heavy duty, 240 volt, 3 pole NEMA 1 fusible, 60 amp	1.000	Ea.	197	330	527
Grounding, wire ground bare armored, #8-1 conductor	.200	C.L.F.	17	76	93
Grounding, clamp, bronze, 3/4" diameter	1.000	Ea.	3.87	23.50	27.37
Grounding, rod, copper clad, 8' long, 3/4" diameter	1.000	Ea.	31.50	143	174.50
Wireway w/fittings, 2-1/2" x 2-1/2"	1.000	L.F.	17.45	16.85	34.30
TOTAL			563.72	1,234.95	1,798.67

D5010 120	Overhead Electric Service, 3 Phase - 4 Wire	COST EACH		
		MAT.	INST.	TOTAL
0200	Service installation, includes breakers, metering, 20' conduit & wire			
0220	3 phase, 4 wire, 120/208 volts, 60 A	565	1,225	1,790
0240	100 A	725	1,400	2,125
0245	100 A w/circuit breaker	1,600	1,725	3,325
0280	200 A	1,300	1,925	3,225
0285	200 A w/circuit breaker	3,375	2,425	5,800
0320	400 A	2,600	3,825	6,425
0325	400 A w/circuit breaker	6,075	4,775	10,850
0360	600 A	4,525	5,650	10,175
0365	600 A, w/switchboard	9,925	7,175	17,100
0400	800 A	6,525	6,775	13,300
0405	800 A, w/switchboard	11,900	8,500	20,400
0440	1000 A	7,975	8,350	16,325
0445	1000 A, w/switchboard	14,500	10,300	24,800
0480	1200 A	10,800	9,525	20,325
0485	1200 A, w/groundfault switchboard	38,800	11,800	50,600
0520	1600 A	13,200	12,500	25,700
0525	1600 A, w/groundfault switchboard	44,100	14,900	59,000

D50 Electrical

D5010 Electrical Service/Distribution

D5010 120	Overhead Electric Service, 3 Phase - 4 Wire	COST EACH		
		MAT.	INST.	TOTAL
0560	2000 A	18,000	15,100	33,100
0565	2000 A, w/groundfault switchboard	52,500	18,600	71,100
0610	1 phase, 3 wire, 120/240 volts, 100 A (no safety switch)	202	680	882
0615	100 A w/load center	485	1,300	1,785
0620	200 A	440	940	1,380
0625	200 A w/load center	1,025	2,000	3,025

For customer support on your Electrical Costs with RSMeans Data, call 800.448.8182.

D5010 Electrical Service/Distribution

Safety Switch

Underground service conductor

Meter and service equipment

Ground rod

System Components	QUANTITY	UNIT	COST EACH		
			MAT.	INST.	TOTAL
SYSTEM D5010 130 1000					
2000 AMP UNDERGROUND ELECTRIC SERVICE WITH SAFETY SWITCH					
INCLUDING EXCAVATION, BACKFILL AND COMPACTION					
Excavate Trench	44.440	B.C.Y.		335.08	335.08
4 inch conduit bank	108.000	L.F.	1,123.20	3,024	4,147.20
4 inch fitting	8.000	Ea.	324	504	828
4 inch bells	4.000	Ea.	20.40	252	272.40
Concrete material	16.580	C.Y.	2,271.46		2,271.46
Concrete placement	16.580	C.Y.		464.24	464.24
Backfill trench	33.710	L.C.Y.		109.90	109.90
Compact fill material in trench	25.930	E.C.Y.		142.10	142.10
Dispose of excess fill material on-site	24.070	L.C.Y.		314.36	314.36
500 kcmil power cable	18.000	C.L.F.	24,750	8,550	33,300
Wire, 600 volt, type THW, copper, stranded, 1/0	6.000	C.L.F.	1,890	1,374	3,264
Saw cutting, concrete walls, plain, per inch of depth	64.000	L.F.	2.56	757.12	759.68
Meter centers and sockets, single pos, 4 terminal, 400 amp	5.000	Ea.	4,625	2,225	6,850
Safety switch, 400 amp	5.000	Ea.	6,625	4,200	10,825
Ground rod clamp	6.000	Ea.	23.22	141	164.22
Wireway	1.000	L.F.	17.45	16.85	34.30
600 volt stranded copper wire, 1/0	1.200	C.L.F.	378	274.80	652.80
Flexible metallic conduit	120.000	L.F.	82.80	567.60	650.40
Ground rod	6.000	Ea.	417	1,134	1,551
TOTAL			42,550.09	24,386.05	66,936.14

D5010 130	Underground Electric Service	COST EACH		
		MAT.	INST.	TOTAL
0950	Underground electric service including excavation, backfill and compaction			
1000	3 phase, 4 wire, 277/480 volts, 2000 A	42,600	24,400	67,000
1050	2000 A w/groundfault switchboard	76,000	27,100	103,100
1100	1600 A	33,700	19,700	53,400
1150	1600 A w/groundfault switchboard	64,500	22,100	86,600
1200	1200 A	25,800	15,700	41,500
1250	1200 A w/groundfault switchboard	54,000	18,000	72,000
1400	800 A	15,000	13,000	28,000

D50 Electrical

D5010 Electrical Service/Distribution

D5010 130	Underground Electric Service	COST EACH		
		MAT.	INST.	TOTAL
1450	800 A w/switchboard	20,400	14,700	35,100
1500	600 A	10,900	12,000	22,900
1550	600 A w/switchboard	16,300	13,500	29,800
1600	1 phase, 3 wire, 120/240 volts, 200 A	3,425	4,225	7,650
1650	200 A w/load center	4,000	5,275	9,275
1700	100 A	2,575	3,475	6,050
1750	100 A w/load center	3,150	4,525	7,675

D5010 Electrical Service/Distribution

System Components			COST PER L.F.		
	QUANTITY	UNIT	MAT.	INST.	TOTAL
SYSTEM D5010 230 0200					
FEEDERS, INCLUDING STEEL CONDUIT & WIRE, 60 A					
Rigid galvanized steel conduit, 3/4", including fittings	1.000	L.F.	3.22	9.45	12.67
Wire 600 volt, type XHHW copper stranded #6	.030	C.L.F.	1.80	3.48	5.28
Wire 600 volt, type XHHW copper stranded #8	.010	C.L.F.	.38	.95	1.33
TOTAL			5.40	13.88	19.28

D5010 230	Feeder Installation	COST PER L.F.		
		MAT.	INST.	TOTAL
0200	Feeder installation 600 V, including RGS conduit and XHHW wire, 60 A	5.40	13.90	19.30
0240	100 A	9.85	18.10	27.95
0280	200 A	19.10	27.50	46.60
0320	400 A	38	55	93
0360	600 A	75.50	89.50	165
0400	800 A	99.50	106	205.50
0440	1000 A	124	138	262
0480	1200 A	151	179	330
0520	1600 A	199	211	410
0560	2000 A	249	277	526
1200	Branch installation 600 V, including EMT conduit and THW wire, 15 A	1.12	6.80	7.92
1240	20 A	1.56	7.20	8.76
1280	30 A	2.03	7.45	9.48
1320	50 A	3.18	10.40	13.58
1360	65 A	4.59	11.05	15.64
1400	85 A	7	12.80	19.80
1440	100 A	8.35	13.05	21.40
1480	130 A	11.90	15.25	27.15
1520	150 A	14.45	17.50	31.95
1560	200 A	17	20	37

D5010 Electrical Service/Distribution

System Components	QUANTITY	UNIT	COST EACH		
			MAT.	INST.	TOTAL
SYSTEM D5010 240 0240					
SWITCHGEAR INSTALLATION, INCL SWBD, PANELS & CIRC BREAKERS, 600 A					
Switchboards, 120/208 V, 600 amp	1.000	Ea.	5,400	1,525	6,925
Aluminum bus bars, 120/208 V, 600 amp	1.000	Ea.	1,725	1,525	3,250
Feeder section, circuit breakers, KA frame, 70 to 225 amp	1.000	Ea.	1,575	237	1,812
Feeder section, circuit breakers, LA frame, 125 to 400 amp	1.000	Ea.	4,675	330	5,005
TOTAL			13,375	3,617	16,992

D5010 240	Switchgear	COST PER EACH		
		MAT.	INST.	TOTAL
0190	Switchgear installation, including switchboard, panels, & circuit breaker			
0200	120/208 V, 3 phase, 400 A	9,825	3,225	13,050
0240	600 A	13,400	3,625	17,025
0280	800 A	15,700	4,150	19,850
0300	1000 A	17,400	4,625	22,025
0320	1200 A	20,100	5,200	25,300
0360	1600 A	32,700	6,025	38,725
0400	2000 A	39,600	6,550	46,150
0500	277/480 V, 3 phase, 400 A	13,500	5,975	19,475
0520	600 A	18,400	6,900	25,300
0540	800 A	20,000	7,600	27,600
0560	1000 A	23,400	8,425	31,825
0580	1200 A	27,100	9,400	36,500
0600	1600 A	41,700	10,400	52,100
0620	2000 A	52,500	11,500	64,000

D5010 Electrical Service/Distribution

System Components			COST EACH		
	QUANTITY	UNIT	MAT.	INST.	TOTAL
SYSTEM D5010 250 1020					
PANELBOARD INSTALLATION, INCLUDES PANELBOARD, CONDUCTOR, & CONDUIT					
Conduit, galvanized steel, 1-1/4"dia.	37.000	L.F.	240.50	466.20	706.70
Panelboards, NQOD, 4 wire, 120/208 volts, 100 amp main, 24 circuits	1.000	Ea.	1,925	1,600	3,525
Wire, 600 volt, type THW, copper, stranded, #3	1.480	C.L.F.	248.64	223.48	472.12
TOTAL			2,414.14	2,289.68	4,703.82

D5010 250	Panelboard	COST EACH		
		MAT.	INST.	TOTAL
0900	Panelboards, NQOD, 4 wire, 120/208 volts w/conductor & conduit			
1000	100 A, 0 stories, 0' horizontal	1,925	1,600	3,525
1020	1 stories, 25' horizontal	2,425	2,300	4,725
1040	5 stories, 50' horizontal	3,375	3,650	7,025
1060	10 stories, 75' horizontal	4,500	5,225	9,725
1080	225A, 0 stories, 0' horizontal	3,750	2,100	5,850
2000	1 stories, 25' horizontal	5,150	3,350	8,500
2020	5 stories, 50' horizontal	7,900	5,800	13,700
2040	10 stories, 75' horizontal	11,100	8,650	19,750
2060	400A, 0 stories, 0' horizontal	5,525	3,150	8,675
2080	1 stories, 25' horizontal	7,150	5,175	12,325
3000	5 stories, 50' horizontal	10,400	9,175	19,575
3020	10 stories, 75' horizontal	19,100	18,000	37,100
3040	600 A, 0 stories, 0' horizontal	8,150	3,775	11,925
3060	1 stories, 25' horizontal	11,100	7,100	18,200
3080	5 stories, 50' horizontal	17,000	13,600	30,600
4000	10 stories, 75' horizontal	32,200	26,800	59,000
4010	Panelboards, NEHB, 4 wire, 277/480 volts w/conductor, conduit, & safety switch			
4020	100 A, 0 stories, 0' horizontal, includes safety switch	3,500	2,200	5,700
4040	1 stories, 25' horizontal	4,000	2,900	6,900
4060	5 stories, 50' horizontal	4,975	4,250	9,225
4080	10 stories, 75' horizontal	6,100	5,825	11,925
5000	225 A, 0 stories, 0' horizontal	5,250	2,675	7,925
5020	1 stories, 25' horizontal	6,650	3,925	10,575
5040	5 stories, 50' horizontal	9,400	6,375	15,775
5060	10 stories, 75' horizontal	12,600	9,225	21,825

D5010 Electrical Service/Distribution

D5010 250	Panelboard	COST EACH		
		MAT.	INST.	TOTAL
5080	400 A, 0 stories, 0' horizontal	8,250	4,150	12,400
6000	1 stories, 25' horizontal	9,875	6,175	16,050
6020	5 stories, 50' horizontal	13,100	10,200	23,300
6040	10 stories, 75' horizontal	21,800	19,000	40,800
6060	600 A, 0 stories, 0' horizontal	12,100	5,225	17,325
6080	1 stories, 25' horizontal	15,100	8,550	23,650
7000	5 stories, 50' horizontal	21,000	15,100	36,100
7020	10 stories, 75' horizontal	36,100	28,300	64,400

405

D5020 Lighting and Branch Wiring

Duplex Receptacle

System Components	QUANTITY	UNIT	COST PER S.F.		
			MAT.	INST.	TOTAL
SYSTEM D5020 110 0200					
RECEPTACLES INCL. PLATE, BOX, CONDUIT, WIRE & TRANS. WHEN REQUIRED					
2.5 PER 1000 S.F., .3 WATTS PER S.F.					
Steel intermediate conduit, (IMC) 1/2" diam	167.000	L.F.	.36	1.26	1.62
Wire 600V type THWN-THHN, copper solid #12	3.340	C.L.F.	.04	.23	.27
Wiring device, receptacle, duplex, 120V grounded, 15 amp	2.500	Ea.		.05	.05
Wall plate, 1 gang, brown plastic	2.500	Ea.		.02	.02
Steel outlet box 4" square	2.500	Ea.	.01	.10	.11
Steel outlet box 4" plaster rings	2.500	Ea.		.03	.03
		TOTAL	.41	1.69	2.10

D5020 110	Receptacle (by Wattage)	COST PER S.F.		
		MAT.	INST.	TOTAL
0190	Receptacles include plate, box, conduit, wire & transformer when required			
0200	2.5 per 1000 S.F., .3 watts per S.F.	.41	1.69	2.10
0240	With transformer	.49	1.77	2.26
0280	4 per 1000 S.F., .5 watts per S.F.	.47	1.97	2.44
0320	With transformer	.59	2.10	2.69
0360	5 per 1000 S.F., .6 watts per S.F.	.13	.78	.91
0400	With transformer	.70	2.49	3.19
0440	8 per 1000 S.F., .9 watts per S.F.	.57	2.57	3.14
0480	With transformer	.81	2.84	3.65
0520	10 per 1000 S.F., 1.2 watts per S.F.	.60	2.79	3.39
0560	With transformer	.90	3.13	4.03
0600	16.5 per 1000 S.F., 2.0 watts per S.F.	.69	3.49	4.18
0640	With transformer	1.19	4.04	5.23
0680	20 per 1000 S.F., 2.4 watts per S.F.	.72	3.80	4.52
0720	With transformer	1.32	4.47	5.79

D5020 Lighting and Branch Wiring

Underfloor Receptacle System

Description: Table D5020 115 includes installed costs of raceways and copper wire from panel to and including receptacle.

National Electrical Code prohibits use of undercarpet system in residential, school or hospital buildings. Can only be used with carpet squares.

Low density = (1) Outlet per 259 S.F. of floor area.

High density = (1) Outlet per 127 S.F. of floor area.

System Components	QUANTITY	UNIT	COST PER S.F.		
			MAT.	INST.	TOTAL
SYSTEM D5020 115 0200					
RECEPTACLE SYSTEMS, UNDERFLOOR DUCT, 5' ON CENTER, LOW DENSITY					
Underfloor duct 3-1/8" x 7/8" w/insert 24" on center	.190	L.F.	4.28	2.05	6.33
Underfloor duct junction box, single duct, 3-1/8"	.003	Ea.	1.49	.57	2.06
Underfloor junction box carpet pan	.003	Ea.	1.23	.03	1.26
Underfloor duct outlet, high tension receptacle	.004	Ea.	.47	.38	.85
Wire 600V type THWN-THHN copper solid #12	.010	C.L.F.	.11	.69	.80
Vertical elbow for underfloor duct, 3-1/8", included					
Underfloor duct conduit adapter, 2" x 1-1/4", included					
TOTAL			7.58	3.72	11.30

D5020 115	Receptacles, Floor	COST PER S.F.		
		MAT.	INST.	TOTAL
0200	Receptacle systems, underfloor duct, 5' on center, low density	7.55	3.72	11.27
0240	High density	8.15	4.78	12.93
0280	7' on center, low density	6	3.19	9.19
0320	High density	6.55	4.26	10.81
0400	Poke thru fittings, low density	1.66	1.81	3.47
0440	High density	3.31	3.61	6.92
0520	Telepoles, using Romex, low density	1.63	1.13	2.76
0560	High density	3.25	2.25	5.50
0600	Using EMT, low density	1.69	1.48	3.17
0640	High density	3.39	2.97	6.36
0720	Conduit system with floor boxes, low density	1.45	1.27	2.72
0760	High density	2.90	2.54	5.44
0840	Undercarpet power system, 3 conductor with 5 conductor feeder, low density	1.59	.46	2.05
0880	High density	3.17	.93	4.10

D5020 Lighting and Branch Wiring

Duplex Receptacle

Wall Switch

System Components

System Components			COST PER S.F.		
	QUANTITY	UNIT	MAT.	INST.	TOTAL
SYSTEM D5020 120 0520					
RECEPTACLES AND WALL SWITCHES					
4 RECEPTACLES PER 400 S.F.					
Steel intermediate conduit, (IMC), 1/2" diam	.220	L.F.	.48	1.66	2.14
Wire, 600 volt, type THWN-THHN, copper, solid #12	.005	C.L.F.	.05	.35	.40
Steel outlet box 4" square	.010	Ea.	.02	.38	.40
Steel outlet box, 4" square, plaster rings	.010	Ea.	.02	.12	.14
Receptacle, duplex, 120 volt grounded, 15 amp	.010	Ea.	.02	.19	.21
Wall plate, 1 gang, brown plastic	.010	Ea.		.09	.09
TOTAL			.59	2.79	3.38

D5020 120	Receptacles & Wall Switches	COST PER S.F.		
		MAT.	INST.	TOTAL
0520	Receptacles and wall switches, 400 S.F., 4 receptacles	.59	2.79	3.38
0560	6 receptacles	.65	3.26	3.91
0600	8 receptacles	.70	3.78	4.48
0640	1 switch	.12	.63	.75
0680	600 S.F., 6 receptacles	.59	2.79	3.38
0720	8 receptacles	.63	3.16	3.79
0760	10 receptacles	.68	3.55	4.23
0800	2 switches	.17	.83	1
0840	1000 S.F., 10 receptacles	.58	2.78	3.36
0880	12 receptacles	.58	2.93	3.51
0920	14 receptacles	.63	3.17	3.80
0960	2 switches	.09	.45	.54
1000	1600 S.F., 12 receptacles	.53	2.47	3
1040	14 receptacles	.56	2.70	3.26
1080	16 receptacles	.58	2.78	3.36
1120	4 switches	.11	.53	.64
1160	2000 S.F., 14 receptacles	.55	2.54	3.09
1200	16 receptacles	.55	2.55	3.10
1240	18 receptacles	.56	2.70	3.26
1280	4 switches	.09	.45	.54
1320	3000 S.F., 12 receptacles	.45	1.94	2.39
1360	18 receptacles	.53	2.39	2.92
1400	24 receptacles	.55	2.55	3.10
1440	6 switches	.09	.45	.54
1480	3600 S.F., 20 receptacles	.51	2.32	2.83
1520	24 receptacles	.52	2.40	2.92
1560	28 receptacles	.54	2.47	3.01
1600	8 switches	.10	.52	.62
1640	4000 S.F., 16 receptacles	.45	1.94	2.39
1680	24 receptacles	.53	2.39	2.92
1720	30 receptacles	.53	2.47	3
1760	8 switches	.09	.45	.54
1800	5000 S.F., 20 receptacles	.45	1.94	2.39
1840	26 receptacles	.53	2.32	2.85
1880	30 receptacles	.53	2.39	2.92
1920	10 switches	.09	.45	.54

D5020 Lighting and Branch Wiring

Duplex Receptacle

Wall Switch

System Components	QUANTITY	UNIT	COST PER EACH		
			MAT.	INST.	TOTAL
SYSTEM D5020 125 0520					
RECEPTACLES AND WALL SWITCHES, RECEPTICLE DUPLEX 120 V GROUNDED, 15 A					
Electric metallic tubing conduit, (EMT), 3/4" diam	22.000	L.F.	24.42	127.60	152.02
Wire, 600 volt, type THWN-THHN, copper, solid #12	.630	C.L.F.	6.68	43.47	50.15
Steel outlet box 4" square	1.000	Ea.	2.42	38	40.42
Steel outlet box, 4" square, plaster rings	1.000	Ea.	1.83	11.85	13.68
Receptacle, duplex, 120 volt grounded, 15 amp	1.000	Ea.	1.52	18.95	20.47
Wall plate, 1 gang, brown plastic	1.000	Ea.	.42	9.45	9.87
TOTAL			37.29	249.32	286.61

D5020 125	Receptacles & Switches by Each	COST PER EACH		
		MAT.	INST.	TOTAL
0460	Receptacles & Switches, with box, plate, 3/4" EMT conduit & wire			
0520	Receptacle duplex 120 V grounded, 15 A	37.50	249	286.50
0560	20 A	48	258	306
0600	Receptacle duplex ground fault interrupting, 15 A	52.50	258	310.50
0640	20 A	97	258	355
0680	Toggle switch single, 15 A	36.50	249	285.50
0720	20 A	41	258	299
0760	3 way switch, 15 A	37.50	263	300.50
0800	20 A	42	272	314
0840	4 way switch, 15 A	48	281	329
0880	20 A	93.50	299	392.50

D50 Electrical

D5020 Lighting and Branch Wiring

Description: Table D5020 130 includes the cost for switch, plate, box, conduit in slab or EMT exposed and copper wire. Add 20% for exposed conduit.

No power required for switches.

Federal energy guidelines recommend the maximum lighting area controlled per switch shall not exceed 1000 S.F. and that areas over 500 S.F. shall be so controlled that total illumination can be reduced by at least 50%.

System Components	QUANTITY	UNIT	COST PER S.F. MAT.	COST PER S.F. INST.	COST PER S.F. TOTAL
SYSTEM D5020 130 0360					
WALL SWITCHES, 5.0 PER 1000 S.F.					
Steel, intermediate conduit (IMC), 1/2" diameter	88.000	L.F.	.19	.66	.85
Wire, 600V type THWN-THHN, copper solid #12	1.710	C.L.F.	.02	.12	.14
Toggle switch, single pole, 15 amp	5.000	Ea.		.09	.09
Wall plate, 1 gang, brown plastic	5.000	Ea.		.05	.05
Steel outlet box 4" plaster rings	5.000	Ea.	.01	.19	.20
Plaster rings	5.000	Ea.	.01	.06	.07
TOTAL			.23	1.17	1.40

D5020 130	Wall Switch by Sq. Ft.	COST PER S.F. MAT.	COST PER S.F. INST.	COST PER S.F. TOTAL
0200	Wall switches, 1.0 per 1000 S.F.	.06	.27	.33
0240	1.2 per 1000 S.F.	.06	.30	.36
0280	2.0 per 1000 S.F.	.08	.44	.52
0320	2.5 per 1000 S.F.	.11	.55	.66
0360	5.0 per 1000 S.F.	.24	1.17	1.41
0400	10.0 per 1000 S.F.	.48	2.37	2.85

D5020 Lighting and Branch Wiring

System D5020 135 includes all wiring and connections.

System Components			COST PER S.F.		
	QUANTITY	UNIT	MAT.	INST.	TOTAL
SYSTEM D5020 135 0200					
MISCELLANEOUS POWER, TO .5 WATTS					
Steel intermediate conduit, (IMC) 1/2″ diam	15.000	L.F.	.03	.11	.14
Wire 600V type THWN-THHN, copper solid #12	.325	C.L.F.		.02	.02
TOTAL			.03	.13	.16

D5020 135	Miscellaneous Power	COST PER S.F.		
		MAT.	INST.	TOTAL
0200	Miscellaneous power, to .5 watts	.04	.14	.18
0240	.8 watts	.05	.19	.24
0280	1 watt	.07	.26	.33
0320	1.2 watts	.08	.30	.38
0360	1.5 watts	.09	.35	.44
0400	1.8 watts	.11	.41	.52
0440	2 watts	.13	.49	.62
0480	2.5 watts	.16	.60	.76
0520	3 watts	.19	.71	.90

411

D50 Electrical

D5020 Lighting and Branch Wiring

System D5020 140 includes all wiring and connections for central air conditioning units.

System Components			COST PER S.F.		
	QUANTITY	UNIT	MAT.	INST.	TOTAL
SYSTEM D5020 140 0200					
CENTRAL AIR CONDITIONING POWER, 1 WATT					
Steel intermediate conduit, 1/2″ diam.	.030	L.F.	.07	.23	.30
Wire 600V type THWN-THHN, copper solid #12	.001	C.L.F.	.01	.07	.08
TOTAL			.08	.30	.38

D5020 140	Central A. C. Power (by Wattage)	COST PER S.F.		
		MAT.	INST.	TOTAL
0200	Central air conditioning power, 1 watt	.08	.30	.38
0220	2 watts	.09	.34	.43
0240	3 watts	.17	.50	.67
0280	4 watts	.23	.51	.74
0320	6 watts	.37	.68	1.05
0360	8 watts	.47	.74	1.21
0400	10 watts	.83	.85	1.68

D50 Electrical

D5020 Lighting and Branch Wiring

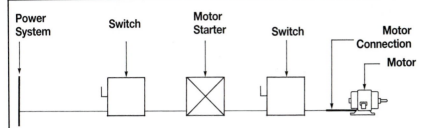

System D5020 145 installed cost of motor wiring using 50' of rigid conduit and copper wire. **Cost and setting of motor not included.**

System Components	QUANTITY	UNIT	COST EACH		
			MAT.	INST.	TOTAL
SYSTEM D5020 145 0200					
MOTOR INST., SINGLE PHASE, 115V, TO AND INCLUDING 1/3 HP MOTOR SIZE					
Wire 600V type THWN-THHN, copper solid #12	1.250	C.L.F.	13.25	86.25	99.50
Steel intermediate conduit, (IMC) 1/2" diam.	50.000	L.F.	108.50	377.50	486
Magnetic FVNR, 115V, 1/3 HP, size 00 starter	1.000	Ea.	193	189	382
Safety switch, fused, heavy duty, 240V 2P 30 amp	1.000	Ea.	88.50	216	304.50
Safety switch, non fused, heavy duty, 600V, 3 phase, 30 A	1.000	Ea.	103	237	340
Flexible metallic conduit, Greenfield 1/2" diam.	1.500	L.F.	.72	5.69	6.41
Connectors for flexible metallic conduit Greenfield 1/2" diam.	1.000	Ea.	2.22	9.45	11.67
Coupling for Greenfield to conduit 1/2" diam. flexible metallic conduit	1.000	Ea.	1.03	15.15	16.18
Fuse cartridge nonrenewable, 250V 30 amp	1.000	Ea.	2.51	15.15	17.66
TOTAL			512.73	1,151.19	1,663.92

D5020 145	Motor Installation	COST EACH		
		MAT.	INST.	TOTAL
0200	Motor installation, single phase, 115V, 1/3 HP motor size	515	1,150	1,665
0240	1 HP motor size	535	1,150	1,685
0280	2 HP motor size	585	1,225	1,810
0320	3 HP motor size	655	1,250	1,905
0360	230V, 1 HP motor size	515	1,175	1,690
0400	2 HP motor size	555	1,175	1,730
0440	3 HP motor size	615	1,250	1,865
0520	Three phase, 200V, 1-1/2 HP motor size	520	1,275	1,795
0560	3 HP motor size	805	1,400	2,205
0600	5 HP motor size	610	1,550	2,160
0640	7-1/2 HP motor size	630	1,600	2,230
0680	10 HP motor size	940	2,000	2,940
0720	15 HP motor size	1,275	2,200	3,475
0760	20 HP motor size	1,725	2,550	4,275
0800	25 HP motor size	1,800	2,550	4,350
0840	30 HP motor size	2,475	3,000	5,475
0880	40 HP motor size	3,200	3,550	6,750
0920	50 HP motor size	5,300	4,150	9,450
0960	60 HP motor size	5,550	4,375	9,925
1000	75 HP motor size	7,275	5,000	12,275
1040	100 HP motor size	15,400	5,900	21,300
1080	125 HP motor size	14,800	6,475	21,275
1120	150 HP motor size	18,600	7,625	26,225
1160	200 HP motor size	19,500	8,975	28,475
1240	230V, 1-1/2 HP motor size	495	1,275	1,770
1280	3 HP motor size	785	1,375	2,160
1320	5 HP motor size	585	1,550	2,135
1360	7-1/2 HP motor size	585	1,550	2,135
1400	10 HP motor size	865	1,875	2,740
1440	15 HP motor size	1,025	2,050	3,075
1480	20 HP motor size	1,600	2,475	4,075
1520	25 HP motor size	1,725	2,550	4,275

413

D5020 Lighting and Branch Wiring

D5020 145	Motor Installation	COST EACH		
		MAT.	INST.	TOTAL
1560	30 HP motor size	1,650	2,550	4,200
1600	40 HP motor size	3,275	3,450	6,725
1640	50 HP motor size	3,875	3,650	7,525
1680	60 HP motor size	5,325	4,150	9,475
1720	75 HP motor size	6,575	4,700	11,275
1760	100 HP motor size	7,875	5,250	13,125
1800	125 HP motor size	16,200	6,100	22,300
1840	150 HP motor size	16,300	6,925	23,225
1880	200 HP motor size	18,400	7,700	26,100
1960	460V, 2 HP motor size	630	1,275	1,905
2000	5 HP motor size	920	1,400	2,320
2040	10 HP motor size	685	1,550	2,235
2080	15 HP motor size	875	1,775	2,650
2120	20 HP motor size	930	1,900	2,830
2160	25 HP motor size	1,000	1,975	2,975
2200	30 HP motor size	1,250	2,125	3,375
2240	40 HP motor size	1,700	2,300	4,000
2280	50 HP motor size	1,975	2,550	4,525
2320	60 HP motor size	2,675	2,975	5,650
2360	75 HP motor size	3,425	3,300	6,725
2400	100 HP motor size	4,200	3,675	7,875
2440	125 HP motor size	5,625	4,175	9,800
2480	150 HP motor size	7,800	4,675	12,475
2520	200 HP motor size	8,725	5,275	14,000
2600	575V, 2 HP motor size	630	1,275	1,905
2640	5 HP motor size	920	1,400	2,320
2680	10 HP motor size	685	1,550	2,235
2720	20 HP motor size	875	1,775	2,650
2760	25 HP motor size	930	1,900	2,830
2800	30 HP motor size	1,250	2,125	3,375
2840	50 HP motor size	1,375	2,225	3,600
2880	60 HP motor size	2,725	2,950	5,675
2920	75 HP motor size	2,675	2,975	5,650
2960	100 HP motor size	3,425	3,300	6,725
3000	125 HP motor size	6,075	4,100	10,175
3040	150 HP motor size	5,625	4,175	9,800
3080	200 HP motor size	7,450	4,725	12,175

D5020 Lighting and Branch Wiring

System Components	QUANTITY	UNIT	COST PER L.F.		
			MAT.	INST.	TOTAL
SYSTEM D5020 155 0200					
MOTOR FEEDER SYSTEMS, SINGLE PHASE, UP TO 115V, 1HP OR 230V, 2HP					
Steel intermediate conduit, (IMC) 1/2" diam	1.000	L.F.	2.17	7.55	9.72
Wire 600V type THWN-THHN, copper solid #12	.020	C.L.F.	.21	1.38	1.59
TOTAL			2.38	8.93	11.31

D5020 155	Motor Feeder	COST PER L.F.		
		MAT.	INST.	TOTAL
0200	Motor feeder systems, single phase, feed up to 115V 1HP or 230V 2 HP	2.38	8.95	11.33
0240	115V 2HP, 230V 3HP	2.51	9.05	11.56
0280	115V 3HP	2.74	9.45	12.19
0360	Three phase, feed to 200V 3HP, 230V 5HP, 460V 10HP, 575V 10HP	2.49	9.60	12.09
0440	200V 5HP, 230V 7.5HP, 460V 15HP, 575V 20HP	2.68	9.80	12.48
0520	200V 10HP, 230V 10HP, 460V 30HP, 575V 30HP	3.03	10.40	13.43
0600	200V 15HP, 230V 15HP, 460V 40HP, 575V 50HP	4.59	11.90	16.49
0680	200V 20HP, 230V 25HP, 460V 50HP, 575V 60HP	5.85	15.10	20.95
0760	200V 25HP, 230V 30HP, 460V 60HP, 575V 75HP	6.35	15.35	21.70
0840	200V 30HP	6.95	15.85	22.80
0920	230V 40HP, 460V 75HP, 575V 100HP	9	17.30	26.30
1000	200V 40HP	9.90	18.50	28.40
1080	230V 50HP, 460V 100HP, 575V 125HP	12.25	20.50	32.75
1160	200V 50HP, 230V 60HP, 460V 125HP, 575V 150HP	13.80	22	35.80
1240	200V 60HP, 460V 150HP	17.05	25.50	42.55
1320	230V 75HP, 575V 200HP	18.55	26.50	45.05
1400	200V 75HP	21	27	48
1480	230V 100HP, 460V 200HP	29	31.50	60.50
1560	200V 100HP	40	39.50	79.50
1640	230V 125HP	40	39.50	79.50
1720	200V 125HP, 230V 150HP	44.50	49	93.50
1800	200V 150HP	53	53	106
1880	200V 200HP	66.50	56.50	123
1960	230V 200HP	70.50	59	129.50

D5020 Lighting and Branch Wiring

Starters are full voltage, type NEMA 1 for general purpose indoor application with motor overload protection and include mounting and wire connections.

System Components	QUANTITY	UNIT	COST EACH		
			MAT.	INST.	TOTAL
SYSTEM D5020 160 0200					
MAGNETIC STARTER, SIZE 00 TO 1/3 HP, 1 PHASE 115V OR 1 HP 230V					
Magnetic starter, size 00, to 1/3 HP, 1 phase, 115V or 1 HP 230V	1.000	Ea.	193	189	382
TOTAL			193	189	382

D5020 160	Magnetic Starter	COST EACH		
		MAT.	INST.	TOTAL
0200	Magnetic starter, size 00, to 1/3 HP, 1 phase, 115V or 1 HP 230V	193	189	382
0280	Size 00, to 1-1/2 HP, 3 phase, 200-230V or 2 HP 460-575V	131	216	347
0360	Size 0, to 1 HP, 1 phase, 115V or 2 HP 230V	215	189	404
0440	Size 0, to 3 HP, 3 phase, 200-230V or 5 HP 460-575V	420	330	750
0520	Size 1, to 2 HP, 1 phase, 115V or 3 HP 230V	247	252	499
0600	Size 1, to 7-1/2 HP, 3 phase, 200-230V or 10 HP 460-575V	186	475	661
0680	Size 2, to 10 HP, 3 phase, 200V, 15 HP-230V or 25 HP 460-575V	365	690	1,055
0760	Size 3, to 25 HP, 3 phase, 200V, 30 HP-230V or 50 HP 460-575V	610	840	1,450
0840	Size 4, to 40 HP, 3 phase, 200V, 50 HP-230V or 100 HP 460-575V	1,375	1,250	2,625
0920	Size 5, to 75 HP, 3 phase, 200V, 100 HP-230V or 200 HP 460-575V	3,250	1,675	4,925
1000	Size 6, to 150 HP, 3 phase, 200V, 200 HP-230V or 400 HP 460-575V	9,200	1,900	11,100

D50 Electrical

D5020 Lighting and Branch Wiring

Safety switches are type NEMA 1 for general purpose indoor application, and include time delay fuses, insulation and wire terminations.

System Components	QUANTITY	UNIT	COST EACH		
			MAT.	INST.	TOTAL
SYSTEM D5020 165 0200					
SAFETY SWITCH, 30A FUSED, 1 PHASE, 115V OR 230V					
Safety switch fused, hvy duty, 240V 2p 30 amp	1.000	Ea.	88.50	216	304.50
Fuse, dual element time delay 250V, 30 amp	2.000	Ea.	21.90	30.30	52.20
TOTAL			110.40	246.30	356.70

D5020 165	Safety Switches	COST EACH		
		MAT.	INST.	TOTAL
0200	Safety switch, 30 A fused, 1 phase, 2 HP 115 V or 3 HP, 230 V	110	246	356
0280	3 phase, 5 HP, 200 V or 7 1/2 HP, 230 V	155	282	437
0360	15 HP, 460 V or 20 HP, 575 V	264	294	558
0440	60 A fused, 3 phase, 15 HP 200 V or 15 HP 230 V	235	375	610
0520	30 HP 460 V or 40 HP 575 V	340	385	725
0600	100 A fused, 3 phase, 20 HP 200 V or 25 HP 230 V	405	455	860
0680	50 HP 460 V or 60 HP 575 V	650	465	1,115
0760	200 A fused, 3 phase, 50 HP 200 V or 60 HP 230 V	750	645	1,395
0840	125 HP 460 V or 150 HP 575 V	1,050	655	1,705
0920	400 A fused, 3 phase, 100 HP 200 V or 125 HP 230 V	1,700	915	2,615
1000	250 HP 460 V or 350 HP 575 V	2,550	935	3,485
1020	600 A fused, 3 phase, 150 HP 200 V or 200 HP 230 V	3,200	1,350	4,550
1040	400 HP 460 V	4,025	1,375	5,400

D5020   Lighting and Branch Wiring

Straight Connector

Angle Connector

Flexible Conduit

Table below includes costs for the flexible conduit. Not included are wire terminations and testing motor for correct rotation.

System Components	QUANTITY	UNIT	COST EACH		
			MAT.	INST.	TOTAL
SYSTEM D5020 170 0200					
MOTOR CONNECTIONS, SINGLE PHASE, 115V/230V UP TO 1 HP					
Motor connection, flexible conduit & fittings, 1 HP motor 115V	1.000	Ea.	6.80	94.50	101.30
TOTAL			6.80	94.50	101.30

D5020 170	Motor Connections	COST EACH		
		MAT.	INST.	TOTAL
0200	Motor connections, single phase, 115/230V, up to 1 HP	6.80	94.50	101.30
0240	Up to 3 HP	11.30	116	127.30
0280	Three phase, 200/230/460/575V, up to 3 HP	7.85	112	119.85
0320	Up to 5 HP	7	138	145
0360	Up to 7-1/2 HP	11.40	164	175.40
0400	Up to 10 HP	18.45	180	198.45
0440	Up to 15 HP	18.45	229	247.45
0480	Up to 25 HP	27.50	280	307.50
0520	Up to 50 HP	56.50	345	401.50
0560	Up to 100 HP	130	505	635

D5020 Lighting and Branch Wiring

For 230/460 Volt A.C., 3 phase, 60 cycle ball bearing squirrel cage induction motors, NEMA Class B standard line. Installation included.

No conduit, wire, or terminations included.

Manual Starter

Magnetic Starter

Induction Motor

System Components	QUANTITY	UNIT	COST EACH MAT.	COST EACH INST.	COST EACH TOTAL
SYSTEM D5020 175 0220					
MOTOR, DRIPPROOF CLASS B INSULATION, 1.15 SERVICE FACTOR, WITH STARTER					
1 H.P., 1200 RPM WITH MANUAL STARTER					
Motor, dripproof, class B insul, 1.15 serv fact, 1,200 RPM, 1 HP	1.000	Ea.	520	168	688
Motor starter, manual, 3 phase, 1 HP motor	1.000	Ea.	222	216	438
TOTAL			742	384	1,126

D5020 175	Motor & Starter	MAT.	INST.	TOTAL
0190	Motor, dripproof, premium efficient, 1.15 service factor			
0200	1 HP, 1200 RPM, motor only	520	168	688
0220	With manual starter	740	385	1,125
0240	With magnetic starter	650	385	1,035
0260	1800 RPM, motor only	279	168	447
0280	With manual starter	500	385	885
0300	With magnetic starter	410	385	795
0320	2 HP, 1200 RPM, motor only	620	168	788
0340	With manual starter	840	385	1,225
0360	With magnetic starter	1,050	500	1,550
0380	1800 RPM, motor only	250	168	418
0400	With manual starter	470	385	855
0420	With magnetic starter	670	500	1,170
0440	3600 RPM, motor only	515	168	683
0460	With manual starter	735	385	1,120
0480	With magnetic starter	935	500	1,435
0500	3 HP, 1200 RPM, motor only	790	168	958
0520	With manual starter	1,000	385	1,385
0540	With magnetic starter	1,200	500	1,700
0560	1800 RPM, motor only	550	168	718
0580	With manual starter	770	385	1,155
0600	With magnetic starter	970	500	1,470
0620	3600 RPM, motor only	605	168	773
0640	With manual starter	825	385	1,210
0660	With magnetic starter	1,025	500	1,525
0680	5 HP, 1200 RPM, motor only	1,025	168	1,193
0700	With manual starter	1,300	550	1,850
0720	With magnetic starter	1,200	645	1,845
0740	1800 RPM, motor only	655	168	823
0760	With manual starter	920	550	1,470
0780	With magnetic starter	840	645	1,485
0800	3600 RPM, motor only	645	168	813

D5020 Lighting and Branch Wiring

D5020 175	Motor & Starter	COST EACH		
		MAT.	**INST.**	**TOTAL**
0820	With manual starter	910	550	1,460
0840	With magnetic starter	830	645	1,475
0860	7.5 HP, 1800 RPM, motor only	975	180	1,155
0880	With manual starter	1,250	560	1,810
0900	With magnetic starter	1,350	870	2,220
0920	10 HP, 1800 RPM, motor only	1,175	189	1,364
0940	With manual starter	1,450	570	2,020
0960	With magnetic starter	1,550	880	2,430
0980	15 HP, 1800 RPM, motor only	1,625	237	1,862
1000	With magnetic starter	2,000	925	2,925
1040	20 HP, 1800 RPM, motor only	2,050	291	2,341
1060	With magnetic starter	2,650	1,125	3,775
1100	25 HP, 1800 RPM, motor only	2,675	305	2,980
1120	With magnetic starter	3,275	1,150	4,425
1160	30 HP, 1800 RPM, motor only	2,925	315	3,240
1180	With magnetic starter	3,525	1,150	4,675
1220	40 HP, 1800 RPM, motor only	3,575	380	3,955
1240	With magnetic starter	4,950	1,625	6,575
1280	50 HP, 1800 RPM, motor only	4,200	475	4,675
1300	With magnetic starter	5,575	1,725	7,300
1340	60 HP, 1800 RPM, motor only	5,225	540	5,765
1360	With magnetic starter	8,475	2,225	10,700
1400	75 HP, 1800 RPM, motor only	6,400	630	7,030
1420	With magnetic starter	9,650	2,300	11,950
1460	100 HP, 1800 RPM, motor only	6,050	840	6,890
1480	With magnetic starter	9,300	2,525	11,825
1520	125 HP, 1800 RPM, motor only	8,350	1,075	9,425
1540	With magnetic starter	17,600	2,975	20,575
1580	150 HP, 1800 RPM, motor only	10,200	1,250	11,450
1600	With magnetic starter	19,400	3,150	22,550
1640	200 HP, 1800 RPM, motor only	10,600	1,525	12,125
1660	With magnetic starter	19,800	3,425	23,225
1680	Totally encl, premium efficient, 1.0 ser. fac., 1HP, 1200 RPM, motor only	555	168	723
1700	With manual starter	775	385	1,160
1720	With magnetic starter	685	385	1,070
1740	1800 RPM, motor only	455	168	623
1760	With manual starter	675	385	1,060
1780	With magnetic starter	585	385	970
1800	2 HP, 1200 RPM, motor only	540	168	708
1820	With manual starter	760	385	1,145
1840	With magnetic starter	960	500	1,460
1860	1800 RPM, motor only	675	168	843
1880	With manual starter	895	385	1,280
1900	With magnetic starter	1,100	500	1,600
1920	3600 RPM, motor only	545	168	713
1940	With manual starter	765	385	1,150
1960	With magnetic starter	965	500	1,465
1980	3 HP, 1200 RPM, motor only	700	168	868
2000	With manual starter	920	385	1,305
2020	With magnetic starter	1,125	500	1,625
2040	1800 RPM, motor only	850	168	1,018
2060	With manual starter	1,075	385	1,460
2080	With magnetic starter	1,275	500	1,775
2100	3600 RPM, motor only	550	168	718
2120	With manual starter	770	385	1,155
2140	With magnetic starter	970	500	1,470
2160	5 HP, 1200 RPM, motor only	1,025	168	1,193
2180	With manual starter	1,300	550	1,850

D5020 Lighting and Branch Wiring

D5020 175	Motor & Starter	COST EACH		
		MAT.	INST.	TOTAL
2200	With magnetic starter	1,200	645	1,845
2220	1800 RPM, motor only	880	168	1,048
2240	With manual starter	1,150	550	1,700
2260	With magnetic starter	1,075	645	1,720
2280	3600 RPM, motor only	715	168	883
2300	With manual starter	980	550	1,530
2320	With magnetic starter	900	645	1,545
2340	7.5 HP, 1800 RPM, motor only	1,050	180	1,230
2360	With manual starter	1,325	560	1,885
2380	With magnetic starter	1,425	870	2,295
2400	10 HP, 1800 RPM, motor only	1,300	189	1,489
2420	With manual starter	1,575	570	2,145
2440	With magnetic starter	1,675	880	2,555
2460	15 HP, 1800 RPM, motor only	1,800	237	2,037
2480	With magnetic starter	2,175	925	3,100
2500	20 HP, 1800 RPM, motor only	2,825	291	3,116
2520	With magnetic starter	3,425	1,125	4,550
2540	25 HP, 1800 RPM, motor only	2,825	305	3,130
2560	With magnetic starter	3,425	1,150	4,575
2580	30 HP, 1800 RPM, motor only	3,725	315	4,040
2600	With magnetic starter	4,325	1,150	5,475
2620	40 HP, 1800 RPM, motor only	4,400	380	4,780
2640	With magnetic starter	5,775	1,625	7,400
2660	50 HP, 1800 RPM, motor only	5,350	475	5,825
2680	With magnetic starter	6,725	1,725	8,450
2700	60 HP, 1800 RPM, motor only	5,600	540	6,140
2720	With magnetic starter	8,850	2,225	11,075
2740	75 HP, 1800 RPM, motor only	7,350	630	7,980
2760	With magnetic starter	10,600	2,300	12,900
2780	100 HP, 1800 RPM, motor only	8,275	840	9,115
2800	With magnetic starter	11,500	2,525	14,025
2820	125 HP, 1800 RPM, motor only	10,200	1,075	11,275
2840	With magnetic starter	19,400	2,975	22,375
2860	150 HP, 1800 RPM, motor only	11,500	1,250	12,750
2880	With magnetic starter	20,700	3,150	23,850
2900	200 HP, 1800 RPM, motor only	15,800	1,525	17,325
2920	With magnetic starter	25,000	3,425	28,425

General: The cost of the lighting portion of the electrical costs is dependent upon:
1. The footcandle requirement of the proposed building.
2. The type of fixtures required.
3. The ceiling heights of the building.
4. Reflectance value of ceilings, walls and floors.
5. Fixture efficiencies and spacing vs. mounting height ratios.

Footcandle Requirements: See Table D5020-204 for Footcandle and Watts per S.F. determination.

Table D5020-201   IESNA* Recommended Illumination Levels in Footcandles

Commercial Buildings			Industrial Buildings		
Type	Description	Footcandles	Type	Description	Footcandles
Bank	Lobby	50	Assembly Areas	Rough bench & machine work	50
	Customer Areas	70		Medium bench & machine work	100
	Teller Stations	150		Fine bench & machine work	500
	Accounting Areas	150	Inspection Areas	Ordinary	50
Offices	Routine Work	100		Difficult	100
	Accounting	150		Highly Difficult	200
	Drafting	200	Material Handling	Loading	20
	Corridors, Halls, Washrooms	30		Stock Picking	30
Schools	Reading or Writing	70		Packing, Wrapping	50
	Drafting, Labs, Shops	100	Stairways	Service Areas	20
	Libraries	70	Washrooms	Service Areas	20
	Auditoriums, Assembly	15	Storage Areas	Inactive	5
	Auditoriums, Exhibition	30		Active, Rough, Bulky	10
Stores	Circulation Areas	30		Active, Medium	20
	Stock Rooms	30		Active, Fine	50
	Merchandise Areas, Service	100	Garages	Active Traffic Areas	20
	Self-Service Areas	200		Service & Repair	100

*IESNA - Illuminating Engineering Society of North America

Table D5020-202 General Lighting Loads by Occupancies

Type of Occupancy	Unit Load per S.F. (Watts)
Armories and Auditoriums	1
Banks	5
Barber Shops and Beauty Parlors	3
Churches	1
Clubs	2
Court Rooms	2
*Dwelling Units	3
Garages — Commercial (storage)	½
Hospitals	2
*Hotels and Motels, including apartment houses without provisions for cooking by tenants	2
Industrial Commercial (Loft) Buildings	2
Lodge Rooms	1½
Office Buildings	5
Restaurants	2
Schools	3
Stores	3
Warehouses (storage)	¼
*In any of the above occupancies except one-family dwellings and individual dwelling units of multi-family dwellings:	
Assembly Halls and Auditoriums	1
Halls, Corridors, Closets	½
Storage Spaces	¼

Table D5020-203 Lighting Limit (Connected Load) for Listed Occupancies: New Building Proposed Energy Conservation Guideline

Type of Use	Maximum Watts per S.F.
Interior	
Category A: Classrooms, office areas, automotive mechanical areas, museums, conference rooms, drafting rooms, clerical areas, laboratories, merchandising areas, kitchens, examining rooms, book stacks, athletic facilities.	3.00
Category B: Auditoriums, waiting areas, spectator areas, restrooms, dining areas, transportation terminals, working corridors in prisons and hospitals, book storage areas, active inventory storage, hospital bedrooms, hotel and motel bedrooms, enclosed shopping mall concourse areas, stairways.	1.00
Category C: Corridors, lobbies, elevators, inactive storage areas.	0.50
Category D: Indoor parking.	0.25
Exterior	
Category E: Building perimeter: wall-wash, facade, canopy.	5.00 (per linear foot)
Category F: Outdoor parking.	0.10

423

Table D5020-204 Procedure for Calculating Footcandles and Watts Per Square Foot

1. Initial footcandles = No. of fixtures × lamps per fixture × lumens per lamp × coefficient of utilization ÷ square feet
2. Maintained footcandles = initial footcandles × maintenance factor
3. Watts per square foot = No. of fixtures × lamps × (lamp watts + ballast watts) ÷ square feet

Example: To find footcandles and watts per S.F. for an office 20′ x 20′ with 11 fluorescent fixtures each having 4–40 watt C.W. lamps.

Based on good reflectance and clean conditions:

Lumens per lamp = 40 watt cool white at 3150 lumens per lamp RD5020-250, Table RD5020-251

Coefficient of utilization = .42 (varies from .62 for light colored areas to .27 for dark)

Maintenance factor = .75 (varies from .80 for clean areas with good maintenance to .50 for poor)

Ballast loss = 8 watts per lamp. (Varies with manufacturer. See manufacturers' catalog.)

1. Initial footcandles:

$$\frac{11 \times 4 \times 3150 \times .42}{400} = \frac{58,212}{400} = 145 \text{ footcandles}$$

2. Maintained footcandles:

$$145 \times .75 = 109 \text{ footcandles}$$

3. Watts per S.F.

$$\frac{11 \times 4 (40 + 8)}{400} = \frac{2,112}{400} = 5.3 \text{ watts per S.F.}$$

Table D5020-205 Approximate Watts Per Square Foot for Popular Fixture Types

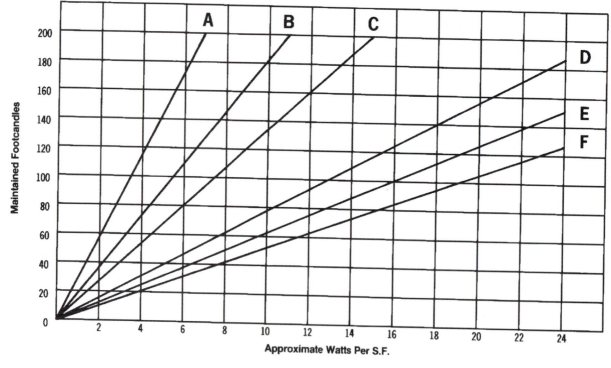

Due to the many variables involved, use for preliminary estimating only:

a. Fluorescent – industrial System D5020 208
b. Fluorescent – lens unit System D5020 208 Fixture types B & C
c. Fluorescent – louvered unit
d. Incandescent – open reflector System D5020 214, Type D
e. Incandescent – lens unit System D5020 214, Type A
f. Incandescent – down light System D5020 214, Type B

D50 Electrical

D5020 Lighting and Branch Wiring

A. Strip Fixture

B. Surface Mounted

C. Recessed

D. Pendent Mounted

Design Assumptions:

1. A 100 footcandle average maintained level of illumination.
2. Ceiling heights range from 9' to 11'.
3. Average reflectance values are assumed for ceilings, walls and floors.
4. Cool white (CW) fluorescent lamps with 3150 lumens for 40 watt lamps and 6300 lumens for 8' slimline lamps.
5. Four 40 watt lamps per 4' fixture and two 8' lamps per 8' fixture.
6. Average fixture efficiency values and spacing to mounting height ratios.
7. Installation labor is average U.S. rate as of January 1.

System Components	QUANTITY	UNIT	COST PER S.F.		
			MAT.	INST.	TOTAL
SYSTEM D5020 208 0520					
FLUORESCENT FIXTURES MOUNTED 9'-11" ABOVE FLOOR, 100 FC					
TYPE A, 8 FIXTURES PER 400 S.F.					
Conduit, steel intermediate, 1/2" diam.	.185	L.F.	.40	1.40	1.80
Wire, 600V, type THWN-THHN, copper, solid, #12	.004	C.L.F.	.04	.26	.30
Fluorescent strip fixture 8' long, surface mounted, two 75W SL	.020	Ea.	1.59	2.44	4.03
Steel outlet box 4" concrete	.020	Ea.	.35	.76	1.11
Steel outlet box plate with stud, 4" concrete	.020	Ea.	.14	.19	.33
Fixture hangers, flexible, 1/2" diameter, 4" long	.040	Ea.	.74	2.52	3.26
Fixture whip, THHN wire, three #12, 3/8" Greenfield	.040	Ea.	.41	.94	1.35
TOTAL			3.67	8.51	12.18

D5020 208	Fluorescent Fixtures (by Type)	COST PER S.F.		
		MAT.	INST.	TOTAL
0520	Fluorescent fixtures, type A, 8 fixtures per 400 S.F.	3.67	8.50	12.17
0560	11 fixtures per 600 S.F.	3.34	7.75	11.09
0600	17 fixtures per 1000 S.F.	3.12	7.20	10.32
0640	23 fixtures per 1600 S.F.	2.61	6.05	8.66
0680	28 fixtures per 2000 S.F.	2.57	5.95	8.52
0720	41 fixtures per 3000 S.F.	2.53	5.85	8.38
0800	53 fixtures per 4000 S.F.	2.42	5.65	8.07
0840	64 fixtures per 5000 S.F.	2.38	5.50	7.88
0880	Type B, 11 fixtures per 400 S.F.	5.50	12.10	17.60
0920	15 fixtures per 600 S.F.	4.95	10.90	15.85
0960	24 fixtures per 1000 S.F.	4.77	10.55	15.32
1000	35 fixtures per 1600 S.F.	4.35	9.60	13.95
1040	42 fixtures per 2000 S.F.	4.16	9.15	13.31
1080	61 fixtures per 3000 S.F.	3.99	8.85	12.84
1160	80 fixtures per 4000 S.F.	2.93	5.60	8.53
1200	98 fixtures per 5000 S.F.	3.92	8.60	12.52
1240	Type C, 11 fixtures per 400 S.F.	4.83	12.60	17.43
1280	14 fixtures per 600 S.F.	4.04	10.55	14.59
1320	23 fixtures per 1000 S.F.	4	10.45	14.45
1360	34 fixtures per 1600 S.F.	3.69	9.65	13.34
1400	43 fixtures per 2000 S.F.	3.79	9.90	13.69
1440	63 fixtures per 3000 S.F.	3.65	9.55	13.20
1520	81 fixtures per 4000 S.F.	3.51	9.20	12.71
1560	101 fixtures per 5000 S.F.	3.48	9.10	12.58
1600	Type D, 8 fixtures per 400 S.F.	4.30	9.50	13.80
1640	12 fixtures per 600 S.F.	4.30	9.50	13.80
1680	19 fixtures per 1000 S.F.	4.09	9.05	13.14
1720	27 fixtures per 1600 S.F.	3.65	8.05	11.70
1760	34 fixtures per 2000 S.F.	3.65	8.05	11.70
1800	48 fixtures per 3000 S.F.	3.44	7.60	11.04
1880	64 fixtures per 4000 S.F.	3.44	7.60	11.04
1920	79 fixtures per 5000 S.F.	3.43	7.60	11.03

D5020   Lighting and Branch Wiring

Type C. Recessed, mounted on grid ceiling suspension system, 2' x 4', four 40 watt lamps, acrylic prismatic diffusers.

5.3 watts per S.F. for 100 footcandles.

3 watts per S.F. for 57 footcandles.

System Components	QUANTITY	UNIT	COST PER S.F.		
			MAT.	INST.	TOTAL
SYSTEM D5020 210 0200					
FLUORESCENT FIXTURES RECESS MOUNTED IN CEILING					
1 WATT PER S.F., 20 FC, 5 FIXTURES PER 1000 S.F.					
Steel intermediate conduit, (IMC) 1/2" diam.	.128	L.F.	.28	.97	1.25
Wire, 600 volt, type THW, copper, solid, #12	.003	C.L.F.	.03	.21	.24
Fluorescent fixture, recessed, 2'x 4', four 40W, w/lens, for grid ceiling	.005	Ea.	.36	.81	1.17
Steel outlet box 4" square	.005	Ea.	.09	.19	.28
Fixture whip, Greenfield w/#12 THHN wire	.005	Ea.	.04	.05	.09
TOTAL			.80	2.23	3.03

D5020 210	Fluorescent Fixtures (by Wattage)	COST PER S.F.		
		MAT.	INST.	TOTAL
0190	Fluorescent fixtures recess mounted in ceiling			
0195	T12, standard 40 watt lamps			
0200	1 watt per S.F., 20 FC, 5 fixtures @40 watts per 1000 S.F.	.80	2.22	3.02
0240	2 watt per S.F., 40 FC, 10 fixtures @40 watt per 1000 S.F.	1.58	4.36	5.94
0280	3 watt per S.F., 60 FC, 15 fixtures @40 watt per 1000 S.F	2.38	6.60	8.98
0320	4 watt per S.F., 80 FC, 20 fixtures @40 watt per 1000 S.F.	3.17	8.70	11.87
0400	5 watt per S.F., 100 FC, 25 fixtures @40 watt per 1000 S.F.	3.96	10.95	14.91
0450	T8, energy saver 32 watt lamps			
0500	0.8 watt per S.F., 20 FC, 5 fixtures @32 watt per 1000 S.F.	.85	2.22	3.07
0520	1.6 watt per S.F., 40 FC, 10 fixtures @32 watt per 1000 S.F.	1.69	4.36	6.05
0540	2.4 watt per S.F., 60 FC, 15 fixtures @ 32 watt per 1000 S.F	2.53	6.60	9.13
0560	3.2 watt per S.F., 80 FC, 20 fixtures @32 watt per 1000 S.F.	3.37	8.70	12.07
0580	4 watt per S.F., 100 FC, 25 fixtures @32 watt per 1000 S.F.	4.22	10.95	15.17

D5020 Lighting and Branch Wiring

Type A. Recessed wide distribution reflector with flat glass lens 150 W.

Maximum spacing = 1.2 x mounting height.

13 watts per S.F. for 100 footcandles.

Type A

Type B

Type C. Recessed PAR–38 flood lamp with concentric louver 150 W.

Maximum spacing = 0.5 x mounting height.

19 watts per S.F. for 100 footcandles.

Type C

Type D

Type B. Recessed reflector down light with baffles 150 W.

Maximum spacing = 0.8 x mounting height.

18 watts per S.F. for 100 footcandles.

Type D. Recessed R–40 flood lamp with reflector skirt.

Maximum spacing = 0.7 x mounting height.

15 watts per S.F. for 100 footcandles.

System Components	QUANTITY	UNIT	COST PER S.F.		
			MAT.	INST.	TOTAL
SYSTEM D5020 214 0400					
INCANDESCENT FIXTURE RECESS MOUNTED, 100 FC					
TYPE A, 34 FIXTURES PER 400 S.F.					
Steel intermediate conduit, (IMC) 1/2" diam	1.060	L.F.	2.30	8	10.30
Wire, 600V, type THWN-THHN, copper, solid, #12	.033	C.L.F.	.35	2.28	2.63
Steel outlet box 4" square	.085	Ea.	1.48	3.23	4.71
Fixture whip, Greenfield w/#12 THHN wire	.085	Ea.	.61	.80	1.41
Incandescent fixture, recessed, w/lens, prewired, square trim, 200W	.085	Ea.	9.18	9.61	18.79
TOTAL			13.92	23.92	37.84

D5020 214	Incandescent Fixture (by Type)	COST PER S.F.		
		MAT.	INST.	TOTAL
0380	Incandescent fixture recess mounted, 100 FC			
0400	Type A, 34 fixtures per 400 S.F.	13.95	24	37.95
0440	49 fixtures per 600 S.F.	13.55	23.50	37.05
0480	63 fixtures per 800 S.F.	13.20	23.50	36.70
0520	90 fixtures per 1200 S.F.	12.75	23	35.75
0560	116 fixtures per 1600 S.F.	12.50	22.50	35
0600	143 fixtures per 2000 S.F.	12.35	22.50	34.85
0640	Type B, 47 fixtures per 400 S.F.	13.35	30	43.35
0680	66 fixtures per 600 S.F.	12.80	29.50	42.30
0720	88 fixtures per 800 S.F.	12.80	29.50	42.30
0760	127 fixtures per 1200 S.F.	12.50	29	41.50
0800	160 fixtures per 1600 S.F.	12.35	29	41.35
0840	206 fixtures per 2000 S.F.	12.30	29	41.30
0880	Type C, 51 fixtures per 400 S.F.	17.65	31.50	49.15
0920	74 fixtures per 600 S.F.	17.20	31	48.20
0960	97 fixtures per 800 S.F.	16.95	31	47.95
1000	142 fixtures per 1200 S.F.	16.65	30.50	47.15
1040	186 fixtures per 1600 S.F.	16.45	30.50	46.95
1080	230 fixtures per 2000 S.F.	16.40	30.50	46.90
1120	Type D, 39 fixtures per 400 S.F.	16.50	25	41.50
1160	57 fixtures per 600 S.F.	16.10	24.50	40.60
1200	75 fixtures per 800 S.F.	16	24.50	40.50
1240	109 fixtures per 1200 S.F.	15.65	24	39.65
1280	143 fixtures per 1600 S.F.	15.35	24	39.35
1320	176 fixtures per 2000 S.F.	15.25	24	39.25

D5020 Lighting and Branch Wiring

Type A. Recessed, wide distribution reflector with flat glass lens.

150 watt inside frost—2500 lumens per lamp.

PS–25 extended service lamp.

Maximum spacing = 1.2 x mounting height.

13 watts per S.F. for 100 footcandles.

System Components	QUANTITY	UNIT	COST PER S.F.		
			MAT.	INST.	TOTAL
SYSTEM D5020 216 0200					
INCANDESCENT FIXTURE RECESS MOUNTED, TYPE A					
1 WATT PER S.F., 8 FC, 6 FIXT PER 1000 S.F.					
Steel intermediate conduit, (IMC) 1/2" diam	.091	L.F.	.20	.69	.89
Wire, 600V, type THWN-THHN, copper, solid, #12	.002	C.L.F.	.02	.14	.16
Incandescent fixture, recessed, w/lens, prewired, square trim, 200W	.006	Ea.	.65	.68	1.33
Steel outlet box 4" square	.006	Ea.	.10	.23	.33
Fixture whip, Greenfield w/#12 THHN wire	.006	Ea.	.04	.06	.10
TOTAL			1.01	1.80	2.81

D5020 216	Incandescent Fixture (by Wattage)	COST PER S.F.		
		MAT.	INST.	TOTAL
0190	Incandescent fixture recess mounted, type A			
0200	1 watt per S.F., 8 FC, 6 fixtures per 1000 S.F.	1.01	1.79	2.80
0240	2 watt per S.F., 16 FC, 12 fixtures per 1000 S.F.	2.03	3.58	5.61
0280	3 watt per S.F., 24 FC, 18 fixtures, per 1000 S.F.	3.03	5.30	8.33
0320	4 watt per S.F., 32 FC, 24 fixtures per 1000 S.F.	4.05	7.10	11.15
0400	5 watt per S.F., 40 FC, 30 fixtures per 1000 S.F.	5.05	8.85	13.90

D5020   Lighting and Branch Wiring

High Bay Fluorescent Fixtures
Four T5 HO/54 watt lamps

System Components	QUANTITY	UNIT	COST PER S.F.		
			MAT.	INST.	TOTAL
SYSTEM D5020 218 0200					
FLUORESCENT HIGH BAY FIXTURE, 4 LAMP, 8'-10' ABOVE WORK PLANE					
0.5 WATT/S.F., 29 FC, 2 FIXTURES/1000 S.F.					
Steel intermediate conduit, (IMC) 1/2" diam	.100	L.F.	.22	.76	.98
Wire, 600V, type THWN-THHN, copper, solid, #10	.002	C.L.F.	.03	.15	.18
Steel outlet box 4" concrete	.002	Ea.	.03	.08	.11
Steel outlet box plate with stud, 4" concrete	.002	Ea.	.01	.02	.03
Flourescent , hi bay, 4 lamp, T-5 fixture	.002	Ea.	.62	.34	.96
TOTAL			.91	1.35	2.26

D5020 218	Fluorescent Fixture, High Bay, 8'-10' (by Wattage)	COST PER S.F.		
		MAT.	INST.	TOTAL
0190	Fluorescent high bay-4 lamp fixture, 8'-10' above work plane			
0200	.5 watt/SF, 29 FC, 2 fixtures per 1000 S.F.	.92	1.34	2.26
0400	1 watt/SF, 59 FC, 4 fixtures per 1000 S.F.	1.81	2.58	4.39
0600	1.5 watt/SF, 103 FC, 7 fixtures per 1000 S.F.	3.03	4.01	7.04
0800	2 watt/SF, 133 FC, 9 fixtures per 1000 S.F.	3.94	5.30	9.24
1000	2.5 watt/SF, 162 FC, 11 fixtures per 1000 S.F.	4.84	6.60	11.44

429

D5020 Lighting and Branch Wiring

HIGH BAY FIXTURES

B. Metal halide 400 watt

C. High pressure sodium 400 watt

E. Metal halide 1000 watt

F. High pressure sodium 1000 watt

G. Metal halide 1000 watt
125,000 lumen lamp

System Components	QUANTITY	UNIT	COST PER S.F.		
			MAT.	INST.	TOTAL
SYSTEM D5020 220 0880					
HIGH INTENSITY DISCHARGE FIXTURE, 8'-10' ABOVE WORK PLANE, 100 FC					
TYPE B, 8 FIXTURES PER 900 S.F.					
Steel intermediate conduit, (IMC) 1/2" diam	.460	L.F.	1	3.47	4.47
Wire, 600V, type THWN-THHN, copper, solid, #10	.009	C.L.F.	.15	.68	.83
Steel outlet box 4" concrete	.009	Ea.	.16	.34	.50
Steel outlet box plate with stud 4" concrete	.009	Ea.	.06	.09	.15
Metal halide, hi bay, aluminum reflector, 400 W lamp	.009	Ea.	4.19	2.97	7.16
TOTAL			5.56	7.55	13.11

D5020 220	H.I.D. Fixture, High Bay, 8'-10' (by Type)	COST PER S.F.		
		MAT.	INST.	TOTAL
0500	High intensity discharge fixture, 8'-10' above work plane, 100 FC			
0880	Type B, 8 fixtures per 900 S.F.	5.55	7.55	13.10
0920	15 fixtures per 1800 S.F.	5.10	7.25	12.35
0960	24 fixtures per 3000 S.F.	5.10	7.25	12.35
1000	31 fixtures per 4000 S.F.	5.10	7.25	12.35
1040	38 fixtures per 5000 S.F.	5.10	7.25	12.35
1080	60 fixtures per 8000 S.F.	5.10	7.25	12.35
1120	72 fixtures per 10000 S.F.	4.56	6.70	11.26
1160	115 fixtures per 16000 S.F.	4.56	6.70	11.26
1200	230 fixtures per 32000 S.F.	4.56	6.70	11.26
1240	Type C, 4 fixtures per 900 S.F.	2.67	4.61	7.28
1280	8 fixtures per 1800 S.F.	2.68	4.63	7.31
1320	13 fixtures per 3000 S.F.	2.67	4.61	7.28
1360	17 fixtures per 4000 S.F.	2.67	4.61	7.28
1400	21 fixtures per 5000 S.F.	2.57	4.23	6.80
1440	33 fixtures per 8000 S.F.	2.56	4.15	6.71
1480	40 fixtures per 10000 S.F.	2.49	3.93	6.42
1520	63 fixtures per 16000 S.F.	2.49	3.93	6.42
1560	126 fixtures per 32000 S.F.	2.49	3.93	6.42

D5020 Lighting and Branch Wiring

HIGH BAY FIXTURES

B. Metal halide 400 watt

C. High pressure sodium 400 watt

E. Metal halide 1000 watt

F. High pressure sodium 1000 watt

G. Metal halide 1000 watt
125,000 lumen lamp

System Components	QUANTITY	UNIT	COST PER S.F.		
			MAT.	INST.	TOTAL
SYSTEM D5020 222 0240					
HIGH INTENSITY DISCHARGE FIXTURE, 8'-10' ABOVE WORK PLANE					
1 WATT/S.F., TYPE B, 29 FC, 2 FIXTURES/1000 S.F.					
Steel intermediate conduit, (IMC) 1/2" diam	.100	L.F.	.22	.76	.98
Wire, 600V, type THWN-THHN, copper, solid, #10	.002	C.L.F.	.03	.15	.18
Steel outlet box 4" concrete	.002	Ea.	.03	.08	.11
Steel outlet box plate with stud, 4" concrete	.002	Ea.	.01	.02	.03
Metal halide, hi bay, aluminum reflector, 400 W lamp	.002	Ea.	.93	.66	1.59
TOTAL			1.22	1.67	2.89

D5020 222	H.I.D. Fixture, High Bay, 8'-10' (by Wattage)	COST PER S.F.		
		MAT.	INST.	TOTAL
0190	High intensity discharge fixture, 8'-10' above work plane			
0240	1 watt/S.F., type B, 29 FC, 2 fixtures/1000 S.F.	1.23	1.66	2.89
0280	Type C, 54 FC, 2 fixtures/1000 S.F.	1.05	1.26	2.31
0400	2 watt/S.F., type B, 59 FC, 4 fixtures/1000 S.F.	2.43	3.22	5.65
0440	Type C, 108 FC, 4 fixtures/1000 S.F.	2.08	2.45	4.53
0560	3 watt/S.F., type B, 103 FC, 7 fixtures/1000 S.F.	4.12	5.15	9.27
0600	Type C, 189 FC, 6 fixtures/1000 S.F.	3.32	3.17	6.49
0720	4 watt/S.F., type B, 133 FC, 9 fixtures/1000 S.F.	5.35	6.75	12.10
0760	Type C, 243 FC, 9 fixtures/1000 S.F.	4.66	5.45	10.11
0880	5 watt/S.F., type B, 162 FC, 11 fixtures/1000 S.F.	6.55	8.35	14.90
0920	Type C, 297 FC, 11 fixtures/1000 S.F.	5.75	6.75	12.50

431

HIGH BAY FIXTURES

B. Metal halide 400 watt

C. High pressure sodium 400 watt

E. Metal halide 1000 watt

F. High pressure sodium 1000 watt

G. Metal halide 1000 watt
   125,000 lumen lamp

System Components	QUANTITY	UNIT	COST PER S.F.		
			MAT.	INST.	TOTAL
SYSTEM D5020 224 1240					
HIGH INTENSITY DISCHARGE FIXTURE, 16′ ABOVE WORK PLANE, 100 FC					
TYPE C, 5 FIXTURES PER 900 S.F.					
Steel intermediate conduit, (IMC) 1/2″ diam	.260	L.F.	.56	1.96	2.52
Wire, 600V, type THWN-THHN, copper, solid, #10	.007	C.L.F.	.12	.53	.65
Steel outlet box 4″ concrete	.006	Ea.	.10	.23	.33
Steel outlet box plate with stud, 4″ concrete	.006	Ea.	.04	.06	.10
High pressure sodium, hi bay, aluminum reflector, 400 W lamp	.006	Ea.	2.58	1.98	4.56
TOTAL			3.40	4.76	8.16

D5020 224	H.I.D. Fixture, High Bay, 16′ (by Type)	COST PER S.F.		
		MAT.	INST.	TOTAL
0510	High intensity discharge fixture, 16′ above work plane, 100 FC			
1240	Type C, 5 fixtures per 900 S.F.	3.41	4.76	8.17
1280	9 fixtures per 1800 S.F.	3.13	4.98	8.11
1320	15 fixtures per 3000 S.F.	3.13	4.98	8.11
1360	18 fixtures per 4000 S.F.	2.97	4.45	7.42
1400	22 fixtures per 5000 S.F.	2.97	4.45	7.42
1440	36 fixtures per 8000 S.F.	2.97	4.45	7.42
1480	42 fixtures per 10,000 S.F.	2.67	4.61	7.28
1520	65 fixtures per 16,000 S.F.	2.67	4.61	7.28
1600	Type G, 4 fixtures per 900 S.F.	4.38	7.50	11.88
1640	6 fixtures per 1800 S.F.	3.66	7	10.66
1720	9 fixtures per 4000 S.F.	3.03	6.80	9.83
1760	11 fixtures per 5000 S.F.	3.03	6.80	9.83
1840	21 fixtures per 10,000 S.F.	2.84	6.15	8.99
1880	33 fixtures per 16,000 S.F.	2.84	6.15	8.99

D5020 Lighting and Branch Wiring

HIGH BAY FIXTURES

B. Metal halide 400 watt

C. High pressure sodium 400 watt

E. Metal halide 1000 watt

F. High pressure sodium 1000 watt

G. Metal halide 1000 watt
125,000 lumen lamp

System Components	QUANTITY	UNIT	COST PER S.F.		
			MAT.	INST.	TOTAL
SYSTEM D5020 226 0240					
HIGH INTENSITY DISCHARGE FIXTURE, 16' ABOVE WORK PLANE					
1 WATT/S.F., TYPE E, 42 FC, 1 FIXTURE/1000 S.F.					
Steel intermediate conduit, (IMC) 1/2" diam	.160	L.F.	.35	1.21	1.56
Wire, 600V, type THWN-THHN, copper, solid, #10	.003	C.L.F.	.05	.23	.28
Steel outlet box 4" concrete	.001	Ea.	.02	.04	.06
Steel outlet box plate with stud, 4" concrete	.001	Ea.	.01	.01	.02
Metal halide, hi bay, aluminum reflector, 1000 W lamp	.001	Ea.	.67	.38	1.05
TOTAL			1.10	1.87	2.97

D5020 226	H.I.D. Fixture, High Bay, 16' (by Wattage)	COST PER S.F.		
		MAT.	INST.	TOTAL
0190	High intensity discharge fixture, 16' above work plane			
0240	1 watt/S.F., type E, 42 FC, 1 fixture/1000 S.F.	1.09	1.86	2.95
0280	Type G, 52 FC, 1 fixture/1000 S.F.	1.09	1.86	2.95
0320	Type C, 54 FC, 2 fixture/1000 S.F.	1.28	2.08	3.36
0440	2 watt/S.F., type E, 84 FC, 2 fixture/1000 S.F.	2.20	3.80	6
0480	Type G, 105 FC, 2 fixture/1000 S.F.	2.20	3.80	6
0520	Type C, 108 FC, 4 fixture/1000 S.F.	2.55	4.17	6.72
0640	3 watt/S.F., type E, 126 FC, 3 fixture/1000 S.F.	3.30	5.65	8.95
0680	Type G, 157 FC, 3 fixture/1000 S.F.	3.30	5.65	8.95
0720	Type C, 162 FC, 6 fixture/1000 S.F.	3.83	6.25	10.08
0840	4 watt/S.F., type E, 168 FC, 4 fixture/1000 S.F.	4.41	7.60	12.01
0880	Type G, 210 FC, 4 fixture/1000 S.F.	4.41	7.60	12.01
0920	Type C, 243 FC, 9 fixture/1000 S.F.	5.55	8.70	14.25
1040	5 watt/S.F., type E, 210 FC, 5 fixture/1000 S.F.	5.50	9.45	14.95
1080	Type G, 262 FC, 5 fixture/1000 S.F.	5.50	9.45	14.95
1120	Type C, 297 FC, 11 fixture/1000 S.F.	6.85	10.80	17.65

433

D5020 Lighting and Branch Wiring

HIGH BAY FIXTURES

B. Metal halide 400 watt

C. High pressure sodium 400 watt

E. Metal halide 1000 watt

F. High pressure sodium 1000 watt

G. Metal halide 1000 watt 125,000 lumen lamp

System Components	QUANTITY	UNIT	COST PER S.F.		
			MAT.	INST.	TOTAL
SYSTEM D5020 228 1240					
HIGH INTENSITY DISCHARGE FIXTURE, 20' ABOVE WORK PLANE, 100 FC					
TYPE C, 6 FIXTURES PER 900 S.F.					
Steel intermediate conduit, (IMC) 1/2" diam	.350	L.F.	.76	2.64	3.40
Wire, 600V, type THWN-THHN, copper, solid, #10.	.011	C.L.F.	.19	.83	1.02
Steel outlet box 4" concrete	.007	Ea.	.12	.27	.39
Steel outlet box plate with stud, 4" concrete	.007	Ea.	.05	.07	.12
High pressure sodium, hi bay, aluminum reflector, 400 W lamp	.007	Ea.	3.01	2.31	5.32
TOTAL			4.13	6.12	10.25

D5020 228	H.I.D. Fixture, High Bay, 20' (by Type)	COST PER S.F.		
		MAT.	INST.	TOTAL
0510	High intensity discharge fixture 20' above work plane, 100 FC			
1240	Type C, 6 fixtures per 900 S.F.	4.13	6.10	10.23
1280	10 fixtures per 1800 S.F.	3.68	5.75	9.43
1320	16 fixtures per 3000 S.F.	3.26	5.50	8.76
1360	20 fixtures per 4000 S.F.	3.26	5.50	8.76
1400	24 fixtures per 5000 S.F.	3.26	5.50	8.76
1440	38 fixtures per 8000 S.F.	3.19	5.30	8.49
1520	68 fixtures per 16000 S.F.	3	5.80	8.80
1560	132 fixtures per 32000 S.F.	3	5.80	8.80
1600	Type G, 4 fixtures per 900 S.F.	4.33	7.35	11.68
1640	6 fixtures per 1800 S.F.	3.65	7	10.65
1680	7 fixtures per 3000 S.F.	3.59	6.80	10.39
1720	10 fixtures per 4000 S.F.	3.57	6.70	10.27
1760	11 fixtures per 5000 S.F.	3.13	7.20	10.33
1800	18 fixtures per 8000 S.F.	3.13	7.20	10.33
1840	22 fixtures per 10000 S.F.	3.13	7.20	10.33
1880	34 fixtures per 16000 S.F.	3.06	6.95	10.01
1920	66 fixtures per 32000 S.F.	2.89	6.35	9.24

D5020 Lighting and Branch Wiring

HIGH BAY FIXTURES

B. Metal halide 400 watt

C. High pressure sodium 400 watt

E. Metal halide 1000 watt

F. High pressure sodium 1000 watt

G. Metal halide 1000 watt
   125,000 lumen lamp

System Components	QUANTITY	UNIT	COST PER S.F.		
			MAT.	INST.	TOTAL
SYSTEM D5020 230 0240					
HIGH INTENSITY DISCHARGE FIXTURE, 20' ABOVE WORK PLANE					
1 WATT/S.F., TYPE E, 40 FC, 1 FIXTURE 1000 S.F.	.160	L.F.	.35	1.21	1.56
Steel intermediate conduit, (IMC) 1/2" diam	.005	C.L.F.	.09	.38	.47
Wire, 600V, type THWN-THHN, copper, solid, #10	.001	Ea.	.02	.04	.06
Steel outlet box 4" concrete	.001	Ea.	.01	.01	.02
Steel outlet box plate with stud, 4" concrete	.001	Ea.	.67	.38	1.05
Metal halide, hi bay, aluminum reflector, 1000 W lamp					
TOTAL			1.14	2.02	3.16

D5020 230	H.I.D. Fixture, High Bay, 20' (by Wattage)	COST PER S.F.		
		MAT.	INST.	TOTAL
0190	High intensity discharge fixture, 20' above work plane			
0240	1 watt/S.F., type E, 40 FC, 1 fixture/1000 S.F.	1.13	2.01	3.14
0280	Type G, 50 FC, 1 fixture/1000 S.F.	1.13	2.01	3.14
0320	Type C, 52 FC, 2 fixtures/1000 S.F.	1.34	2.33	3.67
0440	2 watt/S.F., type E, 81 FC, 2 fixtures/1000 S.F.	2.25	4.03	6.28
0480	Type G, 101 FC, 2 fixtures/1000 S.F.	2.25	4.03	6.28
0520	Type C, 104 FC, 4 fixtures/1000 S.F.	2.66	4.58	7.24
0640	3 watt/S.F., type E, 121 FC, 3 fixtures/1000 S.F.	3.38	6.05	9.43
0680	Type G, 151 FC, 3 fixtures/1000 S.F.	3.38	6.05	9.43
0720	Type C, 155 FC, 6 fixtures/1000 S.F.	4	6.90	10.90
0840	4 watt/S.F., type E, 161 FC, 4 fixtures/1000 S.F.	4.51	8.05	12.56
0880	Type G, 202 FC, 4 fixtures/1000 S.F.	4.51	8.05	12.56
0920	Type C, 233 FC, 9 fixtures/1000 S.F.	5.75	9.55	15.30
1040	5 watt/S.F., type E, 202 FC, 5 fixtures/1000 S.F.	5.65	10.05	15.70
1080	Type G, 252 FC, 5 fixtures/1000 S.F.	5.65	10.05	15.70
1120	Type C, 285 FC, 11 fixtures/1000 S.F.	7.10	11.85	18.95

D5020 Lighting and Branch Wiring

HIGH BAY FIXTURES

B. Metal halide 400 watt

C. High pressure sodium 400 watt

E. Metal halide 1000 watt

F. High pressure sodium 1000 watt

G. Metal halide 1000 watt
125,000 lumen lamp

System Components	QUANTITY	UNIT	COST PER S.F.		
			MAT.	INST.	TOTAL
SYSTEM D5020 232 1240					
HIGH INTENSITY DISCHARGE FIXTURE, 30' ABOVE WORK PLANE, 100 FC					
TYPE F, 4 FIXTURES PER 900 S.F.					
Steel intermediate conduit, (IMC) 1/2" diam	.580	L.F.	1.26	4.38	5.64
Wire, 600V, type THWN-THHN, copper, solid, #10	.018	C.L.F.	.31	1.36	1.67
Steel outlet box 4" concrete	.004	Ea.	.07	.15	.22
Steel outlet box plate with stud, 4" concrete	.004	Ea.	.03	.04	.07
High pressure sodium, hi bay, aluminum, 1000 W lamp	.004	Ea.	2.40	1.52	3.92
TOTAL			4.07	7.45	11.52

D5020 232	H.I.D. Fixture, High Bay, 30' (by Type)	COST PER S.F.		
		MAT.	INST.	TOTAL
0510	High intensity discharge fixture, 30' above work plane, 100 FC			
1240	Type F, 4 fixtures per 900 S.F.	4.07	7.45	11.52
1280	6 fixtures per 1800 S.F.	3.46	7.10	10.56
1320	8 fixtures per 3000 S.F.	3.46	7.10	10.56
1360	9 fixtures per 4000 S.F.	2.86	6.75	9.61
1400	10 fixtures per 5000 S.F.	2.86	6.75	9.61
1440	17 fixtures per 8000 S.F.	2.86	6.75	9.61
1480	18 fixtures per 10,000 S.F.	2.71	6.20	8.91
1520	27 fixtures per 16,000 S.F.	2.71	6.20	8.91
1560	52 fixtures per 32000 S.F.	2.69	6.15	8.84
1600	Type G, 4 fixtures per 900 S.F.	4.48	7.90	12.38
1640	6 fixtures per 1800 S.F.	3.70	7.15	10.85
1680	9 fixtures per 3000 S.F.	3.63	6.95	10.58
1720	11 fixtures per 4000 S.F.	3.57	6.70	10.27
1760	13 fixtures per 5000 S.F.	3.57	6.70	10.27
1800	21 fixtures per 8000 S.F.	3.57	6.70	10.27
1840	23 fixtures per 10,000 S.F.	3.06	7.10	10.16
1880	36 fixtures per 16,000 S.F.	3.06	7.10	10.16
1920	70 fixtures per 32,000 S.F.	3.06	7.10	10.16

D5020 Lighting and Branch Wiring

HIGH BAY FIXTURES

B. Metal halide 400 watt

C. High pressure sodium 400 watt

E. Metal halide 1000 watt

F. High pressure sodium 1000 watt

G. Metal halide 1000 watt
   125,000 lumen lamp

System Components	QUANTITY	UNIT	COST PER S.F.		
			MAT.	INST.	TOTAL
SYSTEM D5020 234 0240					
HIGH INTENSITY DISCHARGE FIXTURE, 30′ ABOVE WORK PLANE					
1 WATT/S.F., TYPE E, 37 FC, 1 FIXTURE/1000 S.F.					
Steel intermediate conduit, (IMC) 1/2″ diam	.196	L.F.	.43	1.48	1.91
Wire, 600V type THWN-THHN, copper, solid, #10	.006	C.L.F.	.10	.45	.55
Steel outlet box 4″ concrete	.001	Ea.	.02	.04	.06
Steel outlet box plate with stud, 4″ concrete	.001	Ea.	.01	.01	.02
Metal halide, hi bay, aluminum reflector, 1000 W lamp	.001	Ea.	.67	.38	1.05
TOTAL			1.23	2.36	3.59

D5020 234	H.I.D. Fixture, High Bay, 30′ (by Wattage)	COST PER S.F.		
		MAT.	INST.	TOTAL
0190	High intensity discharge fixture, 30′ above work plane	1.22	2.36	3.58
0240	1 watt/S.F., type E, 37 FC, 1 fixture/1000 S.F.	1.22	2.36	3.58
0280	Type G, 45 FC., 1 fixture/1000 S.F.	1.01	1.85	2.86
0320	Type F, 50 FC, 1 fixture/1000 S.F.	2.45	4.72	7.17
0440	2 watt/S.F., type E, 74 FC, 2 fixtures/1000 S.F.	2.45	4.72	7.17
0480	Type G, 92 FC, 2 fixtures/1000 S.F.	2.05	3.77	5.82
0520	Type F, 100 FC, 2 fixtures/1000 S.F.	3.68	7.15	10.83
0640	3 watt/S.F., type E, 110 FC, 3 fixtures/1000 S.F.	3.68	7.15	10.83
0680	Type G, 138 FC, 3 fixtures/1000 S.F.	3.06	5.60	8.66
0720	Type F, 150 FC, 3 fixtures/1000 S.F.	4.91	9.50	14.41
0840	4 watt/S.F., type E, 148 FC, 4 fixtures/1000 S.F.	4.91	9.50	14.41
0880	Type G, 185 FC, 4 fixtures/1000 S.F.	4.09	7.55	11.64
0920	Type F, 200 FC, 4 fixtures/1000 S.F.	6.15	11.90	18.05
1040	5 watt/S.F., type E, 185 FC, 5 fixtures/1000 S.F.	6.15	11.90	18.05
1080	Type G, 230 FC, 5 fixtures/1000 S.F.	5.10	9.40	14.50
1120	Type F, 250 FC, 5 fixtures/1000 S.F.			

D50 Electrical

D5020 Lighting and Branch Wiring

LOW BAY FIXTURES

J. Metal halide 250 watt

K. High pressure sodium 150 watt

System Components

System Components	QUANTITY	UNIT	COST PER S.F. MAT.	COST PER S.F. INST.	COST PER S.F. TOTAL
SYSTEM D5020 236 0920					
HIGH INTENSITY DISCHARGE FIXTURE, 8'-10' ABOVE WORK PLANE, 50 FC					
TYPE J, 13 FIXTURES PER 1800 S.F.					
Steel intermediate conduit, (IMC) 1/2" diam	.550	L.F.	1.19	4.15	5.34
Wire, 600V, type THWN-THHN, copper, solid, #10	.012	C.L.F.	.21	.91	1.12
Steel outlet box 4" concrete	.007	Ea.	.12	.27	.39
Steel outlet box plate with stud, 4" concrete	.007	Ea.	.05	.07	.12
Metal halide, lo bay, aluminum reflector, 250 W DX lamp	.007	Ea.	2.80	1.66	4.46
TOTAL			**4.37**	**7.06**	**11.43**

D5020 236	H.I.D. Fixture, Low Bay, 8'-10' (by Type)	COST PER S.F. MAT.	COST PER S.F. INST.	COST PER S.F. TOTAL
0510	High intensity discharge fixture, 8'-10' above work plane, 50 FC			
0880	Type J, 7 fixtures per 900 S.F.	4.74	7.10	11.84
0920	13 fixtures per 1800 S.F.	4.37	7.05	11.42
0960	21 fixtures per 3000 S.F.	4.37	7.05	11.42
1000	28 fixtures per 4000 S.F.	4.37	7.05	11.42
1040	35 fixtures per 5000 S.F.	4.37	7.05	11.42
1120	62 fixtures per 10,000 S.F.	3.95	6.75	10.70
1160	99 fixtures per 16,000 S.F.	3.95	6.75	10.70
1200	199 fixtures per 32,000 S.F.	3.95	6.75	10.70
1240	Type K, 9 fixtures per 900 S.F.	4.84	6.10	10.94
1280	16 fixtures per 1800 S.F.	4.47	5.90	10.37
1320	26 fixtures per 3000 S.F.	4.45	5.80	10.25
1360	31 fixtures per 4000 S.F.	4.10	5.65	9.75
1400	39 fixtures per 5000 S.F.	4.10	5.65	9.75
1440	62 fixtures per 8000 S.F.	4.10	5.65	9.75
1480	78 fixtures per 10,000 S.F.	4.10	5.65	9.75
1520	124 fixtures per 16,000 S.F.	4.06	5.50	9.56
1560	248 fixtures per 32,000 S.F.	3.91	6.15	10.06

D5020 Lighting and Branch Wiring

LOW BAY FIXTURES

J. Metal halide 250 watt

K. High pressure sodium 150 watt

System Components	QUANTITY	UNIT	COST PER S.F.		
			MAT.	INST.	TOTAL
SYSTEM D5020 238 0240					
HIGH INTENSITY DISCHARGE FIXTURE, 8'-10' ABOVE WORK PLANE					
1 WATT/S.F., TYPE J, 30 FC, 4 FIXTURES/1000 S.F.					
Steel intermediate conduit, (IMC) 1/2" diam	.280	L.F.	.61	2.11	2.72
Wire, 600V, type THWN-THHN, copper, solid, #10	.008	C.L.F.	.14	.60	.74
Steel outlet box 4" concrete	.004	Ea.	.07	.15	.22
Steel outlet box plate with stud, 4" concrete	.004	Ea.	.03	.04	.07
Metal halide, lo bay, aluminum reflector, 250 W DX lamp	.004	Ea.	1.60	.95	2.55
TOTAL			2.45	3.85	6.30

D5020 238	H.I.D. Fixture, Low Bay, 8'-10' (by Wattage)	COST PER S.F.		
		MAT.	INST.	TOTAL
0190	High intensity discharge fixture, 8'-10' above work plane			
0240	1 watt/S.F., type J, 30 FC, 4 fixtures/1000 S.F.	2.44	3.86	6.30
0280	Type K, 29 FC, 5 fixtures/1000 S.F.	2.53	3.46	5.99
0400	2 watt/S.F., type J, 52 FC, 7 fixtures/1000 S.F.	4.36	7.05	11.41
0440	Type K, 63 FC, 11 fixtures/1000 S.F.	5.45	7.30	12.75
0560	3 watt/S.F., type J, 81 FC, 11 fixtures/1000 S.F.	6.75	10.65	17.40
0600	Type K, 92 FC, 16 fixtures/1000 S.F.	8	10.75	18.75
0720	4 watt/S.F., type J, 103 FC, 14 fixtures/1000 S.F.	8.70	14.05	22.75
0760	Type K, 127 FC, 22 fixtures/1000 S.F.	10.95	14.55	25.50
0880	5 watt/S.F., type J, 133 FC, 18 fixtures/1000 S.F.	11.10	17.75	28.85
0920	Type K, 155 FC, 27 fixtures/1000 S.F.	13.50	18	31.50

439

For customer support on your Electrical Costs with RSMeans Data, call 800.448.8182.

D5020 Lighting and Branch Wiring

LOW BAY FIXTURES

J. Metal halide 250 watt

K. High pressure sodium 150 watt

System Components	QUANTITY	UNIT	COST PER S.F.		
			MAT.	INST.	TOTAL
SYSTEM D5020 240 0880					
HIGH INTENSITY DISCHARGE FIXTURE, 16' ABOVE WORK PLANE, 50 FC					
TYPE J, 9 FIXTURES PER 900 S.F.					
Steel intermediate conduit, (IMC) 1/2" diam	.630	L.F.	1.37	4.76	6.13
Wire, 600V type, THWN-THHN, copper, solid, #10	.012	C.L.F.	.21	.91	1.12
Steel outlet box 4" concrete	.010	Ea.	.17	.38	.55
Steel outlet box plate with stud, 4" concrete	.010	Ea.	.07	.09	.16
Metal halide, lo bay, aluminum reflector, 250 W DX lamp	.010	Ea.	4	2.37	6.37
TOTAL			5.82	8.51	14.33

D5020 240	H.I.D. Fixture, Low Bay, 16' (by Type)	COST PER S.F.		
		MAT.	INST.	TOTAL
0510	High intensity discharge fixture, 16' above work plane, 50 FC			
0880	Type J, 9 fixtures per 900 S.F.	5.80	8.50	14.30
0920	14 fixtures per 1800 S.F.	4.99	8	12.99
0960	24 fixtures per 3000 S.F.	5.05	8.15	13.20
1000	32 fixtures per 4000 S.F.	5.05	8.15	13.20
1040	35 fixtures per 5000 S.F.	4.63	7.95	12.58
1080	56 fixtures per 8000 S.F.	4.63	7.95	12.58
1120	70 fixtures per 10,000 S.F.	4.62	7.95	12.57
1160	111 fixtures per 16,000 S.F.	4.62	7.95	12.57
1200	222 fixtures per 32,000 S.F.	4.62	7.95	12.57
1240	Type K, 11 fixtures per 900 S.F.	6	7.95	13.95
1280	20 fixtures per 1800 S.F.	5.50	7.35	12.85
1320	29 fixtures per 3000 S.F.	5.15	7.20	12.35
1360	39 fixtures per 4000 S.F.	5.15	7.15	12.30
1400	44 fixtures per 5000 S.F.	4.78	7	11.78
1440	62 fixtures per 8000 S.F.	4.78	7	11.78
1480	87 fixtures per 10,000 S.F.	4.78	7	11.78
1520	138 fixtures per 16,000 S.F.	4.78	7	11.78

D5020 Lighting and Branch Wiring

LOW BAY FIXTURES

J. Metal halide 250 watt

K. High pressure sodium 150 watt

System Components	QUANTITY	UNIT	COST PER S.F.		
			MAT.	INST.	TOTAL
SYSTEM D5020 242 0240					
HIGH INTENSITY DISCHARGE FIXTURE, 16' ABOVE WORK PLANE					
1 WATT/S.F., TYPE J, 28 FC, 4 FIXTURES/1000 S.F.					
Steel intermediate conduit, (IMC) 1/2" diam	.328	L.F.	.71	2.48	3.19
Wire, 600V, type THWN-THHN, copper, solid, #10	.010	C.L.F.	.17	.76	.93
Steel outlet box 4" concrete	.004	Ea.	.07	.15	.22
Steel outlet box plate with stud, 4" concrete	.004	Ea.	.03	.04	.07
Metal halide, lo bay, aluminum reflector, 250 W DX lamp	.004	Ea.	1.60	.95	2.55
TOTAL			2.58	4.38	6.96

D5020 242	H.I.D. Fixture, Low Bay, 16' (by Wattage)	COST PER S.F.		
		MAT.	INST.	TOTAL
0190	High intensity discharge fixture, mounted 16' above work plane			
0240	1 watt/S.F., type J, 28 FC, 4 fixt./1000 S.F.	2.58	4.37	6.95
0280	Type K, 27 FC, 5 fixt./1000 S.F.	2.92	4.90	7.82
0400	2 watt/S.F., type J, 48 FC, 7 fixt/1000 S.F.	4.74	8.45	13.19
0440	Type K, 58 FC, 11 fixt/1000 S.F.	6.20	10	16.20
0560	3 watt/S.F., type J, 75 FC, 11 fixt/1000 S.F.	7.30	12.80	20.10
0600	Type K, 85 FC, 16 fixt/1000 S.F.	9.15	14.90	24.05
0720	4 watt/S.F., type J, 95 FC, 14 fixt/1000 S.F.	9.50	16.90	26.40
0760	Type K, 117 FC, 22 fixt/1000 S.F.	12.45	20	32.45
0880	5 watt/S.F., type J, 122 FC, 18 fixt/1000 S.F.	12.05	21.50	33.55
0920	Type K, 143 FC, 27 fixt/1000 S.F.	15.35	25	40.35

Daylight Dimming System

System Components			COST PER S.F.		
	QUANTITY	UNIT	MAT.	INST.	TOTAL
SYSTEM D5020 290 0800					
DAYLIGHT DIMMING CONTROL SYSTEM					
5 FIXTURES PER 1000 S.F.					
Tray cable, type TC, copper #16-4 conductor	.300	C.L.F.	.02	.03	.05
Wire, 600 volt, type THWN-THHN, copper, solid, #12	.320	C.L.F.		.02	.02
Conduit (EMT) , to 10' H, incl 2 termn,2 elb&11 bm clp per 100', 3/4"	10.000	L.F.	.01	.06	.07
Cabinet, hinged, steel, NEMA 1, 12"W x 12"H x 4"D	.200	Ea.	.01	.03	.04
Lighting control module	.200	Ea.	.07	.08	.15
Dimmable ballast three-lamp	5.000	Ea.	.70	.55	1.25
Daylight level sensor, wall mounted, on/off or dimming	.200	Ea.	.03	.02	.05
Automatic wall switches	.200	Ea.	.02	.01	.03
Remote power pack	.100	Ea.		.01	.01
TOTAL			.86	.81	1.67

D5020 290	Daylight Dimming System	COST PER S.F.		
		MAT.	INST.	TOTAL
0500	Daylight Dimming System (no fixtures or fixture power)			
0800	5 fixtures per 1000 S.F.	.85	.80	1.65
1000	10 fixtures per 1000 S.F.	1.69	1.51	3.20
2000	12 fixtures per 1000 S.F.	2.03	1.80	3.83
3000	15 fixtures per 1000 S.F.	2.53	2.23	4.76
4000	20 fixtures per 1000 S.F.	3.37	2.94	6.31
5000	25 fixtures per 1000 S.F.	4.21	3.66	7.87
6000	50 fixtures per 1000 S.F.	8.40	7.25	15.65

D50 Electrical

D5020 Lighting and Branch Wiring

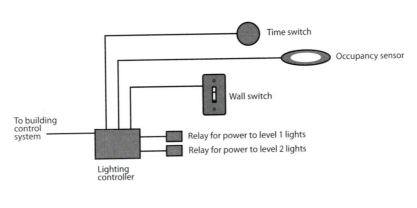

Lighting AC Control System

System Components	QUANTITY	UNIT	COST PER S.F.		
			MAT.	INST.	TOTAL
SYSTEM D5020 295 0800					
LIGHTING ON/OFF CONTROL SYSTEM					
5 FIXTURES PER 1000 S.F.					
Tray cable, type TC, copper #16-4 conductor	.600	C.L.F.	.04	.06	.10
Wire, 600 volt, type THWN-THHN, copper, solid, #12	.320	C.L.F.		.02	.02
Conduit (EMT) , to 10' H, incl 2 termn,2 elb&11 bm clp per 100', 3/4"	10.000	L.F.	.01	.06	.07
Cabinet, hinged, steel, NEMA 1, 12"W x 12"H x 4"D	.400	Ea.	.01	.05	.06
Relays, 120 V or 277 V standard	.800	Ea.	.04	.05	.09
24 hour dial with reserve power	.400	Ea.	.33	.08	.41
Lighting control module	.200	Ea.	.07	.08	.15
Occupancy sensors, passive infrared ceiling mounted	.400	Ea.	.03	.04	.07
Automatic wall switches	.400	Ea.	.03	.01	.04
Remote power pack	.200	Ea.	.01	.02	.03
TOTAL			.57	.47	1.04

D5020 295	Lighting On/Off Control System	COST PER S.F.		
		MAT.	INST.	TOTAL
0500	Includes occupancy and time switching (no fixtures or fixture power)	.57	.47	1.04
0800	5 fixtures per 1000 SF			
1000	10 fixtures per 1000 S.F.	.64	.55	1.19
2000	12 fixtures per 1000 S.F.	.77	.64	1.41
3000	15 fixtures per 1000 S.F.	.95	.79	1.74
4000	20 fixtures per 1000 S.F.	1.27	1.02	2.29
5000	25 fixtures per 1000 S.F.	1.58	1.26	2.84
6000	50 fixtures per 1000 S.F.	3.14	2.43	5.57

443

D5030 Communications and Security

Description: System below includes telephone fitting installed. Does not include cable.

When poke thru fittings and telepoles are used for power, they can also be used for telephones at a negligible additional cost.

System Components	QUANTITY	UNIT	COST PER S.F.		
			MAT.	INST.	TOTAL
SYSTEM D5030 310 0200					
TELEPHONE SYSTEMS, UNDERFLOOR DUCT, 5' ON CENTER, LOW DENSITY					
Underfloor duct 7-1/4" w/insert 2' O.C. 1-3/8" x 7-1/4" super duct	.190	L.F.	7.41	2.88	10.29
Vertical elbow for underfloor superduct, 7-1/4", included					
Underfloor duct conduit adapter, 2" x 1-1/4", included					
Underfloor duct junction box, single duct, 7-1/4" x 3 1/8"	.003	Ea.	1.92	.57	2.49
Underfloor junction box carpet pan	.003	Ea.	1.23	.03	1.26
Underfloor duct outlet, low tension	.004	Ea.	.46	.38	.84
TOTAL			11.02	3.86	14.88

D5030 310	Telephone Systems	COST PER S.F.		
		MAT.	INST.	TOTAL
0200	Telephone systems, underfloor duct, 5' on center, low density	11	3.85	14.85
0240	5' on center, high density	11.50	4.23	15.73
0280	7' on center, low density	8.80	3.20	12
0320	7' on center, high density	9.25	3.58	12.83
0400	Poke thru fittings, low density	1.59	1.36	2.95
0440	High density	3.17	2.71	5.88
0520	Telepoles, low density	1.60	.82	2.42
0560	High density	3.20	1.64	4.84
0640	Conduit system with floor boxes, low density	1.52	1.35	2.87
0680	High density	3.03	2.66	5.69
1020	Telephone wiring for offices & laboratories, 8 jacks/M.S.F.	.37	2.07	2.44

D5030 810	Security & Detection Systems	COST EACH		
		MAT.	INST.	TOTAL
0200	Security system, head end equipment	30,700	5,050	35,750
0240	Security system, door/window contact biased, box, conduit & cable	156	565	721
0280	Security system, door/window contact balanced, box, conduit & cable	101	565	666
0440	Security system, proximity card reader, box, conduit & cable	262	755	1,017
1600	Card control entrance system, to 6 zone, including hardware for 100 doors	43,400	35,100	78,500

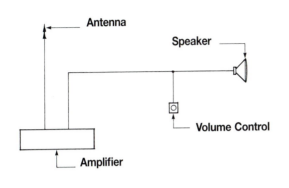

Sound System Includes AM–FM antenna, outlets, rigid conduit, and copper wire.

Fire Detection System Includes pull stations, signals, smoke and heat detectors, rigid conduit, and copper wire.

Intercom System Includes master and remote stations, rigid conduit, and copper wire.

Master Clock System Includes clocks, bells, rigid conduit, and copper wire.

Master TV Antenna Includes antenna, VHF–UHF reception and distribution, rigid conduit, and copper wire.

System Components	QUANTITY	UNIT	COST EACH MAT.	COST EACH INST.	COST EACH TOTAL
SYSTEM D5030 910 0220					
SOUND SYSTEM, INCLUDES OUTLETS, BOXES, CONDUIT & WIRE					
Steel intermediate conduit, (IMC) 1/2" diam	1200.000	L.F.	2,604	9,060	11,664
Wire sound shielded w/drain, #22-2 conductor	15.500	C.L.F.	167.40	1,464.75	1,632.15
Sound system speakers ceiling or wall	12.000	Ea.	1,812	1,134	2,946
Sound system volume control	12.000	Ea.	870	1,134	2,004
Sound system amplifier, 250 W	1.000	Ea.	1,550	755	2,305
Sound system antenna, AM FM	1.000	Ea.	162	189	351
Sound system monitor panel	1.000	Ea.	500	189	689
Sound system cabinet	1.000	Ea.	1,250	755	2,005
Steel outlet box 4" square	12.000	Ea.	29.04	456	485.04
Steel outlet box 4" plaster rings	12.000	Ea.	21.96	142.20	164.16
TOTAL			8,966.40	15,278.95	24,245.35

D5030 910	Communication & Alarm Systems	COST EACH MAT.	COST EACH INST.	COST EACH TOTAL
0200	Communication & alarm systems, includes outlets, boxes, conduit & wire			
0210	Sound system, 6 outlets	6,325	9,550	15,875
0220	12 outlets	8,975	15,300	24,275
0240	30 outlets	15,600	28,900	44,500
0280	100 outlets	46,500	96,500	143,000
0320	Fire detection systems, non-addressable, 12 detectors	3,575	7,925	11,500
0360	25 detectors	6,350	13,400	19,750
0400	50 detectors	12,500	26,600	39,100
0440	100 detectors	23,400	48,400	71,800
0450	Addressable type, 12 detectors	5,000	8,000	13,000
0452	25 detectors	9,050	13,500	22,550
0454	50 detectors	17,400	26,600	44,000
0456	100 detectors	33,700	48,800	82,500
0458	Fire alarm control panel, 8 zone, excluding wire and conduit	460	1,525	1,985
0459	12 zone	2,825	2,275	5,100
0460	Fire alarm command center, addressable without voice, excl. wire & conduit	2,150	1,325	3,475
0462	Addressable with voice	11,100	2,075	13,175
0480	Intercom systems, 6 stations	4,550	6,525	11,075
0520	12 stations	8,425	13,000	21,425
0560	25 stations	14,500	25,000	39,500
0600	50 stations	28,100	47,100	75,200
0640	100 stations	55,500	92,000	147,500
0680	Master clock systems, 6 rooms	5,100	10,700	15,800
0720	12 rooms	8,275	18,200	26,475
0760	20 rooms	11,600	25,800	37,400
0800	30 rooms	19,300	47,500	66,800

445

D50 Electrical

D5030   Communications and Security

<table>
<tr><td rowspan="2">D5030 910</td><td rowspan="2" align="center">Communication & Alarm Systems</td><td colspan="3" align="center">COST EACH</td></tr>
<tr><td>MAT.</td><td>INST.</td><td>TOTAL</td></tr>
<tr><td>0840</td><td>50 rooms</td><td>31,600</td><td>80,000</td><td>111,600</td></tr>
<tr><td>0880</td><td>100 rooms</td><td>61,500</td><td>159,000</td><td>220,500</td></tr>
<tr><td>0920</td><td align="center">Master TV antenna systems, 6 outlets</td><td>2,575</td><td>6,750</td><td>9,325</td></tr>
<tr><td>0960</td><td>12 outlets</td><td>4,825</td><td>12,600</td><td>17,425</td></tr>
<tr><td>1000</td><td>30 outlets</td><td>15,800</td><td>29,100</td><td>44,900</td></tr>
<tr><td>1040</td><td>100 outlets</td><td>56,500</td><td>95,000</td><td>151,500</td></tr>
</table>

<table>
<tr><td rowspan="2">D5030 920</td><td rowspan="2" align="center">Data Communication</td><td colspan="3" align="center">COST PER M.S.F.</td></tr>
<tr><td>MAT.</td><td>INST.</td><td>TOTAL</td></tr>
<tr><td>0100</td><td>Data communication system, incl data/voice outlets, boxes, conduit & cable</td><td></td><td></td><td></td></tr>
<tr><td>0102</td><td align="center">Data and voice system, 2 data/voice outlets per 1000 S.F.</td><td>133</td><td>560</td><td>693</td></tr>
<tr><td>0104</td><td>4 data/voice outlets per 1000 S.F.</td><td>251</td><td>1,100</td><td>1,351</td></tr>
<tr><td>0106</td><td>6 data/voice outlets per 1000 S.F.</td><td>355</td><td>1,625</td><td>1,980</td></tr>
<tr><td>0110</td><td>8 data/voice outlets per 1000 S.F.</td><td>450</td><td>2,150</td><td>2,600</td></tr>
</table>

Description: System below tabulates the installed cost for generators by kW. Included in costs are battery, charger, muffler, and transfer switch.

No conduit, wire, or terminations included.

System Components	QUANTITY	UNIT	COST PER kW		
			MAT.	INST.	TOTAL
SYSTEM D5090 210 0200					
GENERATOR SET, INCL. BATTERY, CHARGER, MUFFLER & TRANSFER SWITCH					
GAS/GASOLINE OPER., 3 PHASE, 4 WIRE, 277/480V, 7.5 kW					
Generator set, gas or gasoline operated, 3 ph 4 W, 277/480 V, 7.5 kW	.133	Ea.	1,106.67	337.06	1,443.73
TOTAL			1,106.67	337.06	1,443.73

D5090 210	Generators (by kW)	COST PER kW		
		MAT.	INST.	TOTAL
0190	Generator sets, include battery, charger, muffler & transfer switch			
0200	Gas/gasoline operated, 3 phase, 4 wire, 277/480 volt, 7.5 kW	1,100	335	1,435
0240	11.5 kW	1,025	256	1,281
0280	20 kW	585	167	752
0320	35 kW	470	109	579
0360	80 kW	325	128	453
0400	100 kW	276	124	400
0440	125 kW	244	117	361
0480	185 kW	435	89	524
0560	Diesel engine with fuel tank, 30 kW	435	127	562
0600	50 kW	555	100	655
0640	75 kW	287	80	367
0680	100 kW	370	67.50	437.50
0720	125 kW	293	58	351
0760	150 kW	178	53.50	231.50
0800	175 kW	282	48	330
0840	200 kW	250	43.50	293.50
0880	250 kW	268	36.50	304.50
0920	300 kW	203	32	235
0960	350 kW	220	30	250
1000	400 kW	168	27.50	195.50
1040	500 kW	215	23.50	238.50
1200	750 kW	272	14.40	286.40
1400	1000 kW	226	15.20	241.20

D5090 Other Electrical Systems

PV power system, stand alone, AC and DC loads

Combiner box 15amp pull out fuses

PV array disconnect 50 amp 2-pole, added lightning protection

Battery disconnect 175 amp

DC to AC inverter

To AC loads

Low voltage disconnect

AC load center

DC loads

PV modules

PV Charge battery temp sensor

C40 charge controller 40 amp, 48 V

Batteries

DC load center

Grounding rod

System Components			COST EACH		
	QUANTITY	UNIT	MAT.	INST.	TOTAL
SYSTEM D5090 420 0100					
PHOTOVOLTAIC POWER SYSTEM, STAND ALONE					
Alternative energy sources, photovoltaic module, 75 watt, 17 V	12.000	Ea.	5,640	1,134	6,774
PV rack system, roof, penetrating surface mount, on wood framing, 1 panel	12.000	Ea.	756	1,488	2,244
DC to AC inverter for, 24 V, 2,500 watt	1.000	Ea.	1,850	189	2,039
PV components, combiner box, 10 lug, NEMA 3R enclosure	1.000	Ea.	296	189	485
Alternative energy sources, PV components, fuse, 15 A for combiner box	10.000	Ea.	235	189.50	424.50
Battery charger controller w/temperature sensor	1.000	Ea.	445	189	634
Digital readout panel, displays hours, volts, amps, etc.	1.000	Ea.	233	189	422
Deep cycle solar battery, 6 V, 180 Ah (C/20)	4.000	Ea.	1,480	378	1,858
Battery interconnection, 15" AWG #2/0, sealed w/copper ring lugs	3.000	Ea.	56.25	142.50	198.75
Battery interconn, 24" AWG #2/0, sealed w/copper ring lugs	2.000	Ea.	58	95	153
Battery interconn, 60" AWG #2/0, sealed w/copper ring lugs	2.000	Ea.	105	95	200
Batt temp computer probe, RJ11 jack, 15' cord	1.000	Ea.	24.50	47.50	72
System disconnect, DC 175 amp circuit breaker	1.000	Ea.	268	94.50	362.50
Conduit box for inverter	1.000	Ea.	69.50	94.50	164
Low voltage disconnect	1.000	Ea.	80	94.50	174.50
Vented battery enclosure, wood	1.000	Ea.	335	325	660
Grounding, rod, copper clad, 8' long, 5/8" diameter	1.000	Ea.	16.50	138	154.50
Grounding, clamp, bronze, 5/8" dia	1.000	Ea.	2.71	23.50	26.21
Bare copper wire, stranded, #8	1.000	C.L.F.	30.50	69	99.50
Wire, 600 volt, type THW, copper, stranded, #12	3.600	C.L.F.	69.12	248.40	317.52
Wire, 600 volt, type THW, copper, stranded, #10	1.050	C.L.F.	32.55	79.28	111.83
Cond to 10' H,incl 2 termn,2 elb&11 bm CLP per 100',galv stl,1/2" dia	120.000	L.F.	346.80	1,008	1,354.80
Conduit,to 10' H,incl 2 termn,2 elb&11 bm clp per 100',(EMT), 1" dia	30.000	L.F.	55.20	198	253.20
Lightning surge suppressor	1.000	Ea.	56	23.50	79.50
General duty 240 volt, 2 pole, nonfusible, NEMA 3R, 60 amp	1.000	Ea.	134	330	464
Load centers, 3 wire, 120/240V, 100 amp main lugs, indoor, 8 circuits	2.000	Ea.	362	1,080	1,442
Circuit breaker, Plug-in, 120/240 volt, to 60 amp, 1 pole	5.000	Ea.	71.50	315	386.50
Fuses, dual element, time delay, 250 volt, 50 amp	2.000	Ea.	24.80	30.30	55.10
TOTAL			13,132.93	8,476.98	21,609.91

D50 Electrical

D5090 Other Electrical Systems

D5090 420	Photovoltaic Power System, Stand Alone	COST EACH		
		MAT.	INST.	TOTAL
0050	24 V capacity, 900 W (~102 SF)			
0100	Roof mounted, on wood framing	13,100	8,475	21,575
0150	with standoff	13,300	8,475	21,775
0200	Ground placement, ballast, fixed	18,300	9,050	27,350
0250	Adjustable	18,700	9,050	27,750
0300	Top of pole, passive tracking	16,500	9,050	25,550
0350	1.8 kW (~204 SF)			
0400	Roof mounted, on wood framing	20,900	13,500	34,400
0450	with standoff	21,200	13,500	34,700
0500	Ground placement, ballast, fixed	30,400	12,700	43,100
0550	Adjustable	31,100	12,700	43,800
0600	Top of pole, passive tracking	31,700	12,700	44,400
0650	3.2 kW (~408 SF)			
0700	Roof mounted, on wood framing	38,700	23,100	61,800
0750	with standoff	39,200	23,100	62,300
0800	Ground placement, ballast, fixed	57,000	19,800	76,800
0850	Adjustable	58,500	19,800	78,300
0900	Top of pole, passive tracking	59,500	19,600	79,100

D5090 Other Electrical Systems

Photovoltaic Modules →

Photovoltaic power system, grid connected 10 kW

Combiner Box

DC Disconnect

Inverter

AC Disconnect

Isolation Transformer

Utility Connection

System Components			COST EACH		
	QUANTITY	UNIT	MAT.	INST.	TOTAL
SYSTEM D5090 430 0100					
PHOTOVOLTAIC POWER SYSTEM, GRID CONNECTED 10 kW					
Photovoltaic module, 150 watt, 33 V	60.000	Ea.	33,300	5,670	38,970
PV rack system, roof, non-penetrating ballast, 1 panel	60.000	Ea.	50,400	2,670	53,070
DC to AC inverter for, 24 V, 4,000 watt	3.000	Ea.	12,375	1,140	13,515
Combiner box 10 lug, NEMA 3R	1.000	Ea.	296	189	485
15 amp fuses	10.000	Ea.	235	189.50	424.50
Safety switch, 60 amp	1.000	Ea.	325	345	670
Safety switch, 100 amp	1.000	Ea.	585	420	1,005
Utility connection, 3 pole breaker	1.000	Ea.	370	122	492
Fuse, 60 A	3.000	Ea.	38.40	45.45	83.85
Fuse, 100 A	3.000	Ea.	93	56.85	149.85
10 kVA isolation tranformer	1.000	Ea.	4,400	945	5,345
EMT Conduit w/fittings & support	180.000	L.F.	331.20	1,188	1,519.20
RGS Conduit w/fittings & support	267.000	L.F.	1,228.20	3,110.55	4,338.75
Enclosure 24" x 24" x 10", NEMA 4	1.000	Ea.	5,300	1,525	6,825
Wire, 600 volt, type THWN-THHN, copper, stranded, #6	8.000	C.L.F.	368	928	1,296
Wire, 600 volt, copper type XLPE-USE(RHW), stranded, #12	18.000	C.L.F.	450	1,242	1,692
Grounding, bare copper wire stranded, 4/0	2.000	C.L.F.	670	532	1,202
Grounding, exothermic weld, 4/0 wire to building steel	4.000	Ea.	12.04	432	444.04
Grounding, brazed connections, #6 wire	2.000	Ea.	35.20	126	161.20
Insulated ground wire, copper, #6	.400	C.L.F.	18.40	46.40	64.80
TOTAL			110,830.44	20,922.75	131,753.19

D5090 430	Photovoltaic Power System, Grid Connected	COST EACH		
		MAT.	INST.	TOTAL
0050	10 kW			
0100	Roof mounted, non-penetrating ballast	111,000	20,900	131,900
0200	Penetrating surface mount, on steel framing	64,500	34,200	98,700
0300	on wood framing	64,000	25,700	89,700
0400	with standoff	65,000	25,800	90,800
0500	Ground placement, ballast, fixed	86,500	20,800	107,300
0600	adjustable	88,500	20,800	109,300
0700	Top of pole, passive tracking	79,500	21,400	100,900
1050	20 kW			
1100	Roof mounted, non-penetrating ballast	215,500	35,000	250,500
1200	Penetrating surface mount, on steel framing	122,000	61,500	183,500
1300	on wood framing	122,000	44,600	166,600
1400	with standoff	123,500	44,600	168,100
1500	Ground placement, ballast, fixed	166,500	33,700	200,200
1600	adjustable	170,000	33,700	203,700
1700	Top of pole, passive tracking	152,000	34,800	186,800

D50 Electrical

D5090 Other Electrical Systems

System Components	QUANTITY	UNIT	COST EACH		
			MAT.	INST.	TOTAL
SYSTEM D5090 480 1100					
ELECTRICAL, SINGLE PHASE, 1 METER					
#18 twisted shielded pair in 1/2" EMT conduit	.050	C.L.F.	7.85	12.60	20.45
Wire, 600 volt, type THW, copper, solid, #12	.150	C.L.F.	1.69	10.35	12.04
Conduit (EMT) , to 10' H, incl 2 termn,2 elb&11 bm clp per 100', 3/4"	5.000	L.F.	5.55	29	34.55
Outlet boxes, pressed steel, handy box	1.000	Ea.	3.12	28	31.12
Outlet boxes, pressed steel, handy box, covers, device	1.000	Ea.	.76	11.85	12.61
Wiring devices, receptacle, duplex, 120 volt, ground, 20 amp	1.000	Ea.	12.10	28	40.10
Single phase, 277 volt, 200 amp	1.000	Ea.	510	86	596
Data recorder, 8 meters	1.000	Ea.	2,500	69	2,569
Software package, per meter, premium	1.000	Ea.	765		765
TOTAL			3,806.07	274.80	4,080.87

D5090 480	Energy Monitoring Systems	COST EACH		
		MAT.	INST.	TOTAL
1000	Electrical			
1100	Single phase, 1 meter	3,800	275	4,075
1110	4 meters	8,350	2,075	10,425
1120	8 meters	16,800	4,275	21,075
1200	Three phase, 1 meter	6,200	350	6,550
1210	5 meters	17,800	3,250	21,050
1220	10 meters	38,300	6,675	44,975
1230	25 meters	84,500	13,300	97,800
2000	Mechanical			
2100	BTU, 1 meter	4,225	990	5,215
2110	w/1 duct sensor	4,425	1,225	5,650
2120	& 1 space sensor	4,900	1,400	6,300
2130	& 5 space sensors	6,775	2,075	8,850
2140	& 10 space sensors	9,950	2,925	12,875
2200	BTU, 3 meters	7,675	2,825	10,500
2210	w/3 duct sensors	8,275	3,550	11,825
2220	& 3 space sensors	9,675	4,050	13,725
2230	& 15 space sensors	15,300	6,075	21,375
2240	& 30 space sensors	23,200	8,625	31,825
9000	Front end display	565	148	713
9100	Computer workstation	1,850	1,975	3,825

451

D5090 Other Electrical Systems

Low Density and Medium Density Baseboard Radiation

The costs shown in Table below are based on the following system considerations:

1. The heat loss per square foot is based on approximately 34 BTU/hr. per S.F. of floor or 10 watts per S.F. of floor.
2. Baseboard radiation is based on the low watt density type rated 187 watts per L.F. and the medium density type rated 250 watts per L.F.
3. Thermostat is not included.
4. Wiring costs include branch circuit wiring.

System Components	QUANTITY	UNIT	COST PER S.F.		
			MAT.	INST.	TOTAL
SYSTEM D5090 510 1000					
ELECTRIC BASEBOARD RADIATION, LOW DENSITY, 900 S.F., 31 MBH, 9 kW					
Electric baseboard radiator, 5′ long 935 watt	.011	Ea.	.63	1.46	2.09
Steel intermediate conduit, (IMC) 1/2″ diam	.170	L.F.	.37	1.28	1.65
Wire 600 volt, type THW, copper, solid, #12	.005	C.L.F.	.06	.35	.41
TOTAL			1.06	3.09	4.15

D5090 510	Electric Baseboard Radiation (Low Density)	COST PER S.F.		
		MAT.	INST.	TOTAL
1000	Electric baseboard radiation, low density, 900 S.F., 31 MBH, 9 kW	1.06	3.09	4.15
1200	1500 S.F., 51 MBH, 15 kW	1.04	3.02	4.06
1400	2100 S.F., 72 MBH, 21 kW	.93	2.70	3.63
1600	3000 S.F., 102 MBH, 30 kW	.85	2.46	3.31
2000	Medium density, 900 S.F., 31 MBH, 9 kW	.94	2.83	3.77
2200	1500 S.F., 51 MBH, 15 kW	.92	2.75	3.67
2400	2100 S.F., 72 MBH, 21 kW	.88	2.57	3.45
2600	3000 S.F., 102 MBH, 30 kW	.79	2.33	3.12

D5090 Other Electrical Systems

Commercial Duty Baseboard Radiation

The costs shown in Table below are based on the following system considerations:

1. The heat loss per square foot is based on approximately 41 BTU/hr. per S.F. of floor or 12 watts per S.F.
2. The baseboard radiation is of the commercial duty type rated 250 watts per L.F. served by 277 volt, single phase power.
3. Thermostat is not included.
4. Wiring costs include branch circuit wiring.

System Components	QUANTITY	UNIT	COST PER S.F.		
			MAT.	INST.	TOTAL
SYSTEM D5090 520 1000					
ELECTRIC BASEBOARD RADIATION, MEDIUM DENSITY, 1230 S.F., 51 MBH, 15 kW					
Electric baseboard radiator, 5' long	.013	Ea.	.75	1.73	2.48
Steel intermediate conduit, (IMC) 1/2" diam	.154	L.F.	.33	1.16	1.49
Wire 600 volt, type THW, copper, solid, #12	.004	C.L.F.	.05	.28	.33
TOTAL			1.13	3.17	4.30

D5090 520	Electric Baseboard Radiation (Medium Density)	COST PER S.F.		
		MAT.	INST.	TOTAL
1000	Electric baseboard radiation, medium density, 1230 SF, 51 MBH, 15 kW	1.13	3.17	4.30
1200	2500 S.F. floor area, 106 MBH, 31 kW	1.05	2.97	4.02
1400	3700 S.F. floor area, 157 MBH, 46 kW	1.03	2.88	3.91
1600	4800 S.F. floor area, 201 MBH, 59 kW	.94	2.67	3.61
1800	11,300 S.F. floor area, 464 MBH, 136 kW	.89	2.45	3.34
2000	30,000 S.F. floor area, 1229 MBH, 360 kW	.87	2.39	3.26

453

Same Data. Simplified.

Enjoy the convenience and efficiency of accessing your costs anywhere:

■ **Skip the multiplier** by setting your location

■ **Quickly search,** edit, favorite and share costs

■ **Stay on top of price changes** with automatic updates

Discover more at rsmeans.com/online

G10 Site Preparation

G1030 Site Earthwork

Trenching Systems are shown on a cost per linear foot basis. The systems include: excavation; backfill and removal of spoil; and compaction for various depths and trench bottom widths. The backfill has been reduced to accommodate a pipe of suitable diameter and bedding.

The slope for trench sides varies from none to 1:1.

The Expanded System Listing shows Trenching Systems that range from 2' to 12' in width. Depths range from 2' to 25'.

System Components	QUANTITY	UNIT	COST PER L.F. EQUIP.	COST PER L.F. LABOR	COST PER L.F. TOTAL
SYSTEM G1030 805 1310					
TRENCHING, COMMON EARTH, NO SLOPE, 2' WIDE, 2' DP, 3/8 C.Y. BUCKET					
Excavation, trench, hyd. backhoe, track mtd., 3/8 C.Y. bucket	.148	B.C.Y.	.24	1.21	1.45
Backfill and load spoil, from stockpile	.153	L.C.Y.	.13	.37	.50
Compaction by vibrating plate, 6" lifts, 4 passes	.118	E.C.Y.	.03	.45	.48
Remove excess spoil, 8 C.Y. dump truck, 2 mile roundtrip	.040	L.C.Y.	.15	.20	.35
TOTAL			.55	2.23	2.78

G1030 805	Trenching Common Earth	COST PER L.F. EQUIP.	COST PER L.F. LABOR	COST PER L.F. TOTAL
1310	Trenching, common earth, no slope, 2' wide, 2' deep, 3/8 C.Y. bucket	.54	2.24	2.78
1320	3' deep, 3/8 C.Y. bucket	.76	3.36	4.12
1330	4' deep, 3/8 C.Y. bucket	.98	4.48	5.46
1340	6' deep, 3/8 C.Y. bucket	1.28	5.80	7.08
1350	8' deep, 1/2 C.Y. bucket	1.66	7.70	9.36
1360	10' deep, 1 C.Y. bucket	3.61	9.20	12.81
1400	4' wide, 2' deep, 3/8 C.Y. bucket	1.28	4.45	5.73
1410	3' deep, 3/8 C.Y. bucket	1.71	6.70	8.41
1420	4' deep, 1/2 C.Y. bucket	1.94	7.45	9.39
1430	6' deep, 1/2 C.Y. bucket	3.20	12	15.20
1440	8' deep, 1/2 C.Y. bucket	6.75	15.50	22.25
1450	10' deep, 1 C.Y. bucket	8.20	19.25	27.45
1460	12' deep, 1 C.Y. bucket	10.55	24.50	35.05
1470	15' deep, 1-1/2 C.Y. bucket	7.70	22	29.70
1480	18' deep, 2-1/2 C.Y. bucket	13	30.50	43.50
1520	6' wide, 6' deep, 5/8 C.Y. bucket w/trench box	9.05	17.95	27
1530	8' deep, 3/4 C.Y. bucket	11.75	23.50	35.25
1540	10' deep, 1 C.Y. bucket	11.70	24.50	36.20
1550	12' deep, 1-1/2 C.Y. bucket	10.40	26	36.40
1560	16' deep, 2-1/2 C.Y. bucket	16.60	32.50	49.10
1570	20' deep, 3-1/2 C.Y. bucket	20.50	39	59.50
1580	24' deep, 3-1/2 C.Y. bucket	24.50	47	71.50
1640	8' wide, 12' deep, 1-1/2 C.Y. bucket w/trench box	14.50	33	47.50
1650	15' deep, 1-1/2 C.Y. bucket	18.80	43.50	62.30
1660	18' deep, 2-1/2 C.Y. bucket	24	43.50	67.50
1680	24' deep, 3-1/2 C.Y. bucket	33	60.50	93.50
1730	10' wide, 20' deep, 3-1/2 C.Y. bucket w/trench box	27	57.50	84.50
1740	24' deep, 3-1/2 C.Y. bucket	39.50	69.50	109
1780	12' wide, 20' deep, 3-1/2 C.Y. bucket w/trench box	42	73	115
1790	25' deep, bucket	52	93	145
1800	1/2 to 1 slope, 2' wide, 2' deep, 3/8 C.Y. bucket	.76	3.36	4.12
1810	3' deep, 3/8 C.Y. bucket	1.25	5.90	7.15

G1030 Site Earthwork

G1030 805	Trenching Common Earth	COST PER L.F.		
		EQUIP.	LABOR	TOTAL
1820	4' deep, 3/8 C.Y. bucket	1.85	8.95	10.80
1840	6' deep, 3/8 C.Y. bucket	3.03	14.55	17.58
1860	8' deep, 1/2 C.Y. bucket	4.78	23	27.78
1880	10' deep, 1 C.Y. bucket	12.45	32.50	44.95
2300	4' wide, 2' deep, 3/8 C.Y. bucket	1.49	5.55	7.04
2310	3' deep, 3/8 C.Y. bucket	2.20	9.20	11.40
2320	4' deep, 1/2 C.Y. bucket	2.72	11.35	14.07
2340	6' deep, 1/2 C.Y. bucket	5.35	21.50	26.85
2360	8' deep, 1/2 C.Y. bucket	13.10	31.50	44.60
2380	10' deep, 1 C.Y. bucket	18.15	44	62.15
2400	12' deep, 1 C.Y. bucket	24	59.50	83.50
2430	15' deep, 1-1/2 C.Y. bucket	22	64	86
2460	18' deep, 2-1/2 C.Y. bucket	45	99.50	144.50
2840	6' wide, 6' deep, 5/8 C.Y. bucket w/trench box	13.15	26.50	39.65
2860	8' deep, 3/4 C.Y. bucket	19.15	39.50	58.65
2880	10' deep, 1 C.Y. bucket	15.60	40.50	56.10
2900	12' deep, 1-1/2 C.Y. bucket	19.75	53	72.75
2940	16' deep, 2-1/2 C.Y. bucket	38	77.50	115.50
2980	20' deep, 3-1/2 C.Y. bucket	51.50	105	156.50
3020	24' deep, 3-1/2 C.Y. bucket	73	143	216
3100	8' wide, 12' deep, 1-1/2 C.Y. bucket w/trench box	24.50	60.50	85
3120	15' deep, 1-1/2 C.Y. bucket	35.50	87.50	123
3140	18' deep, 2-1/2 C.Y. bucket	52.50	104	156.50
3180	24' deep, 3-1/2 C.Y. bucket	81.50	157	238.50
3270	10' wide, 20' deep, 3-1/2 C.Y. bucket w/trench box	52.50	121	173.50
3280	24' deep, 3-1/2 C.Y. bucket	90	171	261
3370	12' wide, 20' deep, 3-1/2 C.Y. bucket w/trench box	75.50	140	215.50
3380	25' deep, 3-1/2 C.Y. bucket	105	200	305
3500	1 to 1 slope, 2' wide, 2' deep, 3/8 C.Y. bucket	.98	4.48	5.46
3520	3' deep, 3/8 C.Y. bucket	3.09	10.25	13.34
3540	4' deep, 3/8 C.Y. bucket	2.72	13.45	16.17
3560	6' deep, 1/2 C.Y. bucket	3.03	14.55	17.58
3580	8' deep, 1/2 C.Y. bucket	5.95	29	34.95
3600	10' deep, 1 C.Y. bucket	21.50	55.50	77
3800	4' wide, 2' deep, 3/8 C.Y. bucket	1.71	6.70	8.41
3820	3' deep, 3/8 C.Y. bucket	2.69	11.75	14.44
3840	4' deep, 1/2 C.Y. bucket	3.50	15.20	18.70
3860	6' deep, 1/2 C.Y. bucket	7.55	30.50	38.05
3880	8' deep, 1/2 C.Y. bucket	19.50	47	66.50
3900	10' deep, 1 C.Y. bucket	28	68.50	96.50
3920	12' deep, 1 C.Y. bucket	41.50	99.50	141
3940	15' deep, 1-1/2 C.Y. bucket	36	106	142
3960	18' deep, 2-1/2 C.Y. bucket	62	137	199
4030	6' wide, 6' deep, 5/8 C.Y. bucket w/trench box	17.30	35.50	52.80
4040	8' deep, 3/4 C.Y. bucket	25	49	74
4050	10' deep, 1 C.Y. bucket	22.50	59.50	82
4060	12' deep, 1-1/2 C.Y. bucket	30	81	111
4070	16' deep, 2-1/2 C.Y. bucket	59.50	123	182.50
4080	20' deep, 3-1/2 C.Y. bucket	83	171	254
4090	24' deep, 3-1/2 C.Y. bucket	121	240	361
4500	8' wide, 12' deep, 1-1/2 C.Y. bucket w/trench box	34.50	88	122.50
4550	15' deep, 1-1/2 C.Y. bucket	52	132	184
4600	18' deep, 2-1/2 C.Y. bucket	79.50	161	240.50
4650	24' deep, 3-1/2 C.Y. bucket	130	253	383
4800	10' wide, 20' deep, 3-1/2 C.Y. bucket w/trench box	78	184	262
4850	24' deep, 3-1/2 C.Y. bucket	138	268	406
4950	12' wide, 20' deep, 3-1/2 C.Y. bucket w/trench box	109	207	316
4980	25' deep, 3-1/2 C.Y. bucket	157	305	462

G1030 Site Earthwork

Trenching Systems are shown on a cost per linear foot basis. The systems include: excavation; backfill and removal of spoil; and compaction for various depths and trench bottom widths. The backfill has been reduced to accommodate a pipe of suitable diameter and bedding.

The slope for trench sides varies from none to 1:1.

The Expanded System Listing shows Trenching Systems that range from 2' to 12' in width. Depths range from 2' to 25'.

System Components	QUANTITY	UNIT	COST PER L.F.		
			EQUIP.	LABOR	TOTAL
SYSTEM G1030 806 1310					
TRENCHING, LOAM & SANDY CLAY, NO SLOPE, 2' WIDE, 2' DP, 3/8 C.Y. BUCKET					
Excavation, trench, hyd. backhoe, track mtd., 3/8 C.Y. bucket	.148	B.C.Y.	.22	1.12	1.34
Backfill and load spoil, from stockpile	.165	L.C.Y.	.14	.40	.54
Compaction by vibrating plate 18" wide, 6" lifts, 4 passes	.118	E.C.Y.	.03	.45	.48
Remove excess spoil, 8 C.Y. dump truck, 2 mile roundtrip	.042	L.C.Y.	.16	.21	.37
TOTAL			.55	2.18	2.73

G1030 806	Trenching Loam & Sandy Clay	COST PER L.F.		
		EQUIP.	LABOR	TOTAL
1310	Trenching, loam & sandy clay, no slope, 2' wide, 2' deep, 3/8 C.Y. bucket	.54	2.19	2.73
1320	3' deep, 3/8 C.Y. bucket	.81	3.59	4.40
1330	4' deep, 3/8 C.Y. bucket	.97	4.37	5.35
1340	6' deep, 3/8 C.Y. bucket	1.82	5.20	7.05
1350	8' deep, 1/2 C.Y. bucket	2.38	6.90	9.25
1360	10' deep, 1 C.Y. bucket	2.70	7.35	10.05
1400	4' wide, 2' deep, 3/8 C.Y. bucket	1.30	4.37	5.65
1410	3' deep, 3/8 C.Y. bucket	1.73	6.55	8.30
1420	4' deep, 1/2 C.Y. bucket	1.96	7.35	9.30
1430	6' deep, 1/2 C.Y. bucket	4.30	10.80	15.10
1440	8' deep, 1/2 C.Y. bucket	6.55	15.30	22
1450	10' deep, 1 C.Y. bucket	6.40	15.65	22
1460	12' deep, 1 C.Y. bucket	8.05	19.50	27.50
1470	15' deep, 1-1/2 C.Y. bucket	8.05	22.50	30.50
1480	18' deep, 2-1/2 C.Y. bucket	11.20	25	36
1520	6' wide, 6' deep, 5/8 C.Y. bucket w/trench box	8.70	17.60	26.50
1530	8' deep, 3/4 C.Y. bucket	11.30	23	34.50
1540	10' deep, 1 C.Y. bucket	10.70	23.50	34
1550	12' deep, 1-1/2 C.Y. bucket	10.25	26	36.50
1560	16' deep, 2-1/2 C.Y. bucket	16.20	33	49
1570	20' deep, 3-1/2 C.Y. bucket	19	39	58
1580	24' deep, 3-1/2 C.Y. bucket	24	47.50	71.50
1640	8' wide, 12' deep, 1-1/4 C.Y. bucket w/trench box	14.45	33.50	47.50
1650	15' deep, 1-1/2 C.Y. bucket	17.70	42.50	60
1660	18' deep, 2-1/2 C.Y. bucket	24.50	47.50	72
1680	24' deep, 3-1/2 C.Y. bucket	32	61.50	93.50
1730	10' wide, 20' deep, 3-1/2 C.Y. bucket w/trench box	32.50	62	94.50
1740	24' deep, 3-1/2 C.Y. bucket	40.50	76	117
1780	12' wide, 20' deep, 3-1/2 C.Y. bucket w/trench box	39	73.50	112
1790	25' deep, 3-1/2 C.Y. bucket	50.50	94.50	145
1800	1/2:1 slope, 2' wide, 2' deep, 3/8 C.Y. bucket	.76	3.28	4.04
1810	3' deep, 3/8 C.Y. bucket	1.24	5.75	7
1820	4' deep, 3/8 C.Y. bucket	1.83	8.75	10.60
1840	6' deep, 3/8 C.Y. bucket	4.37	13.05	17.40

G10 Site Preparation

G1030 Site Earthwork

G1030 806	Trenching Loam & Sandy Clay	COST PER L.F.		
		EQUIP.	LABOR	TOTAL
1860	8' deep, 1/2 C.Y. bucket	6.95	21	28
1880	10' deep, 1 C.Y. bucket	9.25	26	35.50
2300	4' wide, 2' deep, 3/8 C.Y. bucket	1.51	5.45	7
2310	3' deep, 3/8 C.Y. bucket	2.21	9	11.25
2320	4' deep, 1/2 C.Y. bucket	2.74	11.15	13.90
2340	6' deep, 1/2 C.Y. bucket	7.25	19.20	26.50
2360	8' deep, 1/2 C.Y. bucket	12.75	31	43.50
2380	10' deep, 1 C.Y. bucket	14.10	36	50
2400	12' deep, 1 C.Y. bucket	23.50	59.50	83
2430	15' deep, 1-1/2 C.Y. bucket	23	66	89
2460	18' deep, 2-1/2 C.Y. bucket	44.50	101	145
2840	6' wide, 6' deep, 5/8 C.Y. bucket w/trench box	12.70	26.50	39.50
2860	8' deep, 3/4 C.Y. bucket	18.35	39	57.50
2880	10' deep, 1 C.Y. bucket	19.20	44	63
2900	12' deep, 1-1/2 C.Y. bucket	20	53.50	73.50
2940	16' deep, 2-1/2 C.Y. bucket	37	78.50	116
2980	20' deep, 3-1/2 C.Y. bucket	50	106	156
3020	24' deep, 3-1/2 C.Y. bucket	70.50	145	216
3100	8' wide, 12' deep, 1-1/2 C.Y. bucket w/trench box	24	60.50	85
3120	15' deep, 1-1/2 C.Y. bucket	33	85	118
3140	18' deep, 2-1/2 C.Y. bucket	51	105	156
3180	24' deep, 3-1/2 C.Y. bucket	79	159	238
3270	10' wide, 20' deep, 3-1/2 C.Y. bucket w/trench box	63.50	129	192
3280	24' deep, 3-1/2 C.Y. bucket	87.50	174	261
3320	12' wide, 20' deep, 3-1/2 C.Y. bucket w/trench box	70	140	211
3380	25' deep, 3-1/2 C.Y. bucket w/trench box	95.50	188	284
3500	1:1 slope, 2' wide, 2' deep, 3/8 C.Y. bucket	.97	4.38	5.35
3520	3' deep, 3/8 C.Y. bucket	1.72	8.20	9.95
3540	4' deep, 3/8 C.Y. bucket	2.69	13.15	15.80
3560	6' deep, 1/2 C.Y. bucket	4.37	13.05	17.40
3580	8' deep, 1/2 C.Y. bucket	11.45	34.50	46
3600	10' deep, 1 C.Y. bucket	15.85	45	60.50
3800	4' wide, 2' deep, 3/8 C.Y. bucket	1.73	6.55	8.30
3820	3' deep, 1/2 C.Y. bucket	2.69	11.50	14.20
3840	4' deep, 1/2 C.Y. bucket	3.52	15	18.50
3860	6' deep, 1/2 C.Y. bucket	10.25	27.50	38
3880	8' deep, 1/2 C.Y. bucket	18.95	46.50	65.50
3900	10' deep, 1 C.Y. bucket	22	56	78
3920	12' deep, 1 C.Y. bucket	31.50	79.50	111
3940	15' deep, 1-1/2 C.Y. bucket	37.50	110	147
3960	18' deep, 2-1/2 C.Y. bucket	61	139	200
4030	6' wide, 6' deep, 5/8 C.Y. bucket w/trench box	16.70	35.50	52.50
4040	8' deep, 3/4 C.Y. bucket	25.50	55	80.50
4050	10' deep, 1 C.Y. bucket	27.50	64.50	92
4060	12' deep, 1-1/2 C.Y. bucket	30	81	111
4070	16' deep, 2-1/2 C.Y. bucket	58	124	182
4080	20' deep, 3-1/2 C.Y. bucket	81	173	254
4090	24' deep, 3-1/2 C.Y. bucket	118	243	360
4500	8' wide, 12' deep, 1-1/4 C.Y. bucket w/trench box	34	88	122
4550	15' deep, 1-1/2 C.Y. bucket	48	128	176
4600	18' deep, 2-1/2 C.Y. bucket	77.50	163	241
4650	24' deep, 3-1/2 C.Y. bucket	126	257	385
4800	10' wide, 20' deep, 3-1/2 C.Y. bucket w/trench box	94.50	196	291
4850	24' deep, 3-1/2 C.Y. bucket	134	272	405
4950	12' wide, 20' deep, 3-1/2 C.Y. bucket w/trench box	101	208	310
4980	25' deep, 3-1/2 C.Y. bucket	152	305	460

459

G10 Site Preparation

G1030 Site Earthwork

Trenching Systems are shown on a cost per linear foot basis. The systems include: excavation; backfill and removal of spoil; and compaction for various depths and trench bottom widths. The backfill has been reduced to accommodate a pipe of suitable diameter and bedding.

The slope for trench sides varies from none to 1:1.

The Expanded System Listing shows Trenching Systems that range from 2' to 12' in width. Depths range from 2' to 25'.

System Components			COST PER L.F.		
	QUANTITY	UNIT	EQUIP.	LABOR	TOTAL
SYSTEM G1030 807 1310					
TRENCHING, SAND & GRAVEL, NO SLOPE, 2' WIDE, 2' DEEP, 3/8 C.Y. BUCKET					
Excavation, trench, hyd. backhoe, track mtd., 3/8 C.Y. bucket	.148	B.C.Y.	.21	1.10	1.31
Backfill and load spoil, from stockpile	.140	L.C.Y.	.12	.34	.46
Compaction by vibrating plate 18" wide, 6" lifts, 4 passes	.118	E.C.Y.	.03	.45	.48
Remove excess spoil, 8 C.Y. dump truck, 2 mile roundtrip	.035	L.C.Y.	.13	.18	.31
TOTAL			.49	2.07	2.56

G1030 807	Trenching Sand & Gravel	COST PER L.F.		
		EQUIP.	LABOR	TOTAL
1310	Trenching, sand & gravel, no slope, 2' wide, 2' deep, 3/8 C.Y. bucket	.49	2.07	2.56
1320	3' deep, 3/8 C.Y. bucket	.75	3.45	4.20
1330	4' deep, 3/8 C.Y. bucket	.89	4.15	5.05
1340	6' deep, 3/8 C.Y. bucket	1.72	4.97	6.70
1350	8' deep, 1/2 C.Y. bucket	2.26	6.60	8.85
1360	10' deep, 1 C.Y. bucket	2.52	6.95	9.45
1400	4' wide, 2' deep, 3/8 C.Y. bucket	1.16	4.10	5.25
1410	3' deep, 3/8 C.Y. bucket	1.56	6.20	7.75
1420	4' deep, 1/2 C.Y. bucket	1.78	6.95	8.75
1430	6' deep, 1/2 C.Y. bucket	4.07	10.30	14.40
1440	8' deep, 1/2 C.Y. bucket	6.20	14.55	20.50
1450	10' deep, 1 C.Y. bucket	6	14.75	21
1460	12' deep, 1 C.Y. bucket	7.60	18.45	26
1470	15' deep, 1-1/2 C.Y. bucket	7.50	21.50	29
1480	18' deep, 2-1/2 C.Y. bucket	10.55	23.50	34
1520	6' wide, 6' deep, 5/8 C.Y. bucket w/trench box	8.25	16.65	25
1530	8' deep, 3/4 C.Y. bucket	10.75	22	32.50
1540	10' deep, 1 C.Y. bucket	10.05	22	32
1550	12' deep, 1-1/2 C.Y. bucket	9.60	24.50	34.50
1560	16' deep, 2 C.Y. bucket	15.50	31	46.50
1570	20' deep, 3-1/2 C.Y. bucket	17.90	36.50	54.50
1580	24' deep, 3-1/2 C.Y. bucket	22.50	44.50	67
1640	8' wide, 12' deep, 1-1/2 C.Y. bucket w/trench box	13.45	31.50	45
1650	15' deep, 1-1/2 C.Y. bucket	16.40	40	56
1660	18' deep, 2-1/2 C.Y. bucket	23	44.50	68
1680	24' deep, 3-1/2 C.Y. bucket	30.50	57.50	88
1730	10' wide, 20' deep, 3-1/2 C.Y. bucket w/trench box	30.50	58	88
1740	24' deep, 3-1/2 C.Y. bucket	38	71.50	109
1780	12' wide, 20' deep, 3-1/2 C.Y. bucket w/trench box	36.50	68.50	105
1790	25' deep, 3-1/2 C.Y. bucket	47.50	88.50	136
1800	1/2:1 slope, 2' wide, 2' deep, 3/8 C.Y. bucket	.69	3.11	3.80
1810	3' deep, 3/8 C.Y. bucket	1.14	5.45	6.60
1820	4' deep, 3/8 C.Y. bucket	1.68	8.35	10
1840	6' deep, 3/8 C.Y. bucket	4.16	12.45	16.60

G1030 Site Earthwork

G1030 807	Trenching Sand & Gravel	COST PER L.F.		
		EQUIP.	LABOR	TOTAL
1860	8' deep, 1/2 C.Y. bucket	6.60	19.90	26.50
1880	10' deep, 1 C.Y. bucket	8.65	24.50	33
2300	4' wide, 2' deep, 3/8 C.Y. bucket	1.36	5.15	6.50
2310	3' deep, 3/8 C.Y. bucket	2.01	8.55	10.55
2320	4' deep, 1/2 C.Y. bucket	2.49	10.60	13.05
2340	6' deep, 1/2 C.Y. bucket	6.95	18.35	25.50
2360	8' deep, 1/2 C.Y. bucket	12.05	29.50	41.50
2380	10' deep, 1 C.Y. bucket	13.30	34	47
2400	12' deep, 1 C.Y. bucket	22.50	56.50	78.50
2430	15' deep, 1-1/2 C.Y. bucket	21.50	62.50	84
2460	18' deep, 2-1/2 C.Y. bucket	42	95	137
2840	6' wide, 6' deep, 5/8 C.Y. bucket w/trench box	12.05	25.50	37.50
2860	8' deep, 3/4 C.Y. bucket	17.45	37	54.50
2880	10' deep, 1 C.Y. bucket	18.15	41.50	59.50
2900	12' deep, 1-1/2 C.Y. bucket	18.80	50.50	69.50
2940	16' deep, 2 C.Y. bucket	35.50	74	110
2980	20' deep, 3-1/2 C.Y. bucket	47.50	99.50	147
3020	24' deep, 3-1/2 C.Y. bucket	67	137	204
3100	8' wide, 12' deep, 1-1/4 C.Y. bucket w/trench box	22.50	57.50	80
3120	15' deep, 1-1/2 C.Y. bucket	31	80.50	111
3140	18' deep, 2-1/2 C.Y. bucket	48.50	99	148
3180	24' deep, 3-1/2 C.Y. bucket	75	150	224
3270	10' wide, 20' deep, 3-1/2 C.Y. bucket w/trench box	60	121	181
3280	24' deep, 3-1/2 C.Y. bucket	82.50	163	246
3370	12' wide, 20' deep, 3-1/2 C.Y. bucket w/trench box	67	134	201
3380	25' deep, 3-1/2 C.Y. bucket	96	188	284
3500	1:1 slope, 2' wide, 2' deep, 3/8 C.Y. bucket	1.76	5.25	7
3520	3' deep, 3/8 C.Y. bucket	1.58	7.80	9.40
3540	4' deep, 3/8 C.Y. bucket	2.48	12.50	15
3560	6' deep, 3/8 C.Y. bucket	4.16	12.45	16.65
3580	8' deep, 1/2 C.Y. bucket	10.95	33	44
3600	10' deep, 1 C.Y. bucket	14.80	42	57
3800	4' wide, 2' deep, 3/8 C.Y. bucket	1.56	6.20	7.75
3820	3' deep, 3/8 C.Y. bucket	2.45	10.90	13.35
3840	4' deep, 1/2 C.Y. bucket	3.21	14.20	17.40
3860	6' deep, 1/2 C.Y. bucket	9.80	26.50	36
3880	8' deep, 1/2 C.Y. bucket	17.95	44.50	62
3900	10' deep, 1 C.Y. bucket	20.50	53	73.50
3920	12' deep, 1 C.Y. bucket	30	75	105
3940	15' deep, 1-1/2 C.Y. bucket	35	104	139
3960	18' deep, 2-1/2 C.Y. bucket	57.50	131	188
4030	6' wide, 6' deep, 5/8 C.Y. bucket w/trench box	15.85	34	49.50
4040	8' deep, 3/4 C.Y. bucket	24	52.50	76.50
4050	10' deep, 1 C.Y. bucket	26	61	87
4060	12' deep, 1-1/2 C.Y. bucket	28	76.50	104
4070	16' deep, 2 C.Y. bucket	55.50	117	173
4080	20' deep, 3-1/2 C.Y. bucket	76.50	163	239
4090	24' deep, 3-1/2 C.Y. bucket	111	228	340
4500	8' wide, 12' deep, 1-1/2 C.Y. bucket w/trench box	32	83.50	115
4550	15' deep, 1-1/2 C.Y. bucket	45	121	166
4600	18' deep, 2-1/2 C.Y. bucket	73.50	153	227
4650	24' deep, 3-1/2 C.Y. bucket	119	241	360
4800	10' wide, 20' deep, 3-1/2 C.Y. bucket w/trench box	89.50	184	273
4850	24' deep, 3-1/2 C.Y. bucket	127	255	380
4950	12' wide, 20' deep, 3-1/2 C.Y. bucket w/trench box	95.50	195	290
4980	25' deep, 3-1/2 C.Y. bucket	144	288	430

G4010 Electrical Distribution

GROUND SURFACE

BACKFILL

CONCRETE

CONDUIT

CONDUIT SUPPORTS

BEDDING (NATURAL)

UNDERGROUND ELECTRICAL CONDUIT

System Components	QUANTITY	UNIT	COST EACH		
			MAT.	INST.	TOTAL
SYSTEM G4010 312 1000					
2000 AMP UNDERGROUND SERVICE INCLUDING COMMON EARTH EXCAVATION, CONCRETE, BACKFILL AND COMPACTION					
Excavate Trench	44.440	B.C.Y.		335.08	335.08
4 inch conduit bank	108.000	L.F.	1,123.20	3,024	4,147.20
4 inch fitting	8.000	Ea.	324	504	828
4 inch bells	4.000	Ea.	20.40	252	272.40
Concrete material	16.580	C.Y.	2,271.46		2,271.46
Concrete placement	16.580	C.Y.		464.24	464.24
Backfill trench	33.710	L.C.Y.		109.90	109.90
Compact fill material in trench	25.930	E.C.Y.		142.10	142.10
Dispose of excess fill material on-site	24.070	L.C.Y.		314.36	314.36
500 kcmil power cable	18.000	C.L.F.	24,750	8,550	33,300
Wire, 600 volt stranded copper wire, type THW, 1/0	6.000	C.L.F.	1,890	1,374	3,264
TOTAL			30,379.06	15,069.68	45,448.74

G4010 312	Underground Power Feed	COST EACH		
		MAT.	INST.	TOTAL
0900	Underground electrical			
0950	Including common earth excavation, concrete, backfill and compaction			
1000	2000 Amp service, 100' length, 4' depth	30,400	15,100	45,500
1100	1600 Amp service	23,800	11,600	35,400
1200	1200 Amp service	18,500	9,650	28,150
1300	1000 Amp service	21,700	8,825	30,525
1400	800 Amp service	10,000	8,275	18,275
1500	600 Amp service	8,100	7,700	15,800

For customer support on your Electrical Costs with RSMeans Data, call 800.448.8182.

G4010 Electrical Distribution

System Components	QUANTITY	UNIT	COST L.F.		
			MAT.	INST.	TOTAL
SYSTEM G4010 320 1012					
UNDERGROUND ELECTRICAL CONDUIT, 2" DIAMETER, INCLUDING EXCAVATION,					
CONCRETE, BEDDING, BACKFILL AND COMPACTION					
Excavate Trench	.150	B.C.Y.		1.13	1.13
Base spacer	.200	Ea.	.30	2.70	3
Conduit	1.000	L.F.	.70	4.21	4.91
Concrete material	.050	C.Y.	6.85		6.85
Concrete placement	.050	C.Y.		1.41	1.41
Backfill trench	.120	L.C.Y.		.39	.39
Compact fill material in trench	.100	E.C.Y.		.55	.55
Dispose of excess fill material on-site	.120	L.C.Y.		1.56	1.56
Marking tape	.010	C.L.F.	.43	.04	.47
TOTAL			8.28	11.99	20.27

G4010 320	Underground Electrical Conduit	COST L.F.		
		MAT.	INST.	TOTAL
0950	Underground electrical conduit, including common earth excavation,			
0952	Concrete, bedding, backfill and compaction			
1012	2" dia. Schedule 40 PVC, 2' deep	8.30	12	20.30
1014	4' deep	8.30	17	25.30
1016	6' deep	8.30	22	30.30
1022	2 @ 2" dia. Schedule 40 PVC, 2' deep	9.30	18.90	28.20
1024	4' deep	9.30	24	33.30
1026	6' deep	9.30	28.50	37.80
1032	3 @ 2" dia. Schedule 40 PVC, 2' deep	10.30	26	36.30
1034	4' deep	10.30	31	41.30
1036	6' deep	10.30	35.50	45.80
1042	4 @ 2" dia. Schedule 40 PVC, 2' deep	13.95	32.50	46.45
1044	4' deep	13.95	37.50	51.45
1046	6' deep	13.95	42	55.95

463

G4010 320	Underground Electrical Conduit	COST L.F.		
		MAT.	INST.	TOTAL
1062	6 @ 2" dia. Schedule 40 PVC, 2' deep	15.95	46	61.95
1064	4' deep	15.95	51	66.95
1066	6' deep	15.95	56	71.95
1082	8 @ 2" dia. Schedule 40 PVC, 2' deep	17.95	59.50	77.45
1084	4' deep	17.95	64.50	82.45
1086	6' deep	17.95	69.50	87.45
1092	9 @ 2" dia. Schedule 40 PVC, 2' deep	18.90	66	84.90
1094	4' deep	21.50	71	92.50
1096	6' deep	21.50	76	97.50
1212	3" dia. Schedule 40 PVC, 2' deep	10.40	14.75	25.15
1214	4' deep	10.40	19.80	30.20
1216	6' deep	10.40	24.50	34.90
1222	2 @ 3" dia. Schedule 40 PVC, 2' deep	10.80	24	34.80
1224	4' deep	10.80	29	39.80
1226	6' deep	10.80	34	44.80
1232	3 @ 3" dia. Schedule 40 PVC, 2' deep	12.55	33.50	46.05
1234	4' deep	12.55	38.50	51.05
1236	6' deep	12.55	43.50	56.05
1242	4 @ 3" dia. Schedule 40 PVC, 2' deep	18.40	43	61.40
1244	4' deep	18.40	48.50	66.90
1246	6' deep	18.40	53	71.40
1262	6 @ 3" dia. Schedule 40 PVC, 2' deep	20.50	62	82.50
1264	4' deep	20.50	67	87.50
1266	6' deep	20.50	72	92.50
1282	8 @ 3" dia. Schedule 40 PVC, 2' deep	24	81.50	105.50
1284	4' deep	24	86.50	110.50
1286	6' deep	24	92	116
1292	9 @ 3" dia. Schedule 40 PVC, 2' deep	28.50	92	120.50
1294	4' deep	30	96	126
1296	6' deep	30	101	131
1312	4" dia. Schedule 40 PVC, 2' deep	11	18.05	29.05
1314	4' deep	11	23	34
1316	6' deep	11	28	39
1322	2 @ 4" dia. Schedule 40 PVC, 2' deep	13.30	31	44.30
1324	4' deep	13.30	36.50	49.80
1326	6' deep	13.30	41	54.30
1332	3 @ 4" dia. Schedule 40 PVC, 2' deep	14.25	44	58.25
1334	4' deep	14.25	49	63.25
1336	6' deep	14.25	54	68.25
1342	4 @ 4" dia. Schedule 40 PVC, 2' deep	22	57.50	79.50
1344	4' deep	22	62.50	84.50
1346	6' deep	22	67.50	89.50
1362	6 @ 4" dia. Schedule 40 PVC, 2' deep	25.50	83.50	109
1364	4' deep	25.50	88.50	114
1366	6' deep	25.50	93.50	119
1382	8 @ 4" dia. Schedule 40 PVC, 2' deep	31	110	141
1384	4' deep	31	116	147
1386	6' deep	31	122	153
1392	9 @ 4" dia. Schedule 40 PVC, 2' deep	35	124	159
1394	4' deep	35	128	163
1396	6' deep	35	132	167
1412	5" dia. Schedule 40 PVC, 2' deep	14.35	21.50	35.85
1414	4' deep	14.35	26.50	40.85
1416	6' deep	14.35	31.50	45.85
1422	2 @ 5" dia. Schedule 40 PVC, 2' deep	20	38.50	58.50
1424	4' deep	20	43	63
1426	6' deep	20	48	68
1432	3 @ 5" dia. Schedule 40 PVC, 2' deep	25.50	55	80.50

G4010 Electrical Distribution

G4010 320	Underground Electrical Conduit	COST L.F.		
		MAT.	INST.	TOTAL
1434	4' deep	25.50	60.50	86
1436	6' deep	25.50	65.50	91
1442	4 @ 5" dia. Schedule 40 PVC, 2' deep	35.50	71.50	107
1444	4' deep	35.50	76.50	112
1446	6' deep	35.50	81	116.50
1462	6 @ 5" dia. Schedule 40 PVC, 2' deep	45.50	105	150.50
1464	4' deep	45.50	110	155.50
1466	6' deep	45.50	115	160.50
1482	8 @ 5" dia. Schedule 40 PVC, 2' deep	59.50	139	198.50
1484	4' deep	59.50	146	205.50
1486	6' deep	59.50	153	212.50
1492	9 @ 5" dia. Schedule 40 PVC, 2' deep	66.50	157	223.50
1494	4' deep	66.50	160	226.50
1496	6' deep	66.50	165	231.50
1512	6" dia. Schedule 40 PVC, 2' deep	14.05	26.50	40.55
1514	4' deep	14.05	31	45.05
1516	6' deep	14.05	36	50.05
1522	2 @ 6" dia. Schedule 40 PVC, 2' deep	16.70	47.50	64.20
1524	4' deep	16.70	52.50	69.20
1526	6' deep	16.70	57.50	74.20
1532	3 @ 6" dia. Schedule 40 PVC, 2' deep	22	69.50	91.50
1534	4' deep	22	75.50	97.50
1536	6' deep	22	81.50	103.50
1542	4 @ 6" dia. Schedule 40 PVC, 2' deep	29	89.50	118.50
1544	4' deep	29	94.50	123.50
1546	6' deep	29	99.50	128.50
1562	6 @ 6" dia. Schedule 40 PVC, 2' deep	38.50	133	171.50
1564	4' deep	38.50	139	177.50
1566	6' deep	38.50	145	183.50
1582	8 @ 6" dia. Schedule 40 PVC, 2' deep	50.50	177	227.50
1584	4' deep	50.50	185	235.50
1586	6' deep	50.50	192	242.50
1592	9 @ 6" dia. Schedule 40 PVC, 2' deep	56	199	255
1594	4' deep	56	203	259
1596	6' deep	56	208	264

465

For customer support on your Electrical Costs with RSMeans Data, call 800.448.8182.

G4020 Site Lighting

The Site Lighting System includes the complete unit from foundation to electrical fixtures. Each system includes: excavation; concrete base; backfill by hand; compaction with a plate compacter; pole of specified material; all fixtures; and lamps.

The Expanded System Listing shows Site Lighting Systems that use one of two types of lamps: high pressure sodium; and metal halide. Systems are listed for 400-watt and 1000-watt lamps. Pole height varies from 20′ to 40′. There are four types of poles possibly listed: aluminum, fiberglass, steel and wood.

System Components	QUANTITY	UNIT	COST EACH MAT.	COST EACH INST.	COST EACH TOTAL
SYSTEM G4020 110 5820					
SITE LIGHTING, METAL HALIDE, 400 WATT, ALUMINUM POLE, 20′ HIGH					
Excavating, by hand, pits to 6′ deep, heavy soil	2.368	C.Y.		314.94	314.94
Concrete in place incl. forms and reinf. stl. spread footings under 1 CY	.465	C.Y.	106.49	144.65	251.14
Backfill	1.903	C.Y.		91.34	91.34
Compact, vibrating plate	1.903	C.Y.		15.07	15.07
Aluminum light pole, 20′, no concrete base	1.000	Ea.	665	722.50	1,387.50
Road area luminaire wire	.300	C.L.F.	8.55	28.35	36.90
Roadway area luminaire, metal halide 400 W	1.000	Ea.	705	345	1,050
TOTAL			1,485.04	1,661.85	3,146.89

G4020 110	Site Lighting	COST EACH MAT.	COST EACH INST.	COST EACH TOTAL
2320	Site lighting, high pressure sodium, 400 watt, aluminum pole, 20′ high	1,625	1,650	3,275
2340	30′ high	2,725	2,025	4,750
2360	40′ high	3,225	2,625	5,850
2520	Fiberglass pole, 20′ high	2,025	1,450	3,475
2540	30′ high	2,325	1,825	4,150
2560	40′ high	3,500	2,325	5,825
2720	Steel pole, 20′ high	2,200	1,750	3,950
2740	30′ high	2,400	2,150	4,550
2760	40′ high	2,675	2,800	5,475
2920	Wood pole, 20′ high	1,225	1,625	2,850
2940	30′ high	1,350	2,025	3,375
2960	40′ high	1,675	2,500	4,175
3120	1000 watt, aluminum pole, 20′ high	1,750	1,700	3,450
3140	30′ high	2,825	2,075	4,900
3160	40′ high	3,325	2,650	5,975
3320	Fiberglass pole, 20′ high	2,150	1,500	3,650
3340	30′ high	2,425	1,850	4,275
3360	40′ high	3,600	2,350	5,950
3420	Steel pole, 20′ high	2,325	1,775	4,100
3440	30′ high	2,500	2,175	4,675

G4020 Site Lighting

G4020 110	Site Lighting	COST EACH		
		MAT.	INST.	TOTAL
3460	40' high	2,775	2,825	5,600
3520	Wood pole, 20' high	1,350	1,650	3,000
3540	30' high	1,450	2,075	3,525
3560	40' high	1,800	2,525	4,325
5820	Metal halide, 400 watt, aluminum pole, 20' high	1,475	1,650	3,125
5840	30' high	2,575	2,025	4,600
5860	40' high	3,075	2,625	5,700
6120	Fiberglass pole, 20' high	1,900	1,450	3,350
6140	30' high	2,175	1,825	4,000
6160	40' high	3,350	2,325	5,675
6320	Steel pole, 20' high	2,075	1,750	3,825
6340	30' high	2,250	2,150	4,400
6360	40' high	2,525	2,800	5,325
7620	1000 watt, aluminum pole, 20' high	1,625	1,700	3,325
7640	30' high	2,725	2,075	4,800
7660	40' high	3,225	2,650	5,875
7900	Fiberglass pole, 20' high	2,050	1,500	3,550
7920	30' high	2,325	1,850	4,175
7940	40' high	3,500	2,350	5,850
8120	Steel pole, 20' high	2,225	1,775	4,000
8140	30' high	2,400	2,175	4,575
8160	40' high	2,675	2,825	5,500

467

Table G4020 210 Procedure for Calculating Floodlights Required for Various Footcandles
Poles should not be spaced more than 4 times the fixture mounting height for good light distribution.

To maintain 1 footcandle over a large area use these watts per square foot:

Incandescent	0.15
Metal Halide	0.032
Mercury Vapor	0.05
High Pressure Sodium	0.024

Estimating Chart
Select Lamp type.

Determine total square feet.

Chart will show quantity of fixtures to provide 1 footcandle initial, at intersection of lines. Multiply fixture quantity by desired footcandle level.

Chart based on use of wide beam luminaires in an area whose dimensions are large compared to mounting height and is approximate only.

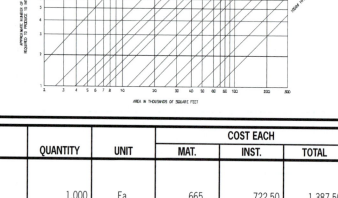

System Components			COST EACH		
	QUANTITY	UNIT	MAT.	INST.	TOTAL
SYSTEM G4020 210 0200					
LIGHT POLES, ALUMINUM, 20' HIGH, 1 ARM BRACKET					
Aluminum light pole, 20', no concrete base	1.000	Ea.	665	722.50	1,387.50
Bracket arm for aluminum light pole	1.000	Ea.	151	94.50	245.50
Excavation by hand, pits to 6' deep, heavy soil or clay	2.368	C.Y.		314.94	314.94
Footing, concrete incl forms, reinforcing, spread, under 1 C.Y.	.465	C.Y.	106.49	144.65	251.14
Backfill by hand	1.903	C.Y.		91.34	91.34
Compaction vibrating plate	1.903	C.Y.		15.07	15.07
TOTAL			922.49	1,383	2,305.49

G4020 210	Light Pole (Installed)	COST EACH		
		MAT.	INST.	TOTAL
0200	Light pole, aluminum, 20' high, 1 arm bracket	920	1,375	2,295
0240	2 arm brackets	1,075	1,375	2,450
0280	3 arm brackets	1,225	1,425	2,650
0320	4 arm brackets	1,375	1,450	2,825
0360	30' high, 1 arm bracket	2,000	1,750	3,750
0400	2 arm brackets	2,150	1,750	3,900
0440	3 arm brackets	2,300	1,800	4,100
0480	4 arm brackets	2,475	1,800	4,275
0680	40' high, 1 arm bracket	2,500	2,325	4,825
0720	2 arm brackets	2,650	2,325	4,975
0760	3 arm brackets	2,800	2,375	5,175
0800	4 arm brackets	2,975	2,400	5,375
0840	Steel, 20' high, 1 arm bracket	1,600	1,475	3,075
0880	2 arm brackets	1,675	1,475	3,150
0920	3 arm brackets	1,625	1,525	3,150
0960	4 arm brackets	1,725	1,525	3,250
1000	30' high, 1 arm bracket	1,775	1,850	3,625
1040	2 arm brackets	1,850	1,850	3,700
1080	3 arm brackets	1,800	1,900	3,700
1120	4 arm brackets	1,900	1,900	3,800
1320	40' high, 1 arm bracket	2,050	2,500	4,550
1360	2 arm brackets	2,125	2,500	4,625
1400	3 arm brackets	2,075	2,550	4,625
1440	4 arm brackets	2,175	2,550	4,725

Reference Section

All the reference information is in one section, making it easy to find what you need to know and easy to use the data set on a daily basis. This section is visually identified by a vertical black bar on the page edges.

In this Reference Section, we've included Equipment Rental Costs, a listing of rental and operating costs; Crew Listings, a full listing of all crews, equipment, and their costs; Historical Cost Indexes for cost comparisons over time; City Cost Indexes and Location Factors for adjusting costs to the region you are in; Reference Tables, where you will find explanations, estimating information and procedures, and technical data; and an explanation of all the Abbreviations in the data set.

Table of Contents

Reference Tables (cont.)

Estimating Tips

- This section contains the average costs to rent and operate hundreds of pieces of construction equipment. This is useful information when one is estimating the time and material requirements of any particular operation in order to establish a unit or total cost. Bare equipment costs shown on a unit cost line include, not only rental, but also operating costs for equipment under normal use.

Rental Costs

- Equipment rental rates are obtained from the following industry sources throughout North America: contractors, suppliers, dealers, manufacturers, and distributors.

- Rental rates vary throughout the country, with larger cities generally having lower rates. Lease plans for new equipment are available for periods in excess of six months, with a percentage of payments applying toward purchase.

- Monthly rental rates vary from 2% to 5% of the purchase price of the equipment depending on the anticipated life of the equipment and its wearing parts.

- Weekly rental rates are about 1/3 of the monthly rates, and daily rental rates are about 1/3 of the weekly rate.

- Rental rates can also be treated as reimbursement costs for contractor-owned equipment. Owned equipment costs include depreciation, loan payments, interest, taxes, insurance, storage, and major repairs.

Operating Costs

- The operating costs include parts and labor for routine servicing, such as the repair and replacement of pumps, filters, and worn lines. Normal operating expendables, such as fuel, lubricants, tires, and electricity (where applicable), are also included.

- Extraordinary operating expendables with highly variable wear patterns, such as diamond bits and blades, are excluded. These costs can be found as material costs in the Unit Price section.

- The hourly operating costs listed do not include the operator's wages.

Equipment Cost/Day

- Any power equipment required by a crew is shown in the Crew Listings with a daily cost.

- This daily cost of equipment needed by a crew includes both the rental cost and the operating cost and is based on dividing the weekly rental rate by 5 (the number of working days in the week), then adding the hourly operating cost multiplied by 8 (the number of hours in a day). This "Equipment Cost/Day" is shown in the far right column of the Equipment Rental section.

- If equipment is needed for only one or two days, it is best to develop your own cost by including components for daily rent and hourly operating costs. This is important when the listed Crew for a task does not contain the equipment needed, such as a crane for lifting mechanical heating/cooling equipment up onto a roof.

- If the quantity of work is less than the crew's Daily Output shown for a Unit Price line item that includes a bare unit equipment cost, the recommendation is to estimate one day's rental cost and operating cost for equipment shown in the Crew Listing for that line item.

- Please note, in some cases the equipment description in the crew is followed by a time period in parenthesis. For example: (daily) or (monthly). In these cases the equipment cost/day is calculated by adding the rental cost per time period to the hourly operating cost multiplied by 8.

Mobilization, Demobilization Costs

- The cost to move construction equipment from an equipment yard or rental company to the job site and back again is not included in equipment rental costs listed in the Reference Section. It is also not included in the bare equipment cost of any Unit Price line item or in any equipment costs shown in the Crew Listings.

- Mobilization (to the site) and demobilization (from the site) costs can be found in the Unit Price section.

- If a piece of equipment is already at the job site, it is not appropriate to utilize mobilization or demobilization costs again in an estimate. ■

Same Data. Simplified.

Enjoy the convenience and efficiency of accessing your costs anywhere:

- **Skip the multiplier** by setting your location
- **Quickly search**, edit, favorite and share costs
- **Stay on top of price changes** with automatic updates

Discover more at rsmeans.com/online

01 54 33 | Equipment Rental

			UNIT	HOURLY OPER. COST	RENT PER DAY	RENT PER WEEK	RENT PER MONTH	EQUIPMENT COST/DAY
10	0010	**CONCRETE EQUIPMENT RENTAL** without operators	R015433 -10							10
	0200	Bucket, concrete lightweight, 1/2 C.Y.		Ea.	.76	38.50	116.16	350	29.30
	0300	1 C.Y.			.99	63.50	190	570	45.90
	0400	1-1/2 C.Y.			1.29	60.50	181.82	545	46.70
	0500	2 C.Y.			1.40	74	222.23	665	55.65
	0580	8 C.Y.			7.01	94.50	282.83	850	112.65
	0600	Cart, concrete, self-propelled, operator walking, 10 C.F.			2.88	157	469.50	1,400	116.95
	0700	Operator riding, 18 C.F.			4.85	167	500.50	1,500	138.95
	0800	Conveyer for concrete, portable, gas, 16" wide, 26' long			10.71	162	484.86	1,450	182.70
	0900	46' long			11.09	177	530.31	1,600	194.80
	1000	56' long			11.26	194	580.82	1,750	206.25
	1100	Core drill, electric, 2-1/2 H.P., 1" to 8" bit diameter			1.58	83.50	250	750	62.65
	1150	11 H.P., 8" to 18" cores			5.45	130	390	1,175	121.60
	1200	Finisher, concrete floor, gas, riding trowel, 96" wide			9.75	155	465.11	1,400	171.05
	1300	Gas, walk-behind, 3 blade, 36" trowel			2.06	101	302.50	910	76.95
	1400	4 blade, 48" trowel			3.10	116	347.50	1,050	94.30
	1500	Float, hand-operated (Bull float), 48" wide			.08	13	39	117	8.45
	1570	Curb builder, 14 H.P., gas, single screw			14.14	256	767.69	2,300	266.65
	1590	Double screw			12.50	256	767.69	2,300	253.55
	1600	Floor grinder, concrete and terrazzo, electric, 22" path			3.07	119	357.50	1,075	96.05
	1700	Edger, concrete, electric, 7" path			1.19	57.50	172.50	520	44.05
	1750	Vacuum pick-up system for floor grinders, wet/dry			1.63	103	309.38	930	74.95
	1800	Mixer, powered, mortar and concrete, gas, 6 C.F., 18 H.P.			7.48	89	267.50	805	113.35
	1900	10 C.F., 25 H.P.			9.08	124	372.50	1,125	147.15
	2000	16 C.F.			9.44	146	436.88	1,300	162.90
	2100	Concrete, stationary, tilt drum, 2 C.Y.			7.98	81	242.43	725	112.35
	2120	Pump, concrete, truck mounted, 4" line, 80' boom			31.24	290	868.70	2,600	423.65
	2140	5" line, 110' boom			39.99	290	868.70	2,600	493.65
	2160	Mud jack, 50 C.F. per hr.			6.50	231	691.93	2,075	190.35
	2180	225 C.F. per hr.			8.61	296	888.91	2,675	246.65
	2190	Shotcrete pump rig, 12 C.Y./hr.			16.73	226	676.78	2,025	269.20
	2200	35 C.Y./hr.			15.91	290	868.70	2,600	301.05
	2600	Saw, concrete, manual, gas, 18 H.P.			5.59	115	345	1,025	113.70
	2650	Self-propelled, gas, 30 H.P.			7.97	82	245.62	735	112.85
	2675	V-groove crack chaser, manual, gas, 6 H.P.			1.65	100	300	900	73.25
	2700	Vibrators, concrete, electric, 60 cycle, 2 H.P.			.47	70	210	630	45.80
	2800	3 H.P.			.57	81	242.50	730	53.05
	2900	Gas engine, 5 H.P.			1.56	17.05	51.21	154	22.70
	3000	8 H.P.			2.10	17.25	51.74	155	27.15
	3050	Vibrating screed, gas engine, 8 H.P.			2.83	108	325	975	87.65
	3120	Concrete transit mixer, 6 x 4, 250 H.P., 8 C.Y., rear discharge			51.08	70.50	212.13	635	451.10
	3200	Front discharge			59.30	136	409.10	1,225	556.20
	3300	6 x 6, 285 H.P., 12 C.Y., rear discharge			58.56	167	500	1,500	568.50
	3400	Front discharge			61.02	167	500	1,500	588.15
20	0010	**EARTHWORK EQUIPMENT RENTAL** without operators	R015433 -10							20
	0040	Aggregate spreader, push type, 8' to 12' wide		Ea.	2.62	65	195	585	59.95
	0045	Tailgate type, 8' wide			2.56	64	191.92	575	58.90
	0055	Earth auger, truck mounted, for fence & sign posts, utility poles			13.95	152	454.55	1,375	202.55
	0060	For borings and monitoring wells			42.95	84	252.53	760	394.15
	0070	Portable, trailer mounted			2.32	92.50	277.50	835	74.05
	0075	Truck mounted, for caissons, water wells			86.00	134	402	1,200	768.40
	0080	Horizontal boring machine, 12" to 36" diameter, 45 H.P.			22.93	110	330	990	249.45
	0090	12" to 48" diameter, 65 H.P.			31.47	130	390	1,175	329.75
	0095	Auger, for fence posts, gas engine, hand held			.45	84.50	254.05	760	54.40
	0100	Excavator, diesel hydraulic, crawler mounted, 1/2 C.Y. cap.			21.92	470	1,408.39	4,225	457
	0120	5/8 C.Y. capacity			29.30	615	1,851.77	5,550	604.75
	0140	3/4 C.Y. capacity			32.95	730	2,190.83	6,575	701.80
	0150	1 C.Y. capacity			41.58	835	2,500	7,500	832.65

01 54 33 | Equipment Rental

Line	Description	UNIT	HOURLY OPER. COST	RENT PER DAY	RENT PER WEEK	RENT PER MONTH	EQUIPMENT COST/DAY
0200	1-1/2 C.Y. capacity	Ea.	49.02	505	1,518	4,550	695.80
0300	2 C.Y. capacity		57.09	810	2,430	7,300	942.70
0320	2-1/2 C.Y. capacity		83.38	1,500	4,500	13,500	1,567
0325	3-1/2 C.Y. capacity		121.20	2,025	6,072	18,200	2,184
0330	4-1/2 C.Y. capacity		152.99	3,700	11,132	33,400	3,450
0335	6 C.Y. capacity		194.11	3,250	9,765.80	29,300	3,506
0340	7 C.Y. capacity		176.77	3,425	10,303.22	30,900	3,475
0342	Excavator attachments, bucket thumbs		3.43	261	783.90	2,350	184.20
0345	Grapples		3.17	225	674.15	2,025	160.20
0346	Hydraulic hammer for boom mounting, 4000 ft lb.		13.60	900	2,702.04	8,100	649.20
0347	5000 ft lb.		16.10	960	2,884.20	8,650	705.60
0348	8000 ft lb.		23.75	1,225	3,643.20	10,900	918.65
0349	12,000 ft lb.		25.95	1,125	3,373	10,100	882.20
0350	Gradall type, truck mounted, 3 ton @ 15' radius, 5/8 C.Y.		43.83	835	2,500	7,500	850.65
0370	1 C.Y. capacity		59.93	840	2,525.30	7,575	984.55
0400	Backhoe-loader, 40 to 45 H.P., 5/8 C.Y. capacity		12.01	286	857.50	2,575	267.55
0450	45 H.P. to 60 H.P., 3/4 C.Y. capacity		18.19	118	353.54	1,050	216.20
0460	80 H.P., 1-1/4 C.Y. capacity		20.54	118	353.54	1,050	235.05
0470	112 H.P., 1-1/2 C.Y. capacity		33.28	620	1,855.22	5,575	637.30
0482	Backhoe-loader attachment, compactor, 20,000 lb.		6.50	157	470.34	1,400	146.05
0485	Hydraulic hammer, 750 ft lb.		3.71	108	324.01	970	94.50
0486	Hydraulic hammer, 1200 ft lb.		6.61	207	621.89	1,875	177.25
0500	Brush chipper, gas engine, 6" cutter head, 35 H.P.		9.25	212	635	1,900	201
0550	Diesel engine, 12" cutter head, 130 H.P.		23.88	292	875	2,625	366.05
0600	15" cutter head, 165 H.P.		26.82	515	1,545	4,625	523.60
0750	Bucket, clamshell, general purpose, 3/8 C.Y.		1.41	92.50	277.78	835	66.85
0800	1/2 C.Y.		1.53	92.50	277.78	835	67.80
0850	3/4 C.Y.		1.65	92.50	277.78	835	68.80
0900	1 C.Y.		1.71	92.50	277.78	835	69.25
0950	1-1/2 C.Y.		2.81	92.50	277.78	835	78.05
1000	2 C.Y.		2.94	95	285	855	80.55
1010	Bucket, dragline, medium duty, 1/2 C.Y.		.83	92.50	277.78	835	62.15
1020	3/4 C.Y.		.79	92.50	277.78	835	61.85
1030	1 C.Y.		.80	92.50	277.78	835	62
1040	1-1/2 C.Y.		1.27	92.50	277.78	835	65.70
1050	2 C.Y.		1.30	92.50	277.78	835	65.95
1070	3 C.Y.		2.10	92.50	277.78	835	72.30
1200	Compactor, manually guided 2-drum vibratory smooth roller, 7.5 H.P.		7.28	181	542.50	1,625	166.75
1250	Rammer/tamper, gas, 8"		2.23	48.50	146.06	440	47
1260	15"		2.65	55.50	167.23	500	54.65
1300	Vibratory plate, gas, 18" plate, 3000 lb. blow		2.15	24.50	73.55	221	31.90
1350	21" plate, 5000 lb. blow		2.64	241	722.50	2,175	165.60
1370	Curb builder/extruder, 14 H.P., gas, single screw		12.50	256	767.69	2,300	253.55
1390	Double screw		12.50	256	767.69	2,300	253.55
1500	Disc harrow attachment, for tractor		.48	83.50	249.80	750	53.80
1810	Feller buncher, shearing & accumulating trees, 100 H.P.		43.75	465	1,393.97	4,175	628.80
1860	Grader, self-propelled, 25,000 lb.		33.65	1,125	3,366.73	10,100	942.50
1910	30,000 lb.		33.15	1,350	4,040.48	12,100	1,073
1920	40,000 lb.		52.35	1,325	4,000	12,000	1,219
1930	55,000 lb.		67.53	1,675	5,000	15,000	1,540
1950	Hammer, pavement breaker, self-propelled, diesel, 1000 to 1250 lb.		28.65	600	1,800	5,400	589.25
2000	1300 to 1500 lb.		43.19	1,025	3,051.51	9,150	955.80
2050	Pile driving hammer, steam or air, 4150 ft lb. @ 225 bpm		12.26	505	1,518	4,550	401.70
2100	8750 ft lb. @ 145 bpm		14.47	710	2,125.20	6,375	540.85
2150	15,000 ft lb. @ 60 bpm		14.81	845	2,530	7,600	624.45
2200	24,450 ft lb. @ 111 bpm		15.83	980	2,934.80	8,800	713.60
2250	Leads, 60' high for pile driving hammers up to 20,000 ft lb.		3.70	305	910.80	2,725	211.80
2300	90' high for hammers over 20,000 ft lb.		5.50	545	1,639.44	4,925	371.85

01 54 33 | Equipment Rental

		UNIT	HOURLY OPER. COST	RENT PER DAY	RENT PER WEEK	RENT PER MONTH	EQUIPMENT COST/DAY
2350	Diesel type hammer, 22,400 ft lb.	Ea.	17.98	495	1,489.40	4,475	441.70
2400	41,300 ft lb.		25.91	625	1,881.35	5,650	583.55
2450	141,000 ft lb.		41.70	995	2,978.80	8,925	929.35
2500	Vib. elec. hammer/extractor, 200 kW diesel generator, 34 H.P.		41.74	725	2,168.78	6,500	767.70
2550	80 H.P.		73.68	1,050	3,135.58	9,400	1,217
2600	150 H.P.		136.35	2,000	6,035.99	18,100	2,298
2700	Hydro excavator w/ext boom 12 C.Y., 1200 gallons		38.15	1,625	4,857.60	14,600	1,277
2800	Log chipper, up to 22" diameter, 600 H.P.		45.00	325	975	2,925	555
2850	Logger, for skidding & stacking logs, 150 H.P.		43.84	940	2,813.18	8,450	913.35
2860	Mulcher, diesel powered, trailer mounted		20.06	310	924.26	2,775	345.35
2900	Rake, spring tooth, with tractor		14.85	375	1,123.58	3,375	343.50
3000	Roller, vibratory, tandem, smooth drum, 20 H.P.		7.88	325	978.60	2,925	258.75
3050	35 H.P.		10.22	275	825	2,475	246.80
3100	Towed type vibratory compactor, smooth drum, 50 H.P.		25.50	525	1,581.85	4,750	520.35
3150	Sheepsfoot, 50 H.P.		25.87	365	1,100	3,300	426.95
3170	Landfill compactor, 220 H.P.		70.51	1,675	5,035.45	15,100	1,571
3200	Pneumatic tire roller, 80 H.P.		13.04	410	1,228.10	3,675	349.90
3250	120 H.P.		19.56	565	1,700	5,100	496.50
3300	Sheepsfoot vibratory roller, 240 H.P.		62.77	1,425	4,303.41	12,900	1,363
3320	340 H.P.		84.58	2,200	6,565.78	19,700	1,990
3350	Smooth drum vibratory roller, 75 H.P.		23.55	660	1,982.18	5,950	584.80
3400	125 H.P.		27.86	750	2,247.17	6,750	672.35
3410	Rotary mower, brush, 60", with tractor		18.96	365	1,097.45	3,300	371.15
3420	Rototiller, walk-behind, gas, 5 H.P.		2.15	60	180	540	53.25
3422	8 H.P.		2.84	116	347.50	1,050	92.20
3440	Scrapers, towed type, 7 C.Y. capacity		6.50	129	386.72	1,150	129.30
3450	10 C.Y. capacity		7.27	172	517.37	1,550	161.65
3500	15 C.Y. capacity		7.46	199	595.76	1,775	178.85
3525	Self-propelled, single engine, 14 C.Y. capacity		134.48	2,250	6,739.92	20,200	2,424
3550	Dual engine, 21 C.Y. capacity		142.64	2,525	7,575.90	22,700	2,656
3600	31 C.Y. capacity		189.53	3,650	10,954.15	32,900	3,707
3640	44 C.Y. capacity		234.76	4,700	14,083.90	42,300	4,695
3650	Elevating type, single engine, 11 C.Y. capacity		62.42	935	2,800	8,400	1,059
3700	22 C.Y. capacity		115.65	1,625	4,850	14,600	1,895
3710	Screening plant, 110 H.P. w/5' x 10' screen		21.32	650	1,952.56	5,850	561.10
3720	5' x 16' screen		26.92	1,225	3,666.67	11,000	948.65
3850	Shovel, crawler-mounted, front-loading, 7 C.Y. capacity		220.61	3,975	11,915.21	35,700	4,148
3855	12 C.Y. capacity		339.92	5,500	16,514.06	49,500	6,022
3860	Shovel/backhoe bucket, 1/2 C.Y.		2.72	74	221.58	665	66.05
3870	3/4 C.Y.		2.69	83	248.76	745	71.25
3880	1 C.Y.		2.78	92	275.42	825	77.35
3890	1-1/2 C.Y.		2.98	108	324.01	970	88.65
3910	3 C.Y.		3.47	146	438.98	1,325	115.55
3950	Stump chipper, 18" deep, 30 H.P.		6.95	223	668	2,000	189.20
4110	Dozer, crawler, torque converter, diesel 80 H.P.		25.48	335	1,010.12	3,025	405.85
4150	105 H.P.		34.64	605	1,818.22	5,450	640.80
4200	140 H.P.		41.64	660	1,980	5,950	729.15
4260	200 H.P.		63.72	1,675	5,050.60	15,200	1,520
4310	300 H.P.		81.45	1,900	5,667.20	17,000	1,785
4360	410 H.P.		107.70	3,250	9,727.46	29,200	2,807
4370	500 H.P.		134.58	3,925	11,788.10	35,400	3,434
4380	700 H.P.		232.23	5,525	16,587.71	49,800	5,175
4400	Loader, crawler, torque conv., diesel, 1-1/2 C.Y., 80 H.P.		29.80	555	1,667.71	5,000	572
4450	1-1/2 to 1-3/4 C.Y., 95 H.P.		30.54	705	2,120	6,350	668.35
4510	1-3/4 to 2-1/4 C.Y., 130 H.P.		48.19	900	2,700	8,100	925.50
4530	2-1/2 to 3-1/4 C.Y., 190 H.P.		58.30	1,125	3,400	10,200	1,146
4560	3-1/2 to 5 C.Y., 275 H.P.		72.05	1,475	4,400	13,200	1,456
4610	Front end loader, 4WD, articulated frame, diesel, 1 to 1-1/4 C.Y., 70 H.P.		16.77	286	857.06	2,575	305.60

20

01 54 33 | Equipment Rental

		UNIT	HOURLY OPER. COST	RENT PER DAY	RENT PER WEEK	RENT PER MONTH	EQUIPMENT COST/DAY
20							
4620	1-1/2 to 1-3/4 C.Y., 95 H.P.	Ea.	20.17	465	1,400	4,200	441.40
4650	1-3/4 to 2 C.Y., 130 H.P.		21.26	380	1,140	3,425	398.05
4710	2-1/2 to 3-1/2 C.Y., 145 H.P.		29.79	665	2,000	6,000	638.30
4730	3 to 4-1/2 C.Y., 185 H.P.		32.38	835	2,500	7,500	759
4760	5-1/4 to 5-3/4 C.Y., 270 H.P.		53.67	900	2,692.98	8,075	967.95
4810	7 to 9 C.Y., 475 H.P.		91.99	2,575	7,744.59	23,200	2,285
4870	9 to 11 C.Y., 620 H.P.		133.10	2,725	8,204.77	24,600	2,706
4880	Skid-steer loader, wheeled, 10 C.F., 30 H.P. gas		9.66	170	511.19	1,525	179.50
4890	1 C.Y., 78 H.P., diesel		18.60	495	1,487.50	4,475	446.30
4892	Skid-steer attachment, auger		.75	149	447.50	1,350	95.50
4893	Backhoe		.75	123	370.35	1,100	80.05
4894	Broom		.71	136	407.50	1,225	87.20
4895	Forks		.16	32	96	288	20.45
4896	Grapple		.73	89.50	267.93	805	59.45
4897	Concrete hammer		1.06	183	550	1,650	118.50
4898	Tree spade		.61	105	313.56	940	67.55
4899	Trencher		.66	102	305	915	66.25
4900	Trencher, chain, boom type, gas, operator walking, 12 H.P.		4.21	208	624.51	1,875	158.60
4910	Operator riding, 40 H.P.		16.83	525	1,580	4,750	450.70
5000	Wheel type, diesel, 4' deep, 12" wide		69.33	975	2,926.54	8,775	1,140
5100	6' deep, 20" wide		88.37	885	2,655	7,975	1,238
5150	Chain type, diesel, 5' deep, 8" wide		16.45	365	1,097.45	3,300	351.05
5200	Diesel, 8' deep, 16" wide		90.47	1,950	5,853.08	17,600	1,894
5202	Rock trencher, wheel type, 6" wide x 18" deep		47.46	91	272.73	820	434.20
5206	Chain type, 18" wide x 7' deep		105.44	286	858.60	2,575	1,015
5210	Tree spade, self-propelled		13.79	232	696.98	2,100	249.70
5250	Truck, dump, 2-axle, 12 ton, 8 C.Y. payload, 220 H.P.		24.16	355	1,071.50	3,225	407.60
5300	Three axle dump, 16 ton, 12 C.Y. payload, 400 H.P.		45.03	365	1,095.41	3,275	579.35
5310	Four axle dump, 25 ton, 18 C.Y. payload, 450 H.P.		50.45	585	1,750	5,250	753.60
5350	Dump trailer only, rear dump, 16-1/2 C.Y.		5.80	153	459.89	1,375	138.35
5400	20 C.Y.		6.26	172	517.37	1,550	153.55
5450	Flatbed, single axle, 1-1/2 ton rating		19.22	74.50	223.67	670	198.50
5500	3 ton rating		25.95	1,075	3,212.18	9,625	850.05
5550	Off highway rear dump, 25 ton capacity		63.42	1,525	4,600	13,800	1,427
5600	35 ton capacity		67.70	675	2,020.24	6,050	945.65
5610	50 ton capacity		84.88	2,175	6,500	19,500	1,979
5620	65 ton capacity		90.64	2,025	6,062.12	18,200	1,938
5630	100 ton capacity		122.70	2,975	8,936.41	26,800	2,769
6000	Vibratory plow, 25 H.P., walking		6.84	305	909.11	2,725	236.55
40							
0010	**GENERAL EQUIPMENT RENTAL** without operators [R015433 -10]	Ea.					
0020	Aerial lift, scissor type, to 20' high, 1200 lb. capacity, electric		3.52	154	462.50	1,400	120.70
0030	To 30' high, 1200 lb. capacity		3.82	235	705	2,125	171.55
0040	Over 30' high, 1500 lb. capacity		5.20	288	865	2,600	214.55
0070	Articulating boom, to 45' high, 500 lb. capacity, diesel [R015433 -15]		10.04	250	750	2,250	230.30
0075	To 60' high, 500 lb. capacity		13.82	305	909.11	2,725	292.40
0080	To 80' high, 500 lb. capacity		16.24	995	2,987.50	8,975	727.45
0085	To 125' high, 500 lb. capacity		18.56	1,675	4,997.50	15,000	1,148
0100	Telescoping boom to 40' high, 500 lb. capacity, diesel		11.37	320	954.56	2,875	281.90
0105	To 45' high, 500 lb. capacity		12.66	325	974.77	2,925	296.20
0110	To 60' high, 500 lb. capacity		16.56	267	800	2,400	292.45
0115	To 80' high, 500 lb. capacity		21.52	360	1,077.80	3,225	387.75
0120	To 100' high, 500 lb. capacity		29.06	985	2,950	8,850	822.45
0125	To 120' high, 500 lb. capacity		29.51	1,525	4,545	13,600	1,145
0195	Air compressor, portable, 6.5 CFM, electric		.91	53.50	160	480	39.30
0196	Gasoline		.66	56	167.50	505	38.80
0200	Towed type, gas engine, 60 CFM		9.55	129	387.50	1,175	153.85
0300	160 CFM		10.60	213	637.50	1,925	212.30

01 54 33 | Equipment Rental

		UNIT	HOURLY OPER. COST	RENT PER DAY	RENT PER WEEK	RENT PER MONTH	EQUIPMENT COST/DAY
0400	Diesel engine, rotary screw, 250 CFM	Ea.	12.23	175	525	1,575	202.85
0500	365 CFM		16.19	355	1,070	3,200	343.55
0550	450 CFM		20.19	300	900	2,700	341.50
0600	600 CFM		34.51	251	752.54	2,250	426.55
0700	750 CFM		35.04	525	1,580	4,750	596.30
0930	Air tools, breaker, pavement, 60 lb.		.57	81.50	245	735	53.60
0940	80 lb.		.57	81	242.50	730	53.05
0950	Drills, hand (jackhammer), 65 lb.		.68	64	192.50	580	43.90
0960	Track or wagon, swing boom, 4" drifter		62.50	1,050	3,135.41	9,400	1,127
0970	5" drifter		62.50	1,050	3,135.41	9,400	1,127
0975	Track mounted quarry drill, 6" diameter drill		102.94	1,900	5,722.33	17,200	1,968
0980	Dust control per drill		1.05	25.50	76.26	229	23.65
0990	Hammer, chipping, 12 lb.		.61	46.50	140	420	32.85
1000	Hose, air with couplings, 50' long, 3/4" diameter		.07	11	33	99	7.15
1100	1" diameter		.08	12.35	37	111	8.05
1200	1-1/2" diameter		.22	35	105	315	22.80
1300	2" diameter		.24	45	135	405	28.95
1400	2-1/2" diameter		.36	57.50	172.50	520	37.40
1410	3" diameter		.42	58.50	175	525	38.35
1450	Drill, steel, 7/8" x 2'		.09	13.05	39.12	117	8.50
1460	7/8" x 6'		.12	19.85	59.58	179	12.85
1520	Moil points		.03	7	21	63	4.40
1525	Pneumatic nailer w/accessories		.48	39.50	119.19	360	27.70
1530	Sheeting driver for 60 lb. breaker		.04	7.85	23.52	70.50	5.05
1540	For 90 lb. breaker		.13	10.65	31.88	95.50	7.45
1550	Spade, 25 lb.		.51	7.50	22.47	67.50	8.55
1560	Tamper, single, 35 lb.		.60	58.50	175	525	39.75
1570	Triple, 140 lb.		.90	62.50	187.09	560	44.60
1580	Wrenches, impact, air powered, up to 3/4" bolt		.43	50	149.75	450	33.40
1590	Up to 1-1/4" bolt		.58	67.50	202.50	610	45.15
1600	Barricades, barrels, reflectorized, 1 to 99 barrels		.03	4	12	36	2.65
1610	100 to 200 barrels		.03	4.46	13.38	40	2.90
1620	Barrels with flashers, 1 to 99 barrels		.04	6.45	19.39	58	4.15
1630	100 to 200 barrels		.03	5.15	15.52	46.50	3.35
1640	Barrels with steady burn type C lights		.05	8.55	25.61	77	5.50
1650	Illuminated board, trailer mounted, with generator		3.32	141	423.30	1,275	111.20
1670	Portable barricade, stock, with flashers, 1 to 6 units		.04	6.45	19.34	58	4.15
1680	25 to 50 units		.03	6	18.03	54	3.85
1685	Butt fusion machine, wheeled, 1.5 HP electric, 2" - 8" diameter pipe		2.66	190	568.50	1,700	134.95
1690	Tracked, 20 HP diesel, 4" - 12" diameter pipe		10.04	565	1,702.05	5,100	420.75
1695	83 HP diesel, 8" - 24" diameter pipe		51.84	1,125	3,358.65	10,100	1,086
1700	Carts, brick, gas engine, 1000 lb. capacity		2.98	65.50	196.97	590	63.20
1800	1500 lb., 7-1/2' lift		2.95	70	210.50	630	65.70
1822	Dehumidifier, medium, 6 lb./hr., 150 CFM		1.20	77.50	232.03	695	56.05
1824	Large, 18 lb./hr., 600 CFM		2.22	585	1,750	5,250	367.75
1830	Distributor, asphalt, trailer mounted, 2000 gal., 38 H.P. diesel		11.13	355	1,071.32	3,225	303.25
1840	3000 gal., 38 H.P. diesel		13.02	385	1,149.71	3,450	334.10
1850	Drill, rotary hammer, electric		1.13	72	216.17	650	52.25
1860	Carbide bit, 1-1/2" diameter, add to electric rotary hammer		.03	41.50	125	375	25.25
1865	Rotary, crawler, 250 H.P.		137.40	2,450	7,375	22,100	2,574
1870	Emulsion sprayer, 65 gal., 5 H.P. gas engine		2.80	108	324.01	970	87.20
1880	200 gal., 5 H.P. engine		7.31	181	543.50	1,625	167.20
1900	Floor auto-scrubbing machine, walk-behind, 28" path		5.69	223	670	2,000	179.55
1930	Floodlight, mercury vapor, or quartz, on tripod, 1000 watt		.46	36.50	110	330	25.70
1940	2000 watt		.60	28.50	85.55	257	21.90
1950	Floodlights, trailer mounted with generator, 1 - 300 watt light		3.59	79.50	238.30	715	76.35
1960	2 - 1000 watt lights		4.54	88.50	265.48	795	89.40
2000	4 - 300 watt lights		4.29	101	303.11	910	95

40

01 54 33 | Equipment Rental

		UNIT	HOURLY OPER. COST	RENT PER DAY	RENT PER WEEK	RENT PER MONTH	EQUIPMENT COST/DAY
2005	Foam spray rig, incl. box trailer, compressor, generator, proportioner	Ea.	25.76	540	1,620.05	4,850	530.15
2015	Forklift, pneumatic tire, rough terr, straight mast, 5000 lb, 12' lift, gas		18.82	221	663.70	2,000	283.30
2025	8000 lb, 12' lift		22.95	365	1,097.45	3,300	403.10
2030	5000 lb, 12' lift, diesel		15.59	247	740.71	2,225	272.85
2035	8000 lb, 12' lift, diesel		16.90	280	839.82	2,525	303.15
2045	All terrain, telescoping boom, diesel, 5000 lb, 10' reach, 19' lift		17.40	200	600	1,800	259.25
2055	6600 lb, 29' reach, 42' lift		21.29	236	707.08	2,125	311.75
2065	10,000 lb, 31' reach, 45' lift		23.30	267	800	2,400	346.45
2070	Cushion tire, smooth floor, gas, 5000 lb capacity		8.32	262	785	2,350	223.60
2075	8000 lb capacity		11.47	320	957.50	2,875	283.25
2085	Diesel, 5000 lb capacity		7.82	257	770	2,300	216.55
2090	12,000 lb capacity		12.16	400	1,206.59	3,625	338.60
2095	20,000 lb capacity		17.41	800	2,405	7,225	620.30
2100	Generator, electric, gas engine, 1.5 kW to 3 kW		2.60	46.50	140	420	48.80
2200	5 kW		3.25	91	272.50	820	80.45
2300	10 kW		5.98	108	322.50	970	112.35
2400	25 kW		7.47	405	1,222.25	3,675	304.25
2500	Diesel engine, 20 kW		9.29	222	665	2,000	207.35
2600	50 kW		16.09	360	1,085	3,250	345.70
2700	100 kW		28.85	485	1,455	4,375	521.85
2800	250 kW		54.84	825	2,472.50	7,425	933.25
2850	Hammer, hydraulic, for mounting on boom, to 500 ft lb.		2.93	94.50	283.25	850	80.05
2860	1000 ft lb.		4.65	141	423.30	1,275	121.85
2900	Heaters, space, oil or electric, 50 MBH		1.48	47	141.42	425	40.10
3000	100 MBH		2.74	47	141.42	425	50.25
3100	300 MBH		8.00	136	409.10	1,225	145.80
3150	500 MBH		13.28	208	625	1,875	231.20
3200	Hose, water, suction with coupling, 20' long, 2" diameter		.02	5.65	17	51	3.55
3210	3" diameter		.03	14.15	42.50	128	8.75
3220	4" diameter		.03	28.50	85.01	255	17.25
3230	6" diameter		.11	41	122.96	370	25.50
3240	8" diameter		.27	54	161.92	485	34.55
3250	Discharge hose with coupling, 50' long, 2" diameter		.01	6.50	19.50	58.50	4
3260	3" diameter		.01	7.35	22	66	4.50
3270	4" diameter		.02	21	63.25	190	12.80
3280	6" diameter		.06	29.50	88.04	264	18.10
3290	8" diameter		.24	38	113.85	340	24.70
3295	Insulation blower		.84	118	353.54	1,050	77.40
3300	Ladders, extension type, 16' to 36' long		.18	41.50	125	375	26.45
3400	40' to 60' long		.64	123	368	1,100	78.75
3405	Lance for cutting concrete		2.22	66.50	200	600	57.80
3407	Lawn mower, rotary, 22", 5 H.P.		1.06	38.50	115	345	31.50
3408	48" self-propelled		2.92	160	481	1,450	119.60
3410	Level, electronic, automatic, with tripod and leveling rod		1.06	43	129.50	390	34.40
3430	Laser type, for pipe and sewer line and grade		2.20	117	350	1,050	87.55
3440	Rotating beam for interior control		.91	64	192.50	580	45.80
3460	Builder's optical transit, with tripod and rod		.10	43	129.50	390	26.70
3500	Light towers, towable, with diesel generator, 2000 watt		4.31	102	307.29	920	95.90
3600	4000 watt		4.55	142	425	1,275	121.40
3700	Mixer, powered, plaster and mortar, 6 C.F., 7 H.P.		2.08	86.50	260	780	68.60
3800	10 C.F., 9 H.P.		2.26	124	372.50	1,125	92.60
3850	Nailer, pneumatic		.48	30.50	91	273	22.10
3900	Paint sprayers complete, 8 CFM		.86	62.50	186.74	560	44.20
4000	17 CFM		1.61	111	333.84	1,000	79.70
4020	Pavers, bituminous, rubber tires, 8' wide, 50 H.P., diesel		32.31	575	1,724.57	5,175	603.40
4030	10' wide, 150 H.P.		96.77	1,975	5,905.34	17,700	1,955
4050	Crawler, 8' wide, 100 H.P., diesel		88.64	2,075	6,245.03	18,700	1,958
4060	10' wide, 150 H.P.		105.21	2,375	7,133.45	21,400	2,268

40

For customer support on your Electrical Costs with RSMeans Data, call 800.448.8182.

47

01 54 33 | Equipment Rental

		UNIT	HOURLY OPER. COST	RENT PER DAY	RENT PER WEEK	RENT PER MONTH	EQUIPMENT COST/DAY
4070	Concrete paver, 12' to 24' wide, 250 H.P.	Ea.	88.67	1,700	5,121.45	15,400	1,734
4080	Placer-spreader-trimmer, 24' wide, 300 H.P.		118.92	2,575	7,760.56	23,300	2,503
4100	Pump, centrifugal gas pump, 1-1/2" diam., 65 GPM		3.97	54.50	164.10	490	64.60
4200	2" diameter, 130 GPM		5.04	44	132.50	400	66.80
4300	3" diameter, 250 GPM		5.18	55	165	495	74.40
4400	6" diameter, 1500 GPM		22.46	92.50	277.78	835	235.25
4500	Submersible electric pump, 1-1/4" diameter, 55 GPM		.40	35	105	315	24.25
4600	1-1/2" diameter, 83 GPM		.45	43.50	130	390	29.60
4700	2" diameter, 120 GPM		1.66	55	165	495	46.30
4800	3" diameter, 300 GPM		3.07	109	327.50	985	90.05
4900	4" diameter, 560 GPM		14.90	76	227.50	685	164.70
5000	6" diameter, 1590 GPM		22.30	65.50	196.97	590	217.80
5100	Diaphragm pump, gas, single, 1-1/2" diameter		1.14	38.50	116.16	350	32.35
5200	2" diameter		4.02	92.50	277.78	835	87.70
5300	3" diameter		4.09	86	258	775	84.35
5400	Double, 4" diameter		6.10	96	287.88	865	106.35
5450	Pressure washer 5 GPM, 3000 psi		3.92	110	330	990	97.35
5460	7 GPM, 3000 psi		5.00	90	270	810	93.95
5470	High pressure water jet 10 KSI		40.02	730	2,185.92	6,550	757.40
5480	40 KSI		28.21	990	2,975.28	8,925	820.75
5500	Trash pump, self-priming, gas, 2" diameter		3.85	109	328.29	985	96.50
5600	Diesel, 4" diameter		6.74	163	489.91	1,475	151.95
5650	Diesel, 6" diameter		17.02	163	489.91	1,475	234.15
5655	Grout pump		18.92	284	851.83	2,550	321.75
5700	Salamanders, L.P. gas fired, 100,000 Btu		2.92	57.50	172.50	520	57.85
5705	50,000 Btu		1.68	23	69	207	27.25
5720	Sandblaster, portable, open top, 3 C.F. capacity		.61	132	395	1,175	83.85
5730	6 C.F. capacity		1.02	133	399	1,200	87.90
5740	Accessories for above		.14	24	72.12	216	15.55
5750	Sander, floor		.78	73.50	220	660	50.20
5760	Edger		.52	35	105	315	25.20
5800	Saw, chain, gas engine, 18" long		1.77	63.50	190	570	52.20
5900	Hydraulic powered, 36" long		.79	59	176.77	530	41.65
5950	60" long		.79	65.50	196.97	590	45.70
6000	Masonry, table mounted, 14" diameter, 5 H.P.		1.34	76.50	230	690	56.70
6050	Portable cut-off, 8 H.P.		1.83	72.50	217.50	655	58.15
6100	Circular, hand held, electric, 7-1/4" diameter		.23	13.85	41.50	125	10.15
6200	12" diameter		.24	41	122.50	370	26.40
6250	Wall saw, w/hydraulic power, 10 H.P.		3.32	99.50	299	895	86.40
6275	Shot blaster, walk-behind, 20" wide		4.79	284	851.83	2,550	208.70
6280	Sidewalk broom, walk-behind		2.27	83.50	250.85	755	68.30
6300	Steam cleaner, 100 gallons per hour		3.38	83.50	250.85	755	77.20
6310	200 gallons per hour		4.38	101	303.11	910	95.70
6340	Tar kettle/pot, 400 gallons		16.65	128	383.85	1,150	209.95
6350	Torch, cutting, acetylene-oxygen, 150' hose, excludes gases		.46	15.50	46.51	140	12.95
6360	Hourly operating cost includes tips and gas		21.17	7.10	21.23	63.50	173.60
6410	Toilet, portable chemical		.13	23.50	70.03	210	15.05
6420	Recycle flush type		.16	29	86.75	260	18.65
6430	Toilet, fresh water flush, garden hose,		.20	34.50	103.47	310	22.25
6440	Hoisted, non-flush, for high rise		.16	28	84.66	254	18.20
6465	Tractor, farm with attachment		17.57	395	1,184.37	3,550	377.45
6480	Trailers, platform, flush deck, 2 axle, 3 ton capacity		1.71	96	287.41	860	71.15
6500	25 ton capacity		6.31	145	433.76	1,300	137.20
6600	40 ton capacity		8.14	206	616.66	1,850	188.45
6700	3 axle, 50 ton capacity		8.82	228	683.33	2,050	207.25
6800	75 ton capacity		11.21	305	907.63	2,725	271.20
6810	Trailer mounted cable reel for high voltage line work		5.94	28.50	85.86	258	64.75
6820	Trailer mounted cable tensioning rig		11.79	28.50	85.86	258	111.50

40

01 54 33 | Equipment Rental

		UNIT	HOURLY OPER. COST	RENT PER DAY	RENT PER WEEK	RENT PER MONTH	EQUIPMENT COST/DAY	
40	6830	Cable pulling rig	Ea.	74.51	28.50	85.86	258	613.30	**40**
	6850	Portable cable/wire puller, 8000 lb max pulling capacity		3.74	121	363.64	1,100	102.65	
	6900	Water tank trailer, engine driven discharge, 5000 gallons		7.24	160	480.79	1,450	154.10	
	6925	10,000 gallons		9.87	218	653.25	1,950	209.60	
	6950	Water truck, off highway, 6000 gallons		72.61	845	2,529.89	7,600	1,087	
	7010	Tram car for high voltage line work, powered, 2 conductor		6.95	29.50	88	264	73.20	
	7020	Transit (builder's level) with tripod		.10	17.75	53.30	160	11.45	
	7030	Trench box, 3000 lb., 6' x 8'		.57	97.50	293.15	880	63.15	
	7040	7200 lb., 6' x 20'		.73	189	566.72	1,700	119.15	
	7050	8000 lb., 8' x 16'		1.09	205	615	1,850	131.70	
	7060	9500 lb., 8' x 20'		1.22	235	705.51	2,125	150.85	
	7065	11,000 lb., 8' x 24'		1.28	221	663.70	2,000	142.95	
	7070	12,000 lb., 10' x 20'		1.51	267	799.57	2,400	172	
	7100	Truck, pickup, 3/4 ton, 2 wheel drive		9.35	62.50	187.09	560	112.20	
	7200	4 wheel drive		9.60	167	500	1,500	176.75	
	7250	Crew carrier, 9 passenger		12.81	109	328.29	985	168.15	
	7290	Flat bed truck, 20,000 lb. GVW		15.44	134	402.40	1,200	204.05	
	7300	Tractor, 4 x 2, 220 H.P.		22.52	218	653.25	1,950	310.80	
	7410	330 H.P.		32.72	298	893.64	2,675	440.45	
	7500	6 x 4, 380 H.P.		36.53	345	1,034.74	3,100	499.15	
	7600	450 H.P.		44.76	420	1,254.23	3,775	608.95	
	7610	Tractor, with A frame, boom and winch, 225 H.P.		25.04	296	886.76	2,650	377.65	
	7620	Vacuum truck, hazardous material, 2500 gallons		12.94	315	940.67	2,825	291.70	
	7625	5,000 gallons		13.18	445	1,332.62	4,000	371.95	
	7650	Vacuum, HEPA, 16 gallon, wet/dry		.86	122	365	1,100	79.90	
	7655	55 gallon, wet/dry		.79	25.50	76.50	230	21.60	
	7660	Water tank, portable		.74	162	485.11	1,450	102.90	
	7690	Sewer/catch basin vacuum, 14 C.Y., 1500 gallons		17.52	670	2,012	6,025	542.60	
	7700	Welder, electric, 200 amp		3.86	33.50	101.01	305	51.10	
	7800	300 amp		5.61	104	313.14	940	107.55	
	7900	Gas engine, 200 amp		9.05	59	176.77	530	107.80	
	8000	300 amp		10.25	111	333.96	1,000	148.75	
	8100	Wheelbarrow, any size		.07	11.15	33.50	101	7.20	
	8200	Wrecking ball, 4000 lb.		2.53	62	186	560	57.45	
50	0010	**HIGHWAY EQUIPMENT RENTAL** without operators [R015433 -10]	Ea.	89.47	1,550	4,677.24	14,000	1,651	**5**
	0050	Asphalt batch plant, portable drum mixer, 100 ton/hr.		103.21	1,675	4,990.80	15,000	1,824	
	0060	200 ton/hr.		121.29	1,950	5,853.08	17,600	2,141	
	0070	300 ton/hr.		.37	26	77.34	232	18.45	
	0100	Backhoe attachment, long stick, up to 185 H.P., 10-1/2' long		.42	29	86.75	260	20.70	
	0140	Up to 250 H.P., 12' long		.57	39.50	118.11	355	28.20	
	0180	Over 250 H.P., 15' long		1.17	80.50	241.44	725	57.65	
	0200	Special dipper arm, up to 100 H.P., 32' long		1.46	101	303.11	910	72.30	
	0240	Over 100 H.P., 33' long		35.81	425	1,280.36	3,850	542.60	
	0280	Catch basin/sewer cleaning truck, 3 ton, 9 C.Y., 1000 gal.		24.47	565	1,698.44	5,100	535.50	
	0300	Concrete batch plant, portable, electric, 200 C.Y./hr.		3.19	64	192.32	575	63.95	
	0520	Grader/dozer attachment, ripper/scarifier, rear mounted, up to 135 H.P.		4.18	97	290.56	870	91.60	
	0540	Up to 180 H.P.		5.92	155	465.11	1,400	140.40	
	0580	Up to 250 H.P.		2.18	59	176.64	530	52.80	
	0700	Pvmt. removal bucket, for hyd. excavator, up to 90 H.P.		2.33	75.50	225.76	675	63.80	
	0740	Up to 200 H.P.		2.55	92.50	276.98	830	75.80	
	0780	Over 200 H.P.		51.21	750	2,247.17	6,750	859.10	
	0900	Aggregate spreader, self-propelled, 187 H.P.		3.20	99.50	299	895	85.40	
	1000	Chemical spreader, 3 C.Y.		68.03	520	1,565.69	4,700	857.40	
	1900	Hammermill, traveling, 250 H.P.		5.47	234	702.03	2,100	184.15	
	2000	Horizontal borer, 3" diameter, 13 H.P. gas driven		27.86	535	1,606.09	4,825	544.10	
	2150	Horizontal directional drill, 20,000 lb. thrust, 78 H.P. diesel		34.24	625	1,868.72	5,600	647.65	
	2160	30,000 lb. thrust, 115 H.P.		49.09	720	2,156.61	6,475	824.05	
	2170	50,000 lb. thrust, 170 H.P.							

01 54 33 | Equipment Rental

		UNIT	HOURLY OPER. COST	RENT PER DAY	RENT PER WEEK	RENT PER MONTH	EQUIPMENT COST/DAY	
50								**50**
2190	Mud trailer for HDD, 1500 gallons, 175 H.P., gas	Ea.	25.76	177	530.31	1,600	312.15	
2200	Hydromulcher, diesel, 3000 gallon, for truck mounting		17.61	193	580	1,750	256.85	
2300	Gas, 600 gallon		7.57	96	287.88	865	118.15	
2400	Joint & crack cleaner, walk behind, 25 H.P.		3.19	46	137.38	410	53	
2500	Filler, trailer mounted, 400 gallons, 20 H.P.		8.43	173	517.50	1,550	170.95	
3000	Paint striper, self-propelled, 40 gallon, 22 H.P.		6.83	123	368.69	1,100	128.35	
3100	120 gallon, 120 H.P.		17.83	385	1,151.54	3,450	372.90	
3200	Post drivers, 6" I-Beam frame, for truck mounting		12.54	325	969.72	2,900	294.25	
3400	Road sweeper, self-propelled, 8' wide, 90 H.P.		36.34	720	2,164.75	6,500	723.65	
3450	Road sweeper, vacuum assisted, 4 C.Y., 220 gallons		58.98	680	2,038.13	6,125	879.45	
4000	Road mixer, self-propelled, 130 H.P.		46.79	835	2,508.46	7,525	876	
4100	310 H.P.		75.91	2,175	6,558.59	19,700	1,919	
4220	Cold mix paver, incl. pug mill and bitumen tank, 165 H.P.		96.11	2,350	7,015.87	21,000	2,172	
4240	Pavement brush, towed		3.47	101	303.11	910	88.35	
4250	Paver, asphalt, wheel or crawler, 130 H.P., diesel		95.37	2,300	6,898.28	20,700	2,143	
4300	Paver, road widener, gas, 1' to 6', 67 H.P.		47.22	985	2,952.67	8,850	968.30	
4400	Diesel, 2' to 14', 88 H.P.		57.06	1,175	3,501.40	10,500	1,157	
4600	Slipform pavers, curb and gutter, 2 track, 75 H.P.		58.53	1,275	3,814.96	11,400	1,231	
4700	4 track, 165 H.P.		36.11	855	2,560.72	7,675	801.05	
4800	Median barrier, 215 H.P.		59.13	1,350	4,076.26	12,200	1,288	
4901	Trailer, low bed, 75 ton capacity		10.84	286	857.06	2,575	258.10	
5000	Road planer, walk behind, 10" cutting width, 10 H.P.		2.48	258	773	2,325	174.45	
5100	Self-propelled, 12" cutting width, 64 H.P.		8.34	192	575.77	1,725	181.85	
5120	Traffic line remover, metal ball blaster, truck mounted, 115 H.P.		47.03	1,025	3,100	9,300	996.25	
5140	Grinder, truck mounted, 115 H.P.		51.41	1,025	3,100	9,300	1,031	
5160	Walk-behind, 11 H.P.		3.59	143	430	1,300	114.75	
5200	Pavement profiler, 4' to 6' wide, 450 H.P.		218.77	1,275	3,838.46	11,500	2,518	
5300	8' to 10' wide, 750 H.P.		334.94	1,350	4,015.23	12,000	3,483	
5400	Roadway plate, steel, 1" x 8' x 20'		.09	61.50	184.69	555	37.65	
5600	Stabilizer, self-propelled, 150 H.P.		41.55	1,050	3,131.37	9,400	958.70	
5700	310 H.P.		76.95	1,325	3,939.47	11,800	1,404	
5800	Striper, truck mounted, 120 gallon paint, 460 H.P.		49.24	350	1,046	3,150	603.10	
5900	Thermal paint heating kettle, 115 gallons		7.78	75	225	675	107.25	
6000	Tar kettle, 330 gallon, trailer mounted		12.40	96	287.50	865	156.70	
7000	Tunnel locomotive, diesel, 8 to 12 ton		30.11	625	1,881.35	5,650	617.20	
7005	Electric, 10 ton		29.60	715	2,142.65	6,425	665.30	
7010	Muck cars, 1/2 C.Y. capacity		2.33	27	81	243	34.80	
7020	1 C.Y. capacity		2.54	35	105.56	315	41.45	
7030	2 C.Y. capacity		2.69	39.50	118.11	355	45.10	
7040	Side dump, 2 C.Y. capacity		2.90	49	146.33	440	52.50	
7050	3 C.Y. capacity		3.90	53.50	160.96	485	63.40	
7060	5 C.Y. capacity		5.69	69.50	207.99	625	87.10	
7100	Ventilating blower for tunnel, 7-1/2 H.P.		2.16	53.50	159.91	480	49.30	
7110	10 H.P.		2.45	55.50	167.23	500	53.05	
7120	20 H.P.		3.58	72.50	217.40	650	72.15	
7140	40 H.P.		6.21	96	287.43	860	107.20	
7160	60 H.P.		8.79	103	308.33	925	132	
7175	75 H.P.		10.49	160	480.79	1,450	180.10	
7180	200 H.P.		21.03	315	945.90	2,850	357.45	
7800	Windrow loader, elevating		54.49	1,700	5,125	15,400	1,461	
60								**60**
0010	**LIFTING AND HOISTING EQUIPMENT RENTAL** without operators							
0150	Crane, flatbed mounted, 3 ton capacity	Ea.	14.58	203	610.30	1,825	238.75	
0200	Crane, climbing, 106' jib, 6000 lb. capacity, 410 fpm		40.13	2,675	8,034	24,100	1,928	
0300	101' jib, 10,250 lb. capacity, 270 fpm		46.90	2,300	6,868.82	20,600	1,749	
0500	Tower, static, 130' high, 106' jib, 6200 lb. capacity at 400 fpm		45.61	2,300	6,916	20,700	1,748	
0520	Mini crawler spider crane, up to 24" wide, 1990 lb. lifting capacity		12.65	555	1,672.31	5,025	435.70	
0525	Up to 30" wide, 6450 lb. lifting capacity		14.70	660	1,985.87	5,950	514.75	
0530	Up to 52" wide, 6680 lb. lifting capacity		23.38	810	2,430.08	7,300	673.05	

Reference boxes: R015433 -10, R312316 -45

01 54 33 | Equipment Rental

Line	Description	UNIT	HOURLY OPER. COST	RENT PER DAY	RENT PER WEEK	RENT PER MONTH	EQUIPMENT COST/DAY
0535	Up to 55" wide, 8920 lb. lifting capacity	Ea.	26.10	895	2,691.37	8,075	747.10
0540	Up to 66" wide, 13,350 lb. lifting capacity		35.34	1,400	4,180.77	12,500	1,119
0600	Crawler mounted, lattice boom, 1/2 C.Y., 15 tons at 12' radius		37.34	855	2,558	7,675	810.30
0700	3/4 C.Y., 20 tons at 12' radius		54.72	960	2,874	8,625	1,013
0800	1 C.Y., 25 tons at 12' radius		68.10	1,025	3,038	9,125	1,152
0900	1-1/2 C.Y., 40 tons at 12' radius		66.98	1,150	3,476	10,400	1,231
1000	2 C.Y., 50 tons at 12' radius		89.66	1,375	4,120	12,400	1,541
1100	3 C.Y., 75 tons at 12' radius		65.65	2,400	7,210	21,600	1,967
1200	100 ton capacity, 60' boom		86.78	2,700	8,080.96	24,200	2,310
1300	165 ton capacity, 60' boom		107.18	3,025	9,091.08	27,300	2,676
1400	200 ton capacity, 70' boom		139.61	3,875	11,616.38	34,800	3,440
1500	350 ton capacity, 80' boom		184.04	4,200	12,626.50	37,900	3,998
1600	Truck mounted, lattice boom, 6 x 4, 20 tons at 10' radius		40.16	2,000	6,026	18,100	1,526
1700	25 tons at 10' radius		43.16	2,400	7,210	21,600	1,787
1800	8 x 4, 30 tons at 10' radius		54.71	2,575	7,725	23,200	1,983
1900	40 tons at 12' radius		54.71	2,825	8,446	25,300	2,127
2000	60 tons at 15' radius		54.23	1,675	5,000.09	15,000	1,434
2050	82 tons at 15' radius		60.03	1,800	5,404.14	16,200	1,561
2100	90 tons at 15' radius		67.06	1,950	5,883.95	17,700	1,713
2200	115 tons at 15' radius		75.66	2,200	6,591.03	19,800	1,923
2300	150 tons at 18' radius		81.91	2,775	8,343	25,000	2,324
2350	165 tons at 18' radius		87.91	2,825	8,498	25,500	2,403
2400	Truck mounted, hydraulic, 12 ton capacity		29.80	395	1,186.89	3,550	475.80
2500	25 ton capacity		36.72	490	1,464.67	4,400	586.70
2550	33 ton capacity		54.71	910	2,727.32	8,175	983.15
2560	40 ton capacity		54.71	910	2,727.32	8,175	983.15
2600	55 ton capacity		54.32	925	2,777.83	8,325	990.15
2700	80 ton capacity		71.13	1,475	4,444.53	13,300	1,458
2720	100 ton capacity		75.72	1,575	4,722.31	14,200	1,550
2740	120 ton capacity		106.34	1,850	5,555.66	16,700	1,962
2760	150 ton capacity		113.80	2,050	6,186.99	18,600	2,148
2800	Self-propelled, 4 x 4, with telescoping boom, 5 ton		15.29	435	1,298	3,900	381.95
2900	12-1/2 ton capacity		21.63	435	1,298	3,900	432.65
3000	15 ton capacity		34.77	455	1,363.66	4,100	550.85
3050	20 ton capacity		22.55	655	1,969.73	5,900	574.35
3100	25 ton capacity		37.06	1,425	4,293.01	12,900	1,155
3150	40 ton capacity		45.35	665	1,994.99	5,975	761.80
3200	Derricks, guy, 20 ton capacity, 60' boom, 75' mast		22.97	1,450	4,378	13,100	1,059
3300	100' boom, 115' mast		36.41	2,050	6,180	18,500	1,527
3400	Stiffleg, 20 ton capacity, 70' boom, 37' mast		25.66	635	1,906	5,725	586.70
3500	100' boom, 47' mast		39.71	685	2,060	6,175	729.70
3550	Helicopter, small, lift to 1250 lb. maximum, w/pilot		100.14	2,175	6,500.12	19,500	2,101
3600	Hoists, chain type, overhead, manual, 3/4 ton		.15	10.25	30.70	92	7.30
3900	10 ton		.80	6.25	18.78	56.50	10.10
4000	Hoist and tower, 5000 lb. cap., portable electric, 40' high		4.82	146	438	1,325	126.15
4100	For each added 10' section, add		.12	32.50	98	294	20.55
4200	Hoist and single tubular tower, 5000 lb. electric, 100' high		7.03	108	325	975	121.20
4300	For each added 6'-6" section, add		.21	39.50	118	355	25.25
4400	Hoist and double tubular tower, 5000 lb., 100' high		7.65	108	325	975	126.20
4500	For each added 6'-6" section, add		.23	42.50	128	385	27.45
4550	Hoist and tower, mast type, 6000 lb., 100' high		9.32	95.50	286.87	860	131.95
4570	For each added 10' section, add		.13	32.50	98	294	20.65
4600	Hoist and tower, personnel, electric, 2000 lb., 100' @ 125 fpm		17.67	25.50	75.76	227	156.55
4700	3000 lb., 100' @ 200 fpm		20.22	25.50	75.76	227	176.95
4800	3000 lb., 150' @ 300 fpm		22.44	25.50	75.76	227	194.70
4900	4000 lb., 100' @ 300 fpm		23.22	25.50	75.76	227	200.90
5000	6000 lb., 100' @ 275 fpm		24.95	25.50	75.76	227	214.75
5100	For added heights up to 500', add	L.F.	.01	3.33	10	30	2.10

60

For customer support on your Electrical Costs with RSMeans Data, call 800.448.8182.

48

01 54 33 | Equipment Rental

		UNIT	HOURLY OPER. COST	RENT PER DAY	RENT PER WEEK	RENT PER MONTH	EQUIPMENT COST/DAY	
60	5200	Jacks, hydraulic, 20 ton	Ea.	.05	19.85	59.60	179	12.30	60
	5500	100 ton		.40	26	78.50	236	18.95	
	6100	Jacks, hydraulic, climbing w/50' jackrods, control console, 30 ton cap.		2.19	32	96	288	36.70	
	6150	For each added 10' jackrod section, add		.05	5	15	45	3.40	
	6300	50 ton capacity		3.52	34.50	103	310	48.75	
	6350	For each added 10' jackrod section, add		.06	5	15	45	3.50	
	6500	125 ton capacity		9.20	53.50	160	480	105.55	
	6550	For each added 10' jackrod section, add		.62	5	15	45	8	
	6600	Cable jack, 10 ton capacity with 200' cable		1.60	36.50	110	330	34.80	
	6650	For each added 50' of cable, add		.22	15.35	46	138	11	
70	0010	**WELLPOINT EQUIPMENT RENTAL** without operators							70
	0020	Based on 2 months rental							
	0100	Combination jetting & wellpoint pump, 60 H.P. diesel	Ea.	15.83	305	922	2,775	311.05	
	0200	High pressure gas jet pump, 200 H.P., 300 psi	"	34.17	278	833.35	2,500	440.05	
	0300	Discharge pipe, 8" diameter	L.F.	.01	1.41	4.24	12.75	.90	
	0350	12" diameter		.01	2.09	6.26	18.80	1.35	
	0400	Header pipe, flows up to 150 GPM, 4" diameter		.01	.74	2.22	6.65	.50	
	0500	400 GPM, 6" diameter		.01	1.08	3.23	9.70	.70	
	0600	800 GPM, 8" diameter		.01	1.41	4.24	12.75	.95	
	0700	1500 GPM, 10" diameter		.01	1.75	5.25	15.75	1.15	
	0800	2500 GPM, 12" diameter		.03	2.09	6.26	18.80	1.45	
	0900	4500 GPM, 16" diameter		.03	2.42	7.27	22	1.70	
	0950	For quick coupling aluminum and plastic pipe, add		.03	9.60	28.84	86.50	6.05	
	1100	Wellpoint, 25' long, with fittings & riser pipe, 1-1/2" or 2" diameter	Ea.	.07	133	399	1,200	80.35	
	1200	Wellpoint pump, diesel powered, 4" suction, 20 H.P.		7.07	152	454.55	1,375	147.50	
	1300	6" suction, 30 H.P.		9.49	140	420	1,250	159.90	
	1400	8" suction, 40 H.P.		12.85	253	757.59	2,275	254.35	
	1500	10" suction, 75 H.P.		18.75	268	803.05	2,400	310.60	
	1600	12" suction, 100 H.P.		27.51	300	904.06	2,700	400.90	
	1700	12" suction, 175 H.P.		39.37	320	959.61	2,875	506.90	
80	0010	**MARINE EQUIPMENT RENTAL** without operators	Ea.						80
	0200	Barge, 400 Ton, 30' wide x 90' long		17.84	1,200	3,632.05	10,900	869.15	
	0240	800 Ton, 45' wide x 90' long		22.41	1,500	4,468.20	13,400	1,073	
	2000	Tugboat, diesel, 100 H.P.		29.92	240	721.18	2,175	383.65	
	2040	250 H.P.		58.10	435	1,306.49	3,925	726.10	
	2080	380 H.P.		126.49	1,300	3,919.48	11,800	1,796	
	3000	Small work boat, gas, 16-foot, 50 H.P.		11.48	48.50	145.01	435	120.85	
	4000	Large, diesel, 48-foot, 200 H.P.		75.57	1,375	4,154.64	12,500	1,436	

R015433-10 (row 0020)

R015433-10 (row 0200)

Crew A-1

Crew A-1	Bare Costs Hr.	Bare Costs Daily	Incl. Subs O&P Hr.	Incl. Subs O&P Daily	Cost Per Labor-Hour Bare Costs	Cost Per Labor-Hour Incl. O&P
1 Building Laborer	$44.40	$355.20	$66.25	$530.00	$44.40	$66.25
1 Concrete Saw, Gas Manual		113.70		125.07	14.21	15.63
8 L.H., Daily Totals		$468.90		$655.07	$58.61	$81.88

Crew A-1A

Crew A-1A	Bare Costs Hr.	Bare Costs Daily	Incl. Subs O&P Hr.	Incl. Subs O&P Daily	Cost Per Labor-Hour Bare Costs	Cost Per Labor-Hour Incl. O&P
1 Skilled Worker	$57.10	$456.80	$85.90	$687.20	$57.10	$85.90
1 Shot Blaster, 20"		208.70		229.57	26.09	28.70
8 L.H., Daily Totals		$665.50		$916.77	$83.19	$114.60

Crew A-1B

Crew A-1B	Bare Costs Hr.	Bare Costs Daily	Incl. Subs O&P Hr.	Incl. Subs O&P Daily	Cost Per Labor-Hour Bare Costs	Cost Per Labor-Hour Incl. O&P
1 Building Laborer	$44.40	$355.20	$66.25	$530.00	$44.40	$66.25
1 Concrete Saw		112.85		124.14	14.11	15.52
8 L.H., Daily Totals		$468.05		$654.13	$58.51	$81.77

Crew A-1C

Crew A-1C	Bare Costs Hr.	Bare Costs Daily	Incl. Subs O&P Hr.	Incl. Subs O&P Daily	Cost Per Labor-Hour Bare Costs	Cost Per Labor-Hour Incl. O&P
1 Building Laborer	$44.40	$355.20	$66.25	$530.00	$44.40	$66.25
1 Chain Saw, Gas, 18"		52.20		57.42	6.53	7.18
8 L.H., Daily Totals		$407.40		$587.42	$50.92	$73.43

Crew A-1D

Crew A-1D	Bare Costs Hr.	Bare Costs Daily	Incl. Subs O&P Hr.	Incl. Subs O&P Daily	Cost Per Labor-Hour Bare Costs	Cost Per Labor-Hour Incl. O&P
1 Building Laborer	$44.40	$355.20	$66.25	$530.00	$44.40	$66.25
1 Vibrating Plate, Gas, 18"		31.90		35.09	3.99	4.39
8 L.H., Daily Totals		$387.10		$565.09	$48.39	$70.64

Crew A-1E

Crew A-1E	Bare Costs Hr.	Bare Costs Daily	Incl. Subs O&P Hr.	Incl. Subs O&P Daily	Cost Per Labor-Hour Bare Costs	Cost Per Labor-Hour Incl. O&P
1 Building Laborer	$44.40	$355.20	$66.25	$530.00	$44.40	$66.25
1 Vibrating Plate, Gas, 21"		165.60		182.16	20.70	22.77
8 L.H., Daily Totals		$520.80		$712.16	$65.10	$89.02

Crew A-1F

Crew A-1F	Bare Costs Hr.	Bare Costs Daily	Incl. Subs O&P Hr.	Incl. Subs O&P Daily	Cost Per Labor-Hour Bare Costs	Cost Per Labor-Hour Incl. O&P
1 Building Laborer	$44.40	$355.20	$66.25	$530.00	$44.40	$66.25
1 Rammer/Tamper, Gas, 8"		47.00		51.70	5.88	6.46
8 L.H., Daily Totals		$402.20		$581.70	$50.27	$72.71

Crew A-1G

Crew A-1G	Bare Costs Hr.	Bare Costs Daily	Incl. Subs O&P Hr.	Incl. Subs O&P Daily	Cost Per Labor-Hour Bare Costs	Cost Per Labor-Hour Incl. O&P
1 Building Laborer	$44.40	$355.20	$66.25	$530.00	$44.40	$66.25
1 Rammer/Tamper, Gas, 15"		54.65		60.12	6.83	7.51
8 L.H., Daily Totals		$409.85		$590.12	$51.23	$73.76

Crew A-1H

Crew A-1H	Bare Costs Hr.	Bare Costs Daily	Incl. Subs O&P Hr.	Incl. Subs O&P Daily	Cost Per Labor-Hour Bare Costs	Cost Per Labor-Hour Incl. O&P
1 Building Laborer	$44.40	$355.20	$66.25	$530.00	$44.40	$66.25
1 Exterior Steam Cleaner		77.20		84.92	9.65	10.62
8 L.H., Daily Totals		$432.40		$614.92	$54.05	$76.86

Crew A-1J

Crew A-1J	Bare Costs Hr.	Bare Costs Daily	Incl. Subs O&P Hr.	Incl. Subs O&P Daily	Cost Per Labor-Hour Bare Costs	Cost Per Labor-Hour Incl. O&P
1 Building Laborer	$44.40	$355.20	$66.25	$530.00	$44.40	$66.25
1 Cultivator, Walk-Behind, 5 H.P.		53.25		58.58	6.66	7.32
8 L.H., Daily Totals		$408.45		$588.58	$51.06	$73.57

Crew A-1K

Crew A-1K	Bare Costs Hr.	Bare Costs Daily	Incl. Subs O&P Hr.	Incl. Subs O&P Daily	Cost Per Labor-Hour Bare Costs	Cost Per Labor-Hour Incl. O&P
1 Building Laborer	$44.40	$355.20	$66.25	$530.00	$44.40	$66.25
1 Cultivator, Walk-Behind, 8 H.P.		92.20		101.42	11.53	12.68
8 L.H., Daily Totals		$447.40		$631.42	$55.92	$78.93

Crew A-1M

Crew A-1M	Bare Costs Hr.	Bare Costs Daily	Incl. Subs O&P Hr.	Incl. Subs O&P Daily	Cost Per Labor-Hour Bare Costs	Cost Per Labor-Hour Incl. O&P
1 Building Laborer	$44.40	$355.20	$66.25	$530.00	$44.40	$66.25
1 Snow Blower, Walk-Behind		68.30		75.13	8.54	9.39
8 L.H., Daily Totals		$423.50		$605.13	$52.94	$75.64

Crew A-2

Crew A-2	Bare Costs Hr.	Bare Costs Daily	Incl. Subs O&P Hr.	Incl. Subs O&P Daily	Cost Per Labor-Hour Bare Costs	Cost Per Labor-Hour Incl. O&P
2 Laborers	$44.40	$710.40	$66.25	$1060.00	$45.87	$68.50
1 Truck Driver (light)	48.80	390.40	73.00	584.00		
1 Flatbed Truck, Gas, 1.5 Ton		198.50		218.35	8.27	9.10
24 L.H., Daily Totals		$1299.30		$1862.35	$54.14	$77.60

Crew A-2A

Crew A-2A	Bare Costs Hr.	Bare Costs Daily	Incl. Subs O&P Hr.	Incl. Subs O&P Daily	Cost Per Labor-Hour Bare Costs	Cost Per Labor-Hour Incl. O&P
2 Laborers	$44.40	$710.40	$66.25	$1060.00	$45.87	$68.50
1 Truck Driver (light)	48.80	390.40	73.00	584.00		
1 Flatbed Truck, Gas, 1.5 Ton		198.50		218.35		
1 Concrete Saw		112.85		124.14	12.97	14.27
24 L.H., Daily Totals		$1412.15		$1986.48	$58.84	$82.77

Crew A-2B

Crew A-2B	Bare Costs Hr.	Bare Costs Daily	Incl. Subs O&P Hr.	Incl. Subs O&P Daily	Cost Per Labor-Hour Bare Costs	Cost Per Labor-Hour Incl. O&P
1 Truck Driver (light)	$48.80	$390.40	$73.00	$584.00	$48.80	$73.00
1 Flatbed Truck, Gas, 1.5 Ton		198.50		218.35	24.81	27.29
8 L.H., Daily Totals		$588.90		$802.35	$73.61	$100.29

Crew A-3A

Crew A-3A	Bare Costs Hr.	Bare Costs Daily	Incl. Subs O&P Hr.	Incl. Subs O&P Daily	Cost Per Labor-Hour Bare Costs	Cost Per Labor-Hour Incl. O&P
1 Equip. Oper. (light)	$55.50	$444.00	$82.70	$661.60	$55.50	$82.70
1 Pickup Truck, 4x4, 3/4 Ton		176.75		194.43	22.09	24.30
8 L.H., Daily Totals		$620.75		$856.02	$77.59	$107.00

Crew A-3B

Crew A-3B	Bare Costs Hr.	Bare Costs Daily	Incl. Subs O&P Hr.	Incl. Subs O&P Daily	Cost Per Labor-Hour Bare Costs	Cost Per Labor-Hour Incl. O&P
1 Equip. Oper. (medium)	$59.00	$472.00	$87.90	$703.20	$55.15	$82.30
1 Truck Driver (heavy)	51.30	410.40	76.70	613.60		
1 Dump Truck, 12 C.Y., 400 H.P.		579.35		637.28		
1 F.E. Loader, W.M., 2.5 C.Y.		638.30		702.13	76.10	83.71
16 L.H., Daily Totals		$2100.05		$2656.22	$131.25	$166.01

Crew A-3C

Crew A-3C	Bare Costs Hr.	Bare Costs Daily	Incl. Subs O&P Hr.	Incl. Subs O&P Daily	Cost Per Labor-Hour Bare Costs	Cost Per Labor-Hour Incl. O&P
1 Equip. Oper. (light)	$55.50	$444.00	$82.70	$661.60	$55.50	$82.70
1 Loader, Skid Steer, 78 H.P.		446.30		490.93	55.79	61.37
8 L.H., Daily Totals		$890.30		$1152.53	$111.29	$144.07

Crew A-3D

Crew A-3D	Bare Costs Hr.	Bare Costs Daily	Incl. Subs O&P Hr.	Incl. Subs O&P Daily	Cost Per Labor-Hour Bare Costs	Cost Per Labor-Hour Incl. O&P
1 Truck Driver (light)	$48.80	$390.40	$73.00	$584.00	$48.80	$73.00
1 Pickup Truck, 4x4, 3/4 Ton		176.75		194.43		
1 Flatbed Trailer, 25 Ton		137.20		150.92	39.24	43.17
8 L.H., Daily Totals		$704.35		$929.35	$88.04	$116.17

Crew A-3E

Crew A-3E	Bare Costs Hr.	Bare Costs Daily	Incl. Subs O&P Hr.	Incl. Subs O&P Daily	Cost Per Labor-Hour Bare Costs	Cost Per Labor-Hour Incl. O&P
1 Equip. Oper. (crane)	$61.45	$491.60	$91.55	$732.40	$56.38	$84.13
1 Truck Driver (heavy)	51.30	410.40	76.70	613.60		
1 Pickup Truck, 4x4, 3/4 Ton		176.75		194.43	11.05	12.15
16 L.H., Daily Totals		$1078.75		$1540.43	$67.42	$96.28

Crew A-3F

Crew A-3F	Bare Costs Hr.	Bare Costs Daily	Incl. Subs O&P Hr.	Incl. Subs O&P Daily	Cost Per Labor-Hour Bare Costs	Cost Per Labor-Hour Incl. O&P
1 Equip. Oper. (crane)	$61.45	$491.60	$91.55	$732.40	$56.38	$84.13
1 Truck Driver (heavy)	51.30	410.40	76.70	613.60		
1 Pickup Truck, 4x4, 3/4 Ton		176.75		194.43		
1 Truck Tractor, 6x4, 380 H.P.		499.15		549.07		
1 Lowbed Trailer, 75 Ton		258.10		283.91	58.38	64.21
16 L.H., Daily Totals		$1836.00		$2373.40	$114.75	$148.34

For customer support on your Electrical Costs with RSMeans Data, call 800.448.8182.

483

Crew No.	Bare Costs		Incl. Subs O&P		Cost Per Labor-Hour	

Crew A-3G

Crew A-3G	Hr.	Daily	Hr.	Daily	Bare Costs	Incl. O&P
1 Equip. Oper. (crane)	$61.45	$491.60	$91.55	$732.40	$56.38	$84.13
1 Truck Driver (heavy)	51.30	410.40	76.70	613.60		
1 Pickup Truck, 4x4, 3/4 Ton		176.75		194.43		
1 Truck Tractor, 6x4, 450 H.P.		608.95		669.85		
1 Lowbed Trailer, 75 Ton		258.10		283.91	65.24	71.76
16 L.H., Daily Totals		$1945.80		$2494.18	$121.61	$155.89

Crew A-3H

Crew A-3H	Hr.	Daily	Hr.	Daily	Bare Costs	Incl. O&P
1 Equip. Oper. (crane)	$61.45	$491.60	$91.55	$732.40	$61.45	$91.55
1 Hyd. Crane, 12 Ton (Daily)		733.15		806.47	91.64	100.81
8 L.H., Daily Totals		$1224.75		$1538.87	$153.09	$192.36

Crew A-3I

Crew A-3I	Hr.	Daily	Hr.	Daily	Bare Costs	Incl. O&P
1 Equip. Oper. (crane)	$61.45	$491.60	$91.55	$732.40	$61.45	$91.55
1 Hyd. Crane, 25 Ton (Daily)		810.50		891.55	101.31	111.44
8 L.H., Daily Totals		$1302.10		$1623.95	$162.76	$202.99

Crew A-3J

Crew A-3J	Hr.	Daily	Hr.	Daily	Bare Costs	Incl. O&P
1 Equip. Oper. (crane)	$61.45	$491.60	$91.55	$732.40	$61.45	$91.55
1 Hyd. Crane, 40 Ton (Daily)		1287.00		1415.70	160.88	176.96
8 L.H., Daily Totals		$1778.60		$2148.10	$222.32	$268.51

Crew A-3K

Crew A-3K	Hr.	Daily	Hr.	Daily	Bare Costs	Incl. O&P
1 Equip. Oper. (crane)	$61.45	$491.60	$91.55	$732.40	$56.98	$84.90
1 Equip. Oper. (oiler)	52.50	420.00	78.25	626.00		
1 Hyd. Crane, 55 Ton (Daily)		1377.00		1514.70		
1 P/U Truck, 3/4 Ton (Daily)		143.85		158.24	95.05	104.56
16 L.H., Daily Totals		$2432.45		$3031.34	$152.03	$189.46

Crew A-3L

Crew A-3L	Hr.	Daily	Hr.	Daily	Bare Costs	Incl. O&P
1 Equip. Oper. (crane)	$61.45	$491.60	$91.55	$732.40	$56.98	$84.90
1 Equip. Oper. (oiler)	52.50	420.00	78.25	626.00		
1 Hyd. Crane, 80 Ton (Daily)		2058.00		2263.80		
1 P/U Truck, 3/4 Ton (Daily)		143.85		158.24	137.62	151.38
16 L.H., Daily Totals		$3113.45		$3780.43	$194.59	$236.28

Crew A-3M

Crew A-3M	Hr.	Daily	Hr.	Daily	Bare Costs	Incl. O&P
1 Equip. Oper. (crane)	$61.45	$491.60	$91.55	$732.40	$56.98	$84.90
1 Equip. Oper. (oiler)	52.50	420.00	78.25	626.00		
1 Hyd. Crane, 100 Ton (Daily)		2253.00		2478.30		
1 P/U Truck, 3/4 Ton (Daily)		143.85		158.24	149.80	164.78
16 L.H., Daily Totals		$3308.45		$3994.93	$206.78	$249.68

Crew A-3N

Crew A-3N	Hr.	Daily	Hr.	Daily	Bare Costs	Incl. O&P
1 Equip. Oper. (crane)	$61.45	$491.60	$91.55	$732.40	$61.45	$91.55
1 Tower Crane (monthly)		1737.00		1910.70	217.13	238.84
8 L.H., Daily Totals		$2228.60		$2643.10	$278.57	$330.39

Crew A-3P

Crew A-3P	Hr.	Daily	Hr.	Daily	Bare Costs	Incl. O&P
1 Equip. Oper. (light)	$55.50	$444.00	$82.70	$661.60	$55.50	$82.70
1 A.T. Forklift, 31' reach, 45' lift		346.45		381.10	43.31	47.64
8 L.H., Daily Totals		$790.45		$1042.69	$98.81	$130.34

Crew A-3Q

Crew A-3Q	Hr.	Daily	Hr.	Daily	Bare Costs	Incl. O&P
1 Equip. Oper. (light)	$55.50	$444.00	$82.70	$661.60	$55.50	$82.70
1 Pickup Truck, 4x4, 3/4 Ton		176.75		194.43		
1 Flatbed Trailer, 3 Ton		71.15		78.27	30.99	34.09
8 L.H., Daily Totals		$691.90		$934.29	$86.49	$116.79

Crew A-3R

Crew A-3R	Hr.	Daily	Hr.	Daily	Bare Costs	Incl. O&P
1 Equip. Oper. (light)	$55.50	$444.00	$82.70	$661.60	$55.50	$82.70
1 Forklift, Smooth Floor, 8,000 Lb.		283.25		311.57	35.41	38.95
8 L.H., Daily Totals		$727.25		$973.17	$90.91	$121.65

Crew A-4

Crew A-4	Hr.	Daily	Hr.	Daily	Bare Costs	Incl. O&P
2 Carpenters	$54.70	$875.20	$81.65	$1306.40	$51.95	$77.40
1 Painter, Ordinary	46.45	371.60	68.90	551.20		
24 L.H., Daily Totals		$1246.80		$1857.60	$51.95	$77.40

Crew A-5

Crew A-5	Hr.	Daily	Hr.	Daily	Bare Costs	Incl. O&P
2 Laborers	$44.40	$710.40	$66.25	$1060.00	$44.89	$67.00
.25 Truck Driver (light)	48.80	97.60	73.00	146.00		
.25 Flatbed Truck, Gas, 1.5 Ton		49.63		54.59	2.76	3.03
18 L.H., Daily Totals		$857.63		$1260.59	$47.65	$70.03

Crew A-6

Crew A-6	Hr.	Daily	Hr.	Daily	Bare Costs	Incl. O&P
1 Instrument Man	$57.10	$456.80	$85.90	$687.20	$54.88	$82.15
1 Rodman/Chainman	52.65	421.20	78.40	627.20		
1 Level, Electronic		34.40		37.84	2.15	2.37
16 L.H., Daily Totals		$912.40		$1352.24	$57.02	$84.52

Crew A-7

Crew A-7	Hr.	Daily	Hr.	Daily	Bare Costs	Incl. O&P
1 Chief of Party	$68.50	$548.00	$102.35	$818.80	$59.42	$88.88
1 Instrument Man	57.10	456.80	85.90	687.20		
1 Rodman/Chainman	52.65	421.20	78.40	627.20		
1 Level, Electronic		34.40		37.84	1.43	1.58
24 L.H., Daily Totals		$1460.40		$2171.04	$60.85	$90.46

Crew A-8

Crew A-8	Hr.	Daily	Hr.	Daily	Bare Costs	Incl. O&P
1 Chief of Party	$68.50	$548.00	$102.35	$818.80	$57.73	$86.26
1 Instrument Man	57.10	456.80	85.90	687.20		
2 Rodmen/Chainmen	52.65	842.40	78.40	1254.40		
1 Level, Electronic		34.40		37.84	1.08	1.18
32 L.H., Daily Totals		$1881.60		$2798.24	$58.80	$87.44

Crew A-9

Crew A-9	Hr.	Daily	Hr.	Daily	Bare Costs	Incl. O&P
1 Asbestos Foreman	$61.45	$491.60	$94.00	$752.00	$61.01	$93.34
7 Asbestos Workers	60.95	3413.20	93.25	5222.00		
64 L.H., Daily Totals		$3904.80		$5974.00	$61.01	$93.34

Crew A-10A

Crew A-10A	Hr.	Daily	Hr.	Daily	Bare Costs	Incl. O&P
1 Asbestos Foreman	$61.45	$491.60	$94.00	$752.00	$61.12	$93.50
2 Asbestos Workers	60.95	975.20	93.25	1492.00		
24 L.H., Daily Totals		$1466.80		$2244.00	$61.12	$93.50

Crew A-10B

Crew A-10B	Hr.	Daily	Hr.	Daily	Bare Costs	Incl. O&P
1 Asbestos Foreman	$61.45	$491.60	$94.00	$752.00	$61.08	$93.44
3 Asbestos Workers	60.95	1462.80	93.25	2238.00		
32 L.H., Daily Totals		$1954.40		$2990.00	$61.08	$93.44

Crew A-10C

Crew A-10C	Hr.	Daily	Hr.	Daily	Bare Costs	Incl. O&P
3 Asbestos Workers	$60.95	$1462.80	$93.25	$2238.00	$60.95	$93.25
1 Flatbed Truck, Gas, 1.5 Ton		198.50		218.35	8.27	9.10
24 L.H., Daily Totals		$1661.30		$2456.35	$69.22	$102.35

Crews - Standard

Crew A-10D

Crew A-10D	Bare Costs Hr.	Bare Costs Daily	Incl. Subs O&P Hr.	Incl. Subs O&P Daily	Cost Per Labor-Hour Bare Costs	Cost Per Labor-Hour Incl. O&P
2 Asbestos Workers	$60.95	$975.20	$93.25	$1492.00	$58.96	$89.08
1 Equip. Oper. (crane)	61.45	491.60	91.55	732.40		
1 Equip. Oper. (oiler)	52.50	420.00	78.25	626.00		
1 Hydraulic Crane, 33 Ton		983.15		1081.46	30.72	33.80
32 L.H., Daily Totals		$2869.95		$3931.86	$89.69	$122.87

Crew A-11

Crew A-11	Bare Costs Hr.	Bare Costs Daily	Incl. Subs O&P Hr.	Incl. Subs O&P Daily	Cost Per Labor-Hour Bare Costs	Cost Per Labor-Hour Incl. O&P
1 Asbestos Foreman	$61.45	$491.60	$94.00	$752.00	$61.01	$93.34
7 Asbestos Workers	60.95	3413.20	93.25	5222.00		
2 Chip. Hammers, 12 Lb., Elec.		65.70		72.27	1.03	1.13
64 L.H., Daily Totals		$3970.50		$6046.27	$62.04	$94.47

Crew A-12

Crew A-12	Bare Costs Hr.	Bare Costs Daily	Incl. Subs O&P Hr.	Incl. Subs O&P Daily	Cost Per Labor-Hour Bare Costs	Cost Per Labor-Hour Incl. O&P
1 Asbestos Foreman	$61.45	$491.60	$94.00	$752.00	$61.01	$93.34
7 Asbestos Workers	60.95	3413.20	93.25	5222.00		
1 Trk-Mtd Vac, 14 CY, 1500 Gal.		542.60		596.86		
1 Flatbed Truck, 20,000 GVW		204.05		224.46	11.67	12.83
64 L.H., Daily Totals		$4651.45		$6795.31	$72.68	$106.18

Crew A-13

Crew A-13	Bare Costs Hr.	Bare Costs Daily	Incl. Subs O&P Hr.	Incl. Subs O&P Daily	Cost Per Labor-Hour Bare Costs	Cost Per Labor-Hour Incl. O&P
1 Equip. Oper. (light)	$55.50	$444.00	$82.70	$661.60	$55.50	$82.70
1 Trk-Mtd Vac, 14 CY, 1500 Gal.		542.60		596.86		
1 Flatbed Truck, 20,000 GVW		204.05		224.46	93.33	102.66
8 L.H., Daily Totals		$1190.65		$1482.92	$148.83	$185.36

Crew B-1

Crew B-1	Bare Costs Hr.	Bare Costs Daily	Incl. Subs O&P Hr.	Incl. Subs O&P Daily	Cost Per Labor-Hour Bare Costs	Cost Per Labor-Hour Incl. O&P
1 Labor Foreman (outside)	$46.40	$371.20	$69.25	$554.00	$45.07	$67.25
2 Laborers	44.40	710.40	66.25	1060.00		
24 L.H., Daily Totals		$1081.60		$1614.00	$45.07	$67.25

Crew B-1A

Crew B-1A	Bare Costs Hr.	Bare Costs Daily	Incl. Subs O&P Hr.	Incl. Subs O&P Daily	Cost Per Labor-Hour Bare Costs	Cost Per Labor-Hour Incl. O&P
1 Labor Foreman (outside)	$46.40	$371.20	$69.25	$554.00	$45.07	$67.25
2 Laborers	44.40	710.40	66.25	1060.00		
2 Cutting Torches		25.90		28.49		
2 Sets of Gases		347.20		381.92	15.55	17.10
24 L.H., Daily Totals		$1454.70		$2024.41	$60.61	$84.35

Crew B-1B

Crew B-1B	Bare Costs Hr.	Bare Costs Daily	Incl. Subs O&P Hr.	Incl. Subs O&P Daily	Cost Per Labor-Hour Bare Costs	Cost Per Labor-Hour Incl. O&P
1 Labor Foreman (outside)	$46.40	$371.20	$69.25	$554.00	$49.16	$73.33
2 Laborers	44.40	710.40	66.25	1060.00		
1 Equip. Oper. (crane)	61.45	491.60	91.55	732.40		
2 Cutting Torches		25.90		28.49		
2 Sets of Gases		347.20		381.92		
1 Hyd. Crane, 12 Ton		475.80		523.38	26.53	29.18
32 L.H., Daily Totals		$2422.10		$3280.19	$75.69	$102.51

Crew B-1C

Crew B-1C	Bare Costs Hr.	Bare Costs Daily	Incl. Subs O&P Hr.	Incl. Subs O&P Daily	Cost Per Labor-Hour Bare Costs	Cost Per Labor-Hour Incl. O&P
1 Labor Foreman (outside)	$46.40	$371.20	$69.25	$554.00	$45.07	$67.25
2 Laborers	44.40	710.40	66.25	1060.00		
1 Telescoping Boom Lift, to 60'		292.45		321.69	12.19	13.40
24 L.H., Daily Totals		$1374.05		$1935.69	$57.25	$80.65

Crew B-1D

Crew B-1D	Bare Costs Hr.	Bare Costs Daily	Incl. Subs O&P Hr.	Incl. Subs O&P Daily	Cost Per Labor-Hour Bare Costs	Cost Per Labor-Hour Incl. O&P
2 Laborers	$44.40	$710.40	$66.25	$1060.00	$44.40	$66.25
1 Small Work Boat, Gas, 50 H.P.		120.85		132.94		
1 Pressure Washer, 7 GPM		93.95		103.35	13.43	14.77
16 L.H., Daily Totals		$925.20		$1296.28	$57.83	$81.02

Crew B-1E

Crew B-1E	Bare Costs Hr.	Bare Costs Daily	Incl. Subs O&P Hr.	Incl. Subs O&P Daily	Cost Per Labor-Hour Bare Costs	Cost Per Labor-Hour Incl. O&P
1 Labor Foreman (outside)	$46.40	$371.20	$69.25	$554.00	$44.90	$67.00
3 Laborers	44.40	1065.60	66.25	1590.00		
1 Work Boat, Diesel, 200 H.P.		1436.00		1579.60		
2 Pressure Washers, 7 GPM		187.90		206.69	50.75	55.82
32 L.H., Daily Totals		$3060.70		$3930.29	$95.65	$122.82

Crew B-1F

Crew B-1F	Bare Costs Hr.	Bare Costs Daily	Incl. Subs O&P Hr.	Incl. Subs O&P Daily	Cost Per Labor-Hour Bare Costs	Cost Per Labor-Hour Incl. O&P
2 Skilled Workers	$57.10	$913.60	$85.90	$1374.40	$52.87	$79.35
1 Laborer	44.40	355.20	66.25	530.00		
1 Small Work Boat, Gas, 50 H.P.		120.85		132.94		
1 Pressure Washer, 7 GPM		93.95		103.35	8.95	9.85
24 L.H., Daily Totals		$1483.60		$2140.68	$61.82	$89.19

Crew B-1G

Crew B-1G	Bare Costs Hr.	Bare Costs Daily	Incl. Subs O&P Hr.	Incl. Subs O&P Daily	Cost Per Labor-Hour Bare Costs	Cost Per Labor-Hour Incl. O&P
2 Laborers	$44.40	$710.40	$66.25	$1060.00	$44.40	$66.25
1 Small Work Boat, Gas, 50 H.P.		120.85		132.94	7.55	8.31
16 L.H., Daily Totals		$831.25		$1192.93	$51.95	$74.56

Crew B-1H

Crew B-1H	Bare Costs Hr.	Bare Costs Daily	Incl. Subs O&P Hr.	Incl. Subs O&P Daily	Cost Per Labor-Hour Bare Costs	Cost Per Labor-Hour Incl. O&P
2 Skilled Workers	$57.10	$913.60	$85.90	$1374.40	$52.87	$79.35
1 Laborer	44.40	355.20	66.25	530.00		
1 Small Work Boat, Gas, 50 H.P.		120.85		132.94	5.04	5.54
24 L.H., Daily Totals		$1389.65		$2037.34	$57.90	$84.89

Crew B-1J

Crew B-1J	Bare Costs Hr.	Bare Costs Daily	Incl. Subs O&P Hr.	Incl. Subs O&P Daily	Cost Per Labor-Hour Bare Costs	Cost Per Labor-Hour Incl. O&P
1 Labor Foreman (inside)	$44.90	$359.20	$67.00	$536.00	$44.65	$66.63
1 Laborer	44.40	355.20	66.25	530.00		
16 L.H., Daily Totals		$714.40		$1066.00	$44.65	$66.63

Crew B-1K

Crew B-1K	Bare Costs Hr.	Bare Costs Daily	Incl. Subs O&P Hr.	Incl. Subs O&P Daily	Cost Per Labor-Hour Bare Costs	Cost Per Labor-Hour Incl. O&P
1 Carpenter Foreman (inside)	$55.20	$441.60	$82.40	$659.20	$54.95	$82.03
1 Carpenter	54.70	437.60	81.65	653.20		
16 L.H., Daily Totals		$879.20		$1312.40	$54.95	$82.03

Crew B-2

Crew B-2	Bare Costs Hr.	Bare Costs Daily	Incl. Subs O&P Hr.	Incl. Subs O&P Daily	Cost Per Labor-Hour Bare Costs	Cost Per Labor-Hour Incl. O&P
1 Labor Foreman (outside)	$46.40	$371.20	$69.25	$554.00	$44.80	$66.85
4 Laborers	44.40	1420.80	66.25	2120.00		
40 L.H., Daily Totals		$1792.00		$2674.00	$44.80	$66.85

Crew B-2A

Crew B-2A	Bare Costs Hr.	Bare Costs Daily	Incl. Subs O&P Hr.	Incl. Subs O&P Daily	Cost Per Labor-Hour Bare Costs	Cost Per Labor-Hour Incl. O&P
1 Labor Foreman (outside)	$46.40	$371.20	$69.25	$554.00	$45.07	$67.25
2 Laborers	44.40	710.40	66.25	1060.00		
1 Telescoping Boom Lift, to 60'		292.45		321.69	12.19	13.40
24 L.H., Daily Totals		$1374.05		$1935.69	$57.25	$80.65

Crew B-3

Crew B-3	Bare Costs Hr.	Bare Costs Daily	Incl. Subs O&P Hr.	Incl. Subs O&P Daily	Cost Per Labor-Hour Bare Costs	Cost Per Labor-Hour Incl. O&P
1 Labor Foreman (outside)	$46.40	$371.20	$69.25	$554.00	$49.47	$73.84
2 Laborers	44.40	710.40	66.25	1060.00		
1 Equip. Oper. (medium)	59.00	472.00	87.90	703.20		
2 Truck Drivers (heavy)	51.30	820.80	76.70	1227.20		
1 Crawler Loader, 3 C.Y.		1146.00		1260.60		
2 Dump Trucks, 12 C.Y., 400 H.P.		1158.70		1274.57	48.01	52.82
48 L.H., Daily Totals		$4679.10		$6079.57	$97.48	$126.66

Crew B-3A

Crew B-3A	Bare Costs Hr.	Bare Costs Daily	Incl. Subs O&P Hr.	Incl. Subs O&P Daily	Cost Per Labor-Hour Bare Costs	Cost Per Labor-Hour Incl. O&P
4 Laborers	$44.40	$1420.80	$66.25	$2120.00	$47.32	$70.58
1 Equip. Oper. (medium)	59.00	472.00	87.90	703.20		
1 Hyd. Excavator, 1.5 C.Y.		695.80		765.38	17.40	19.13
40 L.H., Daily Totals		$2588.60		$3588.58	$64.72	$89.71

For customer support on your Electrical Costs with RSMeans Data, call 800.448.8182.

48.

Crew No.	Bare Costs		Incl. Subs O&P		Cost Per Labor-Hour	
Crew B-3B	Hr.	Daily	Hr.	Daily	Bare Costs	Incl. O&P
2 Laborers	$44.40	$710.40	$66.25	$1060.00	$49.77	$74.28
1 Equip. Oper. (medium)	59.00	472.00	87.90	703.20		
1 Truck Driver (heavy)	51.30	410.40	76.70	613.60		
1 Backhoe Loader, 80 H.P.		235.05		258.56		
1 Dump Truck, 12 C.Y., 400 H.P.		579.35		637.28	25.45	28.00
32 L.H., Daily Totals		$2407.20		$3272.64	$75.22	$102.27
Crew B-3C	Hr.	Daily	Hr.	Daily	Bare Costs	Incl. O&P
3 Laborers	$44.40	$1065.60	$66.25	$1590.00	$48.05	$71.66
1 Equip. Oper. (medium)	59.00	472.00	87.90	703.20		
1 Crawler Loader, 4 C.Y.		1456.00		1601.60	45.50	50.05
32 L.H., Daily Totals		$2993.60		$3894.80	$93.55	$121.71
Crew B-4	Hr.	Daily	Hr.	Daily	Bare Costs	Incl. O&P
1 Labor Foreman (outside)	$46.40	$371.20	$69.25	$554.00	$45.88	$68.49
4 Laborers	44.40	1420.80	66.25	2120.00		
1 Truck Driver (heavy)	51.30	410.40	76.70	613.60		
1 Truck Tractor, 220 H.P.		310.80		341.88		
1 Flatbed Trailer, 40 Ton		188.45		207.29	10.40	11.44
48 L.H., Daily Totals		$2701.65		$3836.78	$56.28	$79.93
Crew B-5	Hr.	Daily	Hr.	Daily	Bare Costs	Incl. O&P
1 Labor Foreman (outside)	$46.40	$371.20	$69.25	$554.00	$48.86	$72.86
4 Laborers	44.40	1420.80	66.25	2120.00		
2 Equip. Oper. (medium)	59.00	944.00	87.90	1406.40		
1 Air Compressor, 250 cfm		202.85		223.13		
2 Breakers, Pavement, 60 lb.		107.20		117.92		
2 -50' Air Hoses, 1.5"		45.60		50.16		
1 Crawler Loader, 3 C.Y.		1146.00		1260.60	26.82	29.50
56 L.H., Daily Totals		$4237.65		$5732.22	$75.67	$102.36
Crew B-5A	Hr.	Daily	Hr.	Daily	Bare Costs	Incl. O&P
1 Labor Foreman (outside)	$46.40	$371.20	$69.25	$554.00	$49.08	$73.22
6 Laborers	44.40	2131.20	66.25	3180.00		
2 Equip. Oper. (medium)	59.00	944.00	87.90	1406.40		
1 Equip. Oper. (light)	55.50	444.00	82.70	661.60		
2 Truck Drivers (heavy)	51.30	820.80	76.70	1227.20		
1 Air Compressor, 365 cfm		343.55		377.90		
2 Breakers, Pavement, 60 lb.		107.20		117.92		
8 -50' Air Hoses, 1"		64.40		70.84		
2 Dump Trucks, 8 C.Y., 220 H.P.		815.20		896.72	13.86	15.24
96 L.H., Daily Totals		$6041.55		$8492.58	$62.93	$88.46
Crew B-5B	Hr.	Daily	Hr.	Daily	Bare Costs	Incl. O&P
1 Powderman	$57.10	$456.80	$85.90	$687.20	$54.83	$81.97
2 Equip. Oper. (medium)	59.00	944.00	87.90	1406.40		
3 Truck Drivers (heavy)	51.30	1231.20	76.70	1840.80		
1 F.E. Loader, W.M., 2.5 C.Y.		638.30		702.13		
3 Dump Trucks, 12 C.Y., 400 H.P.		1738.05		1911.86		
1 Air Compressor, 365 cfm		343.55		377.90	56.66	62.33
48 L.H., Daily Totals		$5351.90		$6926.29	$111.50	$144.30

Crew No.	Bare Costs		Incl. Subs O&P		Cost Per Labor-Hour	
Crew B-5C	Hr.	Daily	Hr.	Daily	Bare Costs	Incl. O&P
3 Laborers	$44.40	$1065.60	$66.25	$1590.00	$51.09	$76.23
1 Equip. Oper. (medium)	59.00	472.00	87.90	703.20		
2 Truck Drivers (heavy)	51.30	820.80	76.70	1227.20		
1 Equip. Oper. (crane)	61.45	491.60	91.55	732.40		
1 Equip. Oper. (oiler)	52.50	420.00	78.25	626.00		
2 Dump Trucks, 12 C.Y., 400 H.P.		1158.70		1274.57		
1 Crawler Loader, 4 C.Y.		1456.00		1601.60		
1 S.P. Crane, 4x4, 25 Ton		1155.00		1270.50	58.90	64.79
64 L.H., Daily Totals		$7039.70		$9025.47	$110.00	$141.02
Crew B-5D	Hr.	Daily	Hr.	Daily	Bare Costs	Incl. O&P
1 Labor Foreman (outside)	$46.40	$371.20	$69.25	$554.00	$49.16	$73.34
4 Laborers	44.40	1420.80	66.25	2120.00		
2 Equip. Oper. (medium)	59.00	944.00	87.90	1406.40		
1 Truck Driver (heavy)	51.30	410.40	76.70	613.60		
1 Air Compressor, 250 cfm		202.85		223.13		
2 Breakers, Pavement, 60 lb.		107.20		117.92		
2 -50' Air Hoses, 1.5"		45.60		50.16		
1 Crawler Loader, 3 C.Y.		1146.00		1260.60		
1 Dump Truck, 12 C.Y., 400 H.P.		579.35		637.28	32.52	35.77
64 L.H., Daily Totals		$5227.40		$6983.10	$81.68	$109.11
Crew B-5E	Hr.	Daily	Hr.	Daily	Bare Costs	Incl. O&P
1 Labor Foreman (outside)	$46.40	$371.20	$69.25	$554.00	$49.16	$73.34
4 Laborers	44.40	1420.80	66.25	2120.00		
2 Equip. Oper. (medium)	59.00	944.00	87.90	1406.40		
1 Truck Driver (heavy)	51.30	410.40	76.70	613.60		
1 Water Tank Trailer, 5000 Gal.		154.10		169.51		
1 High Pressure Water Jet 40 KSI		820.75		902.83		
2 -50' Air Hoses, 1.5"		45.60		50.16		
1 Crawler Loader, 3 C.Y.		1146.00		1260.60		
1 Dump Truck, 12 C.Y., 400 H.P.		579.35		637.28	42.90	47.19
64 L.H., Daily Totals		$5892.20		$7714.38	$92.07	$120.54
Crew B-6	Hr.	Daily	Hr.	Daily	Bare Costs	Incl. O&P
2 Laborers	$44.40	$710.40	$66.25	$1060.00	$48.10	$71.73
1 Equip. Oper. (light)	55.50	444.00	82.70	661.60		
1 Backhoe Loader, 48 H.P.		216.20		237.82	9.01	9.91
24 L.H., Daily Totals		$1370.60		$1959.42	$57.11	$81.64
Crew B-6A	Hr.	Daily	Hr.	Daily	Bare Costs	Incl. O&P
.5 Labor Foreman (outside)	$46.40	$185.60	$69.25	$277.00	$50.64	$75.51
1 Laborer	44.40	355.20	66.25	530.00		
1 Equip. Oper. (medium)	59.00	472.00	87.90	703.20		
1 Vacuum Truck, 5000 Gal.		371.95		409.14	18.60	20.46
20 L.H., Daily Totals		$1384.75		$1919.35	$69.24	$95.97
Crew B-6B	Hr.	Daily	Hr.	Daily	Bare Costs	Incl. O&P
2 Labor Foremen (outside)	$46.40	$742.40	$69.25	$1108.00	$45.07	$67.25
4 Laborers	44.40	1420.80	66.25	2120.00		
1 S.P. Crane, 4x4, 5 Ton		381.95		420.14		
1 Flatbed Truck, Gas, 1.5 Ton		198.50		218.35		
1 Butt Fusion Mach., 4"-12" diam.		420.75		462.82	20.86	22.94
48 L.H., Daily Totals		$3164.40		$4329.32	$65.92	$90.19

Crew No.	Bare Costs		Incl. Subs O&P		Cost Per Labor-Hour	

Crew B-6C	Hr.	Daily	Hr.	Daily	Bare Costs	Incl. O&P
2 Labor Foremen (outside)	$46.40	$742.40	$69.25	$1108.00	$45.07	$67.25
4 Laborers	44.40	1420.80	66.25	2120.00		
1 S.P. Crane, 4x4, 12 Ton		432.65		475.92		
1 Flatbed Truck, Gas, 3 Ton		850.05		935.05		
1 Butt Fusion Mach., 8"-24" diam.		1086.00		1194.60	49.35	54.28
48 L.H., Daily Totals		$4531.90		$5833.57	$94.41	$121.53

Crew B-6D	Hr.	Daily	Hr.	Daily	Bare Costs	Incl. O&P
.5 Labor Foreman (outside)	$46.40	$185.60	$69.25	$277.00	$50.64	$75.51
1 Laborer	44.40	355.20	66.25	530.00		
1 Equip. Oper. (medium)	59.00	472.00	87.90	703.20		
1 Hydro Excavator, 12 C.Y.		1277.00		1404.70	63.85	70.23
20 L.H., Daily Totals		$2289.80		$2914.90	$114.49	$145.75

Crew B-7	Hr.	Daily	Hr.	Daily	Bare Costs	Incl. O&P
1 Labor Foreman (outside)	$46.40	$371.20	$69.25	$554.00	$47.17	$70.36
4 Laborers	44.40	1420.80	66.25	2120.00		
1 Equip. Oper. (medium)	59.00	472.00	87.90	703.20		
1 Brush Chipper, 12", 130 H.P.		366.05		402.65		
1 Crawler Loader, 3 C.Y.		1146.00		1260.60		
2 Chain Saws, Gas, 36" Long		83.30		91.63	33.24	36.56
48 L.H., Daily Totals		$3859.35		$5132.09	$80.40	$106.92

Crew B-7A	Hr.	Daily	Hr.	Daily	Bare Costs	Incl. O&P
2 Laborers	$44.40	$710.40	$66.25	$1060.00	$48.10	$71.73
1 Equip. Oper. (light)	55.50	444.00	82.70	661.60		
1 Rake w/Tractor		343.50		377.85		
2 Chain Saws, Gas, 18"		104.40		114.84	18.66	20.53
24 L.H., Daily Totals		$1602.30		$2214.29	$66.76	$92.26

Crew B-7B	Hr.	Daily	Hr.	Daily	Bare Costs	Incl. O&P
1 Labor Foreman (outside)	$46.40	$371.20	$69.25	$554.00	$47.76	$71.26
4 Laborers	44.40	1420.80	66.25	2120.00		
1 Equip. Oper. (medium)	59.00	472.00	87.90	703.20		
1 Truck Driver (heavy)	51.30	410.40	76.70	613.60		
1 Brush Chipper, 12", 130 H.P.		366.05		402.65		
1 Crawler Loader, 3 C.Y.		1146.00		1260.60		
2 Chain Saws, Gas, 36" Long		83.30		91.63		
1 Dump Truck, 8 C.Y., 220 H.P.		407.60		448.36	35.77	39.34
56 L.H., Daily Totals		$4677.35		$6194.05	$83.52	$110.61

Crew B-7C	Hr.	Daily	Hr.	Daily	Bare Costs	Incl. O&P
1 Labor Foreman (outside)	$46.40	$371.20	$69.25	$554.00	$47.76	$71.26
4 Laborers	44.40	1420.80	66.25	2120.00		
1 Equip. Oper. (medium)	59.00	472.00	87.90	703.20		
1 Truck Driver (heavy)	51.30	410.40	76.70	613.60		
1 Brush Chipper, 12", 130 H.P.		366.05		402.65		
1 Crawler Loader, 3 C.Y.		1146.00		1260.60		
2 Chain Saws, Gas, 36" Long		83.30		91.63		
1 Dump Truck, 12 C.Y., 400 H.P.		579.35		637.28	38.83	42.72
56 L.H., Daily Totals		$4849.10		$6382.97	$86.59	$113.98

Crew B-8	Hr.	Daily	Hr.	Daily	Bare Costs	Incl. O&P
1 Labor Foreman (outside)	$46.40	$371.20	$69.25	$554.00	$51.04	$76.15
2 Laborers	44.40	710.40	66.25	1060.00		
2 Equip. Oper. (medium)	59.00	944.00	87.90	1406.40		
1 Equip. Oper. (oiler)	52.50	420.00	78.25	626.00		
2 Truck Drivers (heavy)	51.30	820.80	76.70	1227.20		
1 Hyd. Crane, 25 Ton		586.70		645.37		
1 Crawler Loader, 3 C.Y.		1146.00		1260.60		
2 Dump Trucks, 12 C.Y., 400 H.P.		1158.70		1274.57	45.18	49.70
64 L.H., Daily Totals		$6157.80		$8054.14	$96.22	$125.85

Crew B-9	Hr.	Daily	Hr.	Daily	Bare Costs	Incl. O&P
1 Labor Foreman (outside)	$46.40	$371.20	$69.25	$554.00	$44.80	$66.85
4 Laborers	44.40	1420.80	66.25	2120.00		
1 Air Compressor, 250 cfm		202.85		223.13		
2 Breakers, Pavement, 60 lb.		107.20		117.92		
2 -50' Air Hoses, 1.5"		45.60		50.16	8.89	9.78
40 L.H., Daily Totals		$2147.65		$3065.22	$53.69	$76.63

Crew B-9A	Hr.	Daily	Hr.	Daily	Bare Costs	Incl. O&P
2 Laborers	$44.40	$710.40	$66.25	$1060.00	$46.70	$69.73
1 Truck Driver (heavy)	51.30	410.40	76.70	613.60		
1 Water Tank Trailer, 5000 Gal.		154.10		169.51		
1 Truck Tractor, 220 H.P.		310.80		341.88		
2 -50' Discharge Hoses, 3"		9.00		9.90	19.75	21.72
24 L.H., Daily Totals		$1594.70		$2194.89	$66.45	$91.45

Crew B-9B	Hr.	Daily	Hr.	Daily	Bare Costs	Incl. O&P
2 Laborers	$44.40	$710.40	$66.25	$1060.00	$46.70	$69.73
1 Truck Driver (heavy)	51.30	410.40	76.70	613.60		
2 -50' Discharge Hoses, 3"		9.00		9.90		
1 Water Tank Trailer, 5000 Gal.		154.10		169.51		
1 Truck Tractor, 220 H.P.		310.80		341.88		
1 Pressure Washer		97.35		107.08	23.80	26.18
24 L.H., Daily Totals		$1692.05		$2301.97	$70.50	$95.92

Crew B-9D	Hr.	Daily	Hr.	Daily	Bare Costs	Incl. O&P
1 Labor Foreman (outside)	$46.40	$371.20	$69.25	$554.00	$44.80	$66.85
4 Common Laborers	44.40	1420.80	66.25	2120.00		
1 Air Compressor, 250 cfm		202.85		223.13		
2 -50' Air Hoses, 1.5"		45.60		50.16		
2 Air Powered Tampers		79.50		87.45	8.20	9.02
40 L.H., Daily Totals		$2119.95		$3034.74	$53.00	$75.87

Crew B-9E	Hr.	Daily	Hr.	Daily	Bare Costs	Incl. O&P
1 Cement Finisher	$51.80	$414.40	$75.90	$607.20	$48.10	$71.08
1 Laborer	44.40	355.20	66.25	530.00		
1 Chip. Hammers, 12 Lb., Elec.		32.85		36.13	2.05	2.26
16 L.H., Daily Totals		$802.45		$1173.34	$50.15	$73.33

Crew B-10	Hr.	Daily	Hr.	Daily	Bare Costs	Incl. O&P
1 Equip. Oper. (medium)	$59.00	$472.00	$87.90	$703.20	$54.13	$80.68
.5 Laborer	44.40	177.60	66.25	265.00		
12 L.H., Daily Totals		$649.60		$968.20	$54.13	$80.68

Crew B-10A	Hr.	Daily	Hr.	Daily	Bare Costs	Incl. O&P
1 Equip. Oper. (medium)	$59.00	$472.00	$87.90	$703.20	$54.13	$80.68
.5 Laborer	44.40	177.60	66.25	265.00		
1 Roller, 2-Drum, W.B., 7.5 H.P.		166.75		183.43	13.90	15.29
12 L.H., Daily Totals		$816.35		$1151.63	$68.03	$95.97

Crew No.	Bare Costs		Incl. Subs O&P		Cost Per Labor-Hour	

Crew B-10B

	Hr.	Daily	Hr.	Daily	Bare Costs	Incl. O&P
1 Equip. Oper. (medium)	$59.00	$472.00	$87.90	$703.20	$54.13	$80.68
.5 Laborer	44.40	177.60	66.25	265.00		
1 Dozer, 200 H.P.		1520.00		1672.00	126.67	139.33
12 L.H., Daily Totals		$2169.60		$2640.20	$180.80	$220.02

Crew B-10C

	Hr.	Daily	Hr.	Daily	Bare Costs	Incl. O&P
1 Equip. Oper. (medium)	$59.00	$472.00	$87.90	$703.20	$54.13	$80.68
.5 Laborer	44.40	177.60	66.25	265.00		
1 Dozer, 200 H.P.		1520.00		1672.00		
1 Vibratory Roller, Towed, 23 Ton		520.35		572.38	170.03	187.03
12 L.H., Daily Totals		$2689.95		$3212.59	$224.16	$267.72

Crew B-10D

	Hr.	Daily	Hr.	Daily	Bare Costs	Incl. O&P
1 Equip. Oper. (medium)	$59.00	$472.00	$87.90	$703.20	$54.13	$80.68
.5 Laborer	44.40	177.60	66.25	265.00		
1 Dozer, 200 H.P.		1520.00		1672.00		
1 Sheepsft. Roller, Towed		426.95		469.64	162.25	178.47
12 L.H., Daily Totals		$2596.55		$3109.84	$216.38	$259.15

Crew B-10E

	Hr.	Daily	Hr.	Daily	Bare Costs	Incl. O&P
1 Equip. Oper. (medium)	$59.00	$472.00	$87.90	$703.20	$54.13	$80.68
.5 Laborer	44.40	177.60	66.25	265.00		
1 Tandem Roller, 5 Ton		258.75		284.63	21.56	23.72
12 L.H., Daily Totals		$908.35		$1252.83	$75.70	$104.40

Crew B-10F

	Hr.	Daily	Hr.	Daily	Bare Costs	Incl. O&P
1 Equip. Oper. (medium)	$59.00	$472.00	$87.90	$703.20	$54.13	$80.68
.5 Laborer	44.40	177.60	66.25	265.00		
1 Tandem Roller, 10 Ton		246.80		271.48	20.57	22.62
12 L.H., Daily Totals		$896.40		$1239.68	$74.70	$103.31

Crew B-10G

	Hr.	Daily	Hr.	Daily	Bare Costs	Incl. O&P
1 Equip. Oper. (medium)	$59.00	$472.00	$87.90	$703.20	$54.13	$80.68
.5 Laborer	44.40	177.60	66.25	265.00		
1 Sheepsfoot Roller, 240 H.P.		1363.00		1499.30	113.58	124.94
12 L.H., Daily Totals		$2012.60		$2467.50	$167.72	$205.63

Crew B-10H

	Hr.	Daily	Hr.	Daily	Bare Costs	Incl. O&P
1 Equip. Oper. (medium)	$59.00	$472.00	$87.90	$703.20	$54.13	$80.68
.5 Laborer	44.40	177.60	66.25	265.00		
1 Diaphragm Water Pump, 2"		87.70		96.47		
1 -20' Suction Hose, 2"		3.55		3.90		
2 -50' Discharge Hoses, 2"		8.00		8.80	8.27	9.10
12 L.H., Daily Totals		$748.85		$1077.38	$62.40	$89.78

Crew B-10I

	Hr.	Daily	Hr.	Daily	Bare Costs	Incl. O&P
1 Equip. Oper. (medium)	$59.00	$472.00	$87.90	$703.20	$54.13	$80.68
.5 Laborer	44.40	177.60	66.25	265.00		
1 Diaphragm Water Pump, 4"		106.35		116.99		
1 -20' Suction Hose, 4"		17.25		18.98		
2 -50' Discharge Hoses, 4"		25.60		28.16	12.43	13.68
12 L.H., Daily Totals		$798.80		$1132.32	$66.57	$94.36

Crew B-10J

	Hr.	Daily	Hr.	Daily	Bare Costs	Incl. O&P
1 Equip. Oper. (medium)	$59.00	$472.00	$87.90	$703.20	$54.13	$80.68
.5 Laborer	44.40	177.60	66.25	265.00		
1 Centrifugal Water Pump, 3"		74.40		81.84		
1 -20' Suction Hose, 3"		8.75		9.63		
2 -50' Discharge Hoses, 3"		9.00		9.90	7.68	8.45
12 L.H., Daily Totals		$741.75		$1069.57	$61.81	$89.13

Crew B-10K

	Hr.	Daily	Hr.	Daily	Bare Costs	Incl. O&P
1 Equip. Oper. (medium)	$59.00	$472.00	$87.90	$703.20	$54.13	$80.68
.5 Laborer	44.40	177.60	66.25	265.00		
1 Centr. Water Pump, 6"		235.25		258.77		
1 -20' Suction Hose, 6"		25.50		28.05		
2 -50' Discharge Hoses, 6"		36.20		39.82	24.75	27.22
12 L.H., Daily Totals		$946.55		$1294.85	$78.88	$107.90

Crew B-10L

	Hr.	Daily	Hr.	Daily	Bare Costs	Incl. O&P
1 Equip. Oper. (medium)	$59.00	$472.00	$87.90	$703.20	$54.13	$80.68
.5 Laborer	44.40	177.60	66.25	265.00		
1 Dozer, 80 H.P.		405.85		446.44	33.82	37.20
12 L.H., Daily Totals		$1055.45		$1414.64	$87.95	$117.89

Crew B-10M

	Hr.	Daily	Hr.	Daily	Bare Costs	Incl. O&P
1 Equip. Oper. (medium)	$59.00	$472.00	$87.90	$703.20	$54.13	$80.68
.5 Laborer	44.40	177.60	66.25	265.00		
1 Dozer, 300 H.P.		1785.00		1963.50	148.75	163.63
12 L.H., Daily Totals		$2434.60		$2931.70	$202.88	$244.31

Crew B-10N

	Hr.	Daily	Hr.	Daily	Bare Costs	Incl. O&P
1 Equip. Oper. (medium)	$59.00	$472.00	$87.90	$703.20	$54.13	$80.68
.5 Laborer	44.40	177.60	66.25	265.00		
1 F.E. Loader, T.M., 1.5 C.Y.		572.00		629.20	47.67	52.43
12 L.H., Daily Totals		$1221.60		$1597.40	$101.80	$133.12

Crew B-10O

	Hr.	Daily	Hr.	Daily	Bare Costs	Incl. O&P
1 Equip. Oper. (medium)	$59.00	$472.00	$87.90	$703.20	$54.13	$80.68
.5 Laborer	44.40	177.60	66.25	265.00		
1 F.E. Loader, T.M., 2.25 C.Y.		925.50		1018.05	77.13	84.84
12 L.H., Daily Totals		$1575.10		$1986.25	$131.26	$165.52

Crew B-10P

	Hr.	Daily	Hr.	Daily	Bare Costs	Incl. O&P
1 Equip. Oper. (medium)	$59.00	$472.00	$87.90	$703.20	$54.13	$80.68
.5 Laborer	44.40	177.60	66.25	265.00		
1 Crawler Loader, 3 C.Y.		1146.00		1260.60	95.50	105.05
12 L.H., Daily Totals		$1795.60		$2228.80	$149.63	$185.73

Crew B-10Q

	Hr.	Daily	Hr.	Daily	Bare Costs	Incl. O&P
1 Equip. Oper. (medium)	$59.00	$472.00	$87.90	$703.20	$54.13	$80.68
.5 Laborer	44.40	177.60	66.25	265.00		
1 Crawler Loader, 4 C.Y.		1456.00		1601.60	121.33	133.47
12 L.H., Daily Totals		$2105.60		$2569.80	$175.47	$214.15

Crew B-10R

	Hr.	Daily	Hr.	Daily	Bare Costs	Incl. O&P
1 Equip. Oper. (medium)	$59.00	$472.00	$87.90	$703.20	$54.13	$80.68
.5 Laborer	44.40	177.60	66.25	265.00		
1 F.E. Loader, W.M., 1 C.Y.		305.60		336.16	25.47	28.01
12 L.H., Daily Totals		$955.20		$1304.36	$79.60	$108.70

Crew B-10S

	Hr.	Daily	Hr.	Daily	Bare Costs	Incl. O&P
1 Equip. Oper. (medium)	$59.00	$472.00	$87.90	$703.20	$54.13	$80.68
.5 Laborer	44.40	177.60	66.25	265.00		
1 F.E. Loader, W.M., 1.5 C.Y.		441.40		485.54	36.78	40.46
12 L.H., Daily Totals		$1091.00		$1453.74	$90.92	$121.15

Crew B-10T

	Hr.	Daily	Hr.	Daily	Bare Costs	Incl. O&P
1 Equip. Oper. (medium)	$59.00	$472.00	$87.90	$703.20	$54.13	$80.68
.5 Laborer	44.40	177.60	66.25	265.00		
1 F.E. Loader, W.M., 2.5 C.Y.		638.30		702.13	53.19	58.51
12 L.H., Daily Totals		$1287.90		$1670.33	$107.33	$139.19

For customer support on your Electrical Costs with RSMeans Data, call 800.448.8182.

Crew No.	Bare Costs		Incl. Subs O&P		Cost Per Labor-Hour	

Crew B-10U

Crew B-10U	Hr.	Daily	Hr.	Daily	Bare Costs	Incl. O&P
1 Equip. Oper. (medium)	$59.00	$472.00	$87.90	$703.20	$54.13	$80.68
.5 Laborer	44.40	177.60	66.25	265.00		
1 F.E. Loader, W.M., 5.5 C.Y.		967.95		1064.74	80.66	88.73
12 L.H., Daily Totals		$1617.55		$2032.94	$134.80	$169.41

Crew B-10V

Crew B-10V	Hr.	Daily	Hr.	Daily	Bare Costs	Incl. O&P
1 Equip. Oper. (medium)	$59.00	$472.00	$87.90	$703.20	$54.13	$80.68
.5 Laborer	44.40	177.60	66.25	265.00		
1 Dozer, 700 H.P.		5175.00		5692.50	431.25	474.38
12 L.H., Daily Totals		$5824.60		$6660.70	$485.38	$555.06

Crew B-10W

Crew B-10W	Hr.	Daily	Hr.	Daily	Bare Costs	Incl. O&P
1 Equip. Oper. (medium)	$59.00	$472.00	$87.90	$703.20	$54.13	$80.68
.5 Laborer	44.40	177.60	66.25	265.00		
1 Dozer, 105 H.P.		640.80		704.88	53.40	58.74
12 L.H., Daily Totals		$1290.40		$1673.08	$107.53	$139.42

Crew B-10X

Crew B-10X	Hr.	Daily	Hr.	Daily	Bare Costs	Incl. O&P
1 Equip. Oper. (medium)	$59.00	$472.00	$87.90	$703.20	$54.13	$80.68
.5 Laborer	44.40	177.60	66.25	265.00		
1 Dozer, 410 H.P.		2807.00		3087.70	233.92	257.31
12 L.H., Daily Totals		$3456.60		$4055.90	$288.05	$337.99

Crew B-10Y

Crew B-10Y	Hr.	Daily	Hr.	Daily	Bare Costs	Incl. O&P
1 Equip. Oper. (medium)	$59.00	$472.00	$87.90	$703.20	$54.13	$80.68
.5 Laborer	44.40	177.60	66.25	265.00		
1 Vibr. Roller, Towed, 12 Ton		584.80		643.28	48.73	53.61
12 L.H., Daily Totals		$1234.40		$1611.48	$102.87	$134.29

Crew B-11A

Crew B-11A	Hr.	Daily	Hr.	Daily	Bare Costs	Incl. O&P
1 Equipment Oper. (med.)	$59.00	$472.00	$87.90	$703.20	$51.70	$77.08
1 Laborer	44.40	355.20	66.25	530.00		
1 Dozer, 200 H.P.		1520.00		1672.00	95.00	104.50
16 L.H., Daily Totals		$2347.20		$2905.20	$146.70	$181.57

Crew B-11B

Crew B-11B	Hr.	Daily	Hr.	Daily	Bare Costs	Incl. O&P
1 Equipment Oper. (light)	$55.50	$444.00	$82.70	$661.60	$49.95	$74.47
1 Laborer	44.40	355.20	66.25	530.00		
1 Air Powered Tamper		39.75		43.73		
1 Air Compressor, 365 cfm		343.55		377.90		
2 -50' Air Hoses, 1.5"		45.60		50.16	26.81	29.49
16 L.H., Daily Totals		$1228.10		$1663.39	$76.76	$103.96

Crew B-11C

Crew B-11C	Hr.	Daily	Hr.	Daily	Bare Costs	Incl. O&P
1 Equipment Oper. (med.)	$59.00	$472.00	$87.90	$703.20	$51.70	$77.08
1 Laborer	44.40	355.20	66.25	530.00		
1 Backhoe Loader, 48 H.P.		216.20		237.82	13.51	14.86
16 L.H., Daily Totals		$1043.40		$1471.02	$65.21	$91.94

Crew B-11J

Crew B-11J	Hr.	Daily	Hr.	Daily	Bare Costs	Incl. O&P
1 Equipment Oper. (med.)	$59.00	$472.00	$87.90	$703.20	$51.70	$77.08
1 Laborer	44.40	355.20	66.25	530.00		
1 Grader, 30,000 Lbs.		1073.00		1180.30		
1 Ripper, Beam & 1 Shank		91.60		100.76	72.79	80.07
16 L.H., Daily Totals		$1991.80		$2514.26	$124.49	$157.14

Crew B-11K

Crew B-11K	Hr.	Daily	Hr.	Daily	Bare Costs	Incl. O&P
1 Equipment Oper. (med.)	$59.00	$472.00	$87.90	$703.20	$51.70	$77.08
1 Laborer	44.40	355.20	66.25	530.00		
1 Trencher, Chain Type, 8' D		1894.00		2083.40	118.38	130.21
16 L.H., Daily Totals		$2721.20		$3316.60	$170.07	$207.29

Crew B-11L

Crew B-11L	Hr.	Daily	Hr.	Daily	Bare Costs	Incl. O&P
1 Equipment Oper. (med.)	$59.00	$472.00	$87.90	$703.20	$51.70	$77.08
1 Laborer	44.40	355.20	66.25	530.00		
1 Grader, 30,000 Lbs.		1073.00		1180.30	67.06	73.77
16 L.H., Daily Totals		$1900.20		$2413.50	$118.76	$150.84

Crew B-11M

Crew B-11M	Hr.	Daily	Hr.	Daily	Bare Costs	Incl. O&P
1 Equipment Oper. (med.)	$59.00	$472.00	$87.90	$703.20	$51.70	$77.08
1 Laborer	44.40	355.20	66.25	530.00		
1 Backhoe Loader, 80 H.P.		235.05		258.56	14.69	16.16
16 L.H., Daily Totals		$1062.25		$1491.76	$66.39	$93.23

Crew B-11N

Crew B-11N	Hr.	Daily	Hr.	Daily	Bare Costs	Incl. O&P
1 Labor Foreman (outside)	$46.40	$371.20	$69.25	$554.00	$52.47	$78.36
2 Equipment Operators (med.)	59.00	944.00	87.90	1406.40		
6 Truck Drivers (heavy)	51.30	2462.40	76.70	3681.60		
1 F.E. Loader, W.M., 5.5 C.Y.		967.95		1064.74		
1 Dozer, 410 H.P.		2807.00		3087.70		
6 Dump Trucks, Off Hwy., 50 Ton		11874.00		13061.40	217.35	239.08
72 L.H., Daily Totals		$19426.55		$22855.85	$269.81	$317.44

Crew B-11Q

Crew B-11Q	Hr.	Daily	Hr.	Daily	Bare Costs	Incl. O&P
1 Equipment Operator (med.)	$59.00	$472.00	$87.90	$703.20	$54.13	$80.68
.5 Laborer	44.40	177.60	66.25	265.00		
1 Dozer, 140 H.P.		729.15		802.07	60.76	66.84
12 L.H., Daily Totals		$1378.75		$1770.27	$114.90	$147.52

Crew B-11R

Crew B-11R	Hr.	Daily	Hr.	Daily	Bare Costs	Incl. O&P
1 Equipment Operator (med.)	$59.00	$472.00	$87.90	$703.20	$54.13	$80.68
.5 Laborer	44.40	177.60	66.25	265.00		
1 Dozer, 200 H.P.		1520.00		1672.00	126.67	139.33
12 L.H., Daily Totals		$2169.60		$2640.20	$180.80	$220.02

Crew B-11S

Crew B-11S	Hr.	Daily	Hr.	Daily	Bare Costs	Incl. O&P
1 Equipment Operator (med.)	$59.00	$472.00	$87.90	$703.20	$54.13	$80.68
.5 Laborer	44.40	177.60	66.25	265.00		
1 Dozer, 300 H.P.		1785.00		1963.50		
1 Ripper, Beam & 1 Shank		91.60		100.76	156.38	172.02
12 L.H., Daily Totals		$2526.20		$3032.46	$210.52	$252.71

Crew B-11T

Crew B-11T	Hr.	Daily	Hr.	Daily	Bare Costs	Incl. O&P
1 Equipment Operator (med.)	$59.00	$472.00	$87.90	$703.20	$54.13	$80.68
.5 Laborer	44.40	177.60	66.25	265.00		
1 Dozer, 410 H.P.		2807.00		3087.70		
1 Ripper, Beam & 2 Shanks		140.40		154.44	245.62	270.18
12 L.H., Daily Totals		$3597.00		$4210.34	$299.75	$350.86

Crew B-11U

Crew B-11U	Hr.	Daily	Hr.	Daily	Bare Costs	Incl. O&P
1 Equipment Operator (med.)	$59.00	$472.00	$87.90	$703.20	$54.13	$80.68
.5 Laborer	44.40	177.60	66.25	265.00		
1 Dozer, 520 H.P.		3434.00		3777.40	286.17	314.78
12 L.H., Daily Totals		$4083.60		$4745.60	$340.30	$395.47

For customer support on your Electrical Costs with RSMeans Data, call 800.448.8182.

489

Crews - Standard

Crew B-11V

	Hr.	Daily	Hr.	Daily	Bare Costs	Incl. O&P
3 Laborers	$44.40	$1065.60	$66.25	$1590.00	$44.40	$66.25
1 Roller, 2-Drum, W.B., 7.5 H.P.		166.75		183.43	6.95	7.64
24 L.H., Daily Totals		$1232.35		$1773.43	$51.35	$73.89

Crew B-11W

	Hr.	Daily	Hr.	Daily	Bare Costs	Incl. O&P
1 Equipment Operator (med.)	$59.00	$472.00	$87.90	$703.20	$51.37	$76.76
1 Common Laborer	44.40	355.20	66.25	530.00		
10 Truck Drivers (heavy)	51.30	4104.00	76.70	6136.00		
1 Dozer, 200 H.P.		1520.00		1672.00		
1 Vibratory Roller, Towed, 23 Ton		520.35		572.38		
10 Dump Trucks, 8 C.Y., 220 H.P.		4076.00		4483.60	63.71	70.08
96 L.H., Daily Totals		$11047.55		$14097.18	$115.08	$146.85

Crew B-11Y

	Hr.	Daily	Hr.	Daily	Bare Costs	Incl. O&P
1 Labor Foreman (outside)	$46.40	$371.20	$69.25	$554.00	$49.49	$73.80
5 Common Laborers	44.40	1776.00	66.25	2650.00		
3 Equipment Operators (med.)	59.00	1416.00	87.90	2109.60		
1 Dozer, 80 H.P.		405.85		446.44		
2 Rollers, 2-Drum, W.B., 7.5 H.P.		333.50		366.85		
4 Vibrating Plates, Gas, 21"		662.40		728.64	19.47	21.42
72 L.H., Daily Totals		$4964.95		$6855.52	$68.96	$95.22

Crew B-12A

	Hr.	Daily	Hr.	Daily	Bare Costs	Incl. O&P
1 Equip. Oper. (crane)	$61.45	$491.60	$91.55	$732.40	$52.92	$78.90
1 Laborer	44.40	355.20	66.25	530.00		
1 Hyd. Excavator, 1 C.Y.		832.65		915.91	52.04	57.24
16 L.H., Daily Totals		$1679.45		$2178.32	$104.97	$136.14

Crew B-12B

	Hr.	Daily	Hr.	Daily	Bare Costs	Incl. O&P
1 Equip. Oper. (crane)	$61.45	$491.60	$91.55	$732.40	$52.92	$78.90
1 Laborer	44.40	355.20	66.25	530.00		
1 Hyd. Excavator, 1.5 C.Y.		695.80		765.38	43.49	47.84
16 L.H., Daily Totals		$1542.60		$2027.78	$96.41	$126.74

Crew B-12C

	Hr.	Daily	Hr.	Daily	Bare Costs	Incl. O&P
1 Equip. Oper. (crane)	$61.45	$491.60	$91.55	$732.40	$52.92	$78.90
1 Laborer	44.40	355.20	66.25	530.00		
1 Hyd. Excavator, 2 C.Y.		942.70		1036.97	58.92	64.81
16 L.H., Daily Totals		$1789.50		$2299.37	$111.84	$143.71

Crew B-12D

	Hr.	Daily	Hr.	Daily	Bare Costs	Incl. O&P
1 Equip. Oper. (crane)	$61.45	$491.60	$91.55	$732.40	$52.92	$78.90
1 Laborer	44.40	355.20	66.25	530.00		
1 Hyd. Excavator, 3.5 C.Y.		2184.00		2402.40	136.50	150.15
16 L.H., Daily Totals		$3030.80		$3664.80	$189.43	$229.05

Crew B-12E

	Hr.	Daily	Hr.	Daily	Bare Costs	Incl. O&P
1 Equip. Oper. (crane)	$61.45	$491.60	$91.55	$732.40	$52.92	$78.90
1 Laborer	44.40	355.20	66.25	530.00		
1 Hyd. Excavator, .5 C.Y.		457.00		502.70	28.56	31.42
16 L.H., Daily Totals		$1303.80		$1765.10	$81.49	$110.32

Crew B-12F

	Hr.	Daily	Hr.	Daily	Bare Costs	Incl. O&P
1 Equip. Oper. (crane)	$61.45	$491.60	$91.55	$732.40	$52.92	$78.90
1 Laborer	44.40	355.20	66.25	530.00		
1 Hyd. Excavator, .75 C.Y.		701.80		771.98	43.86	48.25
16 L.H., Daily Totals		$1548.60		$2034.38	$96.79	$127.15

Crew B-12G

	Hr.	Daily	Hr.	Daily	Bare Costs	Incl. O&P
1 Equip. Oper. (crane)	$61.45	$491.60	$91.55	$732.40	$52.92	$78.90
1 Laborer	44.40	355.20	66.25	530.00		
1 Crawler Crane, 15 Ton		810.30		891.33		
1 Clamshell Bucket, .5 C.Y.		67.80		74.58	54.88	60.37
16 L.H., Daily Totals		$1724.90		$2228.31	$107.81	$139.27

Crew B-12H

	Hr.	Daily	Hr.	Daily	Bare Costs	Incl. O&P
1 Equip. Oper. (crane)	$61.45	$491.60	$91.55	$732.40	$52.92	$78.90
1 Laborer	44.40	355.20	66.25	530.00		
1 Crawler Crane, 25 Ton		1152.00		1267.20		
1 Clamshell Bucket, 1 C.Y.		69.25		76.17	76.33	83.96
16 L.H., Daily Totals		$2068.05		$2605.78	$129.25	$162.86

Crew B-12I

	Hr.	Daily	Hr.	Daily	Bare Costs	Incl. O&P
1 Equip. Oper. (crane)	$61.45	$491.60	$91.55	$732.40	$52.92	$78.90
1 Laborer	44.40	355.20	66.25	530.00		
1 Crawler Crane, 20 Ton		1013.00		1114.30		
1 Dragline Bucket, .75 C.Y.		61.85		68.03	67.18	73.90
16 L.H., Daily Totals		$1921.65		$2444.74	$120.10	$152.80

Crew B-12J

	Hr.	Daily	Hr.	Daily	Bare Costs	Incl. O&P
1 Equip. Oper. (crane)	$61.45	$491.60	$91.55	$732.40	$52.92	$78.90
1 Laborer	44.40	355.20	66.25	530.00		
1 Gradall, 5/8 C.Y.		850.65		935.72	53.17	58.48
16 L.H., Daily Totals		$1697.45		$2198.11	$106.09	$137.38

Crew B-12K

	Hr.	Daily	Hr.	Daily	Bare Costs	Incl. O&P
1 Equip. Oper. (crane)	$61.45	$491.60	$91.55	$732.40	$52.92	$78.90
1 Laborer	44.40	355.20	66.25	530.00		
1 Gradall, 3 Ton, 1 C.Y.		984.55		1083.01	61.53	67.69
16 L.H., Daily Totals		$1831.35		$2345.41	$114.46	$146.59

Crew B-12L

	Hr.	Daily	Hr.	Daily	Bare Costs	Incl. O&P
1 Equip. Oper. (crane)	$61.45	$491.60	$91.55	$732.40	$52.92	$78.90
1 Laborer	44.40	355.20	66.25	530.00		
1 Crawler Crane, 15 Ton		810.30		891.33		
1 F.E. Attachment, .5 C.Y.		66.05		72.66	54.77	60.25
16 L.H., Daily Totals		$1723.15		$2226.39	$107.70	$139.15

Crew B-12M

	Hr.	Daily	Hr.	Daily	Bare Costs	Incl. O&P
1 Equip. Oper. (crane)	$61.45	$491.60	$91.55	$732.40	$52.92	$78.90
1 Laborer	44.40	355.20	66.25	530.00		
1 Crawler Crane, 20 Ton		1013.00		1114.30		
1 F.E. Attachment, .75 C.Y.		71.25		78.38	67.77	74.54
16 L.H., Daily Totals		$1931.05		$2455.07	$120.69	$153.44

Crew B-12N

	Hr.	Daily	Hr.	Daily	Bare Costs	Incl. O&P
1 Equip. Oper. (crane)	$61.45	$491.60	$91.55	$732.40	$52.92	$78.90
1 Laborer	44.40	355.20	66.25	530.00		
1 Crawler Crane, 25 Ton		1152.00		1267.20		
1 F.E. Attachment, 1 C.Y.		77.35		85.08	76.83	84.52
16 L.H., Daily Totals		$2076.15		$2614.68	$129.76	$163.42

Crew B-12O

	Hr.	Daily	Hr.	Daily	Bare Costs	Incl. O&P
1 Equip. Oper. (crane)	$61.45	$491.60	$91.55	$732.40	$52.92	$78.90
1 Laborer	44.40	355.20	66.25	530.00		
1 Crawler Crane, 40 Ton		1231.00		1354.10		
1 F.E. Attachment, 1.5 C.Y.		88.65		97.52	82.48	90.73
16 L.H., Daily Totals		$2166.45		$2714.01	$135.40	$169.63

Crew No.	Bare Costs Hr.	Daily	Incl. Subs O&P Hr.	Daily	Cost Per Labor-Hour Bare Costs	Incl. O&P
Crew B-12P	Hr.	Daily	Hr.	Daily	Bare Costs	Incl. O&P
1 Equip. Oper. (crane)	$61.45	$491.60	$91.55	$732.40	$52.92	$78.90
1 Laborer	44.40	355.20	66.25	530.00		
1 Crawler Crane, 40 Ton		1231.00		1354.10		
1 Dragline Bucket, 1.5 C.Y.		65.70		72.27	81.04	89.15
16 L.H., Daily Totals		$2143.50		$2688.77	$133.97	$168.05
Crew B-12Q	Hr.	Daily	Hr.	Daily	Bare Costs	Incl. O&P
1 Equip. Oper. (crane)	$61.45	$491.60	$91.55	$732.40	$52.92	$78.90
1 Laborer	44.40	355.20	66.25	530.00		
1 Hyd. Excavator, 5/8 C.Y.		604.75		665.23	37.80	41.58
16 L.H., Daily Totals		$1451.55		$1927.63	$90.72	$120.48
Crew B-12S	Hr.	Daily	Hr.	Daily	Bare Costs	Incl. O&P
1 Equip. Oper. (crane)	$61.45	$491.60	$91.55	$732.40	$52.92	$78.90
1 Laborer	44.40	355.20	66.25	530.00		
1 Hyd. Excavator, 2.5 C.Y.		1567.00		1723.70	97.94	107.73
16 L.H., Daily Totals		$2413.80		$2986.10	$150.86	$186.63
Crew B-12T	Hr.	Daily	Hr.	Daily	Bare Costs	Incl. O&P
1 Equip. Oper. (crane)	$61.45	$491.60	$91.55	$732.40	$52.92	$78.90
1 Laborer	44.40	355.20	66.25	530.00		
1 Crawler Crane, 75 Ton		1967.00		2163.70		
1 F.E. Attachment, 3 C.Y.		115.55		127.11	130.16	143.18
16 L.H., Daily Totals		$2929.35		$3553.20	$183.08	$222.08
Crew B-12V	Hr.	Daily	Hr.	Daily	Bare Costs	Incl. O&P
1 Equip. Oper. (crane)	$61.45	$491.60	$91.55	$732.40	$52.92	$78.90
1 Laborer	44.40	355.20	66.25	530.00		
1 Crawler Crane, 75 Ton		1967.00		2163.70		
1 Dragline Bucket, 3 C.Y.		72.30		79.53	127.46	140.20
16 L.H., Daily Totals		$2886.10		$3505.63	$180.38	$219.10
Crew B-12Y	Hr.	Daily	Hr.	Daily	Bare Costs	Incl. O&P
1 Equip. Oper. (crane)	$61.45	$491.60	$91.55	$732.40	$50.08	$74.68
2 Laborers	44.40	710.40	66.25	1060.00		
1 Hyd. Excavator, 3.5 C.Y.		2184.00		2402.40	91.00	100.10
24 L.H., Daily Totals		$3386.00		$4194.80	$141.08	$174.78
Crew B-12Z	Hr.	Daily	Hr.	Daily	Bare Costs	Incl. O&P
1 Equip. Oper. (crane)	$61.45	$491.60	$91.55	$732.40	$50.08	$74.68
2 Laborers	44.40	710.40	66.25	1060.00		
1 Hyd. Excavator, 2.5 C.Y.		1567.00		1723.70	65.29	71.82
24 L.H., Daily Totals		$2769.00		$3516.10	$115.38	$146.50
Crew B-13	Hr.	Daily	Hr.	Daily	Bare Costs	Incl. O&P
1 Labor Foreman (outside)	$46.40	$371.20	$69.25	$554.00	$48.28	$72.01
4 Laborers	44.40	1420.80	66.25	2120.00		
1 Equip. Oper. (crane)	61.45	491.60	91.55	732.40		
1 Equip. Oper. (oiler)	52.50	420.00	78.25	626.00		
1 Hyd. Crane, 25 Ton		586.70		645.37	10.48	11.52
56 L.H., Daily Totals		$3290.30		$4677.77	$58.76	$83.53

Crew No.	Bare Costs Hr.	Daily	Incl. Subs O&P Hr.	Daily	Cost Per Labor-Hour Bare Costs	Incl. O&P
Crew B-13A	Hr.	Daily	Hr.	Daily	Bare Costs	Incl. O&P
1 Labor Foreman (outside)	$46.40	$371.20	$69.25	$554.00	$50.83	$75.85
2 Laborers	44.40	710.40	66.25	1060.00		
2 Equipment Operators (med.)	59.00	944.00	87.90	1406.40		
2 Truck Drivers (heavy)	51.30	820.80	76.70	1227.20		
1 Crawler Crane, 75 Ton		1967.00		2163.70		
1 Crawler Loader, 4 C.Y.		1456.00		1601.60		
2 Dump Trucks, 8 C.Y., 220 H.P.		815.20		896.72	75.68	83.25
56 L.H., Daily Totals		$7084.60		$8909.62	$126.51	$159.10
Crew B-13B	Hr.	Daily	Hr.	Daily	Bare Costs	Incl. O&P
1 Labor Foreman (outside)	$46.40	$371.20	$69.25	$554.00	$48.28	$72.01
4 Laborers	44.40	1420.80	66.25	2120.00		
1 Equip. Oper. (crane)	61.45	491.60	91.55	732.40		
1 Equip. Oper. (oiler)	52.50	420.00	78.25	626.00		
1 Hyd. Crane, 55 Ton		990.15		1089.17	17.68	19.45
56 L.H., Daily Totals		$3693.75		$5121.56	$65.96	$91.46
Crew B-13C	Hr.	Daily	Hr.	Daily	Bare Costs	Incl. O&P
1 Labor Foreman (outside)	$46.40	$371.20	$69.25	$554.00	$48.28	$72.01
4 Laborers	44.40	1420.80	66.25	2120.00		
1 Equip. Oper. (crane)	61.45	491.60	91.55	732.40		
1 Equip. Oper. (oiler)	52.50	420.00	78.25	626.00		
1 Crawler Crane, 100 Ton		2310.00		2541.00	41.25	45.38
56 L.H., Daily Totals		$5013.60		$6573.40	$89.53	$117.38
Crew B-13D	Hr.	Daily	Hr.	Daily	Bare Costs	Incl. O&P
1 Laborer	$44.40	$355.20	$66.25	$530.00	$52.92	$78.90
1 Equip. Oper. (crane)	61.45	491.60	91.55	732.40		
1 Hyd. Excavator, 1 C.Y.		832.65		915.91		
1 Trench Box		119.15		131.07	59.49	65.44
16 L.H., Daily Totals		$1798.60		$2309.38	$112.41	$144.34
Crew B-13E	Hr.	Daily	Hr.	Daily	Bare Costs	Incl. O&P
1 Laborer	$44.40	$355.20	$66.25	$530.00	$52.92	$78.90
1 Equip. Oper. (crane)	61.45	491.60	91.55	732.40		
1 Hyd. Excavator, 1.5 C.Y.		695.80		765.38		
1 Trench Box		119.15		131.07	50.93	56.03
16 L.H., Daily Totals		$1661.75		$2158.84	$103.86	$134.93
Crew B-13F	Hr.	Daily	Hr.	Daily	Bare Costs	Incl. O&P
1 Laborer	$44.40	$355.20	$66.25	$530.00	$52.92	$78.90
1 Equip. Oper. (crane)	61.45	491.60	91.55	732.40		
1 Hyd. Excavator, 3.5 C.Y.		2184.00		2402.40		
1 Trench Box		119.15		131.07	143.95	158.34
16 L.H., Daily Totals		$3149.95		$3795.86	$196.87	$237.24
Crew B-13G	Hr.	Daily	Hr.	Daily	Bare Costs	Incl. O&P
1 Laborer	$44.40	$355.20	$66.25	$530.00	$52.92	$78.90
1 Equip. Oper. (crane)	61.45	491.60	91.55	732.40		
1 Hyd. Excavator, .75 C.Y.		701.80		771.98		
1 Trench Box		119.15		131.07	51.31	56.44
16 L.H., Daily Totals		$1667.75		$2165.45	$104.23	$135.34
Crew B-13H	Hr.	Daily	Hr.	Daily	Bare Costs	Incl. O&P
1 Laborer	$44.40	$355.20	$66.25	$530.00	$52.92	$78.90
1 Equip. Oper. (crane)	61.45	491.60	91.55	732.40		
1 Gradall, 5/8 C.Y.		850.65		935.72		
1 Trench Box		119.15		131.07	60.61	66.67
16 L.H., Daily Totals		$1816.60		$2329.18	$113.54	$145.57

For customer support on your Electrical Costs with RSMeans Data, call 800.448.8182.

49

Crew No.	Hr.	Daily	Hr.	Daily	Bare Costs	Incl. O&P
Crew B-13I						
1 Laborer	$44.40	$355.20	$66.25	$530.00	$52.92	$78.90
1 Equip. Oper. (crane)	61.45	491.60	91.55	732.40		
1 Gradall, 3 Ton, 1 C.Y.		984.55		1083.01		
1 Trench Box		119.15		131.07	68.98	75.88
16 L.H., Daily Totals		$1950.50		$2476.47	$121.91	$154.78
Crew B-13J						
1 Laborer	$44.40	$355.20	$66.25	$530.00	$52.92	$78.90
1 Equip. Oper. (crane)	61.45	491.60	91.55	732.40		
1 Hyd. Excavator, 2.5 C.Y.		1567.00		1723.70		
1 Trench Box		119.15		131.07	105.38	115.92
16 L.H., Daily Totals		$2532.95		$3117.17	$158.31	$194.82
Crew B-13K						
2 Equip. Opers. (crane)	$61.45	$983.20	$91.55	$1464.80	$61.45	$91.55
1 Hyd. Excavator, .75 C.Y.		701.80		771.98		
1 Hyd. Hammer, 4000 ft-lb		649.20		714.12		
1 Hyd. Excavator, .75 C.Y.		701.80		771.98	128.30	141.13
16 L.H., Daily Totals		$3036.00		$3722.88	$189.75	$232.68
Crew B-13L						
2 Equip. Opers. (crane)	$61.45	$983.20	$91.55	$1464.80	$61.45	$91.55
1 Hyd. Excavator, 1.5 C.Y.		695.80		765.38		
1 Hyd. Hammer, 5000 ft-lb		705.60		776.16		
1 Hyd. Excavator, .75 C.Y.		701.80		771.98	131.45	144.60
16 L.H., Daily Totals		$3086.40		$3778.32	$192.90	$236.15
Crew B-13M						
2 Equip. Opers. (crane)	$61.45	$983.20	$91.55	$1464.80	$61.45	$91.55
1 Hyd. Excavator, 2.5 C.Y.		1567.00		1723.70		
1 Hyd. Hammer, 8000 ft-lb		918.65		1010.52		
1 Hyd. Excavator, 1.5 C.Y.		695.80		765.38	198.84	218.72
16 L.H., Daily Totals		$4164.65		$4964.40	$260.29	$310.27
Crew B-13N						
2 Equip. Opers. (crane)	$61.45	$983.20	$91.55	$1464.80	$61.45	$91.55
1 Hyd. Excavator, 3.5 C.Y.		2184.00		2402.40		
1 Hyd. Hammer, 12,000 ft-lb		882.20		970.42		
1 Hyd. Excavator, 1.5 C.Y.		695.80		765.38	235.13	258.64
16 L.H., Daily Totals		$4745.20		$5603.00	$296.57	$350.19
Crew B-14						
1 Labor Foreman (outside)	$46.40	$371.20	$69.25	$554.00	$46.58	$69.49
4 Laborers	44.40	1420.80	66.25	2120.00		
1 Equip. Oper. (light)	55.50	444.00	82.70	661.60		
1 Backhoe Loader, 48 H.P.		216.20		237.82	4.50	4.95
48 L.H., Daily Totals		$2452.20		$3573.42	$51.09	$74.45
Crew B-14A						
1 Equip. Oper. (crane)	$61.45	$491.60	$91.55	$732.40	$55.77	$83.12
.5 Laborer	44.40	177.60	66.25	265.00		
1 Hyd. Excavator, 4.5 C.Y.		3450.00		3795.00	287.50	316.25
12 L.H., Daily Totals		$4119.20		$4792.40	$343.27	$399.37
Crew B-14B						
1 Equip. Oper. (crane)	$61.45	$491.60	$91.55	$732.40	$55.77	$83.12
.5 Laborer	44.40	177.60	66.25	265.00		
1 Hyd. Excavator, 6 C.Y.		3506.00		3856.60	292.17	321.38
12 L.H., Daily Totals		$4175.20		$4854.00	$347.93	$404.50

Crew No.	Hr.	Daily	Hr.	Daily	Bare Costs	Incl. O&P
Crew B-14C						
1 Equip. Oper. (crane)	$61.45	$491.60	$91.55	$732.40	$55.77	$83.12
.5 Laborer	44.40	177.60	66.25	265.00		
1 Hyd. Excavator, 7 C.Y.		3475.00		3822.50	289.58	318.54
12 L.H., Daily Totals		$4144.20		$4819.90	$345.35	$401.66
Crew B-14F						
1 Equip. Oper. (crane)	$61.45	$491.60	$91.55	$732.40	$55.77	$83.12
.5 Laborer	44.40	177.60	66.25	265.00		
1 Hyd. Shovel, 7 C.Y.		4148.00		4562.80	345.67	380.23
12 L.H., Daily Totals		$4817.20		$5560.20	$401.43	$463.35
Crew B-14G						
1 Equip. Oper. (crane)	$61.45	$491.60	$91.55	$732.40	$55.77	$83.12
.5 Laborer	44.40	177.60	66.25	265.00		
1 Hyd. Shovel, 12 C.Y.		6022.00		6624.20	501.83	552.02
12 L.H., Daily Totals		$6691.20		$7621.60	$557.60	$635.13
Crew B-14J						
1 Equip. Oper. (medium)	$59.00	$472.00	$87.90	$703.20	$54.13	$80.68
.5 Laborer	44.40	177.60	66.25	265.00		
1 F.E. Loader, 8 C.Y.		2285.00		2513.50	190.42	209.46
12 L.H., Daily Totals		$2934.60		$3481.70	$244.55	$290.14
Crew B-14K						
1 Equip. Oper. (medium)	$59.00	$472.00	$87.90	$703.20	$54.13	$80.68
.5 Laborer	44.40	177.60	66.25	265.00		
1 F.E. Loader, 10 C.Y.		2706.00		2976.60	225.50	248.05
12 L.H., Daily Totals		$3355.60		$3944.80	$279.63	$328.73
Crew B-15						
1 Equipment Oper. (med.)	$59.00	$472.00	$87.90	$703.20	$52.51	$78.41
.5 Laborer	44.40	177.60	66.25	265.00		
2 Truck Drivers (heavy)	51.30	820.80	76.70	1227.20		
2 Dump Trucks, 12 C.Y., 400 H.P.		1158.70		1274.57		
1 Dozer, 200 H.P.		1520.00		1672.00	95.67	105.23
28 L.H., Daily Totals		$4149.10		$5141.97	$148.18	$183.64
Crew B-16						
1 Labor Foreman (outside)	$46.40	$371.20	$69.25	$554.00	$46.63	$69.61
2 Laborers	44.40	710.40	66.25	1060.00		
1 Truck Driver (heavy)	51.30	410.40	76.70	613.60		
1 Dump Truck, 12 C.Y., 400 H.P.		579.35		637.28	18.10	19.92
32 L.H., Daily Totals		$2071.35		$2864.89	$64.73	$89.53
Crew B-17						
2 Laborers	$44.40	$710.40	$66.25	$1060.00	$48.90	$72.97
1 Equip. Oper. (light)	55.50	444.00	82.70	661.60		
1 Truck Driver (heavy)	51.30	410.40	76.70	613.60		
1 Backhoe Loader, 48 H.P.		216.20		237.82		
1 Dump Truck, 8 C.Y., 220 H.P.		407.60		448.36	19.49	21.44
32 L.H., Daily Totals		$2188.60		$3021.38	$68.39	$94.42
Crew B-17A						
2 Labor Foremen (outside)	$46.40	$742.40	$69.25	$1108.00	$47.54	$71.08
6 Laborers	44.40	2131.20	66.25	3180.00		
1 Skilled Worker Foreman (out)	59.10	472.80	88.90	711.20		
1 Skilled Worker	57.10	456.80	85.90	687.20		
80 L.H., Daily Totals		$3803.20		$5686.40	$47.54	$71.08

Crews - Standard

Crew No.	Bare Costs		Incl. Subs O&P		Cost Per Labor-Hour	
					Bare Costs	Incl. O&P
Crew B-17B	Hr.	Daily	Hr.	Daily	$48.90	$72.97
2 Laborers	$44.40	$710.40	$66.25	$1060.00		
1 Equip. Oper. (light)	55.50	444.00	82.70	661.60		
1 Truck Driver (heavy)	51.30	410.40	76.70	613.60		
1 Backhoe Loader, 48 H.P.		216.20		237.82		
1 Dump Truck, 12 C.Y., 400 H.P.		579.35		637.28	24.86	27.35
32 L.H., Daily Totals		$2360.35		$3210.30	$73.76	$100.32

Crew No.	Bare Costs		Incl. Subs O&P		Cost Per Labor-Hour	
					Bare Costs	Incl. O&P
Crew B-18	Hr.	Daily	Hr.	Daily	$45.07	$67.25
1 Labor Foreman (outside)	$46.40	$371.20	$69.25	$554.00		
2 Laborers	44.40	710.40	66.25	1060.00		
1 Vibrating Plate, Gas, 21"		165.60		182.16	6.90	7.59
24 L.H., Daily Totals		$1247.20		$1796.16	$51.97	$74.84

Crew No.	Bare Costs		Incl. Subs O&P		Cost Per Labor-Hour	
					Bare Costs	Incl. O&P
Crew B-19	Hr.	Daily	Hr.	Daily	$57.11	$87.06
1 Pile Driver Foreman (outside)	$57.90	$463.20	$89.50	$716.00		
4 Pile Drivers	55.90	1788.80	86.40	2764.80		
2 Equip. Oper. (crane)	61.45	983.20	91.55	1464.80		
1 Equip. Oper. (oiler)	52.50	420.00	78.25	626.00		
1 Crawler Crane, 40 Ton		1231.00		1354.10		
1 Lead, 90' High		371.85		409.04		
1 Hammer, Diesel, 22k ft-lb		441.70		485.87	31.95	35.14
64 L.H., Daily Totals		$5699.75		$7820.60	$89.06	$122.20

Crew No.	Bare Costs		Incl. Subs O&P		Cost Per Labor-Hour	
					Bare Costs	Incl. O&P
Crew B-19A	Hr.	Daily	Hr.	Daily	$57.11	$87.06
1 Pile Driver Foreman (outside)	$57.90	$463.20	$89.50	$716.00		
4 Pile Drivers	55.90	1788.80	86.40	2764.80		
2 Equip. Oper. (crane)	61.45	983.20	91.55	1464.80		
1 Equip. Oper. (oiler)	52.50	420.00	78.25	626.00		
1 Crawler Crane, 75 Ton		1967.00		2163.70		
1 Lead, 90' High		371.85		409.04		
1 Hammer, Diesel, 41k ft-lb		583.55		641.90	45.66	50.23
64 L.H., Daily Totals		$6577.60		$8786.24	$102.78	$137.29

Crew No.	Bare Costs		Incl. Subs O&P		Cost Per Labor-Hour	
					Bare Costs	Incl. O&P
Crew B-19B	Hr.	Daily	Hr.	Daily	$57.11	$87.06
1 Pile Driver Foreman (outside)	$57.90	$463.20	$89.50	$716.00		
4 Pile Drivers	55.90	1788.80	86.40	2764.80		
2 Equip. Oper. (crane)	61.45	983.20	91.55	1464.80		
1 Equip. Oper. (oiler)	52.50	420.00	78.25	626.00		
1 Crawler Crane, 40 Ton		1231.00		1354.10		
1 Lead, 90' High		371.85		409.04		
1 Hammer, Diesel, 22k ft-lb		441.70		485.87		
1 Barge, 400 Ton		869.15		956.07	45.53	50.08
64 L.H., Daily Totals		$6568.90		$8776.67	$102.64	$137.14

Crew No.	Bare Costs		Incl. Subs O&P		Cost Per Labor-Hour	
					Bare Costs	Incl. O&P
Crew B-19C	Hr.	Daily	Hr.	Daily	$57.11	$87.06
1 Pile Driver Foreman (outside)	$57.90	$463.20	$89.50	$716.00		
4 Pile Drivers	55.90	1788.80	86.40	2764.80		
2 Equip. Oper. (crane)	61.45	983.20	91.55	1464.80		
1 Equip. Oper. (oiler)	52.50	420.00	78.25	626.00		
1 Crawler Crane, 75 Ton		1967.00		2163.70		
1 Lead, 90' High		371.85		409.04		
1 Hammer, Diesel, 41k ft-lb		583.55		641.90		
1 Barge, 400 Ton		869.15		956.07	59.24	65.17
64 L.H., Daily Totals		$7446.75		$9742.31	$116.36	$152.22

Crew No.	Bare Costs		Incl. Subs O&P		Cost Per Labor-Hour	
					Bare Costs	Incl. O&P
Crew B-20	Hr.	Daily	Hr.	Daily	$49.30	$73.80
1 Labor Foreman (outside)	$46.40	$371.20	$69.25	$554.00		
1 Skilled Worker	57.10	456.80	85.90	687.20		
1 Laborer	44.40	355.20	66.25	530.00		
24 L.H., Daily Totals		$1183.20		$1771.20	$49.30	$73.80

Crew No.	Bare Costs		Incl. Subs O&P		Cost Per Labor-Hour	
					Bare Costs	Incl. O&P
Crew B-20A	Hr.	Daily	Hr.	Daily	$53.16	$79.34
1 Labor Foreman (outside)	$46.40	$371.20	$69.25	$554.00		
1 Laborer	44.40	355.20	66.25	530.00		
1 Plumber	67.70	541.60	101.05	808.40		
1 Plumber Apprentice	54.15	433.20	80.80	646.40		
32 L.H., Daily Totals		$1701.20		$2538.80	$53.16	$79.34

Crew No.	Bare Costs		Incl. Subs O&P		Cost Per Labor-Hour	
					Bare Costs	Incl. O&P
Crew B-21	Hr.	Daily	Hr.	Daily	$51.04	$76.34
1 Labor Foreman (outside)	$46.40	$371.20	$69.25	$554.00		
1 Skilled Worker	57.10	456.80	85.90	687.20		
1 Laborer	44.40	355.20	66.25	530.00		
.5 Equip. Oper. (crane)	61.45	245.80	91.55	366.20		
.5 S.P. Crane, 4x4, 5 Ton		190.97		210.07	6.82	7.50
28 L.H., Daily Totals		$1619.97		$2347.47	$57.86	$83.84

Crew No.	Bare Costs		Incl. Subs O&P		Cost Per Labor-Hour	
					Bare Costs	Incl. O&P
Crew B-21A	Hr.	Daily	Hr.	Daily	$54.82	$81.78
1 Labor Foreman (outside)	$46.40	$371.20	$69.25	$554.00		
1 Laborer	44.40	355.20	66.25	530.00		
1 Plumber	67.70	541.60	101.05	808.40		
1 Plumber Apprentice	54.15	433.20	80.80	646.40		
1 Equip. Oper. (crane)	61.45	491.60	91.55	732.40		
1 S.P. Crane, 4x4, 12 Ton		432.65		475.92	10.82	11.90
40 L.H., Daily Totals		$2625.45		$3747.11	$65.64	$93.68

Crew No.	Bare Costs		Incl. Subs O&P		Cost Per Labor-Hour	
					Bare Costs	Incl. O&P
Crew B-21B	Hr.	Daily	Hr.	Daily	$48.21	$71.91
1 Labor Foreman (outside)	$46.40	$371.20	$69.25	$554.00		
3 Laborers	44.40	1065.60	66.25	1590.00		
1 Equip. Oper. (crane)	61.45	491.60	91.55	732.40		
1 Hyd. Crane, 12 Ton		475.80		523.38	11.90	13.08
40 L.H., Daily Totals		$2404.20		$3399.78	$60.10	$84.99

Crew No.	Bare Costs		Incl. Subs O&P		Cost Per Labor-Hour	
					Bare Costs	Incl. O&P
Crew B-21C	Hr.	Daily	Hr.	Daily	$48.28	$72.01
1 Labor Foreman (outside)	$46.40	$371.20	$69.25	$554.00		
4 Laborers	44.40	1420.80	66.25	2120.00		
1 Equip. Oper. (crane)	61.45	491.60	91.55	732.40		
1 Equip. Oper. (oiler)	52.50	420.00	78.25	626.00		
2 Cutting Torches		25.90		28.49		
2 Sets of Gases		347.20		381.92		
1 Lattice Boom Crane, 90 Ton		1713.00		1884.30	37.25	40.98
56 L.H., Daily Totals		$4789.70		$6327.11	$85.53	$112.98

Crew No.	Bare Costs		Incl. Subs O&P		Cost Per Labor-Hour	
					Bare Costs	Incl. O&P
Crew B-22	Hr.	Daily	Hr.	Daily	$51.73	$77.35
1 Labor Foreman (outside)	$46.40	$371.20	$69.25	$554.00		
1 Skilled Worker	57.10	456.80	85.90	687.20		
1 Laborer	44.40	355.20	66.25	530.00		
.75 Equip. Oper. (crane)	61.45	368.70	91.55	549.30		
.75 S.P. Crane, 4x4, 5 Ton		286.46		315.11	9.55	10.50
30 L.H., Daily Totals		$1838.36		$2635.61	$61.28	$87.85

Crew No.	Bare Costs		Incl. Subs O&P		Cost Per Labor-Hour	
					Bare Costs	Incl. O&P
Crew B-22A	Hr.	Daily	Hr.	Daily	$50.75	$75.84
1 Labor Foreman (outside)	$46.40	$371.20	$69.25	$554.00		
1 Skilled Worker	57.10	456.80	85.90	687.20		
2 Laborers	44.40	710.40	66.25	1060.00		
1 Equipment Operator, Crane	61.45	491.60	91.55	732.40		
1 S.P. Crane, 4x4, 5 Ton		381.95		420.14		
1 Butt Fusion Mach., 4"-12" diam.		420.75		462.82	20.07	22.07
40 L.H., Daily Totals		$2832.70		$3916.57	$70.82	$97.91

For customer support on your Electrical Costs with RSMeans Data, call 800.448.8182.

493

Crew No.	Bare Costs		Incl. Subs O&P		Cost Per Labor-Hour	
Crew B-22B	Hr.	Daily	Hr.	Daily	Bare Costs	Incl. O&P
1 Labor Foreman (outside)	$46.40	$371.20	$69.25	$554.00	$50.75	$75.84
1 Skilled Worker	57.10	456.80	85.90	687.20		
2 Laborers	44.40	710.40	66.25	1060.00		
1 Equip. Oper. (crane)	61.45	491.60	91.55	732.40		
1 S.P. Crane, 4x4, 5 Ton		381.95		420.14		
1 Butt Fusion Mach., 8"-24" diam.		1086.00		1194.60	36.70	40.37
40 L.H., Daily Totals		$3497.95		$4648.35	$87.45	$116.21

Crew No.	Bare Costs		Incl. Subs O&P		Cost Per Labor-Hour	
Crew B-22C	Hr.	Daily	Hr.	Daily	Bare Costs	Incl. O&P
1 Skilled Worker	$57.10	$456.80	$85.90	$687.20	$50.75	$76.08
1 Laborer	44.40	355.20	66.25	530.00		
1 Butt Fusion Mach., 2"-8" diam.		134.95		148.44	8.43	9.28
16 L.H., Daily Totals		$946.95		$1365.65	$59.18	$85.35

Crew No.	Bare Costs		Incl. Subs O&P		Cost Per Labor-Hour	
Crew B-23	Hr.	Daily	Hr.	Daily	Bare Costs	Incl. O&P
1 Labor Foreman (outside)	$46.40	$371.20	$69.25	$554.00	$44.80	$66.85
4 Laborers	44.40	1420.80	66.25	2120.00		
1 Drill Rig, Truck-Mounted		768.40		845.24		
1 Flatbed Truck, Gas, 3 Ton		850.05		935.05	40.46	44.51
40 L.H., Daily Totals		$3410.45		$4454.30	$85.26	$111.36

Crew No.	Bare Costs		Incl. Subs O&P		Cost Per Labor-Hour	
Crew B-23A	Hr.	Daily	Hr.	Daily	Bare Costs	Incl. O&P
1 Labor Foreman (outside)	$46.40	$371.20	$69.25	$554.00	$49.93	$74.47
1 Laborer	44.40	355.20	66.25	530.00		
1 Equip. Oper. (medium)	59.00	472.00	87.90	703.20		
1 Drill Rig, Truck-Mounted		768.40		845.24		
1 Pickup Truck, 3/4 Ton		112.20		123.42	36.69	40.36
24 L.H., Daily Totals		$2079.00		$2755.86	$86.63	$114.83

Crew No.	Bare Costs		Incl. Subs O&P		Cost Per Labor-Hour	
Crew B-23B	Hr.	Daily	Hr.	Daily	Bare Costs	Incl. O&P
1 Labor Foreman (outside)	$46.40	$371.20	$69.25	$554.00	$49.93	$74.47
1 Laborer	44.40	355.20	66.25	530.00		
1 Equip. Oper. (medium)	59.00	472.00	87.90	703.20		
1 Drill Rig, Truck-Mounted		768.40		845.24		
1 Pickup Truck, 3/4 Ton		112.20		123.42		
1 Centr. Water Pump, 6"		235.25		258.77	46.49	51.14
24 L.H., Daily Totals		$2314.25		$3014.64	$96.43	$125.61

Crew No.	Bare Costs		Incl. Subs O&P		Cost Per Labor-Hour	
Crew B-24	Hr.	Daily	Hr.	Daily	Bare Costs	Incl. O&P
1 Cement Finisher	$51.80	$414.40	$75.90	$607.20	$50.30	$74.60
1 Laborer	44.40	355.20	66.25	530.00		
1 Carpenter	54.70	437.60	81.65	653.20		
24 L.H., Daily Totals		$1207.20		$1790.40	$50.30	$74.60

Crew No.	Bare Costs		Incl. Subs O&P		Cost Per Labor-Hour	
Crew B-25	Hr.	Daily	Hr.	Daily	Bare Costs	Incl. O&P
1 Labor Foreman (outside)	$46.40	$371.20	$69.25	$554.00	$48.56	$72.43
7 Laborers	44.40	2486.40	66.25	3710.00		
3 Equip. Oper. (medium)	59.00	1416.00	87.90	2109.60		
1 Asphalt Paver, 130 H.P.		2143.00		2357.30		
1 Tandem Roller, 10 Ton		246.80		271.48		
1 Roller, Pneum. Whl., 12 Ton		349.90		384.89	31.13	34.25
88 L.H., Daily Totals		$7013.30		$9387.27	$79.70	$106.67

Crew No.	Bare Costs		Incl. Subs O&P		Cost Per Labor-Hour	
Crew B-25B	Hr.	Daily	Hr.	Daily	Bare Costs	Incl. O&P
1 Labor Foreman (outside)	$46.40	$371.20	$69.25	$554.00	$49.43	$73.72
7 Laborers	44.40	2486.40	66.25	3710.00		
4 Equip. Oper. (medium)	59.00	1888.00	87.90	2812.80		
1 Asphalt Paver, 130 H.P.		2143.00		2357.30		
2 Tandem Rollers, 10 Ton		493.60		542.96		
1 Roller, Pneum. Whl., 12 Ton		349.90		384.89	31.11	34.22
96 L.H., Daily Totals		$7732.10		$10361.95	$80.54	$107.94

Crew No.	Bare Costs		Incl. Subs O&P		Cost Per Labor-Hour	
Crew B-25C	Hr.	Daily	Hr.	Daily	Bare Costs	Incl. O&P
1 Labor Foreman (outside)	$46.40	$371.20	$69.25	$554.00	$49.60	$73.97
3 Laborers	44.40	1065.60	66.25	1590.00		
2 Equip. Oper. (medium)	59.00	944.00	87.90	1406.40		
1 Asphalt Paver, 130 H.P.		2143.00		2357.30		
1 Tandem Roller, 10 Ton		246.80		271.48	49.79	54.77
48 L.H., Daily Totals		$4770.60		$6179.18	$99.39	$128.73

Crew No.	Bare Costs		Incl. Subs O&P		Cost Per Labor-Hour	
Crew B-25D	Hr.	Daily	Hr.	Daily	Bare Costs	Incl. O&P
1 Labor Foreman (outside)	$46.40	$371.20	$69.25	$554.00	$49.82	$74.30
3 Laborers	44.40	1065.60	66.25	1590.00		
2.125 Equip. Oper. (medium)	59.00	1003.00	87.90	1494.30		
.125 Truck Driver (heavy)	51.30	51.30	76.70	76.70		
.125 Truck Tractor, 6x4, 380 H.P.		62.39		68.63		
.125 Dist. Tanker, 3000 Gallon		41.76		45.94		
1 Asphalt Paver, 130 H.P.		2143.00		2357.30		
1 Tandem Roller, 10 Ton		246.80		271.48	49.88	54.87
50 L.H., Daily Totals		$4985.06		$6458.35	$99.70	$129.17

Crew No.	Bare Costs		Incl. Subs O&P		Cost Per Labor-Hour	
Crew B-25E	Hr.	Daily	Hr.	Daily	Bare Costs	Incl. O&P
1 Labor Foreman (outside)	$46.40	$371.20	$69.25	$554.00	$50.03	$74.61
3 Laborers	44.40	1065.60	66.25	1590.00		
2.250 Equip. Oper. (medium)	59.00	1062.00	87.90	1582.20		
.25 Truck Driver (heavy)	51.30	102.60	76.70	153.40		
.25 Truck Tractor, 6x4, 380 H.P.		124.79		137.27		
.25 Dist. Tanker, 3000 Gallon		83.53		91.88		
1 Asphalt Paver, 130 H.P.		2143.00		2357.30		
1 Tandem Roller, 10 Ton		246.80		271.48	49.96	54.96
52 L.H., Daily Totals		$5199.51		$6737.52	$99.99	$129.57

Crew No.	Bare Costs		Incl. Subs O&P		Cost Per Labor-Hour	
Crew B-26	Hr.	Daily	Hr.	Daily	Bare Costs	Incl. O&P
1 Labor Foreman (outside)	$46.40	$371.20	$69.25	$554.00	$49.23	$73.32
6 Laborers	44.40	2131.20	66.25	3180.00		
2 Equip. Oper. (medium)	59.00	944.00	87.90	1406.40		
1 Rodman (reinf.)	58.90	471.20	88.05	704.40		
1 Cement Finisher	51.80	414.40	75.90	607.20		
1 Grader, 30,000 Lbs.		1073.00		1180.30		
1 Paving Mach. & Equip.		2503.00		2753.30	40.64	44.70
88 L.H., Daily Totals		$7908.00		$10385.60	$89.86	$118.02

Crew No.	Bare Costs		Incl. Subs O&P		Cost Per Labor-Hour	
Crew B-26A	Hr.	Daily	Hr.	Daily	Bare Costs	Incl. O&P
1 Labor Foreman (outside)	$46.40	$371.20	$69.25	$554.00	$49.23	$73.32
6 Laborers	44.40	2131.20	66.25	3180.00		
2 Equip. Oper. (medium)	59.00	944.00	87.90	1406.40		
1 Rodman (reinf.)	58.90	471.20	88.05	704.40		
1 Cement Finisher	51.80	414.40	75.90	607.20		
1 Grader, 30,000 Lbs.		1073.00		1180.30		
1 Paving Mach. & Equip.		2503.00		2753.30		
1 Concrete Saw		112.85		124.14	41.92	46.11
88 L.H., Daily Totals		$8020.85		$10509.74	$91.15	$119.43

Crew No.	Bare Costs		Incl. Subs O&P		Cost Per Labor-Hour	
Crew B-26B	Hr.	Daily	Hr.	Daily	Bare Costs	Incl. O&P
1 Labor Foreman (outside)	$46.40	$371.20	$69.25	$554.00	$50.04	$74.53
6 Laborers	44.40	2131.20	66.25	3180.00		
3 Equip. Oper. (medium)	59.00	1416.00	87.90	2109.60		
1 Rodman (reinf.)	58.90	471.20	88.05	704.40		
1 Cement Finisher	51.80	414.40	75.90	607.20		
1 Grader, 30,000 Lbs.		1073.00		1180.30		
1 Paving Mach. & Equip.		2503.00		2753.30		
1 Concrete Pump, 110' Boom		493.65		543.01	42.39	46.63
96 L.H., Daily Totals		$8873.65		$11631.82	$92.43	$121.16

Crew B-26C

Crew No.	Bare Costs Hr.	Bare Costs Daily	Incl. Subs O&P Hr.	Incl. Subs O&P Daily	Cost Per Labor-Hour Bare Costs	Cost Per Labor-Hour Incl. O&P
1 Labor Foreman (outside)	$46.40	$371.20	$69.25	$554.00	$48.25	$71.86
6 Laborers	44.40	2131.20	66.25	3180.00		
1 Equip. Oper. (medium)	59.00	472.00	87.90	703.20		
1 Rodman (reinf.)	58.90	471.20	88.05	704.40		
1 Cement Finisher	51.80	414.40	75.90	607.20		
1 Paving Mach. & Equip.		2503.00		2753.30		
1 Concrete Saw		112.85		124.14	32.70	35.97
80 L.H., Daily Totals		$6475.85		$8626.24	$80.95	$107.83

Crew B-27

Crew No.	Bare Costs Hr.	Bare Costs Daily	Incl. Subs O&P Hr.	Incl. Subs O&P Daily	Cost Per Labor-Hour Bare Costs	Cost Per Labor-Hour Incl. O&P
1 Labor Foreman (outside)	$46.40	$371.20	$69.25	$554.00	$44.90	$67.00
3 Laborers	44.40	1065.60	66.25	1590.00		
1 Berm Machine		253.55		278.90	7.92	8.72
32 L.H., Daily Totals		$1690.35		$2422.91	$52.82	$75.72

Crew B-28

Crew No.	Bare Costs Hr.	Bare Costs Daily	Incl. Subs O&P Hr.	Incl. Subs O&P Daily	Cost Per Labor-Hour Bare Costs	Cost Per Labor-Hour Incl. O&P
2 Carpenters	$54.70	$875.20	$81.65	$1306.40	$51.27	$76.52
1 Laborer	44.40	355.20	66.25	530.00		
24 L.H., Daily Totals		$1230.40		$1836.40	$51.27	$76.52

Crew B-29

Crew No.	Bare Costs Hr.	Bare Costs Daily	Incl. Subs O&P Hr.	Incl. Subs O&P Daily	Cost Per Labor-Hour Bare Costs	Cost Per Labor-Hour Incl. O&P
1 Labor Foreman (outside)	$46.40	$371.20	$69.25	$554.00	$48.28	$72.01
4 Laborers	44.40	1420.80	66.25	2120.00		
1 Equip. Oper. (crane)	61.45	491.60	91.55	732.40		
1 Equip. Oper. (oiler)	52.50	420.00	78.25	626.00		
1 Gradall, 5/8 C.Y.		850.65		935.72	15.19	16.71
56 L.H., Daily Totals		$3554.25		$4968.11	$63.47	$88.72

Crew B-30

Crew No.	Bare Costs Hr.	Bare Costs Daily	Incl. Subs O&P Hr.	Incl. Subs O&P Daily	Cost Per Labor-Hour Bare Costs	Cost Per Labor-Hour Incl. O&P
1 Equip. Oper. (medium)	$59.00	$472.00	$87.90	$703.20	$53.87	$80.43
2 Truck Drivers (heavy)	51.30	820.80	76.70	1227.20		
1 Hyd. Excavator, 1.5 C.Y.		695.80		765.38		
2 Dump Trucks, 12 C.Y., 400 H.P.		1158.70		1274.57	77.27	85.00
24 L.H., Daily Totals		$3147.30		$3970.35	$131.14	$165.43

Crew B-31

Crew No.	Bare Costs Hr.	Bare Costs Daily	Incl. Subs O&P Hr.	Incl. Subs O&P Daily	Cost Per Labor-Hour Bare Costs	Cost Per Labor-Hour Incl. O&P
1 Labor Foreman (outside)	$46.40	$371.20	$69.25	$554.00	$46.86	$69.93
3 Laborers	44.40	1065.60	66.25	1590.00		
1 Carpenter	54.70	437.60	81.65	653.20		
1 Air Compressor, 250 cfm		202.85		223.13		
1 Sheeting Driver		7.45		8.20		
2 -50' Air Hoses, 1.5"		45.60		50.16	6.40	7.04
40 L.H., Daily Totals		$2130.30		$3078.69	$53.26	$76.97

Crew B-32

Crew No.	Bare Costs Hr.	Bare Costs Daily	Incl. Subs O&P Hr.	Incl. Subs O&P Daily	Cost Per Labor-Hour Bare Costs	Cost Per Labor-Hour Incl. O&P
1 Laborer	$44.40	$355.20	$66.25	$530.00	$55.35	$82.49
3 Equip. Oper. (medium)	59.00	1416.00	87.90	2109.60		
1 Grader, 30,000 Lbs.		1073.00		1180.30		
1 Tandem Roller, 10 Ton		246.80		271.48		
1 Dozer, 200 H.P.		1520.00		1672.00	88.74	97.62
32 L.H., Daily Totals		$4611.00		$5763.38	$144.09	$180.11

Crew B-32A

Crew No.	Bare Costs Hr.	Bare Costs Daily	Incl. Subs O&P Hr.	Incl. Subs O&P Daily	Cost Per Labor-Hour Bare Costs	Cost Per Labor-Hour Incl. O&P
1 Laborer	$44.40	$355.20	$66.25	$530.00	$54.13	$80.68
2 Equip. Oper. (medium)	59.00	944.00	87.90	1406.40		
1 Grader, 30,000 Lbs.		1073.00		1180.30		
1 Roller, Vibratory, 25 Ton		672.35		739.59	72.72	80.00
24 L.H., Daily Totals		$3044.55		$3856.28	$126.86	$160.68

Crew B-32B

Crew No.	Bare Costs Hr.	Bare Costs Daily	Incl. Subs O&P Hr.	Incl. Subs O&P Daily	Bare Costs	Incl. O&P
1 Laborer	$44.40	$355.20	$66.25	$530.00	$54.13	$80.68
2 Equip. Oper. (medium)	59.00	944.00	87.90	1406.40		
1 Dozer, 200 H.P.		1520.00		1672.00		
1 Roller, Vibratory, 25 Ton		672.35		739.59	91.35	100.48
24 L.H., Daily Totals		$3491.55		$4347.98	$145.48	$181.17

Crew B-32C

Crew No.	Bare Costs Hr.	Bare Costs Daily	Incl. Subs O&P Hr.	Incl. Subs O&P Daily	Bare Costs	Incl. O&P
1 Labor Foreman (outside)	$46.40	$371.20	$69.25	$554.00	$52.03	$77.58
2 Laborers	44.40	710.40	66.25	1060.00		
3 Equip. Oper. (medium)	59.00	1416.00	87.90	2109.60		
1 Grader, 30,000 Lbs.		1073.00		1180.30		
1 Tandem Roller, 10 Ton		246.80		271.48		
1 Dozer, 200 H.P.		1520.00		1672.00	59.16	65.08
48 L.H., Daily Totals		$5337.40		$6847.38	$111.20	$142.65

Crew B-33A

Crew No.	Bare Costs Hr.	Bare Costs Daily	Incl. Subs O&P Hr.	Incl. Subs O&P Daily	Bare Costs	Incl. O&P
1 Equip. Oper. (medium)	$59.00	$472.00	$87.90	$703.20	$54.83	$81.71
.5 Laborer	44.40	177.60	66.25	265.00		
.25 Equip. Oper. (medium)	59.00	118.00	87.90	175.80		
1 Scraper, Towed, 7 C.Y.		129.30		142.23		
1.25 Dozers, 300 H.P.		2231.25		2454.38	168.61	185.47
14 L.H., Daily Totals		$3128.15		$3740.61	$223.44	$267.19

Crew B-33B

Crew No.	Bare Costs Hr.	Bare Costs Daily	Incl. Subs O&P Hr.	Incl. Subs O&P Daily	Bare Costs	Incl. O&P
1 Equip. Oper. (medium)	$59.00	$472.00	$87.90	$703.20	$54.83	$81.71
.5 Laborer	44.40	177.60	66.25	265.00		
.25 Equip. Oper. (medium)	59.00	118.00	87.90	175.80		
1 Scraper, Towed, 10 C.Y.		161.65		177.82		
1.25 Dozers, 300 H.P.		2231.25		2454.38	170.92	188.01
14 L.H., Daily Totals		$3160.50		$3776.19	$225.75	$269.73

Crew B-33C

Crew No.	Bare Costs Hr.	Bare Costs Daily	Incl. Subs O&P Hr.	Incl. Subs O&P Daily	Bare Costs	Incl. O&P
1 Equip. Oper. (medium)	$59.00	$472.00	$87.90	$703.20	$54.83	$81.71
.5 Laborer	44.40	177.60	66.25	265.00		
.25 Equip. Oper. (medium)	59.00	118.00	87.90	175.80		
1 Scraper, Towed, 15 C.Y.		178.85		196.74		
1.25 Dozers, 300 H.P.		2231.25		2454.38	172.15	189.37
14 L.H., Daily Totals		$3177.70		$3795.11	$226.98	$271.08

Crew B-33D

Crew No.	Bare Costs Hr.	Bare Costs Daily	Incl. Subs O&P Hr.	Incl. Subs O&P Daily	Bare Costs	Incl. O&P
1 Equip. Oper. (medium)	$59.00	$472.00	$87.90	$703.20	$54.83	$81.71
.5 Laborer	44.40	177.60	66.25	265.00		
.25 Equip. Oper. (medium)	59.00	118.00	87.90	175.80		
1 S.P. Scraper, 14 C.Y.		2424.00		2666.40		
.25 Dozer, 300 H.P.		446.25		490.88	205.02	225.52
14 L.H., Daily Totals		$3637.85		$4301.27	$259.85	$307.23

Crew B-33E

Crew No.	Bare Costs Hr.	Bare Costs Daily	Incl. Subs O&P Hr.	Incl. Subs O&P Daily	Bare Costs	Incl. O&P
1 Equip. Oper. (medium)	$59.00	$472.00	$87.90	$703.20	$54.83	$81.71
.5 Laborer	44.40	177.60	66.25	265.00		
.25 Equip. Oper. (medium)	59.00	118.00	87.90	175.80		
1 S.P. Scraper, 21 C.Y.		2656.00		2921.60		
.25 Dozer, 300 H.P.		446.25		490.88	221.59	243.75
14 L.H., Daily Totals		$3869.85		$4556.48	$276.42	$325.46

For customer support on your Electrical Costs with RSMeans Data, call 800.448.8182.

495

Crews - Standard

Crew B-33F

	Bare Costs Hr.	Daily	Incl. Subs O&P Hr.	Daily	Cost Per Labor-Hour Bare Costs	Incl. O&P
1 Equip. Oper. (medium)	$59.00	$472.00	$87.90	$703.20	$54.83	$81.71
.5 Laborer	44.40	177.60	66.25	265.00		
.25 Equip. Oper. (medium)	59.00	118.00	87.90	175.80		
1 Elev. Scraper, 11 C.Y.		1059.00		1164.90		
.25 Dozer, 300 H.P.		446.25		490.88	107.52	118.27
14 L.H., Daily Totals		$2272.85		$2799.78	$162.35	$199.98

Crew B-33G

	Bare Costs Hr.	Daily	Incl. Subs O&P Hr.	Daily	Cost Per Labor-Hour Bare Costs	Incl. O&P
1 Equip. Oper. (medium)	$59.00	$472.00	$87.90	$703.20	$54.83	$81.71
.5 Laborer	44.40	177.60	66.25	265.00		
.25 Equip. Oper. (medium)	59.00	118.00	87.90	175.80		
1 Elev. Scraper, 22 C.Y.		1895.00		2084.50		
.25 Dozer, 300 H.P.		446.25		490.88	167.23	183.96
14 L.H., Daily Totals		$3108.85		$3719.38	$222.06	$265.67

Crew B-33H

	Bare Costs Hr.	Daily	Incl. Subs O&P Hr.	Daily	Cost Per Labor-Hour Bare Costs	Incl. O&P
.5 Laborer	$44.40	$177.60	$66.25	$265.00	$54.83	$81.71
1 Equipment Operator (med.)	59.00	472.00	87.90	703.20		
.25 Equipment Operator (med.)	59.00	118.00	87.90	175.80		
1 S.P. Scraper, 44 C.Y.		4695.00		5164.50		
.25 Dozer, 410 H.P.		701.75		771.92	385.48	424.03
14 L.H., Daily Totals		$6164.35		$7080.43	$440.31	$505.74

Crew B-33J

	Bare Costs Hr.	Daily	Incl. Subs O&P Hr.	Daily	Cost Per Labor-Hour Bare Costs	Incl. O&P
1 Equipment Operator (med.)	$59.00	$472.00	$87.90	$703.20	$59.00	$87.90
1 S.P. Scraper, 14 C.Y.		2424.00		2666.40	303.00	333.30
8 L.H., Daily Totals		$2896.00		$3369.60	$362.00	$421.20

Crew B-33K

	Bare Costs Hr.	Daily	Incl. Subs O&P Hr.	Daily	Cost Per Labor-Hour Bare Costs	Incl. O&P
1 Equipment Operator (med.)	$59.00	$472.00	$87.90	$703.20	$54.83	$81.71
.25 Equipment Operator (med.)	59.00	118.00	87.90	175.80		
.5 Laborer	44.40	177.60	66.25	265.00		
1 S.P. Scraper, 31 C.Y.		3707.00		4077.70		
.25 Dozer, 410 H.P.		701.75		771.92	314.91	346.40
14 L.H., Daily Totals		$5176.35		$5993.63	$369.74	$428.12

Crew B-34A

	Bare Costs Hr.	Daily	Incl. Subs O&P Hr.	Daily	Cost Per Labor-Hour Bare Costs	Incl. O&P
1 Truck Driver (heavy)	$51.30	$410.40	$76.70	$613.60	$51.30	$76.70
1 Dump Truck, 8 C.Y., 220 H.P.		407.60		448.36	50.95	56.05
8 L.H., Daily Totals		$818.00		$1061.96	$102.25	$132.75

Crew B-34B

	Bare Costs Hr.	Daily	Incl. Subs O&P Hr.	Daily	Cost Per Labor-Hour Bare Costs	Incl. O&P
1 Truck Driver (heavy)	$51.30	$410.40	$76.70	$613.60	$51.30	$76.70
1 Dump Truck, 12 C.Y., 400 H.P.		579.35		637.28	72.42	79.66
8 L.H., Daily Totals		$989.75		$1250.89	$123.72	$156.36

Crew B-34C

	Bare Costs Hr.	Daily	Incl. Subs O&P Hr.	Daily	Cost Per Labor-Hour Bare Costs	Incl. O&P
1 Truck Driver (heavy)	$51.30	$410.40	$76.70	$613.60	$51.30	$76.70
1 Truck Tractor, 6x4, 380 H.P.		499.15		549.07		
1 Dump Trailer, 16.5 C.Y.		138.35		152.19	79.69	87.66
8 L.H., Daily Totals		$1047.90		$1314.85	$130.99	$164.36

Crew B-34D

	Bare Costs Hr.	Daily	Incl. Subs O&P Hr.	Daily	Cost Per Labor-Hour Bare Costs	Incl. O&P
1 Truck Driver (heavy)	$51.30	$410.40	$76.70	$613.60	$51.30	$76.70
1 Truck Tractor, 6x4, 380 H.P.		499.15		549.07		
1 Dump Trailer, 20 C.Y.		153.55		168.91	81.59	89.75
8 L.H., Daily Totals		$1063.10		$1331.57	$132.89	$166.45

Crew B-34E

	Bare Costs Hr.	Daily	Incl. Subs O&P Hr.	Daily	Cost Per Labor-Hour Bare Costs	Incl. O&P
1 Truck Driver (heavy)	$51.30	$410.40	$76.70	$613.60	$51.30	$76.70
1 Dump Truck, Off Hwy., 25 Ton		1427.00		1569.70	178.38	196.21
8 L.H., Daily Totals		$1837.40		$2183.30	$229.68	$272.91

Crew B-34F

	Bare Costs Hr.	Daily	Incl. Subs O&P Hr.	Daily	Cost Per Labor-Hour Bare Costs	Incl. O&P
1 Truck Driver (heavy)	$51.30	$410.40	$76.70	$613.60	$51.30	$76.70
1 Dump Truck, Off Hwy., 35 Ton		945.65		1040.21	118.21	130.03
8 L.H., Daily Totals		$1356.05		$1653.82	$169.51	$206.73

Crew B-34G

	Bare Costs Hr.	Daily	Incl. Subs O&P Hr.	Daily	Cost Per Labor-Hour Bare Costs	Incl. O&P
1 Truck Driver (heavy)	$51.30	$410.40	$76.70	$613.60	$51.30	$76.70
1 Dump Truck, Off Hwy., 50 Ton		1979.00		2176.90	247.38	272.11
8 L.H., Daily Totals		$2389.40		$2790.50	$298.68	$348.81

Crew B-34H

	Bare Costs Hr.	Daily	Incl. Subs O&P Hr.	Daily	Cost Per Labor-Hour Bare Costs	Incl. O&P
1 Truck Driver (heavy)	$51.30	$410.40	$76.70	$613.60	$51.30	$76.70
1 Dump Truck, Off Hwy., 65 Ton		1938.00		2131.80	242.25	266.48
8 L.H., Daily Totals		$2348.40		$2745.40	$293.55	$343.18

Crew B-34I

	Bare Costs Hr.	Daily	Incl. Subs O&P Hr.	Daily	Cost Per Labor-Hour Bare Costs	Incl. O&P
1 Truck Driver (heavy)	$51.30	$410.40	$76.70	$613.60	$51.30	$76.70
1 Dump Truck, 18 C.Y., 450 H.P.		753.60		828.96	94.20	103.62
8 L.H., Daily Totals		$1164.00		$1442.56	$145.50	$180.32

Crew B-34J

	Bare Costs Hr.	Daily	Incl. Subs O&P Hr.	Daily	Cost Per Labor-Hour Bare Costs	Incl. O&P
1 Truck Driver (heavy)	$51.30	$410.40	$76.70	$613.60	$51.30	$76.70
1 Dump Truck, Off Hwy., 100 Ton		2769.00		3045.90	346.13	380.74
8 L.H., Daily Totals		$3179.40		$3659.50	$397.43	$457.44

Crew B-34K

	Bare Costs Hr.	Daily	Incl. Subs O&P Hr.	Daily	Cost Per Labor-Hour Bare Costs	Incl. O&P
1 Truck Driver (heavy)	$51.30	$410.40	$76.70	$613.60	$51.30	$76.70
1 Truck Tractor, 6x4, 450 H.P.		608.95		669.85		
1 Lowbed Trailer, 75 Ton		258.10		283.91	108.38	119.22
8 L.H., Daily Totals		$1277.45		$1567.36	$159.68	$195.92

Crew B-34L

	Bare Costs Hr.	Daily	Incl. Subs O&P Hr.	Daily	Cost Per Labor-Hour Bare Costs	Incl. O&P
1 Equip. Oper. (light)	$55.50	$444.00	$82.70	$661.60	$55.50	$82.70
1 Flatbed Truck, Gas, 1.5 Ton		198.50		218.35	24.81	27.29
8 L.H., Daily Totals		$642.50		$879.95	$80.31	$109.99

Crew B-34M

	Bare Costs Hr.	Daily	Incl. Subs O&P Hr.	Daily	Cost Per Labor-Hour Bare Costs	Incl. O&P
1 Equip. Oper. (light)	$55.50	$444.00	$82.70	$661.60	$55.50	$82.70
1 Flatbed Truck, Gas, 3 Ton		850.05		935.05	106.26	116.88
8 L.H., Daily Totals		$1294.05		$1596.66	$161.76	$199.58

Crew B-34N

	Bare Costs Hr.	Daily	Incl. Subs O&P Hr.	Daily	Cost Per Labor-Hour Bare Costs	Incl. O&P
1 Truck Driver (heavy)	$51.30	$410.40	$76.70	$613.60	$55.15	$82.30
1 Equip. Oper. (medium)	59.00	472.00	87.90	703.20		
1 Truck Tractor, 6x4, 380 H.P.		499.15		549.07		
1 Flatbed Trailer, 40 Ton		188.45		207.29	42.98	47.27
16 L.H., Daily Totals		$1570.00		$2073.16	$98.13	$129.57

Crew B-34P

	Bare Costs Hr.	Daily	Incl. Subs O&P Hr.	Daily	Cost Per Labor-Hour Bare Costs	Incl. O&P
1 Pipe Fitter	$68.35	$546.80	$102.00	$816.00	$58.72	$87.63
1 Truck Driver (light)	48.80	390.40	73.00	584.00		
1 Equip. Oper. (medium)	59.00	472.00	87.90	703.20		
1 Flatbed Truck, Gas, 3 Ton		850.05		935.05		
1 Backhoe Loader, 48 H.P.		216.20		237.82	44.43	48.87
24 L.H., Daily Totals		$2475.45		$3276.07	$103.14	$136.50

Crew No.	Bare Costs		Incl. Subs O&P		Cost Per Labor-Hour	
Crew B-34Q	Hr.	Daily	Hr.	Daily	Bare Costs	Incl. O&P
1 Pipe Fitter	$68.35	$546.80	$102.00	$816.00	$59.53	$88.85
1 Truck Driver (light)	48.80	390.40	73.00	584.00		
1 Equip. Oper. (crane)	61.45	491.60	91.55	732.40		
1 Flatbed Trailer, 25 Ton		137.20		150.92		
1 Dump Truck, 8 C.Y., 220 H.P.		407.60		448.36		
1 Hyd. Crane, 25 Ton		586.70		645.37	47.15	51.86
24 L.H., Daily Totals		$2560.30		$3377.05	$106.68	$140.71
Crew B-34R	Hr.	Daily	Hr.	Daily	Bare Costs	Incl. O&P
1 Pipe Fitter	$68.35	$546.80	$102.00	$816.00	$59.53	$88.85
1 Truck Driver (light)	48.80	390.40	73.00	584.00		
1 Equip. Oper. (crane)	61.45	491.60	91.55	732.40		
1 Flatbed Trailer, 25 Ton		137.20		150.92		
1 Dump Truck, 8 C.Y., 220 H.P.		407.60		448.36		
1 Hyd. Crane, 25 Ton		586.70		645.37		
1 Hyd. Excavator, 1 C.Y.		832.65		915.91	81.84	90.02
24 L.H., Daily Totals		$3392.95		$4292.97	$141.37	$178.87
Crew B-34S	Hr.	Daily	Hr.	Daily	Bare Costs	Incl. O&P
2 Pipe Fitters	$68.35	$1093.60	$102.00	$1632.00	$62.36	$93.06
1 Truck Driver (heavy)	51.30	410.40	76.70	613.60		
1 Equip. Oper. (crane)	61.45	491.60	91.55	732.40		
1 Flatbed Trailer, 40 Ton		188.45		207.29		
1 Truck Tractor, 6x4, 380 H.P.		499.15		549.07		
1 Hyd. Crane, 80 Ton		1458.00		1603.80		
1 Hyd. Excavator, 2 C.Y.		942.70		1036.97	96.51	106.16
32 L.H., Daily Totals		$5083.90		$6375.13	$158.87	$199.22
Crew B-34T	Hr.	Daily	Hr.	Daily	Bare Costs	Incl. O&P
2 Pipe Fitters	$68.35	$1093.60	$102.00	$1632.00	$62.36	$93.06
1 Truck Driver (heavy)	51.30	410.40	76.70	613.60		
1 Equip. Oper. (crane)	61.45	491.60	91.55	732.40		
1 Flatbed Trailer, 40 Ton		188.45		207.29		
1 Truck Tractor, 6x4, 380 H.P.		499.15		549.07		
1 Hyd. Crane, 80 Ton		1458.00		1603.80	67.05	73.75
32 L.H., Daily Totals		$4141.20		$5338.16	$129.41	$166.82
Crew B-34U	Hr.	Daily	Hr.	Daily	Bare Costs	Incl. O&P
1 Truck Driver (heavy)	$51.30	$410.40	$76.70	$613.60	$53.40	$79.70
1 Equip. Oper. (light)	55.50	444.00	82.70	661.60		
1 Truck Tractor, 220 H.P.		310.80		341.88		
1 Flatbed Trailer, 25 Ton		137.20		150.92	28.00	30.80
16 L.H., Daily Totals		$1302.40		$1768.00	$81.40	$110.50
Crew B-34V	Hr.	Daily	Hr.	Daily	Bare Costs	Incl. O&P
1 Truck Driver (heavy)	$51.30	$410.40	$76.70	$613.60	$56.08	$83.65
1 Equip. Oper. (crane)	61.45	491.60	91.55	732.40		
1 Equip. Oper. (light)	55.50	444.00	82.70	661.60		
1 Truck Tractor, 6x4, 450 H.P.		608.95		669.85		
1 Equipment Trailer, 50 Ton		207.25		227.97		
1 Pickup Truck, 4x4, 3/4 Ton		176.75		194.43	41.37	45.51
24 L.H., Daily Totals		$2338.95		$3099.84	$97.46	$129.16

Crew No.	Bare Costs		Incl. Subs O&P		Cost Per Labor-Hour	
Crew B-34W	Hr.	Daily	Hr.	Daily	Bare Costs	Incl. O&P
5 Truck Drivers (heavy)	$51.30	$2052.00	$76.70	$3068.00	$53.92	$80.50
2 Equip. Opers. (crane)	61.45	983.20	91.55	1464.80		
1 Equip. Oper. (mechanic)	61.50	492.00	91.65	733.20		
1 Laborer	44.40	355.20	66.25	530.00		
4 Truck Tractors, 6x4, 380 H.P.		1996.60		2196.26		
2 Equipment Trailers, 50 Ton		414.50		455.95		
2 Flatbed Trailers, 40 Ton		376.90		414.59		
1 Pickup Truck, 4x4, 3/4 Ton		176.75		194.43		
1 S.P. Crane, 4x4, 20 Ton		574.35		631.78	49.15	54.07
72 L.H., Daily Totals		$7421.50		$9689.01	$103.08	$134.57
Crew B-35	Hr.	Daily	Hr.	Daily	Bare Costs	Incl. O&P
1 Labor Foreman (outside)	$46.40	$371.20	$69.25	$554.00	$56.75	$84.76
1 Skilled Worker	57.10	456.80	85.90	687.20		
2 Welders	67.70	1083.20	101.05	1616.80		
1 Laborer	44.40	355.20	66.25	530.00		
1 Equip. Oper. (crane)	61.45	491.60	91.55	732.40		
1 Equip. Oper. (oiler)	52.50	420.00	78.25	626.00		
2 Welder, Electric, 300 amp		215.10		236.61		
1 Hyd. Excavator, .75 C.Y.		701.80		771.98	16.37	18.01
56 L.H., Daily Totals		$4094.90		$5754.99	$73.12	$102.77
Crew B-35A	Hr.	Daily	Hr.	Daily	Bare Costs	Incl. O&P
1 Labor Foreman (outside)	$46.40	$371.20	$69.25	$554.00	$53.42	$79.79
2 Laborers	44.40	710.40	66.25	1060.00		
1 Skilled Worker	57.10	456.80	85.90	687.20		
1 Welder (plumber)	67.70	541.60	101.05	808.40		
1 Equip. Oper. (crane)	61.45	491.60	91.55	732.40		
1 Equip. Oper. (oiler)	52.50	420.00	78.25	626.00		
1 Welder, Gas Engine, 300 amp		148.75		163.63		
1 Crawler Crane, 75 Ton		1967.00		2163.70	37.78	41.56
56 L.H., Daily Totals		$5107.35		$6795.32	$91.20	$121.35
Crew B-36	Hr.	Daily	Hr.	Daily	Bare Costs	Incl. O&P
1 Labor Foreman (outside)	$46.40	$371.20	$69.25	$554.00	$50.64	$75.51
2 Laborers	44.40	710.40	66.25	1060.00		
2 Equip. Oper. (medium)	59.00	944.00	87.90	1406.40		
1 Dozer, 200 H.P.		1520.00		1672.00		
1 Aggregate Spreader		59.95		65.94		
1 Tandem Roller, 10 Ton		246.80		271.48	45.67	50.24
40 L.H., Daily Totals		$3852.35		$5029.82	$96.31	$125.75
Crew B-36A	Hr.	Daily	Hr.	Daily	Bare Costs	Incl. O&P
1 Labor Foreman (outside)	$46.40	$371.20	$69.25	$554.00	$53.03	$79.05
2 Laborers	44.40	710.40	66.25	1060.00		
4 Equip. Oper. (medium)	59.00	1888.00	87.90	2812.80		
1 Dozer, 200 H.P.		1520.00		1672.00		
1 Aggregate Spreader		59.95		65.94		
1 Tandem Roller, 10 Ton		246.80		271.48		
1 Roller, Pneum. Whl., 12 Ton		349.90		384.89	38.87	42.76
56 L.H., Daily Totals		$5146.25		$6821.11	$91.90	$121.81

For customer support on your Electrical Costs with RSMeans Data, call 800.448.8182.

497

Crew No.	Bare Costs		Incl. Subs O&P		Cost Per Labor-Hour	
Crew B-36B	Hr.	Daily	Hr.	Daily	Bare Costs	Incl. O&P
1 Labor Foreman (outside)	$46.40	$371.20	$69.25	$554.00	$52.81	$78.76
2 Laborers	44.40	710.40	66.25	1060.00		
4 Equip. Oper. (medium)	59.00	1888.00	87.90	2812.80		
1 Truck Driver (heavy)	51.30	410.40	76.70	613.60		
1 Grader, 30,000 Lbs.		1073.00		1180.30		
1 F.E. Loader, Crl, 1.5 C.Y.		668.35		735.18		
1 Dozer, 300 H.P.		1785.00		1963.50		
1 Roller, Vibratory, 25 Ton		672.35		739.59		
1 Truck Tractor, 6x4, 450 H.P.		608.95		669.85		
1 Water Tank Trailer, 5000 Gal.		154.10		169.51	77.53	85.28
64 L.H., Daily Totals		$8341.75		$10498.33	$130.34	$164.04

Crew No.	Bare Costs		Incl. Subs O&P		Cost Per Labor-Hour	
Crew B-36C	Hr.	Daily	Hr.	Daily	Bare Costs	Incl. O&P
1 Labor Foreman (outside)	$46.40	$371.20	$69.25	$554.00	$54.94	$81.93
3 Equip. Oper. (medium)	59.00	1416.00	87.90	2109.60		
1 Truck Driver (heavy)	51.30	410.40	76.70	613.60		
1 Grader, 30,000 Lbs.		1073.00		1180.30		
1 Dozer, 300 H.P.		1785.00		1963.50		
1 Roller, Vibratory, 25 Ton		672.35		739.59		
1 Truck Tractor, 6x4, 450 H.P.		608.95		669.85		
1 Water Tank Trailer, 5000 Gal.		154.10		169.51	107.33	118.07
40 L.H., Daily Totals		$6491.00		$7999.94	$162.28	$200.00

Crew No.	Bare Costs		Incl. Subs O&P		Cost Per Labor-Hour	
Crew B-36D	Hr.	Daily	Hr.	Daily	Bare Costs	Incl. O&P
1 Labor Foreman (outside)	$46.40	$371.20	$69.25	$554.00	$55.85	$83.24
3 Equip. Oper. (medium)	59.00	1416.00	87.90	2109.60		
1 Grader, 30,000 Lbs.		1073.00		1180.30		
1 Dozer, 300 H.P.		1785.00		1963.50		
1 Roller, Vibratory, 25 Ton		672.35		739.59	110.32	121.36
32 L.H., Daily Totals		$5317.55		$6546.98	$166.17	$204.59

Crew No.	Bare Costs		Incl. Subs O&P		Cost Per Labor-Hour	
Crew B-37	Hr.	Daily	Hr.	Daily	Bare Costs	Incl. O&P
1 Labor Foreman (outside)	$46.40	$371.20	$69.25	$554.00	$46.58	$69.49
4 Laborers	44.40	1420.80	66.25	2120.00		
1 Equip. Oper. (light)	55.50	444.00	82.70	661.60		
1 Tandem Roller, 5 Ton		258.75		284.63	5.39	5.93
48 L.H., Daily Totals		$2494.75		$3620.22	$51.97	$75.42

Crew No.	Bare Costs		Incl. Subs O&P		Cost Per Labor-Hour	
Crew B-37A	Hr.	Daily	Hr.	Daily	Bare Costs	Incl. O&P
2 Laborers	$44.40	$710.40	$66.25	$1060.00	$45.87	$68.50
1 Truck Driver (light)	48.80	390.40	73.00	584.00		
1 Flatbed Truck, Gas, 1.5 Ton		198.50		218.35		
1 Tar Kettle, T.M.		156.70		172.37	14.80	16.28
24 L.H., Daily Totals		$1456.00		$2034.72	$60.67	$84.78

Crew No.	Bare Costs		Incl. Subs O&P		Cost Per Labor-Hour	
Crew B-37B	Hr.	Daily	Hr.	Daily	Bare Costs	Incl. O&P
3 Laborers	$44.40	$1065.60	$66.25	$1590.00	$45.50	$67.94
1 Truck Driver (light)	48.80	390.40	73.00	584.00		
1 Flatbed Truck, Gas, 1.5 Ton		198.50		218.35		
1 Tar Kettle, T.M.		156.70		172.37	11.10	12.21
32 L.H., Daily Totals		$1811.20		$2564.72	$56.60	$80.15

Crew No.	Bare Costs		Incl. Subs O&P		Cost Per Labor-Hour	
Crew B-37C	Hr.	Daily	Hr.	Daily	Bare Costs	Incl. O&P
2 Laborers	$44.40	$710.40	$66.25	$1060.00	$46.60	$69.63
2 Truck Drivers (light)	48.80	780.80	73.00	1168.00		
2 Flatbed Trucks, Gas, 1.5 Ton		397.00		436.70		
1 Tar Kettle, T.M.		156.70		172.37	17.30	19.03
32 L.H., Daily Totals		$2044.90		$2837.07	$63.90	$88.66

Crew No.	Bare Costs		Incl. Subs O&P		Cost Per Labor-Hour	
Crew B-37D	Hr.	Daily	Hr.	Daily	Bare Costs	Incl. O&P
1 Laborer	$44.40	$355.20	$66.25	$530.00	$46.60	$69.63
1 Truck Driver (light)	48.80	390.40	73.00	584.00		
1 Pickup Truck, 3/4 Ton		112.20		123.42	7.01	7.71
16 L.H., Daily Totals		$857.80		$1237.42	$53.61	$77.34

Crew No.	Bare Costs		Incl. Subs O&P		Cost Per Labor-Hour	
Crew B-37E	Hr.	Daily	Hr.	Daily	Bare Costs	Incl. O&P
3 Laborers	$44.40	$1065.60	$66.25	$1590.00	$49.33	$73.62
1 Equip. Oper. (light)	55.50	444.00	82.70	661.60		
1 Equip. Oper. (medium)	59.00	472.00	87.90	703.20		
2 Truck Drivers (light)	48.80	780.80	73.00	1168.00		
4 Barrels w/ Flasher		16.60		18.26		
1 Concrete Saw		112.85		124.14		
1 Rotary Hammer Drill		52.25		57.48		
1 Hammer Drill Bit		25.25		27.77		
1 Loader, Skid Steer, 30 H.P.		179.50		197.45		
1 Conc. Hammer Attach.		118.50		130.35		
1 Vibrating Plate, Gas, 18"		31.90		35.09		
2 Flatbed Trucks, Gas, 1.5 Ton		397.00		436.70	16.68	18.34
56 L.H., Daily Totals		$3696.25		$5150.03	$66.00	$91.96

Crew No.	Bare Costs		Incl. Subs O&P		Cost Per Labor-Hour	
Crew B-37F	Hr.	Daily	Hr.	Daily	Bare Costs	Incl. O&P
3 Laborers	$44.40	$1065.60	$66.25	$1590.00	$45.50	$67.94
1 Truck Driver (light)	48.80	390.40	73.00	584.00		
4 Barrels w/ Flasher		16.60		18.26		
1 Concrete Mixer, 10 C.F.		147.15		161.87		
1 Air Compressor, 60 cfm		153.85		169.24		
1 -50' Air Hose, 3/4"		7.15		7.87		
1 Spade (Chipper)		8.55		9.40		
1 Flatbed Truck, Gas, 1.5 Ton		198.50		218.35	16.62	18.28
32 L.H., Daily Totals		$1987.80		$2758.98	$62.12	$86.22

Crew No.	Bare Costs		Incl. Subs O&P		Cost Per Labor-Hour	
Crew B-37G	Hr.	Daily	Hr.	Daily	Bare Costs	Incl. O&P
1 Labor Foreman (outside)	$46.40	$371.20	$69.25	$554.00	$46.58	$69.49
4 Laborers	44.40	1420.80	66.25	2120.00		
1 Equip. Oper. (light)	55.50	444.00	82.70	661.60		
1 Berm Machine		253.55		278.90		
1 Tandem Roller, 5 Ton		258.75		284.63	10.67	11.74
48 L.H., Daily Totals		$2748.30		$3899.13	$57.26	$81.23

Crew No.	Bare Costs		Incl. Subs O&P		Cost Per Labor-Hour	
Crew B-37H	Hr.	Daily	Hr.	Daily	Bare Costs	Incl. O&P
1 Labor Foreman (outside)	$46.40	$371.20	$69.25	$554.00	$46.58	$69.49
4 Laborers	44.40	1420.80	66.25	2120.00		
1 Equip. Oper. (light)	55.50	444.00	82.70	661.60		
1 Tandem Roller, 5 Ton		258.75		284.63		
1 Flatbed Truck, Gas, 1.5 Ton		198.50		218.35		
1 Tar Kettle, T.M.		156.70		172.37	12.79	14.07
48 L.H., Daily Totals		$2849.95		$4010.95	$59.37	$83.56

For customer support on your Electrical Costs with RSMeans Data, call 800.448.8182.

Crew No.	Bare Costs		Incl. Subs O&P		Cost Per Labor-Hour	

Crew B-37I	Hr.	Daily	Hr.	Daily	Bare Costs	Incl. O&P
3 Laborers	$44.40	$1065.60	$66.25	$1590.00	$49.33	$73.62
1 Equip. Oper. (light)	55.50	444.00	82.70	661.60		
1 Equip. Oper. (medium)	59.00	472.00	87.90	703.20		
2 Truck Drivers (light)	48.80	780.80	73.00	1168.00		
4 Barrels w/ Flasher		16.60		18.26		
1 Concrete Saw		112.85		124.14		
1 Rotary Hammer Drill		52.25		57.48		
1 Hammer Drill Bit		25.25		27.77		
1 Air Compressor, 60 cfm		153.85		169.24		
1 -50' Air Hose, 3/4"		7.15		7.87		
1 Spade (Chipper)		8.55		9.40		
1 Loader, Skid Steer, 30 H.P.		179.50		197.45		
1 Conc. Hammer Attach.		118.50		130.35		
1 Concrete Mixer, 10 C.F.		147.15		161.87		
1 Vibrating Plate, Gas, 18"		31.90		35.09		
2 Flatbed Trucks, Gas, 1.5 Ton		397.00		436.70	22.33	24.56
56 L.H., Daily Totals		$4012.95		$5498.40	$71.66	$98.19

Crew B-37J	Hr.	Daily	Hr.	Daily	Bare Costs	Incl. O&P
1 Labor Foreman (outside)	$46.40	$371.20	$69.25	$554.00	$46.58	$69.49
4 Laborers	44.40	1420.80	66.25	2120.00		
1 Equip. Oper. (light)	55.50	444.00	82.70	661.60		
1 Air Compressor, 60 cfm		153.85		169.24		
1 -50' Air Hose, 3/4"		7.15		7.87		
2 Concrete Mixers, 10 C.F.		294.30		323.73		
2 Flatbed Trucks, Gas, 1.5 Ton		397.00		436.70		
1 Shot Blaster, 20"		208.70		229.57	22.10	24.31
48 L.H., Daily Totals		$3297.00		$4502.70	$68.69	$93.81

Crew B-37K	Hr.	Daily	Hr.	Daily	Bare Costs	Incl. O&P
1 Labor Foreman (outside)	$46.40	$371.20	$69.25	$554.00	$46.58	$69.49
4 Laborers	44.40	1420.80	66.25	2120.00		
1 Equip. Oper. (light)	55.50	444.00	82.70	661.60		
1 Air Compressor, 60 cfm		153.85		169.24		
1 -50' Air Hose, 3/4"		7.15		7.87		
2 Flatbed Trucks, Gas, 1.5 Ton		397.00		436.70		
1 Shot Blaster, 20"		208.70		229.57	15.97	17.57
48 L.H., Daily Totals		$3002.70		$4178.97	$62.56	$87.06

Crew B-38	Hr.	Daily	Hr.	Daily	Bare Costs	Incl. O&P
1 Labor Foreman (outside)	$46.40	$371.20	$69.25	$554.00	$49.94	$74.47
2 Laborers	44.40	710.40	66.25	1060.00		
1 Equip. Oper. (light)	55.50	444.00	82.70	661.60		
1 Equip. Oper. (medium)	59.00	472.00	87.90	703.20		
1 Backhoe Loader, 48 H.P.		216.20		237.82		
1 Hyd. Hammer (1200 lb.)		177.25		194.97		
1 F.E. Loader, W.M., 4 C.Y.		759.00		834.90		
1 Pvmt. Rem. Bucket		63.80		70.18	30.41	33.45
40 L.H., Daily Totals		$3213.85		$4316.68	$80.35	$107.92

Crew B-39	Hr.	Daily	Hr.	Daily	Bare Costs	Incl. O&P
1 Labor Foreman (outside)	$46.40	$371.20	$69.25	$554.00	$46.58	$69.49
4 Laborers	44.40	1420.80	66.25	2120.00		
1 Equip. Oper. (light)	55.50	444.00	82.70	661.60		
1 Air Compressor, 250 cfm		202.85		223.13		
2 Breakers, Pavement, 60 lb.		107.20		117.92		
2 -50' Air Hoses, 1.5"		45.60		50.16	7.41	8.15
48 L.H., Daily Totals		$2591.65		$3726.82	$53.99	$77.64

Crew B-40	Hr.	Daily	Hr.	Daily	Bare Costs	Incl. O&P
1 Pile Driver Foreman (outside)	$57.90	$463.20	$89.50	$716.00	$57.11	$87.06
4 Pile Drivers	55.90	1788.80	86.40	2764.80		
2 Equip. Oper. (crane)	61.45	983.20	91.55	1464.80		
1 Equip. Oper. (oiler)	52.50	420.00	78.25	626.00		
1 Crawler Crane, 40 Ton		1231.00		1354.10		
1 Vibratory Hammer & Gen.		2298.00		2527.80	55.14	60.65
64 L.H., Daily Totals		$7184.20		$9453.50	$112.25	$147.71

Crew B-40B	Hr.	Daily	Hr.	Daily	Bare Costs	Incl. O&P
1 Labor Foreman (outside)	$46.40	$371.20	$69.25	$554.00	$48.92	$72.97
3 Laborers	44.40	1065.60	66.25	1590.00		
1 Equip. Oper. (crane)	61.45	491.60	91.55	732.40		
1 Equip. Oper. (oiler)	52.50	420.00	78.25	626.00		
1 Lattice Boom Crane, 40 Ton		2127.00		2339.70	44.31	48.74
48 L.H., Daily Totals		$4475.40		$5842.10	$93.24	$121.71

Crew B-41	Hr.	Daily	Hr.	Daily	Bare Costs	Incl. O&P
1 Labor Foreman (outside)	$46.40	$371.20	$69.25	$554.00	$45.91	$68.49
4 Laborers	44.40	1420.80	66.25	2120.00		
.25 Equip. Oper. (crane)	61.45	122.90	91.55	183.10		
.25 Equip. Oper. (oiler)	52.50	105.00	78.25	156.50		
.25 Crawler Crane, 40 Ton		307.75		338.52	6.99	7.69
44 L.H., Daily Totals		$2327.65		$3352.13	$52.90	$76.18

Crew B-42	Hr.	Daily	Hr.	Daily	Bare Costs	Incl. O&P
1 Labor Foreman (outside)	$46.40	$371.20	$69.25	$554.00	$49.78	$74.67
4 Laborers	44.40	1420.80	66.25	2120.00		
1 Equip. Oper. (crane)	61.45	491.60	91.55	732.40		
1 Equip. Oper. (oiler)	52.50	420.00	78.25	626.00		
1 Welder	60.30	482.40	93.30	746.40		
1 Hyd. Crane, 25 Ton		586.70		645.37		
1 Welder, Gas Engine, 300 amp		148.75		163.63		
1 Horz. Boring Csg. Mch.		329.75		362.73	16.64	18.31
64 L.H., Daily Totals		$4251.20		$5950.52	$66.43	$92.98

Crew B-43	Hr.	Daily	Hr.	Daily	Bare Costs	Incl. O&P
1 Labor Foreman (outside)	$46.40	$371.20	$69.25	$554.00	$48.92	$72.97
3 Laborers	44.40	1065.60	66.25	1590.00		
1 Equip. Oper. (crane)	61.45	491.60	91.55	732.40		
1 Equip. Oper. (oiler)	52.50	420.00	78.25	626.00		
1 Drill Rig, Truck-Mounted		768.40		845.24	16.01	17.61
48 L.H., Daily Totals		$3116.80		$4347.64	$64.93	$90.58

Crew B-44	Hr.	Daily	Hr.	Daily	Bare Costs	Incl. O&P
1 Pile Driver Foreman (outside)	$57.90	$463.20	$89.50	$716.00	$56.10	$85.56
4 Pile Drivers	55.90	1788.80	86.40	2764.80		
2 Equip. Oper. (crane)	61.45	983.20	91.55	1464.80		
1 Laborer	44.40	355.20	66.25	530.00		
1 Crawler Crane, 40 Ton		1231.00		1354.10		
1 Lead, 60' High		211.80		232.98		
1 Hammer, Diesel, 15K ft.-lbs.		624.45		686.89	32.30	35.53
64 L.H., Daily Totals		$5657.65		$7749.57	$88.40	$121.09

Crew B-45	Hr.	Daily	Hr.	Daily	Bare Costs	Incl. O&P
1 Equip. Oper. (medium)	$59.00	$472.00	$87.90	$703.20	$55.15	$82.30
1 Truck Driver (heavy)	51.30	410.40	76.70	613.60		
1 Dist. Tanker, 3000 Gallon		334.10		367.51		
1 Truck Tractor, 6x4, 380 H.P.		499.15		549.07	52.08	57.29
16 L.H., Daily Totals		$1715.65		$2233.38	$107.23	$139.59

For customer support on your Electrical Costs with RSMeans Data, call 800.448.8182.

499

Crew No.	Bare Costs		Incl. Subs O&P		Cost Per Labor-Hour	
Crew B-46	Hr.	Daily	Hr.	Daily	Bare Costs	Incl. O&P
1 Pile Driver Foreman (outside)	$57.90	$463.20	$89.50	$716.00	$50.48	$76.84
2 Pile Drivers	55.90	894.40	86.40	1382.40		
3 Laborers	44.40	1065.60	66.25	1590.00		
1 Chain Saw, Gas, 36" Long		41.65		45.81	.87	.95
48 L.H., Daily Totals		$2464.85		$3734.22	$51.35	$77.80

Crew No.	Bare Costs		Incl. Subs O&P		Cost Per Labor-Hour	
Crew B-47	Hr.	Daily	Hr.	Daily	Bare Costs	Incl. O&P
1 Blast Foreman (outside)	$46.40	$371.20	$69.25	$554.00	$48.77	$72.73
1 Driller	44.40	355.20	66.25	530.00		
1 Equip. Oper. (light)	55.50	444.00	82.70	661.60		
1 Air Track Drill, 4"		1127.00		1239.70		
1 Air Compressor, 600 cfm		426.55		469.20		
2 -50' Air Hoses, 3"		76.70		84.37	67.93	74.72
24 L.H., Daily Totals		$2800.65		$3538.88	$116.69	$147.45

Crew No.	Bare Costs		Incl. Subs O&P		Cost Per Labor-Hour	
Crew B-47A	Hr.	Daily	Hr.	Daily	Bare Costs	Incl. O&P
1 Drilling Foreman (outside)	$46.40	$371.20	$69.25	$554.00	$53.45	$79.68
1 Equip. Oper. (heavy)	61.45	491.60	91.55	732.40		
1 Equip. Oper. (oiler)	52.50	420.00	78.25	626.00		
1 Air Track Drill, 5"		1127.00		1239.70	46.96	51.65
24 L.H., Daily Totals		$2409.80		$3152.10	$100.41	$131.34

Crew No.	Bare Costs		Incl. Subs O&P		Cost Per Labor-Hour	
Crew B-47C	Hr.	Daily	Hr.	Daily	Bare Costs	Incl. O&P
1 Laborer	$44.40	$355.20	$66.25	$530.00	$49.95	$74.47
1 Equip. Oper. (light)	55.50	444.00	82.70	661.60		
1 Air Compressor, 750 cfm		596.30		655.93		
2 -50' Air Hoses, 3"		76.70		84.37		
1 Air Track Drill, 4"		1127.00		1239.70	112.50	123.75
16 L.H., Daily Totals		$2599.20		$3171.60	$162.45	$198.22

Crew No.	Bare Costs		Incl. Subs O&P		Cost Per Labor-Hour	
Crew B-47E	Hr.	Daily	Hr.	Daily	Bare Costs	Incl. O&P
1 Labor Foreman (outside)	$46.40	$371.20	$69.25	$554.00	$44.90	$67.00
3 Laborers	44.40	1065.60	66.25	1590.00		
1 Flatbed Truck, Gas, 3 Ton		850.05		935.05	26.56	29.22
32 L.H., Daily Totals		$2286.85		$3079.05	$71.46	$96.22

Crew No.	Bare Costs		Incl. Subs O&P		Cost Per Labor-Hour	
Crew B-47G	Hr.	Daily	Hr.	Daily	Bare Costs	Incl. O&P
1 Labor Foreman (outside)	$46.40	$371.20	$69.25	$554.00	$47.67	$71.11
2 Laborers	44.40	710.40	66.25	1060.00		
1 Equip. Oper. (light)	55.50	444.00	82.70	661.60		
1 Air Track Drill, 4"		1127.00		1239.70		
1 Air Compressor, 600 cfm		426.55		469.20		
2 -50' Air Hoses, 3"		76.70		84.37		
1 Gunite Pump Rig		321.75		353.93	61.00	67.10
32 L.H., Daily Totals		$3477.60		$4422.80	$108.68	$138.21

Crew No.	Bare Costs		Incl. Subs O&P		Cost Per Labor-Hour	
Crew B-47H	Hr.	Daily	Hr.	Daily	Bare Costs	Incl. O&P
1 Skilled Worker Foreman (out)	$59.10	$472.80	$88.90	$711.20	$57.60	$86.65
3 Skilled Workers	57.10	1370.40	85.90	2061.60		
1 Flatbed Truck, Gas, 3 Ton		850.05		935.05	26.56	29.22
32 L.H., Daily Totals		$2693.25		$3707.86	$84.16	$115.87

Crew No.	Bare Costs		Incl. Subs O&P		Cost Per Labor-Hour	
Crew B-48	Hr.	Daily	Hr.	Daily	Bare Costs	Incl. O&P
1 Labor Foreman (outside)	$46.40	$371.20	$69.25	$554.00	$49.86	$74.36
3 Laborers	44.40	1065.60	66.25	1590.00		
1 Equip. Oper. (crane)	61.45	491.60	91.55	732.40		
1 Equip. Oper. (oiler)	52.50	420.00	78.25	626.00		
1 Equip. Oper. (light)	55.50	444.00	82.70	661.60		
1 Centr. Water Pump, 6"		235.25		258.77		
1 -20' Suction Hose, 6"		25.50		28.05		
1 -50' Discharge Hose, 6"		18.10		19.91		
1 Drill Rig, Truck-Mounted		768.40		845.24	18.70	20.57
56 L.H., Daily Totals		$3839.65		$5315.98	$68.57	$94.93

Crew No.	Bare Costs		Incl. Subs O&P		Cost Per Labor-Hour	
Crew B-49	Hr.	Daily	Hr.	Daily	Bare Costs	Incl. O&P
1 Labor Foreman (outside)	$46.40	$371.20	$69.25	$554.00	$52.25	$78.46
3 Laborers	44.40	1065.60	66.25	1590.00		
2 Equip. Oper. (crane)	61.45	983.20	91.55	1464.80		
2 Equip. Oper. (oilers)	52.50	840.00	78.25	1252.00		
1 Equip. Oper. (light)	55.50	444.00	82.70	661.60		
2 Pile Drivers	55.90	894.40	86.40	1382.40		
1 Hyd. Crane, 25 Ton		586.70		645.37		
1 Centr. Water Pump, 6"		235.25		258.77		
1 -20' Suction Hose, 6"		25.50		28.05		
1 -50' Discharge Hose, 6"		18.10		19.91		
1 Drill Rig, Truck-Mounted		768.40		845.24	18.57	20.42
88 L.H., Daily Totals		$6232.35		$8702.15	$70.82	$98.89

Crew No.	Bare Costs		Incl. Subs O&P		Cost Per Labor-Hour	
Crew B-50	Hr.	Daily	Hr.	Daily	Bare Costs	Incl. O&P
2 Pile Driver Foremen (outside)	$57.90	$926.40	$89.50	$1432.00	$54.27	$82.68
6 Pile Drivers	55.90	2683.20	86.40	4147.20		
2 Equip. Oper. (crane)	61.45	983.20	91.55	1464.80		
1 Equip. Oper. (oiler)	52.50	420.00	78.25	626.00		
3 Laborers	44.40	1065.60	66.25	1590.00		
1 Crawler Crane, 40 Ton		1231.00		1354.10		
1 Lead, 60' High		211.80		232.98		
1 Hammer, Diesel, 15K ft.-lbs.		624.45		686.89		
1 Air Compressor, 600 cfm		426.55		469.20		
2 -50' Air Hoses, 3"		76.70		84.37		
1 Chain Saw, Gas, 36" Long		41.65		45.81	23.32	25.66
112 L.H., Daily Totals		$8690.55		$12133.37	$77.59	$108.33

Crew No.	Bare Costs		Incl. Subs O&P		Cost Per Labor-Hour	
Crew B-51	Hr.	Daily	Hr.	Daily	Bare Costs	Incl. O&P
1 Labor Foreman (outside)	$46.40	$371.20	$69.25	$554.00	$45.47	$67.88
4 Laborers	44.40	1420.80	66.25	2120.00		
1 Truck Driver (light)	48.80	390.40	73.00	584.00		
1 Flatbed Truck, Gas, 1.5 Ton		198.50		218.35	4.14	4.55
48 L.H., Daily Totals		$2380.90		$3476.35	$49.60	$72.42

Crew No.	Bare Costs		Incl. Subs O&P		Cost Per Labor-Hour	
Crew B-52	Hr.	Daily	Hr.	Daily	Bare Costs	Incl. O&P
1 Carpenter Foreman (outside)	$56.70	$453.60	$84.60	$676.80	$50.76	$75.55
1 Carpenter	54.70	437.60	81.65	653.20		
3 Laborers	44.40	1065.60	66.25	1590.00		
1 Cement Finisher	51.80	414.40	75.90	607.20		
.5 Rodman (reinf.)	58.90	235.60	88.05	352.20		
.5 Equip. Oper. (medium)	59.00	236.00	87.90	351.60		
.5 Crawler Loader, 3 C.Y.		573.00		630.30	10.23	11.26
56 L.H., Daily Totals		$3415.80		$4861.30	$61.00	$86.81

Crew No.	Bare Costs		Incl. Subs O&P		Cost Per Labor-Hour	
Crew B-53	Hr.	Daily	Hr.	Daily	Bare Costs	Incl. O&P
1 Equip. Oper. (light)	$55.50	$444.00	$82.70	$661.60	$55.50	$82.70
1 Trencher, Chain, 12 H.P.		158.60		174.46	19.82	21.81
8 L.H., Daily Totals		$602.60		$836.06	$75.33	$104.51

For customer support on your Electrical Costs with RSMeans Data, call 800.448.8182.

Crews - Standard

Crew B-54

Crew B-54	Hr.	Daily	Hr.	Daily	Bare Costs	Incl. O&P
1 Equip. Oper. (light)	$55.50	$444.00	$82.70	$661.60	$55.50	$82.70
1 Trencher, Chain, 40 H.P.		450.70		495.77	56.34	61.97
8 L.H., Daily Totals		$894.70		$1157.37	$111.84	$144.67

Crew B-54A

Crew B-54A	Hr.	Daily	Hr.	Daily	Bare Costs	Incl. O&P
.17 Labor Foreman (outside)	$46.40	$63.10	$69.25	$94.18	$57.17	$85.19
1 Equipment Operator (med.)	59.00	472.00	87.90	703.20		
1 Wheel Trencher, 67 H.P.		1140.00		1254.00	121.79	133.97
9.36 L.H., Daily Totals		$1675.10		$2051.38	$178.96	$219.16

Crew B-54B

Crew B-54B	Hr.	Daily	Hr.	Daily	Bare Costs	Incl. O&P
.25 Labor Foreman (outside)	$46.40	$92.80	$69.25	$138.50	$56.48	$84.17
1 Equipment Operator (med.)	59.00	472.00	87.90	703.20		
1 Wheel Trencher, 150 H.P.		1238.00		1361.80	123.80	136.18
10 L.H., Daily Totals		$1802.80		$2203.50	$180.28	$220.35

Crew B-54C

Crew B-54C	Hr.	Daily	Hr.	Daily	Bare Costs	Incl. O&P
1 Laborer	$44.40	$355.20	$66.25	$530.00	$51.70	$77.08
1 Equipment Operator (med.)	59.00	472.00	87.90	703.20		
1 Wheel Trencher, 67 H.P.		1140.00		1254.00	71.25	78.38
16 L.H., Daily Totals		$1967.20		$2487.20	$122.95	$155.45

Crew B-54D

Crew B-54D	Hr.	Daily	Hr.	Daily	Bare Costs	Incl. O&P
1 Laborer	$44.40	$355.20	$66.25	$530.00	$51.70	$77.08
1 Equipment Operator (med.)	59.00	472.00	87.90	703.20		
1 Rock Trencher, 6" Width		434.20		477.62	27.14	29.85
16 L.H., Daily Totals		$1261.40		$1710.82	$78.84	$106.93

Crew B-54E

Crew B-54E	Hr.	Daily	Hr.	Daily	Bare Costs	Incl. O&P
1 Laborer	$44.40	$355.20	$66.25	$530.00	$51.70	$77.08
1 Equipment Operator (med.)	59.00	472.00	87.90	703.20		
1 Rock Trencher, 18" Width		1015.00		1116.50	63.44	69.78
16 L.H., Daily Totals		$1842.20		$2349.70	$115.14	$146.86

Crew B-55

Crew B-55	Hr.	Daily	Hr.	Daily	Bare Costs	Incl. O&P
2 Laborers	$44.40	$710.40	$66.25	$1060.00	$45.87	$68.50
1 Truck Driver (light)	48.80	390.40	73.00	584.00		
1 Truck-Mounted Earth Auger		394.15		433.57		
1 Flatbed Truck, Gas, 3 Ton		850.05		935.05	51.84	57.03
24 L.H., Daily Totals		$2345.00		$3012.62	$97.71	$125.53

Crew B-56

Crew B-56	Hr.	Daily	Hr.	Daily	Bare Costs	Incl. O&P
1 Laborer	$44.40	$355.20	$66.25	$530.00	$49.95	$74.47
1 Equip. Oper. (light)	55.50	444.00	82.70	661.60		
1 Air Track Drill, 4"		1127.00		1239.70		
1 Air Compressor, 600 cfm		426.55		469.20		
1 -50' Air Hose, 3"		38.35		42.19	99.49	109.44
16 L.H., Daily Totals		$2391.10		$2942.69	$149.44	$183.92

Crew B-57

Crew B-57	Hr.	Daily	Hr.	Daily	Bare Costs	Incl. O&P
1 Labor Foreman (outside)	$46.40	$371.20	$69.25	$554.00	$50.77	$75.71
2 Laborers	44.40	710.40	66.25	1060.00		
1 Equip. Oper. (crane)	61.45	491.60	91.55	732.40		
1 Equip. Oper. (light)	55.50	444.00	82.70	661.60		
1 Equip. Oper. (oiler)	52.50	420.00	78.25	626.00		
1 Crawler Crane, 25 Ton		1152.00		1267.20		
1 Clamshell Bucket, 1 C.Y.		69.25		76.17		
1 Centr. Water Pump, 6"		235.25		258.77		
1 -20' Suction Hose, 6"		25.50		28.05		
20 -50' Discharge Hoses, 6"		362.00		398.20	38.42	42.26
48 L.H., Daily Totals		$4281.20		$5662.40	$89.19	$117.97

Crew B-58

Crew B-58	Hr.	Daily	Hr.	Daily	Bare Costs	Incl. O&P
2 Laborers	$44.40	$710.40	$66.25	$1060.00	$48.10	$71.73
1 Equip. Oper. (light)	55.50	444.00	82.70	661.60		
1 Backhoe Loader, 48 H.P.		216.20		237.82		
1 Small Helicopter, w/ Pilot		2101.00		2311.10	96.55	106.21
24 L.H., Daily Totals		$3471.60		$4270.52	$144.65	$177.94

Crew B-59

Crew B-59	Hr.	Daily	Hr.	Daily	Bare Costs	Incl. O&P
1 Truck Driver (heavy)	$51.30	$410.40	$76.70	$613.60	$51.30	$76.70
1 Truck Tractor, 220 H.P.		310.80		341.88		
1 Water Tank Trailer, 5000 Gal.		154.10		169.51	58.11	63.92
8 L.H., Daily Totals		$875.30		$1124.99	$109.41	$140.62

Crew B-59A

Crew B-59A	Hr.	Daily	Hr.	Daily	Bare Costs	Incl. O&P
2 Laborers	$44.40	$710.40	$66.25	$1060.00	$46.70	$69.73
1 Truck Driver (heavy)	51.30	410.40	76.70	613.60		
1 Water Tank Trailer, 5000 Gal.		154.10		169.51		
1 Truck Tractor, 220 H.P.		310.80		341.88	19.37	21.31
24 L.H., Daily Totals		$1585.70		$2184.99	$66.07	$91.04

Crew B-60

Crew B-60	Hr.	Daily	Hr.	Daily	Bare Costs	Incl. O&P
1 Labor Foreman (outside)	$46.40	$371.20	$69.25	$554.00	$51.45	$76.71
2 Laborers	44.40	710.40	66.25	1060.00		
1 Equip. Oper. (crane)	61.45	491.60	91.55	732.40		
2 Equip. Oper. (light)	55.50	888.00	82.70	1323.20		
1 Equip. Oper. (oiler)	52.50	420.00	78.25	626.00		
1 Crawler Crane, 40 Ton		1231.00		1354.10		
1 Lead, 60' High		211.80		232.98		
1 Hammer, Diesel, 15K ft.-lbs.		624.45		686.89		
1 Backhoe Loader, 48 H.P.		216.20		237.82	40.78	44.85
56 L.H., Daily Totals		$5164.65		$6807.40	$92.23	$121.56

Crew B-61

Crew B-61	Hr.	Daily	Hr.	Daily	Bare Costs	Incl. O&P
1 Labor Foreman (outside)	$46.40	$371.20	$69.25	$554.00	$47.02	$70.14
3 Laborers	44.40	1065.60	66.25	1590.00		
1 Equip. Oper. (light)	55.50	444.00	82.70	661.60		
1 Cement Mixer, 2 C.Y.		112.35		123.58		
1 Air Compressor, 160 cfm		212.30		233.53	8.12	8.93
40 L.H., Daily Totals		$2205.45		$3162.72	$55.14	$79.07

Crew B-62

Crew B-62	Hr.	Daily	Hr.	Daily	Bare Costs	Incl. O&P
2 Laborers	$44.40	$710.40	$66.25	$1060.00	$48.10	$71.73
1 Equip. Oper. (light)	55.50	444.00	82.70	661.60		
1 Loader, Skid Steer, 30 H.P.		179.50		197.45	7.48	8.23
24 L.H., Daily Totals		$1333.90		$1919.05	$55.58	$79.96

For customer support on your Electrical Costs with RSMeans Data, call 800.448.8182.

501

Crew No.	Bare Costs		Incl. Subs O&P		Cost Per Labor-Hour	

Crew B-62A

	Hr.	Daily	Hr.	Daily	Bare Costs	Incl. O&P
2 Laborers	$44.40	$710.40	$66.25	$1060.00	$48.10	$71.73
1 Equip. Oper. (light)	55.50	444.00	82.70	661.60		
1 Loader, Skid Steer, 30 H.P.		179.50		197.45		
1 Trencher Attachment		66.25		72.88	10.24	11.26
24 L.H., Daily Totals		$1400.15		$1991.93	$58.34	$83.00

Crew B-63

	Hr.	Daily	Hr.	Daily	Bare Costs	Incl. O&P
4 Laborers	$44.40	$1420.80	$66.25	$2120.00	$46.62	$69.54
1 Equip. Oper. (light)	55.50	444.00	82.70	661.60		
1 Loader, Skid Steer, 30 H.P.		179.50		197.45	4.49	4.94
40 L.H., Daily Totals		$2044.30		$2979.05	$51.11	$74.48

Crew B-63B

	Hr.	Daily	Hr.	Daily	Bare Costs	Incl. O&P
1 Labor Foreman (inside)	$44.90	$359.20	$67.00	$536.00	$47.30	$70.55
2 Laborers	44.40	710.40	66.25	1060.00		
1 Equip. Oper. (light)	55.50	444.00	82.70	661.60		
1 Loader, Skid Steer, 78 H.P.		446.30		490.93	13.95	15.34
32 L.H., Daily Totals		$1959.90		$2748.53	$61.25	$85.89

Crew B-64

	Hr.	Daily	Hr.	Daily	Bare Costs	Incl. O&P
1 Laborer	$44.40	$355.20	$66.25	$530.00	$46.60	$69.63
1 Truck Driver (light)	48.80	390.40	73.00	584.00		
1 Power Mulcher (small)		201.00		221.10		
1 Flatbed Truck, Gas, 1.5 Ton		198.50		218.35	24.97	27.47
16 L.H., Daily Totals		$1145.10		$1553.45	$71.57	$97.09

Crew B-65

	Hr.	Daily	Hr.	Daily	Bare Costs	Incl. O&P
1 Laborer	$44.40	$355.20	$66.25	$530.00	$46.60	$69.63
1 Truck Driver (light)	48.80	390.40	73.00	584.00		
1 Power Mulcher (Large)		345.35		379.88		
1 Flatbed Truck, Gas, 1.5 Ton		198.50		218.35	33.99	37.39
16 L.H., Daily Totals		$1289.45		$1712.23	$80.59	$107.01

Crew B-66

	Hr.	Daily	Hr.	Daily	Bare Costs	Incl. O&P
1 Equip. Oper. (light)	$55.50	$444.00	$82.70	$661.60	$55.50	$82.70
1 Loader-Backhoe, 40 H.P.		267.55		294.31	33.44	36.79
8 L.H., Daily Totals		$711.55		$955.90	$88.94	$119.49

Crew B-67

	Hr.	Daily	Hr.	Daily	Bare Costs	Incl. O&P
1 Millwright	$58.75	$470.00	$84.90	$679.20	$57.13	$83.80
1 Equip. Oper. (light)	55.50	444.00	82.70	661.60		
1 R.T. Forklift, 5,000 Lb., diesel		272.85		300.13	17.05	18.76
16 L.H., Daily Totals		$1186.85		$1640.93	$74.18	$102.56

Crew B-67B

	Hr.	Daily	Hr.	Daily	Bare Costs	Incl. O&P
1 Millwright Foreman (inside)	$59.25	$474.00	$85.60	$684.80	$59.00	$85.25
1 Millwright	58.75	470.00	84.90	679.20		
16 L.H., Daily Totals		$944.00		$1364.00	$59.00	$85.25

Crew B-68

	Hr.	Daily	Hr.	Daily	Bare Costs	Incl. O&P
2 Millwrights	$58.75	$940.00	$84.90	$1358.40	$57.67	$84.17
1 Equip. Oper. (light)	55.50	444.00	82.70	661.60		
1 R.T. Forklift, 5,000 Lb., diesel		272.85		300.13	11.37	12.51
24 L.H., Daily Totals		$1656.85		$2320.14	$69.04	$96.67

Crew B-68A

	Hr.	Daily	Hr.	Daily	Bare Costs	Incl. O&P
1 Millwright Foreman (inside)	$59.25	$474.00	$85.60	$684.80	$58.92	$85.13
2 Millwrights	58.75	940.00	84.90	1358.40		
1 Forklift, Smooth Floor, 8,000 Lb.		283.25		311.57	11.80	12.98
24 L.H., Daily Totals		$1697.25		$2354.78	$70.72	$98.12

Crew B-68B

	Hr.	Daily	Hr.	Daily	Bare Costs	Incl. O&P
1 Millwright Foreman (inside)	$59.25	$474.00	$85.60	$684.80	$62.79	$92.40
2 Millwrights	58.75	940.00	84.90	1358.40		
2 Electricians	63.70	1019.20	94.65	1514.40		
2 Plumbers	67.70	1083.20	101.05	1616.80		
1 R.T. Forklift, 5,000 Lb., gas		283.30		311.63	5.06	5.56
56 L.H., Daily Totals		$3799.70		$5486.03	$67.85	$97.96

Crew B-68C

	Hr.	Daily	Hr.	Daily	Bare Costs	Incl. O&P
1 Millwright Foreman (inside)	$59.25	$474.00	$85.60	$684.80	$62.35	$91.55
1 Millwright	58.75	470.00	84.90	679.20		
1 Electrician	63.70	509.60	94.65	757.20		
1 Plumber	67.70	541.60	101.05	808.40		
1 R.T. Forklift, 5,000 Lb., gas		283.30		311.63	8.85	9.74
32 L.H., Daily Totals		$2278.50		$3241.23	$71.20	$101.29

Crew B-68D

	Hr.	Daily	Hr.	Daily	Bare Costs	Incl. O&P
1 Labor Foreman (inside)	$44.90	$359.20	$67.00	$536.00	$48.27	$71.98
1 Laborer	44.40	355.20	66.25	530.00		
1 Equip. Oper. (light)	55.50	444.00	82.70	661.60		
1 R.T. Forklift, 5,000 Lb., gas		283.30		311.63	11.80	12.98
24 L.H., Daily Totals		$1441.70		$2039.23	$60.07	$84.97

Crew B-68E

	Hr.	Daily	Hr.	Daily	Bare Costs	Incl. O&P
1 Struc. Steel Foreman (inside)	$60.80	$486.40	$94.10	$752.80	$60.40	$93.46
3 Struc. Steel Workers	60.30	1447.20	93.30	2239.20		
1 Welder	60.30	482.40	93.30	746.40		
1 Forklift, Smooth Floor, 8,000 Lb.		283.25		311.57	7.08	7.79
40 L.H., Daily Totals		$2699.25		$4049.97	$67.48	$101.25

Crew B-68F

	Hr.	Daily	Hr.	Daily	Bare Costs	Incl. O&P
1 Skilled Worker Foreman (out)	$59.10	$472.80	$88.90	$711.20	$57.77	$86.90
2 Skilled Workers	57.10	913.60	85.90	1374.40		
1 R.T. Forklift, 5,000 Lb., gas		283.30		311.63	11.80	12.98
24 L.H., Daily Totals		$1669.70		$2397.23	$69.57	$99.88

Crew B-68G

	Hr.	Daily	Hr.	Daily	Bare Costs	Incl. O&P
2 Structural Steel Workers	$60.30	$964.80	$93.30	$1492.80	$60.30	$93.30
1 R.T. Forklift, 5,000 Lb., gas		283.30		311.63	17.71	19.48
16 L.H., Daily Totals		$1248.10		$1804.43	$78.01	$112.78

Crew B-69

	Hr.	Daily	Hr.	Daily	Bare Costs	Incl. O&P
1 Labor Foreman (outside)	$46.40	$371.20	$69.25	$554.00	$48.92	$72.97
3 Laborers	44.40	1065.60	66.25	1590.00		
1 Equip. Oper. (crane)	61.45	491.60	91.55	732.40		
1 Equip. Oper. (oiler)	52.50	420.00	78.25	626.00		
1 Hyd. Crane, 80 Ton		1458.00		1603.80	30.38	33.41
48 L.H., Daily Totals		$3806.40		$5106.20	$79.30	$106.38

Crew B-69A

	Hr.	Daily	Hr.	Daily	Bare Costs	Incl. O&P
1 Labor Foreman (outside)	$46.40	$371.20	$69.25	$554.00	$48.40	$71.97
3 Laborers	44.40	1065.60	66.25	1590.00		
1 Equip. Oper. (medium)	59.00	472.00	87.90	703.20		
1 Concrete Finisher	51.80	414.40	75.90	607.20		
1 Curb/Gutter Paver, 2-Track		1231.00		1354.10	25.65	28.21
48 L.H., Daily Totals		$3554.20		$4808.50	$74.05	$100.18

For customer support on your Electrical Costs with RSMeans Data, call 800.448.8182.

Crew No.	Bare Costs Hr.	Bare Costs Daily	Incl. Subs O&P Hr.	Incl. Subs O&P Daily	Cost Per Labor-Hour Bare Costs	Cost Per Labor-Hour Incl. O&P
Crew B-69B					Bare Costs	Incl. O&P
1 Labor Foreman (outside)	$46.40	$371.20	$69.25	$554.00	$48.40	$71.97
3 Laborers	44.40	1065.60	66.25	1590.00		
1 Equip. Oper. (medium)	59.00	472.00	87.90	703.20		
1 Cement Finisher	51.80	414.40	75.90	607.20		
1 Curb/Gutter Paver, 4-Track		801.05		881.15	16.69	18.36
48 L.H., Daily Totals		$3124.25		$4335.56	$65.09	$90.32
Crew B-70					Bare Costs	Incl. O&P
1 Labor Foreman (outside)	$46.40	$371.20	$69.25	$554.00	$50.94	$75.96
3 Laborers	44.40	1065.60	66.25	1590.00		
3 Equip. Oper. (medium)	59.00	1416.00	87.90	2109.60		
1 Grader, 30,000 Lbs.		1073.00		1180.30		
1 Ripper, Beam & 1 Shank		91.60		100.76		
1 Road Sweeper, S.P., 8' wide		723.65		796.01		
1 F.E. Loader, W.M., 1.5 C.Y.		441.40		485.54	41.60	45.76
56 L.H., Daily Totals		$5182.45		$6816.22	$92.54	$121.72
Crew B-70A					Bare Costs	Incl. O&P
1 Laborer	$44.40	$355.20	$66.25	$530.00	$56.08	$83.57
4 Equip. Oper. (medium)	59.00	1888.00	87.90	2812.80		
1 Grader, 40,000 Lbs.		1219.00		1340.90		
1 F.E. Loader, W.M., 2.5 C.Y.		638.30		702.13		
1 Dozer, 80 H.P.		405.85		446.44		
1 Roller, Pneum. Whl., 12 Ton		349.90		384.89	65.33	71.86
40 L.H., Daily Totals		$4856.25		$6217.15	$121.41	$155.43
Crew B-71					Bare Costs	Incl. O&P
1 Labor Foreman (outside)	$46.40	$371.20	$69.25	$554.00	$50.94	$75.96
3 Laborers	44.40	1065.60	66.25	1590.00		
3 Equip. Oper. (medium)	59.00	1416.00	87.90	2109.60		
1 Pvmt. Profiler, 750 H.P.		3483.00		3831.30		
1 Road Sweeper, S.P., 8' wide		723.65		796.01		
1 F.E. Loader, W.M., 1.5 C.Y.		441.40		485.54	83.00	91.30
56 L.H., Daily Totals		$7500.85		$9366.45	$133.94	$167.26
Crew B-72					Bare Costs	Incl. O&P
1 Labor Foreman (outside)	$46.40	$371.20	$69.25	$554.00	$51.95	$77.45
3 Laborers	44.40	1065.60	66.25	1590.00		
4 Equip. Oper. (medium)	59.00	1888.00	87.90	2812.80		
1 Pvmt. Profiler, 750 H.P.		3483.00		3831.30		
1 Hammermill, 250 H.P.		857.40		943.14		
1 Windrow Loader		1461.00		1607.10		
1 Mix Paver, 165 H.P.		2172.00		2389.20		
1 Roller, Pneum. Whl., 12 Ton		349.90		384.89	130.05	143.06
64 L.H., Daily Totals		$11648.10		$14112.43	$182.00	$220.51
Crew B-73					Bare Costs	Incl. O&P
1 Labor Foreman (outside)	$46.40	$371.20	$69.25	$554.00	$53.77	$80.16
2 Laborers	44.40	710.40	66.25	1060.00		
5 Equip. Oper. (medium)	59.00	2360.00	87.90	3516.00		
1 Road Mixer, 310 H.P.		1919.00		2110.90		
1 Tandem Roller, 10 Ton		246.80		271.48		
1 Hammermill, 250 H.P.		857.40		943.14		
1 Grader, 30,000 Lbs.		1073.00		1180.30		
.5 F.E. Loader, W.M., 1.5 C.Y.		220.70		242.77		
.5 Truck Tractor, 220 H.P.		155.40		170.94		
.5 Water Tank Trailer, 5000 Gal.		77.05		84.75	71.08	78.19
64 L.H., Daily Totals		$7990.95		$10134.29	$124.86	$158.35

Crew No.	Bare Costs Hr.	Bare Costs Daily	Incl. Subs O&P Hr.	Incl. Subs O&P Daily	Cost Per Labor-Hour Bare Costs	Cost Per Labor-Hour Incl. O&P
Crew B-74					$53.67	$80.06
1 Labor Foreman (outside)	$46.40	$371.20	$69.25	$554.00		
1 Laborer	44.40	355.20	66.25	530.00		
4 Equip. Oper. (medium)	59.00	1888.00	87.90	2812.80		
2 Truck Drivers (heavy)	51.30	820.80	76.70	1227.20		
1 Grader, 30,000 Lbs.		1073.00		1180.30		
1 Ripper, Beam & 1 Shank		91.60		100.76		
2 Stabilizers, 310 H.P.		2808.00		3088.80		
1 Flatbed Truck, Gas, 3 Ton		850.05		935.05		
1 Chem. Spreader, Towed		85.40		93.94		
1 Roller, Vibratory, 25 Ton		672.35		739.59		
1 Water Tank Trailer, 5000 Gal.		154.10		169.51		
1 Truck Tractor, 220 H.P.		310.80		341.88	94.46	103.90
64 L.H., Daily Totals		$9480.50		$11773.83	$148.13	$183.97
Crew B-75					$54.01	$80.54
1 Labor Foreman (outside)	$46.40	$371.20	$69.25	$554.00		
1 Laborer	44.40	355.20	66.25	530.00		
4 Equip. Oper. (medium)	59.00	1888.00	87.90	2812.80		
1 Truck Driver (heavy)	51.30	410.40	76.70	613.60		
1 Grader, 30,000 Lbs.		1073.00		1180.30		
1 Ripper, Beam & 1 Shank		91.60		100.76		
2 Stabilizers, 310 H.P.		2808.00		3088.80		
1 Dist. Tanker, 3000 Gallon		334.10		367.51		
1 Truck Tractor, 6x4, 380 H.P.		499.15		549.07		
1 Roller, Vibratory, 25 Ton		672.35		739.59	97.83	107.61
56 L.H., Daily Totals		$8503.00		$10536.42	$151.84	$188.15
Crew B-76					$56.98	$86.98
1 Dock Builder Foreman (outside)	$57.90	$463.20	$89.50	$716.00		
5 Dock Builders	55.90	2236.00	86.40	3456.00		
2 Equip. Oper. (crane)	61.45	983.20	91.55	1464.80		
1 Equip. Oper. (oiler)	52.50	420.00	78.25	626.00		
1 Crawler Crane, 50 Ton		1541.00		1695.10		
1 Barge, 400 Ton		869.15		956.07		
1 Hammer, Diesel, 15K ft.-lbs.		624.45		686.89		
1 Lead, 60' High		211.80		232.98		
1 Air Compressor, 600 cfm		426.55		469.20		
2 -50' Air Hoses, 3"		76.70		84.37	52.08	57.29
72 L.H., Daily Totals		$7852.05		$10387.42	$109.06	$144.27
Crew B-76A					$47.79	$71.29
1 Labor Foreman (outside)	$46.40	$371.20	$69.25	$554.00		
5 Laborers	44.40	1776.00	66.25	2650.00		
1 Equip. Oper. (crane)	61.45	491.60	91.55	732.40		
1 Equip. Oper. (oiler)	52.50	420.00	78.25	626.00		
1 Crawler Crane, 50 Ton		1541.00		1695.10		
1 Barge, 400 Ton		869.15		956.07	37.66	41.42
64 L.H., Daily Totals		$5468.95		$7213.56	$85.45	$112.71
Crew B-77					$45.68	$68.20
1 Labor Foreman (outside)	$46.40	$371.20	$69.25	$554.00		
3 Laborers	44.40	1065.60	66.25	1590.00		
1 Truck Driver (light)	48.80	390.40	73.00	584.00		
1 Crack Cleaner, 25 H.P.		53.00		58.30		
1 Crack Filler, Trailer Mtd.		170.95		188.04		
1 Flatbed Truck, Gas, 3 Ton		850.05		935.05	26.85	29.54
40 L.H., Daily Totals		$2901.20		$3909.40	$72.53	$97.73

Crew B-78

Crew No.	Bare Costs Hr.	Daily	Incl. Subs O&P Hr.	Daily	Cost Per LH Bare Costs	Incl. O&P
1 Labor Foreman (outside)	$46.40	$371.20	$69.25	$554.00	$45.47	$67.88
4 Laborers	44.40	1420.80	66.25	2120.00		
1 Truck Driver (light)	48.80	390.40	73.00	584.00		
1 Paint Striper, S.P., 40 Gallon		128.35		141.19		
1 Flatbed Truck, Gas, 3 Ton		850.05		935.05		
1 Pickup Truck, 3/4 Ton		112.20		123.42	22.72	24.99
48 L.H., Daily Totals		$3273.00		$4457.66	$68.19	$92.87

Crew B-78A

Crew No.	Bare Costs Hr.	Daily	Incl. Subs O&P Hr.	Daily	Cost Per LH Bare Costs	Incl. O&P
1 Equip. Oper. (light)	$55.50	$444.00	$82.70	$661.60	$55.50	$82.70
1 Line Rem. (Metal Balls) 115 H.P.		996.25		1095.88	124.53	136.98
8 L.H., Daily Totals		$1440.25		$1757.47	$180.03	$219.68

Crew B-78B

Crew No.	Bare Costs Hr.	Daily	Incl. Subs O&P Hr.	Daily	Cost Per LH Bare Costs	Incl. O&P
2 Laborers	$44.40	$710.40	$66.25	$1060.00	$45.63	$68.08
.25 Equip. Oper. (light)	55.50	111.00	82.70	165.40		
1 Pickup Truck, 3/4 Ton		112.20		123.42		
1 Line Rem.,11 H.P.,Walk Behind		114.75		126.22		
.25 Road Sweeper, S.P., 8' wide		180.91		199.00	22.66	24.92
18 L.H., Daily Totals		$1229.26		$1674.05	$68.29	$93.00

Crew B-78C

Crew No.	Bare Costs Hr.	Daily	Incl. Subs O&P Hr.	Daily	Cost Per LH Bare Costs	Incl. O&P
1 Labor Foreman (outside)	$46.40	$371.20	$69.25	$554.00	$45.47	$67.88
4 Laborers	44.40	1420.80	66.25	2120.00		
1 Truck Driver (light)	48.80	390.40	73.00	584.00		
1 Paint Striper, T.M., 120 Gal.		603.10		663.41		
1 Flatbed Truck, Gas, 3 Ton		850.05		935.05		
1 Pickup Truck, 3/4 Ton		112.20		123.42	32.61	35.87
48 L.H., Daily Totals		$3747.75		$4979.89	$78.08	$103.75

Crew B-78D

Crew No.	Bare Costs Hr.	Daily	Incl. Subs O&P Hr.	Daily	Cost Per LH Bare Costs	Incl. O&P
2 Labor Foremen (outside)	$46.40	$742.40	$69.25	$1108.00	$45.24	$67.53
7 Laborers	44.40	2486.40	66.25	3710.00		
1 Truck Driver (light)	48.80	390.40	73.00	584.00		
1 Paint Striper, T.M., 120 Gal.		603.10		663.41		
1 Flatbed Truck, Gas, 3 Ton		850.05		935.05		
3 Pickup Trucks, 3/4 Ton		336.60		370.26		
1 Air Compressor, 60 cfm		153.85		169.24		
1 -50' Air Hose, 3/4"		7.15		7.87		
1 Breaker, Pavement, 60 lb.		53.60		58.96	25.05	27.56
80 L.H., Daily Totals		$5623.55		$7606.78	$70.29	$95.08

Crew B-78E

Crew No.	Bare Costs Hr.	Daily	Incl. Subs O&P Hr.	Daily	Cost Per LH Bare Costs	Incl. O&P
2 Labor Foremen (outside)	$46.40	$742.40	$69.25	$1108.00	$45.10	$67.31
9 Laborers	44.40	3196.80	66.25	4770.00		
1 Truck Driver (light)	48.80	390.40	73.00	584.00		
1 Paint Striper, T.M., 120 Gal.		603.10		663.41		
1 Flatbed Truck, Gas, 3 Ton		850.05		935.05		
4 Pickup Trucks, 3/4 Ton		448.80		493.68		
2 Air Compressors, 60 cfm		307.70		338.47		
2 -50' Air Hoses, 3/4"		14.30		15.73		
2 Breakers, Pavement, 60 lb.		107.20		117.92	24.28	26.71
96 L.H., Daily Totals		$6660.75		$9026.26	$69.38	$94.02

Crew B-78F

Crew No.	Bare Costs Hr.	Daily	Incl. Subs O&P Hr.	Daily	Cost Per LH Bare Costs	Incl. O&P
2 Labor Foremen (outside)	$46.40	$742.40	$69.25	$1108.00	$45.00	$67.16
11 Laborers	44.40	3907.20	66.25	5830.00		
1 Truck Driver (light)	48.80	390.40	73.00	584.00		
1 Paint Striper, T.M., 120 Gal.		603.10		663.41		
1 Flatbed Truck, Gas, 3 Ton		850.05		935.05		
7 Pickup Trucks, 3/4 Ton		785.40		863.94		
3 Air Compressors, 60 cfm		461.55		507.70		
3 -50' Air Hoses, 3/4"		21.45		23.59		
3 Breakers, Pavement, 60 lb.		160.80		176.88	25.74	28.31
112 L.H., Daily Totals		$7922.35		$10692.58	$70.74	$95.47

Crew B-79

Crew No.	Bare Costs Hr.	Daily	Incl. Subs O&P Hr.	Daily	Cost Per LH Bare Costs	Incl. O&P
1 Labor Foreman (outside)	$46.40	$371.20	$69.25	$554.00	$45.68	$68.20
3 Laborers	44.40	1065.60	66.25	1590.00		
1 Truck Driver (light)	48.80	390.40	73.00	584.00		
1 Paint Striper, T.M., 120 Gal.		603.10		663.41		
1 Heating Kettle, 115 Gallon		107.25		117.97		
1 Flatbed Truck, Gas, 3 Ton		850.05		935.05		
2 Pickup Trucks, 3/4 Ton		224.40		246.84	44.62	49.08
40 L.H., Daily Totals		$3612.00		$4691.28	$90.30	$117.28

Crew B-79A

Crew No.	Bare Costs Hr.	Daily	Incl. Subs O&P Hr.	Daily	Cost Per LH Bare Costs	Incl. O&P
1.5 Equip. Oper. (light)	$55.50	$666.00	$82.70	$992.40	$55.50	$82.70
.5 Line Remov. (Grinder) 115 H.P.		515.50		567.05		
1 Line Rem. (Metal Balls) 115 H.P.		996.25		1095.88	125.98	138.58
12 L.H., Daily Totals		$2177.75		$2655.32	$181.48	$221.28

Crew B-79B

Crew No.	Bare Costs Hr.	Daily	Incl. Subs O&P Hr.	Daily	Cost Per LH Bare Costs	Incl. O&P
1 Laborer	$44.40	$355.20	$66.25	$530.00	$44.40	$66.25
1 Set of Gases		173.60		190.96	21.70	23.87
8 L.H., Daily Totals		$528.80		$720.96	$66.10	$90.12

Crew B-79C

Crew No.	Bare Costs Hr.	Daily	Incl. Subs O&P Hr.	Daily	Cost Per LH Bare Costs	Incl. O&P
1 Labor Foreman (outside)	$46.40	$371.20	$69.25	$554.00	$45.31	$67.64
5 Laborers	44.40	1776.00	66.25	2650.00		
1 Truck Driver (light)	48.80	390.40	73.00	584.00		
1 Paint Striper, T.M., 120 Gal.		603.10		663.41		
1 Heating Kettle, 115 Gallon		107.25		117.97		
1 Flatbed Truck, Gas, 3 Ton		850.05		935.05		
3 Pickup Trucks, 3/4 Ton		336.60		370.26		
1 Air Compressor, 60 cfm		153.85		169.24		
1 -50' Air Hose, 3/4"		7.15		7.87		
1 Breaker, Pavement, 60 lb.		53.60		58.96	37.71	41.48
56 L.H., Daily Totals		$4649.20		$6110.76	$83.02	$109.12

Crew B-79D

Crew No.	Bare Costs Hr.	Daily	Incl. Subs O&P Hr.	Daily	Cost Per LH Bare Costs	Incl. O&P
2 Labor Foremen (outside)	$46.40	$742.40	$69.25	$1108.00	$45.45	$67.84
5 Laborers	44.40	1776.00	66.25	2650.00		
1 Truck Driver (light)	48.80	390.40	73.00	584.00		
1 Paint Striper, T.M., 120 Gal.		603.10		663.41		
1 Heating Kettle, 115 Gallon		107.25		117.97		
1 Flatbed Truck, Gas, 3 Ton		850.05		935.05		
4 Pickup Trucks, 3/4 Ton		448.80		493.68		
1 Air Compressor, 60 cfm		153.85		169.24		
1 -50' Air Hose, 3/4"		7.15		7.87		
1 Breaker, Pavement, 60 lb.		53.60		58.96	34.75	38.22
64 L.H., Daily Totals		$5132.60		$6788.18	$80.20	$106.07

For customer support on your Electrical Costs with RSMeans Data, call 800.448.8182.

Crew No.	Bare Costs		Incl. Subs O&P		Cost Per Labor-Hour	

Crew B-79E

	Hr.	Daily	Hr.	Daily	Bare Costs	Incl. O&P
2 Labor Foremen (outside)	$46.40	$742.40	$69.25	$1108.00	$45.24	$67.53
7 Laborers	44.40	2486.40	66.25	3710.00		
1 Truck Driver (light)	48.80	390.40	73.00	584.00		
1 Paint Striper, T.M., 120 Gal.		603.10		663.41		
1 Heating Kettle, 115 Gallon		107.25		117.97		
1 Flatbed Truck, Gas, 3 Ton		850.05		935.05		
5 Pickup Trucks, 3/4 Ton		561.00		617.10		
2 Air Compressors, 60 cfm		307.70		338.47		
2 -50' Air Hoses, 3/4"		14.30		15.73		
2 Breakers, Pavement, 60 lb.		107.20		117.92	31.88	35.07
80 L.H., Daily Totals		$6169.80		$8207.66	$77.12	$102.60

Crew B-80

	Hr.	Daily	Hr.	Daily	Bare Costs	Incl. O&P
1 Labor Foreman (outside)	$46.40	$371.20	$69.25	$554.00	$48.77	$72.80
1 Laborer	44.40	355.20	66.25	530.00		
1 Truck Driver (light)	48.80	390.40	73.00	584.00		
1 Equip. Oper. (light)	55.50	444.00	82.70	661.60		
1 Flatbed Truck, Gas, 3 Ton		850.05		935.05		
1 Earth Auger, Truck-Mtd.		202.55		222.81	32.89	36.18
32 L.H., Daily Totals		$2613.40		$3487.46	$81.67	$108.98

Crew B-80A

	Hr.	Daily	Hr.	Daily	Bare Costs	Incl. O&P
3 Laborers	$44.40	$1065.60	$66.25	$1590.00	$44.40	$66.25
1 Flatbed Truck, Gas, 3 Ton		850.05		935.05	35.42	38.96
24 L.H., Daily Totals		$1915.65		$2525.05	$79.82	$105.21

Crew B-80B

	Hr.	Daily	Hr.	Daily	Bare Costs	Incl. O&P
3 Laborers	$44.40	$1065.60	$66.25	$1590.00	$47.17	$70.36
1 Equip. Oper. (light)	55.50	444.00	82.70	661.60		
1 Crane, Flatbed Mounted, 3 Ton		238.75		262.63	7.46	8.21
32 L.H., Daily Totals		$1748.35		$2514.22	$54.64	$78.57

Crew B-80C

	Hr.	Daily	Hr.	Daily	Bare Costs	Incl. O&P
2 Laborers	$44.40	$710.40	$66.25	$1060.00	$45.87	$68.50
1 Truck Driver (light)	48.80	390.40	73.00	584.00		
1 Flatbed Truck, Gas, 1.5 Ton		198.50		218.35		
1 Manual Fence Post Auger, Gas		54.40		59.84	10.54	11.59
24 L.H., Daily Totals		$1353.70		$1922.19	$56.40	$80.09

Crew B-81

	Hr.	Daily	Hr.	Daily	Bare Costs	Incl. O&P
1 Laborer	$44.40	$355.20	$66.25	$530.00	$51.57	$76.95
1 Equip. Oper. (medium)	59.00	472.00	87.90	703.20		
1 Truck Driver (heavy)	51.30	410.40	76.70	613.60		
1 Hydromulcher, T.M., 3000 Gal.		256.85		282.54		
1 Truck Tractor, 220 H.P.		310.80		341.88	23.65	26.02
24 L.H., Daily Totals		$1805.25		$2471.22	$75.22	$102.97

Crew B-81A

	Hr.	Daily	Hr.	Daily	Bare Costs	Incl. O&P
1 Laborer	$44.40	$355.20	$66.25	$530.00	$46.60	$69.63
1 Truck Driver (light)	48.80	390.40	73.00	584.00		
1 Hydromulcher, T.M., 600 Gal.		118.15		129.97		
1 Flatbed Truck, Gas, 3 Ton		850.05		935.05	60.51	66.56
16 L.H., Daily Totals		$1713.80		$2179.02	$107.11	$136.19

Crew B-82

	Hr.	Daily	Hr.	Daily	Bare Costs	Incl. O&P
1 Laborer	$44.40	$355.20	$66.25	$530.00	$49.95	$74.47
1 Equip. Oper. (light)	55.50	444.00	82.70	661.60		
1 Horiz. Borer, 6 H.P.		184.15		202.57	11.51	12.66
16 L.H., Daily Totals		$983.35		$1394.17	$61.46	$87.14

Crew B-82A

	Hr.	Daily	Hr.	Daily	Bare Costs	Incl. O&P
2 Laborers	$44.40	$710.40	$66.25	$1060.00	$49.95	$74.47
2 Equip. Opers. (light)	55.50	888.00	82.70	1323.20		
2 Dump Truck, 8 C.Y., 220 H.P.		815.20		896.72		
1 Flatbed Trailer, 25 Ton		137.20		150.92		
1 Horiz. Dir. Drill, 20k lb. Thrust		544.10		598.51		
1 Mud Trailer for HDD, 1500 Gal.		312.15		343.37		
1 Pickup Truck, 4x4, 3/4 Ton		176.75		194.43		
1 Flatbed Trailer, 3 Ton		71.15		78.27		
1 Loader, Skid Steer, 78 H.P.		446.30		490.93	78.21	86.04
32 L.H., Daily Totals		$4101.25		$5136.34	$128.16	$160.51

Crew B-82B

	Hr.	Daily	Hr.	Daily	Bare Costs	Incl. O&P
2 Laborers	$44.40	$710.40	$66.25	$1060.00	$49.95	$74.47
2 Equip. Opers. (light)	55.50	888.00	82.70	1323.20		
2 Dump Truck, 8 C.Y., 220 H.P.		815.20		896.72		
1 Flatbed Trailer, 25 Ton		137.20		150.92		
1 Horiz. Dir. Drill, 30k lb. Thrust		647.65		712.41		
1 Mud Trailer for HDD, 1500 Gal.		312.15		343.37		
1 Pickup Truck, 4x4, 3/4 Ton		176.75		194.43		
1 Flatbed Trailer, 3 Ton		71.15		78.27		
1 Loader, Skid Steer, 78 H.P.		446.30		490.93	81.45	89.59
32 L.H., Daily Totals		$4204.80		$5250.24	$131.40	$164.07

Crew B-82C

	Hr.	Daily	Hr.	Daily	Bare Costs	Incl. O&P
2 Laborers	$44.40	$710.40	$66.25	$1060.00	$49.95	$74.47
2 Equip. Opers. (light)	55.50	888.00	82.70	1323.20		
2 Dump Truck, 8 C.Y., 220 H.P.		815.20		896.72		
1 Flatbed Trailer, 25 Ton		137.20		150.92		
1 Horiz. Dir. Drill, 50k lb. Thrust		824.05		906.46		
1 Mud Trailer for HDD, 1500 Gal.		312.15		343.37		
1 Pickup Truck, 4x4, 3/4 Ton		176.75		194.43		
1 Flatbed Trailer, 3 Ton		71.15		78.27		
1 Loader, Skid Steer, 78 H.P.		446.30		490.93	86.96	95.66
32 L.H., Daily Totals		$4381.20		$5444.28	$136.91	$170.13

Crew B-82D

	Hr.	Daily	Hr.	Daily	Bare Costs	Incl. O&P
1 Equip. Oper. (light)	$55.50	$444.00	$82.70	$661.60	$55.50	$82.70
1 Mud Trailer for HDD, 1500 Gal.		312.15		343.37	39.02	42.92
8 L.H., Daily Totals		$756.15		$1004.97	$94.52	$125.62

Crew B-83

	Hr.	Daily	Hr.	Daily	Bare Costs	Incl. O&P
1 Tugboat Captain	$59.00	$472.00	$87.90	$703.20	$51.70	$77.08
1 Tugboat Hand	44.40	355.20	66.25	530.00		
1 Tugboat, 250 H.P.		726.10		798.71	45.38	49.92
16 L.H., Daily Totals		$1553.30		$2031.91	$97.08	$126.99

Crew B-84

	Hr.	Daily	Hr.	Daily	Bare Costs	Incl. O&P
1 Equip. Oper. (medium)	$59.00	$472.00	$87.90	$703.20	$59.00	$87.90
1 Rotary Mower/Tractor		371.15		408.26	46.39	51.03
8 L.H., Daily Totals		$843.15		$1111.46	$105.39	$138.93

Crew B-85

	Hr.	Daily	Hr.	Daily	Bare Costs	Incl. O&P
3 Laborers	$44.40	$1065.60	$66.25	$1590.00	$48.70	$72.67
1 Equip. Oper. (medium)	59.00	472.00	87.90	703.20		
1 Truck Driver (heavy)	51.30	410.40	76.70	613.60		
1 Telescoping Boom Lift, to 80'		387.75		426.52		
1 Brush Chipper, 12", 130 H.P.		366.05		402.65		
1 Pruning Saw, Rotary		26.40		29.04	19.50	21.46
40 L.H., Daily Totals		$2728.20		$3765.02	$68.20	$94.13

For customer support on your Electrical Costs with RSMeans Data, call 800.448.8182.

505

Crew No.	Bare Costs		Incl. Subs O&P		Cost Per Labor-Hour	

Crew B-86	Hr.	Daily	Hr.	Daily	Bare Costs	Incl. O&P
1 Equip. Oper. (medium)	$59.00	$472.00	$87.90	$703.20	$59.00	$87.90
1 Stump Chipper, S.P.		189.20		208.12	23.65	26.02
8 L.H., Daily Totals		$661.20		$911.32	$82.65	$113.92

Crew B-86A	Hr.	Daily	Hr.	Daily	Bare Costs	Incl. O&P
1 Equip. Oper. (medium)	$59.00	$472.00	$87.90	$703.20	$59.00	$87.90
1 Grader, 30,000 Lbs.		1073.00		1180.30	134.13	147.54
8 L.H., Daily Totals		$1545.00		$1883.50	$193.13	$235.44

Crew B-86B	Hr.	Daily	Hr.	Daily	Bare Costs	Incl. O&P
1 Equip. Oper. (medium)	$59.00	$472.00	$87.90	$703.20	$59.00	$87.90
1 Dozer, 200 H.P.		1520.00		1672.00	190.00	209.00
8 L.H., Daily Totals		$1992.00		$2375.20	$249.00	$296.90

Crew B-87	Hr.	Daily	Hr.	Daily	Bare Costs	Incl. O&P
1 Laborer	$44.40	$355.20	$66.25	$530.00	$56.08	$83.57
4 Equip. Oper. (medium)	59.00	1888.00	87.90	2812.80		
2 Feller Bunchers, 100 H.P.		1257.60		1383.36		
1 Log Chipper, 22" Tree		555.00		610.50		
1 Dozer, 105 H.P.		640.80		704.88		
1 Chain Saw, Gas, 36" Long		41.65		45.81	62.38	68.61
40 L.H., Daily Totals		$4738.25		$6087.35	$118.46	$152.18

Crew B-88	Hr.	Daily	Hr.	Daily	Bare Costs	Incl. O&P
1 Laborer	$44.40	$355.20	$66.25	$530.00	$56.91	$84.81
6 Equip. Oper. (medium)	59.00	2832.00	87.90	4219.20		
2 Feller Bunchers, 100 H.P.		1257.60		1383.36		
1 Log Chipper, 22" Tree		555.00		610.50		
2 Log Skidders, 50 H.P.		1826.70		2009.37		
1 Dozer, 105 H.P.		640.80		704.88		
1 Chain Saw, Gas, 36" Long		41.65		45.81	77.17	84.89
56 L.H., Daily Totals		$7508.95		$9503.13	$134.09	$169.70

Crew B-89	Hr.	Daily	Hr.	Daily	Bare Costs	Incl. O&P
1 Equip. Oper. (light)	$55.50	$444.00	$82.70	$661.60	$52.15	$77.85
1 Truck Driver (light)	48.80	390.40	73.00	584.00		
1 Flatbed Truck, Gas, 3 Ton		850.05		935.05		
1 Concrete Saw		112.85		124.14		
1 Water Tank, 65 Gal.		102.90		113.19	66.61	73.27
16 L.H., Daily Totals		$1900.20		$2417.98	$118.76	$151.12

Crew B-89A	Hr.	Daily	Hr.	Daily	Bare Costs	Incl. O&P
1 Skilled Worker	$57.10	$456.80	$85.90	$687.20	$50.75	$76.08
1 Laborer	44.40	355.20	66.25	530.00		
1 Core Drill (Large)		121.60		133.76	7.60	8.36
16 L.H., Daily Totals		$933.60		$1350.96	$58.35	$84.44

Crew B-89B	Hr.	Daily	Hr.	Daily	Bare Costs	Incl. O&P
1 Equip. Oper. (light)	$55.50	$444.00	$82.70	$661.60	$52.15	$77.85
1 Truck Driver (light)	48.80	390.40	73.00	584.00		
1 Wall Saw, Hydraulic, 10 H.P.		86.40		95.04		
1 Generator, Diesel, 100 kW		521.85		574.03		
1 Water Tank, 65 Gal.		102.90		113.19		
1 Flatbed Truck, Gas, 3 Ton		850.05		935.05	97.58	107.33
16 L.H., Daily Totals		$2395.60		$2962.92	$149.72	$185.18

Crew B-89C	Hr.	Daily	Hr.	Daily	Bare Costs	Incl. O&P
1 Cement Finisher	$51.80	$414.40	$75.90	$607.20	$51.80	$75.90
1 Masonry cut-off saw, gas		58.15		63.97	7.27	8.00
8 L.H., Daily Totals		$472.55		$671.16	$59.07	$83.90

Crew B-90	Hr.	Daily	Hr.	Daily	Bare Costs	Incl. O&P
1 Labor Foreman (outside)	$46.40	$371.20	$69.25	$554.00	$49.15	$73.35
3 Laborers	44.40	1065.60	66.25	1590.00		
2 Equip. Oper. (light)	55.50	888.00	82.70	1323.20		
2 Truck Drivers (heavy)	51.30	820.80	76.70	1227.20		
1 Road Mixer, 310 H.P.		1919.00		2110.90		
1 Dist. Truck, 2000 Gal.		303.25		333.57	34.72	38.19
64 L.H., Daily Totals		$5367.85		$7138.88	$83.87	$111.54

Crew B-90A	Hr.	Daily	Hr.	Daily	Bare Costs	Incl. O&P
1 Labor Foreman (outside)	$46.40	$371.20	$69.25	$554.00	$53.03	$79.05
2 Laborers	44.40	710.40	66.25	1060.00		
4 Equip. Oper. (medium)	59.00	1888.00	87.90	2812.80		
2 Graders, 30,000 Lbs.		2146.00		2360.60		
1 Tandem Roller, 10 Ton		246.80		271.48		
1 Roller, Pneum. Whl., 12 Ton		349.90		384.89	48.98	53.87
56 L.H., Daily Totals		$5712.30		$7443.77	$102.01	$132.92

Crew B-90B	Hr.	Daily	Hr.	Daily	Bare Costs	Incl. O&P
1 Labor Foreman (outside)	$46.40	$371.20	$69.25	$554.00	$52.03	$77.58
2 Laborers	44.40	710.40	66.25	1060.00		
3 Equip. Oper. (medium)	59.00	1416.00	87.90	2109.60		
1 Roller, Pneum. Whl., 12 Ton		349.90		384.89		
1 Road Mixer, 310 H.P.		1919.00		2110.90	47.27	52.00
48 L.H., Daily Totals		$4766.50		$6219.39	$99.30	$129.57

Crew B-90C	Hr.	Daily	Hr.	Daily	Bare Costs	Incl. O&P
1 Labor Foreman (outside)	$46.40	$371.20	$69.25	$554.00	$50.45	$75.28
4 Laborers	44.40	1420.80	66.25	2120.00		
3 Equip. Oper. (medium)	59.00	1416.00	87.90	2109.60		
3 Truck Drivers (heavy)	51.30	1231.20	76.70	1840.80		
3 Road Mixers, 310 H.P.		5757.00		6332.70	65.42	71.96
88 L.H., Daily Totals		$10196.20		$12957.10	$115.87	$147.24

Crew B-90D	Hr.	Daily	Hr.	Daily	Bare Costs	Incl. O&P
1 Labor Foreman (outside)	$46.40	$371.20	$69.25	$554.00	$49.52	$73.89
6 Laborers	44.40	2131.20	66.25	3180.00		
3 Equip. Oper. (medium)	59.00	1416.00	87.90	2109.60		
3 Truck Drivers (heavy)	51.30	1231.20	76.70	1840.80		
3 Road Mixers, 310 H.P.		5757.00		6332.70	55.36	60.89
104 L.H., Daily Totals		$10906.60		$14017.10	$104.87	$134.78

Crew B-90E	Hr.	Daily	Hr.	Daily	Bare Costs	Incl. O&P
1 Labor Foreman (outside)	$46.40	$371.20	$69.25	$554.00	$50.26	$74.96
4 Laborers	44.40	1420.80	66.25	2120.00		
3 Equip. Oper. (medium)	59.00	1416.00	87.90	2109.60		
1 Truck Driver (heavy)	51.30	410.40	76.70	613.60		
1 Road Mixer, 310 H.P.		1919.00		2110.90	26.65	29.32
72 L.H., Daily Totals		$5537.40		$7508.10	$76.91	$104.28

Crew B-91	Hr.	Daily	Hr.	Daily	Bare Costs	Incl. O&P
1 Labor Foreman (outside)	$46.40	$371.20	$69.25	$554.00	$52.81	$78.76
2 Laborers	44.40	710.40	66.25	1060.00		
4 Equip. Oper. (medium)	59.00	1888.00	87.90	2812.80		
1 Truck Driver (heavy)	51.30	410.40	76.70	613.60		
1 Dist. Tanker, 3000 Gallon		334.10		367.51		
1 Truck Tractor, 6x4, 380 H.P.		499.15		549.07		
1 Aggreg. Spreader, S.P.		859.10		945.01		
1 Roller, Pneum. Whl., 12 Ton		349.90		384.89		
1 Tandem Roller, 10 Ton		246.80		271.48	35.77	39.34
64 L.H., Daily Totals		$5669.05		$7558.35	$88.58	$118.10

Crew No.	Bare Costs Hr.	Daily	Incl. Subs O&P Hr.	Daily	Cost Per Labor-Hour Bare Costs	Incl. O&P
Crew B-91B	Hr.	Daily	Hr.	Daily	Bare Costs	Incl. O&P
1 Laborer	$44.40	$355.20	$66.25	$530.00	$51.70	$77.08
1 Equipment Oper. (med.)	59.00	472.00	87.90	703.20		
1 Road Sweeper, Vac. Assist.		879.45		967.39	54.97	60.46
16 L.H., Daily Totals		$1706.65		$2200.59	$106.67	$137.54
Crew B-91C	Hr.	Daily	Hr.	Daily	Bare Costs	Incl. O&P
1 Laborer	$44.40	$355.20	$66.25	$530.00	$46.60	$69.63
1 Truck Driver (light)	48.80	390.40	73.00	584.00		
1 Catch Basin Cleaning Truck		542.60		596.86	33.91	37.30
16 L.H., Daily Totals		$1288.20		$1710.86	$80.51	$106.93
Crew B-91D	Hr.	Daily	Hr.	Daily	Bare Costs	Incl. O&P
1 Labor Foreman (outside)	$46.40	$371.20	$69.25	$554.00	$51.23	$76.42
5 Laborers	44.40	1776.00	66.25	2650.00		
5 Equip. Oper. (medium)	59.00	2360.00	87.90	3516.00		
2 Truck Drivers (heavy)	51.30	820.80	76.70	1227.20		
1 Aggreg. Spreader, S.P.		859.10		945.01		
2 Truck Tractors, 6x4, 380 H.P.		998.30		1098.13		
2 Dist. Tankers, 3000 Gallon		668.20		735.02		
2 Pavement Brushes, Towed		176.70		194.37		
2 Rollers Pneum. Whl., 12 Ton		699.80		769.78	32.71	35.98
104 L.H., Daily Totals		$8730.10		$11689.51	$83.94	$112.40
Crew B-92	Hr.	Daily	Hr.	Daily	Bare Costs	Incl. O&P
1 Labor Foreman (outside)	$46.40	$371.20	$69.25	$554.00	$44.90	$67.00
3 Laborers	44.40	1065.60	66.25	1590.00		
1 Crack Cleaner, 25 H.P.		53.00		58.30		
1 Air Compressor, 60 cfm		153.85		169.24		
1 Tar Kettle, T.M.		156.70		172.37		
1 Flatbed Truck, Gas, 3 Ton		850.05		935.05	37.92	41.72
32 L.H., Daily Totals		$2650.40		$3478.96	$82.83	$108.72
Crew B-93	Hr.	Daily	Hr.	Daily	Bare Costs	Incl. O&P
1 Equip. Oper. (medium)	$59.00	$472.00	$87.90	$703.20	$59.00	$87.90
1 Feller Buncher, 100 H.P.		628.80		691.68	78.60	86.46
8 L.H., Daily Totals		$1100.80		$1394.88	$137.60	$174.36
Crew B-94A	Hr.	Daily	Hr.	Daily	Bare Costs	Incl. O&P
1 Laborer	$44.40	$355.20	$66.25	$530.00	$44.40	$66.25
1 Diaphragm Water Pump, 2"		87.70		96.47		
1 -20' Suction Hose, 2"		3.55		3.90		
2 -50' Discharge Hoses, 2"		8.00		8.80	12.41	13.65
8 L.H., Daily Totals		$454.45		$639.17	$56.81	$79.90
Crew B-94B	Hr.	Daily	Hr.	Daily	Bare Costs	Incl. O&P
1 Laborer	$44.40	$355.20	$66.25	$530.00	$44.40	$66.25
1 Diaphragm Water Pump, 4"		106.35		116.99		
1 -20' Suction Hose, 4"		17.25		18.98		
2 -50' Discharge Hoses, 4"		25.60		28.16	18.65	20.52
8 L.H., Daily Totals		$504.40		$694.12	$63.05	$86.77
Crew B-94C	Hr.	Daily	Hr.	Daily	Bare Costs	Incl. O&P
1 Laborer	$44.40	$355.20	$66.25	$530.00	$44.40	$66.25
1 Centrifugal Water Pump, 3"		74.40		81.84		
1 -20' Suction Hose, 3"		8.75		9.63		
2 -50' Discharge Hoses, 3"		9.00		9.90	11.52	12.67
8 L.H., Daily Totals		$447.35		$631.37	$55.92	$78.92
Crew B-94D	Hr.	Daily	Hr.	Daily	Bare Costs	Incl. O&P
1 Laborer	$44.40	$355.20	$66.25	$530.00	$44.40	$66.25
1 Centr. Water Pump, 6"		235.25		258.77		
1 -20' Suction Hose, 6"		25.50		28.05		
2 -50' Discharge Hoses, 6"		36.20		39.82	37.12	40.83
8 L.H., Daily Totals		$652.15		$856.64	$81.52	$107.08
Crew C-1	Hr.	Daily	Hr.	Daily	Bare Costs	Incl. O&P
3 Carpenters	$54.70	$1312.80	$81.65	$1959.60	$52.13	$77.80
1 Laborer	44.40	355.20	66.25	530.00		
32 L.H., Daily Totals		$1668.00		$2489.60	$52.13	$77.80
Crew C-2	Hr.	Daily	Hr.	Daily	Bare Costs	Incl. O&P
1 Carpenter Foreman (outside)	$56.70	$453.60	$84.60	$676.80	$53.32	$79.58
4 Carpenters	54.70	1750.40	81.65	2612.80		
1 Laborer	44.40	355.20	66.25	530.00		
48 L.H., Daily Totals		$2559.20		$3819.60	$53.32	$79.58
Crew C-2A	Hr.	Daily	Hr.	Daily	Bare Costs	Incl. O&P
1 Carpenter Foreman (outside)	$56.70	$453.60	$84.60	$676.80	$52.83	$78.62
3 Carpenters	54.70	1312.80	81.65	1959.60		
1 Cement Finisher	51.80	414.40	75.90	607.20		
1 Laborer	44.40	355.20	66.25	530.00		
48 L.H., Daily Totals		$2536.00		$3773.60	$52.83	$78.62
Crew C-3	Hr.	Daily	Hr.	Daily	Bare Costs	Incl. O&P
1 Rodman Foreman (outside)	$60.90	$487.20	$91.05	$728.40	$55.10	$82.31
4 Rodmen (reinf.)	58.90	1884.80	88.05	2817.60		
1 Equip. Oper. (light)	55.50	444.00	82.70	661.60		
2 Laborers	44.40	710.40	66.25	1060.00		
3 Stressing Equipment		56.85		62.53		
.5 Grouting Equipment		123.33		135.66	2.82	3.10
64 L.H., Daily Totals		$3706.57		$5465.79	$57.92	$85.40
Crew C-4	Hr.	Daily	Hr.	Daily	Bare Costs	Incl. O&P
1 Rodman Foreman (outside)	$60.90	$487.20	$91.05	$728.40	$59.40	$88.80
3 Rodmen (reinf.)	58.90	1413.60	88.05	2113.20		
3 Stressing Equipment		56.85		62.53	1.78	1.95
32 L.H., Daily Totals		$1957.65		$2904.14	$61.18	$90.75
Crew C-4A	Hr.	Daily	Hr.	Daily	Bare Costs	Incl. O&P
2 Rodmen (reinf.)	$58.90	$942.40	$88.05	$1408.80	$58.90	$88.05
4 Stressing Equipment		75.80		83.38	4.74	5.21
16 L.H., Daily Totals		$1018.20		$1492.18	$63.64	$93.26
Crew C-5	Hr.	Daily	Hr.	Daily	Bare Costs	Incl. O&P
1 Rodman Foreman (outside)	$60.90	$487.20	$91.05	$728.40	$58.64	$87.58
4 Rodmen (reinf.)	58.90	1884.80	88.05	2817.60		
1 Equip. Oper. (crane)	61.45	491.60	91.55	732.40		
1 Equip. Oper. (oiler)	52.50	420.00	78.25	626.00		
1 Hyd. Crane, 25 Ton		586.70		645.37	10.48	11.52
56 L.H., Daily Totals		$3870.30		$5549.77	$69.11	$99.10
Crew C-6	Hr.	Daily	Hr.	Daily	Bare Costs	Incl. O&P
1 Labor Foreman (outside)	$46.40	$371.20	$69.25	$554.00	$45.97	$68.36
4 Laborers	44.40	1420.80	66.25	2120.00		
1 Cement Finisher	51.80	414.40	75.90	607.20		
2 Gas Engine Vibrators		54.30		59.73	1.13	1.24
48 L.H., Daily Totals		$2260.70		$3340.93	$47.10	$69.60

For customer support on your Electrical Costs with RSMeans Data, call 800.448.8182.

507

Crew No.	Bare Costs		Incl. Subs O&P		Cost Per Labor-Hour	
Crew C-6A	Hr.	Daily	Hr.	Daily	Bare Costs	Incl. O&P
2 Cement Finishers	$51.80	$828.80	$75.90	$1214.40	$51.80	$75.90
1 Concrete Vibrator, Elec, 2 HP		45.80		50.38	2.86	3.15
16 L.H., Daily Totals		$874.60		$1264.78	$54.66	$79.05
Crew C-7	Hr.	Daily	Hr.	Daily	Bare Costs	Incl. O&P
1 Labor Foreman (outside)	$46.40	$371.20	$69.25	$554.00	$47.97	$71.39
5 Laborers	44.40	1776.00	66.25	2650.00		
1 Cement Finisher	51.80	414.40	75.90	607.20		
1 Equip. Oper. (medium)	59.00	472.00	87.90	703.20		
1 Equip. Oper. (oiler)	52.50	420.00	78.25	626.00		
2 Gas Engine Vibrators		54.30		59.73		
1 Concrete Bucket, 1 C.Y.		45.90		50.49		
1 Hyd. Crane, 55 Ton		990.15		1089.17	15.14	16.66
72 L.H., Daily Totals		$4543.95		$6339.78	$63.11	$88.05
Crew C-7A	Hr.	Daily	Hr.	Daily	Bare Costs	Incl. O&P
1 Labor Foreman (outside)	$46.40	$371.20	$69.25	$554.00	$46.38	$69.24
5 Laborers	44.40	1776.00	66.25	2650.00		
2 Truck Drivers (heavy)	51.30	820.80	76.70	1227.20		
2 Conc. Transit Mixers		1176.30		1293.93	18.38	20.22
64 L.H., Daily Totals		$4144.30		$5725.13	$64.75	$89.46
Crew C-7B	Hr.	Daily	Hr.	Daily	Bare Costs	Incl. O&P
1 Labor Foreman (outside)	$46.40	$371.20	$69.25	$554.00	$47.79	$71.29
5 Laborers	44.40	1776.00	66.25	2650.00		
1 Equipment Operator, Crane	61.45	491.60	91.55	732.40		
1 Equipment Oiler	52.50	420.00	78.25	626.00		
1 Conc. Bucket, 2 C.Y.		55.65		61.22		
1 Lattice Boom Crane, 165 Ton		2403.00		2643.30	38.42	42.26
64 L.H., Daily Totals		$5517.45		$7266.92	$86.21	$113.55
Crew C-7C	Hr.	Daily	Hr.	Daily	Bare Costs	Incl. O&P
1 Labor Foreman (outside)	$46.40	$371.20	$69.25	$554.00	$48.30	$72.04
5 Laborers	44.40	1776.00	66.25	2650.00		
2 Equipment Operators (med.)	59.00	944.00	87.90	1406.40		
2 F.E. Loaders, W.M., 4 C.Y.		1518.00		1669.80	23.72	26.09
64 L.H., Daily Totals		$4609.20		$6280.20	$72.02	$98.13
Crew C-7D	Hr.	Daily	Hr.	Daily	Bare Costs	Incl. O&P
1 Labor Foreman (outside)	$46.40	$371.20	$69.25	$554.00	$46.77	$69.77
5 Laborers	44.40	1776.00	66.25	2650.00		
1 Equip. Oper. (medium)	59.00	472.00	87.90	703.20		
1 Concrete Conveyer		206.25		226.88	3.68	4.05
56 L.H., Daily Totals		$2825.45		$4134.07	$50.45	$73.82
Crew C-8	Hr.	Daily	Hr.	Daily	Bare Costs	Incl. O&P
1 Labor Foreman (outside)	$46.40	$371.20	$69.25	$554.00	$48.89	$72.53
3 Laborers	44.40	1065.60	66.25	1590.00		
2 Cement Finishers	51.80	828.80	75.90	1214.40		
1 Equip. Oper. (medium)	59.00	472.00	87.90	703.20		
1 Concrete Pump (Small)		423.65		466.01	7.57	8.32
56 L.H., Daily Totals		$3161.25		$4527.61	$56.45	$80.85
Crew C-8A	Hr.	Daily	Hr.	Daily	Bare Costs	Incl. O&P
1 Labor Foreman (outside)	$46.40	$371.20	$69.25	$554.00	$47.20	$69.97
3 Laborers	44.40	1065.60	66.25	1590.00		
2 Cement Finishers	51.80	828.80	75.90	1214.40		
48 L.H., Daily Totals		$2265.60		$3358.40	$47.20	$69.97

Crew No.	Bare Costs		Incl. Subs O&P		Cost Per Labor-Hour	
Crew C-8B	Hr.	Daily	Hr.	Daily	Bare Costs	Incl. O&P
1 Labor Foreman (outside)	$46.40	$371.20	$69.25	$554.00	$47.72	$71.18
3 Laborers	44.40	1065.60	66.25	1590.00		
1 Equip. Oper. (medium)	59.00	472.00	87.90	703.20		
1 Vibrating Power Screed		87.65		96.42		
1 Roller, Vibratory, 25 Ton		672.35		739.59		
1 Dozer, 200 H.P.		1520.00		1672.00	57.00	62.70
40 L.H., Daily Totals		$4188.80		$5355.20	$104.72	$133.88
Crew C-8C	Hr.	Daily	Hr.	Daily	Bare Costs	Incl. O&P
1 Labor Foreman (outside)	$46.40	$371.20	$69.25	$554.00	$48.40	$71.97
3 Laborers	44.40	1065.60	66.25	1590.00		
1 Cement Finisher	51.80	414.40	75.90	607.20		
1 Equip. Oper. (medium)	59.00	472.00	87.90	703.20		
1 Shotcrete Rig, 12 C.Y./hr		269.20		296.12		
1 Air Compressor, 160 cfm		212.30		233.53		
4 -50' Air Hoses, 1"		32.20		35.42		
4 -50' Air Hoses, 2"		115.80		127.38	13.11	14.43
48 L.H., Daily Totals		$2952.70		$4146.85	$61.51	$86.39
Crew C-8D	Hr.	Daily	Hr.	Daily	Bare Costs	Incl. O&P
1 Labor Foreman (outside)	$46.40	$371.20	$69.25	$554.00	$49.52	$73.53
1 Laborer	44.40	355.20	66.25	530.00		
1 Cement Finisher	51.80	414.40	75.90	607.20		
1 Equipment Oper. (light)	55.50	444.00	82.70	661.60		
1 Air Compressor, 250 cfm		202.85		223.13		
2 -50' Air Hoses, 1"		16.10		17.71	6.84	7.53
32 L.H., Daily Totals		$1803.75		$2593.65	$56.37	$81.05
Crew C-8E	Hr.	Daily	Hr.	Daily	Bare Costs	Incl. O&P
1 Labor Foreman (outside)	$46.40	$371.20	$69.25	$554.00	$47.82	$71.10
3 Laborers	44.40	1065.60	66.25	1590.00		
1 Cement Finisher	51.80	414.40	75.90	607.20		
1 Equipment Oper. (light)	55.50	444.00	82.70	661.60		
1 Shotcrete Rig, 35 C.Y./hr.		301.05		331.15		
1 Air Compressor, 250 cfm		202.85		223.13		
4 -50' Air Hoses, 1"		32.20		35.42		
4 -50' Air Hoses, 2"		115.80		127.38	13.58	14.94
48 L.H., Daily Totals		$2947.10		$4129.89	$61.40	$86.04
Crew C-9	Hr.	Daily	Hr.	Daily	Bare Costs	Incl. O&P
1 Cement Finisher	$51.80	$414.40	$75.90	$607.20	$49.02	$72.78
2 Laborers	44.40	710.40	66.25	1060.00		
1 Equipment Oper. (light)	55.50	444.00	82.70	661.60		
1 Grout Pump, 50 C.F./hr.		190.35		209.38		
1 Air Compressor, 160 cfm		212.30		233.53		
2 -50' Air Hoses, 1"		16.10		17.71		
2 -50' Air Hoses, 2"		57.90		63.69	14.90	16.38
32 L.H., Daily Totals		$2045.45		$2853.11	$63.92	$89.16
Crew C-10	Hr.	Daily	Hr.	Daily	Bare Costs	Incl. O&P
1 Laborer	$44.40	$355.20	$66.25	$530.00	$49.33	$72.68
2 Cement Finishers	51.80	828.80	75.90	1214.40		
24 L.H., Daily Totals		$1184.00		$1744.40	$49.33	$72.68
Crew C-10B	Hr.	Daily	Hr.	Daily	Bare Costs	Incl. O&P
3 Laborers	$44.40	$1065.60	$66.25	$1590.00	$47.36	$70.11
2 Cement Finishers	51.80	828.80	75.90	1214.40		
1 Concrete Mixer, 10 C.F.		147.15		161.87		
2 Trowels, 48" Walk-Behind		188.60		207.46	8.39	9.23
40 L.H., Daily Totals		$2230.15		$3173.72	$55.75	$79.34

Crew No.	Bare Costs Hr.	Bare Costs Daily	Incl. Subs O&P Hr.	Incl. Subs O&P Daily	Cost Per Labor-Hour Bare Costs	Cost Per Labor-Hour Incl. O&P
Crew C-10C					Bare Costs	Incl. O&P
1 Laborer	$44.40	$355.20	$66.25	$530.00	$49.33	$72.68
2 Cement Finishers	51.80	828.80	75.90	1214.40		
1 Trowel, 48" Walk-Behind		94.30		103.73	3.93	4.32
24 L.H., Daily Totals		$1278.30		$1848.13	$53.26	$77.01
Crew C-10D					Bare Costs	Incl. O&P
1 Laborer	$44.40	$355.20	$66.25	$530.00	$49.33	$72.68
2 Cement Finishers	51.80	828.80	75.90	1214.40		
1 Vibrating Power Screed		87.65		96.42		
1 Trowel, 48" Walk-Behind		94.30		103.73	7.58	8.34
24 L.H., Daily Totals		$1365.95		$1944.55	$56.91	$81.02
Crew C-10E					Bare Costs	Incl. O&P
1 Laborer	$44.40	$355.20	$66.25	$530.00	$49.33	$72.68
2 Cement Finishers	51.80	828.80	75.90	1214.40		
1 Vibrating Power Screed		87.65		96.42		
1 Cement Trowel, 96" Ride-On		171.05		188.16	10.78	11.86
24 L.H., Daily Totals		$1442.70		$2028.97	$60.11	$84.54
Crew C-10F					Bare Costs	Incl. O&P
1 Laborer	$44.40	$355.20	$66.25	$530.00	$49.33	$72.68
2 Cement Finishers	51.80	828.80	75.90	1214.40		
1 Telescoping Boom Lift, to 60'		292.45		321.69	12.19	13.40
24 L.H., Daily Totals		$1476.45		$2066.09	$61.52	$86.09
Crew C-11					Bare Costs	Incl. O&P
1 Struc. Steel Foreman (outside)	$62.30	$498.40	$96.40	$771.20	$59.78	$91.78
6 Struc. Steel Workers	60.30	2894.40	93.30	4478.40		
1 Equip. Oper. (crane)	61.45	491.60	91.55	732.40		
1 Equip. Oper. (oiler)	52.50	420.00	78.25	626.00		
1 Lattice Boom Crane, 150 Ton		2324.00		2556.40	32.28	35.51
72 L.H., Daily Totals		$6628.40		$9164.40	$92.06	$127.28
Crew C-12					Bare Costs	Incl. O&P
1 Carpenter Foreman (outside)	$56.70	$453.60	$84.60	$676.80	$54.44	$81.22
3 Carpenters	54.70	1312.80	81.65	1959.60		
1 Laborer	44.40	355.20	66.25	530.00		
1 Equip. Oper. (crane)	61.45	491.60	91.55	732.40		
1 Hyd. Crane, 12 Ton		475.80		523.38	9.91	10.90
48 L.H., Daily Totals		$3089.00		$4422.18	$64.35	$92.13
Crew C-13					Bare Costs	Incl. O&P
1 Struc. Steel Worker	$60.30	$482.40	$93.30	$746.40	$58.43	$89.42
1 Welder	60.30	482.40	93.30	746.40		
1 Carpenter	54.70	437.60	81.65	653.20		
1 Welder, Gas Engine, 300 amp		148.75		163.63	6.20	6.82
24 L.H., Daily Totals		$1551.15		$2309.63	$64.63	$96.23
Crew C-14					Bare Costs	Incl. O&P
1 Carpenter Foreman (outside)	$56.70	$453.60	$84.60	$676.80	$53.39	$79.54
5 Carpenters	54.70	2188.00	81.65	3266.00		
4 Laborers	44.40	1420.80	66.25	2120.00		
4 Rodmen (reinf.)	58.90	1884.80	88.05	2817.60		
2 Cement Finishers	51.80	828.80	75.90	1214.40		
1 Equip. Oper. (crane)	61.45	491.60	91.55	732.40		
1 Equip. Oper. (oiler)	52.50	420.00	78.25	626.00		
1 Hyd. Crane, 80 Ton		1458.00		1603.80	10.13	11.14
144 L.H., Daily Totals		$9145.60		$13057.00	$63.51	$90.67

Crew No.	Bare Costs Hr.	Bare Costs Daily	Incl. Subs O&P Hr.	Incl. Subs O&P Daily	Cost Per Labor-Hour Bare Costs	Cost Per Labor-Hour Incl. O&P
Crew C-14A					Bare Costs	Incl. O&P
1 Carpenter Foreman (outside)	$56.70	$453.60	$84.60	$676.80	$54.68	$81.58
16 Carpenters	54.70	7001.60	81.65	10451.20		
4 Rodmen (reinf.)	58.90	1884.80	88.05	2817.60		
2 Laborers	44.40	710.40	66.25	1060.00		
1 Cement Finisher	51.80	414.40	75.90	607.20		
1 Equip. Oper. (medium)	59.00	472.00	87.90	703.20		
1 Gas Engine Vibrator		27.15		29.86		
1 Concrete Pump (Small)		423.65		466.01	2.25	2.48
200 L.H., Daily Totals		$11387.60		$16811.88	$56.94	$84.06
Crew C-14B					Bare Costs	Incl. O&P
1 Carpenter Foreman (outside)	$56.70	$453.60	$84.60	$676.80	$54.57	$81.36
16 Carpenters	54.70	7001.60	81.65	10451.20		
4 Rodmen (reinf.)	58.90	1884.80	88.05	2817.60		
2 Laborers	44.40	710.40	66.25	1060.00		
2 Cement Finishers	51.80	828.80	75.90	1214.40		
1 Equip. Oper. (medium)	59.00	472.00	87.90	703.20		
1 Gas Engine Vibrator		27.15		29.86		
1 Concrete Pump (Small)		423.65		466.01	2.17	2.38
208 L.H., Daily Totals		$11802.00		$17419.08	$56.74	$83.75
Crew C-14C					Bare Costs	Incl. O&P
1 Carpenter Foreman (outside)	$56.70	$453.60	$84.60	$676.80	$52.29	$77.96
6 Carpenters	54.70	2625.60	81.65	3919.20		
2 Rodmen (reinf.)	58.90	942.40	88.05	1408.80		
4 Laborers	44.40	1420.80	66.25	2120.00		
1 Cement Finisher	51.80	414.40	75.90	607.20		
1 Gas Engine Vibrator		27.15		29.86	.24	.27
112 L.H., Daily Totals		$5883.95		$8761.86	$52.54	$78.23
Crew C-14D					Bare Costs	Incl. O&P
1 Carpenter Foreman (outside)	$56.70	$453.60	$84.60	$676.80	$54.35	$81.07
18 Carpenters	54.70	7876.80	81.65	11757.60		
2 Rodmen (reinf.)	58.90	942.40	88.05	1408.80		
2 Laborers	44.40	710.40	66.25	1060.00		
1 Cement Finisher	51.80	414.40	75.90	607.20		
1 Equip. Oper. (medium)	59.00	472.00	87.90	703.20		
1 Gas Engine Vibrator		27.15		29.86		
1 Concrete Pump (Small)		423.65		466.01	2.25	2.48
200 L.H., Daily Totals		$11320.40		$16709.48	$56.60	$83.55
Crew C-14E					Bare Costs	Incl. O&P
1 Carpenter Foreman (outside)	$56.70	$453.60	$84.60	$676.80	$53.34	$79.52
2 Carpenters	54.70	875.20	81.65	1306.40		
4 Rodmen (reinf.)	58.90	1884.80	88.05	2817.60		
3 Laborers	44.40	1065.60	66.25	1590.00		
1 Cement Finisher	51.80	414.40	75.90	607.20		
1 Gas Engine Vibrator		27.15		29.86	.31	.34
88 L.H., Daily Totals		$4720.75		$7027.86	$53.64	$79.86
Crew C-14F					Bare Costs	Incl. O&P
1 Labor Foreman (outside)	$46.40	$371.20	$69.25	$554.00	$49.56	$73.02
2 Laborers	44.40	710.40	66.25	1060.00		
6 Cement Finishers	51.80	2486.40	75.90	3643.20		
1 Gas Engine Vibrator		27.15		29.86	.38	.41
72 L.H., Daily Totals		$3595.15		$5287.06	$49.93	$73.43

For customer support on your Electrical Costs with RSMeans Data, call 800.448.8182.

509

Crew C-14G

Crew No.	Bare Costs Hr.	Daily	Incl. Subs O&P Hr.	Daily	Cost Per Labor-Hour Bare Costs	Incl. O&P
1 Labor Foreman (outside)	$46.40	$371.20	$69.25	$554.00	$48.91	$72.19
2 Laborers	44.40	710.40	66.25	1060.00		
4 Cement Finishers	51.80	1657.60	75.90	2428.80		
1 Gas Engine Vibrator		27.15		29.86	.48	.53
56 L.H., Daily Totals		$2766.35		$4072.67	$49.40	$72.73

Crew C-14H

Crew No.	Bare Costs Hr.	Daily	Incl. Subs O&P Hr.	Daily	Cost Per Labor-Hour Bare Costs	Incl. O&P
1 Carpenter Foreman (outside)	$56.70	$453.60	$84.60	$676.80	$53.53	$79.68
2 Carpenters	54.70	875.20	81.65	1306.40		
1 Rodman (reinf.)	58.90	471.20	88.05	704.40		
1 Laborer	44.40	355.20	66.25	530.00		
1 Cement Finisher	51.80	414.40	75.90	607.20		
1 Gas Engine Vibrator		27.15		29.86	.57	.62
48 L.H., Daily Totals		$2596.75		$3854.67	$54.10	$80.31

Crew C-14L

Crew No.	Bare Costs Hr.	Daily	Incl. Subs O&P Hr.	Daily	Cost Per Labor-Hour Bare Costs	Incl. O&P
1 Carpenter Foreman (outside)	$56.70	$453.60	$84.60	$676.80	$51.19	$76.28
6 Carpenters	54.70	2625.60	81.65	3919.20		
4 Laborers	44.40	1420.80	66.25	2120.00		
1 Cement Finisher	51.80	414.40	75.90	607.20		
1 Gas Engine Vibrator		27.15		29.86	.28	.31
96 L.H., Daily Totals		$4941.55		$7353.06	$51.47	$76.59

Crew C-14M

Crew No.	Bare Costs Hr.	Daily	Incl. Subs O&P Hr.	Daily	Cost Per Labor-Hour Bare Costs	Incl. O&P
1 Carpenter Foreman (outside)	$56.70	$453.60	$84.60	$676.80	$53.08	$79.03
2 Carpenters	54.70	875.20	81.65	1306.40		
1 Rodman (reinf.)	58.90	471.20	88.05	704.40		
2 Laborers	44.40	710.40	66.25	1060.00		
1 Cement Finisher	51.80	414.40	75.90	607.20		
1 Equip. Oper. (medium)	59.00	472.00	87.90	703.20		
1 Gas Engine Vibrator		27.15		29.86		
1 Concrete Pump (Small)		423.65		466.01	7.04	7.75
64 L.H., Daily Totals		$3847.60		$5553.88	$60.12	$86.78

Crew C-15

Crew No.	Bare Costs Hr.	Daily	Incl. Subs O&P Hr.	Daily	Cost Per Labor-Hour Bare Costs	Incl. O&P
1 Carpenter Foreman (outside)	$56.70	$453.60	$84.60	$676.80	$51.31	$76.28
2 Carpenters	54.70	875.20	81.65	1306.40		
3 Laborers	44.40	1065.60	66.25	1590.00		
2 Cement Finishers	51.80	828.80	75.90	1214.40		
1 Rodman (reinf.)	58.90	471.20	88.05	704.40		
72 L.H., Daily Totals		$3694.40		$5492.00	$51.31	$76.28

Crew C-16

Crew No.	Bare Costs Hr.	Daily	Incl. Subs O&P Hr.	Daily	Cost Per Labor-Hour Bare Costs	Incl. O&P
1 Labor Foreman (outside)	$46.40	$371.20	$69.25	$554.00	$48.89	$72.53
3 Laborers	44.40	1065.60	66.25	1590.00		
2 Cement Finishers	51.80	828.80	75.90	1214.40		
1 Equip. Oper. (medium)	59.00	472.00	87.90	703.20		
1 Gunite Pump Rig		321.75		353.93		
2 -50' Air Hoses, 3/4"		14.30		15.73		
2 -50' Air Hoses, 2"		57.90		63.69	7.03	7.74
56 L.H., Daily Totals		$3131.55		$4494.94	$55.92	$80.27

Crew C-16A

Crew No.	Bare Costs Hr.	Daily	Incl. Subs O&P Hr.	Daily	Cost Per Labor-Hour Bare Costs	Incl. O&P
1 Laborer	$44.40	$355.20	$66.25	$530.00	$51.75	$76.49
2 Cement Finishers	51.80	828.80	75.90	1214.40		
1 Equip. Oper. (medium)	59.00	472.00	87.90	703.20		
1 Gunite Pump Rig		321.75		353.93		
2 -50' Air Hoses, 3/4"		14.30		15.73		
2 -50' Air Hoses, 2"		57.90		63.69		
1 Telescoping Boom Lift, to 60'		292.45		321.69	21.45	23.59
32 L.H., Daily Totals		$2342.40		$3202.64	$73.20	$100.08

Crew C-17

Crew No.	Bare Costs Hr.	Daily	Incl. Subs O&P Hr.	Daily	Cost Per Labor-Hour Bare Costs	Incl. O&P
2 Skilled Worker Foremen (out)	$59.10	$945.60	$88.90	$1422.40	$57.50	$86.50
8 Skilled Workers	57.10	3654.40	85.90	5497.60		
80 L.H., Daily Totals		$4600.00		$6920.00	$57.50	$86.50

Crew C-17A

Crew No.	Bare Costs Hr.	Daily	Incl. Subs O&P Hr.	Daily	Cost Per Labor-Hour Bare Costs	Incl. O&P
2 Skilled Worker Foremen (out)	$59.10	$945.60	$88.90	$1422.40	$57.55	$86.56
8 Skilled Workers	57.10	3654.40	85.90	5497.60		
.125 Equip. Oper. (crane)	61.45	61.45	91.55	91.55		
.125 Hyd. Crane, 80 Ton		182.25		200.47	2.25	2.48
81 L.H., Daily Totals		$4843.70		$7212.02	$59.80	$89.04

Crew C-17B

Crew No.	Bare Costs Hr.	Daily	Incl. Subs O&P Hr.	Daily	Cost Per Labor-Hour Bare Costs	Incl. O&P
2 Skilled Worker Foremen (out)	$59.10	$945.60	$88.90	$1422.40	$57.60	$86.62
8 Skilled Workers	57.10	3654.40	85.90	5497.60		
.25 Equip. Oper. (crane)	61.45	122.90	91.55	183.10		
.25 Hyd. Crane, 80 Ton		364.50		400.95		
.25 Trowel, 48" Walk-Behind		23.57		25.93	4.73	5.21
82 L.H., Daily Totals		$5110.98		$7529.98	$62.33	$91.83

Crew C-17C

Crew No.	Bare Costs Hr.	Daily	Incl. Subs O&P Hr.	Daily	Cost Per Labor-Hour Bare Costs	Incl. O&P
2 Skilled Worker Foremen (out)	$59.10	$945.60	$88.90	$1422.40	$57.64	$86.68
8 Skilled Workers	57.10	3654.40	85.90	5497.60		
.375 Equip. Oper. (crane)	61.45	184.35	91.55	274.65		
.375 Hyd. Crane, 80 Ton		546.75		601.42	6.59	7.25
83 L.H., Daily Totals		$5331.10		$7796.07	$64.23	$93.93

Crew C-17D

Crew No.	Bare Costs Hr.	Daily	Incl. Subs O&P Hr.	Daily	Cost Per Labor-Hour Bare Costs	Incl. O&P
2 Skilled Worker Foremen (out)	$59.10	$945.60	$88.90	$1422.40	$57.69	$86.74
8 Skilled Workers	57.10	3654.40	85.90	5497.60		
.5 Equip. Oper. (crane)	61.45	245.80	91.55	366.20		
.5 Hyd. Crane, 80 Ton		729.00		801.90	8.68	9.55
84 L.H., Daily Totals		$5574.80		$8088.10	$66.37	$96.29

Crew C-17E

Crew No.	Bare Costs Hr.	Daily	Incl. Subs O&P Hr.	Daily	Cost Per Labor-Hour Bare Costs	Incl. O&P
2 Skilled Worker Foremen (out)	$59.10	$945.60	$88.90	$1422.40	$57.50	$86.50
8 Skilled Workers	57.10	3654.40	85.90	5497.60		
1 Hyd. Jack with Rods		36.70		40.37	.46	.50
80 L.H., Daily Totals		$4636.70		$6960.37	$57.96	$87.00

Crew C-18

Crew No.	Bare Costs Hr.	Daily	Incl. Subs O&P Hr.	Daily	Cost Per Labor-Hour Bare Costs	Incl. O&P
.125 Labor Foreman (outside)	$46.40	$46.40	$69.25	$69.25	$44.62	$66.58
1 Laborer	44.40	355.20	66.25	530.00		
1 Concrete Cart, 10 C.F.		116.95		128.65	12.99	14.29
9 L.H., Daily Totals		$518.55		$727.89	$57.62	$80.88

Crew C-19

Crew No.	Bare Costs Hr.	Daily	Incl. Subs O&P Hr.	Daily	Cost Per Labor-Hour Bare Costs	Incl. O&P
.125 Labor Foreman (outside)	$46.40	$46.40	$69.25	$69.25	$44.62	$66.58
1 Laborer	44.40	355.20	66.25	530.00		
1 Concrete Cart, 18 C.F.		138.95		152.85	15.44	16.98
9 L.H., Daily Totals		$540.55		$752.10	$60.06	$83.57

Crew C-20

Crew No.	Bare Costs Hr.	Daily	Incl. Subs O&P Hr.	Daily	Cost Per Labor-Hour Bare Costs	Incl. O&P
1 Labor Foreman (outside)	$46.40	$371.20	$69.25	$554.00	$47.40	$70.54
5 Laborers	44.40	1776.00	66.25	2650.00		
1 Cement Finisher	51.80	414.40	75.90	607.20		
1 Equip. Oper. (medium)	59.00	472.00	87.90	703.20		
2 Gas Engine Vibrators		54.30		59.73		
1 Concrete Pump (Small)		423.65		466.01	7.47	8.21
64 L.H., Daily Totals		$3511.55		$5040.15	$54.87	$78.75

For customer support on your Electrical Costs with RSMeans Data, call 800.448.8182.

Crew No.	Bare Costs		Incl. Subs O&P		Cost Per Labor-Hour	

Crew C-21

	Hr.	Daily	Hr.	Daily	Bare Costs	Incl. O&P
1 Labor Foreman (outside)	$46.40	$371.20	$69.25	$554.00	$47.40	$70.54
5 Laborers	44.40	1776.00	66.25	2650.00		
1 Cement Finisher	51.80	414.40	75.90	607.20		
1 Equip. Oper. (medium)	59.00	472.00	87.90	703.20		
2 Gas Engine Vibrators		54.30		59.73		
1 Concrete Conveyer		206.25		226.88	4.07	4.48
64 L.H., Daily Totals		$3294.15		$4801.01	$51.47	$75.02

Crew C-22

	Hr.	Daily	Hr.	Daily	Bare Costs	Incl. O&P
1 Rodman Foreman (outside)	$60.90	$487.20	$91.05	$728.40	$59.19	$88.47
4 Rodmen (reinf.)	58.90	1884.80	88.05	2817.60		
.125 Equip. Oper. (crane)	61.45	61.45	91.55	91.55		
.125 Equip. Oper. (oiler)	52.50	52.50	78.25	78.25		
.125 Hyd. Crane, 25 Ton		73.34		80.67	1.75	1.92
42 L.H., Daily Totals		$2559.29		$3796.47	$60.94	$90.39

Crew C-23

	Hr.	Daily	Hr.	Daily	Bare Costs	Incl. O&P
2 Skilled Worker Foremen (out)	$59.10	$945.60	$88.90	$1422.40	$57.48	$86.30
6 Skilled Workers	57.10	2740.80	85.90	4123.20		
1 Equip. Oper. (crane)	61.45	491.60	91.55	732.40		
1 Equip. Oper. (oiler)	52.50	420.00	78.25	626.00		
1 Lattice Boom Crane, 90 Ton		1713.00		1884.30	21.41	23.55
80 L.H., Daily Totals		$6311.00		$8788.30	$78.89	$109.85

Crew C-23A

	Hr.	Daily	Hr.	Daily	Bare Costs	Incl. O&P
1 Labor Foreman (outside)	$46.40	$371.20	$69.25	$554.00	$49.83	$74.31
2 Laborers	44.40	710.40	66.25	1060.00		
1 Equip. Oper. (crane)	61.45	491.60	91.55	732.40		
1 Equip. Oper. (oiler)	52.50	420.00	78.25	626.00		
1 Crawler Crane, 100 Ton		2310.00		2541.00		
3 Conc. Buckets, 8 C.Y.		337.95		371.75	66.20	72.82
40 L.H., Daily Totals		$4641.15		$5885.15	$116.03	$147.13

Crew C-24

	Hr.	Daily	Hr.	Daily	Bare Costs	Incl. O&P
2 Skilled Worker Foremen (out)	$59.10	$945.60	$88.90	$1422.40	$57.48	$86.30
6 Skilled Workers	57.10	2740.80	85.90	4123.20		
1 Equip. Oper. (crane)	61.45	491.60	91.55	732.40		
1 Equip. Oper. (oiler)	52.50	420.00	78.25	626.00		
1 Lattice Boom Crane, 150 Ton		2324.00		2556.40	29.05	31.95
80 L.H., Daily Totals		$6922.00		$9460.40	$86.53	$118.26

Crew C-25

	Hr.	Daily	Hr.	Daily	Bare Costs	Incl. O&P
2 Rodmen (reinf.)	$58.90	$942.40	$88.05	$1408.80	$47.42	$73.15
2 Rodmen Helpers	35.95	575.20	58.25	932.00		
32 L.H., Daily Totals		$1517.60		$2340.80	$47.42	$73.15

Crew C-27

	Hr.	Daily	Hr.	Daily	Bare Costs	Incl. O&P
2 Cement Finishers	$51.80	$828.80	$75.90	$1214.40	$51.80	$75.90
1 Concrete Saw		112.85		124.14	7.05	7.76
16 L.H., Daily Totals		$941.65		$1338.54	$58.85	$83.66

Crew C-28

	Hr.	Daily	Hr.	Daily	Bare Costs	Incl. O&P
1 Cement Finisher	$51.80	$414.40	$75.90	$607.20	$51.80	$75.90
1 Portable Air Compressor, Gas		38.80		42.68	4.85	5.34
8 L.H., Daily Totals		$453.20		$649.88	$56.65	$81.23

Crew C-29

	Hr.	Daily	Hr.	Daily	Bare Costs	Incl. O&P
1 Laborer	$44.40	$355.20	$66.25	$530.00	$44.40	$66.25
1 Pressure Washer		97.35		107.08	12.17	13.39
8 L.H., Daily Totals		$452.55		$637.09	$56.57	$79.64

Crew C-30

	Hr.	Daily	Hr.	Daily	Bare Costs	Incl. O&P
1 Laborer	$44.40	$355.20	$66.25	$530.00	$44.40	$66.25
1 Concrete Mixer, 10 C.F.		147.15		161.87	18.39	20.23
8 L.H., Daily Totals		$502.35		$691.87	$62.79	$86.48

Crew C-31

	Hr.	Daily	Hr.	Daily	Bare Costs	Incl. O&P
1 Cement Finisher	$51.80	$414.40	$75.90	$607.20	$51.80	$75.90
1 Grout Pump		321.75		353.93	40.22	44.24
8 L.H., Daily Totals		$736.15		$961.13	$92.02	$120.14

Crew C-32

	Hr.	Daily	Hr.	Daily	Bare Costs	Incl. O&P
1 Cement Finisher	$51.80	$414.40	$75.90	$607.20	$48.10	$71.08
1 Laborer	44.40	355.20	66.25	530.00		
1 Crack Chaser Saw, Gas, 6 H.P.		73.25		80.58		
1 Vacuum Pick-Up System		74.95		82.44	9.26	10.19
16 L.H., Daily Totals		$917.80		$1300.22	$57.36	$81.26

Crew D-1

	Hr.	Daily	Hr.	Daily	Bare Costs	Incl. O&P
1 Bricklayer	$53.70	$429.60	$80.90	$647.20	$48.70	$73.38
1 Bricklayer Helper	43.70	349.60	65.85	526.80		
16 L.H., Daily Totals		$779.20		$1174.00	$48.70	$73.38

Crew D-2

	Hr.	Daily	Hr.	Daily	Bare Costs	Incl. O&P
3 Bricklayers	$53.70	$1288.80	$80.90	$1941.60	$50.15	$75.50
2 Bricklayer Helpers	43.70	699.20	65.85	1053.60		
.5 Carpenter	54.70	218.80	81.65	326.60		
44 L.H., Daily Totals		$2206.80		$3321.80	$50.15	$75.50

Crew D-3

	Hr.	Daily	Hr.	Daily	Bare Costs	Incl. O&P
3 Bricklayers	$53.70	$1288.80	$80.90	$1941.60	$49.94	$75.20
2 Bricklayer Helpers	43.70	699.20	65.85	1053.60		
.25 Carpenter	54.70	109.40	81.65	163.30		
42 L.H., Daily Totals		$2097.40		$3158.50	$49.94	$75.20

Crew D-4

	Hr.	Daily	Hr.	Daily	Bare Costs	Incl. O&P
1 Bricklayer	$53.70	$429.60	$80.90	$647.20	$49.15	$73.83
2 Bricklayer Helpers	43.70	699.20	65.85	1053.60		
1 Equip. Oper. (light)	55.50	444.00	82.70	661.60		
1 Grout Pump, 50 C.F./hr.		190.35		209.38	5.95	6.54
32 L.H., Daily Totals		$1763.15		$2571.78	$55.10	$80.37

Crew D-5

	Hr.	Daily	Hr.	Daily	Bare Costs	Incl. O&P
1 Bricklayer	53.70	429.60	80.90	647.20	53.70	80.90
8 L.H., Daily Totals		$429.60		$647.20	$53.70	$80.90

Crew D-6

	Hr.	Daily	Hr.	Daily	Bare Costs	Incl. O&P
3 Bricklayers	$53.70	$1288.80	$80.90	$1941.60	$48.94	$73.71
3 Bricklayer Helpers	43.70	1048.80	65.85	1580.40		
.25 Carpenter	54.70	109.40	81.65	163.30		
50 L.H., Daily Totals		$2447.00		$3685.30	$48.94	$73.71

Crew D-7

	Hr.	Daily	Hr.	Daily	Bare Costs	Incl. O&P
1 Tile Layer	$51.70	$413.60	$75.55	$604.40	$46.65	$68.17
1 Tile Layer Helper	41.60	332.80	60.80	486.40		
16 L.H., Daily Totals		$746.40		$1090.80	$46.65	$68.17

Crew D-8

	Hr.	Daily	Hr.	Daily	Bare Costs	Incl. O&P
3 Bricklayers	$53.70	$1288.80	$80.90	$1941.60	$49.70	$74.88
2 Bricklayer Helpers	43.70	699.20	65.85	1053.60		
40 L.H., Daily Totals		$1988.00		$2995.20	$49.70	$74.88

Crew No.	Bare Costs		Incl. Subs O&P		Cost Per Labor-Hour	

Crew D-9

	Hr.	Daily	Hr.	Daily	Bare Costs	Incl. O&P
3 Bricklayers	$53.70	$1288.80	$80.90	$1941.60	$48.70	$73.38
3 Bricklayer Helpers	43.70	1048.80	65.85	1580.40		
48 L.H., Daily Totals		$2337.60		$3522.00	$48.70	$73.38

Crew D-10

	Hr.	Daily	Hr.	Daily	Bare Costs	Incl. O&P
1 Bricklayer Foreman (outside)	$55.70	$445.60	$83.90	$671.20	$53.64	$80.55
1 Bricklayer	53.70	429.60	80.90	647.20		
1 Bricklayer Helper	43.70	349.60	65.85	526.80		
1 Equip. Oper. (crane)	61.45	491.60	91.55	732.40		
1 S.P. Crane, 4x4, 12 Ton		432.65		475.92	13.52	14.87
32 L.H., Daily Totals		$2149.05		$3053.51	$67.16	$95.42

Crew D-11

	Hr.	Daily	Hr.	Daily	Bare Costs	Incl. O&P
1 Bricklayer Foreman (outside)	$55.70	$445.60	$83.90	$671.20	$51.03	$76.88
1 Bricklayer	53.70	429.60	80.90	647.20		
1 Bricklayer Helper	43.70	349.60	65.85	526.80		
24 L.H., Daily Totals		$1224.80		$1845.20	$51.03	$76.88

Crew D-12

	Hr.	Daily	Hr.	Daily	Bare Costs	Incl. O&P
1 Bricklayer Foreman (outside)	$55.70	$445.60	$83.90	$671.20	$49.20	$74.13
1 Bricklayer	53.70	429.60	80.90	647.20		
2 Bricklayer Helpers	43.70	699.20	65.85	1053.60		
32 L.H., Daily Totals		$1574.40		$2372.00	$49.20	$74.13

Crew D-13

	Hr.	Daily	Hr.	Daily	Bare Costs	Incl. O&P
1 Bricklayer Foreman (outside)	$55.70	$445.60	$83.90	$671.20	$52.16	$78.28
1 Bricklayer	53.70	429.60	80.90	647.20		
2 Bricklayer Helpers	43.70	699.20	65.85	1053.60		
1 Carpenter	54.70	437.60	81.65	653.20		
1 Equip. Oper. (crane)	61.45	491.60	91.55	732.40		
1 S.P. Crane, 4x4, 12 Ton		432.65		475.92	9.01	9.91
48 L.H., Daily Totals		$2936.25		$4233.52	$61.17	$88.20

Crew D-14

	Hr.	Daily	Hr.	Daily	Bare Costs	Incl. O&P
3 Bricklayers	$53.70	$1288.80	$80.90	$1941.60	$51.20	$77.14
1 Bricklayer Helper	43.70	349.60	65.85	526.80		
32 L.H., Daily Totals		$1638.40		$2468.40	$51.20	$77.14

Crew E-1

	Hr.	Daily	Hr.	Daily	Bare Costs	Incl. O&P
1 Welder Foreman (outside)	$62.30	$498.40	$96.40	$771.20	$59.37	$90.80
1 Welder	60.30	482.40	93.30	746.40		
1 Equip. Oper. (light)	55.50	444.00	82.70	661.60		
1 Welder, Gas Engine, 300 amp		148.75		163.63	6.20	6.82
24 L.H., Daily Totals		$1573.55		$2342.82	$65.56	$97.62

Crew E-2

	Hr.	Daily	Hr.	Daily	Bare Costs	Incl. O&P
1 Struc. Steel Foreman (outside)	$62.30	$498.40	$96.40	$771.20	$59.64	$91.34
4 Struc. Steel Workers	60.30	1929.60	93.30	2985.60		
1 Equip. Oper. (crane)	61.45	491.60	91.55	732.40		
1 Equip. Oper. (oiler)	52.50	420.00	78.25	626.00		
1 Lattice Boom Crane, 90 Ton		1713.00		1884.30	30.59	33.65
56 L.H., Daily Totals		$5052.60		$6999.50	$90.22	$124.99

Crew E-3

	Hr.	Daily	Hr.	Daily	Bare Costs	Incl. O&P
1 Struc. Steel Foreman (outside)	$62.30	$498.40	$96.40	$771.20	$60.97	$94.33
1 Struc. Steel Worker	60.30	482.40	93.30	746.40		
1 Welder	60.30	482.40	93.30	746.40		
1 Welder, Gas Engine, 300 amp		148.75		163.63	6.20	6.82
24 L.H., Daily Totals		$1611.95		$2427.63	$67.16	$101.15

Crew E-3A

	Hr.	Daily	Hr.	Daily	Bare Costs	Incl. O&P
1 Struc. Steel Foreman (outside)	$62.30	$498.40	$96.40	$771.20	$60.97	$94.33
1 Struc. Steel Worker	60.30	482.40	93.30	746.40		
1 Welder	60.30	482.40	93.30	746.40		
1 Welder, Gas Engine, 300 amp		148.75		163.63		
1 Telescoping Boom Lift, to 40'		281.90		310.09	17.94	19.74
24 L.H., Daily Totals		$1893.85		$2737.72	$78.91	$114.07

Crew E-4

	Hr.	Daily	Hr.	Daily	Bare Costs	Incl. O&P
1 Struc. Steel Foreman (outside)	$62.30	$498.40	$96.40	$771.20	$60.80	$94.08
3 Struc. Steel Workers	60.30	1447.20	93.30	2239.20		
1 Welder, Gas Engine, 300 amp		148.75		163.63	4.65	5.11
32 L.H., Daily Totals		$2094.35		$3174.03	$65.45	$99.19

Crew E-5

	Hr.	Daily	Hr.	Daily	Bare Costs	Incl. O&P
2 Struc. Steel Foremen (outside)	$62.30	$996.80	$96.40	$1542.40	$60.03	$92.24
5 Struc. Steel Workers	60.30	2412.00	93.30	3732.00		
1 Equip. Oper. (crane)	61.45	491.60	91.55	732.40		
1 Welder	60.30	482.40	93.30	746.40		
1 Equip. Oper. (oiler)	52.50	420.00	78.25	626.00		
1 Lattice Boom Crane, 90 Ton		1713.00		1884.30		
1 Welder, Gas Engine, 300 amp		148.75		163.63	23.27	25.60
80 L.H., Daily Totals		$6664.55		$9427.13	$83.31	$117.84

Crew E-6

	Hr.	Daily	Hr.	Daily	Bare Costs	Incl. O&P
3 Struc. Steel Foremen (outside)	$62.30	$1495.20	$96.40	$2313.60	$59.96	$92.17
9 Struc. Steel Workers	60.30	4341.60	93.30	6717.60		
1 Equip. Oper. (crane)	61.45	491.60	91.55	732.40		
1 Welder	60.30	482.40	93.30	746.40		
1 Equip. Oper. (oiler)	52.50	420.00	78.25	626.00		
1 Equip. Oper. (light)	55.50	444.00	82.70	661.60		
1 Lattice Boom Crane, 90 Ton		1713.00		1884.30		
1 Welder, Gas Engine, 300 amp		148.75		163.63		
1 Air Compressor, 160 cfm		212.30		233.53		
2 Impact Wrenches		90.30		99.33	16.91	18.60
128 L.H., Daily Totals		$9839.15		$14178.39	$76.87	$110.77

Crew E-7

	Hr.	Daily	Hr.	Daily	Bare Costs	Incl. O&P
1 Struc. Steel Foreman (outside)	$62.30	$498.40	$96.40	$771.20	$60.03	$92.24
4 Struc. Steel Workers	60.30	1929.60	93.30	2985.60		
1 Equip. Oper. (crane)	61.45	491.60	91.55	732.40		
1 Equip. Oper. (oiler)	52.50	420.00	78.25	626.00		
1 Welder Foreman (outside)	62.30	498.40	96.40	771.20		
2 Welders	60.30	964.80	93.30	1492.80		
1 Lattice Boom Crane, 90 Ton		1713.00		1884.30		
2 Welder, Gas Engine, 300 amp		297.50		327.25	25.13	27.64
80 L.H., Daily Totals		$6813.30		$9590.75	$85.17	$119.88

Crew E-8

	Hr.	Daily	Hr.	Daily	Bare Costs	Incl. O&P
1 Struc. Steel Foreman (outside)	$62.30	$498.40	$96.40	$771.20	$59.73	$91.67
4 Struc. Steel Workers	60.30	1929.60	93.30	2985.60		
1 Welder Foreman (outside)	62.30	498.40	96.40	771.20		
4 Welders	60.30	1929.60	93.30	2985.60		
1 Equip. Oper. (crane)	61.45	491.60	91.55	732.40		
1 Equip. Oper. (oiler)	52.50	420.00	78.25	626.00		
1 Equip. Oper. (light)	55.50	444.00	82.70	661.60		
1 Lattice Boom Crane, 90 Ton		1713.00		1884.30		
4 Welder, Gas Engine, 300 amp		595.00		654.50	22.19	24.41
104 L.H., Daily Totals		$8519.60		$12072.40	$81.92	$116.08

Crew E-9	Hr.	Daily	Hr.	Daily	Bare Costs	Incl. O&P
2 Struc. Steel Foremen (outside)	$62.30	$996.80	$96.40	$1542.40	$59.96	$92.17
5 Struc. Steel Workers	60.30	2412.00	93.30	3732.00		
1 Welder Foreman (outside)	62.30	498.40	96.40	771.20		
5 Welders	60.30	2412.00	93.30	3732.00		
1 Equip. Oper. (crane)	61.45	491.60	91.55	732.40		
1 Equip. Oper. (oiler)	52.50	420.00	78.25	626.00		
1 Equip. Oper. (light)	55.50	444.00	82.70	661.60		
1 Lattice Boom Crane, 90 Ton		1713.00		1884.30		
5 Welder, Gas Engine, 300 amp		743.75		818.13	19.19	21.11
128 L.H., Daily Totals		$10131.55		$14500.03	$79.15	$113.28

Crew E-10	Hr.	Daily	Hr.	Daily	Bare Costs	Incl. O&P
1 Welder Foreman (outside)	$62.30	$498.40	$96.40	$771.20	$61.30	$94.85
1 Welder	60.30	482.40	93.30	746.40		
1 Welder, Gas Engine, 300 amp		148.75		163.63		
1 Flatbed Truck, Gas, 3 Ton		850.05		935.05	62.42	68.67
16 L.H., Daily Totals		$1979.60		$2616.28	$123.72	$163.52

Crew E-11	Hr.	Daily	Hr.	Daily	Bare Costs	Incl. O&P
2 Painters, Struc. Steel	$47.20	$755.20	$75.80	$1212.80	$48.58	$75.14
1 Building Laborer	44.40	355.20	66.25	530.00		
1 Equip. Oper. (light)	55.50	444.00	82.70	661.60		
1 Air Compressor, 250 cfm		202.85		223.13		
1 Sandblaster, Portable, 3 C.F.		83.85		92.23		
1 Set Sand Blasting Accessories		15.55		17.11	9.45	10.39
32 L.H., Daily Totals		$1856.65		$2736.88	$58.02	$85.53

Crew E-11A	Hr.	Daily	Hr.	Daily	Bare Costs	Incl. O&P
2 Painters, Struc. Steel	$47.20	$755.20	$75.80	$1212.80	$48.58	$75.14
1 Building Laborer	44.40	355.20	66.25	530.00		
1 Equip. Oper. (light)	55.50	444.00	82.70	661.60		
1 Air Compressor, 250 cfm		202.85		223.13		
1 Sandblaster, Portable, 3 C.F.		83.85		92.23		
1 Set Sand Blasting Accessories		15.55		17.11		
1 Telescoping Boom Lift, to 60'		292.45		321.69	18.58	20.44
32 L.H., Daily Totals		$2149.10		$3058.57	$67.16	$95.58

Crew E-11B	Hr.	Daily	Hr.	Daily	Bare Costs	Incl. O&P
2 Painters, Struc. Steel	$47.20	$755.20	$75.80	$1212.80	$46.27	$72.62
1 Building Laborer	44.40	355.20	66.25	530.00		
2 Paint Sprayer, 8 C.F.M.		88.40		97.24		
1 Telescoping Boom Lift, to 60'		292.45		321.69	15.87	17.46
24 L.H., Daily Totals		$1491.25		$2161.74	$62.14	$90.07

Crew E-12	Hr.	Daily	Hr.	Daily	Bare Costs	Incl. O&P
1 Welder Foreman (outside)	$62.30	$498.40	$96.40	$771.20	$58.90	$89.55
1 Equip. Oper. (light)	55.50	444.00	82.70	661.60		
1 Welder, Gas Engine, 300 amp		148.75		163.63	9.30	10.23
16 L.H., Daily Totals		$1091.15		$1596.43	$68.20	$99.78

Crew E-13	Hr.	Daily	Hr.	Daily	Bare Costs	Incl. O&P
1 Welder Foreman (outside)	$62.30	$498.40	$96.40	$771.20	$60.03	$91.83
.5 Equip. Oper. (light)	55.50	222.00	82.70	330.80		
1 Welder, Gas Engine, 300 amp		148.75		163.63	12.40	13.64
12 L.H., Daily Totals		$869.15		$1265.63	$72.43	$105.47

Crew E-14	Hr.	Daily	Hr.	Daily	Bare Costs	Incl. O&P
1 Welder Foreman (outside)	$62.30	$498.40	$96.40	$771.20	$62.30	$96.40
1 Welder, Gas Engine, 300 amp		148.75		163.63	18.59	20.45
8 L.H., Daily Totals		$647.15		$934.83	$80.89	$116.85

Crew E-16	Hr.	Daily	Hr.	Daily	Bare Costs	Incl. O&P
1 Welder Foreman (outside)	$62.30	$498.40	$96.40	$771.20	$61.30	$94.85
1 Welder	60.30	482.40	93.30	746.40		
1 Welder, Gas Engine, 300 amp		148.75		163.63	9.30	10.23
16 L.H., Daily Totals		$1129.55		$1681.22	$70.60	$105.08

Crew E-17	Hr.	Daily	Hr.	Daily	Bare Costs	Incl. O&P
1 Struc. Steel Foreman (outside)	$62.30	$498.40	$96.40	$771.20	$61.30	$94.85
1 Structural Steel Worker	60.30	482.40	93.30	746.40		
16 L.H., Daily Totals		$980.80		$1517.60	$61.30	$94.85

Crew E-18	Hr.	Daily	Hr.	Daily	Bare Costs	Incl. O&P
1 Struc. Steel Foreman (outside)	$62.30	$498.40	$96.40	$771.20	$60.44	$92.84
3 Structural Steel Workers	60.30	1447.20	93.30	2239.20		
1 Equipment Operator (med.)	59.00	472.00	87.90	703.20		
1 Lattice Boom Crane, 20 Ton		1526.00		1678.60	38.15	41.97
40 L.H., Daily Totals		$3943.60		$5392.20	$98.59	$134.81

Crew E-19	Hr.	Daily	Hr.	Daily	Bare Costs	Incl. O&P
1 Struc. Steel Foreman (outside)	$62.30	$498.40	$96.40	$771.20	$59.37	$90.80
1 Structural Steel Worker	60.30	482.40	93.30	746.40		
1 Equip. Oper. (light)	55.50	444.00	82.70	661.60		
1 Lattice Boom Crane, 20 Ton		1526.00		1678.60	63.58	69.94
24 L.H., Daily Totals		$2950.80		$3857.80	$122.95	$160.74

Crew E-20	Hr.	Daily	Hr.	Daily	Bare Costs	Incl. O&P
1 Struc. Steel Foreman (outside)	$62.30	$498.40	$96.40	$771.20	$59.72	$91.59
5 Structural Steel Workers	60.30	2412.00	93.30	3732.00		
1 Equip. Oper. (crane)	61.45	491.60	91.55	732.40		
1 Equip. Oper. (oiler)	52.50	420.00	78.25	626.00		
1 Lattice Boom Crane, 40 Ton		2127.00		2339.70	33.23	36.56
64 L.H., Daily Totals		$5949.00		$8201.30	$92.95	$128.15

Crew E-22	Hr.	Daily	Hr.	Daily	Bare Costs	Incl. O&P
1 Skilled Worker Foreman (out)	$59.10	$472.80	$88.90	$711.20	$57.77	$86.90
2 Skilled Workers	57.10	913.60	85.90	1374.40		
24 L.H., Daily Totals		$1386.40		$2085.60	$57.77	$86.90

Crew E-24	Hr.	Daily	Hr.	Daily	Bare Costs	Incl. O&P
3 Structural Steel Workers	$60.30	$1447.20	$93.30	$2239.20	$59.98	$91.95
1 Equipment Operator (med.)	59.00	472.00	87.90	703.20		
1 Hyd. Crane, 25 Ton		586.70		645.37	18.33	20.17
32 L.H., Daily Totals		$2505.90		$3587.77	$78.31	$112.12

Crew E-25	Hr.	Daily	Hr.	Daily	Bare Costs	Incl. O&P
1 Welder Foreman (outside)	$62.30	$498.40	$96.40	$771.20	$62.30	$96.40
1 Cutting Torch		12.95		14.24	1.62	1.78
8 L.H., Daily Totals		$511.35		$785.45	$63.92	$98.18

Crew E-26	Hr.	Daily	Hr.	Daily	Bare Costs	Incl. O&P
1 Struc. Steel Foreman (outside)	$62.30	$498.40	$96.40	$771.20	$61.64	$94.84
1 Struc. Steel Worker	60.30	482.40	93.30	746.40		
1 Welder	60.30	482.40	93.30	746.40		
.25 Electrician	63.70	127.40	94.65	189.30		
.25 Plumber	67.70	135.40	101.05	202.10		
1 Welder, Gas Engine, 300 amp		148.75		163.63	5.31	5.84
28 L.H., Daily Totals		$1874.75		$2819.03	$66.96	$100.68

For customer support on your Electrical Costs with RSMeans Data, call 800.448.8182.

513

Crew No.	Bare Costs		Incl. Subs O&P		Cost Per Labor-Hour	

Crew E-27

Crew E-27	Hr.	Daily	Hr.	Daily	Bare Costs	Incl. O&P
1 Struc. Steel Foreman (outside)	$62.30	$498.40	$96.40	$771.20	$59.72	$91.59
5 Struc. Steel Workers	60.30	2412.00	93.30	3732.00		
1 Equip. Oper. (crane)	61.45	491.60	91.55	732.40		
1 Equip. Oper. (oiler)	52.50	420.00	78.25	626.00		
1 Hyd. Crane, 12 Ton		475.80		523.38		
1 Hyd. Crane, 80 Ton		1458.00		1603.80	30.22	33.24
64 L.H., Daily Totals		$5755.80		$7988.78	$89.93	$124.82

Crew F-3	Hr.	Daily	Hr.	Daily	Bare Costs	Incl. O&P
4 Carpenters	$54.70	$1750.40	$81.65	$2612.80	$56.05	$83.63
1 Equip. Oper. (crane)	61.45	491.60	91.55	732.40		
1 Hyd. Crane, 12 Ton		475.80		523.38	11.90	13.08
40 L.H., Daily Totals		$2717.80		$3868.58	$67.94	$96.71

Crew F-4	Hr.	Daily	Hr.	Daily	Bare Costs	Incl. O&P
4 Carpenters	$54.70	$1750.40	$81.65	$2612.80	$55.46	$82.73
1 Equip. Oper. (crane)	61.45	491.60	91.55	732.40		
1 Equip. Oper. (oiler)	52.50	420.00	78.25	626.00		
1 Hyd. Crane, 55 Ton		990.15		1089.17	20.63	22.69
48 L.H., Daily Totals		$3652.15		$5060.36	$76.09	$105.42

Crew F-5	Hr.	Daily	Hr.	Daily	Bare Costs	Incl. O&P
1 Carpenter Foreman (outside)	$56.70	$453.60	$84.60	$676.80	$55.20	$82.39
3 Carpenters	54.70	1312.80	81.65	1959.60		
32 L.H., Daily Totals		$1766.40		$2636.40	$55.20	$82.39

Crew F-6	Hr.	Daily	Hr.	Daily	Bare Costs	Incl. O&P
2 Carpenters	$54.70	$875.20	$81.65	$1306.40	$51.93	$77.47
2 Building Laborers	44.40	710.40	66.25	1060.00		
1 Equip. Oper. (crane)	61.45	491.60	91.55	732.40		
1 Hyd. Crane, 12 Ton		475.80		523.38	11.90	13.08
40 L.H., Daily Totals		$2553.00		$3622.18	$63.83	$90.55

Crew F-7	Hr.	Daily	Hr.	Daily	Bare Costs	Incl. O&P
2 Carpenters	$54.70	$875.20	$81.65	$1306.40	$49.55	$73.95
2 Building Laborers	44.40	710.40	66.25	1060.00		
32 L.H., Daily Totals		$1585.60		$2366.40	$49.55	$73.95

Crew G-1	Hr.	Daily	Hr.	Daily	Bare Costs	Incl. O&P
1 Roofer Foreman (outside)	$50.20	$401.60	$81.35	$650.80	$44.99	$72.92
4 Roofers Composition	48.20	1542.40	78.15	2500.80		
2 Roofer Helpers	35.95	575.20	58.25	932.00		
1 Application Equipment		194.80		214.28		
1 Tar Kettle/Pot		209.95		230.94		
1 Crew Truck		168.15		184.97	10.23	11.25
56 L.H., Daily Totals		$3092.10		$4713.79	$55.22	$84.17

Crew G-2	Hr.	Daily	Hr.	Daily	Bare Costs	Incl. O&P
1 Plasterer	$49.85	$398.80	$74.25	$594.00	$46.27	$68.95
1 Plasterer Helper	44.55	356.40	66.35	530.80		
1 Building Laborer	44.40	355.20	66.25	530.00		
1 Grout Pump, 50 C.F./hr.		190.35		209.38	7.93	8.72
24 L.H., Daily Totals		$1300.75		$1864.18	$54.20	$77.67

Crew G-2A	Hr.	Daily	Hr.	Daily	Bare Costs	Incl. O&P
1 Roofer Composition	$48.20	$385.60	$78.15	$625.20	$42.85	$67.55
1 Roofer Helper	35.95	287.60	58.25	466.00		
1 Building Laborer	44.40	355.20	66.25	530.00		
1 Foam Spray Rig, Trailer-Mtd.		530.15		583.16		
1 Pickup Truck, 3/4 Ton		112.20		123.42	26.76	29.44
24 L.H., Daily Totals		$1670.75		$2327.78	$69.61	$96.99

Crew G-3	Hr.	Daily	Hr.	Daily	Bare Costs	Incl. O&P
2 Sheet Metal Workers	$65.45	$1047.20	$98.70	$1579.20	$54.92	$82.47
2 Building Laborers	44.40	710.40	66.25	1060.00		
32 L.H., Daily Totals		$1757.60		$2639.20	$54.92	$82.47

Crew G-4	Hr.	Daily	Hr.	Daily	Bare Costs	Incl. O&P
1 Labor Foreman (outside)	$46.40	$371.20	$69.25	$554.00	$45.07	$67.25
2 Building Laborers	44.40	710.40	66.25	1060.00		
1 Flatbed Truck, Gas, 1.5 Ton		198.50		218.35		
1 Air Compressor, 160 cfm		212.30		233.53	17.12	18.83
24 L.H., Daily Totals		$1492.40		$2065.88	$62.18	$86.08

Crew G-5	Hr.	Daily	Hr.	Daily	Bare Costs	Incl. O&P
1 Roofer Foreman (outside)	$50.20	$401.60	$81.35	$650.80	$43.70	$70.83
2 Roofers Composition	48.20	771.20	78.15	1250.40		
2 Roofer Helpers	35.95	575.20	58.25	932.00		
1 Application Equipment		194.80		214.28	4.87	5.36
40 L.H., Daily Totals		$1942.80		$3047.48	$48.57	$76.19

Crew G-6A	Hr.	Daily	Hr.	Daily	Bare Costs	Incl. O&P
2 Roofers Composition	$48.20	$771.20	$78.15	$1250.40	$48.20	$78.15
1 Small Compressor, Electric		39.30		43.23		
2 Pneumatic Nailers		55.40		60.94	5.92	6.51
16 L.H., Daily Totals		$865.90		$1354.57	$54.12	$84.66

Crew G-7	Hr.	Daily	Hr.	Daily	Bare Costs	Incl. O&P
1 Carpenter	$54.70	$437.60	$81.65	$653.20	$54.70	$81.65
1 Small Compressor, Electric		39.30		43.23		
1 Pneumatic Nailer		27.70		30.47	8.38	9.21
8 L.H., Daily Totals		$504.60		$726.90	$63.08	$90.86

Crew H-1	Hr.	Daily	Hr.	Daily	Bare Costs	Incl. O&P
2 Glaziers	$52.65	$842.40	$78.40	$1254.40	$56.48	$85.85
2 Struc. Steel Workers	60.30	964.80	93.30	1492.80		
32 L.H., Daily Totals		$1807.20		$2747.20	$56.48	$85.85

Crew H-2	Hr.	Daily	Hr.	Daily	Bare Costs	Incl. O&P
2 Glaziers	$52.65	$842.40	$78.40	$1254.40	$49.90	$74.35
1 Building Laborer	44.40	355.20	66.25	530.00		
24 L.H., Daily Totals		$1197.60		$1784.40	$49.90	$74.35

Crew H-3	Hr.	Daily	Hr.	Daily	Bare Costs	Incl. O&P
1 Glazier	$52.65	$421.20	$78.40	$627.20	$47.35	$71.03
1 Helper	42.05	336.40	63.65	509.20		
16 L.H., Daily Totals		$757.60		$1136.40	$47.35	$71.03

Crew H-4	Hr.	Daily	Hr.	Daily	Bare Costs	Incl. O&P
1 Carpenter	$54.70	$437.60	$81.65	$653.20	$51.44	$77.05
1 Carpenter Helper	42.05	336.40	63.65	509.20		
.5 Electrician	63.70	254.80	94.65	378.60		
20 L.H., Daily Totals		$1028.80		$1541.00	$51.44	$77.05

For customer support on your Electrical Costs with RSMeans Data, call 800.448.8182.

Crews - Standard

Crew J-1

Crew No.	Bare Costs Hr.	Daily	Incl. Subs O&P Hr.	Daily	Cost Per Labor-Hour Bare Costs	Incl. O&P
3 Plasterers	$49.85	$1196.40	$74.25	$1782.00	$47.73	$71.09
2 Plasterer Helpers	44.55	712.80	66.35	1061.60		
1 Mixing Machine, 6 C.F.		113.35		124.69	2.83	3.12
40 L.H., Daily Totals		$2022.55		$2968.28	$50.56	$74.21

Crew J-2

Crew No.	Bare Costs Hr.	Daily	Incl. Subs O&P Hr.	Daily	Cost Per Labor-Hour Bare Costs	Incl. O&P
3 Plasterers	$49.85	$1196.40	$74.25	$1782.00	$48.79	$72.41
2 Plasterer Helpers	44.55	712.80	66.35	1061.60		
1 Lather	54.10	432.80	79.00	632.00		
1 Mixing Machine, 6 C.F.		113.35		124.69	2.36	2.60
48 L.H., Daily Totals		$2455.35		$3600.28	$51.15	$75.01

Crew J-3

Crew No.	Bare Costs Hr.	Daily	Incl. Subs O&P Hr.	Daily	Cost Per Labor-Hour Bare Costs	Incl. O&P
1 Terrazzo Worker	$51.75	$414.00	$75.60	$604.80	$47.65	$69.63
1 Terrazzo Helper	43.55	348.40	63.65	509.20		
1 Floor Grinder, 22" Path		96.05		105.66		
1 Terrazzo Mixer		162.90		179.19	16.18	17.80
16 L.H., Daily Totals		$1021.35		$1398.85	$63.83	$87.43

Crew J-4

Crew No.	Bare Costs Hr.	Daily	Incl. Subs O&P Hr.	Daily	Cost Per Labor-Hour Bare Costs	Incl. O&P
2 Cement Finishers	$51.80	$828.80	$75.90	$1214.40	$49.33	$72.68
1 Laborer	44.40	355.20	66.25	530.00		
1 Floor Grinder, 22" Path		96.05		105.66		
1 Floor Edger, 7" Path		44.05		48.45		
1 Vacuum Pick-Up System		74.95		82.44	8.96	9.86
24 L.H., Daily Totals		$1399.05		$1980.95	$58.29	$82.54

Crew J-4A

Crew No.	Bare Costs Hr.	Daily	Incl. Subs O&P Hr.	Daily	Cost Per Labor-Hour Bare Costs	Incl. O&P
2 Cement Finishers	$51.80	$828.80	$75.90	$1214.40	$48.10	$71.08
2 Laborers	44.40	710.40	66.25	1060.00		
1 Floor Grinder, 22" Path		96.05		105.66		
1 Floor Edger, 7" Path		44.05		48.45		
1 Vacuum Pick-Up System		74.95		82.44		
1 Floor Auto Scrubber		179.55		197.51	12.33	13.56
32 L.H., Daily Totals		$1933.80		$2708.46	$60.43	$84.64

Crew J-4B

Crew No.	Bare Costs Hr.	Daily	Incl. Subs O&P Hr.	Daily	Cost Per Labor-Hour Bare Costs	Incl. O&P
1 Laborer	$44.40	$355.20	$66.25	$530.00	$44.40	$66.25
1 Floor Auto Scrubber		179.55		197.51	22.44	24.69
8 L.H., Daily Totals		$534.75		$727.51	$66.84	$90.94

Crew J-6

Crew No.	Bare Costs Hr.	Daily	Incl. Subs O&P Hr.	Daily	Cost Per Labor-Hour Bare Costs	Incl. O&P
2 Painters	$46.45	$743.20	$68.90	$1102.40	$48.20	$71.69
1 Building Laborer	44.40	355.20	66.25	530.00		
1 Equip. Oper. (light)	55.50	444.00	82.70	661.60		
1 Air Compressor, 250 cfm		202.85		223.13		
1 Sandblaster, Portable, 3 C.F.		83.85		92.23		
1 Set Sand Blasting Accessories		15.55		17.11	9.45	10.39
32 L.H., Daily Totals		$1844.65		$2626.47	$57.65	$82.08

Crew J-7

Crew No.	Bare Costs Hr.	Daily	Incl. Subs O&P Hr.	Daily	Cost Per Labor-Hour Bare Costs	Incl. O&P
2 Painters	$46.45	$743.20	$68.90	$1102.40	$46.45	$68.90
1 Floor Belt Sander		50.20		55.22		
1 Floor Sanding Edger		25.20		27.72	4.71	5.18
16 L.H., Daily Totals		$818.60		$1185.34	$51.16	$74.08

Crew K-1

Crew No.	Bare Costs Hr.	Daily	Incl. Subs O&P Hr.	Daily	Cost Per Labor-Hour Bare Costs	Incl. O&P
1 Carpenter	$54.70	$437.60	$81.65	$653.20	$51.75	$77.33
1 Truck Driver (light)	48.80	390.40	73.00	584.00		
1 Flatbed Truck, Gas, 3 Ton		850.05		935.05	53.13	58.44
16 L.H., Daily Totals		$1678.05		$2172.26	$104.88	$135.77

Crew K-2

Crew No.	Bare Costs Hr.	Daily	Incl. Subs O&P Hr.	Daily	Cost Per Labor-Hour Bare Costs	Incl. O&P
1 Struc. Steel Foreman (outside)	$62.30	$498.40	$96.40	$771.20	$57.13	$87.57
1 Struc. Steel Worker	60.30	482.40	93.30	746.40		
1 Truck Driver (light)	48.80	390.40	73.00	584.00		
1 Flatbed Truck, Gas, 3 Ton		850.05		935.05	35.42	38.96
24 L.H., Daily Totals		$2221.25		$3036.66	$92.55	$126.53

Crew L-1

Crew No.	Bare Costs Hr.	Daily	Incl. Subs O&P Hr.	Daily	Cost Per Labor-Hour Bare Costs	Incl. O&P
1 Electrician	$63.70	$509.60	$94.65	$757.20	$65.70	$97.85
1 Plumber	67.70	541.60	101.05	808.40		
16 L.H., Daily Totals		$1051.20		$1565.60	$65.70	$97.85

Crew L-2

Crew No.	Bare Costs Hr.	Daily	Incl. Subs O&P Hr.	Daily	Cost Per Labor-Hour Bare Costs	Incl. O&P
1 Carpenter	$54.70	$437.60	$81.65	$653.20	$48.38	$72.65
1 Carpenter Helper	42.05	336.40	63.65	509.20		
16 L.H., Daily Totals		$774.00		$1162.40	$48.38	$72.65

Crew L-3

Crew No.	Bare Costs Hr.	Daily	Incl. Subs O&P Hr.	Daily	Cost Per Labor-Hour Bare Costs	Incl. O&P
1 Carpenter	$54.70	$437.60	$81.65	$653.20	$59.64	$89.16
.5 Electrician	63.70	254.80	94.65	378.60		
.5 Sheet Metal Worker	65.45	261.80	98.70	394.80		
16 L.H., Daily Totals		$954.20		$1426.60	$59.64	$89.16

Crew L-3A

Crew No.	Bare Costs Hr.	Daily	Incl. Subs O&P Hr.	Daily	Cost Per Labor-Hour Bare Costs	Incl. O&P
1 Carpenter Foreman (outside)	$56.70	$453.60	$84.60	$676.80	$59.62	$89.30
.5 Sheet Metal Worker	65.45	261.80	98.70	394.80		
12 L.H., Daily Totals		$715.40		$1071.60	$59.62	$89.30

Crew L-4

Crew No.	Bare Costs Hr.	Daily	Incl. Subs O&P Hr.	Daily	Cost Per Labor-Hour Bare Costs	Incl. O&P
2 Skilled Workers	$57.10	$913.60	$85.90	$1374.40	$52.08	$78.48
1 Helper	42.05	336.40	63.65	509.20		
24 L.H., Daily Totals		$1250.00		$1883.60	$52.08	$78.48

Crew L-5

Crew No.	Bare Costs Hr.	Daily	Incl. Subs O&P Hr.	Daily	Cost Per Labor-Hour Bare Costs	Incl. O&P
1 Struc. Steel Foreman (outside)	$62.30	$498.40	$96.40	$771.20	$60.75	$93.49
5 Struc. Steel Workers	60.30	2412.00	93.30	3732.00		
1 Equip. Oper. (crane)	61.45	491.60	91.55	732.40		
1 Hyd. Crane, 25 Ton		586.70		645.37	10.48	11.52
56 L.H., Daily Totals		$3988.70		$5880.97	$71.23	$105.02

Crew L-5A

Crew No.	Bare Costs Hr.	Daily	Incl. Subs O&P Hr.	Daily	Cost Per Labor-Hour Bare Costs	Incl. O&P
1 Struc. Steel Foreman (outside)	$62.30	$498.40	$96.40	$771.20	$61.09	$93.64
2 Structural Steel Workers	60.30	964.80	93.30	1492.80		
1 Equip. Oper. (crane)	61.45	491.60	91.55	732.40		
1 S.P. Crane, 4x4, 25 Ton		1155.00		1270.50	36.09	39.70
32 L.H., Daily Totals		$3109.80		$4266.90	$97.18	$133.34

For customer support on your Electrical Costs with RSMeans Data, call 800.448.8182.

515

Crew No.	Bare Costs		Incl. Subs O&P		Cost Per Labor-Hour	
Crew L-5B	Hr.	Daily	Hr.	Daily	Bare Costs	Incl. O&P
1 Struc. Steel Foreman (outside)	$62.30	$498.40	$96.40	$771.20	$62.33	$94.01
2 Structural Steel Workers	60.30	964.80	93.30	1492.80		
2 Electricians	63.70	1019.20	94.65	1514.40		
2 Steamfitters/Pipefitters	68.35	1093.60	102.00	1632.00		
1 Equip. Oper. (crane)	61.45	491.60	91.55	732.40		
1 Equip. Oper. (oiler)	52.50	420.00	78.25	626.00		
1 Hyd. Crane, 80 Ton		1458.00		1603.80	20.25	22.27
72 L.H., Daily Totals		$5945.60		$8372.60	$82.58	$116.29
Crew L-6	Hr.	Daily	Hr.	Daily	Bare Costs	Incl. O&P
1 Plumber	$67.70	$541.60	$101.05	$808.40	$66.37	$98.92
.5 Electrician	63.70	254.80	94.65	378.60		
12 L.H., Daily Totals		$796.40		$1187.00	$66.37	$98.92
Crew L-7	Hr.	Daily	Hr.	Daily	Bare Costs	Incl. O&P
2 Carpenters	$54.70	$875.20	$81.65	$1306.40	$53.04	$79.11
1 Building Laborer	44.40	355.20	66.25	530.00		
.5 Electrician	63.70	254.80	94.65	378.60		
28 L.H., Daily Totals		$1485.20		$2215.00	$53.04	$79.11
Crew L-8	Hr.	Daily	Hr.	Daily	Bare Costs	Incl. O&P
2 Carpenters	$54.70	$875.20	$81.65	$1306.40	$57.30	$85.53
.5 Plumber	67.70	270.80	101.05	404.20		
20 L.H., Daily Totals		$1146.00		$1710.60	$57.30	$85.53
Crew L-9	Hr.	Daily	Hr.	Daily	Bare Costs	Incl. O&P
1 Labor Foreman (inside)	$44.90	$359.20	$67.00	$536.00	$50.19	$75.58
2 Building Laborers	44.40	710.40	66.25	1060.00		
1 Struc. Steel Worker	60.30	482.40	93.30	746.40		
.5 Electrician	63.70	254.80	94.65	378.60		
36 L.H., Daily Totals		$1806.80		$2721.00	$50.19	$75.58
Crew L-10	Hr.	Daily	Hr.	Daily	Bare Costs	Incl. O&P
1 Struc. Steel Foreman (outside)	$62.30	$498.40	$96.40	$771.20	$61.35	$93.75
1 Structural Steel Worker	60.30	482.40	93.30	746.40		
1 Equip. Oper. (crane)	61.45	491.60	91.55	732.40		
1 Hyd. Crane, 12 Ton		475.80		523.38	19.82	21.81
24 L.H., Daily Totals		$1948.20		$2773.38	$81.17	$115.56
Crew L-11	Hr.	Daily	Hr.	Daily	Bare Costs	Incl. O&P
2 Wreckers	$44.40	$710.40	$67.40	$1078.40	$51.44	$77.26
1 Equip. Oper. (crane)	61.45	491.60	91.55	732.40		
1 Equip. Oper. (light)	55.50	444.00	82.70	661.60		
1 Hyd. Excavator, 2.5 C.Y.		1567.00		1723.70		
1 Loader, Skid Steer, 78 H.P.		446.30		490.93	62.92	69.21
32 L.H., Daily Totals		$3659.30		$4687.03	$114.35	$146.47
Crew M-1	Hr.	Daily	Hr.	Daily	Bare Costs	Incl. O&P
3 Elevator Constructors	$90.30	$2167.20	$133.80	$3211.20	$85.79	$127.11
1 Elevator Apprentice	72.25	578.00	107.05	856.40		
5 Hand Tools		50.50		55.55	1.58	1.74
32 L.H., Daily Totals		$2795.70		$4123.15	$87.37	$128.85

Crew No.	Bare Costs		Incl. Subs O&P		Cost Per Labor-Hour	
Crew M-3	Hr.	Daily	Hr.	Daily	Bare Costs	Incl. O&P
1 Electrician Foreman (outside)	$65.70	$525.60	$97.65	$781.20	$67.62	$100.41
1 Common Laborer	44.40	355.20	66.25	530.00		
.25 Equipment Operator (med.)	59.00	118.00	87.90	175.80		
1 Elevator Constructor	90.30	722.40	133.80	1070.40		
1 Elevator Apprentice	72.25	578.00	107.05	856.40		
.25 S.P. Crane, 4x4, 20 Ton		143.59		157.95	4.22	4.65
34 L.H., Daily Totals		$2442.79		$3571.75	$71.85	$105.05
Crew M-4	Hr.	Daily	Hr.	Daily	Bare Costs	Incl. O&P
1 Electrician Foreman (outside)	$65.70	$525.60	$97.65	$781.20	$66.92	$99.38
1 Common Laborer	44.40	355.20	66.25	530.00		
.25 Equipment Operator, Crane	61.45	122.90	91.55	183.10		
.25 Equip. Oper. (oiler)	52.50	105.00	78.25	156.50		
1 Elevator Constructor	90.30	722.40	133.80	1070.40		
1 Elevator Apprentice	72.25	578.00	107.05	856.40		
.25 S.P. Crane, 4x4, 40 Ton		190.45		209.50	5.29	5.82
36 L.H., Daily Totals		$2599.55		$3787.09	$72.21	$105.20
Crew Q-1	Hr.	Daily	Hr.	Daily	Bare Costs	Incl. O&P
1 Plumber	$67.70	$541.60	$101.05	$808.40	$60.92	$90.92
1 Plumber Apprentice	54.15	433.20	80.80	646.40		
16 L.H., Daily Totals		$974.80		$1454.80	$60.92	$90.92
Crew Q-1A	Hr.	Daily	Hr.	Daily	Bare Costs	Incl. O&P
.25 Plumber Foreman (outside)	$69.70	$139.40	$104.00	$208.00	$68.10	$101.64
1 Plumber	67.70	541.60	101.05	808.40		
10 L.H., Daily Totals		$681.00		$1016.40	$68.10	$101.64
Crew Q-1C	Hr.	Daily	Hr.	Daily	Bare Costs	Incl. O&P
1 Plumber	$67.70	$541.60	$101.05	$808.40	$60.28	$89.92
1 Plumber Apprentice	54.15	433.20	80.80	646.40		
1 Equip. Oper. (medium)	59.00	472.00	87.90	703.20		
1 Trencher, Chain Type, 8' D		1894.00		2083.40	78.92	86.81
24 L.H., Daily Totals		$3340.80		$4241.40	$139.20	$176.72
Crew Q-2	Hr.	Daily	Hr.	Daily	Bare Costs	Incl. O&P
2 Plumbers	$67.70	$1083.20	$101.05	$1616.80	$63.18	$94.30
1 Plumber Apprentice	54.15	433.20	80.80	646.40		
24 L.H., Daily Totals		$1516.40		$2263.20	$63.18	$94.30
Crew Q-3	Hr.	Daily	Hr.	Daily	Bare Costs	Incl. O&P
1 Plumber Foreman (inside)	$68.20	$545.60	$101.80	$814.40	$64.44	$96.17
2 Plumbers	67.70	1083.20	101.05	1616.80		
1 Plumber Apprentice	54.15	433.20	80.80	646.40		
32 L.H., Daily Totals		$2062.00		$3077.60	$64.44	$96.17
Crew Q-4	Hr.	Daily	Hr.	Daily	Bare Costs	Incl. O&P
1 Plumber Foreman (inside)	$68.20	$545.60	$101.80	$814.40	$64.44	$96.17
1 Plumber	67.70	541.60	101.05	808.40		
1 Welder (plumber)	67.70	541.60	101.05	808.40		
1 Plumber Apprentice	54.15	433.20	80.80	646.40		
1 Welder, Electric, 300 amp		107.55		118.31	3.36	3.70
32 L.H., Daily Totals		$2169.55		$3195.91	$67.80	$99.87
Crew Q-5	Hr.	Daily	Hr.	Daily	Bare Costs	Incl. O&P
1 Steamfitter	$68.35	$546.80	$102.00	$816.00	$61.52	$91.83
1 Steamfitter Apprentice	54.70	437.60	81.65	653.20		
16 L.H., Daily Totals		$984.40		$1469.20	$61.52	$91.83

For customer support on your **Electrical Costs** with RSMeans Data, call 800.448.8182.

Crew Q-6

	Bare Costs Hr.	Daily	Incl. Subs O&P Hr.	Daily	Bare Costs	Incl. O&P
2 Steamfitters	$68.35	$1093.60	$102.00	$1632.00	$63.80	$95.22
1 Steamfitter Apprentice	54.70	437.60	81.65	653.20		
24 L.H., Daily Totals		$1531.20		$2285.20	$63.80	$95.22

Crew Q-7

	Bare Costs Hr.	Daily	Incl. Subs O&P Hr.	Daily	Bare Costs	Incl. O&P
1 Steamfitter Foreman (inside)	$68.85	$550.80	$102.75	$822.00	$65.06	$97.10
2 Steamfitters	68.35	1093.60	102.00	1632.00		
1 Steamfitter Apprentice	54.70	437.60	81.65	653.20		
32 L.H., Daily Totals		$2082.00		$3107.20	$65.06	$97.10

Crew Q-8

	Bare Costs Hr.	Daily	Incl. Subs O&P Hr.	Daily	Bare Costs	Incl. O&P
1 Steamfitter Foreman (inside)	$68.85	$550.80	$102.75	$822.00	$65.06	$97.10
1 Steamfitter	68.35	546.80	102.00	816.00		
1 Welder (steamfitter)	68.35	546.80	102.00	816.00		
1 Steamfitter Apprentice	54.70	437.60	81.65	653.20		
1 Welder, Electric, 300 amp		107.55		118.31	3.36	3.70
32 L.H., Daily Totals		$2189.55		$3225.51	$68.42	$100.80

Crew Q-9

	Bare Costs Hr.	Daily	Incl. Subs O&P Hr.	Daily	Bare Costs	Incl. O&P
1 Sheet Metal Worker	$65.45	$523.60	$98.70	$789.60	$58.90	$88.83
1 Sheet Metal Apprentice	52.35	418.80	78.95	631.60		
16 L.H., Daily Totals		$942.40		$1421.20	$58.90	$88.83

Crew Q-10

	Bare Costs Hr.	Daily	Incl. Subs O&P Hr.	Daily	Bare Costs	Incl. O&P
2 Sheet Metal Workers	$65.45	$1047.20	$98.70	$1579.20	$61.08	$92.12
1 Sheet Metal Apprentice	52.35	418.80	78.95	631.60		
24 L.H., Daily Totals		$1466.00		$2210.80	$61.08	$92.12

Crew Q-11

	Bare Costs Hr.	Daily	Incl. Subs O&P Hr.	Daily	Bare Costs	Incl. O&P
1 Sheet Metal Foreman (inside)	$65.95	$527.60	$99.50	$796.00	$62.30	$93.96
2 Sheet Metal Workers	65.45	1047.20	98.70	1579.20		
1 Sheet Metal Apprentice	52.35	418.80	78.95	631.60		
32 L.H., Daily Totals		$1993.60		$3006.80	$62.30	$93.96

Crew Q-12

	Bare Costs Hr.	Daily	Incl. Subs O&P Hr.	Daily	Bare Costs	Incl. O&P
1 Sprinkler Installer	$66.50	$532.00	$99.35	$794.80	$59.85	$89.42
1 Sprinkler Apprentice	53.20	425.60	79.50	636.00		
16 L.H., Daily Totals		$957.60		$1430.80	$59.85	$89.42

Crew Q-13

	Bare Costs Hr.	Daily	Incl. Subs O&P Hr.	Daily	Bare Costs	Incl. O&P
1 Sprinkler Foreman (inside)	$67.00	$536.00	$100.10	$800.80	$63.30	$94.58
2 Sprinkler Installers	66.50	1064.00	99.35	1589.60		
1 Sprinkler Apprentice	53.20	425.60	79.50	636.00		
32 L.H., Daily Totals		$2025.60		$3026.40	$63.30	$94.58

Crew Q-14

	Bare Costs Hr.	Daily	Incl. Subs O&P Hr.	Daily	Bare Costs	Incl. O&P
1 Asbestos Worker	$60.95	$487.60	$93.25	$746.00	$54.85	$83.90
1 Asbestos Apprentice	48.75	390.00	74.55	596.40		
16 L.H., Daily Totals		$877.60		$1342.40	$54.85	$83.90

Crew Q-15

	Bare Costs Hr.	Daily	Incl. Subs O&P Hr.	Daily	Bare Costs	Incl. O&P
1 Plumber	$67.70	$541.60	$101.05	$808.40	$60.92	$90.92
1 Plumber Apprentice	54.15	433.20	80.80	646.40		
1 Welder, Electric, 300 amp		107.55		118.31	6.72	7.39
16 L.H., Daily Totals		$1082.35		$1573.11	$67.65	$98.32

Crew Q-16

	Bare Costs Hr.	Daily	Incl. Subs O&P Hr.	Daily	Bare Costs	Incl. O&P
2 Plumbers	$67.70	$1083.20	$101.05	$1616.80	$63.18	$94.30
1 Plumber Apprentice	54.15	433.20	80.80	646.40		
1 Welder, Electric, 300 amp		107.55		118.31	4.48	4.93
24 L.H., Daily Totals		$1623.95		$2381.51	$67.66	$99.23

Crew Q-17

	Bare Costs Hr.	Daily	Incl. Subs O&P Hr.	Daily	Bare Costs	Incl. O&P
1 Steamfitter	$68.35	$546.80	$102.00	$816.00	$61.52	$91.83
1 Steamfitter Apprentice	54.70	437.60	81.65	653.20		
1 Welder, Electric, 300 amp		107.55		118.31	6.72	7.39
16 L.H., Daily Totals		$1091.95		$1587.51	$68.25	$99.22

Crew Q-17A

	Bare Costs Hr.	Daily	Incl. Subs O&P Hr.	Daily	Bare Costs	Incl. O&P
1 Steamfitter	$68.35	$546.80	$102.00	$816.00	$61.50	$91.73
1 Steamfitter Apprentice	54.70	437.60	81.65	653.20		
1 Equip. Oper. (crane)	61.45	491.60	91.55	732.40		
1 Hyd. Crane, 12 Ton		475.80		523.38		
1 Welder, Electric, 300 amp		107.55		118.31	24.31	26.74
24 L.H., Daily Totals		$2059.35		$2843.28	$85.81	$118.47

Crew Q-18

	Bare Costs Hr.	Daily	Incl. Subs O&P Hr.	Daily	Bare Costs	Incl. O&P
2 Steamfitters	$68.35	$1093.60	$102.00	$1632.00	$63.80	$95.22
1 Steamfitter Apprentice	54.70	437.60	81.65	653.20		
1 Welder, Electric, 300 amp		107.55		118.31	4.48	4.93
24 L.H., Daily Totals		$1638.75		$2403.51	$68.28	$100.15

Crew Q-19

	Bare Costs Hr.	Daily	Incl. Subs O&P Hr.	Daily	Bare Costs	Incl. O&P
1 Steamfitter	$68.35	$546.80	$102.00	$816.00	$62.25	$92.77
1 Steamfitter Apprentice	54.70	437.60	81.65	653.20		
1 Electrician	63.70	509.60	94.65	757.20		
24 L.H., Daily Totals		$1494.00		$2226.40	$62.25	$92.77

Crew Q-20

	Bare Costs Hr.	Daily	Incl. Subs O&P Hr.	Daily	Bare Costs	Incl. O&P
1 Sheet Metal Worker	$65.45	$523.60	$98.70	$789.60	$59.86	$89.99
1 Sheet Metal Apprentice	52.35	418.80	78.95	631.60		
.5 Electrician	63.70	254.80	94.65	378.60		
20 L.H., Daily Totals		$1197.20		$1799.80	$59.86	$89.99

Crew Q-21

	Bare Costs Hr.	Daily	Incl. Subs O&P Hr.	Daily	Bare Costs	Incl. O&P
2 Steamfitters	$68.35	$1093.60	$102.00	$1632.00	$63.77	$95.08
1 Steamfitter Apprentice	54.70	437.60	81.65	653.20		
1 Electrician	63.70	509.60	94.65	757.20		
32 L.H., Daily Totals		$2040.80		$3042.40	$63.77	$95.08

Crew Q-22

	Bare Costs Hr.	Daily	Incl. Subs O&P Hr.	Daily	Bare Costs	Incl. O&P
1 Plumber	$67.70	$541.60	$101.05	$808.40	$60.92	$90.92
1 Plumber Apprentice	54.15	433.20	80.80	646.40		
1 Hyd. Crane, 12 Ton		475.80		523.38	29.74	32.71
16 L.H., Daily Totals		$1450.60		$1978.18	$90.66	$123.64

Crew Q-22A

	Bare Costs Hr.	Daily	Incl. Subs O&P Hr.	Daily	Bare Costs	Incl. O&P
1 Plumber	$67.70	$541.60	$101.05	$808.40	$56.92	$84.91
1 Plumber Apprentice	54.15	433.20	80.80	646.40		
1 Laborer	44.40	355.20	66.25	530.00		
1 Equip. Oper. (crane)	61.45	491.60	91.55	732.40		
1 Hyd. Crane, 12 Ton		475.80		523.38	14.87	16.36
32 L.H., Daily Totals		$2297.40		$3240.58	$71.79	$101.27

For customer support on your Electrical Costs with RSMeans Data, call 800.448.8182.

517

Crews - Standard

Crew No.	Bare Costs Hr.	Daily	Incl. Subs O&P Hr.	Daily	Cost Per Labor-Hour Bare Costs	Incl. O&P
Crew Q-23	**Hr.**	**Daily**	**Hr.**	**Daily**	**Bare Costs**	**Incl. O&P**
1 Plumber Foreman (outside)	$69.70	$557.60	$104.00	$832.00	$65.47	$97.65
1 Plumber	67.70	541.60	101.05	808.40		
1 Equip. Oper. (medium)	59.00	472.00	87.90	703.20		
1 Lattice Boom Crane, 20 Ton		1526.00		1678.60	63.58	69.94
24 L.H., Daily Totals		$3097.20		$4022.20	$129.05	$167.59
Crew R-1	**Hr.**	**Daily**	**Hr.**	**Daily**	**Bare Costs**	**Incl. O&P**
1 Electrician Foreman	$64.20	$513.60	$95.40	$763.20	$59.53	$88.46
3 Electricians	63.70	1528.80	94.65	2271.60		
2 Electrician Apprentices	50.95	815.20	75.70	1211.20		
48 L.H., Daily Totals		$2857.60		$4246.00	$59.53	$88.46
Crew R-1A	**Hr.**	**Daily**	**Hr.**	**Daily**	**Bare Costs**	**Incl. O&P**
1 Electrician	$63.70	$509.60	$94.65	$757.20	$57.33	$85.17
1 Electrician Apprentice	50.95	407.60	75.70	605.60		
16 L.H., Daily Totals		$917.20		$1362.80	$57.33	$85.17
Crew R-1B	**Hr.**	**Daily**	**Hr.**	**Daily**	**Bare Costs**	**Incl. O&P**
1 Electrician	$63.70	$509.60	$94.65	$757.20	$55.20	$82.02
2 Electrician Apprentices	50.95	815.20	75.70	1211.20		
24 L.H., Daily Totals		$1324.80		$1968.40	$55.20	$82.02
Crew R-1C	**Hr.**	**Daily**	**Hr.**	**Daily**	**Bare Costs**	**Incl. O&P**
2 Electricians	$63.70	$1019.20	$94.65	$1514.40	$57.33	$85.17
2 Electrician Apprentices	50.95	815.20	75.70	1211.20		
1 Portable cable puller, 8000 lb.		102.65		112.92	3.21	3.53
32 L.H., Daily Totals		$1937.05		$2838.51	$60.53	$88.70
Crew R-2	**Hr.**	**Daily**	**Hr.**	**Daily**	**Bare Costs**	**Incl. O&P**
1 Electrician Foreman	$64.20	$513.60	$95.40	$763.20	$59.81	$88.90
3 Electricians	63.70	1528.80	94.65	2271.60		
2 Electrician Apprentices	50.95	815.20	75.70	1211.20		
1 Equip. Oper. (crane)	61.45	491.60	91.55	732.40		
1 S.P. Crane, 4x4, 5 Ton		381.95		420.14	6.82	7.50
56 L.H., Daily Totals		$3731.15		$5398.55	$66.63	$96.40
Crew R-3	**Hr.**	**Daily**	**Hr.**	**Daily**	**Bare Costs**	**Incl. O&P**
1 Electrician Foreman	$64.20	$513.60	$95.40	$763.20	$63.45	$94.33
1 Electrician	63.70	509.60	94.65	757.20		
.5 Equip. Oper. (crane)	61.45	245.80	91.55	366.20		
.5 S.P. Crane, 4x4, 5 Ton		190.97		210.07	9.55	10.50
20 L.H., Daily Totals		$1459.97		$2096.67	$73.00	$104.83
Crew R-4	**Hr.**	**Daily**	**Hr.**	**Daily**	**Bare Costs**	**Incl. O&P**
1 Struc. Steel Foreman (outside)	$62.30	$498.40	$96.40	$771.20	$61.38	$94.19
3 Struc. Steel Workers	60.30	1447.20	93.30	2239.20		
1 Electrician	63.70	509.60	94.65	757.20		
1 Welder, Gas Engine, 300 amp		148.75		163.63	3.72	4.09
40 L.H., Daily Totals		$2603.95		$3931.22	$65.10	$98.28

Crew No.	Bare Costs Hr.	Daily	Incl. Subs O&P Hr.	Daily	Cost Per Labor-Hour Bare Costs	Incl. O&P
Crew R-5	**Hr.**	**Daily**	**Hr.**	**Daily**	**Bare Costs**	**Incl. O&P**
1 Electrician Foreman	$64.20	$513.60	$95.40	$763.20	$55.87	$83.45
4 Electrician Linemen	63.70	2038.40	94.65	3028.80		
2 Electrician Operators	63.70	1019.20	94.65	1514.40		
4 Electrician Groundmen	42.05	1345.60	63.65	2036.80		
1 Crew Truck		168.15		184.97		
1 Flatbed Truck, 20,000 GVW		204.05		224.46		
1 Pickup Truck, 3/4 Ton		112.20		123.42		
.2 Hyd. Crane, 55 Ton		198.03		217.83		
.2 Hyd. Crane, 12 Ton		95.16		104.68		
.2 Earth Auger, Truck-Mtd.		40.51		44.56		
1 Tractor w/Winch		377.65		415.42	13.59	14.95
88 L.H., Daily Totals		$6112.55		$8658.52	$69.46	$98.39
Crew R-6	**Hr.**	**Daily**	**Hr.**	**Daily**	**Bare Costs**	**Incl. O&P**
1 Electrician Foreman	$64.20	$513.60	$95.40	$763.20	$55.87	$83.45
4 Electrician Linemen	63.70	2038.40	94.65	3028.80		
2 Electrician Operators	63.70	1019.20	94.65	1514.40		
4 Electrician Groundmen	42.05	1345.60	63.65	2036.80		
1 Crew Truck		168.15		184.97		
1 Flatbed Truck, 20,000 GVW		204.05		224.46		
1 Pickup Truck, 3/4 Ton		112.20		123.42		
.2 Hyd. Crane, 55 Ton		198.03		217.83		
.2 Hyd. Crane, 12 Ton		95.16		104.68		
.2 Earth Auger, Truck-Mtd.		40.51		44.56		
1 Tractor w/Winch		377.65		415.42		
3 Cable Trailers		194.25		213.68		
.5 Tensioning Rig		55.75		61.33		
.5 Cable Pulling Rig		306.65		337.32	19.91	21.91
88 L.H., Daily Totals		$6669.20		$9270.84	$75.79	$105.35
Crew R-7	**Hr.**	**Daily**	**Hr.**	**Daily**	**Bare Costs**	**Incl. O&P**
1 Electrician Foreman	$64.20	$513.60	$95.40	$763.20	$45.74	$68.94
5 Electrician Groundmen	42.05	1682.00	63.65	2546.00		
1 Crew Truck		168.15		184.97	3.50	3.85
48 L.H., Daily Totals		$2363.75		$3494.17	$49.24	$72.80
Crew R-8	**Hr.**	**Daily**	**Hr.**	**Daily**	**Bare Costs**	**Incl. O&P**
1 Electrician Foreman	$64.20	$513.60	$95.40	$763.20	$56.57	$84.44
3 Electrician Linemen	63.70	1528.80	94.65	2271.60		
2 Electrician Groundmen	42.05	672.80	63.65	1018.40		
1 Pickup Truck, 3/4 Ton		112.20		123.42		
1 Crew Truck		168.15		184.97	5.84	6.42
48 L.H., Daily Totals		$2995.55		$4361.59	$62.41	$90.87
Crew R-9	**Hr.**	**Daily**	**Hr.**	**Daily**	**Bare Costs**	**Incl. O&P**
1 Electrician Foreman	$64.20	$513.60	$95.40	$763.20	$52.94	$79.24
1 Electrician Lineman	63.70	509.60	94.65	757.20		
2 Electrician Operators	63.70	1019.20	94.65	1514.40		
4 Electrician Groundmen	42.05	1345.60	63.65	2036.80		
1 Pickup Truck, 3/4 Ton		112.20		123.42		
1 Crew Truck		168.15		184.97	4.38	4.82
64 L.H., Daily Totals		$3668.35		$5379.98	$57.32	$84.06
Crew R-10	**Hr.**	**Daily**	**Hr.**	**Daily**	**Bare Costs**	**Incl. O&P**
1 Electrician Foreman	$64.20	$513.60	$95.40	$763.20	$60.17	$89.61
4 Electrician Linemen	63.70	2038.40	94.65	3028.80		
1 Electrician Groundman	42.05	336.40	63.65	509.20		
1 Crew Truck		168.15		184.97		
3 Tram Cars		219.60		241.56	8.08	8.89
48 L.H., Daily Totals		$3276.15		$4727.73	$68.25	$98.49

Crew No.	Bare Costs		Incl. Subs O&P		Cost Per Labor-Hour	

Crew R-11	Hr.	Daily	Hr.	Daily	Bare Costs	Incl. O&P
1 Electrician Foreman	$64.20	$513.60	$95.40	$763.20	$60.69	$90.26
4 Electricians	63.70	2038.40	94.65	3028.80		
1 Equip. Oper. (crane)	61.45	491.60	91.55	732.40		
1 Common Laborer	44.40	355.20	66.25	530.00		
1 Crew Truck		168.15		184.97		
1 Hyd. Crane, 12 Ton		475.80		523.38	11.50	12.65
56 L.H., Daily Totals		$4042.75		$5762.74	$72.19	$102.91

Crew R-12	Hr.	Daily	Hr.	Daily	Bare Costs	Incl. O&P
1 Carpenter Foreman (inside)	$55.20	$441.60	$82.40	$659.20	$51.90	$77.75
4 Carpenters	54.70	1750.40	81.65	2612.80		
4 Common Laborers	44.40	1420.80	66.25	2120.00		
1 Equip. Oper. (medium)	59.00	472.00	87.90	703.20		
1 Steel Worker	60.30	482.40	93.30	746.40		
1 Dozer, 200 H.P.		1520.00		1672.00		
1 Pickup Truck, 3/4 Ton		112.20		123.42	18.55	20.40
88 L.H., Daily Totals		$6199.40		$8637.02	$70.45	$98.15

Crew R-13	Hr.	Daily	Hr.	Daily	Bare Costs	Incl. O&P
1 Electrician Foreman	$64.20	$513.60	$95.40	$763.20	$61.55	$91.52
3 Electricians	63.70	1528.80	94.65	2271.60		
.25 Equip. Oper. (crane)	61.45	122.90	91.55	183.10		
1 Equipment Oiler	52.50	420.00	78.25	626.00		
.25 Hydraulic Crane, 33 Ton		245.79		270.37	5.85	6.44
42 L.H., Daily Totals		$2831.09		$4114.27	$67.41	$97.96

Crew R-15	Hr.	Daily	Hr.	Daily	Bare Costs	Incl. O&P
1 Electrician Foreman	$64.20	$513.60	$95.40	$763.20	$62.42	$92.78
4 Electricians	63.70	2038.40	94.65	3028.80		
1 Equipment Oper. (light)	55.50	444.00	82.70	661.60		
1 Telescoping Boom Lift, to 40'		281.90		310.09	5.87	6.46
48 L.H., Daily Totals		$3277.90		$4763.69	$68.29	$99.24

Crew R-15A	Hr.	Daily	Hr.	Daily	Bare Costs	Incl. O&P
1 Electrician Foreman	$64.20	$513.60	$95.40	$763.20	$55.98	$83.32
2 Electricians	63.70	1019.20	94.65	1514.40		
2 Common Laborers	44.40	710.40	66.25	1060.00		
1 Equip. Oper. (light)	55.50	444.00	82.70	661.60		
1 Telescoping Boom Lift, to 40'		281.90		310.09	5.87	6.46
48 L.H., Daily Totals		$2969.10		$4309.29	$61.86	$89.78

Crew R-18	Hr.	Daily	Hr.	Daily	Bare Costs	Incl. O&P
.25 Electrician Foreman	$64.20	$128.40	$95.40	$190.80	$55.89	$83.05
1 Electrician	63.70	509.60	94.65	757.20		
2 Electrician Apprentices	50.95	815.20	75.70	1211.20		
26 L.H., Daily Totals		$1453.20		$2159.20	$55.89	$83.05

Crew R-19	Hr.	Daily	Hr.	Daily	Bare Costs	Incl. O&P
.5 Electrician Foreman	$64.20	$256.80	$95.40	$381.60	$63.80	$94.80
2 Electricians	63.70	1019.20	94.65	1514.40		
20 L.H., Daily Totals		$1276.00		$1896.00	$63.80	$94.80

Crew R-21	Hr.	Daily	Hr.	Daily	Bare Costs	Incl. O&P
1 Electrician Foreman	$64.20	$513.60	$95.40	$763.20	$63.71	$94.67
3 Electricians	63.70	1528.80	94.65	2271.60		
.1 Equip. Oper. (medium)	59.00	47.20	87.90	70.32		
.1 S.P. Crane, 4x4, 25 Ton		115.50		127.05	3.52	3.87
32.8 L.H., Daily Totals		$2205.10		$3232.17	$67.23	$98.54

Crew R-22	Hr.	Daily	Hr.	Daily	Bare Costs	Incl. O&P
.66 Electrician Foreman	$64.20	$338.98	$95.40	$503.71	$58.30	$86.62
2 Electricians	63.70	1019.20	94.65	1514.40		
2 Electrician Apprentices	50.95	815.20	75.70	1211.20		
37.28 L.H., Daily Totals		$2173.38		$3229.31	$58.30	$86.62

Crew R-30	Hr.	Daily	Hr.	Daily	Bare Costs	Incl. O&P
.25 Electrician Foreman (outside)	$65.70	$131.40	$97.65	$195.30	$51.98	$77.40
1 Electrician	63.70	509.60	94.65	757.20		
2 Laborers (Semi-Skilled)	44.40	710.40	66.25	1060.00		
26 L.H., Daily Totals		$1351.40		$2012.50	$51.98	$77.40

Crew R-31	Hr.	Daily	Hr.	Daily	Bare Costs	Incl. O&P
1 Electrician	$63.70	$509.60	$94.65	$757.20	$63.70	$94.65
1 Core Drill, Electric, 2.5 H.P.		62.65		68.92	7.83	8.61
8 L.H., Daily Totals		$572.25		$826.12	$71.53	$103.26

Crew W-41E	Hr.	Daily	Hr.	Daily	Bare Costs	Incl. O&P
.5 Plumber Foreman (outside)	$69.70	$278.80	$104.00	$416.00	$58.78	$87.72
1 Plumber	67.70	541.60	101.05	808.40		
1 Laborer	44.40	355.20	66.25	530.00		
20 L.H., Daily Totals		$1175.60		$1754.40	$58.78	$87.72

Historical Cost Indexes

The table below lists both the RSMeans® historical cost index based on Jan. 1, 1993 = 100 as well as the computed value of an index based on Jan. 1, 2021 costs. Since the Jan. 1, 2021 figure is estimated, space is left to write in the actual index figures as they become available through the quarterly *RSMeans Construction Cost Indexes*.

To compute the actual index based on Jan. 1, 2021 = 100, divide the historical cost index for a particular year by the actual Jan. 1, 2021 construction cost index. Space has been left to advance the index figures as the year progresses.

Year	Historical Cost Index Jan. 1, 1993 = 100		Current Index Based on Jan. 1, 2021 = 100		Year	Historical Cost Index Jan. 1, 1993 = 100	Current Index Based on Jan. 1, 2021 = 100		Year	Historical Cost Index Jan. 1, 1993 = 100	Current Index Based on Jan. 1, 2021 = 100	
	Est.	Actual	Est.	Actual		Actual	Est.	Actual		Actual	Est.	Actual
Oct 2021					July 2004	143.7	60.7		1984	82.0	34.6	
July 2021					2003	132.0	55.8		1983	80.2	33.9	
April 2021					2002	128.7	54.4		1982	76.1	32.2	
Jan 2021	236.7		100.0		2001	125.1	52.9		1981	70.0	29.6	
2020		234.6	99.1		2000	120.9	51.1		1980	62.9	26.6	
2019		232.2	98.1		1999	117.6	49.7		1979	57.8	24.4	
2018		222.9	94.2		1998	115.1	48.6		1978	53.5	22.6	
2017		213.6	90.2		1997	112.8	47.7		1977	49.5	20.9	
2016		207.3	87.6		1996	110.2	46.6		1976	46.9	19.8	
2015		206.2	87.1		1995	107.6	45.5		1975	44.8	18.9	
2014		204.9	86.6		1994	104.4	44.1		1974	41.4	17.5	
2013		201.2	85.0		1993	101.7	43.0		1973	37.7	15.9	
2012		194.6	82.2		1992	99.4	42.0		1972	34.8	14.7	
2011		191.2	80.8		1991	96.8	40.9		1971	32.1	13.6	
2010		183.5	77.5		1990	94.3	39.8		1970	28.7	12.1	
2009		180.1	76.1		1989	92.1	38.9		1969	26.9	11.4	
2008		180.4	76.2		1988	89.9	38.0					
2007		169.4	71.6		1987	87.7	37.0					
2006		162.0	68.4		1986	84.2	35.6					
2005		151.6	64.0		1985	82.6	34.9					

Adjustments to Costs

The "Historical Cost Index" can be used to convert national average building costs at a particular time to the approximate building costs for some other time.

Example:

Estimate and compare construction costs for different years in the same city.

To estimate the national average construction cost of a building in 1970, knowing that it cost $900,000 in 2021:

INDEX in 1970 = 28.7

INDEX in 2021 = 236.7

Note: The city cost indexes for Canada can be used to convert U.S. national averages to local costs in Canadian dollars.

Time Adjustment Using the Historical Cost Indexes:

$$\frac{\text{Index for Year A}}{\text{Index for Year B}} \times \text{Cost in Year B} = \text{Cost in Year A}$$

$$\frac{\text{INDEX 1970}}{\text{INDEX 2021}} \times \text{Cost 2021} = \text{Cost 1970}$$

$$\frac{28.7}{236.7} \times \$900,000 = .121 \times \$900,000 = \$108,900$$

The construction cost of the building in 1970 was $108,900.

Example:

To estimate and compare the cost of a building in Toronto, ON in 2021 with the known cost of $600,000 (US$) in New York, NY in 2021:

INDEX Toronto = 112.5

INDEX New York = 134.1

$$\frac{\text{INDEX Toronto}}{\text{INDEX New York}} \times \text{Cost New York} = \text{Cost Toronto}$$

$$\frac{112.5}{134.1} \times \$600,000 = .839 \times \$600,000 = \$503,400$$

The construction cost of the building in Toronto is $503,400 (CN$).

*Historical Cost Index updates and other resources are provided on the following website:
rsmeans.com/2021books

How to Use the City Cost Indexes

What you should know before you begin

RSMeans City Cost Indexes (CCI) are an extremely useful tool for when you want to compare costs from city to city and region to region.

This publication contains average construction cost indexes for 731 U.S. and Canadian cities covering over 930 three-digit zip code locations, as listed directly under each city.

Keep in mind that a City Cost Index number is a percentage ratio of a specific city's cost to the national average cost of the same item at a stated time period.

In other words, these index figures represent relative construction factors (or, if you prefer, multipliers) for material and installation costs, as well as the weighted average for Total In Place costs for each CSI MasterFormat division. Installation costs include both labor and equipment rental costs. When estimating equipment rental rates only for a specific location, use 01 54 33 EQUIPMENT RENTAL COSTS in the Reference Section.

The 30 City Average Index is the average of 30 major U.S. cities and serves as a national average. This national average represents the baseline from which all locations are calculated.

Index figures for both material and installation are based on the 30 major city average of 100 and represent the cost relationship as of July 1, 2020. The index for each division is computed from representative material and labor quantities for that division. The weighted average for each city is a weighted total of the components listed above it. It does not include relative productivity between trades or cities.

As changes occur in local material prices, labor rates, and equipment rental rates (including fuel costs), the impact of these changes should be accurately measured by the change in the City Cost Index for each particular city (as compared to the 30 city average).

Therefore, if you know (or have estimated) building costs in one city today, you can easily convert those costs to expected building costs in another city.

In addition, by using the Historical Cost Index, you can easily convert national average building costs at a particular time to the approximate building costs for some other time. The City Cost Indexes can then be applied to calculate the costs for a particular city.

Quick calculations

Location Adjustment Using the City Cost Indexes:

$$\frac{\text{Index for City A}}{\text{Index for City B}} \times \text{Cost in City B} = \text{Cost in City A}$$

Time Adjustment for the National Average Using the Historical Cost Index:

$$\frac{\text{Index for Year A}}{\text{Index for Year B}} \times \text{Cost in Year B} = \text{Cost in Year A}$$

Adjustment from the National Average:

$$\frac{\text{Index for City A}}{100} \times \text{National Average Cost} = \text{Cost in City A}$$

Since each of the other RSMeans data sets contains many different items, any *one* item multiplied by the particular city index may give incorrect results. However, the larger the number of items compiled, the closer the results should be to actual costs for that particular city.

The City Cost Indexes for Canadian cities are calculated using Canadian material and equipment prices and labor rates in Canadian dollars. When compared to the baseline mentioned previously, the resulting index enables the user to convert baseline costs to local costs. Therefore, indexes for Canadian cities can be used to convert U.S. national average prices to local costs in Canadian dollars.

How to use this section

1. Compare costs from city to city.

In using the RSMeans Indexes, remember that an index number is not a fixed number but a ratio: It's a percentage ratio of a building component's cost at any stated time to the national average cost of that same component at the same time period. Put in the form of an equation:

$$\frac{\text{Specific City Cost}}{\text{National Average Cost}} \times 100 = \text{City Index Number}$$

Therefore, when making cost comparisons between cities, do not subtract one city's index number from the index number of another city and read the result as a percentage difference. Instead, divide one city's index number by that of the other city. The resulting number may then be used as a multiplier to calculate cost differences from city to city.

The formula used to find cost differences between cities for the purpose of comparison is as follows:

$$\frac{\text{City A Index}}{\text{City B Index}} \times \text{City B Cost (Known)} = \text{City A Cost (Unknown)}$$

In addition, you can use RSMeans CCI to calculate and compare costs division by division between cities using the same basic formula. (Just be sure that you're comparing similar divisions.)

2. Compare a specific city's construction costs with the national average.

When you're studying construction location feasibility, it's advisable to compare a prospective project's cost index with an index of the national average cost.

For example, divide the weighted average index of construction costs of a specific city by that of the 30 City Average, which = 100.

$$\frac{\text{City Index}}{100} = \% \text{ of National Average}$$

As a result, you get a ratio that indicates the relative cost of construction in that city in comparison with the national average.

3. Convert U.S. national average to actual costs in Canadian City.

$$\frac{\text{Index for Canadian City}}{100} \times \text{National Average Cost} = \text{Cost in Canadian City in \$ CAN}$$

For customer support on your Electrical Costs with RSMeans Data, call 800.448.8182.

52

4. Adjust construction cost data based on a national average.

When you use a source of construction cost data which is based on a national average (such as RSMeans cost data), it is necessary to adjust those costs to a specific location.

$$\frac{\text{City Index}}{100} \times \frac{\text{Cost Based on}}{\text{National Average Costs}} = \frac{\text{City Cost}}{\text{(Unknown)}}$$

5. When applying the City Cost Indexes to demolition projects, use the appropriate division installation index. For example, for removal of existing doors and windows, use the Division 8 (Openings) index.

What you might like to know about how we developed the Indexes

The information presented in the CCI is organized according to the Construction Specifications Institute (CSI) MasterFormat 2018 classification system.

To create a reliable index, RSMeans researched the building type most often constructed in the United States and Canada. Because it was concluded that no one type of building completely represented the building construction industry, nine different types of buildings were combined to create a composite model.

The exact material, labor, and equipment quantities are based on detailed analyses of these nine building types, and then each quantity is weighted in proportion to expected usage. These various material items, labor hours, and equipment rental rates are thus combined to form a composite building representing as closely as possible the actual usage of materials, labor, and equipment in the North American building construction industry.

The following structures were chosen to make up that composite model:

1. Factory, 1 story
2. Office, 2–4 stories
3. Store, Retail
4. Town Hall, 2–3 stories
5. High School, 2–3 stories
6. Hospital, 4–8 stories
7. Garage, Parking
8. Apartment, 1–3 stories
9. Hotel/Motel, 2–3 stories

For the purposes of ensuring the timeliness of the data, the components of the index for the composite model have been streamlined. They currently consist of:

- specific quantities of 66 commonly used construction materials;
- specific labor-hours for 21 building construction trades; and
- specific days of equipment rental for 6 types of construction equipment (normally used to install the 66 material items by the 21 trades.) Fuel costs and routine maintenance costs are included in the equipment cost.

Material and equipment price quotations are gathered quarterly from cities in the United States and Canada. These prices and the latest negotiated labor wage rates for 21 different building trades are used to compile the quarterly update of the City Cost Index.

The 30 major U.S. cities used to calculate the national average are:

Atlanta, GA	Memphis, TN
Baltimore, MD	Milwaukee, WI
Boston, MA	Minneapolis, MN
Buffalo, NY	Nashville, TN
Chicago, IL	New Orleans, LA
Cincinnati, OH	New York, NY
Cleveland, OH	Philadelphia, PA
Columbus, OH	Phoenix, AZ
Dallas, TX	Pittsburgh, PA
Denver, CO	St. Louis, MO
Detroit, MI	San Antonio, TX
Houston, TX	San Diego, CA
Indianapolis, IN	San Francisco, CA
Kansas City, MO	Seattle, WA
Los Angeles, CA	Washington, DC

What the CCI does not indicate

The weighted average for each city is a total of the divisional components weighted to reflect typical usage. It does not include the productivity variations between trades or cities.

In addition, the CCI does not take into consideration factors such as the following:

- managerial efficiency
- competitive conditions
- automation
- restrictive union practices
- unique local requirements
- regional variations due to specific building codes

ALABAMA

DIVISION		UNITED STATES 30 CITY AVERAGE			ANNISTON 362			BIRMINGHAM 350-352			BUTLER 369			DECATUR 356			DOTHAN 363		
		MAT.	INST.	TOTAL	MAT.	INST.	TOTAL	MAT.	INST.	TOTAL	MAT.	INST.	TOTAL	MAT.	INST.	TOTAL	MAT.	INST.	TOTAL
015433	CONTRACTOR EQUIPMENT		100.0	100.0		104.5	104.5		104.7	104.7		102.1	102.1		104.5	104.5		102.1	102.1
0241, 31 - 34	SITE & INFRASTRUCTURE, DEMOLITION	100.0	100.0	100.0	90.9	92.7	92.1	92.0	93.1	92.7	105.4	88.1	93.3	85.9	92.0	90.2	102.9	88.7	93.1
0310	Concrete Forming & Accessories	100.0	100.0	100.0	86.5	65.3	68.5	92.0	66.5	70.3	83.0	65.9	68.5	93.0	61.4	66.2	91.4	68.2	71.7
0320	Concrete Reinforcing	100.0	100.0	100.0	86.9	72.3	79.5	96.0	72.3	83.9	91.9	72.9	82.2	89.8	68.7	79.1	91.9	72.3	82.0
0330	Cast-in-Place Concrete	100.0	100.0	100.0	87.8	66.3	79.8	102.6	68.4	89.8	85.6	66.6	78.5	100.7	66.1	87.8	85.6	68.0	79.0
03	CONCRETE	100.0	100.0	100.0	92.3	68.6	81.6	94.4	69.9	83.4	93.3	69.1	82.4	94.3	66.1	81.6	92.5	70.6	82.6
04	MASONRY	100.0	100.0	100.0	85.1	69.1	75.2	86.0	67.7	74.7	89.3	69.1	76.8	84.1	68.6	74.5	90.6	60.2	71.8
05	METALS	100.0	100.0	100.0	98.1	94.5	97.0	98.1	95.0	97.2	97.0	95.5	96.6	100.2	93.2	98.1	97.1	95.4	96.6
06	WOOD, PLASTICS & COMPOSITES	100.0	100.0	100.0	84.9	65.9	74.6	94.0	65.9	78.7	79.4	65.9	72.1	98.1	60.4	77.6	91.4	68.7	79.1
07	THERMAL & MOISTURE PROTECTION	100.0	100.0	100.0	95.5	63.9	81.9	96.0	67.3	83.6	95.5	67.3	83.4	95.1	66.4	82.8	95.5	65.0	82.4
08	OPENINGS	100.0	100.0	100.0	96.8	67.7	90.0	103.3	67.7	95.1	96.8	67.8	90.1	109.4	63.7	98.8	96.9	69.3	90.5
0920	Plaster & Gypsum Board	100.0	100.0	100.0	90.5	65.5	74.0	92.0	65.5	74.5	87.1	65.5	72.9	100.4	59.8	73.6	97.8	68.3	78.4
0950, 0980	Ceilings & Acoustic Treatment	100.0	100.0	100.0	80.8	65.5	70.4	82.1	65.5	70.8	80.8	65.5	70.4	82.4	59.8	67.1	80.8	68.3	72.3
0960	Flooring	100.0	100.0	100.0	83.8	80.6	82.9	94.8	76.3	89.4	88.6	80.6	86.3	85.6	80.6	84.2	94.0	82.2	90.6
0970, 0990	Wall Finishes & Painting/Coating	100.0	100.0	100.0	90.3	58.9	71.7	88.5	60.1	71.7	90.3	48.0	65.3	80.4	61.4	69.2	90.3	80.9	84.8
09	FINISHES	100.0	100.0	100.0	80.0	67.8	73.4	86.5	67.4	76.1	83.0	66.7	74.1	82.3	64.7	72.7	85.9	72.3	78.4
COVERS	DIVS. 10 - 14, 25, 28, 41, 43, 44, 46	100.0	100.0	100.0	100.0	70.2	93.4	100.0	83.4	96.3	100.0	77.0	94.9	100.0	69.5	93.3	100.0	72.0	93.8
21, 22, 23	FIRE SUPPRESSION, PLUMBING & HVAC	100.0	100.0	100.0	101.2	50.4	80.8	100.0	66.4	86.6	98.7	65.7	85.5	100.1	64.7	85.9	98.7	65.4	85.3
26, 27, 3370	ELECTRICAL, COMMUNICATIONS & UTIL.	100.0	100.0	100.0	96.4	58.4	77.0	97.2	61.2	78.8	98.4	65.4	81.5	93.0	64.0	78.2	97.2	75.3	86.0
MF2018	WEIGHTED AVERAGE	100.0	100.0	100.0	95.6	67.4	83.5	97.0	71.7	86.2	95.9	71.6	85.5	97.0	70.1	85.4	96.0	72.9	86.1

ALABAMA

DIVISION		EVERGREEN 364			GADSDEN 359			HUNTSVILLE 357-358			JASPER 355			MOBILE 365-366			MONTGOMERY 360-361		
		MAT.	INST.	TOTAL	MAT.	INST.	TOTAL	MAT.	INST.	TOTAL	MAT.	INST.	TOTAL	MAT.	INST.	TOTAL	MAT.	INST.	TOTAL
015433	CONTRACTOR EQUIPMENT		102.1	102.1		104.5	104.5		104.5	104.5		104.5	104.5		102.1	102.1		102.1	102.1
0241, 31 - 34	SITE & INFRASTRUCTURE, DEMOLITION	105.8	88.0	93.4	91.3	92.7	92.3	85.6	92.6	90.3	91.3	92.8	92.4	98.0	89.1	91.8	96.0	88.5	90.8
0310	Concrete Forming & Accessories	80.0	66.7	68.7	85.2	65.8	68.7	93.0	65.2	69.4	90.1	66.1	69.7	90.5	66.7	70.3	93.7	65.9	70.1
0320	Concrete Reinforcing	92.0	72.8	82.2	95.4	72.3	83.6	89.8	73.7	81.6	89.8	72.4	81.0	89.6	72.2	80.8	97.6	71.8	84.5
0330	Cast-in-Place Concrete	85.6	66.3	78.4	100.7	67.3	88.3	98.0	67.0	86.5	111.6	67.1	95.1	89.7	66.3	81.0	86.8	67.6	79.7
03	CONCRETE	93.6	69.3	82.7	98.8	69.2	85.4	93.0	69.0	82.2	102.5	69.2	87.5	88.8	69.2	80.0	87.2	69.2	79.1
04	MASONRY	89.3	68.7	76.5	82.6	65.8	72.3	85.2	65.4	73.0	80.4	70.3	74.1	87.9	58.8	69.9	85.2	59.2	69.2
05	METALS	97.0	95.4	96.5	98.1	95.0	97.2	100.2	95.3	98.7	98.1	94.9	97.1	99.1	94.7	97.7	98.1	94.4	97.0
06	WOOD, PLASTICS & COMPOSITES	76.0	67.4	71.3	88.5	65.9	76.3	98.1	65.3	80.3	95.2	65.9	79.3	89.9	67.4	77.7	93.3	65.9	78.5
07	THERMAL & MOISTURE PROTECTION	95.5	66.5	83.0	95.4	66.8	83.1	95.1	66.6	82.8	95.4	62.4	81.2	95.1	64.9	82.1	93.5	65.1	81.3
08	OPENINGS	96.8	68.6	90.3	105.9	67.7	97.0	109.1	67.7	99.5	105.9	67.7	97.0	99.5	68.6	92.3	98.8	67.7	91.6
0920	Plaster & Gypsum Board	86.5	67.0	73.7	92.6	65.5	74.7	100.4	64.8	77.0	97.2	65.5	76.3	94.2	67.0	76.3	89.9	65.5	73.8
0950, 0980	Ceilings & Acoustic Treatment	80.8	67.0	71.4	79.0	65.5	69.8	84.3	64.8	71.1	79.0	65.5	69.8	86.1	67.0	73.1	84.8	65.5	71.7
0960	Flooring	86.4	80.6	84.7	82.2	77.3	80.8	85.6	77.3	83.2	84.1	80.6	83.1	93.5	63.7	84.8	94.0	63.7	85.2
0970, 0990	Wall Finishes & Painting/Coating	90.3	48.0	65.3	80.4	58.9	67.7	80.4	67.3	72.6	80.4	58.9	67.7	93.6	49.9	67.7	91.3	58.9	72.1
09	FINISHES	82.3	67.4	74.2	80.0	67.1	72.9	82.7	67.5	74.4	81.1	68.1	74.0	85.8	64.2	74.0	86.1	64.4	74.2
COVERS	DIVS. 10 - 14, 25, 28, 41, 43, 44, 46	100.0	70.1	93.4	100.0	82.8	96.2	100.0	82.6	96.2	100.0	71.4	93.7	100.0	83.7	96.4	100.0	83.0	96.2
21, 22, 23	FIRE SUPPRESSION, PLUMBING & HVAC	98.7	65.5	85.4	102.0	62.8	86.3	100.1	63.5	85.4	102.0	66.2	87.6	100.0	62.8	85.1	100.0	65.0	85.9
26, 27, 3370	ELECTRICAL, COMMUNICATIONS & UTIL.	95.7	65.4	80.2	93.0	61.2	76.7	93.9	65.7	79.5	92.7	58.4	75.2	99.1	58.1	78.1	99.7	73.9	86.5
MF2018	WEIGHTED AVERAGE	95.6	71.4	85.2	97.1	70.6	85.7	96.9	71.4	86.0	97.5	71.0	86.2	96.4	68.8	84.5	95.8	71.4	85.3

ALABAMA / ALASKA

DIVISION		PHENIX CITY 368			SELMA 367			TUSCALOOSA 354			ANCHORAGE 995-996			FAIRBANKS 997			JUNEAU 998		
		MAT.	INST.	TOTAL	MAT.	INST.	TOTAL	MAT.	INST.	TOTAL	MAT.	INST.	TOTAL	MAT.	INST.	TOTAL	MAT.	INST.	TOTAL
015433	CONTRACTOR EQUIPMENT		102.1	102.1		102.1	102.1		104.5	104.5		115.9	115.9		115.9	115.9		115.9	115.9
0241, 31 - 34	SITE & INFRASTRUCTURE, DEMOLITION	109.7	88.5	95.0	102.8	88.5	92.9	86.2	92.7	90.8	122.1	130.7	128.1	121.2	130.7	127.8	140.2	130.7	133.6
0310	Concrete Forming & Accessories	86.5	66.1	69.1	84.2	65.9	68.7	93.0	66.0	70.1	115.1	117.7	117.3	119.9	117.9	118.2	123.8	117.7	118.6
0320	Concrete Reinforcing	91.9	69.2	80.4	91.9	72.4	82.0	89.8	72.3	80.9	152.4	121.3	136.6	141.8	121.3	131.4	127.6	121.3	124.4
0330	Cast-in-Place Concrete	85.6	67.7	78.9	85.6	66.6	78.5	102.1	67.6	89.3	97.6	117.2	104.9	108.1	117.6	111.7	120.4	117.2	119.2
03	CONCRETE	96.5	68.9	84.1	92.0	69.0	81.7	94.9	69.4	83.4	116.9	117.3	117.1	108.4	117.6	112.5	124.8	117.3	121.4
04	MASONRY	89.2	69.1	76.8	92.5	69.1	78.1	84.3	66.3	73.2	183.5	119.5	144.0	185.0	119.8	144.7	157.6	119.5	134.1
05	METALS	97.0	94.2	96.1	97.0	95.1	96.4	99.4	95.0	98.1	125.0	104.8	118.9	125.2	104.9	119.1	119.7	104.8	115.2
06	WOOD, PLASTICS & COMPOSITES	84.6	65.8	74.4	81.6	65.9	73.1	98.1	65.9	80.6	114.0	115.8	115.0	125.8	115.8	120.4	123.4	115.8	119.3
07	THERMAL & MOISTURE PROTECTION	95.9	67.8	83.8	95.4	67.3	83.3	95.2	67.0	83.1	169.5	117.6	147.2	181.1	118.7	154.2	185.1	117.6	156.0
08	OPENINGS	96.8	66.8	89.8	96.8	67.7	90.0	109.1	67.7	99.5	130.6	116.5	127.3	135.8	116.5	131.4	130.4	116.5	127.2
0920	Plaster & Gypsum Board	91.7	65.3	74.3	89.9	65.5	73.8	100.4	65.5	77.4	142.8	116.1	125.2	172.8	116.1	135.4	155.4	116.1	129.5
0950, 0980	Ceilings & Acoustic Treatment	80.8	65.3	70.3	80.8	65.5	70.4	84.3	65.5	71.5	132.6	116.1	121.4	127.9	116.1	119.9	138.9	116.1	123.4
0960	Flooring	90.6	80.6	87.7	89.1	80.6	86.6	85.6	77.3	83.2	125.1	122.6	124.4	122.1	122.6	122.2	129.4	122.6	127.4
0970, 0990	Wall Finishes & Painting/Coating	90.3	81.0	84.8	90.3	58.9	71.7	80.4	58.9	67.7	108.7	113.2	111.3	113.7	117.7	116.1	110.8	113.2	112.2
09	FINISHES	84.7	70.2	76.8	83.2	67.8	74.8	82.6	67.2	74.2	125.7	118.3	121.7	128.6	118.9	123.3	129.3	118.3	123.3
COVERS	DIVS. 10 - 14, 25, 28, 41, 43, 44, 46	100.0	82.9	96.2	100.0	70.7	93.6	100.0	82.9	96.2	100.0	109.3	102.0	100.0	109.4	102.1	100.0	109.3	102.0
21, 22, 23	FIRE SUPPRESSION, PLUMBING & HVAC	98.7	65.0	85.2	98.7	64.3	84.9	100.1	66.5	86.6	100.5	106.1	102.8	100.4	110.3	104.3	100.5	106.1	102.8
26, 27, 3370	ELECTRICAL, COMMUNICATIONS & UTIL.	97.8	65.7	81.4	96.9	73.9	85.1	93.5	61.2	77.0	112.5	108.8	110.7	120.6	108.8	114.6	108.9	108.8	108.9
MF2018	WEIGHTED AVERAGE	96.5	72.0	86.0	95.7	72.4	85.7	97.0	71.5	86.0	119.3	113.6	116.9	120.3	114.7	117.9	119.0	113.6	116.7

For customer support on your Electrical Costs with RSMeans Data, call 800.448.8182.

523

DIVISION		ALASKA KETCHIKAN 999			ARIZONA CHAMBERS 865			FLAGSTAFF 860			GLOBE 855			KINGMAN 864			MESA/TEMPE 852		
		MAT.	INST.	TOTAL	MAT.	INST.	TOTAL	MAT.	INST.	TOTAL	MAT.	INST.	TOTAL	MAT.	INST.	TOTAL	MAT.	INST.	TOTAL
015433	CONTRACTOR EQUIPMENT		115.9	115.9		91.1	91.1		91.1	91.1		92.2	92.2		91.1	91.1		92.2	92.2
0241, 31 - 34	SITE & INFRASTRUCTURE, DEMOLITION	176.2	130.7	144.5	71.0	95.0	87.7	89.5	95.0	93.3	102.0	95.9	97.8	70.9	95.0	87.7	92.1	95.9	94.8
0310	Concrete Forming & Accessories	111.6	117.7	116.7	98.2	68.0	72.5	103.7	69.2	74.4	93.9	68.0	71.9	96.3	67.9	72.2	97.1	68.1	72.5
0320	Concrete Reinforcing	108.4	121.3	114.9	103.2	79.1	91.0	103.0	79.0	90.8	109.0	79.1	93.8	103.4	79.1	91.0	109.7	79.1	94.2
0330	Cast-in-Place Concrete	218.8	117.2	181.0	89.7	69.4	82.2	89.8	69.4	82.2	84.1	69.0	78.5	89.5	69.4	82.0	84.8	69.1	79.0
03	CONCRETE	186.7	117.3	155.4	92.7	70.3	82.7	112.5	70.9	93.8	104.5	70.3	89.1	92.4	70.3	82.5	95.5	70.4	84.2
04	MASONRY	192.7	119.5	147.5	95.2	60.7	73.9	95.3	59.0	72.9	100.4	60.6	75.8	95.2	60.7	73.9	100.6	60.6	75.9
05	METALS	125.4	104.8	119.1	101.9	72.4	92.9	102.4	72.3	93.3	100.7	73.2	92.0	102.6	72.3	93.4	97.7	73.3	90.3
06	WOOD, PLASTICS & COMPOSITES	116.0	115.8	115.9	102.8	69.3	84.6	109.3	70.9	88.4	94.2	69.4	80.8	97.5	69.3	82.2	98.1	69.4	82.5
07	THERMAL & MOISTURE PROTECTION	186.0	117.6	156.5	94.1	70.1	83.8	96.1	69.6	84.7	100.2	68.7	86.6	94.1	69.9	83.7	100.0	68.4	86.4
08	OPENINGS	131.5	116.5	128.0	108.2	66.4	98.5	108.3	72.2	99.9	97.2	66.7	90.2	108.4	68.7	99.2	97.3	68.8	90.7
0920	Plaster & Gypsum Board	158.3	116.1	130.5	92.6	68.6	76.8	96.2	70.2	79.1	88.0	68.6	75.2	84.6	68.6	74.1	90.8	68.6	76.2
0950, 0980	Ceilings & Acoustic Treatment	121.3	116.1	117.8	107.2	68.6	81.1	108.1	70.2	82.4	93.5	68.6	76.6	108.1	68.6	81.3	93.5	68.6	76.6
0960	Flooring	122.1	122.6	122.2	89.9	58.8	80.9	92.1	58.8	82.4	104.0	58.8	90.9	88.7	58.8	80.0	105.6	58.8	92.0
0970, 0990	Wall Finishes & Painting/Coating	113.7	113.2	113.4	88.5	68.3	76.6	88.5	68.3	76.6	92.8	68.3	78.3	88.5	68.3	76.6	92.8	68.3	78.3
09	FINISHES	129.1	118.3	123.2	90.1	66.1	77.0	93.2	67.0	78.9	96.5	66.2	80.0	88.7	66.1	76.4	96.3	66.2	79.9
COVERS	DIVS. 10 - 14, 25, 28, 41, 43, 44, 46	100.0	109.3	102.0	100.0	83.0	96.3	100.0	83.1	96.3	100.0	83.2	96.3	100.0	82.9	96.2	100.0	83.2	96.3
21, 22, 23	FIRE SUPPRESSION, PLUMBING & HVAC	98.6	106.1	101.6	98.3	77.0	89.7	100.2	77.2	91.0	97.6	77.0	89.3	98.3	77.0	89.7	100.0	77.2	90.9
26, 27, 3370	ELECTRICAL, COMMUNICATIONS & UTIL.	120.6	108.8	114.6	102.8	83.6	93.0	101.8	62.2	81.5	96.9	64.4	80.3	102.9	64.4	83.2	94.0	64.4	78.9
MF2018	WEIGHTED AVERAGE	131.0	113.6	123.5	98.3	74.4	88.0	102.0	71.7	89.0	98.7	71.8	87.2	98.2	71.8	86.9	97.7	72.0	86.7

DIVISION		ARIZONA PHOENIX 850,853			PRESCOTT 863			SHOW LOW 859			TUCSON 856 - 857			ARKANSAS BATESVILLE 725			CAMDEN 717		
		MAT.	INST.	TOTAL	MAT.	INST.	TOTAL	MAT.	INST.	TOTAL	MAT.	INST.	TOTAL	MAT.	INST.	TOTAL	MAT.	INST.	TOTAL
015433	CONTRACTOR EQUIPMENT		93.2	93.2		91.1	91.1		92.2	92.2		92.2	92.2		91.9	91.9		91.9	91.9
0241, 31 - 34	SITE & INFRASTRUCTURE, DEMOLITION	92.6	94.5	93.9	77.5	95.0	89.7	104.2	95.9	98.4	87.7	95.9	93.4	76.6	89.7	85.7	79.6	89.7	86.6
0310	Concrete Forming & Accessories	101.8	71.6	76.1	99.8	68.0	72.8	101.3	68.1	73.1	97.6	69.3	73.6	79.5	56.4	59.9	79.8	56.3	59.8
0320	Concrete Reinforcing	107.8	79.2	93.2	103.0	79.1	90.9	109.7	79.1	94.2	91.0	79.0	84.9	93.3	65.3	79.1	98.6	65.0	81.5
0330	Cast-in-Place Concrete	84.5	69.3	78.8	89.8	69.4	82.2	84.2	69.1	78.5	87.3	69.1	80.5	74.8	76.4	75.4	78.1	76.4	77.5
03	CONCRETE	98.1	72.1	86.4	98.1	70.3	85.6	107.0	70.3	90.5	94.2	70.9	83.7	76.3	65.7	71.5	79.1	65.7	73.0
04	MASONRY	93.5	62.7	74.5	95.3	60.7	73.9	100.4	60.6	75.8	87.5	58.9	69.8	92.1	42.3	61.4	104.9	36.9	62.9
05	METALS	99.1	75.2	91.8	102.4	72.3	93.3	97.2	73.2	89.9	98.4	73.2	90.7	100.2	72.8	91.9	96.2	72.4	89.0
06	WOOD, PLASTICS & COMPOSITES	101.1	72.9	85.7	104.4	69.5	85.4	102.7	69.4	84.6	98.3	71.0	83.5	92.9	57.2	73.5	91.0	57.2	72.6
07	THERMAL & MOISTURE PROTECTION	101.3	69.2	87.5	94.7	70.2	84.2	100.4	69.0	86.9	101.1	68.1	86.9	102.7	57.4	83.2	93.5	55.8	77.3
08	OPENINGS	98.6	72.8	92.6	108.3	66.7	98.7	96.5	68.8	90.1	93.7	72.3	88.7	102.0	53.1	90.7	105.6	54.9	93.8
0920	Plaster & Gypsum Board	98.0	71.9	80.8	93.0	68.8	77.1	92.6	68.6	76.8	95.3	70.2	78.8	78.6	56.4	64.0	87.9	56.4	67.1
0950, 0980	Ceilings & Acoustic Treatment	104.1	71.9	82.3	106.4	68.8	80.9	93.5	68.6	76.6	94.3	70.2	78.0	85.7	56.4	65.8	80.0	56.4	64.0
0960	Flooring	107.7	58.8	93.5	90.7	58.8	81.5	107.3	58.8	93.2	96.1	58.8	85.3	86.3	53.0	76.6	92.4	35.5	75.9
0970, 0990	Wall Finishes & Painting/Coating	97.5	65.0	78.3	88.5	68.3	76.6	92.8	68.3	78.3	93.3	68.3	78.5	90.7	53.0	68.7	94.5	53.5	70.2
09	FINISHES	100.9	68.5	83.2	90.7	66.2	77.3	98.3	66.2	80.8	94.3	67.2	79.5	76.9	54.9	64.9	80.4	51.7	64.7
COVERS	DIVS. 10 - 14, 25, 28, 41, 43, 44, 46	100.0	84.7	96.6	100.0	82.9	96.2	100.0	83.2	96.3	100.0	83.4	96.3	100.0	65.9	92.5	100.0	65.9	92.5
21, 22, 23	FIRE SUPPRESSION, PLUMBING & HVAC	99.9	78.6	91.3	100.2	77.0	90.9	97.6	77.0	89.3	100.0	77.1	90.8	97.6	52.2	79.4	97.5	56.9	81.2
26, 27, 3370	ELECTRICAL, COMMUNICATIONS & UTIL.	99.6	62.2	80.5	101.5	64.4	82.5	94.3	64.4	79.0	96.1	59.5	77.4	96.5	60.7	78.2	94.4	57.2	75.4
MF2018	WEIGHTED AVERAGE	99.0	73.0	87.9	99.6	71.7	87.6	98.9	71.9	87.3	96.6	71.4	85.8	93.6	60.0	79.2	94.1	59.6	79.3

DIVISION		ARKANSAS FAYETTEVILLE 727			FORT SMITH 729			HARRISON 726			HOT SPRINGS 719			JONESBORO 724			LITTLE ROCK 720 - 722		
		MAT.	INST.	TOTAL	MAT.	INST.	TOTAL	MAT.	INST.	TOTAL	MAT.	INST.	TOTAL	MAT.	INST.	TOTAL	MAT.	INST.	TOTAL
015433	CONTRACTOR EQUIPMENT		91.9	91.9		91.9	91.9		91.9	91.9		91.9	91.9		115.7	115.7		91.9	91.9
0241, 31 - 34	SITE & INFRASTRUCTURE, DEMOLITION	76.0	89.7	85.6	81.2	89.8	87.2	81.5	89.7	87.2	82.7	89.7	87.6	100.5	107.5	105.4	89.1	90.1	89.8
0310	Concrete Forming & Accessories	75.3	56.5	59.3	93.6	61.7	66.5	83.8	56.3	60.4	77.4	56.5	59.6	82.7	56.5	60.5	90.3	62.0	66.3
0320	Concrete Reinforcing	93.4	62.6	77.7	94.5	65.9	80.0	92.9	65.4	78.9	96.7	65.3	80.8	90.1	65.5	77.6	100.0	67.9	83.7
0330	Cast-in-Place Concrete	74.8	76.5	75.4	85.6	76.7	82.2	83.0	76.4	80.5	79.9	76.5	78.6	81.5	77.7	80.1	82.9	76.8	80.6
03	CONCRETE	76.0	65.3	71.2	83.7	68.3	76.7	83.1	65.7	75.2	82.4	65.8	74.9	80.6	67.3	74.6	84.5	68.9	77.4
04	MASONRY	83.5	41.8	57.8	90.1	62.1	72.8	92.6	43.3	62.1	78.8	42.0	56.1	84.8	39.9	57.1	86.0	62.1	71.2
05	METALS	100.2	72.0	91.6	102.5	74.6	94.0	101.4	72.8	92.7	96.2	72.9	89.1	96.7	87.8	94.0	102.8	75.6	94.5
06	WOOD, PLASTICS & COMPOSITES	89.2	57.2	71.8	110.0	63.6	84.8	98.9	57.2	76.2	88.4	57.2	71.4	97.1	57.5	75.6	96.5	63.6	78.6
07	THERMAL & MOISTURE PROTECTION	103.5	57.2	83.6	103.9	63.6	86.5	103.0	57.7	83.5	93.7	57.3	78.0	108.4	56.6	86.1	98.0	63.9	83.3
08	OPENINGS	102.0	54.9	91.1	104.1	60.1	93.8	102.8	52.0	91.0	105.5	54.8	93.8	107.8	56.0	95.8	97.2	59.7	88.5
0920	Plaster & Gypsum Board	78.0	56.4	63.8	84.8	63.0	70.4	83.9	56.4	65.8	86.6	56.4	66.7	89.9	56.4	67.8	94.5	63.0	73.8
0950, 0980	Ceilings & Acoustic Treatment	85.7	56.4	65.8	87.4	63.0	70.9	87.4	56.4	66.4	80.0	56.4	64.0	90.7	56.4	67.4	89.5	63.0	71.5
0960	Flooring	83.4	51.9	74.3	92.9	73.3	87.2	88.6	53.0	78.3	91.3	51.9	79.8	62.6	47.0	58.1	94.0	82.1	90.5
0970, 0990	Wall Finishes & Painting/Coating	90.7	53.5	68.7	90.7	53.5	68.7	90.7	53.5	68.7	94.5	53.5	70.2	80.6	53.5	64.6	91.2	53.5	68.2
09	FINISHES	76.1	54.5	64.3	80.3	63.1	70.9	79.0	54.7	65.8	80.2	54.7	66.3	74.9	53.8	63.4	86.3	64.9	74.6
COVERS	DIVS. 10 - 14, 25, 28, 41, 43, 44, 46	100.0	65.9	92.5	100.0	78.4	95.2	100.0	67.0	92.7	100.0	65.9	92.5	100.0	64.9	92.3	100.0	78.4	95.3
21, 22, 23	FIRE SUPPRESSION, PLUMBING & HVAC	97.6	61.2	83.0	100.1	49.1	79.7	97.6	48.5	77.9	97.5	52.5	79.5	100.4	52.2	81.1	99.8	52.6	80.9
26, 27, 3370	ELECTRICAL, COMMUNICATIONS & UTIL.	90.8	53.1	71.5	94.0	58.4	75.8	95.2	58.9	76.6	96.3	66.8	81.2	100.2	60.9	80.1	101.1	66.8	83.6
MF2018	WEIGHTED AVERAGE	92.5	60.7	78.9	95.9	63.6	82.0	95.0	59.1	79.6	93.5	61.0	79.6	95.5	62.7	81.4	96.2	65.9	83.2

For customer support on your Electrical Costs with RSMeans Data, call 800.448.8182.

City Cost Indexes

ARKANSAS / CALIFORNIA

DIVISION		PINE BLUFF 716			RUSSELLVILLE 728			TEXARKANA 718			WEST MEMPHIS 723			ALHAMBRA 917 - 918			ANAHEIM 928		
		MAT.	INST.	TOTAL	MAT.	INST.	TOTAL	MAT.	INST.	TOTAL	MAT.	INST.	TOTAL	MAT.	INST.	TOTAL	MAT.	INST.	TOTAL
015433	CONTRACTOR EQUIPMENT		91.9	91.9		91.9	91.9		93.2	93.2		115.7	115.7		97.7	97.7		101.8	101.8
0241, 31 - 34	SITE & INFRASTRUCTURE, DEMOLITION	84.7	89.8	88.3	77.8	89.7	86.1	95.5	91.9	93.0	107.3	107.7	107.6	97.1	110.3	106.3	100.2	110.6	107.4
0310	Concrete Forming & Accessories	77.1	61.9	64.2	80.2	56.3	59.9	82.8	56.4	60.4	87.9	56.8	61.5	116.4	140.0	136.4	101.6	140.1	134.3
0320	Concrete Reinforcing	98.4	67.9	82.9	94.0	65.2	79.4	98.0	65.3	81.4	90.1	63.7	76.7	102.5	130.1	116.5	95.4	129.9	112.9
0330	Cast-in-Place Concrete	79.9	76.7	78.7	78.4	76.4	77.6	87.1	76.4	83.2	85.5	77.8	82.6	83.4	129.8	100.7	85.2	133.2	103.1
03	CONCRETE	83.1	68.8	76.6	79.3	65.6	73.1	81.4	65.7	74.4	87.3	67.1	78.2	95.2	133.3	112.3	96.3	134.6	113.5
04	MASONRY	110.4	62.1	80.5	89.4	40.9	59.4	92.9	41.3	61.0	73.6	39.8	52.7	111.7	141.9	130.4	78.5	140.2	116.6
05	METALS	97.0	75.4	90.4	100.2	72.6	91.8	89.7	72.8	84.6	95.7	87.4	93.2	90.3	111.7	96.5	107.1	111.7	108.5
06	WOOD, PLASTICS & COMPOSITES	88.0	63.6	74.8	94.2	57.2	74.1	95.6	57.2	74.7	103.0	57.5	78.3	102.8	137.1	121.4	97.9	137.4	119.3
07	THERMAL & MOISTURE PROTECTION	93.8	64.0	81.0	103.7	57.0	83.6	94.4	57.1	78.4	108.9	56.6	86.4	96.2	133.2	112.1	105.0	136.3	118.5
08	OPENINGS	106.8	60.5	96.0	102.0	54.1	90.9	111.8	54.1	98.4	105.3	54.8	93.5	89.8	135.7	100.4	106.2	135.9	113.1
0920	Plaster & Gypsum Board	86.3	63.0	71.0	78.6	56.4	64.0	89.2	56.4	67.6	92.1	56.4	68.6	96.3	138.3	124.0	111.2	138.3	129.1
0950, 0980	Ceilings & Acoustic Treatment	80.0	63.0	68.5	85.7	56.4	65.8	83.4	56.4	65.1	88.8	56.4	66.8	112.6	138.3	130.0	112.6	138.3	130.0
0960	Flooring	91.1	73.3	85.9	85.9	53.0	76.3	93.2	44.8	79.1	64.9	47.0	59.7	100.0	120.4	105.9	97.3	120.4	104.0
0970, 0990	Wall Finishes & Painting/Coating	94.5	53.5	70.2	90.7	53.5	68.7	94.5	53.5	70.2	80.6	53.5	64.6	105.4	121.1	114.7	90.7	121.1	108.7
09	FINISHES	80.2	63.1	70.9	77.1	54.8	64.9	82.4	53.4	66.6	76.2	54.0	64.1	100.8	134.3	119.1	98.1	134.5	118.0
COVERS	DIVS. 10 - 14, 25, 28, 41, 43, 44, 46	100.0	78.4	95.2	100.0	65.9	92.5	100.0	65.9	92.5	100.0	66.7	92.7	100.0	119.6	104.3	100.0	120.1	104.4
21, 22, 23	FIRE SUPPRESSION, PLUMBING & HVAC	100.0	52.6	81.0	97.6	51.9	79.3	100.0	55.6	82.2	97.9	64.2	84.4	97.5	131.7	111.2	100.0	131.8	112.7
26, 27, 3370	ELECTRICAL, COMMUNICATIONS & UTIL.	94.5	58.9	76.3	94.0	58.9	76.1	96.2	58.9	77.1	101.7	62.7	81.7	115.3	130.5	123.1	90.3	111.0	100.9
MF2018	WEIGHTED AVERAGE	95.9	64.6	82.4	93.7	59.6	79.1	94.9	60.4	80.1	95.3	65.5	82.5	98.2	129.4	111.6	99.3	126.9	111.2

CALIFORNIA

DIVISION		BAKERSFIELD 932 - 933			BERKELEY 947			EUREKA 955			FRESNO 936 - 938			INGLEWOOD 903 - 905			LONG BEACH 906 - 908		
		MAT.	INST.	TOTAL	MAT.	INST.	TOTAL	MAT.	INST.	TOTAL	MAT.	INST.	TOTAL	MAT.	INST.	TOTAL	MAT.	INST.	TOTAL
015433	CONTRACTOR EQUIPMENT		99.8	99.8		102.8	102.8		99.6	99.6		99.8	99.8		99.0	99.0		99.0	99.0
0241, 31 - 34	SITE & INFRASTRUCTURE, DEMOLITION	97.1	108.4	105.0	111.1	110.7	110.8	111.5	107.4	108.7	101.1	107.8	105.8	91.4	106.7	102.1	98.7	106.7	104.3
0310	Concrete Forming & Accessories	101.0	138.9	133.2	112.1	167.2	158.9	109.5	152.9	146.4	98.2	152.1	144.0	106.9	140.3	135.3	101.6	140.3	134.5
0320	Concrete Reinforcing	103.5	129.9	116.9	89.1	131.8	110.8	103.9	131.3	117.8	86.5	130.8	109.0	103.0	130.1	116.8	102.1	130.1	116.4
0330	Cast-in-Place Concrete	89.3	132.3	105.3	111.5	133.6	119.7	92.5	129.4	106.3	96.8	129.4	108.9	89.7	134.4	109.8	99.2	134.4	115.0
03	CONCRETE	89.8	133.7	109.5	101.4	147.1	122.0	106.8	139.3	121.5	93.5	138.7	113.9	89.7	134.4	109.8	99.2	134.4	115.0
04	MASONRY	92.5	139.6	121.6	124.9	156.3	144.3	103.7	152.4	133.8	97.2	146.3	127.5	75.6	142.1	116.6	84.7	142.1	120.1
05	METALS	105.1	111.1	106.9	108.5	114.0	110.2	106.9	113.6	108.9	105.3	112.1	107.4	99.7	112.1	103.5	99.6	112.1	103.4
06	WOOD, PLASTICS & COMPOSITES	92.8	136.2	116.4	106.7	172.6	142.5	111.8	156.0	135.8	98.8	156.0	129.9	96.9	137.5	119.0	90.7	137.5	116.1
07	THERMAL & MOISTURE PROTECTION	100.7	128.2	112.5	99.8	157.2	124.5	108.6	150.6	126.7	92.1	134.4	110.3	100.3	134.9	115.2	100.5	134.9	115.3
08	OPENINGS	96.1	133.1	104.7	92.7	159.5	108.2	105.5	141.6	113.9	97.0	144.0	108.0	89.4	135.9	100.2	89.4	135.9	100.2
0920	Plaster & Gypsum Board	97.4	137.0	123.5	113.1	174.1	153.3	116.2	157.4	143.3	94.3	157.4	135.8	105.2	138.3	127.0	101.3	138.3	125.7
0950, 0980	Ceilings & Acoustic Treatment	104.7	137.0	126.6	98.5	174.1	149.8	116.0	157.4	144.0	103.9	157.4	140.1	110.8	138.3	129.5	110.8	138.3	129.5
0960	Flooring	102.9	120.4	108.0	115.9	142.0	123.5	101.1	142.0	113.0	102.1	118.0	106.7	101.8	120.4	107.2	98.9	120.4	105.1
0970, 0990	Wall Finishes & Painting/Coating	92.1	109.3	102.3	113.3	162.7	142.5	92.1	137.6	119.0	104.1	123.9	115.8	105.2	121.1	114.6	105.2	121.1	114.6
09	FINISHES	94.0	132.5	115.0	107.6	163.7	138.3	103.1	150.9	129.2	93.5	145.3	121.8	104.5	134.6	120.9	103.5	134.6	120.4
COVERS	DIVS. 10 - 14, 25, 28, 41, 43, 44, 46	100.0	120.2	104.4	100.0	129.9	106.6	100.0	126.7	105.9	100.0	126.7	105.9	100.0	120.4	104.5	100.0	120.4	104.5
21, 22, 23	FIRE SUPPRESSION, PLUMBING & HVAC	100.1	129.0	111.7	97.7	168.0	125.9	97.5	129.1	110.2	100.2	130.3	112.2	97.1	131.8	111.0	97.1	131.8	111.0
26, 27, 3370	ELECTRICAL, COMMUNICATIONS & UTIL.	105.1	110.7	108.0	103.0	161.0	132.7	97.0	124.3	111.0	94.9	108.0	101.6	101.0	130.5	116.1	100.7	130.5	116.0
MF2018	WEIGHTED AVERAGE	98.7	125.3	110.1	102.7	151.2	123.5	102.7	133.2	115.8	98.4	129.3	111.7	96.0	129.6	110.4	97.6	129.6	111.3

CALIFORNIA

DIVISION		LOS ANGELES 900 - 902			MARYSVILLE 959			MODESTO 953			MOJAVE 935			OAKLAND 946			OXNARD 930		
		MAT.	INST.	TOTAL	MAT.	INST.	TOTAL	MAT.	INST.	TOTAL	MAT.	INST.	TOTAL	MAT.	INST.	TOTAL	MAT.	INST.	TOTAL
015433	CONTRACTOR EQUIPMENT		105.2	105.2		99.6	99.6		99.6	99.6		99.8	99.8		102.8	102.8		98.7	98.7
0241, 31 - 34	SITE & INFRASTRUCTURE, DEMOLITION	97.2	112.4	107.8	108.0	107.3	107.5	102.7	107.3	105.9	94.2	108.4	104.1	116.0	110.7	112.3	101.4	106.5	104.9
0310	Concrete Forming & Accessories	104.8	140.6	135.3	99.7	152.5	144.5	96.2	152.3	143.9	108.4	138.8	134.3	101.3	167.2	157.3	100.4	140.2	134.2
0320	Concrete Reinforcing	103.8	130.2	117.2	103.9	130.6	117.5	107.6	130.5	119.3	104.1	129.9	117.2	91.1	131.8	111.8	102.3	129.8	116.3
0330	Cast-in-Place Concrete	82.1	133.4	101.2	103.3	129.5	113.1	92.5	129.4	106.2	83.7	132.3	101.8	105.8	133.6	116.2	97.5	132.5	110.6
03	CONCRETE	96.8	135.0	114.0	107.3	138.9	121.5	98.3	138.8	116.5	85.6	133.6	107.2	102.1	147.1	122.1	93.8	134.3	112.0
04	MASONRY	91.0	142.1	122.5	104.7	141.7	127.5	102.1	141.7	126.6	95.4	139.6	122.7	132.8	156.3	147.3	98.3	138.0	122.8
05	METALS	106.1	114.5	108.6	106.4	112.3	108.1	103.4	112.1	106.0	102.3	111.0	104.9	104.1	114.0	107.1	100.2	111.3	103.6
06	WOOD, PLASTICS & COMPOSITES	99.9	137.7	120.4	98.8	156.0	129.9	94.3	156.0	127.8	98.4	136.2	118.9	94.6	172.6	137.0	93.3	137.4	117.3
07	THERMAL & MOISTURE PROTECTION	96.5	134.8	113.0	108.0	139.1	121.4	107.6	135.5	119.6	97.7	125.6	109.7	98.6	157.2	123.8	101.1	134.3	115.4
08	OPENINGS	98.9	136.1	107.5	104.7	144.9	114.1	103.5	144.9	113.1	92.0	133.1	101.6	92.8	159.5	108.3	94.6	135.9	104.2
0920	Plaster & Gypsum Board	100.5	138.3	125.4	108.7	157.4	140.8	111.2	157.4	141.6	103.8	137.0	125.7	107.2	174.1	151.3	97.8	138.3	124.5
0950, 0980	Ceilings & Acoustic Treatment	117.7	138.3	131.7	115.1	157.4	143.8	112.6	157.4	143.0	106.1	137.0	127.1	101.3	174.1	150.6	107.7	138.3	128.4
0960	Flooring	100.0	120.4	105.9	96.8	123.5	104.6	97.2	129.9	106.7	102.3	120.4	107.5	109.6	142.0	119.0	94.9	120.4	102.3
0970, 0990	Wall Finishes & Painting/Coating	101.9	121.1	113.2	92.1	135.6	119.0	92.1	137.6	119.0	88.9	109.3	100.9	113.3	162.7	142.5	88.9	115.6	104.7
09	FINISHES	105.5	134.8	121.5	100.1	147.7	126.1	99.6	148.8	126.5	93.8	132.5	114.9	105.9	163.7	137.5	91.3	133.9	114.6
COVERS	DIVS. 10 - 14, 25, 28, 41, 43, 44, 46	100.0	121.1	104.6	100.0	126.7	105.9	100.0	126.7	105.9	100.0	120.2	104.4	100.0	129.9	106.6	100.0	120.4	104.5
21, 22, 23	FIRE SUPPRESSION, PLUMBING & HVAC	99.9	131.9	112.7	97.5	130.3	110.7	100.0	130.3	112.1	97.6	129.0	110.2	100.2	168.0	127.4	100.1	131.8	112.8
26, 27, 3370	ELECTRICAL, COMMUNICATIONS & UTIL.	99.4	130.5	115.3	93.8	122.4	108.4	96.1	108.4	102.4	93.4	110.7	102.3	102.2	161.0	132.3	99.2	117.5	108.5
MF2018	WEIGHTED AVERAGE	100.2	130.4	113.2	101.9	131.3	114.5	100.6	129.4	112.9	95.6	125.2	108.3	102.8	151.2	123.5	97.9	127.1	110.4

For customer support on your Electrical Costs with RSMeans Data, call 800.448.8182.

525

City Cost Indexes

CALIFORNIA

DIVISION		PALM SPRINGS 922			PALO ALTO 943			PASADENA 910 - 912			REDDING 960			RICHMOND 948			RIVERSIDE 925		
		MAT.	INST.	TOTAL	MAT.	INST.	TOTAL	MAT.	INST.	TOTAL	MAT.	INST.	TOTAL	MAT.	INST.	TOTAL	MAT.	INST.	TOTAL
015433	CONTRACTOR EQUIPMENT		100.7	100.7		102.8	102.8		97.7	97.7		99.6	99.6		102.8	102.8		100.7	100.7
0241, 31 - 34	SITE & INFRASTRUCTURE, DEMOLITION	91.8	108.6	103.5	107.4	110.7	109.7	93.9	110.3	105.3	126.7	107.3	113.2	115.2	110.6	112.0	98.7	108.6	105.6
0310	Concrete Forming & Accessories	98.4	140.1	133.8	99.5	167.2	157.0	104.4	140.0	134.6	103.2	152.7	145.3	114.9	166.8	159.0	102.3	140.1	134.4
0320	Concrete Reinforcing	109.3	129.9	119.8	89.1	131.1	110.5	103.4	130.1	116.9	132.1	130.6	131.3	89.1	131.3	110.5	106.2	129.9	118.2
0330	Cast-in-Place Concrete	81.2	133.2	100.6	94.4	133.6	109.0	79.1	129.8	98.0	101.6	129.5	112.0	108.5	133.5	117.8	88.4	133.2	105.0
03	CONCRETE	90.5	134.6	110.3	91.4	147.0	116.4	90.7	133.3	109.9	114.8	139.0	125.7	103.6	146.8	123.1	96.5	134.6	113.6
04	MASONRY	76.7	139.8	115.7	105.9	146.3	130.8	97.9	141.9	125.1	132.2	141.7	138.1	124.6	150.4	140.6	77.7	139.8	116.1
05	METALS	107.8	111.7	109.0	101.7	113.3	105.2	90.4	110.5	96.5	107.4	112.3	108.8	101.7	113.2	105.2	107.3	111.7	108.6
06	WOOD, PLASTICS & COMPOSITES	92.6	137.4	116.9	92.2	172.6	135.9	88.1	137.1	114.7	110.1	156.3	135.2	110.0	172.6	144.0	97.9	137.4	119.3
07	THERMAL & MOISTURE PROTECTION	104.2	136.2	118.0	97.8	154.4	122.2	95.8	133.2	111.9	131.6	139.2	134.8	98.5	154.5	122.6	105.1	136.2	118.5
08	OPENINGS	102.1	135.9	109.9	92.8	159.5	108.3	89.8	135.7	100.4	118.1	143.8	124.0	92.8	159.5	108.3	104.8	135.9	112.0
0920	Plaster & Gypsum Board	106.2	138.3	127.4	105.7	174.1	150.7	90.3	138.3	121.9	108.9	157.8	141.1	114.2	174.1	153.7	110.5	138.3	128.8
0950, 0980	Ceilings & Acoustic Treatment	109.2	138.3	128.9	99.4	174.1	150.0	112.6	138.3	130.0	148.5	157.8	154.8	99.4	174.1	150.0	118.7	138.3	132.0
0960	Flooring	99.4	120.4	105.5	108.2	142.0	118.0	94.2	120.4	101.8	93.5	132.9	105.0	118.2	142.0	125.1	100.9	120.4	106.6
0970, 0990	Wall Finishes & Painting/Coating	89.1	121.1	108.1	113.3	162.7	142.5	105.4	121.1	114.7	100.0	137.6	122.3	113.3	162.7	142.5	89.1	121.1	108.1
09	FINISHES	96.6	134.5	117.3	104.3	163.7	136.8	98.0	134.3	117.8	106.4	149.4	129.8	109.0	163.7	138.9	99.8	134.5	118.7
COVERS	DIVS. 10 - 14, 25, 28, 41, 43, 44, 46	100.0	120.2	104.4	100.0	129.9	106.6	100.0	119.6	104.3	100.0	126.8	105.9	100.0	129.6	106.6	100.0	120.2	104.4
21, 22, 23	FIRE SUPPRESSION, PLUMBING & HVAC	97.5	131.8	111.2	97.7	168.7	126.2	97.5	131.7	111.2	100.4	130.3	112.4	97.7	159.7	122.6	100.0	131.8	112.7
26, 27, 3370	ELECTRICAL, COMMUNICATIONS & UTIL.	93.3	113.5	103.6	102.1	169.9	136.8	112.3	130.5	121.6	101.9	122.4	112.4	102.7	136.5	120.0	90.1	113.5	102.1
MF2018	WEIGHTED AVERAGE	97.5	127.1	110.2	98.8	151.4	121.4	96.3	129.4	110.5	108.8	131.5	118.6	102.0	145.2	120.5	99.3	127.1	111.2

CALIFORNIA

DIVISION		SACRAMENTO 942,956 - 958			SALINAS 939			SAN BERNARDINO 923 - 924			SAN DIEGO 919 - 921			SAN FRANCISCO 940 - 941			SAN JOSE 951		
		MAT.	INST.	TOTAL	MAT.	INST.	TOTAL	MAT.	INST.	TOTAL	MAT.	INST.	TOTAL	MAT.	INST.	TOTAL	MAT.	INST.	TOTAL
015433	CONTRACTOR EQUIPMENT		101.6	101.6		99.8	99.8		100.7	100.7		102.5	102.5		110.9	110.9		101.3	101.3
0241, 31 - 34	SITE & INFRASTRUCTURE, DEMOLITION	97.4	116.3	110.6	114.7	107.9	110.0	78.2	108.6	99.4	101.9	107.8	106.0	116.9	114.2	115.0	133.8	102.3	111.9
0310	Concrete Forming & Accessories	99.6	155.0	146.7	103.6	155.3	147.5	106.0	140.1	135.0	105.9	128.0	124.6	105.2	167.7	158.3	102.1	166.9	157.2
0320	Concrete Reinforcing	84.6	130.9	108.2	102.8	131.0	117.2	106.2	129.9	118.2	100.1	129.8	115.2	103.7	132.6	118.4	94.9	131.4	113.5
0330	Cast-in-Place Concrete	89.5	130.5	104.7	96.3	129.8	108.8	61.0	133.2	87.9	90.6	123.5	102.9	118.1	133.8	123.9	107.5	132.7	116.9
03	CONCRETE	93.0	140.1	114.2	102.7	140.4	119.7	71.1	134.6	99.7	101.2	125.8	112.3	112.7	148.1	128.6	104.3	147.0	123.5
04	MASONRY	106.8	147.2	131.8	95.4	147.3	127.5	84.0	139.8	118.5	89.8	132.3	116.0	142.9	153.6	149.5	132.3	150.4	143.5
05	METALS	99.8	107.5	102.1	105.0	113.2	107.5	107.3	111.7	108.6	107.4	113.7	109.3	109.4	122.6	113.4	101.4	119.0	106.8
06	WOOD, PLASTICS & COMPOSITES	88.9	159.3	127.1	98.1	159.0	131.2	101.6	137.4	121.0	99.4	124.9	113.2	97.0	172.6	138.1	106.5	172.4	142.3
07	THERMAL & MOISTURE PROTECTION	108.4	141.0	122.4	98.3	145.2	118.5	103.3	136.2	117.5	100.6	120.3	109.1	103.3	155.7	125.8	104.1	155.6	126.3
08	OPENINGS	105.8	146.7	115.3	95.7	152.0	108.8	102.1	135.9	110.0	98.3	126.1	104.7	98.5	159.8	112.7	94.6	159.4	109.7
0920	Plaster & Gypsum Board	103.2	160.5	140.9	98.9	160.5	139.5	111.8	138.3	129.3	96.3	125.3	115.4	105.3	174.1	150.6	107.3	174.1	151.3
0950, 0980	Ceilings & Acoustic Treatment	99.4	160.5	140.8	106.1	160.5	143.0	112.6	138.3	130.0	123.1	125.3	124.6	108.6	174.1	153.0	113.6	174.1	154.6
0960	Flooring	108.2	127.6	113.9	96.9	140.6	109.6	102.8	120.8	108.0	103.6	120.4	108.4	110.0	142.0	119.3	90.8	142.0	105.7
0970, 0990	Wall Finishes & Painting/Coating	109.9	137.6	126.3	89.7	162.7	132.9	89.1	121.1	108.1	103.0	121.1	113.7	111.2	179.4	151.6	92.4	162.7	134.0
09	FINISHES	103.2	150.4	128.9	93.4	155.2	127.1	98.0	134.5	117.9	105.5	125.5	116.4	107.5	165.6	139.2	98.9	163.6	134.2
COVERS	DIVS. 10 - 14, 25, 28, 41, 43, 44, 46	100.0	127.6	106.1	100.0	127.1	106.0	100.0	120.2	104.4	100.0	117.0	103.7	100.0	130.0	106.6	100.0	129.4	106.5
21, 22, 23	FIRE SUPPRESSION, PLUMBING & HVAC	100.1	129.9	112.1	97.6	136.3	113.1	97.5	131.8	111.2	99.9	128.4	111.3	100.1	185.2	134.2	100.0	168.6	127.5
26, 27, 3370	ELECTRICAL, COMMUNICATIONS & UTIL.	97.7	122.4	110.3	94.4	127.6	111.4	93.3	111.8	102.7	98.7	102.7	100.8	102.3	176.3	140.1	99.1	169.9	135.3
MF2018	WEIGHTED AVERAGE	100.1	132.7	114.1	99.1	135.7	114.8	95.2	126.6	108.8	101.0	120.8	109.5	106.3	158.1	128.5	102.6	151.7	123.6

CALIFORNIA

DIVISION		SAN LUIS OBISPO 934			SAN MATEO 944			SAN RAFAEL 949			SANTA ANA 926 - 927			SANTA BARBARA 931			SANTA CRUZ 950		
		MAT.	INST.	TOTAL	MAT.	INST.	TOTAL	MAT.	INST.	TOTAL	MAT.	INST.	TOTAL	MAT.	INST.	TOTAL	MAT.	INST.	TOTAL
015433	CONTRACTOR EQUIPMENT		99.8	99.8		102.8	102.8		102.0	102.0		100.7	100.7		99.8	99.8		101.3	101.3
0241, 31 - 34	SITE & INFRASTRUCTURE, DEMOLITION	106.9	108.4	108.0	113.1	110.7	111.4	110.0	116.2	114.3	90.3	108.6	103.0	101.3	108.4	106.3	133.4	102.2	111.6
0310	Concrete Forming & Accessories	110.1	140.2	135.7	105.4	167.1	157.8	110.7	167.0	158.6	106.1	140.1	135.0	101.0	140.0	134.3	102.1	155.6	147.6
0320	Concrete Reinforcing	104.1	129.8	117.2	89.1	131.6	110.7	89.7	131.6	111.0	109.8	129.9	120.1	102.3	129.9	116.3	117.4	131.1	124.3
0330	Cast-in-Place Concrete	103.6	132.4	114.3	105.0	133.6	115.7	122.2	132.6	126.1	77.9	133.2	98.5	97.2	132.4	110.3	106.7	131.6	116.0
03	CONCRETE	101.2	134.3	116.1	100.3	147.0	121.3	121.8	146.4	132.9	88.0	134.6	109.0	93.7	134.3	111.9	106.6	141.4	122.3
04	MASONRY	97.1	137.3	121.9	124.3	153.6	142.4	101.5	150.6	133.7	73.6	140.2	114.7	95.7	137.3	121.4	136.1	147.5	143.1
05	METALS	103.0	111.2	105.5	101.6	113.7	105.3	102.7	110.0	104.9	107.4	111.7	108.7	100.7	111.2	103.9	109.0	117.6	111.6
06	WOOD, PLASTICS & COMPOSITES	100.5	137.4	120.6	99.7	172.6	139.3	97.4	172.4	138.1	103.5	137.4	121.9	93.3	137.4	117.3	106.5	159.2	135.1
07	THERMAL & MOISTURE PROTECTION	98.5	133.7	113.6	98.2	155.9	123.1	102.1	156.2	125.4	104.6	136.3	118.2	97.9	133.7	113.3	103.6	147.5	122.5
08	OPENINGS	93.9	133.8	103.2	92.8	159.5	108.3	103.4	159.4	116.4	101.4	135.9	109.4	95.5	135.9	104.9	95.9	152.1	108.9
0920	Plaster & Gypsum Board	104.7	138.3	126.8	110.8	174.1	152.5	112.7	174.1	153.2	113.3	138.3	129.8	97.8	138.3	124.5	115.3	160.5	145.1
0950, 0980	Ceilings & Acoustic Treatment	106.1	138.3	127.9	99.4	174.1	150.0	106.2	174.1	152.2	112.6	138.3	130.0	107.7	138.3	128.4	113.4	160.5	145.3
0960	Flooring	103.1	120.4	108.1	111.8	142.0	120.6	123.4	140.6	128.4	103.4	120.4	108.3	96.0	120.4	103.1	94.8	140.6	108.1
0970, 0990	Wall Finishes & Painting/Coating	88.9	115.6	104.7	113.3	162.7	142.5	109.3	169.9	145.2	89.1	121.1	108.1	88.9	115.6	104.7	92.5	162.7	134.0
09	FINISHES	95.1	133.9	116.3	106.5	163.7	137.8	109.1	164.1	139.2	99.5	134.5	118.6	91.8	133.9	114.8	101.2	155.2	130.7
COVERS	DIVS. 10 - 14, 25, 28, 41, 43, 44, 46	100.0	126.4	105.8	100.0	129.9	106.6	100.0	129.2	106.4	100.0	120.2	104.4	100.0	119.5	104.3	100.0	127.4	106.0
21, 22, 23	FIRE SUPPRESSION, PLUMBING & HVAC	97.6	131.8	111.3	97.7	163.6	124.1	97.7	185.1	132.7	97.5	131.8	111.2	100.1	131.8	112.8	100.0	136.7	114.7
26, 27, 3370	ELECTRICAL, COMMUNICATIONS & UTIL.	93.4	113.1	103.5	102.1	163.0	133.3	98.9	124.3	111.9	93.3	112.8	103.3	92.4	114.2	103.5	98.3	127.6	113.3
MF2018	WEIGHTED AVERAGE	98.4	126.6	110.5	101.2	150.2	122.2	103.9	149.3	123.4	97.2	127.0	110.0	97.2	126.7	109.8	104.5	136.0	118.0

City Cost Indexes

CALIFORNIA / COLORADO

Division		SANTA ROSA 954			STOCKTON 952			SUSANVILLE 961			VALLEJO 945			VAN NUYS 913–916			ALAMOSA 811		
		MAT.	INST.	TOTAL	MAT.	INST.	TOTAL	MAT.	INST.	TOTAL	MAT.	INST.	TOTAL	MAT.	INST.	TOTAL	MAT.	INST.	TOTAL
015433	CONTRACTOR EQUIPMENT		100.1	100.1		99.6	99.6		99.6	99.6		102.0	102.0		97.7	97.7		93.1	93.1
0241, 31 - 34	SITE & INFRASTRUCTURE, DEMOLITION	104.2	107.3	106.4	102.5	107.3	105.8	134.4	107.3	115.5	98.7	116.1	110.8	111.4	110.3	110.6	102.0	62.7	104.1
0310	Concrete Forming & Accessories	99.3	166.2	156.2	100.1	154.6	146.4	104.4	152.5	145.3	101.3	165.9	156.2	103.4	130.1	116.9	111.5	69.8	90.3
0320	Concrete Reinforcing	104.8	131.5	118.4	107.6	130.5	119.3	132.1	130.4	131.2	90.8	131.4	111.5	83.5	129.8	100.7	100.1	73.5	90.2
0330	Cast-in-Place Concrete	101.5	131.1	112.5	90.1	129.5	104.8	92.5	129.5	106.3	97.4	131.7	110.2	105.2	133.3	117.9	111.4	68.4	92.0
03	CONCRETE	107.3	145.9	124.7	97.4	139.9	116.5	118.4	138.9	127.6	97.6	145.6	119.2	105.2	133.3	117.9	111.4	68.4	92.0
04	MASONRY	102.3	152.5	133.3	102.0	141.7	126.5	130.7	139.2	136.0	77.4	152.5	123.8	111.7	141.9	130.4	129.7	60.6	87.0
05	METALS	107.7	115.1	109.9	103.6	112.2	106.2	106.3	112.1	108.0	102.7	109.3	104.7	89.5	110.5	95.9	104.3	78.4	96.4
06	WOOD, PLASTICS & COMPOSITES	94.1	172.2	136.5	99.9	159.0	132.0	111.9	156.0	135.9	86.6	172.4	133.2	97.2	137.1	118.9	99.6	63.0	79.7
07	THERMAL & MOISTURE PROTECTION	105.0	152.9	125.6	108.0	138.1	121.0	133.4	138.4	135.6	100.3	153.6	123.2	96.9	133.2	112.5	104.9	68.2	89.1
08	OPENINGS	103.0	159.3	116.1	103.5	146.6	113.5	119.0	144.9	125.0	105.1	159.4	117.7	89.6	135.7	100.3	98.5	66.4	91.0
0920	Plaster & Gypsum Board	108.2	174.1	151.6	111.2	160.5	143.7	109.6	157.4	141.1	106.9	174.1	151.2	94.4	138.3	123.3	81.5	61.9	68.6
0950, 0980	Ceilings & Acoustic Treatment	112.6	174.1	154.3	122.1	160.5	148.1	140.0	157.4	151.8	108.1	174.1	152.8	109.2	138.3	128.9	103.8	61.9	75.4
0960	Flooring	99.9	133.5	109.7	97.2	129.9	106.7	94.0	132.9	105.3	117.5	142.0	124.6	97.3	120.4	104.0	108.4	71.3	97.6
0970, 0990	Wall Finishes & Painting/Coating	89.1	169.9	136.9	92.1	137.6	119.0	100.0	137.6	122.3	110.3	169.9	145.6	105.4	121.1	114.7	102.7	60.8	77.9
09	FINISHES	98.6	162.5	133.5	101.5	150.6	128.3	106.0	149.3	129.6	105.8	164.1	137.6	100.1	134.3	118.8	98.4	63.5	79.4
COVERS	DIVS. 10 - 14, 25, 28, 41, 43, 44, 46	100.0	128.3	106.2	100.0	127.0	106.0	100.0	126.7	105.9	100.0	128.9	106.4	100.0	119.6	104.3	100.0	84.8	96.6
21, 22, 23	FIRE SUPPRESSION, PLUMBING & HVAC	97.5	184.5	132.4	100.0	130.3	112.1	97.9	130.3	110.9	100.2	147.9	119.3	97.5	131.7	111.2	97.5	66.0	84.9
26, 27, 3370	ELECTRICAL, COMMUNICATIONS & UTIL.	93.6	124.3	109.3	96.1	114.3	105.4	102.2	122.4	112.6	95.1	127.9	111.9	112.3	130.5	121.6	97.0	64.8	80.5
MF2018	WEIGHTED AVERAGE	101.4	148.5	121.6	100.7	130.8	113.6	108.8	131.2	118.4	99.5	141.5	117.5	99.3	129.4	112.3	103.5	68.8	88.6

COLORADO

Division		BOULDER 803			COLORADO SPRINGS 808–809			DENVER 800–802			DURANGO 813			FORT COLLINS 805			FORT MORGAN 807		
		MAT.	INST.	TOTAL	MAT.	INST.	TOTAL	MAT.	INST.	TOTAL	MAT.	INST.	TOTAL	MAT.	INST.	TOTAL	MAT.	INST.	TOTAL
015433	CONTRACTOR EQUIPMENT		95.2	95.2		93.1	93.1		99.4	99.4		93.1	93.1		95.2	95.2		95.2	95.2
0241, 31 - 34	SITE & INFRASTRUCTURE, DEMOLITION	98.0	95.8	96.5	99.9	91.3	93.9	104.9	102.3	103.1	133.3	88.5	102.1	110.8	95.7	100.2	100.5	95.5	97.0
0310	Concrete Forming & Accessories	101.7	73.6	77.8	92.3	65.1	69.2	101.0	64.8	70.3	108.4	62.5	69.4	99.3	68.6	73.3	102.3	62.4	68.4
0320	Concrete Reinforcing	98.3	69.8	83.8	97.5	69.9	83.5	97.5	69.9	83.5	111.5	69.7	90.3	98.3	69.7	83.8	98.5	69.7	83.9
0330	Cast-in-Place Concrete	112.5	72.8	97.7	115.5	73.9	100.0	120.5	71.3	102.2	115.1	73.4	99.6	127.0	71.4	106.3	103.1	67.9	87.3
03	CONCRETE	104.6	73.0	90.4	108.2	69.6	90.8	111.0	68.6	91.9	113.6	68.2	93.2	115.9	70.3	95.4	116.7	60.9	82.2
04	MASONRY	101.3	63.3	77.8	103.2	63.6	78.7	104.4	63.8	79.3	104.3	78.2	96.4	98.2	77.3	91.9	96.8	77.3	90.8
05	METALS	97.0	77.4	91.1	100.0	78.1	93.4	102.4	78.6	95.2	109.5	63.0	84.2	104.2	70.9	86.1	106.7	62.5	82.7
06	WOOD, PLASTICS & COMPOSITES	106.7	76.3	90.2	96.1	65.5	79.5	104.3	65.3	83.1	104.9	68.2	89.1	96.5	70.4	85.3	96.1	69.6	84.7
07	THERMAL & MOISTURE PROTECTION	96.2	72.7	86.1	96.9	71.4	85.9	94.7	70.9	84.5	104.9	68.2	89.1	96.5	70.4	85.3	97.4	66.1	90.1
08	OPENINGS	97.5	73.4	91.9	101.7	67.8	93.8	104.5	67.7	96.0	105.4	66.4	96.4	97.4	70.4	91.2	97.4	66.1	90.1
0920	Plaster & Gypsum Board	116.5	76.0	89.8	100.0	64.8	76.8	111.7	64.8	80.8	95.3	61.9	73.3	110.7	70.5	84.2	116.5	61.9	80.5
0950, 0980	Ceilings & Acoustic Treatment	86.2	76.0	79.3	93.0	64.8	73.9	97.5	64.8	75.3	103.8	61.9	75.4	86.2	70.5	75.5	86.2	61.9	69.7
0960	Flooring	108.3	71.3	97.6	99.7	71.3	91.5	105.9	71.3	95.9	113.4	71.3	101.2	104.6	71.3	95.0	108.7	71.3	96.5
0970, 0990	Wall Finishes & Painting/Coating	99.4	63.0	77.8	99.1	63.0	77.7	106.2	63.0	80.6	101.0	63.5	80.6	99.1	68.2	82.2	100.0	63.2	80.2
09	FINISHES	100.5	72.2	85.0	97.0	65.5	79.8	102.3	65.3	82.1	101.0	65.3	80.6	99.1	68.2	82.2	100.0	63.2	80.2
COVERS	DIVS. 10 - 14, 25, 28, 41, 43, 44, 46	100.0	86.1	96.9	100.0	84.7	96.6	100.0	84.2	96.5	100.0	84.7	96.6	100.0	84.7	96.6	100.0	83.8	96.4
21, 22, 23	FIRE SUPPRESSION, PLUMBING & HVAC	97.6	74.6	88.4	100.2	73.8	89.6	100.0	73.6	89.4	97.5	57.2	81.3	100.1	73.3	89.3	97.6	73.3	87.9
26, 27, 3370	ELECTRICAL, COMMUNICATIONS & UTIL.	96.5	80.0	88.1	99.6	74.9	87.0	101.3	80.2	90.5	96.5	51.5	73.5	96.5	77.2	86.6	96.8	80.0	88.2
MF2018	WEIGHTED AVERAGE	98.9	75.9	89.0	101.0	73.0	89.0	102.7	74.4	90.6	103.9	65.0	87.2	102.1	73.7	89.9	99.5	72.9	88.1

COLORADO

Division		GLENWOOD SPRINGS 816			GOLDEN 804			GRAND JUNCTION 815			GREELEY 806			MONTROSE 814			PUEBLO 810		
		MAT.	INST.	TOTAL	MAT.	INST.	TOTAL	MAT.	INST.	TOTAL	MAT.	INST.	TOTAL	MAT.	INST.	TOTAL	MAT.	INST.	TOTAL
015433	CONTRACTOR EQUIPMENT		96.3	96.3		95.2	95.2		96.3	96.3		95.2	95.2		94.7	94.7		93.1	93.1
0241, 31 - 34	SITE & INFRASTRUCTURE, DEMOLITION	148.9	96.4	112.3	111.2	95.8	100.5	132.8	96.6	107.6	96.7	95.7	96.0	142.3	92.5	107.6	124.4	88.9	99.7
0310	Concrete Forming & Accessories	98.7	52.4	59.4	94.6	63.4	68.1	107.2	73.0	78.2	97.3	72.6	76.3	98.2	62.7	68.0	104.4	65.4	71.2
0320	Concrete Reinforcing	110.3	69.7	89.7	98.5	69.7	83.9	110.7	69.8	89.9	98.3	69.7	83.8	110.2	69.8	89.6	106.6	69.8	87.9
0330	Cast-in-Place Concrete	100.1	72.3	89.7	110.4	74.1	96.9	110.8	73.7	97.0	106.1	72.7	93.7	100.1	72.9	90.0	99.4	74.2	90.0
03	CONCRETE	116.4	63.3	92.5	113.4	68.9	93.4	110.1	73.2	93.5	99.6	72.6	87.4	107.4	68.2	89.8	100.3	69.9	86.6
04	MASONRY	101.6	60.8	76.4	119.7	63.3	84.9	136.8	64.0	91.8	112.5	60.9	80.7	109.5	60.6	79.3	97.8	61.6	75.4
05	METALS	104.0	77.9	96.0	96.9	77.3	91.0	105.6	78.2	97.3	98.2	77.4	91.9	103.0	78.3	95.5	107.2	79.0	98.7
06	WOOD, PLASTICS & COMPOSITES	94.6	49.2	69.9	98.5	62.5	79.0	107.2	75.7	90.1	101.3	76.2	87.7	95.6	62.8	77.8	102.3	65.8	82.5
07	THERMAL & MOISTURE PROTECTION	104.8	66.7	88.4	97.3	71.3	86.1	103.8	71.6	89.9	95.8	71.0	85.1	104.9	68.2	89.1	103.3	69.5	88.5
08	OPENINGS	104.4	58.8	93.8	97.5	66.1	90.2	105.1	73.1	97.7	97.4	73.7	91.9	105.6	66.3	96.5	100.3	67.9	92.8
0920	Plaster & Gypsum Board	122.5	48.0	73.4	108.0	61.9	77.6	135.9	75.2	96.0	109.2	76.0	87.3	80.6	61.9	68.3	84.9	64.8	71.7
0950, 0980	Ceilings & Acoustic Treatment	103.0	48.0	65.7	86.2	61.9	69.7	103.0	75.2	84.2	86.2	76.0	79.3	103.8	61.9	75.4	109.8	64.8	79.3
0960	Flooring	107.5	71.3	97.0	102.3	71.3	93.3	112.8	71.3	100.7	103.5	71.3	94.2	110.8	71.3	99.4	109.5	75.5	99.7
0970, 0990	Wall Finishes & Painting/Coating	102.7	60.8	77.9	99.4	63.0	77.8	102.7	63.0	79.2	99.4	60.8	76.5	102.7	60.8	77.9	102.7	60.8	77.9
09	FINISHES	104.3	55.4	77.6	98.9	64.1	79.9	106.0	71.8	87.3	97.8	71.3	83.3	99.0	63.4	79.6	98.6	66.3	80.9
COVERS	DIVS. 10 - 14, 25, 28, 41, 43, 44, 46	100.0	82.6	96.2	100.0	84.6	96.6	100.0	86.3	97.0	100.0	85.3	96.8	100.0	84.4	96.6	100.0	85.4	96.8
21, 22, 23	FIRE SUPPRESSION, PLUMBING & HVAC	97.5	67.0	85.3	97.6	74.6	88.4	99.9	75.8	90.2	100.1	73.3	89.3	97.5	75.8	88.8	97.0	64.8	80.5
26, 27, 3370	ELECTRICAL, COMMUNICATIONS & UTIL.	94.0	51.5	72.3	96.8	80.0	88.2	96.2	51.5	73.3	96.5	72.7	84.8	96.2	51.5	73.3	101.4	71.4	88.5
MF2018	WEIGHTED AVERAGE	103.7	65.4	87.2	101.0	73.7	89.3	105.5	72.2	91.2	99.3	74.7	88.7	102.5	69.3	88.3	101.4	71.4	88.5

For customer support on your Electrical Costs with RSMeans Data, call 800.448.8182.

City Cost Indexes

COLORADO / CONNECTICUT

DIVISION		SALIDA 812 MAT	INST	TOTAL	BRIDGEPORT 066 MAT	INST	TOTAL	BRISTOL 060 MAT	INST	TOTAL	HARTFORD 061 MAT	INST	TOTAL	MERIDEN 064 MAT	INST	TOTAL	NEW BRITAIN 060 MAT	INST	TOTAL
015433	CONTRACTOR EQUIPMENT		94.7	94.7		97.9	97.9		97.9	97.9		97.9	97.9		98.3	98.3		97.9	97.9
0241, 31-34	SITE & INFRASTRUCTURE, DEMOLITION	133.0	92.4	104.8	103.3	102.3	102.6	102.4	102.3	102.3	97.8	102.3	100.9	100.2	103.1	102.2	102.6	102.3	102.4
0310	Concrete Forming & Accessories	107.1	52.5	60.7	98.6	119.5	116.3	98.6	119.3	116.2	99.9	119.3	116.4	98.3	119.3	116.2	99.0	119.3	116.3
0320	Concrete Reinforcing	109.9	69.7	89.5	125.5	145.1	135.4	125.5	145.1	135.4	120.3	145.1	132.9	125.5	145.1	135.4	99.0	119.3	116.3
0330	Cast-in-Place Concrete	114.7	72.8	99.1	102.1	128.8	112.1	95.7	128.7	112.1	100.2	128.7	110.8	92.1	128.7	105.7	125.5	145.1	135.4
03	CONCRETE	108.7	63.5	88.4	98.6	126.2	111.0	95.5	126.1	109.3	95.9	126.1	109.5	93.8	126.1	108.3	97.2	128.7	108.9
04	MASONRY	138.1	60.6	90.2	116.6	130.9	125.4	107.4	130.9	121.9	106.8	130.9	121.7	107.0	130.9	121.8	96.3	126.1	109.7
05	METALS	102.7	78.1	95.2	91.2	115.7	98.6	91.2	115.6	98.6	95.7	115.6	101.7	88.5	115.6	96.7	109.7	130.9	122.8
06	WOOD, PLASTICS & COMPOSITES	103.6	49.3	74.1	100.8	117.5	109.9	100.8	117.5	109.9	91.0	117.5	105.4	100.8	117.5	109.9	87.7	115.6	96.1
07	THERMAL & MOISTURE PROTECTION	103.8	66.7	87.8	98.0	125.8	110.0	98.1	121.9	108.4	102.8	121.9	111.0	98.1	121.9	108.4	100.8	117.5	109.9
08	OPENINGS	98.6	58.8	89.3	98.4	122.7	104.0	98.4	122.7	104.0	96.5	122.7	102.5	100.6	122.7	105.7	98.1	122.2	108.5
0920	Plaster & Gypsum Board	80.9	48.0	59.2	116.7	117.7	117.4	116.7	117.7	117.4	104.7	117.7	113.3	117.8	117.7	117.7	98.4	122.7	104.0
0950, 0980	Ceilings & Acoustic Treatment	103.8	48.0	66.0	97.9	117.7	111.3	97.9	117.7	111.3	94.1	117.7	110.1	101.1	117.7	112.4	116.7	117.7	117.4
0960	Flooring	116.1	71.3	103.1	90.6	131.2	102.4	90.6	128.7	101.6	94.9	131.2	105.4	90.6	128.7	101.6	90.6	128.7	101.6
0970, 0990	Wall Finishes & Painting/Coating	102.7	60.8	77.9	89.8	128.8	112.9	89.8	128.8	112.9	94.3	128.8	114.8	89.8	128.8	112.9	89.8	128.8	112.9
09	FINISHES	99.5	55.5	75.5	90.9	122.2	108.0	90.9	121.8	107.8	92.0	122.2	108.5	91.7	121.8	108.1	89.8	121.8	107.8
COVERS	DIVS. 10-14, 25, 28, 41, 43, 44, 46	100.0	83.0	96.3	100.0	113.4	102.9	100.0	113.4	102.9	100.0	113.4	102.9	100.0	113.4	102.9	100.0	113.4	102.9
21, 22, 23	FIRE SUPPRESSION, PLUMBING & HVAC	97.5	65.0	84.5	100.1	118.5	107.5	100.1	118.5	107.5	100.0	118.5	107.4	97.6	118.5	106.0	100.1	118.5	107.5
26, 27, 3370	ELECTRICAL, COMMUNICATIONS & UTIL.	96.4	64.8	80.2	97.4	107.6	102.6	97.4	106.2	101.9	97.0	109.3	103.3	97.3	108.4	103.0	97.4	106.2	101.9
MF2018	WEIGHTED AVERAGE	103.1	66.6	87.5	98.1	118.4	106.8	97.2	118.1	106.2	97.8	118.5	106.7	96.2	118.4	105.7	96.9	118.1	106.0

CONNECTICUT

DIVISION		NEW HAVEN 065 MAT	INST	TOTAL	NEW LONDON 063 MAT	INST	TOTAL	NORWALK 068 MAT	INST	TOTAL	STAMFORD 069 MAT	INST	TOTAL	WATERBURY 067 MAT	INST	TOTAL	WILLIMANTIC 062 MAT	INST	TOTAL
015433	CONTRACTOR EQUIPMENT		98.3	98.3		98.3	98.3		97.9	97.9		97.9	97.9		97.9	97.9		97.9	97.9
0241, 31-34	SITE & INFRASTRUCTURE, DEMOLITION	102.5	103.1	102.9	94.9	102.8	100.4	103.1	102.3	102.5	103.7	102.3	102.8	103.1	102.3	102.6	103.0	102.1	102.3
0310	Concrete Forming & Accessories	98.3	119.4	116.2	98.3	119.3	116.1	98.6	119.5	116.4	98.6	119.8	116.6	98.6	119.4	116.2	98.6	119.2	116.1
0320	Concrete Reinforcing	125.5	145.1	135.4	98.3	145.1	122.1	125.5	145.2	135.5	125.5	145.3	135.5	125.5	145.1	135.4	125.5	145.1	135.4
0330	Cast-in-Place Concrete	98.9	128.7	110.0	84.3	128.7	100.8	100.5	130.0	111.5	102.1	130.1	112.6	102.1	128.7	112.0	95.4	127.0	107.1
03	CONCRETE	108.9	126.1	116.6	84.3	126.0	103.1	97.8	126.6	110.8	98.6	126.8	111.3	98.6	126.1	111.0	95.4	125.4	108.9
04	MASONRY	107.7	130.9	122.0	106.0	130.9	121.4	107.1	132.2	122.6	108.0	132.2	122.9	108.0	130.9	122.1	107.3	130.9	121.9
05	METALS	87.9	115.6	96.3	87.6	115.6	96.1	91.2	115.9	98.7	91.2	116.2	98.8	91.2	115.6	98.6	90.9	115.4	98.4
06	WOOD, PLASTICS & COMPOSITES	100.8	117.5	109.9	100.8	117.5	109.9	100.8	117.5	109.9	100.8	117.5	109.9	100.8	117.5	109.9	100.8	115.4	109.9
07	THERMAL & MOISTURE PROTECTION	98.2	122.5	108.7	98.0	121.9	108.3	98.2	126.4	110.3	98.1	126.4	110.3	98.1	122.5	108.6	98.3	121.3	108.2
08	OPENINGS	98.4	122.7	104.0	100.8	122.7	105.9	98.4	122.7	104.0	98.4	122.7	104.0	98.4	122.7	104.0	100.8	122.7	105.9
0920	Plaster & Gypsum Board	116.7	117.7	117.4	116.7	117.7	117.4	116.7	117.7	117.4	116.7	117.7	117.4	116.7	117.7	117.4	116.7	117.7	117.4
0950, 0980	Ceilings & Acoustic Treatment	97.9	117.7	111.3	96.0	117.7	110.7	97.9	117.7	111.3	116.7	117.7	117.4	116.7	117.7	117.4	116.7	117.7	117.4
0960	Flooring	90.6	131.2	102.4	90.6	128.7	101.6	90.6	128.7	101.6	97.9	117.7	111.3	97.9	117.7	111.3	96.0	117.7	110.7
0970, 0990	Wall Finishes & Painting/Coating	89.8	128.8	112.9	89.8	128.8	112.9	89.8	128.8	112.9	90.6	131.2	102.4	90.6	131.2	102.4	90.6	125.2	100.6
09	FINISHES	91.0	122.2	108.0	90.1	121.8	107.4	90.9	121.8	107.8	91.0	122.2	108.0	90.8	122.2	107.9	90.6	121.2	107.3
COVERS	DIVS. 10-14, 25, 28, 41, 43, 44, 46	100.0	113.4	102.9	100.0	113.4	102.9	100.0	113.4	102.9	100.0	113.6	103.0	100.0	113.4	103.0	100.0	113.4	102.9
21, 22, 23	FIRE SUPPRESSION, PLUMBING & HVAC	100.1	118.5	107.5	97.6	118.5	106.0	100.1	118.5	107.5	100.1	118.6	107.5	100.1	118.5	107.5	100.0	113.4	102.9
26, 27, 3370	ELECTRICAL, COMMUNICATIONS & UTIL.	97.3	108.6	103.1	94.5	108.4	101.6	97.4	107.4	102.5	97.4	153.8	126.2	96.9	109.7	103.5	100.1	118.4	107.4
MF2018	WEIGHTED AVERAGE	98.4	118.5	107.0	94.3	118.4	104.7	97.5	118.6	106.6	97.7	125.2	109.5	97.6	118.6	106.6	97.4	118.2	106.4

D.C. / DELAWARE / FLORIDA

DIVISION		WASHINGTON 200-205 MAT	INST	TOTAL	DOVER 199 MAT	INST	TOTAL	NEWARK 197 MAT	INST	TOTAL	WILMINGTON 198 MAT	INST	TOTAL	DAYTONA BEACH 321 MAT	INST	TOTAL	FORT LAUDERDALE 333 MAT	INST	TOTAL
015433	CONTRACTOR EQUIPMENT		105.6	105.6		121.3	121.3		121.3	121.3		121.5	121.5		102.1	102.1		95.2	95.2
0241, 31-34	SITE & INFRASTRUCTURE, DEMOLITION	105.2	95.9	98.7	100.5	113.6	109.7	102.7	113.6	110.3	105.0	114.0	111.2	113.2	88.9	96.3	92.1	76.8	81.4
0310	Concrete Forming & Accessories	97.0	74.0	77.4	96.8	101.9	101.2	96.5	101.9	101.1	95.0	101.9	100.9	97.1	62.7	67.9	92.0	63.3	67.6
0320	Concrete Reinforcing	100.9	88.7	94.7	109.1	118.4	113.8	103.3	118.4	111.0	107.4	118.4	113.0	93.3	62.7	77.7	94.5	61.4	77.7
0330	Cast-in-Place Concrete	112.6	80.6	100.7	100.6	107.5	103.1	90.0	107.5	96.5	102.3	107.5	104.3	88.1	65.9	79.8	90.1	65.9	79.8
03	CONCRETE	109.9	79.8	96.4	96.3	107.7	101.4	92.4	107.7	99.3	96.8	107.7	101.7	87.5	65.6	77.6	90.1	64.2	78.4
04	MASONRY	100.4	80.9	88.3	97.5	100.1	99.1	101.3	100.1	100.5	97.1	100.1	98.9	88.4	59.5	70.6	93.3	62.3	74.1
05	METALS	99.1	95.8	98.1	92.8	123.7	102.2	97.5	123.7	105.5	97.4	123.7	105.4	101.5	89.0	97.7	99.3	89.2	96.3
06	WOOD, PLASTICS & COMPOSITES	95.0	72.7	82.9	88.7	100.2	95.0	91.1	100.2	96.0	81.7	100.2	91.8	96.4	61.8	77.6	80.2	65.2	72.1
07	THERMAL & MOISTURE PROTECTION	103.7	84.9	95.6	104.3	112.7	107.9	107.0	112.7	109.5	103.9	112.7	107.7	99.8	65.7	85.1	98.5	62.5	83.0
08	OPENINGS	99.3	75.9	93.9	90.4	110.4	95.1	93.4	110.4	97.4	88.8	110.4	93.9	95.1	60.9	87.2	96.4	62.4	88.5
0920	Plaster & Gypsum Board	105.7	71.9	83.5	97.8	100.2	99.4	95.9	100.2	98.7	96.8	100.2	99.1	95.0	61.2	72.8	109.2	64.7	79.9
0950, 0980	Ceilings & Acoustic Treatment	105.6	71.9	82.8	99.1	100.2	99.9	96.6	100.2	99.1	90.3	100.2	97.0	83.7	61.2	66.0	86.1	64.7	71.6
0960	Flooring	96.4	78.1	91.1	90.4	109.0	95.8	87.7	109.0	93.9	91.5	109.0	96.6	101.0	56.7	88.2	86.8	76.7	91.6
0970, 0990	Wall Finishes & Painting/Coating	99.1	75.0	84.8	88.6	122.2	108.5	84.6	122.2	106.8	86.8	122.2	107.7	97.1	63.1	77.0	88.4	58.6	70.8
09	FINISHES	97.9	74.0	84.9	91.5	104.5	98.6	87.7	104.5	96.9	90.1	104.5	97.9	90.4	61.0	74.4	90.5	65.1	76.7
COVERS	DIVS. 10-14, 25, 28, 41, 43, 44, 46	100.0	96.3	99.2	100.0	105.4	101.2	100.0	105.4	101.2	100.0	105.4	101.2	100.0	82.7	96.2	100.0	82.7	96.2
21, 22, 23	FIRE SUPPRESSION, PLUMBING & HVAC	100.1	90.3	96.2	100.0	105.4	101.2	100.2	122.0	108.8	100.1	122.0	108.9	100.0	82.7	96.2	100.0	82.7	96.2
26, 27, 3370	ELECTRICAL, COMMUNICATIONS & UTIL.	101.3	102.4	101.9	96.7	112.4	104.7	99.2	114.4	106.0	98.0	112.4	105.4	94.4	60.3	77.0	91.7	69.1	80.1
MF2018	WEIGHTED AVERAGE	101.2	87.7	95.4	96.3	112.3	103.1	97.2	112.3	103.7	97.0	112.3	103.6	96.6	70.3	85.3	96.0	69.5	84.6

For customer support on your Electrical Costs with RSMeans Data, call 800.448.8182.

City Cost Indexes

FLORIDA

Code	Division	FORT MYERS 339, 341			GAINESVILLE 326, 344			JACKSONVILLE 320, 322			LAKELAND 338			MELBOURNE 329			MIAMI 330-332, 340		
		MAT.	INST.	TOTAL	MAT.	INST.	TOTAL	MAT.	INST.	TOTAL	MAT.	INST.	TOTAL	MAT.	INST.	TOTAL	MAT.	INST.	TOTAL
015433	CONTRACTOR EQUIPMENT		102.1	102.1		102.1	102.1		102.1	102.1		102.1	102.1		102.1	102.1		95.2	95.2
0241, 31 - 34	SITE & INFRASTRUCTURE, DEMOLITION	103.8	89.0	93.5	122.4	88.7	99.0	113.1	88.9	96.3	105.7	88.9	94.0	120.9	89.1	98.8	93.5	77.1	82.1
0310	Concrete Forming & Accessories	88.1	62.1	66.0	92.3	56.9	62.3	96.8	60.1	65.6	84.8	61.5	65.0	93.5	62.9	67.5	97.3	63.3	68.4
0320	Concrete Reinforcing	95.5	78.9	87.1	99.0	60.7	79.5	93.3	60.8	76.8	97.9	78.3	87.9	94.4	67.4	80.7	101.4	60.9	80.8
0330	Cast-in-Place Concrete	93.9	63.4	82.6	101.2	63.0	87.0	89.0	64.0	79.7	96.1	65.2	84.6	106.2	66.0	91.3	87.0	63.2	78.2
03	CONCRETE	90.5	67.1	80.0	98.3	61.6	81.8	87.9	63.4	76.9	92.2	67.4	81.0	98.8	66.5	84.2	88.3	64.6	77.6
04	MASONRY	87.1	64.8	73.3	100.7	55.9	73.0	88.1	56.5	68.6	101.9	58.3	74.9	85.0	59.6	69.3	92.6	54.7	69.2
05	METALS	101.5	95.2	99.6	100.5	87.9	96.7	100.1	88.1	96.4	101.4	94.2	99.3	110.2	90.8	104.3	99.7	88.3	96.2
06	WOOD, PLASTICS & COMPOSITES	77.1	63.2	69.6	90.0	56.2	71.6	96.4	60.3	76.8	72.4	61.0	66.2	91.8	61.8	75.5	90.3	65.2	76.7
07	THERMAL & MOISTURE PROTECTION	98.3	64.4	83.7	100.2	60.7	83.2	100.1	61.5	83.5	98.2	63.0	83.1	100.3	64.3	84.8	98.4	61.3	82.4
08	OPENINGS	97.7	65.2	90.2	94.8	57.4	86.1	95.1	59.7	86.9	97.7	64.0	89.8	94.4	62.0	86.9	99.9	62.4	91.2
0920	Plaster & Gypsum Board	105.3	62.7	77.2	91.8	55.4	67.8	95.0	59.7	71.8	101.6	60.4	74.5	91.8	61.2	71.7	95.5	64.7	75.2
0950, 0980	Ceilings & Acoustic Treatment	81.6	62.7	68.8	78.1	55.4	62.7	83.7	59.7	67.4	81.6	60.4	67.3	82.9	61.2	68.2	84.9	64.7	71.2
0960	Flooring	93.7	79.8	89.7	98.4	56.7	86.3	101.0	58.5	88.7	91.6	58.4	82.0	98.6	57.8	86.8	98.8	58.4	87.1
0970, 0990	Wall Finishes & Painting/Coating	92.6	63.6	75.4	97.1	63.6	77.3	97.1	63.6	77.3	92.6	63.6	75.4	97.1	83.0	88.8	88.9	58.6	70.9
09	FINISHES	90.3	65.3	76.6	89.1	56.9	71.6	90.5	59.7	73.7	89.2	60.6	73.6	89.8	63.4	75.4	89.1	61.6	74.1
COVERS	DIVS. 10 - 14, 25, 28, 41, 43, 44, 46	100.0	79.4	95.5	100.0	81.1	95.8	100.0	78.2	95.2	100.0	80.4	95.7	100.0	82.8	96.2	100.0	83.0	96.2
21, 22, 23	FIRE SUPPRESSION, PLUMBING & HVAC	98.7	57.5	82.2	98.8	62.3	84.2	99.9	62.3	84.9	98.7	59.4	82.9	99.9	75.3	90.1	100.0	62.5	85.0
26, 27, 3370	ELECTRICAL, COMMUNICATIONS & UTIL.	93.5	61.5	77.1	94.7	57.2	75.5	94.2	63.4	78.5	91.9	60.0	75.6	95.5	64.3	79.5	95.1	72.3	83.5
MF2018	WEIGHTED AVERAGE	96.3	68.3	84.3	98.2	64.9	83.9	96.4	66.5	83.6	97.0	67.1	84.2	99.4	71.2	87.3	96.5	67.4	84.0

FLORIDA

Code	Division	ORLANDO 327-328, 347			PANAMA CITY 324			PENSACOLA 325			SARASOTA 342			ST. PETERSBURG 337			TALLAHASSEE 323		
		MAT.	INST.	TOTAL	MAT.	INST.	TOTAL	MAT.	INST.	TOTAL	MAT.	INST.	TOTAL	MAT.	INST.	TOTAL	MAT.	INST.	TOTAL
015433	CONTRACTOR EQUIPMENT		102.1	102.1		102.1	102.1		102.1	102.1		102.1	102.1		102.1	102.1		102.1	102.1
0241, 31 - 34	SITE & INFRASTRUCTURE, DEMOLITION	113.6	89.2	96.6	126.3	88.2	99.8	126.4	88.6	100.1	116.9	89.0	97.4	107.3	88.5	94.2	106.7	88.7	94.1
0310	Concrete Forming & Accessories	101.5	62.5	68.4	96.1	60.3	65.7	94.2	63.6	68.2	93.8	61.6	66.5	91.4	59.3	64.1	98.6	56.9	63.1
0320	Concrete Reinforcing	98.5	67.4	82.7	97.5	69.8	83.4	100.0	69.2	84.3	93.7	78.3	85.8	97.9	78.2	87.9	99.1	60.7	79.6
0330	Cast-in-Place Concrete	102.6	65.8	88.9	93.5	58.8	80.6	115.6	63.9	96.4	103.7	65.4	89.5	97.1	61.0	83.7	93.1	62.8	81.9
03	CONCRETE	94.0	66.2	81.5	96.4	63.2	81.5	106.2	66.4	88.2	96.0	67.5	83.2	93.7	64.9	80.8	90.7	61.5	77.6
04	MASONRY	92.3	59.5	72.1	92.4	57.4	70.8	113.1	57.4	78.7	90.1	58.8	70.7	138.6	52.1	85.2	90.1	55.4	68.7
05	METALS	99.2	90.4	96.5	101.3	91.4	98.3	102.4	90.8	98.9	102.8	93.6	100.0	102.4	93.9	99.8	99.0	88.0	95.7
06	WOOD, PLASTICS & COMPOSITES	93.5	61.8	76.3	95.2	64.2	78.4	93.1	64.2	77.4	94.4	61.0	76.3	81.5	61.0	70.4	95.5	56.2	74.1
07	THERMAL & MOISTURE PROTECTION	101.9	65.7	86.3	100.5	58.8	82.5	100.4	61.8	83.8	98.5	63.2	83.3	98.5	58.8	81.4	100.6	61.2	83.6
08	OPENINGS	99.4	62.0	90.7	93.2	63.7	86.4	93.2	63.7	86.4	100.0	63.6	91.5	96.5	64.0	88.9	100.3	57.4	90.3
0920	Plaster & Gypsum Board	96.0	61.2	73.1	94.3	63.7	74.1	102.8	63.7	77.0	100.9	60.4	74.2	107.7	60.4	76.6	99.3	55.4	70.4
0950, 0980	Ceilings & Acoustic Treatment	90.0	61.2	70.5	82.9	63.7	69.9	82.9	63.7	69.9	88.7	60.4	69.6	83.5	60.4	67.9	88.3	55.4	66.0
0960	Flooring	97.2	58.4	85.9	100.6	75.0	93.2	96.4	56.7	84.9	103.5	56.7	89.9	95.7	56.7	84.4	99.0	56.7	86.7
0970, 0990	Wall Finishes & Painting/Coating	93.5	61.9	74.8	97.1	63.6	77.3	97.1	63.6	77.3	98.2	63.6	77.7	92.6	63.6	75.4	92.3	63.6	75.3
09	FINISHES	92.1	61.2	75.2	91.5	63.5	76.2	91.1	62.1	75.3	95.8	60.4	76.5	91.7	58.9	73.8	91.2	56.8	72.4
COVERS	DIVS. 10 - 14, 25, 28, 41, 43, 44, 46	100.0	82.7	96.2	100.0	76.5	94.8	100.0	81.0	95.8	100.0	80.6	95.7	100.0	77.1	95.0	100.0	78.7	95.3
21, 22, 23	FIRE SUPPRESSION, PLUMBING & HVAC	100.1	57.1	82.8	99.9	51.5	80.5	99.9	63.2	85.2	99.9	58.8	83.4	100.0	55.7	82.2	100.0	66.6	86.6
26, 27, 3370	ELECTRICAL, COMMUNICATIONS & UTIL.	98.9	64.7	81.4	93.3	57.2	74.8	96.7	51.9	73.8	96.1	60.0	77.6	91.9	62.1	76.7	98.4	57.2	77.3
MF2018	WEIGHTED AVERAGE	98.3	67.1	84.9	98.0	64.2	83.5	100.6	66.4	86.0	99.1	67.0	85.3	99.6	65.2	84.8	97.5	65.7	83.9

FLORIDA / GEORGIA

Code	Division	TAMPA 335-336, 346			WEST PALM BEACH 334, 349			ALBANY 317, 398			ATHENS 306			ATLANTA 300-303, 399			AUGUSTA 308-309		
		MAT.	INST.	TOTAL	MAT.	INST.	TOTAL	MAT.	INST.	TOTAL	MAT.	INST.	TOTAL	MAT.	INST.	TOTAL	MAT.	INST.	TOTAL
015433	CONTRACTOR EQUIPMENT		102.1	102.1		95.2	95.2		96.1	96.1		94.6	94.6		96.5	96.5		94.6	94.6
0241, 31 - 34	SITE & INFRASTRUCTURE, DEMOLITION	107.8	89.0	94.7	88.9	76.8	80.4	102.4	79.4	86.4	101.3	94.9	96.9	98.3	95.3	96.2	93.4	72.7	75.8
0310	Concrete Forming & Accessories	94.1	61.9	66.7	95.2	62.9	67.7	87.8	67.7	70.7	92.3	44.3	51.5	96.9	71.9	75.7	99.3	68.0	83.4
0320	Concrete Reinforcing	94.5	78.3	86.2	97.1	59.2	77.8	90.4	73.3	81.7	98.9	65.4	81.9	98.2	73.4	85.6	103.5	70.1	91.1
0330	Cast-in-Place Concrete	94.9	65.8	84.1	85.6	61.8	76.8	85.7	67.7	79.0	109.6	68.5	94.3	113.1	71.6	97.6	103.5	70.1	91.1
03	CONCRETE	92.4	67.8	81.3	86.7	63.6	76.3	86.8	70.4	79.4	101.4	57.7	81.7	103.6	72.7	89.7	94.2	71.6	84.0
04	MASONRY	94.2	59.3	72.6	92.8	53.3	68.4	91.9	67.7	77.0	76.6	77.8	77.3	89.9	69.7	77.4	90.1	69.2	77.2
05	METALS	101.4	94.3	99.2	98.4	87.7	95.1	101.3	97.5	100.1	94.9	79.5	90.2	95.7	84.6	92.4	96.5	75.4	85.0
06	WOOD, PLASTICS & COMPOSITES	85.3	61.0	72.1	85.0	65.2	74.3	79.5	68.4	73.5	95.0	36.8	63.4	99.4	72.4	87.8	97.7	71.3	86.3
07	THERMAL & MOISTURE PROTECTION	98.7	63.5	83.6	98.2	61.2	82.3	87.9	69.5	85.7	93.0	50.9	83.3	97.6	73.4	92.0	93.1	73.3	88.5
08	OPENINGS	97.7	64.0	89.8	96.0	62.0	88.1	87.6	70.5	83.6	97.4	35.5	56.6	97.6	73.4	92.0	98.7	75.1	83.1
0920	Plaster & Gypsum Board	110.1	60.4	77.4	113.8	64.7	81.5	103.8	68.0	80.2	97.4	35.5	56.6	99.6	72.6	81.8	102.2	75.1	83.8
0950, 0980	Ceilings & Acoustic Treatment	86.1	60.4	68.7	81.6	64.7	70.2	81.9	68.0	72.5	101.2	35.5	56.7	94.0	72.6	79.5	94.1	65.9	85.9
0960	Flooring	96.8	58.4	85.6	98.6	56.7	86.5	102.3	65.9	91.8	94.0	84.7	91.3	97.0	68.0	88.6	94.1	65.9	85.9
0970, 0990	Wall Finishes & Painting/Coating	92.6	63.6	75.4	88.4	58.6	70.8	94.5	96.5	95.6	101.8	96.5	98.7	106.2	96.5	100.5	101.8	80.6	89.2
09	FINISHES	93.0	60.8	75.4	90.6	61.1	74.5	92.4	70.0	80.2	97.7	55.2	74.5	97.8	73.5	84.5	97.4	72.5	83.8
COVERS	DIVS. 10 - 14, 25, 28, 41, 43, 44, 46	100.0	80.8	95.8	100.0	82.7	96.2	100.0	83.3	96.3	100.0	79.6	95.5	100.0	86.4	97.0	100.0	83.9	96.5
21, 22, 23	FIRE SUPPRESSION, PLUMBING & HVAC	100.0	60.0	84.0	98.7	59.9	83.1	99.9	69.9	87.9	97.6	67.3	85.5	99.5	72.6	85.7	100.6	69.8	84.8
26, 27, 3370	ELECTRICAL, COMMUNICATIONS & UTIL.	91.6	64.7	77.8	92.7	69.1	80.6	96.1	63.4	79.4	96.7	65.4	82.3	98.7	75.6	88.8	96.9	72.7	86.5
MF2018	WEIGHTED AVERAGE	97.4	68.1	84.8	95.1	66.0	82.6	95.8	72.6	85.8	96.7	67.9	84.4	96.7	77.9	88.8	96.9	72.7	86.5

For customer support on your Electrical Costs with RSMeans Data, call 800.448.8182.

529

DIVISION		GEORGIA																
		COLUMBUS 318 - 319			DALTON 307			GAINESVILLE 305			MACON 310 - 312			SAVANNAH 313 - 314			STATESBORO 304		
		MAT.	INST.	TOTAL	MAT.	INST.	TOTAL	MAT.	INST.	TOTAL	MAT.	INST.	TOTAL	MAT.	INST.	TOTAL	MAT.	INST.	TOTAL
015433	CONTRACTOR EQUIPMENT		96.1	96.1		110.1	110.1		94.6	94.6		105.9	105.9		97.0	97.0		97.7	97.7
0241, 31 - 34	SITE & INFRASTRUCTURE, DEMOLITION	102.3	79.5	86.5	101.0	100.3	100.5	101.2	94.9	96.8	104.0	94.6	97.5	102.2	80.9	87.4	102.1	81.4	87.7
0310	Concrete Forming & Accessories	87.7	67.8	70.8	84.9	65.5	68.4	96.0	41.8	49.9	87.6	67.8	70.8	91.3	70.4	73.6	79.7	52.1	56.2
0320	Concrete Reinforcing	90.3	73.3	81.7	98.4	62.5	80.2	98.7	65.3	81.7	91.5	73.3	82.3	97.3	68.0	82.4	98.0	36.6	66.8
0330	Cast-in-Place Concrete	85.4	68.0	78.8	106.5	67.2	91.8	115.2	68.3	97.7	84.2	69.7	78.8	89.0	67.6	81.0	109.4	67.2	93.7
03	CONCRETE	86.6	70.5	79.4	100.5	67.4	85.6	103.4	56.4	82.2	86.2	71.2	79.5	88.0	70.7	80.2	100.9	57.0	81.1
04	MASONRY	91.9	67.8	77.0	77.4	77.1	77.2	85.4	77.8	80.7	104.2	67.9	81.8	86.8	67.7	75.0	79.4	77.8	78.4
05	METALS	100.9	97.9	100.0	96.1	93.8	95.4	94.2	78.9	89.6	96.5	98.0	96.9	97.5	95.3	96.9	99.7	84.0	94.9
06	WOOD, PLASTICS & COMPOSITES	79.5	68.4	73.5	78.2	67.1	72.1	98.9	34.1	63.7	85.0	68.4	76.0	88.2	72.3	79.6	71.5	48.5	59.0
07	THERMAL & MOISTURE PROTECTION	99.5	69.4	86.5	100.0	71.7	87.8	97.9	69.2	85.5	98.0	71.5	86.6	97.5	68.9	85.2	98.5	68.1	85.5
08	OPENINGS	87.6	70.5	83.6	94.2	66.9	87.9	93.0	49.4	82.9	87.4	70.5	83.5	95.1	71.3	89.6	95.1	49.5	84.5
0920	Plaster & Gypsum Board	103.8	68.0	80.2	85.0	66.6	72.9	99.9	32.7	55.6	108.4	68.0	81.8	102.7	72.0	82.5	85.8	47.5	60.6
0950, 0980	Ceilings & Acoustic Treatment	81.9	68.0	72.5	117.4	66.6	83.0	101.2	32.7	54.8	77.1	68.0	70.9	88.3	72.0	77.3	111.8	47.5	68.2
0960	Flooring	102.3	65.9	91.8	94.5	84.7	91.6	95.5	84.7	92.3	81.0	65.9	76.6	98.3	65.9	88.9	111.2	84.7	103.5
0970, 0990	Wall Finishes & Painting/Coating	94.5	96.5	95.6	91.6	71.9	80.0	101.8	96.5	98.7	96.5	96.5	96.5	91.5	86.3	88.4	100.0	71.9	83.4
09	FINISHES	92.3	70.0	80.1	107.5	70.2	87.2	98.3	53.6	73.9	82.3	70.0	75.6	92.3	71.2	80.8	110.2	59.4	82.5
COVERS	DIVS. 10 - 14, 25, 28, 41, 43, 44, 46	100.0	83.3	96.3	100.0	28.4	84.2	100.0	36.2	85.9	100.0	83.4	96.3	100.0	83.3	96.3	100.0	42.2	87.3
21, 22, 23	FIRE SUPPRESSION, PLUMBING & HVAC	100.0	66.2	86.5	97.7	62.0	83.4	97.6	67.1	85.4	100.0	69.3	87.7	100.1	67.1	86.8	98.1	67.8	86.0
26, 27, 3370	ELECTRICAL, COMMUNICATIONS & UTIL.	96.2	69.4	82.5	109.3	63.7	86.0	100.0	69.3	84.3	94.8	64.2	79.1	99.8	70.2	84.7	100.3	58.6	79.0
MF2018	WEIGHTED AVERAGE	95.7	72.7	85.8	98.6	71.0	86.7	97.3	66.5	84.1	94.6	74.0	85.8	96.3	73.1	86.3	98.7	65.8	84.6

DIVISION		GEORGIA						HAWAII									IDAHO		
		VALDOSTA 316			WAYCROSS 315			HILO 967			HONOLULU 968			STATES & POSS., GUAM 969			BOISE 836 - 837		
		MAT.	INST.	TOTAL	MAT.	INST.	TOTAL	MAT.	INST.	TOTAL	MAT.	INST.	TOTAL	MAT.	INST.	TOTAL	MAT.	INST.	TOTAL
015433	CONTRACTOR EQUIPMENT		96.1	96.1		96.1	96.1		100.4	100.4		100.4	100.4		165.0	165.0		97.6	97.6
0241, 31 - 34	SITE & INFRASTRUCTURE, DEMOLITION	111.6	79.4	89.2	108.5	80.6	89.1	147.0	107.5	119.5	155.4	107.5	122.0	191.2	103.5	130.1	86.5	97.0	93.8
0310	Concrete Forming & Accessories	79.0	42.0	47.6	80.9	64.7	67.1	106.5	123.7	121.1	119.6	123.7	123.1	108.9	53.0	61.4	98.7	79.8	82.7
0320	Concrete Reinforcing	92.4	61.2	76.5	92.4	61.4	76.7	137.5	126.6	131.9	159.9	126.6	142.9	237.2	28.2	131.0	105.1	78.8	91.8
0330	Cast-in-Place Concrete	83.9	67.5	77.8	94.8	67.4	84.6	170.9	125.6	154.1	137.5	125.6	133.1	148.2	98.5	129.7	90.6	83.1	87.8
03	CONCRETE	91.8	56.7	76.0	94.6	66.7	82.0	144.8	123.7	135.3	138.4	123.7	131.8	146.1	65.7	109.9	96.8	81.0	89.7
04	MASONRY	97.2	77.8	85.2	98.0	77.8	85.5	148.6	120.0	130.9	135.5	120.0	125.9	205.8	35.5	100.6	125.2	87.2	101.7
05	METALS	100.4	93.1	98.2	99.5	88.8	96.3	114.4	106.6	112.0	128.1	106.6	121.6	147.7	76.6	126.1	108.5	82.1	100.5
06	WOOD, PLASTICS & COMPOSITES	68.1	33.9	49.5	69.6	65.2	67.2	113.9	124.5	119.6	131.3	124.5	127.6	126.6	55.4	87.9	94.6	78.5	85.9
07	THERMAL & MOISTURE PROTECTION	99.8	67.5	85.9	99.9	70.3	87.0	131.8	120.2	126.8	149.6	120.2	136.9	154.9	59.8	114.0	94.9	85.5	90.9
08	OPENINGS	84.2	48.2	75.9	84.6	62.9	79.5	116.3	123.0	117.9	130.9	123.0	129.1	120.7	45.0	103.1	99.2	73.1	93.1
0920	Plaster & Gypsum Board	95.8	32.5	54.1	95.8	64.7	75.4	112.1	124.9	120.6	155.7	124.9	135.5	228.9	43.7	106.9	94.2	77.9	83.5
0950, 0980	Ceilings & Acoustic Treatment	80.2	32.5	47.9	78.3	64.7	69.1	137.5	124.9	129.0	144.3	124.9	131.2	254.9	43.7	111.8	101.2	77.9	85.4
0960	Flooring	96.1	86.3	93.2	97.3	84.7	93.6	110.8	140.7	119.5	130.2	140.7	133.3	130.1	40.8	104.2	93.2	79.6	89.3
0970, 0990	Wall Finishes & Painting/Coating	94.5	96.5	95.6	94.5	71.9	81.1	94.7	140.6	121.9	105.1	140.6	126.1	100.0	32.4	60.0	93.4	40.5	62.1
09	FINISHES	90.0	53.8	70.2	89.6	69.3	78.5	110.5	129.0	120.6	125.7	129.0	127.5	183.1	49.4	110.1	91.4	75.7	82.8
COVERS	DIVS. 10 - 14, 25, 28, 41, 43, 44, 46	100.0	79.1	95.4	100.0	49.8	89.0	100.0	112.6	102.8	100.0	112.6	102.8	100.0	67.3	92.8	100.0	88.6	97.5
21, 22, 23	FIRE SUPPRESSION, PLUMBING & HVAC	100.0	69.3	87.7	98.5	64.2	84.7	100.4	112.0	105.1	100.4	112.0	105.1	100.4	34.3	75.5	100.0	74.8	89.9
26, 27, 3370	ELECTRICAL, COMMUNICATIONS & UTIL.	94.6	52.9	73.3	98.3	54.0	75.6	109.0	124.3	116.9	110.7	124.3	117.7	158.3	36.8	96.1	97.6	69.8	83.4
MF2018	WEIGHTED AVERAGE	96.0	66.2	83.2	96.2	68.4	84.2	115.9	118.3	116.9	120.4	118.3	119.5	137.8	52.6	101.3	100.7	79.4	91.5

DIVISION		IDAHO															ILLINOIS		
		COEUR D'ALENE 838			IDAHO FALLS 834			LEWISTON 835			POCATELLO 832			TWIN FALLS 833			BLOOMINGTON 617		
		MAT.	INST.	TOTAL	MAT.	INST.	TOTAL	MAT.	INST.	TOTAL	MAT.	INST.	TOTAL	MAT.	INST.	TOTAL	MAT.	INST.	TOTAL
015433	CONTRACTOR EQUIPMENT		92.5	92.5		97.6	97.6		92.5	92.5		97.6	97.6		97.6	97.6		104.7	104.7
0241, 31 - 34	SITE & INFRASTRUCTURE, DEMOLITION	84.5	91.0	89.0	84.5	96.8	93.1	91.2	91.9	91.7	87.7	97.0	94.2	94.5	96.7	96.0	97.0	100.8	99.6
0310	Concrete Forming & Accessories	106.0	82.0	85.6	92.4	78.6	80.7	111.3	83.4	87.6	98.9	79.5	82.4	100.1	78.5	81.7	83.5	119.2	113.8
0320	Concrete Reinforcing	113.2	99.0	106.0	107.2	77.4	92.0	113.2	99.0	106.0	105.6	78.8	92.0	107.6	77.3	92.2	96.1	105.3	100.8
0330	Cast-in-Place Concrete	98.0	84.1	92.8	86.2	83.3	85.2	101.8	84.5	95.4	93.1	83.0	89.4	95.6	83.3	91.0	98.1	117.0	105.2
03	CONCRETE	103.0	85.7	95.2	88.8	80.2	85.0	106.6	86.5	97.5	96.0	86.0	89.1	103.6	80.2	93.1	93.9	116.7	104.1
04	MASONRY	126.8	85.0	100.9	120.2	84.6	98.2	127.2	86.5	102.0	122.7	87.2	100.8	125.4	84.6	100.2	110.9	124.2	119.2
05	METALS	101.8	89.0	97.9	117.0	81.4	106.2	101.3	89.2	97.6	117.1	81.7	106.3	117.1	81.2	106.2	97.0	123.8	105.1
06	WOOD, PLASTICS & COMPOSITES	97.6	82.1	89.2	88.1	78.5	82.9	103.7	82.1	92.0	94.6	78.5	85.9	95.7	78.5	86.3	83.4	116.3	101.3
07	THERMAL & MOISTURE PROTECTION	147.9	84.1	120.4	94.1	75.4	86.1	148.3	85.5	121.3	94.7	76.4	86.8	95.5	82.7	90.0	98.7	115.2	105.8
08	OPENINGS	116.4	78.1	107.5	102.1	72.8	95.3	110.1	82.0	103.6	99.9	68.1	92.5	102.9	63.6	93.8	93.3	122.1	100.0
0920	Plaster & Gypsum Board	170.4	81.8	112.0	80.8	77.9	78.9	172.2	81.8	112.7	82.2	77.9	79.4	84.4	77.9	80.1	88.8	116.9	107.3
0950, 0980	Ceilings & Acoustic Treatment	136.3	81.8	99.4	104.7	77.9	86.6	136.3	81.8	99.4	109.8	77.9	88.2	107.2	77.9	87.4	89.7	116.9	108.1
0960	Flooring	131.9	79.6	116.7	93.1	79.6	89.2	135.1	79.6	119.0	96.5	79.6	91.6	97.7	79.6	92.5	87.3	124.9	98.2
0970, 0990	Wall Finishes & Painting/Coating	112.5	74.5	90.0	93.4	40.5	62.1	112.5	73.2	89.3	93.3	40.5	62.1	93.4	40.5	62.1	88.7	138.3	118.0
09	FINISHES	155.0	80.7	114.4	89.7	75.0	81.7	156.3	81.2	115.3	92.3	75.7	83.2	93.1	75.0	83.2	86.9	122.5	106.3
COVERS	DIVS. 10 - 14, 25, 28, 41, 43, 44, 46	100.0	93.6	98.6	100.0	87.7	97.3	100.0	94.6	98.8	100.0	88.6	97.5	100.0	87.7	97.3	100.0	106.7	101.5
21, 22, 23	FIRE SUPPRESSION, PLUMBING & HVAC	99.5	81.7	92.4	100.8	73.4	89.8	100.7	86.9	95.2	99.9	74.8	89.8	99.9	69.6	87.7	97.5	107.7	101.6
26, 27, 3370	ELECTRICAL, COMMUNICATIONS & UTIL.	89.3	83.5	86.4	89.0	69.4	79.0	87.5	80.1	83.7	94.9	67.6	81.0	90.3	69.4	79.6	94.4	92.5	93.4
MF2018	WEIGHTED AVERAGE	107.7	84.4	97.7	100.2	78.2	90.8	107.8	85.7	98.3	101.7	78.6	91.8	102.9	77.2	91.9	96.1	112.2	103.0

ILLINOIS

DIVISION		CARBONDALE 629 MAT.	INST.	TOTAL	CENTRALIA 628 MAT.	INST.	TOTAL	CHAMPAIGN 618-619 MAT.	INST.	TOTAL	CHICAGO 606-608 MAT.	INST.	TOTAL	DECATUR 625 MAT.	INST.	TOTAL	EAST ST. LOUIS 620-622 MAT.	INST.	TOTAL
015433	CONTRACTOR EQUIPMENT		113.3	113.3		113.3	113.3		105.6	105.6		98.9	98.9		105.6	105.6		113.3	113.3
0241, 31 - 34	SITE & INFRASTRUCTURE, DEMOLITION	98.6	101.5	100.7	99.0	103.4	102.0	106.1	101.6	103.0	98.1	162.0	152.4	90.4	115.9	112.1	85.7	111.5	107.6
0310	Concrete Forming & Accessories	88.3	108.2	105.2	89.8	114.7	110.9	96.1	100.6	98.4	100.2	154.7	128.0	88.5	100.9	94.8	90.1	107.0	98.7
0320	Concrete Reinforcing	90.2	106.3	98.4	90.2	107.0	98.8	96.1	100.6	98.4	126.4	154.6	136.9	97.0	113.7	103.2	90.8	121.5	102.3
0330	Cast-in-Place Concrete	88.9	102.3	93.8	89.3	121.6	101.3	113.8	111.9	113.1	109.2	157.3	130.8	91.6	113.1	101.3	82.6	115.7	97.5
03	CONCRETE	81.1	107.3	92.9	81.6	117.1	97.6	106.4	112.8	109.3	109.2	157.3	130.8	91.6	113.1	101.3	76.3	125.5	106.7
04	MASONRY	76.0	112.4	98.5	76.0	125.5	106.6	133.9	121.8	126.4	104.5	166.2	142.6	71.8	120.2	101.7	97.8	133.5	108.6
05	METALS	96.6	130.7	107.0	96.6	133.8	107.9	97.0	119.7	103.9	99.0	146.6	113.5	100.3	119.2	106.0	85.8	106.9	97.3
06	WOOD, PLASTICS & COMPOSITES	88.7	105.4	97.8	91.2	110.9	101.9	90.4	114.9	103.7	101.1	161.7	134.0	89.8	114.9	103.4	95.1	112.5	102.6
07	THERMAL & MOISTURE PROTECTION	95.1	102.3	98.2	95.1	113.2	102.9	99.5	114.7	106.0	104.4	170.9	119.9	100.8	113.0	106.1	89.3	117.7	95.9
08	OPENINGS	89.2	116.9	95.6	89.2	120.0	96.3	93.9	116.8	99.2	104.4	170.9	119.9	100.3	116.9	104.2	93.8	107.2	102.6
0920	Plaster & Gypsum Board	95.3	105.7	102.2	96.6	111.3	106.3	90.6	115.4	107.0	94.8	163.5	140.1	97.8	115.4	109.4	84.6	107.2	99.9
0950, 0980	Ceilings & Acoustic Treatment	84.6	105.7	98.9	84.6	111.3	102.7	89.7	115.4	107.1	99.8	163.5	143.0	90.5	115.4	107.4	110.2	127.7	115.3
0960	Flooring	111.2	113.8	112.0	112.0	127.7	116.5	90.4	114.7	97.4	98.7	164.8	117.9	99.5	115.2	104.1	109.1	114.6	112.3
0970, 0990	Wall Finishes & Painting/Coating	109.1	107.7	108.3	109.1	114.6	112.3	88.7	115.4	104.5	104.9	162.3	138.9	100.8	114.4	108.9	92.4	114.4	104.4
09	FINISHES	92.8	108.1	101.1	93.2	117.1	106.3	88.8	116.6	104.0	98.4	163.5	133.9	92.8	116.2	105.6	100.0	103.8	100.8
COVERS	DIVS. 10 - 14, 25, 28, 41, 43, 44, 46	100.0	102.3	100.5	100.0	104.3	100.9	100.0	105.4	101.2	100.0	128.1	106.2	100.0	104.8	101.1	99.9	99.3	99.7
21, 22, 23	FIRE SUPPRESSION, PLUMBING & HVAC	97.4	106.5	101.1	97.4	98.8	98.0	97.5	106.3	101.0	99.8	138.0	115.1	99.9	96.4	98.5	92.6	102.1	97.5
26, 27, 3370	ELECTRICAL, COMMUNICATIONS & UTIL.	91.7	108.8	100.5	93.0	108.9	101.1	97.5	92.6	95.0	101.4	146.1	120.5	96.4	107.0	101.0	93.7	111.3	101.2
MF2018	WEIGHTED AVERAGE	92.6	109.8	100.0	92.8	113.0	101.5	99.6	109.8	104.0	101.4	146.1	120.5	96.4	107.0	101.0	93.7	111.3	101.2

ILLINOIS

DIVISION		EFFINGHAM 624 MAT.	INST.	TOTAL	GALESBURG 614 MAT.	INST.	TOTAL	JOLIET 604 MAT.	INST.	TOTAL	KANKAKEE 609 MAT.	INST.	TOTAL	LA SALLE 613 MAT.	INST.	TOTAL	NORTH SUBURBAN 600-603 MAT.	INST.	TOTAL
015433	CONTRACTOR EQUIPMENT		105.6	105.6		104.7	104.7		95.4	95.4		95.4	95.4		104.7	104.7		95.4	95.4
0241, 31 - 34	SITE & INFRASTRUCTURE, DEMOLITION	97.9	100.4	99.7	99.4	99.8	99.7	98.9	101.7	100.8	92.7	101.3	98.7	103.9	126.0	122.6	96.7	158.8	149.5
0310	Concrete Forming & Accessories	94.9	116.6	113.3	89.8	119.5	115.0	97.4	161.1	151.5	90.8	146.0	137.7	105.9	137.5	116.9	100.2	153.3	127.2
0320	Concrete Reinforcing	91.4	96.2	93.9	95.5	108.6	102.1	100.2	140.6	120.3	101.0	138.9	120.3	100.9	125.0	109.9	115.0	154.5	129.7
0330	Cast-in-Place Concrete	96.7	109.2	101.3	101.0	111.5	105.2	115.0	148.1	127.3	107.1	138.1	118.6	97.6	128.2	111.4	101.9	155.4	126.0
03	CONCRETE	92.3	111.0	100.7	96.8	115.4	105.2	102.0	151.9	124.5	95.8	141.3	116.3	111.1	145.6	132.4	100.7	162.1	138.6
04	MASONRY	80.2	114.1	101.2	111.1	123.7	118.9	103.7	160.0	138.5	100.0	148.4	129.9	97.1	142.8	110.9	98.1	142.0	111.4
05	METALS	97.4	114.5	102.6	97.0	124.1	105.2	96.9	135.6	108.7	96.9	133.8	108.1	106.5	123.3	115.6	98.1	157.7	130.5
06	WOOD, PLASTICS & COMPOSITES	92.1	117.3	105.8	90.3	118.7	105.7	99.2	162.3	133.5	92.1	144.6	120.6	99.1	128.9	111.9	103.8	148.9	123.2
07	THERMAL & MOISTURE PROTECTION	100.2	109.8	104.4	98.9	112.1	104.6	103.5	148.2	122.7	102.5	141.1	119.1	99.3	135.2	103.0	102.0	168.4	117.4
08	OPENINGS	95.3	116.2	100.2	93.3	118.6	99.2	101.9	166.5	116.9	94.8	156.8	109.2	93.3	135.2	103.0	90.9	159.5	136.1
0920	Plaster & Gypsum Board	97.8	117.9	111.0	90.6	119.3	109.5	87.3	164.1	137.9	85.5	146.0	125.3	98.3	124.1	115.2	99.4	159.5	140.1
0950, 0980	Ceilings & Acoustic Treatment	84.6	117.9	107.1	89.7	119.3	109.8	99.4	164.1	143.3	99.4	146.0	131.0	89.7	124.1	113.0	99.4	159.5	140.1
0960	Flooring	100.5	116.9	105.3	90.2	124.9	100.3	96.7	154.8	113.5	93.7	155.7	111.7	96.7	122.9	104.3	97.2	164.8	116.8
0970, 0990	Wall Finishes & Painting/Coating	100.8	112.2	107.6	88.7	101.1	96.0	97.0	167.3	138.6	97.0	138.3	121.4	88.7	136.5	117.0	99.0	168.2	139.9
09	FINISHES	91.9	117.4	105.8	88.1	119.4	105.2	93.7	162.0	131.0	92.2	147.5	122.4	91.1	125.3	109.8	94.5	161.8	131.2
COVERS	DIVS. 10 - 14, 25, 28, 41, 43, 44, 46	100.0	77.2	95.0	100.0	101.4	100.3	100.0	125.0	105.5	100.0	117.8	103.9	100.0	102.7	100.6	100.0	121.3	104.7
21, 22, 23	FIRE SUPPRESSION, PLUMBING & HVAC	97.5	104.7	100.4	97.5	105.7	100.8	99.9	133.0	113.1	97.4	132.4	111.4	97.5	127.0	109.3	99.8	136.4	114.5
26, 27, 3370	ELECTRICAL, COMMUNICATIONS & UTIL.	93.4	108.8	101.3	95.3	83.8	89.4	97.2	136.5	117.3	92.5	133.8	113.7	92.5	133.9	113.7	97.0	135.7	116.8
MF2018	WEIGHTED AVERAGE	95.2	109.2	101.2	96.8	109.5	102.3	99.4	142.2	117.8	96.3	136.1	113.3	97.0	128.9	110.7	99.4	144.1	118.8

ILLINOIS

DIVISION		PEORIA 615-616 MAT.	INST.	TOTAL	QUINCY 623 MAT.	INST.	TOTAL	ROCK ISLAND 612 MAT.	INST.	TOTAL	ROCKFORD 610-611 MAT.	INST.	TOTAL	SOUTH SUBURBAN 605 MAT.	INST.	TOTAL	SPRINGFIELD 626-627 MAT.	INST.	TOTAL
015433	CONTRACTOR EQUIPMENT		104.7	104.7		105.6	105.6		104.7	104.7		104.7	104.7		95.4	95.4		105.6	105.6
0241, 31 - 34	SITE & INFRASTRUCTURE, DEMOLITION	99.8	100.7	100.4	96.7	101.2	99.9	97.5	99.6	99.0	99.2	101.8	101.0	98.2	101.7	100.6	98.4	101.6	100.6
0310	Concrete Forming & Accessories	92.8	119.2	115.2	92.8	116.3	112.8	91.4	101.9	100.3	97.2	131.3	126.2	96.7	158.8	149.4	91.6	116.6	112.8
0320	Concrete Reinforcing	93.1	107.7	100.5	91.0	88.5	89.7	95.5	101.8	98.7	88.1	138.0	113.5	100.2	153.3	127.2	92.1	111.1	99.2
0330	Cast-in-Place Concrete	98.0	119.0	105.8	96.8	109.6	101.6	98.8	101.3	99.8	100.3	130.5	111.5	115.0	154.4	129.6	89.7	112.5	100.0
03	CONCRETE	94.0	117.7	104.7	91.9	109.7	99.9	94.7	102.7	98.3	94.8	132.6	111.8	100.7	162.1	138.6	81.7	121.6	106.4
04	MASONRY	110.8	124.0	119.0	100.2	114.1	108.8	110.9	102.4	105.6	86.3	145.3	122.8	98.1	142.0	111.4	97.8	119.3	104.4
05	METALS	99.7	125.0	107.3	97.5	113.0	102.2	97.0	119.6	103.9	99.7	142.8	112.8	98.1	157.7	130.5	91.0	114.9	104.0
06	WOOD, PLASTICS & COMPOSITES	98.8	116.5	108.4	89.6	117.3	104.6	92.0	100.4	96.5	98.8	127.2	114.2	103.8	148.9	123.2	103.1	114.6	108.1
07	THERMAL & MOISTURE PROTECTION	99.7	115.6	106.5	100.2	108.5	103.8	98.9	101.2	99.9	102.0	132.6	115.2	102.0	168.4	117.4	100.8	117.4	104.6
08	OPENINGS	98.8	123.0	104.4	96.0	114.9	100.4	93.3	106.3	96.3	98.8	140.5	108.5	90.9	159.5	136.1	99.5	115.4	110.0
0920	Plaster & Gypsum Board	95.1	117.0	109.5	96.6	117.9	110.6	90.6	100.5	97.1	95.1	128.0	116.8	90.9	159.5	136.1	91.0	115.4	107.5
0950, 0980	Ceilings & Acoustic Treatment	94.8	117.0	109.8	84.6	117.9	107.1	89.7	100.5	97.0	94.8	128.0	117.3	99.4	159.5	140.1	102.0	115.2	105.8
0960	Flooring	93.7	123.5	102.3	99.5	111.5	103.0	91.3	97.8	93.2	93.7	129.1	103.9	97.2	164.8	116.8	102.9	113.1	108.9
0970, 0990	Wall Finishes & Painting/Coating	88.7	138.3	118.0	100.8	114.4	108.9	88.7	98.0	94.2	88.7	146.5	122.9	94.5	161.8	131.2	97.3	116.4	107.7
09	FINISHES	90.8	122.2	107.9	91.4	116.5	105.1	88.4	100.4	94.9	90.8	132.2	113.4	100.0	121.3	104.7	100.0	105.3	101.2
COVERS	DIVS. 10 - 14, 25, 28, 41, 43, 44, 46	100.0	106.6	101.4	100.0	79.8	95.5	100.0	99.5	99.9	100.0	116.5	103.6	100.0	121.3	104.7	99.9	102.4	100.9
21, 22, 23	FIRE SUPPRESSION, PLUMBING & HVAC	99.9	102.4	100.9	97.5	103.5	99.9	97.5	99.7	98.4	96.5	128.7	113.0	97.0	135.7	116.8	97.9	84.2	90.9
26, 27, 3370	ELECTRICAL, COMMUNICATIONS & UTIL.	96.3	93.7	95.0	91.1	84.2	87.6	87.8	95.4	91.7	97.5	128.7	110.9	99.4	144.0	118.6	97.1	107.7	101.7
MF2018	WEIGHTED AVERAGE	98.5	111.5	104.0	95.8	105.1	99.8	95.8	102.1	98.5	97.5	128.7	110.9	99.4	144.0	118.6	97.1	107.7	101.7

For customer support on your Electrical Costs with RSMeans Data, call 800.448.8182.

531

City Cost Indexes

INDIANA

DIVISION		ANDERSON 460 MAT.	INST.	TOTAL	BLOOMINGTON 474 MAT.	INST.	TOTAL	COLUMBUS 472 MAT.	INST.	TOTAL	EVANSVILLE 476-477 MAT.	INST.	TOTAL	FORT WAYNE 467-468 MAT.	INST.	TOTAL	GARY 463-464 MAT.	INST.	TOTAL
015433	CONTRACTOR EQUIPMENT		97.4	97.4		83.9	83.9		83.9	83.9		113.7	113.7		97.4	97.4		97.4	97.4
0241, 31-34	SITE & INFRASTRUCTURE, DEMOLITION	99.1	93.7	95.4	87.0	92.1	90.6	83.7	91.9	89.4	92.4	121.2	112.5	100.2	93.6	95.6	99.8	97.3	98.0
0310	Concrete Forming & Accessories	95.8	79.6	82.0	100.6	80.1	83.2	94.5	79.9	82.1	93.8	80.8	82.7	93.7	75.7	78.4	95.8	112.6	110.0
0320	Concrete Reinforcing	95.6	85.4	90.4	89.4	85.1	87.2	89.8	87.2	88.5	97.9	83.4	90.5	95.6	80.2	87.8	95.6	112.7	104.3
0330	Cast-in-Place Concrete	103.7	77.1	93.8	99.3	76.2	90.7	98.8	74.1	89.6	94.9	85.7	91.5	110.2	76.2	97.6	108.4	113.7	110.4
03	CONCRETE	92.9	80.3	87.2	98.1	79.2	89.6	97.3	78.7	89.0	98.1	83.1	91.3	95.9	77.4	87.5	95.1	112.7	103.0
04	MASONRY	87.3	75.6	80.1	88.1	72.0	78.1	87.9	73.1	78.7	83.7	78.9	80.7	90.4	72.9	79.6	88.7	109.2	101.3
05	METALS	99.1	89.1	96.1	99.5	74.2	91.8	99.5	74.4	91.9	92.3	84.0	89.8	99.1	87.2	95.5	99.1	105.1	100.9
06	WOOD, PLASTICS & COMPOSITES	97.5	79.8	87.9	112.3	80.6	95.1	106.6	80.4	92.4	93.2	79.9	86.0	97.2	76.0	85.6	94.7	111.0	103.5
07	THERMAL & MOISTURE PROTECTION	113.0	77.2	97.6	98.6	77.7	89.6	98.0	78.1	89.4	103.0	84.0	94.8	112.8	78.9	98.2	111.4	106.1	109.1
08	OPENINGS	95.3	78.0	91.3	99.7	78.6	94.8	95.9	79.0	92.0	93.8	78.2	90.2	95.3	73.2	90.1	95.3	114.2	99.7
0920	Plaster & Gypsum Board	104.6	79.5	88.1	98.6	80.7	86.8	95.5	80.6	85.7	94.0	79.0	84.1	103.7	75.6	85.2	97.3	111.6	106.7
0950, 0980	Ceilings & Acoustic Treatment	91.9	79.5	83.5	79.6	80.7	80.4	79.6	80.6	80.3	83.4	79.0	80.4	91.9	75.6	80.8	91.9	111.6	105.2
0960	Flooring	95.1	77.5	90.0	99.1	83.1	94.4	94.0	83.1	90.9	94.0	73.8	88.1	95.1	73.5	88.9	95.1	114.3	100.7
0970, 0990	Wall Finishes & Painting/Coating	94.0	67.6	78.4	85.4	79.5	81.9	85.4	79.5	81.9	91.0	86.6	88.4	94.0	72.1	81.0	94.0	121.4	110.2
09	FINISHES	91.8	77.9	84.2	90.1	80.7	85.0	88.2	80.6	84.0	88.8	79.9	84.0	91.5	75.1	82.5	90.6	113.9	103.3
COVERS	DIVS. 10-14, 25, 28, 41, 43, 44, 46	100.0	89.2	97.6	100.0	85.8	96.9	100.0	85.8	96.9	100.0	93.1	98.5	100.0	89.1	97.6	100.0	105.1	101.1
21, 22, 23	FIRE SUPPRESSION, PLUMBING & HVAC	99.9	78.7	91.4	99.8	78.9	91.4	97.3	78.6	89.8	100.0	80.4	92.1	99.9	73.8	89.5	99.9	105.1	102.0
26, 27, 3370	ELECTRICAL, COMMUNICATIONS & UTIL.	86.1	85.2	85.6	98.7	86.6	92.5	97.9	87.0	92.3	94.7	83.4	88.9	86.8	76.7	81.6	97.5	110.7	104.2
MF2018	WEIGHTED AVERAGE	96.1	81.8	90.0	97.8	80.3	90.3	96.4	80.4	89.5	95.5	84.8	90.9	96.7	78.2	88.8	97.5	108.3	102.1

INDIANA

DIVISION		INDIANAPOLIS 461-462 MAT.	INST.	TOTAL	KOKOMO 469 MAT.	INST.	TOTAL	LAFAYETTE 479 MAT.	INST.	TOTAL	LAWRENCEBURG 470 MAT.	INST.	TOTAL	MUNCIE 473 MAT.	INST.	TOTAL	NEW ALBANY 471 MAT.	INST.	TOTAL
015433	CONTRACTOR EQUIPMENT		85.3	85.3		97.4	97.4		83.9	83.9		103.6	103.6		95.5	95.5		93.3	93.3
0241, 31-34	SITE & INFRASTRUCTURE, DEMOLITION	100.0	90.9	93.6	95.5	93.6	94.2	84.5	92.0	89.8	82.0	107.3	99.7	86.9	92.6	90.9	79.2	94.5	89.8
0310	Concrete Forming & Accessories	100.2	85.1	87.3	99.1	78.0	81.2	91.9	80.9	82.5	90.9	76.6	78.8	91.6	79.1	81.0	89.8	77.4	79.2
0320	Concrete Reinforcing	95.6	87.6	91.5	86.4	87.3	86.9	89.4	85.3	87.3	88.7	77.2	82.9	98.9	85.3	92.0	90.0	83.0	86.4
0330	Cast-in-Place Concrete	100.6	84.7	94.7	102.7	80.9	94.6	99.4	80.0	92.2	92.9	74.4	86.0	104.3	76.2	93.9	95.9	74.3	87.9
03	CONCRETE	97.9	84.7	92.0	90.0	81.2	86.0	97.6	80.9	90.1	90.7	76.6	84.4	96.4	79.7	88.9	96.1	77.5	87.7
04	MASONRY	88.0	79.1	82.5	87.0	73.6	78.7	93.3	75.5	82.3	73.2	71.8	72.3	89.4	75.7	81.1	79.8	68.1	72.6
05	METALS	97.0	75.7	90.5	95.6	89.6	93.8	97.9	74.4	90.8	94.4	84.4	91.4	101.3	88.9	97.5	96.5	82.0	92.1
06	WOOD, PLASTICS & COMPOSITES	98.6	86.0	91.8	100.8	77.4	88.1	103.6	81.5	91.6	91.3	76.5	83.3	104.9	79.4	91.0	93.7	78.6	85.5
07	THERMAL & MOISTURE PROTECTION	103.3	81.4	93.9	111.9	76.5	96.6	98.0	79.4	90.0	104.0	76.2	92.0	101.2	78.5	91.4	89.6	72.2	82.1
08	OPENINGS	104.1	82.2	99.0	90.3	77.3	87.3	94.4	78.9	90.8	95.8	73.8	90.7	93.0	77.8	89.5	93.4	77.5	89.7
0920	Plaster & Gypsum Board	96.4	85.8	89.4	109.5	77.1	88.1	93.3	81.7	85.7	71.9	76.5	74.9	94.0	79.5	84.5	91.9	78.5	83.0
0950, 0980	Ceilings & Acoustic Treatment	94.9	85.8	88.7	91.9	77.1	81.9	76.2	81.7	80.0	86.0	76.5	79.5	79.6	79.5	79.6	83.4	78.5	80.1
0960	Flooring	96.9	81.8	92.5	99.0	91.3	96.8	93.0	81.8	89.8	69.0	82.9	73.1	93.4	77.5	88.8	91.4	55.8	81.1
0970, 0990	Wall Finishes & Painting/Coating	99.3	79.5	87.6	94.0	69.7	79.6	85.4	83.5	84.3	86.2	74.5	79.3	85.4	67.6	74.8	91.0	68.0	77.4
09	FINISHES	96.0	84.2	89.6	93.4	79.6	85.8	86.9	81.5	83.9	78.6	77.7	78.1	87.4	77.6	82.0	91.0	68.0	77.4
COVERS	DIVS. 10-14, 25, 28, 41, 43, 44, 46	100.0	90.6	97.9	100.0	86.4	97.0	100.0	88.4	97.5	100.0	84.8	96.6	100.0	88.1	97.4	88.0	72.3	79.5
21, 22, 23	FIRE SUPPRESSION, PLUMBING & HVAC	99.9	79.7	91.8	97.5	78.8	90.0	97.3	77.9	89.5	98.2	74.5	88.7	99.8	78.6	91.3	100.0	83.5	96.4
26, 27, 3370	ELECTRICAL, COMMUNICATIONS & UTIL.	102.6	87.0	94.6	90.3	79.1	84.6	97.4	80.5	88.8	92.7	72.9	82.6	90.5	75.9	83.0	97.5	76.0	88.9
MF2018	WEIGHTED AVERAGE	99.1	83.0	92.2	94.5	81.0	88.7	96.1	80.1	89.2	92.9	78.5	86.8	96.3	80.3	89.4	94.2	77.2	86.9

INDIANA / IOWA

DIVISION		SOUTH BEND 465-466 MAT.	INST.	TOTAL	TERRE HAUTE 478 MAT.	INST.	TOTAL	WASHINGTON 475 MAT.	INST.	TOTAL	BURLINGTON 526 MAT.	INST.	TOTAL	CARROLL 514 MAT.	INST.	TOTAL	CEDAR RAPIDS 522-524 MAT.	INST.	TOTAL
015433	CONTRACTOR EQUIPMENT		108.9	108.9		113.7	113.7		113.7	113.7		101.4	101.4		101.4	101.4		98.2	98.2
0241, 31-34	SITE & INFRASTRUCTURE, DEMOLITION	98.8	94.3	95.7	94.0	121.6	113.3	94.0	121.8	113.4	98.9	96.5	97.2	88.2	97.3	94.5	100.5	96.0	97.4
0310	Concrete Forming & Accessories	95.2	78.1	80.6	94.5	79.2	81.5	95.6	82.1	84.1	94.3	96.8	96.4	81.6	79.8	80.1	100.4	86.3	88.4
0320	Concrete Reinforcing	95.8	86.0	90.8	97.9	85.4	91.6	90.5	84.9	87.7	94.4	100.6	97.6	95.1	88.3	91.6	95.1	82.5	88.7
0330	Cast-in-Place Concrete	103.6	79.9	94.8	91.8	80.6	87.7	99.9	86.0	94.7	107.9	56.4	88.7	107.9	84.4	99.2	108.2	85.9	99.9
03	CONCRETE	92.5	81.7	87.6	100.9	81.0	91.9	106.7	84.1	96.5	96.0	83.7	90.5	94.8	83.7	89.8	96.1	86.1	91.6
04	MASONRY	92.9	75.6	82.2	91.0	74.8	81.0	83.9	78.6	80.7	99.7	73.4	83.5	101.3	74.0	84.4	105.3	81.7	90.7
05	METALS	102.1	104.8	102.9	93.1	85.2	90.7	87.7	85.2	86.9	88.3	99.9	91.8	88.3	95.8	90.6	90.6	93.9	91.6
06	WOOD, PLASTICS & COMPOSITES	95.8	77.2	85.7	95.5	78.9	86.5	95.9	81.5	88.1	93.0	101.0	97.4	78.7	82.9	81.0	100.4	86.1	92.6
07	THERMAL & MOISTURE PROTECTION	104.5	80.4	94.1	103.0	81.4	93.7	102.9	84.2	94.9	104.1	79.2	93.4	104.4	76.3	92.3	105.2	82.3	95.3
08	OPENINGS	95.8	77.7	91.6	94.3	77.5	90.4	91.2	79.4	88.4	96.5	98.6	97.0	101.0	82.2	96.6	101.5	83.6	97.3
0920	Plaster & Gypsum Board	94.5	76.9	82.9	94.0	78.0	83.4	94.3	80.6	85.3	105.0	101.2	102.5	100.5	82.6	88.7	110.1	86.0	94.3
0950, 0980	Ceilings & Acoustic Treatment	90.3	76.9	81.2	83.4	78.0	79.7	79.2	80.6	80.1	95.0	101.2	99.2	95.0	82.6	86.6	97.5	86.0	89.3
0960	Flooring	93.6	89.1	92.3	94.0	78.2	89.4	94.9	79.3	90.3	95.3	71.4	88.4	89.6	82.2	87.5	109.6	87.1	103.0
0970, 0990	Wall Finishes & Painting/Coating	88.9	84.7	86.4	91.0	81.9	85.6	91.0	86.6	88.4	92.3	87.5	89.5	92.3	85.8	88.4	93.9	72.6	81.3
09	FINISHES	90.0	80.5	84.9	88.8	79.2	83.6	88.6	81.9	85.0	92.4	92.4	92.4	88.7	81.2	84.6	97.8	85.1	90.9
COVERS	DIVS. 10-14, 25, 28, 41, 43, 44, 46	100.0	90.3	97.9	100.0	90.5	97.9	100.0	93.3	98.5	100.0	92.1	98.3	100.0	85.8	96.9	100.0	94.0	98.7
21, 22, 23	FIRE SUPPRESSION, PLUMBING & HVAC	99.9	76.7	90.8	100.0	78.2	91.2	97.5	80.4	90.7	97.7	85.7	92.9	97.7	72.6	87.6	100.1	83.0	93.3
26, 27, 3370	ELECTRICAL, COMMUNICATIONS & UTIL.	95.9	86.0	90.8	93.1	85.5	89.2	93.5	83.5	88.4	98.7	74.0	86.0	99.3	78.8	88.8	96.4	82.0	89.1
MF2018	WEIGHTED AVERAGE	97.5	83.6	91.5	96.3	83.8	90.9	94.9	85.4	90.8	96.0	86.3	91.8	95.7	81.3	89.6	98.1	85.8	92.8

For customer support on your Electrical Costs with RSMeans Data, call 800.448.8182.

IOWA

DIVISION		COUNCIL BLUFFS 515			CRESTON 508			DAVENPORT 527 - 528			DECORAH 521			DES MOINES 500 - 503, 509			DUBUQUE 520		
		MAT.	INST.	TOTAL	MAT.	INST.	TOTAL	MAT.	INST.	TOTAL	MAT.	INST.	TOTAL	MAT.	INST.	TOTAL	MAT.	INST.	TOTAL
015433	CONTRACTOR EQUIPMENT		97.5	97.5		101.4	101.4		101.4	101.4		101.4	101.4		103.2	103.2		97.0	97.0
0241, 31 - 34	SITE & INFRASTRUCTURE, DEMOLITION	104.3	93.0	96.4	95.3	98.4	97.4	99.1	99.4	99.3	97.5	96.3	96.6	101.2	99.9	100.3	98.3	93.2	94.8
0310	Concrete Forming & Accessories	80.9	75.6	76.4	75.8	82.6	81.6	99.9	98.6	98.8	91.8	73.7	76.5	92.7	86.1	87.1	82.4	83.8	83.6
0320	Concrete Reinforcing	97.0	82.1	89.4	94.5	88.3	91.3	95.1	101.9	98.6	94.4	88.0	91.1	99.0	85.7	92.3	93.7	82.3	87.9
0330	Cast-in-Place Concrete	112.5	81.4	100.9	114.7	62.5	95.3	104.2	98.0	101.9	105.0	78.2	95.0	97.6	90.5	95.0	105.9	85.4	98.3
03	CONCRETE	98.3	79.7	89.9	98.3	77.3	88.8	94.2	99.4	96.5	93.9	78.7	87.1	91.3	88.3	90.0	92.9	84.8	89.3
04	MASONRY	106.7	77.8	88.8	99.3	81.6	88.3	102.1	94.1	97.2	120.9	76.2	93.3	87.3	85.5	86.2	106.3	70.3	84.1
05	METALS	95.7	93.5	95.1	92.1	96.3	93.4	90.6	106.4	95.4	88.4	95.4	90.5	98.1	97.2	97.8	89.3	93.1	90.4
06	WOOD, PLASTICS & COMPOSITES	77.5	74.2	75.7	70.6	82.9	77.3	100.4	98.1	99.1	89.8	72.9	80.7	87.7	85.3	86.4	79.3	84.6	82.2
07	THERMAL & MOISTURE PROTECTION	104.5	76.9	92.6	105.4	81.5	95.1	104.6	94.2	100.1	104.3	73.4	91.0	97.0	85.9	92.2	104.8	78.6	93.5
08	OPENINGS	100.5	77.6	95.2	111.0	84.0	104.7	101.5	99.5	101.0	99.5	80.2	95.0	98.9	88.2	96.4	100.5	84.8	96.9
0920	Plaster & Gypsum Board	100.5	73.9	82.9	96.1	82.6	87.2	110.1	98.2	102.3	103.8	72.3	83.1	93.5	85.0	87.9	100.5	84.5	90.0
0950, 0980	Ceilings & Acoustic Treatment	95.0	73.9	80.7	88.5	82.6	84.5	97.5	98.2	98.0	95.0	72.3	79.6	91.9	85.0	87.3	95.0	84.5	87.9
0960	Flooring	88.5	87.1	88.1	83.9	71.4	80.3	97.7	92.4	96.2	94.9	71.4	88.1	97.9	94.8	97.0	100.8	71.4	92.3
0970, 0990	Wall Finishes & Painting/Coating	89.1	72.6	79.3	83.2	76.2	79.1	92.3	92.9	92.6	92.3	85.8	88.4	88.8	85.8	87.0	93.1	80.5	85.7
09	FINISHES	89.4	77.1	82.7	82.5	79.9	81.1	94.3	96.8	95.7	92.0	74.4	82.4	90.2	87.7	88.8	93.1	81.2	86.6
COVERS	DIVS. 10 - 14, 25, 28, 41, 43, 44, 46	100.0	89.4	97.7	100.0	88.4	97.4	100.0	97.6	99.5	100.0	86.2	97.0	100.0	94.4	98.8	100.0	92.9	98.4
21, 22, 23	FIRE SUPPRESSION, PLUMBING & HVAC	100.1	74.4	89.8	97.6	80.2	90.6	100.1	96.0	98.5	97.7	75.3	88.7	99.8	84.7	93.8	100.0	77.3	91.0
26, 27, 3370	ELECTRICAL, COMMUNICATIONS & UTIL.	101.3	82.3	91.6	92.5	78.8	85.5	94.4	89.7	92.0	96.4	48.4	71.9	103.6	84.4	93.8	100.0	78.8	89.1
MF2018	WEIGHTED AVERAGE	98.8	80.8	91.1	96.7	83.1	90.9	97.1	96.9	97.0	96.7	75.9	87.8	97.3	88.4	93.5	97.1	81.9	90.6

IOWA

DIVISION		FORT DODGE 505			MASON CITY 504			OTTUMWA 525			SHENANDOAH 516			SIBLEY 512			SIOUX CITY 510 - 511		
		MAT.	INST.	TOTAL	MAT.	INST.	TOTAL	MAT.	INST.	TOTAL	MAT.	INST.	TOTAL	MAT.	INST.	TOTAL	MAT.	INST.	TOTAL
015433	CONTRACTOR EQUIPMENT		101.4	101.4		101.4	101.4		97.0	97.0		97.5	97.5		101.4	101.4		101.4	101.4
0241, 31 - 34	SITE & INFRASTRUCTURE, DEMOLITION	103.6	95.1	97.7	103.6	96.2	98.4	98.8	91.0	93.4	102.6	93.1	96.0	108.6	96.2	100.0	100.4	74.4	78.3
0310	Concrete Forming & Accessories	76.4	74.1	74.4	80.2	73.3	74.3	89.6	73.7	76.1	82.8	79.2	79.7	83.2	38.7	45.4	100.4	74.4	78.3
0320	Concrete Reinforcing	94.5	88.0	91.2	94.3	88.0	91.1	94.4	100.6	97.5	97.0	88.3	92.6	97.0	87.9	92.4	95.1	79.3	87.1
0330	Cast-in-Place Concrete	107.6	43.0	83.6	107.6	73.3	94.9	108.6	65.6	92.6	95.8	83.6	90.3	94.7	55.5	77.1	95.5	75.3	86.4
03	CONCRETE	93.6	66.7	81.5	93.8	76.8	86.2	95.5	76.6	87.0	106.2	76.0	87.6	124.9	58.8	84.1	99.4	68.4	80.2
04	MASONRY	98.1	58.6	73.7	110.9	75.0	88.7	102.6	54.0	72.6	106.2	76.0	87.6	124.9	58.8	84.1	99.4	68.4	80.2
05	METALS	92.2	95.0	93.1	92.3	95.4	93.2	88.2	100.1	91.8	94.7	96.3	95.2	88.5	94.3	90.3	90.6	91.4	90.9
06	WOOD, PLASTICS & COMPOSITES	71.1	82.9	77.5	75.0	72.9	73.9	86.9	79.7	83.0	79.3	80.8	80.1	80.1	35.1	55.7	100.4	74.3	86.2
07	THERMAL & MOISTURE PROTECTION	104.7	67.9	88.9	104.1	74.0	91.2	105.0	68.7	89.4	103.7	74.4	91.1	104.0	55.8	83.2	104.6	72.6	90.8
08	OPENINGS	104.8	73.7	97.6	96.8	80.2	93.0	101.0	83.9	97.0	92.1	78.7	89.0	99.7	47.3	86.0	101.5	75.5	95.4
0920	Plaster & Gypsum Board	96.1	82.6	87.2	96.1	72.3	80.5	101.4	79.5	87.0	100.5	80.6	87.4	100.5	33.5	56.3	110.1	73.8	86.2
0950, 0980	Ceilings & Acoustic Treatment	88.5	82.6	84.5	88.5	72.3	77.6	95.0	79.5	84.5	95.0	80.6	85.2	95.0	33.5	53.3	97.5	73.8	81.4
0960	Flooring	85.2	71.4	81.2	87.2	71.4	82.7	103.8	71.4	94.4	89.2	76.4	85.5	90.5	71.4	85.0	97.7	74.0	90.9
0970, 0990	Wall Finishes & Painting/Coating	83.2	83.4	83.3	83.2	85.8	84.7	93.1	85.8	88.8	89.1	85.8	87.1	92.3	85.8	88.4	92.3	68.2	78.0
09	FINISHES	84.3	75.5	79.5	84.9	74.1	79.0	94.3	74.8	83.6	89.5	79.3	83.9	91.7	47.7	67.7	95.7	73.5	83.6
COVERS	DIVS. 10 - 14, 25, 28, 41, 43, 44, 46	100.0	82.4	96.1	100.0	85.8	96.9	100.0	82.8	96.2	100.0	85.8	96.9	100.0	76.3	94.8	100.0	91.1	98.0
21, 22, 23	FIRE SUPPRESSION, PLUMBING & HVAC	97.6	69.5	86.3	97.6	78.6	89.9	97.7	72.6	87.6	97.7	85.7	92.9	97.7	67.7	85.6	100.1	78.1	91.3
26, 27, 3370	ELECTRICAL, COMMUNICATIONS & UTIL.	98.4	71.1	84.4	97.6	48.4	72.4	98.5	72.7	85.3	96.4	80.4	88.2	96.4	48.4	71.9	96.4	75.8	85.9
MF2018	WEIGHTED AVERAGE	96.4	74.1	86.8	96.2	76.2	87.6	96.6	76.3	87.9	96.4	83.8	91.0	97.0	63.3	82.5	97.8	78.7	89.6

DIVISION		IOWA SPENCER 513			WATERLOO 506 - 507			KANSAS BELLEVILLE 669			COLBY 677			DODGE CITY 678			EMPORIA 668		
		MAT.	INST.	TOTAL	MAT.	INST.	TOTAL	MAT.	INST.	TOTAL	MAT.	INST.	TOTAL	MAT.	INST.	TOTAL	MAT.	INST.	TOTAL
015433	CONTRACTOR EQUIPMENT		101.4	101.4		101.4	101.4		106.5	106.5		106.5	106.5		106.5	106.5		104.7	104.7
0241, 31 - 34	SITE & INFRASTRUCTURE, DEMOLITION	108.6	94.9	99.1	108.9	97.4	100.9	109.2	96.7	100.5	107.5	97.4	100.4	109.9	96.4	100.5	101.7	94.2	96.5
0310	Concrete Forming & Accessories	89.5	38.5	46.2	91.4	70.6	73.7	97.0	55.7	61.9	99.3	62.9	68.4	92.5	62.7	67.2	87.5	67.4	70.4
0320	Concrete Reinforcing	97.0	87.9	92.3	95.1	82.6	88.7	105.6	103.6	104.6	105.2	103.6	104.4	102.5	103.6	103.1	104.1	103.8	103.9
0330	Cast-in-Place Concrete	106.6	67.4	92.0	115.3	86.5	104.6	119.6	86.0	107.1	116.6	89.8	106.6	118.7	89.5	107.8	115.7	89.9	106.1
03	CONCRETE	95.2	59.0	78.9	99.7	79.3	90.5	112.6	75.9	96.1	109.9	80.6	96.7	111.2	80.3	97.3	105.1	82.6	95.0
04	MASONRY	124.9	58.8	84.1	98.9	76.7	85.2	90.4	59.7	71.5	99.5	66.0	78.8	108.0	65.9	82.0	96.0	67.4	78.3
05	METALS	88.4	94.1	90.2	94.5	93.8	94.3	96.0	95.3	97.0	93.2	100.2	95.3	94.7	99.1	96.0	95.7	100.3	97.1
06	WOOD, PLASTICS & COMPOSITES	86.8	35.1	58.7	88.7	66.2	76.5	97.0	52.6	72.9	101.1	59.0	78.2	92.5	59.0	74.3	87.3	64.8	75.1
07	THERMAL & MOISTURE PROTECTION	105.0	58.9	85.2	104.4	79.3	93.6	93.8	63.9	80.9	101.0	67.2	86.5	101.0	69.9	87.6	91.9	76.8	85.4
08	OPENINGS	109.0	47.3	94.7	97.3	75.1	92.1	96.2	63.7	88.7	100.2	71.0	93.4	100.1	67.2	92.5	94.2	72.8	89.2
0920	Plaster & Gypsum Board	101.4	33.5	56.7	104.9	65.4	78.9	89.2	51.4	64.3	99.3	58.0	72.1	92.9	58.0	69.9	86.2	63.9	71.5
0950, 0980	Ceilings & Acoustic Treatment	95.0	33.5	53.3	91.1	65.4	73.7	85.4	51.4	62.4	82.9	58.0	66.0	82.9	58.0	66.0	85.4	63.9	70.8
0960	Flooring	93.3	71.4	86.9	92.4	81.6	89.3	90.5	69.3	84.4	86.2	69.3	81.3	82.4	69.3	78.6	86.1	69.3	81.2
0970, 0990	Wall Finishes & Painting/Coating	92.3	58.7	72.4	83.2	83.4	83.3	92.0	56.5	71.0	92.8	56.5	71.3	92.8	56.5	71.3	83.2	65.4	73.5
09	FINISHES	92.6	43.5	65.8	88.4	73.0	80.0	85.9	56.7	70.0	86.0	62.0	72.9	84.1	62.0	72.0	83.2	65.4	73.5
COVERS	DIVS. 10 - 14, 25, 28, 41, 43, 44, 46	100.0	76.2	94.8	100.0	91.5	98.1	100.0	83.6	96.4	100.0	86.4	97.0	100.0	86.4	97.0	100.0	87.0	97.1
21, 22, 23	FIRE SUPPRESSION, PLUMBING & HVAC	97.7	69.7	86.4	100.0	81.9	92.8	97.4	71.3	87.0	97.6	72.1	87.4	100.1	72.1	88.9	97.4	74.9	88.4
26, 27, 3370	ELECTRICAL, COMMUNICATIONS & UTIL.	98.0	48.4	72.6	94.3	63.1	78.3	97.5	64.2	80.5	97.4	69.5	83.2	94.5	69.5	81.7	94.9	68.5	81.4
MF2018	WEIGHTED AVERAGE	98.5	63.6	83.6	97.5	79.3	89.7	98.1	72.1	87.0	98.4	75.7	88.7	99.3	75.4	89.1	96.4	77.2	88.2

For customer support on your Electrical Costs with RSMeans Data, call 800.448.8182.

533

City Cost Indexes

KANSAS

DIVISION		FORT SCOTT 667 MAT.	INST.	TOTAL	HAYS 676 MAT.	INST.	TOTAL	HUTCHINSON 675 MAT.	INST.	TOTAL	INDEPENDENCE 673 MAT.	INST.	TOTAL	KANSAS CITY 660 - 662 MAT.	INST.	TOTAL	LIBERAL 679 MAT.	INST.	TOTAL
015433	CONTRACTOR EQUIPMENT		105.6	105.6		106.5	106.5		106.5	106.5		106.5	106.5		103.1	103.1		106.5	106.5
0241, 31 - 34	SITE & INFRASTRUCTURE, DEMOLITION	98.5	94.2	95.5	111.8	96.9	101.5	91.3	97.4	95.5	110.8	97.4	101.5	93.3	94.5	94.1	111.7	97.0	101.5
0310	Concrete Forming & Accessories	105.6	82.1	85.6	96.7	60.5	65.9	87.1	58.1	62.5	107.9	69.6	75.4	101.7	97.2	97.8	92.9	60.4	65.3
0320	Concrete Reinforcing	103.4	103.7	103.6	102.5	103.6	103.1	102.5	103.6	103.1	101.9	103.7	102.8	100.3	105.3	102.9	104.0	103.6	103.8
0330	Cast-in-Place Concrete	107.2	84.8	98.9	92.0	86.2	89.8	85.2	89.4	86.8	119.2	89.6	108.2	91.8	98.6	94.4	92.0	86.0	89.7
03	CONCRETE	100.5	87.5	94.6	101.1	78.2	90.8	84.3	78.2	81.5	112.5	83.5	99.4	92.5	99.5	95.6	103.0	78.1	91.8
04	MASONRY	97.0	57.3	72.5	107.5	59.8	78.0	99.5	65.9	78.7	96.7	65.9	77.7	98.1	98.0	98.1	105.6	59.7	77.3
05	METALS	95.6	99.7	96.9	92.8	99.9	94.9	92.6	99.0	94.6	92.5	99.6	94.7	102.7	105.2	103.4	93.1	99.2	94.9
06	WOOD, PLASTICS & COMPOSITES	108.2	88.9	97.7	97.8	59.0	76.7	87.2	52.9	68.5	112.3	67.8	88.1	103.8	96.8	100.0	93.2	59.0	74.6
07	THERMAL & MOISTURE PROTECTION	92.9	74.4	85.0	101.4	64.6	85.5	99.7	66.2	85.3	101.1	77.0	90.7	92.6	99.8	95.7	101.5	64.0	85.4
08	OPENINGS	94.2	86.1	92.3	100.1	67.2	92.4	100.1	63.8	91.6	98.1	71.9	92.0	95.4	96.4	95.6	100.1	67.2	92.5
0920	Plaster & Gypsum Board	91.6	88.7	89.7	96.5	58.0	71.1	91.6	51.6	65.3	107.8	67.0	80.9	85.5	96.8	92.9	93.8	58.0	70.2
0950, 0980	Ceilings & Acoustic Treatment	85.4	88.7	87.7	82.9	58.0	66.0	82.9	51.6	61.7	82.9	67.0	72.1	85.4	96.8	93.1	82.9	58.0	66.0
0960	Flooring	99.7	69.1	90.9	85.0	69.3	80.4	79.7	69.3	76.6	90.4	69.1	84.2	80.9	97.5	85.7	82.7	68.1	78.5
0970, 0990	Wall Finishes & Painting/Coating	93.6	79.5	85.3	92.8	56.5	71.3	92.8	56.5	71.3	92.8	56.5	71.3	99.6	100.9	100.3	92.8	56.5	71.3
09	FINISHES	88.2	80.5	84.0	85.8	60.5	72.0	81.6	58.4	68.9	88.6	67.7	77.2	83.7	97.5	91.3	85.0	60.2	71.5
COVERS	DIVS. 10 - 14, 25, 28, 41, 43, 44, 46	100.0	86.7	97.1	100.0	84.3	96.5	100.0	85.7	96.8	100.0	87.4	97.2	100.0	94.5	98.8	100.0	84.3	96.5
21, 22, 23	FIRE SUPPRESSION, PLUMBING & HVAC	97.4	68.3	85.8	97.6	68.9	86.1	97.6	70.7	86.8	97.6	72.0	87.3	99.8	98.8	99.4	97.6	70.3	86.7
26, 27, 3370	ELECTRICAL, COMMUNICATIONS & UTIL.	94.2	69.5	81.6	96.4	69.5	82.7	91.8	62.4	76.8	93.8	74.6	84.0	99.5	97.9	98.7	94.5	69.5	81.7
MF2018	WEIGHTED AVERAGE	96.3	78.2	88.5	97.6	73.5	87.3	93.7	73.1	84.8	98.3	77.9	89.5	97.2	98.6	97.8	97.5	73.7	87.3

KANSAS / KENTUCKY

DIVISION		SALINA 674 MAT.	INST.	TOTAL	TOPEKA 664 - 666 MAT.	INST.	TOTAL	WICHITA 670 - 672 MAT.	INST.	TOTAL	ASHLAND 411 - 412 MAT.	INST.	TOTAL	BOWLING GREEN 421 - 422 MAT.	INST.	TOTAL	CAMPTON 413 - 414 MAT.	INST.	TOTAL
015433	CONTRACTOR EQUIPMENT		106.5	106.5		104.7	104.7		106.5	106.5		100.4	100.4		93.3	93.3		99.7	99.7
0241, 31 - 34	SITE & INFRASTRUCTURE, DEMOLITION	100.4	96.9	98.0	96.4	94.0	94.7	96.4	98.1	97.6	114.6	83.0	92.6	79.2	94.2	89.6	88.2	95.5	93.3
0310	Concrete Forming & Accessories	88.9	55.8	60.8	98.7	67.8	72.4	94.6	55.0	61.0	86.3	92.2	91.3	86.0	77.8	79.0	88.6	83.2	84.0
0320	Concrete Reinforcing	101.9	103.4	102.7	99.9	102.7	101.3	100.0	101.0	100.5	91.4	96.9	94.2	88.7	79.0	83.8	89.6	95.7	92.7
0330	Cast-in-Place Concrete	103.0	86.2	96.8	96.5	87.9	93.3	96.3	74.6	88.2	87.3	97.5	91.1	86.6	71.5	81.0	96.6	69.7	86.6
03	CONCRETE	98.2	76.0	88.2	94.2	82.0	88.7	93.8	71.2	83.7	91.7	95.9	93.6	90.3	76.2	84.0	93.9	80.9	88.1
04	MASONRY	122.3	56.0	81.4	91.8	65.5	75.6	96.6	50.5	68.1	89.6	94.0	92.3	91.6	70.6	78.6	89.0	55.9	68.5
05	METALS	94.5	99.6	96.0	99.7	100.4	99.9	96.5	97.8	96.9	95.5	108.8	99.6	97.2	83.8	93.1	96.5	90.7	94.7
06	WOOD, PLASTICS & COMPOSITES	88.7	52.6	69.1	102.0	66.9	82.9	98.6	52.9	73.8	75.3	90.2	83.4	87.9	78.4	82.7	86.3	90.9	88.8
07	THERMAL & MOISTURE PROTECTION	100.4	62.8	84.2	96.7	76.1	87.8	99.3	59.4	82.1	94.0	90.9	92.6	89.6	79.6	85.3	103.1	69.7	88.7
08	OPENINGS	100.1	63.7	91.6	103.6	75.2	97.0	104.5	63.3	94.9	92.6	91.6	92.4	93.4	80.5	90.4	94.6	89.9	93.5
0920	Plaster & Gypsum Board	91.6	51.4	65.1	97.9	66.0	76.9	94.8	51.6	66.4	60.0	90.2	79.9	87.3	78.2	81.3	87.3	90.2	89.2
0950, 0980	Ceilings & Acoustic Treatment	82.9	51.4	61.6	91.7	66.0	74.3	86.7	51.6	63.0	78.7	90.2	86.5	83.4	78.2	79.9	83.4	90.2	88.0
0960	Flooring	81.0	69.3	77.6	95.7	69.3	88.0	92.8	69.3	86.0	74.3	84.0	77.1	89.4	64.0	82.0	91.5	66.5	84.3
0970, 0990	Wall Finishes & Painting/Coating	92.8	56.5	71.3	96.1	69.5	80.4	91.5	56.5	70.8	92.9	92.8	92.8	91.0	90.4	90.7	91.0	55.3	69.9
09	FINISHES	82.8	56.7	68.6	92.5	67.2	78.7	89.6	56.3	71.5	75.9	90.7	84.0	86.7	76.3	81.0	87.4	78.1	82.4
COVERS	DIVS. 10 - 14, 25, 28, 41, 43, 44, 46	100.0	84.3	96.5	100.0	81.6	95.9	100.0	83.7	96.4	100.0	90.0	97.8	100.0	85.6	96.8	100.0	48.8	88.7
21, 22, 23	FIRE SUPPRESSION, PLUMBING & HVAC	100.1	70.9	88.4	100.0	74.7	89.9	99.8	68.5	87.3	97.3	86.9	93.2	100.0	78.8	91.5	97.5	78.1	90.3
26, 27, 3370	ELECTRICAL, COMMUNICATIONS & UTIL.	94.3	71.9	82.8	98.4	73.6	85.7	98.2	71.8	84.7	91.1	90.4	90.7	93.6	76.5	84.9	91.1	90.3	90.7
MF2018	WEIGHTED AVERAGE	98.0	72.8	87.2	98.3	77.7	89.5	97.8	70.8	86.2	93.6	92.0	92.9	94.7	78.9	87.9	94.8	79.9	88.4

KENTUCKY

DIVISION		CORBIN 407 - 409 MAT.	INST.	TOTAL	COVINGTON 410 MAT.	INST.	TOTAL	ELIZABETHTOWN 427 MAT.	INST.	TOTAL	FRANKFORT 406 MAT.	INST.	TOTAL	HAZARD 417 - 418 MAT.	INST.	TOTAL	HENDERSON 424 MAT.	INST.	TOTAL
015433	CONTRACTOR EQUIPMENT		99.7	99.7		103.6	103.6		93.3	93.3		99.7	99.7		99.7	99.7		113.7	113.7
0241, 31 - 34	SITE & INFRASTRUCTURE, DEMOLITION	90.8	96.0	94.4	83.5	107.3	100.1	73.6	93.9	87.7	89.1	96.5	94.2	85.9	96.6	93.4	82.2	120.8	109.1
0310	Concrete Forming & Accessories	84.8	78.1	79.1	84.0	72.7	74.4	80.6	74.0	75.0	101.4	79.3	82.7	85.0	84.1	84.3	91.6	79.1	81.0
0320	Concrete Reinforcing	92.1	92.3	92.2	88.3	80.9	84.5	89.2	84.3	86.7	96.9	82.2	89.5	90.0	92.5	91.3	88.8	81.1	84.9
0330	Cast-in-Place Concrete	92.8	74.3	85.9	92.4	82.1	88.6	78.3	69.0	74.8	94.1	76.4	87.6	92.9	71.7	85.0	76.5	87.0	80.4
03	CONCRETE	88.1	79.6	84.3	92.2	78.2	85.9	82.3	74.5	78.8	91.7	79.2	86.1	90.7	81.5	86.6	87.3	82.5	85.2
04	MASONRY	81.0	61.3	68.8	103.0	73.5	84.8	76.5	63.0	68.2	79.5	71.8	74.7	87.5	58.0	69.3	94.8	80.4	85.9
05	METALS	94.1	89.3	92.6	94.3	88.9	92.7	96.3	84.9	92.9	94.5	86.0	91.9	96.5	89.7	94.4	87.4	85.8	86.9
06	WOOD, PLASTICS & COMPOSITES	75.0	79.8	77.6	83.7	70.2	76.4	82.5	76.1	79.0	102.1	79.8	90.0	82.9	90.9	87.2	90.8	78.0	83.8
07	THERMAL & MOISTURE PROTECTION	107.0	71.5	91.7	104.3	73.7	91.1	89.0	70.5	81.1	105.6	75.7	92.7	103.0	71.5	89.4	102.2	85.1	94.9
08	OPENINGS	89.2	71.9	85.2	96.6	73.9	91.4	93.4	74.7	89.0	96.7	79.6	92.7	95.0	89.1	93.6	91.6	79.4	88.7
0920	Plaster & Gypsum Board	93.1	78.9	83.7	68.6	70.0	69.5	86.1	75.8	79.3	95.2	78.9	84.4	86.1	90.2	88.8	91.0	77.0	81.8
0950, 0980	Ceilings & Acoustic Treatment	79.8	78.9	79.2	86.0	70.0	75.2	83.4	75.8	78.3	90.1	78.9	82.5	83.4	90.2	88.0	79.2	77.0	77.7
0960	Flooring	86.0	66.5	80.3	66.6	86.3	72.3	86.8	76.3	83.8	95.1	66.5	86.8	89.8	66.5	83.0	93.0	78.6	88.9
0970, 0990	Wall Finishes & Painting/Coating	90.7	63.6	74.7	86.2	73.0	78.4	91.0	69.9	78.6	97.5	94.1	95.5	91.0	55.3	69.9	91.0	89.0	89.9
09	FINISHES	81.3	74.6	77.6	77.6	74.8	76.1	85.3	73.7	79.0	89.1	78.9	83.5	86.6	78.7	82.3	86.8	80.1	83.2
COVERS	DIVS. 10 - 14, 25, 28, 41, 43, 44, 46	100.0	91.7	98.2	100.0	87.7	97.3	100.0	71.5	93.7	100.0	58.6	90.9	100.0	49.6	88.9	100.0	56.9	90.5
21, 22, 23	FIRE SUPPRESSION, PLUMBING & HVAC	97.7	74.9	88.5	98.2	77.7	90.0	97.6	77.8	89.7	100.0	78.7	91.5	97.5	79.2	90.2	97.6	78.1	89.8
26, 27, 3370	ELECTRICAL, COMMUNICATIONS & UTIL.	90.9	90.3	90.6	94.6	74.2	84.2	90.9	78.2	84.4	102.9	73.2	87.7	91.1	90.4	90.7	93.1	75.5	84.1
MF2018	WEIGHTED AVERAGE	92.4	79.6	86.9	94.7	79.8	88.3	91.7	76.7	85.3	96.1	78.7	88.6	94.2	80.6	88.4	92.5	82.4	88.2

For customer support on your Electrical Costs with RSMeans Data, call 800.448.8182.

KENTUCKY

DIVISION	LEXINGTON 403-405 MAT.	INST.	TOTAL	LOUISVILLE 400-402 MAT.	INST.	TOTAL	OWENSBORO 423 MAT.	INST.	TOTAL	PADUCAH 420 MAT.	INST.	TOTAL	PIKEVILLE 415-416 MAT.	INST.	TOTAL	SOMERSET 425-426 MAT.	INST.	TOTAL
015433 CONTRACTOR EQUIPMENT		99.7	99.7		93.3	93.3		113.7	113.7		113.7	113.7		100.4	100.4		99.7	99.7
0241, 31-34 SITE & INFRASTRUCTURE, DEMOLITION	92.9	97.8	96.3	88.0	94.1	92.3	92.3	121.4	112.6	84.9	120.9	110.0	125.9	82.1	95.4	78.6	96.0	90.7
0310 Concrete Forming & Accessories	97.0	75.6	78.8	94.0	76.9	79.5	90.1	79.0	80.6	88.0	80.8	81.9	91.9	96.8	94.4	89.2	62.1	75.4
0320 Concrete Reinforcing	101.2	82.8	91.8	96.9	82.6	89.6	88.8	79.5	84.1	89.4	82.9	86.1	96.0	92.1	94.6	76.5	90.1	81.6
0330 Cast-in-Place Concrete	95.0	86.9	92.0	89.9	70.6	82.7	89.3	86.4	88.2	81.6	83.1	82.1	105.1	92.2	99.3	77.6	77.8	77.7
03 CONCRETE	91.0	81.2	86.6	89.2	76.1	83.3	99.1	82.0	91.4	91.8	82.3	87.5	87.9	85.0	86.1	82.4	64.0	71.1
04 MASONRY	79.6	72.6	75.3	75.7	68.0	70.9	87.8	80.3	83.1	90.6	80.1	84.1	95.4	108.6	99.4	96.4	78.3	90.9
05 METALS	96.5	86.1	93.4	100.5	84.6	95.7	88.9	85.6	87.9	85.9	87.6	86.4	85.7	90.2	88.2	83.4	73.6	78.1
06 WOOD, PLASTICS & COMPOSITES	91.3	73.6	81.7	87.9	78.4	82.7	88.5	78.0	82.8	86.0	80.7	83.1	94.8	82.8	89.6	102.2	71.8	89.1
07 THERMAL & MOISTURE PROTECTION	107.4	75.5	93.7	103.4	72.8	90.2	103.0	78.8	92.6	102.3	84.3	94.6	93.2	88.2	92.1	94.1	68.0	88.0
08 OPENINGS	89.4	74.0	86.3	87.4	75.5	84.6	91.6	79.0	88.7	90.9	81.3	88.7	63.7	90.2	81.2	86.1	72.5	77.1
0920 Plaster & Gypsum Board	102.7	72.5	82.8	95.9	78.2	84.3	89.5	77.0	81.2	88.5	79.8	82.8	78.7	90.2	86.5	83.4	72.5	76.0
0950, 0980 Ceilings & Acoustic Treatment	83.2	72.5	76.0	88.3	78.2	81.5	79.2	77.0	77.7	79.2	79.8	79.6	78.7	90.2	86.5	83.4	72.5	76.0
0960 Flooring	90.8	66.5	83.7	93.5	64.0	84.9	92.5	64.0	84.2	91.4	78.6	87.7	78.4	66.5	74.9	90.1	66.5	83.2
0970, 0990 Wall Finishes & Painting/Coating	90.7	81.8	85.4	96.9	69.9	80.9	91.0	89.0	89.9	91.0	75.1	81.6	92.9	94.1	93.6	91.0	69.9	78.6
09 FINISHES	84.7	74.1	78.9	88.2	73.8	80.4	87.0	76.9	81.5	86.2	79.7	82.7	78.4	85.0	82.0	86.0	71.6	78.1
COVERS DIVS. 10-14, 25, 28, 41, 43, 44, 46	100.0	90.3	97.9	100.0	87.5	97.2	100.0	94.9	98.9	100.0	86.3	97.0	100.0	49.2	88.8	100.0	88.9	97.6
21, 22, 23 FIRE SUPPRESSION, PLUMBING & HVAC	100.1	78.0	91.3	100.0	77.2	90.9	100.0	78.8	91.5	97.6	78.4	89.9	97.3	84.1	92.0	97.6	74.5	88.4
26, 27, 3370 ELECTRICAL, COMMUNICATIONS & UTIL.	93.4	76.5	84.8	96.0	76.5	86.0	93.1	75.5	84.1	95.3	77.5	86.2	93.9	90.4	92.1	92.1	77.8	86.0
MF2018 WEIGHTED AVERAGE	94.4	79.6	88.0	94.6	77.7	87.4	94.7	83.1	89.7	92.7	83.8	88.9	96.0	87.6	92.4	92.1	77.8	86.0

LOUISIANA

DIVISION	ALEXANDRIA 713-714 MAT.	INST.	TOTAL	BATON ROUGE 707-708 MAT.	INST.	TOTAL	HAMMOND 704 MAT.	INST.	TOTAL	LAFAYETTE 705 MAT.	INST.	TOTAL	LAKE CHARLES 706 MAT.	INST.	TOTAL	MONROE 712 MAT.	INST.	TOTAL
015433 CONTRACTOR EQUIPMENT		93.2	93.2		90.4	90.4		90.9	90.9		90.9	90.9		90.4	90.4		93.2	93.2
0241, 31-34 SITE & INFRASTRUCTURE, DEMOLITION	101.6	92.1	95.0	103.5	90.4	94.4	100.0	89.6	92.7	101.0	91.8	94.6	101.8	91.0	94.3	101.6	92.0	94.9
0310 Concrete Forming & Accessories	78.9	61.6	64.2	95.1	73.4	76.7	77.4	56.2	59.4	94.6	69.5	73.3	93.0	55.9	74.1	98.8	55.6	76.8
0320 Concrete Reinforcing	99.9	55.6	77.4	90.5	55.6	72.8	91.7	55.3	73.2	93.0	55.6	74.0	97.2	67.9	82.3	91.1	65.9	81.7
0330 Cast-in-Place Concrete	91.1	68.1	82.5	97.0	75.6	89.1	92.8	64.2	82.1	92.3	67.8	83.2	92.4	67.0	81.0	86.3	62.5	75.6
03 CONCRETE	86.5	63.5	76.2	91.0	71.7	82.3	89.0	59.6	75.8	90.0	66.9	79.6	93.9	68.2	78.0	104.1	60.3	77.1
04 MASONRY	109.4	61.6	79.9	91.6	64.0	74.5	94.4	53.6	69.2	94.4	64.0	75.6	89.3	69.7	83.3	88.4	70.8	83.1
05 METALS	88.5	71.0	83.2	99.8	74.3	92.1	90.0	68.7	83.5	89.3	69.3	83.2	89.3	69.7	83.3	90.4	60.5	74.2
06 WOOD, PLASTICS & COMPOSITES	91.0	60.5	74.5	99.7	75.1	86.3	83.4	55.8	68.4	104.2	70.0	85.6	102.4	70.0	84.8	94.9	65.5	82.3
07 THERMAL & MOISTURE PROTECTION	94.9	66.3	82.6	97.7	68.7	85.2	96.1	62.4	81.6	96.7	68.1	84.4	96.4	69.1	84.6	113.4	58.7	100.7
08 OPENINGS	113.4	60.3	101.1	101.2	72.3	94.5	98.2	56.5	88.5	101.9	63.9	93.1	101.9	63.9	93.1	86.5	59.8	68.9
0920 Plaster & Gypsum Board	86.8	59.8	69.0	97.1	74.9	82.4	97.9	55.0	69.6	106.1	69.6	82.0	106.1	69.6	82.0	80.9	59.8	66.6
0950, 0980 Ceilings & Acoustic Treatment	80.9	59.8	66.6	95.3	74.9	81.5	98.8	55.0	69.1	97.1	69.6	78.4	98.0	69.6	78.7	90.9	56.9	81.0
0960 Flooring	91.3	63.5	83.2	92.1	59.0	82.5	88.7	57.1	79.6	97.2	62.9	87.3	97.2	65.9	88.1	94.5	61.3	74.9
0970, 0990 Wall Finishes & Painting/Coating	94.5	61.3	74.9	91.5	61.3	73.7	95.5	61.4	75.3	95.5	61.3	75.3	95.5	61.3	75.3	94.5	61.3	74.9
09 FINISHES	81.5	61.6	70.6	87.8	69.9	78.0	89.0	56.4	71.2	92.3	67.6	78.8	92.5	68.2	79.2	81.4	59.9	69.7
COVERS DIVS. 10-14, 25, 28, 41, 43, 44, 46	100.0	80.7	95.8	100.0	85.5	96.8	100.0	80.1	95.6	100.0	85.0	96.7	100.0	85.0	96.7	100.0	80.6	95.7
21, 22, 23 FIRE SUPPRESSION, PLUMBING & HVAC	100.0	64.7	85.9	100.0	65.5	86.2	97.7	62.3	83.5	100.1	65.6	86.3	100.1	65.9	86.4	100.0	63.3	85.3
26, 27, 3370 ELECTRICAL, COMMUNICATIONS & UTIL.	93.1	63.0	77.7	99.7	59.5	79.1	95.5	71.3	83.1	96.5	65.3	80.6	96.1	67.6	81.5	94.7	58.7	76.3
MF2018 WEIGHTED AVERAGE	96.0	66.6	83.4	97.6	69.7	85.7	94.4	64.5	81.6	96.0	68.9	84.4	96.2	69.8	84.9	95.8	65.1	82.6

LOUISIANA / MAINE

DIVISION	NEW ORLEANS 700-701 MAT.	INST.	TOTAL	SHREVEPORT 710-711 MAT.	INST.	TOTAL	THIBODAUX 703 MAT.	INST.	TOTAL	AUGUSTA 043 MAT.	INST.	TOTAL	BANGOR 044 MAT.	INST.	TOTAL	BATH 045 MAT.	INST.	TOTAL
015433 CONTRACTOR EQUIPMENT		86.7	86.7		93.2	93.2		90.9	90.9		97.9	97.9		97.9	97.9		97.9	97.9
0241, 31-34 SITE & INFRASTRUCTURE, DEMOLITION	100.9	92.9	95.4	103.9	92.0	95.6	102.3	91.6	94.8	87.8	98.5	95.2	90.7	99.8	97.0	88.6	98.5	95.5
0310 Concrete Forming & Accessories	96.8	70.6	74.5	94.6	61.6	66.6	88.6	65.8	69.2	101.1	78.7	82.1	95.2	78.0	80.6	90.6	78.7	80.5
0320 Concrete Reinforcing	94.7	56.0	75.0	100.1	55.4	77.4	91.7	55.1	73.1	105.3	82.9	93.9	94.5	83.1	88.7	93.5	82.6	88.0
0330 Cast-in-Place Concrete	93.1	70.7	84.7	94.0	66.1	83.6	99.7	65.8	87.1	90.5	110.3	97.8	73.2	111.1	87.3	73.2	110.2	87.0
03 CONCRETE	95.6	68.0	83.2	90.1	62.9	77.8	94.0	64.5	80.7	90.3	90.9	90.5	115.5	93.5	101.9	122.6	85.9	99.9
04 MASONRY	98.4	65.6	78.2	94.7	57.7	71.8	118.4	52.8	77.9	102.9	90.3	95.2	89.1	92.2	90.0	87.5	90.8	88.5
05 METALS	99.4	63.1	88.4	92.3	70.7	85.7	90.0	68.4	83.4	99.9	91.1	97.2	94.4	75.1	83.9	88.4	77.6	82.6
06 WOOD, PLASTICS & COMPOSITES	99.4	73.0	85.0	98.0	61.3	78.0	90.9	67.4	78.1	95.3	77.6	85.7	102.5	82.9	94.0	102.4	75.4	90.8
07 THERMAL & MOISTURE PROTECTION	93.9	69.7	83.5	93.2	64.9	81.0	96.2	63.8	82.2	104.8	76.0	92.4	102.5	82.9	94.0	102.4	75.4	90.8
08 OPENINGS	100.8	66.0	92.7	107.2	58.2	95.8	102.8	58.6	92.5	100.3	80.7	95.8	98.8	79.3	94.3	98.8	80.7	94.6
0920 Plaster & Gypsum Board	96.9	72.6	80.9	95.4	60.6	72.5	99.1	66.9	77.9	114.0	76.6	89.4	117.1	74.0	88.7	112.3	76.6	88.8
0950, 0980 Ceilings & Acoustic Treatment	102.1	72.6	82.1	87.6	60.6	69.3	98.8	66.9	77.2	105.6	76.6	86.0	90.9	74.0	79.5	90.0	76.6	80.9
0960 Flooring	102.3	65.7	91.7	97.6	56.9	85.8	94.6	39.4	78.6	88.2	47.3	76.3	80.9	102.1	87.0	79.4	47.3	70.0
0970, 0990 Wall Finishes & Painting/Coating	100.2	62.2	77.7	89.8	61.3	72.9	96.7	61.4	75.8	95.0	91.3	92.4	86.4	83.1	84.6	85.1	72.9	78.4
09 FINISHES	98.7	69.0	82.5	88.2	60.4	73.1	91.1	60.4	74.4	92.1	72.9	81.6	90.1	83.1	86.1	88.0	73.0	79.0
COVERS DIVS. 10-14, 25, 28, 41, 43, 44, 46	100.0	85.8	96.9	100.0	84.0	96.5	100.0	83.4	96.4	100.0	99.3	99.9	100.0	101.8	100.4	100.0	99.3	99.9
21, 22, 23 FIRE SUPPRESSION, PLUMBING & HVAC	100.1	63.8	85.5	99.9	64.1	85.6	97.7	63.8	84.1	100.0	74.1	89.6	100.1	71.6	88.7	97.7	74.1	88.2
26, 27, 3370 ELECTRICAL, COMMUNICATIONS & UTIL.	100.7	71.3	85.6	100.4	67.1	83.4	94.2	71.3	82.5	98.8	78.6	88.5	97.9	71.1	84.2	96.0	78.6	87.1
MF2018 WEIGHTED AVERAGE	99.2	69.5	86.5	97.0	66.4	83.9	96.8	66.4	83.8	98.0	83.1	91.6	95.4	83.5	90.3	94.4	82.6	89.4

For customer support on your Electrical Costs with RSMeans Data, call 800.448.8182.

535

MAINE

DIVISION		HOULTON 047			KITTERY 039			LEWISTON 042			MACHIAS 046			PORTLAND 040 - 041			ROCKLAND 048		
		MAT.	INST.	TOTAL	MAT.	INST.	TOTAL	MAT.	INST.	TOTAL	MAT.	INST.	TOTAL	MAT.	INST.	TOTAL	MAT.	INST.	TOTAL
015433	CONTRACTOR EQUIPMENT		97.9	97.9		97.9	97.9		97.9	97.9		97.9	97.9		97.9	97.9		97.9	97.9
0241, 31 - 34	SITE & INFRASTRUCTURE, DEMOLITION	90.4	98.5	96.0	81.5	98.5	93.3	88.4	99.8	96.3	89.8	98.5	95.8	88.6	99.8	96.4	86.4	98.5	94.8
0310	Concrete Forming & Accessories	99.4	78.5	81.6	90.8	79.0	80.7	101.2	78.0	81.5	96.2	78.4	81.1	102.8	78.0	81.7	97.3	78.7	81.5
0320	Concrete Reinforcing	94.5	81.1	87.7	89.3	82.6	85.9	116.1	83.1	99.3	94.5	81.1	87.7	104.7	83.1	93.7	94.5	82.9	88.6
0330	Cast-in-Place Concrete	73.2	110.2	86.9	71.9	110.4	86.2	74.7	111.1	88.2	73.2	110.1	86.9	90.8	111.1	98.4	74.7	110.3	87.9
03	CONCRETE	84.7	90.4	87.3	78.8	91.0	84.3	84.1	90.9	87.2	84.2	90.4	87.0	90.1	90.9	90.4	81.9	90.8	85.9
04	MASONRY	98.2	61.9	75.8	113.2	85.9	96.4	98.8	93.5	95.5	98.2	61.9	75.8	102.5	93.5	96.9	92.3	90.3	91.1
05	METALS	87.7	88.8	88.1	78.8	91.1	82.5	92.3	92.2	92.2	87.8	88.8	88.1	98.2	92.2	96.4	87.6	91.1	88.7
06	WOOD, PLASTICS & COMPOSITES	98.6	77.6	87.2	91.1	77.6	83.8	100.5	75.1	86.7	95.3	77.6	85.7	97.5	75.1	85.3	96.2	77.6	86.1
07	THERMAL & MOISTURE PROTECTION	102.6	69.1	88.2	105.4	74.5	92.1	102.3	82.9	93.9	102.5	68.4	87.8	106.5	82.9	96.3	102.2	76.4	91.1
08	OPENINGS	98.8	80.7	94.6	98.4	84.3	95.2	101.8	79.3	96.6	98.8	80.7	94.6	100.9	79.3	95.9	98.8	80.7	94.6
0920	Plaster & Gypsum Board	119.6	76.6	91.3	108.0	76.6	87.4	122.2	74.0	90.5	117.8	76.6	90.7	111.0	74.0	86.6	117.8	76.6	90.7
0950, 0980	Ceilings & Acoustic Treatment	90.0	76.6	80.9	101.6	76.6	84.7	100.3	74.0	82.5	90.0	76.6	80.9	104.7	74.0	83.9	90.0	76.6	80.9
0960	Flooring	82.1	44.2	71.1	85.2	49.3	74.8	83.5	102.1	88.9	81.3	44.2	70.6	87.7	102.1	91.9	81.7	44.2	70.8
0970, 0990	Wall Finishes & Painting/Coating	90.0	91.3	90.7	83.3	104.8	96.1	90.0	91.3	90.7	90.0	91.3	90.7	93.6	91.3	92.2	90.0	91.3	90.7
09	FINISHES	87.0	72.4	79.0	87.9	74.7	80.7	89.4	83.1	85.9	86.4	72.4	78.8	91.8	83.1	87.0	86.2	72.4	78.7
COVERS	DIVS. 10 - 14, 25, 28, 41, 43, 44, 46	100.0	99.3	99.8	100.0	99.3	99.9	100.0	101.8	100.4	100.0	99.3	99.9	100.0	101.8	100.4	100.0	99.3	99.9
21, 22, 23	FIRE SUPPRESSION, PLUMBING & HVAC	97.7	74.0	88.2	97.6	81.6	91.2	100.1	71.6	88.7	97.7	74.0	88.2	100.0	71.6	88.6	97.7	74.1	88.2
26, 27, 3370	ELECTRICAL, COMMUNICATIONS & UTIL.	99.8	78.6	89.0	92.5	78.6	85.4	99.8	75.3	87.3	99.8	78.6	89.0	102.8	75.3	88.7	99.7	78.6	88.9
MF2018	WEIGHTED AVERAGE	94.1	79.7	87.9	91.7	84.6	88.7	95.9	84.1	90.8	94.0	79.7	87.8	98.2	84.1	92.2	93.3	83.1	88.9

MAINE / MARYLAND

DIVISION		WATERVILLE 049			ANNAPOLIS 214			BALTIMORE 210 - 212			COLLEGE PARK 207 - 208			CUMBERLAND 215			EASTON 216		
		MAT.	INST.	TOTAL	MAT.	INST.	TOTAL	MAT.	INST.	TOTAL	MAT.	INST.	TOTAL	MAT.	INST.	TOTAL	MAT.	INST.	TOTAL
015433	CONTRACTOR EQUIPMENT		97.9	97.9		104.4	104.4		104.4	104.4		110.1	110.1		104.4	104.4		104.4	104.4
0241, 31 - 34	SITE & INFRASTRUCTURE, DEMOLITION	90.3	98.5	96.0	104.6	93.4	96.8	100.2	97.5	98.3	102.5	98.0	99.3	94.2	94.3	94.3	101.5	91.5	94.6
0310	Concrete Forming & Accessories	90.1	78.7	80.4	99.0	74.3	78.0	99.9	78.4	81.6	83.3	77.1	78.0	91.7	84.2	85.3	89.8	74.9	77.1
0320	Concrete Reinforcing	94.5	82.9	88.6	113.0	89.7	101.2	100.7	89.8	95.2	100.9	96.7	98.7	95.4	90.0	92.6	94.6	89.6	92.0
0330	Cast-in-Place Concrete	73.2	110.3	87.0	124.7	73.7	105.7	114.1	79.0	101.0	119.0	77.4	103.5	102.7	85.2	96.2	114.1	66.5	96.4
03	CONCRETE	85.0	90.8	87.6	105.7	78.2	93.3	110.9	81.5	97.7	109.5	82.1	97.2	94.4	86.7	90.9	102.9	75.8	90.7
04	MASONRY	109.3	90.3	97.6	96.6	69.1	79.6	99.9	74.6	84.1	113.9	75.1	89.9	97.9	87.3	91.3	111.7	59.8	79.7
05	METALS	87.7	91.1	88.7	102.7	103.1	102.8	99.0	98.3	98.8	86.0	108.8	92.9	97.0	104.7	99.4	97.3	99.9	98.1
06	WOOD, PLASTICS & COMPOSITES	87.8	77.6	82.3	96.2	75.9	85.2	102.9	79.0	89.9	78.1	76.3	77.2	88.4	83.3	85.6	86.2	82.3	84.1
07	THERMAL & MOISTURE PROTECTION	102.6	76.0	91.1	102.1	78.5	92.0	103.1	82.0	94.0	104.5	83.0	95.3	100.4	84.1	93.4	100.6	75.0	89.6
08	OPENINGS	98.8	80.7	94.6	100.1	81.4	95.7	98.8	83.1	95.1	94.7	83.4	92.1	99.7	87.1	96.8	98.1	84.9	95.0
0920	Plaster & Gypsum Board	112.3	76.6	88.8	103.0	75.6	84.9	104.8	78.4	87.4	96.7	75.8	83.0	105.0	83.1	90.6	105.0	82.2	89.9
0950, 0980	Ceilings & Acoustic Treatment	90.0	76.6	80.9	87.9	75.6	79.6	105.5	78.4	87.2	109.0	75.8	86.5	103.9	83.1	89.9	103.9	82.2	89.2
0960	Flooring	79.0	47.3	69.8	89.3	78.2	86.1	101.2	78.2	94.5	88.6	79.1	85.8	89.1	96.2	91.2	88.2	78.2	85.3
0970, 0990	Wall Finishes & Painting/Coating	90.0	91.3	90.7	91.7	76.6	82.7	104.1	77.3	88.2	100.0	74.4	84.9	96.9	86.9	91.0	96.9	74.4	83.6
09	FINISHES	85.1	72.9	78.5	86.5	74.8	80.1	101.7	77.9	88.7	94.6	76.5	84.8	94.9	86.7	90.4	95.1	76.1	84.7
COVERS	DIVS. 10 - 14, 25, 28, 41, 43, 44, 46	100.0	99.2	99.8	100.0	86.1	96.9	100.0	88.6	97.5	100.0	89.0	97.9	100.0	90.5	97.9	100.0	69.8	93.3
21, 22, 23	FIRE SUPPRESSION, PLUMBING & HVAC	97.7	74.1	88.2	100.0	82.7	93.1	100.0	83.8	93.5	97.7	88.1	93.8	97.4	73.6	87.9	97.4	74.4	88.2
26, 27, 3370	ELECTRICAL, COMMUNICATIONS & UTIL.	99.7	78.6	88.9	100.9	88.2	94.4	101.1	89.0	94.9	100.8	101.0	100.9	98.6	81.0	89.6	98.1	63.8	80.6
MF2018	WEIGHTED AVERAGE	94.4	83.1	89.6	100.2	83.1	92.8	101.4	85.0	94.3	98.1	88.6	94.0	97.3	85.5	92.2	99.0	76.0	89.1

MARYLAND / MASSACHUSETTS

DIVISION		ELKTON 219			HAGERSTOWN 217			SALISBURY 218			SILVER SPRING 209			WALDORF 206			BOSTON 020 - 022, 024		
		MAT.	INST.	TOTAL	MAT.	INST.	TOTAL	MAT.	INST.	TOTAL	MAT.	INST.	TOTAL	MAT.	INST.	TOTAL	MAT.	INST.	TOTAL
015433	CONTRACTOR EQUIPMENT		104.4	104.4		104.4	104.4		104.4	104.4		101.8	101.8		101.8	101.8		102.0	102.0
0241, 31 - 34	SITE & INFRASTRUCTURE, DEMOLITION	88.1	93.0	91.5	92.4	94.4	93.8	101.4	91.5	94.5	90.5	89.6	89.9	97.1	89.6	91.9	94.9	101.3	99.4
0310	Concrete Forming & Accessories	95.7	94.0	94.2	90.8	80.9	82.4	104.3	51.6	59.5	91.5	78.6	78.6	98.6	76.3	79.7	102.5	137.0	131.8
0320	Concrete Reinforcing	94.6	115.3	105.1	95.4	90.0	92.6	94.6	67.0	80.6	99.7	96.5	98.1	100.3	96.5	98.1	105.8	152.0	129.3
0330	Cast-in-Place Concrete	92.3	73.1	85.2	97.8	85.2	93.1	114.1	64.6	95.7	121.8	77.8	105.4	136.5	77.8	114.6	91.1	141.0	109.7
03	CONCRETE	86.7	91.2	88.7	90.5	85.2	88.2	103.9	60.7	84.4	107.5	81.6	95.8	118.7	81.7	102.0	98.4	140.3	117.3
04	MASONRY	96.6	68.4	79.2	103.4	87.3	93.4	111.4	56.3	77.4	113.4	75.4	89.9	97.4	75.4	83.8	108.3	148.1	132.9
05	METALS	97.3	112.4	101.9	97.2	104.7	99.5	97.3	91.3	95.5	90.3	104.6	94.7	90.3	104.6	94.7	97.0	133.0	107.9
06	WOOD, PLASTICS & COMPOSITES	94.1	103.0	98.9	87.5	78.9	82.8	105.2	52.8	76.8	84.9	75.6	79.9	92.1	75.6	83.2	103.2	136.5	121.3
07	THERMAL & MOISTURE PROTECTION	100.1	81.1	91.9	100.8	85.2	94.1	100.9	70.1	87.6	108.1	88.0	99.5	108.7	88.0	99.8	110.5	138.6	122.6
08	OPENINGS	98.1	103.0	99.2	98.0	83.1	94.6	98.3	62.5	90.0	86.7	83.0	86.4	108.7	88.0	99.8	110.5	138.6	122.6
0920	Plaster & Gypsum Board	108.9	103.5	105.3	105.0	78.6	87.6	116.2	51.8	73.8	103.1	75.8	85.1	106.4	75.8	86.3	97.8	143.1	108.4
0950, 0980	Ceilings & Acoustic Treatment	103.9	103.5	103.6	104.9	78.6	87.1	103.9	51.8	68.6	117.5	75.8	89.3	117.5	75.8	89.3	107.2	137.4	127.1
0960	Flooring	90.5	78.2	87.0	88.7	96.2	90.9	94.1	78.2	89.5	94.6	79.1	90.1	98.2	79.1	92.6	106.3	137.4	127.4
0970, 0990	Wall Finishes & Painting/Coating	96.9	74.4	83.6	96.9	74.4	83.6	96.9	74.4	83.6	107.2	74.4	87.8	107.2	74.4	87.8	93.4	169.1	115.4
09	FINISHES	95.5	90.4	92.7	94.8	82.7	88.2	98.5	57.8	76.2	95.1	75.9	84.6	96.9	76.0	85.5	97.4	153.2	130.4
COVERS	DIVS. 10 - 14, 25, 28, 41, 43, 44, 46	100.0	57.7	90.7	100.0	90.0	97.8	100.0	78.0	95.2	100.0	84.6	96.6	100.0	84.6	96.6	101.6	145.2	125.4
21, 22, 23	FIRE SUPPRESSION, PLUMBING & HVAC	97.4	80.4	90.6	99.9	84.9	93.9	97.4	72.3	87.3	97.7	87.9	93.7	97.7	87.9	93.7	100.0	117.3	103.8
26, 27, 3370	ELECTRICAL, COMMUNICATIONS & UTIL.	99.8	89.0	94.3	98.4	81.0	89.5	96.9	61.6	78.8	97.9	101.0	99.4	95.5	101.0	98.3	100.1	127.5	111.1
MF2018	WEIGHTED AVERAGE	96.2	87.6	92.5	97.5	87.0	93.0	99.5	68.5	86.2	97.3	87.4	93.0	98.1	87.4	93.5	100.0	133.3	114.3

City Cost Indexes

MASSACHUSETTS

DIVISION		BROCKTON 023 MAT.	INST.	TOTAL	BUZZARDS BAY 025 MAT.	INST.	TOTAL	FALL RIVER 027 MAT.	INST.	TOTAL	FITCHBURG 014 MAT.	INST.	TOTAL	FRAMINGHAM 017 MAT.	INST.	TOTAL	GREENFIELD 013 MAT.	INST.	TOTAL
015433	CONTRACTOR EQUIPMENT		100.3	100.3		100.3	100.3		101.2	101.2		97.9	97.9		99.7	99.7		97.9	97.9
0241, 31 - 34	SITE & INFRASTRUCTURE, DEMOLITION	91.8	102.7	99.4	82.4	102.9	96.6	99.2	124.0	120.3	94.2	121.6	117.5	102.3	124.5	121.2	92.4	120.8	116.5
0310	Concrete Forming & Accessories	99.2	124.3	120.5	96.9	123.9	119.8	104.9	125.8	115.5	88.3	146.6	118.0	88.3	146.9	118.1	92.0	123.8	108.1
0320	Concrete Reinforcing	104.9	146.8	126.2	84.1	125.8	105.3	104.9	125.8	115.5	88.3	146.6	118.0	88.3	146.9	118.1	80.9	119.4	95.2
0330	Cast-in-Place Concrete	84.6	133.5	102.8	70.3	133.3	93.7	81.8	133.7	101.2	78.7	133.3	99.0	78.7	135.3	99.7	80.4	120.1	98.3
03	CONCRETE	89.7	130.8	108.2	76.1	126.9	98.9	88.4	127.1	105.8	77.1	129.2	100.6	79.7	131.5	103.0	80.4	120.1	98.3
04	MASONRY	101.9	138.0	124.2	94.8	138.0	121.5	102.5	137.9	124.4	100.3	134.7	121.6	106.8	139.0	126.7	105.0	121.6	115.2
05	METALS	91.7	126.4	102.2	86.7	117.7	96.1	91.7	118.0	99.7	88.3	122.8	98.8	88.3	126.5	99.9	90.6	111.6	97.0
06	WOOD, PLASTICS & COMPOSITES	99.2	123.1	112.2	95.9	123.1	110.7	99.2	123.4	112.3	96.4	119.9	109.2	103.2	122.9	113.9	94.4	123.1	110.0
07	THERMAL & MOISTURE PROTECTION	102.4	129.8	114.2	101.2	127.0	112.3	102.2	126.2	112.5	101.7	122.7	110.7	101.7	129.9	113.9	101.7	112.1	106.2
08	OPENINGS	99.5	132.1	107.0	95.4	124.2	101.7	99.5	121.8	104.6	101.5	130.3	108.2	91.8	132.0	101.1	101.7	122.1	106.4
0920	Plaster & Gypsum Board	90.2	123.4	112.1	86.0	123.4	112.1	90.2	123.4	112.1	110.9	120.1	117.0	114.0	123.4	120.2	111.8	123.4	119.9
0950, 0980	Ceilings & Acoustic Treatment	106.5	123.4	118.0	90.2	123.4	112.7	106.5	123.4	118.0	94.5	120.1	111.9	94.5	123.4	114.1	103.0	123.4	116.9
0960	Flooring	88.9	167.4	111.6	86.7	167.4	110.1	87.9	167.4	110.9	85.8	167.4	109.5	87.2	167.4	110.5	85.0	139.6	100.9
0970, 0990	Wall Finishes & Painting/Coating	91.0	142.9	121.7	91.0	142.9	121.7	91.0	142.9	121.7	87.2	142.9	120.1	88.1	142.9	120.5	87.2	112.2	102.2
09	FINISHES	90.8	134.6	114.7	85.8	134.6	112.4	90.5	134.8	114.7	86.5	132.7	111.7	87.2	134.4	113.0	88.4	124.4	108.2
COVERS	DIVS. 10 - 14, 25, 28, 41, 43, 44, 46	100.0	111.9	102.6	100.0	112.0	102.6	100.0	112.5	102.7	100.0	106.4	101.4	100.0	111.6	102.5	100.0	105.0	101.1
21, 22, 23	FIRE SUPPRESSION, PLUMBING & HVAC	100.3	106.1	102.6	97.8	105.8	101.0	100.3	105.8	102.5	97.8	106.8	101.4	97.8	123.7	108.2	99.3	98.5	98.9
26, 27, 3370	ELECTRICAL, COMMUNICATIONS & UTIL.	98.7	100.1	99.4	96.0	102.6	99.4	98.6	100.1	99.4	99.3	106.6	103.1	96.1	127.4	112.1	99.3	98.5	98.9
MF2018	WEIGHTED AVERAGE	96.5	119.2	106.2	91.7	117.7	102.8	96.3	117.4	105.3	98.7	118.7	104.3	92.8	127.0	107.4	94.7	110.9	101.0

MASSACHUSETTS

DIVISION		HYANNIS 026 MAT.	INST.	TOTAL	LAWRENCE 019 MAT.	INST.	TOTAL	LOWELL 018 MAT.	INST.	TOTAL	NEW BEDFORD 027 MAT.	INST.	TOTAL	PITTSFIELD 012 MAT.	INST.	TOTAL	SPRINGFIELD 010-011 MAT.	INST.	TOTAL
015433	CONTRACTOR EQUIPMENT		100.3	100.3		100.3	100.3		97.9	97.9		101.2	101.2		97.9	97.9		97.9	97.9
0241, 31 - 34	SITE & INFRASTRUCTURE, DEMOLITION	88.3	102.9	98.5	93.1	102.7	99.8	92.0	102.8	99.5	89.4	102.8	98.7	93.0	101.3	98.8	92.5	101.7	98.9
0310	Concrete Forming & Accessories	90.9	123.9	118.9	103.5	124.6	121.4	99.7	125.7	121.8	99.2	124.1	120.3	99.7	105.7	104.8	99.9	120.7	117.3
0320	Concrete Reinforcing	84.1	125.8	105.3	109.3	138.5	124.2	110.2	138.3	124.5	104.9	125.9	115.6	91.3	117.4	104.6	110.2	123.8	117.1
0330	Cast-in-Place Concrete	77.2	133.3	98.1	91.0	133.6	106.9	82.8	135.3	102.3	72.1	133.8	95.1	86.5	111.2	97.6	86.2	119.4	98.1
03	CONCRETE	81.7	126.9	102.0	92.9	129.5	109.4	85.1	130.4	105.5	83.7	127.2	103.3	86.5	111.2	97.6	86.8	120.1	101.0
04	MASONRY	100.9	138.0	123.8	112.5	138.8	128.7	99.3	137.8	123.1	100.8	138.6	124.1	99.9	113.8	108.5	99.6	121.6	113.0
05	METALS	88.2	117.7	97.1	91.0	123.4	100.8	91.0	120.0	99.8	91.7	118.1	99.7	90.8	108.6	96.2	93.5	111.5	99.0
06	WOOD, PLASTICS & COMPOSITES	89.0	123.1	107.5	103.7	123.1	114.2	102.9	123.1	113.9	99.2	123.4	112.3	102.9	105.5	104.3	102.9	123.1	113.0
07	THERMAL & MOISTURE PROTECTION	101.7	127.0	112.6	102.4	130.0	114.3	102.2	130.3	114.3	102.2	126.4	112.6	102.2	107.2	104.4	102.2	112.1	106.3
08	OPENINGS	96.0	122.4	102.1	95.5	129.9	103.5	102.8	129.9	109.0	99.5	126.6	105.8	102.8	110.7	104.6	102.8	122.1	107.4
0920	Plaster & Gypsum Board	81.4	123.4	109.1	116.4	123.4	121.0	116.4	123.4	121.0	90.2	123.4	112.1	116.4	105.3	109.1	116.4	123.4	121.0
0950, 0980	Ceilings & Acoustic Treatment	99.7	123.4	115.8	104.9	123.4	117.5	104.9	123.4	117.5	106.5	123.4	118.0	104.9	105.3	105.2	104.9	123.4	117.5
0960	Flooring	84.2	167.4	108.3	87.8	167.4	110.9	87.8	167.4	110.9	87.9	167.4	110.9	88.2	134.2	101.5	87.2	139.6	102.0
0970, 0990	Wall Finishes & Painting/Coating	91.0	142.9	121.7	87.3	142.9	120.2	87.2	142.9	120.1	91.0	142.9	121.7	87.2	112.2	102.0	88.0	112.2	102.0
09	FINISHES	86.6	134.6	112.8	90.4	134.6	114.6	90.4	135.4	115.0	90.4	134.8	114.6	90.5	111.7	102.1	90.3	124.4	108.0
COVERS	DIVS. 10 - 14, 25, 28, 41, 43, 44, 46	100.0	112.0	102.6	100.0	112.1	102.7	100.0	113.1	102.9	100.0	112.5	102.7	100.0	101.5	100.3	100.0	105.0	101.1
21, 22, 23	FIRE SUPPRESSION, PLUMBING & HVAC	100.3	106.5	102.8	100.1	121.9	108.8	100.1	123.4	109.4	100.3	105.7	102.4	100.1	96.3	98.5	100.1	100.6	100.0
26, 27, 3370	ELECTRICAL, COMMUNICATIONS & UTIL.	96.4	100.1	98.3	98.4	125.9	112.5	98.9	122.2	110.8	99.5	100.1	99.8	98.9	98.5	98.7	98.9	95.2	97.0
MF2018	WEIGHTED AVERAGE	93.8	117.5	103.9	96.1	125.8	109.2	96.0	125.4	108.6	95.7	117.6	105.1	96.2	105.1	100.0	96.6	110.4	102.0

MASSACHUSETTS / MICHIGAN

DIVISION		WORCESTER 015-016 MAT.	INST.	TOTAL	ANN ARBOR 481 MAT.	INST.	TOTAL	BATTLE CREEK 490 MAT.	INST.	TOTAL	BAY CITY 487 MAT.	INST.	TOTAL	DEARBORN 481 MAT.	INST.	TOTAL	DETROIT 482 MAT.	INST.	TOT
015433	CONTRACTOR EQUIPMENT		97.9	97.9		111.9	111.9		99.0	99.0		111.9	111.9		111.9	111.9		96.4	96
0241, 31 - 34	SITE & INFRASTRUCTURE, DEMOLITION	92.4	102.6	99.5	78.5	96.7	91.2	93.4	86.0	88.2	70.2	96.2	88.3	78.2	96.8	91.1	93.1	100.7	98
0310	Concrete Forming & Accessories	100.3	121.5	118.3	96.5	104.3	103.1	96.3	79.5	82.0	96.5	83.7	85.7	96.4	104.9	103.6	100.8	108.2	102
0320	Concrete Reinforcing	110.2	146.6	128.7	102.3	106.5	104.5	94.8	83.3	88.9	102.3	105.7	104.0	102.3	106.6	104.5	103.1	108.3	105
0330	Cast-in-Place Concrete	85.7	133.3	103.4	81.0	99.2	87.8	82.0	93.8	86.4	77.5	85.9	80.6	79.2	100.0	87.0	86.1	101.5	91
03	CONCRETE	86.6	129.2	105.7	85.3	103.7	93.6	84.2	84.9	84.5	83.6	89.7	86.4	84.4	104.3	93.4	91.5	105.1	92
04	MASONRY	99.1	134.7	121.1	99.9	98.4	99.0	97.3	80.1	86.7	99.5	79.3	87.0	99.7	100.3	100.1	102.0	101.9	102
05	METALS	93.6	122.7	102.4	100.0	116.6	105.0	100.7	84.5	95.8	100.6	114.9	104.9	100.1	116.7	105.1	99.2	109.8	105
06	WOOD, PLASTICS & COMPOSITES	103.4	119.9	112.3	92.4	106.0	99.8	92.1	77.8	84.3	92.4	84.3	88.0	92.4	106.0	99.8	106.7	106.2	106
07	THERMAL & MOISTURE PROTECTION	102.2	122.7	111.0	110.4	100.6	106.1	100.2	81.3	92.1	107.5	83.2	97.1	108.5	103.1	106.2	98.7	104.1	99
08	OPENINGS	102.8	130.3	109.2	96.3	101.6	97.5	88.4	76.3	85.6	96.3	85.9	93.9	96.3	101.9	97.6	97.3	109.9	100
0920	Plaster & Gypsum Board	116.4	120.1	118.8	100.4	105.8	104.0	90.2	73.8	79.4	100.4	83.5	89.3	100.4	105.8	104.0	97.3	109.9	105
0950, 0980	Ceilings & Acoustic Treatment	104.9	120.1	115.2	85.1	105.8	99.1	77.8	73.8	75.1	86.1	83.5	84.3	85.1	105.8	99.1	95.9	109.9	105
0960	Flooring	87.8	160.1	108.8	89.4	110.8	95.7	89.7	72.8	84.8	89.4	79.5	86.6	89.0	105.0	93.6	93.7	106.0	97
0970, 0990	Wall Finishes & Painting/Coating	87.2	142.9	120.1	83.5	98.0	92.1	86.8	78.3	81.7	83.5	81.2	82.1	83.5	96.3	91.1	91.4	97.9	10
09	FINISHES	90.4	131.2	112.7	89.4	105.1	98.0	83.3	77.4	80.1	89.2	82.2	85.4	89.2	104.3	97.5	89.2	103.4	10
COVERS	DIVS. 10 - 14, 25, 28, 41, 43, 44, 46	100.0	106.4	101.4	100.0	95.2	98.9	100.0	95.7	99.0	100.0	89.8	97.8	100.0	95.6	99.0	100.0	103.7	10
21, 22, 23	FIRE SUPPRESSION, PLUMBING & HVAC	100.1	106.8	102.7	100.1	93.4	97.4	100.1	84.5	93.8	100.1	80.4	92.2	100.1	102.3	101.0	100.9	101.3	10
26, 27, 3370	ELECTRICAL, COMMUNICATIONS & UTIL.	98.9	106.6	102.9	96.9	105.5	101.3	93.4	79.6	86.3	96.0	84.5	90.1	96.9	98.4	97.7	99.1	103.0	10
MF2018	WEIGHTED AVERAGE	96.6	118.5	106.0	96.4	101.7	98.6	94.6	82.5	89.4	95.8	87.5	92.2	96.6	101.7	98.6	99.1	103.0	10

MICHIGAN

DIVISION		FLINT 484 - 485			GAYLORD 497			GRAND RAPIDS 493, 495			IRON MOUNTAIN 498 - 499			JACKSON 492			KALAMAZOO 491		
		MAT.	INST.	TOTAL	MAT.	INST.	TOTAL	MAT.	INST.	TOTAL	MAT.	INST.	TOTAL	MAT.	INST.	TOTAL	MAT.	INST.	TOTAL
015433	CONTRACTOR EQUIPMENT		111.9	111.9		106.8	106.8		99.0	99.0		93.8	93.8		106.8	106.8		99.0	99.0
0241, 31 - 34	SITE & INFRASTRUCTURE, DEMOLITION	68.3	96.3	87.8	88.0	82.8	84.4	92.3	85.9	87.9	96.6	93.2	94.2	110.9	85.5	93.2	93.7	86.0	88.3
0310	Concrete Forming & Accessories	99.5	84.1	86.4	95.2	74.3	77.4	95.1	78.0	80.5	87.1	79.7	80.8	91.7	84.6	85.7	96.3	79.3	81.9
0320	Concrete Reinforcing	102.3	106.1	104.2	88.4	89.7	89.0	99.5	83.0	91.1	88.1	87.3	87.7	85.8	106.1	96.1	94.8	78.4	86.5
0330	Cast-in-Place Concrete	81.6	88.3	84.1	81.8	80.6	81.3	86.1	92.0	88.3	96.8	69.7	86.7	81.6	92.5	85.7	83.6	93.7	87.4
03	CONCRETE	85.7	90.8	88.0	81.1	80.9	81.0	88.0	83.6	86.0	89.3	78.2	84.3	76.4	92.3	83.6	87.3	84.0	85.8
04	MASONRY	99.9	87.7	92.4	107.9	74.2	87.1	90.6	75.7	81.4	93.7	79.4	84.8	87.9	88.8	88.4	95.9	80.1	86.1
05	METALS	100.1	115.5	104.8	102.3	111.6	105.1	97.8	84.0	93.6	101.6	91.8	98.6	102.4	113.9	105.9	100.7	82.7	95.3
06	WOOD, PLASTICS & COMPOSITES	96.2	83.0	89.1	85.5	73.8	79.1	95.8	76.3	85.2	81.7	79.8	80.7	84.2	82.1	83.1	92.1	77.8	84.3
07	THERMAL & MOISTURE PROTECTION	107.8	86.9	98.8	98.8	70.6	86.6	102.1	72.7	89.4	102.7	76.2	91.3	98.0	90.8	94.9	100.2	81.3	92.1
08	OPENINGS	96.3	85.0	93.7	87.9	80.1	86.1	99.0	76.0	93.6	94.6	71.1	89.2	87.1	88.0	87.3	88.4	75.6	85.4
0920	Plaster & Gypsum Board	102.6	82.2	89.1	90.0	72.1	78.2	97.1	72.2	80.7	48.2	79.8	69.0	88.5	80.6	83.3	90.2	73.8	79.4
0950, 0980	Ceilings & Acoustic Treatment	85.1	82.2	83.1	77.0	72.1	73.6	89.6	72.2	77.8	74.0	79.8	77.9	77.0	80.6	79.4	77.8	73.8	75.1
0960	Flooring	89.4	88.9	89.3	83.0	88.5	84.6	93.6	78.7	89.3	101.9	90.3	98.5	81.7	81.3	81.6	89.7	72.8	84.8
0970, 0990	Wall Finishes & Painting/Coating	83.5	80.4	81.7	83.1	79.5	81.0	93.6	78.9	84.9	100.5	71.5	83.3	83.1	96.3	90.9	86.8	78.3	81.7
09	FINISHES	88.9	84.0	86.2	82.7	76.9	79.6	89.5	77.6	83.0	82.7	80.7	81.6	83.8	84.3	84.1	83.3	77.4	80.1
COVERS	DIVS. 10 - 14, 25, 28, 41, 43, 44, 46	100.0	90.6	97.9	100.0	79.2	95.4	100.0	95.4	99.0	100.0	87.5	97.3	100.0	93.5	98.6	100.0	95.7	99.0
21, 22, 23	FIRE SUPPRESSION, PLUMBING & HVAC	100.1	85.1	94.1	97.8	80.6	90.9	100.1	81.4	92.6	97.7	84.8	92.5	97.8	87.6	93.7	100.1	80.0	92.0
26, 27, 3370	ELECTRICAL, COMMUNICATIONS & UTIL.	96.9	91.3	94.0	91.4	74.3	82.7	98.6	85.0	91.6	97.4	82.6	89.8	95.1	105.5	100.4	93.2	75.9	84.4
MF2018	WEIGHTED AVERAGE	96.1	90.8	93.8	94.0	81.3	88.5	96.5	81.6	90.1	95.8	83.0	90.3	93.4	93.0	93.2	95.0	80.7	88.8

MICHIGAN / MINNESOTA

DIVISION		LANSING 488 - 489			MUSKEGON 494			ROYAL OAK 480, 483			SAGINAW 486			TRAVERSE CITY 496			BEMIDJI 566		
		MAT.	INST.	TOTAL	MAT.	INST.	TOTAL	MAT.	INST.	TOTAL	MAT.	INST.	TOTAL	MAT.	INST.	TOTAL	MAT.	INST.	TOTAL
015433	CONTRACTOR EQUIPMENT		111.9	111.9		99.0	99.0		91.9	91.9		111.9	111.9		93.8	93.8		100.3	100.3
0241, 31 - 34	SITE & INFRASTRUCTURE, DEMOLITION	89.5	96.5	94.4	91.4	85.9	87.6	82.3	97.2	92.7	71.1	96.2	88.6	81.9	91.1	88.3	95.9	99.8	98.6
0310	Concrete Forming & Accessories	93.5	77.0	79.5	96.6	79.2	81.8	92.0	105.1	103.1	96.5	81.9	84.1	87.1	72.3	74.5	85.8	84.3	84.5
0320	Concrete Reinforcing	105.3	105.9	105.6	95.4	83.0	89.1	92.7	89.8	91.2	102.3	105.7	104.0	89.5	82.5	85.9	97.1	101.4	99.3
0330	Cast-in-Place Concrete	99.5	87.2	95.0	81.7	91.9	85.5	70.9	100.3	81.8	80.0	85.8	82.2	75.5	76.9	76.0	102.4	105.7	103.6
03	CONCRETE	93.2	87.2	90.5	82.7	84.1	83.3	73.4	100.2	85.5	84.8	88.8	86.6	73.4	76.5	74.8	91.2	96.0	93.3
04	MASONRY	94.7	86.3	89.5	94.5	75.7	82.9	93.1	99.4	97.0	101.4	79.3	87.7	91.9	75.3	81.6	99.1	102.4	101.2
05	METALS	98.2	115.0	103.3	98.4	83.8	94.0	103.5	91.0	99.7	100.1	114.7	104.5	101.6	89.4	97.9	91.6	116.8	99.2
06	WOOD, PLASTICS & COMPOSITES	90.1	73.3	81.0	89.2	78.1	83.2	88.1	106.0	97.8	89.1	82.0	85.2	81.7	72.2	76.6	69.9	79.2	75.0
07	THERMAL & MOISTURE PROTECTION	106.8	85.5	97.6	99.1	73.2	87.9	106.0	102.7	104.6	108.8	83.1	97.8	101.7	67.9	87.1	106.0	92.2	100.1
08	OPENINGS	100.5	79.4	95.6	87.7	77.6	85.3	96.1	101.3	97.3	94.4	84.6	92.1	94.6	66.4	88.0	101.9	102.3	102.0
0920	Plaster & Gypsum Board	96.3	72.2	80.4	72.6	74.1	73.6	97.9	105.8	103.1	100.4	81.1	87.7	48.2	72.1	63.9	104.3	79.0	87.7
0950, 0980	Ceilings & Acoustic Treatment	84.9	72.2	76.3	77.8	74.1	75.3	84.4	105.8	98.9	85.1	81.1	82.4	74.0	72.1	72.7	121.3	79.0	92.6
0960	Flooring	96.2	81.3	91.9	88.4	74.7	84.4	86.9	105.0	92.1	89.4	79.5	86.6	101.9	88.5	98.0	92.3	88.4	91.2
0970, 0990	Wall Finishes & Painting/Coating	95.0	78.9	85.5	85.0	78.9	81.4	85.1	96.3	91.7	83.5	81.2	82.1	100.5	40.5	65.0	89.1	106.5	99.4
09	FINISHES	88.8	76.8	82.2	79.8	77.2	78.4	88.3	104.2	97.0	89.1	80.9	84.6	81.6	71.8	76.2	96.0	86.9	91.1
COVERS	DIVS. 10 - 14, 25, 28, 41, 43, 44, 46	100.0	90.7	97.9	100.0	95.5	99.0	100.0	101.6	100.1	100.0	89.5	97.7	100.0	85.6	96.8	100.0	93.9	98.6
21, 22, 23	FIRE SUPPRESSION, PLUMBING & HVAC	100.0	85.5	94.2	99.9	81.4	92.5	97.8	103.6	100.1	100.1	80.0	92.1	97.7	80.0	90.6	100.0	82.0	91.5
26, 27, 3370	ELECTRICAL, COMMUNICATIONS & UTIL.	97.5	89.8	93.5	93.6	74.7	84.0	98.9	101.3	100.1	95.1	86.6	90.7	93.0	75.8	84.2	102.3	101.1	101.7
MF2018	WEIGHTED AVERAGE	97.4	88.7	93.6	93.5	80.3	87.8	94.6	100.6	97.2	95.7	87.3	92.1	92.8	78.3	86.6	96.9	95.4	96.3

MINNESOTA

DIVISION		BRAINERD 564			DETROIT LAKES 565			DULUTH 556 - 558			MANKATO 560			MINNEAPOLIS 553 - 555			ROCHESTER 559		
		MAT.	INST.	TOTAL	MAT.	INST.	TOTAL	MAT.	INST.	TOTAL	MAT.	INST.	TOTAL	MAT.	INST.	TOTAL	MAT.	INST.	TOTAL
015433	CONTRACTOR EQUIPMENT		102.9	102.9		100.3	100.3		103.2	103.2		102.9	102.9		107.6	107.6		103.2	103.2
0241, 31 - 34	SITE & INFRASTRUCTURE, DEMOLITION	97.5	104.8	102.6	94.1	100.0	98.2	97.8	102.9	101.4	94.1	104.4	101.3	95.8	108.4	104.6	97.1	102.5	100.8
0310	Concrete Forming & Accessories	86.1	85.0	85.1	82.5	84.2	83.9	97.9	97.0	97.1	94.5	93.7	93.8	99.6	117.1	114.4	98.4	97.2	97.4
0320	Concrete Reinforcing	95.9	101.8	98.9	97.1	101.4	99.3	96.3	101.6	99.0	95.8	109.0	102.5	97.5	112.5	105.1	94.1	108.7	101.5
0330	Cast-in-Place Concrete	111.4	110.3	111.0	99.4	108.6	102.9	99.3	99.6	99.4	102.5	99.3	101.3	103.3	117.7	108.7	99.1	98.7	98.7
03	CONCRETE	95.1	98.0	96.4	88.7	96.9	92.4	94.4	99.8	96.8	90.7	99.4	94.6	100.8	117.1	108.1	92.0	100.6	95.9
04	MASONRY	123.4	115.3	118.4	123.3	108.0	113.8	101.6	107.9	105.5	111.9	109.3	110.3	112.2	122.0	118.3	103.3	105.7	104.8
05	METALS	92.7	117.1	100.1	91.5	116.2	99.0	97.5	119.0	104.1	92.5	120.6	101.1	99.7	126.4	107.8	98.0	122.2	105.3
06	WOOD, PLASTICS & COMPOSITES	86.0	76.4	80.8	66.8	76.6	72.1	93.8	94.4	94.1	95.9	91.6	93.6	96.7	113.9	106.0	97.8	95.6	96.6
07	THERMAL & MOISTURE PROTECTION	104.2	103.9	104.1	105.8	101.2	103.8	105.6	104.4	105.0	104.6	93.6	99.9	104.2	118.5	110.3	109.9	92.5	102.4
08	OPENINGS	88.3	100.7	91.2	101.8	100.8	101.6	106.3	106.5	106.3	92.9	110.8	97.1	101.6	125.0	107.0	100.6	113.8	103.6
0920	Plaster & Gypsum Board	91.4	76.3	81.5	103.4	76.3	85.5	94.4	94.8	94.6	96.0	92.0	93.3	97.4	114.4	108.6	103.5	95.9	98.5
0950, 0980	Ceilings & Acoustic Treatment	56.8	76.3	70.0	121.3	76.3	90.8	93.2	94.8	94.2	56.8	92.0	80.6	98.2	114.4	109.2	93.3	95.9	95.1
0960	Flooring	91.1	88.4	90.3	91.0	88.4	90.3	96.5	119.7	103.3	92.9	84.6	90.5	99.7	118.5	105.2	93.2	84.6	90.7
0970, 0990	Wall Finishes & Painting/Coating	83.6	106.5	97.1	89.1	85.6	87.0	89.5	107.2	100.0	94.4	102.5	99.2	97.4	127.1	115.0	85.3	100.8	94.5
09	FINISHES	80.8	87.0	84.2	95.4	84.5	89.4	91.5	101.6	97.0	82.5	93.0	88.2	100.5	118.2	110.2	91.2	95.8	93.7
COVERS	DIVS. 10 - 14, 25, 28, 41, 43, 44, 46	100.0	95.5	99.0	100.0	95.3	99.0	100.0	96.2	99.2	100.0	95.1	98.9	100.0	107.3	101.6	100.0	100.3	100.1
21, 22, 23	FIRE SUPPRESSION, PLUMBING & HVAC	97.1	85.7	92.5	97.8	84.7	92.6	99.8	96.5	98.5	97.1	84.8	92.2	99.9	113.0	105.2	99.9	93.5	97.4
26, 27, 3370	ELECTRICAL, COMMUNICATIONS & UTIL.	100.1	101.8	101.0	102.1	70.4	85.9	95.8	101.8	98.9	106.2	92.4	99.1	100.2	113.0	106.8	99.1	92.4	95.6
MF2018	WEIGHTED AVERAGE	95.9	98.6	97.0	97.6	92.2	95.3	98.5	102.7	100.3	96.0	98.0	96.8	100.8	116.5	107.6	98.3	100.3	99.1

For customer support on your Electrical Costs with RSMeans Data, call 800.448.8182.

DIVISION		MINNESOTA															MISSISSIPPI		
		SAINT PAUL 550 - 551			ST. CLOUD 563			THIEF RIVER FALLS 567			WILLMAR 562			WINDOM 561			BILOXI 395		
		MAT.	INST.	TOTAL	MAT.	INST.	TOTAL	MAT.	INST.	TOTAL	MAT.	INST.	TOTAL	MAT.	INST.	TOTAL	MAT.	INST.	TOTAL
015433	CONTRACTOR EQUIPMENT		103.2	103.2		102.9	102.9		100.3	100.3		102.9	102.9		102.9	102.9		102.7	102.7
0241, 31 - 34	SITE & INFRASTRUCTURE, DEMOLITION	95.0	103.8	101.1	92.7	105.8	101.8	94.8	99.6	98.1	92.1	104.7	100.9	87.6	84.3	84.8	92.0	65.7	69.6
0310	Concrete Forming & Accessories	97.7	116.8	113.9	83.4	112.9	108.5	86.4	83.4	83.8	83.2	89.1	88.2	87.6	84.3	84.8	86.0	50.0	67.7
0320	Concrete Reinforcing	95.5	109.2	102.5	95.9	109.2	102.7	97.5	101.2	99.4	95.5	109.0	102.4	95.5	107.9	101.8	86.0	50.0	67.7
0330	Cast-in-Place Concrete	100.5	116.4	106.4	98.2	115.1	104.5	101.5	82.7	94.5	99.8	81.7	93.0	86.3	85.6	86.0	120.0	66.8	100.2
03	CONCRETE	94.1	115.9	103.9	86.7	113.6	98.8	89.9	87.5	88.8	86.6	91.2	88.7	77.7	90.2	83.3	102.9	65.1	85.9
04	MASONRY	102.5	122.0	114.6	107.8	121.0	115.9	99.1	102.4	101.1	112.2	115.3	114.1	122.7	88.3	101.4	97.8	60.2	74.6
05	METALS	97.4	124.1	105.5	93.3	122.0	102.0	91.7	115.5	98.9	92.5	120.4	101.0	92.4	118.4	100.3	92.8	84.2	90.2
06	WOOD, PLASTICS & COMPOSITES	95.4	113.6	105.3	83.1	109.7	97.5	71.0	79.2	75.5	82.8	82.2	82.5	87.4	82.2	84.6	95.3	66.4	79.6
07	THERMAL & MOISTURE PROTECTION	104.1	119.3	110.7	104.4	113.1	108.1	106.8	90.3	99.7	104.1	103.4	103.8	104.1	84.5	95.7	99.6	63.2	83.9
08	OPENINGS	98.8	123.7	104.6	93.4	121.6	99.9	101.9	102.3	102.0	90.3	104.7	93.6	93.9	104.7	96.4	98.6	56.6	88.9
0920	Plaster & Gypsum Board	99.0	114.4	109.2	91.4	110.5	104.0	104.0	79.0	87.6	91.4	82.3	85.4	91.4	82.3	85.4	105.2	65.9	79.3
0950, 0980	Ceilings & Acoustic Treatment	96.2	114.4	108.5	56.8	110.5	93.2	121.3	79.0	92.6	56.8	82.3	74.0	56.8	82.3	74.0	94.2	65.9	75.0
0960	Flooring	93.1	124.6	102.2	87.9	118.5	96.8	91.9	88.4	90.9	89.3	88.4	89.1	91.8	88.4	90.8	93.5	56.2	82.7
0970, 0990	Wall Finishes & Painting/Coating	91.3	127.0	112.5	94.4	127.0	113.7	89.1	85.6	87.0	89.1	85.6	87.0	89.1	102.5	97.0	83.0	48.1	62.4
09	FINISHES	92.0	119.3	106.9	80.3	115.5	99.5	95.9	84.6	89.7	80.3	87.7	84.4	80.5	86.8	83.9	88.0	62.4	74.0
COVERS	DIVS. 10 - 14, 25, 28, 41, 43, 44, 46	100.0	106.6	101.5	100.0	101.1	100.2	100.0	93.7	98.6	100.0	96.1	99.1	100.0	92.4	98.3	100.0	73.4	94.1
21, 22, 23	FIRE SUPPRESSION, PLUMBING & HVAC	99.9	112.7	105.0	99.5	110.9	104.1	97.8	81.7	91.4	97.1	103.7	99.7	97.1	82.0	91.0	100.0	56.5	82.6
26, 27, 3370	ELECTRICAL, COMMUNICATIONS & UTIL.	102.2	113.0	107.7	100.1	113.0	106.7	99.6	70.4	84.6	100.1	85.8	92.8	106.2	92.3	99.1	101.7	55.0	77.8
MF2018	WEIGHTED AVERAGE	98.4	115.8	105.9	95.1	114.0	103.2	96.5	89.4	93.4	96.5	99.9	96.6	94.5	92.3	93.5	98.3	64.6	83.8

DIVISION		MISSISSIPPI																	
		CLARKSDALE 386			COLUMBUS 397			GREENVILLE 387			GREENWOOD 389			JACKSON 390 - 392			LAUREL 394		
		MAT.	INST.	TOTAL	MAT.	INST.	TOTAL	MAT.	INST.	TOTAL	MAT.	INST.	TOTAL	MAT.	INST.	TOTAL	MAT.	INST.	TOTAL
015433	CONTRACTOR EQUIPMENT		102.7	102.7		102.7	102.7		102.7	102.7		102.7	102.7		102.7	102.7		102.7	102.7
0241, 31 - 34	SITE & INFRASTRUCTURE, DEMOLITION	99.7	88.5	91.9	106.0	89.4	94.4	105.6	89.9	94.7	102.8	88.2	92.7	101.9	89.9	93.6	111.3	88.6	95.5
0310	Concrete Forming & Accessories	83.4	44.8	50.6	80.8	47.2	52.3	80.2	63.8	66.2	92.5	45.0	52.2	91.0	66.1	69.9	81.0	60.7	63.7
0320	Concrete Reinforcing	101.0	66.0	83.2	92.1	66.4	79.0	101.6	67.3	84.2	101.0	66.1	83.3	95.6	51.9	73.4	92.7	32.8	62.3
0330	Cast-in-Place Concrete	102.8	50.0	83.1	122.4	55.3	97.4	105.9	56.6	87.5	110.5	49.7	87.9	104.5	65.9	90.2	119.7	51.2	94.2
03	CONCRETE	94.4	52.4	75.5	104.3	55.5	82.3	99.9	63.6	83.6	100.8	52.4	79.0	96.8	65.3	82.6	106.9	54.4	83.3
04	MASONRY	90.6	40.2	59.4	123.7	46.2	75.8	134.1	68.2	93.4	91.3	40.0	59.6	103.4	59.1	76.1	119.8	42.9	72.3
05	METALS	93.0	86.8	91.1	89.9	89.6	89.8	94.0	90.8	93.0	93.0	86.5	91.0	99.6	85.0	95.2	90.0	75.5	85.6
06	WOOD, PLASTICS & COMPOSITES	80.8	45.7	61.7	80.6	46.5	62.1	77.4	64.6	70.5	93.8	45.7	67.7	94.9	67.8	80.2	81.8	66.4	73.4
07	THERMAL & MOISTURE PROTECTION	96.2	48.6	75.7	99.5	53.5	79.7	96.7	63.4	82.3	96.6	50.8	76.9	97.8	62.6	82.6	99.7	54.4	80.2
08	OPENINGS	97.2	49.2	86.0	98.2	49.6	87.0	96.9	60.0	88.3	97.2	49.2	86.0	99.9	58.2	90.2	95.4	53.3	85.6
0920	Plaster & Gypsum Board	93.4	44.6	61.3	96.6	45.5	63.0	93.1	64.1	74.0	104.6	44.6	65.1	90.0	67.3	75.1	96.6	65.9	76.4
0950, 0980	Ceilings & Acoustic Treatment	86.2	44.6	58.0	88.1	45.5	59.2	90.0	64.1	72.5	86.2	44.6	58.0	94.8	67.3	76.2	88.1	65.9	73.1
0960	Flooring	97.7	46.5	82.9	87.1	52.2	76.9	96.1	46.5	81.7	103.5	46.5	87.0	91.7	57.9	81.9	85.7	46.5	74.3
0970, 0990	Wall Finishes & Painting/Coating	93.8	48.1	66.8	83.0	48.1	62.4	93.8	49.1	67.3	93.8	48.1	66.8	85.0	49.1	63.8	83.0	48.1	62.4
09	FINISHES	89.9	45.2	65.5	83.8	47.5	63.9	90.7	59.0	73.4	93.5	45.2	67.1	87.0	63.2	74.0	83.9	58.1	69.8
COVERS	DIVS. 10 - 14, 25, 28, 41, 43, 44, 46	100.0	48.9	88.8	100.0	50.0	89.0	100.0	72.6	94.0	100.0	48.9	88.8	100.0	72.9	94.0	100.0	34.9	85.7
21, 22, 23	FIRE SUPPRESSION, PLUMBING & HVAC	98.8	52.4	80.2	98.7	54.2	80.9	100.0	58.2	83.3	98.8	52.8	80.4	100.0	59.0	83.6	98.7	48.0	78.4
26, 27, 3370	ELECTRICAL, COMMUNICATIONS & UTIL.	96.0	42.6	68.7	99.2	55.0	76.6	96.0	56.1	75.6	96.0	39.9	67.3	103.3	56.1	79.2	100.7	57.7	78.7
MF2018	WEIGHTED AVERAGE	95.7	54.4	78.0	98.1	58.4	81.1	99.1	66.0	84.9	97.0	54.2	78.6	99.0	65.4	84.6	98.3	56.9	80.5

DIVISION		MISSISSIPPI									MISSOURI								
		MCCOMB 396			MERIDIAN 393			TUPELO 388			BOWLING GREEN 633			CAPE GIRARDEAU 637			CHILLICOTHE 646		
		MAT.	INST.	TOTAL	MAT.	INST.	TOTAL	MAT.	INST.	TOTAL	MAT.	INST.	TOTAL	MAT.	INST.	TOTAL	MAT.	INST.	TOTAL
015433	CONTRACTOR EQUIPMENT		102.7	102.7		102.7	102.7		102.7	102.7		110.5	110.5		110.5	110.5		104.5	104.5
0241, 31 - 34	SITE & INFRASTRUCTURE, DEMOLITION	98.4	88.4	91.4	102.7	90.0	93.9	97.3	88.4	91.1	90.5	94.7	93.4	92.0	94.6	93.8	104.1	92.8	96.2
0310	Concrete Forming & Accessories	80.8	46.3	51.5	78.2	64.2	66.3	80.8	46.9	52.0	96.9	90.2	91.2	89.2	80.7	82.0	83.3	96.1	94.2
0320	Concrete Reinforcing	93.3	34.5	63.4	92.1	51.9	71.7	98.9	66.1	82.2	96.5	99.3	98.0	97.7	78.2	87.8	100.7	105.6	103.2
0330	Cast-in-Place Concrete	106.2	49.5	85.1	113.7	66.5	96.1	102.8	69.7	90.5	90.0	97.0	92.6	89.1	86.8	88.2	97.2	87.8	93.7
03	CONCRETE	93.1	47.5	72.6	97.9	64.6	82.9	94.0	60.2	78.8	91.9	95.7	93.6	90.9	84.3	87.9	102.0	95.6	99.1
04	MASONRY	125.1	39.6	72.3	97.3	59.6	74.0	123.0	42.7	73.3	110.9	97.4	102.5	107.5	80.2	90.6	99.2	92.9	95.3
05	METALS	90.1	74.8	85.5	91.0	84.9	89.2	92.9	86.8	91.0	96.4	118.9	103.2	97.4	109.0	101.0	84.0	111.2	92.3
06	WOOD, PLASTICS & COMPOSITES	80.6	48.1	62.9	77.8	64.6	70.6	78.0	46.5	60.9	101.3	89.8	95.1	93.1	78.7	85.3	93.9	97.5	95.9
07	THERMAL & MOISTURE PROTECTION	99.0	50.9	78.3	99.2	62.6	83.4	96.1	52.1	77.2	103.2	99.5	101.6	102.6	85.3	95.2	93.8	94.2	94.0
08	OPENINGS	98.3	42.5	85.3	98.0	56.1	88.3	97.2	52.5	86.8	100.7	97.0	99.8	100.7	75.4	94.8	87.9	97.7	90.2
0920	Plaster & Gypsum Board	96.6	47.1	64.0	96.6	64.1	75.2	93.1	45.5	61.7	97.7	89.9	92.6	97.1	78.4	84.8	98.8	97.3	97.8
0950, 0980	Ceilings & Acoustic Treatment	88.1	47.1	60.3	90.0	64.1	72.5	86.2	45.5	58.6	91.9	89.9	90.5	91.9	78.4	82.8	88.9	97.3	94.6
0960	Flooring	87.1	46.5	75.3	85.7	56.2	77.1	96.3	46.5	81.9	96.9	100.1	97.9	93.6	87.4	91.8	99.1	103.0	100.2
0970, 0990	Wall Finishes & Painting/Coating	83.0	48.1	62.4	83.0	48.1	62.4	93.8	46.4	65.8	97.0	109.1	104.1	97.0	68.8	80.3	93.8	105.9	101.0
09	FINISHES	83.1	46.5	63.1	83.4	61.2	71.3	89.4	46.1	65.8	97.4	93.5	95.3	96.3	79.5	87.1	98.1	98.8	98.5
COVERS	DIVS. 10 - 14, 25, 28, 41, 43, 44, 46	100.0	51.9	89.4	100.0	73.0	94.0	100.0	50.0	89.0	100.0	81.4	95.9	100.0	91.8	98.2	100.0	82.4	96.1
21, 22, 23	FIRE SUPPRESSION, PLUMBING & HVAC	98.7	51.8	79.9	100.0	58.7	83.5	98.9	53.8	80.8	97.6	100.4	98.7	100.0	99.8	99.9	97.7	101.5	99.2
26, 27, 3370	ELECTRICAL, COMMUNICATIONS & UTIL.	97.7	56.7	76.7	100.7	57.0	78.3	95.8	54.9	74.9	100.4	78.7	89.3	100.4	100.9	100.6	92.7	77.7	85.0
MF2018	WEIGHTED AVERAGE	96.5	54.5	78.5	96.6	65.1	83.0	97.0	58.2	80.4	98.1	96.0	97.2	98.5	91.8	95.6	94.8	95.4	95.0

For customer support on your Electrical Costs with RSMeans Data, call 800.448.8182.

5

City Cost Indexes

MISSOURI

DIVISION		COLUMBIA 652			FLAT RIVER 636			HANNIBAL 634			HARRISONVILLE 647			JEFFERSON CITY 650 - 651			JOPLIN 648		
		MAT.	INST.	TOTAL	MAT.	INST.	TOTAL	MAT.	INST.	TOTAL	MAT.	INST.	TOTAL	MAT.	INST.	TOTAL	MAT.	INST.	TOTAL
015433	CONTRACTOR EQUIPMENT		113.3	113.3		110.5	110.5		110.5	110.5		104.5	104.5		113.3	113.3		108.0	108.0
0241, 31 - 34	SITE & INFRASTRUCTURE, DEMOLITION	103.2	98.4	99.9	93.2	94.5	94.1	88.2	94.6	92.7	95.6	94.1	94.6	102.6	98.4	99.7	105.0	97.6	99.8
0310	Concrete Forming & Accessories	81.6	80.1	80.3	103.7	89.2	91.4	95.0	84.8	86.4	80.6	101.7	98.5	93.1	80.0	82.0	94.1	76.3	79.0
0320	Concrete Reinforcing	90.7	96.0	93.4	97.7	105.9	101.9	99.3	99.3	99.3	100.3	111.1	105.8	98.0	96.0	97.0	103.9	86.8	95.2
0330	Cast-in-Place Concrete	87.0	84.6	86.1	93.0	93.3	93.1	85.2	96.3	89.4	99.6	102.4	100.6	92.2	84.6	89.4	105.3	77.8	95.0
03	CONCRETE	83.8	86.3	84.9	94.9	95.1	95.0	88.2	93.1	90.4	97.8	104.2	100.6	90.8	86.2	88.7	100.7	79.8	91.3
04	MASONRY	139.0	86.5	106.6	107.9	78.0	89.4	103.0	96.6	99.0	93.9	101.5	98.6	97.7	86.5	90.8	93.0	81.9	86.1
05	METALS	100.6	116.3	105.4	96.3	120.4	103.6	96.4	118.5	103.1	84.4	114.7	93.6	100.2	116.2	105.0	86.9	99.8	90.8
06	WOOD, PLASTICS & COMPOSITES	80.9	77.6	79.1	110.7	90.0	99.4	99.3	83.5	90.7	90.4	101.6	96.5	95.8	77.6	85.9	106.0	75.7	89.5
07	THERMAL & MOISTURE PROTECTION	93.1	86.1	90.1	103.5	93.3	99.1	103.1	96.1	100.1	93.0	103.2	97.4	100.2	86.1	94.2	92.8	83.5	88.8
08	OPENINGS	96.5	82.3	93.2	100.7	99.1	100.3	100.7	86.2	97.3	87.6	104.8	91.6	96.9	82.3	93.5	88.9	78.5	86.5
0920	Plaster & Gypsum Board	84.6	77.1	79.7	104.1	90.0	94.8	97.4	83.4	88.2	94.8	101.5	99.2	93.9	77.1	82.8	105.7	74.9	85.4
0950, 0980	Ceilings & Acoustic Treatment	97.8	77.1	83.8	91.9	90.0	90.6	91.9	83.4	86.1	88.9	101.5	97.5	97.3	77.1	83.6	89.8	74.9	79.7
0960	Flooring	90.6	96.3	92.2	100.2	87.4	96.5	96.3	100.1	97.4	94.3	104.2	97.2	98.0	73.9	91.0	125.7	74.6	110.8
0970, 0990	Wall Finishes & Painting/Coating	91.6	80.2	84.9	97.0	73.6	83.2	97.0	96.6	96.7	98.1	105.6	102.5	88.4	80.2	83.6	93.4	77.9	84.2
09	FINISHES	84.9	82.2	83.4	99.4	86.3	92.3	96.9	87.8	92.0	95.7	102.6	99.4	90.8	78.3	84.0	105.1	76.6	89.6
COVERS	DIVS. 10 - 14, 25, 28, 41, 43, 44, 46	100.0	94.6	98.8	100.0	92.3	98.3	100.0	80.5	95.7	100.0	84.5	96.6	100.0	94.6	98.8	100.0	81.2	95.9
21, 22, 23	FIRE SUPPRESSION, PLUMBING & HVAC	99.9	98.1	99.2	97.6	99.0	98.1	97.6	99.9	98.5	97.6	103.0	99.8	99.9	98.1	99.2	100.1	72.7	89.1
26, 27, 3370	ELECTRICAL, COMMUNICATIONS & UTIL.	94.4	81.3	87.7	105.1	100.9	103.0	99.1	78.7	88.7	98.9	101.0	100.0	99.6	81.3	90.3	90.7	67.7	79.0
MF2018	WEIGHTED AVERAGE	97.6	91.3	94.9	99.1	96.2	97.9	97.0	94.0	95.7	94.2	102.6	97.8	97.7	90.8	94.8	95.9	79.7	89.0

MISSOURI

DIVISION		KANSAS CITY 640 - 641			KIRKSVILLE 635			POPLAR BLUFF 639			ROLLA 654 - 655			SEDALIA 653			SIKESTON 638		
		MAT.	INST.	TOTAL	MAT.	INST.	TOTAL	MAT.	INST.	TOTAL	MAT.	INST.	TOTAL	MAT.	INST.	TOTAL	MAT.	INST.	TOTAL
015433	CONTRACTOR EQUIPMENT		105.5	105.5		100.4	100.4		102.7	102.7		113.3	113.3		103.1	103.1		102.7	102.7
0241, 31 - 34	SITE & INFRASTRUCTURE, DEMOLITION	96.9	98.2	97.8	93.2	89.5	90.6	79.8	93.3	89.2	102.3	98.9	99.9	100.6	93.6	95.8	83.4	93.9	90.7
0310	Concrete Forming & Accessories	93.9	102.2	100.9	87.2	80.6	81.6	87.4	78.8	80.1	88.7	96.6	95.4	86.6	80.4	81.3	88.5	78.9	80.3
0320	Concrete Reinforcing	98.7	111.2	105.0	96.7	85.7	91.1	99.7	78.2	88.7	91.2	96.1	93.7	89.9	110.5	100.3	99.0	78.2	88.4
0330	Cast-in-Place Concrete	103.2	103.9	103.5	92.9	84.8	89.9	71.5	86.0	76.9	89.1	95.5	92.2	93.0	83.5	89.5	76.4	86.0	80.0
03	CONCRETE	100.7	104.8	102.5	107.0	84.1	96.7	81.8	82.4	82.1	85.5	98.2	91.2	99.8	87.8	94.4	85.6	82.5	84.2
04	MASONRY	99.7	103.7	102.2	114.4	86.6	97.3	105.8	75.7	87.2	112.7	87.2	96.9	119.0	83.3	96.9	105.6	75.7	87.1
05	METALS	93.5	113.4	99.5	96.0	102.1	97.8	96.6	98.7	97.2	100.0	116.9	105.2	98.7	112.2	102.8	96.9	98.8	97.5
06	WOOD, PLASTICS & COMPOSITES	101.3	101.5	101.4	85.8	78.8	82.0	84.8	78.7	81.5	88.9	98.4	94.0	82.3	79.0	80.5	86.5	78.7	82.3
07	THERMAL & MOISTURE PROTECTION	93.2	104.9	98.3	109.9	92.9	102.6	108.1	84.1	97.7	93.4	94.5	93.9	99.0	89.7	95.0	108.2	84.1	97.8
08	OPENINGS	94.4	104.8	96.8	106.2	81.1	100.4	107.2	75.4	99.8	96.5	93.7	95.9	101.7	88.8	98.7	107.2	75.4	99.8
0920	Plaster & Gypsum Board	101.3	101.5	101.5	92.2	78.5	83.2	92.6	78.4	83.3	87.0	98.4	94.5	80.6	78.5	79.2	94.4	78.4	83.9
0950, 0980	Ceilings & Acoustic Treatment	92.1	101.5	98.5	90.2	78.5	82.3	91.9	78.4	82.8	97.8	98.4	98.2	97.8	78.5	84.7	91.9	78.4	82.8
0960	Flooring	100.3	104.2	101.4	75.2	99.7	82.3	88.5	87.4	88.1	94.1	99.7	95.7	72.9	99.7	80.7	89.0	87.4	88.6
0970, 0990	Wall Finishes & Painting/Coating	96.8	105.9	102.2	92.8	80.4	85.4	92.0	68.8	78.3	91.6	92.6	92.2	91.6	105.9	100.1	92.0	68.8	78.3
09	FINISHES	99.0	102.8	101.1	96.4	83.4	89.3	95.9	78.6	86.4	86.4	96.7	92.0	86.3	85.5	84.7	96.6	78.9	86.9
COVERS	DIVS. 10 - 14, 25, 28, 41, 43, 44, 46	100.0	99.0	99.8	100.0	80.0	95.6	100.0	90.1	97.8	100.0	98.7	99.7	100.0	89.7	97.7	100.0	90.3	97.9
21, 22, 23	FIRE SUPPRESSION, PLUMBING & HVAC	100.0	103.0	101.2	97.6	99.3	98.3	97.6	97.0	97.4	97.5	100.8	98.8	97.4	97.0	97.3	97.6	97.0	97.4
26, 27, 3370	ELECTRICAL, COMMUNICATIONS & UTIL.	100.5	101.0	100.8	99.3	78.6	88.8	99.6	100.8	100.2	93.0	81.3	87.0	94.0	101.0	97.6	98.6	100.8	99.7
MF2018	WEIGHTED AVERAGE	98.2	103.6	100.5	100.6	88.8	95.5	96.9	89.2	93.6	95.9	96.6	96.2	98.2	93.7	96.3	97.4	89.4	94.0

MISSOURI / MONTANA

DIVISION		SPRINGFIELD 656 - 658			ST. JOSEPH 644 - 645			ST. LOUIS 630 - 631			BILLINGS 590 - 591			BUTTE 597			GREAT FALLS 594		
		MAT.	INST.	TOTAL	MAT.	INST.	TOTAL	MAT.	INST.	TOTAL	MAT.	INST.	TOTAL	MAT.	INST.	TOTAL	MAT.	INST.	TOTAL
015433	CONTRACTOR EQUIPMENT		105.6	105.6		104.5	104.5		111.0	111.0		100.5	100.5		100.3	100.3		100.3	100.3
0241, 31 - 34	SITE & INFRASTRUCTURE, DEMOLITION	102.8	96.1	98.1	99.0	91.9	94.1	97.6	99.2	98.7	96.6	97.5	97.2	103.4	96.3	98.4	107.2	97.2	100.2
0310	Concrete Forming & Accessories	94.3	77.5	80.0	93.0	91.5	91.8	100.6	102.6	102.3	98.3	66.4	71.2	84.9	66.5	69.3	98.2	65.7	70.6
0320	Concrete Reinforcing	87.4	95.6	91.6	97.5	110.6	104.1	89.3	107.2	98.4	90.7	80.6	85.6	98.4	81.1	89.6	90.8	80.6	85.6
0330	Cast-in-Place Concrete	94.6	75.9	87.6	97.8	98.2	97.9	100.5	103.5	101.6	115.6	69.8	98.6	127.2	69.8	105.9	134.5	69.7	110.4
03	CONCRETE	95.8	81.2	89.2	96.8	98.0	97.3	97.9	104.8	101.0	95.8	71.0	84.7	99.5	71.1	86.7	104.7	70.6	89.4
04	MASONRY	90.6	81.6	85.0	95.7	90.5	92.5	92.8	109.7	103.2	121.7	73.8	92.1	117.6	82.2	95.7	121.8	73.8	92.1
05	METALS	105.1	102.5	104.3	90.1	113.4	97.2	102.2	121.0	107.9	105.3	87.1	99.8	99.5	87.8	95.9	102.9	87.1	98.1
06	WOOD, PLASTICS & COMPOSITES	90.3	77.2	83.2	106.0	91.2	98.0	102.7	100.8	101.6	92.6	63.3	76.7	78.5	63.3	70.3	93.9	62.5	76.8
07	THERMAL & MOISTURE PROTECTION	97.4	77.7	88.9	93.4	90.5	92.1	101.0	106.0	103.1	105.6	69.7	90.2	105.4	73.3	91.6	106.1	71.0	91.0
08	OPENINGS	104.0	86.7	100.0	91.9	98.8	93.5	99.3	105.8	100.8	100.8	65.0	92.5	99.0	65.0	91.1	101.9	64.6	93.3
0920	Plaster & Gypsum Board	88.0	76.6	80.5	107.1	90.9	96.4	102.7	101.1	101.6	114.5	62.7	80.4	112.1	62.7	79.6	125.5	61.8	82.5
0950, 0980	Ceilings & Acoustic Treatment	97.8	76.6	83.5	96.6	90.9	92.7	93.6	101.1	98.7	90.6	62.7	71.7	96.5	62.7	73.6	98.2	61.8	73.5
0960	Flooring	93.0	73.9	87.5	104.0	102.4	103.5	99.3	100.1	99.5	92.9	78.0	88.6	90.8	94.0	91.7	97.5	78.0	91.8
0970, 0990	Wall Finishes & Painting/Coating	86.1	102.0	95.5	93.8	105.6	100.8	98.0	109.1	104.5	88.8	69.5	77.4	88.0	69.5	77.1	88.0	69.5	77.1
09	FINISHES	88.5	79.9	83.8	101.4	95.0	97.9	99.9	102.4	101.3	90.5	68.3	78.4	90.4	71.5	80.1	94.5	67.7	79.9
COVERS	DIVS. 10 - 14, 25, 28, 41, 43, 44, 46	100.0	92.4	98.3	100.0	96.0	99.1	100.0	102.3	100.5	100.0	90.3	97.9	100.0	90.3	97.9	100.0	90.2	97.9
21, 22, 23	FIRE SUPPRESSION, PLUMBING & HVAC	100.0	71.0	88.4	100.1	89.4	95.8	100.0	106.1	102.5	100.0	73.2	89.3	100.1	71.5	88.6	100.1	69.6	87.9
26, 27, 3370	ELECTRICAL, COMMUNICATIONS & UTIL.	98.1	70.5	84.0	99.0	77.7	88.1	103.0	100.9	101.9	96.9	69.9	83.1	103.3	70.9	86.7	96.2	69.1	82.4
MF2018	WEIGHTED AVERAGE	99.1	81.0	91.3	96.8	92.8	95.1	100.0	105.8	102.5	100.4	75.0	89.5	100.3	76.2	89.9	101.8	74.0	89.9

For customer support on your Electrical Costs with RSMeans Data, call 800.448.8182.

City Cost Indexes

MONTANA

DIVISION		HAVRE 595 MAT.	INST.	TOTAL	HELENA 596 MAT.	INST.	TOTAL	KALISPELL 599 MAT.	INST.	TOTAL	MILES CITY 593 MAT.	INST.	TOTAL	MISSOULA 598 MAT.	INST.	TOTAL	WOLF POINT 592 MAT.	INST.	TOTAL
015433	CONTRACTOR EQUIPMENT		100.3	100.3		100.3	100.3		100.3	100.3		100.3	100.3		100.3	100.3		100.3	100.3
0241, 31 - 34	SITE & INFRASTRUCTURE, DEMOLITION	111.0	96.4	100.8	96.8	96.3	96.4	93.1	96.2	95.3	99.7	96.3	97.3	85.4	96.0	92.8	117.7	96.4	102.8
0310	Concrete Forming & Accessories	77.6	64.4	68.1	99.4	66.5	71.5	88.2	67.6	70.7	96.7	66.4	71.1	88.2	66.5	69.7	89.0	65.9	69.4
0320	Concrete Reinforcing	99.2	81.1	90.0	104.7	79.9	92.1	101.0	85.9	93.3	98.8	81.2	89.8	100.1	85.9	92.9	100.2	80.4	90.2
0330	Cast-in-Place Concrete	137.1	69.7	112.0	99.5	69.8	88.4	110.5	69.8	95.3	120.9	69.8	101.9	93.7	69.7	84.8	135.5	69.8	111.1
03	CONCRETE	107.1	71.0	90.9	91.9	70.7	82.4	89.4	72.4	81.8	96.5	71.1	85.1	78.4	71.9	75.4	110.4	70.7	92.5
04	MASONRY	118.6	82.2	96.1	110.3	82.2	92.9	116.4	82.2	95.2	123.5	82.2	98.0	140.6	82.2	104.5	124.7	82.2	98.4
05	METALS	95.8	87.7	93.4	101.7	87.4	97.3	95.7	89.4	93.8	95.0	87.9	92.9	96.2	89.3	94.1	95.1	87.6	92.9
06	WOOD, PLASTICS & COMPOSITES	69.7	63.3	66.2	93.3	63.3	77.0	82.1	64.7	72.6	90.8	63.3	75.9	82.1	63.3	71.9	81.5	62.5	71.2
07	THERMAL & MOISTURE PROTECTION	105.8	67.3	89.2	100.4	73.3	88.8	104.9	74.0	91.6	105.3	68.8	89.6	104.5	72.8	90.8	106.4	68.7	90.1
08	OPENINGS	99.5	65.0	91.5	99.8	64.8	91.7	99.5	66.9	91.9	99.0	65.0	91.1	99.0	66.1	91.4	99.0	64.4	91.0
0920	Plaster & Gypsum Board	107.5	62.7	78.0	108.5	62.7	78.3	112.1	64.0	80.5	121.8	62.7	82.9	112.1	62.7	79.6	115.7	61.8	80.2
0950, 0980	Ceilings & Acoustic Treatment	96.5	62.7	73.6	100.6	62.7	74.9	96.5	64.0	74.5	94.8	62.7	73.0	96.5	62.7	73.6	94.8	61.8	72.4
0960	Flooring	88.4	94.0	90.0	101.6	94.0	99.4	92.5	94.0	92.9	97.4	94.0	96.4	92.5	94.0	92.9	93.9	94.0	93.9
0970, 0990	Wall Finishes & Painting/Coating	88.0	69.5	77.1	93.5	69.5	79.3	88.0	69.5	77.1	88.0	69.5	77.1	88.0	55.0	68.5	88.0	69.5	77.1
09	FINISHES	89.7	71.5	79.8	97.5	71.4	83.2	90.3	72.3	80.5	93.2	71.5	81.4	89.8	69.9	78.9	92.7	71.0	80.8
COVERS	DIVS. 10 - 14, 25, 28, 41, 43, 44, 46	100.0	90.3	97.9	100.0	90.3	97.9	100.0	90.5	97.9	100.0	90.3	97.9	100.0	90.3	97.9	100.0	90.3	97.9
21, 22, 23	FIRE SUPPRESSION, PLUMBING & HVAC	97.6	70.4	86.7	100.1	71.5	88.6	97.6	69.7	86.4	97.6	76.9	89.3	100.1	69.7	87.9	96.2	76.3	86.0
26, 27, 3370	ELECTRICAL, COMMUNICATIONS & UTIL.	96.2	70.9	83.3	102.8	70.9	86.5	100.3	69.8	84.7	96.2	76.3	86.0	101.3	69.7	85.1	96.2	76.3	86.0
MF2018	WEIGHTED AVERAGE	99.5	75.7	89.3	99.7	76.1	89.6	97.2	76.2	88.2	98.4	77.9	89.6	97.5	78.1	88.1	100.5	77.8	90.8

NEBRASKA

DIVISION		ALLIANCE 693 MAT.	INST.	TOTAL	COLUMBUS 686 MAT.	INST.	TOTAL	GRAND ISLAND 688 MAT.	INST.	TOTAL	HASTINGS 689 MAT.	INST.	TOTAL	LINCOLN 683 - 685 MAT.	INST.	TOTAL	MCCOOK 690 MAT.	INST.	TOTAL
015433	CONTRACTOR EQUIPMENT		97.8	97.8		104.7	104.7		104.7	104.7		104.7	104.7		104.7	104.7		104.7	104.7
0241, 31 - 34	SITE & INFRASTRUCTURE, DEMOLITION	100.9	100.8	100.8	105.6	95.0	98.2	110.6	95.2	99.9	109.3	95.0	99.3	99.0	95.2	96.4	103.9	94.9	97.7
0310	Concrete Forming & Accessories	86.6	56.6	61.1	97.5	77.0	80.1	97.1	70.7	74.6	100.5	74.5	78.4	92.5	77.4	79.7	92.1	57.3	62.6
0320	Concrete Reinforcing	110.9	89.9	100.2	100.7	88.4	94.5	100.1	78.3	89.0	100.1	78.3	89.1	101.2	78.6	89.7	103.3	78.5	90.7
0330	Cast-in-Place Concrete	108.2	83.9	99.1	109.4	84.0	99.9	115.8	79.7	102.4	115.8	75.9	101.0	89.3	84.3	87.4	106.0	69.1	89.3
03	CONCRETE	116.4	72.5	96.7	101.3	82.5	92.8	106.1	76.4	92.7	106.3	76.8	93.0	88.5	81.1	85.1	102.9	78.9	88.1
04	MASONRY	107.1	79.0	89.7	112.2	78.9	91.7	105.0	75.3	86.7	113.4	75.4	89.9	92.5	79.8	84.6	95.7	94.0	95.2
05	METALS	101.0	85.1	96.1	95.9	98.2	96.6	97.7	94.1	96.6	98.6	93.9	97.2	99.6	94.6	98.1	95.7	94.0	95.2
06	WOOD, PLASTICS & COMPOSITES	84.6	50.1	65.9	96.1	77.0	85.7	95.2	68.2	80.5	99.1	74.2	85.6	96.3	77.0	85.8	92.8	51.1	70.1
07	THERMAL & MOISTURE PROTECTION	104.3	67.7	88.5	103.4	81.7	94.1	103.6	79.6	93.3	103.6	79.0	93.0	100.4	81.8	92.4	99.3	77.4	89.9
08	OPENINGS	92.8	59.8	85.1	93.2	75.9	89.2	93.2	68.7	87.5	93.2	71.8	88.2	102.5	70.8	95.1	93.3	57.4	84.9
0920	Plaster & Gypsum Board	80.1	48.9	59.5	94.5	76.5	82.6	93.6	67.4	76.3	95.4	73.5	81.0	103.7	76.5	85.8	90.6	49.8	63.8
0950, 0980	Ceilings & Acoustic Treatment	93.7	48.9	63.3	88.3	76.5	80.3	88.3	67.4	74.1	88.3	73.5	78.3	92.2	76.5	81.6	89.7	49.8	62.7
0960	Flooring	92.6	90.3	92.0	85.8	90.3	87.1	85.5	81.5	84.4	86.8	84.0	86.0	99.5	89.5	96.6	91.1	90.3	90.9
0970, 0990	Wall Finishes & Painting/Coating	157.3	53.0	95.6	75.6	61.7	67.3	75.6	64.1	68.8	75.6	61.7	67.3	97.7	81.4	88.1	88.7	46.2	63.5
09	FINISHES	91.7	60.9	74.9	84.8	77.8	81.0	84.9	71.4	77.5	85.5	74.8	79.7	94.5	80.1	86.6	89.1	60.8	73.6
COVERS	DIVS. 10 - 14, 25, 28, 41, 43, 44, 46	100.0	85.6	96.8	100.0	85.2	96.7	100.0	87.8	97.3	100.0	84.9	96.7	100.0	88.7	97.5	100.0	85.8	96.9
21, 22, 23	FIRE SUPPRESSION, PLUMBING & HVAC	97.6	76.7	89.2	97.6	77.0	89.3	100.0	81.3	92.5	97.6	76.2	89.0	99.9	81.3	92.5	97.5	76.9	89.2
26, 27, 3370	ELECTRICAL, COMMUNICATIONS & UTIL.	92.4	67.2	79.5	95.1	83.4	89.1	93.7	67.3	80.2	93.1	81.4	87.1	104.4	67.3	85.4	94.4	67.3	80.5
MF2018	WEIGHTED AVERAGE	99.8	74.7	89.0	97.3	82.7	91.0	98.4	78.5	89.9	98.4	80.0	90.5	98.3	81.1	90.9	97.5	74.8	87.8

NEBRASKA / NEVADA

DIVISION		NORFOLK 687 MAT.	INST.	TOTAL	NORTH PLATTE 691 MAT.	INST.	TOTAL	OMAHA 680 - 681 MAT.	INST.	TOTAL	VALENTINE 692 MAT.	INST.	TOTAL	CARSON CITY 897 MAT.	INST.	TOTAL	ELKO 898 MAT.	INST.	TOTAL
015433	CONTRACTOR EQUIPMENT		94.1	94.1		104.7	104.7		94.1	94.1		97.6	97.6		97.6	97.6		97.6	97.6
0241, 31 - 34	SITE & INFRASTRUCTURE, DEMOLITION	86.5	94.1	91.8	105.1	94.7	97.9	90.3	94.3	93.1	88.1	99.7	96.2	86.4	97.7	94.3	69.7	96.3	88.2
0310	Concrete Forming & Accessories	83.4	76.1	77.2	94.6	77.2	79.8	97.1	75.5	78.7	83.0	54.4	58.7	105.7	78.3	82.4	110.8	99.1	100.8
0320	Concrete Reinforcing	100.8	68.4	84.3	102.8	78.4	90.4	97.7	78.5	88.0	103.3	68.2	85.5	108.1	119.5	113.9	116.1	117.9	117.0
0330	Cast-in-Place Concrete	109.9	73.1	96.2	117.1	62.0	96.6	92.6	79.8	87.9	103.2	55.3	85.4	97.0	83.6	92.0	93.9	74.7	86.7
03	CONCRETE	99.8	74.2	88.3	106.1	73.2	91.3	90.2	78.0	84.7	103.2	58.2	82.9	96.0	87.3	92.1	93.7	93.4	93.6
04	MASONRY	117.8	78.9	93.8	91.1	78.4	83.2	95.6	80.2	86.1	102.5	78.3	87.6	118.0	70.2	88.5	125.0	69.1	90.7
05	METALS	99.4	80.4	93.7	94.8	93.2	94.4	99.6	84.6	95.1	107.1	79.2	98.6	109.6	95.7	105.4	113.1	94.7	107.5
06	WOOD, PLASTICS & COMPOSITES	80.0	76.6	78.1	94.8	79.1	86.3	94.5	74.5	83.6	79.4	48.8	62.8	92.3	76.4	83.7	105.4	105.3	105.3
07	THERMAL & MOISTURE PROTECTION	103.2	79.7	93.1	99.2	79.5	90.7	99.3	80.5	91.2	100.0	74.9	89.2	108.6	79.7	96.2	104.8	75.7	92.3
08	OPENINGS	94.6	70.7	89.1	92.6	74.2	88.3	100.6	75.3	94.7	94.8	55.0	85.6	99.4	79.7	94.8	104.5	95.3	102.4
0920	Plaster & Gypsum Board	94.3	76.5	82.6	90.6	78.6	82.7	101.6	74.3	83.6	91.9	47.9	62.9	103.2	75.8	85.2	109.5	105.5	106.9
0950, 0980	Ceilings & Acoustic Treatment	103.1	76.5	85.1	89.7	78.6	82.2	104.1	74.3	83.9	106.1	47.9	66.7	102.3	75.8	84.4	102.7	105.5	104.6
0960	Flooring	106.9	90.3	102.1	92.2	81.5	89.1	101.8	90.3	98.4	118.5	90.3	110.3	97.8	69.4	89.6	101.1	69.4	91.9
0970, 0990	Wall Finishes & Painting/Coating	132.0	61.7	90.4	88.7	60.4	72.0	101.1	74.0	85.1	157.4	62.6	101.3	96.7	81.4	87.7	93.9	81.4	86.5
09	FINISHES	101.4	77.4	88.3	89.4	76.9	82.6	98.0	77.8	87.0	109.7	61.0	83.1	95.2	76.1	84.8	93.2	93.1	93.1
COVERS	DIVS. 10 - 14, 25, 28, 41, 43, 44, 46	100.0	84.8	96.7	100.0	62.4	91.7	100.0	87.5	97.2	100.0	57.2	90.6	100.0	102.0	100.4	100.0	92.9	98.4
21, 22, 23	FIRE SUPPRESSION, PLUMBING & HVAC	97.3	76.0	88.7	99.9	75.3	90.1	99.9	76.8	90.6	97.1	75.2	88.3	100.1	78.7	91.5	98.8	78.7	90.7
26, 27, 3370	ELECTRICAL, COMMUNICATIONS & UTIL.	94.0	83.4	88.6	92.6	67.3	79.7	103.3	83.4	93.1	90.0	67.3	78.4	99.6	90.4	94.9	96.9	90.4	93.6
MF2018	WEIGHTED AVERAGE	98.6	79.2	90.3	97.2	77.3	88.7	98.4	80.8	90.9	99.7	70.8	87.3	101.3	84.2	93.9	101.3	87.4	95.3

For customer support on your Electrical Costs with RSMeans Data, call 800.448.8182.

541

City Cost Indexes

NEVADA / NEW HAMPSHIRE

DIVISION		NEVADA ELY 893			NEVADA LAS VEGAS 889-891			NEVADA RENO 894-895			NEW HAMPSHIRE CHARLESTON 036			NEW HAMPSHIRE CLAREMONT 037			NEW HAMPSHIRE CONCORD 032-033		
		MAT.	INST.	TOTAL	MAT.	INST.	TOTAL	MAT.	INST.	TOTAL	MAT.	INST.	TOTAL	MAT.	INST.	TOTAL	MAT.	INST.	TOTAL
015433	CONTRACTOR EQUIPMENT		97.6	97.6		97.6	97.6		97.6	97.6		97.9	97.9		97.9	97.9		97.9	97.9
0241, 31 - 34	SITE & INFRASTRUCTURE, DEMOLITION	75.5	97.5	90.8	78.9	100.1	93.7	75.3	97.7	90.9	83.5	100.4	95.3	77.6	100.4	93.5	90.8	100.6	97.6
0310	Concrete Forming & Accessories	103.7	104.5	104.3	104.4	107.0	106.6	99.8	78.3	81.6	88.1	85.0	85.4	94.4	85.0	86.4	99.5	94.5	95.3
0320	Concrete Reinforcing	114.8	118.1	116.5	105.8	122.0	114.0	108.5	120.9	114.8	89.3	86.2	87.7	89.3	86.2	87.7	104.0	86.6	95.2
0330	Cast-in-Place Concrete	100.8	98.1	99.8	97.6	107.9	101.4	106.6	83.6	98.0	86.1	113.0	96.1	79.1	113.0	91.7	105.0	113.8	108.3
03	CONCRETE	101.9	104.1	102.9	97.2	109.3	102.7	101.6	87.6	95.3	87.2	95.0	90.7	80.2	95.0	86.9	96.1	99.7	97.7
04	MASONRY	130.4	76.4	97.0	117.2	97.1	104.8	124.1	70.2	90.8	93.8	94.8	94.4	94.2	94.8	94.6	101.1	101.3	101.2
05	METALS	113.1	97.6	108.4	122.0	101.3	115.7	114.8	96.2	109.2	85.1	89.5	86.5	85.1	89.5	86.5	91.5	90.4	91.2
06	WOOD, PLASTICS & COMPOSITES	95.2	107.5	101.9	93.0	105.3	99.7	88.9	76.4	82.1	89.4	81.7	85.2	95.8	81.7	88.1	93.8	93.6	93.7
07	THERMAL & MOISTURE PROTECTION	105.4	94.8	100.8	118.9	98.9	110.3	104.8	79.7	94.0	104.9	104.5	104.8	104.7	104.5	104.6	111.4	109.4	110.5
08	OPENINGS	104.5	96.5	102.6	103.4	110.0	105.0	102.2	80.0	97.1	98.4	81.8	94.5	99.5	81.8	95.4	94.4	88.4	93.0
0920	Plaster & Gypsum Board	104.1	107.7	106.5	97.1	105.5	102.6	92.6	75.8	81.5	108.0	80.8	90.1	109.2	80.8	90.5	107.7	93.1	98.1
0950, 0980	Ceilings & Acoustic Treatment	102.7	107.7	106.1	108.6	105.5	106.5	106.1	75.8	85.6	101.6	80.8	87.5	101.6	80.8	87.5	101.7	93.1	95.8
0960	Flooring	98.8	69.4	90.2	90.5	99.9	93.2	95.5	69.4	88.0	83.7	104.3	89.7	86.0	104.3	91.3	93.2	104.3	96.5
0970, 0990	Wall Finishes & Painting/Coating	93.9	115.0	106.4	96.4	115.0	107.4	93.9	81.4	86.5	83.3	92.4	88.7	83.3	92.4	88.7	91.1	92.4	91.8
09	FINISHES	92.2	99.8	96.4	90.1	106.6	99.1	90.1	76.1	82.5	86.6	88.5	87.6	87.0	88.5	87.8	90.1	95.8	93.2
COVERS	DIVS. 10 - 14, 25, 28, 41, 43, 44, 46	100.0	72.3	93.9	100.0	101.8	100.4	100.0	102.0	100.4	100.0	90.3	97.9	100.0	90.3	97.9	100.0	104.6	101.0
21, 22, 23	FIRE SUPPRESSION, PLUMBING & HVAC	98.8	98.3	98.6	100.1	101.3	100.6	100.0	78.7	91.5	97.6	83.9	92.1	97.6	83.9	92.1	100.0	86.3	94.5
26, 27, 3370	ELECTRICAL, COMMUNICATIONS & UTIL.	97.1	95.2	96.2	101.3	110.4	106.0	97.5	90.4	93.9	93.7	77.1	85.2	93.7	77.1	85.2	95.4	77.1	86.1
MF2018	WEIGHTED AVERAGE	102.6	95.6	99.6	103.7	104.1	103.9	102.4	84.2	94.6	92.9	88.6	91.1	92.1	88.6	90.6	96.4	92.4	94.7

NEW HAMPSHIRE / NEW JERSEY

DIVISION		NEW HAMPSHIRE KEENE 034			NEW HAMPSHIRE LITTLETON 035			NEW HAMPSHIRE MANCHESTER 031			NEW HAMPSHIRE NASHUA 030			NEW HAMPSHIRE PORTSMOUTH 038			NEW JERSEY ATLANTIC CITY 082, 084		
		MAT.	INST.	TOTAL	MAT.	INST.	TOTAL	MAT.	INST.	TOTAL	MAT.	INST.	TOTAL	MAT.	INST.	TOTAL	MAT.	INST.	TOTAL
015433	CONTRACTOR EQUIPMENT		97.9	97.9		97.9	97.9		97.9	97.9		97.9	97.9		97.9	97.9		95.4	95.4
0241, 31 - 34	SITE & INFRASTRUCTURE, DEMOLITION	90.5	100.4	97.4	77.8	99.3	92.8	88.6	100.6	96.9	92.0	100.5	97.9	85.7	101.0	96.3	88.7	104.0	99.3
0310	Concrete Forming & Accessories	92.9	85.2	86.3	107.0	79.0	83.2	98.9	94.7	95.4	101.8	94.7	95.7	89.7	94.2	93.5	113.8	141.1	137.0
0320	Concrete Reinforcing	89.3	86.2	87.7	90.0	86.2	88.1	104.4	86.6	95.4	111.5	86.6	98.9	89.3	86.6	87.9	80.2	141.7	114.4
0330	Cast-in-Place Concrete	86.5	113.1	96.4	77.6	104.0	87.4	98.0	116.0	104.7	81.8	115.9	94.5	77.6	115.8	91.8	81.4	139.4	103.0
03	CONCRETE	87.0	95.1	90.6	79.9	89.2	84.0	92.8	100.5	96.3	87.0	100.4	93.0	79.7	100.3	89.0	84.3	139.0	108.9
04	MASONRY	96.4	94.8	95.4	106.4	79.0	89.5	99.9	101.3	100.8	98.9	101.3	100.4	94.4	100.5	98.1	105.2	137.8	125.3
05	METALS	85.8	89.7	87.0	85.8	89.5	86.9	92.3	90.7	91.8	91.1	90.5	90.9	87.3	91.4	88.6	92.9	114.5	99.5
06	WOOD, PLASTICS & COMPOSITES	94.0	81.7	87.3	105.9	81.7	92.8	93.9	93.6	93.7	104.1	93.6	98.4	90.7	93.6	92.2	122.6	141.9	133.1
07	THERMAL & MOISTURE PROTECTION	105.5	104.5	105.1	104.9	97.3	101.6	107.1	109.7	108.2	106.0	109.7	107.6	105.5	108.5	106.8	100.7	134.5	115.2
08	OPENINGS	97.0	85.4	94.3	100.4	81.8	96.1	99.3	92.0	97.6	101.4	91.3	99.1	102.0	81.7	97.3	98.7	137.3	107.7
0920	Plaster & Gypsum Board	108.3	80.8	90.2	123.2	80.8	95.3	105.3	93.1	97.2	117.7	93.1	101.5	108.0	93.1	98.2	117.5	142.7	134.1
0950, 0980	Ceilings & Acoustic Treatment	101.6	80.8	87.5	101.6	80.8	87.5	102.6	93.1	96.1	113.6	93.1	99.7	102.6	93.1	96.1	90.9	142.7	126.0
0960	Flooring	85.7	104.3	91.1	96.0	103.2	98.1	94.1	108.6	98.3	89.3	104.3	93.7	83.8	104.3	89.8	92.5	164.3	113.4
0970, 0990	Wall Finishes & Painting/Coating	83.3	106.1	96.8	83.3	92.4	88.7	94.3	106.1	101.3	83.3	104.9	96.1	83.3	104.9	96.1	81.1	145.5	119.2
09	FINISHES	88.4	90.0	89.3	91.9	84.3	87.8	91.9	98.2	95.3	93.1	97.1	95.3	87.5	96.9	92.6	87.7	147.1	120.1
COVERS	DIVS. 10 - 14, 25, 28, 41, 43, 44, 46	100.0	96.6	99.3	100.0	97.5	99.5	100.0	104.6	101.0	100.0	104.6	101.0	100.0	104.3	100.9	100.0	116.6	103.7
21, 22, 23	FIRE SUPPRESSION, PLUMBING & HVAC	97.6	84.0	92.2	97.6	75.7	88.8	100.0	86.3	94.5	100.1	86.3	94.6	100.1	85.7	94.3	99.8	129.2	111.6
26, 27, 3370	ELECTRICAL, COMMUNICATIONS & UTIL.	93.7	77.1	85.2	94.6	50.3	72.0	93.5	77.1	85.1	95.8	77.1	86.2	94.1	77.1	85.4	99.8	134.0	113.6
MF2018	WEIGHTED AVERAGE	93.4	89.2	91.6	93.4	80.1	87.7	96.4	93.0	94.9	96.1	92.8	94.7	93.5	92.2	93.0	95.2	132.6	111.3

NEW JERSEY

DIVISION		CAMDEN 081			DOVER 078			ELIZABETH 072			HACKENSACK 076			JERSEY CITY 073			LONG BRANCH 077		
		MAT.	INST.	TOTAL	MAT.	INST.	TOTAL	MAT.	INST.	TOTAL	MAT.	INST.	TOTAL	MAT.	INST.	TOTAL	MAT.	INST.	TOTAL
015433	CONTRACTOR EQUIPMENT		95.4	95.4		97.9	97.9		97.9	97.9		97.9	97.9		95.4	95.4		95.1	95.1
0241, 31 - 34	SITE & INFRASTRUCTURE, DEMOLITION	89.8	104.0	99.7	105.9	105.9	105.9	110.8	105.9	107.4	106.9	106.1	106.3	97.3	106.0	103.4	102.2	105.2	104.3
0310	Concrete Forming & Accessories	103.9	141.0	135.4	100.9	150.2	142.7	114.1	150.2	144.8	100.9	150.1	142.7	105.9	150.2	143.6	106.6	142.7	137.3
0320	Concrete Reinforcing	105.2	132.2	118.9	81.2	151.8	117.1	81.2	151.8	117.1	81.2	151.8	117.1	105.3	151.8	129.0	81.2	151.4	116.9
0330	Cast-in-Place Concrete	78.9	137.5	100.7	82.4	134.8	101.9	70.8	138.0	95.8	80.6	137.9	101.9	64.3	134.9	90.6	71.5	134.7	96.0
03	CONCRETE	84.6	136.6	108.0	83.2	143.5	110.3	80.0	144.6	109.0	81.7	144.5	110.0	77.8	143.3	107.3	81.4	140.7	108.1
04	MASONRY	96.5	137.8	122.0	90.5	148.2	126.1	107.5	148.2	132.6	94.4	148.2	127.6	86.7	148.2	124.7	100.0	138.3	123.7
05	METALS	98.3	111.1	102.2	85.8	123.3	97.2	87.3	123.3	98.2	85.9	123.2	97.2	91.2	120.3	100.1	85.9	119.2	96.0
06	WOOD, PLASTICS & COMPOSITES	109.9	141.9	127.2	98.6	150.2	126.6	114.4	150.2	133.9	98.6	150.2	126.6	99.5	150.2	127.1	100.6	143.3	123.8
07	THERMAL & MOISTURE PROTECTION	100.5	135.5	115.6	111.2	145.5	126.0	111.5	146.0	126.3	111.0	138.4	122.8	110.7	145.5	125.7	110.9	132.4	120.1
08	OPENINGS	100.8	135.1	108.8	104.1	145.7	113.8	102.5	145.7	112.5	101.9	145.7	112.1	100.5	145.7	111.0	96.5	141.2	106.9
0920	Plaster & Gypsum Board	112.5	142.7	132.4	108.5	151.3	136.7	116.1	151.3	139.3	108.5	151.3	136.7	111.3	151.3	137.6	110.3	144.3	132.7
0950, 0980	Ceilings & Acoustic Treatment	99.5	142.7	128.8	87.3	151.3	130.7	89.2	151.3	131.3	87.3	151.3	130.7	96.9	151.3	133.8	110.3	144.3	132.7
0960	Flooring	88.9	164.3	110.8	81.5	190.7	113.2	86.3	190.7	116.6	81.5	190.7	113.2	82.5	190.7	113.9	82.7	182.9	111.8
0970, 0990	Wall Finishes & Painting/Coating	81.1	145.5	119.2	83.0	145.5	120.0	83.0	145.5	120.0	83.0	145.5	120.0	83.1	145.5	120.1	83.1	145.5	120.1
09	FINISHES	87.4	147.1	120.0	84.4	157.8	124.4	87.8	157.8	126.0	84.2	157.8	124.4	86.7	158.1	125.7	85.3	151.3	121.3
COVERS	DIVS. 10 - 14, 25, 28, 41, 43, 44, 46	100.0	116.6	103.7	100.0	133.7	107.4	100.0	133.7	107.4	100.0	133.8	107.4	100.0	133.8	107.4	100.0	116.8	103.7
21, 22, 23	FIRE SUPPRESSION, PLUMBING & HVAC	99.9	129.1	111.6	99.9	138.3	115.3	100.1	137.9	115.2	99.9	138.3	115.3	100.1	138.3	115.4	100.0	116.8	103.7
26, 27, 3370	ELECTRICAL, COMMUNICATIONS & UTIL.	97.7	135.3	116.9	98.5	141.9	120.7	99.1	141.9	121.0	98.5	142.5	121.0	103.4	142.5	123.4	98.1	133.1	116.0
MF2018	WEIGHTED AVERAGE	96.3	130.7	111.1	94.6	139.6	113.9	95.7	139.7	114.6	94.4	139.6	113.8	94.8	139.4	113.9	94.0	133.6	111.0

City Cost Indexes

NEW JERSEY

DIVISION		NEW BRUNSWICK 088-089 MAT.	INST.	TOTAL	NEWARK 070-071 MAT.	INST.	TOTAL	PATERSON 074-075 MAT.	INST.	TOTAL	POINT PLEASANT 087 MAT.	INST.	TOTAL	SUMMIT 079 MAT.	INST.	TOTAL	TRENTON 085-086 MAT.	INST.	TOTAL
015433	CONTRACTOR EQUIPMENT		95.1	95.1		97.9	97.9		97.9	97.9		95.1	95.1		97.9	97.9		95.1	95.1
0241, 31 - 34	SITE & INFRASTRUCTURE, DEMOLITION	101.3	105.6	104.3	110.5	105.9	107.3	108.8	106.1	106.9	102.8	105.2	104.5	108.3	105.9	106.6	88.2	105.2	100.0
0310	Concrete Forming & Accessories	107.4	149.9	143.5	104.9	150.2	143.4	102.8	150.1	143.0	101.2	142.7	136.4	103.8	150.2	143.2	102.8	142.5	136.5
0320	Concrete Reinforcing	81.1	151.5	116.9	103.6	151.8	128.1	105.3	151.8	129.0	81.1	151.4	116.8	81.2	151.8	117.1	106.5	119.6	113.1
0330	Cast-in-Place Concrete	100.7	140.4	115.4	91.0	138.0	108.5	82.0	137.9	102.8	100.7	139.5	115.1	68.5	138.0	94.3	99.1	137.4	113.3
03	CONCRETE	100.0	145.0	120.3	90.4	144.6	114.8	86.0	144.5	112.3	99.6	141.4	118.4	77.3	144.6	107.6	94.4	135.2	112.8
04	MASONRY	104.4	143.1	128.3	93.1	148.2	127.2	90.9	148.2	126.3	93.9	138.3	121.3	93.5	148.2	127.3	97.1	138.3	122.6
05	METALS	93.0	119.6	101.1	92.9	123.3	102.1	87.1	123.2	98.0	93.0	119.2	100.9	85.8	123.3	97.2	98.0	108.1	101.1
06	WOOD, PLASTICS & COMPOSITES	115.3	150.2	134.3	99.4	150.2	127.0	101.3	150.2	127.9	107.0	143.3	126.7	102.5	150.2	128.4	99.0	143.3	123.1
07	THERMAL & MOISTURE PROTECTION	101.0	141.6	118.5	112.8	146.0	127.1	111.3	138.4	123.0	101.0	134.8	115.5	111.7	146.0	126.4	101.5	136.5	116.6
08	OPENINGS	93.8	145.7	105.9	100.3	145.7	110.8	107.2	145.7	116.1	95.6	141.2	106.2	108.3	145.7	117.0	100.0	133.8	107.8
0920	Plaster & Gypsum Board	114.8	151.3	138.8	104.9	151.3	135.4	111.3	151.3	137.6	109.6	144.3	132.4	110.3	151.3	137.3	103.6	144.3	130.4
0950, 0980	Ceilings & Acoustic Treatment	90.9	151.3	131.8	95.8	151.3	133.4	96.9	151.3	133.8	90.9	144.3	127.1	87.3	151.3	130.7	94.9	144.3	128.4
0960	Flooring	90.2	190.7	119.4	94.0	190.7	122.1	82.5	190.7	113.9	87.8	164.3	110.0	82.7	190.7	114.0	94.9	182.9	120.4
0970, 0990	Wall Finishes & Painting/Coating	81.1	145.5	119.2	88.0	145.5	122.0	83.0	145.5	120.0	81.1	145.5	119.2	83.0	145.5	120.0	88.5	145.5	122.2
09	FINISHES	87.7	157.7	125.9	90.4	157.8	127.2	86.8	157.8	125.5	86.2	148.1	120.0	85.3	157.8	124.9	89.7	151.3	123.3
COVERS	DIVS. 10 - 14, 25, 28, 41, 43, 44, 46	100.0	133.6	107.4	100.0	133.8	107.4	100.0	133.8	107.4	100.0	113.9	103.1	100.0	133.7	107.4	100.0	116.8	103.7
21, 22, 23	FIRE SUPPRESSION, PLUMBING & HVAC	99.8	136.5	114.5	100.1	137.9	115.2	100.1	138.3	115.4	99.8	134.0	113.5	99.9	137.9	115.1	100.1	133.7	113.6
26, 27, 3370	ELECTRICAL, COMMUNICATIONS & UTIL.	94.0	137.0	116.0	107.7	142.5	125.5	103.4	141.9	123.1	93.4	133.1	113.7	99.1	141.9	121.0	101.6	131.8	117.0
MF2018	WEIGHTED AVERAGE	97.0	137.7	114.5	98.0	139.7	115.9	96.3	139.5	114.9	96.4	133.3	112.2	94.7	139.7	114.0	97.9	131.4	112.3

NEW JERSEY / NEW MEXICO

DIVISION		VINELAND 080, 083 MAT.	INST.	TOTAL	ALBUQUERQUE 870-872 MAT.	INST.	TOTAL	CARRIZOZO 883 MAT.	INST.	TOTAL	CLOVIS 881 MAT.	INST.	TOTAL	FARMINGTON 874 MAT.	INST.	TOTAL	GALLUP 873 MAT.	INST.	TOTAL
015433	CONTRACTOR EQUIPMENT		95.4	95.4		110.9	110.9		110.9	110.9		110.9	110.9		110.9	110.9		110.9	110.9
0241, 31 - 34	SITE & INFRASTRUCTURE, DEMOLITION	92.9	104.0	100.6	90.3	102.3	98.6	108.6	102.3	104.2	96.1	102.3	100.4	96.7	102.3	100.6	105.5	102.3	103.3
0310	Concrete Forming & Accessories	98.1	141.2	134.7	100.2	66.9	71.9	97.0	66.9	71.4	96.9	66.8	71.3	100.2	66.9	71.9	100.3	66.9	71.9
0320	Concrete Reinforcing	80.2	135.0	108.0	108.9	71.9	90.1	113.5	71.9	92.3	114.7	71.9	92.9	119.1	71.9	95.1	113.9	71.9	92.6
0330	Cast-in-Place Concrete	87.8	137.6	106.3	91.4	69.1	83.1	94.8	69.1	85.2	94.7	69.1	85.2	92.3	69.1	83.7	86.8	69.1	80.2
03	CONCRETE	88.5	137.3	110.4	93.9	69.8	83.1	115.0	69.8	94.7	103.3	69.8	88.2	97.4	69.8	85.0	103.6	69.8	88.4
04	MASONRY	93.5	138.3	121.2	105.7	58.7	76.7	106.9	58.7	77.1	106.9	58.7	77.1	114.3	58.7	80.0	100.9	58.7	74.9
05	METALS	92.9	112.7	98.9	110.1	90.6	104.2	105.7	90.6	101.1	105.3	90.5	100.8	107.8	90.6	102.5	106.9	90.6	101.9
06	WOOD, PLASTICS & COMPOSITES	103.5	141.9	124.3	104.0	68.8	84.9	94.6	68.8	80.6	94.6	68.8	80.6	104.1	68.8	84.9	104.1	68.8	84.9
07	THERMAL & MOISTURE PROTECTION	100.4	134.7	115.2	101.3	70.0	87.8	100.1	70.0	87.1	98.8	70.0	86.4	101.6	70.0	88.0	102.8	70.0	88.7
08	OPENINGS	95.2	136.1	104.7	100.9	67.7	93.2	98.7	67.7	91.5	98.8	67.7	91.6	103.3	67.7	95.1	103.4	67.7	95.1
0920	Plaster & Gypsum Board	108.1	142.7	130.9	107.8	67.7	81.4	80.9	67.7	72.2	80.9	67.7	72.2	98.8	67.7	78.3	98.8	67.7	78.3
0950, 0980	Ceilings & Acoustic Treatment	90.9	142.7	126.0	103.0	67.7	79.1	103.8	67.7	79.4	103.8	67.7	79.4	102.6	67.7	79.0	102.6	67.7	79.0
0960	Flooring	86.9	164.3	109.4	87.8	68.0	82.0	97.5	68.0	88.9	97.5	68.0	88.9	89.2	68.0	83.1	89.2	68.0	83.1
0970, 0990	Wall Finishes & Painting/Coating	81.1	145.5	119.2	96.1	51.9	69.9	93.4	51.9	68.8	93.4	51.9	68.8	90.8	51.9	67.7	90.8	51.9	67.7
09	FINISHES	85.0	147.2	119.0	89.8	65.5	76.5	92.9	65.5	78.0	91.7	65.5	77.4	88.8	65.5	76.1	90.0	65.5	76.6
COVERS	DIVS. 10 - 14, 25, 28, 41, 43, 44, 46	100.0	116.8	103.7	100.0	85.1	96.7	100.0	85.1	96.7	100.0	85.1	96.7	100.0	85.1	96.7	100.0	85.1	96.7
21, 22, 23	FIRE SUPPRESSION, PLUMBING & HVAC	99.8	129.4	111.7	100.1	68.7	87.5	98.4	68.7	86.5	98.4	68.4	86.4	99.9	68.7	87.4	98.4	68.7	86.5
26, 27, 3370	ELECTRICAL, COMMUNICATIONS & UTIL.	93.4	143.6	119.1	89.5	85.8	87.6	90.8	85.8	88.3	88.6	85.8	87.2	87.5	85.8	86.7	86.8	85.8	86.3
MF2018	WEIGHTED AVERAGE	94.6	132.3	110.8	99.2	75.0	88.8	101.3	75.0	90.0	99.1	74.9	88.7	99.8	75.0	89.2	99.7	75.0	89.1

NEW MEXICO

DIVISION		LAS CRUCES 880 MAT.	INST.	TOTAL	LAS VEGAS 877 MAT.	INST.	TOTAL	ROSWELL 882 MAT.	INST.	TOTAL	SANTA FE 875 MAT.	INST.	TOTAL	SOCORRO 878 MAT.	INST.	TOTAL	TRUTH/CONSEQUENCES 879 MAT.	INST.	TOTAL
015433	CONTRACTOR EQUIPMENT		86.2	86.2		110.9	110.9		110.9	110.9		110.9	110.9		110.9	110.9		86.2	86.2
0241, 31 - 34	SITE & INFRASTRUCTURE, DEMOLITION	96.8	81.1	85.9	95.9	102.3	100.4	98.3	102.3	101.1	100.0	102.3	101.6	92.3	102.3	99.3	111.7	81.2	90.4
0310	Concrete Forming & Accessories	93.8	65.8	70.0	100.2	66.9	71.9	97.0	66.9	71.4	99.1	66.9	71.8	100.3	66.9	71.9	98.2	65.8	70.7
0320	Concrete Reinforcing	110.5	71.7	90.8	115.9	71.9	93.5	114.7	71.9	92.9	108.3	71.9	89.8	118.2	71.9	94.6	111.1	71.7	91.1
0330	Cast-in-Place Concrete	89.5	61.9	79.2	89.7	69.1	82.1	94.7	69.1	85.2	95.3	69.1	85.5	87.9	69.1	80.9	96.4	61.9	83.5
03	CONCRETE	82.0	66.4	75.0	95.0	69.8	83.7	104.1	69.8	88.7	94.2	69.8	83.2	93.9	69.8	83.1	87.8	66.4	78.2
04	MASONRY	102.3	58.4	75.2	101.2	58.7	75.0	118.1	58.7	81.4	94.3	58.7	72.3	101.1	58.7	74.9	98.2	58.4	73.6
05	METALS	104.1	82.8	97.6	106.6	90.6	101.7	106.6	90.6	101.7	103.6	90.6	99.7	106.9	90.6	101.9	106.4	82.8	99.3
06	WOOD, PLASTICS & COMPOSITES	83.9	67.8	75.1	104.1	68.8	84.9	94.6	68.8	80.6	99.1	68.8	82.7	104.1	68.8	84.9	95.4	67.8	80.4
07	THERMAL & MOISTURE PROTECTION	86.2	65.3	77.2	101.1	70.0	87.7	99.0	70.0	86.5	104.0	70.0	89.3	101.1	70.0	87.7	89.0	65.3	78.8
08	OPENINGS	94.1	67.1	87.9	99.8	67.7	92.3	98.6	67.7	91.5	100.2	67.7	92.6	99.6	67.7	92.2	92.9	67.1	86.9
0920	Plaster & Gypsum Board	79.0	67.7	71.6	98.8	67.7	78.3	80.9	67.7	72.2	111.2	67.7	82.6	98.8	67.7	78.3	99.7	67.7	78.7
0950, 0980	Ceilings & Acoustic Treatment	87.0	67.7	73.9	102.6	67.7	79.0	103.8	67.7	79.4	101.4	67.7	78.6	102.6	67.7	79.0	86.9	67.7	73.9
0960	Flooring	127.8	68.0	110.4	89.2	68.0	83.1	97.5	68.0	88.9	99.4	68.0	90.3	89.2	68.0	83.1	117.2	68.0	103.0
0970, 0990	Wall Finishes & Painting/Coating	82.6	51.9	64.4	90.8	51.9	67.7	93.4	51.9	68.8	98.2	51.9	70.8	90.8	51.9	67.7	83.3	51.9	64.7
09	FINISHES	100.5	64.7	81.0	88.6	65.5	76.0	91.8	65.5	77.4	96.1	65.5	79.4	88.6	65.5	76.0	98.7	64.7	80.2
COVERS	DIVS. 10 - 14, 25, 28, 41, 43, 44, 46	100.0	82.6	96.2	100.0	85.1	96.7	100.0	85.1	96.7	100.0	85.1	96.7	100.0	85.1	96.7	100.0	82.6	96.2
21, 22, 23	FIRE SUPPRESSION, PLUMBING & HVAC	100.3	68.5	87.5	98.4	68.7	86.5	99.9	68.7	87.4	100.0	68.7	87.5	98.4	68.7	86.5	98.3	68.5	86.4
26, 27, 3370	ELECTRICAL, COMMUNICATIONS & UTIL.	90.7	85.8	88.2	89.0	85.8	87.4	89.9	85.8	87.8	102.4	85.8	93.9	87.3	85.8	86.5	91.1	85.8	88.4
MF2018	WEIGHTED AVERAGE	96.5	71.7	85.9	98.1	75.0	88.2	100.5	75.0	89.6	99.7	75.0	89.1	97.7	75.0	87.9	97.3	71.7	86.3

For customer support on your Electrical Costs with RSMeans Data, call 800.448.8182.

543

NEW MEXICO / NEW YORK

DIVISION		TUCUMCARI 884			ALBANY 120 - 122			BINGHAMTON 137 - 139			BRONX 104			BROOKLYN 112			BUFFALO 140 - 142		
		MAT.	INST.	TOTAL	MAT.	INST.	TOTAL	MAT.	INST.	TOTAL	MAT.	INST.	TOTAL	MAT.	INST.	TOTAL	MAT.	INST.	TOTAL
015433	CONTRACTOR EQUIPMENT		110.9	110.9		115.2	115.2		120.3	120.3		106.8	106.8		111.8	111.8		99.5	99.5
0241, 31 - 34	SITE & INFRASTRUCTURE, DEMOLITION	95.7	102.3	100.3	82.5	104.6	97.9	95.7	93.4	94.1	96.2	114.2	108.7	120.0	125.6	123.9	106.0	101.8	103.1
0310	Concrete Forming & Accessories	96.9	66.8	71.3	100.5	107.8	106.7	99.9	94.0	94.9	100.5	193.1	179.2	105.9	193.1	180.0	100.1	118.9	116.1
0320	Concrete Reinforcing	112.5	71.9	91.8	106.2	116.1	111.3	102.8	109.4	106.1	99.4	179.6	140.2	104.1	246.0	176.2	109.1	115.0	112.1
0330	Cast-in-Place Concrete	94.7	69.1	85.2	80.0	116.1	93.4	104.1	108.3	105.7	77.3	175.4	113.8	106.1	173.8	131.3	115.0	124.3	118.5
03	CONCRETE	102.6	69.8	87.8	84.6	113.1	97.4	93.4	103.8	98.1	85.9	183.7	129.9	104.5	193.3	144.4	108.1	119.5	113.2
04	MASONRY	117.9	58.7	81.3	87.4	116.9	105.6	106.7	106.2	106.4	87.5	189.3	150.4	117.1	189.2	161.7	123.5	123.6	123.5
05	METALS	105.3	90.5	100.8	98.3	125.9	106.7	90.7	134.4	104.0	85.9	173.7	112.6	98.2	174.0	121.2	98.9	107.1	101.4
06	WOOD, PLASTICS & COMPOSITES	94.6	68.8	80.6	95.6	104.7	100.5	103.7	90.7	96.6	103.5	194.1	152.7	105.9	193.8	153.6	99.6	118.2	109.7
07	THERMAL & MOISTURE PROTECTION	98.7	70.0	86.4	107.8	111.5	109.4	108.3	96.2	103.1	104.9	171.4	133.5	108.0	170.6	135.0	104.0	113.6	108.1
08	OPENINGS	98.6	67.7	91.4	97.2	105.0	99.0	92.4	95.3	93.1	94.0	202.1	119.1	89.8	201.9	115.8	99.1	112.0	102.1
0920	Plaster & Gypsum Board	80.9	67.7	72.2	103.7	104.7	104.4	105.9	90.2	95.5	95.9	196.6	162.2	102.3	196.6	164.4	103.6	118.6	113.5
0950, 0980	Ceilings & Acoustic Treatment	103.8	67.7	79.4	99.5	104.7	103.1	96.5	90.2	92.2	94.3	196.6	163.6	92.3	196.6	162.9	93.3	118.6	110.4
0960	Flooring	97.5	68.0	88.9	93.5	111.3	98.7	101.8	103.9	102.4	96.1	188.7	123.0	108.8	188.7	132.0	98.3	120.6	104.8
0970, 0990	Wall Finishes & Painting/Coating	93.4	51.9	68.8	93.4	103.9	99.6	87.8	106.9	99.1	104.8	166.6	141.3	114.1	166.6	145.1	98.9	118.6	110.6
09	FINISHES	91.6	65.5	77.4	90.4	107.6	99.8	91.7	96.5	94.3	97.5	190.5	148.2	104.7	190.2	151.4	98.7	119.9	110.3
COVERS	DIVS. 10 - 14, 25, 28, 41, 43, 44, 46	100.0	85.1	96.7	100.0	102.5	100.6	100.0	98.0	99.6	100.0	138.0	108.4	100.0	137.2	108.2	100.0	106.4	101.4
21, 22, 23	FIRE SUPPRESSION, PLUMBING & HVAC	98.4	68.4	86.4	100.1	111.8	104.8	100.5	98.7	99.8	100.2	176.4	130.7	99.8	176.3	130.4	100.0	99.7	99.9
26, 27, 3370	ELECTRICAL, COMMUNICATIONS & UTIL.	90.8	85.8	88.3	98.1	107.6	102.9	101.1	100.9	101.0	94.1	186.7	141.4	101.0	186.7	144.8	102.4	102.8	102.6
MF2018	WEIGHTED AVERAGE	99.8	74.9	89.1	95.7	111.5	102.5	96.9	102.8	99.5	94.1	176.8	129.6	101.2	178.9	134.6	102.2	110.0	105.6

NEW YORK

DIVISION		ELMIRA 148 - 149			FAR ROCKAWAY 116			FLUSHING 113			GLENS FALLS 128			HICKSVILLE 115, 117, 118			JAMAICA 114		
		MAT.	INST.	TOTAL	MAT.	INST.	TOTAL	MAT.	INST.	TOTAL	MAT.	INST.	TOTAL	MAT.	INST.	TOTAL	MAT.	INST.	TOTAL
015433	CONTRACTOR EQUIPMENT		123.0	123.0		111.8	111.8		111.8	111.8		115.2	115.2		111.8	111.8		111.8	111.8
0241, 31 - 34	SITE & INFRASTRUCTURE, DEMOLITION	104.9	93.8	97.2	123.3	125.6	124.9	123.3	125.6	124.9	73.1	104.2	94.8	112.9	124.4	120.9	117.5	125.6	123.1
0310	Concrete Forming & Accessories	83.9	97.4	95.3	92.2	193.1	177.9	96.1	193.1	178.5	84.1	100.7	98.2	88.5	161.5	150.5	96.1	193.1	178.5
0320	Concrete Reinforcing	108.5	108.2	108.4	104.1	246.0	176.2	105.9	246.0	177.1	101.8	116.0	109.0	104.1	179.4	142.4	104.1	246.0	176.2
0330	Cast-in-Place Concrete	103.2	107.8	104.9	114.8	173.8	136.7	114.8	173.8	136.7	77.1	114.0	90.9	97.4	166.2	123.0	106.1	173.8	131.3
03	CONCRETE	91.7	105.2	97.8	110.6	193.3	147.8	111.1	193.3	148.1	78.5	109.1	92.3	96.4	166.2	127.8	103.9	193.3	144.1
04	MASONRY	110.3	106.6	108.2	121.8	189.2	163.4	115.8	189.2	161.2	93.3	113.6	105.8	111.6	176.9	152.0	119.9	189.2	162.7
05	METALS	94.6	136.7	107.4	98.2	174.0	121.2	98.2	174.0	121.2	91.9	125.3	102.0	99.7	173.6	122.2	98.2	174.0	121.2
06	WOOD, PLASTICS & COMPOSITES	84.8	95.6	90.7	88.8	193.8	145.8	93.6	193.8	148.0	85.0	97.0	91.5	85.3	158.4	125.0	93.6	193.8	148.0
07	THERMAL & MOISTURE PROTECTION	103.7	96.4	100.6	107.9	170.6	134.9	108.0	170.6	135.0	100.6	109.2	104.3	107.6	161.9	131.0	107.8	170.6	134.8
08	OPENINGS	98.6	97.7	98.4	88.5	201.9	114.8	88.5	201.9	114.8	92.0	100.9	94.0	88.9	183.2	110.8	88.5	201.9	114.8
0920	Plaster & Gypsum Board	98.0	95.4	96.3	91.1	196.6	160.6	93.6	196.6	161.4	94.8	96.8	96.1	90.8	160.2	136.5	93.6	196.6	161.4
0950, 0980	Ceilings & Acoustic Treatment	99.9	95.4	96.9	81.3	196.6	159.4	81.3	196.6	159.4	86.5	96.8	93.5	80.3	160.2	134.5	81.3	196.6	159.4
0960	Flooring	88.8	103.9	93.2	103.7	188.7	128.3	105.2	188.7	129.4	84.9	111.3	92.6	102.6	185.7	126.7	105.2	188.7	129.4
0970, 0990	Wall Finishes & Painting/Coating	98.0	94.0	95.6	114.1	166.6	145.1	114.1	166.6	145.1	86.9	102.0	95.9	114.1	166.6	145.1	114.1	166.6	145.1
09	FINISHES	91.7	98.0	95.1	99.9	190.2	149.2	100.7	190.2	149.6	83.0	102.0	93.4	98.5	165.8	135.2	100.3	190.2	149.4
COVERS	DIVS. 10 - 14, 25, 28, 41, 43, 44, 46	100.0	99.1	99.8	100.0	137.2	108.2	100.0	137.2	108.2	100.0	98.1	99.6	100.0	134.8	107.7	100.0	137.2	108.2
21, 22, 23	FIRE SUPPRESSION, PLUMBING & HVAC	97.7	97.2	97.5	97.3	176.3	128.9	97.3	176.3	128.9	97.7	108.9	102.2	99.8	163.2	125.2	97.3	176.3	128.9
26, 27, 3370	ELECTRICAL, COMMUNICATIONS & UTIL.	100.4	104.3	102.4	107.9	186.7	148.2	107.9	186.7	148.2	92.8	107.6	100.4	100.4	144.4	122.9	99.4	186.7	144.1
MF2018	WEIGHTED AVERAGE	97.4	103.9	100.2	101.7	178.9	134.8	101.6	178.9	134.8	91.5	108.7	98.9	99.2	160.5	125.5	99.8	178.9	133.8

NEW YORK

DIVISION		JAMESTOWN 147			KINGSTON 124			LONG ISLAND CITY 111			MONTICELLO 127			MOUNT VERNON 105			NEW ROCHELLE 108		
		MAT.	INST.	TOTAL	MAT.	INST.	TOTAL	MAT.	INST.	TOTAL	MAT.	INST.	TOTAL	MAT.	INST.	TOTAL	MAT.	INST.	TOTAL
015433	CONTRACTOR EQUIPMENT		92.7	92.7		111.8	111.8		111.8	111.8		111.8	111.8		106.8	106.8		106.8	106.8
0241, 31 - 34	SITE & INFRASTRUCTURE, DEMOLITION	106.5	94.2	97.9	143.9	121.2	128.1	121.1	125.6	124.2	139.1	121.1	126.6	101.1	110.1	107.3	100.7	110.0	107.2
0310	Concrete Forming & Accessories	83.9	92.2	90.9	85.3	133.6	126.3	100.5	193.1	179.2	92.7	133.6	127.5	90.0	141.8	134.0	106.2	138.2	133.4
0320	Concrete Reinforcing	108.7	114.4	111.6	102.3	164.9	134.1	104.1	246.0	176.2	101.4	164.9	133.7	98.1	182.1	140.8	98.3	181.9	140.8
0330	Cast-in-Place Concrete	107.0	106.6	106.8	104.1	147.5	120.2	109.5	173.8	133.4	97.5	147.5	116.2	86.3	151.4	110.5	86.3	151.1	110.4
03	CONCRETE	94.7	101.0	97.5	98.7	143.2	118.7	107.0	193.3	145.8	94.0	143.2	116.1	94.6	152.2	120.5	94.0	150.5	119.4
04	MASONRY	120.0	104.7	110.6	107.8	154.7	136.8	113.9	189.2	160.5	100.1	154.7	133.8	92.8	156.2	132.0	92.8	156.2	132.0
05	METALS	92.0	102.5	95.2	100.0	136.5	111.1	98.2	174.0	121.2	99.9	136.7	111.1	85.7	165.6	110.0	86.0	164.9	109.9
06	WOOD, PLASTICS & COMPOSITES	83.4	88.7	86.3	86.2	127.6	108.7	99.9	193.8	150.9	93.1	127.6	111.8	92.4	136.3	116.2	111.5	132.4	122.9
07	THERMAL & MOISTURE PROTECTION	103.2	96.7	100.4	124.7	146.8	134.2	107.9	170.6	134.9	124.3	146.8	134.0	105.8	148.9	124.4	105.9	146.3	123.3
08	OPENINGS	98.4	95.7	97.8	93.7	146.1	105.8	88.5	201.9	114.8	89.5	146.0	102.6	94.0	169.4	111.5	94.1	167.2	111.1
0920	Plaster & Gypsum Board	85.7	88.3	87.4	96.9	128.5	117.7	98.5	196.6	163.1	97.5	128.5	117.9	90.9	137.1	121.3	104.3	133.1	123.3
0950, 0980	Ceilings & Acoustic Treatment	96.5	88.3	90.9	77.1	128.5	112.0	81.3	196.6	159.4	77.1	128.5	111.9	92.6	137.1	122.7	92.6	133.1	120.0
0960	Flooring	91.4	103.9	95.0	102.2	166.0	120.7	106.8	188.7	130.6	104.5	166.0	122.4	87.9	185.7	116.3	95.3	169.5	116.9
0970, 0990	Wall Finishes & Painting/Coating	99.6	101.0	100.4	113.6	130.5	123.6	114.1	166.6	145.1	113.6	126.6	121.3	103.2	166.6	140.7	103.2	166.6	140.7
09	FINISHES	90.2	94.3	92.5	96.5	138.5	119.4	101.6	190.2	150.0	96.9	138.0	119.3	94.5	151.4	125.5	98.4	145.8	124.3
COVERS	DIVS. 10 - 14, 25, 28, 41, 43, 44, 46	100.0	98.5	99.7	100.0	119.7	104.3	100.0	137.2	108.2	100.0	120.6	104.5	100.0	131.4	106.9	100.0	114.8	103.3
21, 22, 23	FIRE SUPPRESSION, PLUMBING & HVAC	97.6	95.3	96.6	97.6	136.7	113.3	99.8	176.3	130.4	97.6	141.0	115.0	97.8	144.9	116.7	97.8	144.6	116.6
26, 27, 3370	ELECTRICAL, COMMUNICATIONS & UTIL.	99.2	97.2	98.2	94.2	119.2	107.0	99.9	186.7	144.3	94.2	119.2	107.0	92.2	170.1	132.1	92.2	145.2	119.3
MF2018	WEIGHTED AVERAGE	97.5	97.8	97.7	99.8	136.0	115.3	100.8	178.9	134.4	98.4	136.9	114.9	94.4	151.3	118.8	94.9	146.0	116.8

NEW YORK

	DIVISION	NEW YORK 100 - 102			NIAGARA FALLS 143			PLATTSBURGH 129			POUGHKEEPSIE 125 - 126			QUEENS 110			RIVERHEAD 119		
		MAT.	INST.	TOTAL	MAT.	INST.	TOTAL	MAT.	INST.	TOTAL	MAT.	INST.	TOTAL	MAT.	INST.	TOTAL	MAT.	INST.	TOTAL
015433	CONTRACTOR EQUIPMENT		104.8	104.8		92.7	92.7		96.4	96.4		111.8	111.8		111.8	111.8		111.8	111.8
0241, 31 - 34	SITE & INFRASTRUCTURE, DEMOLITION	105.0	111.5	109.5	108.9	95.2	99.4	112.6	101.3	104.7	140.2	120.3	126.3	116.2	125.6	122.7	114.0	123.9	120.9
0310	Concrete Forming & Accessories	110.3	192.3	180.0	83.9	116.3	111.4	90.5	93.9	93.4	85.3	172.3	159.2	88.8	193.1	177.4	93.3	160.5	150.4
0320	Concrete Reinforcing	105.2	183.4	144.9	107.3	115.2	111.3	106.3	115.3	110.8	102.3	164.7	134.0	105.9	246.0	177.1	106.1	221.8	164.9
0330	Cast-in-Place Concrete	89.3	174.6	121.0	110.7	127.7	117.0	94.3	103.5	97.7	100.8	139.3	115.2	100.8	173.8	128.0	99.1	164.4	123.4
03	CONCRETE	99.1	183.5	137.1	97.0	119.3	107.0	91.3	100.6	95.5	96.2	157.8	123.9	98.6	193.3	141.7	97.1	171.0	130.3
04	MASONRY	101.2	187.1	154.2	128.3	127.8	128.0	87.7	99.4	94.9	100.1	142.5	126.3	108.1	189.2	158.2	117.3	175.8	153.5
05	METALS	96.9	173.1	120.0	94.6	103.1	97.2	96.0	100.0	97.2	100.0	136.3	111.0	98.2	174.0	121.2	100.1	161.6	118.8
06	WOOD, PLASTICS & COMPOSITES	109.2	194.2	155.4	83.4	111.4	98.6	91.7	90.7	91.2	86.2	184.5	139.6	85.4	193.8	144.3	90.5	158.4	127.4
07	THERMAL & MOISTURE PROTECTION	107.9	170.7	134.9	103.3	113.1	107.5	119.5	99.9	111.0	124.6	147.6	134.5	107.5	170.6	134.7	108.6	158.8	130.2
08	OPENINGS	97.1	203.0	121.7	98.5	107.7	100.6	99.4	97.4	98.9	93.7	175.5	112.7	88.5	201.9	114.8	88.9	176.6	109.3
0920	Plaster & Gypsum Board	101.2	196.6	164.0	85.7	111.6	102.8	114.7	90.0	98.4	96.9	187.0	156.2	90.8	196.6	160.5	91.9	160.2	136.9
0950, 0980	Ceilings & Acoustic Treatment	109.6	196.6	168.5	96.5	111.6	106.7	106.5	90.0	95.3	77.1	187.0	151.6	81.3	196.6	159.4	81.2	160.2	134.7
0960	Flooring	97.3	188.7	123.8	91.4	115.8	98.5	107.4	108.6	107.8	102.2	164.0	120.1	102.6	188.7	127.6	103.7	172.5	123.6
0970, 0990	Wall Finishes & Painting/Coating	101.4	166.6	140.0	99.6	114.4	108.4	108.3	97.6	102.0	113.6	126.6	121.3	114.1	166.6	145.1	114.1	166.6	145.1
09	FINISHES	102.3	190.0	150.2	90.4	116.2	104.5	94.9	96.3	95.7	96.3	169.0	136.0	98.9	190.2	148.7	99.1	162.7	133.8
COVERS	DIVS. 10 - 14, 25, 28, 41, 43, 44, 46	100.0	143.1	109.5	100.0	104.8	101.1	100.0	94.6	98.8	100.0	118.8	104.1	100.0	137.2	108.2	100.0	122.6	105.0
21, 22, 23	FIRE SUPPRESSION, PLUMBING & HVAC	100.1	175.3	130.2	97.6	107.5	101.5	97.6	102.6	99.6	97.6	123.0	107.8	99.8	176.3	130.4	99.9	157.9	123.2
26, 27, 3370	ELECTRICAL, COMMUNICATIONS & UTIL.	100.9	186.7	144.8	97.9	101.3	99.6	90.5	94.0	92.3	94.2	124.2	109.6	100.4	186.7	144.6	102.0	138.2	120.5
MF2018	WEIGHTED AVERAGE	99.8	176.2	132.6	98.6	110.2	103.5	96.5	99.0	97.5	99.0	140.2	116.7	99.2	178.9	133.4	100.0	156.8	124.3

NEW YORK

	DIVISION	ROCHESTER 144 - 146			SCHENECTADY 123			STATEN ISLAND 103			SUFFERN 109			SYRACUSE 130 - 132			UTICA 133 - 135		
		MAT.	INST.	TOTAL	MAT.	INST.	TOTAL	MAT.	INST.	TOTAL	MAT.	INST.	TOTAL	MAT.	INST.	TOTAL	MAT.	INST.	TOTAL
015433	CONTRACTOR EQUIPMENT		119.9	119.9		115.2	115.2		106.8	106.8		106.8	106.8		115.2	115.2		115.2	115.2
0241, 31 - 34	SITE & INFRASTRUCTURE, DEMOLITION	92.9	108.7	103.9	83.2	104.6	98.1	105.1	114.2	111.5	98.0	108.8	105.5	94.3	103.2	100.5	72.8	102.7	93.6
0310	Concrete Forming & Accessories	106.0	101.9	102.5	100.9	107.7	106.7	89.4	193.4	177.7	99.0	144.7	137.8	99.0	91.6	92.7	100.0	88.7	90.4
0320	Concrete Reinforcing	110.9	108.0	109.4	100.8	116.1	108.6	99.4	222.2	161.8	98.3	156.4	127.8	103.8	109.2	106.5	103.8	103.2	103.5
0330	Cast-in-Place Concrete	101.1	105.3	102.7	93.1	116.0	101.7	86.3	175.6	119.5	83.7	143.0	105.8	96.6	105.3	99.9	88.4	103.9	94.2
03	CONCRETE	96.3	105.4	100.4	91.8	113.0	101.3	96.4	190.3	138.6	91.4	145.1	115.6	96.0	100.8	98.1	93.9	98.0	95.7
04	MASONRY	101.4	105.6	104.0	90.2	116.9	106.7	99.5	189.3	155.0	92.7	145.2	125.1	99.4	103.2	101.7	90.5	101.5	97.3
05	METALS	100.4	121.5	106.8	95.9	125.8	105.0	84.2	173.9	111.4	84.2	132.9	98.9	94.0	120.6	102.1	92.1	118.3	100.1
06	WOOD, PLASTICS & COMPOSITES	108.4	101.4	104.6	103.5	104.7	104.1	90.9	194.1	147.0	103.0	146.6	126.7	100.3	88.5	93.9	100.3	85.1	92.0
07	THERMAL & MOISTURE PROTECTION	104.8	101.8	103.5	102.3	111.5	106.2	105.2	171.4	133.7	105.8	144.3	122.4	102.7	96.8	100.2	91.1	96.7	93.5
08	OPENINGS	102.1	101.3	101.9	97.4	105.0	99.1	94.0	202.1	119.1	94.1	154.5	108.1	94.5	92.1	94.0	97.4	88.8	95.4
0920	Plaster & Gypsum Board	103.7	101.6	102.3	104.1	104.7	104.5	91.0	196.6	160.5	95.2	147.7	129.8	95.9	88.1	90.8	95.9	84.6	88.5
0950, 0980	Ceilings & Acoustic Treatment	102.4	101.6	101.8	93.2	104.7	101.0	94.3	196.6	163.6	92.6	147.7	129.9	96.5	88.1	90.8	96.5	84.6	88.4
0960	Flooring	90.0	108.5	95.4	92.1	111.3	97.7	91.5	188.7	119.7	91.4	177.3	116.3	90.7	95.4	92.0	88.5	95.5	90.6
0970, 0990	Wall Finishes & Painting/Coating	97.1	101.0	99.4	86.9	103.9	97.0	104.8	166.6	141.3	103.2	130.3	119.2	90.4	101.3	96.9	84.2	101.3	94.3
09	FINISHES	94.1	103.4	99.2	88.4	107.6	98.8	96.2	190.5	147.7	95.8	150.1	125.4	90.5	92.6	91.6	88.9	90.4	89.7
COVERS	DIVS. 10 - 14, 25, 28, 41, 43, 44, 46	100.0	100.4	100.1	100.0	102.5	100.6	100.0	138.0	108.4	100.0	117.0	103.8	100.0	97.0	99.3	100.0	91.6	98.1
21, 22, 23	FIRE SUPPRESSION, PLUMBING & HVAC	100.0	90.1	96.0	100.1	107.7	103.2	100.2	176.4	130.7	97.8	126.5	109.3	100.2	96.8	98.8	100.2	94.6	98.0
26, 27, 3370	ELECTRICAL, COMMUNICATIONS & UTIL.	104.7	92.1	98.3	96.9	107.6	102.4	94.1	186.7	141.5	98.9	117.4	108.4	101.0	103.5	102.3	98.9	103.5	101.3
MF2018	WEIGHTED AVERAGE	99.9	101.3	100.5	96.0	110.6	102.3	95.7	177.8	130.9	94.6	133.5	111.3	97.3	100.8	98.8	95.4	98.9	96.9

NEW YORK / NORTH CAROLINA

	DIVISION	WATERTOWN 136			WHITE PLAINS 106			YONKERS 107			ASHEVILLE 287 - 288			CHARLOTTE 281 - 282			DURHAM 277		
		MAT.	INST.	TOTAL	MAT.	INST.	TOTAL	MAT.	INST.	TOTAL	MAT.	INST.	TOTAL	MAT.	INST.	TOTAL	MAT.	INST.	TOTAL
015433	CONTRACTOR EQUIPMENT		115.2	115.2		106.8	106.8		106.8	106.8		100.9	100.9		100.9	100.9		106.2	106.2
0241, 31 - 34	SITE & INFRASTRUCTURE, DEMOLITION	80.5	103.2	96.3	95.9	110.1	105.8	102.7	110.1	107.8	95.9	81.1	85.6	98.1	81.3	86.4	99.7	90.1	93.0
0310	Concrete Forming & Accessories	84.8	95.6	94.0	104.4	152.6	145.4	104.7	149.2	142.5	88.7	62.1	66.1	96.0	61.7	66.9	94.0	62.1	66.9
0320	Concrete Reinforcing	104.5	109.2	106.9	98.3	182.1	140.9	102.2	182.0	142.8	92.3	67.8	79.9	96.5	68.6	82.3	95.8	68.6	82.0
0330	Cast-in-Place Concrete	103.0	107.7	104.7	76.7	151.4	104.5	85.8	151.4	110.2	104.8	69.7	91.7	106.8	69.0	92.7	104.2	69.7	91.3
03	CONCRETE	106.3	103.4	105.0	85.6	157.1	117.8	94.1	155.6	121.8	96.8	67.4	83.6	96.5	67.1	83.3	96.6	67.5	83.5
04	MASONRY	91.7	107.0	101.2	92.2	156.2	131.7	96.3	156.2	133.3	86.5	62.4	71.6	90.2	61.0	72.2	85.6	62.4	71.3
05	METALS	92.2	120.6	100.8	85.7	165.6	110.0	93.5	165.5	115.4	98.1	90.1	95.6	98.9	90.5	96.3	114.7	90.4	107.3
06	WOOD, PLASTICS & COMPOSITES	81.7	92.5	87.6	109.1	151.0	131.9	109.0	146.6	129.4	87.4	60.7	72.9	92.3	60.7	75.1	90.3	60.7	74.2
07	THERMAL & MOISTURE PROTECTION	91.3	99.9	95.0	105.6	150.4	124.9	105.9	150.0	124.9	100.9	63.9	85.0	95.0	63.3	81.4	105.4	63.9	87.6
08	OPENINGS	97.4	96.5	97.2	94.1	177.5	113.5	97.8	173.0	115.2	90.9	61.6	84.1	97.8	61.7	89.4	97.4	61.7	89.1
0920	Plaster & Gypsum Board	86.7	92.2	90.4	98.2	152.2	133.7	101.4	147.7	131.9	107.3	59.5	75.8	104.4	59.5	74.8	94.1	59.5	71.3
0950, 0980	Ceilings & Acoustic Treatment	96.5	92.2	93.6	92.6	152.2	132.9	108.7	147.7	135.1	81.6	59.5	66.6	87.3	59.5	68.5	84.3	59.5	67.5
0960	Flooring	81.8	95.5	85.8	93.7	185.7	120.4	93.2	188.7	120.9	94.7	62.1	85.2	95.2	62.1	85.6	98.3	62.1	87.8
0970, 0990	Wall Finishes & Painting/Coating	84.2	95.4	90.9	103.2	166.6	140.7	103.2	166.6	140.7	103.7	58.6	77.0	97.4	58.6	74.4	98.9	58.6	75.1
09	FINISHES	86.2	95.1	91.1	96.6	160.0	131.2	100.5	158.0	131.9	88.6	61.5	73.8	88.7	61.1	73.6	87.0	61.5	73.1
COVERS	DIVS. 10 - 14, 25, 28, 41, 43, 44, 46	100.0	98.4	99.6	100.0	129.6	106.5	100.0	126.8	105.9	100.0	83.4	96.3	100.0	83.0	96.3	100.0	83.4	96.3
21, 22, 23	FIRE SUPPRESSION, PLUMBING & HVAC	100.2	89.8	96.0	100.3	144.9	118.2	100.3	144.8	118.1	100.5	61.5	84.9	99.9	61.9	84.7	100.5	61.5	84.9
26, 27, 3370	ELECTRICAL, COMMUNICATIONS & UTIL.	101.0	92.4	96.6	92.2	170.1	132.1	99.0	159.8	130.1	101.8	59.3	80.0	100.9	61.8	80.9	95.9	58.8	76.9
MF2018	WEIGHTED AVERAGE	97.0	99.2	98.0	94.1	153.5	119.6	98.1	151.2	120.9	96.9	67.0	84.1	97.6	67.3	84.6	99.8	67.7	86.0

For customer support on your Electrical Costs with RSMeans Data, call 800.448.8182.

545

City Cost Indexes

NORTH CAROLINA

DIVISION		ELIZABETH CITY 279 MAT.	INST.	TOTAL	FAYETTEVILLE 283 MAT.	INST.	TOTAL	GASTONIA 280 MAT.	INST.	TOTAL	GREENSBORO 270, 272-274 MAT.	INST.	TOTAL	HICKORY 286 MAT.	INST.	TOTAL	KINSTON 285 MAT.	INST.	TOTAL
015433	CONTRACTOR EQUIPMENT		110.3	110.3		106.2	106.2		100.9	100.9		106.2	106.2		106.2	106.2		106.2	106.2
0241, 31-34	SITE & INFRASTRUCTURE, DEMOLITION	104.1	91.8	95.5	95.2	90.0	91.6	95.7	81.4	85.8	99.6	90.1	93.0	94.8	88.8	90.6	93.7	88.7	90.2
0310	Concrete Forming & Accessories	80.3	62.5	65.2	88.3	61.6	65.6	94.8	62.4	67.2	93.8	62.1	66.9	85.3	62.1	65.6	81.8	61.6	64.6
0320	Concrete Reinforcing	93.9	71.7	82.6	95.9	68.6	82.0	92.7	68.6	80.5	94.7	68.6	81.4	92.3	68.6	80.3	91.9	68.6	80.0
0330	Cast-in-Place Concrete	104.4	71.4	92.1	109.9	68.9	94.6	100.0	67.7	90.3	103.4	69.7	90.8	104.7	69.7	91.7	101.1	68.9	89.1
03	CONCRETE	96.8	68.8	84.2	98.6	67.0	84.4	95.3	67.7	82.9	96.1	67.5	83.2	96.5	67.5	83.5	93.5	67.0	81.6
04	MASONRY	96.8	61.0	74.7	89.6	61.0	72.0	90.5	62.4	73.1	82.9	62.4	70.2	75.9	62.4	67.6	82.6	61.0	69.3
05	METALS	101.1	92.6	98.5	118.2	90.4	109.7	98.7	90.6	96.2	107.4	90.3	102.2	98.1	90.4	95.8	96.9	90.3	94.9
06	WOOD, PLASTICS & COMPOSITES	75.0	61.4	67.6	86.6	60.7	72.5	95.3	60.7	76.5	90.1	60.7	74.1	82.3	60.7	70.6	79.3	60.7	69.2
07	THERMAL & MOISTURE PROTECTION	104.6	63.2	86.8	100.4	63.3	84.4	101.1	63.9	85.1	105.2	63.9	87.4	101.3	63.9	85.2	101.1	63.3	84.8
08	OPENINGS	94.7	62.7	87.3	91.0	61.7	84.2	94.3	61.7	86.7	97.4	61.7	89.1	90.9	61.7	84.2	91.0	61.7	84.2
0920	Plaster & Gypsum Board	88.8	59.5	69.5	111.2	59.5	77.2	113.2	59.5	77.9	95.5	59.5	71.8	107.3	59.5	75.8	107.1	59.5	75.8
0950, 0980	Ceilings & Acoustic Treatment	84.3	59.5	67.5	83.3	59.5	67.2	85.0	59.5	67.7	84.3	59.5	67.5	81.6	59.5	66.6	85.0	59.5	67.7
0960	Flooring	90.4	79.8	87.3	94.8	62.1	85.3	98.0	79.8	92.7	98.3	62.1	87.8	94.6	79.8	90.3	91.7	79.8	88.3
0970, 0990	Wall Finishes & Painting/Coating	98.9	58.6	75.1	103.7	58.6	77.0	103.7	58.6	77.0	98.9	58.6	75.1	103.7	58.6	77.0	103.7	58.6	77.0
09	FINISHES	84.3	65.2	73.9	89.5	61.1	74.0	91.0	65.1	76.8	87.2	61.5	73.2	88.8	65.1	75.8	88.4	64.7	75.5
COVERS	DIVS. 10-14, 25, 28, 41, 43, 44, 46	100.0	81.7	96.0	100.0	83.0	96.2	100.0	83.4	96.4	100.0	80.7	95.7	100.0	83.4	96.3	100.0	82.9	96.2
21, 22, 23	FIRE SUPPRESSION, PLUMBING & HVAC	98.0	59.9	82.7	100.3	60.8	84.5	100.5	60.4	84.4	100.5	61.5	84.9	98.0	60.4	83.0	98.0	59.7	82.7
26, 27, 3370	ELECTRICAL, COMMUNICATIONS & UTIL.	95.6	66.2	80.6	101.1	58.8	79.6	101.2	61.9	81.1	95.1	59.3	76.7	99.4	61.9	80.2	99.2	57.2	77.7
MF2018	WEIGHTED AVERAGE	96.9	69.2	85.0	100.5	67.3	86.2	97.6	67.7	84.8	98.3	67.7	85.2	95.5	68.3	83.8	95.2	67.1	83.1

NORTH CAROLINA / NORTH DAKOTA

DIVISION		MURPHY 289 MAT.	INST.	TOTAL	RALEIGH 275-276 MAT.	INST.	TOTAL	ROCKY MOUNT 278 MAT.	INST.	TOTAL	WILMINGTON 284 MAT.	INST.	TOTAL	WINSTON-SALEM 271 MAT.	INST.	TOTAL	BISMARCK 585 MAT.	INST.	TOTAL
015433	CONTRACTOR EQUIPMENT		100.9	100.9		106.2	106.2		106.2	106.2		100.9	100.9		106.2	106.2		100.3	100.3
0241, 31-34	SITE & INFRASTRUCTURE, DEMOLITION	97.1	79.6	84.9	99.6	90.1	93.0	102.1	90.1	93.7	97.0	81.0	85.8	99.9	90.1	93.1	101.1	98.4	99.2
0310	Concrete Forming & Accessories	95.4	61.5	66.6	95.2	61.6	66.7	86.3	61.8	65.5	90.1	61.6	65.9	95.7	62.1	67.2	103.0	73.4	77.9
0320	Concrete Reinforcing	91.9	66.1	78.8	96.8	68.6	82.5	93.9	68.6	81.1	93.1	68.6	80.6	94.7	68.6	81.4	103.8	94.5	99.1
0330	Cast-in-Place Concrete	108.4	68.9	93.7	107.1	68.9	92.9	102.1	69.0	89.8	104.3	68.9	91.2	106.0	69.7	92.5	102.2	83.9	95.4
03	CONCRETE	99.7	66.5	84.8	95.1	67.0	82.5	97.5	67.1	83.8	96.7	67.0	83.4	97.5	67.5	84.0	94.4	81.4	88.5
04	MASONRY	78.9	61.0	67.9	81.6	61.0	68.9	75.3	61.0	66.5	76.5	61.0	66.9	83.2	62.4	70.3	98.4	82.4	88.5
05	METALS	95.9	89.5	94.0	99.8	90.4	96.9	100.3	90.5	97.3	97.6	90.4	95.4	104.6	90.4	100.3	98.7	92.6	96.8
06	WOOD, PLASTICS & COMPOSITES	96.0	60.6	76.7	90.4	60.7	74.3	81.8	60.7	70.4	89.2	60.7	73.7	90.1	60.7	74.1	102.1	69.5	84.4
07	THERMAL & MOISTURE PROTECTION	101.1	63.3	84.8	99.5	63.3	83.9	105.2	63.3	87.1	100.9	63.3	84.7	105.2	63.9	87.4	105.1	83.2	95.7
08	OPENINGS	90.9	61.1	84.0	98.7	61.7	90.1	93.9	61.7	86.5	91.0	61.7	84.2	97.4	61.7	89.1	106.3	79.3	100.0
0920	Plaster & Gypsum Board	112.5	59.3	77.5	91.1	59.5	70.3	90.3	59.5	70.0	109.2	59.5	76.5	95.5	59.5	71.8	100.0	69.0	79.6
0950, 0980	Ceilings & Acoustic Treatment	81.6	59.3	66.5	82.2	59.5	66.8	82.6	59.5	67.0	83.3	59.5	67.2	84.3	59.5	67.5	109.7	69.0	82.1
0960	Flooring	98.2	79.8	92.9	93.3	62.1	84.3	93.9	79.8	89.8	95.4	62.1	85.7	98.3	62.1	87.8	91.3	51.4	79.8
0970, 0990	Wall Finishes & Painting/Coating	103.7	58.6	77.0	92.1	58.6	72.3	98.9	58.6	75.1	103.7	58.6	77.0	98.9	58.6	75.1	93.7	59.0	73.2
09	FINISHES	90.6	64.6	76.4	85.6	61.1	72.3	85.3	64.7	74.1	89.4	61.1	74.0	87.2	61.5	73.2	95.6	68.3	80.7
COVERS	DIVS. 10-14, 25, 28, 41, 43, 44, 46	100.0	82.9	96.2	100.0	83.0	96.2	100.0	83.0	96.2	100.0	83.0	96.2	100.0	83.4	96.3	100.0	89.1	97.6
21, 22, 23	FIRE SUPPRESSION, PLUMBING & HVAC	98.0	59.6	82.6	100.0	60.8	84.3	98.0	59.7	82.6	100.5	60.8	84.6	100.5	61.5	84.9	100.0	89.1	97.6
26, 27, 3370	ELECTRICAL, COMMUNICATIONS & UTIL.	102.6	57.2	79.4	98.3	57.7	77.5	97.5	58.8	77.7	102.0	57.2	79.1	95.0	59.3	76.7	99.9	75.5	90.1
MF2018	WEIGHTED AVERAGE	96.3	66.2	83.4	96.9	67.1	84.1	96.1	67.5	83.8	96.5	66.3	83.6	98.0	67.8	85.0	99.1	80.0	90.9

NORTH DAKOTA

DIVISION		DEVILS LAKE 583 MAT.	INST.	TOTAL	DICKINSON 586 MAT.	INST.	TOTAL	FARGO 580-581 MAT.	INST.	TOTAL	GRAND FORKS 582 MAT.	INST.	TOTAL	JAMESTOWN 584 MAT.	INST.	TOTAL	MINOT 587 MAT.	INST.	TOTAL
015433	CONTRACTOR EQUIPMENT		100.3	100.3		100.3	100.3		100.3	100.3		100.3	100.3		100.3	100.3		100.3	100.3
0241, 31-34	SITE & INFRASTRUCTURE, DEMOLITION	106.5	98.5	100.9	114.1	98.4	103.2	97.7	98.4	98.2	110.0	98.4	101.9	105.5	98.5	100.6	107.6	98.4	101.2
0310	Concrete Forming & Accessories	101.9	73.5	77.7	90.9	73.4	76.0	97.2	73.5	77.1	94.9	73.4	76.6	92.6	73.4	76.3	90.6	73.4	76.0
0320	Concrete Reinforcing	100.2	94.4	97.3	101.1	94.5	97.8	94.6	94.8	94.7	98.7	94.4	96.5	100.7	94.8	97.7	102.1	94.5	98.2
0330	Cast-in-Place Concrete	122.1	84.0	107.9	110.4	83.9	100.5	99.9	83.9	93.9	110.4	83.9	100.5	120.5	84.0	106.0	110.4	83.9	100.5
03	CONCRETE	103.8	81.4	93.8	102.8	81.4	93.1	95.9	81.5	89.4	100.2	81.4	91.7	102.4	81.5	93.0	98.8	81.4	91.0
04	MASONRY	114.6	80.9	93.8	116.6	82.4	95.5	103.2	87.1	93.3	108.9	80.9	91.6	127.0	87.2	102.4	107.2	82.4	91.9
05	METALS	95.2	92.7	94.4	95.1	92.5	94.3	97.9	93.0	96.4	95.1	92.4	94.3	95.2	93.1	94.5	95.4	92.5	94.6
06	WOOD, PLASTICS & COMPOSITES	99.9	69.5	83.4	86.2	69.5	77.1	94.9	69.5	81.1	91.0	69.5	79.3	88.3	69.5	78.0	85.9	69.5	77.0
07	THERMAL & MOISTURE PROTECTION	106.5	82.8	96.3	107.1	83.2	96.8	106.5	85.0	97.2	106.8	82.8	96.4	106.3	85.0	97.2	106.5	83.2	96.5
08	OPENINGS	103.1	79.3	97.6	103.1	79.3	97.6	100.2	79.3	95.4	101.7	79.3	96.5	103.1	79.3	97.6	101.9	79.3	96.6
0920	Plaster & Gypsum Board	114.9	69.0	84.7	105.2	69.0	81.3	98.2	69.0	78.9	106.4	69.0	81.7	106.1	69.0	81.6	105.2	69.0	81.3
0950, 0980	Ceilings & Acoustic Treatment	101.8	69.0	79.6	101.8	69.0	79.6	95.2	69.0	77.4	101.8	69.0	79.6	101.8	69.0	79.6	101.8	69.0	79.6
0960	Flooring	95.3	51.4	82.6	88.8	51.4	77.9	100.0	51.4	85.9	90.6	51.4	79.3	89.5	51.4	78.4	88.5	51.4	77.8
0970, 0990	Wall Finishes & Painting/Coating	88.1	57.8	70.2	88.1	57.8	70.2	92.8	69.1	78.8	88.1	67.8	76.1	88.1	57.8	70.2	88.1	59.0	70.9
09	FINISHES	93.8	68.0	79.7	91.4	68.0	78.6	94.9	69.2	80.9	91.6	69.1	79.3	90.8	68.0	78.4	90.6	68.1	78.3
COVERS	DIVS. 10-14, 25, 28, 41, 43, 44, 46	100.0	89.1	97.6	100.0	89.1	97.6	100.0	89.1	97.6	100.0	89.1	97.6	100.0	89.1	97.6	100.0	89.1	97.6
21, 22, 23	FIRE SUPPRESSION, PLUMBING & HVAC	97.7	80.0	90.6	97.7	74.9	88.5	99.9	75.6	90.2	100.2	73.8	89.6	97.7	73.8	88.1	100.0	89.1	97.6
26, 27, 3370	ELECTRICAL, COMMUNICATIONS & UTIL.	94.4	69.9	81.9	102.2	69.5	85.5	101.2	71.1	85.8	97.5	69.9	83.4	94.4	69.9	81.9	100.2	73.6	89.5
MF2018	WEIGHTED AVERAGE	99.5	80.2	91.2	100.1	79.1	91.1	99.1	80.3	91.0	99.3	79.0	90.6	99.5	79.6	91.0	99.2	79.6	90.8

For customer support on your Electrical Costs with RSMeans Data, call 800.448.8182.

DIVISION		NORTH DAKOTA			OHIO														
		WILLISTON			AKRON			ATHENS			CANTON			CHILLICOTHE			CINCINNATI		
		588			442 - 443			457			446 - 447			456			451 - 452		
		MAT.	INST.	TOTAL	MAT.	INST.	TOTAL	MAT.	INST.	TOTAL	MAT.	INST.	TOTAL	MAT.	INST.	TOTAL	MAT.	INST.	TOTAL
015433	CONTRACTOR EQUIPMENT		100.3	100.3		91.4	91.4		87.5	87.5		91.4	91.4		98.4	98.4		96.0	96.0
0241, 31 - 34	SITE & INFRASTRUCTURE, DEMOLITION	108.0	96.2	99.8	99.8	99.1	99.4	116.0	89.8	97.8	100.0	98.8	99.2	102.4	100.0	100.8	97.9	99.0	98.7
0310	Concrete Forming & Accessories	96.8	73.1	76.7	100.8	85.2	87.6	92.9	79.2	81.3	100.8	75.7	79.5	95.7	83.9	85.6	99.1	82.1	84.7
0320	Concrete Reinforcing	103.1	94.5	98.7	95.4	93.7	94.6	90.7	91.8	91.2	95.4	77.2	86.2	88.0	91.4	89.7	93.3	80.0	86.6
0330	Cast-in-Place Concrete	110.4	83.8	100.5	107.0	91.7	101.3	111.5	98.3	106.6	108.0	89.0	100.9	101.2	95.2	99.0	96.6	77.5	89.5
03	CONCRETE	100.2	81.2	91.6	102.2	88.4	96.0	100.6	87.7	94.8	102.7	80.3	92.6	95.1	89.4	92.5	94.0	80.3	87.8
04	MASONRY	102.1	82.4	89.9	94.2	91.3	92.4	72.3	98.5	88.5	94.9	80.9	86.2	78.7	92.4	87.2	81.3	81.0	81.1
05	METALS	95.3	92.2	94.4	97.9	81.1	92.8	101.6	81.4	95.5	97.9	74.4	90.8	93.6	90.8	92.7	95.8	82.4	91.7
06	WOOD, PLASTICS & COMPOSITES	92.7	69.5	80.1	108.4	83.5	94.8	87.7	73.9	80.2	108.7	73.9	89.8	100.7	80.3	89.6	100.7	82.3	90.7
07	THERMAL & MOISTURE PROTECTION	106.7	83.2	96.6	105.3	93.7	100.3	101.9	93.5	98.3	106.4	89.0	98.9	104.0	90.9	98.4	101.9	82.0	93.3
08	OPENINGS	103.2	79.3	97.6	109.5	85.1	103.8	99.1	76.8	93.9	103.1	71.3	95.8	90.3	80.3	88.0	98.4	78.1	93.7
0920	Plaster & Gypsum Board	106.4	69.0	81.7	101.8	82.9	89.3	96.9	73.0	81.2	102.7	73.0	83.2	100.3	80.1	87.0	99.9	82.2	88.2
0950, 0980	Ceilings & Acoustic Treatment	101.8	69.0	79.6	94.9	82.9	86.8	110.6	73.0	85.1	94.9	73.0	80.1	103.2	80.1	87.6	97.0	82.2	87.0
0960	Flooring	91.6	51.4	79.9	98.1	86.1	94.6	121.9	77.8	109.1	98.3	76.6	92.0	99.5	77.8	93.2	100.8	81.2	95.1
0970, 0990	Wall Finishes & Painting/Coating	88.1	57.8	70.2	95.1	93.3	94.0	103.6	91.5	96.4	95.1	70.9	80.8	100.8	91.5	95.3	100.4	73.2	84.3
09	FINISHES	91.8	68.0	78.8	99.0	86.3	92.1	101.9	79.4	89.7	99.2	74.6	85.7	99.3	82.7	90.3	98.5	81.2	89.1
COVERS	DIVS. 10 - 14, 25, 28, 41, 43, 44, 46	100.0	89.0	97.6	100.0	92.2	98.3	100.0	89.2	97.6	100.0	89.8	97.7	100.0	88.7	97.5	100.0	89.8	97.8
21, 22, 23	FIRE SUPPRESSION, PLUMBING & HVAC	97.7	74.8	88.5	100.0	89.8	95.9	97.6	82.6	91.6	100.0	79.6	91.8	98.2	94.5	96.7	100.0	77.2	90.9
26, 27, 3370	ELECTRICAL, COMMUNICATIONS & UTIL.	97.9	70.4	83.8	94.9	83.9	89.3	97.7	91.9	94.7	94.2	87.1	90.6	97.2	91.9	94.5	96.1	72.9	84.3
MF2018	WEIGHTED AVERAGE	98.6	79.0	90.2	100.3	88.3	95.1	98.6	86.4	93.4	99.7	81.4	91.8	95.7	90.8	93.6	97.0	80.7	90.0

DIVISION		OHIO																	
		CLEVELAND			COLUMBUS			DAYTON			HAMILTON			LIMA			LORAIN		
		441			430 - 432			453 - 454			450			458			440		
		MAT.	INST.	TOTAL	MAT.	INST.	TOTAL	MAT.	INST.	TOTAL	MAT.	INST.	TOTAL	MAT.	INST.	TOTAL	MAT.	INST.	TOTAL
015433	CONTRACTOR EQUIPMENT		92.5	92.5		92.0	92.0		91.6	91.6		98.4	98.4		90.9	90.9		91.4	91.4
0241, 31 - 34	SITE & INFRASTRUCTURE, DEMOLITION	98.4	97.4	97.7	101.3	92.5	95.2	98.0	99.3	98.9	97.9	99.5	99.1	109.5	89.5	95.5	99.1	99.2	99.2
0310	Concrete Forming & Accessories	100.2	89.2	90.9	98.7	77.9	81.0	97.5	79.4	82.2	97.6	79.6	82.3	92.8	78.5	80.7	100.8	76.6	80.2
0320	Concrete Reinforcing	95.9	91.8	93.8	99.2	81.0	90.0	93.3	81.4	87.3	93.3	79.5	86.3	90.7	81.6	86.1	95.4	94.0	94.7
0330	Cast-in-Place Concrete	103.7	98.7	101.8	100.6	82.4	93.8	86.7	83.0	85.3	92.9	83.4	89.4	102.4	92.2	98.6	101.8	94.2	99.0
03	CONCRETE	102.3	92.4	97.9	96.3	80.0	89.0	86.3	80.9	83.9	89.3	81.4	85.7	93.5	83.8	89.2	99.7	85.4	93.3
04	MASONRY	100.7	97.9	99.0	84.3	86.9	85.9	77.6	78.0	77.9	78.0	82.9	81.0	97.4	79.8	86.5	90.8	95.0	93.4
05	METALS	99.4	83.7	94.6	96.8	80.8	92.0	95.1	77.6	89.7	95.1	86.5	92.5	101.6	81.0	95.4	98.5	81.8	93.4
06	WOOD, PLASTICS & COMPOSITES	102.3	86.7	93.8	97.8	76.7	86.3	104.6	79.4	90.9	103.4	79.4	90.3	87.6	77.3	82.0	108.4	72.1	88.7
07	THERMAL & MOISTURE PROTECTION	102.4	97.8	100.4	96.7	84.9	91.6	108.3	81.7	96.8	104.1	82.2	94.7	101.5	86.0	94.8	106.3	93.6	100.9
08	OPENINGS	97.7	86.3	95.0	99.3	74.4	93.5	97.5	76.8	92.7	95.2	77.0	91.0	99.1	75.9	93.7	103.1	78.9	97.5
0920	Plaster & Gypsum Board	101.4	86.1	91.3	97.3	76.1	83.4	102.0	79.2	87.0	102.0	79.2	87.0	96.9	76.5	83.5	101.8	71.2	81.7
0950, 0980	Ceilings & Acoustic Treatment	90.4	86.1	87.5	91.4	76.1	81.1	105.0	79.2	87.5	104.1	79.2	87.2	109.7	76.5	87.2	94.9	71.2	78.9
0960	Flooring	99.5	90.5	96.9	94.2	77.8	89.5	103.2	74.8	95.0	100.6	81.2	95.0	120.8	79.8	108.9	98.3	92.8	96.7
0970, 0990	Wall Finishes & Painting/Coating	97.7	93.3	95.1	95.8	81.7	87.5	100.8	72.6	84.1	100.8	73.2	84.5	103.7	78.5	88.8	95.1	93.3	94.0
09	FINISHES	93.9	89.6	94.0	94.7	77.9	85.5	100.7	77.6	88.1	99.7	79.2	88.5	101.0	78.2	88.5	99.0	80.3	88.8
COVERS	DIVS. 10 - 14, 25, 28, 41, 43, 44, 46	100.0	98.0	99.6	100.0	89.5	97.7	100.0	86.9	97.1	100.0	87.1	97.2	100.0	86.2	97.0	100.0	92.8	98.4
21, 22, 23	FIRE SUPPRESSION, PLUMBING & HVAC	100.0	92.1	96.8	100.0	85.9	94.4	100.8	83.0	93.7	100.6	76.6	91.0	97.6	93.8	96.1	100.0	90.4	96.2
26, 27, 3370	ELECTRICAL, COMMUNICATIONS & UTIL.	94.6	93.0	93.8	99.0	82.2	90.4	94.8	77.5	86.0	95.1	77.3	86.0	98.0	77.5	87.5	94.4	80.1	87.1
MF2018	WEIGHTED AVERAGE	99.4	92.2	96.3	97.6	83.2	91.4	96.1	81.3	89.7	96.1	81.5	89.8	98.7	83.8	92.3	99.2	86.8	93.9

DIVISION		OHIO																	
		MANSFIELD			MARION			SPRINGFIELD			STEUBENVILLE			TOLEDO			YOUNGSTOWN		
		448 - 449			433			455			439			434 - 436			444 - 445		
		MAT.	INST.	TOTAL	MAT.	INST.	TOTAL	MAT.	INST.	TOTAL	MAT.	INST.	TOTAL	MAT.	INST.	TOTAL	MAT.	INST.	TOTAL
015433	CONTRACTOR EQUIPMENT		91.4	91.4		91.8	91.8		91.6	91.6		95.6	95.6		95.6	95.6		91.4	91.4
0241, 31 - 34	SITE & INFRASTRUCTURE, DEMOLITION	95.4	99.4	97.9	95.5	95.8	95.7	98.3	99.2	99.0	140.6	103.8	115.0	99.3	95.9	96.9	99.7	98.9	99.2
0310	Concrete Forming & Accessories	89.7	75.6	77.7	96.6	79.4	82.0	97.5	79.3	82.0	97.6	81.2	83.7	100.4	86.7	88.8	100.8	78.8	82.1
0320	Concrete Reinforcing	86.9	81.3	84.1	91.4	81.3	86.3	93.3	81.4	87.3	89.2	97.6	93.4	99.2	85.1	92.0	95.4	87.7	91.5
0330	Cast-in-Place Concrete	99.0	90.4	95.8	87.2	90.9	88.6	89.0	82.8	86.7	94.7	92.2	93.7	95.5	92.3	94.3	105.9	89.9	100.0
03	CONCRETE	94.2	81.5	88.5	83.6	83.6	83.6	87.5	80.7	84.4	87.6	87.3	87.5	91.2	88.4	89.9	101.7	83.8	93.6
04	MASONRY	93.3	90.8	91.7	86.5	91.8	89.7	77.8	77.5	77.6	76.8	91.4	85.8	91.7	90.2	90.8	94.5	88.5	90.8
05	METALS	98.7	76.7	92.0	95.8	79.8	91.0	95.0	77.6	89.7	92.3	82.7	89.4	96.6	86.2	93.5	97.9	78.6	92.1
06	WOOD, PLASTICS & COMPOSITES	94.8	72.1	82.5	95.0	76.8	85.1	106.0	79.4	91.5	90.2	78.6	83.9	99.3	86.4	92.3	108.4	76.8	91.2
07	THERMAL & MOISTURE PROTECTION	104.4	90.9	98.6	93.4	91.0	92.4	108.1	81.5	96.7	105.0	89.5	98.3	93.6	92.2	93.0	106.6	90.7	99.8
08	OPENINGS	104.2	73.1	97.0	93.1	75.6	89.0	95.6	76.8	91.3	93.3	80.7	90.4	95.7	83.7	92.9	103.1	77.4	97.2
0920	Plaster & Gypsum Board	94.6	71.2	79.2	98.5	76.1	83.8	102.0	79.2	87.0	96.7	77.6	84.1	100.3	86.0	90.9	101.8	76.1	84.8
0950, 0980	Ceilings & Acoustic Treatment	95.7	71.2	79.1	97.6	76.1	83.1	105.0	79.2	87.5	94.7	77.6	83.1	97.6	86.0	89.8	94.9	76.1	82.1
0960	Flooring	93.3	95.8	94.0	93.6	95.8	94.3	103.2	74.8	95.0	121.5	93.3	113.3	94.0	93.8	93.9	98.3	93.0	96.8
0970, 0990	Wall Finishes & Painting/Coating	95.1	79.9	86.1	100.2	79.9	88.2	100.8	72.6	84.1	113.6	89.4	99.3	100.2	89.3	93.8	95.1	82.1	87.4
09	FINISHES	96.4	78.8	86.8	96.1	81.8	88.3	100.7	75.5	88.0	113.9	83.5	97.3	96.8	88.0	92.0	99.1	81.1	89.3
COVERS	DIVS. 10 - 14, 25, 28, 41, 43, 44, 46	100.0	90.2	97.8	100.0	87.4	97.2	100.0	86.7	97.1	100.0	88.0	97.4	100.0	92.9	98.4	100.0	90.4	97.9
21, 22, 23	FIRE SUPPRESSION, PLUMBING & HVAC	97.6	89.4	94.3	97.6	95.2	96.6	100.8	82.8	93.6	98.0	93.4	96.1	100.1	95.1	98.1	100.0	85.8	94.3
26, 27, 3370	ELECTRICAL, COMMUNICATIONS & UTIL.	92.2	93.0	92.6	93.1	93.0	93.1	94.8	82.2	88.3	88.0	109.1	98.8	99.2	106.0	102.7	94.4	76.2	85.0
MF2018	WEIGHTED AVERAGE	97.5	86.4	92.7	94.0	88.5	91.6	96.1	81.8	89.9	95.9	92.1	94.3	97.0	92.8	95.2	99.5	84.0	92.7

For customer support on your Electrical Costs with RSMeans Data, call 800.448.8182.

547

City Cost Indexes

Table 1

DIVISION		OHIO ZANESVILLE 437 - 438			OKLAHOMA ARDMORE 734			OKLAHOMA CLINTON 736			OKLAHOMA DURANT 747			OKLAHOMA ENID 737			OKLAHOMA GUYMON 739		
		MAT.	INST.	TOTAL	MAT.	INST.	TOTAL	MAT.	INST.	TOTAL	MAT.	INST.	TOTAL	MAT.	INST.	TOTAL	MAT.	INST.	TOTAL
015433	CONTRACTOR EQUIPMENT		91.8	91.8		82.7	82.7		81.8	81.8		81.8	81.8		81.8	81.8		81.8	81.8
0241, 31 - 34	SITE & INFRASTRUCTURE, DEMOLITION	98.6	95.5	96.4	97.9	96.0	96.6	99.2	94.5	95.9	93.8	93.0	93.2	100.9	94.5	96.4	103.5	95.0	97.6
0310	Concrete Forming & Accessories	93.6	78.0	80.4	85.9	56.6	61.0	84.6	56.6	60.8	82.4	56.4	60.3	88.0	56.8	61.5	90.9	56.7	61.8
0320	Concrete Reinforcing	90.9	95.2	93.1	87.7	67.2	77.3	88.2	67.2	77.5	93.2	65.5	79.1	87.6	67.3	77.3	88.2	65.1	76.4
0330	Cast-in-Place Concrete	91.9	88.8	90.7	87.0	72.4	81.6	84.1	72.4	79.7	90.2	72.3	83.5	84.1	72.4	79.8	84.1	72.1	79.6
03	CONCRETE	87.2	84.6	86.0	82.8	63.8	74.3	82.8	63.8	74.0	85.5	63.4	75.5	82.8	63.9	74.3	85.4	63.3	75.4
04	MASONRY	83.5	86.2	85.2	94.6	57.5	71.7	118.1	57.5	80.6	87.3	63.4	72.5	100.2	57.5	73.8	97.1	56.6	72.1
05	METALS	97.3	85.1	93.6	100.5	61.2	88.5	100.6	61.2	88.6	92.6	60.7	82.9	102.0	61.3	89.6	101.0	59.7	88.5
06	WOOD, PLASTICS & COMPOSITES	90.3	76.8	82.9	99.6	56.0	75.9	98.7	56.0	75.5	89.3	56.0	71.2	102.0	56.0	77.0	105.4	56.0	78.5
07	THERMAL & MOISTURE PROTECTION	93.5	85.8	90.2	101.7	64.5	85.7	101.9	64.5	85.8	97.9	66.1	84.2	102.0	64.5	85.8	102.3	64.0	85.8
08	OPENINGS	93.1	79.7	90.0	105.9	56.0	94.3	105.9	56.0	94.3	98.1	55.5	88.2	107.1	56.7	95.4	106.0	55.5	94.3
0920	Plaster & Gypsum Board	94.5	76.1	82.4	93.7	55.3	68.4	93.4	55.3	68.3	79.6	55.3	63.6	94.3	55.3	68.6	94.3	55.3	68.6
0950, 0980	Ceilings & Acoustic Treatment	97.6	76.1	83.1	83.2	55.3	64.3	83.2	55.3	64.3	79.4	55.3	63.1	83.2	55.3	64.3	83.2	55.3	64.3
0960	Flooring	91.8	77.8	87.8	87.3	59.2	79.1	86.2	59.2	78.3	91.4	53.4	80.4	87.8	73.1	83.6	89.3	59.2	80.6
0970, 0990	Wall Finishes & Painting/Coating	100.2	91.5	95.0	84.6	45.1	61.2	84.6	45.1	61.2	92.9	45.1	64.6	84.6	45.1	61.2	84.6	43.4	60.2
09	FINISHES	95.2	78.5	86.1	81.0	54.9	66.8	80.8	54.9	66.7	81.4	53.8	66.3	81.5	57.8	68.6	82.4	56.1	68.0
COVERS	DIVS. 10 - 14, 25, 28, 41, 43, 44, 46	100.0	84.5	96.6	100.0	79.0	95.4	100.0	79.0	95.4	100.0	79.0	95.4	100.0	79.0	95.4	100.0	79.0	95.4
21, 22, 23	FIRE SUPPRESSION, PLUMBING & HVAC	97.6	91.5	95.1	97.7	67.8	85.7	97.7	67.8	85.7	97.8	67.7	85.7	100.2	67.8	87.2	97.7	65.0	84.6
26, 27, 3370	ELECTRICAL, COMMUNICATIONS & UTIL.	93.3	91.9	92.6	94.0	70.9	82.2	95.1	70.9	82.7	96.3	70.9	83.3	95.1	70.9	82.7	96.7	67.2	81.6
MF2018	WEIGHTED AVERAGE	94.6	87.1	91.4	95.7	66.2	83.0	96.9	66.1	83.6	93.5	66.3	81.8	97.1	66.5	84.0	96.8	64.8	83.1

Table 2 — OKLAHOMA

DIVISION		LAWTON 735			MCALESTER 745			MIAMI 743			MUSKOGEE 744			OKLAHOMA CITY 730 - 731			PONCA CITY 746		
		MAT.	INST.	TOTAL	MAT.	INST.	TOTAL	MAT.	INST.	TOTAL	MAT.	INST.	TOTAL	MAT.	INST.	TOTAL	MAT.	INST.	TOTAL
015433	CONTRACTOR EQUIPMENT		82.7	82.7		81.8	81.8		93.2	93.2		93.2	93.2		83.0	83.0		81.8	81.8
0241, 31 - 34	SITE & INFRASTRUCTURE, DEMOLITION	97.1	96.0	96.4	87.2	95.1	92.7	88.4	92.4	91.2	88.8	92.2	91.2	95.3	96.5	96.2	94.4	95.4	95.1
0310	Concrete Forming & Accessories	90.8	56.8	61.9	80.5	42.5	48.2	93.4	56.6	62.2	97.8	56.3	62.6	91.0	63.5	67.6	89.0	56.8	61.6
0320	Concrete Reinforcing	87.8	67.3	77.4	92.9	65.1	78.7	91.4	67.3	79.1	92.3	64.7	78.2	97.1	67.3	81.9	92.3	67.2	79.5
0330	Cast-in-Place Concrete	81.3	72.4	78.0	79.1	72.0	76.4	82.9	73.4	79.3	83.8	73.2	79.9	81.3	74.3	78.7	92.6	72.4	85.1
03	CONCRETE	79.5	63.9	72.5	76.5	56.9	67.7	81.1	65.2	73.9	82.6	64.5	74.5	82.5	67.6	75.8	87.6	63.9	76.9
04	MASONRY	97.0	57.5	72.5	104.4	57.4	75.3	90.1	57.5	70.0	106.1	50.2	71.6	99.6	57.5	73.5	82.9	57.5	67.2
05	METALS	105.8	61.3	92.3	92.5	59.6	82.5	92.5	76.4	87.6	94.0	74.4	88.0	98.7	61.4	87.4	92.5	61.3	83.0
06	WOOD, PLASTICS & COMPOSITES	104.5	56.0	78.1	86.8	37.2	59.9	102.1	56.1	77.1	106.6	56.1	79.2	92.9	65.0	77.7	97.5	56.0	75.0
07	THERMAL & MOISTURE PROTECTION	101.7	64.5	85.7	97.4	62.4	82.4	97.9	65.4	83.9	98.2	62.8	83.0	93.4	65.8	81.5	98.1	64.7	83.7
08	OPENINGS	108.9	56.7	96.8	98.1	45.2	85.8	98.1	56.1	88.3	99.3	55.6	89.1	103.4	61.6	93.7	98.1	56.0	88.3
0920	Plaster & Gypsum Board	95.8	55.3	69.1	78.7	36.0	50.6	85.7	55.3	65.7	87.5	55.3	66.3	100.2	64.5	76.7	84.5	55.3	65.3
0950, 0980	Ceilings & Acoustic Treatment	90.0	55.3	66.5	79.4	36.0	50.0	79.4	55.3	63.1	87.9	55.3	65.8	91.5	64.5	73.2	79.4	55.3	63.1
0960	Flooring	89.6	73.1	84.8	90.3	59.2	81.3	97.4	53.4	84.6	99.9	34.3	80.8	91.1	73.1	85.9	94.5	59.2	84.2
0970, 0990	Wall Finishes & Painting/Coating	84.6	45.1	61.2	92.9	43.4	63.6	92.9	43.4	63.6	92.9	43.4	63.6	88.6	45.1	62.9	92.9	45.1	64.6
09	FINISHES	83.1	57.8	69.3	80.4	43.7	60.3	83.3	53.7	67.1	86.0	50.3	66.5	86.5	63.1	73.7	83.1	55.9	68.2
COVERS	DIVS. 10 - 14, 25, 28, 41, 43, 44, 46	100.0	79.0	95.4	100.0	77.0	94.9	100.0	77.9	95.1	100.0	77.9	95.1	100.0	80.0	95.6	100.0	79.0	95.4
21, 22, 23	FIRE SUPPRESSION, PLUMBING & HVAC	100.2	67.8	87.2	97.8	65.0	84.6	97.8	65.0	84.7	100.2	62.7	85.2	100.1	67.8	87.1	97.8	65.0	84.6
26, 27, 3370	ELECTRICAL, COMMUNICATIONS & UTIL.	96.7	70.9	83.5	94.8	68.6	81.4	96.2	68.7	82.1	94.4	68.8	81.3	102.2	70.9	86.2	94.4	67.2	80.5
MF2018	WEIGHTED AVERAGE	97.6	66.6	84.3	92.8	61.9	79.5	93.2	66.5	81.7	95.2	64.4	82.0	96.8	68.2	84.5	93.6	65.2	81.4

Table 3

DIVISION		OKLAHOMA POTEAU 749			OKLAHOMA SHAWNEE 748			OKLAHOMA TULSA 740 - 741			OKLAHOMA WOODWARD 738			OREGON BEND 977			OREGON EUGENE 974		
		MAT.	INST.	TOTAL	MAT.	INST.	TOTAL	MAT.	INST.	TOTAL	MAT.	INST.	TOTAL	MAT.	INST.	TOTAL	MAT.	INST.	TOTAL
015433	CONTRACTOR EQUIPMENT		91.9	91.9		81.8	81.8		93.2	93.2		81.8	81.8		99.8	99.8		99.8	99.8
0241, 31 - 34	SITE & INFRASTRUCTURE, DEMOLITION	75.7	87.9	84.2	97.4	95.4	96.0	95.0	91.4	92.5	99.6	94.5	96.0	114.1	103.2	106.5	103.7	103.2	103.4
0310	Concrete Forming & Accessories	86.9	56.4	61.0	82.2	56.7	60.6	98.0	57.5	63.6	84.7	44.6	50.6	106.6	96.8	98.3	103.0	96.8	97.7
0320	Concrete Reinforcing	93.3	67.2	80.1	92.3	67.2	79.5	92.5	67.3	79.7	87.6	67.2	77.3	100.4	111.7	106.1	104.8	111.7	108.3
0330	Cast-in-Place Concrete	82.9	73.3	79.3	95.6	72.4	87.0	91.3	75.3	85.4	84.1	72.3	79.7	115.1	99.9	109.4	111.4	99.9	107.1
03	CONCRETE	83.1	65.0	75.0	89.2	63.9	77.8	87.8	66.3	78.1	82.6	58.3	71.7	109.4	99.9	105.2	100.9	99.9	100.4
04	MASONRY	90.4	57.5	70.1	105.8	57.5	75.9	91.0	57.5	70.3	90.0	57.5	69.9	101.6	101.6	101.6	98.5	101.6	100.4
05	METALS	92.5	76.1	87.6	92.4	61.3	83.0	97.0	76.5	90.8	100.6	61.1	88.6	96.0	94.9	95.7	96.7	94.9	96.2
06	WOOD, PLASTICS & COMPOSITES	94.1	56.1	73.5	89.1	56.0	71.1	105.8	57.2	79.4	98.8	39.8	66.8	101.1	96.1	98.4	96.9	96.1	96.5
07	THERMAL & MOISTURE PROTECTION	98.0	65.4	84.0	98.1	64.0	83.4	98.3	65.6	84.2	101.9	62.9	85.1	110.2	98.7	105.3	109.3	101.7	106.0
08	OPENINGS	98.1	56.1	88.3	98.1	56.0	88.3	101.0	57.3	90.9	105.9	47.1	92.2	99.2	99.3	99.2	99.5	99.3	99.5
0920	Plaster & Gypsum Board	82.6	55.3	64.6	79.6	55.3	63.6	87.5	56.4	67.0	93.4	38.7	57.3	119.9	95.8	104.0	118.1	95.8	103.4
0950, 0980	Ceilings & Acoustic Treatment	79.4	55.3	63.1	79.4	55.3	63.1	87.9	56.4	66.6	83.2	38.7	53.0	95.9	95.8	95.9	96.9	95.8	96.2
0960	Flooring	93.9	53.4	82.1	91.4	59.2	82.2	98.7	63.7	88.5	86.2	56.5	77.5	97.8	104.2	99.7	96.3	104.2	98.6
0970, 0990	Wall Finishes & Painting/Coating	92.9	45.1	64.6	92.9	43.4	63.6	92.9	52.6	69.1	84.6	45.1	61.2	96.1	71.4	81.5	96.1	71.4	81.5
09	FINISHES	81.1	53.9	66.3	81.6	54.7	66.9	85.9	57.4	70.3	80.9	44.8	61.2	96.5	95.2	95.8	94.9	95.2	95.1
COVERS	DIVS. 10 - 14, 25, 28, 41, 43, 44, 46	100.0	79.3	95.4	100.0	79.0	95.4	100.0	78.0	95.2	100.0	77.3	95.0	100.0	98.3	99.6	100.0	98.3	99.6
21, 22, 23	FIRE SUPPRESSION, PLUMBING & HVAC	97.8	65.0	84.6	97.8	67.8	85.8	100.2	65.1	86.2	97.7	67.8	85.7	97.6	107.5	101.5	100.0	100.8	100.3
26, 27, 3370	ELECTRICAL, COMMUNICATIONS & UTIL.	94.5	68.7	81.3	96.4	70.9	83.4	96.3	68.7	82.2	96.5	70.9	83.4	100.6	93.7	97.1	99.3	93.7	96.4
MF2018	WEIGHTED AVERAGE	92.8	66.1	81.3	95.0	66.1	82.6	96.1	67.1	83.7	95.7	63.4	81.8	100.4	99.8	100.1	99.3	98.5	99.0

OREGON

DIVISION		KLAMATH FALLS 976			MEDFORD 975			PENDLETON 978			PORTLAND 970 - 972			SALEM 973			VALE 979		
		MAT.	INST.	TOTAL	MAT.	INST.	TOTAL	MAT.	INST.	TOTAL	MAT.	INST.	TOTAL	MAT.	INST.	TOTAL	MAT.	INST.	TOTAL
015433	CONTRACTOR EQUIPMENT		99.8	99.8		99.8	99.8		97.3	97.3		99.8	99.8		99.8	99.8		97.3	97.3
0241, 31 - 34	SITE & INFRASTRUCTURE, DEMOLITION	118.4	103.2	107.8	111.8	103.2	105.8	111.9	96.3	101.0	106.6	103.2	104.2	97.8	103.2	101.6	98.5	96.2	96.9
0310	Concrete Forming & Accessories	99.4	96.6	97.0	98.5	96.6	96.9	100.2	97.0	97.5	104.2	96.8	97.9	105.3	96.8	98.1	106.8	95.7	97.4
0320	Concrete Reinforcing	100.4	111.6	106.1	102.1	111.6	106.9	99.5	111.7	105.7	105.5	111.7	108.7	113.5	111.7	112.6	97.1	111.5	104.4
0330	Cast-in-Place Concrete	115.1	99.8	109.4	115.0	99.8	109.4	115.9	100.3	110.1	114.5	99.9	109.0	105.2	99.9	103.2	91.2	100.6	94.7
03	CONCRETE	112.1	99.8	106.6	106.9	99.8	103.7	94.0	100.2	96.8	102.5	99.9	101.4	98.3	99.9	99.0	78.2	99.6	87.9
04	MASONRY	114.5	101.6	106.5	95.4	101.6	99.2	105.6	101.7	103.2	100.0	101.6	101.0	103.4	101.6	102.3	103.4	101.7	102.3
05	METALS	96.0	94.6	95.6	96.3	94.6	95.8	103.1	95.5	100.8	97.8	94.9	97.0	103.8	94.9	101.1	103.0	94.2	100.3
06	WOOD, PLASTICS & COMPOSITES	91.9	96.1	94.2	90.8	96.1	93.7	94.0	96.2	95.2	98.1	96.1	97.0	93.1	96.1	94.7	102.6	96.2	99.1
07	THERMAL & MOISTURE PROTECTION	110.5	95.3	103.9	110.1	95.3	103.7	102.5	97.0	100.2	109.3	98.8	104.8	105.9	99.1	103.0	101.9	92.8	98.0
08	OPENINGS	99.2	99.3	99.3	102.0	99.3	101.4	95.5	99.4	96.4	97.4	99.3	97.9	101.2	99.3	100.7	95.4	89.2	94.0
0920	Plaster & Gypsum Board	114.8	95.8	102.3	114.2	95.8	102.1	100.3	95.8	97.4	118.0	95.8	103.4	113.0	95.8	101.7	106.4	95.8	99.4
0950, 0980	Ceilings & Acoustic Treatment	103.6	95.8	98.3	110.2	95.8	100.5	68.4	95.8	87.0	98.8	95.8	96.8	103.9	95.8	98.4	68.4	95.8	87.0
0960	Flooring	95.0	104.2	97.6	94.4	104.2	97.2	65.5	104.2	76.7	94.2	104.2	97.1	99.1	104.2	100.6	67.5	104.2	78.1
0970, 0990	Wall Finishes & Painting/Coating	96.1	67.0	78.9	96.1	67.0	78.9	87.6	76.2	80.9	95.9	76.2	84.3	93.4	75.1	82.6	87.6	71.4	78.0
09	FINISHES	96.9	94.7	95.7	97.1	94.7	95.8	68.8	95.8	83.5	94.7	95.7	95.2	95.1	95.6	95.4	69.3	95.3	83.5
COVERS	DIVS. 10 - 14, 25, 28, 41, 43, 44, 46	100.0	98.2	99.6	100.0	98.2	99.6	100.0	98.2	99.6	100.0	98.3	99.6	100.0	98.3	99.6	100.0	98.5	99.7
21, 22, 23	FIRE SUPPRESSION, PLUMBING & HVAC	97.6	107.5	101.5	100.0	107.5	103.0	99.3	114.5	105.4	100.0	104.7	101.9	100.0	107.5	103.0	99.3	72.6	88.6
26, 27, 3370	ELECTRICAL, COMMUNICATIONS & UTIL.	99.4	80.5	89.7	102.8	80.5	91.4	91.7	99.2	95.5	99.5	101.5	100.5	107.2	93.7	100.3	91.7	69.7	80.4
MF2018	WEIGHTED AVERAGE	101.3	97.8	99.8	100.8	97.8	99.5	96.4	101.7	98.7	99.6	100.4	100.0	101.1	99.9	100.6	94.1	87.9	91.4

PENNSYLVANIA

DIVISION		ALLENTOWN 181			ALTOONA 166			BEDFORD 155			BRADFORD 167			BUTLER 160			CHAMBERSBURG 172		
		MAT.	INST.	TOTAL	MAT.	INST.	TOTAL	MAT.	INST.	TOTAL	MAT.	INST.	TOTAL	MAT.	INST.	TOTAL	MAT.	INST.	TOTAL
015433	CONTRACTOR EQUIPMENT		115.2	115.2		115.2	115.2		113.2	113.2		115.2	115.2		115.2	115.2		114.4	114.4
0241, 31 - 34	SITE & INFRASTRUCTURE, DEMOLITION	93.0	101.8	99.1	95.9	101.6	99.9	100.4	99.3	99.6	92.0	100.5	97.9	87.2	102.4	97.8	88.0	96.0	96.2
0310	Concrete Forming & Accessories	98.3	111.2	109.3	83.2	84.9	84.6	81.9	83.5	83.2	85.4	99.2	97.1	84.7	96.7	94.9	84.5	78.9	79.8
0320	Concrete Reinforcing	103.8	116.6	110.3	100.6	108.8	104.8	99.5	108.8	104.2	102.8	108.8	105.9	101.3	122.8	112.2	103.4	115.5	109.5
0330	Cast-in-Place Concrete	87.6	102.8	93.2	97.6	89.3	94.5	94.0	88.3	91.9	93.2	93.3	93.3	86.2	98.9	90.9	95.1	96.5	95.6
03	CONCRETE	90.7	109.9	99.3	86.6	91.9	89.0	95.9	90.9	93.6	92.5	99.6	95.7	78.8	103.0	89.7	96.1	92.9	94.7
04	MASONRY	94.2	95.1	94.8	97.9	85.1	90.0	109.3	81.9	92.4	95.3	83.7	88.1	99.9	94.9	96.8	95.1	80.8	86.3
05	METALS	94.2	121.3	102.5	88.5	115.6	96.8	97.7	114.1	102.7	92.1	114.0	98.8	88.3	122.5	98.7	88.1	119.6	97.7
06	WOOD, PLASTICS & COMPOSITES	99.6	114.8	107.8	78.0	84.1	81.3	80.6	84.1	82.5	84.2	104.2	95.0	79.5	96.8	88.9	85.2	77.8	81.1
07	THERMAL & MOISTURE PROTECTION	102.7	110.8	106.2	101.5	92.5	97.6	101.5	89.8	96.5	102.5	91.0	97.6	101.0	97.1	99.3	93.2	85.4	89.8
08	OPENINGS	94.5	110.4	98.2	88.2	88.1	88.2	95.5	88.1	93.7	94.5	98.0	95.3	88.2	102.2	91.4	91.2	83.8	89.5
0920	Plaster & Gypsum Board	94.2	115.1	108.0	85.6	83.6	84.3	100.1	83.6	89.2	86.7	104.2	98.2	85.6	96.6	92.9	102.2	77.1	85.7
0950, 0980	Ceilings & Acoustic Treatment	88.8	115.1	106.6	91.4	83.6	86.1	101.3	83.6	89.3	91.5	104.2	100.1	92.3	96.6	95.2	92.1	77.1	81.9
0960	Flooring	90.7	98.3	92.9	84.2	102.3	89.4	96.7	107.2	99.7	85.0	107.2	91.5	85.1	108.9	92.0	88.1	80.6	85.9
0970, 0990	Wall Finishes & Painting/Coating	90.4	104.5	98.8	86.0	109.0	99.6	94.8	108.5	102.9	90.4	108.5	101.1	86.0	109.0	99.6	91.5	103.4	98.5
09	FINISHES	88.8	108.9	99.8	86.4	90.0	88.4	99.0	89.9	94.1	86.6	102.2	95.2	86.4	100.0	93.8	85.9	81.1	83.3
COVERS	DIVS. 10 - 14, 25, 28, 41, 43, 44, 46	100.0	101.6	100.4	100.0	96.4	99.2	100.0	95.1	98.9	100.0	98.1	99.6	100.0	99.3	99.8	100.0	94.6	98.8
21, 22, 23	FIRE SUPPRESSION, PLUMBING & HVAC	100.2	117.1	107.0	99.8	85.5	94.0	97.5	86.3	93.0	97.8	92.6	95.7	97.3	97.6	97.4	97.7	91.7	95.3
26, 27, 3370	ELECTRICAL, COMMUNICATIONS & UTIL.	100.3	98.1	99.2	90.7	110.3	100.8	98.7	110.3	104.7	94.3	110.3	102.5	91.3	110.4	101.0	91.7	87.9	89.8
MF2018	WEIGHTED AVERAGE	96.2	108.4	101.4	92.9	95.1	93.8	98.2	94.3	96.5	94.6	99.5	96.7	91.2	103.1	96.3	93.3	91.6	92.6

PENNSYLVANIA

DIVISION		DOYLESTOWN 189			DUBOIS 158			ERIE 164 - 165			GREENSBURG 156			HARRISBURG 170 - 171			HAZLETON 182		
		MAT.	INST.	TOTAL	MAT.	INST.	TOTAL	MAT.	INST.	TOTAL	MAT.	INST.	TOTAL	MAT.	INST.	TOTAL	MAT.	INST.	TOTAL
015433	CONTRACTOR EQUIPMENT		93.4	93.4		113.2	113.2		115.2	115.2		113.2	113.2		114.4	114.4		115.2	115.2
0241, 31 - 34	SITE & INFRASTRUCTURE, DEMOLITION	106.8	89.8	95.0	105.1	99.7	101.3	92.8	101.9	99.2	96.6	101.2	99.8	88.1	99.9	96.3	86.3	101.6	97.0
0310	Concrete Forming & Accessories	82.4	130.2	123.0	81.4	86.0	85.3	97.6	88.2	89.6	88.2	91.0	90.6	95.8	87.9	89.1	80.0	91.3	89.6
0320	Concrete Reinforcing	100.4	154.7	128.0	98.8	122.8	111.0	102.8	110.4	106.7	98.8	122.5	110.9	113.1	114.2	113.7	100.8	116.3	108.7
0330	Cast-in-Place Concrete	82.7	133.4	101.6	90.6	96.8	92.9	95.9	91.2	94.2	87.2	98.4	91.4	93.7	98.7	95.5	82.7	98.3	88.5
03	CONCRETE	86.3	134.6	108.0	98.0	97.3	97.7	85.7	94.4	89.6	91.4	100.2	95.3	91.7	97.5	94.3	83.5	99.3	90.6
04	MASONRY	97.6	128.4	116.6	109.8	95.4	100.9	86.9	88.7	88.0	120.0	90.9	102.1	89.9	85.6	87.3	107.2	91.4	97.4
05	METALS	91.9	123.9	101.6	97.7	119.6	104.3	88.7	116.3	97.1	97.6	120.5	104.6	94.2	119.3	101.9	94.0	120.5	102.0
06	WOOD, PLASTICS & COMPOSITES	79.8	129.4	106.7	79.5	84.1	82.0	96.0	86.8	91.0	86.8	89.6	88.3	98.0	87.6	92.4	78.3	89.5	84.4
07	THERMAL & MOISTURE PROTECTION	99.9	132.7	114.0	101.8	95.5	99.1	102.1	91.0	97.4	101.4	94.8	98.5	96.1	102.6	98.9	102.0	103.5	102.7
08	OPENINGS	96.9	137.0	106.2	95.5	91.2	94.5	88.3	90.5	88.9	95.4	98.3	96.1	98.0	88.9	95.9	95.0	96.0	95.3
0920	Plaster & Gypsum Board	84.7	130.2	114.7	98.9	83.6	88.8	94.2	86.4	89.1	101.1	89.3	93.4	108.1	87.2	94.3	85.1	89.2	87.8
0950, 0980	Ceilings & Acoustic Treatment	88.1	130.2	116.6	101.3	83.6	89.3	88.8	86.4	87.2	100.5	89.3	92.9	99.7	87.2	91.2	89.8	89.2	89.4
0960	Flooring	75.2	144.0	95.2	96.4	107.2	99.5	91.1	102.3	94.3	100.3	80.6	94.6	92.0	93.5	92.4	82.3	92.3	85.2
0970, 0990	Wall Finishes & Painting/Coating	90.0	147.5	124.0	94.8	109.0	103.2	96.5	94.8	95.5	94.8	109.0	103.2	96.6	87.0	90.9	90.4	106.8	100.1
09	FINISHES	80.5	134.3	109.9	99.2	91.4	95.0	89.8	91.0	90.5	99.7	90.6	94.8	91.1	88.2	89.5	84.9	92.8	89.2
COVERS	DIVS. 10 - 14, 25, 28, 41, 43, 44, 46	100.0	115.3	103.4	100.0	97.1	99.4	100.0	97.9	99.5	100.0	98.4	99.6	100.0	96.9	99.3	100.0	97.4	99.4
21, 22, 23	FIRE SUPPRESSION, PLUMBING & HVAC	97.3	135.6	112.6	97.5	89.5	94.3	99.8	95.7	98.1	97.5	89.3	94.2	100.1	93.7	97.5	97.8	97.7	97.8
26, 27, 3370	ELECTRICAL, COMMUNICATIONS & UTIL.	93.7	135.6	115.1	99.3	110.3	105.0	92.4	95.4	93.9	99.3	110.3	105.0	98.9	87.9	93.3	95.1	92.2	93.6
MF2018	WEIGHTED AVERAGE	93.8	129.2	109.0	98.7	98.3	98.5	92.8	96.2	94.3	98.2	98.7	98.4	96.1	94.9	95.6	94.1	98.3	95.9

For customer support on your Electrical Costs with RSMeans Data, call 800.448.8182.

549

PENNSYLVANIA

DIVISION		INDIANA 157			JOHNSTOWN 159			KITTANNING 162			LANCASTER 175 - 176			LEHIGH VALLEY 180			MONTROSE 188		
		MAT.	INST.	TOTAL	MAT.	INST.	TOTAL	MAT.	INST.	TOTAL	MAT.	INST.	TOTAL	MAT.	INST.	TOTAL	MAT.	INST.	TOTAL
015433	CONTRACTOR EQUIPMENT		113.2	113.2		113.2	113.2		115.2	115.2		114.4	114.4		115.2	115.2		115.2	115.2
0241, 31 - 34	SITE & INFRASTRUCTURE, DEMOLITION	94.7	100.3	98.6	100.8	100.5	100.6	89.8	102.0	98.3	80.0	100.1	94.0	90.3	102.0	98.5	89.0	101.7	97.9
0310	Concrete Forming & Accessories	82.4	95.5	93.5	81.4	84.8	84.3	84.7	92.2	91.0	86.5	88.8	88.4	91.8	111.6	108.6	80.9	91.8	90.2
0320	Concrete Reinforcing	98.1	122.9	110.7	99.5	123.3	111.1	101.3	122.6	112.2	103.0	114.2	108.7	100.8	111.1	106.0	105.6	119.6	112.7
0330	Cast-in-Place Concrete	85.5	97.1	89.8	94.8	89.1	92.7	89.6	97.4	92.5	80.8	99.5	87.8	89.6	102.0	94.2	87.8	95.7	90.7
03	CONCRETE	88.9	101.7	94.7	96.8	94.0	95.5	81.3	100.4	89.9	84.4	98.2	90.6	89.8	108.9	98.4	88.5	99.2	93.3
04	MASONRY	105.6	95.5	99.3	106.4	85.5	93.5	102.5	90.7	95.2	100.9	88.6	93.3	94.3	98.5	96.9	94.2	93.5	93.7
05	METALS	97.7	120.7	104.7	97.7	118.7	104.1	88.3	121.4	98.4	88.1	119.5	97.7	93.9	119.7	101.7	92.2	121.2	101.0
06	WOOD, PLASTICS & COMPOSITES	81.3	96.7	89.7	79.5	84.1	82.0	79.5	91.9	86.2	88.1	87.6	87.8	90.9	113.5	103.2	79.1	89.5	84.8
07	THERMAL & MOISTURE PROTECTION	101.2	96.9	99.4	101.5	91.0	97.0	101.2	95.0	98.5	92.6	103.8	97.4	102.4	112.8	106.9	102.0	93.5	98.4
08	OPENINGS	95.5	98.2	96.1	95.5	91.2	94.5	88.2	99.5	90.8	91.2	88.9	90.7	95.0	108.5	98.1	91.8	94.5	92.4
0920	Plaster & Gypsum Board	100.4	96.6	97.9	98.7	83.6	88.8	85.6	91.6	89.6	104.6	87.2	93.1	87.5	113.8	104.8	85.5	89.2	87.9
0950, 0980	Ceilings & Acoustic Treatment	101.3	96.6	98.1	100.5	83.6	89.1	92.3	91.6	91.8	92.1	87.2	88.8	89.8	113.8	106.1	91.5	89.2	89.9
0960	Flooring	97.2	107.2	100.1	96.4	80.6	91.8	85.1	107.2	91.5	89.1	98.9	91.9	87.7	97.3	90.5	83.0	107.2	90.0
0970, 0990	Wall Finishes & Painting/Coating	94.8	109.0	103.2	94.8	109.0	103.2	86.0	109.0	99.6	91.5	88.9	90.0	90.4	103.0	97.8	90.4	106.8	100.1
09	FINISHES	98.8	98.9	98.9	98.6	86.3	91.9	86.5	96.1	91.7	86.0	89.9	88.1	87.0	108.3	98.6	85.6	95.4	91.0
COVERS	DIVS. 10 - 14, 25, 28, 41, 43, 44, 46	100.0	98.4	99.7	100.0	96.3	99.2	100.0	98.1	99.6	100.0	97.6	99.5	100.0	104.3	101.0	100.0	98.1	99.6
21, 22, 23	FIRE SUPPRESSION, PLUMBING & HVAC	97.5	87.3	93.4	97.5	80.7	90.8	97.3	92.7	95.5	97.7	94.8	96.5	97.8	118.6	106.2	97.8	96.9	97.4
26, 27, 3370	ELECTRICAL, COMMUNICATIONS & UTIL.	99.3	110.3	105.0	99.3	110.3	105.0	90.7	110.3	100.8	93.0	95.6	94.3	95.1	132.5	114.3	94.3	97.6	96.0
MF2018	WEIGHTED AVERAGE	97.1	100.1	98.4	98.2	94.2	96.5	91.7	100.4	95.4	92.1	96.8	94.1	94.7	113.6	102.8	93.6	99.2	96.0

PENNSYLVANIA

DIVISION		NEW CASTLE 161			NORRISTOWN 194			OIL CITY 163			PHILADELPHIA 190 - 191			PITTSBURGH 150 - 152			POTTSVILLE 179		
		MAT.	INST.	TOTAL	MAT.	INST.	TOTAL	MAT.	INST.	TOTAL	MAT.	INST.	TOTAL	MAT.	INST.	TOTAL	MAT.	INST.	TOTAL
015433	CONTRACTOR EQUIPMENT		115.2	115.2		99.0	99.0		115.2	115.2		99.3	99.3		102.0	102.0		114.4	114.4
0241, 31 - 34	SITE & INFRASTRUCTURE, DEMOLITION	87.6	102.3	97.8	95.7	99.3	98.2	86.1	100.8	96.4	98.6	100.0	99.6	100.5	99.9	100.1	82.9	100.3	95.0
0310	Concrete Forming & Accessories	84.7	96.2	94.4	82.6	128.7	121.7	84.7	95.3	93.7	100.4	140.0	134.0	98.3	97.2	97.3	78.3	90.1	88.3
0320	Concrete Reinforcing	100.1	103.1	101.6	101.9	154.7	128.7	101.3	97.7	99.5	105.2	126.5	116.0	100.0	123.1	111.8	102.2	117.4	109.9
0330	Cast-in-Place Concrete	87.0	96.9	90.7	85.0	131.0	102.1	84.5	97.1	89.2	93.3	135.3	108.9	93.5	99.2	95.6	86.1	99.1	91.0
03	CONCRETE	79.1	98.7	88.0	86.9	133.0	107.7	77.6	97.5	86.6	101.3	134.8	116.4	97.5	102.3	99.7	87.9	99.3	93.0
04	MASONRY	99.6	94.6	96.5	108.7	124.6	118.5	98.8	90.6	93.7	98.9	131.3	118.9	101.3	99.6	100.2	94.6	89.4	91.4
05	METALS	88.3	115.0	96.4	93.5	123.8	102.7	88.3	113.9	96.1	98.1	113.5	102.8	99.1	107.4	101.6	88.3	121.2	98.3
06	WOOD, PLASTICS & COMPOSITES	79.5	96.8	88.9	76.1	129.3	105.0	79.5	96.8	88.9	96.8	141.8	121.2	99.1	107.4	101.6	77.8	87.6	83.1
07	THERMAL & MOISTURE PROTECTION	101.1	94.6	98.3	106.5	131.1	117.1	100.9	93.3	97.6	99.7	136.4	115.5	101.8	98.4	100.4	102.2	94.5	99.0
08	OPENINGS	88.2	94.1	89.5	88.7	137.0	99.9	88.2	97.7	90.4	101.3	137.3	109.6	97.8	102.5	98.9	92.7	102.4	96.9
0920	Plaster & Gypsum Board	85.6	96.6	92.9	83.8	130.2	114.3	85.6	96.6	92.9	97.8	142.8	127.4	98.7	96.9	97.5	99.1	87.2	91.3
0950, 0980	Ceilings & Acoustic Treatment	92.3	96.6	95.2	91.5	130.2	117.7	92.3	96.6	95.2	101.6	142.8	129.5	95.1	96.9	96.3	92.1	87.2	88.8
0960	Flooring	85.1	108.9	92.0	83.2	144.0	100.8	85.1	107.2	91.5	97.5	145.7	111.5	104.9	107.5	105.7	92.1	87.2	88.8
0970, 0990	Wall Finishes & Painting/Coating	86.0	109.0	99.6	85.8	147.5	122.3	86.0	109.0	99.6	91.8	152.1	127.5	97.5	111.4	105.7	85.2	107.2	91.6
09	FINISHES	86.4	99.8	93.7	83.3	133.3	110.6	86.3	98.9	93.2	98.4	142.8	122.7	100.4	100.2	100.3	91.5	106.8	100.6
COVERS	DIVS. 10 - 14, 25, 28, 41, 43, 44, 46	100.0	99.3	99.8	100.0	111.3	102.5	100.0	98.6	99.7	100.0	120.0	104.4	100.4	100.2	100.3	84.3	93.9	89.6
21, 22, 23	FIRE SUPPRESSION, PLUMBING & HVAC	97.3	99.0	98.0	97.6	133.6	112.0	97.3	96.3	96.9	100.2	140.0	116.1	99.9	98.3	99.2	100.0	99.5	99.9
26, 27, 3370	ELECTRICAL, COMMUNICATIONS & UTIL.	91.3	100.6	96.1	94.1	145.5	120.4	93.0	110.3	101.9	101.2	160.5	131.5	101.7	110.8	106.4	97.7	97.9	97.8
MF2018	WEIGHTED AVERAGE	91.2	100.3	95.1	94.0	130.0	109.4	91.1	100.3	95.1	99.9	135.3	115.1	99.6	102.2	100.7	91.3	93.5	92.4

PENNSYLVANIA

DIVISION		READING 195 - 196			SCRANTON 184 - 185			STATE COLLEGE 168			STROUDSBURG 183			SUNBURY 178			UNIONTOWN 154		
		MAT.	INST.	TOTAL	MAT.	INST.	TOTAL	MAT.	INST.	TOTAL	MAT.	INST.	TOTAL	MAT.	INST.	TOTAL	MAT.	INST.	TOTAL
015433	CONTRACTOR EQUIPMENT		121.3	121.3		115.2	115.2		114.4	114.4		115.2	115.2		115.2	115.2		113.2	113.2
0241, 31 - 34	SITE & INFRASTRUCTURE, DEMOLITION	99.9	112.5	108.7	93.5	101.8	99.3	84.2	100.2	95.4	88.0	101.9	97.7	94.7	101.2	99.3	95.3	101.1	99.3
0310	Concrete Forming & Accessories	98.2	88.5	90.0	98.5	89.0	90.5	84.4	85.1	85.0	86.2	93.1	92.0	89.0	87.1	87.4	75.9	96.5	93.4
0320	Concrete Reinforcing	103.3	152.9	128.5	103.8	119.7	111.9	102.0	108.9	105.5	104.1	119.8	112.1	105.0	115.5	110.9	98.8	122.9	111.1
0330	Cast-in-Place Concrete	76.2	99.7	84.9	91.4	95.8	93.1	88.2	89.4	88.6	86.2	97.2	90.3	94.2	97.6	95.4	85.5	98.4	90.7
03	CONCRETE	85.9	104.7	94.4	92.5	98.0	95.0	92.7	92.1	92.4	87.3	100.3	93.1	91.6	97.1	94.1	88.6	102.7	94.9
04	MASONRY	97.7	93.4	95.0	94.6	95.8	95.4	100.2	85.7	91.2	92.4	98.3	96.1	94.9	82.7	87.4	122.0	97.4	106.8
05	METALS	93.8	133.7	105.9	96.2	121.3	103.8	92.0	115.8	99.2	94.0	121.7	102.4	88.1	120.2	97.8	97.4	121.0	104.6
06	WOOD, PLASTICS & COMPOSITES	93.7	84.8	88.8	99.6	85.7	92.1	86.1	84.1	85.0	85.1	89.5	87.5	86.0	87.6	86.9	73.7	96.7	86.2
07	THERMAL & MOISTURE PROTECTION	106.9	106.2	106.6	102.6	93.9	98.8	101.7	100.4	101.1	102.2	93.9	98.6	93.6	95.3	94.4	101.1	97.8	99.7
08	OPENINGS	93.2	102.3	95.3	94.5	92.4	94.0	91.6	88.1	90.8	95.1	97.3	95.6	91.3	92.5	91.6	95.4	102.2	97.0
0920	Plaster & Gypsum Board	93.2	84.3	87.4	95.9	85.3	88.9	87.4	83.6	84.9	86.3	89.2	88.2	98.4	87.2	91.0	96.9	96.6	96.7
0950, 0980	Ceilings & Acoustic Treatment	82.9	84.3	83.9	96.5	85.3	88.9	88.9	83.6	85.3	88.1	89.2	88.8	88.7	87.2	87.7	96.9	96.6	96.7
0960	Flooring	87.7	101.2	91.6	90.7	96.0	92.2	88.1	101.2	91.9	85.5	97.2	88.9	86.0	107.2	92.1	93.5	107.2	97.9
0970, 0990	Wall Finishes & Painting/Coating	84.6	104.5	96.4	90.4	111.1	102.7	90.4	109.0	101.4	90.4	106.8	100.1	91.5	106.8	100.6	94.8	112.9	105.5
09	FINISHES	84.5	91.0	88.0	90.5	91.9	91.3	86.2	89.8	88.2	85.8	94.5	90.5	85.1	92.2	89.0	97.1	99.8	98.6
COVERS	DIVS. 10 - 14, 25, 28, 41, 43, 44, 46	100.0	97.6	99.5	100.0	96.5	99.2	100.0	96.5	99.2	100.0	99.2	99.8	100.0	94.9	98.9	100.0	99.1	99.8
21, 22, 23	FIRE SUPPRESSION, PLUMBING & HVAC	100.2	109.2	103.8	100.2	98.2	99.4	97.8	93.4	96.0	97.8	100.1	98.7	97.7	89.3	94.3	97.5	93.7	96.0
26, 27, 3370	ELECTRICAL, COMMUNICATIONS & UTIL.	100.2	93.6	96.8	100.3	97.6	98.9	93.5	110.3	102.1	95.1	145.2	120.8	91.7	94.1	92.9	96.3	110.3	103.5
MF2018	WEIGHTED AVERAGE	95.4	104.0	99.1	96.9	98.9	97.8	94.2	97.0	95.4	94.1	107.3	99.8	92.9	95.0	93.8	97.2	102.2	99.4

For customer support on your Electrical Costs with RSMeans Data, call 800.448.8182.

PENNSYLVANIA

DIVISION		WASHINGTON 153			WELLSBORO 169			WESTCHESTER 193			WILKES-BARRE 186 - 187			WILLIAMSPORT 177			YORK 173 - 174		
		MAT.	INST.	TOTAL	MAT.	INST.	TOTAL	MAT.	INST.	TOTAL	MAT.	INST.	TOTAL	MAT.	INST.	TOTAL	MAT.	INST.	TOTAL
015433	CONTRACTOR EQUIPMENT		113.2	113.2		115.2	115.2		99.0	99.0	86.0	101.8	97.0	85.7	101.4	96.6	83.8	100.1	95.1
0241, 31 - 34	SITE & INFRASTRUCTURE, DEMOLITION	95.3	101.4	99.6	95.5	101.2	99.4	101.8	100.3	100.8	89.1	87.6	87.9	85.7	88.0	87.6	81.6	88.7	87.6
0310	Concrete Forming & Accessories	82.6	96.6	94.5	84.9	86.5	86.3	89.2	130.0	123.9	102.8	119.7	111.4	104.2	114.1	109.3	105.0	114.2	109.7
0320	Concrete Reinforcing	98.8	123.0	111.1	102.0	119.5	110.9	101.0	154.6	128.3	102.8	119.7	111.4	79.7	92.5	84.5	86.5	99.5	91.3
0330	Cast-in-Place Concrete	85.5	98.6	90.4	92.4	90.7	91.8	94.1	133.1	108.7	82.7	95.6	87.5	79.9	95.4	86.9	89.0	98.2	93.1
03	CONCRETE	89.0	102.8	95.2	94.9	95.0	94.9	94.7	134.4	112.5	84.4	97.3	90.2	79.9	95.4	86.9	95.9	88.6	91.3
04	MASONRY	105.3	97.5	100.5	101.4	83.1	90.1	103.2	128.4	118.8	107.5	95.5	100.1	86.9	88.0	87.6	89.8	119.4	98.8
05	METALS	97.4	121.4	104.7	92.0	120.5	100.7	93.5	123.9	102.7	92.2	121.2	101.0	88.1	118.9	97.5	81.3	87.6	84.7
06	WOOD, PLASTICS & COMPOSITES	81.4	96.7	89.7	83.6	86.8	85.4	82.7	129.3	108.0	87.6	84.1	85.7	82.5	87.6	85.3	81.3	87.6	84.7
07	THERMAL & MOISTURE PROTECTION	101.2	97.8	99.7	102.8	88.4	96.6	107.0	128.0	116.0	102.0	93.4	98.3	93.0	90.7	92.0	92.8	103.8	97.5
08	OPENINGS	95.4	102.2	96.9	94.5	92.6	94.0	88.7	137.0	99.9	91.8	91.5	91.7	91.3	92.2	91.5	91.2	88.9	90.7
0920	Plaster & Gypsum Board	100.2	96.6	97.8	86.1	86.4	86.3	84.4	130.2	114.5	87.0	83.6	84.7	99.1	87.2	91.3	99.5	87.2	91.4
0950, 0980	Ceilings & Acoustic Treatment	100.5	96.6	97.9	88.9	86.4	87.2	91.5	130.2	117.7	91.5	83.6	86.1	92.1	87.2	88.8	91.1	87.2	88.5
0960	Flooring	97.3	107.2	100.2	84.7	107.2	91.2	86.0	144.0	102.8	86.4	96.0	89.2	84.9	89.7	86.3	86.5	98.9	90.1
0970, 0990	Wall Finishes & Painting/Coating	94.8	112.9	105.5	90.4	106.8	100.1	85.8	147.5	122.3	90.4	104.5	98.8	91.5	104.5	99.2	91.5	87.0	88.8
09	FINISHES	98.7	99.8	99.3	86.3	92.2	89.5	84.6	134.3	111.7	86.6	90.1	88.5	84.8	89.2	87.2	84.6	89.6	87.3
COVERS	DIVS. 10 - 14, 25, 28, 41, 43, 44, 46	100.0	99.1	99.8	100.0	98.2	99.6	100.0	117.6	103.9	100.0	96.2	99.2	100.0	97.4	99.4	100.0	97.6	99.5
21, 22, 23	FIRE SUPPRESSION, PLUMBING & HVAC	97.5	95.3	96.6	97.8	89.3	94.4	97.6	135.5	112.8	97.8	98.8	98.2	97.7	94.3	96.3	100.1	94.8	98.0
26, 27, 3370	ELECTRICAL, COMMUNICATIONS & UTIL.	98.7	110.4	104.7	94.3	97.6	96.0	94.0	135.6	115.3	95.1	92.2	93.6	92.2	85.4	88.7	93.0	84.9	88.9
MF2018	WEIGHTED AVERAGE	96.9	102.7	99.4	95.2	95.2	95.2	95.0	129.9	110.0	93.8	97.9	95.6	90.8	94.6	92.4	93.2	95.3	94.1

PUERTO RICO / RHODE ISLAND / SOUTH CAROLINA

DIVISION		PUERTO RICO — SAN JUAN 009			RHODE ISLAND — NEWPORT 028			PROVIDENCE 029			AIKEN 298			BEAUFORT 299			CHARLESTON 294		
		MAT.	INST.	TOTAL	MAT.	INST.	TOTAL	MAT.	INST.	TOTAL	MAT.	INST.	TOTAL	MAT.	INST.	TOTAL	MAT.	INST.	TOTAL
015433	CONTRACTOR EQUIPMENT		86.5	86.5		100.1	100.1		100.1	100.1		105.8	105.8		105.8	105.8		105.8	105.8
0241, 31 - 34	SITE & INFRASTRUCTURE, DEMOLITION	129.2	89.2	101.4	86.7	102.2	97.5	86.8	102.2	97.5	126.7	89.0	100.4	121.9	88.4	98.5	107.3	88.7	94.3
0310	Concrete Forming & Accessories	91.5	21.3	31.9	99.1	121.2	117.9	99.5	121.2	117.9	91.3	63.7	67.9	90.5	39.1	46.8	89.7	63.7	67.6
0320	Concrete Reinforcing	198.3	18.3	106.8	104.9	118.9	112.0	105.8	118.9	112.5	92.7	62.4	77.3	91.8	26.5	58.6	91.6	67.7	79.4
0330	Cast-in-Place Concrete	94.9	32.2	71.6	68.9	120.7	88.1	88.1	120.7	100.2	91.9	66.1	82.3	91.8	65.7	82.1	107.8	66.1	92.3
03	CONCRETE	96.2	25.6	64.4	82.1	120.1	99.2	87.7	120.1	102.3	104.5	66.0	87.2	101.7	48.8	77.9	97.6	66.9	83.8
04	MASONRY	87.7	20.4	46.1	96.0	124.0	113.3	100.4	124.0	115.0	78.2	65.3	70.3	92.4	65.3	75.7	93.8	65.3	76.2
05	METALS	109.8	38.4	88.1	91.6	112.3	97.9	94.1	112.3	99.6	99.1	89.4	96.2	99.1	79.7	93.3	101.0	91.1	98.0
06	WOOD, PLASTICS & COMPOSITES	97.6	20.7	55.8	99.2	120.3	110.7	97.7	120.3	110.0	87.9	65.6	75.8	86.9	33.3	57.8	97.3	64.5	83.2
07	THERMAL & MOISTURE PROTECTION	129.8	25.4	84.9	101.8	119.5	109.4	102.9	119.5	110.0	98.6	64.9	84.1	95.8	56.9	80.5	99.4	63.8	91.1
08	OPENINGS	155.2	19.0	123.6	99.5	120.2	104.3	100.8	120.2	105.3	95.8	62.6	88.1	95.8	38.8	82.6	96.7	64.6	75.5
0920	Plaster & Gypsum Board	170.9	18.6	70.6	89.5	120.6	109.9	99.7	120.6	113.4	92.0	64.6	73.9	85.1	64.6	48.7	85.1	64.6	71.2
0950, 0980	Ceilings & Acoustic Treatment	254.2	18.6	94.5	93.6	120.6	111.9	98.6	120.6	113.5	82.6	64.6	70.4	85.1	31.4	48.7	99.3	79.8	93.7
0960	Flooring	197.7	20.8	146.4	87.9	130.5	100.2	91.6	130.5	102.9	98.2	92.8	96.6	99.7	78.4	93.5	97.0	67.8	79.7
0970, 0990	Wall Finishes & Painting/Coating	186.1	22.2	89.1	91.0	118.7	107.4	95.3	118.7	109.1	97.0	67.8	79.7	97.0	58.4	74.1	88.7	67.5	77.1
09	FINISHES	191.9	21.4	98.8	87.9	123.1	107.1	89.2	123.1	107.7	89.6	69.7	78.7	90.6	47.1	66.9	88.7	67.5	77.1
COVERS	DIVS. 10 - 14, 25, 28, 41, 43, 44, 46	100.0	19.5	82.3	100.0	106.1	101.3	100.0	106.1	101.3	100.0	70.6	93.5	100.0	81.4	95.9	100.0	70.6	93.5
21, 22, 23	FIRE SUPPRESSION, PLUMBING & HVAC	103.5	17.9	69.2	100.3	113.2	105.5	100.1	113.2	105.3	98.0	56.0	81.2	98.0	56.0	81.2	100.5	58.3	83.6
26, 27, 3370	ELECTRICAL, COMMUNICATIONS & UTIL.	122.6	16.9	68.6	99.5	97.9	98.7	98.1	97.9	98.0	95.1	63.4	78.9	98.5	31.6	64.3	97.0	58.3	77.2
MF2018	WEIGHTED AVERAGE	118.1	27.2	79.1	95.0	113.7	103.0	96.3	113.7	103.8	97.7	67.9	84.9	98.3	56.0	80.1	98.4	67.6	85.2

SOUTH CAROLINA / SOUTH DAKOTA

DIVISION		COLUMBIA 290 - 292			FLORENCE 295			GREENVILLE 296			ROCK HILL 297			SPARTANBURG 293			SOUTH DAKOTA — ABERDEEN 574		
		MAT.	INST.	TOTAL	MAT.	INST.	TOTAL	MAT.	INST.	TOTAL	MAT.	INST.	TOTAL	MAT.	INST.	TOTAL	MAT.	INST.	TOTAL
015433	CONTRACTOR EQUIPMENT		105.8	105.8		105.8	105.8		105.8	105.8		105.8	105.8		105.8	105.8		100.3	100.3
0241, 31 - 34	SITE & INFRASTRUCTURE, DEMOLITION	106.5	88.7	94.1	116.2	88.5	96.9	111.3	88.8	95.6	109.1	88.0	94.4	111.0	88.8	95.5	96.6	77.8	80.6
0310	Concrete Forming & Accessories	91.7	63.7	67.9	78.7	63.6	65.9	89.2	63.7	67.5	87.4	63.4	67.0	91.7	63.7	67.9	96.6	77.8	80.6
0320	Concrete Reinforcing	96.7	67.7	81.9	91.2	67.7	79.3	91.1	66.0	78.4	91.9	67.7	79.6	91.1	67.7	79.2	105.3	72.4	88.5
0330	Cast-in-Place Concrete	109.0	66.1	93.1	91.8	65.9	82.2	91.8	66.1	82.2	91.8	66.0	82.2	95.1	66.9	82.4	112.8	81.2	101.1
03	CONCRETE	95.8	66.9	82.8	96.3	66.7	83.0	94.9	66.6	82.2	92.9	66.7	81.1	95.1	66.9	82.4	101.9	78.6	91.4
04	MASONRY	87.6	65.3	73.8	78.3	65.3	70.3	76.0	65.3	69.4	99.5	65.3	78.4	78.3	65.3	70.3	114.9	74.9	90.2
05	METALS	98.1	91.1	96.0	99.9	90.5	97.1	99.9	91.1	97.3	99.1	90.8	96.6	99.9	91.1	97.3	97.6	82.4	93.0
06	WOOD, PLASTICS & COMPOSITES	90.0	65.6	76.7	73.6	65.6	69.3	85.3	65.6	74.6	97.4	60.2	81.4	97.6	64.5	83.4	103.9	79.4	93.4
07	THERMAL & MOISTURE PROTECTION	93.4	64.5	80.9	97.7	64.5	83.4	97.6	64.5	83.4	97.4	60.2	81.4	95.8	63.8	88.4	99.5	63.8	91.2
08	OPENINGS	98.1	63.8	90.1	95.9	63.8	88.4	95.8	63.8	88.4	95.8	63.8	88.4	95.8	63.8	88.4	108.3	76.1	87.1
0920	Plaster & Gypsum Board	97.3	64.6	75.7	86.4	64.6	72.0	91.1	64.6	73.6	90.8	64.6	73.5	93.9	64.6	74.6	108.3	76.1	87.1
0950, 0980	Ceilings & Acoustic Treatment	87.6	64.6	72.0	83.4	64.6	70.6	82.6	64.6	70.4	82.6	64.6	70.4	82.6	64.6	70.4	93.9	76.1	81.8
0960	Flooring	91.4	78.5	87.6	90.7	78.4	87.1	97.0	78.4	91.6	95.9	78.4	90.8	98.4	78.4	92.6	91.9	45.3	78.4
0970, 0990	Wall Finishes & Painting/Coating	94.9	67.8	78.9	97.0	67.8	79.7	97.0	67.8	79.7	97.0	67.8	79.7	97.0	67.8	79.7	87.8	34.3	56.1
09	FINISHES	86.7	67.2	76.1	85.6	67.2	75.6	87.5	67.2	76.4	86.9	67.2	76.1	88.3	67.2	76.8	89.1	67.9	77.5
COVERS	DIVS. 10 - 14, 25, 28, 41, 43, 44, 46	100.0	70.6	93.5	100.0	70.6	93.5	100.0	70.6	93.5	100.0	70.6	93.5	100.0	70.6	93.5	100.0	86.5	97.0
21, 22, 23	FIRE SUPPRESSION, PLUMBING & HVAC	100.0	57.7	83.0	100.5	57.7	83.3	100.0	57.7	83.3	98.0	56.1	81.2	100.5	57.7	83.3	100.1	55.4	82.2
26, 27, 3370	ELECTRICAL, COMMUNICATIONS & UTIL.	97.3	61.1	78.8	95.0	61.1	77.7	97.0	61.3	78.8	97.0	61.4	78.8	96.9	67.9	84.5	99.6	71.5	87.6
MF2018	WEIGHTED AVERAGE	97.0	67.9	84.5	96.7	67.8	84.3	96.7	67.9	84.3	96.8	67.3	84.1	96.9	67.9	84.5			

For customer support on your Electrical Costs with RSMeans Data, call 800.448.8182.

551

SOUTH DAKOTA

DIVISION		MITCHELL 573			MOBRIDGE 576			PIERRE 575			RAPID CITY 577			SIOUX FALLS 570 - 571			WATERTOWN 572		
		MAT.	INST.	TOTAL	MAT.	INST.	TOTAL	MAT.	INST.	TOTAL	MAT.	INST.	TOTAL	MAT.	INST.	TOTAL	MAT.	INST.	TOTAL
015433	CONTRACTOR EQUIPMENT		100.3	100.3		100.3	100.3		100.3	100.3		100.3	100.3		100.3	100.3		100.3	100.3
0241, 31 - 34	SITE & INFRASTRUCTURE, DEMOLITION	97.6	97.9	97.8	97.5	98.0	97.8	100.5	97.4	98.3	99.1	97.4	97.9	91.3	99.7	97.2	97.4	98.0	97.8
0310	Concrete Forming & Accessories	95.7	45.4	53.0	85.8	45.8	51.8	97.2	46.6	54.3	104.9	58.1	65.1	100.7	78.4	81.7	82.2	76.6	77.4
0320	Concrete Reinforcing	104.7	72.6	88.4	107.4	72.5	88.4	96.7	96.7	102.0	98.4	96.9	97.6	105.2	97.1	101.1	101.8	72.5	86.9
0330	Cast-in-Place Concrete	109.7	53.8	88.9	109.7	79.5	98.4	110.2	79.1	98.6	108.8	79.4	97.9	87.1	80.3	84.6	109.7	79.4	98.4
03	CONCRETE	99.5	54.5	79.3	99.2	63.6	83.2	98.3	67.8	84.6	98.9	73.1	87.3	87.5	82.8	85.4	98.3	77.5	88.9
04	MASONRY	102.5	77.1	86.8	111.2	74.4	88.4	113.0	74.1	89.0	110.9	75.6	89.1	93.2	77.1	83.2	138.5	75.3	99.4
05	METALS	96.6	82.2	92.3	96.7	82.7	92.4	100.3	88.9	96.8	99.3	89.8	96.4	99.0	92.0	96.9	96.6	82.6	92.4
06	WOOD, PLASTICS & COMPOSITES	96.5	34.6	62.9	84.5	34.4	57.3	102.2	34.9	65.6	102.1	49.3	73.4	98.8	76.4	86.6	80.4	76.4	78.2
07	THERMAL & MOISTURE PROTECTION	103.7	74.9	91.3	103.7	77.2	92.3	104.9	74.2	91.7	104.4	80.1	93.9	103.2	84.2	95.0	103.5	78.8	92.9
08	OPENINGS	98.6	40.8	85.2	101.2	40.1	87.0	100.7	58.1	90.8	103.4	66.1	94.7	102.3	81.0	97.3	98.6	63.3	90.4
0920	Plaster & Gypsum Board	106.8	33.2	58.3	100.6	32.9	56.0	102.8	33.4	57.1	107.5	48.3	68.5	98.5	76.1	83.7	98.5	76.1	83.7
0950, 0980	Ceilings & Acoustic Treatment	89.7	33.2	51.4	93.9	32.9	52.6	91.4	33.4	52.1	94.8	48.3	63.3	92.8	76.1	81.5	89.7	76.1	80.5
0960	Flooring	91.6	45.3	78.1	87.3	48.0	75.9	97.5	32.3	78.6	91.3	75.1	86.6	97.0	77.5	91.4	86.0	45.3	74.2
0970, 0990	Wall Finishes & Painting/Coating	87.8	37.5	58.0	87.8	38.5	58.6	95.5	96.2	95.9	87.8	96.2	92.8	99.9	96.2	97.7	87.8	34.3	56.1
09	FINISHES	87.8	42.3	63.0	86.6	42.8	62.7	92.3	46.2	67.2	89.0	63.4	75.0	93.6	79.8	86.1	85.0	66.6	74.9
COVERS	DIVS. 10 - 14, 25, 28, 41, 43, 44, 46	100.0	83.7	96.4	100.0	83.6	96.4	100.0	83.7	96.4	100.0	85.3	96.8	100.0	88.3	97.4	100.0	87.6	97.3
21, 22, 23	FIRE SUPPRESSION, PLUMBING & HVAC	97.6	50.1	78.6	97.6	70.6	86.8	100.0	78.4	91.3	100.1	78.5	91.4	100.0	71.6	88.6	97.6	53.2	79.8
26, 27, 3370	ELECTRICAL, COMMUNICATIONS & UTIL.	96.7	64.9	80.5	98.3	40.7	68.8	101.4	48.9	74.5	95.0	48.9	71.4	99.4	65.0	81.8	95.9	65.0	80.1
MF2018	WEIGHTED AVERAGE	97.5	62.4	82.5	98.1	64.4	83.7	100.2	69.4	87.0	99.4	73.3	88.2	97.5	79.2	89.6	98.6	70.8	86.7

TENNESSEE

DIVISION		CHATTANOOGA 373 - 374			COLUMBIA 384			COOKEVILLE 385			JACKSON 383			JOHNSON CITY 376			KNOXVILLE 377 - 379		
		MAT.	INST.	TOTAL	MAT.	INST.	TOTAL	MAT.	INST.	TOTAL	MAT.	INST.	TOTAL	MAT.	INST.	TOTAL	MAT.	INST.	TOTAL
015433	CONTRACTOR EQUIPMENT		107.6	107.6		102.1	102.1		102.1	102.1		108.3	108.3		101.4	101.4		101.4	101.4
0241, 31 - 34	SITE & INFRASTRUCTURE, DEMOLITION	104.2	98.6	100.3	89.9	88.8	89.1	95.6	85.6	88.6	99.1	98.7	98.8	110.4	85.2	92.9	90.8	88.0	88.9
0310	Concrete Forming & Accessories	96.5	58.9	64.5	80.4	64.0	66.5	80.5	32.5	39.7	87.2	41.9	48.7	82.3	59.0	62.5	94.6	63.6	68.1
0320	Concrete Reinforcing	95.2	67.0	80.9	88.7	66.8	77.6	88.7	66.4	77.4	88.7	67.3	77.8	95.9	63.6	79.5	95.2	63.6	79.1
0330	Cast-in-Place Concrete	98.5	63.3	85.4	91.6	62.0	80.6	103.9	40.5	80.3	101.4	67.8	88.9	79.2	58.7	71.6	92.5	64.8	82.2
03	CONCRETE	94.0	63.6	80.3	92.9	65.5	80.6	103.1	43.7	76.4	93.9	57.6	77.5	102.8	61.4	84.2	91.4	65.5	79.8
04	MASONRY	100.1	57.5	73.8	112.7	56.7	78.1	107.9	40.3	66.1	113.1	42.5	69.5	112.9	44.1	70.4	77.9	51.7	61.7
05	METALS	98.2	88.8	95.3	94.0	89.5	92.6	94.1	87.9	92.2	96.4	89.2	94.2	95.4	86.9	92.8	98.7	87.1	95.2
06	WOOD, PLASTICS & COMPOSITES	106.5	57.9	80.1	69.1	65.2	67.0	69.3	30.3	48.1	84.4	40.8	60.7	78.8	64.4	71.0	93.3	64.4	77.6
07	THERMAL & MOISTURE PROTECTION	96.1	62.1	81.5	94.2	62.9	80.7	94.7	50.4	75.6	96.6	55.3	78.8	92.3	54.6	76.1	90.0	61.2	77.6
08	OPENINGS	104.1	59.2	93.6	93.1	54.9	84.2	93.1	36.5	79.9	99.8	45.2	87.2	100.3	61.3	91.3	97.6	56.8	88.1
0920	Plaster & Gypsum Board	83.1	57.2	66.1	91.2	64.7	73.8	91.2	28.8	50.1	93.6	39.6	58.1	101.7	63.9	76.8	109.5	63.9	79.5
0950, 0980	Ceilings & Acoustic Treatment	107.2	57.2	73.3	80.2	64.7	69.7	80.2	28.8	45.4	89.6	39.6	55.8	102.6	63.9	76.4	103.4	63.9	76.7
0960	Flooring	98.0	57.1	86.1	82.0	55.8	74.4	82.0	51.7	73.2	81.3	56.1	74.0	93.3	45.4	79.4	98.3	51.3	84.7
0970, 0990	Wall Finishes & Painting/Coating	97.7	61.1	76.0	84.9	57.9	69.0	84.9	57.9	69.0	86.6	57.9	69.6	94.4	60.3	74.2	94.4	59.1	73.5
09	FINISHES	96.6	58.2	75.6	85.9	61.9	72.8	86.4	37.2	59.6	86.2	44.6	63.5	100.5	56.9	76.7	93.5	60.7	75.6
COVERS	DIVS. 10 - 14, 25, 28, 41, 43, 44, 46	100.0	69.2	93.2	100.0	68.7	93.1	100.0	36.6	86.0	100.0	62.5	91.7	100.0	77.3	95.0	100.0	74.2	95.3
21, 22, 23	FIRE SUPPRESSION, PLUMBING & HVAC	100.1	60.6	84.3	98.3	75.3	89.1	98.3	68.5	86.4	100.1	60.0	84.1	99.9	56.8	82.6	100.0	81.0	95.8
26, 27, 3370	ELECTRICAL, COMMUNICATIONS & UTIL.	100.7	84.9	92.6	93.3	51.8	72.1	94.8	62.8	78.4	99.7	52.8	75.7	91.4	41.8	66.0	99.9	61.8	84.7
MF2018	WEIGHTED AVERAGE	99.3	69.6	86.5	95.2	67.9	83.5	96.6	57.3	79.8	97.9	59.7	81.5	99.3	59.9	82.4	96.0	65.2	82.8

TENNESSEE / TEXAS

DIVISION		TENNESSEE McKENZIE 382			TENNESSEE MEMPHIS 375, 380 - 381			TENNESSEE NASHVILLE 370 - 372			TEXAS ABILENE 795 - 796			TEXAS AMARILLO 790 - 791			TEXAS AUSTIN 786 - 787		
		MAT.	INST.	TOTAL	MAT.	INST.	TOTAL	MAT.	INST.	TOTAL	MAT.	INST.	TOTAL	MAT.	INST.	TOTAL	MAT.	INST.	TOTAL
015433	CONTRACTOR EQUIPMENT		102.1	102.1		101.5	101.5		105.2	105.2		93.2	93.2		93.2	93.2		92.7	92.7
0241, 31 - 34	SITE & INFRASTRUCTURE, DEMOLITION	95.3	85.8	88.7	88.1	94.9	92.8	101.6	97.5	98.7	93.7	91.8	92.4	92.8	91.1	91.6	95.9	90.8	92.4
0310	Concrete Forming & Accessories	88.2	34.8	42.9	95.1	66.3	70.7	96.3	64.9	69.6	95.6	60.8	66.0	96.9	53.0	59.6	94.4	54.4	60.4
0320	Concrete Reinforcing	88.8	67.4	77.9	102.8	67.9	85.1	101.3	67.0	83.9	95.6	52.7	73.8	106.3	51.0	78.2	90.0	48.5	68.9
0330	Cast-in-Place Concrete	101.6	53.2	83.6	95.5	78.0	89.0	91.1	65.6	81.6	84.8	66.8	78.1	81.6	66.7	76.1	94.6	66.9	84.3
03	CONCRETE	101.7	49.3	78.1	96.5	71.8	85.4	94.1	66.8	81.8	82.8	62.3	73.6	84.6	58.4	72.8	89.5	58.7	75.6
04	MASONRY	111.8	46.1	71.2	95.5	57.5	72.0	87.7	57.0	68.8	94.3	61.9	74.3	93.1	61.4	73.5	92.9	61.9	73.8
05	METALS	94.1	88.5	92.4	99.9	84.1	95.1	99.4	85.8	95.3	108.2	70.7	96.8	102.3	69.8	92.4	100.1	68.8	90.6
06	WOOD, PLASTICS & COMPOSITES	78.2	32.6	53.4	97.0	67.1	80.7	97.6	65.6	80.2	102.6	62.5	80.8	100.8	52.3	74.5	93.2	54.0	71.9
07	THERMAL & MOISTURE PROTECTION	94.7	52.6	76.6	89.1	68.6	80.3	94.8	63.9	81.5	98.9	65.2	84.4	98.7	63.4	83.5	94.8	65.0	81.9
08	OPENINGS	93.1	38.2	80.3	100.5	63.0	91.8	98.9	64.4	90.9	102.9	58.2	92.5	103.7	52.2	91.8	104.0	50.9	91.7
0920	Plaster & Gypsum Board	94.5	31.2	52.8	94.1	66.4	75.8	93.8	64.7	74.7	88.6	61.9	71.0	95.9	51.4	66.6	90.4	50.9	65.8
0950, 0980	Ceilings & Acoustic Treatment	80.2	31.2	47.0	98.9	66.4	76.9	98.7	64.7	75.7	90.4	61.9	71.1	97.5	51.4	66.3	89.4	53.1	64.8
0960	Flooring	84.9	55.8	76.4	97.2	57.1	85.6	97.4	57.1	85.7	92.5	69.9	85.9	96.5	65.9	87.6	92.3	65.9	84.6
0970, 0990	Wall Finishes & Painting/Coating	84.9	45.3	61.5	94.0	60.9	74.4	99.5	69.1	81.5	94.4	53.0	69.9	89.4	53.0	67.9	94.7	45.4	65.5
09	FINISHES	87.6	38.6	60.8	98.1	63.6	79.3	98.3	63.5	79.3	82.7	61.8	71.3	89.7	55.0	70.8	87.8	55.1	70.0
COVERS	DIVS. 10 - 14, 25, 28, 41, 43, 44, 46	100.0	24.5	83.4	100.0	81.9	96.0	100.0	81.7	96.0	100.0	80.5	95.7	100.0	73.8	94.2	100.0	77.8	95.1
21, 22, 23	FIRE SUPPRESSION, PLUMBING & HVAC	98.3	59.9	82.9	100.0	71.9	88.7	100.0	75.5	90.2	100.3	53.3	81.5	100.0	52.2	80.8	100.0	60.0	84.0
26, 27, 3370	ELECTRICAL, COMMUNICATIONS & UTIL.	94.6	56.7	75.2	101.4	65.1	82.9	94.8	62.8	78.4	97.9	55.3	76.1	100.6	60.3	80.0	96.1	58.7	76.9
MF2018	WEIGHTED AVERAGE	96.7	56.0	79.3	98.8	71.1	86.9	97.7	71.1	86.3	97.6	62.8	82.7	97.6	61.1	81.9	97.1	62.7	82.3

TEXAS

DIVISION		BEAUMONT 776 - 777			BROWNWOOD 768			BRYAN 778			CHILDRESS 792			CORPUS CHRISTI 783 - 784			DALLAS 752 - 753		
		MAT.	INST.	TOTAL	MAT.	INST.	TOTAL	MAT.	INST.	TOTAL	MAT.	INST.	TOTAL	MAT.	INST.	TOTAL	MAT.	INST.	TOTAL
015433	CONTRACTOR EQUIPMENT		97.3	97.3		93.2	93.2		97.3	97.3		93.2	93.2		103.7	103.7		106.9	106.9
0241, 31 - 34	SITE & INFRASTRUCTURE, DEMOLITION	87.7	96.0	93.5	101.1	91.7	94.5	78.9	96.2	90.9	104.3	90.1	94.4	141.5	87.3	103.7	105.7	96.8	99.5
0310	Concrete Forming & Accessories	99.5	56.1	62.7	98.5	60.4	66.1	79.5	59.8	62.8	94.0	60.2	65.3	99.2	53.6	60.5	97.8	62.9	68.2
0320	Concrete Reinforcing	90.3	63.8	76.8	86.5	52.6	69.3	92.5	52.6	72.2	95.8	52.5	73.8	82.4	52.4	67.1	92.0	53.8	72.6
0330	Cast-in-Place Concrete	97.8	67.0	86.3	93.0	60.3	80.8	78.3	61.2	71.9	87.0	60.2	77.0	119.7	69.1	100.9	93.0	72.5	85.3
03	CONCRETE	95.0	62.2	80.2	89.3	59.8	76.0	78.9	60.0	70.4	90.6	59.7	76.7	99.6	60.8	82.1	93.8	66.2	81.4
04	MASONRY	94.3	63.0	75.0	124.4	61.9	85.8	123.8	61.9	85.6	98.3	61.4	75.5	84.5	62.0	70.6	98.2	60.7	75.0
05	METALS	96.9	76.5	90.7	97.2	70.2	89.0	96.8	72.3	89.4	105.7	70.1	94.9	95.7	85.3	92.5	101.8	84.0	96.4
06	WOOD, PLASTICS & COMPOSITES	111.1	55.6	81.0	109.6	62.5	84.0	76.8	61.0	68.2	101.9	62.5	80.5	118.0	53.0	82.7	98.8	64.7	80.2
07	THERMAL & MOISTURE PROTECTION	103.3	66.1	87.2	93.2	63.4	80.4	94.9	65.6	82.3	99.6	61.9	83.3	101.7	63.9	85.4	88.6	67.1	79.4
08	OPENINGS	97.1	56.6	87.7	102.4	58.2	92.1	97.9	57.6	88.5	98.2	58.2	88.9	103.7	51.5	91.6	97.5	60.4	88.9
0920	Plaster & Gypsum Board	104.4	54.8	71.7	86.6	61.9	70.3	91.4	60.4	71.0	88.2	61.9	70.9	99.4	51.9	68.1	94.2	63.7	74.1
0950, 0980	Ceilings & Acoustic Treatment	99.7	54.8	69.3	79.3	61.9	67.5	92.7	60.4	70.8	88.7	61.9	70.5	92.4	51.9	65.0	89.6	63.7	72.1
0960	Flooring	116.6	77.2	105.2	83.6	56.4	75.7	87.4	69.2	82.1	90.9	56.4	80.9	104.8	70.0	94.7	98.1	70.4	90.0
0970, 0990	Wall Finishes & Painting/Coating	92.7	55.3	70.6	91.7	53.0	68.8	90.0	58.9	71.6	94.4	53.0	69.9	112.2	45.4	72.7	101.4	55.2	74.0
09	FINISHES	94.3	59.4	75.3	78.1	59.1	67.7	82.4	61.1	70.7	83.1	59.1	70.0	97.7	55.4	74.6	95.8	63.5	78.2
COVERS	DIVS. 10 - 14, 25, 28, 41, 43, 44, 46	100.0	79.3	95.4	100.0	74.9	94.5	100.0	78.5	95.3	100.0	74.9	94.5	100.0	79.4	95.5	100.0	81.5	95.9
21, 22, 23	FIRE SUPPRESSION, PLUMBING & HVAC	100.1	64.4	85.8	97.6	50.6	78.8	97.7	63.8	84.1	97.8	52.9	79.8	100.1	49.8	80.0	100.0	60.7	84.2
26, 27, 3370	ELECTRICAL, COMMUNICATIONS & UTIL.	97.3	64.9	80.7	93.8	47.3	70.0	95.6	64.9	79.9	97.9	57.1	77.0	92.7	65.0	78.5	98.2	61.3	79.3
MF2018	WEIGHTED AVERAGE	97.5	67.1	84.4	96.6	60.2	81.0	94.6	66.4	82.5	97.5	61.8	82.2	99.2	63.0	83.7	98.5	67.7	85.3

TEXAS

DIVISION		DEL RIO 788			DENTON 762			EASTLAND 764			EL PASO 798 - 799, 885			FORT WORTH 760 - 761			GALVESTON 775		
		MAT.	INST.	TOTAL	MAT.	INST.	TOTAL	MAT.	INST.	TOTAL	MAT.	INST.	TOTAL	MAT.	INST.	TOTAL	MAT.	INST.	TOTAL
015433	CONTRACTOR EQUIPMENT		92.7	92.7		102.8	102.8		93.2	93.2		93.2	93.2		93.2	93.2		110.3	110.3
0241, 31 - 34	SITE & INFRASTRUCTURE, DEMOLITION	122.0	90.1	99.8	101.1	83.7	89.0	104.0	89.8	94.1	92.4	92.5	92.5	97.0	92.5	93.9	103.6	94.8	97.5
0310	Concrete Forming & Accessories	95.8	52.5	59.0	106.5	60.5	67.4	99.4	60.2	66.1	96.5	58.2	63.9	98.3	61.0	66.6	88.7	57.3	62.0
0320	Concrete Reinforcing	82.9	48.9	65.6	87.9	52.5	69.9	86.7	52.5	69.4	98.3	52.5	75.1	93.1	52.3	72.4	92.0	61.4	76.5
0330	Cast-in-Place Concrete	128.9	60.2	103.3	71.9	61.4	68.0	98.3	60.2	84.1	75.9	69.0	73.3	85.7	66.9	78.7	104.0	62.2	88.4
03	CONCRETE	120.1	55.5	91.1	70.3	61.3	66.2	93.7	59.7	78.4	80.8	61.8	72.3	85.2	62.3	74.9	96.5	61.8	80.9
04	MASONRY	99.5	61.8	76.2	133.1	60.6	88.3	94.2	61.9	74.2	85.5	64.0	72.2	93.8	60.5	73.2	89.5	62.0	72.5
05	METALS	95.6	68.0	87.2	96.8	85.5	93.4	97.0	70.0	88.8	103.7	70.5	93.6	98.8	70.7	90.2	98.5	91.8	96.5
06	WOOD, PLASTICS & COMPOSITES	98.2	51.9	73.0	122.3	62.7	89.9	117.0	62.5	87.4	90.5	58.6	73.2	101.2	62.5	80.2	93.5	57.5	73.9
07	THERMAL & MOISTURE PROTECTION	98.2	63.8	83.4	91.0	63.9	79.4	93.5	63.4	80.6	93.3	65.6	81.3	90.1	65.5	79.5	93.9	66.2	82.0
08	OPENINGS	98.9	49.2	87.3	120.9	57.9	106.2	71.7	58.2	68.6	97.8	54.3	87.7	102.0	58.1	91.8	101.9	57.7	91.6
0920	Plaster & Gypsum Board	96.0	50.9	66.3	91.3	61.9	71.9	86.6	61.9	70.3	97.2	57.8	71.2	93.6	61.9	72.7	98.4	56.6	70.9
0950, 0980	Ceilings & Acoustic Treatment	88.8	50.9	63.2	84.4	61.9	69.1	79.3	61.9	67.5	91.1	57.8	68.5	89.3	61.9	70.7	97.0	56.6	69.6
0960	Flooring	89.9	56.4	80.2	79.2	56.4	72.6	106.4	56.4	91.9	93.1	72.1	87.0	104.6	66.6	93.6	102.7	69.2	93.0
0970, 0990	Wall Finishes & Painting/Coating	99.5	43.7	66.5	101.9	50.2	71.3	93.1	53.0	69.4	99.7	50.9	70.8	97.2	53.2	71.1	101.6	55.1	74.1
09	FINISHES	90.3	51.8	69.3	77.0	58.9	67.1	85.1	59.1	70.9	88.4	59.7	72.7	90.7	61.2	74.6	91.9	58.6	73.7
COVERS	DIVS. 10 - 14, 25, 28, 41, 43, 44, 46	100.0	75.2	94.5	100.0	80.8	95.8	100.0	73.2	94.1	100.0	77.9	95.1	100.0	80.5	95.7	100.0	79.8	95.6
21, 22, 23	FIRE SUPPRESSION, PLUMBING & HVAC	97.6	60.7	82.8	97.6	57.1	81.4	97.6	50.6	78.8	99.9	65.9	86.3	100.9	56.6	82.6	97.6	63.9	84.1
26, 27, 3370	ELECTRICAL, COMMUNICATIONS & UTIL.	94.6	61.3	77.6	96.1	60.1	77.7	93.7	60.0	76.5	97.6	51.7	74.1	95.3	60.0	77.2	97.1	65.0	80.7
MF2018	WEIGHTED AVERAGE	100.2	62.0	83.8	96.7	64.4	82.9	93.2	61.8	79.7	95.7	64.7	82.4	96.3	64.1	82.5	97.4	68.1	84.8

TEXAS

DIVISION		GIDDINGS 789			GREENVILLE 754			HOUSTON 770 - 772			HUNTSVILLE 773			LAREDO 780			LONGVIEW 756		
		MAT.	INST.	TOTAL	MAT.	INST.	TOTAL	MAT.	INST.	TOTAL	MAT.	INST.	TOTAL	MAT.	INST.	TOTAL	MAT.	INST.	TOTAL
015433	CONTRACTOR EQUIPMENT		92.7	92.7		103.6	103.6		102.2	102.2		97.3	97.3		92.7	92.7		94.6	94.6
0241, 31 - 34	SITE & INFRASTRUCTURE, DEMOLITION	106.8	90.4	95.4	96.6	86.8	89.8	102.5	94.6	97.0	94.2	95.9	95.4	100.1	90.9	93.7	94.6	93.3	93.7
0310	Concrete Forming & Accessories	93.0	52.6	58.7	88.9	58.7	63.2	93.9	58.6	63.9	86.5	55.6	60.2	95.8	53.6	60.0	84.7	60.4	64.1
0320	Concrete Reinforcing	83.4	49.8	66.4	92.4	52.6	72.1	91.8	52.8	72.0	92.7	52.6	72.3	82.9	52.4	67.4	91.5	52.2	71.5
0330	Cast-in-Place Concrete	109.1	60.5	91.0	92.5	61.4	80.9	96.5	63.3	84.2	107.7	61.1	90.4	92.1	67.0	82.7	107.6	60.3	90.0
03	CONCRETE	96.4	55.9	78.2	87.5	60.4	75.3	95.3	60.6	79.7	103.0	58.0	82.8	90.4	59.0	76.3	103.4	59.7	83.7
04	MASONRY	106.7	61.9	79.1	154.1	60.5	96.3	88.7	62.9	72.8	124.6	61.9	85.9	92.5	61.9	73.6	150.1	60.5	94.7
05	METALS	95.1	68.9	87.1	99.3	84.1	94.7	101.2	77.8	94.1	96.8	72.1	89.3	98.1	70.1	89.6	92.1	69.1	85.1
06	WOOD, PLASTICS & COMPOSITES	97.4	51.9	72.7	87.7	60.2	72.7	98.9	59.0	77.2	86.1	55.6	69.5	98.2	52.9	73.6	81.6	62.6	71.3
07	THERMAL & MOISTURE PROTECTION	98.7	64.5	84.0	88.8	63.0	77.7	95.8	67.3	83.6	96.0	65.0	82.7	97.5	64.8	83.4	90.4	62.5	78.4
08	OPENINGS	98.0	50.1	86.9	94.1	56.9	85.4	105.1	56.5	93.8	97.8	54.0	87.7	99.0	51.2	87.9	85.1	57.8	78.7
0920	Plaster & Gypsum Board	95.1	50.9	66.0	84.9	59.2	68.0	97.7	58.0	71.5	96.0	54.8	68.9	96.7	51.9	67.2	84.0	61.9	69.4
0950, 0980	Ceilings & Acoustic Treatment	88.8	50.9	63.2	85.0	59.2	67.5	93.9	58.0	69.6	92.7	54.8	67.0	92.2	51.9	64.9	80.8	61.9	68.0
0960	Flooring	89.5	56.4	79.9	91.8	56.4	81.5	103.8	72.1	94.6	91.7	56.4	81.5	89.8	65.9	82.8	96.9	56.4	85.1
0970, 0990	Wall Finishes & Painting/Coating	99.5	45.4	67.5	95.8	53.0	70.5	99.4	58.9	75.5	90.0	58.9	71.6	99.5	45.4	67.5	86.0	50.2	64.8
09	FINISHES	89.0	52.0	68.8	90.3	57.7	72.5	97.9	60.9	77.7	85.1	55.6	69.0	89.4	54.4	70.3	95.5	58.8	75.5
COVERS	DIVS. 10 - 14, 25, 28, 41, 43, 44, 46	100.0	71.8	93.8	100.0	78.1	95.2	100.0	81.9	96.0	100.0	71.2	93.7	100.0	77.4	95.0	100.0	80.7	95.7
21, 22, 23	FIRE SUPPRESSION, PLUMBING & HVAC	97.7	62.8	83.7	97.7	59.3	82.3	100.1	65.5	86.2	97.7	63.8	84.1	100.1	60.2	84.1	97.6	58.8	82.1
26, 27, 3370	ELECTRICAL, COMMUNICATIONS & UTIL.	91.9	57.0	74.0	95.0	60.1	77.1	99.1	68.1	83.2	95.6	60.6	77.7	94.7	58.5	76.2	95.3	52.1	73.2
MF2018	WEIGHTED AVERAGE	96.7	61.9	81.8	98.0	64.5	83.6	99.3	67.9	85.8	98.3	64.3	83.7	96.6	62.7	82.1	98.0	62.6	82.8

For customer support on your Electrical Costs with RSMeans Data, call 800.448.8182.

553

City Cost Indexes

TEXAS

DIVISION		LUBBOCK 793-794 MAT.	INST.	TOTAL	LUFKIN 759 MAT.	INST.	TOTAL	MCALLEN 785 MAT.	INST.	TOTAL	MCKINNEY 750 MAT.	INST.	TOTAL	MIDLAND 797 MAT.	INST.	TOTAL	ODESSA 797 MAT.	INST.	TOTAL
015433	CONTRACTOR EQUIPMENT		105.6	105.6		94.6	94.6		103.8	103.8		103.6	103.6		105.6	105.6		93.2	93.2
0241, 31 - 34	SITE & INFRASTRUCTURE, DEMOLITION	118.2	91.0	99.3	89.6	94.8	93.2	146.3	87.3	105.2	93.1	86.8	88.7	121.2	90.0	99.5	94.0	92.5	92.9
0310	Concrete Forming & Accessories	94.8	54.5	60.5	87.9	55.7	60.5	99.7	52.6	59.7	88.0	58.7	63.2	98.6	60.7	66.4	95.6	60.8	66.0
0320	Concrete Reinforcing	96.9	52.8	74.4	93.1	68.7	80.7	82.5	52.3	67.2	92.4	52.6	72.2	97.9	52.7	74.9	95.6	52.6	73.7
0330	Cast-in-Place Concrete	85.0	70.2	79.5	96.2	60.1	82.8	130.0	61.6	104.6	86.7	61.4	77.3	90.3	67.8	81.9	84.8	66.8	78.1
03	CONCRETE	81.6	61.6	72.6	95.6	60.2	79.7	107.6	57.7	85.1	82.8	60.4	72.7	85.6	63.6	75.7	82.8	62.3	73.6
04	MASONRY	93.7	62.3	74.3	115.0	61.8	82.1	99.2	62.0	76.2	165.6	60.5	100.7	110.1	61.5	80.1	94.3	61.4	74.0
05	METALS	112.1	86.3	104.3	99.4	74.0	91.7	95.5	85.1	92.3	99.2	84.2	94.6	110.4	85.8	102.9	107.5	70.6	96.3
06	WOOD, PLASTICS & COMPOSITES	102.2	53.6	75.8	89.0	55.7	70.9	116.6	52.0	81.5	86.6	60.2	72.2	106.9	62.7	82.9	102.6	62.5	80.8
07	THERMAL & MOISTURE PROTECTION	88.1	65.3	78.3	90.1	61.6	77.8	101.4	64.7	85.6	88.5	63.0	77.6	88.4	65.1	78.4	98.9	64.2	84.0
08	OPENINGS	110.9	53.8	97.6	64.5	58.2	63.1	102.7	50.0	90.5	94.1	56.9	85.4	112.4	58.3	99.8	102.9	58.2	92.5
0920	Plaster & Gypsum Board	88.8	52.6	64.9	83.1	54.8	64.5	100.4	50.9	67.8	84.9	59.2	68.0	90.6	61.9	71.7	88.6	61.9	71.0
0950, 0980	Ceilings & Acoustic Treatment	91.3	52.6	65.0	76.5	54.8	61.8	92.2	50.9	64.3	85.0	59.2	67.5	89.6	61.9	70.8	90.4	61.9	71.1
0960	Flooring	86.4	71.3	82.0	129.5	56.4	108.3	104.4	75.2	95.9	91.4	56.4	81.2	87.6	65.9	81.3	92.5	65.9	84.8
0970, 0990	Wall Finishes & Painting/Coating	104.9	55.2	75.5	86.0	53.0	66.5	112.2	43.7	71.7	95.8	53.0	70.5	104.9	53.0	74.2	94.4	53.0	69.9
09	FINISHES	85.1	57.1	69.8	103.7	55.5	77.4	98.1	55.7	74.9	89.9	57.7	72.3	85.8	61.1	72.3	82.7	61.0	70.9
COVERS	DIVS. 10 - 14, 25, 28, 41, 43, 44, 46	100.0	79.9	95.6	100.0	71.5	93.7	100.0	77.5	95.0	100.0	80.8	95.8	100.0	75.3	94.6	100.0	75.0	94.5
21, 22, 23	FIRE SUPPRESSION, PLUMBING & HVAC	99.7	54.3	81.5	97.6	59.4	82.3	97.7	49.8	78.5	97.7	61.6	83.2	97.3	48.5	77.7	100.3	53.7	81.6
26, 27, 3370	ELECTRICAL, COMMUNICATIONS & UTIL.	96.6	61.6	78.7	96.5	62.9	79.3	92.5	32.4	61.8	95.1	60.1	77.2	96.6	60.4	78.1	98.0	60.3	78.7
MF2018	WEIGHTED AVERAGE	99.1	64.4	84.2	95.1	64.2	81.9	100.3	57.9	82.1	97.8	65.1	83.8	99.8	63.7	84.3	97.5	63.3	82.8

TEXAS

DIVISION		PALESTINE 758 MAT.	INST.	TOTAL	SAN ANGELO 769 MAT.	INST.	TOTAL	SAN ANTONIO 781-782 MAT.	INST.	TOTAL	TEMPLE 765 MAT.	INST.	TOTAL	TEXARKANA 755 MAT.	INST.	TOTAL	TYLER 757 MAT.	INST.	TOTAL
015433	CONTRACTOR EQUIPMENT		94.6	94.6		93.2	93.2		93.5	93.5		93.2	93.2		94.6	94.6		94.6	94.6
0241, 31 - 34	SITE & INFRASTRUCTURE, DEMOLITION	94.9	93.6	94.0	97.5	92.1	93.7	100.0	93.5	95.5	86.0	92.0	90.2	84.3	96.0	92.5	93.9	93.4	93.6
0310	Concrete Forming & Accessories	78.7	60.4	63.2	98.9	52.9	59.8	97.1	53.9	60.4	102.4	52.6	60.1	95.5	61.0	66.2	89.7	60.5	64.9
0320	Concrete Reinforcing	90.8	52.5	71.3	86.4	52.6	69.2	92.1	49.1	70.2	86.6	52.3	69.2	90.6	52.3	71.2	91.5	52.2	71.5
0330	Cast-in-Place Concrete	87.9	60.3	77.7	87.8	66.8	80.0	93.7	69.4	84.7	71.9	60.3	67.6	88.6	66.9	80.5	105.7	60.3	88.8
03	CONCRETE	97.0	59.8	80.3	85.2	58.7	73.3	94.3	59.3	78.5	72.8	56.2	65.3	88.6	62.3	76.8	102.9	59.8	83.5
04	MASONRY	109.4	60.5	79.2	121.0	61.9	84.5	95.5	62.0	74.8	132.2	61.9	88.8	169.9	60.5	102.3	159.7	60.5	98.4
05	METALS	99.2	69.2	90.1	97.3	70.6	89.2	99.5	66.7	89.6	97.0	69.7	88.7	92.0	69.7	85.3	99.0	69.2	90.0
06	WOOD, PLASTICS & COMPOSITES	79.5	62.6	70.3	109.9	51.9	78.4	99.5	53.0	74.3	119.9	51.9	83.0	94.1	62.6	77.0	90.8	62.6	75.5
07	THERMAL & MOISTURE PROTECTION	90.7	63.5	79.0	93.0	64.4	80.6	93.8	66.7	82.1	92.5	63.4	80.0	89.9	64.8	79.1	90.5	63.5	78.9
08	OPENINGS	64.5	58.3	63.0	102.4	52.3	90.8	101.7	50.8	89.9	68.3	52.3	64.5	85.0	58.2	78.8	64.4	58.2	63.0
0920	Plaster & Gypsum Board	81.3	61.9	68.5	86.6	50.9	63.1	98.5	51.9	67.9	86.6	50.9	63.1	88.6	61.9	71.0	83.1	61.9	69.1
0950, 0980	Ceilings & Acoustic Treatment	76.5	61.9	66.6	79.3	50.9	60.1	91.3	51.9	64.6	79.3	50.9	60.1	80.8	61.9	68.0	76.5	61.9	66.6
0960	Flooring	120.4	56.4	101.8	83.7	56.4	75.8	98.5	70.0	90.2	108.0	56.4	93.0	105.3	65.9	93.8	131.8	56.4	110.0
0970, 0990	Wall Finishes & Painting/Coating	86.0	53.0	66.5	91.7	53.0	68.8	99.4	45.4	67.5	93.1	45.4	64.9	86.0	53.0	66.5	86.0	53.0	66.5
09	FINISHES	101.4	59.2	78.3	77.8	52.8	64.2	99.0	55.4	75.2	84.3	52.0	66.6	98.0	61.1	77.8	104.7	59.2	79.8
COVERS	DIVS. 10 - 14, 25, 28, 41, 43, 44, 46	100.0	75.1	94.5	100.0	77.6	95.1	100.0	79.3	95.4	100.0	73.8	94.2	100.0	80.7	95.8	100.0	80.7	95.7
21, 22, 23	FIRE SUPPRESSION, PLUMBING & HVAC	97.6	58.0	81.7	97.6	53.4	79.9	100.1	61.1	84.5	97.6	54.4	80.3	97.6	56.8	81.2	97.6	59.5	82.3
26, 27, 3370	ELECTRICAL, COMMUNICATIONS & UTIL.	92.8	49.6	70.7	97.8	55.9	76.4	96.2	61.2	78.3	94.9	53.3	73.6	96.4	55.9	75.7	95.3	55.0	74.7
MF2018	WEIGHTED AVERAGE	94.5	62.0	80.6	96.3	60.8	81.1	98.6	63.5	83.5	91.7	60.0	78.1	97.2	63.6	82.8	98.2	63.2	83.2

TEXAS / UTAH

DIVISION		VICTORIA 779 MAT.	INST.	TOTAL	WACO 766-767 MAT.	INST.	TOTAL	WAXAHACHIE 751 MAT.	INST.	TOTAL	WHARTON 774 MAT.	INST.	TOTAL	WICHITA FALLS 763 MAT.	INST.	TOTAL	LOGAN 843 MAT.	INST.	TOTAL
015433	CONTRACTOR EQUIPMENT		109.0	109.0		93.2	93.2		103.6	103.6		110.3	110.3		93.2	93.2		96.9	96.9
0241, 31 - 34	SITE & INFRASTRUCTURE, DEMOLITION	108.6	92.0	97.1	94.4	92.4	93.0	94.9	87.0	89.4	113.6	94.2	100.1	95.2	92.5	93.3	98.5	94.1	95.4
0310	Concrete Forming & Accessories	87.9	52.7	58.0	100.9	60.9	66.9	88.0	60.7	64.8	83.4	54.5	58.9	100.9	60.9	66.9	101.8	67.5	72.7
0320	Concrete Reinforcing	88.4	52.4	70.1	86.3	48.6	67.1	92.4	52.6	72.2	91.9	52.4	71.8	86.3	51.1	68.4	105.6	86.4	95.8
0330	Cast-in-Place Concrete	116.8	62.1	96.5	77.8	66.8	73.7	91.5	61.6	80.4	120.0	61.1	98.1	83.1	66.8	77.1	86.4	75.2	82.2
03	CONCRETE	104.6	58.2	83.7	78.9	61.6	71.1	86.5	61.4	75.2	108.7	58.6	86.2	81.4	62.1	72.7	104.0	74.0	90.5
04	MASONRY	104.0	62.0	78.0	92.5	61.9	73.6	154.7	60.6	96.5	90.9	61.9	73.0	92.9	61.4	73.5	108.1	67.8	83.2
05	METALS	96.9	88.2	94.3	99.3	68.8	90.0	99.3	84.4	94.8	98.5	87.8	95.2	99.2	70.7	90.6	108.4	83.6	100.8
06	WOOD, PLASTICS & COMPOSITES	96.5	52.0	72.3	118.0	62.5	87.8	86.6	62.8	73.6	86.6	54.2	69.0	118.0	62.5	87.8	87.0	67.8	76.6
07	THERMAL & MOISTURE PROTECTION	97.8	62.8	82.8	93.4	65.5	81.4	88.7	63.2	77.7	94.3	65.5	81.9	93.4	64.5	81.0	97.0	71.7	86.1
08	OPENINGS	101.7	52.5	90.3	79.7	57.2	74.5	94.1	58.3	85.8	101.9	53.8	90.7	79.7	58.2	74.7	94.4	66.8	88.0
0920	Plaster & Gypsum Board	94.6	50.9	65.9	87.0	61.9	70.4	85.3	61.9	69.9	93.5	53.3	67.0	87.0	61.9	70.4	81.1	66.9	71.8
0950, 0980	Ceilings & Acoustic Treatment	97.8	50.9	66.1	81.0	61.9	68.0	86.7	61.9	69.9	97.0	53.3	67.4	81.0	61.9	68.0	104.7	66.9	79.1
0960	Flooring	101.4	56.4	88.3	107.3	65.9	95.3	91.4	56.4	81.2	100.0	68.8	90.9	108.0	74.3	98.3	97.5	59.7	86.5
0970, 0990	Wall Finishes & Painting/Coating	101.9	58.9	76.5	93.1	53.0	69.4	95.8	53.0	70.5	101.6	57.2	75.3	95.1	53.0	70.2	93.4	55.7	71.1
09	FINISHES	89.4	53.5	69.8	84.9	61.0	71.9	90.5	59.3	73.4	91.0	56.8	72.3	85.3	62.7	73.0	92.4	64.4	77.1
COVERS	DIVS. 10 - 14, 25, 28, 41, 43, 44, 46	100.0	71.1	93.6	100.0	80.5	95.7	100.0	81.0	95.8	100.0	71.3	93.7	100.0	74.9	94.5	100.0	83.7	96.4
21, 22, 23	FIRE SUPPRESSION, PLUMBING & HVAC	97.7	63.7	84.1	100.1	60.1	84.0	97.7	59.3	82.3	97.6	63.3	83.9	100.1	53.3	81.3	99.9	69.9	87.9
26, 27, 3370	ELECTRICAL, COMMUNICATIONS & UTIL.	101.9	55.3	78.1	98.0	55.4	76.2	95.0	60.1	77.2	100.8	60.6	80.2	99.7	55.5	77.0	95.8	70.5	82.8
MF2018	WEIGHTED AVERAGE	99.4	64.4	84.4	93.3	63.9	80.7	97.9	65.1	83.8	99.5	65.8	85.0	93.8	62.7	80.5	100.4	73.1	88.7

For customer support on your Electrical Costs with RSMeans Data, call 800.448.8182.

UTAH / VERMONT

DIVISION		OGDEN 842,844 MAT.	INST.	TOTAL	PRICE 845 MAT.	INST.	TOTAL	PROVO 846-847 MAT.	INST.	TOTAL	SALT LAKE CITY 840-841 MAT.	INST.	TOTAL	BELLOWS FALLS 051 MAT.	INST.	TOTAL	BENNINGTON 052 MAT.	INST.	TOTAL
015433	CONTRACTOR EQUIPMENT		96.9	96.9		95.9	95.9		95.9	95.9		96.8	96.8		97.9	97.9		97.9	97.9
0241, 31-34	SITE & INFRASTRUCTURE, DEMOLITION	86.7	94.1	91.8	95.9	92.2	93.3	94.7	92.5	93.2	86.4	94.0	91.7	89.8	100.7	97.4	89.2	100.7	97.2
0310	Concrete Forming & Accessories	101.8	67.5	72.7	104.4	66.0	71.8	103.6	67.5	73.0	104.1	67.5	73.0	97.0	104.2	103.1	94.4	104.1	102.6
0320	Concrete Reinforcing	105.2	86.4	95.7	113.2	86.4	99.6	114.2	86.4	100.1	107.5	86.4	96.8	85.4	84.3	84.9	85.4	84.3	84.8
0330	Cast-in-Place Concrete	87.7	75.2	83.1	86.5	73.8	81.8	86.5	75.2	82.3	95.8	75.2	88.2	91.3	112.8	99.3	91.3	112.7	99.3
03	CONCRETE	93.6	74.0	84.7	105.1	72.8	90.5	103.5	73.9	90.2	112.7	74.0	95.3	86.3	103.4	94.0	86.2	103.3	93.9
04	MASONRY	103.0	67.8	81.3	113.9	66.5	84.6	114.0	67.8	85.5	116.7	67.8	86.5	101.6	88.0	93.2	112.4	88.0	97.3
05	METALS	108.9	83.6	101.2	105.5	83.5	98.8	106.5	83.6	99.5	112.4	83.6	103.6	85.9	89.6	87.0	85.9	89.4	86.9
06	WOOD, PLASTICS & COMPOSITES	87.0	67.8	76.6	90.3	65.9	77.1	88.6	67.8	77.3	89.0	67.8	77.5	106.9	109.8	108.5	103.8	109.8	107.1
07	THERMAL & MOISTURE PROTECTION	95.8	71.7	85.5	98.9	70.2	86.5	98.9	71.7	87.2	103.4	71.7	89.8	98.9	90.7	95.4	98.8	90.7	95.3
08	OPENINGS	94.4	66.8	88.0	98.3	78.5	93.7	98.3	66.8	91.0	96.2	66.8	89.4	102.4	99.3	101.7	102.4	99.3	101.7
0920	Plaster & Gypsum Board	81.1	66.9	71.8	83.9	65.0	71.4	81.7	66.9	72.0	93.1	66.9	75.9	104.9	109.8	108.1	103.4	109.8	107.6
0950, 0980	Ceilings & Acoustic Treatment	104.7	66.9	79.1	104.7	65.0	77.8	104.7	66.9	79.1	96.1	66.9	76.3	97.0	109.8	105.7	97.0	109.8	105.7
0960	Flooring	95.4	59.7	85.0	98.6	58.0	86.8	98.4	59.7	87.1	99.3	59.7	87.8	91.2	102.2	94.4	90.4	102.2	93.8
0970, 0990	Wall Finishes & Painting/Coating	93.4	55.7	71.1	93.4	55.7	71.1	93.4	55.7	71.1	96.5	56.8	73.0	87.1	104.2	97.2	87.1	104.2	97.2
09	FINISHES	90.6	64.4	76.3	93.6	62.8	76.8	93.1	64.4	77.4	92.5	64.5	77.2	86.3	104.8	96.4	85.9	104.8	96.2
COVERS	DIVS. 10-14, 25, 28, 41, 43, 44, 46	100.0	83.7	96.4	100.0	83.4	96.3	100.0	83.7	96.4	100.0	83.7	96.4	100.0	93.7	98.6	100.0	93.6	98.6
21, 22, 23	FIRE SUPPRESSION, PLUMBING & HVAC	99.9	69.9	87.9	98.6	64.6	85.0	99.9	69.9	87.9	100.1	69.9	88.0	97.7	89.2	94.3	97.7	89.2	94.3
26, 27, 3370	ELECTRICAL, COMMUNICATIONS & UTIL.	96.1	70.5	83.0	101.0	70.5	85.4	96.4	70.3	83.0	98.7	70.5	84.3	99.2	81.5	90.1	99.2	55.8	77.0
MF2018	WEIGHTED AVERAGE	98.5	73.1	87.6	101.1	71.7	88.5	100.8	72.9	88.8	102.9	73.1	90.1	94.4	93.6	94.1	94.8	90.0	92.8

VERMONT

DIVISION		BRATTLEBORO 053 MAT.	INST.	TOTAL	BURLINGTON 054 MAT.	INST.	TOTAL	GUILDHALL 059 MAT.	INST.	TOTAL	MONTPELIER 056 MAT.	INST.	TOTAL	RUTLAND 057 MAT.	INST.	TOTAL	ST. JOHNSBURY 058 MAT.	INST.	TOTAL
015433	CONTRACTOR EQUIPMENT		97.9	97.9		97.9	97.9		97.9	97.9		97.9	97.9		97.9	97.9		97.9	97.9
0241, 31-34	SITE & INFRASTRUCTURE, DEMOLITION	90.7	100.7	97.6	92.7	100.9	98.4	88.7	96.2	93.9	94.8	100.6	98.8	93.1	100.6	98.3	88.8	99.6	96.4
0310	Concrete Forming & Accessories	97.2	104.2	103.2	99.8	81.9	84.6	95.0	97.5	97.1	97.7	103.6	102.7	97.5	81.9	84.3	93.6	97.9	97.2
0320	Concrete Reinforcing	84.5	84.3	84.4	104.8	84.2	94.3	86.2	84.1	85.1	104.9	84.2	94.4	106.8	84.2	95.3	84.5	84.2	84.3
0330	Cast-in-Place Concrete	94.2	112.8	101.1	110.1	112.5	111.0	88.5	103.8	94.2	107.8	112.6	109.6	89.5	112.5	98.0	88.5	104.0	94.3
03	CONCRETE	88.3	103.4	95.1	98.6	93.2	96.1	83.9	97.1	89.9	97.4	103.0	99.9	88.7	93.1	90.7	83.7	97.4	89.8
04	MASONRY	112.0	88.0	97.2	113.8	88.1	97.9	112.3	73.0	88.0	104.3	88.1	94.3	92.5	88.1	89.8	140.6	73.0	98.8
05	METALS	85.9	89.6	87.0	93.5	88.8	92.1	85.9	88.5	86.7	91.0	88.9	90.3	91.5	88.7	90.7	85.9	88.9	86.8
06	WOOD, PLASTICS & COMPOSITES	107.2	109.8	108.6	100.8	80.6	89.8	102.8	109.8	106.6	96.5	109.8	103.8	107.4	80.6	92.8	97.5	109.8	104.2
07	THERMAL & MOISTURE PROTECTION	99.0	90.7	95.4	106.0	87.6	98.0	98.7	83.8	92.3	105.9	90.8	99.4	99.1	87.6	94.1	98.6	83.8	92.2
08	OPENINGS	102.4	99.6	101.8	104.1	79.6	98.4	102.4	95.7	100.9	106.0	95.7	103.6	105.6	79.6	99.5	102.4	95.7	100.9
0920	Plaster & Gypsum Board	104.9	109.8	108.1	106.4	79.7	88.8	110.7	109.8	110.1	106.7	109.8	108.7	105.3	79.7	88.4	112.2	109.8	110.6
0950, 0980	Ceilings & Acoustic Treatment	97.0	109.8	105.7	101.0	79.7	86.6	97.0	109.8	105.7	106.9	109.8	108.9	101.5	79.7	86.7	97.0	109.8	105.7
0960	Flooring	91.3	102.2	94.5	98.1	102.2	99.3	94.1	102.2	96.5	100.4	102.2	100.9	91.2	102.2	94.4	97.4	102.2	98.8
0970, 0990	Wall Finishes & Painting/Coating	87.1	104.2	97.2	96.4	90.7	93.0	87.1	90.7	89.2	95.3	90.7	92.5	87.1	90.7	89.2	87.1	90.7	89.2
09	FINISHES	86.5	104.8	96.5	92.5	86.1	89.0	87.9	99.5	94.3	94.2	103.4	99.2	87.4	86.1	86.7	89.1	99.5	94.8
COVERS	DIVS. 10-14, 25, 28, 41, 43, 44, 46	100.0	93.7	98.6	100.0	90.4	97.9	100.0	88.5	97.5	100.0	93.6	98.6	100.0	90.4	97.9	100.0	88.5	97.5
21, 22, 23	FIRE SUPPRESSION, PLUMBING & HVAC	97.7	89.2	94.3	100.1	69.4	87.8	97.7	61.5	83.2	97.6	69.4	86.3	100.2	69.4	87.9	97.7	61.5	83.2
26, 27, 3370	ELECTRICAL, COMMUNICATIONS & UTIL.	99.2	81.5	90.1	99.5	55.0	76.7	99.2	55.8	77.0	98.1	55.0	76.1	99.3	55.0	76.6	99.2	55.8	77.0
MF2018	WEIGHTED AVERAGE	95.2	94.0	94.5	99.2	80.6	91.2	94.7	80.2	88.5	97.8	85.3	92.4	96.3	80.5	89.5	96.1	80.5	89.4

VERMONT / VIRGINIA

DIVISION		WHITE RIVER JCT. 050 MAT.	INST.	TOTAL	ALEXANDRIA 223 MAT.	INST.	TOTAL	ARLINGTON 222 MAT.	INST.	TOTAL	BRISTOL 242 MAT.	INST.	TOTAL	CHARLOTTESVILLE 229 MAT.	INST.	TOTAL	CULPEPER 227 MAT.	INST.	TOTAL
015433	CONTRACTOR EQUIPMENT		97.9	97.9		107.5	107.5		106.2	106.2		106.2	106.2		110.4	110.4		106.2	106.2
0241, 31-34	SITE & INFRASTRUCTURE, DEMOLITION	93.1	99.7	97.7	114.2	92.2	98.9	124.3	89.9	100.4	108.9	89.2	95.2	113.4	91.7	98.3	112.0	89.9	96.6
0310	Concrete Forming & Accessories	91.6	98.4	97.4	90.4	70.2	73.2	89.1	70.1	73.0	85.2	65.6	68.5	83.6	47.6	53.0	80.7	72.9	74.0
0320	Concrete Reinforcing	85.4	84.2	84.8	83.9	87.4	85.7	94.7	87.4	91.0	94.7	69.8	82.0	94.1	70.1	81.9	94.7	87.4	91.0
0330	Cast-in-Place Concrete	94.2	104.8	98.2	110.1	76.5	97.6	107.1	76.5	95.7	106.7	45.4	83.9	111.0	76.9	98.3	109.7	76.4	97.3
03	CONCRETE	89.8	97.9	93.4	101.3	76.8	90.3	105.6	76.8	92.6	102.0	61.0	83.5	102.7	63.8	85.2	99.6	78.0	89.9
04	MASONRY	126.0	74.4	94.1	89.4	72.3	78.9	103.1	72.3	84.1	92.7	46.3	64.1	116.9	55.7	79.1	104.6	72.3	84.7
05	METALS	85.9	88.9	86.8	102.9	100.4	102.1	101.5	100.4	101.2	100.3	92.2	97.9	100.6	94.1	98.6	100.7	100.0	100.5
06	WOOD, PLASTICS & COMPOSITES	100.6	109.8	105.6	91.8	68.4	79.1	88.3	68.4	77.5	80.9	71.8	76.0	79.4	40.8	58.5	78.1	72.1	74.8
07	THERMAL & MOISTURE PROTECTION	99.0	84.5	92.8	103.0	79.8	93.0	105.1	79.8	94.2	104.4	59.5	85.0	103.9	68.0	88.5	104.1	79.7	93.6
08	OPENINGS	102.4	95.7	100.9	97.4	73.6	91.9	95.6	73.6	90.5	98.4	65.7	90.8	96.7	54.6	86.9	97.0	75.9	92.1
0920	Plaster & Gypsum Board	102.2	109.8	107.2	101.7	67.4	79.1	98.3	67.4	77.9	94.8	70.9	79.1	94.8	38.4	57.7	94.9	71.2	79.3
0950, 0980	Ceilings & Acoustic Treatment	97.0	109.8	105.7	93.4	67.4	75.8	91.7	67.4	75.2	90.8	70.9	77.4	90.8	38.4	55.3	91.7	71.2	77.8
0960	Flooring	89.3	102.2	93.0	97.0	76.9	91.2	95.5	76.9	90.1	92.1	57.6	82.1	90.6	57.6	81.0	90.6	76.9	86.7
0970, 0990	Wall Finishes & Painting/Coating	87.1	90.7	89.2	115.2	73.6	90.6	115.2	75.1	91.4	101.5	55.6	74.4	101.5	58.9	76.3	115.2	75.1	91.4
09	FINISHES	85.7	99.9	93.5	93.1	70.9	81.0	93.1	71.1	81.1	89.8	64.0	75.7	89.4	48.6	67.1	90.1	73.3	80.9
COVERS	DIVS. 10-14, 25, 28, 41, 43, 44, 46	100.0	89.0	97.6	100.0	87.3	97.2	100.0	84.9	96.7	100.0	76.0	94.7	100.0	80.4	95.7	100.0	85.3	96.8
21, 22, 23	FIRE SUPPRESSION, PLUMBING & HVAC	97.7	62.2	83.5	100.4	87.2	95.1	100.4	87.2	95.1	98.0	46.8	77.5	98.0	69.1	86.4	98.0	87.2	93.6
26, 27, 3370	ELECTRICAL, COMMUNICATIONS & UTIL.	99.2	55.7	76.9	96.9	96.7	96.8	94.6	99.3	97.0	96.5	37.2	66.2	96.5	70.0	83.0	98.9	99.3	99.1
MF2018	WEIGHTED AVERAGE	96.0	81.0	89.5	99.5	84.2	92.9	100.3	84.3	93.4	98.3	59.4	81.6	99.5	68.0	86.0	98.8	84.9	92.8

For customer support on your Electrical Costs with RSMeans Data, call 800.448.8182.

555

VIRGINIA

DIVISION		FAIRFAX 220 - 221			FARMVILLE 239			FREDERICKSBURG 224 - 225			GRUNDY 246			HARRISONBURG 228			LYNCHBURG 245		
		MAT.	INST.	TOTAL	MAT.	INST.	TOTAL	MAT.	INST.	TOTAL	MAT.	INST.	TOTAL	MAT.	INST.	TOTAL	MAT.	INST.	TOTAL
015433	CONTRACTOR EQUIPMENT		106.2	106.2		110.4	110.4		106.2	106.2		106.2	106.2		106.2	106.2		106.2	106.2
0241, 31 - 34	SITE & INFRASTRUCTURE, DEMOLITION	123.0	89.9	100.0	109.6	90.7	96.4	111.5	89.7	96.3	106.7	88.5	94.1	120.0	88.3	97.9	107.5	90.0	95.3
0310	Concrete Forming & Accessories	83.6	70.1	72.1	98.5	45.2	53.2	83.6	67.1	69.6	88.3	38.1	45.7	79.7	38.7	44.8	85.2	71.5	73.6
0320	Concrete Reinforcing	94.7	87.4	91.0	90.0	69.7	79.7	95.4	82.5	88.8	93.4	45.3	69.0	94.7	58.9	76.5	94.1	70.6	82.1
0330	Cast-in-Place Concrete	107.1	76.5	95.7	106.2	86.0	98.7	108.7	76.2	96.6	106.7	51.0	86.0	107.1	53.8	87.3	106.7	75.7	95.2
03	CONCRETE	105.3	76.7	92.4	96.4	65.7	82.6	99.3	74.5	88.2	100.6	46.0	76.0	103.0	49.8	79.1	100.4	74.3	88.7
04	MASONRY	103.0	72.3	84.0	101.3	51.0	70.2	103.7	68.2	81.7	94.2	54.6	69.7	101.5	59.1	75.3	109.0	64.4	81.5
05	METALS	100.8	100.3	100.6	97.0	90.6	95.1	100.7	98.4	100.0	100.3	75.1	92.7	100.6	87.2	96.6	100.5	94.2	98.6
06	WOOD, PLASTICS & COMPOSITES	80.9	68.4	74.1	94.1	39.5	64.4	80.9	66.5	73.1	83.8	31.4	55.3	77.1	34.9	54.2	80.9	73.2	76.7
07	THERMAL & MOISTURE PROTECTION	104.8	75.6	92.2	104.3	54.9	83.0	104.1	77.1	92.5	104.3	47.4	79.8	104.6	62.9	86.6	104.1	71.1	89.9
08	OPENINGS	95.6	73.9	90.6	96.2	44.4	84.2	96.7	70.5	90.6	98.4	30.9	82.7	97.0	44.3	84.8	97.0	66.4	89.9
0920	Plaster & Gypsum Board	94.9	67.4	76.8	103.8	37.0	59.8	94.9	65.4	75.5	94.8	29.4	51.7	94.8	33.0	54.1	94.8	72.3	80.0
0950, 0980	Ceilings & Acoustic Treatment	91.7	67.4	75.2	86.3	37.0	52.9	91.7	65.4	73.9	90.8	29.4	49.2	90.8	33.0	51.6	90.8	72.3	78.3
0960	Flooring	92.4	77.7	88.1	92.5	52.2	80.8	92.4	74.8	87.3	93.4	29.2	74.8	90.5	76.9	86.5	92.1	67.6	85.0
0970, 0990	Wall Finishes & Painting/Coating	115.2	67.7	86.6	101.8	58.7	76.3	115.2	58.5	81.6	101.5	32.1	60.4	115.2	49.6	76.4	101.5	55.6	74.4
09	FINISHES	91.7	71.5	80.7	88.7	46.3	65.6	92.0	66.9	77.7	90.0	34.4	59.6	90.6	45.3	65.8	89.6	69.2	78.5
COVERS	DIVS. 10 - 14, 25, 28, 41, 43, 44, 46	100.0	84.9	96.7	100.0	76.2	94.8	100.0	79.8	95.6	100.0	74.3	94.4	100.0	61.2	91.5	100.0	79.5	95.5
21, 22, 23	FIRE SUPPRESSION, PLUMBING & HVAC	98.0	87.2	93.6	98.0	53.1	80.0	98.0	81.7	91.4	98.0	66.1	85.2	98.0	66.8	85.5	98.0	69.4	86.5
26, 27, 3370	ELECTRICAL, COMMUNICATIONS & UTIL.	97.6	99.3	98.5	90.5	70.0	80.0	94.8	94.0	94.4	96.5	70.0	82.9	96.7	88.5	92.5	97.6	69.4	83.2
MF2018	WEIGHTED AVERAGE	99.6	84.2	93.0	96.7	62.8	82.2	98.4	80.6	90.7	98.2	59.2	81.5	99.1	66.0	84.9	98.9	73.7	88.1

VIRGINIA

DIVISION		NEWPORT NEWS 236			NORFOLK 233 - 235			PETERSBURG 238			PORTSMOUTH 237			PULASKI 243			RICHMOND 230 - 232		
		MAT.	INST.	TOTAL	MAT.	INST.	TOTAL	MAT.	INST.	TOTAL	MAT.	INST.	TOTAL	MAT.	INST.	TOTAL	MAT.	INST.	TOTAL
015433	CONTRACTOR EQUIPMENT		110.4	110.4		111.0	111.0		110.4	110.4		110.3	110.3		106.2	106.2		110.4	110.4
0241, 31 - 34	SITE & INFRASTRUCTURE, DEMOLITION	108.4	91.7	96.8	105.1	92.7	96.5	112.1	91.7	97.9	107.0	91.0	95.8	106.0	89.0	94.2	106.0	91.7	96.0
0310	Concrete Forming & Accessories	97.7	61.3	66.8	103.7	61.4	67.8	90.9	62.5	66.8	86.7	48.0	53.8	88.3	41.2	48.3	95.3	80.7	82.9
0320	Concrete Reinforcing	89.8	67.3	78.3	98.0	67.3	82.4	89.4	70.7	79.9	89.4	66.6	77.8	93.4	86.2	89.7	98.8	70.7	84.5
0330	Cast-in-Place Concrete	103.2	76.2	93.1	108.0	77.2	96.6	109.6	77.4	97.6	102.2	61.2	86.9	106.7	84.6	98.5	97.1	77.5	89.8
03	CONCRETE	93.7	69.3	82.7	96.1	69.7	84.2	98.8	70.8	86.2	92.4	58.0	76.9	100.6	66.2	85.1	90.5	79.1	85.3
04	MASONRY	96.2	63.2	75.8	98.4	63.2	76.6	109.6	63.2	80.9	101.9	48.4	68.9	88.1	55.4	67.9	90.5	63.2	73.6
05	METALS	99.3	93.5	97.5	99.6	93.6	97.8	97.1	95.1	96.5	98.2	91.9	96.3	100.4	96.2	99.1	102.4	95.2	100.2
06	WOOD, PLASTICS & COMPOSITES	92.8	60.3	75.1	101.3	60.3	79.0	84.0	61.2	71.6	80.2	46.1	61.7	83.8	32.9	56.2	93.1	85.7	89.1
07	THERMAL & MOISTURE PROTECTION	104.3	70.9	89.9	102.3	71.1	88.8	104.2	72.1	90.4	104.2	60.8	85.6	104.3	57.9	84.3	101.5	74.8	90.0
08	OPENINGS	96.6	59.0	87.9	98.3	60.3	89.5	96.2	65.8	89.0	96.7	48.8	85.6	98.4	44.1	85.8	100.1	79.4	95.3
0920	Plaster & Gypsum Board	104.7	58.4	74.2	99.9	58.4	72.5	98.2	59.4	72.6	98.6	43.8	62.5	94.8	30.9	52.7	101.7	84.6	90.4
0950, 0980	Ceilings & Acoustic Treatment	90.5	58.4	68.8	94.8	58.4	70.1	87.1	59.4	68.3	90.5	43.8	58.9	90.8	30.9	50.2	92.7	84.6	87.2
0960	Flooring	92.5	67.6	85.3	90.6	67.6	83.9	88.7	72.7	84.0	85.8	57.6	77.6	93.4	57.6	83.0	93.6	72.7	87.5
0970, 0990	Wall Finishes & Painting/Coating	101.8	58.9	76.4	101.5	58.9	76.3	101.8	58.9	76.4	101.8	58.9	76.4	101.5	49.6	70.8	99.6	58.9	75.5
09	FINISHES	89.5	61.7	74.3	89.0	61.7	74.1	87.1	63.3	74.1	86.7	49.5	66.4	90.0	42.0	63.8	91.6	77.8	84.1
COVERS	DIVS. 10 - 14, 25, 28, 41, 43, 44, 46	100.0	79.5	95.5	100.0	79.5	95.5	100.0	82.2	96.1	100.0	65.2	92.3	100.0	74.6	94.4	100.0	84.8	96.7
21, 22, 23	FIRE SUPPRESSION, PLUMBING & HVAC	100.5	66.5	86.9	100.5	66.0	86.4	98.0	68.5	86.2	100.5	63.2	85.5	98.0	67.5	85.8	100.0	68.5	87.4
26, 27, 3370	ELECTRICAL, COMMUNICATIONS & UTIL.	93.1	64.7	78.6	95.5	61.9	78.3	93.2	70.0	81.3	91.5	61.9	76.4	96.5	85.4	90.8	97.6	70.0	83.5
MF2018	WEIGHTED AVERAGE	97.4	69.7	85.5	98.1	70.0	86.1	97.5	72.4	86.8	96.9	63.1	82.4	97.9	68.3	85.2	98.0	76.4	88.7

		VIRGINIA									WASHINGTON								
DIVISION		ROANOKE 240 - 241			STAUNTON 244			WINCHESTER 226			CLARKSTON 994			EVERETT 982			OLYMPIA 985		
		MAT.	INST.	TOTAL	MAT.	INST.	TOTAL	MAT.	INST.	TOTAL	MAT.	INST.	TOTAL	MAT.	INST.	TOTAL	MAT.	INST.	TOTAL
015433	CONTRACTOR EQUIPMENT		106.2	106.2		110.4	110.4		106.2	106.2		92.5	92.5		102.5	102.5		102.5	102.5
0241, 31 - 34	SITE & INFRASTRUCTURE, DEMOLITION	106.8	89.9	95.1	110.0	90.6	96.5	118.7	89.8	98.6	102.2	90.4	94.0	93.8	110.8	105.7	93.3	110.8	105.4
0310	Concrete Forming & Accessories	94.6	71.6	75.1	88.0	49.4	55.2	82.0	68.1	70.2	102.2	66.0	71.5	111.2	101.2	102.7	100.2	101.1	100.9
0320	Concrete Reinforcing	94.5	70.6	82.3	94.1	86.1	90.0	94.1	82.5	88.2	104.8	98.7	101.7	111.1	111.6	111.4	118.0	111.6	114.8
0330	Cast-in-Place Concrete	121.2	85.7	108.0	111.0	86.5	101.9	107.1	65.6	91.7	81.1	85.3	82.7	99.4	108.9	103.0	87.2	108.8	95.2
03	CONCRETE	105.2	77.8	92.9	102.0	70.5	87.8	102.6	71.3	88.5	96.1	78.7	88.3	90.4	105.1	97.0	86.2	105.0	94.7
04	MASONRY	95.6	64.4	76.3	104.5	54.4	73.6	99.0	71.2	81.8	99.6	93.6	95.9	117.9	103.8	109.2	103.0	103.8	103.5
05	METALS	102.6	94.3	100.1	100.6	95.9	99.1	100.7	98.3	100.0	93.7	88.0	92.0	104.4	96.0	101.8	103.1	95.6	100.8
06	WOOD, PLASTICS & COMPOSITES	92.7	73.2	82.1	83.8	44.7	62.5	79.4	66.7	72.5	103.6	60.2	80.0	107.5	99.8	103.3	93.6	99.8	97.0
07	THERMAL & MOISTURE PROTECTION	104.1	74.5	91.4	103.9	56.3	83.4	104.7	76.7	92.6	157.7	85.2	126.5	111.4	106.8	109.4	110.1	104.3	107.6
08	OPENINGS	97.4	66.4	90.2	97.0	50.9	86.3	98.5	70.3	91.9	119.0	66.9	106.9	108.2	103.6	107.2	110.7	102.1	108.7
0920	Plaster & Gypsum Board	101.7	72.3	82.4	94.8	42.4	60.3	94.9	65.6	75.6	151.9	58.9	90.7	109.4	99.9	103.1	103.2	99.9	101.0
0950, 0980	Ceilings & Acoustic Treatment	93.4	72.3	79.1	90.8	42.4	58.0	91.7	65.6	74.0	106.8	58.9	74.4	109.1	99.9	102.8	103.6	99.9	101.1
0960	Flooring	97.0	67.6	88.5	93.0	34.2	75.9	91.9	76.9	87.5	88.3	80.1	85.9	105.2	99.3	103.5	94.0	99.3	95.6
0970, 0990	Wall Finishes & Painting/Coating	101.5	55.6	74.4	101.5	31.5	60.1	115.2	79.6	94.1	83.8	74.0	78.0	91.1	95.6	93.8	92.3	95.6	94.2
09	FINISHES	92.1	69.2	79.6	89.8	43.7	64.6	91.1	70.3	79.7	107.3	67.7	85.7	103.8	99.7	101.5	92.3	99.7	96.3
COVERS	DIVS. 10 - 14, 25, 28, 41, 43, 44, 46	100.0	79.4	95.5	100.0	77.2	95.0	100.0	84.3	96.6	100.0	91.2	98.1	100.0	100.4	100.1	100.0	100.8	100.2
21, 22, 23	FIRE SUPPRESSION, PLUMBING & HVAC	100.4	65.2	86.3	98.0	59.2	82.4	98.0	83.2	92.0	98.0	82.4	91.8	100.1	102.8	101.2	100.0	102.8	101.1
26, 27, 3370	ELECTRICAL, COMMUNICATIONS & UTIL.	96.6	56.1	75.9	95.5	70.1	82.5	95.2	94.0	94.6	87.4	97.9	92.8	103.5	101.5	102.5	100.9	100.1	100.5
MF2018	WEIGHTED AVERAGE	100.0	71.6	87.8	98.7	65.6	84.5	99.0	81.3	91.4	100.9	83.9	93.6	102.1	102.6	102.3	99.9	102.3	100.9

City Cost Indexes

WASHINGTON

DIVISION		RICHLAND 993			SEATTLE 980 - 981, 987			SPOKANE 990 - 992			TACOMA 983 - 984			VANCOUVER 986			WENATCHEE 988		
		MAT.	INST.	TOTAL	MAT.	INST.	TOTAL	MAT.	INST.	TOTAL	MAT.	INST.	TOTAL	MAT.	INST.	TOTAL	MAT.	INST.	TOTAL
015433	CONTRACTOR EQUIPMENT		92.5	92.5		104.0	104.0		92.5	92.5		102.5	102.5		98.8	98.8		102.5	102.5
0241, 31 - 34	SITE & INFRASTRUCTURE, DEMOLITION	104.2	90.6	94.7	100.9	110.0	107.3	103.7	90.5	94.5	97.1	110.8	106.6	107.3	98.1	100.9	106.6	108.5	107.9
0310	Concrete Forming & Accessories	102.3	82.2	85.2	107.0	103.6	104.1	107.5	81.7	85.5	101.4	101.1	101.1	101.6	93.2	94.5	103.3	78.6	82.3
0320	Concrete Reinforcing	100.5	98.1	99.3	114.5	111.7	113.0	101.2	98.0	99.6	109.7	111.5	110.6	110.7	110.8	110.7	104.4	93.6	100.4
0330	Cast-in-Place Concrete	81.3	86.9	83.4	107.8	108.7	108.2	84.5	86.7	85.3	102.2	108.8	104.7	114.2	99.0	108.6	99.0	87.3	93.8
03	CONCRETE	95.8	86.5	91.6	99.2	106.3	102.4	97.8	86.1	92.6	92.0	105.0	97.9	101.5	98.1	99.9	99.0	87.3	93.8
04	MASONRY	100.6	85.5	91.3	125.2	103.8	112.0	101.2	85.5	91.5	116.6	102.4	107.8	116.6	90.6	100.6	120.0	96.1	105.2
05	METALS	94.1	87.8	92.2	105.8	99.1	103.8	96.5	87.3	93.7	106.2	95.5	102.9	103.6	96.0	101.3	103.6	88.2	98.9
06	WOOD, PLASTICS & COMPOSITES	103.8	80.2	91.0	101.3	102.7	102.1	112.7	80.2	95.0	97.0	99.8	98.5	89.5	92.2	91.0	98.8	75.5	86.1
07	THERMAL & MOISTURE PROTECTION	158.9	86.7	127.8	107.7	106.0	106.9	155.2	87.0	125.8	111.1	104.1	108.1	111.1	96.9	105.0	110.6	90.8	102.1
08	OPENINGS	117.0	76.9	107.7	108.4	105.2	107.7	117.6	76.8	108.1	109.0	102.1	107.4	105.0	96.6	103.0	108.5	75.1	100.7
0920	Plaster & Gypsum Board	151.9	79.5	104.2	107.8	102.9	104.5	144.5	79.5	101.7	109.2	99.9	103.0	107.4	92.4	97.5	112.5	74.9	87.7
0950, 0980	Ceilings & Acoustic Treatment	113.5	79.5	90.4	120.3	102.9	108.5	108.7	79.5	88.9	112.5	99.9	103.9	109.6	92.4	97.9	104.7	74.9	84.5
0960	Flooring	88.6	80.1	86.1	105.3	99.3	103.5	87.7	80.1	85.5	98.2	99.3	98.5	103.9	81.7	97.5	101.0	80.1	94.9
0970, 0990	Wall Finishes & Painting/Coating	83.8	77.0	79.8	101.9	95.6	98.2	83.9	78.4	80.7	91.1	95.6	93.8	93.8	73.8	82.0	91.1	72.8	80.3
09	FINISHES	108.9	80.4	93.3	109.6	101.4	105.1	106.8	80.5	92.4	102.5	99.7	101.0	100.5	88.5	93.9	103.2	77.1	89.0
COVERS	DIVS. 10 - 14, 25, 28, 41, 43, 44, 46	100.0	94.1	98.7	100.0	101.2	100.3	100.0	94.1	98.7	100.0	100.8	100.2	100.0	96.3	99.2	100.0	93.1	98.5
21, 22, 23	FIRE SUPPRESSION, PLUMBING & HVAC	100.5	110.9	104.7	100.1	112.5	105.0	100.5	85.5	94.5	100.1	102.8	101.2	100.2	100.0	100.1	97.7	94.1	96.2
26, 27, 3370	ELECTRICAL, COMMUNICATIONS & UTIL.	85.2	95.1	90.3	103.7	115.4	109.7	83.5	78.9	81.2	103.3	100.1	101.7	108.6	98.5	103.4	104.2	96.8	100.4
MF2018	WEIGHTED AVERAGE	101.4	92.2	97.5	104.3	107.4	105.6	101.7	84.5	94.3	102.5	102.1	102.3	103.4	96.2	100.3	102.9	91.0	97.8

WASHINGTON / WEST VIRGINIA

DIVISION		YAKIMA 989			BECKLEY 258 - 259			BLUEFIELD 247 - 248			BUCKHANNON 262			CHARLESTON 250 - 253			CLARKSBURG 263 - 264		
		MAT.	INST.	TOTAL	MAT.	INST.	TOTAL	MAT.	INST.	TOTAL	MAT.	INST.	TOTAL	MAT.	INST.	TOTAL	MAT.	INST.	TOTAL
015433	CONTRACTOR EQUIPMENT		102.5	102.5		106.2	106.2		106.2	106.2		106.2	106.2		106.2	106.2		106.2	106.2
0241, 31 - 34	SITE & INFRASTRUCTURE, DEMOLITION	99.6	109.3	106.4	97.9	91.3	93.3	100.6	91.3	94.1	106.8	91.2	95.9	97.5	92.3	93.9	107.6	91.2	96.2
0310	Concrete Forming & Accessories	102.0	96.1	97.0	84.4	89.7	88.9	85.2	89.5	88.8	84.6	85.9	85.7	96.5	89.6	90.7	82.0	86.0	85.4
0320	Concrete Reinforcing	110.2	98.2	104.1	88.8	89.4	89.1	93.0	84.9	88.9	93.7	84.7	89.1	96.0	89.4	92.7	93.7	98.0	95.9
0330	Cast-in-Place Concrete	109.2	85.6	100.4	103.1	95.1	100.1	104.4	95.0	100.9	104.0	94.4	100.5	99.6	96.1	98.3	114.0	90.0	105.1
03	CONCRETE	96.5	92.4	94.7	94.6	92.4	93.7	96.4	91.6	94.2	99.4	89.7	95.0	93.3	92.8	93.1	103.6	90.5	97.7
04	MASONRY	109.2	82.8	92.9	88.7	90.9	90.1	89.7	90.9	90.5	101.4	86.5	92.2	85.8	92.3	89.8	105.0	86.5	93.6
05	METALS	104.3	88.2	99.4	101.8	103.3	102.3	100.6	101.7	100.9	100.8	101.6	101.0	100.3	103.5	101.3	78.5	85.4	82.2
06	WOOD, PLASTICS & COMPOSITES	97.4	99.8	98.7	81.7	89.9	86.2	82.8	89.9	86.7	82.1	85.4	83.9	93.6	88.8	91.0	104.3	88.3	97.4
07	THERMAL & MOISTURE PROTECTION	111.3	87.7	101.1	106.5	88.4	98.7	104.1	88.4	97.3	104.4	88.6	97.6	102.8	88.9	96.8	98.9	84.4	95.6
08	OPENINGS	108.5	99.1	106.3	96.5	84.9	93.8	98.9	83.9	95.4	98.9	81.5	94.9	96.9	84.3	93.9	98.9	84.4	95.6
0920	Plaster & Gypsum Board	108.8	99.9	102.9	97.3	89.5	92.2	94.1	89.5	91.1	94.5	84.9	88.2	101.7	88.4	92.9	91.7	84.9	87.2
0950, 0980	Ceilings & Acoustic Treatment	106.6	99.9	102.0	81.9	89.5	87.1	89.1	89.5	89.4	90.8	84.9	86.8	93.8	88.4	90.1	90.8	84.9	86.8
0960	Flooring	99.0	80.1	93.5	91.3	101.6	94.3	89.9	101.6	93.3	89.6	96.4	91.6	97.3	101.6	98.5	88.6	96.4	90.9
0970, 0990	Wall Finishes & Painting/Coating	91.1	78.4	83.6	92.3	94.6	93.7	101.5	92.2	96.0	101.5	90.3	94.9	90.7	94.6	93.0	101.5	90.3	94.9
09	FINISHES	101.7	91.3	96.0	86.2	92.7	89.7	88.1	92.4	90.5	89.0	88.2	88.5	92.4	92.4	92.4	88.2	88.2	88.2
COVERS	DIVS. 10 - 14, 25, 28, 41, 43, 44, 46	100.0	97.3	99.4	100.0	92.1	98.3	100.0	92.1	98.3	100.0	90.8	98.0	100.0	92.1	98.3	100.0	90.8	98.0
21, 22, 23	FIRE SUPPRESSION, PLUMBING & HVAC	100.1	109.8	104.0	98.2	93.2	96.2	98.0	85.9	93.1	98.0	91.7	95.4	100.1	94.3	97.8	96.9	91.5	94.1
26, 27, 3370	ELECTRICAL, COMMUNICATIONS & UTIL.	106.3	95.1	100.6	93.7	86.2	89.9	95.5	86.2	90.7	96.9	91.5	94.1	97.5	92.5	95.4	99.1	91.4	95.8
MF2018	WEIGHTED AVERAGE	102.7	96.6	100.1	96.6	92.1	94.6	97.2	90.2	94.2	98.5	90.7	95.1	97.5	92.5	95.4	99.1	91.4	95.8

WEST VIRGINIA

DIVISION		GASSAWAY 266			HUNTINGTON 255 - 257			LEWISBURG 249			MARTINSBURG 254			MORGANTOWN 265			PARKERSBURG 261		
		MAT.	INST.	TOTAL	MAT.	INST.	TOTAL	MAT.	INST.	TOTAL	MAT.	INST.	TOTAL	MAT.	INST.	TOTAL	MAT.	INST.	TOTAL
015433	CONTRACTOR EQUIPMENT		106.2	106.2		106.2	106.2		106.2	106.2		106.2	106.2		106.2	106.2		106.2	106.2
0241, 31 - 34	SITE & INFRASTRUCTURE, DEMOLITION	104.3	91.2	95.2	102.6	92.3	95.5	116.2	91.2	98.8	101.7	92.0	94.9	101.6	92.0	94.9	109.8	92.2	97.5
0310	Concrete Forming & Accessories	83.9	89.2	88.4	96.3	91.7	92.4	82.3	89.2	88.1	84.4	81.7	82.1	93.7	98.0	95.9	93.0	84.7	88.8
0320	Concrete Reinforcing	93.7	94.2	94.0	90.1	95.8	93.0	93.7	84.9	89.2	88.8	97.9	93.4	104.0	94.5	100.5	106.3	91.4	100.8
0330	Cast-in-Place Concrete	108.9	94.4	103.5	112.4	96.5	106.5	104.4	94.9	100.9	108.0	90.8	101.6	96.1	92.1	94.3	101.5	89.8	96.2
03	CONCRETE	99.9	92.8	96.7	100.0	95.0	97.8	106.0	91.4	99.4	98.3	88.8	94.0	96.1	92.1	94.3	101.5	89.8	96.2
04	MASONRY	105.9	89.7	95.9	89.0	95.4	92.9	93.6	90.9	91.9	90.5	85.7	87.5	122.8	86.5	100.4	80.0	87.0	84.3
05	METALS	100.7	104.9	102.0	104.1	105.9	104.7	100.7	101.4	100.9	102.2	105.0	103.0	100.8	106.2	102.4	101.4	101.6	101.5
06	WOOD, PLASTICS & COMPOSITES	81.0	89.9	85.8	92.8	91.2	91.9	78.8	89.9	84.8	81.7	80.9	81.3	78.8	85.4	82.4	104.2	89.1	95.7
07	THERMAL & MOISTURE PROTECTION	104.1	90.0	98.0	106.6	90.4	99.7	105.0	88.4	97.9	106.8	83.0	96.6	104.1	84.4	96.5	97.8	82.7	94.3
08	OPENINGS	97.2	86.0	94.6	95.7	87.0	93.7	98.9	83.9	95.4	97.4	80.2	86.1	91.7	84.9	87.2	94.8	87.6	90.1
0920	Plaster & Gypsum Board	93.2	89.5	90.8	104.1	90.8	95.4	91.7	89.5	90.3	97.4	80.2	86.1	91.7	84.9	87.2	94.8	87.6	90.1
0950, 0980	Ceilings & Acoustic Treatment	90.8	89.5	89.9	82.8	90.8	88.2	90.8	89.5	89.9	82.8	80.2	81.0	90.8	84.9	86.8	90.8	87.6	88.7
0960	Flooring	89.4	101.6	93.0	98.6	106.7	100.9	88.7	101.6	92.4	91.3	102.3	94.5	88.7	96.4	90.9	92.5	95.0	93.2
0970, 0990	Wall Finishes & Painting/Coating	101.5	94.6	97.4	92.3	92.2	92.3	101.5	66.2	80.6	92.3	85.9	88.6	101.5	90.3	94.9	101.5	85.9	92.3
09	FINISHES	88.4	92.4	90.6	89.4	94.7	92.3	89.2	89.6	89.4	86.6	85.4	86.0	87.8	88.2	88.0	89.9	89.3	89.6
COVERS	DIVS. 10 - 14, 25, 28, 41, 43, 44, 46	100.0	91.3	98.1	100.0	92.5	98.3	100.0	66.7	92.7	100.0	83.0	96.3	100.0	85.2	96.7	100.0	91.5	98.1
21, 22, 23	FIRE SUPPRESSION, PLUMBING & HVAC	98.0	85.3	92.9	100.7	92.6	97.5	98.0	93.1	96.0	98.2	82.7	92.0	98.0	91.8	95.5	100.4	90.3	96.3
26, 27, 3370	ELECTRICAL, COMMUNICATIONS & UTIL.	96.9	86.2	91.4	97.4	91.8	94.6	93.1	86.2	89.5	99.5	78.7	88.9	97.1	91.5	94.2	97.0	90.4	93.6
MF2018	WEIGHTED AVERAGE	98.4	90.5	95.0	99.0	94.3	97.0	98.8	90.5	95.2	98.1	86.1	92.9	99.0	91.5	95.8	98.4	90.7	95.1

For customer support on your Electrical Costs with RSMeans Data, call 800.448.8182.

557

City Cost Indexes

DIVISION		WEST VIRGINIA PETERSBURG 268			ROMNEY 267			WHEELING 260			WISCONSIN BELOIT 535			EAU CLAIRE 547			GREEN BAY 541 - 543		
		MAT.	INST.	TOTAL	MAT.	INST.	TOTAL	MAT.	INST.	TOTAL	MAT.	INST.	TOTAL	MAT.	INST.	TOTAL	MAT.	INST.	TOTAL
015433	CONTRACTOR EQUIPMENT		106.2	106.2		106.2	106.2		106.2	106.2		101.8	101.8		102.6	102.6		100.3	100.3
0241, 31 - 34	SITE & INFRASTRUCTURE, DEMOLITION	100.9	92.2	94.8	103.8	92.2	95.7	110.5	92.0	97.6	96.0	107.0	103.6	96.7	104.0	101.8	100.1	100.2	100.2
0310	Concrete Forming & Accessories	85.6	88.8	88.3	81.6	89.1	88.0	88.3	86.1	86.4	97.5	98.9	98.7	96.1	97.4	97.2	106.4	100.3	101.2
0320	Concrete Reinforcing	93.0	94.0	93.5	93.7	98.2	96.0	92.5	98.0	95.3	96.4	141.1	119.2	93.9	117.5	105.9	91.9	112.6	102.4
0330	Cast-in-Place Concrete	104.0	94.3	100.4	108.9	85.7	100.3	106.3	94.5	101.9	103.9	103.2	103.6	100.5	97.6	99.4	103.9	102.9	103.5
03	CONCRETE	96.3	92.5	94.6	99.7	90.4	95.5	101.5	92.1	97.3	94.9	107.6	100.6	93.1	101.1	96.7	96.3	103.4	99.5
04	MASONRY	96.7	86.5	90.4	94.0	89.7	91.4	104.2	86.4	93.2	98.5	102.4	100.9	91.3	98.1	95.5	122.4	99.9	108.5
05	METALS	100.9	104.2	101.9	100.9	106.2	102.5	101.6	106.2	103.0	97.9	113.0	102.5	94.4	106.7	98.2	96.8	105.1	99.3
06	WOOD, PLASTICS & COMPOSITES	83.1	89.9	86.8	77.9	89.9	84.4	83.8	85.4	84.7	98.6	96.9	97.6	102.0	97.0	99.3	109.6	100.0	104.4
07	THERMAL & MOISTURE PROTECTION	104.2	84.5	95.7	104.3	86.1	96.4	104.6	88.8	97.8	98.1	96.1	97.2	103.6	97.2	100.9	105.8	100.2	103.4
08	OPENINGS	100.1	86.0	96.8	100.0	86.9	97.0	98.6	84.4	95.3	100.4	111.4	103.0	105.4	102.2	104.7	101.3	104.2	102.0
0920	Plaster & Gypsum Board	94.5	89.5	91.2	91.4	89.5	90.2	94.8	84.9	88.3	93.7	97.2	96.0	105.7	97.2	100.1	104.2	100.4	101.7
0950, 0980	Ceilings & Acoustic Treatment	90.8	89.5	89.9	90.8	89.5	89.9	90.8	84.9	86.8	82.5	97.2	92.5	87.3	97.2	94.0	80.7	100.4	94.0
0960	Flooring	90.5	96.4	92.2	88.5	102.3	92.5	93.4	96.4	94.3	94.1	122.4	102.3	81.9	110.1	90.1	97.6	118.7	103.7
0970, 0990	Wall Finishes & Painting/Coating	101.5	90.3	94.9	101.5	90.3	94.9	101.5	90.3	94.9	94.3	106.3	101.4	83.0	78.4	80.3	91.5	82.1	85.9
09	FINISHES	88.7	90.9	89.9	87.9	92.1	90.2	90.1	88.2	89.1	90.0	103.7	97.5	86.0	97.7	92.4	90.2	102.0	96.6
COVERS	DIVS. 10 - 14, 25, 28, 41, 43, 44, 46	100.0	56.0	90.3	100.0	91.2	98.1	100.0	86.0	96.9	100.0	100.7	100.2	100.0	96.1	99.1	100.0	97.7	99.5
21, 22, 23	FIRE SUPPRESSION, PLUMBING & HVAC	98.0	85.6	93.0	98.0	85.4	92.9	100.4	91.8	97.0	100.1	97.6	99.1	100.1	90.2	96.1	100.3	85.3	94.3
26, 27, 3370	ELECTRICAL, COMMUNICATIONS & UTIL.	100.0	78.8	89.2	99.3	78.8	88.8	94.3	91.5	92.9	99.2	86.3	92.6	101.6	86.5	93.9	96.6	83.2	89.7
MF2018	WEIGHTED AVERAGE	98.2	87.7	93.7	98.4	89.2	94.4	99.5	91.5	96.1	98.0	101.4	99.5	97.5	96.5	97.1	99.4	95.8	97.9

DIVISION		WISCONSIN KENOSHA 531			LA CROSSE 546			LANCASTER 538			MADISON 537			MILWAUKEE 530, 532			NEW RICHMOND 540		
		MAT.	INST.	TOTAL	MAT.	INST.	TOTAL	MAT.	INST.	TOTAL	MAT.	INST.	TOTAL	MAT.	INST.	TOTAL	MAT.	INST.	TOTAL
015433	CONTRACTOR EQUIPMENT		99.8	99.8		102.6	102.6		101.8	101.8		101.8	101.8		89.6	89.6		102.9	102.9
0241, 31 - 34	SITE & INFRASTRUCTURE, DEMOLITION	101.8	103.6	103.0	90.4	104.0	99.9	95.1	106.3	102.9	93.7	106.9	102.9	94.0	95.2	94.9	94.8	104.7	101.7
0310	Concrete Forming & Accessories	107.7	109.1	108.9	82.6	97.2	95.0	96.8	97.8	97.7	101.4	98.3	98.8	100.6	110.5	109.0	91.2	95.1	94.5
0320	Concrete Reinforcing	96.2	113.7	105.1	93.5	109.3	101.5	97.8	109.3	103.7	100.6	109.5	105.1	99.9	113.9	107.0	91.0	115.1	101.1
0330	Cast-in-Place Concrete	113.0	105.8	110.4	90.3	99.4	93.7	103.3	102.7	103.1	101.3	102.5	101.7	90.5	110.4	97.9	104.6	105.8	105.0
03	CONCRETE	99.9	108.5	103.8	84.5	100.2	91.6	94.6	101.5	97.7	94.8	101.7	97.9	92.1	110.1	100.2	91.1	102.9	96.4
04	MASONRY	96.2	110.3	104.9	90.4	98.1	95.2	98.6	102.3	100.9	97.1	101.2	99.6	100.6	113.0	108.3	117.5	100.5	107.0
05	METALS	98.7	104.2	100.4	94.3	103.7	97.2	95.4	101.2	97.2	99.9	102.1	100.6	99.2	96.5	98.4	94.6	106.3	98.2
06	WOOD, PLASTICS & COMPOSITES	106.0	108.8	107.5	86.2	97.0	92.1	97.7	96.9	97.3	98.4	96.9	97.5	101.3	109.2	105.6	90.8	92.8	91.9
07	THERMAL & MOISTURE PROTECTION	98.6	106.9	102.2	103.0	96.1	100.0	97.9	96.3	97.3	94.3	102.8	97.9	101.5	109.0	104.7	104.4	103.1	103.8
08	OPENINGS	94.7	108.5	97.9	105.4	94.7	102.9	96.2	94.6	95.8	103.5	101.7	103.0	101.4	109.5	103.3	90.6	93.0	91.1
0920	Plaster & Gypsum Board	82.7	109.5	100.4	100.7	97.2	98.4	93.0	97.2	95.8	105.1	97.2	99.9	98.0	109.5	105.6	92.1	93.2	92.8
0950, 0980	Ceilings & Acoustic Treatment	82.5	109.5	100.8	86.5	97.2	93.8	79.1	97.2	91.4	89.5	97.2	94.7	89.7	109.5	103.1	55.9	93.2	81.2
0960	Flooring	111.8	118.7	113.8	76.0	115.4	87.4	93.7	114.7	99.8	93.4	114.7	99.6	100.9	114.7	104.9	91.2	118.4	99.1
0970, 0990	Wall Finishes & Painting/Coating	104.4	121.9	114.8	83.0	83.9	83.5	94.3	100.3	97.8	93.6	106.3	101.1	106.1	122.7	115.9	94.4	83.9	88.1
09	FINISHES	94.4	112.6	104.3	83.0	99.3	91.9	89.1	101.3	95.7	90.2	102.1	96.7	99.4	112.8	106.7	81.4	98.0	90.5
COVERS	DIVS. 10 - 14, 25, 28, 41, 43, 44, 46	100.0	99.4	99.9	100.0	94.7	98.8	100.0	86.0	96.9	100.0	98.6	99.7	100.0	102.7	100.6	100.0	97.1	99.4
21, 22, 23	FIRE SUPPRESSION, PLUMBING & HVAC	100.3	98.1	99.4	100.1	90.1	96.1	97.6	91.2	95.1	100.0	97.4	98.9	100.0	105.6	102.2	100.0	97.1	99.4
26, 27, 3370	ELECTRICAL, COMMUNICATIONS & UTIL.	99.6	99.5	99.5	101.9	86.5	94.0	98.9	86.9	92.8	99.0	96.3	97.6	100.0	101.6	100.8	97.1	91.0	94.6
MF2018	WEIGHTED AVERAGE	98.7	104.6	101.2	95.9	95.9	95.9	96.4	96.7	96.5	98.4	100.3	99.2	98.9	105.9	101.9	95.6	97.0	96.2

DIVISION		WISCONSIN OSHKOSH 549			PORTAGE 539			RACINE 534			RHINELANDER 545			SUPERIOR 548			WAUSAU 544		
		MAT.	INST.	TOTAL	MAT.	INST.	TOTAL	MAT.	INST.	TOTAL	MAT.	INST.	TOTAL	MAT.	INST.	TOTAL	MAT.	INST.	TOTAL
015433	CONTRACTOR EQUIPMENT		100.3	100.3		101.8	101.8		101.8	101.8		100.3	100.3		102.9	102.9		100.3	100.3
0241, 31 - 34	SITE & INFRASTRUCTURE, DEMOLITION	91.8	100.1	97.6	86.0	106.2	100.0	95.8	107.4	103.9	104.0	100.1	101.3	91.6	104.4	100.5	87.7	100.7	96.7
0310	Concrete Forming & Accessories	88.6	97.4	96.1	88.7	98.4	97.0	97.8	109.1	107.4	86.2	97.3	95.6	89.0	91.0	90.7	88.0	100.1	98.3
0320	Concrete Reinforcing	92.1	112.5	102.4	97.9	109.5	103.8	96.4	113.7	105.2	92.3	109.5	101.0	89.0	91.0	90.7	92.3	109.3	100.9
0330	Cast-in-Place Concrete	96.3	100.9	98.0	88.7	101.3	93.4	101.9	105.6	103.3	109.3	100.9	106.1	98.5	101.2	99.5	89.8	95.2	91.8
03	CONCRETE	86.3	101.4	93.1	82.7	101.5	91.1	94.0	108.4	100.4	98.1	100.8	99.3	85.8	98.2	91.4	81.5	100.1	89.9
04	MASONRY	104.9	99.8	101.8	97.4	102.3	100.4	98.5	110.3	105.8	122.2	99.7	108.3	116.7	102.5	107.9	104.3	99.7	101.5
05	METALS	94.8	104.4	97.8	96.1	105.3	98.9	99.6	104.2	101.0	94.7	103.3	97.3	95.5	104.2	98.2	94.6	103.8	97.4
06	WOOD, PLASTICS & COMPOSITES	87.9	97.0	92.8	87.1	96.9	92.4	98.9	108.8	104.3	85.3	97.0	91.6	89.0	88.2	88.6	87.3	100.0	94.2
07	THERMAL & MOISTURE PROTECTION	104.8	85.6	96.5	97.2	103.0	99.7	98.1	105.6	101.3	105.7	85.4	97.0	104.0	98.6	101.7	104.6	97.8	101.7
08	OPENINGS	97.7	99.0	98.0	96.4	102.6	97.8	100.4	108.5	102.3	97.7	94.6	97.0	90.0	93.8	90.9	104.6	97.8	98.4
0920	Plaster & Gypsum Board	90.8	97.2	95.0	85.6	97.2	93.3	93.7	109.5	104.1	90.8	97.2	95.0	92.0	88.5	89.7	90.8	100.4	97.1
0950, 0980	Ceilings & Acoustic Treatment	80.7	97.2	91.9	81.6	97.2	92.2	82.5	109.5	100.8	80.7	97.2	91.9	56.8	88.5	78.3	80.7	100.4	94.0
0960	Flooring	89.3	118.7	97.8	90.0	118.4	98.3	94.1	118.7	101.2	88.7	118.4	97.3	92.3	127.2	102.4	89.2	118.4	97.7
0970, 0990	Wall Finishes & Painting/Coating	89.1	107.0	99.7	94.3	100.3	97.8	94.3	119.0	108.9	89.1	82.1	84.9	83.6	108.6	98.3	89.1	85.4	86.9
09	FINISHES	85.1	102.9	94.8	86.9	102.3	95.3	90.0	112.3	102.1	86.0	100.1	93.7	80.8	99.2	90.8	84.8	102.3	94.4
COVERS	DIVS. 10 - 14, 25, 28, 41, 43, 44, 46	100.0	87.3	97.2	100.0	87.4	97.2	100.0	99.4	99.9	100.0	87.6	97.3	100.0	95.8	99.1	100.0	97.7	99.5
21, 22, 23	FIRE SUPPRESSION, PLUMBING & HVAC	97.8	84.9	92.7	97.6	98.0	97.8	100.1	98.2	99.3	97.8	90.6	94.9	97.1	93.0	95.4	97.8	90.8	95.0
26, 27, 3370	ELECTRICAL, COMMUNICATIONS & UTIL.	100.4	81.1	90.5	102.5	96.3	99.4	99.1	100.4	99.8	99.8	80.0	89.7	104.5	101.0	102.7	101.5	80.0	90.5
MF2018	WEIGHTED AVERAGE	95.6	94.2	95.0	94.8	100.5	97.3	98.1	105.0	101.1	98.2	94.6	96.6	95.3	98.7	96.8	95.0	95.8	95.3

For customer support on your Electrical Costs with RSMeans Data, call 800.448.8182.

DIVISION		WYOMING																	
		CASPER 826			CHEYENNE 820			NEWCASTLE 827			RAWLINS 823			RIVERTON 825			ROCK SPRINGS 829 - 831		
		MAT.	INST.	TOTAL	MAT.	INST.	TOTAL	MAT.	INST.	TOTAL	MAT.	INST.	TOTAL	MAT.	INST.	TOTAL	MAT.	INST.	TOTAL
015433	CONTRACTOR EQUIPMENT		97.6	97.6		97.6	97.6		97.6	97.6		97.6	97.6		97.6	97.6		97.6	97.6
0241, 31 - 34	SITE & INFRASTRUCTURE, DEMOLITION	95.6	94.7	94.9	90.9	94.7	93.5	84.1	94.2	91.1	97.2	94.2	95.1	91.0	94.1	93.2	88.1	94.2	92.3
0310	Concrete Forming & Accessories	99.8	54.3	61.1	101.2	67.2	72.3	91.8	72.5	75.4	96.1	72.7	76.2	90.8	61.9	66.2	98.3	62.8	68.2
0320	Concrete Reinforcing	109.6	84.9	97.0	105.0	84.9	94.8	112.9	84.7	98.6	112.6	84.8	98.4	113.6	84.7	98.9	113.6	83.0	98.0
0330	Cast-in-Place Concrete	104.4	76.0	93.8	98.6	76.0	90.2	99.6	75.0	90.4	99.7	75.1	90.5	99.6	75.0	90.5	99.6	75.0	90.4
03	CONCRETE	101.9	67.9	86.6	99.5	73.7	87.9	99.6	75.7	88.9	113.3	75.9	96.5	107.8	71.0	91.2	100.1	71.1	87.1
04	MASONRY	99.3	64.7	77.9	102.6	63.8	78.5	98.9	60.5	75.2	98.9	60.5	75.2	98.9	60.5	75.2	159.3	56.3	95.7
05	METALS	103.4	80.9	96.6	105.9	80.9	98.3	102.2	80.5	95.6	102.2	80.7	95.7	102.3	80.5	95.7	103.1	79.7	96.0
06	WOOD, PLASTICS & COMPOSITES	94.5	49.0	69.8	93.4	66.5	78.8	83.7	74.8	78.9	87.7	74.8	80.7	82.6	60.4	70.6	92.9	61.7	75.9
07	THERMAL & MOISTURE PROTECTION	100.4	66.2	85.7	95.4	67.9	83.6	97.0	65.5	83.4	98.5	76.8	89.2	97.9	69.4	85.6	97.1	68.3	84.7
08	OPENINGS	106.0	60.0	95.3	109.6	69.7	100.3	114.0	74.3	104.7	113.6	74.3	104.5	113.8	66.3	102.8	114.4	66.7	103.3
0920	Plaster & Gypsum Board	103.3	47.6	66.6	92.9	65.6	74.9	89.9	74.1	79.5	90.2	74.1	79.6	89.9	59.4	69.8	102.4	60.7	74.9
0950, 0980	Ceilings & Acoustic Treatment	105.4	47.6	66.2	99.3	65.6	76.5	101.2	74.1	82.9	101.2	74.1	82.9	101.2	59.4	72.9	101.2	60.7	73.7
0960	Flooring	104.2	72.0	94.8	102.2	72.0	93.5	96.1	46.0	81.6	98.9	46.0	83.6	95.6	62.3	85.9	101.2	46.0	85.2
0970, 0990	Wall Finishes & Painting/Coating	94.7	59.9	74.1	98.9	59.9	75.8	95.5	62.1	75.7	95.5	62.1	75.7	95.5	62.1	75.7	95.5	62.1	75.7
09	FINISHES	96.7	56.6	74.8	93.7	67.0	79.1	89.2	66.5	76.8	91.3	66.5	77.8	89.9	61.3	74.3	92.5	58.8	74.1
COVERS	DIVS. 10 - 14, 25, 28, 41, 43, 44, 46	100.0	93.7	98.6	100.0	89.0	97.6	100.0	93.8	98.6	100.0	93.8	98.6	100.0	79.0	95.4	100.0	84.3	96.5
21, 22, 23	FIRE SUPPRESSION, PLUMBING & HVAC	100.0	71.3	88.5	99.9	71.3	88.5	98.6	70.5	87.3	98.6	70.5	87.3	98.6	70.5	87.3	99.9	70.5	88.1
26, 27, 3370	ELECTRICAL, COMMUNICATIONS & UTIL.	101.2	61.8	81.0	103.1	62.7	82.4	101.9	63.9	82.4	101.9	63.9	82.4	101.9	60.6	80.7	100.0	78.5	89.0
MF2018	WEIGHTED AVERAGE	101.1	69.5	87.5	101.4	72.1	88.8	100.1	72.3	88.2	102.3	72.7	89.6	101.4	69.7	87.8	103.8	71.6	89.9

DIVISION		WYOMING												CANADA					
		SHERIDAN 828			WHEATLAND 822			WORLAND 824			YELLOWSTONE NAT'L PA 821			BARRIE, ONTARIO			BATHURST, NEW BRUNSWICK		
		MAT.	INST.	TOTAL	MAT.	INST.	TOTAL	MAT.	INST.	TOTAL	MAT.	INST.	TOTAL	MAT.	INST.	TOTAL	MAT.	INST.	TOTAL
015433	CONTRACTOR EQUIPMENT		97.6	97.6		97.6	97.6		97.6	97.6		97.6	97.6		103.0	103.0		102.7	102.7
0241, 31 - 34	SITE & INFRASTRUCTURE, DEMOLITION	91.2	94.7	93.6	88.6	94.2	92.5	86.1	94.2	91.7	86.2	94.2	91.7	118.0	99.7	105.3	105.8	95.5	98.6
0310	Concrete Forming & Accessories	98.9	62.7	68.1	93.7	49.7	56.3	93.8	62.7	67.4	93.8	62.8	67.5	124.4	84.6	90.6	102.5	59.4	65.9
0320	Concrete Reinforcing	113.6	84.8	98.9	112.9	84.2	98.3	113.6	84.7	98.9	115.6	84.7	99.9	177.7	89.4	132.8	144.5	59.2	101.1
0330	Cast-in-Place Concrete	103.0	76.0	92.9	104.0	74.9	93.2	99.6	75.0	90.4	99.6	75.0	90.4	156.6	84.6	129.8	111.1	57.9	91.3
03	CONCRETE	108.0	71.6	91.7	104.7	65.3	86.9	99.8	71.3	87.0	100.1	71.3	87.1	139.3	85.8	115.2	110.5	59.9	87.7
04	MASONRY	99.2	62.0	76.2	99.2	52.2	70.2	98.9	60.5	75.2	98.9	60.5	75.2	165.9	91.2	119.7	160.0	58.7	97.4
05	METALS	105.7	80.7	98.1	102.1	79.6	95.3	102.3	80.4	95.7	102.9	79.9	95.9	110.9	92.4	105.3	120.3	74.3	106.3
06	WOOD, PLASTICS & COMPOSITES	95.0	60.4	76.2	85.5	44.2	63.1	85.6	61.7	72.6	85.6	61.7	72.6	118.5	83.2	99.3	99.7	59.6	77.9
07	THERMAL & MOISTURE PROTECTION	98.2	66.7	84.7	97.2	59.5	81.0	97.0	66.4	83.9	96.4	66.4	83.5	114.4	88.5	103.2	109.9	60.1	88.5
08	OPENINGS	114.5	66.3	103.3	112.5	57.4	99.7	114.2	67.0	103.2	106.8	66.7	97.5	92.6	83.2	90.4	85.8	52.7	78.1
0920	Plaster & Gypsum Board	112.7	59.4	77.6	89.9	42.7	58.8	89.9	60.7	70.7	90.1	60.7	70.7	151.9	82.7	106.3	121.5	58.4	79.9
0950, 0980	Ceilings & Acoustic Treatment	104.0	59.4	73.8	101.2	42.7	61.6	101.2	60.7	73.7	102.0	60.7	74.0	95.1	82.7	86.7	115.9	58.4	76.9
0960	Flooring	99.5	62.3	88.7	97.6	45.2	82.4	97.6	46.0	82.6	97.6	46.0	82.6	118.9	90.2	110.6	100.1	42.9	83.5
0970, 0990	Wall Finishes & Painting/Coating	97.7	59.9	75.3	95.5	59.9	74.5	95.5	62.1	75.7	95.5	62.1	75.7	105.0	86.0	93.7	110.7	49.2	74.3
09	FINISHES	96.5	61.7	77.5	89.9	48.1	67.1	89.7	58.8	72.8	89.8	58.9	72.9	109.7	85.9	96.7	105.7	55.4	78.2
COVERS	DIVS. 10 - 14, 25, 28, 41, 43, 44, 46	100.0	94.9	98.9	100.0	83.2	96.3	100.0	77.8	95.1	100.0	77.8	95.1	139.2	66.4	123.2	131.1	59.1	115.3
21, 22, 23	FIRE SUPPRESSION, PLUMBING & HVAC	98.6	71.3	87.7	98.6	70.5	87.3	98.6	70.5	87.3	98.6	70.5	87.3	103.8	95.6	100.5	103.7	66.2	88.7
26, 27, 3370	ELECTRICAL, COMMUNICATIONS & UTIL.	104.4	60.6	82.0	101.9	78.5	89.9	101.9	78.5	89.9	100.7	88.2	94.3	115.3	84.9	99.8	108.7	57.2	82.4
MF2018	WEIGHTED AVERAGE	102.9	70.6	89.1	100.8	68.1	86.8	100.3	71.8	88.1	99.5	73.2	88.2	116.5	89.3	104.8	111.2	63.9	90.9

DIVISION		CANADA																	
		BRANDON, MANITOBA			BRANTFORD, ONTARIO			BRIDGEWATER, NOVA SCOTIA			CALGARY, ALBERTA			CAP-DE-LA-MADELEINE, QUEBEC			CHARLESBOURG, QUEBEC		
		MAT.	INST.	TOTAL	MAT.	INST.	TOTAL	MAT.	INST.	TOTAL	MAT.	INST.	TOTAL	MAT.	INST.	TOTAL	MAT.	INST.	TOTAL
015433	CONTRACTOR EQUIPMENT		104.7	104.7		103.1	103.1		102.7	102.7		132.6	132.6		103.5	103.5		103.5	103.5
0241, 31 - 34	SITE & INFRASTRUCTURE, DEMOLITION	128.1	98.0	107.1	117.5	100.1	105.4	102.0	97.4	98.8	126.7	125.7	126.0	97.8	98.9	98.6	97.8	98.9	98.6
0310	Concrete Forming & Accessories	140.8	68.0	78.9	122.8	91.7	96.3	95.7	69.8	73.7	117.7	97.4	100.4	128.7	81.4	88.5	128.7	81.4	88.5
0320	Concrete Reinforcing	184.3	56.1	119.1	171.2	88.1	128.9	145.8	49.0	96.6	136.9	84.2	110.1	145.8	74.6	109.6	145.8	74.6	109.6
0330	Cast-in-Place Concrete	104.8	72.4	92.7	127.3	105.0	119.0	130.8	69.0	107.8	137.0	107.0	125.8	102.7	90.2	98.1	102.7	90.2	98.1
03	CONCRETE	118.4	68.4	95.9	122.4	95.9	110.5	120.6	66.9	96.5	125.2	99.3	113.5	108.0	83.8	97.1	108.0	83.8	97.1
04	MASONRY	213.4	62.0	119.8	166.5	95.4	122.6	162.1	67.0	103.3	204.0	88.8	132.8	162.8	77.9	110.4	162.8	77.9	110.4
05	METALS	135.2	79.1	118.2	118.0	93.1	110.4	117.3	76.9	105.0	129.6	104.4	122.0	116.4	86.0	107.2	116.4	86.0	107.2
06	WOOD, PLASTICS & COMPOSITES	154.7	68.8	108.0	121.8	90.3	104.7	90.4	69.4	79.0	99.9	97.2	98.4	133.5	81.3	105.2	133.5	81.3	105.2
07	THERMAL & MOISTURE PROTECTION	127.7	70.6	103.1	119.2	94.0	108.4	113.7	69.6	94.7	125.7	96.5	113.1	112.5	86.5	101.3	112.5	86.5	101.3
08	OPENINGS	102.4	61.6	92.9	89.7	89.2	89.6	84.2	63.1	79.3	82.9	86.0	83.7	91.1	74.4	87.2	91.1	74.4	87.2
0920	Plaster & Gypsum Board	113.7	67.5	83.3	113.1	90.0	97.9	118.4	68.4	85.5	125.1	96.0	106.0	142.5	80.7	101.8	142.5	80.7	101.8
0950, 0980	Ceilings & Acoustic Treatment	122.4	67.5	85.2	104.7	90.0	94.7	104.7	68.4	80.1	150.3	96.0	113.5	104.7	80.7	88.4	104.7	80.7	88.4
0960	Flooring	131.3	64.0	111.8	113.6	90.2	106.8	96.1	60.9	85.9	121.3	86.7	111.2	113.6	88.8	106.5	113.6	88.8	106.5
0970, 0990	Wall Finishes & Painting/Coating	117.6	55.5	80.8	109.8	94.4	100.7	109.8	60.9	80.8	115.0	109.2	111.6	109.8	85.8	95.6	109.8	85.8	95.6
09	FINISHES	119.9	66.2	90.6	106.6	92.0	98.6	101.7	67.5	83.0	124.6	97.2	109.7	110.2	83.6	95.7	110.2	83.6	95.7
COVERS	DIVS. 10 - 14, 25, 28, 41, 43, 44, 46	131.1	61.4	115.8	131.1	68.4	117.3	131.1	62.1	115.9	131.1	94.0	123.0	131.1	77.2	119.3	131.1	77.2	119.3
21, 22, 23	FIRE SUPPRESSION, PLUMBING & HVAC	103.9	80.7	94.6	103.8	98.5	101.7	103.8	81.3	94.8	104.4	91.6	99.3	104.1	86.1	96.9	104.1	86.1	96.9
26, 27, 3370	ELECTRICAL, COMMUNICATIONS & UTIL.	110.6	64.7	87.1	107.6	84.5	95.8	111.9	60.1	85.4	102.1	96.2	99.1	106.6	66.8	86.2	106.6	66.8	86.2
MF2018	WEIGHTED AVERAGE	121.8	72.4	100.6	113.7	93.1	104.9	111.9	72.2	94.8	118.3	97.6	109.4	111.3	82.1	98.8	111.3	82.1	98.8

For customer support on your Electrical Costs with RSMeans Data, call 800.448.8182.

559

City Cost Indexes

CANADA

DIVISION		CHARLOTTETOWN, PRINCE EDWARD ISLAND			CHICOUTIMI, QUEBEC			CORNER BROOK, NEWFOUNDLAND			CORNWALL, ONTARIO			DALHOUSIE, NEW BRUNSWICK			DARTMOUTH, NOVA SCOTIA		
		MAT.	INST.	TOTAL	MAT.	INST.	TOTAL	MAT.	INST.	TOTAL	MAT.	INST.	TOTAL	MAT.	INST.	TOTAL	MAT.	INST.	TOTAL
015433	CONTRACTOR EQUIPMENT		120.9	120.9		103.3	103.3		103.2	103.2		103.1	103.1		103.0	103.0		102.0	102.0
0241, 31 - 34	SITE & INFRASTRUCTURE, DEMOLITION	132.9	108.0	115.5	102.7	98.4	99.7	132.3	95.4	106.6	115.7	99.6	104.5	101.1	95.7	97.3	119.5	97.0	103.8
0310	Concrete Forming & Accessories	115.9	54.4	63.6	131.2	89.2	95.5	117.6	77.7	83.7	120.5	84.7	90.1	101.7	59.7	66.0	107.4	69.8	75.4
0320	Concrete Reinforcing	164.4	47.9	105.2	109.0	93.9	101.3	169.0	50.2	108.6	171.2	87.8	128.8	146.3	59.2	102.1	176.7	49.0	111.8
0330	Cast-in-Place Concrete	147.6	58.3	114.4	104.7	94.9	101.0	122.0	65.5	101.0	114.5	95.7	107.5	106.9	58.1	88.7	117.7	68.8	99.5
03	CONCRETE	134.2	56.5	99.3	102.0	92.1	97.6	149.1	69.4	113.3	116.1	89.5	104.1	112.7	60.0	89.0	132.0	66.8	102.7
04	MASONRY	176.6	56.0	102.1	162.3	88.4	116.7	209.3	75.9	126.9	165.5	87.4	117.2	163.7	58.7	98.8	223.5	67.0	126.8
05	METALS	144.2	81.5	125.2	120.1	91.2	111.3	135.7	76.0	117.5	117.8	91.8	109.9	110.2	74.6	99.4	136.4	76.6	118.2
06	WOOD, PLASTICS & COMPOSITES	103.7	53.8	76.6	135.4	89.8	110.6	131.2	83.8	105.5	120.2	84.0	100.5	97.6	59.6	76.9	118.4	69.3	91.7
07	THERMAL & MOISTURE PROTECTION	126.0	58.4	96.9	110.4	96.0	104.2	132.2	67.8	104.5	119.0	88.6	105.9	118.1	60.1	93.2	129.2	69.5	103.5
08	OPENINGS	85.3	47.0	76.4	89.7	76.1	86.6	108.6	68.5	99.3	91.1	82.9	89.2	87.3	52.7	79.3	93.2	63.0	86.2
0920	Plaster & Gypsum Board	118.6	52.0	74.8	141.4	89.3	107.1	145.6	83.2	104.6	168.5	83.5	112.5	126.5	58.4	81.6	140.7	68.4	93.1
0950, 0980	Ceilings & Acoustic Treatment	128.3	52.0	76.6	115.1	89.3	97.6	123.2	83.2	96.1	107.2	83.5	91.2	107.0	58.4	74.1	130.0	68.4	88.3
0960	Flooring	116.7	57.1	99.4	115.8	88.8	108.0	113.2	51.2	95.2	113.6	88.9	106.5	102.3	65.6	91.7	108.1	60.9	94.4
0970, 0990	Wall Finishes & Painting/Coating	116.4	40.9	71.7	110.7	103.2	106.3	117.4	57.9	82.2	109.8	87.9	96.8	113.4	49.2	75.4	117.4	60.9	84.0
09	FINISHES	118.8	53.8	83.3	112.6	91.6	101.1	119.8	72.3	93.8	115.1	86.0	99.2	106.5	60.0	81.1	117.6	67.5	90.2
COVERS	DIVS. 10 - 14, 25, 28, 41, 43, 44, 46	131.1	59.3	115.3	131.1	80.9	120.1	131.1	61.9	115.9	131.1	66.0	116.8	131.1	59.1	115.3	131.1	62.1	115.9
21, 22, 23	FIRE SUPPRESSION, PLUMBING & HVAC	104.0	60.3	86.5	103.7	80.0	94.2	103.9	67.4	89.3	104.1	96.4	101.1	103.8	66.2	88.7	103.9	81.3	94.8
26, 27, 3370	ELECTRICAL, COMMUNICATIONS & UTIL.	108.1	48.6	77.7	105.4	83.1	94.0	108.0	52.8	79.8	107.8	85.4	96.4	108.6	54.1	80.7	112.1	60.1	85.5
MF2018	WEIGHTED AVERAGE	121.0	61.8	95.6	111.1	87.3	100.9	125.8	70.2	101.9	113.7	89.6	103.4	110.4	64.1	90.5	122.6	72.1	100.9

CANADA

DIVISION		EDMONTON, ALBERTA			FORT MCMURRAY, ALBERTA			FREDERICTON, NEW BRUNSWICK			GATINEAU, QUEBEC			GRANBY, QUEBEC			HALIFAX, NOVA SCOTIA		
		MAT.	INST.	TOTAL	MAT.	INST.	TOTAL	MAT.	INST.	TOTAL	MAT.	INST.	TOTAL	MAT.	INST.	TOTAL	MAT.	INST.	TOTAL
015433	CONTRACTOR EQUIPMENT		132.2	132.2		105.3	105.3		120.6	120.6		103.5	103.5		103.5	103.5		119.0	119.0
0241, 31 - 34	SITE & INFRASTRUCTURE, DEMOLITION	123.9	125.5	125.0	122.4	100.7	107.3	111.8	109.3	110.1	97.6	98.9	98.5	98.1	98.9	98.7	103.7	109.4	107.7
0310	Concrete Forming & Accessories	119.5	97.4	100.7	120.5	89.8	94.4	120.7	60.2	69.3	128.7	81.3	88.4	128.7	81.3	88.4	116.2	83.3	88.3
0320	Concrete Reinforcing	137.3	84.2	110.3	158.5	84.1	120.6	161.2	59.5	109.5	154.2	74.5	113.7	154.2	74.5	113.7	155.1	72.1	112.9
0330	Cast-in-Place Concrete	131.0	106.9	122.0	169.0	101.1	143.7	108.4	58.7	89.9	101.3	90.2	97.2	104.7	90.2	99.3	102.5	84.6	95.8
03	CONCRETE	122.5	99.2	112.0	140.6	93.0	119.2	115.4	61.3	91.1	108.3	83.7	97.3	109.9	83.7	98.1	110.9	83.1	98.4
04	MASONRY	199.0	88.8	130.9	206.9	85.5	131.9	183.6	60.0	107.2	162.7	77.9	110.3	163.0	77.9	110.4	181.0	85.3	121.9
05	METALS	136.5	104.2	126.7	147.2	91.6	130.3	144.5	86.5	126.9	116.4	85.9	107.1	116.6	85.8	107.3	144.4	97.6	130.2
06	WOOD, PLASTICS & COMPOSITES	98.8	97.2	97.9	116.0	89.3	101.5	109.0	59.8	82.3	133.5	81.3	105.2	133.5	81.3	105.2	104.1	82.5	92.4
07	THERMAL & MOISTURE PROTECTION	133.4	96.5	117.5	127.6	92.0	112.3	127.9	60.5	98.9	112.5	86.5	101.3	112.5	85.0	100.6	130.5	84.9	110.9
08	OPENINGS	82.0	86.0	82.9	91.1	81.6	88.9	86.7	51.7	78.6	91.1	69.9	86.2	91.1	69.9	86.2	87.1	75.2	84.4
0920	Plaster & Gypsum Board	122.4	96.0	105.0	113.6	88.6	97.1	121.7	58.4	80.0	112.4	80.7	91.5	112.4	80.7	91.5	111.5	81.7	91.9
0950, 0980	Ceilings & Acoustic Treatment	145.4	96.0	112.0	112.3	88.6	96.3	132.9	58.4	82.4	104.7	80.7	88.4	104.7	80.7	88.4	142.1	81.7	101.2
0960	Flooring	122.0	86.7	111.7	113.6	86.7	105.8	118.0	68.5	103.6	113.6	88.8	106.5	113.6	88.8	106.5	112.2	81.9	102.7
0970, 0990	Wall Finishes & Painting/Coating	114.8	109.2	111.5	109.9	90.6	98.5	116.0	62.7	84.4	109.8	85.8	95.6	109.8	85.8	95.6	119.2	90.0	101.9
09	FINISHES	119.6	97.2	107.4	109.5	89.6	98.6	120.1	62.1	88.5	105.8	83.6	93.7	105.8	83.6	93.7	115.1	84.2	98.2
COVERS	DIVS. 10 - 14, 25, 28, 41, 43, 44, 46	131.1	94.0	123.0	131.1	90.4	122.2	131.1	59.7	115.4	131.1	77.2	119.3	131.1	77.2	119.3	131.1	67.4	117.1
21, 22, 23	FIRE SUPPRESSION, PLUMBING & HVAC	104.3	91.6	99.2	104.2	95.3	100.6	104.1	75.3	92.6	104.1	86.1	96.9	103.8	86.1	96.7	104.2	80.9	94.9
26, 27, 3370	ELECTRICAL, COMMUNICATIONS & UTIL.	105.8	96.2	100.9	102.0	79.4	90.4	106.6	71.4	88.6	106.6	66.8	86.2	107.2	66.8	86.5	108.6	84.4	96.2
MF2018	WEIGHTED AVERAGE	118.9	97.5	109.7	122.9	90.2	108.9	118.7	71.2	98.4	111.0	81.9	98.5	111.3	81.8	98.6	117.7	85.8	104.0

CANADA

DIVISION		HAMILTON, ONTARIO			HULL, QUEBEC			JOLIETTE, QUEBEC			KAMLOOPS, BRITISH COLUMBIA			KINGSTON, ONTARIO			KITCHENER, ONTARIO		
		MAT.	INST.	TOTAL	MAT.	INST.	TOTAL	MAT.	INST.	TOTAL	MAT.	INST.	TOTAL	MAT.	INST.	TOTAL	MAT.	INST.	TOTAL
015433	CONTRACTOR EQUIPMENT		120.9	120.9		103.5	103.5		103.5	103.5		106.7	106.7		105.3	105.3		105.5	105.5
0241, 31 - 34	SITE & INFRASTRUCTURE, DEMOLITION	109.2	114.9	113.2	97.6	98.9	98.5	98.2	98.9	98.7	120.3	102.6	108.0	115.7	103.4	107.2	95.9	104.9	102.2
0310	Concrete Forming & Accessories	120.5	94.0	98.0	128.7	81.3	88.4	128.7	81.4	88.5	119.4	84.6	89.8	120.6	84.8	90.2	112.3	86.8	90.7
0320	Concrete Reinforcing	151.4	98.6	124.5	154.2	74.5	113.7	145.8	74.6	109.6	114.3	78.0	95.9	171.2	87.8	128.8	105.9	98.4	102.1
0330	Cast-in-Place Concrete	113.9	102.2	109.5	101.3	90.2	97.2	105.5	90.2	99.9	91.4	94.7	92.6	114.5	95.7	107.5	111.0	97.3	105.9
03	CONCRETE	113.1	98.3	106.4	108.3	83.7	97.3	109.3	83.8	97.8	117.4	87.4	103.9	118.0	89.5	105.1	111.0	97.3	105.9
04	MASONRY	167.0	99.7	125.4	162.7	77.9	110.3	163.1	77.9	110.5	170.0	85.7	117.9	172.4	87.5	119.9	143.0	96.9	114.5
05	METALS	143.5	105.7	132.0	116.6	85.9	107.3	116.6	86.0	107.3	118.4	88.2	109.3	119.1	91.8	110.8	131.7	96.3	120.9
06	WOOD, PLASTICS & COMPOSITES	102.6	93.1	97.4	133.5	81.3	105.2	133.5	81.3	105.2	103.4	83.0	92.3	120.2	84.1	100.6	108.5	84.8	95.6
07	THERMAL & MOISTURE PROTECTION	121.5	100.0	112.2	112.5	86.5	101.3	112.5	86.5	101.3	128.8	84.1	109.6	119.0	89.8	106.4	113.6	96.8	106.4
08	OPENINGS	83.5	91.9	85.4	91.1	69.9	86.2	91.1	74.4	87.2	87.9	80.8	86.3	91.1	82.6	89.1	81.8	85.6	82.7
0920	Plaster & Gypsum Board	118.6	92.3	101.3	112.4	80.7	91.5	142.5	80.7	101.8	98.6	82.1	87.8	171.0	83.6	113.5	102.6	84.4	90.6
0950, 0980	Ceilings & Acoustic Treatment	134.8	92.3	106.0	104.7	80.7	88.4	104.7	80.7	88.4	104.7	82.1	89.4	119.1	83.6	95.1	107.6	84.4	91.8
0960	Flooring	116.4	95.6	110.4	113.6	88.8	106.5	113.6	88.8	106.5	112.8	50.8	94.8	113.6	88.9	106.5	101.4	95.6	99.7
0970, 0990	Wall Finishes & Painting/Coating	110.4	103.5	106.3	109.8	85.8	95.6	109.8	85.8	95.6	109.8	78.4	91.2	109.8	81.3	92.9	103.8	93.3	97.6
09	FINISHES	115.6	95.3	104.5	105.8	83.6	93.7	110.2	83.6	95.7	106.1	78.0	90.8	117.8	85.4	100.1	99.6	88.7	93.6
COVERS	DIVS. 10 - 14, 25, 28, 41, 43, 44, 46	131.1	91.8	122.5	131.1	77.2	119.3	131.1	77.2	119.3	131.1	86.2	121.2	131.1	66.0	116.8	131.1	89.0	121.9
21, 22, 23	FIRE SUPPRESSION, PLUMBING & HVAC	104.4	90.5	98.8	103.8	86.1	96.7	103.8	86.1	96.7	103.8	89.5	98.1	104.1	96.6	101.1	103.7	89.2	97.9
26, 27, 3370	ELECTRICAL, COMMUNICATIONS & UTIL.	106.4	99.6	102.9	108.5	66.8	87.2	107.2	66.8	86.5	110.4	76.0	92.8	107.9	84.2	95.7	108.2	97.1	102.5
MF2018	WEIGHTED AVERAGE	116.6	98.1	108.6	111.2	81.9	98.6	111.5	82.1	98.9	113.6	85.7	101.6	114.7	89.7	104.0	109.9	93.4	102.8

For customer support on your Electrical Costs with RSMeans Data, call 800.448.8182.

City Cost Indexes

CANADA

DIVISION		LAVAL, QUEBEC			LETHBRIDGE, ALBERTA			LLOYDMINSTER, ALBERTA			LONDON, ONTARIO			MEDICINE HAT, ALBERTA			MONCTON, NEW BRUNSWICK		
		MAT.	INST.	TOTAL	MAT.	INST.	TOTAL	MAT.	INST.	TOTAL	MAT.	INST.	TOTAL	MAT.	INST.	TOTAL	MAT.	INST.	TOTAL
015433	CONTRACTOR EQUIPMENT		103.5	103.5		105.3	105.3		105.3	105.3		122.4	122.4		105.3	105.3		102.7	102.7
0241, 31 - 34	SITE & INFRASTRUCTURE, DEMOLITION	98.0	98.9	98.6	115.1	101.3	105.5	115.0	100.7	105.0	103.8	114.7	111.4	113.8	100.8	104.7	105.2	95.8	98.7
0310	Concrete Forming & Accessories	128.9	81.3	88.5	122.0	89.9	94.7	120.1	80.3	86.3	120.7	88.4	93.2	121.9	80.3	86.6	102.5	62.0	68.1
0320	Concrete Reinforcing	154.2	74.5	113.7	158.5	84.1	120.6	158.5	84.1	120.6	151.4	97.3	123.9	158.5	84.0	120.6	144.5	70.8	107.0
0330	Cast-in-Place Concrete	104.7	90.2	99.3	126.8	101.1	117.2	117.6	97.5	110.1	120.4	100.1	112.8	117.6	97.5	110.1	107.0	68.2	92.6
03	CONCRETE	109.9	83.7	98.1	120.6	93.1	108.2	116.1	87.5	103.2	116.2	94.9	106.6	116.2	87.5	103.3	108.6	66.9	89.8
04	MASONRY	162.9	77.9	110.4	180.7	85.5	121.9	162.0	79.2	110.9	182.7	97.9	130.3	162.0	79.2	110.9	159.7	60.1	98.2
05	METALS	116.5	85.9	107.2	141.9	91.6	126.6	118.4	91.5	110.2	143.5	107.0	132.4	118.6	91.4	110.4	120.3	84.9	109.6
06	WOOD, PLASTICS & COMPOSITES	133.6	81.3	105.2	119.5	89.3	103.1	116.0	79.7	96.3	109.2	86.3	96.7	119.5	79.7	97.9	99.7	61.5	78.9
07	THERMAL & MOISTURE PROTECTION	113.0	86.5	101.6	124.8	92.0	110.7	121.3	87.5	106.7	122.7	97.6	111.9	128.1	87.5	110.6	114.6	64.2	92.9
08	OPENINGS	91.1	69.9	86.2	91.1	81.6	88.9	91.1	76.4	87.7	80.2	87.0	81.8	91.1	76.4	87.7	85.8	60.2	79.8
0920	Plaster & Gypsum Board	112.7	80.7	91.6	103.7	88.6	93.7	99.9	78.8	86.0	123.9	85.3	98.5	102.1	78.8	86.7	121.5	60.3	81.2
0950, 0980	Ceilings & Acoustic Treatment	104.7	80.7	88.4	112.3	88.6	96.3	104.7	78.8	87.1	142.7	85.3	103.8	104.7	78.8	87.1	115.9	60.3	78.2
0960	Flooring	113.6	88.8	106.5	113.6	86.7	105.8	113.6	86.7	105.8	112.8	95.6	107.8	113.6	86.7	105.8	100.1	67.2	90.5
0970, 0990	Wall Finishes & Painting/Coating	109.8	85.8	95.6	109.7	99.0	103.4	109.9	77.1	90.5	113.8	100.1	105.7	109.7	77.1	90.4	110.7	50.4	75.1
09	FINISHES	105.9	83.6	93.7	107.2	90.5	98.1	105.2	80.9	92.0	118.5	90.6	103.3	105.4	80.9	92.0	105.7	61.7	81.6
COVERS	DIVS. 10 - 14, 25, 28, 41, 43, 44, 46	131.1	77.2	119.3	131.1	90.4	122.2	131.1	87.3	121.5	131.1	90.7	122.2	131.1	87.3	121.5	131.1	60.7	115.6
21, 22, 23	FIRE SUPPRESSION, PLUMBING & HVAC	103.8	86.1	96.7	104.0	92.1	99.2	104.1	92.0	99.3	104.4	87.8	97.7	103.8	88.8	97.8	103.7	68.6	89.6
26, 27, 3370	ELECTRICAL, COMMUNICATIONS & UTIL.	105.6	66.8	85.8	103.4	79.4	91.1	101.2	79.4	90.0	103.6	97.1	100.3	101.2	79.4	90.0	112.3	59.0	85.0
MF2018	WEIGHTED AVERAGE	111.1	81.9	98.6	118.0	89.7	105.8	112.2	86.5	101.2	117.2	95.6	108.0	112.4	85.8	101.0	111.4	68.1	92.8

CANADA

DIVISION		MONTREAL, QUEBEC			MOOSE JAW, SASKATCHEWAN			NEW GLASGOW, NOVA SCOTIA			NEWCASTLE, NEW BRUNSWICK			NORTH BAY, ONTARIO			OSHAWA, ONTARIO		
		MAT.	INST.	TOTAL	MAT.	INST.	TOTAL	MAT.	INST.	TOTAL	MAT.	INST.	TOTAL	MAT.	INST.	TOTAL	MAT.	INST.	TOTAL
015433	CONTRACTOR EQUIPMENT		121.9	121.9		101.7	101.7		102.0	102.0		102.7	102.7		102.4	102.4		105.5	105.5
0241, 31 - 34	SITE & INFRASTRUCTURE, DEMOLITION	113.4	112.2	112.6	114.9	95.0	101.0	113.7	97.0	102.1	105.8	95.5	98.6	128.5	98.7	107.8	106.6	103.9	104.7
0310	Concrete Forming & Accessories	124.5	89.7	94.9	105.4	56.1	63.5	107.4	69.8	75.4	102.5	59.7	66.1	141.7	82.3	91.2	117.6	87.0	91.6
0320	Concrete Reinforcing	134.8	94.0	114.0	111.8	63.2	87.1	169.0	49.0	108.0	144.5	59.2	101.2	199.7	87.3	142.6	167.8	93.3	129.9
0330	Cast-in-Place Concrete	123.6	96.6	113.5	114.1	65.7	96.1	117.7	68.8	99.5	111.1	58.0	91.4	113.4	82.6	101.9	128.3	85.4	112.3
03	CONCRETE	116.0	93.6	105.9	101.9	61.7	83.8	131.0	66.8	102.2	110.5	60.0	87.8	133.0	83.7	110.8	121.0	87.9	106.1
04	MASONRY	168.0	88.4	118.9	160.7	57.6	97.0	208.9	67.0	121.2	160.0	58.7	97.4	215.0	83.6	133.8	146.0	89.9	111.4
05	METALS	151.6	102.6	136.7	115.4	75.9	103.4	133.8	76.6	116.4	120.3	74.5	106.4	134.5	91.2	121.3	121.1	95.0	113.2
06	WOOD, PLASTICS & COMPOSITES	108.8	90.2	98.7	100.7	54.7	75.7	118.4	69.3	91.7	99.7	59.6	77.9	157.8	82.6	117.0	115.5	85.9	99.4
07	THERMAL & MOISTURE PROTECTION	119.6	96.2	109.5	111.8	61.8	90.3	129.2	69.5	103.5	114.6	60.1	91.1	135.7	85.0	113.9	114.5	89.3	103.7
08	OPENINGS	85.7	78.2	83.9	87.1	52.9	79.2	93.2	63.0	86.2	85.8	52.7	78.1	100.8	80.5	96.1	86.9	87.1	86.9
0920	Plaster & Gypsum Board	117.7	89.3	99.0	96.1	53.4	68.0	139.1	68.4	92.6	121.5	58.4	79.9	136.4	82.1	100.6	105.7	85.4	92.4
0950, 0980	Ceilings & Acoustic Treatment	140.2	89.3	105.7	104.7	53.4	70.0	122.4	68.4	85.8	115.9	58.4	76.9	122.4	82.1	95.1	103.3	85.4	91.2
0960	Flooring	114.0	91.0	107.3	103.9	55.5	89.8	108.1	60.9	94.4	100.1	65.6	90.0	131.3	88.9	119.0	104.3	98.1	102.5
0970, 0990	Wall Finishes & Painting/Coating	112.2	103.2	106.9	109.8	62.6	81.9	117.4	60.9	84.0	110.7	49.2	74.3	117.4	87.3	99.6	103.8	107.5	106.0
09	FINISHES	116.5	92.2	103.2	101.7	56.4	76.9	115.9	67.5	89.4	105.7	59.9	80.7	122.9	84.3	101.9	100.7	91.0	95.4
COVERS	DIVS. 10 - 14, 25, 28, 41, 43, 44, 46	131.1	81.8	120.3	131.1	58.7	115.2	131.1	62.1	115.9	131.1	59.1	115.3	131.1	64.7	116.5	131.1	88.6	121.8
21, 22, 23	FIRE SUPPRESSION, PLUMBING & HVAC	104.6	80.1	94.8	104.1	71.9	91.2	103.9	81.3	94.8	103.7	66.2	88.7	103.9	94.5	100.1	103.7	100.0	102.2
26, 27, 3370	ELECTRICAL, COMMUNICATIONS & UTIL.	106.0	83.1	94.3	109.5	57.7	83.0	108.3	60.1	83.7	108.1	57.2	82.1	109.0	85.3	96.9	109.2	87.6	98.2
MF2018	WEIGHTED AVERAGE	118.7	89.9	106.3	109.7	65.6	90.8	120.7	72.1	99.9	111.3	64.5	91.2	123.7	87.4	108.1	112.0	92.9	103.8

CANADA

DIVISION		OTTAWA, ONTARIO			OWEN SOUND, ONTARIO			PETERBOROUGH, ONTARIO			PORTAGE LA PRAIRIE, MANITOBA			PRINCE ALBERT, SASKATCHEWAN			PRINCE GEORGE, BRITISH COLUMBIA		
		MAT.	INST.	TOTAL	MAT.	INST.	TOTAL	MAT.	INST.	TOTAL	MAT.	INST.	TOTAL	MAT.	INST.	TOTAL	MAT.	INST.	TOTAL
015433	CONTRACTOR EQUIPMENT		123.1	123.1		103.0	103.0		103.1	103.1		105.3	105.3		101.7	101.7		106.7	106.7
0241, 31 - 34	SITE & INFRASTRUCTURE, DEMOLITION	108.2	114.7	112.7	118.0	99.6	105.2	117.5	99.5	105.0	116.2	98.4	103.8	110.6	95.2	99.9	123.7	102.6	109.0
0310	Concrete Forming & Accessories	121.9	88.4	93.4	124.4	80.9	87.4	122.8	83.3	89.2	122.1	67.6	75.8	105.4	55.9	63.4	110.7	79.7	84.3
0320	Concrete Reinforcing	135.1	97.3	115.9	177.7	89.4	132.8	171.2	87.8	128.8	158.5	56.1	106.4	116.8	63.1	89.5	114.3	78.0	95.9
0330	Cast-in-Place Concrete	121.6	102.9	114.6	156.6	78.9	127.7	127.3	84.3	111.3	117.6	71.9	100.6	103.4	65.6	89.3	114.5	94.7	107.1
03	CONCRETE	114.9	96.0	106.4	139.3	82.2	113.6	122.4	84.9	105.5	110.1	68.1	91.2	97.4	61.6	81.3	127.9	85.2	108.7
04	MASONRY	164.0	100.4	124.7	165.9	88.9	118.3	166.5	89.9	119.2	165.2	61.0	100.8	159.9	57.6	96.7	172.0	85.7	118.7
05	METALS	144.3	108.4	133.4	110.9	92.2	105.2	118.0	91.9	110.1	118.6	79.3	106.7	115.4	75.7	103.4	118.4	88.2	109.3
06	WOOD, PLASTICS & COMPOSITES	107.8	85.2	95.5	118.5	79.5	97.3	121.8	81.4	99.8	119.5	68.8	91.9	100.7	54.7	75.7	103.4	76.3	88.7
07	THERMAL & MOISTURE PROTECTION	131.9	98.8	117.7	114.4	85.8	102.1	119.2	90.6	106.9	112.3	70.2	94.2	111.7	60.7	89.7	122.5	83.3	105.6
08	OPENINGS	88.5	86.3	88.0	92.6	79.9	89.6	89.7	82.3	88.0	91.1	61.7	84.2	86.1	52.9	78.4	87.9	77.1	85.4
0920	Plaster & Gypsum Board	120.1	84.1	96.4	151.9	78.9	103.8	113.1	80.8	91.8	101.7	67.5	79.2	96.1	53.4	68.0	98.6	75.2	83.2
0950, 0980	Ceilings & Acoustic Treatment	136.4	84.1	101.0	95.1	78.9	84.1	104.7	80.8	88.5	104.7	67.5	79.5	104.7	53.4	70.0	104.7	75.2	84.7
0960	Flooring	110.7	91.1	105.0	118.9	90.2	110.6	113.6	88.9	106.5	113.6	64.0	99.2	109.3	55.5	89.8	109.2	69.7	97.7
0970, 0990	Wall Finishes & Painting/Coating	110.6	94.9	101.3	105.0	86.0	93.7	109.8	89.4	97.7	109.9	55.5	77.7	109.8	53.4	76.4	109.8	78.4	91.2
09	FINISHES	117.0	89.1	101.8	109.7	83.2	95.2	106.6	85.0	94.8	105.3	66.0	83.8	101.7	55.4	76.4	105.0	77.3	89.9
COVERS	DIVS. 10 - 14, 25, 28, 41, 43, 44, 46	131.1	89.3	121.9	139.2	65.2	122.9	131.1	66.2	116.8	131.1	61.1	115.7	131.1	58.7	115.2	131.1	85.5	121.1
21, 22, 23	FIRE SUPPRESSION, PLUMBING & HVAC	104.4	89.5	98.4	103.8	94.4	100.0	103.8	97.9	101.4	103.8	80.1	94.3	104.1	64.8	88.4	103.8	89.5	98.1
26, 27, 3370	ELECTRICAL, COMMUNICATIONS & UTIL.	103.5	97.9	100.6	116.7	94.2	100.1	107.6	84.9	96.0	109.2	55.8	81.8	109.5	57.7	83.0	107.6	76.0	91.4
MF2018	WEIGHTED AVERAGE	117.4	96.4	108.4	116.6	87.6	104.2	113.7	89.4	103.3	112.2	70.9	94.5	108.9	63.9	89.6	114.5	85.0	101.9

CANADA

	DIVISION	QUEBEC CITY, QUEBEC			RED DEER, ALBERTA			REGINA, SASKATCHEWAN			RIMOUSKI, QUEBEC			ROUYN-NORANDA, QUEBEC			SAINT HYACINTHE, QUEBEC		
		MAT.	INST.	TOTAL	MAT.	INST.	TOTAL	MAT.	INST.	TOTAL	MAT.	INST.	TOTAL	MAT.	INST.	TOTAL	MAT.	INST.	TOTAL
015433	CONTRACTOR EQUIPMENT		122.8	122.8		105.3	105.3		127.7	127.7		103.5	103.5		103.5	103.5		103.5	103.5
0241, 31 - 34	SITE & INFRASTRUCTURE, DEMOLITION	115.8	112.1	113.2	113.8	100.8	104.7	126.6	119.3	121.5	97.8	98.5	98.3	97.6	98.9	98.5	98.0	98.9	98.6
0310	Concrete Forming & Accessories	121.8	89.8	94.6	134.6	80.3	88.5	123.8	93.8	98.3	128.7	89.3	95.2	128.7	81.3	88.4	128.7	81.3	88.4
0320	Concrete Reinforcing	135.0	94.0	114.2	158.5	84.0	120.6	137.5	91.4	114.1	110.0	93.9	101.8	154.2	74.5	113.7	154.2	74.5	113.7
0330	Cast-in-Place Concrete	118.2	96.3	110.1	117.6	97.5	110.1	138.1	99.2	123.6	106.6	94.9	102.2	101.3	90.2	97.2	104.7	90.2	99.3
03	CONCRETE	113.3	93.7	104.5	117.0	87.5	103.7	124.6	96.1	111.7	105.5	92.1	99.5	108.3	83.7	97.3	109.9	83.7	98.1
04	MASONRY	162.6	88.4	116.8	162.0	79.2	110.9	192.6	89.2	128.7	162.5	88.4	116.7	162.7	77.9	110.3	162.9	77.9	110.4
05	METALS	144.6	104.4	132.4	118.6	91.4	110.4	153.6	103.1	138.2	116.1	91.3	108.6	116.6	85.9	107.3	116.6	85.9	107.3
06	WOOD, PLASTICS & COMPOSITES	113.7	90.1	100.9	119.5	79.7	97.9	110.3	94.6	101.8	133.5	89.8	109.7	133.5	81.3	105.2	133.5	81.3	105.2
07	THERMAL & MOISTURE PROTECTION	119.2	96.2	109.3	138.5	87.5	116.5	139.3	88.5	117.4	112.5	96.0	105.4	112.5	86.5	101.3	112.7	86.5	101.4
08	OPENINGS	87.4	85.6	87.0	91.1	76.4	87.7	86.3	81.8	85.3	90.7	76.1	87.3	91.1	69.9	86.2	91.1	69.9	86.2
0920	Plaster & Gypsum Board	121.7	89.3	100.4	102.1	78.8	86.7	131.2	93.8	106.6	142.3	89.3	107.4	112.2	80.7	91.4	112.2	80.7	91.4
0950, 0980	Ceilings & Acoustic Treatment	139.5	89.3	105.5	104.7	78.8	87.1	147.1	93.8	111.0	103.8	89.3	94.0	103.8	80.7	88.1	103.8	80.7	88.1
0960	Flooring	112.3	91.0	106.1	116.3	86.7	107.7	123.9	95.6	115.7	115.0	88.8	107.4	113.6	88.8	106.5	113.6	88.8	106.5
0970, 0990	Wall Finishes & Painting/Coating	122.0	103.2	110.9	109.7	77.1	90.4	115.3	91.4	101.2	113.0	103.2	107.2	109.8	85.8	95.6	109.8	85.8	95.6
09	FINISHES	116.5	92.2	103.2	106.1	80.9	92.4	128.0	94.9	109.9	110.7	91.6	100.2	105.7	83.6	93.6	105.7	83.6	93.6
COVERS	DIVS. 10 - 14, 25, 28, 41, 43, 44, 46	131.1	81.6	120.2	131.1	87.3	121.5	131.1	70.8	117.8	131.1	80.9	120.1	131.1	77.2	119.3	131.1	77.2	119.3
21, 22, 23	FIRE SUPPRESSION, PLUMBING & HVAC	104.4	80.1	94.6	103.8	88.8	97.8	104.1	88.8	98.0	103.8	80.0	94.2	103.8	86.1	96.7	100.4	86.1	94.6
26, 27, 3370	ELECTRICAL, COMMUNICATIONS & UTIL.	105.7	83.1	94.2	101.2	79.4	90.0	104.3	93.7	98.9	107.2	83.1	94.9	107.2	66.8	86.5	107.8	66.8	86.8
MF2018	WEIGHTED AVERAGE	117.1	90.3	105.6	112.8	85.8	101.2	122.8	94.2	110.5	111.0	87.3	100.8	111.0	81.9	98.5	110.5	81.9	98.2

CANADA

	DIVISION	SAINT JOHN, NEW BRUNSWICK			SARNIA, ONTARIO			SASKATOON, SASKATCHEWAN			SAULT STE MARIE, ONTARIO			SHERBROOKE, QUEBEC			SOREL, QUEBEC		
		MAT.	INST.	TOTAL	MAT.	INST.	TOTAL	MAT.	INST.	TOTAL	MAT.	INST.	TOTAL	MAT.	INST.	TOTAL	MAT.	INST.	TOTAL
015433	CONTRACTOR EQUIPMENT		102.7	102.7		103.1	103.1		101.7	101.7		103.1	103.1		103.5	103.5		103.5	103.5
0241, 31 - 34	SITE & INFRASTRUCTURE, DEMOLITION	105.9	97.2	99.8	116.1	99.6	104.6	112.3	97.9	102.3	106.1	99.2	101.3	98.1	98.9	98.7	98.2	98.9	98.7
0310	Concrete Forming & Accessories	122.6	65.0	73.6	121.5	90.1	94.8	105.3	93.2	95.0	110.8	86.2	89.9	128.7	81.3	88.4	128.7	81.4	88.5
0320	Concrete Reinforcing	144.5	71.4	107.3	122.2	89.2	105.4	117.1	91.3	104.0	109.8	87.9	98.7	154.2	74.5	113.7	145.8	74.6	109.6
0330	Cast-in-Place Concrete	109.5	70.0	94.8	117.4	97.2	109.9	109.7	96.0	104.6	105.5	83.0	97.1	104.7	90.2	99.3	105.5	90.2	99.9
03	CONCRETE	111.2	68.9	92.2	111.7	92.6	103.1	101.2	93.8	97.9	97.9	85.8	92.5	109.9	83.7	98.1	109.3	83.8	97.8
04	MASONRY	181.3	70.0	112.6	178.2	92.5	125.2	172.6	89.2	121.1	163.4	91.9	119.2	163.0	77.9	110.4	163.1	77.9	110.5
05	METALS	120.2	86.2	109.9	117.9	92.3	110.1	111.2	89.6	104.6	117.1	94.0	110.1	116.4	85.9	107.1	116.6	86.0	107.3
06	WOOD, PLASTICS & COMPOSITES	123.8	63.2	90.9	121.0	89.3	103.8	98.0	94.0	95.8	107.8	87.1	96.6	133.5	81.3	105.2	133.5	81.3	105.2
07	THERMAL & MOISTURE PROTECTION	114.9	71.0	96.0	119.3	94.1	108.4	114.4	87.4	102.8	118.0	89.4	105.7	112.5	86.5	101.3	112.5	86.5	101.3
08	OPENINGS	85.7	60.0	79.7	92.4	86.0	90.9	87.1	81.4	85.8	84.2	85.3	84.4	91.1	69.9	86.2	91.1	74.4	87.2
0920	Plaster & Gypsum Board	135.3	62.1	87.1	137.6	89.0	105.6	113.0	93.8	100.4	104.4	86.8	92.8	112.2	80.7	91.4	142.3	80.7	101.7
0950, 0980	Ceilings & Acoustic Treatment	121.0	62.1	81.1	108.9	89.0	95.4	124.0	93.8	103.6	104.7	86.8	92.5	103.8	80.7	88.1	103.8	80.7	88.1
0960	Flooring	111.6	67.2	98.7	113.6	97.2	108.9	106.8	95.6	103.6	106.7	94.2	103.1	113.6	88.8	106.5	113.6	88.8	106.5
0970, 0990	Wall Finishes & Painting/Coating	110.7	82.7	94.1	109.8	101.1	104.7	113.4	91.4	100.4	109.8	93.7	100.3	109.8	85.8	95.6	109.8	85.8	95.6
09	FINISHES	112.1	66.8	87.4	111.0	92.9	101.1	109.9	94.4	101.4	102.7	88.7	95.0	105.7	83.6	93.6	110.0	83.6	95.6
COVERS	DIVS. 10 - 14, 25, 28, 41, 43, 44, 46	131.1	61.8	115.9	131.1	67.5	117.1	131.1	69.6	117.6	131.1	87.7	121.6	131.1	77.2	119.3	131.1	77.2	119.3
21, 22, 23	FIRE SUPPRESSION, PLUMBING & HVAC	103.7	76.5	92.8	103.8	103.8	103.8	104.0	88.7	97.8	103.8	92.3	99.1	104.1	86.1	96.9	103.8	86.1	96.7
26, 27, 3370	ELECTRICAL, COMMUNICATIONS & UTIL.	114.8	83.7	98.9	110.5	87.1	98.5	107.5	93.7	100.4	109.2	85.3	97.0	107.2	66.8	86.5	107.2	66.8	86.5
MF2018	WEIGHTED AVERAGE	113.8	75.6	97.4	113.8	93.7	105.2	109.9	90.8	101.7	109.3	90.0	101.0	111.3	81.9	98.7	111.5	82.1	98.9

CANADA

	DIVISION	ST. CATHARINES, ONTARIO			ST JEROME, QUEBEC			ST. JOHN'S, NEWFOUNDLAND			SUDBURY, ONTARIO			SUMMERSIDE, PRINCE EDWARD ISLAND			SYDNEY, NOVA SCOTIA		
		MAT.	INST.	TOTAL	MAT.	INST.	TOTAL	MAT.	INST.	TOTAL	MAT.	INST.	TOTAL	MAT.	INST.	TOTAL	MAT.	INST.	TOTAL
015433	CONTRACTOR EQUIPMENT		103.4	103.4		103.5	103.5		122.7	122.7		103.4	103.4		102.0	102.0		102.0	102.0
0241, 31 - 34	SITE & INFRASTRUCTURE, DEMOLITION	96.0	101.4	99.8	97.6	98.9	98.5	118.7	111.9	113.9	96.1	101.0	99.5	123.5	94.3	103.2	109.8	97.0	100.9
0310	Concrete Forming & Accessories	110.3	93.0	95.6	128.7	81.3	88.4	119.4	86.4	91.3	106.3	88.1	90.9	107.7	54.0	62.0	107.4	69.8	75.4
0320	Concrete Reinforcing	106.8	98.4	102.5	154.2	74.5	113.7	168.2	84.7	125.8	107.7	102.2	104.9	166.8	47.8	106.3	169.0	49.0	108.0
0330	Cast-in-Place Concrete	106.0	99.1	103.4	101.3	90.2	97.2	132.0	99.6	119.9	106.9	95.8	102.8	111.8	56.4	91.2	90.9	68.8	82.7
03	CONCRETE	96.8	96.2	96.5	108.3	83.7	97.3	128.0	91.7	111.7	97.1	93.4	95.4	139.1	54.9	101.2	118.2	66.8	95.1
04	MASONRY	142.6	99.2	115.8	162.7	77.9	110.3	187.8	91.7	128.5	142.7	95.6	113.6	208.2	56.0	114.2	206.6	67.0	120.3
05	METALS	121.0	96.3	113.5	116.6	85.9	107.3	147.1	100.2	132.9	120.5	95.8	113.0	133.8	70.0	114.4	133.8	76.6	116.4
06	WOOD, PLASTICS & COMPOSITES	106.0	92.5	98.7	133.5	81.3	105.2	105.1	84.3	93.8	101.9	87.2	93.9	119.0	53.4	83.3	118.4	69.3	91.7
07	THERMAL & MOISTURE PROTECTION	113.5	100.9	108.1	112.5	86.5	101.3	141.8	96.5	122.3	113.0	95.6	105.5	128.6	58.9	98.6	129.2	69.5	103.5
08	OPENINGS	81.3	90.2	83.4	91.1	69.9	86.2	85.0	75.4	82.8	82.0	85.7	82.8	104.7	46.8	91.2	93.2	63.0	86.2
0920	Plaster & Gypsum Board	94.2	92.3	92.9	112.2	80.7	91.4	138.3	83.2	102.0	95.6	86.8	89.8	140.0	52.0	82.1	139.1	68.4	92.6
0950, 0980	Ceilings & Acoustic Treatment	103.3	92.3	95.8	103.8	80.7	88.1	136.9	83.2	100.6	98.2	86.8	90.5	122.4	52.0	74.7	122.4	68.4	85.8
0960	Flooring	100.2	92.4	98.0	113.6	88.8	106.5	117.5	53.0	98.8	98.4	94.2	97.2	108.1	57.1	93.3	108.1	60.9	94.4
0970, 0990	Wall Finishes & Painting/Coating	103.8	103.5	103.6	109.8	85.8	95.6	118.0	101.2	108.0	103.8	94.4	98.2	117.4	40.9	72.1	117.4	60.9	84.0
09	FINISHES	97.2	94.1	95.5	105.7	83.6	93.6	123.2	81.7	100.5	95.9	89.7	92.5	117.0	53.4	82.3	115.9	67.5	89.4
COVERS	DIVS. 10 - 14, 25, 28, 41, 43, 44, 46	131.1	69.2	117.5	131.1	77.2	119.3	131.1	68.6	117.4	131.1	89.0	121.9	131.1	58.2	115.1	131.1	62.1	115.9
21, 22, 23	FIRE SUPPRESSION, PLUMBING & HVAC	103.7	89.1	97.9	103.8	86.1	96.7	104.2	84.8	96.4	103.6	88.6	97.6	103.9	60.2	86.4	103.9	81.3	94.8
26, 27, 3370	ELECTRICAL, COMMUNICATIONS & UTIL.	109.8	97.8	103.7	107.8	66.8	86.8	106.0	83.4	94.4	109.7	98.3	103.9	107.4	48.6	77.3	108.3	60.1	83.7
MF2018	WEIGHTED AVERAGE	107.7	94.3	102.0	111.1	81.9	98.6	121.5	88.8	107.4	107.6	93.1	101.4	123.1	59.4	95.8	119.0	72.1	98.8

CANADA

DIVISION		THUNDER BAY, ONTARIO			TIMMINS, ONTARIO			TORONTO, ONTARIO			TROIS RIVIERES, QUEBEC			TRURO, NOVA SCOTIA			VANCOUVER, BRITISH COLUMBIA		
		MAT.	INST.	TOTAL	MAT.	INST.	TOTAL	MAT.	INST.	TOTAL	MAT.	INST.	TOTAL	MAT.	INST.	TOTAL	MAT.	INST.	TOTAL
015433	CONTRACTOR EQUIPMENT		103.4	103.4		103.1	103.1		122.5	122.5		102.9	102.9		102.7	102.7		142.3	142.3
0241, 31 - 34	SITE & INFRASTRUCTURE, DEMOLITION	100.6	101.3	101.1	117.5	99.1	104.7	115.6	114.5	114.8	110.1	98.4	102.0	102.2	97.4	98.9	114.4	130.8	125.8
0310	Concrete Forming & Accessories	117.5	91.3	95.3	122.8	82.3	88.4	119.4	99.1	102.2	149.5	81.4	91.6	95.7	69.8	73.7	120.6	92.8	96.9
0320	Concrete Reinforcing	95.5	97.9	96.7	171.2	87.3	128.5	151.4	100.3	125.4	169.0	74.5	121.0	145.8	49.0	96.6	122.5	93.4	107.7
0330	Cast-in-Place Concrete	116.7	98.4	109.9	127.3	82.8	110.7	113.9	111.9	113.1	94.1	90.1	92.6	132.2	66.9	108.7	110.9	92.3	104.0
03	CONCRETE	104.1	95.0	100.0	122.4	83.8	105.0	113.0	104.4	109.1	120.6	83.7	104.0	121.3	66.9	96.8	110.7	94.0	103.2
04	MASONRY	143.2	99.4	116.1	166.5	83.6	115.3	166.8	106.1	129.3	210.6	77.9	128.6	162.2	67.0	103.3	161.7	86.3	115.1
05	METALS	121.0	95.3	113.2	118.0	91.5	109.9	144.1	108.9	133.4	133.1	85.6	118.7	117.3	76.9	105.0	138.6	111.8	130.4
06	WOOD, PLASTICS & COMPOSITES	115.5	89.9	101.6	121.8	82.6	100.5	105.7	96.9	100.9	173.4	81.3	123.4	90.4	69.4	79.0	103.3	92.8	97.6
07	THERMAL & MOISTURE PROTECTION	113.8	98.2	107.1	119.2	85.0	104.5	129.3	106.4	119.4	127.9	86.5	110.1	113.7	69.6	94.7	134.4	88.0	114.4
08	OPENINGS	80.6	88.1	82.4	89.7	80.5	87.6	82.1	96.0	85.4	102.4	74.4	95.9	84.2	63.1	79.3	85.0	89.7	86.1
0920	Plaster & Gypsum Board	120.6	89.6	100.2	113.1	82.1	92.7	116.8	96.2	103.2	169.8	80.7	111.1	118.4	68.4	85.5	118.0	91.5	100.5
0950, 0980	Ceilings & Acoustic Treatment	98.2	89.6	92.4	104.7	82.1	89.4	133.9	96.2	108.4	120.7	80.7	93.6	104.7	68.4	80.1	143.7	91.5	108.3
0960	Flooring	104.3	99.0	102.8	113.6	88.9	106.5	109.1	101.4	106.9	131.3	88.8	119.0	96.1	60.9	85.9	127.9	90.7	117.1
0970, 0990	Wall Finishes & Painting/Coating	103.8	95.3	98.8	109.8	87.3	96.5	107.0	107.5	107.3	117.4	85.8	98.7	109.8	60.9	80.8	115.0	87.6	98.8
09	FINISHES	101.6	93.2	97.0	106.6	84.4	94.5	112.2	100.4	105.7	126.7	83.5	103.1	101.7	67.5	83.0	122.8	92.2	106.1
COVERS	DIVS. 10 - 14, 25, 28, 41, 43, 44, 46	131.1	69.3	117.5	131.1	64.8	116.5	131.1	93.9	122.9	131.1	77.2	119.3	131.1	62.1	115.9	131.1	91.7	122.5
21, 22, 23	FIRE SUPPRESSION, PLUMBING & HVAC	103.7	89.3	98.0	103.8	94.5	100.0	104.3	97.2	101.5	103.9	86.1	96.8	103.8	81.3	94.8	104.4	77.1	93.5
26, 27, 3370	ELECTRICAL, COMMUNICATIONS & UTIL.	108.2	96.5	102.2	109.2	85.3	97.0	104.8	100.0	102.3	108.6	66.8	87.2	106.9	60.1	83.0	101.8	80.5	90.9
MF2018	WEIGHTED AVERAGE	109.0	93.7	102.4	113.9	87.5	102.6	116.4	102.4	110.4	121.6	82.0	104.6	111.4	72.2	94.6	115.9	91.6	105.5

CANADA

DIVISION		VICTORIA, BRITISH COLUMBIA			WHITEHORSE, YUKON			WINDSOR, ONTARIO			WINNIPEG, MANITOBA			YARMOUTH, NOVA SCOTIA			YELLOWKNIFE, NWT		
		MAT.	INST.	TOTAL	MAT.	INST.	TOTAL	MAT.	INST.	TOTAL	MAT.	INST.	TOTAL	MAT.	INST.	TOTAL	MAT.	INST.	TOTAL
015433	CONTRACTOR EQUIPMENT		109.9	109.9		132.7	132.7		103.4	103.4		126.9	126.9		102.0	102.0		131.8	131.8
0241, 31 - 34	SITE & INFRASTRUCTURE, DEMOLITION	123.6	106.8	111.9	139.0	120.4	126.0	92.3	101.3	98.5	115.6	115.2	115.3	113.5	97.0	102.0	147.7	125.6	132.3
0310	Concrete Forming & Accessories	110.4	85.9	89.6	128.7	58.1	68.7	117.6	89.6	93.8	124.9	65.6	74.5	107.4	69.8	75.4	130.8	77.7	85.7
0320	Concrete Reinforcing	114.5	93.2	103.7	153.0	65.1	108.3	104.7	97.2	100.9	140.0	61.1	99.9	169.0	49.0	108.0	150.2	67.7	108.2
0330	Cast-in-Place Concrete	113.7	88.9	104.5	159.0	71.9	126.6	108.6	99.9	105.3	138.6	72.6	114.0	116.5	68.8	98.8	157.1	89.1	131.8
03	CONCRETE	129.1	88.5	110.8	143.6	66.0	108.7	98.3	94.7	96.7	125.2	68.8	99.8	130.5	66.8	101.8	142.5	81.3	114.9
04	MASONRY	175.6	86.2	120.4	240.3	59.2	128.4	142.7	98.4	115.3	188.9	64.9	112.3	208.8	67.0	121.2	235.0	70.3	133.2
05	METALS	112.2	91.4	105.9	151.7	91.8	133.5	121.0	95.8	113.4	153.6	89.3	134.1	133.8	76.6	116.4	149.9	94.7	133.1
06	WOOD, PLASTICS & COMPOSITES	101.8	84.6	92.5	123.9	56.7	87.4	115.5	88.2	100.7	110.1	66.2	86.3	118.4	69.3	91.7	132.9	79.2	103.7
07	THERMAL & MOISTURE PROTECTION	124.5	87.6	108.6	140.0	64.2	107.4	113.6	97.6	106.7	132.5	69.9	105.5	129.2	69.5	103.5	140.7	80.3	114.7
08	OPENINGS	88.6	80.9	86.8	102.8	54.4	91.5	80.4	87.6	82.1	85.6	59.8	79.6	93.2	63.0	86.2	97.1	67.4	90.2
0920	Plaster & Gypsum Board	106.4	83.8	91.5	172.9	54.6	95.0	106.6	87.8	94.2	124.1	64.2	84.6	139.1	68.4	92.6	179.7	77.7	112.5
0950, 0980	Ceilings & Acoustic Treatment	106.2	83.8	91.0	168.6	54.6	91.4	98.2	87.8	91.2	147.1	64.2	90.9	122.4	68.4	85.8	164.1	77.7	105.5
0960	Flooring	112.2	69.7	99.9	130.8	57.0	109.4	104.3	96.4	102.0	118.6	68.8	104.2	108.1	60.9	94.4	124.1	75.7	112.9
0970, 0990	Wall Finishes & Painting/Coating	113.4	87.6	98.1	121.7	54.5	82.0	103.8	96.3	99.4	111.5	53.4	77.1	117.4	60.9	84.0	124.4	77.9	96.9
09	FINISHES	108.4	83.7	94.9	145.3	57.2	97.2	99.2	91.5	95.0	123.1	65.2	91.5	115.9	67.5	89.4	142.1	78.7	107.5
COVERS	DIVS. 10 - 14, 25, 28, 41, 43, 44, 46	131.1	66.1	116.8	131.1	61.3	115.8	131.1	68.8	117.4	131.1	63.8	116.3	131.1	62.1	115.9	131.1	64.8	116.5
21, 22, 23	FIRE SUPPRESSION, PLUMBING & HVAC	103.8	75.2	92.3	104.2	73.2	91.8	103.7	89.2	97.9	104.4	64.3	88.3	103.9	81.3	94.8	104.7	90.4	99.0
26, 27, 3370	ELECTRICAL, COMMUNICATIONS & UTIL.	107.4	80.0	93.4	127.3	58.5	92.1	113.1	97.9	105.3	107.1	63.2	84.6	108.3	60.1	83.7	121.9	79.3	100.1
MF2018	WEIGHTED AVERAGE	114.2	84.2	101.3	133.0	70.5	106.2	108.3	93.5	101.9	122.1	71.2	100.3	120.7	72.1	99.8	131.3	85.0	111.4

For customer support on your Electrical Costs with RSMeans Data, call 800.448.8182.

563

Location Factors - Commercial

Costs shown in RSMeans cost data publications are based on national averages for materials and installation. To adjust these costs to a specific location, simply multiply the base cost by the factor and divide by 100 for that city. The data is arranged alphabetically by state and postal zip code numbers. For a city not listed, use the factor for a nearby city with similar economic characteristics.

STATE/ZIP	CITY	MAT.	INST.	TOTAL
ALABAMA				
350-352	Birmingham	98.2	70.9	86.4
354	Tuscaloosa	96.9	70.5	85.5
355	Jasper	97.6	69.6	85.5
356	Decatur	96.9	70.2	85.4
357-358	Huntsville	96.9	69.9	85.2
359	Gadsden	97.2	70.8	85.8
360-361	Montgomery	98.0	71.8	86.6
362	Anniston	97.1	66.6	83.9
363	Dothan	97.2	71.6	86.1
364	Evergreen	96.8	67.8	84.3
365-366	Mobile	97.8	68.7	85.2
367	Selma	96.9	71.2	85.8
368	Phenix City	97.7	70.8	86.1
369	Butler	97.1	69.7	85.2
ALASKA				
995-996	Anchorage	119.7	110.4	115.7
997	Fairbanks	120.2	111.0	116.2
998	Juneau	119.7	110.4	115.7
999	Ketchikan	130.1	110.9	121.8
ARIZONA				
850,853	Phoenix	98.4	72.1	87.0
851,852	Mesa/Tempe	98.0	71.1	86.4
855	Globe	98.2	70.7	86.3
856-857	Tucson	96.9	71.8	86.1
859	Show Low	98.4	70.8	86.5
860	Flagstaff	102.4	70.4	88.6
863	Prescott	99.9	71.1	87.5
864	Kingman	98.2	69.8	85.9
865	Chambers	98.2	70.4	86.2
ARKANSAS				
716	Pine Bluff	96.2	62.8	81.8
717	Camden	94.1	64.1	81.1
718	Texarkana	95.1	63.7	81.5
719	Hot Springs	93.4	62.5	80.1
720-722	Little Rock	96.0	63.5	81.9
723	West Memphis	94.2	67.5	82.7
724	Jonesboro	94.8	65.7	82.2
725	Batesville	92.4	62.9	79.6
726	Harrison	93.7	62.3	80.1
727	Fayetteville	91.3	64.2	79.6
728	Russellville	92.5	61.3	79.0
729	Fort Smith	95.1	61.9	80.8
CALIFORNIA				
900-902	Los Angeles	98.4	129.3	111.8
903-905	Inglewood	93.6	127.4	108.2
906-908	Long Beach	95.2	127.2	109.0
910-912	Pasadena	95.4	127.0	109.0
913-916	Van Nuys	98.5	127.0	110.8
917-918	Alhambra	97.4	127.0	110.2
919-921	San Diego	100.3	121.3	109.4
922	Palm Springs	97.3	121.9	108.0
923-924	San Bernardino	95.0	125.4	108.1
925	Riverside	99.4	126.0	110.9
926-927	Santa Ana	97.1	125.7	109.5
928	Anaheim	99.5	125.9	110.9
930	Oxnard	98.2	125.8	110.1
931	Santa Barbara	97.4	125.0	109.4
932-933	Bakersfield	99.0	124.8	110.1
934	San Luis Obispo	98.4	125.5	110.1
935	Mojave	95.6	123.6	107.7
936-938	Fresno	98.6	131.2	112.7
939	Salinas	99.2	137.8	115.9
940-941	San Francisco	107.6	159.0	129.8
942,956-958	Sacramento	101.0	133.1	114.9
943	Palo Alto	99.4	150.8	121.6
944	San Mateo	101.9	156.6	125.5
945	Vallejo	100.5	143.4	119.0
946	Oakland	103.9	153.2	125.2
947	Berkeley	103.3	151.9	124.3
948	Richmond	102.8	148.2	122.4
949	San Rafael	105.0	150.3	124.6
950	Santa Cruz	104.9	138.1	119.2

STATE/ZIP	CITY	MAT.	INST.	TOTAL
CALIFORNIA (CONT'D)				
951	San Jose	103.0	158.4	126.9
952	Stockton	101.0	131.9	114.3
953	Modesto	100.8	132.1	114.3
954	Santa Rosa	101.2	149.8	122.2
955	Eureka	102.6	137.6	117.7
959	Marysville	101.8	130.7	114.3
960	Redding	107.8	132.6	118.6
961	Susanville	107.3	131.0	117.5
COLORADO				
800-802	Denver	104.0	75.1	91.5
803	Boulder	99.2	73.7	88.2
804	Golden	101.5	73.0	89.2
805	Fort Collins	102.8	73.2	90.0
806	Greeley	99.9	73.3	88.4
807	Fort Morgan	99.9	72.5	88.0
808-809	Colorado Springs	101.8	72.3	89.1
810	Pueblo	101.6	70.7	88.2
811	Alamosa	103.3	69.2	88.6
812	Salida	102.9	69.8	88.6
813	Durango	103.6	65.5	87.1
814	Montrose	102.3	66.3	86.7
815	Grand Junction	105.6	70.0	90.2
816	Glenwood Springs	103.6	66.6	87.6
CONNECTICUT				
060	New Britain	99.7	116.9	107.1
061	Hartford	101.7	118.7	109.0
062	Willimantic	100.3	116.9	107.5
063	New London	96.7	116.6	105.3
064	Meriden	98.8	116.4	106.4
065	New Haven	101.5	117.0	108.2
066	Bridgeport	100.9	116.7	107.7
067	Waterbury	100.5	117.2	107.7
068	Norwalk	100.4	116.9	107.5
069	Stamford	100.5	123.9	110.6
D.C.				
200-205	Washington	101.3	88.0	95.5
DELAWARE				
197	Newark	98.7	110.6	103.8
198	Wilmington	99.1	110.7	104.1
199	Dover	99.5	110.4	104.2
FLORIDA				
320,322	Jacksonville	97.0	66.9	84.0
321	Daytona Beach	97.2	69.0	85.0
323	Tallahassee	98.1	66.7	84.6
324	Panama City	98.7	67.2	85.1
325	Pensacola	101.2	65.3	85.7
326,344	Gainesville	98.3	63.8	83.4
327-328,347	Orlando	97.9	66.3	84.3
329	Melbourne	99.9	70.6	87.2
330-332,340	Miami	96.2	70.4	85.1
333	Fort Lauderdale	96.5	72.2	86.0
334,349	West Palm Beach	95.1	69.0	83.8
335-336,346	Tampa	98.0	67.4	84.8
337	St. Petersburg	100.4	65.8	85.4
338	Lakeland	97.1	66.3	83.8
339,341	Fort Myers	96.4	66.7	83.6
342	Sarasota	99.1	65.3	84.5
GEORGIA				
300-303,399	Atlanta	98.7	76.7	89.2
304	Statesboro	97.9	67.1	84.6
305	Gainesville	96.5	64.3	82.6
306	Athens	95.9	63.8	82.0
307	Dalton	97.7	68.0	84.9
308-309	Augusta	96.4	73.5	86.5
310-312	Macon	95.3	74.4	86.3
313-314	Savannah	97.3	72.8	86.7
315	Waycross	96.6	66.9	83.8
316	Valdosta	96.7	65.4	83.2
317,398	Albany	96.5	72.2	86.0
318-319	Columbus	96.4	73.4	86.4

Location Factors - Commercial

STATE/ZIP	CITY	MAT.	INST.	TOTAL
HAWAII				
967	Hilo	115.6	117.2	116.3
968	Honolulu	120.3	117.8	119.2
STATES & POSS.				
969	Guam	138.0	63.4	105.7
IDAHO				
832	Pocatello	101.8	77.8	91.4
833	Twin Falls	102.9	76.9	91.7
834	Idaho Falls	100.2	76.9	90.1
835	Lewiston	108.4	84.9	98.2
836-837	Boise	100.9	78.2	91.1
838	Coeur d'Alene	108.1	81.5	96.6
ILLINOIS				
600-603	North Suburban	98.5	137.3	115.3
604	Joliet	98.5	140.3	116.5
605	South Suburban	98.5	137.2	115.3
606-608	Chicago	100.5	144.4	119.5
609	Kankakee	95.1	129.5	110.0
610-611	Rockford	96.3	126.3	109.3
612	Rock Island	94.5	95.5	94.9
613	La Salle	95.8	122.2	107.2
614	Galesburg	95.5	104.8	99.5
615-616	Peoria	97.4	109.9	102.8
617	Bloomington	94.8	105.4	99.4
618-619	Champaign	98.4	106.3	101.8
620-622	East St. Louis	94.1	109.8	100.9
623	Quincy	96.1	101.1	98.3
624	Effingham	95.3	104.9	99.4
625	Decatur	96.7	107.6	101.4
626-627	Springfield	97.8	110.4	103.2
628	Centralia	92.9	106.7	98.9
629	Carbondale	92.7	105.2	98.1
INDIANA				
460	Anderson	96.8	80.4	89.7
461-462	Indianapolis	99.5	82.9	92.3
463-464	Gary	98.3	105.8	101.5
465-466	South Bend	99.6	83.4	92.6
467-468	Fort Wayne	97.5	75.9	88.2
469	Kokomo	94.9	79.8	88.4
470	Lawrenceburg	93.0	78.3	86.6
471	New Albany	94.3	76.3	86.5
472	Columbus	96.5	79.4	89.1
473	Muncie	96.6	78.9	89.0
474	Bloomington	98.2	79.7	90.2
475	Washington	95.1	84.8	90.7
476-477	Evansville	96.0	83.2	90.5
478	Terre Haute	96.9	82.9	90.8
479	Lafayette	96.2	79.9	89.1
IOWA				
500-503,509	Des Moines	97.3	90.0	94.2
504	Mason City	95.0	71.2	84.7
505	Fort Dodge	95.0	71.0	84.6
506-507	Waterloo	96.5	78.4	88.7
508	Creston	95.2	81.1	89.1
510-511	Sioux City	98.0	80.3	90.3
512	Sibley	96.9	60.1	81.0
513	Spencer	98.3	59.9	81.7
514	Carroll	95.5	79.3	88.5
515	Council Bluffs	98.8	81.9	91.5
516	Shenandoah	96.1	78.7	88.6
520	Dubuque	97.3	80.6	90.1
521	Decorah	96.7	71.0	85.6
522-524	Cedar Rapids	98.3	86.6	93.2
525	Ottumwa	96.4	75.7	87.5
526	Burlington	95.8	81.1	89.5
527-528	Davenport	97.3	96.6	97.0
KANSAS				
660-662	Kansas City	96.4	99.2	97.6
664-666	Topeka	98.1	78.0	89.4
667	Fort Scott	95.2	73.9	86.0
668	Emporia	95.2	74.0	86.0
669	Belleville	96.9	69.1	84.9
670-672	Wichita	96.5	72.2	86.0
673	Independence	96.4	73.6	86.6
674	Salina	96.8	73.3	86.6
675	Hutchinson	91.9	69.7	82.3
676	Hays	96.0	70.4	84.9
677	Colby	96.5	71.7	85.7

STATE/ZIP	CITY	MAT.	INST.	TOTAL
KANSAS (CONT'D)				
678	Dodge City	98.0	74.8	88.0
679	Liberal	96.0	69.5	84.6
KENTUCKY				
400-402	Louisville	94.9	79.6	88.3
403-405	Lexington	94.2	78.8	87.5
406	Frankfort	96.8	78.6	88.9
407-409	Corbin	91.8	75.0	84.6
410	Covington	94.9	75.0	86.3
411-412	Ashland	93.6	86.3	90.5
413-414	Campton	94.9	76.3	86.9
415-416	Pikeville	96.1	82.1	90.1
417-418	Hazard	94.3	77.1	86.9
420	Paducah	93.0	80.0	87.4
421-422	Bowling Green	95.2	78.4	87.9
423	Owensboro	95.3	81.8	89.4
424	Henderson	92.7	78.7	86.7
425-426	Somerset	92.1	76.6	85.4
427	Elizabethtown	91.8	73.4	83.8
LOUISIANA				
700-701	New Orleans	97.7	68.3	85.0
703	Thibodaux	95.5	65.0	82.3
704	Hammond	93.2	63.4	80.3
705	Lafayette	95.1	65.9	82.5
706	Lake Charles	95.3	68.4	83.7
707-708	Baton Rouge	97.0	68.3	84.6
710-711	Shreveport	97.7	66.8	84.4
712	Monroe	96.0	64.6	82.4
713-714	Alexandria	96.1	64.7	82.5
MAINE				
039	Kittery	94.4	84.9	90.3
040-041	Portland	100.6	85.6	94.1
042	Lewiston	98.0	84.9	92.4
043	Augusta	101.6	86.0	94.9
044	Bangor	97.5	84.5	91.9
045	Bath	96.2	84.6	91.2
046	Machias	95.8	84.5	90.9
047	Houlton	95.9	84.4	91.0
048	Rockland	95.0	84.6	90.5
049	Waterville	96.2	84.5	91.2
MARYLAND				
206	Waldorf	97.4	84.8	92.0
207-208	College Park	97.5	85.8	92.5
209	Silver Spring	96.7	84.9	91.6
210-212	Baltimore	102.3	83.1	94.0
214	Annapolis	101.4	83.4	93.6
215	Cumberland	97.1	83.1	91.0
216	Easton	98.9	72.6	87.5
217	Hagerstown	97.7	86.2	92.8
218	Salisbury	99.3	65.0	84.5
219	Elkton	96.1	83.0	90.5
MASSACHUSETTS				
010-011	Springfield	99.1	110.8	104.2
012	Pittsfield	98.6	105.1	101.4
013	Greenfield	96.7	109.5	102.3
014	Fitchburg	95.4	113.2	103.1
015-016	Worcester	99.1	118.6	107.5
017	Framingham	94.9	122.4	106.8
018	Lowell	98.4	126.4	110.5
019	Lawrence	99.4	124.9	110.4
020-022, 024	Boston	100.2	132.9	114.3
023	Brockton	98.9	117.8	107.1
025	Buzzards Bay	93.3	114.1	102.3
026	Hyannis	96.0	116.9	105.1
027	New Bedford	98.1	116.3	106.0
MICHIGAN				
480,483	Royal Oak	94.1	96.9	95.3
481	Ann Arbor	96.3	102.0	98.7
482	Detroit	99.5	100.9	100.1
484-485	Flint	95.9	90.9	93.8
486	Saginaw	95.6	86.7	91.8
487	Bay City	95.7	85.3	91.2
488-489	Lansing	98.1	90.5	94.8
490	Battle Creek	95.7	79.4	88.6
491	Kalamazoo	96.0	80.9	89.5
492	Jackson	94.1	88.0	91.4
493,495	Grand Rapids	98.9	82.2	91.6
494	Muskegon	94.5	79.7	88.1

For customer support on your Electrical Costs with RSMeans Data, call 800.448.8182.

56.

Location Factors - Commercial

STATE/ZIP	CITY	MAT.	INST.	TOTAL
MICHIGAN (CONT'D)				
496	Traverse City	93.5	75.2	85.6
497	Gaylord	94.6	78.1	87.5
498-499	Iron Mountain	96.5	78.9	88.9
MINNESOTA				
550-551	Saint Paul	99.3	119.2	107.9
553-555	Minneapolis	99.7	116.6	107.0
556-558	Duluth	99.9	103.9	101.6
559	Rochester	98.4	106.1	101.7
560	Mankato	95.5	105.0	99.6
561	Windom	94.0	91.7	93.0
562	Willmar	93.7	96.6	94.9
563	St. Cloud	94.9	115.9	104.0
564	Brainerd	95.5	97.9	96.5
565	Detroit Lakes	97.6	90.3	94.4
566	Bemidji	96.8	94.6	95.8
567	Thief River Falls	96.4	89.4	93.4
MISSISSIPPI				
386	Clarksdale	95.7	53.8	77.6
387	Greenville	99.4	65.2	84.6
388	Tupelo	97.2	56.4	79.6
389	Greenwood	97.0	53.5	78.2
390-392	Jackson	98.6	67.9	85.3
393	Meridian	95.4	67.8	83.5
394	Laurel	96.8	56.4	79.4
395	Biloxi	97.1	66.6	83.9
396	McComb	95.1	53.8	77.3
397	Columbus	96.7	57.1	79.6
MISSOURI				
630-631	St. Louis	97.5	104.6	100.6
633	Bowling Green	95.6	92.3	94.2
634	Hannibal	94.5	88.5	91.9
635	Kirksville	98.0	83.6	91.8
636	Flat River	96.6	90.6	94.0
637	Cape Girardeau	96.4	90.6	93.9
638	Sikeston	94.9	84.4	90.4
639	Poplar Bluff	94.4	84.3	90.0
640-641	Kansas City	97.5	101.8	99.3
644-645	St. Joseph	96.0	93.3	94.9
646	Chillicothe	93.7	90.7	92.4
647	Harrisonville	93.2	97.4	95.0
648	Joplin	95.2	80.1	88.7
650-651	Jefferson City	95.7	90.0	93.2
652	Columbia	95.7	94.5	95.2
653	Sedalia	95.6	88.1	92.4
654-655	Rolla	93.4	90.6	92.2
656-658	Springfield	96.5	81.4	90.0
MONTANA				
590-591	Billings	101.4	77.2	90.9
592	Wolf Point	101.3	74.5	89.7
593	Miles City	99.1	74.6	88.5
594	Great Falls	102.8	77.1	91.7
595	Havre	100.2	72.5	88.3
596	Helena	101.2	74.7	89.7
597	Butte	101.3	74.4	89.7
598	Missoula	98.8	74.8	88.5
599	Kalispell	98.0	74.1	87.7
NEBRASKA				
680-681	Omaha	98.6	83.0	91.9
683-685	Lincoln	99.8	79.3	90.9
686	Columbus	96.0	77.7	88.1
687	Norfolk	97.4	74.5	87.5
688	Grand Island	97.4	76.2	88.2
689	Hastings	97.1	75.2	87.7
690	McCook	95.9	70.2	84.8
691	North Platte	96.1	74.9	86.9
692	Valentine	98.0	67.4	84.8
693	Alliance	98.4	70.4	86.3
NEVADA				
889-891	Las Vegas	104.2	107.0	105.4
893	Ely	102.9	89.5	97.1
894-895	Reno	102.8	85.6	95.4
897	Carson City	103.2	85.6	95.6
898	Elko	101.5	81.8	93.0
NEW HAMPSHIRE				
030	Nashua	99.3	92.3	96.3
031	Manchester	100.5	93.0	97.3

STATE/ZIP	CITY	MAT.	INST.	TOTAL
NEW HAMPSHIRE (CONT'D)				
032-033	Concord	100.8	91.3	96.7
034	Keene	96.1	86.2	91.8
035	Littleton	96.1	77.7	88.1
036	Charleston	95.6	85.6	91.3
037	Claremont	94.8	85.7	90.8
038	Portsmouth	96.5	91.5	94.3
NEW JERSEY				
070-071	Newark	101.8	137.9	117.4
072	Elizabeth	98.9	137.6	115.6
073	Jersey City	97.7	137.1	114.7
074-075	Paterson	99.3	137.3	115.7
076	Hackensack	97.4	137.1	114.6
077	Long Branch	97.1	126.6	109.9
078	Dover	97.6	137.5	114.8
079	Summit	97.7	136.9	114.6
080,083	Vineland	97.2	126.6	109.9
081	Camden	98.9	133.7	114.0
082,084	Atlantic City	97.8	132.6	112.8
085-086	Trenton	100.5	124.4	110.8
087	Point Pleasant	99.1	126.5	110.9
088-089	New Brunswick	99.7	132.2	113.8
NEW MEXICO				
870-872	Albuquerque	98.9	71.6	87.1
873	Gallup	99.2	71.6	87.3
874	Farmington	99.3	71.6	87.3
875	Santa Fe	100.5	72.2	88.3
877	Las Vegas	97.4	71.6	86.3
878	Socorro	97.1	71.6	86.1
879	Truth/Consequences	96.8	68.6	84.6
880	Las Cruces	96.8	70.0	85.2
881	Clovis	99.1	71.5	87.2
882	Roswell	100.7	71.6	88.1
883	Carrizozo	101.4	71.6	88.5
884	Tucumcari	99.7	71.5	87.6
NEW YORK				
100-102	New York	100.3	174.2	132.2
103	Staten Island	96.5	171.9	129.1
104	Bronx	94.8	171.2	127.8
105	Mount Vernon	94.9	147.0	117.4
106	White Plains	94.9	149.8	118.6
107	Yonkers	99.1	150.1	121.1
108	New Rochelle	95.3	141.4	115.2
109	Suffern	94.9	125.4	108.1
110	Queens	100.5	173.1	131.8
111	Long Island City	102.2	173.1	132.8
112	Brooklyn	102.7	173.1	133.1
113	Flushing	102.5	173.1	133.0
114	Jamaica	100.9	173.1	132.1
115,117,118	Hicksville	100.5	151.7	122.6
116	Far Rockaway	102.7	173.1	133.1
119	Riverhead	101.3	156.9	125.3
120-122	Albany	98.3	109.4	103.1
123	Schenectady	97.9	108.6	102.5
124	Kingston	101.2	128.8	113.1
125-126	Poughkeepsie	100.3	131.7	113.9
127	Monticello	99.8	129.5	112.7
128	Glens Falls	92.8	103.2	97.3
129	Plattsburgh	97.7	92.8	95.6
130-132	Syracuse	98.7	101.6	99.9
133-135	Utica	96.8	102.5	99.3
136	Watertown	98.4	98.9	98.6
137-139	Binghamton	98.5	100.5	99.3
140-142	Buffalo	100.9	109.8	104.8
143	Niagara Falls	98.0	103.4	100.3
144-146	Rochester	100.8	101.2	101.0
147	Jamestown	97.1	92.3	95.0
148-149	Elmira	96.9	100.4	98.4
NORTH CAROLINA				
270,272-274	Greensboro	99.7	65.3	84.8
271	Winston-Salem	99.5	65.3	84.7
275-276	Raleigh	99.0	65.3	84.4
277	Durham	101.2	65.2	85.6
278	Rocky Mount	97.1	64.5	83.0
279	Elizabeth City	98.1	66.6	84.5
280	Gastonia	97.7	68.5	85.1
281-282	Charlotte	98.4	72.1	87.0
283	Fayetteville	100.5	64.0	84.7
284	Wilmington	96.6	63.1	82.1
285	Kinston	94.9	63.5	81.3

For customer support on your Electrical Costs with RSMeans Data, call 800.448.8182.

STATE/ZIP	CITY	MAT.	INST.	TOTAL
NORTH CAROLINA (CONT'D)				
286	Hickory	95.2	69.0	83.9
287-288	Asheville	97.0	64.6	83.0
289	Murphy	96.1	62.6	81.6
NORTH DAKOTA				
580-581	Fargo	99.6	78.9	90.6
582	Grand Forks	98.8	81.0	91.1
583	Devils Lake	98.6	77.8	89.6
584	Jamestown	98.7	76.9	89.3
585	Bismarck	99.1	83.1	92.2
586	Dickinson	99.2	75.7	89.0
587	Minot	98.7	76.5	89.1
588	Williston	97.8	80.7	90.4
OHIO				
430-432	Columbus	99.2	83.2	92.3
433	Marion	95.5	83.6	90.3
434-436	Toledo	98.9	87.7	94.1
437-438	Zanesville	96.1	83.0	90.4
439	Steubenville	97.1	87.4	92.9
440	Lorain	98.7	81.8	91.4
441	Cleveland	99.4	91.5	96.0
442-443	Akron	99.7	83.0	92.5
444-445	Youngstown	99.1	79.7	90.7
446-447	Canton	99.2	77.5	89.8
448-449	Mansfield	96.6	81.1	89.9
450	Hamilton	96.9	76.7	88.2
451-452	Cincinnati	98.0	79.2	89.9
453-454	Dayton	96.9	76.6	88.1
455	Springfield	96.9	77.2	88.4
456	Chillicothe	96.3	85.3	91.5
457	Athens	99.2	81.8	91.7
458	Lima	99.6	79.0	90.7
OKLAHOMA				
730-731	Oklahoma City	97.6	67.8	84.8
734	Ardmore	95.9	63.8	82.0
735	Lawton	98.1	63.9	83.3
736	Clinton	97.3	63.7	82.8
737	Enid	97.8	64.2	83.3
738	Woodward	95.9	61.2	80.9
739	Guymon	97.0	61.7	81.7
740-741	Tulsa	96.9	65.5	83.3
743	Miami	93.5	63.0	80.3
744	Muskogee	96.1	65.7	83.0
745	McAlester	93.2	58.3	78.1
746	Ponca City	93.9	61.8	80.0
747	Durant	93.8	62.7	80.4
748	Shawnee	95.3	63.3	81.5
749	Poteau	93.0	62.6	79.9
OREGON				
970-972	Portland	102.0	104.7	103.2
973	Salem	103.3	102.8	103.0
974	Eugene	101.6	103.7	102.5
975	Medford	103.1	99.3	101.4
976	Klamath Falls	103.3	97.8	100.9
977	Bend	102.3	100.5	101.5
978	Pendleton	98.3	101.0	99.5
979	Vale	96.0	87.8	92.4
PENNSYLVANIA				
150-152	Pittsburgh	99.3	102.6	100.7
153	Washington	96.6	101.9	98.9
154	Uniontown	97.1	99.2	98.0
155	Bedford	98.0	90.1	94.6
156	Greensburg	98.0	95.8	97.1
157	Indiana	96.8	96.2	96.5
158	Dubois	98.5	94.0	96.6
159	Johnstown	98.0	97.2	97.7
160	Butler	92.1	98.7	95.0
161	New Castle	92.2	98.9	95.1
162	Kittanning	92.6	96.6	94.3
163	Oil City	92.1	94.4	93.1
164-165	Erie	94.0	94.6	94.3
166	Altoona	94.2	97.5	95.6
167	Bradford	95.6	95.0	95.3
168	State College	95.2	99.1	96.9
169	Wellsboro	96.2	89.3	93.2
170-171	Harrisburg	98.5	97.0	97.9
172	Chambersburg	95.4	86.0	91.3
173-174	York	96.2	94.6	95.5
175-176	Lancaster	94.2	96.5	95.2

STATE/ZIP	CITY	MAT.	INST.	TOTAL
PENNSYLVANIA (CONT'D)				
177	Williamsport	92.8	94.2	93.4
178	Sunbury	95.0	89.3	92.5
179	Pottsville	94.0	92.5	93.3
180	Lehigh Valley	95.6	107.5	100.8
181	Allentown	97.5	108.1	102.1
182	Hazleton	95.1	92.5	94.0
183	Stroudsburg	95.0	100.9	97.6
184-185	Scranton	98.3	96.8	97.7
186-187	Wilkes-Barre	94.9	96.8	95.7
188	Montrose	94.6	93.3	94.0
189	Doylestown	94.7	122.6	106.7
190-191	Philadelphia	99.7	136.9	115.8
193	Westchester	96.1	123.2	107.8
194	Norristown	95.1	123.4	107.3
195-196	Reading	97.3	104.7	100.5
PUERTO RICO				
009	San Juan	123.0	28.3	82.1
RHODE ISLAND				
028	Newport	97.0	112.6	103.8
029	Providence	101.0	113.3	106.3
SOUTH CAROLINA				
290-292	Columbia	98.8	67.6	85.3
293	Spartanburg	97.7	70.2	85.8
294	Charleston	99.4	67.0	85.4
295	Florence	97.4	67.0	84.2
296	Greenville	97.5	70.0	85.6
297	Rock Hill	97.3	68.2	84.7
298	Aiken	98.1	65.2	83.9
299	Beaufort	98.9	61.5	82.7
SOUTH DAKOTA				
570-571	Sioux Falls	99.3	82.1	91.8
572	Watertown	97.7	66.6	84.3
573	Mitchell	96.6	59.6	80.6
574	Aberdeen	99.1	66.7	85.1
575	Pierre	100.2	68.9	86.7
576	Mobridge	97.1	60.7	81.4
577	Rapid City	98.7	71.0	86.7
TENNESSEE				
370-372	Nashville	100.0	74.5	89.0
373-374	Chattanooga	99.0	68.7	85.9
375,380-381	Memphis	96.5	72.5	86.1
376	Johnson City	99.2	60.5	82.5
377-379	Knoxville	95.7	66.0	82.9
382	McKenzie	96.7	54.7	78.6
383	Jackson	98.1	60.0	81.6
384	Columbia	95.2	64.6	82.0
385	Cookeville	96.6	56.1	79.1
TEXAS				
750	McKinney	99.5	61.8	83.2
751	Waxahachie	99.5	62.0	83.3
752-753	Dallas	100.1	67.4	86.0
754	Greenville	99.6	61.3	83.1
755	Texarkana	99.1	60.4	82.4
756	Longview	99.8	59.2	82.3
757	Tyler	100.1	60.2	82.8
758	Palestine	95.9	58.9	80.0
759	Lufkin	96.6	60.5	81.0
760-761	Fort Worth	99.0	63.9	83.9
762	Denton	98.9	61.6	82.8
763	Wichita Falls	96.5	62.5	81.8
764	Eastland	95.7	58.9	79.8
765	Temple	94.1	56.4	77.8
766-767	Waco	95.8	63.2	81.7
768	Brownwood	99.1	57.2	81.0
769	San Angelo	98.7	57.3	80.8
770-772	Houston	101.3	67.3	86.6
773	Huntsville	99.9	60.7	83.0
774	Wharton	101.0	62.1	84.2
775	Galveston	98.8	64.0	83.8
776-777	Beaumont	99.2	64.9	84.4
778	Bryan	96.3	62.4	81.7
779	Victoria	101.0	60.8	83.6
780	Laredo	97.6	60.9	81.8
781-782	San Antonio	98.9	63.8	83.7
783-784	Corpus Christi	100.2	64.6	84.8
785	McAllen	100.9	55.2	81.2
786-787	Austin	98.6	62.3	82.9

For customer support on your Electrical Costs with RSMeans Data, call 800.448.8182.

567

Location Factors - Commercial

STATE/ZIP	CITY	MAT.	INST.	TOTAL
TEXAS (CONT'D)				
788	Del Rio	100.7	58.2	82.3
789	Giddings	97.3	58.3	80.5
790-791	Amarillo	99.9	61.8	83.5
792	Childress	98.7	59.4	81.7
793-794	Lubbock	100.4	60.7	83.3
795-796	Abilene	99.0	61.6	82.9
797	Midland	100.9	61.7	84.0
798-799,885	El Paso	98.3	63.3	83.2
UTAH				
840-841	Salt Lake City	103.3	73.7	90.5
842,844	Ogden	98.6	74.0	88.0
843	Logan	100.7	74.0	89.2
845	Price	101.1	68.6	87.1
846-847	Provo	101.1	74.0	89.4
VERMONT				
050	White River Jct.	98.7	79.6	90.4
051	Bellows Falls	97.1	90.6	94.3
052	Bennington	97.5	87.2	93.0
053	Brattleboro	97.8	90.7	94.7
054	Burlington	103.1	82.3	94.1
056	Montpelier	101.2	86.0	94.6
057	Rutland	99.4	81.6	91.7
058	St. Johnsbury	98.7	79.2	90.3
059	Guildhall	97.3	79.2	89.5
VIRGINIA				
220-221	Fairfax	99.9	81.7	92.0
222	Arlington	101.0	81.6	92.6
223	Alexandria	100.0	84.8	93.4
224-225	Fredericksburg	98.6	79.9	90.5
226	Winchester	99.2	76.4	89.3
227	Culpeper	99.1	82.0	91.7
228	Harrisonburg	99.4	75.6	89.1
229	Charlottesville	99.8	74.2	88.7
230-232	Richmond	99.6	74.6	88.8
233-235	Norfolk	100.7	68.3	86.7
236	Newport News	99.5	69.5	86.5
237	Portsmouth	99.0	67.4	85.3
238	Petersburg	99.4	74.4	88.6
239	Farmville	98.5	66.0	84.5
240-241	Roanoke	100.5	69.1	87.0
242	Bristol	98.5	59.4	81.6
243	Pulaski	98.1	69.7	85.8
244	Staunton	99.0	67.8	85.5
245	Lynchburg	99.1	73.1	87.9
246	Grundy	98.4	63.2	83.2
WASHINGTON				
980-981,987	Seattle	105.4	108.4	106.7
982	Everett	104.5	104.2	104.4
983-984	Tacoma	104.9	105.7	105.2
985	Olympia	103.9	105.2	104.5
986	Vancouver	106.2	102.2	104.4
988	Wenatchee	105.0	85.6	96.6
989	Yakima	105.0	101.6	103.5
990-992	Spokane	101.0	86.8	94.9
993	Richland	100.7	93.0	97.4
994	Clarkston	99.9	79.5	91.1
WEST VIRGINIA				
247-248	Bluefield	97.3	85.2	92.1
249	Lewisburg	99.0	86.8	93.8
250-253	Charleston	96.6	91.1	94.2
254	Martinsburg	96.5	80.8	89.7
255-257	Huntington	97.8	91.7	95.1
258-259	Beckley	95.2	87.7	91.9
260	Wheeling	100.1	90.9	96.1
261	Parkersburg	98.9	90.0	95.1
262	Buckhannon	98.7	90.5	95.1
263-264	Clarksburg	99.3	91.0	95.7
265	Morgantown	99.3	91.4	95.9
266	Gassaway	98.7	87.3	93.7
267	Romney	98.5	85.3	92.8
268	Petersburg	98.3	85.6	92.8
WISCONSIN				
530,532	Milwaukee	97.9	111.7	103.9
531	Kenosha	99.2	106.2	102.2
534	Racine	98.5	106.2	101.8
535	Beloit	97.8	97.3	97.6
537	Madison	98.6	106.4	101.9

STATE/ZIP	CITY	MAT.	INST.	TOTAL
WISCONSIN (CONT'D)				
538	Lancaster	95.8	92.9	94.6
539	Portage	94.2	96.5	95.2
540	New Richmond	94.9	92.9	94.0
541-543	Green Bay	99.4	103.1	101.0
544	Wausau	94.5	92.9	93.8
545	Rhinelander	98.0	91.1	95.0
546	La Crosse	95.7	100.3	97.7
547	Eau Claire	97.4	100.4	98.7
548	Superior	94.6	94.6	94.6
549	Oshkosh	95.2	92.7	94.1
WYOMING				
820	Cheyenne	101.0	72.4	88.7
821	Yellowstone Nat'l Park	98.9	71.4	87.0
822	Wheatland	100.1	69.9	87.0
823	Rawlins	101.9	71.3	88.7
824	Worland	99.5	71.9	87.6
825	Riverton	100.8	71.4	88.1
826	Casper	102.5	72.3	89.4
827	Newcastle	99.4	71.3	87.3
828	Sheridan	102.4	71.1	88.9
829-831	Rock Springs	103.3	70.6	89.1
CANADIAN FACTORS (reflect Canadian currency)				
ALBERTA				
	Calgary	123.3	91.7	109.7
	Edmonton	125.3	91.9	110.9
	Fort McMurray	123.9	86.2	107.6
	Lethbridge	118.9	85.7	104.6
	Lloydminster	113.6	82.6	100.2
	Medicine Hat	113.7	81.9	100.0
	Red Deer	114.2	81.9	100.2
BRITISH COLUMBIA				
	Kamloops	115.6	81.6	100.9
	Prince George	116.5	81.0	101.2
	Vancouver	121.7	91.7	108.7
	Victoria	116.8	86.4	103.7
MANITOBA				
	Brandon	122.8	68.9	99.5
	Portage la Prairie	113.6	67.5	93.6
	Winnipeg	125.8	68.4	101.0
NEW BRUNSWICK				
	Bathurst	112.7	60.8	90.3
	Dalhousie	112.4	61.0	90.2
	Fredericton	121.9	67.8	98.5
	Moncton	113.0	73.2	95.8
	Newcastle	112.8	61.4	90.5
	St. John	115.6	72.9	97.1
NEWFOUNDLAND				
	Corner Brook	127.3	66.7	101.1
	St. Johns	126.2	85.2	108.5
NORTHWEST TERRITORIES				
	Yellowknife	135.0	81.1	111.7
NOVA SCOTIA				
	Bridgewater	113.3	68.7	94.0
	Dartmouth	124.0	68.6	100.1
	Halifax	120.5	84.9	105.1
	New Glasgow	121.9	68.6	98.9
	Sydney	120.1	68.6	97.9
	Truro	112.8	68.7	93.7
	Yarmouth	121.9	68.6	98.8
ONTARIO				
	Barrie	118.0	85.2	103.8
	Brantford	115.2	88.8	103.8
	Cornwall	115.3	85.5	102.4
	Hamilton	119.2	94.5	108.6
	Kingston	116.5	85.5	103.1
	Kitchener	109.9	91.2	101.8
	London	119.2	92.9	107.8
	North Bay	125.0	83.4	107.0
	Oshawa	112.4	93.6	104.3
	Ottawa	120.6	93.0	108.7
	Owen Sound	118.1	83.5	103.2
	Peterborough	115.2	85.2	102.3
	Sarnia	115.5	89.4	104.2

STATE/ZIP	CITY	MAT.	INST.	TOTAL
ONTARIO (CONT'D)				
	Sault Ste. Marie	110.6	86.4	100.1
	St. Catharines	108.0	92.1	101.1
	Sudbury	107.7	90.3	100.2
	Thunder Bay	109.3	91.6	101.6
	Timmins	115.4	83.4	101.6
	Toronto	119.3	99.9	111.0
	Windsor	108.6	91.2	101.1
PRINCE EDWARD ISLAND				
	Charlottetown	124.1	58.7	95.8
	Summerside	124.4	56.4	95.0
QUEBEC				
	Cap-de-la-Madeleine	112.8	78.2	97.8
	Charlesbourg	112.8	78.2	97.8
	Chicoutimi	112.5	86.9	101.4
	Gatineau	112.5	78.0	97.6
	Granby	112.7	77.9	97.7
	Hull	112.6	78.0	97.7
	Joliette	113.0	78.2	97.9
	Laval	112.6	78.6	97.9
	Montreal	122.1	89.4	108.0
	Quebec City	122.2	89.9	108.2
	Rimouski	112.3	86.9	101.4
	Rouyn-Noranda	112.4	78.0	97.6
	Saint-Hyacinthe	111.8	78.0	97.2
	Sherbrooke	112.8	78.0	97.7
	Sorel	112.9	78.2	97.9
	Saint-Jerome	112.5	78.0	97.6
	Trois-Rivieres	122.8	78.1	103.5
SASKATCHEWAN				
	Moose Jaw	111.0	62.4	90.0
	Prince Albert	110.2	60.7	88.8
	Regina	127.0	88.5	110.3
	Saskatoon	112.0	85.1	100.4
YUKON				
	Whitehorse	136.0	68.6	106.9

R011105-05   Tips for Accurate Estimating

1. Use pre-printed or columnar forms for orderly sequence of dimensions and locations and for recording telephone quotations.

2. Use only the front side of each paper or form except for certain pre-printed summary forms.

3. Be consistent in listing dimensions: For example, length x width x height. This helps in rechecking to ensure that the total length of partitions is appropriate for the building area.

4. Use printed (rather than measured) dimensions where given.

5. Add up multiple printed dimensions for a single entry where possible.

6. Measure all other dimensions carefully.

7. Use each set of dimensions to calculate multiple related quantities.

8. Convert foot and inch measurements to decimal feet when listing. Memorize decimal equivalents to .01 parts of a foot (1/8" equals approximately .01').

9. Do not "round off" quantities until the final summary.

10. Mark drawings with different colors as items are taken off.

11. Keep similar items together, different items separate.

12. Identify location and drawing numbers to aid in future checking for completeness.

13. Measure or list everything on the drawings or mentioned in the specifications.

14. It may be necessary to list items not called for to make the job complete.

15. Be alert for: Notes on plans such as N.T.S. (not to scale); changes in scale throughout the drawings; reduced size drawings; discrepancies between the specifications and the drawings.

16. Develop a consistent pattern of performing an estimate. For example:
    a. Start the quantity takeoff at the lower floor and move to the next higher floor.
    b. Proceed from the main section of the building to the wings.
    c. Proceed from south to north or vice versa, clockwise or counterclockwise.
    d. Take off floor plan quantities first, elevations next, then detail drawings.

17. List all gross dimensions that can be either used again for different quantities, or used as a rough check of other quantities for verification (exterior perimeter, gross floor area, individual floor areas, etc.).

18. Utilize design symmetry or repetition (repetitive floors, repetitive wings, symmetrical design around a center line, similar room layouts, etc.). Note: Extreme caution is needed here so as not to omit or duplicate an area.

19. Do not convert units until the final total is obtained. For instance, when estimating concrete work, keep all units to the nearest cubic foot, then summarize and convert to cubic yards.

20. When figuring alternatives, it is best to total all items involved in the basic system, then total all items involved in the alternates. Therefore, you work with positive numbers in all cases. When adds and deducts are used, it is often confusing whether to add or subtract a portion of an item; especially on a complicated or involved alternate.

R011105-10   Unit Gross Area Requirements

The figures in the table below indicate typical ranges in square feet as a function of the "occupant" unit. This table is best used in the preliminary design stages to help determine the probable size requirement for the total project.

Building Type	Unit	Gross Area in S.F.		
		1/4	Median	3/4
Apartments	Unit	660	860	1,100
Auditorium & Play Theaters	Seat	18	25	38
Bowling Alleys	Lane		940	
Churches & Synagogues	Seat	20	28	39
Dormitories	Bed	200	230	275
Fraternity & Sorority Houses	Bed	220	315	370
Garages, Parking	Car	325	355	385
Hospitals	Bed	685	850	1,075
Hotels	Rental Unit	475	600	710
Housing for the elderly	Unit	515	635	755
Housing, Public	Unit	700	875	1,030
Ice Skating Rinks	Total	27,000	30,000	36,000
Motels	Rental Unit	360	465	620
Nursing Homes	Bed	290	350	450
Restaurants	Seat	23	29	39
Schools, Elementary	Pupil	65	77	90
Junior High & Middle		85	110	129
Senior High		102	130	145
Vocational		110	135	195
Shooting Ranges	Point		450	
Theaters & Movies	Seat		15	

R011105-50 Metric Conversion Factors

Description: This table is primarily for converting customary U.S. units in the left hand column to SI metric units in the right hand column. In addition, conversion factors for some commonly encountered Canadian and non-SI metric units are included.

If You Know		Multiply By		To Find
Length	Inches	x 25.4[a]	=	Millimeters
	Feet	x 0.3048[a]	=	Meters
	Yards	x 0.9144[a]	=	Meters
	Miles (statute)	x 1.609	=	Kilometers
Area	Square inches	x 645.2	=	Square millimeters
	Square feet	x 0.0929	=	Square meters
	Square yards	x 0.8361	=	Square meters
Volume (Capacity)	Cubic inches	x 16,387	=	Cubic millimeters
	Cubic feet	x 0.02832	=	Cubic meters
	Cubic yards	x 0.7646	=	Cubic meters
	Gallons (U.S. liquids)[b]	x 0.003785	=	Cubic meters[c]
	Gallons (Canadian liquid)[b]	x 0.004546	=	Cubic meters[c]
	Ounces (U.S. liquid)[b]	x 29.57	=	Milliliters[c, d]
	Quarts (U.S. liquid)[b]	x 0.9464	=	Liters[c, d]
	Gallons (U.S. liquid)[b]	x 3.785	=	Liters[c, d]
Force	Kilograms force[d]	x 9.807	=	Newtons
	Pounds force	x 4.448	=	Newtons
	Pounds force	x 0.4536	=	Kilograms force[d]
	Kips	x 4448	=	Newtons
	Kips	x 453.6	=	Kilograms force[d]
Pressure, Stress, Strength (Force per unit area)	Kilograms force per square centimeter[d]	x 0.09807	=	Megapascals
	Pounds force per square inch (psi)	x 0.006895	=	Megapascals
	Kips per square inch	x 6.895	=	Megapascals
	Pounds force per square inch (psi)	x 0.07031	=	Kilograms force per square centimeter[d]
	Pounds force per square foot	x 47.88	=	Pascals
	Pounds force per square foot	x 4.882	=	Kilograms force per square meter[d]
Flow	Cubic feet per minute	x 0.4719	=	Liters per second
	Gallons per minute	x 0.0631	=	Liters per second
	Gallons per hour	x 1.05	=	Milliliters per second
Bending Moment Or Torque	Inch-pounds force	x 0.01152	=	Meter-kilograms force[d]
	Inch-pounds force	x 0.1130	=	Newton-meters
	Foot-pounds force	x 0.1383	=	Meter-kilograms force[d]
	Foot-pounds force	x 1.356	=	Newton-meters
	Meter-kilograms force[d]	x 9.807	=	Newton-meters
Mass	Ounces (avoirdupois)	x 28.35	=	Grams
	Pounds (avoirdupois)	x 0.4536	=	Kilograms
	Tons (metric)	x 1000	=	Kilograms
	Tons, short (2000 pounds)	x 907.2	=	Kilograms
	Tons, short (2000 pounds)	x 0.9072	=	Megagrams[e]
Mass per Unit Volume	Pounds mass per cubic foot	x 16.02	=	Kilograms per cubic meter
	Pounds mass per cubic yard	x 0.5933	=	Kilograms per cubic meter
	Pounds mass per gallon (U.S. liquid)[b]	x 119.8	=	Kilograms per cubic meter
	Pounds mass per gallon (Canadian liquid)[b]	x 99.78	=	Kilograms per cubic meter
Temperature	Degrees Fahrenheit	(F-32)/1.8	=	Degrees Celsius
	Degrees Fahrenheit	(F+459.67)/1.8	=	Degrees Kelvin
	Degrees Celsius	C+273.15	=	Degrees Kelvin

[a]The factor given is exact
[b]One U.S. gallon = 0.8327 Canadian gallon
[c]1 liter = 1000 milliliters = 1000 cubic centimeters
1 cubic decimeter = 0.001 cubic meter

[d]Metric but not SI unit
[e]Called "tonne" in England and
"metric ton" in other metric countries

R011105-60   Weights and Measures

Measures of Length
1 Mile = 1760 Yards = 5280 Feet
1 Yard = 3 Feet = 36 inches
1 Foot = 12 Inches
1 Mil = 0.001 Inch
1 Fathom = 2 Yards = 6 Feet
1 Rod = 5.5 Yards = 16.5 Feet
1 Hand = 4 Inches
1 Span = 9 Inches
1 Micro-inch = One Millionth Inch or 0.000001 Inch
1 Micron = One Millionth Meter + 0.00003937 Inch

Surveyor's Measure
1 Mile = 8 Furlongs = 80 Chains
1 Furlong = 10 Chains = 220 Yards
1 Chain = 4 Rods = 22 Yards = 66 Feet = 100 Links
1 Link = 7.92 Inches

Square Measure
1 Square Mile = 640 Acres = 6400 Square Chains
1 Acre = 10 Square Chains = 4840 Square Yards = 43,560 Sq. Ft.
1 Square Chain = 16 Square Rods = 484 Square Yards = 4356 Sq. Ft.
1 Square Rod = 30.25 Square Yards = 272.25 Square Feet = 625 Square Lines
1 Square Yard = 9 Square Feet
1 Square Foot = 144 Square Inches
An Acre equals a Square 208.7 Feet per Side

Cubic Measure
1 Cubic Yard = 27 Cubic Feet
1 Cubic Foot = 1728 Cubic Inches
1 Cord of Wood = 4 x 4 x 8 Feet = 128 Cubic Feet
1 Perch of Masonry = 16½ x 1½ x 1 Foot = 24.75 Cubic Feet

Avoirdupois or Commercial Weight
1 Gross or Long Ton = 2240 Pounds
1 Net or Short Ton = 2000 Pounds
1 Pound = 16 Ounces = 7000 Grains
1 Ounce = 16 Drachms = 437.5 Grains
1 Stone = 14 Pounds

Power
1 British Thermal Unit per Hour = 0.2931 Watts
1 Ton (Refrigeration) = 3.517 Kilowatts
1 Horsepower (Boiler) = 9.81 Kilowatts
1 Horsepower (550 ft-lb/s) = 0.746 Kilowatts

Shipping Measure
For Measuring Internal Capacity of a Vessel:
1 Register Ton = 100 Cubic Feet

For Measurement of Cargo:
Approximately 40 Cubic Feet of Merchandise is considered a Shipping Ton, unless that bulk would weigh more than 2000 Pounds, in which case Freight Charge may be based upon weight.
40 Cubic Feet = 32.143 U.S. Bushels = 31.16 Imp. Bushels

Liquid Measure
1 Imperial Gallon = 1.2009 U.S. Gallon = 277.42 Cu. In.
1 Cubic Foot = 7.48 U.S. Gallons

R011110-30   Engineering Fees

Typical **Structural Engineering Fees** based on type of construction and total project size. These fees are included in Architectural Fees.

Type of Construction	Total Project Size (in thousands of dollars)			
	$500	$500-$1,000	$1,000-$5,000	Over $5000
Industrial buildings, factories & warehouses	Technical payroll times 2.0 to 2.5	1.60%	1.25%	1.00%
Hotels, apartments, offices, dormitories, hospitals, public buildings, food stores		2.00%	1.70%	1.20%
Museums, banks, churches and cathedrals		2.00%	1.75%	1.25%
Thin shells, prestressed concrete, earthquake resistive		2.00%	1.75%	1.50%
Parking ramps, auditoriums, stadiums, convention halls, hangars & boiler houses		2.50%	2.00%	1.75%
Special buildings, major alterations, underpinning & future expansion		Add to above 0.5%	Add to above 0.5%	Add to above 0.5%

For complex reinforced concrete or unusually complicated structures, add 20% to 50%.

Typical **Mechanical and Electrical Engineering Fees** are based on the size of the subcontract. The fee structure for both is shown below. These fees are included in Architectural Fees.

Type of Construction	Subcontract Size							
	$25,000	$50,000	$100,000	$225,000	$350,000	$500,000	$750,000	$1,000,000
Simple structures	6.4%	5.7%	4.8%	4.5%	4.4%	4.3%	4.2%	4.1%
Intermediate structures	8.0	7.3	6.5	5.6	5.1	5.0	4.9	4.8
Complex structures	10.1	9.0	9.0	8.0	7.5	7.5	7.0	7.0

For renovations, add 15% to 25% to applicable fee.

R012153-10   Repair and Remodeling

Cost figures are based on new construction utilizing the most cost-effective combination of labor, equipment and material with the work scheduled in proper sequence to allow the various trades to accomplish their work in an efficient manner.

The costs for repair and remodeling work must be modified due to the following factors that may be present in any given repair and remodeling project.

1. Equipment usage curtailment due to the physical limitations of the project, with only hand-operated equipment being used.

2. Increased requirement for shoring and bracing to hold up the building while structural changes are being made and to allow for temporary storage of construction materials on above-grade floors.

3. Material handling becomes more costly due to having to move within the confines of an enclosed building. For multi-story construction, low capacity elevators and stairwells may be the only access to the upper floors.

4. Large amounts of cutting and patching and attempting to match the existing construction is required. It is often more economical to remove entire walls rather than create many new door and window openings. This sort of trade-off has to be carefully analyzed.

5. Cost of protection of completed work is increased since the usual sequence of construction usually cannot be accomplished.

6. Economies of scale usually associated with new construction may not be present. If small quantities of components must be custom fabricated due to job requirements, unit costs will naturally increase. Also, if only small work areas are available at a given time, job scheduling between trades becomes difficult and subcontractor quotations may reflect the excessive start-up and shut-down phases of the job.

7. Work may have to be done on other than normal shifts and may have to be done around an existing production facility which has to stay in production during the course of the repair and remodeling.

8. Dust and noise protection of adjoining non-construction areas can involve substantial special protection and alter usual construction methods.

9. Job may be delayed due to unexpected conditions discovered during demolition or removal. These delays ultimately increase construction costs.

10. Piping and ductwork runs may not be as simple as for new construction. Wiring may have to be snaked through walls and floors.

11. Matching "existing construction" may be impossible because materials may no longer be manufactured. Substitutions may be expensive.

12. Weather protection of existing structure requires additional temporary structures to protect building at openings.

13. On small projects, because of local conditions, it may be necessary to pay a tradesman for a minimum of four hours for a task that is completed in one hour.

All of the above areas can contribute to increased costs for a repair and remodeling project. Each of the above factors should be considered in the planning, bidding and construction stage in order to minimize the increased costs associated with repair and remodeling jobs.

R012153-60   Security Factors

Contractors entering, working in, and exiting secure facilities often lose productive time during a normal workday. The recommended allowances in this section are intended to provide for the loss of productivity by increasing labor costs. Note that different costs are associated with searches upon entry only and searches upon entry and exit. Time spent in a queue is unpredictable and not part of these allowances. Contractors should plan ahead for this situation.

Security checkpoints are designed to reflect the level of security required to gain access or egress. An extreme example is when contractors, along with any materials, tools, equipment, and vehicles, must be physically searched and have all materials, tools, equipment, and vehicles inventoried and documented prior to both entry and exit.

Physical searches without going through the documentation process represent the next level and take up less time.

Electronic searches—passing through a detector or x-ray machine with no documentation of materials, tools, equipment, and vehicles—take less time than physical searches.

Visual searches of materials, tools, equipment, and vehicles represent the next level of security.

Finally, access by means of an ID card or displayed sticker takes the least amount of time.

Another consideration is if the searches described above are performed each and every day, or if they are performed only on the first day with access granted by ID card or displayed sticker for the remainder of the project. The figures for this situation have been calculated to represent the initial check-in as described and subsequent entry by ID card or displayed sticker for up to 20 days on site. For the situation described above, where the time period is beyond 20 days, the impact on labor cost is negligible.

There are situations where tradespeople must be accompanied by an escort and observed during the work day. The loss of freedom of movement will slow down productivity for the tradesperson. Costs for the observer have not been included. Those costs are normally born by the owner.

R012153-65   Infectious Disease Precautions

Contractors entering and working on job sites may be required to take precautions to prevent the spread of infectious diseases. Those precautions will reduce the amount of productive time during a normal workday. The recommended allowances in this section are intended to provide for the loss of productive time by increasing labor costs.

The estimator should be aware that one, many or none of these precautions apply to the line items in an estimate. Job site requirements and sound judgement must be applied.

Labor cost implications are based upon:

Temperature checks	Once per day
Donning/doffing masks and gloves	Four times per day (2 times each)
Washing hands (additional over normal)	Eight times per day
Informational meetings	Once per day
Maintaining social distance	Throughout the day
Disinfecting tools or equipment	Throughout the day

R012909-80   Sales Tax by State

State sales tax on materials is tabulated below (5 states have no sales tax). Many states allow local jurisdictions, such as a county or city, to levy additional sales tax.

Some projects may be sales tax exempt, particularly those constructed with public funds.

State	Tax (%)	State	Tax (%)	State	Tax (%)	State	Tax (%)
Alabama	4	Illinois	6.25	Montana	0	Rhode Island	7
Alaska	0	Indiana	7	Nebraska	5.5	South Carolina	6
Arizona	5.6	Iowa	6	Nevada	6.85	South Dakota	4.5
Arkansas	6.5	Kansas	6.5	New Hampshire	0	Tennessee	7
California	7.25	Kentucky	6	New Jersey	6.625	Texas	6.25
Colorado	2.9	Louisiana	4.45	New Mexico	5.125	Utah	6.10
Connecticut	6.35	Maine	5.5	New York	4	Vermont	6
Delaware	0	Maryland	6	North Carolina	4.75	Virginia	5.3
District of Columbia	6	Massachusetts	6.25	North Dakota	5	Washington	6.5
Florida	6	Michigan	6	Ohio	5.75	West Virginia	6
Georgia	4	Minnesota	6.875	Oklahoma	4.5	Wisconsin	5
Hawaii	4	Mississippi	7	Oregon	0	Wyoming	4
Idaho	6	Missouri	4.225	Pennsylvania	6	Average	5.11 %

Sales Tax by Province (Canada)

GST - a value-added tax, which the government imposes on most goods and services provided in or imported into Canada. PST - a retail sales tax, which five of the provinces impose on the prices of most goods and some

services. QST - a value-added tax, similar to the federal GST, which Quebec imposes. HST - Three provinces have combined their retail sales taxes with the federal GST into one harmonized tax.

Province	PST (%)	QST(%)	GST(%)	HST(%)
Alberta	7	0	5	0
British Columbia	7	0	5	0
Manitoba	7	0	5	0
New Brunswick	0	0	0	15
Newfoundland	0	0	0	15
Northwest Territories	0	0	5	0
Nova Scotia	0	0	0	15
Ontario	0	0	0	13
Prince Edward Island	0	0	0	15
Quebec	9.975	0	5	0
Saskatchewan	6	0	5	0
Yukon	0	0	5	0

R012909-85   Unemployment Taxes and Social Security Taxes

State unemployment tax rates vary not only from state to state, but also with the experience rating of the contractor. The federal unemployment tax rate is 6.2% of the first $7,000 of wages. This is reduced by a credit of up to 5.4% for timely payment to the state. The minimum federal unemployment tax is 0.6% after all credits.

Social security (FICA) for 2021 is estimated at time of publication to be 7.65% of wages up to $137,700.

R012909-86   Unemployment Tax by State

Information is from the U.S. Department of Labor, state unemployment tax rates.

State	Tax (%)	State	Tax (%)	State	Tax (%)	State	Tax (%)
Alabama	6.80	Illinois	6.93	Montana	6.30	Rhode Island	9.49
Alaska	5.4	Indiana	7.4	Nebraska	5.4	South Carolina	5.46
Arizona	12.76	Iowa	7.5	Nevada	5.4	South Dakota	9.35
Arkansas	14.3	Kansas	7.6	New Hampshire	7.5	Tennessee	10.0
California	6.2	Kentucky	9.3	New Jersey	5.8	Texas	6.5
Colorado	8.15	Louisiana	6.2	New Mexico	5.4	Utah	7.1
Connecticut	6.8	Maine	5.46	New York	9.1	Vermont	7.7
Delaware	8.20	Maryland	7.50	North Carolina	5.76	Virginia	6.21
District of Columbia	7	Massachusetts	12.65	North Dakota	10.74	Washington	7.73
Florida	5.4	Michigan	10.3	Ohio	9	West Virginia	8.5
Georgia	5.4	Minnesota	9.1	Oklahoma	5.5	Wisconsin	12.0
Hawaii	5.6	Mississippi	5.6	Oregon	5.4	Wyoming	8.8
Idaho	5.4	Missouri	8.37	Pennsylvania	11.03	Median	7.40%

R012909-90   Overtime

One way to improve the completion date of a project or eliminate negative float from a schedule is to compress activity duration times. This can be achieved by increasing the crew size or working overtime with the proposed crew.

To determine the costs of working overtime to compress activity duration times, consider the following examples. Below is an overtime efficiency and cost chart based on a five, six, or seven day week with an eight through twelve hour day. Payroll percentage increases for time and one half and double times are shown for the various working days.

Days per Week	Hours per Day	Production Efficiency					Payroll Cost Factors	
		1st Week	2nd Week	3rd Week	4th Week	Average 4 Weeks	@ 1-1/2 Times	@ 2 Times
5	8	100%	100%	100%	100%	100%	1.000	1.000
	9	100	100	95	90	96	1.056	1.111
	10	100	95	90	85	93	1.100	1.200
	11	95	90	75	65	81	1.136	1.273
	12	90	85	70	60	76	1.167	1.333
6	8	100	100	95	90	96	1.083	1.167
	9	100	95	90	85	93	1.130	1.259
	10	95	90	85	80	88	1.167	1.333
	11	95	85	70	65	79	1.197	1.394
	12	90	80	65	60	74	1.222	1.444
7	8	100	95	85	75	89	1.143	1.286
	9	95	90	80	70	84	1.183	1.365
	10	90	85	75	65	79	1.214	1.429
	11	85	80	65	60	73	1.240	1.481
	12	85	75	60	55	69	1.262	1.524

R013113-40   Builder's Risk Insurance

Builder's risk insurance is insurance on a building during construction. Premiums are paid by the owner or the contractor. Blasting, collapse and underground insurance would raise total insurance costs.

R013113-50   General Contractor's Overhead

There are two distinct types of overhead on a construction project: Project overhead and main office overhead. Project overhead includes those costs at a construction site not directly associated with the installation of construction materials. Examples of project overhead costs include the following:

1. Superintendent
2. Construction office and storage trailers
3. Temporary sanitary facilities
4. Temporary utilities
5. Security fencing
6. Photographs
7. Cleanup
8. Performance and payment bonds

The above project overhead items are also referred to as general requirements and therefore are estimated in Division 1. Division 1 is the first division listed in the CSI MasterFormat but it is usually the last division estimated. The sum of the costs in Divisions 1 through 49 is referred to as the sum of the direct costs.

All construction projects also include indirect costs. The primary components of indirect costs are the contractor's main office overhead and profit. The amount of the main office overhead expense varies depending on the following:

1. Owner's compensation
2. Project managers' and estimators' wages
3. Clerical support wages
4. Office rent and utilities
5. Corporate legal and accounting costs
6. Advertising
7. Automobile expenses
8. Association dues
9. Travel and entertainment expenses

These costs are usually calculated as a percentage of annual sales volume. This percentage can range from 35% for a small contractor doing less than $500,000 to 5% for a large contractor with sales in excess of $100 million.

R013113-55   Installing Contractor's Overhead

Installing contractors (subcontractors) also incur costs for general requirements and main office overhead.

Included within the total incl. overhead and profit costs is a percent mark-up for overhead that includes:

1. Compensation and benefits for office staff and project managers
2. Office rent, utilities, business equipment, and maintenance
3. Corporate legal and accounting costs

4. Advertising
5. Vehicle expenses (for office staff and project managers)
6. Association dues
7. Travel, entertainment
8. Insurance
9. Small tools and equipment

R013113-60   Workers' Compensation Insurance Rates by Trade

The table below tabulates the national averages for workers' compensation insurance rates by trade and type of building. The average "Insurance Rate" is multiplied by the "% of Building Cost" for each trade. This produces the "Workers' Compensation" cost by % of total labor cost, to be added for each trade by building type to determine the weighted average workers' compensation rate for the building types analyzed.

Trade	Insurance Rate (% Labor Cost) Range		Average	% of Building Cost Office Bldgs.	Schools & Apts.	Mfg.	Workers' Compensation Office Bldgs.	Schools & Apts.	Mfg.
Excavation, Grading, etc.	2.1 % to	16.5%	9.3%	4.8%	4.9%	4.5%	0.45%	0.46%	0.42%
Piles & Foundations	3.1 to	28.0	15.6	7.1	5.2	8.7	1.11	0.81	1.36
Concrete	2.5 to	24.9	13.7	5.0	14.8	3.7	0.69	2.03	0.51
Masonry	3.2 to	53.5	28.4	6.9	7.5	1.9	1.96	2.13	0.54
Structural Steel	3.3 to	30.8	17.1	10.7	3.9	17.6	1.83	0.67	3.01
Miscellaneous & Ornamental Metals	2.3 to	22.8	12.5	2.8	4.0	3.6	0.35	0.50	0.45
Carpentry & Millwork	2.6 to	29.1	15.8	3.7	4.0	0.5	0.58	0.63	0.08
Metal or Composition Siding	4.2 to	135.3	69.7	2.3	0.3	4.3	1.60	0.21	3.00
Roofing	4.2 to	105.1	54.7	2.3	2.6	3.1	1.26	1.42	1.70
Doors & Hardware	3.1 to	29.1	16.1	0.9	1.4	0.4	0.14	0.23	0.06
Sash & Glazing	2.6 to	20.5	11.5	3.5	4.0	1.0	0.40	0.46	0.12
Lath & Plaster	2.1 to	33.7	17.9	3.3	6.9	0.8	0.59	1.24	0.14
Tile, Marble & Floors	1.9 to	17.4	9.7	2.6	3.0	0.5	0.25	0.29	0.05
Acoustical Ceilings	1.7 to	21.5	11.6	2.4	0.2	0.3	0.28	0.02	0.03
Painting	2.8 to	37.7	20.3	1.5	1.6	1.6	0.30	0.32	0.32
Interior Partitions	2.6 to	29.1	15.8	3.9	4.3	4.4	0.62	0.68	0.70
Miscellaneous Items	1.6 to	97.7	49.7	5.2	3.7	9.7	2.58	1.84	4.82
Elevators	1.0 to	8.9	5.0	2.1	1.1	2.2	0.11	0.06	0.11
Sprinklers	1.4 to	14.2	7.8	0.5	—	2.0	0.04	—	0.16
Plumbing	1.4 to	13.2	7.3	4.9	7.2	5.2	0.36	0.53	0.38
Heat., Vent., Air Conditioning	2.8 to	15.8	9.3	13.5	11.0	12.9	1.26	1.02	1.20
Electrical	1.2 to	10.6	5.9	10.1	8.4	11.1	0.60	0.50	0.65
Total	1.0 % to	135.3%	68.1	100.0%	100.0%	100.0%	17.36%	16.05%	19.81%
			Overall Weighted Average 17.74%						

Workers' Compensation Insurance Rates by States

The table below lists the weighted average Workers' Compensation base rate for each state with a factor comparing this with the national average of 15.1%.

State	Weighted Average	Factor	State	Weighted Average	Factor	State	Weighted Average	Factor
Alabama	17.2%	159	Kentucky	17.5%	162	North Dakota	8.3%	77
Alaska	11.9	110	Louisiana	25.5	236	Ohio	6.4	59
Arizona	11.1	103	Maine	13.5	125	Oklahoma	13.3	123
Arkansas	7.3	68	Maryland	15.2	141	Oregon	10.8	100
California	32.6	302	Massachusetts	12.0	111	Pennsylvania	26.1	242
Colorado	8.1	75	Michigan	5.8	54	Rhode Island	11.0	102
Connecticut	19.8	183	Minnesota	20.7	192	South Carolina	24.6	228
Delaware	18.2	169	Mississippi	13.7	127	South Dakota	11.3	105
District of Columbia	11.3	105	Missouri	21.4	198	Tennessee	8.8	81
Florida	12.5	116	Montana	12.2	113	Texas	8.0	74
Georgia	50.0	463	Nebraska	16.6	154	Utah	9.2	85
Hawaii	14.0	130	Nevada	12.0	111	Vermont	13.6	126
Idaho	12.9	119	New Hampshire	12.7	118	Virginia	9.9	92
Illinois	27.9	258	New Jersey	21.5	199	Washington	10.4	96
Indiana	4.9	45	New Mexico	20.7	192	West Virginia	6.7	62
Iowa	15.1	140	New York	25.9	240	Wisconsin	16.0	148
Kansas	8.8	81	North Carolina	17.4	161	Wyoming	7.0	65
			Weighted Average for U.S. is 15.1% of payroll = 100%					

The weighted average skilled worker rate for 35 trades is 17.74%. For bidding purposes, apply the full value of Workers' Compensation directly to total labor costs, or if labor is 38%, materials 42% and overhead and profit 20% of total cost, carry 38/80 x 17.74% = 8.43% of cost (before overhead and profit) into overhead. Rates vary not only from state to state but also with the experience rating of the contractor.

Rates are the most current available at the time of publication.

R013113-80   Performance Bond

This table shows the cost of a Performance Bond for a construction job scheduled to be completed in 12 months. Add 1% of the premium cost per month for jobs requiring more than 12 months to complete. The rates are "standard" rates offered to contractors that the bonding company considers financially sound and capable of doing the work. Preferred rates are offered by some bonding companies based upon financial strength of the contractor. Actual rates vary from contractor to contractor and from bonding company to bonding company. Contractors should prequalify through a bonding agency before submitting a bid on a contract that requires a bond.

Contract Amount		Building Construction Class B Projects			Highways & Bridges						
					Class A New Construction			Class A-1 Highway Resurfacing			
First	$ 100,000 bid	$25.00 per M			$15.00 per M			$9.40 per M			
Next	400,000 bid	$ 2,500	plus	$15.00 per M	$ 1,500	plus	$10.00 per M	$ 940	plus	$7.20 per M
Next	2,000,000 bid	8,500	plus	10.00 per M	5,500	plus	7.00 per M	3,820	plus	5.00 per M
Next	2,500,000 bid	28,500	plus	7.50 per M	19,500	plus	5.50 per M	15,820	plus	4.50 per M
Next	2,500,000 bid	47,250	plus	7.00 per M	33,250	plus	5.00 per M	28,320	plus	4.50 per M
Over	7,500,000 bid	64,750	plus	6.00 per M	45,750	plus	4.50 per M	39,570	plus	4.00 per M

R015113-65   Temporary Power Equipment

Cost data for the temporary equipment was developed utilizing the following information.

1) Re-usable material-services, transformers, equipment and cords are based on new purchase and prorated to three projects.
2) PVC feeder includes trench and backfill.
3) Connections include disconnects and fuses.
4) Labor units include an allowance for removal.
5) No utility company charges or fees are included.
6) Concrete pads or vaults are not included.
7) Utility company conduits not included.

R015423-10   Steel Tubular Scaffolding

On new construction, tubular scaffolding is efficient up to 60' high or five stories. Above this it is usually better to use hung scaffolding if construction permits. Swing scaffolding operations may interfere with tenants. In this case, the tubular is more practical at all heights.

In repairing or cleaning the front of an existing building the cost of tubular scaffolding per S.F. of building front goes up as the height increases above the first tier. The first tier cost is relatively high due to leveling and alignment.

The minimum efficient crew for erecting and dismantling is three workers. They can set up and remove 18 frame sections per day up to 5 stories high. For 6 to 12 stories high, a crew of four is most efficient. Use two or more on top and two on the bottom for handing up or hoisting. They can

also set up and remove 18 frame sections per day. At 7' horizontal spacing, this will run about 800 S.F. per day of erecting and dismantling. Time for placing and removing planks must be added to the above. A crew of three can place and remove 72 planks per day up to 5 stories. For over 5 stories, a crew of four can place and remove 80 planks per day.

The table below shows the number of pieces required to erect tubular steel scaffolding for 1000 S.F. of building frontage. This area is made up of a scaffolding system that is 12 frames (11 bays) long by 2 frames high.

For jobs under twenty-five frames, add 50% to the rental cost. Rental rates will be lower for jobs over three months duration. Large quantities for long periods can reduce rental rates by 20%.

Description of Component	Number of Pieces for 1000 S.F. of Building Front	Unit
5' Wide Standard Frame, 6'-4" High	24	Ea.
Leveling Jack & Plate	24	
Cross Brace	44	
Side Arm Bracket, 21"	12	
Guardrail Post	12	
Guardrail, 7' section	22	
Stairway Section	2	
Stairway Starter Bar	1	
Stairway Inside Handrail	2	
Stairway Outside Handrail	2	
Walk-Through Frame Guardrail	2	

Scaffolding is often used as falsework over 15' high during construction of cast-in-place concrete beams and slabs. Two foot wide scaffolding is generally used for heavy beam construction. The span between frames depends upon the load to be carried with a maximum span of 5'.

Heavy duty shoring frames with a capacity of 10,000#/leg can be spaced up to 10' O.C. depending upon form support design and loading.

Scaffolding used as horizontal shoring requires less than half the material required with conventional shoring.

On new construction, erection is done by carpenters.

Rolling towers supporting horizontal shores can reduce labor and speed the job. For maintenance work, catwalks with spans up to 70' can be supported by the rolling towers.

R015433-10   Contractor Equipment

Rental Rates shown elsewhere in the data set pertain to late model high quality machines in excellent working condition, rented from equipment dealers. Rental rates from contractors may be substantially lower than the rental rates from equipment dealers depending upon economic conditions; for older, less productive machines, reduce rates by a maximum of 15%. Any overtime must be added to the base rates. For shift work, rates are lower. Usual rule of thumb is 150% of one shift rate for two shifts; 200% for three shifts.

For periods of less than one week, operated equipment is usually more economical to rent than renting bare equipment and hiring an operator.

Costs to move equipment to a job site (mobilization) or from a job site (demobilization) are not included in rental rates, nor in any Equipment costs on any Unit Price line items or crew listings. These costs can be found elsewhere. If a piece of equipment is already at a job site, it is not appropriate to utilize mob/demob costs in an estimate again.

Rental rates vary throughout the country with larger cities generally having lower rates. Lease plans for new equipment are available for periods in excess of six months with a percentage of payments applying toward purchase.

Rental rates can also be treated as reimbursement costs for contractor-owned equipment. Owned equipment costs include depreciation, loan payments, interest, taxes, insurance, storage, and major repairs.

Monthly rental rates vary from 2% to 5% of the cost of the equipment depending on the anticipated life of the equipment and its wearing parts. Weekly rates are about 1/3 the monthly rates and daily rental rates are about 1/3 the weekly rates.

The hourly operating costs for each piece of equipment include costs to the user such as fuel, oil, lubrication, normal expendables for the equipment, and a percentage of the mechanic's wages chargeable to maintenance. The hourly operating costs listed do not include the operator's wages.

The daily cost for equipment used in the standard crews is figured by dividing the weekly rate by five, then adding eight times the hourly operating cost to give the total daily equipment cost, not including the operator. This figure is in the right hand column of the Equipment listings under Equipment Cost/Day.

Pile Driving rates shown for the pile hammer and extractor do not include leads, cranes, boilers or compressors. Vibratory pile driving requires an added field specialist during set-up and pile driving operation for the electric model. The hydraulic model requires a field specialist for set-up only. Up to 125 reuses of sheet piling are possible using vibratory drivers. For normal conditions, crane capacity for hammer type and size is as follows.

Crane Capacity	Hammer Type and Size		
	Air or Steam	Diesel	Vibratory
25 ton	to 8,750 ft.-lb.		70 H.P.
40 ton	15,000 ft.-lb.	to 32,000 ft.-lb.	170 H.P.
60 ton	25,000 ft.-lb.		300 H.P.
100 ton		112,000 ft.-lb.	

Cranes should be specified for the job by size, building and site characteristics, availability, performance characteristics, and duration of time required.

Backhoes & Shovels rent for about the same as equivalent size cranes but maintenance and operating expenses are higher. The crane operator's rate must be adjusted for high boom heights. Average adjustments: for 150' boom add 2% per hour; over 185', add 4% per hour; over 210', add 6% per hour; over 250', add 8% per hour and over 295', add 12% per hour.

Tower Cranes of the climbing or static type have jibs from 50' to 200' and capacities at maximum reach range from 4,000 to 14,000 pounds. Lifting capacities increase up to maximum load as the hook radius decreases.

Typical rental rates, based on purchase price, are about 2% to 3% per month.

Erection and dismantling run between 500 and 2000 labor hours. Climbing operation takes 10 labor hours per 20' climb. Crane dead time is about 5 hours per 40' climb. If crane is bolted to side of the building add cost of ties and extra mast sections. Climbing cranes have from 80' to 180' of mast while static cranes have 80' to 800' of mast.

Truck Cranes can be converted to tower cranes by using tower attachments. Mast heights over 400' have been used.

A single 100' high material **Hoist and Tower** can be erected and dismantled in about 400 labor hours; a double 100' high hoist and tower in about 600 labor hours. Erection times for additional heights are 3 and 4 labor hours

per vertical foot respectively up to 150', and 4 to 5 labor hours per vertical foot over 150' high. A 40' high portable Buck hoist takes about 160 labor hours to erect and dismantle. Additional heights take 2 labor hours per vertical foot to 80' and 3 labor hours per vertical foot for the next 100'. Most material hoists do not meet local code requirements for carrying personnel.

A 150' high **Personnel Hoist** requires about 500 to 800 labor hours to erect and dismantle. Budget erection time at 5 labor hours per vertical foot for all trades. Local code requirements or labor scarcity requiring overtime can add up to 50% to any of the above erection costs.

Earthmoving Equipment: The selection of earthmoving equipment depends upon the type and quantity of material, moisture content, haul distance, haul road, time available, and equipment available. Short haul cut and fill operations may require dozers only, while another operation may require excavators, a fleet of trucks, and spreading and compaction equipment. Stockpiled material and granular material are easily excavated with front end loaders. Scrapers are most economically used with hauls between 300' and 1-1/2 miles if adequate haul roads can be maintained. Shovels are often used for blasted rock and any material where a vertical face of 8' or more can be excavated. Special conditions may dictate the use of draglines, clamshells, or backhoes. Spreading and compaction equipment must be matched to the soil characteristics, the compaction required and the rate the fill is being supplied.

R015433-15   Heavy Lifting

Hydraulic Climbing Jacks

The use of hydraulic heavy lift systems is an alternative to conventional type crane equipment. The lifting, lowering, pushing, or pulling mechanism is a hydraulic climbing jack moving on a square steel jackrod from 1-5/8" to 4" square, or a steel cable. The jackrod or cable can be vertical or horizontal, stationary or movable, depending on the individual application. When the jackrod is stationary, the climbing jack will climb the rod and push or pull the load along with itself. When the climbing jack is stationary, the jackrod is movable with the load attached to the end and the climbing jack will lift or lower the jackrod with the attached load. The heavy lift system is normally operated by a single control lever located at the hydraulic pump.

The system is flexible in that one or more climbing jacks can be applied wherever a load support point is required, and the rate of lift synchronized.

Economic benefits have been demonstrated on projects such as: erection of ground assembled roofs and floors, complete bridge spans, girders and trusses, towers, chimney liners and steel vessels, storage tanks, and heavy machinery. Other uses are raising and lowering offshore work platforms, caissons, tunnel sections and pipelines.

R015436-50   Mobilization

Costs to move rented construction equipment to a job site from an equipment dealer's or contractor's yard (mobilization) or off the job site (demobilization) are not included in the rental or operating rates, nor in the equipment cost on a unit price line or in a crew listing. These costs can be found consolidated in the Mobilization section of the data and elsewhere in particular site work sections. If a piece of equipment is already on the job site, it is not appropriate to include mob/demob costs in a new estimate that requires use of that equipment. The following table identifies approximate sizes of rented construction equipment that would be hauled on a towed trailer. Because this listing is not all-encompassing, the user can infer as to what size trailer might be required for a piece of equipment not listed.

3-ton Trailer	20-ton Trailer	40-ton Trailer	50-ton Trailer
20 H.P. Excavator	110 H.P. Excavator	200 H.P. Excavator	270 H.P. Excavator
50 H.P. Skid Steer	165 H.P. Dozer	300 H.P. Dozer	Small Crawler Crane
35 H.P. Roller	150 H.P. Roller	400 H.P. Scraper	500 H.P. Scraper
40 H.P. Trencher	Backhoe	450 H.P. Art. Dump Truck	500 H.P. Art. Dump Truck

R024119-10   Demolition Defined

Whole Building Demolition - Demolition of the whole building with no concern for any particular building element, component, or material type being demolished. This type of demolition is accomplished with large pieces of construction equipment that break up the structure, load it into trucks and haul it to a disposal site, but disposal or dump fees are not included. Demolition of below-grade foundation elements, such as footings, foundation walls, grade beams, slabs on grade, etc., is not included. Certain mechanical equipment containing flammable liquids or ozone-depleting refrigerants, electric lighting elements, communication equipment components, and other building elements may contain hazardous waste, and must be removed, either selectively or carefully, as hazardous waste before the building can be demolished.

Foundation Demolition - Demolition of below-grade foundation footings, foundation walls, grade beams, and slabs on grade. This type of demolition is accomplished by hand or pneumatic hand tools, and does not include saw cutting, or handling, loading, hauling, or disposal of the debris.

Gutting - Removal of building interior finishes and electrical/mechanical systems down to the load-bearing and sub-floor elements of the rough building frame, with no concern for any particular building element, component, or material type being demolished. This type of demolition is accomplished by hand or pneumatic hand tools, and includes loading into trucks, but not hauling, disposal or dump fees, scaffolding, or shoring. Certain mechanical equipment containing flammable liquids or ozone-depleting refrigerants, electric lighting elements, communication equipment components, and other building elements may contain hazardous waste, and must be removed, either selectively or carefully, as hazardous waste, before the building is gutted.

Selective Demolition - Demolition of a selected building element, component, or finish, with some concern for surrounding or adjacent elements, components, or finishes (see the first Subdivision (s) at the beginning of appropriate Divisions). This type of demolition is accomplished by hand or pneumatic hand tools, and does not include handling, loading,

storing, hauling, or disposal of the debris, scaffolding, or shoring. "Gutting" methods may be used in order to save time, but damage that is caused to surrounding or adjacent elements, components, or finishes may have to be repaired at a later time.

Careful Removal - Removal of a piece of service equipment, building element or component, or material type, with great concern for both the removed item and surrounding or adjacent elements, components or finishes. The purpose of careful removal may be to protect the removed item for later re-use, preserve a higher salvage value of the removed item, or replace an item while taking care to protect surrounding or adjacent elements, components, connections, or finishes from cosmetic and/or structural damage. An approximation of the time required to perform this type of removal is 1/3 to 1/2 the time it would take to install a new item of like kind (see Reference Number R260105-30). This type of removal is accomplished by hand or pneumatic hand tools, and does not include loading, hauling, or storing the removed item, scaffolding, shoring, or lifting equipment.

Cutout Demolition - Demolition of a small quantity of floor, wall, roof, or other assembly, with concern for the appearance and structural integrity of the surrounding materials. This type of demolition is accomplished by hand or pneumatic hand tools, and does not include saw cutting, handling, loading, hauling, or disposal of debris, scaffolding, or shoring.

Rubbish Handling - Work activities that involve handling, loading or hauling of debris. Generally, the cost of rubbish handling must be added to the cost of all types of demolition, with the exception of whole building demolition.

Minor Site Demolition - Demolition of site elements outside the footprint of a building. This type of demolition is accomplished by hand or pneumatic hand tools, or with larger pieces of construction equipment, and may include loading a removed item onto a truck (check the Crew for equipment used). It does not include saw cutting, hauling or disposal of debris, and, sometimes, handling or loading.

R024119-20   Dumpsters

Dumpster rental costs on construction sites are presented in two ways.

The cost per week rental includes the delivery of the dumpster; its pulling or emptying once per week, and its final removal. The assumption is made that the dumpster contractor could choose to empty a dumpster by simply bringing in an empty unit and removing the full one. These costs also include the disposal of the materials in the dumpster.

The Alternate Pricing can be used when actual planned conditions are not approximated by the weekly numbers. For example, these lines can be used when a dumpster is needed for 4 weeks and will need to be emptied 2 or 3 times per week. Conversely the Alternate Pricing lines can be used when a dumpster will be rented for several weeks or months but needs to be emptied only a few times over this period.

Reference Tables

R053100-10   Decking Descriptions

General - All Deck Products

A steel deck is made by cold forming structural grade sheet steel into a repeating pattern of parallel ribs. The strength and stiffness of the panels are the result of the ribs and the material properties of the steel. Deck lengths can be varied to suit job conditions, but because of shipping considerations, are usually less than 40 feet. Standard deck width varies with the product used but full sheets are usually 12″, 18″, 24″, 30″, or 36″. The deck is typically furnished in a standard width with the ends cut square. Any cutting for width, such as at openings or for angular fit, is done at the job site.

The deck is typically attached to the building frame with arc puddle welds, self-drilling screws, or powder or pneumatically driven pins. Sheet to sheet fastening is done with screws, button punching (crimping), or welds.

Composite Floor Deck

After installation and adequate fastening, a floor deck serves several purposes. It (a) acts as a working platform, (b) stabilizes the frame, (c) serves as a concrete form for the slab, and (d) reinforces the slab to carry the design loads applied during the life of the building. Composite decks are distinguished by the presence of shear connector devices as part of the deck. These devices are designed to mechanically lock the concrete and deck together so that the concrete and the deck work together to carry subsequent floor loads. These shear connector devices can be rolled-in embossments, lugs, holes, or wires welded to the panels. The deck profile can also be used to interlock concrete and steel.

Composite deck finishes are either galvanized (zinc coated) or phosphatized/painted. Galvanized deck has a zinc coating on both the top and bottom surfaces. The phosphatized/painted deck has a bare (phosphatized) top surface that will come into contact with the concrete. This bare top surface can be expected to develop rust before the concrete is placed. The bottom side of the deck has a primer coat of paint.

A composite floor deck is normally installed so the panel ends do not overlap on the supporting beams. Shear lugs or panel profile shapes often prevent a tight metal to metal fit if the panel ends overlap; the air gap caused by overlapping will prevent proper fusion with the structural steel supports when the panel end laps are shear stud welded.

Adequate end bearing of the deck must be obtained as shown on the drawings. If bearing is actually less in the field than shown on the drawings, further investigation is required.

Roof Deck

A roof deck is not designed to act compositely with other materials. A roof deck acts alone in transferring horizontal and vertical loads into the building frame. Roof deck rib openings are usually narrower than floor deck rib openings. This provides adequate support of the rigid thermal insulation board.

A roof deck is typically installed to endlap approximately 2″ over supports. However, it can be butted (or lapped more than 2″) to solve field fit problems. Since designers frequently use the installed deck system as part of the horizontal bracing system (the deck as a diaphragm), any fastening substitution or change should be approved by the designer. Continuous perimeter support of the deck is necessary to limit edge deflection in the finished roof and may be required for diaphragm shear transfer.

Standard roof deck finishes are galvanized or primer painted. The standard factory applied paint for roof decks is a primer paint and is not intended to weather for extended periods of time. Field painting or touching up of abrasions and deterioration of the primer coat or other protective finishes is the responsibility of the contractor.

Cellular Deck

A cellular deck is made by attaching a bottom steel sheet to a roof deck or composite floor deck panel. A cellular deck can be used in the same manner as a floor deck. Electrical, telephone, and data wires are easily run through the chase created between the deck panel and the bottom sheet.

When used as part of the electrical distribution system, the cellular deck must be installed so that the ribs line up and create a smooth cell transition at abutting ends. The joint that occurs at butting cell ends must be taped or otherwise sealed to prevent wet concrete from seeping into the cell. Cell interiors must be free of welding burrs, or other sharp intrusions, to prevent damage to wires.

When used as a roof deck, the bottom flat plate is usually left exposed to view. Care must be maintained during erection to keep good alignment and prevent damage.

A cellular deck is sometimes used with the flat plate on the top side to provide a flat working surface. Installation of the deck for this purpose requires special methods for attachment to the frame because the flat plate, now on the top, can prevent direct access to the deck material that is bearing on the structural steel. It may be advisable to treat the flat top surface to prevent slipping.

A cellular deck is always furnished galvanized or painted over galvanized.

Form Deck

A form deck can be any floor or roof deck product used as a concrete form. Connections to the frame are by the same methods used to anchor floor and roof decks. Welding washers are recommended when welding a deck that is less than 20 gauge thickness.

A form deck is furnished galvanized, prime painted, or uncoated. A galvanized deck must be used for those roof deck systems where a form deck is used to carry a lightweight insulating concrete fill.

R078413-30   Firestopping

Firestopping is the sealing of structural, mechanical, electrical and other penetrations through fire-rated assemblies. The basic components of firestop systems are safing insulation and firestop sealant on both sides of wall penetrations and the top side of floor penetrations.

Pipe penetrations are assumed to be through concrete, grout, or joint compound and can be sleeved or unsleeved. Costs for the penetrations and sleeves are not included. An annular space of 1″ is assumed. Escutcheons are not included.

A metallic pipe is assumed to be copper, aluminum, cast iron or similar metallic material. An insulated metallic pipe is assumed to be covered with a thermal insulating jacket of varying thickness and materials.

A non-metallic pipe is assumed to be PVC, CPVC, FR Polypropylene or similar plastic piping material. Intumescent firestop sealant or wrap strips are included. Collars on both sides of wall penetrations and a sheet metal plate on the underside of floor penetrations are included.

Ductwork is assumed to be sheet metal, stainless steel or similar metallic material. Duct penetrations are assumed to be through concrete, grout or joint compound. Costs for penetrations and sleeves are not included. An annular space of 1/2″ is assumed.

Multi-trade openings include costs for sheet metal forms, firestop mortar, wrap strips, collars and sealants as necessary.

Structural penetrations joints are assumed to be 1/2″ or less. CMU walls are assumed to be within 1-1/2″ of the metal deck. Drywall walls are assumed to be tight to the underside of metal decking.

Metal panel, glass or curtain wall systems include a spandrel area of 5′ filled with mineral wool foil-faced insulation. Fasteners and stiffeners are included.

R224000-10 Water Consumption Rates

Fixture Type	Water Supply Fixture Unit Value		
	Hot Water	Cold Water	Combined
Bathtub	1.0	1.0	1.4
Clothes Washer	1.0	1.0	1.4
Dishwasher	1.4	0.0	1.4
Kitchen Sink	1.0	1.0	1.4
Laundry Tub	1.0	1.0	1.4
Lavatory	0.5	0.5	0.7
Shower Stall	1.0	1.0	1.4
Water Closet (tank type)	0.0	2.2	2.2
Full Bath Group			
w/bathtub or shower stall	1.5	2.7	3.6
Half Bath Group			
w/W.C. and Lavatory	0.5	2.5	2.6
Kitchen Group			
w/Dishwasher and Sink	1.9	1.0	2.5
Laundry Group			
w/Clothes Washer and			
Laundry Tub	1.8	1.8	2.5
Hose bibb (sillcock)	0.0	2.5	2.5

Notes:
Typically, WSFU = 1GPM

Supply loads in the building water-distribution system shall be determined by total load on the pipe
being sized, in terms of water supply fixture units (WSFU) and gallons per minute (GPM) flow rates.
For fixtures not listed, choose a WSFU value of a fixture with similar flow characteristics. Water Fixture
Supply Units determined the required water supply to fixtures and their service systems. Fixture units
are equal to one (1) cubic foot of water drained in a 1-1/4" pipe per minute. It is not a flow rate unit but
a design factor.

Excerpted from the 2012 *International Plumbing Code*, Copyright 2011.
Washington, D.C.: International Code Council. Reproduced with permission.
All rights reserved. www.ICCSAFE.org

R224000-20 Fixture Demands in Gallons per Fixture per Hour

The table below is based on a 140°F final temperature except for dishwashers in public places where 180°F water is mandatory.

Supply Systems for Flush Tanks			Supply Systems for Flushometer Valves		
Load	Demand		Load	Demand	
WSFU	GPM	CU. FT.	WSFU	GPM	CU. FT.
1.0	3.0	0.041040	–	–	–
2.0	5.0	0.068400	–	–	–
3.0	6.5	0.868920	–	–	–
4.0	8.0	1.069440	–	–	–
5.0	9.4	1.256592	5.0	15	2.0052
6.0	10.7	1.430376	6.0	17.4	2.326032
7.0	11.8	1.577424	7.0	19.8	2.646364
8.0	12.8	1.711104	8.0	22.2	2.967696
9.0	13.7	1.831416	9.0	24.6	3.288528
10.0	14.6	1.951728	10.0	27	3.60936
11.0	15.4	2.058672	11.0	27.8	3.716304
12.0	16.0	2.138880	12.0	28.6	3.823248
13.0	16.5	2.205720	13.0	29.4	3.930192
14.0	17.0	2.272560	14.0	30.2	4.037136
15.0	17.5	2.339400	15.0	31	4.14408
16.0	18.0	2.906240	16.0	31.8	4.241024
17.0	18.4	2.459712	17.0	32.6	4.357968
18.0	18.8	2.513184	18.0	33.4	4.464912
19.0	19.2	2.566656	19.0	34.2	4.571856
20.0	19.6	2.620218	20.0	35.0	4.678800
25.0	21.5	2.874120	25.0	38.0	5.079840
30.0	23.3	3.114744	30.0	42.0	5.611356
35.0	24.9	3.328632	35.0	44.0	5.881920
40.0	26.3	3.515784	40.0	46.0	6.149280
45.0	27.7	3.702936	45.0	48.0	6.416640
50.0	29.1	3.890088	50.0	50.0	6.684000

Notes:

When designing a plumbing system that utilizes fixtures other than, or in addition to, water closets, use the data provided in the Supply Systems for Flush Tanks section of the above table.

To obtain the probable maximum demand, multiply the total demands for the fixtures (gal./fixture/hour) by the demand factor. The heater should have a heating capacity in gallons per hour equal to this maximum. The storage tank should have a capacity in gallons equal to the probable maximum demand multiplied by the storage capacity factor.

For customer support on your Electrical Costs with RSMeans Data, call 800.448.8182.

585

Fig. R238313-11

R238313-10   Heat Trace Systems

Before you can determine the cost of a HEAT TRACE installation, the method of attachment must be established. There are four (4) common methods:

1. Cable is simply attached to the pipe with polyester tape every 12'.
2. Cable is attached with a continuous cover of 2" wide aluminum tape.
3. Cable is attached with factory extruded heat transfer cement and covered with metallic raceway with clips every 10'.
4. Cable is attached between layers of pipe insulation using either clips or polyester tape.

Example: Components for method 3 must include:
   A. Heat trace cable by voltage and watts per linear foot.
   B. Heat transfer cement, 1 gallon per 60 linear feet of cover.
   C. Metallic raceway by size and type.
   D. Raceway clips by size of pipe.

When taking off linear foot lengths of cable add the following for each valve in the system. (E)

In all of the above methods each component of the system must be priced individually.

SCREWED OR WELDED VALVE:		
1/2"	=	6"
3/4"	=	9"
1"	=	1' -0"
1-1/2"	=	1' -6"
2"	=	2'
2-1/2"	=	2' -6"
3"	=	2' -6"
4"	=	4' -0"
6"	=	7' -0"
8"	=	9' -6"
10"	=	12' -6"
12"	=	15' -0"
14"	=	18' -0"
16"	=	21' -6"
18"	=	25' -6"
20"	=	28' -6"
24"	=	34' -0"
30"	=	40' -0"

FLANGED VALVE:		
1/2"	=	1' -0"
3/4"	=	1' -6"
1"	=	2' -0"
1-1/2"	=	2' -6"
2"	=	2' -6"
2-1/2"	=	3' -0"
3"	=	3' -6"
4"	=	4' -0"
6"	=	8' -0"
8"	=	11' -0"
10"	=	14' -0"
12"	=	16' -6"
14"	=	19' -6"
16"	=	23' -0"
18"	=	27' -0"
20"	=	30' -0"
24"	=	36' -0"
30"	=	42' -0"

BUTTERFLY VALVES:		
1/2"	=	0'
3/4"	=	0'
1"	=	1' -0"
1-1/2"	=	1' -6"
2"	=	2' -0"
2-1/2"	=	2' -6"
3"	=	2' -6"
4"	=	3' -0"
6"	=	3' -6"
8"	=	4' -0"
10"	=	4' -0"
12"	=	5' -0"
14"	=	5' -6"
16"	=	6' -0"
18"	=	6' -6"
20"	=	7' -0"
24"	=	8' -0"
30"	=	10' -0"

R238313-10 Heat Trace Systems (cont.)

Add the following quantities of heat transfer cement to linear foot totals for each valve:

Nominal Valve Size	Gallons of Cement per Valve
1/2"	0.14
3/4"	0.21
1"	0.29
1-1/2"	0.36
2"	0.43
2-1/2"	0.70
3"	0.71
4"	1.00
6"	1.43
8"	1.48
10"	1.50
12"	1.60
14"	1.75
16"	2.00
18"	2.25
20"	2.50
24"	3.00
30"	3.75

The following must be added to the list of components to accurately price HEAT TRACE systems:

1. Expediter fitting and clamp fasteners (F)
2. Junction box and nipple connected to expediter fitting (G)
3. Field installed terminal blocks within junction box
4. Ground lugs
5. Piping from power source to expediter fitting
6. Controls
7. Thermostats
8. Branch wiring
9. Cable splices
10. End of cable terminations
11. Branch piping fittings and boxes

Deduct the following percentages from labor if cable lengths in the same area exceed:

150' to 250'	10%	351' to 500'	20%
251' to 350'	15%	Over 500'	25%

Add the following percentages to labor for elevated installations:

15' to 20' high	10%	31' to 35' high	40%
21' to 25' high	20%	36' to 40' high	50%
26' to 30' high	30%	Over 40' high	60%

R238313-20 Spiral-Wrapped Heat Trace Cable (Pitch Table)

In order to increase the amount of heat, occasionally heat trace cable is wrapped in a spiral fashion around a pipe, increasing the number of feet of heater cable per linear foot of pipe.

Engineers first determine the heat loss per foot of pipe (based on the insulating material, its thickness, and the temperature differential across it). A ratio is then calculated by the formula:

$$\text{Feet of Heat Trace per Foot of Pipe} = \frac{\text{Watts/Foot of Heat Loss}}{\text{Watts/Foot of the Cable}}$$

The linear distance between wraps (pitch) is then taken from a chart or table. Generally, the pitch is listed on a drawing leaving the estimator to calculate the total length of heat tape required. An approximation may be taken from this table.

Feet of Heat Trace Per Foot of Pipe																	
	Nominal Pipe Size in Inches																
Pitch In Inches	1	1¼	1½	2	2½	3	4	6	8	10	12	14	16	18	20	24
3.5	1.80															
4	1.65															
5	1.46	1.60	1.80													
6	1.34	1.45	1.55	1.75												
7	1.25	1.35	1.43	1.57	1.75											
8	1.20	1.28	1.34	1.45	1.60	1.80										
9	1.16	1.23	1.28	1.37	1.51	1.68										
10	1.13	1.19	1.24	1.32	1.44	1.57	1.82									
15	1.06	1.08	1.10	1.15	1.21	1.29	1.42	1.78								
20	1.04	1.05	1.06	1.08	1.13	1.17	1.25	1.49	1.73							
25		1.04	1.04	1.06	1.08	1.11	1.17	1.33	1.51	1.72						
30				1.04	1.05	1.07	1.12	1.24	1.37	1.54	1.70					
35					1.06	1.06	1.09	1.17	1.28	1.42	1.54	1.64	1.78			
40						1.05	1.07	1.14	1.22	1.33	1.44	1.52	1.64	1.75		
50							1.05	1.09	1.15	1.22	1.29	1.35	1.44	1.53	1.64	1.83
60								1.06	1.11	1.16	1.21	1.25	1.31	1.39	1.46	1.62
70								1.05	1.08	1.12	1.17	1.19	1.24	1.30	1.35	1.47
80									1.06	1.09	1.13	1.15	1.19	1.24	1.28	1.38
90									1.04	1.06	1.10	1.13	1.16	1.19	1.23	1.32
100										1.05	1.08	1.10	1.13	1.15	1.19	1.23

Note: Common practice would normally limit the lower end of the table to 5% of additional heat. Above 80% an engineer would likely opt for two (2) parallel cables.

For customer support on your Electrical Costs with RSMeans Data, call 800.448.8182.

587

R260102-20　Labor Adjustment Factors

Labor Adjustment Factors are provided for Divisions 26, 27, and 28 to assist the mechanical estimator account for the various complexities and special conditions of any particular project. While a single percentage has been entered on each line of Division 26 01 02.20, it should be understood that these are just suggested midpoints of ranges of values commonly used by mechanical estimators. They may be increased or decreased depending on the severity of the special conditions.

The group for "existing occupied buildings" has been the subject of requests for explanation. Actually there are two stages to this group: buildings that are existing and "finished" but unoccupied, and those that also are occupied. Buildings that are "finished" may result in higher labor costs due to the workers having to be more careful not to damage finished walls, ceilings, floors, etc. and may necessitate special protective coverings and barriers. Also corridor bends and doorways may not accommodate long pieces of pipe or larger pieces of equipment. Work above an already hung ceiling can be very time consuming. The addition of occupants may force the work to be done on premium time (nights and/or weekends), eliminate the possible use of some preferred tools such as pneumatic drivers, powder charged drivers, etc. The estimator should evaluate the access to the work area and just how the work is going to be accomplished to arrive at an increase in labor costs over "normal" new construction productivity.

R260105-30　Electrical Demolition (Removal for Replacement)

The purpose of this reference number is to provide a guide to users for electrical "removal for replacement" by applying the rule of thumb: 1/3 of new installation time (typical range from 20% to 50%) for removal. Remember to use reasonable judgment when applying the suggested percentage factor. For example:

Contractors have been requested to remove an existing fluorescent lighting fixture and replace with a new fixture utilizing energy saver lamps and electronic ballast:

In order to fully understand the extent of the project, contractors should visit the job site and estimate the time to perform the renovation work in accordance with applicable national, state and local regulation and codes.

The contractor may need to add extra labor hours to his estimate if he discovers unknown concealed conditions such as: contaminated asbestos ceiling, broken acoustical ceiling tile and need to repair, patch and touch-up paint all the damaged or disturbed areas, tasks normally assigned to general contractors. In addition, the owner could request that the contractors salvage the materials removed and turn over the materials to the owner or dispose of the materials to a reclamation station. The normal removal item is 0.5 labor-hour for a lighting fixture and 1.5 labor-hours for new installation time. Revise the estimate times from 2 labor-hours work up to a minimum 4 labor-hours work for just a fluorescent lighting fixture.

For removal of large concentrations of lighting fixtures in the same area, apply an "economy of scale" to reduce estimating labor hours.

Figure R260110-10   Typical Overhead Service Entrance

Figure R260110-11   Typical Commercial Electric System

A = 1" Conduit w/1-#1/0 Wire XHHW
B = 2-3" Conduits w/4-350 kcmil XHHW in Each
C = 1¼" Conduit w/4 #3 THHN
D = 1¼" Conduit w/3 #1 XHHW
E = 1" Conduit w/4 #6 THHN
F = ½" Conduit w/2 #12 THHN

For customer support on your Electrical Costs with RSMeans Data, call 800.448.8182.

589

Figure R260110-12 Preliminary Procedure

1. Determine building size and use.
2. Develop total load in watts.
   a. Lighting
   b. Receptacle
   c. Air Conditioning
   d. Elevator
   e. Other power requirements
3. Determine best voltage available from utility company.
4. Determine cost from tables for loads (a) through (e) above.
5. Determine size of service from formulas (R260110-17).
6. Determine costs for service, panels, and feeders from tables.

Figure R260110-13 Office Building 90′ x 210′, 3 story, w/garage

Garage Area = 18,900 S.F.
Office Area = 56,700 S.F.
Elevator = 2 @ 125 FPM

Tables	Power Required	Watts
D5010-1151	Garage Lighting .5 Watt/S.F.	9,450
	Office Lighting 3 Watts/S.F.	170,100
D5010-1151	Office Receptacles 2 Watts/S.F.	113,400
D5010-1151, R262213-27	Low Rise Office A.C. 4.3 Watts/S.F.	243,810
D5010-1152, 1153	Elevators - 2 @ 20 HP = 2 @ 17,404 Watts/Ea.	34,808
D5010-1151	Misc. Motors + Power 1.2 Watts/S.F.	68,040
	Total	639,608 Watts

Voltage Available
277/480V, 3 Phase, 4 Wire

Formula

D5010-116
$$\text{Amperes} = \frac{\text{Watts}}{\text{Volts} \times \text{Power Factor} \times 1.73} = \frac{639,608}{480V \times .8 \times 1.73} = 963 \text{ Amps}$$

Use 1200 Amp Service

System	Description	Unit Cost	Unit	Total
D5020 210 0200	Garage Lighting (Interpolated)	$3.02	S.F.	$ 57,078
D5020 210 0540	Office Lighting (Using 32 W lamps)	9.13	S.F.	517,671
D5020 115 0880	Receptacle-Undercarpet	4.10	S.F.	232,470
D5020 140 0280	Air Conditioning	0.74	S.F.	41,958
D5020 135 0320	Misc. Pwr.	0.38	S.F.	21,546
D5020 145 2120	Elevators - 2 @ 20HP	2,830.00	Ea.	5,660
D5010 130 1250	Underground Electric Service-1200 Amp	72,000.00	Ea.	72,000
D5010 230 0480	Feeder - Assume 200 feet	330.00	L.F.	66,000
D5010 240 0580	Panels - 277/480V, 1200 Amp	36,500.00	Ea.	36,500
D5030 910 0400	Fire Detection	39,100.00	Ea.	39,100
		Total		$1,089,983
		or		$1,090,000

Table D5010-1151 Nominal Watts Per S.F. for Electric Systems for Various Building Types

Type Construction	1. Lighting	2. Devices	3. HVAC	4. Misc.	5. Elevator	Total Watts
Apartment, luxury high rise	2	2.2	3	1		
Apartment, low rise	2	2	3	1		
Auditorium	2.5	1	3.3	.8		
Bank, branch office	3	2.1	5.7	1.4		
Bank, main office	2.5	1.5	5.7	1.4		
Church	1.8	.8	3.3	.8		
College, science building	3	3	5.3	1.3		
College, library	2.5	.8	5.7	1.4		
College, physical education center	2	1	4.5	1.1		
Department store	2.5	.9	4	1		
Dormitory, college	1.5	1.2	4	1		
Drive-in doughnut shop	3	4	6.8	1.7		
Garage, commercial	.5	.5	0	.5		
Hospital, general	2	4.5	5	1.3		
Hospital, pediatric	3	3.8	5	1.3		
Hotel, airport	2	1	5	1.3		
Housing for the elderly	2	1.2	4	1		
Manufacturing, food processing	3	1	4.5	1.1		
Manufacturing, apparel	2	1	4.5	1.1		
Manufacturing, tools	4	1	4.5	1.1		
Medical clinic	2.5	1.5	3.2	1		
Nursing home	2	1.6	4	1		
Office building, high rise	3	2	4.7	1.2		
Office building, low rise	3	2	4.3	1.2		
Radio-TV studio	3.8	2.2	7.6	1.9		
Restaurant	2.5	2	6.8	1.7		
Retail store	2.5	.9	5.5	1.4		
School, elementary	3	1.9	5.3	1.3		
School, junior high	3	1.5	5.3	1.3		
School, senior high	2.3	1.7	5.3	1.3		
Supermarket	3	1	4	1		
Telephone exchange	1	.6	4.5	1.1		
Theater	2.5	1	3.3	.8		
Town Hall	2	1.9	5.3	1.3		
U.S. Post Office	3	2	5	1.3		
Warehouse, grocery	1	.6	0	.5		

Rule of Thumb: 1 KVA = 1 HP (Single Phase)

Three Phase:

Watts = 1.73 x Volts x Current x Power Factor x Efficiency

$$\text{Horsepower} = \frac{\text{Volts x Current x 1.73 x Power Factor}}{746 \text{ Watts}}$$

Table D5010-1152 Horsepower Requirements for Elevators with 3 Phase Motors

Type	Maximum Travel Height in Ft.	Travel Speeds in FPM	Capacity of Cars in Lbs.		
			1200	1500	1800
Hydraulic	70	70	10	15	15
		85	15	15	15
		100	15	15	20
		110	20	20	20
		125	20	20	20
		150	25	25	25
		175	25	30	30
		200	30	30	40
Geared Traction	300	200			
		350			
			2000	2500	3000
Hydraulic	70	70	15	20	20
		85	20	20	25
		100	20	25	30
		110	20	25	30
		125	25	30	40
		150	30	40	50
		175	40	50	50
		200	40	50	60
Geared Traction	300	200	10	10	15
		350	15	15	23
			3500	4000	4500
Hydraulic	70	70	20	25	30
		85	25	30	30
		100	30	40	40
		110	40	40	50
		125	40	50	50
		150	50	50	60
		175	60		
		200	60		
Geared Traction	300	200	15		23
		350	23		35

The power factor of electric motors varies from 80% to 90% in larger size motors. The efficiency likewise varies from 80% on a small motor to 90% on a large motor.

Table D5010-1153 Watts per Motor

90% Power Factor & Efficiency @ 200 or 460V			
HP	Watts	HP	Watts
10	9024	30	25784
15	13537	40	33519
20	17404	50	41899
25	21916	60	49634

For customer support on your Electrical Costs with RSMeans Data, call 800.448.8182.

591

Table R260110-17 Electrical Formulas

Ohm's Law

Ohm's Law is a method of explaining the relation between voltage, current, and resistance in an electrical circuit. It is practically the basis of all electrical calculations. The term "electromotive force" is often used to designate pressure in volts. This formula can be expressed in various forms.

To find the current in amperes:

$$\text{Current} = \frac{\text{Voltage}}{\text{Resistance}} \quad \text{or} \quad \text{Amperes} = \frac{\text{Volts}}{\text{Ohms}} \quad \text{or} \quad I = \frac{E}{R}$$

The flow of current in amperes through any circuit is equal to the voltage or electromotive force divided by the resistance of that circuit.

To find the pressure or voltage:

$$\text{Voltage} = \text{Current} \times \text{Resistance} \quad \text{or} \quad \text{Volts} = \text{Amperes} \times \text{Ohms}$$
$$\text{or} \quad E = I \times R$$

The voltage required to force a current through a circuit is equal to the resistance of the circuit multiplied by the current.

To find the resistance:

$$\text{Resistance} = \frac{\text{Voltage}}{\text{Current}} \quad \text{or} \quad \text{Ohms} = \frac{\text{Volts}}{\text{Amperes}} \quad \text{or} \quad R = \frac{E}{I}$$

The resistance of a circuit is equal to the voltage divided by the current flowing through that circuit.

Power Formulas

One horsepower = 746 watts   One kilowatt = 1000 watts

The power factor of electric motors varies from 80% to 90% in the larger size motors.

Single-Phase Alternating Current Circuits

Power in Watts = Volts x Amperes x Power Factor

To find current in amperes:

$$\text{Current} = \frac{\text{Watts}}{\text{Volts} \times \text{Power Factor}} \quad \text{or}$$

$$\text{Amperes} = \frac{\text{Watts}}{\text{Volts} \times \text{Power Factor}} \quad \text{or} \quad I = \frac{W}{E \times PF}$$

To find current of a motor, single phase:

$$\text{Current} = \frac{\text{Horsepower} \times 746}{\text{Volts} \times \text{Power Factor} \times \text{Efficiency}} \quad \text{or}$$

$$I = \frac{HP \times 746}{E \times PF \times Eff.}$$

To find horsepower of a motor, single phase:

$$\text{Horsepower} = \frac{\text{Volts} \times \text{Current} \times \text{Power Factor} \times \text{Efficiency}}{746 \text{ Watts}}$$

$$HP = \frac{E \times I \times PF \times Eff.}{746}$$

To find power in watts of a motor, single phase:

Watts = Volts x Current x Power Factor x Efficiency or
Watts = E x I x PF x Eff.

To find single phase kVA:

$$\text{1 Phase kVA} = \frac{\text{Volts} \times \text{Amps}}{1000}$$

Three-Phase Alternating Current Circuits

Power in Watts = Volts x Amperes x Power Factor x 1.73

To find current in amperes in each wire:

$$\text{Current} = \frac{\text{Watts}}{\text{Voltage} \times \text{Power Factor} \times 1.73} \quad \text{or}$$

$$\text{Amperes} = \frac{\text{Watts}}{\text{Volts} \times \text{Power Factor} \times 1.73} \quad \text{or} \quad I = \frac{W}{E \times PF \times 1.73}$$

To find current of a motor, 3 phase:

$$\text{Current} = \frac{\text{Horsepower} \times 746}{\text{Volts} \times \text{Power Factor} \times \text{Efficiency} \times 1.73} \quad \text{or}$$

$$I = \frac{HP \times 746}{E \times PF \times Eff. \times 1.73}$$

To find horsepower of a motor, 3 phase:

$$\text{Horsepower} = \frac{\text{Volts} \times \text{Current} \times 1.73 \times \text{Power Factor}}{746 \text{ Watts}}$$

$$HP = \frac{E \times I \times 1.73 \times PF}{746}$$

To find power in watts of a motor, 3 phase:

Watts = Volts x Current x 1.73 x Power Factor x Efficiency or
Watts = E x I x 1.73 x PF x Eff.

To find 3 phase kVA:

$$\text{3 phase kVA} = \frac{\text{Volts} \times \text{Amps} \times 1.73}{1000} \quad \text{or}$$

$$kVA = \frac{V \times A \times 1.73}{1000}$$

Power Factor (PF) is the percentage ratio of the measured watts (effective power) to the volt-amperes (apparent watts)

$$\text{Power Factor} = \frac{\text{Watts}}{\text{Volts} \times \text{Amperes}} \times 100\%$$

A Conceptual Estimate of the costs for a building when final drawings are not available can be quickly figured by using **Table R260110-18 Cost Per S.F. for Electrical Systems for Various Building Types.** The following definitions apply to this table.

1. **Service and Distribution:** This system includes the incoming primary feeder from the power company, main building transformer, metering arrangement, switchboards, distribution panel boards, stepdown transformers, and power and lighting panels. Items marked (*) include the cost of the primary feeder and transformer. In all other projects the cost of the primary feeder and transformer is paid for by the local power company.
2. **Lighting:** Includes all interior fixtures for decor, illumination, exit, and emergency lighting. Fixtures for exterior building lighting are included, but parking area lighting is not included unless mentioned. See also Section D5020 for a detailed analysis of lighting requirements and costs.
3. **Devices:** Includes all outlet boxes, receptacles, switches for lighting control, dimmers, and cover plates.
4. **Equipment Connections:** Includes all materials and equipment for making connections for heating, ventilating, and air conditioning, as well as food service and other motorized items requiring connections.
5. **Basic Materials:** This category includes all disconnect power switches not part of service equipment, raceways for wires, pull boxes, junction boxes, supports, fittings, grounding materials, wireways, busways, and wire and cable systems.
6. **Special Systems:** Includes installed equipment only for the particular system such as fire detection and alarm, sound, emergency generator, and others as listed in the table.

Table R260110-18 Cost per S.F. for Electric Systems for Various Building Types

System	1. Service & Distribution	2. Lighting	3. Devices	4. Equipment Connections	5. Basic Materials	6a. Fire Alarm & Detection	6b. Telephone	6c. Television Distribution
Auto Sales with Repair	$5.13	$3.99	$0.86	$0.86	$9.41	$3.14	$1.43	$0.00
Banking Institutions	8.19	5.85	0.59	1.76	12.87	7.02	4.68	0.00
Detention Centers	10.89	9.08	1.21	1.21	15.13	6.05	1.21	0.00
Fire Stations	6.80	4.80	1.20	0.80	11.60	4.00	1.20	0.80
Hospitals	11.16	6.20	1.86	1.86	17.36	4.96	2.48	0.62
Industrial Buildings	14.10	10.58	2.12	2.12	20.45	2.82	1.41	1.41
Medical Clinics & Offices	6.08	4.56	1.14	0.76	8.74	3.04	2.28	0.76
Mixed Use	11.24	6.42	2.14	1.61	10.70	3.21	2.14	2.14
Multi-Family Housing	3.60	3.60	0.90	0.90	6.30	1.35	1.35	0.68
Nursing Home & Assisted Living	3.76	2.82	0.71	0.94	4.70	1.88	1.41	1.18
Office Buildings	5.10	4.76	1.02	1.36	9.18	3.40	2.72	0.68
Parking Garage	1.03	0.90	0.13	0.26	1.29	1.35	0.32	0.00
Parking Garage/Mixed Use	3.96	3.08	0.44	0.44	5.06	1.76	0.88	0.44
Police Stations	4.88	3.66	0.92	1.22	7.63	3.36	0.92	0.61
Public Assembly Buildings	5.81	6.64	1.25	2.08	10.38	3.32	1.25	0.83
Recreational	6.40	5.60	1.20	1.60	10.80	3.20	1.20	1.60
Restaurants	8.73	6.79	1.46	1.94	14.55	5.82	0.97	2.43
Retail	5.04	5.67	0.32	1.26	10.71	2.84	0.95	0.00
Schools	5.28	4.62	0.99	1.32	7.26	2.64	0.99	0.99
University, College & Private School Classroom & Admin Buildings	5.70	5.03	1.01	1.34	8.38	2.68	0.67	1.01
University, College & Private School Dormitories	4.88	4.88	0.92	1.22	7.63	2.44	0.61	0.92
University, College & Private School Science, Eng. & Lab Buildings	6.93	6.16	1.16	1.54	10.78	3.08	1.16	0.77
Warehouses	5.36	4.69	1.01	1.68	8.04	4.02	1.34	0.67

COST ASSUMPTIONS:

Each of the projects analyzed in Table R260110-18 was bid within the last 10 years in the northeastern part of the United States. Bid prices have been adjusted to Jan. 1 levels. The list of projects is by no means all-inclusive, yet by carefully examining the various systems for a particular building type, certain cost relationships will emerge. The use of Square Foot/Cubic Costs, in the Reference Section, along with the Project Size Modifier, should produce a budget S.F. cost for the electrical portion of a job that is consistent with the amount of design information normally available at the conceptual estimate stage.

Table R260110-18 Cost per S.F. for Electric Systems for Various Building Types (cont.)

System	6d. Intercom	6e. Sound System	6f. Video Monitoring	6g. Internet	6h. Emergency Generator	6i. Access Control	6j. Master Clock System
Auto Sales with Repair	$0.00	$0.29	$1.14	$1.43	$0.57	$0.29	$0.00
Banking Institutions	0.00	0.00	5.85	5.27	1.17	5.27	0.00
Detention Centers	1.82	1.82	2.42	0.00	3.63	6.05	0.00
Fire Stations	1.20	1.20	1.20	1.20	1.60	1.60	0.80
Hospitals	1.86	1.24	1.24	1.24	7.44	1.86	0.62
Industrial Buildings	1.41	1.41	1.41	4.23	2.82	2.12	2.12
Medical Clinics & Offices	1.14	0.76	1.90	3.42	1.52	1.90	0.00
Mixed Use	1.61	1.07	2.68	3.21	3.21	2.14	0.00
Multi-Family Housing	0.45	0.00	0.45	2.25	0.00	0.68	0.00
Nursing Home & Assisted Living	0.94	0.47	1.18	2.12	0.94	0.47	0.00
Office Buildings	0.68	0.68	1.02	1.36	1.02	1.02	0.00
Parking Garage	0.00	0.00	0.52	0.13	0.19	0.32	0.00
Parking Garage/Mixed Use	0.44	0.44	1.76	1.32	0.88	1.10	0.00
Police Stations	0.92	0.92	1.22	1.22	1.83	1.22	0.00
Public Assembly Buildings	0.00	4.98	1.25	1.25	1.25	1.25	0.00
Recreational	0.00	1.60	1.20	2.00	2.00	1.20	0.40
Restaurants	0.00	2.43	0.97	0.97	1.46	0.00	0.00
Retail	0.63	0.32	1.58	0.32	0.95	0.95	0.00
Schools	0.99	1.65	1.65	1.32	1.32	1.32	0.66
University, College & Private School Classroom & Admin Buildings	1.01	1.01	0.67	1.68	1.01	1.68	0.67
University, College & Private School Dormitories	1.22	0.61	0.92	1.53	1.22	0.92	0.61
University, College & Private School Science, Eng. & Lab Buildings	1.16	0.39	0.77	1.54	1.16	1.16	0.77
Warehouses	0.67	0.34	1.68	1.01	1.34	1.68	0.00

Table R260120-10 ASHRAE Climate Zone Map

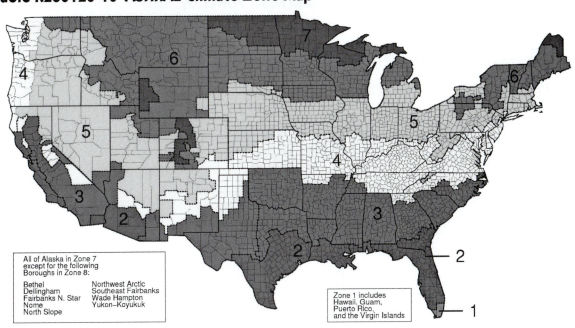

All of Alaska in Zone 7
except for the following
Boroughs in Zone 8:

Bethel   Northwest Arctic
Dellingham   Southeast Fairbanks
Fairbanks N. Star   Wade Hampton
Nome   Yukon–Koyukuk
North Slope

Zone 1 includes
Hawaii, Guam,
Puerto Rico,
and the Virgin Islands

Table R260120-20 Watts per Square Foot for Lighting and Misc. Power by Building Type & Space

Building / Space	Percent	Lighting	Misc
Apartment, Low: Area = 20,000 SF; Floors = 2; Heating = DX Coils; Cooling = DX Coils			
Room	85%	0.50	0.10
Storage	8%	1.19	0.00
Laundry	7%	1.28	0.15
Weighted Average		**0.61**	**0.10**
Apartment, Mid: Area = 40,000 SF; Floors = 5; Heating = DX Coils; Cooling = DX Coils			
Room	71%	0.50	0.30
Corridor	16%	0.57	0.00
Storage	7%	1.19	0.00
Laundry	6%	1.28	0.15
Average		**0.61**	**0.22**
Apartment, High: Area = 72,000 SF; Floors = 8; Heating = DX Coils; Cooling = DX Coils			
Room	71%	0.50	0.30
Corridor	16%	0.57	0.00
Storage	7%	1.19	0.00
Laundry	6%	1.28	0.15
Weighted Average		**0.61**	**0.22**
Assisted Living: Area = 10,000 SF; Floors = 1; Heating = Furnace; Cooling = DX Coils			
Room	70%	1.46	0.25
Dining	15%	1.85	0.10
Food Prep	10%	1.19	1.00
Care	5%	1.34	1.00
Weighted Average		**1.49**	**0.34**

Building / Space	Percent	Lighting	Misc
Auditorium: Area = 70,000 SF; Floors = 1; Heating = Hot Water Coils; Cooling = Chilled Water Coils			
Theater	70%	0.90	
Lobby	15%	1.50	
Office	5%	1.30	
Corridor	5%	0.60	
MEP	5%	0.70	
Weighted Average		**0.99**	
Bank: Area = 25,000 SF; Floors = 2; Heating = Furnace; Cooling = DX Coils			
Bank	60%	2.33	0.45
Corridor	10%	0.57	0.00
Office	10%	1.49	0.75
Lobby	5%	1.52	0.25
Restrooms	5%	0.77	0.10
Conference	4%	0.92	0.10
MEP	4%	0.81	0.10
Production	2%	1.64	0.70
Weighted Average		**1.82**	**0.38**
Community Center: Area = 35,000 SF; Floors = 1; Heating = Furnace; Cooling = DX Coils			
Convention	50%	1.27	
Dining	15%	1.85	
Food Prep	15%	1.19	
Storage	10%	1.19	
Corridor	5%	0.57	
Restrooms	5%	0.77	
Weighted Average		**1.28**	
Convention Center: Area = 150,000 SF; Floors = 2; Heating = Hot Water Coils; Cooling = Chilled Water Coils			
Convention	65%	1.27	0.25
Conference	20%	0.92	0.10
Dining	5%	1.85	0.10
Corridor	4%	0.57	0.00
Food Prep	3%	1.19	1.00
Restrooms	3%	0.77	0.10
Weighted Average		**1.18**	**0.22**

Building / Space	Percent	Lighting	Misc
Day Care: Area = 2,500 SF; Floors = 1; Heating = Furnace; Cooling = DX Coils			
Classroom	45%	1.45	
Corridor	15%	0.57	
Library	15%	1.67	
Food Prep	10%	1.19	
Restrooms	7%	0.77	
Office	5%	1.49	
Storage	3%	1.19	
Weighted Average		**1.27**	
Fitness Center: Area = 20,000 SF; Floors = 1; Heating = Furnace; Cooling = DX Coils			
Exercise	80%	1.13	0.10
Corridor	5%	0.57	0.00
MEP	5%	0.81	0.10
Office	5%	1.24	0.75
Restrooms	7%	0.77	0.10
Weighted Average		**1.07**	**0.13**
Hospital: Area = 250,000 SF; Floors = 4; Heating = Hot Water Coils; Cooling = Chilled Water Coils			
Medical	60%	1.34	1.00
Laboratory	15%	3.19	1.00
Corridor	10%	0.57	0.00
Laundry	5%	1.28	0.15
MEP	5%	0.81	0.10
Restrooms	5%	0.77	0.10
Weighted Average		**1.48**	**0.77**
Hotel: Area = 180,000 SF; Floors = 10; Heating = Furnace; Cooling = DX Coils			
Room	75%	1.46	
Meeting Center	15%	1.27	
Corridor	8%	0.57	
MEP	2%	0.81	
Weighted Average		**1.35**	

Steps:
1. Determine climate zone using R260120-10
2. Use R260120-20 to determine lighting and miscellaneous power for appropriate building type
3. Use corresponding building type and climate zone in R260120-30 to determine other systems

*Values for R260120-20 & -30 were derived by performing an energy analysis using the eQUEST® software from James J. Hirsch & Associates. eQUEST is a trademark of James J. Hirsch
**Additional systems may be necessary

Table R260120-20 Watts per Square Foot for Lighting and Misc. Power by Building Type & Space (cont)

Building / Space	Percent	Lighting	Misc
Manufacturing, General: Area = 100,000 SF; Floors = 10; Heating = Furnace; Cooling = DX Coils			
Work	60%	2.66	1.00
Storage	25%	1.19	0.00
Office	10%	1.24	0.75
Restrooms	3%	0.77	0.10
MEP	2%	0.81	0.10
Weighted Average		2.06	0.68
Manufacturing, High Tech: Area = 200,000 SF; Floors = 1; Heating = Furnace; Cooling = DX Coils			
Work	45%	3.19	1.00
Office	27%	1.24	0.75
Corridor	18%	0.57	0.00
Dining	3%	1.85	0.10
Conference	2%	0.92	0.10
Server	2%	1.24	5.00
MEP	2%	0.81	0.10
Food Prep	1%	1.19	1.00
Weighted Average		2.00	0.77
Medical Clinic: Area = 20,000 SF; Floors = 1; Heating = Furnace; Cooling = DX Coils			
Clinic	50%	1.40	
General Office	15%	1.30	
Private Office	10%	1.30	
Laboratory	10%	1.30	
Storage	5%	0.60	
Restrooms	5%	0.60	
MEP	5%	0.70	
Weighted Average		1.25	
Motel: Area = 10,000 SF; Floors = 1; Heating = DX Coils; Cooling = DX Coils			
Room	95%	1.46	0.25
Office	3%	1.24	0.75
MEP	2%	0.81	0.10
Weighted Average		1.44	0.26
Museum: Area = 20,000 SF; Floors = 1; Heating = Hot Water Coils; Cooling = Chilled Water Coils			
Exhibit Area	65%	1.03	0.25
Theater	15%	1.02	0.25
Lobby	5%	1.77	0.25
MEP	5%	0.81	0.10
Office	5%	1.49	0.75
Restrooms	5%	0.77	0.10
Weighted Average		1.06	0.26
Office, Low: Area = 25,000 SF; Floors = 2; Heating = Furnace; Cooling = DX Coils			
Office	70%	1.30	0.75
Corridor	10%	0.60	0.00
Lobby	5%	1.10	0.25
Restrooms	5%	0.60	0.10
Conference	4%	1.60	0.10
MEP	4%	0.70	0.10
Production	2%	1.50	0.70
Weighted Average		1.18	0.56
Office, Mid: Area = 125,000 SF; Floors = 4; Heating = Hot Water Coils; Cooling = Chilled Water Coils			
Office Open	40%	1.24	0.75
Office Private	30%	1.49	0.75
Corridor	10%	0.57	0.00
Lobby	5%	1.52	0.25
Restrooms	5%	0.77	0.10
Conference	4%	0.92	0.10
MEP	4%	0.81	0.10
Production	2%	1.64	0.70
Weighted Average		1.22	0.56
Office, High: Area = 250,000 SF; Floors = 8; Heating = Hot Water Coils; Cooling = Chilled Water Coils			
Office Open	45%	1.24	0.75
Office Private	25%	1.49	0.75
Corridor	10%	0.57	0.00
Lobby	5%	1.52	0.25
Conference	4%	0.92	0.10
Production	2%	1.64	0.70
Restrooms	5%	0.77	0.10
MEP	4%	0.81	0.10
Weighted Average		1.20	0.56

Building / Space	Percent	Lighting	Misc
Restaurant, Bar/Lounge: Area = 4,000 SF; Floors = 1; Heating = Furnace; Cooling = DX Coils			
Lounge	35%	1.85	0.10
Dining	30%	1.85	0.10
Food Prep	25%	1.19	1.00
MEP	5%	0.81	0.10
Restrooms	5%	0.77	0.10
Weighted Average		1.58	0.33
Restaurant, Full: Area = 5,000 SF; Floors = 1; Heating = Furnace; Cooling = DX Coils			
Dining	60%	1.85	0.10
Food Prep	30%	1.19	1.00
Restrooms	5%	0.77	0.10
MEP	5%	0.81	0.10
Weighted Average		1.55	0.37
Restaurant, Quick: Area = 2,500 SF; Floors = 1; Heating = Furnace; Cooling = DX Coils			
Dining	50%	1.85	0.10
Food Prep	35%	1.19	1.00
Restrooms	10%	0.77	0.10
MEP	5%	0.81	0.10
Weighted Average		1.46	0.42
Store, Convenience: Area = 3,000 SF; Floors = 1; Heating = Furnace; Cooling = DX Coils			
Retail	80%	2.38	0.25
Office	10%	1.49	0.75
Restrooms	10%	0.77	0.10
Weighted Average		2.13	0.29
Store, Department: Area = 120,000 SF; Floors = 2; Heating = Hot Water Coils; Cooling = Chilled Water Coils			
Retail	55%	2.38	0.25
Exhibit Area	15%	1.03	0.25
Storage	16%	1.19	0.00
Office	5%	1.24	0.75
Smoking Lounge	2%	1.85	0.10
Restrooms	2%	0.77	0.10
MEP	5%	0.81	0.10
Weighted Average		1.81	0.22
Store, Large: Area = 70,000 SF; Floors = 1; Heating = Furnace; Cooling = DX Coils			
Retail	65%	2.38	0.25
Storage	20%	1.19	0.00
Office	5%	1.24	0.75
Dining	3%	1.85	0.10
Food Prep	1%	1.19	1.00
Lounge	2%	1.85	0.10
Restrooms	2%	0.77	0.10
MEP	2%	0.81	0.10
Weighted Average		1.98	0.22
Store, Small: Area = 5,000 SF; Floors = 1; Heating = Furnace; Cooling = DX Coils			
Retail	58%	2.38	0.25
Exhibit Area	15%	1.03	0.25
Storage	15%	1.19	0.00
Office	5%	1.24	0.75
Restrooms	5%	0.77	0.10
MEP	2%	0.81	0.10
Weighted Average		1.83	0.23
Store, Strip Mall: Area = 15,000 SF; Floors = 1; Heating = Furnace; Cooling = DX Coils			
Retail	38%	2.38	0.25
Atrium	20%	1.55	0.13
Exhibit Area	15%	1.03	0.25
Storage	15%	1.19	0.00
Office	5%	1.24	0.75
Restrooms	5%	0.77	0.10
MEP	2%	0.81	0.10
Weighted Average		1.66	0.20

Building / Space	Percent	Lighting	Misc
Store, Warehouse: Area = 150,000 SF; Floors = 1; Heating = Furnace; Cooling = DX Coils			
Retail	75%	2.38	0.25
Storage	20%	1.19	0.00
Office	4%	1.24	0.75
Restrooms	1%	0.77	0.10
Weighted Average		2.08	0.22
School, Elementary: Area = 35,000 SF; Floors = 1; Heating = Furnace; Cooling = DX Coils			
Classroom	50%	1.45	0.50
Corridor	15%	0.57	0.00
Gym	10%	1.13	0.10
Dining	5%	1.85	0.10
Food Prep	5%	1.19	1.00
Library	5%	1.78	0.25
Auditorium	5%	1.62	0.25
Restrooms	5%	0.77	0.10
Weighted Average		1.28	0.35
School, Middle: Area = 75,000 SF; Floors = 1; Heating = Furnace; Cooling = DX Coils			
Classroom	50%	1.45	0.50
Corridor	14%	0.57	0.00
Gym	9%	1.13	0.10
Auditorium	7%	1.62	0.25
Dining	6%	1.85	0.10
Food Prep	5%	1.19	1.00
Library	5%	1.78	0.25
Restrooms	4%	0.77	0.10
Weighted Average		1.31	0.35
School, High: Area = 150,000 SF; Floors = 2; Heating = Furnace; Cooling = DX Coils			
Classroom	45%	1.45	0.50
Corridor	10%	0.57	0.00
Gym	10%	1.13	0.10
Dining	10%	1.85	0.10
Food Prep	5%	1.19	1.00
Library	5%	1.78	0.25
Auditorium	10%	1.62	0.25
Restrooms	5%	0.77	0.10
Weighted Average		1.36	0.34
School, College: Area = 250,000 SF; Floors = 4; Heating = Furnace; Cooling = DX Coils			
Classroom	50%	1.45	0.50
Office	15%	1.49	0.75
Corridor	12%	0.57	0.00
Gym	10%	1.13	0.10
Library	5%	1.78	0.25
Dining	3%	1.85	0.10
Restrooms	3%	0.77	0.10
Food Prep	2%	1.19	1.00
Weighted Average		1.32	0.41
Service Station: Area = 2,000 SF; Floors = 1; Heating = Furnace; Cooling = DX Coils			
Workshop	70%	1.31	0.50
Office	10%	1.49	0.75
Restrooms	10%	0.77	0.10
Retail	10%	2.38	0.25
Weighted Average		1.38	0.46
Storage, High Bay: Area = 200,000 SF; Floors = 1; Heating = Furnace; Cooling = DX Coils			
Storage	85%	1.19	0.00
Office	10%	1.24	0.75
Restrooms	3%	0.77	0.10
MEP	2%	0.81	0.10
Weighted Average		1.17	0.08
Storage, Low Bay: Area = 70,000 SF; Floors = 1; Heating = Furnace; Cooling = DX Coils			
Storage	85%	1.19	0.00
Office	10%	1.24	0.75
Restrooms	3%	0.77	0.10
MEP	2%	0.81	0.10
Weighted Average		1.17	0.08

596

For customer support on your Electrical Costs with RSMeans Data, call 800.448.8182.

Table R260120-30 Watts per S.F. for Various Systems by Building Type & Climate Zone

Building / Zone	Heating	Cooling	Heat rejection	Auxiliary (pumps)	Vent fan	Supp heat pump	Building / Zone	Heating	Cooling	Heat rejection	Auxiliary (pumps)	Vent fan	Supp heat pump
Apartment, Low							**Community Center**						
Zone 1	0.42	3.65	0.00	0.00	0.12	1.47	Zone 1	0.00	2.70	0.00	0.00	0.44	0.00
Zone 2	0.63	3.53	0.00	0.00	0.12	4.01	Zone 2	0.00	2.66	0.00	0.00	0.44	0.00
Zone 3	0.64	3.45	0.00	0.00	0.13	4.63	Zone 3	0.00	2.66	0.00	0.00	0.45	0.00
Zone 4	0.65	3.09	0.00	0.00	0.13	7.41	Zone 4	0.00	2.44	0.00	0.00	0.45	0.00
Zone 5	0.67	2.88	0.00	0.00	0.12	10.30	Zone 5	0.00	2.42	0.00	0.00	0.44	0.00
Zone 6	0.69	2.66	0.00	0.00	0.13	10.68	Zone 6	0.00	2.16	0.00	0.00	0.44	0.00
Zone 7	0.85	2.11	0.00	0.00	0.12	11.76	Zone 7	0.00	1.91	0.00	0.00	0.44	0.00
Average	**0.65**	**3.05**	**0.00**	**0.00**	**0.13**	**7.18**	**Average**	**0.00**	**2.42**	**0.00**	**0.00**	**0.45**	**0.00**
Apartment, Mid							**Convention Center**						
Zone 1	0.00	1.90	0.00	0.00	0.11	0.00	Zone 1	0.00	1.46	0.25	0.27	0.31	0.00
Zone 2	0.43	1.82	0.00	0.00	0.11	2.12	Zone 2	0.00	1.44	0.25	0.27	0.29	0.00
Zone 3	0.48	1.69	0.00	0.00	0.11	2.32	Zone 3	0.00	1.43	0.24	0.26	0.30	0.00
Zone 4	0.67	1.67	0.00	0.00	0.12	4.25	Zone 4	0.00	1.42	0.23	0.26	0.28	0.00
Zone 5	0.70	1.54	0.00	0.00	0.12	4.79	Zone 5	0.00	1.30	0.29	0.31	0.22	0.00
Zone 6	0.72	1.45	0.00	0.00	0.12	5.07	Zone 6	0.00	1.17	0.12	0.29	0.16	0.00
Zone 7	0.93	1.39	0.00	0.00	0.12	5.11	Zone 7	0.00	0.90	0.20	0.31	0.16	0.00
Average	**0.56**	**1.63**	**0.00**	**0.00**	**0.12**	**3.38**	**Average**	**0.00**	**1.30**	**0.22**	**0.28**	**0.25**	**0.00**
Apartment, High							**Day Care**						
Zone 1	0.00	2.11	0.00	0.00	0.14	0.00	Zone 1	0.00	3.47	0.00	0.02	0.29	0.00
Zone 2	0.00	2.03	0.00	0.00	0.13	0.00	Zone 2	0.00	3.42	0.00	0.02	0.28	0.00
Zone 3	0.48	1.88	0.00	0.00	0.14	1.75	Zone 3	0.00	3.40	0.00	0.02	0.29	0.00
Zone 4	0.57	1.77	0.00	0.00	0.14	4.22	Zone 4	0.00	3.29	0.00	0.02	0.29	0.00
Zone 5	0.66	1.64	0.00	0.00	0.14	5.40	Zone 5	0.00	3.08	0.00	0.02	0.29	0.00
Zone 6	0.71	1.63	0.00	0.00	0.15	5.64	Zone 6	0.00	2.29	0.00	0.02	0.29	0.00
Zone 7	0.91	1.18	0.00	0.00	0.14	5.75	Zone 7	0.00	2.12	0.00	0.02	0.24	0.00
Average	**0.48**	**1.75**	**0.00**	**0.00**	**0.14**	**3.25**	**Average**	**0.05**	**3.01**	**0.05**	**0.07**	**0.29**	**0.05**
Assisted Living							**Fitness Center**						
Zone 1	0.00	2.65	0.00	0.04	0.31	0.00	Zone 1	0.00	4.97	0.00	0.00	0.59	0.00
Zone 2	0.00	2.63	0.00	0.04	0.31	0.00	Zone 2	0.00	4.94	0.00	0.00	0.59	0.00
Zone 3	0.00	2.57	0.00	0.04	0.31	0.00	Zone 3	0.00	4.80	0.00	0.00	0.60	0.00
Zone 4	0.00	2.50	0.00	0.04	0.31	0.00	Zone 4	0.00	4.76	0.00	0.00	0.60	0.00
Zone 5	0.00	2.42	0.00	0.04	0.30	0.00	Zone 5	0.00	4.75	0.00	0.00	0.59	0.00
Zone 6	0.00	1.99	0.00	0.04	0.32	0.00	Zone 6	0.00	4.13	0.00	0.00	0.60	0.00
Zone 7	0.00	1.90	0.00	0.04	0.30	0.00	Zone 7	0.00	3.27	0.00	0.00	0.60	0.00
Average	**0.00**	**2.38**	**0.00**	**0.04**	**0.31**	**0.00**	**Average**	**0.00**	**4.52**	**0.00**	**0.00**	**0.59**	**0.00**
Auditorium							**Hospital**						
Zone 1	0.00	2.32	0.34	0.37	1.03	0.00	Zone 1	0.00	1.54	0.23	0.26	0.80	0.00
Zone 2	0.00	2.32	0.34	0.38	1.03	0.00	Zone 2	0.00	1.47	0.22	0.25	0.80	0.00
Zone 3	0.00	2.29	0.33	0.38	1.04	0.00	Zone 3	0.00	1.44	0.22	0.25	0.81	0.00
Zone 4	0.00	2.26	0.29	0.35	1.05	0.00	Zone 4	0.00	1.42	0.22	0.25	0.81	0.00
Zone 5	0.00	2.20	0.34	0.39	1.04	0.00	Zone 5	0.00	1.37	0.23	0.27	0.81	0.00
Zone 6	0.00	1.71	0.14	0.34	1.05	0.00	Zone 6	0.00	1.26	0.15	0.25	0.82	0.00
Zone 7	0.00	1.38	0.23	0.36	1.04	0.00	Zone 7	0.00	1.07	0.19	0.26	0.82	0.00
Average	**0.00**	**2.07**	**0.29**	**0.37**	**1.04**	**0.00**	**Average**	**0.00**	**1.36**	**0.21**	**0.26**	**0.81**	**0.00**
Bank							**Hotel**						
Zone 1	0.00	3.32	0.00	0.00	0.60	0.00	Zone 1	0.00	1.77	0.00	0.01	0.26	0.00
Zone 2	0.00	3.28	0.00	0.00	0.61	0.00	Zone 2	0.00	1.64	0.00	0.01	0.26	0.00
Zone 3	0.00	3.28	0.00	0.00	0.61	0.00	Zone 3	0.00	1.62	0.00	0.01	0.27	0.00
Zone 4	0.00	3.21	0.00	0.00	0.60	0.00	Zone 4	0.00	1.62	0.00	0.01	0.27	0.00
Zone 5	0.00	3.06	0.00	0.00	0.59	0.00	Zone 5	0.00	1.51	0.00	0.01	0.27	0.00
Zone 6	0.00	2.82	0.00	0.00	0.61	0.00	Zone 6	0.00	1.25	0.00	0.01	0.28	0.00
Zone 7	0.00	2.66	0.00	0.00	0.60	0.00	Zone 7	0.00	1.23	0.00	0.01	0.28	0.00
Average	**0.00**	**3.00**	**0.00**	**0.00**	**0.60**	**0.00**	**Average**	**0.00**	**1.52**	**0.00**	**0.01**	**0.27**	**0.00**

For customer support on your Electrical Costs with RSMeans Data, call 800.448.8182.

597

Table R260120-30 Watts per S.F. for Various Systems by Building Type & Climate Zone (cont.)

Building / Zone	Heating	Cooling	Heat rejection	Auxiliary (pumps)	Vent fan	Supp heat pump	Building / Zone	Heating	Cooling	Heat rejection	Auxiliary (pumps)	Vent fan	Supp heat pump
Manufacturing, General							**Office, Mid**						
Zone 1	0.00	1.46	0.00	0.00	0.13	0.00	Zone 1	0.00	1.61	0.31	0.33	0.45	0.00
Zone 2	0.00	1.45	0.00	0.00	0.13	0.00	Zone 2	0.00	1.60	0.29	0.32	0.46	0.00
Zone 3	0.00	1.43	0.00	0.00	0.13	0.00	Zone 3	0.00	1.59	0.28	0.34	0.42	0.00
Zone 4	0.00	1.39	0.00	0.00	0.13	0.00	Zone 4	0.00	1.56	0.27	0.33	0.43	0.00
Zone 5	0.00	1.39	0.00	0.00	0.13	0.00	Zone 5	0.00	1.55	0.34	0.36	0.37	0.00
Zone 6	0.00	1.21	0.00	0.00	0.13	0.00	Zone 6	0.00	1.29	0.16	0.34	0.31	0.00
Zone 7	0.00	1.00	0.00	0.00	0.13	0.00	Zone 7	0.00	1.10	0.22	0.35	0.32	0.00
Average	**0.00**	**1.33**	**0.00**	**0.00**	**0.13**	**0.00**	**Average**	**0.00**	**1.47**	**0.27**	**0.34**	**0.40**	**0.00**
Manufacturing, High Tech							**Office, High**						
Zone 1	0.00	2.15	0.00	0.00	0.17	0.00	Zone 1	0.00	1.70	0.30	0.33	0.44	0.00
Zone 2	0.00	2.12	0.00	0.00	0.17	0.00	Zone 2	0.00	1.59	0.29	0.32	0.41	0.00
Zone 3	0.00	2.04	0.00	0.00	0.17	0.00	Zone 3	0.00	1.59	0.32	0.34	0.42	0.00
Zone 4	0.00	2.01	0.00	0.00	0.17	0.00	Zone 4	0.00	1.58	0.28	0.35	0.42	0.00
Zone 5	0.00	1.88	0.00	0.00	0.17	0.00	Zone 5	0.00	1.58	0.32	0.36	0.41	0.00
Zone 6	0.00	1.63	0.00	0.00	0.17	0.00	Zone 6	0.00	1.11	0.16	0.29	0.33	0.00
Zone 7	0.00	1.50	0.00	0.00	0.17	0.00	Zone 7	0.00	0.86	0.23	0.31	0.37	0.00
Average	**0.00**	**1.90**	**0.00**	**0.00**	**0.17**	**0.00**	**Average**	**0.00**	**1.43**	**0.27**	**0.33**	**0.40**	**0.00**
Medical Clinic							**Restaurant, Bar/Lounge**						
Zone 1	0.00	3.15	0.00	0.00	0.54	0.00	Zone 1	0.00	3.07	0.00	0.06	0.24	0.00
Zone 2	0.00	3.09	0.00	0.00	0.53	0.00	Zone 2	0.00	2.87	0.00	0.06	0.25	0.00
Zone 3	0.00	3.07	0.00	0.00	0.54	0.00	Zone 3	0.00	2.85	0.00	0.06	0.24	0.00
Zone 4	0.00	3.02	0.00	0.00	0.54	0.00	Zone 4	0.00	2.83	0.00	0.06	0.24	0.00
Zone 5	0.00	2.99	0.00	0.00	0.54	0.00	Zone 5	0.00	2.80	0.00	0.06	0.24	0.00
Zone 6	0.00	2.58	0.00	0.00	0.55	0.00	Zone 6	0.00	1.88	0.00	0.06	0.24	0.00
Zone 7	0.00	2.45	0.00	0.00	0.52	0.00	Zone 7	0.00	1.87	0.00	0.06	0.23	0.00
Average	**0.00**	**2.91**	**0.00**	**0.00**	**0.54**	**0.00**	**Average**	**0.00**	**2.59**	**0.00**	**0.06**	**0.24**	**0.00**
Motel							**Restaurant, Full**						
Zone 1	0.97	2.68	0.00	0.00	0.19	2.15	Zone 1	0.00	3.37	0.00	0.05	0.38	0.00
Zone 2	1.19	2.52	0.00	0.00	0.18	4.43	Zone 2	0.00	3.30	0.00	0.05	0.39	0.00
Zone 3	1.19	2.52	0.00	0.00	0.19	4.81	Zone 3	0.00	3.25	0.00	0.05	0.39	0.00
Zone 4	1.19	2.42	0.00	0.00	0.21	7.39	Zone 4	0.00	3.20	0.00	0.05	0.39	0.00
Zone 5	1.28	2.26	0.00	0.00	0.21	7.82	Zone 5	0.00	3.03	0.00	0.05	0.38	0.00
Zone 6	1.29	2.07	0.00	0.00	0.23	8.89	Zone 6	0.00	2.35	0.00	0.05	0.39	0.00
Zone 7	1.39	1.97	0.00	0.00	0.23	9.51	Zone 7	0.00	2.33	0.00	0.05	0.38	0.00
Average	**1.21**	**2.35**	**0.00**	**0.00**	**0.21**	**6.43**	**Average**	**0.00**	**2.98**	**0.00**	**0.05**	**0.38**	**0.00**
Museum							**Restaurant, Quick**						
Zone 1	0.00	3.77	0.00	0.11	0.67	0.00	Zone 1	0.00	3.78	0.00	0.08	0.48	0.00
Zone 2	0.00	3.71	0.00	0.12	0.77	0.00	Zone 2	0.00	3.72	0.00	0.08	0.50	0.00
Zone 3	0.00	3.59	0.00	0.12	0.72	0.00	Zone 3	0.00	3.64	0.00	0.08	0.50	0.00
Zone 4	0.00	3.47	0.00	0.14	0.74	0.00	Zone 4	0.00	3.60	0.00	0.08	0.50	0.00
Zone 5	0.00	3.42	0.00	0.16	0.51	0.00	Zone 5	0.00	3.42	0.00	0.08	0.49	0.00
Zone 6	0.00	2.39	0.00	0.15	0.23	0.00	Zone 6	0.00	2.74	0.00	0.08	0.51	0.00
Zone 7	0.00	1.69	0.00	0.17	0.24	0.00	Zone 7	0.00	2.69	0.00	0.08	0.49	0.00
Average	**0.00**	**3.15**	**0.00**	**0.14**	**0.55**	**0.00**	**Average**	**0.00**	**3.37**	**0.00**	**0.08**	**0.50**	**0.00**
Office, Low							**Store, Convenience**						
Zone 1	0.00	2.90	0.00	0.02	0.29	0.00	Zone 1	0.00	4.08	0.00	0.02	0.29	0.00
Zone 2	0.00	2.90	0.00	0.02	0.30	0.00	Zone 2	0.00	3.66	0.00	0.02	0.30	0.00
Zone 3	0.00	2.81	0.00	0.02	0.30	0.00	Zone 3	0.00	3.60	0.00	0.02	0.30	0.00
Zone 4	0.00	2.70	0.00	0.02	0.29	0.00	Zone 4	0.00	3.46	0.00	0.02	0.29	0.00
Zone 5	0.00	2.66	0.00	0.02	0.29	0.00	Zone 5	0.00	3.28	0.00	0.02	0.29	0.00
Zone 6	0.00	2.04	0.00	0.02	0.30	0.00	Zone 6	0.00	2.66	0.00	0.02	0.27	0.00
Zone 7	0.00	1.82	0.00	0.02	0.29	0.00	Zone 7	0.00	2.50	0.00	0.02	0.24	0.00
Average	**0.00**	**2.55**	**0.00**	**0.02**	**0.29**	**0.00**	**Average**	**0.00**	**3.32**	**0.00**	**0.02**	**0.28**	**0.00**

Table R260120-30 Watts per S.F. for Various Systems by Building Type & Climate Zone (cont.)

Building / Zone	Heating	Cooling	Heat rejection	Auxiliary (pumps)	Vent fan	Supp heat pump	Building / Zone	Heating	Cooling	Heat rejection	Auxiliary (pumps)	Vent fan	Supp heat pump
Store, Department							**School, Middle**						
Zone 1	0.00	1.65	0.20	0.23	0.63	0.00	Zone 1	0.00	2.48	0.00	0.00	0.17	0.00
Zone 2	0.00	1.61	0.20	0.23	0.63	0.00	Zone 2	0.00	2.45	0.00	0.00	0.17	0.00
Zone 3	0.00	1.59	0.19	0.22	0.64	0.00	Zone 3	0.00	2.41	0.00	0.00	0.18	0.00
Zone 4	0.00	1.52	0.19	0.23	0.64	0.00	Zone 4	0.00	2.35	0.00	0.00	0.17	0.00
Zone 5	0.00	1.51	0.21	0.24	0.63	0.00	Zone 5	0.00	2.26	0.00	0.00	0.17	0.00
Zone 6	0.00	1.32	0.13	0.21	0.63	0.00	Zone 6	0.00	1.61	0.00	0.00	0.17	0.00
Zone 7	0.00	0.98	0.17	0.23	0.63	0.00	Zone 7	0.00	1.49	0.00	0.00	0.17	0.00
Average	**0.00**	**1.45**	**0.18**	**0.23**	**0.63**	**0.00**	**Average**	**0.00**	**2.15**	**0.00**	**0.00**	**0.17**	**0.00**
Store, Large							**School, High**						
Zone 1	0.00	2.68	0.00	0.00	0.16	0.00	Zone 1	0.00	3.24	0.00	0.01	0.22	0.00
Zone 2	0.00	2.61	0.00	0.00	0.17	0.00	Zone 2	0.00	3.08	0.00	0.01	0.22	0.00
Zone 3	0.00	2.61	0.00	0.00	0.17	0.00	Zone 3	0.00	3.05	0.00	0.01	0.23	0.00
Zone 4	0.00	2.56	0.00	0.00	0.16	0.00	Zone 4	0.00	3.00	0.00	0.01	0.23	0.00
Zone 5	0.00	2.53	0.00	0.00	0.16	0.00	Zone 5	0.00	2.94	0.00	0.01	0.23	0.00
Zone 6	0.00	1.84	0.00	0.00	0.16	0.00	Zone 6	0.00	1.98	0.00	0.01	0.23	0.00
Zone 7	0.00	1.50	0.00	0.00	0.16	0.00	Zone 7	0.00	1.81	0.00	0.01	0.23	0.00
Average	**0.00**	**2.33**	**0.00**	**0.00**	**0.16**	**0.00**	**Average**	**0.00**	**2.73**	**0.00**	**0.01**	**0.23**	**0.00**
Store, Small							**School, College**						
Zone 1	0.00	3.60	0.00	0.01	0.25	0.00	Zone 1	0.00	2.92	0.00	0.00	0.20	0.00
Zone 2	0.00	3.51	0.00	0.01	0.26	0.00	Zone 2	0.00	2.91	0.00	0.00	0.20	0.00
Zone 3	0.00	3.38	0.00	0.01	0.25	0.00	Zone 3	0.00	2.78	0.00	0.00	0.20	0.00
Zone 4	0.00	3.32	0.00	0.01	0.25	0.00	Zone 4	0.00	2.72	0.00	0.00	0.20	0.00
Zone 5	0.00	3.23	0.00	0.01	0.24	0.00	Zone 5	0.00	2.43	0.00	0.00	0.20	0.00
Zone 6	0.00	2.49	0.00	0.01	0.23	0.00	Zone 6	0.00	1.77	0.00	0.00	0.20	0.00
Zone 7	0.00	2.17	0.00	0.01	0.21	0.00	Zone 7	0.00	1.51	0.00	0.00	0.19	0.00
Average	**0.00**	**3.10**	**0.00**	**0.01**	**0.24**	**0.00**	**Average**	**0.00**	**2.43**	**0.00**	**0.00**	**0.20**	**0.00**
Store, Strip Mall							**Service Station**						
Zone 1	0.00	3.36	0.00	0.02	0.25	0.00	Zone 1	0.00	10.42	0.00	0.03	0.47	0.00
Zone 2	0.00	3.30	0.00	0.02	0.25	0.00	Zone 2	0.00	9.01	0.00	0.03	0.46	0.00
Zone 3	0.00	3.14	0.00	0.02	0.25	0.00	Zone 3	0.00	7.82	0.00	0.03	0.49	0.00
Zone 4	0.00	3.14	0.00	0.02	0.25	0.00	Zone 4	0.00	7.02	0.00	0.03	0.47	0.00
Zone 5	0.00	2.24	0.00	0.02	0.24	0.00	Zone 5	0.00	6.13	0.00	0.03	0.46	0.00
Zone 6	0.00	1.96	0.00	0.02	0.25	0.00	Zone 6	0.00	4.32	0.00	0.03	0.45	0.00
Zone 7	0.00	1.64	0.00	0.02	0.24	0.00	Zone 7	0.00	4.18	0.00	0.03	0.40	0.00
Average	**0.00**	**2.52**	**0.00**	**0.02**	**0.25**	**0.00**	**Average**	**0.00**	**6.99**	**0.00**	**0.03**	**0.46**	**0.00**
Store, Warehouse							**Storage, High Bay**						
Zone 1	0.00	2.99	0.00	0.00	0.21	0.00	Zone 1	0.00	1.98	0.00	0.00	0.17	0.00
Zone 2	0.00	2.97	0.00	0.00	0.21	0.00	Zone 2	0.00	1.97	0.00	0.00	0.18	0.00
Zone 3	0.00	2.94	0.00	0.00	0.21	0.00	Zone 3	0.00	1.90	0.00	0.00	0.18	0.00
Zone 4	0.00	2.83	0.00	0.00	0.21	0.00	Zone 4	0.00	1.89	0.00	0.00	0.18	0.00
Zone 5	0.00	2.70	0.00	0.00	0.21	0.00	Zone 5	0.00	1.85	0.00	0.00	0.17	0.00
Zone 6	0.00	2.10	0.00	0.00	0.21	0.00	Zone 6	0.00	1.29	0.00	0.00	0.17	0.00
Zone 7	0.00	1.78	0.00	0.00	0.21	0.00	Zone 7	0.00	1.10	0.00	0.00	0.17	0.00
Average	**0.00**	**2.61**	**0.00**	**0.00**	**0.21**	**0.00**	**Average**	**0.00**	**1.71**	**0.00**	**0.00**	**0.17**	**0.00**
School, Elementary							**Storage, Low Bay**						
Zone 1	0.00	2.73	0.00	0.01	0.21	0.00	Zone 1	0.00	1.87	0.00	0.00	0.16	0.00
Zone 2	0.00	2.69	0.00	0.01	0.21	0.00	Zone 2	0.00	1.85	0.00	0.00	0.16	0.00
Zone 3	0.00	2.63	0.00	0.01	0.21	0.00	Zone 3	0.00	1.82	0.00	0.00	0.16	0.00
Zone 4	0.00	2.61	0.00	0.01	0.20	0.00	Zone 4	0.00	1.77	0.00	0.00	0.16	0.00
Zone 5	0.00	2.51	0.00	0.01	0.20	0.00	Zone 5	0.00	1.76	0.00	0.00	0.16	0.00
Zone 6	0.00	1.71	0.00	0.01	0.20	0.00	Zone 6	0.00	1.20	0.00	0.00	0.16	0.00
Zone 7	0.00	1.64	0.00	0.01	0.20	0.00	Zone 7	0.00	1.00	0.00	0.00	0.16	0.00
Average	**0.00**	**2.36**	**0.00**	**0.01**	**0.20**	**0.00**	**Average**	**0.00**	**1.61**	**0.00**	**0.00**	**0.16**	**0.00**

R260519-20 Armored Cable

Armored Cable – Quantities are taken off in the same manner as wire.

Bx Type Cable – Productivities are based on an average run of 50' before terminating at a box fixture, etc. Each 50' section includes field preparation of (2) ends with hacksaw, identification and tagging of wire. Set up is open coil type without reels attaching cable to snake and pulling across a suspended ceiling or open face wood or steel studding, price does not include drilling of studs.

Cable in Tray – Productivities are based on an average run of 100 L.F. with set up of pulling equipment for (2) 90° bends, attaching cable to pull-in means, identification and tagging of wires, set up of reels. Wire termination to breakers equipment, etc., are not included.

Job Conditions – Productivities are based on new construction to a height of 10' in an unobstructed area. Material staging is assumed to be within 100' of work being performed.

R260519-80 Undercarpet Systems

Takeoff Procedure for Power Systems: List components for each fitting type, tap, splice, and bend on your quantity takeoff sheet. Each component must be priced separately. Start at the power supply transition fittings and survey each circuit for the components needed. List the quantities of each component under a specific circuit number. Use the floor plan layout scale to get cable footage.

Reading across the list, combine the totals of each component in each circuit and list the total quantity in the last column. Calculate approximately 5% for scrap for items such as cable, top shield, tape, and spray adhesive. Also provide for final variations that may occur on-site.

Suggested guidelines are:
1. Equal amounts of cable and top shield should be priced.
2. For each roll of cable, price a set of cable splices.
3. For every 1 ft. of cable, price 2-1/2 ft. of hold-down tape.
4. For every 3 rolls of hold-down tape, price 1 can of spray adhesive.

Adjust final figures wherever possible to accommodate standard packaging of the product. This information is available from the distributor.

Each transition fitting requires:
1. 1 base
2. 1 cover
3. 1 transition block

Each floor fitting requires:
1. 1 frame/base kit
2. 1 transition block
3. 2 covers (duplex/blank)

Each tap requires:
1. 1 tap connector for each conductor
2. 1 pair insulating patches
3. 2 top shield connectors

Each splice requires:
1. 1 splice connector for each conductor
2. 1 pair insulating patches
3. 3 top shield connectors

Each cable bend requires:
1. 2 top shield connectors

Each cable dead end (outside of transition block) requires:
1. 1 pair insulating patches

Labor does not include:
1. Patching or leveling uneven floors.
2. Filling in holes or removing projections from concrete slabs.
3. Sealing porous floors.
4. Sweeping and vacuuming floors.
5. Removal of existing carpeting.
6. Carpet square cut-outs.
7. Installation of carpet squares.

Takeoff Procedures for Telephone Systems: After reviewing floor plans identify each transition. Number or letter each cable run from that fitting.

Start at the transition fitting and survey each circuit for the components needed. List the cable type, terminations, cable length, and floor fitting type under the specific circuit number. Use the floor plan layout scale to get the cable footage. Add some extra length (next higher increment of 5 feet) to preconnectorized cable.

Transition fittings require:
1. 1 base plate
2. 1 cover
3. 1 transition block

Floor fittings require:
1. 1 frame/base kit
2. 2 covers
3. Modular jacks

Reading across the list, combine the list of components in each circuit and list the total quantity in the last column. Calculate the necessary scrap factors for such items as tape, bottom shield and spray adhesive. Also provide for final variations that may occur on-site.

Adjust final figures whenever possible to accommodate standard packaging. Check that items such as transition fittings, floor boxes, and floor fittings that are to utilize both power and telephone have been priced as combination fittings, so as to avoid duplication.

Make sure to include marking of floors and drilling of fasteners if fittings specified are not the adhesive type.

Labor does not include:
1. Conduit or raceways before transition of floor boxes
2. Telephone cable before transition boxes
3. Terminations before transition boxes
4. Floor preparation as described in power section

Be sure to include all cable folds when pricing labor.

Takeoff Procedure for Data Systems: Start at the transition fittings and take off quantities in the same manner as the telephone system, keeping in mind that data cable does not require top or bottom shields.

The data cable is simply cross-taped on the cable run to the floor fitting.

Data cable can be purchased in either bulk form in which case coaxial connector material and labor must be priced, or in preconnectorized cut lengths.

Data cable cannot be folded and must be notched at 1 inch intervals. A count of all turns must be added to the labor portion of the estimate. (Note: Some manufacturers have prenotched cable.)

Notching required:
1. 90 degree turn requires 8 notches per side.
2. 180 degree turn requires 16 notches per side.

Floor boxes, transition boxes, and fittings are the same as described in the power and telephone procedures.

Since undercarpet systems require special hand tools, be sure to include this cost in proportion to number of crews involved in the installation.

Job Conditions: Productivity is based on new construction in an unobstructed area. Staging area is assumed to be within 200' of work being performed.

R260519-90   Wire

Wire quantities are taken off by either measuring each cable run or by extending the conduit and raceway quantities times the number of conductors in the raceway. Ten percent should be added for waste and tie-ins. Keep in mind that the unit of measure of wire is C.L.F. not L.F. as in raceways so the formula would read:

$$\frac{\text{(L.F. Raceway x No. of Conductors) x 1.10}}{100} = \text{C.L.F.}$$

Price per C.L.F. of wire includes:
1. Setting up wire coils or spools on racks
2. Attaching wire to pull in means
3. Measuring and cutting wire
4. Pulling wire into a raceway
5. Identifying and tagging

Price does not include:
1. Connections to breakers, panelboards, or equipment
2. Splices

Job Conditions: Productivity is based on new construction to a height of 10′ in an unobstructed area. Material staging is assumed to be within 100′ of work being performed.

Economy of Scale: If more than three wires at a time are being pulled, deduct the following percentages from the labor of that grouping:

4-5 wires	25%
6-10 wires	30%
11-15 wires	35%
over 15	40%

If a wire pull is less than 100′ in length and is interrupted several times by boxes, lighting outlets, etc., it may be necessary to add the following lengths to each wire being pulled:

Junction box to junction box	2 L.F.
Lighting panel to junction box	6 L.F.
Distribution panel to sub panel	8 L.F.
Switchboard to distribution panel	12 L.F.
Switchboard to motor control center	20 L.F.
Switchboard to cable tray	40 L.F.

Measure of Drops and Riser: It is important when taking off wire quantities to include the wire for drops to electrical equipment. If heights of electrical equipment are not clearly stated, use the following guide:

	Bottom A.F.F.	Top A.F.F.	Inside Cabinet
Safety switch to 100A	5′	6′	2′
Safety switch 400 to 600A	4′	6′	3′
100A panel 12 to 30 circuit	4′	6′	3′
42 circuit panel	3′	6′	4′
Switch box	3′	3′6″	1′
Switchgear	0′	8′	8′
Motor control centers	0′	8′	8′
Transformers - wall mount	4′	8′	2′
Transformers - floor mount	0′	12′	4′

R260519-92   Minimum Copper and Aluminum Wire Size Allowed for Various Types of Insulation

Minimum Wire Sizes

Amperes	Copper THW THWN or XHHW	Copper THHN XHHW *	Aluminum THW XHHW	Aluminum THHN XHHW *	Amperes	Copper THW THWN or XHHW	Copper THHN XHHW *	Aluminum THW XHHW	Aluminum THHN XHHW *
15A	#14	#14	#12	#12	195	3/0	2/0	250kcmil	4/0
20	#12	#12	#10	#10	200	3/0	3/0	250kcmil	4/0
25	#10	#10	#10	#10	205	4/0	3/0	250kcmil	4/0
30	#10	#10	#8	#8	225	4/0	3/0	300kcmil	250kcmil
40	#8	#8	#8	#8	230	4/0	4/0	300kcmil	250kcmil
45	#8	#8	#6	#8	250	250kcmil	4/0	350kcmil	300kcmil
50	#8	#8	#6	#6	255	250kcmil	4/0	400kcmil	300kcmil
55	#6	#8	#4	#6	260	300kcmil	4/0	400kcmil	350kcmil
60	#6	#6	#4	#6	270	300kcmil	250kcmil	400kcmil	350kcmil
65	#6	#6	#4	#4	280	300kcmil	250kcmil	500kcmil	350kcmil
75	#4	#6	#3	#4	285	300kcmil	250kcmil	500kcmil	400kcmil
85	#4	#4	#2	#3	290	350kcmil	250kcmil	500kcmil	400kcmil
90	#3	#4	#2	#2	305	350kcmil	300kcmil	500kcmil	400kcmil
95	#3	#4	#1	#2	310	350kcmil	300kcmil	500kcmil	500kcmil
100	#3	#3	#1	#2	320	400kcmil	300kcmil	600kcmil	500kcmil
110	#2	#3	1/0	#1	335	400kcmil	350kcmil	600kcmil	500kcmil
115	#2	#2	1/0	#1	340	500kcmil	350kcmil	600kcmil	500kcmil
120	#1	#2	1/0	1/0	350	500kcmil	350kcmil	700kcmil	500kcmil
130	#1	#2	2/0	1/0	375	500kcmil	400kcmil	700kcmil	600kcmil
135	1/0	#1	2/0	1/0	380	500kcmil	400kcmil	750kcmil	600kcmil
150	1/0	#1	3/0	2/0	385	600kcmil	500kcmil	750kcmil	600kcmil
155	2/0	1/0	3/0	3/0	420	600kcmil	500kcmil		700kcmil
170	2/0	1/0	4/0	3/0	430		500kcmil		750kcmil
175	2/0	2/0	4/0	3/0	435		600kcmil		750kcmil
180	3/0	2/0	4/0	4/0	475		600kcmil		

*Dry Locations Only

Notes:
1. Size #14 to 4/0 is in AWG units (American Wire Gauge).
2. Size 250 to 750 is in kcmil units (Thousand Circular Mils).
3. Use next higher ampere value if exact value is not listed in table.
4. For loads that operate continuously increase ampere value by 25% to obtain proper wire size.
5. Table R260519-92 has been written for estimating only. It is based on an ambient temperature of 30°C (86° F). For an ambient temperature other than 30°C (86° F), ampacity correction factors will be applied.

For customer support on your Electrical Costs with RSMeans Data, call 800.448.8182.

R260519-93   Metric Equivalent, Wire

United States		European	
Size AWG or kcmil	Area Cir. Mils. (cmil) mm²	Size mm²	Area Cir. Mils.
18	1620/.82	.75	1480
16	2580/1.30	1.0	1974
14	4110/2.08	1.5	2961
12	6530/3.30	2.5	4935
10	10,380/5.25	4	7896
8	16,510/8.36	6	11,844
6	26,240/13.29	10	19,740
4	41,740/21.14	16	31,584
3	52,620/26.65	25	49,350
2	66,360/33.61	–	–
1	83,690/42.39	35	69,090
1/0	105,600/53.49	50	98,700
2/0	133,100/67.42	–	–
3/0	167,800/85.00	70	138,180
4/0	211,600/107.19	95	187,530
250	250,000/126.64	120	236,880
300	300,000/151.97	150	296,100
350	350,000/177.30	–	–
400	400,000/202.63	185	365,190
500	500,000/253.29	240	473,760
600	600,000/303.95	300	592,200
700	700,000/354.60	–	–
750	750,000/379.93	–	–

U.S. vs. European Wire – Approximate Equivalents

R260519-94   Size Required and Weight (Lbs./1000 L.F.) of Aluminum and Copper THW Wire by Ampere Load

Amperes	Copper Size	Aluminum Size	Copper Weight	Aluminum Weight
15	14	12	24	11
20	12	10	33	17
30	10	8	48	39
45	8	6	77	52
65	6	4	112	72
85	4	2	167	101
100	3	1	205	136
115	2	1/0	252	162
130	1	2/0	324	194
150	1/0	3/0	397	233
175	2/0	4/0	491	282
200	3/0	250	608	347
230	4/0	300	753	403
255	250	400	899	512
285	300	500	1068	620
310	350	500	1233	620
335	400	600	1396	772
380	500	750	1732	951

For customer support on your Electrical Costs with RSMeans Data, call 800.448.8182.

603

R260526-80   Grounding

Grounding When taking off grounding systems, identify separately the type and size of wire.

Example:

Bare copper & size

Bare aluminum & size

Insulated copper & size

Insulated aluminum & size

Count the number of ground rods and their size.

Example:

1.   8′ grounding rod – 5/8″ dia. 20 Ea.
2.   10′ grounding rod – 5/8″ dia. 12 Ea.
3.   15′ grounding rod – 3/4″ dia.   4 Ea.

Count the number of connections; the size of the largest wire will determine the productivity.

Example:

Braze a #2 wire to a #4/0 cable

The 4/0 cable will determine the L.H. and cost to be used.

Include individual connections to:

1. Ground rods
2. Building steel
3. Equipment
4. Raceways

Price does not include:

1. Excavation
2. Backfill
3. Sleeves or raceways used to protect grounding wires
4. Wall penetrations
5. Floor cutting
6. Core drilling

Job Conditions: Productivity is based on a ground floor area, using cable reels in an unobstructed area. The material staging area is assumed to be within 100′ of work being performed.

R260533-20   Conduit To 10′ High

List conduit by quantity, size, and type. Do not deduct for lengths occupied by fittings, since this will be allowance for scrap. Example:

A. Aluminum — size
B. Rigid galvanized — size
C. Steel intermediate (IMC) — size
D. Rigid steel, plastic-coated 20 Mil. — size
E. Rigid steel, plastic-coated 40 Mil. — size
F. Electric metallic tubing (EMT) — size
G. PVC Schedule 40 — size

Types (A) through (E) listed above contain the following per 100 L.F.:

1. (11) Threaded couplings
2. (11) Beam-type hangers
3. (2) Factory sweeps
4. (2) Fiber bushings
5. (4) Locknuts
6. (2) Field threaded pipe terminations
7. (2) Removal of concentric knockouts

Type (F) contains per 100 L.F.:

1. (11) Set screw couplings
2. (11) Beam clamps
3. (2) Field bends on 1/2″ and 3/4″ diameter
4. (2) Factory sweeps for 1″ and above
5. (2) Set screw steel connectors
6. (2) Removal of concentric knockouts

Type (G) contains per 100 L.F.:

1. (11) Field cemented couplings
2. (34) Beam clamps

3. (2) Factory sweeps
4. (2) Adapters
5. (2) Locknuts
6. (2) Removal of concentric knockouts

Labor-hours for all conduit to 10′ include:

1. Unloading by hand
2. Hauling by hand to an area up to 200′ from loading dock
3. Installation of conduit and fittings as described in Conduit models (A) through (G)

Not included in the material and labor are:

1. Staging rental or purchase
2. Structural modifications
3. Wire
4. Junction boxes
5. Fittings in excess of those described in conduit models (A) through (G)
6. Painting of conduit

Fittings

Only those fittings listed above are included in the linear foot totals, although they should be listed separately from conduit lengths, without prices, to ensure proper quantities for material procurement.

If the fittings required exceed the quantities included in the model conduit runs, then material and labor costs must be added to the difference. If actual needs per 100 L.F. of conduit are: (2) sweeps, (4) LBs and (1) field bend, then, (4) LBs and (1) field bend must be priced additionally.

For customer support on your Electrical Costs with RSMeans Data, call 800.448.8182.

R260533-21 Hangers

It is sometimes desirable to substitute an alternate style of hanger if the support being used is not the type described in the conduit models.

One approach is the substitution method:

1. Find the cost of the type hanger described in the conduit model.
2. Calculate the cost of the desired type hanger (it may be necessary to calculate individual components such as drilling, expansion shields, etc.).
3. Calculate the cost difference (delta) between the two types of hangers.
4. Multiply the cost delta by the number of hangers in the model.
5. Divide the total delta cost for hangers in the model by the length of the model to find the delta cost per L.F. for that model.
6. Modify the given unit costs per L.F. for the model by the delta cost per L.F.

Another approach to hanger configurations would be to start with the conduit only and add all the supports and any other items as separate lines. This procedure is most useful if the project involves racking many runs of conduit on a single hanger; for instance, a trapeze type hanger.

Example: Five (5) 2″ RGS conduits, 50 L.F. each, are to be run on trapeze hangers from one pull box to another. The run includes one 90° bend.

1. List the hangers' components to create an assembly cost for each 2′-wide trapeze.
2. List the components for the 50′ conduit run, noting that 6 trapeze supports will be required.

Job Conditions: Productivities are based on new construction to 10′ high, using scaffolding, scissor lift, bucket lift in an unobstructed area. Material storage is assumed to be within 100′ of work being performed.

Add to labor for elevated installations:

10′ to 14.5′ High–10%	30′ to 35′ High–40%
15′ to 20′ High–20%	35′ to 40′ High–50%
20′ to 25′ High–25%	Over 40′ High–55%
25′ to 30′ High–35%	

Add these percentages to the L.F. labor cost, but not to fittings. Add these percentages only to quantities exceeding the different height levels, rather than the total conduit quantities.

Linear foot price for labor does not include penetrations in walls or floors and must be added to the estimate.

R260533-22   Conductors in Conduit

The table below lists the maximum number of conductors for various sized conduit using THW, TW or THWN insulations.

Copper Wire Size	Conduit Size																											
	½″			¾″			1″			1-¼″			1-½″			2″			2-½″			3″		3-½″		4″		
	TW	THW	THWN	TW	THW	THWN	TW	THW	THWN	TW	THW	THWN	TW	THW	THWN	TW	THW	THWN	TW	THW	THWN	THW	THWN	THW	THWN	THW	THWN
#14	9	6	13	15	10	24	25	16	39	44	29	69	60	40	94	99	65	154	142	93		143		192			
#12	7	4	10	12	8	18	19	13	29	35	24	51	47	32	70	78	53	114	111	76	164	117		157			
#10	5	4	6	9	6	11	15	11	18	26	19	32	36	26	44	60	43	73	85	61	104	95	160	127		163	
#8	2	1	3	4	3	5	7	5	9	12	10	16	17	13	22	28	22	36	40	32	51	49	79	66	106	85	136
#6		1	1		2	4		4	6		7	11		10	15		16	26		23	37	36	57	48	76	62	98
#4		1	1		1	2		3	4		5	7		7	9		12	16		17	22	27	35	36	47	47	60
#3		1	1		1	1		2	3		4	6		6	8		10	13		15	19	23	29	31	39	40	51
#2		1	1		1	1		2	3		4	5		5	7		9	11		13	16	20	25	27	33	34	43
#1					1	1		1	1		3	3		4	5		6	8		9	12	14	18	19	25	25	32
1/0					1	1		1	1		2	3		3	4		5	7		8	10	12	15	16	21	21	27
2/0					1	1		1	1		1	2		3	3		5	6		7	8	10	13	14	17	18	22
3/0					1	1		1	1		1	1		2	3		4	5		6	7	9	11	12	14	15	18
4/0						1		1	1		1	1		1	2		3	4		5	6	7	9	10	12	13	15
250 kcmil								1	1		1	1		1	1		2	3		4	4	6	7	8	10	10	12
300								1	1		1	1		1	1		2	3		3	4	5	6	7	8	9	11
350									1		1	1		1	1		1	2		3	3	4	5	6	7	8	9
400											1	1		1	1		1	1		2	3	4	5	5	6	7	8
500											1	1		1	1		1	1		1	2	3	4	4	5	6	7
600														1	1		1	1		1	1	3	3	4	4	5	5
700														1	1		1	1		1	1	2	3	3	4	4	5
750														1	1		1	1		1	1	2	2	3	3	4	4

For customer support on your Electrical Costs with RSMeans Data, call 800.448.8182.

605

R260533-25 Conduit Weight Comparisons (Lbs. per 100 ft.) with Maximum Cable Fill*

Type	1/2"	3/4"	1"	1-1/4"	1-1/2"	2"	2-1/2"	3"	3-1/2"	4"	5"	6"
Rigid Galvanized Steel (RGS)	104	140	235	358	455	721	1022	1451	1749	2148	3083	4343
Intermediate Steel (IMC)	84	113	186	293	379	611	883	1263	1501	1830		
Electrical Metallic Tubing (EMT)	54	116	183	296	368	445	641	930	1215	1540		

*Conduit & Heaviest Conductor Combination

R260533-30 Labor for Couplings and Fittings

The labor included in the unit price lines for couplings is for their installation separate from a fitting. However, the labor included in the unit price lines for fittings covers their complete installation, which may include the labor installation of the coupling(s). If coupling(s) are required to complete a fitting installation, only the material for the required number of couplings needs to be added to the appropriate unit price lines.

R260533-60   Wireway

When "taking off" Wireway, list by size and type.

Example:
1. Screw cover, unflanged + size
2. Screw cover, flanged + size
3. Hinged cover, flanged + size
4. Hinged cover, unflanged + size

Each 10' length on Wireway contains:
1. 10' of cover either screw or hinged type
2. (1) Coupling or flange gasket
3. (1) Wall type mount

All fittings must be priced separately.

Substitution of hanger types is done the same as described in R260533-21, "HANGERS," keeping in mind that the wireway model is based on 10' sections instead of a 100' conduit run.

Labor-hours for wireway include:
1. Unloading by hand
2. Hauling by hand up to 100' from loading dock
3. Measuring and marking
4. Mounting wall bracket using two (2) anchor type lead fasteners
5. Installing wireway on brackets, to 10' high (for higher elevations use factors in R260533-20)

Job Conditions: Productivity is based on new construction, to a height of 10' in an unobstructed area.

Material staging area is assumed to be within 100' of work being performed.

R260533-65   Outlet Boxes

Outlet boxes should be included on the same takeoff sheet as branch piping or devices to better explain what is included in each circuit.

Each unit price in this section is a stand alone item and contains no other component unless specified. For example, to estimate a duplex outlet, components that must be added are:
1. 4" square box
2. 4" plaster ring
3. Duplex receptacle
4. Device cover

The method of mounting outlet boxes is two (2) plastic shield fasteners.

Outlet boxes plastic, labor-hours include:
1. Marking box location on wood studding
2. Mounting box

Economy of Scale – For large concentrations of plastic boxes in the same area deduct the following percentages from labor-hour totals:

1	to	10	0%
11	to	25	20%
26	to	50	25%
51	to	100	30%
	over	100	35%

Note: It is important to understand that these percentages are not used on the total job quantities, but only areas where concentrations exceed the levels specified.

R260533-70   Pull Boxes and Cabinets

List cabinets and pull boxes by NEMA type and size.

Example:	TYPE	SIZE
	NEMA 1	6"W x 6"H x 4"D
	NEMA 3R	6"W x 6"H x 4"D

Labor-hours for wall mount (indoor or outdoor) installations include:
1. Unloading and uncrating
2. Handling of enclosures up to 200' from loading dock using a dolly or pipe rollers
3. Measuring and marking
4. Drilling four (4) anchor type lead fasteners using a hammer drill
5. Mounting and leveling boxes

Note: A plywood backboard is not included.

Labor-hours for ceiling mounting include:
1. Unloading and uncrating
2. Handling boxes up to 100' from loading dock

3. Measuring and marking
4. Drilling four (4) anchor type lead fasteners using a hammer drill
5. Installing and leveling boxes to a height of 10' using rolling staging

Labor-hours for free standing cabinets include:
1. Unloading and uncrating
2. Handling of cabinets up to 200' from loading dock using a dolly or pipe rollers
3. Marking of floor
4. Drilling four (4) anchor type lead fasteners using a hammer drill
5. Leveling and shimming

Labor-hours for telephone cabinets include:
1. Unloading and uncrating
2. Handling cabinets up to 200' using a dolly or pipe rollers
3. Measuring and marking
4. Mounting and leveling, using four (4) lead anchor type fasteners

Fig. R260536-11

R260536-10 Cable Tray

Cable Tray - When taking off cable tray it is important to identify separately the different types and sizes involved in the system being estimated. (Fig. R260536-11)

  A. – Ladder Type, galvanized or aluminum
  B. – Trough Type, galvanized or aluminum
  C. – Solid Bottom, galvanized or aluminum

The unit of measure is calculated in linear feet. Do not deduct from this footage any length occupied by fittings. This will be the only allowance for scrap. Be sure to include all vertical drops to panels, switch gear, etc.

Hangers – Included in the linear footage of cable tray is

  D. – 1 – Pair of connector plates per 12 L.F.
  E. – 1 – Pair clamp type hangers and 4' of 3/8" threaded rod per 12 L.F.

Not included are structural supports, which must be priced in addition to the hangers.

Fittings – Identify separately the different types of fittings

  A. – Ladder Type, galvanized or aluminum
  B. – Trough Type, galvanized or aluminum
  C. – Solid Bottom Type, galvanized or aluminum

The configuration, radius and rung spacing must also be listed. The unit of measure is "Ea."

  D. – Elbow, vertical Ea.
  E. – Elbow, horizontal Ea.
  F. – Tee, vertical Ea.
  G. – Cross, horizontal Ea.
  H. – Wye, horizontal Ea.
  I. – Tee, horizontal Ea.
  J. – Reducing fitting Ea.

Depending on the use of the system other examples of units which must be included are:

  K. – Divider strip L.F.
  L. – Drop-outs Ea.

  M. – End caps Ea.
  N. – Panel connectors Ea.

Wire and cable are not included and should be taken off separately. See Unit Price sections.

Job Conditions – Unit prices are based on a new installation to a work plane of 10' in an unobstructed area.

Add to labor for elevated installations:

10' to 14.5' High	10%
15' to 20' High	20%
20' to 25' High	25%
25' to 30' High	35%
30' to 35' High	40%
35' to 40' High	50%
Over 40' High	55%

Add these percentages for L.F. totals but not to fittings. Add percentages to only those quantities that fall in the different elevations. In other words, if the total quantity of cable tray is 200' but only 75' is above 15' then the 10% is added to the 75' only.

Linear foot costs do not include penetrations through walls and floors which must be added to the estimate.

Cable Tray Covers

Covers – Cable tray covers are taken off in the same manner as the tray itself, making distinctions as to the type of cover. (Fig. R260536-11)

  O. – Vented, galvanized or aluminum
  P. – Solid, galvanized or aluminum

Cover configurations are taken off separately noting type, specific radius and widths.

Note: Care should be taken to identify from plans and specifications exactly what is being covered. In many systems only vertical fittings are covered to retain wire and cable.

R260539-30 Conduit In Concrete Slab

List conduit by quantity, size and type.

Example:

A. Rigid galvanized steel + size
B. P.V.C. + size

Rigid galvanized steel (A) contains per 100 L.F.:

1. (20) Ties to slab reinforcing
2. (11) Threaded steel couplings
3. (2) Factory sweeps
4. (2) Field threaded conduit terminations
5. (2) Fiber bushings + locknuts
6. (2) Removal of concentric knockouts

P.V.C. (B) contains per 100 L.F.:

1. (20) Ties to slab reinforcing
2. (11) Field cemented couplings
3. (2) Factory sweeps
4. (2) Adapters
5. (2) Removal of concentric knockouts

R260539-40 Conduit In Trench

Conduit in trench is galvanized steel and contains per 100 L.F.:

1. (11) Threaded couplings
2. (2) Factory sweeps
3. (2) Fiber bushings + (4) locknuts
4. (2) Field threaded conduit terminations
5. (2) Removal of concentric knockouts

Note:

Conduit in Unit Price sections do not include:

1. Floor cutting
2. Excavation or backfill
3. Grouting or patching

Conduit fittings in excess of those listed in the above Conduit model must be added. (Refer to R260533-20 for Procedure example.)

Fig. R260543-51

R260543-50 Underfloor Duct

When pricing Underfloor Duct it is important to identify and list each component, since costs vary significantly from one type of fitting to another. Do not deduct boxes or fittings from linear foot totals; this will be your allowance for scrap.

The first step is to identify the system as either:

FIG. R260543-51 Single Level

FIG. R260543-52 Dual Level

Single Level System

Include on your "takeoff sheet" the following unit price items, making sure to distinquish between Standard and Super duct:

A. Feeder duct (blank) in L.F.
B. Distribution duct (Inserts 2' on center) in L.F.
C. Elbows (Vertical) Ea.
D. Elbows (Horizontal) Ea.
E. Cabinet connector Ea.
F. Single duct junction box Ea.
G. Double duct junction box Ea.
H. Triple duct junction box Ea.
I. Support, single cell Ea.
J. Support, double cell Ea.
K. Support, triple cell Ea.
L. Carpet pan Ea.
M. Terrazzo pan Ea.
N. Insert to conduit adapter Ea.
O. Conduit adapter Ea.
P. Low tension outlet Ea.
Q. High tension outlet Ea.
R. Galvanized nipple Ea.
S. Wire per C.L.F.
T. Offset (Duct type) Ea.

Fig. R260543-52

R260543-50 Underfloor Duct (cont.)

Dual Level System + Labor

Dual Level

Include the following when "taking off" Dual Level systems:

Distinguish between Standard and Super duct.

A. Feeder duct (blank) in L.F.
B. Distribution duct (Inserts 2′ on center) in L.F.
C. Elbows (Vertical), Ea.
D. Elbows (Horizontal), Ea.
E. Cabinet connector, Ea.
F. Single duct, 2 level, junction box, Ea.
G. Double duct, 2 level, junction box, Ea.
H. Support, single cell, Ea.
I. Support, double cell, Ea.
J. Support, triple cell, Ea.
K. Carpet pan, Ea.
L. Terrazzo pan, Ea.
M. Insert to conduit adapter, Ea.
N. Conduit adapter, Ea.
O. Low tension outlet, Ea.
P. High tension outlet, Ea.
Q. Wire per C.L.F.

Note: Make sure to include risers in linear foot totals. High tension outlets include box, receptacle, covers and related mounting hardware.

Labor-hours for both Single and Dual Level systems include:

1. Unloading and uncrating
2. Hauling up to 200′ from loading dock

3. Measuring and marking
4. Setting raceway and fittings in slab or on grade
5. Leveling raceway and fittings

Labor-hours do not include:

1. Floor cutting
2. Excavation or backfill
3. Concrete pour
4. Grouting or patching
5. Wire or wire pulls
6. Additional outlets after concrete is poured
7. Piping to or from Underfloor Duct

Note: Installation is based on installing up to 150′ of duct. If quantities exceed this, deduct the following percentages:

1. 150′ to 250′ - 10%
2. 250′ to 350′ - 15%
3. 350′ to 500′ - 20%
4. over 500′ - 25%

Deduct these percentages from labor only.

Deduct these percentages from straight sections only.

Do not deduct from fittings or junction boxes.

Job Conditions: Productivity is based on new construction.

Underfloor duct to be installed on first three floors.

The material staging area is within 100′ of work being performed.

The area unobstructed and duct are not subject to physical damage.

For customer support on your Electrical Costs with RSMeans Data, call 800.448.8182.

611

R260580-75 Motor Connections

Motor connections should be listed by size and type of motor.
Included in the material and labor cost are:

1. (2) Flex connectors
2. 18" of flexible metallic wireway
3. Wire identification and termination
4. Test for rotation
5. (2) or (3) Conductors

Price does not include:

1. Mounting of motor
2. Disconnect switch
3. Motor starter
4. Controls
5. Conduit or wire ahead of flex

Note: When "taking off" motor connections, it is advisable to list connections on the same quantity sheet as motors, motor starters and controls.

R260913-80 Switchboard Instruments

Switchboard instruments are added to the price of switchboards according to job specifications. This equipment is usually included when ordering "gear" from the manufacturer and will arrive factory installed.

Included in the labor cost are:

1. Internal wiring connections
2. Wire identification
3. Wire tagging

Transition sections include:

1. Uncrating
2. Hauling sections up to 100' from loading dock
3. Positioning sections
4. Leveling sections

5. Bolting enclosures
6. Bolting vertical bus bars

Price does not include:

1. Equipment pads
2. Steel channels embedded or grouted in concrete
3. Special knockouts
4. Rigging

Job Conditions: Productivity is based on new construction and equipment to be installed on the first floor within 200' of the loading dock.

R262213-10 Electric Circuit Voltages

General: The following method provides the user with a simple non-technical means of obtaining comparative costs of wiring circuits. The circuits considered serve the electrical loads of motors, electric heating, lighting and transformers, for example, that require low voltage 60 Hertz alternating current.

The method used here is suitable only for obtaining estimated costs. It is **not** intended to be used as a substitute for electrical engineering design applications.

Conduit and wire circuits can represent from twenty to thirty percent of the total building electrical cost. By following the described steps and using the tables the user can translate the various types of electric circuits into estimated costs.

Wire Size: Wire size is a function of the electric load which is usually listed in one of the following units:

1. Amperes (A)
2. Watts (W)
3. Kilowatts (kW)
4. Volt amperes (VA)
5. Kilovolt amperes (kVA)
6. Horsepower (HP)

These units of electric load must be converted to amperes in order to obtain the size of wire necessary to carry the load. To convert electric load units to amperes one must have an understanding of the voltage classification of the power source and the voltage characteristics of the electrical equipment or load to be energized. The seven A.C. circuits commonly used are illustrated in Figures R262213-11 through R262213-17 showing the tranformer load voltage and the point of use voltage at the point on the circuit where the load is connected. The difference between the source and point of use voltage is attributed to the circuit voltage drop and is considered to be approximately 4%.

Motor Voltages: Motor voltages are listed by their point of use voltage and not the power source voltage.

For example: 460 volts instead of 480 volts
200 instead of 208 volts
115 volts instead of 120 volts

Lighting and Heating Voltages: Lighting and heating equipment voltages are listed by the power source voltage and not the point of wire voltage.

For example: 480, 277, 120 volt lighting
480 volt heating or air conditioning unit
208 volt heating unit

Transformer Voltages: Transformer primary (input) and secondary (output) voltages are listed by the power source voltage.

For example: Single phase 10 kVA
Primary 240/480 volts
Secondary 120/240 volts

In this case, the primary voltage may be 240 volts with a 120 volts secondary or may be 480 volts with either a 120V or a 240V secondary.

For example: Three phase 10 kVA
Primary 480 volts
Secondary 208Y/120 volts

In this case the transformer is suitable for connection to a circuit with a 3 phase, 3 wire or a 3 phase, 4 wire circuit with a 480 voltage. This application will provide a secondary circuit of 3 phase, 4 wire with 208 volts between phase wires and 120 volts between any phase wire and the neutral (white) wire.

R262213-11

3 Wire, 1 Phase, 120/240 Volt System

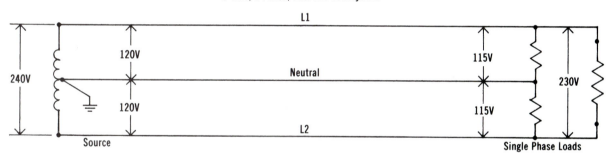

R262213-12

4 Wire, 3 Phase, 208Y/120 Volt System

For customer support on your Electrical Costs with RSMeans Data, call 800.448.8182.

613

R262213-13　Electric Circuit Voltages (cont.)

3 Wire, 3 Phase 240 Volt System

R262213-14

4 Wire, 3 Phase, 240/120 Volt System

R262213-15

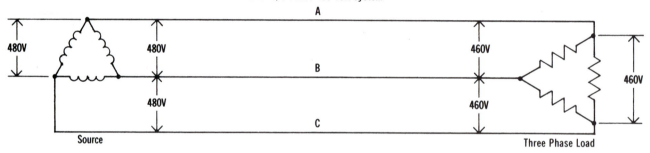

3 Wire, 3 Phase 480 Volt System

R262213-16

4 Wire, 3 Phase, 480Y/277 Volt System

R262213-17   Electric Circuit Voltages (cont.)

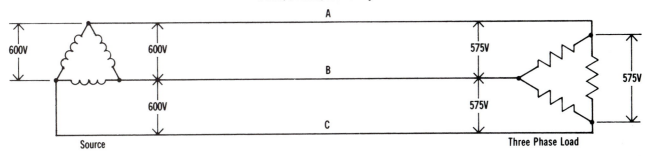

3 Wire, 3 Phase, 600 Volt System

R262213-20   kW Value/Cost Determination

General: Lighting and electric heating loads are expressed in watts and kilowatts.

Cost Determination:

The proper ampere values can be obtained as follows:

1. Convert watts to kilowatts
   (watts 1000 ÷ kilowatts)
2. Determine voltage rating of equipment.
3. Determine whether equipment is single phase or three phase.
4. Refer to Table R262213-21 to find ampere value from kW, Ton and BTU/hr. values.
5. Determine type of wire insulation – TW, THW, THWN.
6. Determine if wire is copper or aluminum.
7. Refer to Table R260519-92 to obtain copper or aluminum wire size from ampere values.
8. Next refer to Table R260533-22 for the proper conduit size to accommodate the number and size of wires in each particular case.
9. Next refer to unit cost data for the per linear foot cost of the conduit.
10. Next refer to unit cost data for the per linear foot cost of the wire. Multiply cost of wire per L.F. x number of wires in the circuits to obtain total wire cost per L.F.
11. Add values obtained in Step 9 and 10 for total cost per linear foot for conduit and wire x length of circuit = Total Cost.

Notes:

1. 1 Phase refers to single phase, 2 wire circuits.
2. 3 Phase refers to three phase, 3 wire circuits.
3. For circuits which operate continuously for 3 hours or more, multiply the ampere values by 1.25 for a given kW requirement.
4. For kW ratings not listed, add ampere values.

   For example: Find the ampere value of
   9 kW at 208 volt, single phase.

$$\begin{array}{rl} 4 \text{ kW} = & 19.2\text{A} \\ 5 \text{ kW} = & 24.0\text{A} \\ \hline 9 \text{ kW} = & 43.2\text{A} \end{array}$$

5. "Length of Circuit" refers to the one way distance of the run, not to the total sum of wire lengths.

R262213-21   Ampere Values as Determined by kW Requirements, BTU/HR or Ton, Voltage and Phase Values

			Ampere Values						
			120V	208V		240V		277V	480V
kW	Ton	BTU/HR	1 Phase	1 Phase	3 Phase	1 Phase	3 Phase	1 Phase	3 Phase
0.5	.1422	1,707	4.2A	2.4A	1.4A	2.1A	1.2A	1.8A	0.6A
0.75	.2133	2,560	6.2	3.6	2.1	3.1	1.9	2.7	.9
1.0	.2844	3,413	8.3	4.9	2.8	4.2	2.4	3.6	1.2
1.25	.3555	4,266	10.4	6.0	3.5	5.2	3.0	4.5	1.5
1.5	.4266	5,120	12.5	7.2	4.2	6.3	3.1	5.4	1.8
2.0	.5688	6,826	16.6	9.7	5.6	8.3	4.8	7.2	2.4
2.5	.7110	8,533	20.8	12.0	7.0	10.4	6.1	9.1	3.1
3.0	.8532	10,239	25.0	14.4	8.4	12.5	7.2	10.8	3.6
4.0	1.1376	13,652	33.4	19.2	11.1	16.7	9.6	14.4	4.8
5.0	1.4220	17,065	41.6	24.0	13.9	20.8	12.1	18.1	6.1
7.5	2.1331	25,598	62.4	36.0	20.8	31.2	18.8	27.0	9.0
10.0	2.8441	34,130	83.2	48.0	27.7	41.6	24.0	36.5	12.0
12.5	3.5552	42,663	104.2	60.1	35.0	52.1	30.0	45.1	15.0
15.0	4.2662	51,195	124.8	72.0	41.6	62.4	37.6	54.0	18.0
20.0	5.6883	68,260	166.4	96.0	55.4	83.2	48.0	73.0	24.0
25.0	7.1104	85,325	208.4	120.2	70.0	104.2	60.0	90.2	30.0
30.0	8.5325	102,390		144.0	83.2	124.8	75.2	108.0	36.0
35.0	9.9545	119,455		168.0	97.1	145.6	87.3	126.0	42.1
40.0	11.3766	136,520		192.0	110.8	166.4	96.0	146.0	48.0
45.0	12.7987	153,585			124.8	187.5	112.8	162.0	54.0
50.0	14.2208	170,650			140.0	208.4	120.0	180.4	60.0
60.0	17.0650	204,780			166.4		150.4	216.0	72.0
70.0	19.9091	238,910			194.2		174.6		84.2
80.0	22.7533	273,040			221.6		192.0		96.0
90.0	25.5975	307,170					225.6		108.0
100.0	28.4416	341,300							120.0

R262213-25 kVA Value/Cost Determination

General: Control transformers are listed in VA. Step-down and power transformers are listed in kVA.

Cost Determination:

1. Convert VA to kVA. Volt amperes (VA) ÷ 1000 = Kilovolt amperes (kVA).
2. Determine voltage rating of equipment.
3. Determine whether equipment is single phase or three phase.
4. Refer to Table R262213-26 to find ampere value from kVA value.
5. Determine type of wire insulation – TW, THW, THWN.
6. Determine if wire is copper or aluminum.
7. Refer to Table R260519-92 to obtain copper or aluminum wire size from ampere values.

8. Next refer to Table R260533-22 for the proper conduit size to accommodate the number and size of wires in each particular case.
9. Next refer to unit price data for the per linear foot cost of the conduit.
10. Next refer to unit price data for the per linear foot cost of the wire. Multiply cost of wire per L.F. x number of wires in the circuits to obtain total wire cost.
11. Add values obtained in Step 9 and 10 for total cost per linear foot for conduit and wire x length of circuit = Total Cost.

Example: A transformer rated 10 kVA 480 volts primary, 240 volts secondary, 3 phase has the capacity to furnish the following:

1. Primary amperes = 10 kVA x 1.20 = 12 amperes
   (from Table R262213-26)
2. Secondary amperes = 10 kVA x 2.40 = 24 amperes
   (from Table R262213-26)

Note: Transformers can deliver generally 125% of their rated kVA. For instance, a 10 kVA rated transformer can safely deliver 12.5 kVA.

R262213-26 Multiplier Values for kVA to Amperes Determined by Voltage and Phase Values

Volts	Multiplier for Circuits	
	2 Wire, 1 Phase	3 Wire, 3 Phase
115	8.70	
120	8.30	
230	4.30	2.51
240	4.16	2.40
200	5.00	2.89
208	4.80	2.77
265	3.77	2.18
277	3.60	2.08
460	2.17	1.26
480	2.08	1.20
575	1.74	1.00
600	1.66	0.96

R262213-60   Oil Filled Transformers

Transformers in this section include:
1. Rigging (as required)
2. Rental of crane and operator
3. Setting of oil filled transformer
4. (4) Anchor bolts, nuts and washers in concrete pad

Price does not include:
1. Primary and secondary terminations
2. Transformer pad
3. Equipment grounding
4. Cable
5. Conduit locknuts or bushings

Transformers in Unit Price sections for dry type, back-boost and isolating transformers include:
1. Unloading and uncrating
2. Hauling transformer to within 200' of loading dock

3. Setting in place
4. Wall mounting hardware
5. Testing

Price does not include:
1. Structural supports
2. Suspension systems
3. Welding or fabrication
4. Primary & secondary terminations

Add the following percentages to the labor for ceiling mounted transformers:

10' to 15'	= + 15%
15' to 25'	= + 30%
Over 25'	= + 35%

Job Conditions: Productivities are based on new construction. Installation is assumed to be on the first floor, in an obstructed area to a height of 10'. Material staging area is within 100' of final transformer location.

For customer support on your Electrical Costs with RSMeans Data, call 800.448.8182.

R262416-50   Load Centers and Panelboards

When pricing load centers list panels by size and type. List breakers in a separate column of the "quantity sheet," and define by phase and ampere rating.

Material and labor prices include breakers; for example: a 100A, 3-Wire, 102/240V, 18 circuit panel w/ main breaker, as described in the unit cost section, contains 18 single pole 20A breakers.

If you do not choose to include a full panel of single pole breakers, use the following method to adjust material and labor costs.

Example: In an 18 circuit panel only 16 single pole breakers are desired, requiring that the cost of 2 breakers be subtracted from the panel cost.

1. Go to the appropriate unit cost section of the data set to find the unit prices of the given circuit breaker type.
2. Modify those costs as follows:
    Bare material price x 0.50
    Bare labor cost x 0.60.
3. Multiply those modified bare costs by 2 (breakers in this example).
4. Subtract the modified costs for the 2 breakers from the given cost of the panel.

Labor-hours for load center installation includes:

1. Unloading, uncrating, and handling enclosures 200' from unloading area

2. Measuring and marking
3. Drilling (4) lead anchor type fasteners using a hammer drill
4. Mounting and leveling panel to a height of 6'
5. Preparation and termination of feeder cable to lugs or main breaker
6. Branch circuit identification
7. Lacing using tie wraps
8. Testing and load balancing
9. Marking panel directory

Not included in the material and labor are:

1. Modifications to enclosure
2. Structural supports
3. Additional lugs
4. Plywood backboards
5. Painting or lettering

Note: Knockouts are included in the price of terminating pipe runs and need not be added to the load center costs.

Job Conditions: Productivity is based on new construction to a height of 6', in an unobstructed area. Material staging area is assumed to be within 100' of work being performed.

Fig. R262419-61

R262419-60 Motor Control Centers

When taking off motor control centers, list the size, type and height of structures.

Example:
1. 600A, 22,000 RMS, 72″ high
2. 600A, back to back, 72″ high

Next take off individual starters; the number of structures can also be determined by adding the height in inches of starters divided by the height of the structure, and list on the same quantity sheet as the structures. Identify starters by type, horsepower rating, size and height in inches.

Example:
A. Class I, Type B, FVNR starter, 25 H.P., 18″ high.
   Add to the list with starters, factory installed controls.

Example:

B. Pilot lights

C. Push buttons

D. Auxiliary contacts

Identify starters and structures as either copper or aluminum and by the NEMA type of enclosure.

When pricing starters and structures, be sure to add or deduct adjustments, using lines in the unit price section.

Included in the cost of motor control structures are:
1. Uncrating
2. Hauling to location within 100′ of loading dock
3. Setting structures
4. Leveling
5. Aligning
6. Bolting together structure frames
7. Bolting horizontal bus bars

Labor-hours do not include:
1. Equipment pad
2. Steel channels embedded or grouted in concrete
3. Pull boxes
4. Special knockouts
5. Main switchboard section
6. Transition section
7. Instrumentation
8. External control wiring
9. Conduit or wire

Material for starters include:
1. Circuit breaker or fused disconnect
2. Magnetic motor starter
3. Control transformer
4. Control fuse and fuse block

Labor-hours for starters include:
1. Handling
2. Installing starter within structure
3. Internal wiring connections
4. Lacing within enclosure
5. Testing
6. Phasing

Job Conditions: Productivity is based on new construction. The motor control location is assumed to be on first floor in an unobstructed area. The material staging area is within 100′ of the final location.

Note: Additional labor-hours must be added if M.C.C. is to be installed on floors other than the first, or if rigging is required.

R262419-65   Motor Starters and Controls

Motor starters should be listed on the same "quantity sheet" as motors and motor connections. Identify each starter by:

1. Size
2. Voltage
3. Type

Example:

  A. FVNR, 480V, 2HP, size 00
  B. FVNR, 480V, 5HP, size 0, combination type
  C. FVR, 480V, size 2

The NEMA type of enclosure should also be identified.

Included in the labor-hours are:

1. Unloading, uncrating and handling of starters up to 200' from loading area
2. Measuring and marking
3. Drilling (4) anchor type lead fasteners using a hammer drill
4. Mounting and leveling starter
5. Connecting wire or cable to line and load sides of starter (when already lugged)
6. Installation of (3) thermal type heaters
7. Testing

The following are not included unless specified in the unit price description.

1. Control transformer
2. Controls, either factory or field installed
3. Conduit and wire to or from starter
4. Plywood backboard
5. Cable terminations

The following material and labor were included for combination type starters:

1. Unloading, uncrating and handling of starter up to 200' of loading dock
2. Measuring and marking
3. Drilling (4) anchor type lead fasteners using a hammer drill
4. Mounting and leveling
5. Connecting prepared cable conductors
6. Installation of (3) dual element cartridge type fuses
7. Installation of (3) thermal type heaters
8. Test for rotation

MOTOR CONTROLS

When pricing motor controls make sure you consider the type of control system being utilized. If the controls are factory installed and located in the enclosure itself, then you would add to your starter price the items 5200 through 5800 in the Unit Price section. If control voltage is different from line voltage, add the material and labor cost of a control transformer.

For external control of starters include the following items:

1. Raceways
2. Wire & terminations
3. Control enclosures
4. Fittings
5. Push button stations
6. Indicators

Job Conditions: Productivity is based on new construction to a height of 10'.

Material staging area is assumed to be within 100' of work being performed.

Fig. R262419-81

R262419-80   Distribution Section

After "taking off" the switchboard section, include on the same "quantity sheet" the distribution section; identify by:

1. Voltage
2. Ampere rating
3. Type

Example:

(Fig. R262419-81) (B) Distribution Section

Included in the labor costs of the distribution section are:

1. Uncrating
2. Hauling to 200' of loading dock
3. Setting of distribution panel
4. Leveling & shimming

5. Anchoring of equipment to pad or floor
6. Bolting of horizontal bus bars between
7. Testing of equipment

Not included in the Distribution section are:

1. Breakers (C)
2. Equipment pads
3. Steel channels embedded or grouted in concrete
4. Pull boxes
5. Special knockouts
6. Transition section
7. Conduit or wire

R262419-82   Feeder Section

List quantities on the same sheet as the distribution section. Identify breakers by: (Fig. R262419-81)

1. Frame type
2. Number of poles
3. Ampere rating

Installation includes:

1. Handling

2. Placing breakers in distribution panel
3. Preparing wire or cable
4. Lacing wire
5. Marking each phase with colored tape
6. Marking panel legend
7. Testing
8. Balancing

R262419-84 Switchgear

It is recommended that "switchgear" or those items contained in the following sections be quoted from equipment manufacturers as a package price.

> Switchboard instruments
>
> Switchboard distribution sections
>
> Switchboard feeder sections
>
> Switchboard Service Disc.
>
> Switchboard In-plant Dist.

Included in these sections are the most common types and sizes of factory assembled equipment.

The recommended procedure for low voltage switchgear would be to price (Fig. R262419-81) (A) Main Switchboard

Identify by:

1. Voltage
2. Amperage
3. Type

Example:

1. 120/208V, 4-wire, 600A, nonfused
2. 277/480V, 4-wire, 600A, nonfused
3. 120/208V, 4-wire, 400A w/fused switch & CT compartment
4. 277/480V, 4-wire, 400A, w/fused switch & CT compartment
5. 120/208V, 4-wire, 800A, w/pressure switch & CT compartment
6. 277/480V, 4-wire, 800A, w/molded CB & CT compartment

Included in the labor costs for switchboards are:

1. Uncrating
2. Hauling to 200' of loading dock
3. Setting equipment
4. Leveling and shimming
5. Anchoring
6. Cable identification
7. Testing of equipment

Not included in the switchboard price are:

1. Rigging
2. Equipment pads
3. Steel channels embedded or grouted in concrete
4. Special knockouts
5. Transition or Auxiliary sections
6. Instrumentation
7. External control
8. Conduit and wire
9. Conductor terminations

For customer support on your Electrical Costs with RSMeans Data, call 800.448.8182.

623

Fig. R16450-101

R262513-10 Aluminum Bus Duct

When taking off bus duct identify the system as either:
1. Aluminum
2. Copper

List straight lengths by type and size
- A. Feeder — 800 A
- B. Plug in — 800 A

Do not measure through fittings as you would on conduit, since there is no allowance for scrap in bus duct systems.

If upon taking off linear foot quantities of bus duct you find your quantities are not divisible evenly by 10 ft., then the remainder must be priced as a special item and quoted from the manufacturer. Do not use the bare material cost per L.F. for these special items. You can, however, safely use the bare labor cost per L.F. for the entire length.

Identify fittings by type and ampere rating.

Example:

C. Switchboard stub, 800 A

D. Elbows, 800 A

E. End box, 800 A

F. Cable tap box, 800 A

G. Tee Fittings, 800 A

H. Hangers

Plug-in Units – List separately plug-in units and identify by type and ampere rating

I. Plug-in switches 600 Volt 3 phase 60 A

J. Plug-in molded case C.B. 60 A

K. Combination starter FVNR NEMA 1

L. Combination contactor & fused switch NEMA 1

M. Combination fusible switch & lighting control 60 A

Labor-hours for feeder and plug-in sections include:
1. Unloading and uncrating
2. Hauling up to 200 ft. from loading dock
3. Measuring and marking
4. Installing hangers
5. Hanging and bolting sections
6. Aligning and leveling
7. Testing

Labor-hours do not include:
1. Modifications to existing structure for hanger supports
2. Threaded rod in excess of 2 ft.
3. Welding
4. Penetrations through walls
5. Staging rental

Deduct the following percentages from labor only:

150 ft. to 250 ft. —	10%
251 ft. to 350 ft. —	15%
351 ft. to 500 ft. —	20%
Over 500 ft. —	25%

Deduct percentage only if runs are contained in the same area.

Example: If the job entails running 100 ft. in 5 different locations do not deduct 20%, but if the duct is being run in 1 area and the quantity is 500 ft. then you would deduct 20%.

Deduct only from straight lengths, not fittings or plug-in units.

R262513-10   Aluminum Bus Duct (cont.)

Add to labor for elevated installations:

10 ft. to 14.5 ft. high	10%
15 ft. to 20 ft. high	20%
21 ft. to 25 ft. high	25%
26 ft. to 30 ft. high	35%
31 ft. to 35 ft. high	40%
36 ft. to 40 ft. high	50%
Over 40 ft. high	55%

Bus Duct Fittings:

Labor-hours for fittings include:
1. Unloading and uncrating
2. Hauling up to 200 ft. from loading dock
3. Installing, fitting, and bolting all ends to in-place sections

Plug-in units include:
1. Unloading and uncrating
2. Hauling up to 200 ft. from loading dock
3. Installing plug-in into in-place duct
4. Connecting load wire to lugs
5. Marking wire
6. Checking phase rotation

Labor-hours for plug-ins do not include:
1. Conduit runs from plug-in
2. Wire from plug-in
3. Conduit termination

Economy of Scale – For large concentrations of plug-in units in the same area deduct the following percentages:

11	to	25	15%
26	to	50	20%
51	to	75	25%
76	to	100	30%
100	and	over	35%

Job Conditions: Productivities are based on new construction in an unobstructed first floor area to a height of 10' in an unobstructed area.

Material staging area is within 100 ft. of work being performed.

Add to the duct fittings and hangers:
Plug-in switches (fused)
Plug-in breakers
Combination starters
Combination contactors
Combination fusible switch and lighting control

Labor-hours for duct and fittings include:
1. Unloading and uncrating
2. Hauling up to 200 ft. from loading dock
3. Measuring and marking
4. Installing duct runs
5. Leveling
6. Sound testing

Labor-hours for plug-ins include:
1. Unloading and uncrating
2. Hauling up to 200 ft. from loading dock
3. Installing plug-ins
4. Preparing wire
5. Wire connections and marking

R262716-40 Standard Electrical Enclosure Types

NEMA Enclosures

Electrical enclosures serve two basic purposes: they protect people from accidental contact with enclosed electrical devices and connections, and they protect the enclosed devices and connections from specified external conditions. The National Electrical Manufacturers Association (NEMA) has established the following standards. Because these descriptions are not intended to be complete representations of NEMA listings, consultation of NEMA literature is advised for detailed information.

The following definitions and descriptions pertain to NONHAZARDOUS locations.

NEMA Type 1: General purpose enclosures intended for use indoors, primarily to prevent accidental contact of personnel with the enclosed equipment in areas that do not involve unusual conditions.

NEMA Type 2: Dripproof indoor enclosures intended to protect the enclosed equipment against dripping noncorrosive liquids and falling dirt.

NEMA Type 3: Dustproof, raintight and sleet-resistant (ice-resistant) enclosures intended for use outdoors to protect the enclosed equipment against wind-blown dust, rain, sleet, and external ice formation.

NEMA Type 3R: Rainproof and sleet-resistant (ice-resistant) enclosures intended for use outdoors to protect the enclosed equipment against rain. These enclosures are constructed so that the accumulation and melting of sleet (ice) will not damage the enclosure and its internal mechanisms.

NEMA Type 3S: Enclosures intended for outdoor use to provide limited protection against wind-blown dust, rain, and sleet (ice) and to allow operation of external mechanisms when ice-laden.

NEMA Type 4: Watertight and dust-tight enclosures intended for use indoors and out – to protect the enclosed equipment against splashing water, seepage of water, falling or hose-directed water, and severe external condensation.

NEMA Type 4X: Watertight, dust-tight, and corrosion-resistant indoor and outdoor enclosures featuring the same provisions as Type 4 enclosures, plus corrosion resistance.

NEMA Type 5: Indoor enclosures intended primarily to provide limited protection against dust and falling dirt.

NEMA Type 6: Enclosures intended for indoor and outdoor use – primarily to provide limited protection against the entry of water during occasional temporary submersion at a limited depth.

NEMA Type 6R: Enclosures intended for indoor and outdoor use – primarily to provide limited protection against the entry of water during prolonged submersion at a limited depth.

NEMA Type 11: Enclosures intended for indoor use – primarily to provide, by means of oil immersion, limited protection to enclosed equipment against the corrosive effects of liquids and gases.

NEMA Type 12: Dust-tight and driptight indoor enclosures intended for use indoors in industrial locations to protect the enclosed equipment against fibers, flyings, lint, dust, and dirt, as well as light splashing, seepage, dripping, and external condensation of noncorrosive liquids.

NEMA Type 13: Oil-tight and dust-tight indoor enclosures intended primarily to house pilot devices, such as limit switches, foot switches, push buttons, selector switches, and pilot lights, and to protect these devices against lint and dust, seepage, external condensation, and sprayed water, oil, and noncorrosive coolant.

The following definitions and descriptions pertain to HAZARDOUS, or CLASSIFIED, locations:

NEMA Type 7: Enclosures intended to use in indoor locations classified as Class 1, Groups A, B, C, or D, as defined in the National Electrical Code.

NEMA Type 9: Enclosures intended for use in indoor locations classified as Class 2, Groups E, F, or G, as defined in the National Electrical Code.

R262726-90 Wiring Devices

Wiring devices should be priced on a separate takeoff form which includes boxes, covers, conduit and wire.

Labor-hours for devices include:
1. Stripping of wire
2. Attaching wire to device using terminators on the device itself, lugs, set screws etc.
3. Mounting of device in box

Labor-hours do not include:
1. Conduit
2. Wire
3. Boxes
4. Plates

Economy of Scale – for large concentrations of devices in the same area deduct the following percentages from labor-hours:

1	to	10	0%
11	to	25	20%
26	to	50	25%
51	to	100	30%
	over	100	35%

R262726-90 Wiring Devices (cont.)

NEMA No.	15 R	20 R	30 R	50 R	60 R
1 125V 2 Pole, 2 Wire					
2 250V 2 Pole, 2 Wire					
5 125V 2 Pole, 3 Wire					
6 250V 2 Pole, 3 Wire					
7 277V, AC 2 Pole, 3 Wire					
10 125/250V 3 Pole, 3 Wire					
11 3 Phase 250V 3 Pole, 3 Wire					
14 125/250V 3 Pole, 4 Wire					
15 3 Phase 250V 3 Pole, 4 Wire					
18 3 Phase 208Y/120V 4 Pole, 4 Wire					

For customer support on your Electrical Costs with RSMeans Data, call 800.448.8182.

627

R262726-90   Wiring Devices (cont.)

NEMA No.	15 R	20 R	30 R	NEMA No.	15 R	20 R	30 R
L 1 125V 2 Pole, 2 Wire				**L 13** 3 Phase 600V 3 Pole, 3 Wire			
L 2 250V 2 Pole, 2 Wire		 15A		**L 14** 125/250V 3 Pole, 4 Wire			
L 5 125 V 2 Pole, 3 Wire				**L 15** 3 Phase 250 V 3 Pole, 4 Wire			
L 6 250 V 2 Pole, 3 Wire				**L 16** 3 Phase 480V 3 Pole, 4 Wire			
L 7 227 V, AC 2 Pole, 3 Wire				**L 17** 3 Phase 600 V 3 Pole, 4 Wire			
L 8 480 V 2 Pole, 3 Wire				**L 18** 3 Phase 208Y/120V 4 Pole, 4 Wire			
L 9 600 V 2 Pole, 3 Wire				**L 19** 3 Phase 480Y/277V 4 Pole, 4 Wire			
L 10 125/250V 3 Pole, 3 Wire				**L 20** 3 Phase 600Y/347V 4 Pole, 4 Wire			
L 11 3 Phase 250 V 3 Pole, 3 Wire				**L 21** 3 Phase 208Y/120V 4 Pole, 5 Wire			
L 12 3 Phase 480 V 3 Pole, 3 Wire				**L 22** 3 Phase 480Y/277V 4 Pole, 5 Wire			
				L 23 3 Phase 600Y/347V 4 Pole, 5 Wire			

R262816-80   Safety Switches

List each safety switch by type, ampere rating, voltage, single or three phase, fused or nonfused.

Example:

  A. General duty, 240V, 3-pole, fused
  B. Heavy Duty, 600V, 3-pole, nonfused
  C. Heavy Duty, 240V, 2-pole, fused
  D. Heavy Duty, 600V, 3-pole, fused

Also include NEMA enclosure type and identify as:

1. Indoor
2. Weatherproof
3. Explosionproof

Installation of safety switches includes:

1. Unloading and uncrating
2. Handling disconnects up to 200' from loading dock
3. Measuring and marking location
4. Drilling (4) anchor type lead fasteners using a hammer drill
5. Mounting and leveling safety switch
6. Installing (3) fuses
7. Phasing and tagging line and load wires

Price does not include:

1. Modifications to enclosure
2. Plywood backboard
3. Conduit or wire
4. Fuses
5. Termination of wires

Job Conditions: Productivities are based on new construction to an installed height of 6' above finished floor.

Material staging area is assumed to be within 100' of work in progress.

For customer support on your Electrical Costs with RSMeans Data, call 800.448.8182.

629

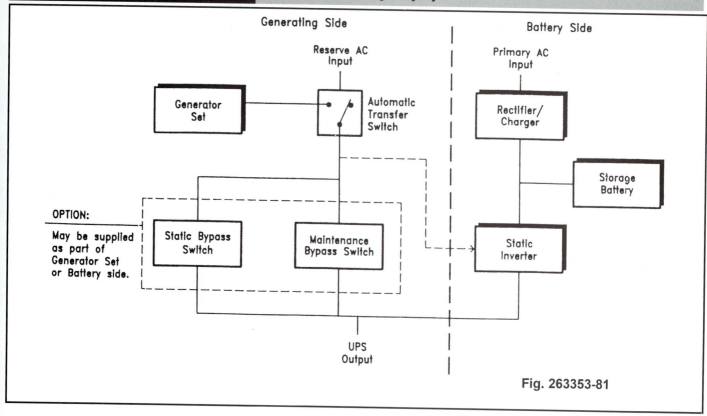

Fig. 263353-81

R263353-80 Uninterruptible Power Supply Systems

General: Uninterruptible Power Supply (UPS) Systems are used to provide power for legally required standby systems. They are also designed to protect computers and provide optional additional coverage for any or all loads in a facility. Figure R263353-81 shows a typical configuration for a UPS System with a generator set.

Cost Determination:

It is recommended that UPS System material costs be obtained from equipment manufacturers as a package. Installation costs should be obtained from a vendor or contractor.

The recommended procedure for pricing UPS Systems would be to identify:
1. Frequency – by Hertz (Hz.)
   For example: 50, 60, and 400/415 Hz.
2. Apparent power and real power
   For example: 200 kVA/170 kW
3. Input/Output voltage
   For example: 120V, 208V or 240V

4. Phase – either single or three-phase
5. Options and accessories – For extended run time more batteries would be required. Other accessories include battery cabinets, battery racks, remote control panel, power distribution unit (PDU), and warranty enhancement plans.

Note:
1. Larger systems can be configured by paralleling two or more standard small size single modules.
   For example: two 15 kVA modules combined together and configured to 30 kVA.
2. Maximum input current during battery recharge is typically 15% higher than normal current.
3. UPS Systems weights vary depending on options purchased and power rating in kVA.

R263413-30 HP Value/Cost Determination

General: Motors can be powered by any of the seven systems shown in Figure R262213-11 through Figure R262213-17 provided the motor voltage characteristics are compatible with the power system characteristics.

Cost Determination:

Motor Amperes for the various size H.P. and voltage are listed in Table R263413-31. To find the amperes, locate the required H.P. rating and locate the amperes under the appropriate circuit characteristics.

For example:

A. 100 H.P., 3 phase, 460 volt motor = 124 amperes (Table R263413-31)

B. 10 H.P., 3 phase, 200 volt motor = 32.2 amperes (Table R263413-31)

Motor Wire Size: After the amperes are found in Table R263413-31 the amperes must be increased 25% to compensate for power losses. Next refer to Table R260519-92. Find the appropriate insulation column for copper or aluminum wire to determine the proper wire size.

For example:

A. 100 H.P., 3 phase, 460 volt motor has an ampere value of 124 amperes from Table R263413-31

B. 124A x 1.25 = 155 amperes

C. Refer to Table R260519-92 for THW or THWN wire insulations to find the proper wire size. For a 155 ampere load using copper wire a size 2/0 wire is needed.

D. For the 3 phase motor three wires of 2/0 size are required.

Conduit Size: To obtain the proper conduit size for the wires and type of insulation used, refer to Table R260533-22.

For example: For the 100 H.P., 460V, 3 phase motor, it was determined that three 2/0 wires are required. Assuming THWN insulated copper wire, use Table R260533-22 to determine that three 2/0 wires require 1-1/2″ conduit.

Material Cost of the conduit and wire system depends on:
1. Wire size required
2. Copper or aluminum wire
3. Wire insulation type selected
4. Steel or plastic conduit
5. Type of conduit raceway selected.

Labor Cost of the conduit and wire system depends on:
1. Type and size of conduit
2. Type and size of wires installed
3. Location and height of installation in building or depth of trench
4. Support system for conduit.

Magnetic starters, switches, and motor connection:

To complete the cost picture from H.P. to Costs additional items must be added to the cost of the conduit and wire system to arrive at a total cost.
1. Assembly Table D5020-160 Magnetic Starters Installed Cost lists the various size starters for single phase and three phase motors.
2. Assembly Table D5020-165 Heavy Duty Safety Switches Installed Cost lists safety switches required at the beginning of a motor circuit and also one required in the vicinity of the motor location.
3. Assembly Table D5020-170 Motor Connection lists the various costs for single and three phase motors.

Worksheet to obtain total motor wiring costs:

It is assumed that the motors or motor driven equipment are furnished and installed under other sections for this estimate and the following work is done under this section:
1. Conduit
2. Wire (add 10% for additional wire beyond conduit ends for connections to switches, boxes, starters, etc.)
3. Starters
4. Safety switches
5. Motor connections

R263413-31   Ampere Values Determined by Horsepower, Voltage and Phase Values

H.P.	Amperes							
	Single Phase			Three Phase				
	115V	208V	230V	200V	208V	230V	460V	575V
1/6	4.4A	2.4A	2.2A					
1/4	5.8	3.2	2.9					
1/3	7.2	4.0	3.6					
1/2	9.8	5.4	4.9	2.5A	2.4A	2.2A	1.1A	0.9A
3/4	13.8	7.6	6.9	3.7	3.5	3.2	1.6	1.3
1	16	8.8	8	4.8	4.6	4.2	2.1	1.7
1-1/2	20	11	10	6.9	6.6	6.0	3.0	2.4
2	24	13.2	12	7.8	7.5	6.8	3.4	2.7
3	34	18.7	17	11.0	10.6	9.6	4.8	3.9
5	56	30.8	28	17.5	16.7	15.2	7.6	6.1
7-1/2	80	44	40	25.3	24.2	22	11	9
10	100	55	50	32.2	30.8	28	14	11
15				48.3	46.2	42	21	17
20				62.1	59.4	54	27	22
25				78.2	74.8	68	34	27
30				92	88	80	40	32
40				120	114	104	52	41
50				150	143	130	65	52
60				177	169	154	77	62
75				221	211	192	96	77
100				285	273	248	124	99
125				359	343	312	156	125
150				414	396	360	180	144
200				552	528	480	240	192
250							302	242
300							361	289
350							414	336
400							477	382

632

For customer support on your Electrical Costs with RSMeans Data, call 800.448.8182.

Figure R263413-32

Worksheet for Motor Circuits					
				Cost	
Item	Type	Size	Quantity	Unit	Total
Wire					
Conduit					
Switch					
Starter					
Switch					
Motor Connection					
Other					
Total Cost					

R263413-33 Maximum Horsepower for Starter Size by Voltage

Starter	Maximum HP (3φ)			
Size	208V	240V	480V	600V
00	1½	1½	2	2
0	3	3	5	5
1	7½	7½	10	10
2	10	15	25	25
3	25	30	50	50
4	40	50	100	100
5		100	200	200
6		200	300	300
7		300	600	600
8		450	900	900
8L		700	1500	1500

For customer support on your Electrical Costs with RSMeans Data, call 800.448.8182.

633

R263623-60   Automatic Transfer Switches

When taking off automatic transfer switches identify by voltage, amperage and number of poles.

Example: Automatic transfer switch, 480V, 3 phase, 30A, NEMA 1 enclosure

Labor-hours for transfer switches include:

1. Unloading, uncrating and handling switches up to 200′ from loading dock
2. Measuring and marking
3. Drilling (4) lead type anchors using a hammer drill
4. Mounting and leveling
5. Circuit identification
6. Testing and load balancing

Labor-hours do not include:

1. Modifications in enclosure
2. Structural supports
3. Additional lugs
4. Plywood backboard
5. Painting or lettering
6. Conduit runs to or from transfer switch
7. Wire
8. Termination of wires

R265113-40   Interior Lighting Fixtures

When taking off interior lighting fixtures, it is advisable to set up your quantity worksheet to conform to the lighting schedule as it appears on the print. Include the alpha-numeric code plus the symbol on your worksheet.

Take off a particular section or floor of the building and count each type of fixture before going on to another type. It would also be advantageous to include on the same worksheet the pipe, wire, fittings and circuit number associated with each type of lighting fixture. This will help you identify the costs associated with any particular lighting system and in turn make material purchases more specific as to when and how much to order under the classification of lighting.

By taking off lighting first you can get a complete "walk through" of the job. This will become helpful when doing other phases of the project.

Materials for a recessed fixture include:

1. Fixture
2. Lamps
3. 6′ of jack chain
4. (2) S hooks
5. (2) Wire nuts

Labor for interior recessed fixtures include:

1. Unloading by hand
2. Hauling by hand to an area up to 200′ from loading dock
3. Uncrating
4. Layout
5. Installing fixture
6. Attaching jack chain & S hooks
7. Connecting circuit power
8. Reassembling fixture
9. Installing lamps
10. Testing

Material for surface mounted fixtures include:

1. Fixture
2. Lamps
3. Either (4) lead type anchors, (4) toggle bolts, or (4) ceiling grid clips
4. (2) Wire nuts

Material for pendent mounted fixtures include:

1. Fixture
2. Lamps
3. (2) Wire nuts
4. Rigid pendents as required by type of fixtures

5. Canopies as required by type of fixture

Labor hours include the following for both surface and pendent fixtures:

1. Unloading by hand
2. Hauling by hand to an area up to 200′ from loading dock
3. Uncrating
4. Layout and marking
5. Drilling (4) holes for either lead anchors or toggle bolts using a hammer drill
6. Installing fixture
7. Leveling fixture
8. Connecting circuit power
9. Installing lamps
10. Testing

Labor for surface or pendent fixtures does not include:

1. Conduit
2. Boxes or covers
3. Connectors
4. Fixture whips
5. Special support
6. Switching
7. Wire

Economy of Scale: For large concentrations of lighting fixtures in the same area deduct the following percentages from labor:

25	to	50	fixtures	15%
51	to	75	fixtures	20%
76	to	100	fixtures	25%
101 and over				30%

Job Conditions: Productivity is based on new construction in an unobstructed first floor location, to 10′ high.

Material staging is assumed to be within 100′ of work being performed.

Add the following percentages to labor for elevated installations:

10′	to	14.5′	high	10%
15′	to	20′	high	20%
21′	to	25′	high	25%
26′	to	30′	high	35%
31′	to	35′	high	40%
36′	to	40′	high	50%
41′ and over				55%

R265723-10 For Other than Regular Cool White (CW) Lamps

	Multiply Material Costs as Follows:				
Regular Lamps	Cool white deluxe (CWX)	x 1.35	Energy Saving Lamps	Cool white (CW/ES)	x 1.35
	Warm white deluxe (WWX)	x 1.35		Cool white deluxe (CWX/ES)	x 1.65
	Warm white (WW)	x 1.30		Warm white (WW/ES)	x 1.55
	Natural (N)	x 2.05		Warm white deluxe (WWX/ES)	x 1.65

R271323-40 Fiber Optics

Fiber optic systems use optical fiber such as plastic, glass, or fused silica, a transparent material, to transmit radiant power (i.e., light) for control, communication, and signaling applications. The types of fiber optic cables can be nonconductive, conductive, or composite. The composite cables contain fiber optics and current-carrying electrical conductors. The configuration for one of the fiber optic systems is as follows:

The transceiver module acts as transmitting and receiving equipment in a common house, which converts electrical energy to light energy or vice versa.

Pricing the fiber optic system is not an easy task. The performance of the whole system will affect the cost significantly. New specialized tools and techniques decrease the installing cost tremendously. In the fiber optic section of Means Electrical Cost Data, a benchmark for labor-hours and material costs is set up so that users can adjust their costs according to unique project conditions.

Units for Measure: Fiber optic cable is measured in hundred linear feet (C.L.F.) or industry units of measure - meter (m) or kilometer (km). The connectors are counted as units (EA.)

Material Units: Generally, the material costs include only the cable. All the accessories shall be priced separately.

Labor Units: The following procedures are generally included for the installation of fiber optic cables:

- Receiving
- Material handling
- Setting up pulling equipment
- Measuring and cutting cable
- Pulling cable

These additional items are listed and extended: Terminations

Takeoff Procedure: Cable should be taken off by type, size, number of fibers, and number of terminations required. List the lengths of each type of cable on the takeoff sheets. Total and add 10% for waste. Transfer the figures to a cost analysis sheet and extend.

R271513-75 High Performance Cable

There are several categories used to describe high performance cable. The following information includes a description of categories CAT 3, 5, 5e, 6, and 7, and details classifications of frequency and specific standards. The category standards have evolved under the sponsorship of organizations such as the Telecommunication Industry Association (TIA), the Electronic Industries Alliance (EIA), the American National Standards Institute (ANSI), the International Organization for Standardization (ISO), and the International Electrotechnical Commission (IEC), all of which have catered to the increasing complexities of modern network technology. For network cabling, users must comply with national or international standards. A breakdown of these categories is as follows:

Category 3: Designed to handle frequencies up to 16 MHz.

Category 5: (TIA/EIA 568A) Designed to handle frequencies up to 100 MHz.

Category 5e: Additional transmission performance to exceed Category 5.

Category 6: Development by TIA and other international groups to handle frequencies of 250 MHz.

Category 7: Development to handle a frequency range up to 600 MHz.

R312316-45   Excavating Equipment

The table below lists theoretical hourly production in C.Y./hr. bank measure for some typical excavation equipment. Figures assume 50 minute hours, 83% job efficiency, 100% operator efficiency, 90° swing and properly sized hauling units, which must be modified for adverse digging and loading conditions. Actual production costs in the front of the data set average about 50% of the theoretical values listed here.

Equipment	Soil Type	B.C.Y. Weight	% Swell	1 C.Y.	1-1/2 C.Y.	2 C.Y.	2-1/2 C.Y.	3 C.Y.	3-1/2 C.Y.	4 C.Y.
Hydraulic Excavator "Backhoe" 15' Deep Cut	Moist loam, sandy clay	3400 lb.	40%	165	195	200	275	330	385	440
	Sand and gravel	3100	18	140	170	225	240	285	330	380
	Common earth	2800	30	150	180	230	250	300	350	400
	Clay, hard, dense	3000	33	120	140	190	200	240	260	320
Power Shovel Optimum Cut (Ft.)	Moist loam, sandy clay	3400	40	170 (6.0)	245 (7.0)	295 (7.8)	335 (8.4)	385 (8.8)	435 (9.1)	475 (9.4)
	Sand and gravel	3100	18	165 (6.0)	225 (7.0)	275 (7.8)	325 (8.4)	375 (8.8)	420 (9.1)	460 (9.4)
	Common earth	2800	30	145 (7.8)	200 (9.2)	250 (10.2)	295 (11.2)	335 (12.1)	375 (13.0)	425 (13.8)
	Clay, hard, dense	3000	33	120 (9.0)	175 (10.7)	220 (12.2)	255 (13.3)	300 (14.2)	335 (15.1)	375 (16.0)
Drag Line Optimum Cut (Ft.)	Moist loam, sandy clay	3400	40	130 (6.6)	180 (7.4)	220 (8.0)	250 (8.5)	290 (9.0)	325 (9.5)	385 (10.0)
	Sand and gravel	3100	18	130 (6.6)	175 (7.4)	210 (8.0)	245 (8.5)	280 (9.0)	315 (9.5)	375 (10.0)
	Common earth	2800	30	110 (8.0)	160 (9.0)	190 (9.9)	220 (10.5)	250 (11.0)	280 (11.5)	310 (12.0)
	Clay, hard, dense	3000	33	90 (9.3)	130 (10.7)	160 (11.8)	190 (12.3)	225 (12.8)	250 (13.3)	280 (12.0)

Equipment	Soil Type	B.C.Y. Weight	% Swell	Wheel Loaders				Track Loaders		
				3 C.Y.	4 C.Y.	6 C.Y.	8 C.Y.	2-1/4 C.Y.	3 C.Y.	4 C.Y.
Loading Tractors	Moist loam, sandy clay	3400	40	260	340	510	690	135	180	250
	Sand and gravel	3100	18	245	320	480	650	130	170	235
	Common earth	2800	30	230	300	460	620	120	155	220
	Clay, hard, dense	3000	33	200	270	415	560	110	145	200
	Rock, well-blasted	4000	50	180	245	380	520	100	130	180

For customer support on your Electrical Costs with RSMeans Data, call 800.448.8182.

637

R337116-60 Average Transmission Line Material Requirements (Per Mile)

Conductor: 397,500 – Cir. Mil., 26/7–ACSR

Terrain:		Flat				Rolling				Mountain			
Item		69kV	161kV	161kV	500kV	69kV	161kV	161kV	500kV	69kV	161kV	161kV	500kV
Pole Type	Unit	Wood	Wood	Steel	Steel	Wood	Wood	Steel	Steel	Wood	Wood	Steel	Steel
Structures	Ea.	12[1]				9[2]				7[2]			
Poles	Ea.	12				18				14			
Crossarms	Ea.	24[3]				9[4]				7[4]			
Conductor	Ft.	15,990				15,990				15,990			
Insulators[5]	Ea.	180				135				105			
Ground Wire	Ft.	5,330				10,660				10,660			

Conductor: 636,000 – Cir. Mil., 26/7–ACSR

Item	Unit	69kV Wood	161kV Wood	161kV Steel	500kV Steel	69kV Wood	161kV Wood	161kV Steel	500kV Steel	69kV Wood	161kV Wood	161kV Steel	500kV Steel
Structures	Ea.	13[1]	11[6]			10[2]	9[2]	6[9]		8[2]	8[2]	6[9]	
Excavation	C.Y.	–	–			–	–	120		–	–	120	
Concrete	C.Y.	–	–			–	–	10		–	–	10	
Steel Towers	Tons	–	–			–	–	32		–	–	32	
Poles	Ea.	13	11			20	18	–		16	16	–	
Crossarms	Ea.	26[3]	33[7]			10[4]	9[8]	–		8[4]	8[8]	–	
Conductor	Ft.	15,990	15,990			15,990	15,990	15,990		15,990	15,990	15,990	
Insulators[5]	Ea.	195	165			150	297	297		120	264	297	
Ground Wire	Ft.	5330	5330			10,660	10,660	10,660		10,660	10,660	10,660	

Conductor: 954,000 – Cir. Mil., 45/7–ACSR

Item	Unit	69kV Wood	161kV Wood	161kV Steel	500kV Steel	69kV Wood	161kV Wood	161kV Steel	500kV Steel	69kV Wood	161kV Wood	161kV Steel	500kV Steel
Structures	Ea.	14[1]	12[6]		4[11]	10[2]	9[2]	6[9]	4[11]	8[2]	8[2]	6[9]	4[13]
Excavation	C.Y.	–	–		200	–	–	125	214	–	–	125	233
Concrete	C.Y.	–	–		20	–	–	10	21	–	–	10	21
Steel Towers	Tons	–	–		57	–	–	33	57	–	–	33	63
Poles	Ea.	14	12		–	20	18	–	–	16	16	–	–
Crossarms	Ea.	28[10]	36[7]		–	10[4]	9[8]	–	–	8[4]	8[8]	–	–
Conductor	Ft.	15,990	15,990		47,970[12]	15,990	15,990	15,990	47,970[12]	15,990	15,990	15,990	47,970[12]
Insulators[5]	Ea.	210	180		288	150	297	297	288	120	264	297	576
Ground Wire	Ft.	5330	5330		10,660	10,660	10,660	10,660	10,660	10,660	10,660	10,660	10,660

Conductor: 1,351,500 – Cir. Mil., 45/7–ACSR

Item	Unit	69kV Wood	161kV Wood	161kV Steel	500kV Steel	69kV Wood	161kV Wood	161kV Steel	500kV Steel	69kV Wood	161kV Wood	161kV Steel	500kV Steel
Structures	Ea.			8[14]				8[14]					
Excavation	C.Y.			220				220					
Concrete	C.Y.			28				28					
Steel Towers	Ton			46				46					
Conductor	Ft.			31,680[15]				31,680[15]					
Insulators	Ea.			528				528					
Ground Wire	Ft.			10,660				10,660					

1. Single pole two-arm suspension type construction
2. Two-pole wood H-frame construction
3. 4¾" x 5¾" x 8' and 4¾" x 5¾" x 10' wood crossarms
4. 6" x 8" x 26'-0" wood crossarm
5. 5¾" x 10" disc insulator
6. Single pole construction with 3 fiberglass crossarms (5 fog-type insulators per phase)
7. 7'-0" fiberglass crossarms
8. 6" x 10" x 35'-0" wood crossarm
9. Laced steel tower, single circuit construction
10. 5" x 7" x 8' and 5" x 7" x 10' wood crossarms
11. Laced steel tower, single circuit 500-kV construction
12. Bundled conductor (3 sub-conductors per phase)
13. Laced steel tower, single circuit restrained phases (500-kV)
14. Laced steel tower, double circuit construction
15. Both sides of double circuit strung

Note: To allow for sagging, a mile (5280 Ft.) of transmission line uses 5330 Ft. of conductor per wire (called a wire mile).

R337119-30 Concrete for Conduit Encasement

The table below lists C.Y. of concrete for 100 L.F. of trench. Conduits separation center to center should meet 7.5" (N.E.C.).

Number of Conduits	1	2	3	4	6	8	9	Number of Conduits
Trench Dimension	11.5" x 11.5"	11.5" x 19"	11.5" x 27"	19" x 19"	19" x 27"	19" x 38"	27" x 27"	Trench Dimension
Conduit Diameter 2.0"	3.29	5.39	7.64	8.83	12.51	17.66	17.72	Conduit Diameter 2.0"
2.5"	3.23	5.29	7.49	8.62	12.19	17.23	17.25	2.5"
3.0"	3.15	5.13	7.24	8.29	11.71	16.59	16.52	3.0"
3.5"	3.08	4.97	7.02	7.99	11.26	15.98	15.84	3.5"
4.0"	2.99	4.80	6.76	7.65	10.74	15.30	15.07	4.0"
5.0"	2.78	4.37	6.11	6.78	9.44	13.57	13.12	5.0"
6.0"	2.52	3.84	5.33	5.74	7.87	11.48	10.77	6.0"
	▪	▪▪	▪▪▪	▪▪/▪▪	▪▪▪/▪▪▪	▪▪▪▪/▪▪▪▪	▪▪▪/▪▪▪/▪▪▪	

For customer support on your Electrical Costs with RSMeans Data, call 800.448.8182.

639

Change Orders

Change Order Considerations

A change order is a written document usually prepared by the design professional and signed by the owner, the architect/engineer, and the contractor. A change order states the agreement of the parties to: an addition, deletion, or revision in the work; an adjustment in the contract sum, if any; or an adjustment in the contract time, if any. Change orders, or "extras", in the construction process occur after execution of the construction contract and impact architects/ engineers, contractors, and owners.

Change orders that are properly recognized and managed can ensure orderly, professional, and profitable progress for everyone involved in the project. There are many causes for change orders and change order requests. In all cases, change orders or change order requests should be addressed promptly and in a precise and prescribed manner. The following paragraphs include information regarding change order pricing and procedures.

The Causes of Change Orders

Reasons for issuing change orders include:

- Unforeseen field conditions that require a change in the work
- Correction of design discrepancies, errors, or omissions in the contract documents
- Owner-requested changes, either by design criteria, scope of work, or project objectives
- Completion date changes for reasons unrelated to the construction process
- Changes in building code interpretations, or other public authority requirements that require a change in the work
- Changes in availability of existing or new materials and products

Procedures

Properly written contract documents must include the correct change order procedures for all parties—owners, design professionals, and contractors—to follow in order to avoid costly delays and litigation.

Being "in the right" is not always a sufficient or acceptable defense. The contract provisions requiring notification and documentation must be adhered to within a defined or reasonable time frame.

The appropriate method of handling change orders is by a written proposal and acceptance by all parties involved. Prior to starting work on a project, all parties should identify their authorized agents who may sign and accept change orders, as well as any limits placed on their authority.

Time may be a critical factor when the need for a change arises. For such cases, the contractor might be directed to proceed on a "time and materials" basis, rather than wait for all paperwork to be processed—a delay that could impede progress. In this situation, the contractor must still follow the prescribed change order procedures including, but not limited to, notification and documentation.

Lack of documentation can be very costly, especially if legal judgments are to be made, and if certain field personnel are no longer available. For time and material change orders, the contractor should keep accurate daily records of all labor and material allocated to the change.

Owners or awarding authorities who do considerable and continual building construction (such as the federal government) realize the inevitability of change orders for numerous reasons, both predictable and unpredictable. As a result, the federal government, the American Institute of Architects (AIA), the Engineers Joint Contract Documents Committee (EJCDC), and other contractor, legal, and technical organizations have developed standards and procedures to be followed by all parties to achieve contract continuance and timely completion, while being financially fair to all concerned.

Pricing Change Orders

When pricing change orders, regardless of their cause, the most significant factor is when the change occurs. The need for a change may be perceived in the field or requested by the architect/engineer *before* any of the actual installation has begun, or may evolve or appear *during* construction when the item of work in question is partially installed. In the latter cases, the original sequence of construction is disrupted, along with all contiguous and supporting systems. Change orders cause the greatest impact when they occur *after* the installation has been completed and must be uncovered, or even replaced. Post-completion changes may be caused by necessary design changes, product failure, or changes in the owner's requirements that are not discovered until the building or the systems begin to function.

Specified procedures of notification and record keeping must be adhered to and enforced regardless of the stage of construction: *before, during,* or *after* installation. Some bidding documents anticipate change orders by requiring that unit prices including overhead and profit percentages—for additional as well as deductible changes—be listed. Generally these unit prices do not fully take into account the ripple effect, or impact on other trades, and should be used for general guidance only.

When pricing change orders, it is important to classify the time frame in which the change occurs. There are two basic time frames for change orders: *pre-installation change orders,* which occur before the start of construction, and *post-installation change orders,* which involve reworking after the original installation. Change orders that occur between these stages may be priced according to the extent of work completed using a combination of techniques developed for pricing *pre-* and *post-installation* changes.

Factors To Consider When Pricing Change Orders

As an estimator begins to prepare a change order, the following questions should be reviewed to determine their impact on the final price.

General

- *Is the change order work* pre-installation *or* post-installation?

  Change order work costs vary according to how much of the installation has been completed. Once workers have the project scoped in their minds, even though they have not started, it can be difficult to refocus. Consequently they may spend more than the normal amount of time understanding the change. Also, modifications to work in place, such as trimming or refitting, usually take more time than was initially estimated. The greater the amount of work in place, the more reluctant workers are to change it. Psychologically they may resent the change and as a result the rework takes longer than normal. Post-installation change order estimates must include demolition of existing work as required to accomplish the change. If the work is performed at a later time, additional obstacles, such as building finishes, may be present which must be protected. Regardless of whether the change occurs

For customer support on your Electrical Costs with RSMeans Data, call 800.448.8182.

pre-installation or post-installation, attempt to isolate the identifiable factors and price them separately. For example, add shipping costs that may be required pre-installation or any demolition required post-installation. Then analyze the potential impact on productivity of psychological and/or learning curve factors and adjust the output rates accordingly. One approach is to break down the typical workday into segments and quantify the impact on each segment.

Change Order Installation Efficiency

The labor-hours expressed (for new construction) are based on average installation time, using an efficiency level. For change order situations, adjustments to this efficiency level should reflect the daily labor-hour allocation for that particular occurrence.

- *Will the change substantially delay the original completion date?*

A significant change in the project may cause the original completion date to be extended. The extended schedule may subject the contractor to new wage rates dictated by relevant labor contracts. Project supervision and other project overhead must also be extended beyond the original completion date. The schedule extension may also put installation into a new weather season. For example, underground piping scheduled for October installation was delayed until January. As a result, frost penetrated the trench area, thereby changing the degree of difficulty of the task. Changes and delays may have a ripple effect throughout the project. This effect must be analyzed and negotiated with the owner.

- *What is the net effect of a deduct change order?*

In most cases, change orders resulting in a deduction or credit reflect only bare costs. The contractor may retain the overhead and profit based on the original bid.

Materials

- *Will you have to pay more or less for the new material, required by the change order, than you paid for the original purchase?*

The same material prices or discounts will usually apply to materials purchased for change orders as new construction. In some instances, however, the contractor may forfeit the advantages of competitive pricing for change orders. Consider the following example:

A contractor purchased over $20,000 worth of fan coil units for an installation and obtained the maximum discount. Some time later it was determined the project required an additional matching unit. The contractor has to purchase this unit from the original supplier to ensure a match. The supplier at this time may not discount the unit because of the small quantity, and he is no longer in a competitive situation. The impact of quantity on purchase can add between 0% and 25% to material prices and/or subcontractor quotes.

- *If materials have been ordered or delivered to the job site, will they be subject to a cancellation charge or restocking fee?*

Check with the supplier to determine if ordered materials are subject to a cancellation charge. Delivered materials not used as a result of a change order may be subject to a restocking fee if returned to the supplier. Common restocking charges run between 20% and 40%. Also, delivery charges to return the goods to the supplier must be added.

Labor

- *How efficient is the existing crew at the actual installation?*

Is the same crew that performed the initial work going to do the change order? Possibly the change consists of the installation of a unit identical to one already installed; therefore, the change should take less time. Be sure to consider this potential productivity increase and modify the productivity rates accordingly.

- *If the crew size is increased, what impact will that have on supervision requirements?*

Under most bargaining agreements or management practices, there is a point at which a working foreman is replaced by a nonworking foreman. This replacement increases project overhead by adding a nonproductive worker. If additional workers are added to accelerate the project or to perform changes while maintaining the schedule, be sure to add additional supervision time if warranted. Calculate the hours involved and the additional cost directly if possible.

- *What are the other impacts of increased crew size?*

The larger the crew, the greater the potential for productivity to decrease. Some of the factors that cause this productivity loss are: overcrowding (producing restrictive conditions in the working space) and possibly a shortage of any special tools and equipment required. Such factors affect not only the crew working on the elements directly involved in the change order, but other crews whose movements may also be hampered. As the crew increases, check its basic composition for changes by the addition or deletion of apprentices or nonworking foreman, and quantify the potential effects of equipment shortages or other logistical factors.

- *As new crews, unfamiliar with the project, are brought onto the site, how long will it take them to become oriented to the project requirements?*

The orientation time for a new crew to become 100% effective varies with the site and type of project. Orientation is easiest at a new construction site and most difficult at existing, very restrictive renovation sites. The type of work also affects orientation time. When all elements of the work are exposed, such as concrete or masonry work, orientation is decreased. When the work is concealed or less visible, such as existing electrical systems, orientation takes longer. Usually orientation can be accomplished in one day or less. Costs for added orientation should be itemized and added to the total estimated cost.

- *How much actual production can be gained by working overtime?*

Short term overtime can be used effectively to accomplish more work in a day. However, as overtime is scheduled to run beyond several weeks, studies have shown marked decreases in output. The following chart shows the effect of long term overtime on worker efficiency. If the anticipated change requires extended overtime to keep the job on schedule, these factors can be used as a guide to predict the impact on time and cost. Add project overhead, particularly supervision, that may also be incurred.

Days per Week	Hours per Day	Production Efficiency					Payroll Cost Factors	
		1st Week	2nd Week	3rd Week	4th Week	Average 4 Weeks	@ 1-1/2 Times	@ 2 Times
5	8	100%	100%	100%	100%	100%	100%	100%
	9	100	100	95	90	96	1.056	1.111
	10	100	95	90	85	93	1.100	1.200
	11	95	90	75	65	81	1.136	1.273
	12	90	85	70	60	76	1.167	1.333
6	8	100	100	95	90	96	1.083	1.167
	9	100	95	90	85	93	1.130	1.259
	10	95	90	85	80	88	1.167	1.333
	11	95	85	70	65	79	1.197	1.394
	12	90	80	65	60	74	1.222	1.444
7	8	100	95	85	75	89	1.143	1.286
	9	95	90	80	70	84	1.183	1.365
	10	90	85	75	65	79	1.214	1.429
	11	85	80	65	60	73	1.240	1.481
	12	85	75	60	55	69	1.262	1.524

Effects of Overtime

Caution: Under many labor agreements, Sundays and holidays are paid at a higher premium than the normal overtime rate.

The use of long-term overtime is counterproductive on almost any construction job; that is, the longer the period of overtime, the lower the actual production rate. Numerous studies have been conducted, and while they have resulted in slightly different numbers, all reach the same conclusion. The figure above tabulates the effects of overtime work on efficiency.

As illustrated, there can be a difference between the *actual* payroll cost per hour and the *effective* cost per hour for overtime work. This is due to the reduced production efficiency with the increase in weekly hours beyond 40. This difference between actual and effective cost results from overtime work over a prolonged period. Short-term overtime work does not result in as great a reduction in efficiency and, in such cases, effective cost may not vary significantly from the actual payroll cost. As the total hours per week are increased on a regular basis, more time is lost due to fatigue, lowered morale, and an increased accident rate.

As an example, assume a project where workers are working 6 days a week, 10 hours per day. From the figure above (based on productivity studies), the average effective productive hours over a 4-week period are:

$$0.875 \times 60 = 52.5$$

Depending upon the locale and day of week, overtime hours may be paid at time and a half or double time. For time and a half, the overall (average) *actual* payroll cost (including regular and overtime hours) is determined as follows:

$$\frac{40 \text{ reg. hrs.} + (20 \text{ overtime hrs.} \times 1.5)}{60 \text{ hrs.}} = 1.167$$

Based on 60 hours, the payroll cost per hour will be 116.7% of the normal rate at 40 hours per week. However, because the effective production (efficiency) for 60 hours is reduced to the equivalent of 52.5 hours, the effective cost of overtime is calculated as follows:

For time and a half:

$$\frac{40 \text{ reg. hrs.} + (20 \text{ overtime hrs.} \times 1.5)}{52.5 \text{ hrs.}} = 1.33$$

The installed cost will be 133% of the normal rate (for labor).

Thus, when figuring overtime, the actual cost per unit of work will be higher than the apparent overtime payroll dollar increase, due to the reduced productivity of the longer work week. These efficiency calculations are true only for those cost factors determined by hours worked. Costs that are applied weekly or monthly, such as equipment rentals, will not be similarly affected.

Equipment

- *What equipment is required to complete the change order?*

Change orders may require extending the rental period of equipment already on the job site, or the addition of special equipment brought in to accomplish the change work. In either case, the additional rental charges and operator labor charges must be added.

Summary

The preceding considerations and others you deem appropriate should be analyzed and applied to a change order estimate. The impact of each should be quantified and listed on the estimate to form an audit trail.

Change orders that are properly identified, documented, and managed help to ensure the orderly, professional, and profitable progress of the work. They also minimize potential claims or disputes at the end of the project.

Back by customer demand!

You asked and we listened. For customer convenience and estimating ease, we have made the 2021 Project Costs available for download at **RSMeans.com/2021books**. You will also find sample estimates, an RSMeans data overview video, and a book registration form to receive quarterly data updates throughout 2021.

Estimating Tips

- The cost figures available in the download were derived from hundreds of projects contained in the RSMeans database of completed construction projects. They include the contractor's overhead and profit. The figures have been adjusted to January of the current year.

- These projects were located throughout the U.S. and reflect a tremendous variation in square foot (S.F.) costs. This is due to differences, not only in labor and material costs, but also in individual owners' requirements. For instance, a bank in a large city would have different features than one in a rural area. This is true of all the different types of buildings analyzed. Therefore, caution should be exercised when using these Project Costs. For example, for courthouses, costs in the database are local courthouse costs and will not apply to the larger, more elaborate federal courthouses.

- None of the figures "go with" any others. All individual cost items were computed and tabulated separately. Thus, the sum of the median figures for plumbing, HVAC, and electrical will not normally total up to the total mechanical and electrical costs arrived at by separate analysis and tabulation of the projects.

- Each building was analyzed as to total and component costs and percentages. The figures were arranged in ascending order with the results tabulated as shown. The 1/4 column shows that 25% of the projects had lower costs and 75% had higher. The 3/4 column shows that 75% of the projects had lower costs and 25% had higher. The median column shows that 50% of the projects had lower costs and 50% had higher.

- Project Costs are useful in the conceptual stage when no details are available. As soon as details become available in the project design, the square foot approach should be discontinued and the project should be priced as to its particular components. When more precision is required, or for estimating the replacement cost of specific buildings, the current edition of *Square Foot Costs with RSMeans data* should be used.

- In using the figures in this section, it is recommended that the median column be used for preliminary figures if no additional information is available. The median figures, when multiplied by the total city construction cost index figures (see City Cost Indexes) and then multiplied by the project size modifier at the end of this section, should present a fairly accurate base figure, which would then have to be adjusted in view of the estimator's experience, local economic conditions, code requirements, and the owner's particular requirements. There is no need to factor in the percentage figures, as these should remain constant from city to city.

- The editors of this data would greatly appreciate receiving cost figures on one or more of your recent projects, which would then be included in the averages for next year. All cost figures received will be kept confidential, except that they will be averaged with other similar projects to arrive at square foot cost figures for next year.

See the website above for details and the discount available for submitting one or more of your projects.

Same Data. Simplified.

Enjoy the convenience and efficiency of accessing your costs anywhere:

- **Skip the multiplier** by setting your location
- **Quickly search,** edit, favorite and share costs
- **Stay on top of price changes** with automatic updates

Discover more at rsmeans.com/online

50 17 00 | Project Costs

			UNIT	UNIT COSTS			% OF TOTAL		
				1/4	MEDIAN	3/4	1/4	MEDIAN	3/4
01	0000	**Auto Sales with Repair**	S.F.						
	0100	Architectural		107	120	130	58%	64%	67%
	0200	Plumbing		8.95	9.40	12.55	4.84%	5.20%	6.80%
	0300	Mechanical		12	16.10	17.75	6.40%	8.70%	10.15%
	0400	Electrical		18.45	23	28.50	9.05%	11.70%	15.90%
	0500	Total Project Costs		180	187	193			
02	0000	**Banking Institutions**	S.F.						
	0100	Architectural		161	198	241	59%	65%	69%
	0200	Plumbing		6.50	9.10	12.60	2.12%	3.39%	4.19%
	0300	Mechanical		12.95	17.90	21	4.41%	5.10%	10.75%
	0400	Electrical		31.50	38	58.50	10.45%	13.05%	15.90%
	0500	Total Project Costs		268	300	370			
03	0000	**Court House**	S.F.						
	0100	Architectural		85	167	167	54.50%	58.50%	58.50%
	0200	Plumbing		3.22	3.22	3.22	2.07%	2.07%	2.07%
	0300	Mechanical		20	20	20	12.95%	12.95%	12.95%
	0400	Electrical		26	26	26	16.60%	16.60%	16.60%
	0500	Total Project Costs		155	286	286			
04	0000	**Data Centers**	S.F.						
	0100	Architectural		192	192	192	68%	68%	68%
	0200	Plumbing		10.55	10.55	10.55	3.71%	3.71%	3.71%
	0300	Mechanical		27	27	27	9.45%	9.45%	9.45%
	0400	Electrical		25.50	25.50	25.50	9%	9%	9%
	0500	Total Project Costs		284	284	284			
05	0000	**Detention Centers**	S.F.						
	0100	Architectural		178	188	200	52%	53%	60.50%
	0200	Plumbing		18.80	23	27.50	5.15%	7.10%	7.25%
	0300	Mechanical		24	34	41	7.55%	9.50%	13.80%
	0400	Electrical		39	46.50	60.50	10.90%	14.85%	17.95%
	0500	Total Project Costs		300	320	375			
06	0000	**Fire Stations**	S.F.						
	0100	Architectural		101	130	190	46%	54.50%	61.50%
	0200	Plumbing		10.25	14.05	18.35	4.59%	5.60%	6.30%
	0300	Mechanical		15.10	22	29.50	6.05%	8.25%	10.20%
	0400	Electrical		23.50	30	40	10.75%	12.55%	14.95%
	0500	Total Project Costs		211	241	335			
07	0000	**Gymnasium**	S.F.						
	0100	Architectural		89	118	118	57%	64.50%	64.50%
	0200	Plumbing		2.19	7.15	7.15	1.58%	3.48%	3.48%
	0300	Mechanical		3.35	30	30	2.42%	14.65%	14.65%
	0400	Electrical		11	21.50	21.50	7.95%	10.35%	10.35%
	0500	Total Project Costs		139	206	206			
08	0000	**Hospitals**	S.F.						
	0100	Architectural		108	178	193	43%	47.50%	48%
	0200	Plumbing		7.95	15.15	33	6%	7.45%	7.65%
	0300	Mechanical		52.50	59	77	14.20%	17.95%	23.50%
	0400	Electrical		24	48	62	10.95%	13.75%	16.85%
	0500	Total Project Costs		253	375	405			
09	0000	**Industrial Buildings**	S.F.						
	0100	Architectural		59	116	235	52%	56%	82%
	0200	Plumbing		1.70	4.62	13.35	1.57%	2.11%	6.30%
	0300	Mechanical		4.86	9.25	44	4.77%	5.55%	14.80%
	0400	Electrical		7.40	8.45	70.50	7.85%	13.55%	16.20%
	0500	Total Project Costs		88	133	435			
10	0000	**Medical Clinics & Offices**	S.F.						
	0100	Architectural		90.50	122	163	48.50%	55.50%	62.50%
	0200	Plumbing		9	13.50	21.50	4.47%	6.65%	8.75%
	0300	Mechanical		14.60	22.50	41.50	7.80%	10.90%	16.05%
	0400	Electrical		19.90	24.50	38	8.90%	11.40%	14.05%
	0500	Total Project Costs		168	224	295			

		50 17 00 | Project Costs	UNIT	UNIT COSTS			% OF TOTAL			
				1/4	MEDIAN	3/4	1/4	MEDIAN	3/4	
11	0000	**Mixed Use**	S.F.							**11**
	0100	Architectural		92	130	198	45.50%	52.50%	61.50%	
	0200	Plumbing		6.25	11.45	12.15	3.31%	3.47%	4.18%	
	0300	Mechanical		15.25	25	46	4.68%	13.60%	17.05%	
	0400	Electrical		16.15	36	53.50	8.30%	11.40%	15.65%	
	0500	Total Project Costs		190	335	340				
12	0000	**Multi-Family Housing**	S.F.							**12**
	0100	Architectural		77.50	105	155	54.50%	61.50%	66.50%	
	0200	Plumbing		6.90	11.10	15.10	5.30%	6.85%	8%	
	0300	Mechanical		7.15	9.55	27.50	4.92%	6.90%	10.40%	
	0400	Electrical		10	15.70	22.50	6.20%	8%	10.25%	
	0500	Total Project Costs		128	210	253				
13	0000	**Nursing Home & Assisted Living**	S.F.							**13**
	0100	Architectural		72.50	94.50	119	51.50%	55.50%	63.50%	
	0200	Plumbing		7.80	11.75	12.90	6.25%	7.40%	8.80%	
	0300	Mechanical		6.40	9.45	18.50	4.04%	6.70%	9.55%	
	0400	Electrical		10.60	16.70	23.50	7%	10.75%	13.10%	
	0500	Total Project Costs		123	161	194				
14	0000	**Office Buildings**	S.F.							**14**
	0100	Architectural		93	130	179	54.50%	61%	69%	
	0200	Plumbing		5.15	8.10	15.15	2.70%	3.78%	5.85%	
	0300	Mechanical		10.10	17.15	26.50	5.60%	8.20%	11.10%	
	0400	Electrical		12.90	22	34	7.75%	10%	12.70%	
	0500	Total Project Costs		159	202	285				
15	0000	**Parking Garage**	S.F.							**15**
	0100	Architectural		32	39	40.50	70%	79%	88%	
	0200	Plumbing		1.05	1.10	2.06	2.05%	2.70%	2.83%	
	0300	Mechanical		.82	1.25	4.76	2.11%	3.62%	3.81%	
	0400	Electrical		2.80	3.08	6.45	5.30%	6.35%	7.95%	
	0500	Total Project Costs		39	47.50	51.50				
16	0000	**Parking Garage/Mixed Use**	S.F.							**16**
	0100	Architectural		103	113	115	61%	62%	65.50%	
	0200	Plumbing		3.33	4.36	6.65	2.47%	2.72%	3.66%	
	0300	Mechanical		14.20	16	23	7.80%	13.10%	13.60%	
	0400	Electrical		14.90	21.50	22	8.20%	12.65%	18.15%	
	0500	Total Project Costs		169	176	182				
17	0000	**Police Stations**	S.F.							**17**
	0100	Architectural		117	131	165	49%	56.50%	61%	
	0200	Plumbing		15.45	18.55	18.65	5.05%	5.55%	9.05%	
	0300	Mechanical		35	48.50	50.50	13%	14.55%	16.55%	
	0400	Electrical		26.50	29	30.50	9.15%	12.10%	14%	
	0500	Total Project Costs		219	270	305				
18	0000	**Police/Fire**	S.F.							**18**
	0100	Architectural		114	114	350	55.50%	66%	68%	
	0200	Plumbing		9.15	9.45	35	5.45%	5.50%	5.55%	
	0300	Mechanical		13.95	22	80	8.35%	12.70%	12.80%	
	0400	Electrical		15.90	20.50	91.50	9.50%	11.75%	14.55%	
	0500	Total Project Costs		167	173	630				
19	0000	**Public Assembly Buildings**	S.F.							**19**
	0100	Architectural		116	160	240	57.50%	61.50%	66%	
	0200	Plumbing		6.15	9	13.50	2.60%	3.36%	4.79%	
	0300	Mechanical		12.95	23	35.50	6.55%	8.95%	12.45%	
	0400	Electrical		19.15	26	41.50	8.60%	10.75%	13%	
	0500	Total Project Costs		186	255	375				
20	0000	**Recreational**	S.F.							**20**
	0100	Architectural		111	176	238	49.50%	59%	65.50%	
	0200	Plumbing		8.60	14.85	25.50	3.12%	4.76%	7.40%	
	0300	Mechanical		13.25	20	32	4.98%	6.60%	11.55%	
	0400	Electrical		19.15	27.50	40	7.15%	8.95%	10.75%	
	0500	Total Project Costs		198	296	445				

50 17 00	Project Costs	UNIT	UNIT COSTS			% OF TOTAL		
			1/4	MEDIAN	3/4	1/4	MEDIAN	3/4
21 0000	**Restaurants**	S.F.							**21**
0100	Architectural		127	195	250	59%	60%	63.50%
0200	Plumbing		13.90	32	40.50	7.35%	7.75%	8.95%
0300	Mechanical		15	17.70	37.50	6.50%	8.15%	11.15%
0400	Electrical		14.90	24.50	48.50	7.10%	10.30%	11.60%
0500	Total Project Costs		210	305	420			
22 0000	**Retail**	S.F.							**22**
0100	Architectural		56.50	88.50	182	54.50%	60%	64.50%
0200	Plumbing		7.05	9.85	12.25	4.18%	5.45%	8.45%
0300	Mechanical		6.60	9.35	17.20	4.98%	6.15%	7.05%
0400	Electrical		10.50	21.50	31.50	7.90%	11.25%	12.45%
0500	Total Project Costs		86	153	292			
23 0000	**Schools**	S.F.							**23**
0100	Architectural		98	127	173	51%	56%	60.50%
0200	Plumbing		7.75	10.70	15.65	3.66%	4.59%	7%
0300	Mechanical		18.70	26.50	38.50	8.95%	12%	14.55%
0400	Electrical		18.45	25.50	33	9.45%	11.25%	13.30%
0500	Total Project Costs		178	227	310			
24 0000	**University, College & Private School Classroom & Admin Buildings**	S.F.							**24**
0100	Architectural		126	160	194	50.50%	55%	59.50%
0200	Plumbing		8.50	11.50	19.90	3.02%	4.46%	7.35%
0300	Mechanical		25	37.50	46.50	9.95%	11.95%	14.70%
0400	Electrical		20	28.50	35.50	7.70%	9.90%	11.55%
0500	Total Project Costs		219	293	380			
25 0000	**University, College & Private School Dormitories**	S.F.							**25**
0100	Architectural		81.50	143	152	53%	61.50%	68%
0200	Plumbing		10.80	15.25	17.50	6.35%	6.65%	8.95%
0300	Mechanical		4.84	20.50	32.50	4.13%	9%	11.80%
0400	Electrical		5.75	19.95	30.50	4.75%	7.35%	10.60%
0500	Total Project Costs		121	229	271			
26 0000	**University, College & Private School Science, Eng. & Lab Buildings**	S.F.							**26**
0100	Architectural		140	166	195	48.50%	56.50%	58%
0200	Plumbing		9.65	14.65	24.50	3.29%	3.77%	5%
0300	Mechanical		37	69	71	11.70%	19.40%	23.50%
0400	Electrical		29	38	39	9%	12.05%	13.15%
0500	Total Project Costs		293	315	370			
27 0000	**University, College & Private School Student Union Buildings**	S.F.							**27**
0100	Architectural		111	292	292	54.50%	54.50%	59.50%
0200	Plumbing		16.80	16.80	25	3.13%	4.27%	11.45%
0300	Mechanical		32	51.50	51.50	9.60%	9.60%	14.55%
0400	Electrical		28	48.50	48.50	9.05%	12.80%	13.15%
0500	Total Project Costs		219	535	535			
28 0000	**Warehouses**	S.F.							**28**
0100	Architectural		47.50	72.50	132	60.50%	67%	71.50%
0200	Plumbing		2.48	5.30	10.20	2.82%	3.72%	5%
0300	Mechanical		2.93	16.70	26	4.56%	8.15%	10.70%
0400	Electrical		6.15	20	33.50	7.75%	10.10%	18.30%
0500	Total Project Costs		71	113	228			

Square Foot Project Size Modifier

One factor that affects the S.F. cost of a particular building is the size. In general, for buildings built to the same specifications in the same locality, the larger building will have the lower S.F. cost. This is due mainly to the decreasing contribution of the exterior walls plus the economy of scale usually achievable in larger buildings. The Area Conversion Scale shown below will give a factor to convert costs for the typical size building to an adjusted cost for the particular project.

The Square Foot Base Size lists the median costs, most typical project size in our accumulated data, and the range in size of the projects.

The Size Factor for your project is determined by dividing your project area in S.F. by the typical project size for the particular Building Type. With this factor, enter the Area Conversion Scale at the appropriate Size Factor and determine the appropriate Cost Multiplier for your building size.

Example: Determine the cost per S.F. for a 107,200 S.F. Multi-family housing.

$$\frac{\text{Proposed building area} = 107,200 \text{ S.F.}}{\text{Typical size from below} = 53,600 \text{ S.F.}} = 2.00$$

Enter Area Conversion Scale at 2.0, intersect curve, read horizontally the appropriate cost multiplier of .94. Size adjusted cost becomes .94 x $210.00 = $197.40 based on national average costs.

Note: For Size Factors less than .50, the Cost Multiplier is 1.1
For Size Factors greater than 3.5, the Cost Multiplier is .90

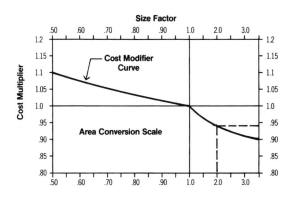

System	Median Cost (Total Project Costs)	Typical Size Gross S.F. (Median of Projects)	Typical Range (Low – High) (Projects)
Auto Sales with Repair	$187.00	24,900	4,680 – 29,253
Banking Institutions	300.00	9,300	3,267 – 38,148
Court House	286.00	47,600	24,680 – 70,500
Data Centers	284.00	14,400	14,369 – 14,369
Detention Centers	320.00	37,800	12,257 – 183,339
Fire Stations	241.00	12,500	6,300 – 49,577
Gymnasium	206.00	52,400	22,844 – 81,992
Hospitals	375.00	87,100	22,428 – 410,273
Industrial Buildings	133.00	21,100	5,146 – 200,625
Medical Clinics & Offices	224.00	22,500	2,273 – 327,000
Mixed Use	335.00	28,500	7,200 – 188,944
Multi-Family Housing	210.00	53,600	2,472 – 1,161,450
Nursing Home & Assisted Living	161.00	38,200	1,515 – 242,555
Office Buildings	202.00	20,600	1,115 – 930,000
Parking Garage	47.5	151,800	99,884 – 287,040
Parking Garage/Mixed Use	176.00	254,200	5,302 – 318,033
Police Stations	270.00	28,500	15,377 – 88,625
Police/Fire	173.00	44,300	8,600 – 50,279
Public Assembly Buildings	255.00	21,000	2,232 – 235,301
Recreational	296.00	27,700	1,000 – 223,787
Restaurants	305.00	6,000	3,975 – 42,000
Retail	153.00	26,000	4,035 – 84,270
Schools	227.00	69,900	1,344 – 410,848
University, College & Private School Classroom & Admin Buildings	293.00	48,200	8,806 – 196,187
University, College & Private School Dormitories	229.00	39,600	1,500 – 126,889
University, College & Private School Science, Eng. & Lab Buildings	315.00	60,000	5,311 – 117,643
University, College & Private School Student Union Buildings	535.00	47,300	42,075 – 50,000
Warehouses	113.00	11,100	640 – 303,750

A	Area Square Feet; Ampere	Brk., brk	Brick	Csc	Cosecant
AAFES	Army and Air Force Exchange Service	brkt	Bracket	C.S.F.	Hundred Square Feet
		Brs.	Brass	CSI	Construction Specifications Institute
ABS	Acrylonitrile Butadiene Stryrene; Asbestos Bonded Steel	Brz.	Bronze		
		Bsn.	Basin	CT	Current Transformer
A.C., AC	Alternating Current; Air-Conditioning; Asbestos Cement; Plywood Grade A & C	Btr.	Better	CTS	Copper Tube Size
		BTU	British Thermal Unit	Cu	Copper, Cubic
		BTUH	BTU per Hour	Cu. Ft.	Cubic Foot
		Bu.	Bushels	cw	Continuous Wave
ACI	American Concrete Institute	BUR	Built-up Roofing	C.W.	Cool White; Cold Water
ACR	Air Conditioning Refrigeration	BX	Interlocked Armored Cable	Cwt.	100 Pounds
ADA	Americans with Disabilities Act	°C	Degree Centigrade	C.W.X.	Cool White Deluxe
AD	Plywood, Grade A & D	c	Conductivity, Copper Sweat	C.Y.	Cubic Yard (27 cubic feet)
Addit.	Additional	C	Hundred; Centigrade	C.Y./Hr.	Cubic Yard per Hour
Adh.	Adhesive	C/C	Center to Center, Cedar on Cedar	Cyl.	Cylinder
Adj.	Adjustable	C-C	Center to Center	d	Penny (nail size)
af	Audio-frequency	Cab	Cabinet	D	Deep; Depth; Discharge
AFFF	Aqueous Film Forming Foam	Cair.	Air Tool Laborer	Dis., Disch.	Discharge
AFUE	Annual Fuel Utilization Efficiency	Cal.	Caliper	Db	Decibel
AGA	American Gas Association	Calc.	Calculated	Dbl.	Double
Agg.	Aggregate	Cap.	Capacity	DC	Direct Current
A.H., Ah	Ampere Hours	Carp.	Carpenter	DDC	Direct Digital Control
A hr.	Ampere-hour	C.B.	Circuit Breaker	Demob.	Demobilization
A.H.U., AHU	Air Handling Unit	C.C.A.	Chromate Copper Arsenate	d.f.t.	Dry Film Thickness
A.I.A.	American Institute of Architects	C.C.F.	Hundred Cubic Feet	d.f.u.	Drainage Fixture Units
AIC	Ampere Interrupting Capacity	cd	Candela	D.H.	Double Hung
Allow.	Allowance	cd/sf	Candela per Square Foot	DHW	Domestic Hot Water
alt., alt	Alternate	CD	Grade of Plywood Face & Back	DI	Ductile Iron
Alum.	Aluminum	CDX	Plywood, Grade C & D, exterior glue	Diag.	Diagonal
a.m.	Ante Meridiem			Diam., Dia	Diameter
Amp.	Ampere	Cefi.	Cement Finisher	Distrib.	Distribution
Anod.	Anodized	Cem.	Cement	Div.	Division
ANSI	American National Standards Institute	CF	Hundred Feet	Dk.	Deck
		C.F.	Cubic Feet	D.L.	Dead Load; Diesel
APA	American Plywood Association	CFM	Cubic Feet per Minute	DLH	Deep Long Span Bar Joist
Approx.	Approximate	CFRP	Carbon Fiber Reinforced Plastic	dlx	Deluxe
Apt.	Apartment	c.g.	Center of Gravity	Do.	Ditto
Asb.	Asbestos	CHW	Chilled Water; Commercial Hot Water	DOP	Dioctyl Phthalate Penetration Test (Air Filters)
A.S.B.C.	American Standard Building Code				
Asbe.	Asbestos Worker	C.I., CI	Cast Iron	Dp., dp	Depth
ASCE	American Society of Civil Engineers	C.I.P., CIP	Cast in Place	D.P.S.T.	Double Pole, Single Throw
A.S.H.R.A.E.	American Society of Heating, Refrig. & AC Engineers	Circ.	Circuit	Dr.	Drive
		C.L.	Carload Lot	DR	Dimension Ratio
ASME	American Society of Mechanical Engineers	CL	Chain Link	Drink.	Drinking
		Clab.	Common Laborer	D.S.	Double Strength
ASTM	American Society for Testing and Materials	Clam	Common Maintenance Laborer	D.S.A.	Double Strength A Grade
		C.L.F.	Hundred Linear Feet	D.S.B.	Double Strength B Grade
Attchmt.	Attachment	CLF	Current Limiting Fuse	Dty.	Duty
Avg., Ave.	Average	CLP	Cross Linked Polyethylene	DWV	Drain Waste Vent
AWG	American Wire Gauge	cm	Centimeter	DX	Deluxe White, Direct Expansion
AWWA	American Water Works Assoc.	CMP	Corr. Metal Pipe	dyn	Dyne
Bbl.	Barrel	CMU	Concrete Masonry Unit	e	Eccentricity
B&B, BB	Grade B and Better; Balled & Burlapped	CN	Change Notice	E	Equipment Only; East; Emissivity
		Col.	Column	Ea.	Each
B&S	Bell and Spigot	CO₂	Carbon Dioxide	EB	Encased Burial
B.&W.	Black and White	Comb.	Combination	Econ.	Economy
b.c.c.	Body-centered Cubic	comm.	Commercial, Communication	E.C.Y	Embankment Cubic Yards
B.C.Y.	Bank Cubic Yards	Compr.	Compressor	EDP	Electronic Data Processing
BE	Bevel End	Conc.	Concrete	EIFS	Exterior Insulation Finish System
B.F.	Board Feet	Cont., cont	Continuous; Continued, Container	E.D.R.	Equiv. Direct Radiation
Bg. cem.	Bag of Cement	Corkbd.	Cork Board	Eq.	Equation
BHP	Boiler Horsepower; Brake Horsepower	Corr.	Corrugated	EL	Elevation
		Cos	Cosine	Elec.	Electrician; Electrical
B.I.	Black Iron	Cot	Cotangent	Elev.	Elevator; Elevating
bidir.	bidirectional	Cov.	Cover	EMT	Electrical Metallic Conduit; Thin Wall Conduit
Bit., Bitum.	Bituminous	C/P	Cedar on Paneling		
Bit., Conc.	Bituminous Concrete	CPA	Control Point Adjustment	Eng.	Engine, Engineered
Bk.	Backed	Cplg.	Coupling	EPDM	Ethylene Propylene Diene Monomer
Bkrs.	Breakers	CPM	Critical Path Method		
Bldg., bldg	Building	CPVC	Chlorinated Polyvinyl Chloride	EPS	Expanded Polystyrene
Blk.	Block	C.Pr.	Hundred Pair	Eqhv.	Equip. Oper., Heavy
Bm.	Beam	CRC	Cold Rolled Channel	Eqlt.	Equip. Oper., Light
Boil.	Boilermaker	Creos.	Creosote	Eqmd.	Equip. Oper., Medium
bpm	Blows per Minute	Crpt.	Carpet & Linoleum Layer	Eqmm.	Equip. Oper., Master Mechanic
BR	Bedroom	CRT	Cathode-ray Tube	Eqol.	Equip. Oper., Oilers
Brg., brng.	Bearing	CS	Carbon Steel, Constant Shear Bar Joist	Equip.	Equipment
Brhe.	Bricklayer Helper			ERW	Electric Resistance Welded
Bric.	Bricklayer				

E.S.	Energy Saver	H	High Henry	Lath.	Lather
Est.	Estimated	HC	High Capacity	Lav.	Lavatory
esu	Electrostatic Units	H.D., HD	Heavy Duty; High Density	lb.; #	Pound
E.W.	Each Way	H.D.O.	High Density Overlaid	L.B., LB	Load Bearing; L Conduit Body
EWT	Entering Water Temperature	HDPE	High Density Polyethylene Plastic	L. & E.	Labor & Equipment
Excav.	Excavation	Hdr.	Header	lb./hr.	Pounds per Hour
excl	Excluding	Hdwe.	Hardware	lb./L.F.	Pounds per Linear Foot
Exp., exp	Expansion, Exposure	H.I.D., HID	High Intensity Discharge	lbf/sq.in.	Pound-force per Square Inch
Ext., ext	Exterior; Extension	Help.	Helper Average	L.C.L.	Less than Carload Lot
Extru.	Extrusion	HEPA	High Efficiency Particulate Air	L.C.Y.	Loose Cubic Yard
f.	Fiber Stress		Filter	Ld.	Load
F	Fahrenheit; Female; Fill	Hg	Mercury	LE	Lead Equivalent
Fab., fab	Fabricated; Fabric	HIC	High Interrupting Capacity	LED	Light Emitting Diode
FBGS	Fiberglass	HM	Hollow Metal	L.F.	Linear Foot
F.C.	Footcandles	HMWPE	High Molecular Weight	L.F. Hdr	Linear Feet of Header
f.c.c.	Face-centered Cubic		Polyethylene	L.F. Nose	Linear Foot of Stair Nosing
f'c.	Compressive Stress in Concrete;	HO	High Output	L.F. Rsr	Linear Foot of Stair Riser
	Extreme Compressive Stress	Horiz.	Horizontal	Lg.	Long; Length; Large
F.E.	Front End	H.P., HP	Horsepower; High Pressure	L & H	Light and Heat
FEP	Fluorinated Ethylene Propylene	H.P.F.	High Power Factor	LH	Long Span Bar Joist
	(Teflon)	Hr.	Hour	L.H.	Labor Hours
F.G.	Flat Grain	Hrs./Day	Hours per Day	L.L., LL	Live Load
F.H.A.	Federal Housing Administration	HSC	High Short Circuit	L.L.D.	Lamp Lumen Depreciation
Fig.	Figure	Ht.	Height	lm	Lumen
Fin.	Finished	Htg.	Heating	lm/sf	Lumen per Square Foot
FIPS	Female Iron Pipe Size	Htrs.	Heaters	lm/W	Lumen per Watt
Fixt.	Fixture	HVAC	Heating, Ventilation & Air-	LOA	Length Over All
FJP	Finger jointed and primed		Conditioning	log	Logarithm
Fl. Oz.	Fluid Ounces	Hvy.	Heavy	L-O-L	Lateralolet
Flr.	Floor	HW	Hot Water	long.	Longitude
Flrs.	Floors	Hyd.; Hydr.	Hydraulic	L.P., LP	Liquefied Petroleum; Low Pressure
FM	Frequency Modulation;	Hz	Hertz (cycles)	L.P.F.	Low Power Factor
	Factory Mutual	I.	Moment of Inertia	LR	Long Radius
Fmg.	Framing	IBC	International Building Code	L.S.	Lump Sum
FM/UL	Factory Mutual/Underwriters Labs	I.C.	Interrupting Capacity	Lt.	Light
Fdn.	Foundation	ID	Inside Diameter	Lt. Ga.	Light Gauge
FNPT	Female National Pipe Thread	I.D.	Inside Dimension; Identification	L.T.L.	Less than Truckload Lot
Fori.	Foreman, Inside	I.F.	Inside Frosted	Lt. Wt.	Lightweight
Foro.	Foreman, Outside	I.M.C.	Intermediate Metal Conduit	L.V.	Low Voltage
Fount.	Fountain	In.	Inch	M	Thousand; Material; Male;
fpm	Feet per Minute	Incan.	Incandescent		Light Wall Copper Tubing
FPT	Female Pipe Thread	Incl.	Included; Including	M²CA	Meters Squared Contact Area
Fr	Frame	Int.	Interior	m/hr.; M.H.	Man-hour
F.R.	Fire Rating	Inst.	Installation	mA	Milliampere
FRK	Foil Reinforced Kraft	Insul., insul	Insulation/Insulated	Mach.	Machine
FSK	Foil/Scrim/Kraft	I.P.	Iron Pipe	Mag. Str.	Magnetic Starter
FRP	Fiberglass Reinforced Plastic	I.P.S., IPS	Iron Pipe Size	Maint.	Maintenance
FS	Forged Steel	IPT	Iron Pipe Threaded	Marb.	Marble Setter
FSC	Cast Body; Cast Switch Box	I.W.	Indirect Waste	Mat; Mat'l.	Material
Ft., ft	Foot; Feet	J	Joule	Max.	Maximum
Ftng.	Fitting	J.I.C.	Joint Industrial Council	MBF	Thousand Board Feet
Ftg.	Footing	K	Thousand; Thousand Pounds;	MBH	Thousand BTU's per hr.
Ft lb.	Foot Pound		Heavy Wall Copper Tubing, Kelvin	MC	Metal Clad Cable
Furn.	Furniture	K.A.H.	Thousand Amp. Hours	MCC	Motor Control Center
FVNR	Full Voltage Non-Reversing	kcmil	Thousand Circular Mils	M.C.F.	Thousand Cubic Feet
FVR	Full Voltage Reversing	KD	Knock Down	MCFM	Thousand Cubic Feet per Minute
FXM	Female by Male	K.D.A.T.	Kiln Dried After Treatment	M.C.M.	Thousand Circular Mils
Fy.	Minimum Yield Stress of Steel	kg	Kilogram	MCP	Motor Circuit Protector
g	Gram	kG	Kilogauss	MD	Medium Duty
G	Gauss	kgf	Kilogram Force	MDF	Medium-density fibreboard
Ga.	Gauge	kHz	Kilohertz	M.D.O.	Medium Density Overlaid
Gal., gal.	Gallon	Kip	1000 Pounds	Med.	Medium
Galv., galv	Galvanized	KJ	Kilojoule	MF	Thousand Feet
GC/MS	Gas Chromatograph/Mass	K.L.	Effective Length Factor	M.F.B.M.	Thousand Feet Board Measure
	Spectrometer	K.L.F.	Kips per Linear Foot	Mfg.	Manufacturing
Gen.	General	Km	Kilometer	Mfrs.	Manufacturers
GFI	Ground Fault Interrupter	KO	Knock Out	mg	Milligram
GFRC	Glass Fiber Reinforced Concrete	K.S.F.	Kips per Square Foot	MGD	Million Gallons per Day
Glaz.	Glazier	K.S.I.	Kips per Square Inch	MGPH	Million Gallons per Hour
GPD	Gallons per Day	kV	Kilovolt	MH, M.H.	Manhole; Metal Halide; Man-Hour
gpf	Gallon per Flush	kVA	Kilovolt Ampere	MHz	Megahertz
GPH	Gallons per Hour	kVAR	Kilovar (Reactance)	Mi.	Mile
gpm, GPM	Gallons per Minute	KW	Kilowatt	MI	Malleable Iron; Mineral Insulated
GR	Grade	KWh	Kilowatt-hour	MIPS	Male Iron Pipe Size
Gran.	Granular	L	Labor Only; Length; Long;	mj	Mechanical Joint
Grnd.	Ground		Medium Wall Copper Tubing	m	Meter
GVW	Gross Vehicle Weight	Lab.	Labor	mm	Millimeter
GWB	Gypsum Wall Board	lat	Latitude	Mill.	Millwright
				Min., min.	Minimum, Minute

Abbreviations

Misc.	Miscellaneous	PCM	Phase Contrast Microscopy	SBS	Styrene Butadiere Styrene
ml	Milliliter, Mainline	PDCA	Painting and Decorating	SC	Screw Cover
M.L.F.	Thousand Linear Feet		Contractors of America	SCFM	Standard Cubic Feet per Minute
Mo.	Month	P.E., PE	Professional Engineer;	Scaf.	Scaffold
Mobil.	Mobilization		Porcelain Enamel;	Sch., Sched.	Schedule
Mog.	Mogul Base		Polyethylene; Plain End	S.C.R.	Modular Brick
MPH	Miles per Hour	P.E.C.I.	Porcelain Enamel on Cast Iron	S.D.	Sound Deadening
MPT	Male Pipe Thread	Perf.	Perforated	SDR	Standard Dimension Ratio
MRGWB	Moisture Resistant Gypsum	PEX	Cross Linked Polyethylene	S.E.	Surfaced Edge
	Wallboard	Ph.	Phase	Sel.	Select
MRT	Mile Round Trip	P.I.	Pressure Injected	SER, SEU	Service Entrance Cable
ms	Millisecond	Pile.	Pile Driver	S.F.	Square Foot
M.S.F.	Thousand Square Feet	Pkg.	Package	S.F.C.A.	Square Foot Contact Area
Mstz.	Mosaic & Terrazzo Worker	Pl.	Plate	S.F. Flr.	Square Foot of Floor
M.S.Y.	Thousand Square Yards	Plah.	Plasterer Helper	S.F.G.	Square Foot of Ground
Mtd., mtd., mtd	Mounted	Plas.	Plasterer	S.F. Hor.	Square Foot Horizontal
Mthe.	Mosaic & Terrazzo Helper	plf	Pounds Per Linear Foot	SFR	Square Feet of Radiation
Mtng.	Mounting	Pluh.	Plumber Helper	S.F. Shlf.	Square Foot of Shelf
Mult.	Multi; Multiply	Plum.	Plumber	S4S	Surface 4 Sides
MUTCD	Manual on Uniform Traffic Control	Ply.	Plywood	Shee.	Sheet Metal Worker
	Devices	p.m.	Post Meridiem	Sin.	Sine
M.V.A.	Million Volt Amperes	Pntd.	Painted	Skwk.	Skilled Worker
M.V.A.R.	Million Volt Amperes Reactance	Pord.	Painter, Ordinary	SL	Saran Lined
MV	Megavolt	pp	Pages	S.L.	Slimline
MW	Megawatt	PP, PPL	Polypropylene	Sldr.	Solder
MXM	Male by Male	P.P.M.	Parts per Million	SLH	Super Long Span Bar Joist
MYD	Thousand Yards	Pr.	Pair	S.N.	Solid Neutral
N	Natural; North	P.E.S.B.	Pre-engineered Steel Building	SO	Stranded with oil resistant inside
nA	Nanoampere	Prefab.	Prefabricated		insulation
NA	Not Available; Not Applicable	Prefin.	Prefinished	S-O-L	Socketolet
N.B.C.	National Building Code	Prop.	Propelled	sp	Standpipe
NC	Normally Closed	PSF, psf	Pounds per Square Foot	S.P.	Static Pressure; Single Pole; Self-
NEMA	National Electrical Manufacturers	PSI, psi	Pounds per Square Inch		Propelled
	Assoc.	PSIG	Pounds per Square Inch Gauge	Spri.	Sprinkler Installer
NEHB	Bolted Circuit Breaker to 600V.	PSP	Plastic Sewer Pipe	spwg	Static Pressure Water Gauge
NFPA	National Fire Protection Association	Pspr.	Painter, Spray	S.P.D.T.	Single Pole, Double Throw
NLB	Non-Load-Bearing	Psst.	Painter, Structural Steel	SPF	Spruce Pine Fir; Sprayed
NM	Non-Metallic Cable	P.T.	Potential Transformer		Polyurethane Foam
nm	Nanometer	P. & T.	Pressure & Temperature	S.P.S.T.	Single Pole, Single Throw
No.	Number	Ptd.	Painted	SPT	Standard Pipe Thread
NO	Normally Open	Ptns.	Partitions	Sq.	Square; 100 Square Feet
N.O.C.	Not Otherwise Classified	Pu	Ultimate Load	Sq. Hd.	Square Head
Nose.	Nosing	PVC	Polyvinyl Chloride	Sq. In.	Square Inch
NPT	National Pipe Thread	Pvmt.	Pavement	S.S.	Single Strength; Stainless Steel
NQOD	Combination Plug-on/Bolt on	PRV	Pressure Relief Valve	S.S.B.	Single Strength B Grade
	Circuit Breaker to 240V.	Pwr.	Power	sst, ss	Stainless Steel
N.R.C., NRC	Noise Reduction Coefficient/	Q	Quantity Heat Flow	Sswk.	Structural Steel Worker
	Nuclear Regulator Commission	Qt.	Quart	Sswl.	Structural Steel Welder
N.R.S.	Non Rising Stem	Quan., Qty.	Quantity	St.; Stl.	Steel
ns	Nanosecond	Q.C.	Quick Coupling	STC	Sound Transmission Coefficient
NTP	Notice to Proceed	r	Radius of Gyration	Std.	Standard
nW	Nanowatt	R	Resistance	Stg.	Staging
OB	Opposing Blade	R.C.P.	Reinforced Concrete Pipe	STK	Select Tight Knot
OC	On Center	Rect.	Rectangle	STP	Standard Temperature & Pressure
OD	Outside Diameter	recpt.	Receptacle	Stpi.	Steamfitter, Pipefitter
O.D.	Outside Dimension	Reg.	Regular	Str.	Strength; Starter; Straight
ODS	Overhead Distribution System	Reinf.	Reinforced	Strd.	Stranded
O.G.	Ogee	Req'd.	Required	Struct.	Structural
O.H.	Overhead	Res.	Resistant	Sty.	Story
O&P	Overhead and Profit	Resi.	Residential	Subj.	Subject
Oper.	Operator	RF	Radio Frequency	Subs.	Subcontractors
Opng.	Opening	RFID	Radio-frequency Identification	Surf.	Surface
Orna.	Ornamental	Rgh.	Rough	Sw.	Switch
OSB	Oriented Strand Board	RGS	Rigid Galvanized Steel	Swbd.	Switchboard
OS&Y	Outside Screw and Yoke	RHW	Rubber, Heat & Water Resistant;	S.Y.	Square Yard
OSHA	Occupational Safety and Health		Residential Hot Water	Syn.	Synthetic
	Act	rms	Root Mean Square	S.Y.P.	Southern Yellow Pine
Ovhd.	Overhead	Rnd.	Round	Sys.	System
OWG	Oil, Water or Gas	Rodm.	Rodman	t.	Thickness
Oz.	Ounce	Rofc.	Roofer, Composition	T	Temperature; Ton
P.	Pole; Applied Load; Projection	Rofp.	Roofer, Precast	Tan	Tangent
p.	Page	Rohe.	Roofer Helpers (Composition)	T.C.	Terra Cotta
Pape.	Paperhanger	Rots.	Roofer, Tile & Slate	T & C	Threaded and Coupled
P.A.P.R.	Powered Air Purifying Respirator	R.O.W.	Right of Way	T.D.	Temperature Difference
PAR	Parabolic Reflector	RPM	Revolutions per Minute	TDD	Telecommunications Device for
P.B., PB	Push Button	R.S.	Rapid Start		the Deaf
Pc., Pcs.	Piece, Pieces	Rsr	Riser	T.E.M.	Transmission Electron Microscopy
P.C.	Portland Cement; Power Connector	RT	Round Trip	temp	Temperature, Tempered, Temporary
P.C.F.	Pounds per Cubic Foot	S.	Suction; Single Entrance; South	TFFN	Nylon Jacketed Wire

For customer support on your Electrical Costs with RSMeans Data, call 800.448.8182.

TFE	Tetrafluoroethylene (Teflon)	U.L., UL	Underwriters Laboratory	w/	With
T. & G.	Tongue & Groove;	Uld.	Unloading	W.C., WC	Water Column; Water Closet
	Tar & Gravel	Unfin.	Unfinished	W.F.	Wide Flange
Th., Thk.	Thick	UPS	Uninterruptible Power Supply	W.G.	Water Gauge
Thn.	Thin	URD	Underground Residential	Wldg.	Welding
Thrded	Threaded		Distribution	W. Mile	Wire Mile
Tilf.	Tile Layer, Floor	US	United States	W-O-L	Weldolet
Tilh.	Tile Layer, Helper	USGBC	U.S. Green Building Council	W.R.	Water Resistant
THHN	Nylon Jacketed Wire	USP	United States Primed	Wrck.	Wrecker
THW.	Insulated Strand Wire	UTMCD	Uniform Traffic Manual For Control	WSFU	Water Supply Fixture Unit
THWN	Nylon Jacketed Wire		Devices	W.S.P.	Water, Steam, Petroleum
T.L., TL	Truckload	UTP	Unshielded Twisted Pair	WT., Wt.	Weight
T.M.	Track Mounted	V	Volt	WWF	Welded Wire Fabric
Tot.	Total	VA	Volt Amperes	XFER	Transfer
T-O-L	Threadolet	VAT	Vinyl Asbestos Tile	XFMR	Transformer
tmpd	Tempered	V.C.T.	Vinyl Composition Tile	XHD	Extra Heavy Duty
TPO	Thermoplastic Polyolefin	VAV	Variable Air Volume	XHHW	Cross-Linked Polyethylene Wire
T.S.	Trigger Start	VC	Veneer Core	XLPE	Insulation
Tr.	Trade	VDC	Volts Direct Current	XLP	Cross-linked Polyethylene
Transf.	Transformer	Vent.	Ventilation	Xport	Transport
Trhv.	Truck Driver, Heavy	Vert.	Vertical	Y	Wye
Trlr	Trailer	V.F.	Vinyl Faced	yd	Yard
Trlt.	Truck Driver, Light	V.G.	Vertical Grain	yr	Year
TTY	Teletypewriter	VHF	Very High Frequency	Δ	Delta
TV	Television	VHO	Very High Output	%	Percent
T.W.	Thermoplastic Water Resistant	Vib.	Vibrating	~	Approximately
	Wire	VLF	Vertical Linear Foot	Ø	Phase; diameter
UCI	Uniform Construction Index	VOC	Volatile Organic Compound	@	At
UF	Underground Feeder	Vol.	Volume	#	Pound; Number
UGND	Underground Feeder	VRP	Vinyl Reinforced Polyester	<	Less Than
UHF	Ultra High Frequency	W	Wire; Watt; Wide; West	>	Greater Than
U.I.	United Inch			Z	Zone

For customer support on your Electrical Costs with RSMeans Data, call 800.448.8182.

651

Index

For customer support on your Electrical Costs with RSMeans Data, call 800.448.8182.

Index

Index

Index

For customer support on your Electrical Costs with RSMeans Data, call 800.448.8182.

Index

Division Notes

	CREW	DAILY OUTPUT	LABOR-HOURS	UNIT	BARE COSTS				TOTAL INCL O&P
					MAT.	LABOR	EQUIP.	TOTAL	

Division Notes

	CREW	DAILY OUTPUT	LABOR-HOURS	UNIT	BARE COSTS				TOTAL INCL O&P
					MAT.	LABOR	EQUIP.	TOTAL	

Division Notes

	CREW	DAILY OUTPUT	LABOR-HOURS	UNIT	BARE COSTS				TOTAL INCL O&P
					MAT.	LABOR	EQUIP.	TOTAL	

Division Notes

	CREW	DAILY OUTPUT	LABOR-HOURS	UNIT	BARE COSTS				TOTAL INCL O&P
					MAT.	LABOR	EQUIP.	TOTAL	

Division Notes

	CREW	DAILY OUTPUT	LABOR-HOURS	UNIT	BARE COSTS				TOTAL INCL O&P
					MAT.	LABOR	EQUIP.	TOTAL	

Other Data & Services

A tradition of excellence in construction cost information and services since 1942

Table of Contents
Cost Data Selection Guide
RSMeans Data Online
Training

For more information visit our website at RSMeans.com

Unit prices according to the latest MasterFormat®

Cost Data Selection Guide

The following table provides definitive information on the content of each cost data publication. The number of lines of data provided in each unit price or assemblies division, as well as the number of crews, is listed for each data set. The presence of other elements such as reference tables, square foot models, equipment rental costs, historical cost indexes, and city cost indexes, is also indicated. You can use the table to help select the RSMeans data set that has the quantity and type of information you most need in your work.

Unit Cost Divisions	Building Construction	Mechanical	Electrical	Commercial Renovation	Square Foot	Site Work Landsc.	Green Building	Interior	Concrete Masonry	Open Shop	Heavy Construction	Light Commercial	Facilities Construction	Plumbing	Residential
1	621	454	475	576	0	543	200	377	505	620	560	322	1092	462	219
2	825	347	157	781	0	972	181	466	221	824	739	550	1268	355	340
3	1745	341	232	1265	0	1537	1043	355	2274	1745	1930	538	2028	317	445
4	960	22	0	920	0	724	180	613	1158	928	614	532	1175	0	446
5	1895	158	155	1099	0	858	1793	1106	735	1895	1031	985	1912	204	752
6	2468	18	18	2127	0	110	589	1544	281	2464	123	2157	2141	22	2677
7	1593	215	128	1633	0	580	761	532	523	1590	26	1326	1693	227	1046
8	2140	80	3	2733	0	255	1138	1813	105	2142	0	2328	2966	0	1552
9	2125	86	45	1943	0	313	464	2216	424	2062	15	1779	2379	54	1544
10	1088	17	10	684	0	232	32	898	136	1088	34	588	1179	237	224
11	1095	199	166	539	0	135	56	923	29	1062	0	229	1115	162	108
12	539	0	2	297	0	219	147	1552	14	506	0	272	1571	23	216
13	745	149	157	252	0	370	124	250	77	721	271	109	761	115	103
14	273	36	0	223	0	0	0	257	0	273	0	12	293	16	6
21	127	0	41	37	0	0	0	293	0	127	0	121	665	685	259
22	1165	7543	160	1226	0	2010	1061	849	20	1154	2109	875	7505	9400	719
23	1263	7010	639	1033	0	250	864	775	38	1246	191	980	5244	2011	579
25	0	0	14	14	0	0	0	0	0	0	0	0	0	0	0
26	1513	491	10473	1293	0	859	645	1160	55	1439	648	1360	10254	399	636
27	95	0	467	105	0	0	0	71	0	95	45	67	389	0	56
28	143	79	223	124	0	0	28	97	0	127	0	70	209	57	41
31	1511	733	610	807	0	3261	284	7	1216	1456	3280	607	1568	660	616
32	906	49	8	944	0	4523	419	418	367	877	1942	496	1803	140	544
33	1319	1100	563	310	0	3142	33	0	253	596	3277	135	1790	2165	161
34	112	0	47	9	0	195	0	0	36	67	226	0	141	0	0
35	18	0	0	0	0	327	0	0	0	18	442	0	84	0	0
41	63	0	0	34	0	8	0	22	0	62	31	0	69	14	0
44	75	79	0	0	0	0	0	0	0	0	0	0	75	75	0
46	23	16	0	0	0	274	261	0	0	23	264	0	33	33	0
48	8	0	36	2	0	0	21	0	0	8	15	8	21	0	8
Totals	26453	19222	14829	21010	0	21697	10324	16594	8467	25215	17813	16446	51423	17833	13297

Assem Div	Building Construction	Mechanical	Electrical	Commercial Renovation	Square Foot	Site Work Landscape	Assemblies	Green Building	Interior	Concrete Masonry	Heavy Construction	Light Commercial	Facilities Construction	Plumbing	Asm Div	Residential
A		15	0	190	165	579	599	0	0	537	573	155	24	0	1	378
B		0	0	848	2567	0	5673	56	329	1979	368	2108	174	0	2	211
C		0	0	647	956	0	1338	0	1645	146	0	846	251	0	3	591
D		1060	945	712	1866	72	2544	330	827	0	0	1353	1108	1090	4	851
E		0	0	85	262	0	302	0	5	0	0	259	5	0	5	391
F		0	0	0	114	0	143	0	0	0	0	114	0	0	6	357
G		527	447	318	312	3378	792	0	0	535	1349	205	293	677	7	307
															8	760
															9	80
															10	0
															11	0
															12	0
Totals		1602	1392	2800	6242	4029	11391	386	2806	3197	2290	5040	1855	1767		3926

Reference Section	Building Construction Costs	Mechanical	Electrical	Commercial Renovation	Square Foot	Site Work Landscape	Assem.	Green Building	Interior	Concrete Masonry	Open Shop	Heavy Construction	Light Commercial	Facilities Construction	Plumbing	Resi.
Reference Tables	yes	yes	yes	yes	no	yes	yes	yes	yes	yes	yes	yes	yes	yes	yes	yes
Models					111			25					50			28
Crews	584	584	584	564		584		584	584	584	562	584	562	564	584	560
Equipment Rental Costs	yes	yes	yes	yes		yes		yes	yes	yes	yes	yes	yes	yes	yes	yes
Historical Cost Indexes	yes	yes	yes	yes	yes	yes	yes	yes	yes	yes	yes	yes	yes	yes	yes	no
City Cost Indexes	yes	yes	yes	yes	yes	yes	yes	yes	yes	yes	yes	yes	yes	yes	yes	yes

673

Virtual classes offered monthly, call for the schedule—see note for details

2021 Training Class Options ☎ 877.620.6245

RSMeans offers training classes in a variety of formats—eLearning training modules, instructor-led classes, on-site training, and virtual training classes. Due to our current Covid-19 situation, RSMeans has adjusted our current training offerings and instructor-led classes for 2021 have not been determined. Please visit our website at https://www.rsmeans.com/products/training/seminars for our current schedule of class offerings.

Training classes that are offered in one or more of our formats are: Building Systems and the Construction Process; Construction Cost Estimating: Concepts & Practice; Introduction to Estimating; Scope of Work for Facilities Estimating; Facilities Construction Estimating; Maintenance & Repair Estimating for Facilities; Mechanical & Electrical Estimating; RSMeans CostWorks CD; and RSMeans Online Training.

If you have any questions or you would like to register for any class or purchase a training module, call us at 877-620.6245.

Facilities Construction Estimating

In this two-day course, professionals working in facilities management can get help with their daily challenges to establish budgets for all phases of a project.

Some of what you'll learn:
- Determining the full scope of a project
- Understanding of RSMeans data and what is included in prices
- Identifying appropriate factors to be included in your estimate
- Creative solutions to estimating issues
- Organizing estimates for presentation and discussion
- Special estimating techniques for repair/remodel and maintenance projects
- Appropriate use of contingency, city cost indexes, and reference notes
- Techniques to get to the correct estimate quickly

Who should attend: facility managers, engineers, contractors, facility tradespeople, planners, and project managers.

Construction Cost Estimating: Concepts and Practice

This one- or two-day introductory course to improve estimating skills and effectiveness starts with the details of interpreting bid documents and ends with the summary of the estimate and bid submission.

Some of what you'll learn:
- Using the plans and specifications to create estimates
- The takeoff process—deriving all tasks with correct quantities
- Developing pricing using various sources; how subcontractor pricing fits in
- Summarizing the estimate to arrive at the final number
- Formulas for area and cubic measure, adding waste and adjusting productivity to specific projects
- Evaluating subcontractors' proposals and prices
- Adding insurance and bonds
- Understanding how labor costs are calculated
- Submitting bids and proposals

Who should attend: project managers, architects, engineers, owners' representatives, contractors, and anyone who's responsible for budgeting or estimating construction projects.

RSMeans data Training

Assessing Scope of Work for Facilities Construction Estimating

This two-day practical training program addresses the vital importance of understanding the scope of projects in order to produce accurate cost estimates for facilities repair and remodeling.

Some of what you'll learn:

- Discussions of site visits, plans/specs, record drawings of facilities, and site-specific lists
- Review of CSI divisions, including means, methods, materials, and the challenges of scoping each topic
- Exercises in scope identification and scope writing for accurate estimating of projects
- Hands-on exercises that require scope, takeoff, and pricing

Who should attend: corporate and government estimators, planners, facility managers, and others who need to produce accurate project estimates.

Maintenance & Repair Estimating for Facilities

This two-day course teaches attendees how to plan, budget, and estimate the cost of ongoing and preventive maintenance and repair for existing buildings and grounds.

Some of what you'll learn:

- The most financially favorable maintenance, repair, and replacement scheduling and estimating
- Auditing and value engineering facilities
- Preventive planning and facilities upgrading
- Determining both in-house and contract-out service costs
- Annual, asset-protecting M&R plan

Who should attend: facility managers, maintenance supervisors, buildings and grounds superintendents, plant managers, planners, estimators, and others involved in facilities planning and budgeting.

Mechanical & Electrical Estimating

This two-day course teaches attendees how to prepare more accurate and complete mechanical/electrical estimates, avoid the pitfalls of omission and double-counting, and understand the composition and rationale within the RSMeans mechanical/electrical database.

Some of what you'll learn:

- The unique way mechanical and electrical systems are interrelated
- M&E estimates—conceptual, planning, budgeting, and bidding stages
- Order of magnitude, square foot, assemblies, and unit price estimating
- Comparative cost analysis of equipment and design alternatives

Who should attend: architects, engineers, facilities managers, mechanical and electrical contractors, and others who need a highly reliable method for developing, understanding, and evaluating mechanical and electrical contracts.

Building Systems and the Construction Process

This one-day course was written to assist novices and those outside the industry in obtaining a solid understanding of the construction process - from both a building systems and construction administration approach.

Some of what you'll learn:

- Various systems used and how components come together to create a building
- Start with foundation and end with the physical systems of the structure such as HVAC and Electrical
- Focus on the process from start of design through project closeout

This training session requires you to bring a laptop computer to class.

Who should attend: building professionals or novices to help make the crossover to the construction industry; suited for anyone responsible for providing high level oversight on construction projects.

Practical Project Management for Construction Professionals

In this two-day course you will acquire the essential knowledge and develop the skills to effectively and efficiently execute the day-to-day responsibilities of the construction project manager.

Some of what you'll learn:

- General conditions of the construction contract
- Contract modifications: change orders and construction change directives
- Negotiations with subcontractors and vendors
- Effective writing: notification and communications
- Dispute resolution: claims and liens

Who should attend: architects, engineers, owners' representatives, and project managers.

RSMeans data Training

675

Training for our Online Estimating Solution

Construction estimating is vital to the decision-making process at each state of every project. Our online solution works the way you do. It's systematic, flexible, and intuitive. In this one-day class you will see how you can estimate any phase of any project faster and better.

Some of what you'll learn:
- Customizing our online estimating solution
- Making the most of RSMeans "Circle Reference" numbers
- How to integrate your cost data
- Generating reports, exporting estimates to MS Excel, sharing, collaborating, and more

Also offered as a self-paced or on-site training program!

Training for our CD Estimating Solution

This one-day course helps users become more familiar with the functionality of the CD. Each menu, icon, screen, and function found in the program is explained in depth. Time is devoted to hands-on estimating exercises.

Some of what you'll learn:
- Searching the database using all navigation methods
- Exporting RSMeans data to your preferred spreadsheet format
- Viewing crews, assembly components, and much more
- Automatically regionalizing the database

This training session requires you to bring a laptop computer to class.

When you register for this course you will receive an outline for your laptop requirements.

Also offered as a self-paced or on-site training program!

Site Work Estimating with RSMeans data

This one-day program focuses directly on site work costs. Accurately scoping, quantifying, and pricing site preparation, underground utility work, and improvements to exterior site elements are often the most difficult estimating tasks on any project. Some of what you'll learn:
- Evaluation of site work and understanding site scope including: site clearing, grading, excavation, disposal and trucking of materials, backfill and compaction, underground utilities, paving, sidewalks, and seeding & planting.
- Unit price site work estimates—Correct use of RSMeans site work cost data to develop a cost estimate.
- Using and modifying assemblies—Save valuable time when estimating site work activities using custom assemblies.

Who should attend: engineers, contractors, estimators, project managers, owners' representatives, and others who are concerned with the proper preparation and/or evaluation of site work estimates.

Please bring a laptop with capability to access the internet.

Facilities Estimating Using the CD

This two-day class combines hands-on skill-building with best estimating practices and real-life problems. You will learn key concepts, tips, pointers, and guidelines to save time and avoid cost oversights and errors.

Some of what you'll learn:
- Estimating process concepts
- Customizing and adapting RSMeans cost data
- Establishing scope of work to account for all known variables
- Budget estimating: when, why, and how
- Site visits: what to look for and what you can't afford to overlook
- How to estimate repair and remodeling variables

This training session requires you to bring a laptop computer to class.

Who should attend: facility managers, architects, engineers, contractors, facility tradespeople, planners, project managers, and anyone involved with JOC, SABRE, or IDIQ.

Registration Information

RSMeans data Training

How to register
By Phone
Register by phone at 877.620.6245
Online
Register online at
RSMeans.com/products/services/training
Note: Purchase Orders or Credits Cards are required to register.

Two-day seminar registration fee - $1,300*.

One-Day Construction Cost Estimating or Building Systems and the Construction Process - $825*.

Two-day virtual training classes - $825*.

Three-day virtual training classes - $995*.

Instructor-led Government pricing
All federal government employees save off the regular seminar price. Other promotional discounts cannot be combined with the government discount. Call 781.422.5115 for government pricing.

CANCELLATION POLICY FOR INSTRUCTOR-LED CLASSES:
If you are unable to attend a seminar, substitutions may be made at any time before the session starts by notifying the seminar registrar at 781.422.5115 or your sales representative.
If you cancel twenty-one (21) days or more prior to the seminar, there will be no penalty and your registration fees will be refunded. These cancellations must be received by the seminar registrar or your sales representative and will be confirmed to be eligible for cancellation.
If you cancel fewer than twenty-one (21) days prior to the seminar, you will forfeit the registration fee.
In the unfortunate event of an RSMeans cancellation, RSMeans will work with you to reschedule your attendance in the same seminar at a later date or will

fully refund your registration fee. RSMeans cannot be responsible for any non-refundable travel expenses incurred by you or another as a result of your registration, attendance at, or cancellation of an RSMeans seminar.
Any on-demand training modules are not eligible for cancellation, substitution, transfer, return, or refund.

AACE approved courses
Many seminars described and offered here have been approved for 14 hours (1.4 recertification credits) of credit by the AACE International Certification Board toward meeting the continuing education requirements for recertification as a Certified Cost Engineer/Certified Cost Consultant.

AIA Continuing Education
We are registered with the AIA Continuing Education System (AIA/CES) and are committed to developing quality learning activities in accordance with the CES criteria. Many seminars meet the AIA/CES criteria for Quality Level 2. AIA members may receive 14 learning units (LUs) for each two-day RSMeans course.

Daily course schedule
The first day of each seminar session begins at 8:30 a.m. and ends at 4:30 p.m. The second day begins at 8:00 a.m. and ends at 4:00 p.m. Participants are urged to bring a hand-held calculator since many actual problems will be worked out in each session.

Continental breakfast
Your registration includes the cost of a continental breakfast and a morning and afternoon refreshment break. These informal segments allow you to discuss topics of mutual interest with other seminar attendees. (You are free to make your own lunch and dinner arrangements.)

Hotel/transportation arrangements
We arrange to hold a block of rooms at most host hotels. To take advantage of special group rates when making your reservation, be sure to mention that you are attending the RSMeans Institute data seminar. You are, of course, free to stay at the lodging place of your choice. (Hotel reservations and transportation arrangements should be made directly by seminar attendees.)

Important
Class sizes are limited, so please register as soon as possible.

*Note: Pricing subject to change.